Medical
Physiology

Medical Physiology

Edited by

ARTHUR M. BROWN, M.D., Ph.D.

Professor and Chairman
Department of Physiology and Biophysics
University of Texas Medical Branch
Galveston, Texas

DONALD W. STUBBS, Ph.D.

Professor
Department of Physiology and Biophysics
University of Texas Medical Branch
Galveston, Texas

A WILEY MEDICAL PUBLICATION
JOHN WILEY & SONS
New York · Chichester · Brisbane · Toronto · Singapore

Cover illustration by Beata Petek

Library of Congress Cataloging in Publication Data
Main entry under title:

Medical physiology.

   (A Wiley medical publication)
   Includes index.
    1. Human physiology. I. Brown, Arthur M., 1932-
II. Stubbs, Donald W. (Donald William) III. Series.
[DNLM: 1. Physiology. QT 104 M4891]
QP34.5.M47   1982   612   82-8585
ISBN 0-471-05207-8   AACR2

Printed in the United States of America

10 9 8 7 6 5 4 3 2

To our students—past, present, and future.

Contributors

R. David Baker, Ph.D.
Associate Professor
Department of Physiology and Biophysics
University of Texas Medical Branch
Galveston, Texas

Malcolm S. Brodwick, Ph.D.
Associate Professor
Department of Physiology and Biophysics
University of Texas Medical Branch
Galveston, Texas

Arthur M. Brown, M.D., Ph.D.
Professor and Chairman
Department of Physiology and Biophysics
University of Texas Medical Branch
Galveston, Texas

Raymond Y. Chan, M.D., Ph.D.
Postdoctoral Fellow
Department of Ophthalmology
Baylor College of Medicine
Houston, Texas

Burgess Christensen, Ph.D.
Associate Professor
Department of Physiology and Biophysics
University of Texas Medical Branch
Galveston, Texas

Douglas C. Eaton, Ph.D.
Associate Professor
Department of Physiology and Biophysics
University of Texas Medical Branch
Galveston, Texas

Wayne R. Giles, Ph.D.
Associate Professor
Department of Physiology and Biophysics
University of Texas Medical Branch
Galveston, Texas

M. Mason Guest, Ph.D.
Ashbel Smith Professor
Department of Physiology and Biophysics
University of Texas Medical Branch
Galveston, Texas

Charles E. Hall, Ph.D.
Professor
Department of Physiology and Biophysics
University of Texas Medical Branch
Galveston, Texas

Diana L. Kunze, Ph.D.
Associate Professor
Department of Physiology and Biophysics
University of Texas Medical Branch
Galveston, Texas

Lee E. Moore, Ph.D.
Professor
Department of Physiology and Biophysics
University of Texas Medical Branch
Galveston, Texas

Ken-Icha Naka, Ph.D.
Visiting Professor
Department of Physiology and Biophysics
University of Texas Medical Branch
Galveston, Texas
Professor
Department of Biology
Institute for Basic Biology
Okazaki, Japan

John E. Remmers, M.D.
Professor
Department of Physiology and Biophysics
    and Department of Medicine
University of Texas Medical Branch
Galveston, Texas

Aileen K. Ritchie, Ph.D.
Assistant Professor
Department of Physiology and Biophysics

University of Texas Medical Branch
Galveston, Texas

John M. Russell, Ph.D.
Associate Professor
Department of Physiology and Biophysics
University of Texas Medical Branch
Galveston, Texas

Donald W. Stubbs, Ph.D.
Professor
Department of Physiology and Biophysics
University of Texas Medical Branch
Galveston, Texas

Daniel L. Traber, Ph.D.
Professor
Department of Physiology and Biophysics
    and Department of Anesthesiology
University of Texas Medical Branch
Galveston, Texas

Preface

The purpose of this book is to provide students of health sciences with a solid understanding of modern medical physiology. Our clinical approach has been to combine expertise in the subject with simple, effective teaching techniques. To achieve this, each major area of subject matter has been written by an author or authors who are actively engaged in research or education in that area.

We also provide students with three valuable tools to assess their understanding of human physiology. Each chapter includes objectives, self-evaluation questions, and fully annotated answers. These tools allow students immediate feedback as well as an efficient system for review.

We have experimented with this approach for nine years in our teaching of health science students (medical, nursing and allied health)—each year furthering its refinement—and each year witnessing new success. We have tested our approach by our own examinations as well as by use of the Flex exam and the National Board of Medical Examiners, Part I in Physiology. Our students have performed well; they have demonstrated a high level of understanding and relative ease in handling physiological principles and clinical applications.

It gives us great satisfaction to present this validated approach to the study of modern medical physiology in book form—to share our efforts in overcoming some of the pitfalls that students encounter in this subject area. We would also like to take this opportunity to acknowledge the efforts of Ms. Dee Kelly, who provided immeasurable help in seeing that individual deadlines were met, in typing and proofreading the text, and in preventing minor revolts that might have taken the pleasure out of writing this book.

Arthur M. Brown
Donald W. Stubbs

Contents

4 SKELETAL MUSCLE   69
R. David Baker and Lee E. Moore

5 HEART MUSCLE   97
Wayne R. Giles and Arthur M. Brown

6 SMOOTH MUSCLE   117
Wayne R. Giles

7 ELECTRICAL EVENTS OF THE CARDIAC CYCLE AND THE ELECTROCARDIOGRAM 131
Daniel L. Traber and Donald W. Stubbs

8 THE NATURE OF THE CARDIAC PUMP 149
Daniel L. Traber

9 THE CARDIAC OUTPUT AND ITS MEASUREMENTS 159
Daniel L. Traber

17 MICROCIRCULATION: CAPILLARY DYNAMICS, LYMPH CIRCULATION, AND LOCAL VASCULAR REGULATION 293

M. Mason Guest

18 THE VOLUME AND ELASTIC PROPERTIES OF THE LUNG 315

John E. Remmers

19 AIRFLOW IN THE LUNGS 329

John E. Remmers

25 CONTROL OF BREATHING   399
John E. Remmers

26 THE PHYSIOLOGY OF EXERCISE   413
Diana L. Kunze

27 GASTROINTESTINAL PHYSIOLOGY: THE CONTROL MECHANISMS   427
R. David Baker

28 MOVEMENTS OF THE DIGESTIVE TRACT   439
R. David Baker

29   SECRETIONS OF THE DIGESTIVE SYSTEM   457
R. David Baker

30   INTESTINAL ABSORPTION OF WATER AND ELECTROLYTES   471
R. David Baker

33 THE PHYSIOLOGY OF WATER AND ELECTROLYTES   537
Donald W. Stubbs

34 THE PHYSIOLOGY OF ACID–BASE BALANCE   567
Donald W. Stubbs

35 HEMOFLUIDITY AND HEMOSTASIS   599
M. Mason Guest

36   ENERGY EXCHANGE AND TEMPERATURE REGULATION   617
M. Mason Guest

37   RECEPTORS   635
Burgess Christensen

38   THE SOMATOSENSORY SYSTEM   649
Burgess Christensen

39   THE MOTOR SYSTEM   665
Burgess Christensen

40 THE LIMBIC SYSTEM   689
Burgess Christensen

41 THE AUTONOMIC NERVOUS SYSTEM   695
Arthur M. Brown

42 THE HUMAN EYE   705
Ken-Ichi Naka and Raymond Y. Chan

43 THE AUDITORY AND VESTIBULAR SYSTEM   721
Ken-Ichi Naka and Raymond Y. Chan

44 TASTE AND OLFACTION   737
Arthur M. Brown

51   THE OVARIES AND FEMALE ENDOCRINE FUNCTION   875
Charles E. Hall

52   PREGNANCY, PARTURITION, AND LACTATION   889
Charles E. Hall

Medical Physiology

1

Movements of Molecules and Ions Across Cell Membranes

ARTHUR M. BROWN
JOHN M. RUSSELL

OBJECTIVES

After completion of this chapter, you should be able to

1. Account for the selective barrier properties of the cell plasma membrane toward ions and molecules.

2. Distinguish between equilibrium states and steady states.

3. Describe and account for each term in Fick's law of diffusion.

4. Relate the concentrations of ions or molecules in intra- and extracellular solutions to their concentrations in the plasma membrane.

5. Relate inward and outward fluxes across the plasma membrane.

6. Describe the factors relating osmotic pressure to solute concentration, including the reflection coefficient.

7. Discuss the relationship between tonicity and osmolarity.

8. Discuss the factors involved in ultrafiltration.

9. Describe each term in the Nernst and constant field equations.

10. Understand how the Gibbs–Donnan equilibrium accounts for the distribution of some ions across some cell membranes.

11. Understand the role of the sodium pump in the regulation of cell volume and in the double Donnan equilibrium system.

12. Discuss the mosaic membrane as it pertains to membrane pores, membrane carriers, and diffusion through the lipid phase.

13. Understand the notion of membrane pores and how their sizes are approximated.

14. Understand how mediated transport differs from simple diffusion.

15. Understand how the energy of adenosine triphosphate (ATP) can be used to cause net movements of substances across biological membranes.

The general objectives of this chapter are (1) to describe the distribution between the extracellular and intracellular compartments of certain physiologically important ions and molecules, (2) to examine the role of the cell membrane in establishing and maintaining these distributions, and (3) to introduce certain functional consequences of these ionic distributions.

INTRODUCTION TO THE CELL MEMBRANE

The cell membrane provides a selective barrier for the entry and egress of material essential to cellular function. It consists of lipids that provide a high energy barrier to the indiscriminate movement of polar substances such as water, ions, sugars, amino acids, and fatty acids and, at the same time, contains specialized proteins that lower energy barriers and control transport of these materials. The regulated operation of such membrane structures provides an optimum cellular environment for enzymes and for special functions, such as excitation, contraction, and secretion. The cell also has to interact with neighboring cells and, in a system such as the human body, with the products of remote cells. To do this, the membrane contains receptor proteins that recognize and react with these substances producing immunological responses, hormonal release, action potentials, and so forth. Specialized membrane functions are more developed in certain cells, which form aggregates or organs; these aspects are dealt with in the appropriate organ system chapter. For the moment, the following three features are emphasized: (1) the cell membrane acts as a barrier to the passive movement of most molecules and ions; (2) this barrier is selective; and (3) the membrane contains specific proteins, which are involved in the transport of some ions and molecules from outside the cell to inside the cell and vice versa.

Table 1 presents some interesting data with respect to features (1) and (2). Sodium, potassium, and chloride ions have similar hydrated diameters, yet they move across membranes much more slowly than they do in water. This characteristic is denoted by their diffusion coefficients, the diffusion coefficient being a constant that indicates rate of movement. It is defined more precisely on page 4. Thus the membrane acts as a barrier; moreover, it appears to be selective, since it is one-tenth to one-hundredth as permeable to sodium ions as it is to potassium ions. It is also true that molecules of approximately the same weight and diameter, such as glycerol and urea, move at entirely different rates through cell membranes. Moreover, the diffusion coefficients of most molecules in water do not differ by more than a factor of 10, but the diffusion coefficients across cell membranes can differ by factors of up to 1 million. For example, water molecules move through cell membranes about 100 times more slowly than in an aqueous solution, whereas calcium ions travel at rates 1/100-millionth their rate in an aqueous solution.

There is no great mystery about the barrier properties of the membrane. The interior of the membrane is hydrophobic or nonpolar, since it is largely lipid in nature and contains little or no electrical charge. The terms hydrophobic, lipophilic, and nonpolar are used interchangeably in this section, as are the terms hydrophilic, lipophobic, and polar. Therefore, polar or charged substances in the extracellular and intracellular fluids, such as water, ions, urea, glucose, and amino acids, encounter energy barriers that restrict translocation. However, protein molecules are embedded in the membrane at numerous locations. The outside surface of these proteins in contact with the lipids of the membrane is mainly lipophilic, but their interior or core can be charged. For appropriate polar substances, it is possible for the energy barriers to be lowered sufficiently by these proteins in order for translocation to occur.

Evidence that membranes transport certain ions into cells and that they transport other ions out of cells is presented in Table 2. Distributions of monovalent ions across the plasma membranes of a number of different animal cells are given. It can be seen that the distributions of sodium and potassium ions are unequal, with large quantities of sodium outside the cell (extracellular) and very little sodium inside (intracellular), and vice versa for potassium.

Since the cell interior is electrically negative with re-

Table 1 Characteristics of the Cell Membrane Barrier

Ion	MW	Hydrated Diameter (Å)	D, $cm^2\ s^{-1}$ Aqueous ($\times 10^{-5}$)	D, $cm^2\ s^{-1}$ Membrane ($\times 10^{-5}$)
Na$^+$	22.98	5.0	1.8	0.0001–0.00001
K$^+$	39.10	4.0	1.8	0.001
Cl$^-$	35.45	4.0	1.8	0.001–0.0001

[a]MW = molecular weight; Å = angstrom = 10^{-10} m, 10^{-8} cm (m = meters, cm = centimeters); D = diffusion coefficients in $cm^2\ s^{-1}$ determined by a method using the Fick equation described on page 00.

Table 2 Monovalent Ion Concentrations in Various Tissues (in mEq/L)

Compartment	Animal Tissue	Na^+	K^+	Cl^-
Blood plasma (extracellular)	Human	142	5.5	103
	Rat	152	5.9	112
	Frog	112	2.5	90
	Squid	350	17	470
Erythrocyte (intracellular)	Human	19	136	78
Muscle (intracellular)	Rat	16	152	5
Nerve (intracellular)	Squid	48	406	110

spect to the exterior, one would expect positively charged particles such as K^+ and Na^+ to be somewhat concentrated inside the cell. Clearly, this is not the case for Na^+; hence we say it is actively extruded from the cell. The precise equilibrium relationship between the cellular electrical potential and ion distribution between the inside and the outside of cells is dealt with more fully under the section entitled the Nernst equation.

In addition to the ion distributions presented in Table 2, other permeant substances, such as amino acids, are also distributed away from equilibrium across the plasma membranes of most cells. However, a variety of molecules and ions, including water, glucose, urea, and many organic acids and bases, are distributed to equilibrium between the intra- and extracellular spaces. A substance such as glycerol, for example, when added to blood, diffuses across the plasma membrane and reaches a concentration within the cell equal to that outside.

To summarize, we find that most molecules and ions move across, or permeate, cell membranes much more slowly than they move in free solution. Moreover, the membrane is a selective barrier in hindering these movements. But we also know that cell membranes cause cells to accumulate or extrude certain molecules and ions actively; in other words, many substances are not at equilibrium across plasma membranes. The sodium ion distribution shown in Table 1 is a classical example. However, the fact that the intracellular sodium concentration is low inside the cell does not arise, because the membranes themselves are *impermeable* to this ion. The membranes are, in fact, *permeable* to this ion, that is, they permit the Na^+ to pass through them, and energy must be used to maintain its low intracellular concentra-

tion. It is held that movement of substances across biological membranes involves one of two processes: (1) passive transport, or (2) energy-dependent active transport. We will consider each of these processes in turn, but before doing so we will digress briefly to more general concepts that will lay the groundwork for an understanding of how the differences between intra- and extracellular fluids are maintained.

CONCEPT OF EQUILIBRIUM STATE VERSUS STEADY STATE

Living cells may be characterized by their ability to transform energy and perform work. Since the composition of the intracellular fluid does not vary significantly with time, living cells are said to be in a *steady state* with respect to their environment. Living cells are *not at equilibrium* with their environment; they must continuously take up energy from the environment in order to remain in the steady state.

The distinction between equilibrium and steady state can be represented graphically, as in Figures 1–3. With the

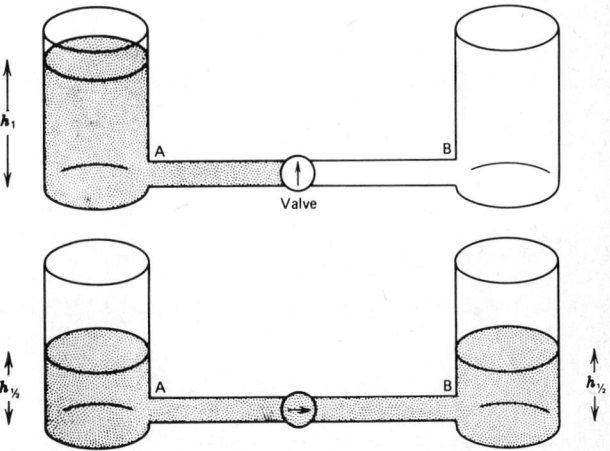

Figures 1 and 2 State of equilibrium for two connected water reservoirs. Note equal fluid levels after valve was opened.

valve shut, the water is at height h_1 in chamber A. When the valve is opened, as in Figure 2, the water runs out of A into B until the two levels are equal. A and B are in *equilibrium* with respect to these water levels. If, however, a pump is connected to B so as to pump water continually from B to A and keep the water levels unequal, but at a steady flow, as in Figure 3, then A and B are in a *steady state* with respect to these water levels. In this system energy is supplied to the pump as electricity and is converted into mechanical work, delivering the water to a higher level. The pump will heat up because of the frictional resistance of its moving parts, and the heat will be transferred to the air or fluid surrounding the pump. Extending the analogy to the living structure, we might expect that in the steady state energy made available from chemical reactions within the cell will appear as work and heat. For the living cell to remain in the steady state, it must therefore take up energy from the environment, do the work required of it, and return heat to the environment.

The constant low level of intracellular Na^+ concentration is the net result of two opposing membrane processes. The Na^+ ion present in much larger concentrations outside cells tends to leak passively into the negatively charged cells. Because this movement is passive, it does not require metabolic energy. This passive process tends to increase the intracellular concentration of Na^+. However, in the living cell the intracellular Na^+ concentration is kept low because the cell membrane actively transports

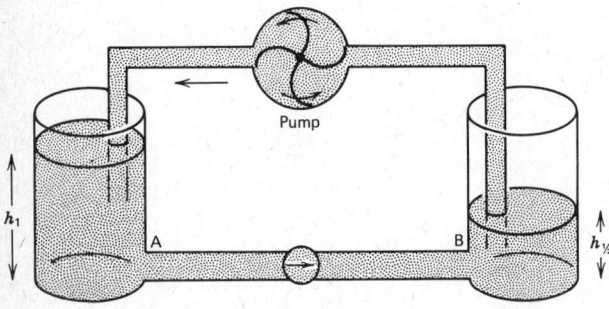

Figure 3 Steady state for two connected reservoirs. Note different levels maintained by the pump's work.

the Na^+ ions that have leaked into the cell back into the extracellular fluid. This movement is an active process, requires metabolic energy, and maintains a steady state of low intracellular Na^+ concentration. These two Na^+ processes are considered in greater detail in a subsequent section.

We now turn to a more general consideration of passive or non-energy-requiring movements of molecules or ions across the cell membranes, and then to mediated movements, which can be passive or active, the latter requiring energy.

PASSIVE MOVEMENT OF MOLECULES AND IONS ACROSS CELL MEMBRANES

The net movement of substances from one location in space to another is governed by a process called diffusion. The basic law of diffusion was first derived by Adolph Eugene Fick and is known as *Fick's law of diffusion.* The Fick principle states that the net movement of a substance, for example, a nonelectrolyte, taken as ds/dt in moles per second (mol/s), is equal to the product of the cross-sectional area A in cm^2 across which the diffusion is occurring, the concentration gradient driving its net movement, dc/dx in mol/cm^4, and a *diffusion coefficient, D* in cm^2/s. Hence

$$\frac{ds}{dt} = DA \frac{dc^*}{dx} \qquad (1)$$

For any substance, the greater its difference in concentration from one place to another, the faster its rate of diffusion; the larger the surface area over which it moves, the greater will be its rate of net movement. This expres-

---

*In order to be mathematically correct, a negative sign $(-)$ should be inserted in front of the right-hand side of the equation to account for the fact that the direction of the net solute movement is the opposite of the direction of concentration gradient, that is, diffusion occurs down the concentration gradient.

sion is for one-dimensional diffusion, across a plane surface. In three-dimensional space, the diffusion equation becomes more complicated.

For gas molecules, the diffusion coefficient D is inversely proportional to the square root of the molecular weight; that is, the larger the molecule, the slower its rate of diffusion. This relationship also holds approximately for the diffusion of some small spherical molecules in solution. For molecules that are large in comparison with the solvent molecules, such as spherical colloidal particles, the diffusion coefficient also becomes smaller as the molecular radius increases or the solution becomes more viscous, but the relationship varies with molecular shape and other factors.

In applying the diffusion law to thin membranes, such as biological membranes, we treat the membrane as a uniform phase and make the simplifying assumption that the concentration gradient *across the membrane, dc/dx* is constant and can therefore be expressed as

$$\frac{dc}{dx} = \frac{C'_o - C'_i}{a} \tag{2}$$

where a is the membrane thickness and C'_o and C'_i are the concentrations of substance C *in the membrane,* near its outer and inner surfaces, respectively. The concentrations of molecules within the membrane differ from those in the aqueous phases on either side of the membrane, partly because of the high lipid and low water concentrations of most membranes. Comparatively high concentrations of a substance in the membrane indicate that the molecule is lipophilic. Lower concentrations indicate that the molecule is lipophobic or hydrophilic (Fig. 4). Neutral anesthetics are highly lipophilic molecules, and their action on nerve cell membranes is attributable in part to this property. The concentration of C in the membrane phase (C') is related to that in the aqueous phase by the *partition coefficient B:*

$$B = \frac{C'_o}{C_o} = \frac{C'_i}{C_i} \tag{3}$$

where C_o and C_i are the concentrations of substance C in the solutions just outside the membrane at its two surfaces. Substitution of expressions (2) and (3) into the diffusion equation gives

$$\frac{ds}{dt} = \frac{DBA}{a}(C_o - C_i) \tag{4}$$

$$\frac{ds}{dt} = PA(C_o - C_i) \tag{5}$$

where DB/a is known as the *permeability constant P* and has the dimensions of velocity, that is, $cm^2 \ s^{-1} \div cm = cm \ s^{-1}$. Note that B is dimensionless.

It is clear from the above that the rate of diffusion across cell membranes will increase for substances having (1) higher partition coefficients (i.e., higher lipid solubilities), (2) larger concentration differences, and (3) higher diffusion coefficients within the cell membrane. In addition, morphological features that increase surface area relative to volume—denoted as the greater surface area/volume (SA/V) ratios—will favor faster diffusion. The structure with the least favorable SA/V is a sphere, and the most favorable is a sheet with folds, such as respiratory alveoli or intestinal villi. Since the respiratory alveoli and intestinal villi have as their main function the exchange of oxygen and nutrients across their respective surfaces, one can appreciate the optimum nature of their design. So far we have not considered the time required for a substance such as CO_2 to diffuse from the center of a cell to the plasma membrane. This time is roughly proportional to the square of the radius. Moreover, we have treated only the simpler case of diffusion across a plane, whereas diffusion in reality occurs in three-dimensional space. Further read-

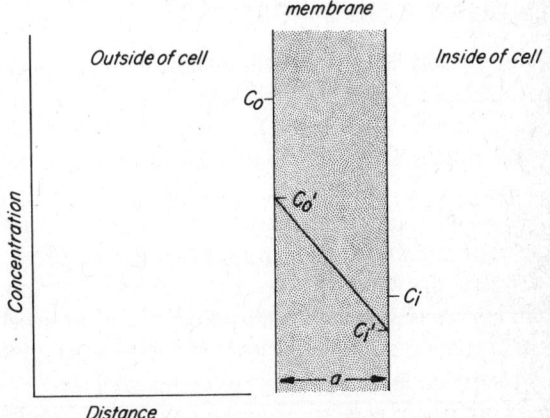

Figure 4 Diffusion profile of a hydrophilic molecule across a membrane.

ing on these aspects of diffusion is provided in the first three references in the Selected Bibliography at the end of this chapter.

GENERALITY OF THE DIFFUSION LAWS

In the foregoing discussion we have considered the basic concepts of diffusion primarily from the point of view of transfer of solute across membranes. It should be apparent, however, that the concepts apply equally well to the movements of particles in a gas or liquid phase. For example, the same principles govern the diffusion of sucrose through a beaker of water, when sucrose crystals are added to pure water. Similarly, these principles can be applied to the transfer of heat in solids; if one end of a metal block is heated, the law of diffusion governs the transfer of heat to the rest of the block. In humans heat exchange with the environment is very important.

On the cellular level, once an amino acid crosses the plasma membrane into the cell, it must diffuse to the ribosomes to be incorporated into proteins. The same diffusion laws we discussed govern this intracellular movement as well. Another important application of the diffusion laws relates to the movements of CO_2 and O_2 in the lungs, in the blood, and in the tissues.

WATER MOVEMENT ACROSS CELL MEMBRANES—OSMOSIS

Water is the solvent in which the chemical processes of life are carried out. It is therefore necessary that we understand the factors responsible for water movement across cell membranes. In general, two kinds of forces cause water molecules to travel across cell membranes. One is simply a hydrostatic pressure difference similar to that which causes water to flow from A to B in Figure 2. In the body the hydrostatic pressure in the capillaries (about 20 mm Hg) causes water to flow across pores in the capillary endothelium. However, hydrostatic pressure differences across animal cells are usually insignificant.

A second force acts on water to cause it to cross cell membranes. This force is called *osmotic pressure,* and the net movement of water in response to this pressure is termed *osmosis.* In order to understand the concepts of

osmotic pressure and osmosis it is helpful to consider first a simple, ideal situation.

Suppose we have a vessel (Fig. 5) that is initially divided into two 1-liter (L) compartments by a piston, the head of which is permeable to water, but not to glucose. The left-hand side initially contains 2 M glucose, whereas the right side contains 1 M glucose. What will be observed is that at equilibrium enough water will have moved from the right-hand compartment to the left-hand compartment to make the glucose concentration 1.5 M on both sides (Fig. 5b). The mechanical force that would have had to been applied to the piston in the left-hand compartment to prevent net water flow into that compartment is the osmotic pressure. The force involved is not insignificant. In the example depicted in Figure 5, it is 19,300 mm Hg, or 25.4 atmospheres. It is important to note explicitly that according to the definition of osmotic pressure and the way in which it is measured, water flows from a region of *low* osmotic pressure to a region of *high* osmotic pressure. However, it is still moving down its concentration gradient from a region of higher to one of lower water concentration.

Osmotic pressure is one of the colligative properties of solutions that depend on the number of particles in the solutions. Other such properties are ability to lower vapor pressure, to depress the freezing point, and to elevate the boiling point, all of which are mathematically related to osmotic pressure. The relationship between osmotic pressure and the number of solute particles was formalized by

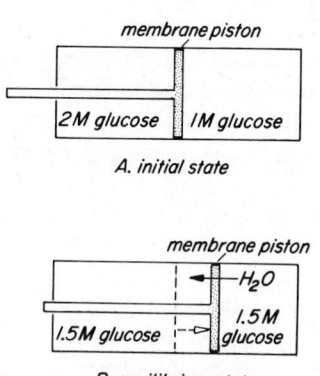

Figure 5 Effects of osmotic pressure. (Redrawn from Davson H: *A Textbook of General Physiology,* ed 3. Boston, Little, Brown, 1964.)

the Dutch chemist Jacobus Hendricus van't Hoff. According to the *van't Hoff equation*

$$\Delta\pi = RT(C_L - C_R) \qquad (6)$$

where $\Delta\pi$ is the osmotic pressure difference (in atmospheres), T is the absolute temperature (°K), R is the universal gas constant,* 0.082 L atm $\cdot$ °K^{-1} $\cdot$ mol^{-1}, and C_L and C_R are the concentrations of the solute (mol/L) on the left and right sides, respectively.

One can also speak of the osmotic pressure of a single solution with reference to pure water.

$$\pi = RTC \qquad (7)$$

Differences in osmotic pressure need not necessarily lead to net movements of water, the process of osmosis, if there are counterbalancing forces. As you will see in studying the circulatory system, such interrelationships between hydrostatic and osmotic pressures are very important.

We have already noted that the osmotic pressure of a solution depends on the number of particles in solution. The identity of these particles is essentially unimportant. Thus, 10 glucose particles will have the same osmotic effect as that of 10 sucrose particles. But remember that some substances, salts, acids, and bases, ionize into separate particles or ions when placed into an aqueous solution and that each of these separate particles or ions contributes to the osmotic pressure of the solution. The unit of concentration of osmotically active particles is called the *Osmole,* and the osmotic strength of a solution is expressed in Osmoles per unit volume of solvent—usually called the osmolarity. The ideal relationship between molarity and osmolarity is straightforward; 1 mol of glucose is 1 Osm. One mole of NaCl is ideally 2 Osm, since it dissociates into Na$^+$ and Cl$^-$ ions. A 1 molar (M) solution of a salt such as CaCl$_2$ is therefore 3 Osm. Thus, the van't Hoff relation is more correctly

$$\pi = RT \cdot \text{Osm} \qquad (8)$$

---

*The universal gas constant R is used because the osmotic pressure of a solution varies with concentration and temperature in the same manner as does pressure of an ideal gas. R can be expressed in several different ways, depending on the units employed for pressure and volume. For example, if pressure is in dyn/cm^2 and the volume in cm^3, R can be given as 8.32×10^7 ergs $\times$ °K^{-1} $\times$ mol^{-1}. Furthermore, because 1 joule = 10^7 ergs, R could also be expressed as 8.32 J $\times$ °K^{-1} $\times$ mol^{-1}.

where Osm is osmolarity. A somewhat more useful unit is the milliosmol (mOsm). A 1-mOsm difference between two solutions can result in an osmotic pressure of 19.3 mm Hg (0.025 atm).

The conditions under which the van't Hoff formula is derived require that in order for it to be absolutely correct, the solution must be quite dilute, certainly less than 100 mM; otherwise errors will be made in calculating the osmotic pressure from the ideal osmolarity. These errors arise because as an aqueous solution becomes more concentrated, the behavior of the individual particles becomes less than ideal. For most substances, the results of the interactions between the water and solute molecules and among the solute molecules themselves are such that the number of *effective* particles is less than that calculated. In the case of ionizable substances such as salts, this could be a case of incomplete ionization, although a complete understanding of this less-than-perfect osmotic behavior is not yet in hand. Nevertheless, it has been possible to measure the osmotic pressure developed by single-solute solutions over a range of concentrations. Thus, by comparing the osmotic pressure actually measured with that predicted, one can get a correction factor that takes into account the nonideal behavior of the solute. This correction factor is called the osmotic coefficient ϕ.

$$\phi = \pi_{\text{actual}} / \pi_{\text{predicted}} \qquad (9)$$

Therefore, to calculate the actual osmotic pressure for a nondilute solution using the van't Hoff formula,

$$\pi_{\text{actual}} = RT\phi\text{Osm} \qquad (10)$$

In order for solute particles to exert an osmotic force at osmotic equilibrium, an osmotic gradient must exist. If the solute molecules can freely cross the membrane and be distributed on both sides of the membrane at the same osmolarity, they will exert no net osmotic force at equilibrium. Remember,

$$\Delta\pi = RT\phi \, (\text{Osm}_1 - \text{Osm}_2) \qquad (11)$$

In the case of a passive membrane (i.e., one that cannot actively transport the solute in question, see page 20) the solute must be totally incapable of permeating the membrane to exert any osmotic force *at equilibrium.* This is because at equilibrium, the concentration of a permeant substance would be the same on both sides, assuming an uncharged solute or no membrane potential. As we will see in a later section of this chapter, there are a number of permeant substances the intracellular concentrations of which are maintained at a steady level by an energy-requiring process called *active transport.* Thus, although these substances are permeant and can cross the membrane, their relative concentrations on the two sides of the membrane are relatively invariant with time, that is, they are in steady state, and can thus induce an osmotic pressure difference. This is particularly important for the sodium ion, as we shall see.

Another complication arises when dealing with osmotic pressure and biological cells, that is, when a substance is neither perfectly impermeant nor actively transported. Such a substance can exert an osmotic force, but its effects are necessarily transient because, with time, it will distribute across the membrane in accordance with the passive forces acting on it. Moreover, water can cross biological membranes faster than any solute, which also reduces the differences in concentration and the osmotic pressure difference. This is associated with a change in volume or pressure, or both. How large the transient difference might be will depend on the relative rates of water and solute transit and on the mechanical properties of the structure separating the two solutions. This problem is taken care of by another correction factor known as the reflection coefficient σ, a property characteristic for a given solute and a given membrane. This factor varies between 1, for a perfectly impermeant solute or an actively transported solute in the steady state, to 0 for a substance as permeant as water. Thus

$$\Delta\pi = \sigma RT\phi \, (\text{Osm}_1 - \text{Osm}_2) \qquad (12)$$

Note that if a substance is permeant, its osmolarity on either side of the membrane will be changing with time because of its diffusion until it is at equilibrium or at steady state, if actively transported.

All body cells are essentially in osmotic equilibrium with blood plasma. Blood plasma has an osmolarity of about 286 mOsm, hence any fluids introduced into the vascular system should usually have an osmolarity of 286 mOsm. Such fluids are said to be iso-osmotic with our cells. The osmolarity of a given solution is easily determined by means of an osmometer. This instrument simply measures the number of osmotically active particles and has no membrane across which an osmotic pressure is

actually developed. The osmometer cannot determine whether these particles will be able to cross our cell membranes. Therefore, knowing the osmolarity of a fluid is only the first step in determining whether it is osmotically compatible with our cells. The next step is to determine whether the solution is also *isotonic,* that is, whether cells bathed by the solution maintain their normal size or retain their normal tone. A solution could be iso-osmotic; yet if the solute can rapidly enter the cells, it will cause water to enter resulting in a cellular swelling. Such a solution acts as though it is more osmotically dilute than the cells; it is termed *hypotonic.* By contrast, a *hypertonic* solution causes water to leave cells, resulting in cell shrinkage. Such tonicity determinations are most conveniently made using red blood cells, the relative volume changes of which can easily be measured as changes in hematocrit values. If swelling is severe enough to cause lysis (hemolysis), the appearance of the red hemoglobin in the solution can be used to make a determination.

If red blood cells are suspended in a solution of 155 mM NaCl, they neither swell nor shrink. This tells us that such a solution is iso-osmotic with the intracellular solution and that it is also isotonic, since no change in cell size occurs when the cells are in contact with this solution. Nevertheless, even though a 280 mM solution of glycerol has the same osmolarity as the 155 mM NaCl solution (286 mOsm), human red blood cells bathed in this medium will rapidly swell and burst. On the other hand, rabbit or bovine erythrocytes are unaffected by the glycerol solution. Thus, the glycerol solution is iso-osmotic to the red blood cell water, but is *hypotonic* with respect to human red blood cells and isotonic to rabbit or bovine red blood cells. This difference is the result of a specialized transport system that produces facilitated diffusion (see page 19), which moves glycerol into human red blood cells, but which is absent from the other two types of erythrocytes. Thus, in human red blood cells glycerol enters the cells, raising the intracellular osmolarity, and thereby forcing water to enter. This process continues until the cell bursts.

Since there are no known instances of active water transport across cell membranes, the only forces available to transport water across membranes in our bodies are hydrostatic pressure (capillary pressure) or osmotic pressure. You will see that osmotic forces play a key role in the absorption of water by the intestine, and in its secretion into the intestine in certain disease states. The kidneys use both types of forces to perform their primary task of maintaining body fluid and electrolyte balance. Water balance across blood capillaries depends critically on the relationship between blood pressure and osmotic pressure; slight imbalances can lead to disastrous results.

ULTRAFILTRATION

We described the osmotic pressure in terms of the mechanical pressure required to prevent fluid from flowing downhill from a high *solvent* concentration (dilute solution) to a low solvent concentration (concentrated solution). If not prevented from doing so by externally applied forces, water will flow from dilute to concentrated solution until the concentrations of the two solutions are equal (Fig. 5*b*). By contrast, water can be made to flow between solutions of equivalent osmotic pressure when hydrostatic pressure is applied to one of the solutions, a process known as *ultrafiltration*. In the absence of an osmotic pressure difference, the rate of ultrafiltration is given by

$$J^{H_2O} = P_H A (HP_L - HP_R)$$

where J^{H_2O} is the net flux of water, in L/s, P_H is the hydraulic permeability constant, and HP_L and HP_R are the hydrostatic pressures on either side of the membrane, and A is the membrane area. As noted earlier, large hydraulic pressure differences are not maintained across the cell membranes of animals. However, ultrafiltration is an important phenomenon in capillary physiology and is involved in the first stage of urine formation in the glomerulus. The force that drives fluid out of the proximal ends of the capillaries is derived from a combination of the pressure–volume characteristics of the circulation and the contraction of the heart and is discussed further in the chapter on hemodynamics of microcirculation (Chapter 10).

PASSIVE MOVEMENTS OF ELECTRICALLY CHARGED PARTICLES

Up to this point we have not considered electrical forces in our discussion of the movement of solutes across membranes. Body fluids are solutions containing large numbers of electrically charged particles or ions. The movement of

these ions across membranes can create electrical potential differences between the compartments separated by these membranes. Conversely, ion movements between compartments will be influenced by the existing electrical potential difference. Let us now consider the simplest case.

DIFFUSION POTENTIALS

Suppose we have a system containing two compartments separated by a porous membrane (Fig. 6). Into one compartment we shall put 1.0 M NaCl and into the other compartment 0.1 M NaCl. Let us now measure the electrical potential difference between the two compartments, assuming that the compartments are large enough so that neither the volumes nor the concentrations of electrolytes change appreciably during the course of the measurements. It turns out that at 20°C we would find the low concentration compartment about 12 mV with respect to negative to the high concentration compartment. This potential difference results from the fact that Cl ions in solution have a mobility about 1.5 times greater than that of Na ions; that is, they tend to diffuse more rapidly. Because particles tend to move from a region of high concentration to one of low concentration, and because Cl ions diffuse more rapidly than do Na ions, there will be a very small net accumulation of Cl ions in excess of Na ions on the low concentration side of the porous membrane. The Na ions lag behind as a small net accumulation of positive ions on the high concentration side of the porous membrane. It should be noted that any difference in concentration of

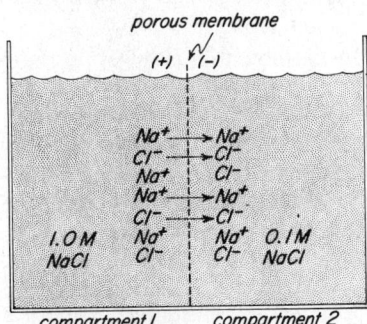

Figure 6 Development of diffusion potential. Note the slight excess of Cl⁻ over Na⁺ ions immediately adjacent to the membrane on side 2 attributable to the greater mobility of Cl⁻ ions. The result is the electrical negativity (−) of side 2.

Na^+ and Cl^- will be limited to the microenvironment immediately adjacent to the membrane; in the bulk of the solutions macroscopic electroneutrality will prevail. The magnitude of the resulting potential difference, known as the *liquid-junction potential* or the *diffusion potential*, can be calculated from the following formula, which is known as the *Henderson equation:*

$$E = \frac{RT}{F}\frac{u - v}{u + v} \ln \frac{[C]_1}{[C]_2} \qquad (13)$$

where E is the diffusion potential (potential in compartment 2 minus potential in compartment 1) in mV; R is the universal gas constant; T is the absolute temperature in °K; F is Faraday's number; u and v = the sodium and chloride *ion mobilities,* respectively; and $[C]_1$ and $[C]_2$ are the salt concentrations in the high concentration and low concentration compartments, respectively. R has dimensions of J/mol/°K, T = °K, and F = C/mol, so that RT/F is joules/coulomb or volts. The mobility and concentration ratios are dimensionless. At 20°C RT/F has a value of about 25 mV. As Cl ions have a mobility 1.5 times that of Na, the mobility factor $(u - v/u + v) = -0.2$. Converting from natural logarithms to common logarithms, we use the factor 2.3, so that E = 25 mV $\times$ $-0.2 \times 2.3 \times \log 10 = -12$ mV. This liquid-junction potential persists only as long as the concentration differences are maintained.

Let us now suppose that instead of a simple porous membrane separating our two compartments we have a membrane that is selectively impermeable to cations. Using the same formula, we now find that u drops out, because Na ions cannot pass across the membrane. Thus the mobility factor reduces to $-v/v = -1$. And in this case, the diffusion potential is given by the *Nernst equation:*

$$E = \frac{-RT}{F}\ln \frac{[C]_1}{[C]_2} = -58 \log \frac{1.0}{0.1}$$
$$= -58 \text{ mV} \qquad (14)$$

The Nernst equation for electrochemical equilibrium of a charged substance is considered in more detail in the following section.

THE NERNST EQUATION

In the absence of active transport (see below), only two physical processes are generally involved in forcing a net movement of ions across cell membranes: (1) diffusion down a concentration gradient, and (2) migration along an electrical potential gradient. From physical chemistry we know that the energy available per mole for diffusion down a chemical gradient from outside to inside is $RT \ln C_o/C_i$, where C_o and C_i are the effective concentrations outside and inside the cell, respectively. The energy available per mole for electromigration (i.e., diffusion down an electrical gradient) from inside to outside is zFE, where z is the valence of the ion. At equilibrium, that is, when there is no net energy available for movement of the particular ion in question

$$RT \ln \frac{C_o}{C_i} = zFE_{eq} \qquad (15)$$

where E_{eq} is the membrane potential during this state of equilibrium. This equation merely points out that when a permeant ion is distributed at equilibrium across a cell membrane, there is no longer any net movement in either direction, because the energy available from the concentration difference is precisely balanced by the energy available from the electrical potential difference. If this equation is solved for the *equilibrium potential E_{eq}*, that is, the specific potential difference that will precisely balance the concentration difference, we obtain

$$E_{eq} = \frac{RT}{zF} \ln \frac{[C]_o}{[C]_i} \text{*} \qquad (16)$$

Equation (14) is known as the Nernst equation. With it we can predict, the concentration ratio at a measured membrane potential or the membrane potential that will balance a known concentration ratio, for any ion we know to be in electrochemical equilibrium across a membrane. However, if we measure both the steady state concentration ratio and membrane potential and find that they are not quantitatively related by the Nernst equation, we can safely conclude that the ion is not distributed at equilibrium and that it is therefore either being actively trans-

---

*At 20°C $\dfrac{RT}{zF} \ln \dfrac{[C]_o}{[C]_i} = \dfrac{58}{z} \log \dfrac{[C]_o}{[C]_i}$,   at 37°C $= \dfrac{61}{z} \log \dfrac{[C]_o}{[C]_i}$

ported, or the membrane is completely impermeable to the ion in question. Taking the first and more interesting case, if Na_o^+ is 140 mM and Na_i^+ is 20 mM for a skeletal muscle cell, then at 37°C the Nernst equation will predict that there must be a membrane potential of +52 mV if Na^+ is passively distributed at equilibrium, that is, the equilibrium potential for Na^+ (E_{Na}) is +52 mV. Now, if the membrane potential is actually measured and found to be, say, −90 mV—a reasonable value for skeletal muscle fibers—it is clear that some process other than diffusion and electromigration is influencing the distribution of Na^+. This process is assumed to be active transport.

The situation with biological membranes is considerably more complicated. Several different ions are present, and cell membranes are rarely perfectly selective for any one of them. A more complicated form of the Henderson equation (13) can be used for diffusion potentials across cell membranes, but more often another equation, derived by Goldman in 1943 and in another form by Hodgkin and Katz in 1949, is employed. This equation assumes that the potential field in the membrane is constant and is known as the *constant field* or *Goldman–Hodgkin–Katz (G–H–K) equation*. If, as is frequently the case, only Na^+, K^+, and Cl^- are contributing to the diffusion potential, the G–H–K equation takes the following form:

$$E = \frac{RT}{F} \ln \frac{P_K[K]_o + P_{Na}[Na]_o + P_{Cl}[Cl]_i}{P_K[K]_i + P_{Na}[Na]_i + P_{Cl}[Cl]_o} \qquad \textbf{(17)}$$

where E is the membrane potential (inside minus outside); K_o, Na_o, and Cl_o are extracellular (outside) concentrations; K_i, Na_i, and Cl_i are intracellular (inside) concentrations; and P_K, P_{Na}, and P_{Cl} are the respective membrane permeabilities. This equation requires only that the system be in a steady state, not in equilibrium. Equation (17) is widely used in electrophysiologic determinations. If the membrane happens to be permeable to only one of these ions, then Eq. 17 reduces to the same form as Eq. (16), as was also the case for the Henderson equation (13).

It should also be clear from examining the G–H–K equation that whenever the permeability to one ion becomes much greater than the permeabilities to the other ions, the actual membrane potential E approaches the equilibrium potential (E_{eq}) for that ion. This phenomenon is of considerable importance for excitable membranes.

GIBBS–DONNAN EQUILIBRIUM

We have discussed osmotic pressure in terms of a system in which two compartments were separated by a membrane permeable only to water. Let us now complicate the situation by changing our semipermeable membrane into an animal cell membrane that is permeable both to small anions and cations as well as to water, but let us keep it impermeable to large protein (Pr) molecules, which carry a net electrical charge, as depicted in Figure 7. Into the compartments on the right we put a solution of 0.1 M KCl, and into one on the left, which can be considered to approximate the intracellular fluid, we put a solution of 0.1 M K proteinate ($= K^+ Pr^-$, where for the sake of convenience we will assume that the proteinate ions are monovalent). The situation is not very different from that which applies to living cells impermeable to proteins. We will want to see what happens at equilibrium when the KCl has had time to diffuse across the membrane. To simplify the situation further, we will apply pressure to the left-hand or internal solution to prevent volume changes as KCl diffuses from the outside to the inside of the cell membrane.

It turns out that the side of the membrane that has the impermeant ion (inside in Fig. 7) also accumulates more of the permeant ions. The difference in Cl^- concentrations leads to a net flux of Cl^- from outside to inside. In order to satisfy the preservation of overall electroneutrality, a phenomenon referred to as "macroscopic electroneutrality," the K^+ concentration on the inside becomes greater

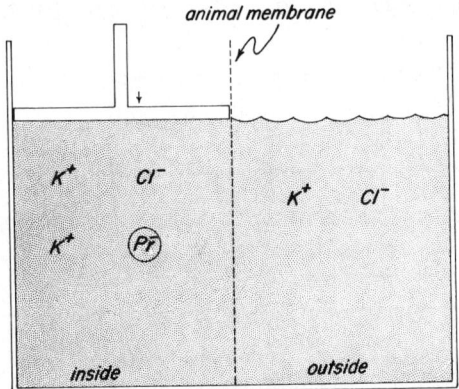

Figure 7 Conditions for Gibbs–Donnan equilibrium.

than that on the outside. Although Cl^- diffuses down its concentration gradient from outside to inside, its concentration at equilibrium is smaller inside than it is outside. As $K_i \neq K_o$ and $Cl_i \neq Cl_o$, a diffusion potential arises between the two compartments, such that applying the Nernst equation gives

$$E_K = \frac{RT}{zF} \ln \frac{[K]_o}{[K]_i} \qquad z \text{ is } +1 \text{ for } K^+ \qquad \textbf{(18)}$$

and

$$E_{Cl} = \frac{RT}{zF} \ln \frac{[Cl]_o}{[Cl]_i} \qquad z \text{ is } -1 \text{ for } Cl^- \qquad \textbf{(19)}$$

However, there can only be a single potential difference across the membrane. Hence, potassium and chloride will continue to move until $E_K = E_{Cl}$. At this point the system is in equilibrium. Therefore

$$\frac{RT}{z_K F} \ln \frac{[K]_o}{[K]_i} = \frac{RT}{z_{Cl} F} \ln \frac{[Cl]_o}{[Cl]_i} \qquad \textbf{(20)}$$

Thus

$$\frac{[K]_o}{[K]_i} = \frac{[Cl]_i}{[Cl]_o} \qquad \textbf{(21)}$$

or

$$[K]_i \times [Cl]_i = [K]_o \times [Cl]_o \qquad \textbf{(22)}$$

This is the *Donnan rule,* which states that at equilibrium, the product of the concentrations of the diffusible ions in one compartment is equal to the product of the concentrations of the diffusible ions in the other compartment. This is also called the Donnan equilibrium. Note that

$$[K]_o = [Cl]_o \qquad \textbf{(23)}$$

and

$$[K]_i > [Cl]_i \qquad \textbf{(24)}$$

Now

$$[K]_i \times [Cl]_i = [K]_o \times [Cl]_o \qquad \textbf{(25)}$$

Therefore

$$[K]_i + [Cl]_i > [K]_o + [Cl]_o \qquad \textbf{(26)}$$

As an example, let

$$K_i = 8 \quad Cl_i = 2 \quad K_o = 4 \quad Cl_o = 4$$

then

$$8 + 2 > 4 + 4$$

Therefore

$$[K]_i + [Cl]_i + [Pr]_i > [K]_o + [Cl]_o \qquad \textbf{(27)}$$

that is, the osmotic concentration inside is greater than that outside the cell. Thus, if the constant cell volume constraint were removed from the system, water would move from the extracellular compartment (o) to the intracellular compartment (i). This would change the ionic concentrations in the two compartments and so disturb the Donnan equilibrium. Hence more potassium chloride would move from the extracellular to the intracellular compartment (o to i), again upsetting the osmotic equilibrium. These processes would continue until the concentration of Pr, the impermeant protein anion inside the cell, was infinitessimal and the concentrations of potassium and chloride were equal throughout the system, or as would be more likely, the cells burst. We prevented water flow in Figure 7 by applying a hydrostatic pressure equal to the osmotic pressure that would develop inside the cell. But as we have seen, animal cells, unlike plant cells, cannot sustain much of a hydrostatic pressure gradient across their membranes without bursting. How do animal cells overcome this problem?

Consider a similar system (Fig. 8) in which the membrane is impermeable to sodium ions and in which some sodium chloride is added to the outside compartment. This solution now approximates the extracellular fluid. This is referred to as a *double Donnan equilibrium* system because an impermeant ion is present in the solution on either side of the membrane, that is, Pr^- and Na^+. Although Na^+ is, in fact, slightly permeant, the Na^+–K^+ active transport process keeps $[Na]_i$ at constant low levels, rendering Na^+ effectively impermeant.

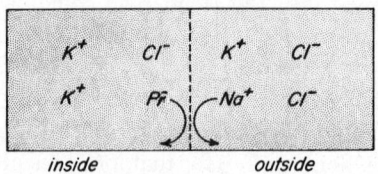

Figure 8 Double Donnan equilibrium system.

Potassium and chloride will move until the Donnan equilibrium is established, that is, until

$$[K]_i \times [Cl]_i = [K]_o \times [Cl]_o \qquad (28)$$

But now $[K]_o$ is less than $[Cl]_o$ and not equal to it. Thus, $[K]_o + [Na]_o = [Cl]_o$. Hence it is possible for the system to be in Donnan equilibrium and osmotic equilibrium at the same time because the osmotic effect of the indiffusible protein anion inside the cell is balanced by that of the effectively indiffusible sodium cation outside the cell. This, of course, requires that the osmotic activities of the two indiffusible molecules are equivalent.

The double Donnan equilibrium hypothesis assumes that the cell membrane is effectively impermeable to sodium ions. Although resting nerve and muscle cells as well as red blood cells, and other cells do take up radioactive sodium ions (e.g., the resting influx of sodium into giant axons of the cuttlefish *Sepia* is about 35×10^{-12} mol $cm^{-2}s^{-1}$), these cells act as if they were impermeable to Na^+, because the Na^+ is pumped back out. If the Na^+ that leaks in were not pumped out, that is, if nothing were done about this influx, the cells would swell and eventually die. In fact, cells prevent sodium influx by means of a continuous extrusion of sodium ions, the mechanism of which is discussed later in this chapter. Since such extrusion must occur against an electrochemical gradient, we would expect this process to be an active one, involving the consumption of metabolic energy. Tissues are most commonly deprived of their energy supply when blood flow to the cells ceases, as occurs in cardiac arrest or stroke. Under these conditions, heart and brain cells swell and die because the energy needed to pump out Na^+ is inadequate. The cellular extrusion of Na^+ is considered in more detail in a subsequent section.

COLLOID OSMOTIC PRESSURE

Blood contains proteins having a net negative charge to which the endothelium of the capillaries is impermeable. These proteins themselves exert an osmotic pressure in addition to exerting a Donnan effect on Na^+ and Cl^- such that the sum of Na and Cl concentrations inside the capillaries is greater than the sum outside. Thus, the total osmotic pressure caused by the plasma proteins is about 27 mm Hg, one-third of which comes from the Donnan effect on Na^+ and Cl^-. The extra osmotic pressure attributable to proteins and the associated diffusible ions is called the *colloid osmotic pressure,* or *oncotic pressure.* This oncotic pressure is offset by the hydrostatic pressure within these vessels and is a very important factor in determining the distribution of fluids between intravascular and interstitial spaces.

SPECIFIC MECHANISMS OF MEMBRANE PERMEATION

We have thus far discussed in general terms the driving *forces* involved in the net movement of substances across membranes: (1) concentration or chemical gradients, or in the case of charged substances, electrochemical gradients; and (2) osmotic gradients.

The problem we now want to consider concerns the specific pathways by which substances actually cross biological membranes. As you recall, biological membranes are composed of proteins and lipids and may be considered to be a mosaic structure as shown in Figure 9. Specific pathways especially for charged or polar molecules make use of proteins that either span the membrane or can shuttle back and forth across it. The specific structures of these proteins remain unknown at the present time. Nonpolar molecules can also dissolve directly into the lipid parts of the membrane. In either case one generalization can be made. The likelihood or probability of membrane permeation by a substance will be increased when less energy is required for the substance to be in the membrane than in the aqueous phases separated by the membrane. With this single generalization we can now turn to specific mechanisms of membrane permeation.

SIMPLE DIFFUSION THROUGH THE LIPID MATRIX OF THE MEMBRANE

When the permeability of biological membranes to a number of different organic molecules, such as alcohols (CH_3–CH_2–CH_2–OH or R–OH), ethers, (R–O–R'), and urea (NH_2–CO–NH_2), are examined in relationship to their molecule structure, a correlation emerges. Molecules such as long chain alcohols and ethers with few polar or ionic groups are found to enter cells rapidly and thus have large

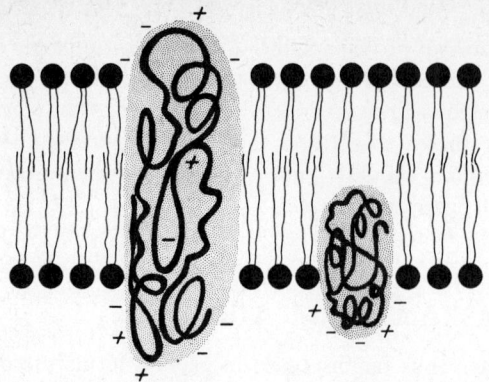

Figure 9 Schematic cross-sectional view of the lipid–globular protein mosaic model of membrane structure. The phospholipids are arranged as a bilayer with their polar heads in contact with water. The integral proteins, with the heavy lines representing the folded polypeptide chains, are shown as globular molecules partially embedded in, and partially protruding from, the membrane. The protruding parts have on their surfaces the ionic residues (− and +) of the protein, whereas the nonpolar residues are largely in the embedded parts; accordingly, the protein molecules are amphipathic. The degree to which the integral proteins are embedded and, in particular, whether they span the entire membrane thickness, depends on the size and structure of the molecules. Note that the protein surfaces spanning the membrane but not in contact with the nonpolar lipids may also carry a charge. These can form aqueous channels lined with charges through which ions may move. However, such channels are usually scarce, occupying only a fraction of the total surface area. (Redrawn from Singer SJ, Nicolson G: *Science* 175:720–731, 1972.)

permeability constants, whereas highly polar or ionized molecules enter very slowly or not at all. The lipid membrane provides a low energy barrier for nonpolar lipid molecules but is a high energy barrier for polar or charged molecules. Propyl alcohol, which is relatively nonpolar and which has the same dimensions as urea, permeates about 1,000 times faster than urea. This suggests that the relatively nonpolar molecules with long hydrocarbon chains are passing through a membrane composed of nonpolar molecules, that is, the lipid interior of the membrane in which other nonpolar molecules can readily dissolve. The solubility of a substance in a nonpolar medium, such as oil, can be measured by shaking an aqueous solution of the substance with a layer of oil. Some of the solute molecules dissolve in the oil phase and some in the water phase. The ratio of the two concentrations C_{oil}/C_{water} is a measure of the solubility of the molecule in oil relative to water. This ratio is the partition coefficient mentioned earlier. If the ratio is 1, the substance is equally soluble in oil and water. A ratio greater than 1 means that the molecule is more soluble in oil than in water, and vice versa. Many biologically important molecules are less soluble in oil than in water, since they usually contain at least a few polar groups. When the permeability constants of a number of different molecules are plotted against their oil/water partition coefficients, a broad linear relationship is found, shown in Figure 10. The larger the partition coefficient, the greater its lipid solubility and the larger its permeability constant. The more readily a molecule dissolves in the membrane lipid, the faster its rate of entry into the cell. If a molecule cannot dissolve in a lipid barrier, it will be excluded from the cell unless some other route of entry exists. Note that certain polar molecules are

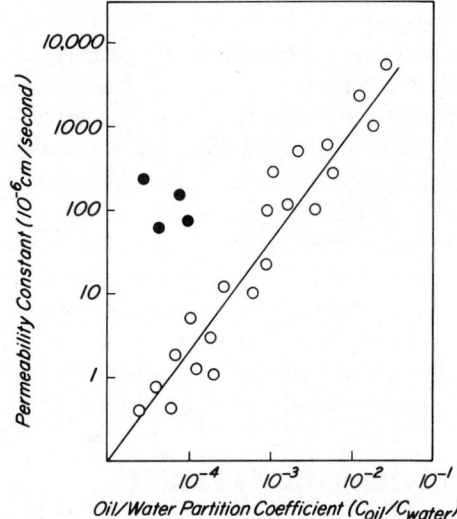

Figure 10 Relationship between lipid solubility (C_{oil}/C_{water}; partition coefficient) and rate at which molecules diffuse across a membrane measured in terms of their permeability constants *P*. Some molecules (e.g., water and ions Na^+, K^+, and Cl^-) are permeable but are not lipid soluble, as indicated by the filled circles in the graph. Filled and open circles represent different types of molecules. (Modified from Vander AJ, Sherman JH, Luciano DS: *Human Physiology.* New York, McGraw-Hill, 1980.)

more permeable than would be expected from their oil-/water partition coefficients. An explanation for this observation is provided in the following section.

PORE-MEDIATED TRANSPORT

How does the permeation of polar molecules occur or, put another way, how are the energy barriers of the membrane for polar molecules lowered? There are two mechanisms for achieving this. One is an appropriately charged water-filled pore spanning the membrane that shields the polar molecule from the hydrophobic lipid interior and the other is a carrier shuttling across the membrane whose interior also effectively shields the polar molecule. The difference between the two is somewhat arbitrary but flux rates help distinguish between the two. Thus the fastest transport rate for a known carrier is about 10^5 ions/s, whereas pore mechanisms for ion transport occur at rates of 10^7 or more ions/s. Water permeates at even faster rates, so again a pore mechanism seems likely. Some pores are highly selective for a single ionic species, such as Na^+, K^+, or Ca^{2+} ions. The pores for Na^+, K^+, and Ca^{2+} are sometimes called the Na^+ channel or K^+ or Ca^{2+} channels, respectively. It now seems likely that the K^+ channel can contain more than one K^+ ion at a time, and that some artificial channels such as the gramicidin channel may contain 5–10 ions and water molecules at a time. The use of ions of various sizes has permitted estimation of the sizes of the Na^+ and K^+ channels and, as might be expected, they are roughly the same size as the ions they have to accommodate. Larger pores are relatively unselective as is the case for the end plate channels that are involved in synaptic transmission at the neuromuscular junction.

What about the pathway whereby water permeates the cell membrane? Consider the fact that even though water moves through the cell membrane at about 1/100 its speed in aqueous solution, it can still move much more rapidly —about 10–100 times—than urea, which is only slightly larger and has only a slightly smaller diffusion coefficient in water. Also, the oil/water partition coefficient of water is 0.0007, which is much less than the similar partition coefficient for methanol, which is 0.01. Nevertheless, water passes across membranes at rates higher than methanol does. Because the rates of water movement are very much greater than the highest rates known to be mediated by carrier molecules, the concept of narrow aqueous channels or pores in plasma membranes through which water can pass has been developed. Further support comes from the following observation, which might at first appear very surprising. The calculated membrane permeabilities to water from a net water flux induced by either a hydrostatic or osmotic pressure gradient are the same and are much larger (up to 100-fold greater) than found when one simply puts radioisotopically labeled water on one side of the membrane and measures its rate of appearance on the other side.

Such a result is to be expected if water moves through the membrane in response to a pressure difference by flowing through pores rather than, or in addition to, molecular diffusion through the membrane matrix. Thus if the pore were to contain several water molecules at one time, the rate of transport of labeled molecules added to the aqueous solutions in trace amounts would be impeded. On the other hand, osmotic or hydrostatic pressures would produce bulk flow of all the molecules in the channel. Some differences might also result if the main delay in the passage of water were the time required for diffusion across an unstirred layer at the membrane surface rather than diffusion through the membrane itself.

Estimates of the dimensions of pores in biological membranes have been made assuming that osmotic flow through the pores is governed by *Poiseuille's equation:*

$$v = \frac{A_o r^2}{l} \cdot \frac{1}{8\eta} \cdot \Delta P_{Osm} \qquad (29)$$

where v is the volume of fluid transferred per unit time, A_o is the total cross-sectional area of all the aqueous channels, r is the radius of each channel, l is the length of each channel, η is the viscosity of water, and ΔP_{Osm} is the osmotic pressure difference. The derivation of this equation, as well as how it applies to blood flow in the vascular system, is more throughly discussed in Chapter 10.) All the terms in Eq (29) are either known or measurable, except $A_o r^2/l$. If it is assumed that l is simply the membrane thickness, then $A_o r^2/l$ can be determined from experiments using radioisotopically labeled water and/or other small molecules of varying radii. Consequently, a value for r can be calculated. Originally values for r of about 4 Å were obtained in human erythrocytes, and this has subsequently been confirmed for various other plasma membranes. It should be recognized, however, that this calculation involves the assumptions that the pores are all identical and are straight cylinders penetrating the membrane at right angles to its surface. Because these assumptions are highly unlikely, we say that the membrane *behaves as though* it had uniform right cylindrical pores with radii of about 4 Å. The term "equivalent pore radius" is used to indicate this abstraction.

These principles can also be applied to ultrafiltration across the capillary endothelium, but ΔP_{osm} is replaced by the hydrostatic pressure difference. In this case, much of the flow appears to occur through "pores" between the endothelial cells. An equivalent pore radius of about 40 Å was found experimentally. It would seem that holes of this size should be demonstrable in the electron microscope, but none has been seen to date in the plasma membrane of the endothelial cell.

SUMMARY OF PORE-MEDIATED ELECTROLYTE TRANSPORT

Table 3 shows that the hydrated radii of water, urea, Cl⁻, K⁺, and Na⁺ are similar. The effective radii are the radii that the ions or molecules appear to possess in diffusion or viscosity measurements. They behave in aqueous solution as though they have the radii listed in Table 3. The effective radii given for ions in solution are larger than their radii in salt crystals because in solution the effective radii include some water molecules held to each ion by electrostatic attraction. Pore radii are thought to be 4.0

Table 3 Hydrated Radii of Permeant Species

Permeant Species	Effective Radius[a] (Å)
Water	1.5
Urea	1.6
Cl⁻	2.0
K⁺	2.0
Na⁺	2.5
Glucose	4.2
Sucrose	5.2

[a]See text for explanation.

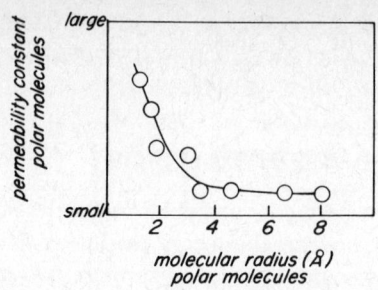

Figure 11 Relationship between the radius of polar molecules and the rate at which the molecules cross the cell membrane. (Modified from Vander AJ, Sherman JH, Luciano DS: *Human Physiology.* New York, McGraw-Hill, 1980.)

Å or less. As noted earlier, the aqueous diffusion coefficients of these molecules are similar, but their permeation through the cell membrane is much slower and shows much greater variation. Still, as shown in Figure 10, these polar molecules permeate faster than would be predicted from their oil/water partition coefficient. The best explanation for electrolyte transport is that they permeate membrane pores having an average radius of about 4 Å (Fig. 11). Smaller pores would let water molecules through, but would be more difficult for hydrated ions to traverse. Divalent and trivalent ions surrounded by large hydration shells might not get through at all. The factor of pore size is related to the steric factor which determines the permeability of membranes to small, uncharged hydrophilic solutes.

Size alone, however, could not be the only property that determines the ionic selectivity of membranes. For example, K⁺ and Cl⁻ ions have almost identically hydrated radii. The mobility of both ions in free solution or across artificial nonselective membranes is almost equal, so that their diffusion down a concentration gradient is not associated with the presence of a diffusion potential; yet the human erythrocyte membrane is 10 to 100 times more permeable to Cl⁻ than to K⁺. This is because the erythrocyte membrane possesses a special carrier protein (described below) that greatly promotes the movement of Cl⁻ across it.

In resting muscle, K⁺ and Cl⁻ ions permeate almost equally well, whereas Na⁺ permeates poorly. It has been suggested that resting muscle consists of a mosaic of pores,

suggested that resting muscle consists of a mosaic of pores, some negatively charged to favor K^+ entry, and vice versa for Cl^-. Why, then, does the negatively charged pore exclude permeability of Na^+ ions? Possibly because such steric factors as size and shape or fit of the hydrated Na^+ ion, or more likely because the interaction between the Na^+ and the net charge of the chemical groups forming the pore is less favorable for Na^+ than for K^+.

In summary, one mechanism whereby ions are able to permeate cell membranes is via pores; the permeability of a given ion is determined by the radius of the hydrated ion, pore size, and the interaction between the charge of the ion and the net charge of the chemical groups that line or form the pore. Some pores are highly selective for a specific ion, whereas others are less specific. Some pores permit only a few ions of similar size and charge to pass, in contrast to those that let a wide range of sizes and opposite-charge ions through. But as we shall see, ionic transport also occurs by another mechanism—one that for many larger polar molecules provides the only means of membrane permeation: carrier-mediated transport.

CARRIER-MEDIATED TRANSPORT ACROSS CELL MEMBRANES

As we have seen, pores might account for the rapid entry of small polar molecules, such as water and certain ions, but they cannot explain the rates at which large polar molecules, such as the sugars and amino acids, have been observed to pass through cell membranes. Amino acids and sugars are polar molecules, most with a radius larger than 4 Å. Clearly, passive diffusion of such molecules through a lipid membrane with pores of the same size would occur very slowly, if at all. Yet the cell requires the entry of large quantities of these molecules to provide the raw materials for protein synthesis and a source of chemical energy. Movements of such molecules across biological membranes appear to be mediated by a series of chemical reactions between these molecules and specialized components of the membrane. We can ascribe a physical signifi-

cance to these membrane components if we assume that they are carriers that can combine reversibly with specific substances in order to transfer them across the membrane. The carriers are probably proteins that have a hydrophilic interior and a lipophilic exterior. Thus the carrier–molecule complex is far more soluble in the membrane than is the molecule alone. Carriers are highly selective and achieve their selectivity using a combination of steric and electrostatic forces. Figure 12 depicts the movement of a substance across a membrane by such a carrier system. Substance S is at a higher concentration at the outside of the membrane.

In the model illustrated in Figure 12, carrier molecule A is assumed to bind rapidly to a substance S at the interface between the membrane and the external medium. The $S–A$ complex then diffuses across the membrane to the inside surface, where the complex dissociates and S is discharged to the cytoplasm. Note that the net flux of S is from a region of high concentration to a region of low concentration, so that no energy has to be injected into the system. The type of transfer illustrated by this model is known as *facilitated diffusion* and is described in a later section. In essence, the effectiveness of the lipid barrier is reduced for the particular substance S for which the carrier is specific. Instead of the diffusion rate being limited by the diffusion of the substance alone, it is now limited by the diffusion of the $S–A$ complex (which might, for example, be much more lipid soluble than is S alone).

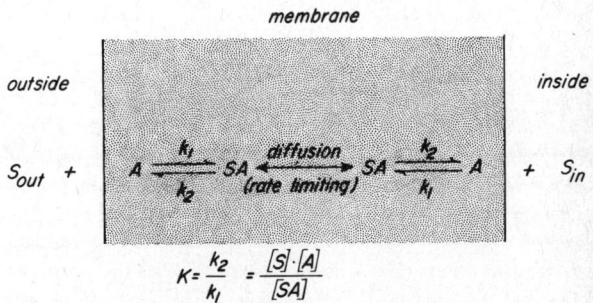

Figure 12 Membrane model for the movement of molecules via carriers.

CRITERIA FOR MEDIATED TRANSPORT PROCESSES

If one compares the rate at which a substance can cross a membrane at any given concentration by simple diffusion with the rate for mediated transport, four differences between the two modes of transmembrane movement can be seen (Fig. 13).

Faster Rate of Mediated Transport

First, we observe that mediated transport permits a more rapid movement of the substance at the lower concentrations of the substance than does simple diffusion. It was this sort of observation that originally led physiologists at the turn of the century to begin to think about specialized transport processes across biological membranes.

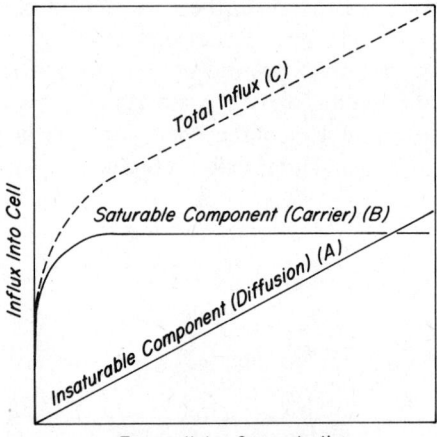

Figure 13 The possible relationships between one-way flux of a solute into a cell and the extracellular concentration of the solute. Curve A represents the general situation, in which the solute crosses the membrane only by simple diffusion. Curve B illustrates the properties of a carrier-mediated transport process showing the much-enhanced transport rate at low concentrations of the transported solute and the saturation of the process at higher concentrations. Curve C represents total transport rate for a system in which the solute enters the cell by both mechanisms.

Saturation Kinetics of Mediated Transport

In simple diffusion, the flux of molecules entering the cell at any time is linearly related to the concentration of the substance outside the cell. A doubling of the concentration will result in a doubling of the flux, as shown by the straight curve in Figure 13, sometimes referred to as the *independence principle.* However, because of the limited number of carrier sites available to transport substances across the membrane, the relationship between flux rate and extracellular concentration is not linear for mediated transport (Fig. 13). As more and more of the substance enters the solution, the chances of an individual molecule "finding" an unoccupied carrier decrease, until finally an extracellular concentration level is reached at which for any given instant in time, all the carrier sites are occupied. Raising the extracellular concentration above this value will result in no further increase of carrier-mediated flux. This is the phenomenon of saturation.

It should be pointed out that substances that cross cell membranes via carrier mechanisms also can cross, to some extent, by diffusion through the lipid matrix and through pores. The result is that the total flux is the sum of all the fluxes, as illustrated in Figure 13.

Specificity for Transported Substances

A third important property of mediated transport processes is that they possess to various degrees an ability to "recognize" certain chemical structures and to preferentially carry one type of chemical rather than another. Some systems are very specific, being able to distinguish between molecules that differ only in their three-dimensional spatial arrangement, that is, stereoisomers. For example, a certain amino acid in the levo (L) configuration might enter a cell 100 times faster than the same amino acid with the dextro (D) configuration.

Competition Among Transported Substances

With mediated transport systems, the phenomenon of *competition* can usually be observed among those molecules or ions capable of being transported by the system. For example, if two molecules *A* and *B,* such as the amino acids glycine and alanine, both of which enter the cell by the same transport system, are present simultaneously outside the cell, they must complete with each other for the available transport site on the membrane (Fig. 14). The presence of B decreases the rate at which A enters the cell because B occupies some of the binding sites that would otherwise be occupied by A. Molecules entering by simple diffusion do not complete with each other because their entry is not restricted to a limited number of sites.

Although the carrier model offers an explanation of how carrier-mediated transport systems operate, the complete identity of actual carrier molecules in the membranes of vertebrates has yet to be demonstrated. The carrier systems that have been most extensively studied appear to involve protein molecules as the carrier structure. This makes sense, as proteins are the only known molecules that possess the high degree of chemical specificity shown by mediated transport systems. Just how the combination of a solute molecule with its carrier leads to an increase in its rate of passage through the membrane is unknown. That such entities exist is exemplified by certain antibiotics, such as nonactin, which act as carriers in membranes. In fact, when a large number of carrier molecules synthe-

sized or harvested from bacteria are incorporated into artificial lipid membranes, they demonstrate the properties we have described above.

There are two general types of mediated transport systems, namely, facilitated diffusion systems, which lead to more rapid *equilibration* across a membrane than would occur by simple diffusion, and active transport systems, which can move substances against a concentration or an electrochemical gradient in order to establish a *steady state* rather than an *equilibrium state*.

FACILITATED DIFFUSION

The final result of this process is no different from the final result of a simple diffusion process; substances move down their concentration or electrochemical gradient until they are in equilibrium across the membrane. That is, the force driving the inward movement of the substance becomes equal to that driving outward movement.

The differences between facilitated diffusion and simple passive diffusion are those of rate, specificity, saturation, and competition, and presumably they reflect quite different mechanisms of translocation. Thus, it is only when the concentration of the transported substance is varied and competing molecules are added to the system that a facilitated diffusion system can be identified.

In facilitated diffusion, all net movement of molecules across the membrane is from a region of high concentration (or electrochemical potential) to one of low concentration (or electrochemical potential). The most important example of facilitated diffusion in the body is the movement of glucose from blood across the membranes of most cells. This relatively large polar molecule would not be expected to diffuse readily through the lipid or pore region of the membrane. Facilitated diffusion through the membrane provides a mechanism for supplying cells with this essential compound. Another very important facilitated diffusion system is the Cl^-/HCO_3^- exchange system in the red blood cell membrane. This mechanism allows for a very rapid exchange of Cl^- and HCO_3^-, which is important for efficient buffering of metabolic CO_2.

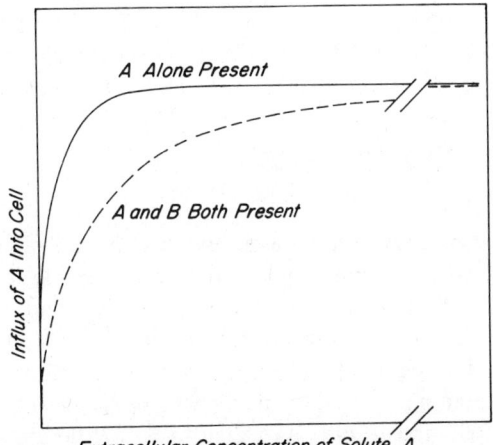

Figure 14 Effect of competition between two molecules, A and B, for the limited number of carrier sites on the rate of inward transport of A. The dashed line represents the situation when a constant amount of solute B is present. We see that the flux of molecule A can reach the same limiting (saturating) value, but a much higher extracellular concentration is necessary.

ACTIVE TRANSPORT

In facilitated diffusion net movements of solute occur due to physical gradients of concentration or electrical potential and no expenditure of cellular metabolic energy is required. However, there are numerous substances that are moved across biological membranes *against* such physical gradients. This uphill process is known as active transport or pumping. Such a process at some point requires the expenditure of metabolic energy. Active transport results not in an equilibrium across the membrane, but in a steady state. That is, the passive forces (i.e., concentration gradients and/or electrical gradients) will favor the net movement of a substance, but no such net movement will occur. Therefore, a steady state exists between fluxes arising from passive driving forces and fluxes caused by active transport, with the result being that at the steady state set point there will be no change in the intra- and extracellular concentrations of the substance. The actual molecular mechanism of biological active transport is unknown. However, they behave experimentally as though there were a carrier mechanism such as that pictured in Figure 15. The most important aspect of this model is that energy, usually in the form of ATP (but see below) is used by the system. The affinity of the carrier for the transported species is altered by the energy. Presumably this reflects some reversible structural change in the carrier molecule. Thus, the carrier exhibits different affinities for the transported substance on either side of the membrane. An active transport system whose function is to maintain higher-than-equilibrium intracellular values of a substance would exhibit a high affinity at the external surface and a low affinity at internal surface. Such an arrangement would result in a net uptake of the substance by the cell until the backleak via passive diffusion would become equal to the active uptake. At this point, a steady state is reached and the intracellular concentration remains unchanging. Table 4 presents a sampling of substances known to be actively transported in various mammalian tissues.

It appears that most active transport systems utilize the energy of the 5'-phosphate bond of ATP. In fact, as we shall see, the best studied pump of all, the Na–K pump, appears to incorporate ATP-splitting enzyme (Na–K ATP-ase). However, there appears to be an alternate method of fueling active transport systems. This method involves using the electrochemical gradient of Na^+, which in turn has been generated by an ATP-requiring pump. Experimentally it has been possible to show that certain sugars, amino acids, and the Ca^{2+} ion (Table 4) can be pumped in the absence of ATP, provided the Na transmembrane electrochemical gradient is artificially maintained.

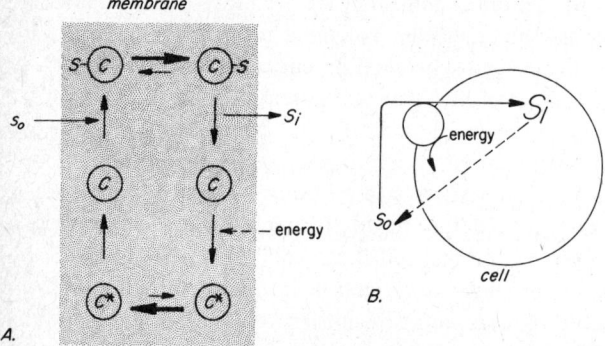

Figure 15 Diagram illustrating the two types of fluxes that maintain the cell in a steady state with respect to solute *S*. The cell uses energy, generally ATP, to effect a net influx into the cell despite the unfavorable chemical gradient represented by the different size of the letter S. The dashed line represents the passive (no cellular energy involved) leak of solute *S* down its concentration gradient from the cellular fluid to the extracellular fluid. A steady state occurs when the pump flux is matched exactly by the leak flux.

Na⁺ EXTRUSION AND K⁺ UPTAKE: THE SODIUM PUMP

The resting nerve membrane, like virtually all other cell membranes, is slightly but definitely permeable to Na^+ and K^+. You might therefore expect Na^+ and K^+ to trickle down their respective concentration (or electrochemical, to be more exact) gradients. Furthermore, during the action potential the nerve becomes transiently very permeable, first to Na^+ and then to K^+. Whereas in the squid axon, for example, this amounts to a loss of only about 1 millionth of the total internal potassium level during each nerve impulse, the effect would be cumulative unless the axon had some means of replenishing its store of K^+. Also, a small amount of Na^+ enters the cell during each action potential, which must be extruded.

Table 4 Examples of Active (Uphill) Transport[a]

Substance	Location	Pump Direction	Energy Source (If Known)
Sugars	Intestine	Lumen to blood	Na^+ gradient
	Renal tubules	Lumen to blood	Na^+ gradient
Amino acids	Most cells	Outside to inside	Na^+ gradient
	Intestine	Lumen to blood	Na^+ gradient
	Renal tubules	Lumen to blood	Na^+ gradient
Bile salts	Intestine	Lumen to blood	Na^+ gradient
Na^+	All cells	Inside to outside	ATP
	Intestine	Lumen to blood	ATP
	Renal tubules	Lumen to blood	ATP
K^+	All cells	Outside to inside	ATP
	Proximal renal	Lumen to blood	ATP
H^+	Parietal cells of stomach	Cell to lumen	?
Ca^{2+}	All cells	Inside to outside	Na^+ gradient
	Intestine	Lumen to blood	Na^+ gradient
	Renal tubules	Lumen to blood	Na^+ gradient?
	Red blood cells	Inside to outside	ATP
	Sarcoplasmic reticulum	Outside to inside	ATP
Cl^-	Parietal cells of stomach	Cell to lumen	ATP
	Thick ascending loop of Henle	Lumen to interstitium	?
I^-	Thyroid cells	Blood to follicles	ATP
HPO_4^{2-}	Renal tubules	Lumen to blood	?

[a]This list is not intended to be complete, but rather gives some of the most important examples.

A system for accomplishing Na^+ extrusion and K^+ uptake has been shown to exist in virtually every cell type studied. This mechanism is commonly referred to as the *sodium pump.* Although it is called the sodium pump it is also directly responsible for K^+ uptake, or pumping, as well. It has been shown that for every cycle of the pump, Na^+ is pumped out and K^+ pumped in. In order for the pump to operate, there are several absolute requirements: (1) the presence of intracellular ATP, and the simultaneous presence of (2) intracellular Na^+, and (3) extracellular K^+. Provided ATP is present, the rate of pumping is most sensitive to $[Na]_i$. Relatively small changes of $[Na]_i$ can cause relatively large changes in pumping rate. The pump can be inhibited in a graded manner up to virtually complete inhibition by cardiac or digitalis glycosides, such as ouabain. A schematic model of the pump based on these observations is presented in Figure 16.

The molecule responsible for this pumping is a membrane-bound enzyme called Na–K ATPase. This enzyme possesses all the properties of the Na^+–K^+ pump and is inhibited by ouabain. Recently, purified Na–K ATPase extracted from animal membranes has been incorporated into artificial membranes with the result that an ATP-dependent active transport of Na^+ and K^+ now takes place across these membranes. This transport is, as expected, inhibited by ouabain.

Electrogenic Na^+ Pumping

Although some Na^+ moves outward and K^+ inward during each cycle of the pump, the movements of the two ions are not perfectly balanced. It has often been shown that for every $3Na^+$ extruded, $2K^+$ are taken up. Thus, every cycle of the pump causes the net loss of a positive charge from the interior of the cell. If there is no simultaneous movement of another cation inward or anion outward, this net loss of internal positivity will cause the inside of

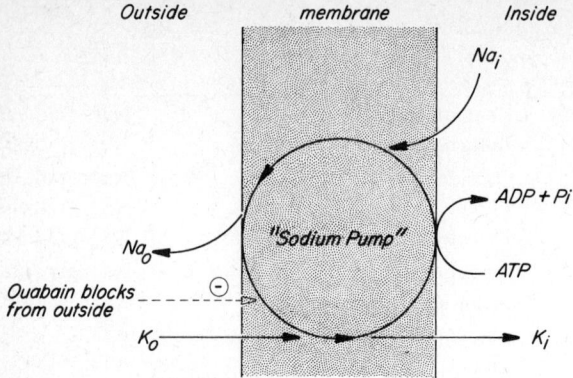

Figure 16 The sodium pump.

the cell to become electrically more negative relative to the outside of the cell. Thus, the Na^+ pump can *directly* influence membrane potential independently of ion gradients established by the pump. The extent to which this "electrogenic" potential is actually realized in any given cell type is determined by the ease with which this electrogenic pump component can be short-circuited by the movement of other ions through pores. For example, if Na^+ could leak back into the cell easily, then almost as fast as the extra Na^+ is pumped out, another Na^+ might leak back in, with very little effect on membrane potential seen. If, on the other hand, the membrane has a very low permeability to ions, then the contribution to the membrane potential by the electrogenic pump can be considerable. Another factor that can increase the electrogenicity of the pump is the fact that under certain conditions, the coupling ratio between Na^+ and K^+ can vary from 3:2. It has been observed to be as great as 5:1 in some circumstances.

The phenomenon of electrogenic sodium pumping has been reported for numerous excitable cell types and the physiological significance of such a contribution to membrane potential will be discussed in relationship to specific cell types. Proton pumping across mitochondrial membranes is electrogenic and is essential for ATP synthesis; this is the basis of the chemiosmotic theory.

Significance of the Sodium Pump

The ubiquitous sodium pump is used to maintain the Na^+ and K^+ concentration gradients that are normally present across most animal cell membranes. It is certainly important in such electrically excitable cells as nerve and muscle, since the electrical activity directly depends on the ionic gradients. Moreover, it might be related to the inotropic action (enhancement of contraction) of certain drugs in cardiac muscle.

The sodium pump is also involved in the regulation of cell volume as already mentioned in relation to the double Donnan equilibrium. By pumping Na^+ out and thereby maintaining a low $[Na]_i$, the sodium pump in effect makes Na^+ an extracellular nonpenetrating cation that offsets the osmotic effects of the intracellular nonpenetrating anions. That this function is important can be seen from the effects of drugs or physical agents that inhibit the Na^+–K^+ pump or that increase the permeability of the membrane to Na^+. Such agents cause cell swelling and perhaps *lysis* (disruption or bursting of cells). Hypoxia, for example, leads to decreased ATP production which, in turn, causes the Na^+–K^+ pump to slow down or stop altogether, making the cells swell. Various *hemolytic agents* cause *hemolysis* (lysis of red blood cells) by inhibiting the Na^+–K^+ pump or by increasing the permeability of the membrane to Na^+.

Metabolism and The Sodium Pump

Discounting the oxygen consumption associated with muscular contraction, perhaps as much as 40% or 50% of the basal O_2 consumption (to use a conservative estimate) of the entire body is associated with the activity of the sodium pump in all the various tissues. For example, although the brain and kidneys together account for only 4% of the total body weight, they use a considerable fraction ($\sim$35%) of the total basal O_2 consumption, and in vitro studies suggest that about 50% of the O_2 utilization by these two tissues is inhibited by cardiac glycosides. You already realize the significance of the sodium pump to neurophysiology. In the kidney, sodium is continually being filtered in the glomerulus and must be reabsorbed by the tubules against an electrochemical gradient. This process is responsible for the reabsorption of about 180 L water per day.

Although the ultimate source of energy for active transport of substances across biological membranes must come from ATP a number of substances are actively transported by systems that do not directly require ATP

(Table 4). Instead they use the energy of the Na^+ electrochemical gradient established by the sodium pump—usually called coupled transport. This mechanism is discussed in a later section.

OTHER ACTIVE TRANSPORT SYSTEMS

The Calcium Pump in Mitochondria and in Sarcoplasmic Reticulum

Sodium and potassium are not the only ions that are not at equilibrium across biological membranes. Furthermore, concentration gradients are not limited to the plasma membrane, but can be found across intracellular membranes as well. For example, the concentration of ionized calcium in the cytoplasm of most animal cells is normally very low. And there is now evidence that mitochondria from a variety of cell types may accumulate calcium from the cytoplasm against a concentration gradient at the expense of ATP breakdown. The significance of this "Ca pump" in mitochondria is not known; however, a similar Ca pump found in sarcoplasmic reticulum plays an important role in muscle physiology. Muscle contraction is associated with a large transient increase in the level of ionized Ca in the sarcoplasm; in fact, the contractile process is dependent on this increase in ionized Ca. But in order for the muscle to relax, the ionized Ca concentration must then be rapidly reduced. An ATP-dependent Ca uptake against a large concentration gradient has been demonstrated in isolated sarcoplasmic reticulum; this Ca transport system is involved in muscle relaxation.

The Proton Pump in Mitochondria

As noted earlier, proton transport in mitochondrial membranes is an important step in ATP synthesis. The *chemiosmotic hypothesis* states that this proton transport is the result of the vectorial arrangement of the electron transport chain enzymes in the mitochondrial membranes. This arrangement results in a net translocation of protons from the mitochondria into the cytoplasm. The resulting pH gradient then provides the energy for ATP synthesis. When this gradient is abolished, by substances such as dinitrophenol, ATP production ceases, although electron transport persists.

COUPLED TRANSPORT: THE ROLE OF SODIUM IN THE TRANSPORT OF OTHER SUBSTANCES

Cotransport: Sugars and Amino Acids

Amino acids are not at equilibrium across the plasma membranes of most animal cells. In addition, amino acids and certain sugars, during their transport across intestinal and renal tubular epithelia, become distributed away from equilibrium. It is of interest that these substances are usually only transported in the presence of sodium ions. For example, when glucose is placed in the medium in which the mucosal surface of hamster or rabbit small intestine is bathed, it is taken up by the mucosal cells against a concentration gradient, but only if sodium is also present in the medium. Furthermore, cardiac glycosides interfere with the uptake of glucose, but this appears to be an indirect consequence of their action on the sodium pump, rather than the result of a direct effect on the glucose uptake process.

A net accumulation (i.e., against a concentration gradient) of the amino acids has been shown to occur in metabolically poisoned cells and in the absence of a chemical energy source, such as ATP. The one definite prerequisite for net accumulation appears to be a sodium concentration gradient in the normal direction (i.e., $[Na]_o > [Na]_i$).

How do we explain the paradoxical buildup of a concentration gradient in the apparent absence of a direct chemical energy source? The answer is that we do, in fact, have an energy source—the sodium battery or electrochemical gradient.

Let us consider a membrane in which we have a carrier molecule X that binds both sodium ions and, say, glucose (G). Let us stipulate that X can move back and forth within the membrane, from inner to outer surface. But X can move only when (1) it is empty (i.e., as X alone), or (2) *both* Na^+ and G are bound to it (i.e., as $Na–X–G$).

It will not move when either $Na–X$ or glucose $X–G$ is present alone (Fig. 17).

We will assume that the ternary complex ($Na–X–G$) has the same dissociation constant at both surfaces of the membrane and that transmembrane glucose movements are virtually limited to this carrier mechanism. It can then

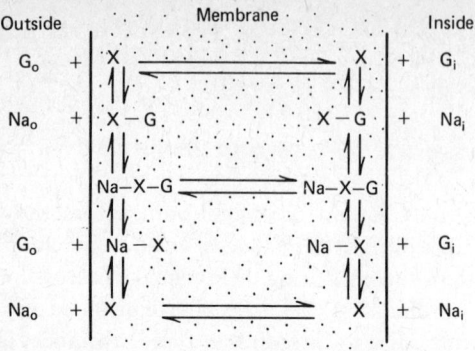

Figure 17 Carrier model for cotransport of glucose (G) with Na^+. (From Goldner AM, Schultz SG, Curran P: *J Gen Physiol* 53:362, 1969.)

be shown that in the steady state the Na and glucose concentrations in the internal and external solutions will be related by the equation

$$\frac{[Na]_o}{[Na]_i} = \frac{[G]_i}{[G]_o} \tag{30}$$

What this means is that with a carrier system, such as the one shown in Figure 17, Na moving into the cell down its concentration gradient can drive another substance (glucose, in this case) up a concentration gradient.

There is considerable evidence that carrier mechanisms of this type are used to concentrate glucose and alanine, and perhaps other amino acids, in the intestine. It has been shown that in metabolically poisoned cancer cells the glycine and sodium concentration ratios vary inversely in accordance with Eq. (30) over a wide range of sodium ratios. The relationship shown in the above equation helps to explain how cardiac glycosides may interfere with glucose and amino acid uptake, although they have no direct effect on the carrier systems per se. Inhibition of the sodium pump will lead to a rise in cell sodium, and the consequences of this are obvious (see above equation).

An understanding of the intestinal coupled carrier systems has been used to advantage in the design of therapy for cholera patients. The primary manifestation of this disease is severe dehydration as a result of massive (5–10 L/day), watery, diarrhea ("rice-water stools"), which carries along copious quantities of electrolytes (especially NaCl). The cause is probably a Na^+, Cl^-, and bicarbonate secretion with an osmotic equivalent of water coming from the blood. In order to replace both the fluid and electrolytes, the standard treatment has been intravenous fluids. However, you can imagine the difficulty involved in trying to obtain sufficient quantities of sterile fluid in remote rural areas of India, for example, during an epidemic. An alternative treatment has now been designed in which moderately concentrated solutions containing both glucose and sodium are administered orally or by Levine (nasogastric) tube. By means of the Na–glucose coupled transport mechanism, both Na^+ and glucose are absorbed *pari passu* in the upper intestine, and the absorption of solute draws solvent water into the blood stream osmotically. Evidence that the coupled transport system(s) is involved is provided by the observation that substitution of other transported sugars or transported amino acids are also effective in helping rehydrate the cholera patients, whereas nontransported sugars are ineffective.

Countertransport of Sodium and Calcium

The squid axon demonstrates that energy from the sodium gradient is apparently used to drive calcium uphill out of the cell against a concentration gradient; in this case, sodium enters the cell (moving "downhill" or "counter") as calcium leaves. There is evidence that a similar mechanism might help regulate cell calcium concentrations in a variety of tissues, including frog skeletal muscle, frog heart, mammalian heart, mammalian vascular smooth muscle, and mammalian liver. This mechanism of sodium–calcium exchange might also be involved in the absorption of calcium in the small intestine; sodium must be present at the submucosal border of the mucosal cells for net transfer of calcium from the intestinal lumen to the serosal surface to occur. (We might point out that calcium gradients play an important role in both contractile and secretory tissues.)

In sum, these observations illustrate the variety of physiological activities in which the sodium gradient—and therefore, indirectly, the sodium pump—has been implicated. And the list continues to grow.

These activities further emphasize the central role of the sodium pump in the economy of the cell and of the body as a whole. The mechanism by which the sodium pump converts chemical energy into useful work is not understood, and this remains a major unsolved problem in physiology.

ENDOCYTOSIS, EXOCYTOSIS, AND BUDDING

Relatively large movements of membranes are involved in the processes of endocytosis, exocytosis, and budding— that is, portions of the membranes themselves undergo transport.

ENDOCYTOSIS

In the transport process of endocytosis, a portion of the plasma membrane invaginates and then pinches off from the surface to form an intracellular vesicle. The plasma membrane seals itself as the vesicle pinches off. In this way the cell ingests whatever extracellular material gets into the invagination. If mainly liquid material is ingested, the process is called *pinocytosis* (cell drinking). If a large proportion of solid material is ingested, the process is called *phagocytosis* (cell eating). Endocytosis is especially well developed in some cells; for example, pinocytosis in capillary endothelial cells and pinocytosis of colloid by thyroid follicular cells; phagocytosis occurs in cells of the reticuloendothelial system. In susceptible cells endocytosis is triggered by attachment of various inducing agents to the outer surface of the membrane. Then, when endocytosis occurs, the inducing agent is transported into the cell. Probably proteins and polysaccharides are the major inducing agents for mammalian cells.

EXOCYTOSIS

Exocytosis is just the reverse of endocytosis. An intracellular vesicle fuses with the plasma membrane and then opens to the outside, allowing its contents to flow out. Exocytosis is an important mechanism of secretion, the one used, for example, by the pancreatic acinar cells, many endocrine cells, and in the release of transmitter from presynaptic nerve endings. In the latter case, the population of synaptic vesicles is maintained by endocytotic incorporation of plasma membranes.

BUDDING

In the process of budding, a portion of the plasma membrane evaginates and then pinches off to form an extracellular droplet of cytoplasm enclosed within a membrane. An example of this type of transport process is the secretion of the lipid components of milk by lactating mammary glands.

SELECTED BIBLIOGRAPHY

Armstrong C: Ionic pores, gates and gating currents. *Q J Biophys* 7:179–210, 1975.

Bull HB: *An Introduction to Physical Biochemistry.* Philadelphia, Davis, 1971.

Crank J: *The Mathematics of Diffusion.* London, Oxford University Press, 1956.

Davson H, Segal MB: *Introduction to Physiology,* vol I. London, Academic Press, 1975.

LeFevre PG: The present state of the carrier hypothesis. *Curr Top Membr Transport* 7:109–215, 1975.

Singer SJ, Nicolson GL: The fluid mosaic model of the structure of cell membranes. *Science* 175:720–731, 1972.

SELF-STUDY QUESTIONS

MULTIPLE CHOICE

Select the best answer(s). (In many cases, more than one answer is correct.)

1. The fact that intracellular levels of Na^+ are low and intracellular K^+ is high depends on
   A. the passive distribution of Cl^-.
   B. the impermeability of the cell membrane to these ions.
   C. the cell gel.
   D. metabolic energy-requiring processes in the membrane.

2. Living cells are in a steady state with their environment when
   A. intracellular Na^+ is low.
   B. intracellular K^+ is high.
   C. their composition changes very little with time.
   D. they take up energy from their environment to maintain a fairly constant composition.

3. When cells die
   A. intracellular Na^+ increases.
   B. the external membrane becomes leaky and nonselective.
   C. their intracellular Cl^- levels rise.
   D. intracellular K^+ increases.

4. Na^+ permeates cell membranes less readily than does K^+. This could be because
   A. it has a smaller hydrated diameter.
   B. it is positively charged.
   C. chloride gets in the way.
   D. it has a larger hydrated diameter.

Questions 5–11. Make the following calculations.

5. Satisfy yourself that $RT(C_L - C_R)$ has units of pressure, using the procedure called dimensional analysis.

6. A 0.9% (0.9 g/100 ml) solution of NaCl is isotonic to mammalian cells. Assuming a reflection coefficient of 1, what is the osmotic coefficient of NaCl? MW NaCl = 58.5.

7. A solution of urea in water is carefully prepared to give an osmolarity of 286 mOsm. It is found that when red blood cells are placed in such a solution, they rapidly burst. Why?

8. Micropipettes are often inserted into cells to measure the potential difference between the inside and outside of the cell. Consider the cell's internal concentration of K^+ to be 0.1 M. If we fill the pipettes with 3 M KCl to enhance electrical conductivity, what will be the liquid junction potential between the pipette and the intracellular fluid? The mobilities of K^+ and Cl^- can be considered equal.

9. $K_i^+ = 100$ mM
   $K_o^+ = 0.01$ M
   A. Calculate the Nernst or equilibrium potential for K^+ at 20°C.
   B. If the cell has a Donnan distribution of K and Cl, what is the value for Cl_i^-, knowing that Cl_o is 100 mM?

10. $K_i = 100$ mM $Na_i = 1$ mM
    $K_o = 1$ mM $Na_o = 100$ mM
    $P_{Na}/P_K = 0.1$ Both Cl_i and Cl_o are zero.
    A. Calculate the membrane potential E_m using the Goldman equation.
    B. Calculate E_m for $P_{Na}/P_K = 1$.

11. The universal gas constant R has units of J/°K/mol. The Faraday F has units of C/mol and T is in °K. What are the units of RT/F? Note that J/C = V.

Again, select the best answer(s). (In many cases, more than one answer is correct.)

12. The sodium pump
    A. pumps K^+ uphill.
    B. maintains cell volume.

C. is inhibited by cardiac glycosides.

D. pumps Na$^+$ downhill.

13. The sodium pump

    A. has a high affinity for K$^+$ on the outside of the membrane.

    B. requires energy to be converted into the high affinity form.

    C. is stimulated by increasing intracellular Na$^+$.

    D. has a high affinity for Na$^+$ along the outside of the cell membrane.

14. The sodium pump

    A. is inhibited by cyanide.

    B. is inhibited by anoxia.

    C. is inhibited by zero K$^+$ extracellularly.

    D. is responsible for a considerable amount of basal O$_2$ consumption.

15. An increase in intracellular sodium would be expected to

    A. stimulate the sodium pump.

    B. raise the levels of intracellular calcium in many tissues.

    C. lower the levels of amino acids in most cells.

    D. cause cellular swelling.

ANSWERS

1. **D.** Intracellular Na$^+$ levels are low and K$^+$ high because of an energy-requiring mechanism in the cell membrane that "pumps" Na$^+$ out of the cell and K$^+$ into it. Radioisotope studies demonstrate that the cell membrane is not impermeable to Na$^+$ or K$^+$. The Cl$^-$ ion is not always passively distributed; it varies among cell types. In virtually every animal cell, however, Na$^+$ is low and K$^+$ is high. Preferential adsorption of K$^+$ and exclusion of Na$^+$ by an intracellular polyelectrolyte gel is a theory of cellular ionic distribution that currently has very few advocates.

2. **A–D.** Selections C and D are part of the general definition of a steady state; A and B are particular aspects of the cellular ionic steady state condition.

3. **A–C.** When cells die, they no longer produce ATP, their metabolic energy substrate. Therefore, the Na–K pump ceases to operate, causing cellular Na$^+$ levels to rise and K$^+$ to fall. This causes the membrane potential to fall toward zero, in turn forcing Cl$^-$ to enter the cell down its electrochemical gradient. In addition, a breakdown of the cellular membrane occurs, making it less and less of a barrier to ionic traffic.

4. **D.** Na$^+$ is believed to leak into cells through small water-filled holes (pores). Na$^+$ with its adherent water molecules is too tight a fit to easily enter through these pores.

5. $R = $ J/mol/°K; $T = $ °K. $C_L - C_R = $ mol/cm^3. $RT(C_L - C_R) = $ J/cm^3 = erg/cm^3. 1 erg = 1 dyn cm. $RT(C_L - C_R) = $ dyn/cm^2, which is pressure, or R can be given in units of L · atm/°K · mol and concentration in terms of mol/L. Therefore L · atm/°K · mol × °K × mol/L reduces to atmosphere, another unit of pressure; 1 atm = 760 mm Hg.

6. Concentration of NaCl 0.9 g/100 ml = 9.0 g/L 9.0 ÷ 58.5 = 0.154 mol = 154 mMol/L. To obtain ideal osmolarity, remember NaCl dissociates in solution to Na$^+$ and Cl$^-$; therefore, 154 mM Na$^+$ + 154 mM Cl$^-$ = 308 mOsm. An isotonic solution is one with an effective osmolarity of 286 mOsm. This means that the 308 "ideal" mOsm of the 0.9% NaCl solution acts as though it were 286 "effective" mOsm. Therefore, the osmotic coefficient = 286/308 = 0.93 [see Eq. (9)].

7. Urea, which is quite permeant relative to the osmotically active particles inside the cells rapidly enters the cells, causing the osmolarity of the cell interior to increase. Thus water enters the cells, causing them to swell and eventually burst.

8. Zero, because both ion mobilities are equal. Hence, movements of K$^+$ and Cl$^-$ are of equal size but of

opposite charge, preventing any net potential difference from being generated.

9. A. -58 mV

   B. 10 mM

10. A. $E_m = 58 \log \dfrac{1 + (0.1)(100)}{100 + 0.1}$

   $= 58 \log \dfrac{11}{100.1} = -55.6$ mV

   B. $E_m = 58 \log \dfrac{1 + 100}{100 + 1} = 0$ mV

11. Volts

12. A–C. The sodium pump, more properly known as the Na–K pump, moves Na^+ out of the cell and K^+ into the cell against their electrochemical gradients. This uphill transport process renders the membrane effectively impermeable to Na^+. An extracellular impermeant ion is thus created to offset the intracellular protein and create the "double Donnan" situation, which is thought to be important in maintaining cell volume. The only known biochemical action of the cardiac glycosides is to inhibit Na/K ATPase, the sodium pump.

13. A–C. The sodium pump has a high affinity for K^+ outside the cell and for Na inside the cell. Energy is required to bring about these high affinity states. The most important physiological stimulant of the sodium pump is an increase in the cellular concentration of sodium.

14. A–D. Cyanide and anoxia both block ATP synthesis. Without ATP the sodium pump cannot work. External K^+ is required in order to activate the pump. As virtually every cell in the body has a sodium pump using up ATP, a considerable portion of basal O_2 utilization is directed toward the synthesis of ATP via oxidative phosphorylation.

15. A–D. The sodium pump is stimulated by an increase in cellular Na levels. The Na-coupled transport systems of calcium and amino acids require a normal Na^+ gradient. An increase in cellular Na^+ decreases the gradient: therefore, lower levels of calcium will be transported out of the cell and lower levels of amino acids will be pumped into it. A rise in intracellular Na^+ levels will, by the Donnan effect, cause more chloride to enter the cell, thereby increasing the number of osmotic particles. In response water will enter the cell, and the cell will swell.

Excitability

MALCOLM S. BRODWICK

OBJECTIVES

After completion of this chapter, you should be able to

1. Understand the passive and active electrical properties of nerve membranes and know the important terminology.

2. Understand how an action potential is propagated along an axon, including the role of local circuits and saltatory conduction in myelinated fibers.

3. Have an elementary understanding of conduction velocity and extracellular recording.

The nervous system is designed to collect information about external and internal events, to integrate this information, and to provide an appropriate response. Other organ systems, particularly the endocrine system, can be thought of as functioning in a similar fashion, but the design is not as local. Nor is the speed of the response as quick or detailed. The nervous system has been developed for speedy transmission of information from one part of the body to another. Nerve cells are the unit structures in which transmission occurs; the process by which transmission is accomplished is electrical. The all-or-none action potential represents a particular specialization of a more general and fundamental property of living systems, excitability. Excitability can be defined broadly as the ability to be activated by and to react to stimuli.

In higher organisms, in which cells have differentiated to perform specialized tasks, the property of excitability is especially enhanced in sensory cells, nerves, muscles, and certain glandular tissues. In these tissues excitability functions to transduce, transmit, and control the responses to a wide range of environmental stimuli. In this chapter we will consider how information is transmitted from one part of a cell to another and will examine the electrical and chemical nature of the membrane changes underlying the action potential. Because the electrical signal decrements with distance, action potential propagation requires a system of local amplification. We will therefore discuss aspects of local circuit theory, as well as the remarkable consequences of nerve myelination. The chapter ends with a discussion of extracellular recording techniques encountered by medical students in the clinical setting. Chapter 3 will show how nerves communicate with each other, while chapters 4–6 will show how excitability controls muscular contraction.

ELECTRICAL PROPERTIES OF MEMBRANES

Because of the central role of electrical processes in excitability we will review some basic electrical concepts relevant to membrane function.

THE MEMBRANE AS A RESISTOR

The property of a material to impede the flow of direct current (DC) is known as resistance. The relationship between current, voltage, and resistance is given by *Ohm's law:*

$$E = IR \tag{1}$$

Potential difference = current × resistance

The electrical resistance of a material is determined by its dimensions as well as by its electrical properties through the following relationship:

$$R = r\frac{d}{A} \tag{1a}$$

Resistance = resistivity (Ωcm)

$$\times \frac{\text{thickness}}{\text{cross-sectional area}}$$

From this equation we see that the electrical resistance across a membrane decreases with increasing membrane area and increases with increasing membrane thickness. Because membrane thickness is generally constant and membrane surface area varies for purposes of comparison among different tissues, the membrane resistance is expressed as areal specific membrane resistance in units of Ωcm^2. For example, the resting squid axon membrane has a specific resistance of about 1,000 Ωcm^2, which means if you had 1 cm^2 of membrane, it would have a resistance of 1,000 Ω. If you had 2 cm^2, the total resistance would be 500. Cellular membranes generally have about 100 to 1,000 times more resistance to current flow than do the electrolyte solutions bathing their surfaces. Therefore, the resistance of the extracellular fluids may be neglected for most purposes. Membrane electrical events are mediated by net ionic fluxes (unidirectional influx minus unidirectional efflux), which constitute the transmembrane ionic

currents. Membrane physiologists often use the membrane conductance g, which is simply the reciprocal of the membrane resistance 1/R, to describe membrane properties.

THE MEMBRANE AS A CAPACITOR

The lipid bilayer that constitutes most of the membrane structure is a thin (40–100 Å) sheet of low dielectric constant separating two aqueous phases of high dielectric constant containing electrolytes. These two fluid phases are analogous to the two plates of a capacitor. Capacitance is the property of a material to store charge when a voltage is applied across the material:

$$q = CV \tag{2}$$

Charge = capacitance × voltage

The capacitance of biological membranes is generally constant and is equal to about 1 μF/cm^2. If we take the derivative of Eq. (2), we get

$$\frac{dq}{dt} = C\frac{dV}{dt}$$

but

$$\frac{dq}{dt} = I$$

where I is current. Therefore,

$$I = C\frac{dV}{dt} \tag{3}$$

The total current flowing across the membrane is the sum of a resistive component and a capacitative component:

$$I_{\text{total}} = \frac{V}{R} + C\frac{dV}{dt} \tag{4}$$

THE MEMBRANE AS AN RC NETWORK

The electrical properties of a membrane can be represented by an *equivalent circuit,* as illustrated in Figure 1, where subscript m denotes the membrane. An equivalent circuit is an electrical analog of the biological material and

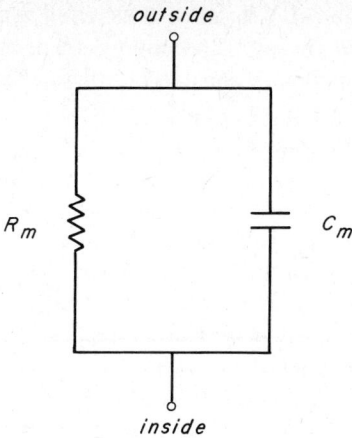

Figure 1 Membrane patch equivalent circuit.

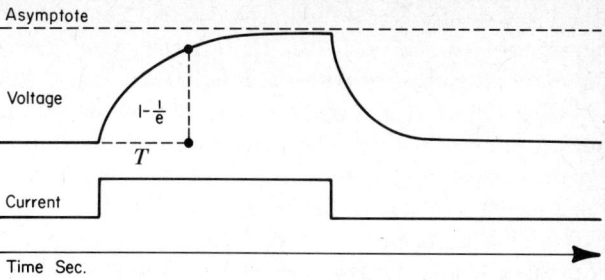

Figure 2 Exponential voltage response to rectangular current pulse.

mimics the current and voltage response to voltage and current stimuli. A section of biological membrane lacking excitable components can be adequately modeled by an equivalent circuit composed of a resistor and capacitor in parallel. We will now consider some of the electrical properties of this membrane and its equivalent circuit.

Time Constant

If a rectangular current pulse is delivered across the parallel RC (resistance–capacitance) network of Figure 1, we get the exponential voltage response illustrated in Figure 2. The form of the rising phase of the voltage response is given by

$$V = V_{asymptote} (1 - e^{-t/\tau}) \tag{5}$$

This is an exponential function of time, and the rate at which it changes is characterized by the symbol τ (Greek letter tau), known as the *time constant*. The time constant is given by

$$\tau = R_m C_m \tag{6}$$

In the time of one time constant, the response has completed $1-(1/e)$ (i.e., about ⅔) of its asymptotic value, which is given by Ohm's law, $E = IR$.

Length Constant

Real cell membranes can be thought of as being composed of many sections of RC membrane units connected in parallel. For some cells, such as vertebrate neurons or skeletal muscle fibers, the RC units may be connected in complex branching networks. The equivalent circuit for a simple section of axon is illustrated in Figure 3. Note that RC units are connected through an internal resistance r_i on the inside and connected directly on the outside. The resistivity of a unit cube of internal fluid is approximately

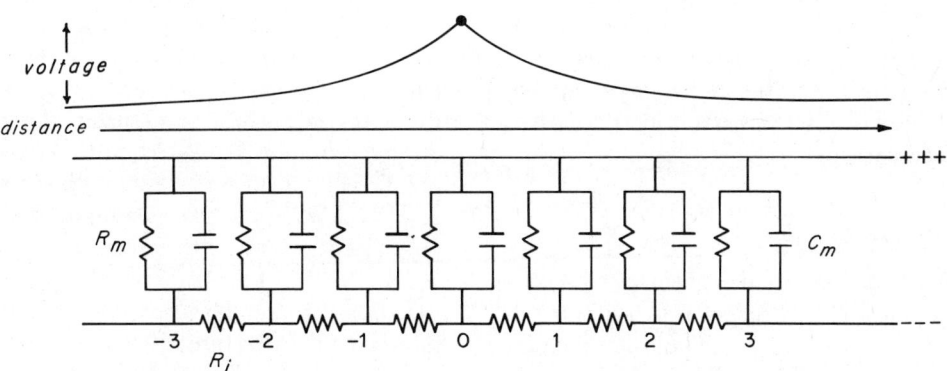

Figure 3 Spatial decrement of voltage to a point current stimulus.

equal to that of external fluid. The external fluid usually fills a large volume such that the external resistance is quite low as compared with the internal resistance. The structure depicted in Figure 3 is known as an infinite one-dimensional *cable*. The system of current–voltage equations governing such structures was first described by Lord Kelvin to describe the behavior of transmitted signals in the trans-Atlantic cable. Cables are dissipative structures. If a voltage is applied at point 0 and the voltage response is measured along the length of the cable, we find that the voltage decrements exponentially with distance as

$$V = V_o \, e^{-x/\lambda} \tag{7}$$

where x is the cable length along the axon, V_o is the voltage at point 0, and

$$\lambda = \left(\frac{r_m}{r_i} \right)^{1/2} \tag{8}$$

where r_m is the membrane resistance per unit length (Ω cm) and r_i is the internal resistance per unit length (Ω /cm).

The *cable properties* of membranes attenuate potential signals both temporally and spatially; the potential responses are referred to as *electrotonic potentials*. If a real nervous system were made of such *passive* components, the distances between "connected" nerve cells would have to be less than one or two space constants. As the distance between the ends of most sensory and motor nerve fibers is hundreds of times longer than this, it would be necessary either to contract the dimensions of the organism or to amplify the signal locally. Signal amplification is accomplished by voltage-dependent changes in membrane properties that produce the *action potential*. These voltage-dependent properties are often described as *active* in contradistinction to the passive cable properties we have just discussed. The active properties underlying excitation should not be confused with active ion transport discussed in the previous chapter.

THE ACTION POTENTIAL

When a nerve or muscle membrane is penetrated with a microelectrode, a potential between -60 to -100 mV is recorded with reference to the potential outside the cell, which is usually set at ground potential, that is, zero mil-livolts. This potential is called the *resting potential*. If we deliver current through a second electrode this potential can be altered. Before considering these effects we must first discuss some conventions regarding current. Current is defined as the flow of positive charge. Current flowing inward across the passive membrane increases the negativity of the intracellular potential and is said to *hyperpolarize* the membrane. Conversely, if a hyperpolarizing potential is applied, current will flow from the outside relative positivity to the inside negativity. If the current is applied in the outward direction the membrane potential becomes less negative, that is, it moves toward zero and is said to *depolarize* the membrane. Figure 4 shows a series of two inward and three outward current pulses being delivered to an excitable membrane. We observe that the voltage response for the two inward current pulses follows Eq. (5) for passive *RC* circuits. The response to the first outward current pulse obeys this relationship as well, being simply the mirror image of the response to the first inward current pulse. However, the second depolarizing response departs from the expected passive response in three important ways: (1) the potential rises more quickly; (2) the steady state potential may be different from that predicted by the product of the resting resistance and the outward current using Ohm's law; and (3) the time course of the potential response when the current pulse is terminated is prolonged. Occasionally, small oscillations are observed as well. These added features brought about by depolarizing stimuli are known as the *local responses*. If the amplitude of the current pulse is increased further the membrane produces the complex, highly nonlinear response known as the *action potential* or *spike*. The membrane potential at which an action potential is produced 50% of the time is known as the *threshold potential*. The threshold itself is not constant but varies with the duration and intensity of the pulse as shown by the strength–duration relationship illustrated in Figure 5. Shorter duration pulses require a stronger stimulus. A natural consequence of the strength–duration relationship is that the *latency* between stimulus onset and the action potential decreases with increasing stimulus strength (Fig. 6).

Let us examine the structure of the action potential in more detail. After threshold potential has been reached, the potential depolarizes steeply during the *rising phase*.

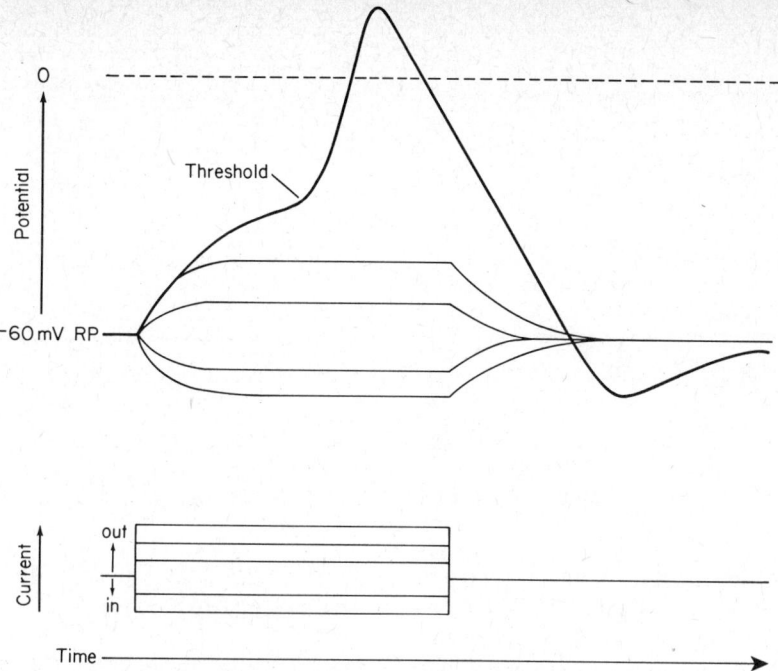

Figure 4 Response of an excitable membrane to current stimuli.

The potential of the rising phase actually *overshoots* zero potential and for nerve axon and muscle approaches the Nernst equilibrium potential for sodium ion, E_{Na}. Following the *peak* of the action potential, the *falling phase* restores the potential, but the potential may actually *undershoot* its resting value. For historical reasons, the undershoot potential, when it occurs, has been called the positive *afterpotential*. In skeletal muscle the action potential has a depolarizing foot on the falling phase known as the *negative afterpotential*. The overshoot potential follows changes in external sodium concentration according to the changes in Nernst potentials for Na ion. The undershoot varies with changes of external potassium concentration but the relationship is not perfectly Nernstian for K ion.

The action potential, when it occurs, is an *all-or-none* event. That is, for a stimulus greater than threshold the resulting action potential is a stereotyped event with respect to amplitude and time course. If the height of the action potential is decreased by reducing external sodium, the resulting action potential, although of smaller amplitude, is still an all-or-none event.

Two action potentials may be evoked by two successive threshold pulses. If the interval between the pulses is decreased, the second pulse will eventually fail to evoke an action potential. The action potential can be reinstated by increasing the magnitude of the stimulus. If the interval is contracted still further, eventually it will be impossible to generate the second action potential at all. The interval wherein an action potential may be evoked by increasing the stimulus intensity is known as the *relative refractory period*. The shorter interval when even increased stimulus intensity fails to evoke an action potential is known as the *absolute refractory period*.

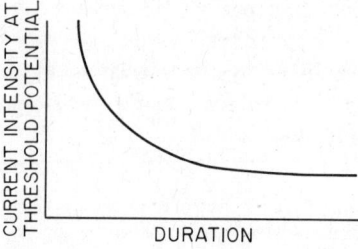

Figure 5 Strength–duration relation for spike threshold.

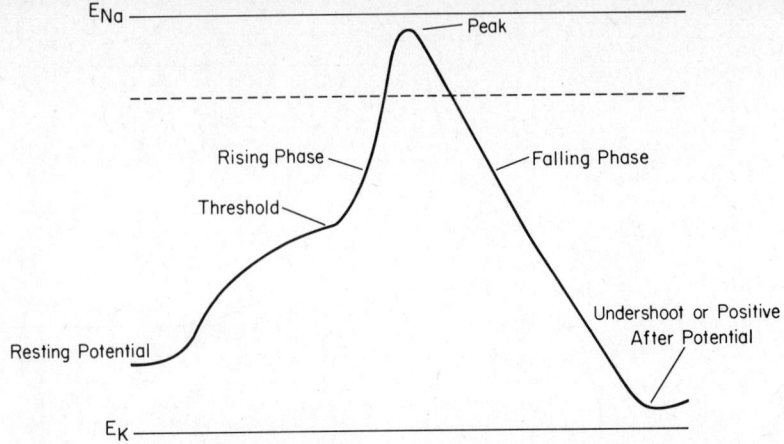

Figure 6 Components of the action potential.

Some excitable tissues fire *repetitive* action potentials in response to prolonged depolarizing current pulses. For a constant current stimulus, the interspike intervals generally get successively longer. This process is known as *adaptation*. *Rapidly adapting* nerves fire only one or two action potentials to prolonged stimuli. Some nerves will fail to fire an action potential if a sufficiently slow ramp stimulus is given. This process is known as *accommodation*. Occasionally, the term "accommodation" is used instead of the term "adaptation" to describe a decelerating spike train, but this is incorrect usage.

If a hyperpolarizing pulse precedes a depolarizing stimulus, the magnitude of the depolarization necessary to evoke an action potential may be decreased. Following a hyperpolarizing pulse of sufficient amplitude, an action potential may result without any depolarizing stimulus at all. This phenomenon is known as *anodal break.* * Because of its stereotyped structure the action potential does not contain information per se. Information in the nervous system is encoded either by the *intervals* between the spikes in a train or by the activation of special nerves referred to as *labeled lines* using the language of communications theory.

---

*The anode is the positive electrode. In the days before intracellular recording, when the term "anodal break" was created, hyperpolarization of the membrane was accomplished with an external anode. This is equivalent to an internal cathode or negative electrode.

THE IONIC BASIS OF THE ACTION POTENTIAL

In 1952 experiments performed on squid giant axon elucidated the ionic basis of the action potential. For this work Hodgkin and Huxley received the Nobel Prize. Some of the antecedents for these experiments will be described before presenting the description of the ionic currents underlying the action potential.

Around the turn of the century, Bernstein hypothesized that the resting membrane was permeable only to potassium and that during the action potential, the membrane became equally permeable to all ions. This may be referred to as the membrane breakdown hypothesis. Later Overton demonstrated that sodium ion was necessary for excitation, but the significance of this observation was somehow overlooked. The original membrane breakdown hypothesis was disproved when it was shown that the action potential had an overshoot. If the membrane became equally permeable to all ions, the potential at the peak of the action potential would approach zero. However, the peak varied with external sodium concentration in a Nernstian fashion. Hence, it was proposed that the depolarizing phase of the action potential was due to an increase in sodium permeability. The next problem was to determine what features of the stimulus and membrane were responsible for activating the all-or-none action potential.

Let us consider the equivalent circuit of Figure 1. The

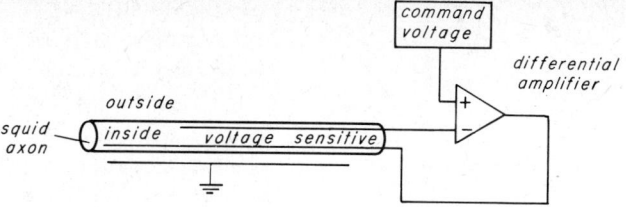

Figure 7 Schematic of voltage clamp technique.

action potential could, in principle, be generated by either a change in membrane resistance (e.g., sodium permeability) or a change in membrane capacitance ($V = q/C$). Thus as the capacitance decreases at constant charge, the voltage increases. Just before the demonstration of the role of Na ions in the action potential overshoot, Cole showed that the action potential was accompanied by a decrease in resistance, while the capacitance remained approximately constant (see Selected Bibliography). When an action potential occurs the change in potential is accompanied by parallel changes both in resistance and current. If the resistance changes are a function of voltage, one cannot use Ohm's law, which assumes a constant resistance to determine membrane resistance; during an action potential all three variables—current, voltage, and resistance—are changing.

To make the voltage "sit still," a technique called *voltage clamp* was used. The membrane was clamped to a given command potential or voltage in much the same way that temperature is controlled by a thermostat. The experimental arrangement is diagrammed in Figure 7. The voltage sensed by an intracellular electrode is fed to one terminal of a differential amplifier. The other terminal of the amplifier is connected to a command voltage. If the membrane potential differs from the command potential, a current proportional to the difference in the potentials is delivered to the axon. The current flows across the resistance of the axon until the voltage of the axon is identical to the command voltage. The general strategy of producing an output to lessen the difference between a set point and the actual operating point of the system is known as *negative feedback*. Negative feedback systems are the essence of homeostasis, one of the central themes of physiology.

Voltage clamping has three advantages. First, ionic cur-

rents can be studied without capacitative currents. Recall $I_{total} = I_{ionic} + I_{capacitative}$ from Eq. (**4**). But $I_c = C \, dv/dt$ and $dv/dt = 0$ for a square pulse of voltage (i.e., the voltage does not change after the instantaneous rise of the square pulse). Second, we can determine whether the conductance changes responsible for the action potential are in fact *voltage dependent* or *current dependent* by changing the ionic solutions. Third, given that the conductances are voltage dependent, we may determine the voltage and time dependencies of various components of the currents provided we have adequate techniques for separating the specific ionic components.

We are now ready to observe the currents obtained under voltage clamp conditions. Figure 8 illustrates a family of currents at several different voltage steps. We note that for hyperpolarizing voltage commands the small currents are inward and flat as expected of a resistor (Fig. 8, M and N). For small depolarizing command steps, the current is outward and flat (Fig. 8, K and L). For larger depolarizing pulses a *transient inward current* is followed by an *outward current* that reaches a steady state. Unlike the current stimulus condition presented earlier in Figure 4, in which the all-or-none action potential was produced by a suprathreshold stimulus, under voltage clamp the currents make smooth transitions in a graded fashion indicating that the conductance changes are continuous, that is, the inward current is not an all-or-none event. Hodgkin and Huxley found that the early transient inward current I_m followed the equation and was carried by sodium ion,

$$I_{Na} = G_{Na} \, (E - E_{Na}) \qquad \textbf{(9)}$$

where G_{Na}, the Na conductance, is a *function* of *voltage and time*, as shown in Figure 9; E_{Na} is the equilibrium potential for sodium ions; and E is the membrane po-

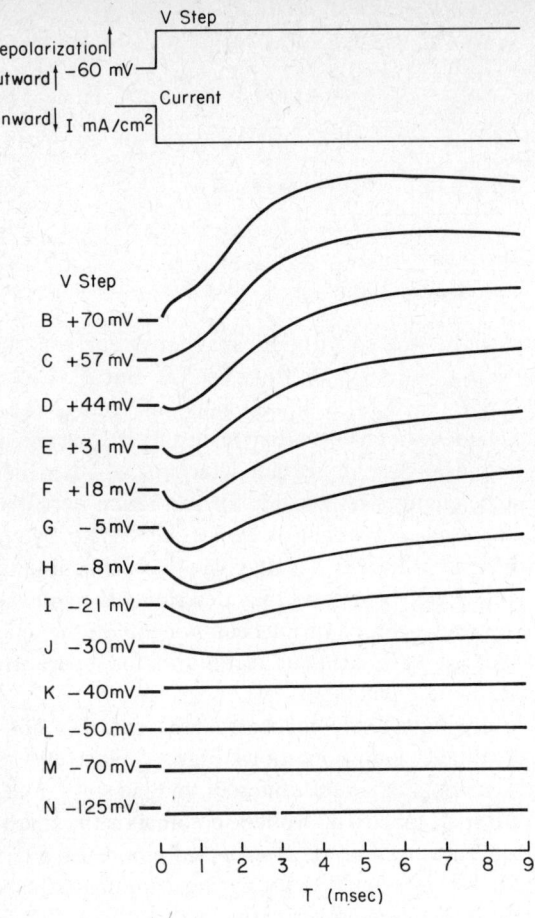

Figure 8 Family of currents produced by different membrane potentials in squid axons.

reversal potential, even though the conductance might be large, no net current is flowing through the sodium system at all, since the driving force $E - E_{Na} = 0$.

The reversal potential for the early current was shown to vary with external sodium in a Nernstian manner, whereas the late current varied with external potassium. By appropriate ionic substitutions, the early sodium and late potassium currents can be separated as shown in Figure 9. The sodium conductance has an early turn-on phase termed *sodium activation,* followed by a turn-off phase termed *sodium inactivation.* The magnitudes and time courses of both phases are voltage dependent. The magnitudes of both phases increase, whereas the time courses speed up with increasing depolarization. The actual observed current flowing through this voltage- and time-dependent current still depends on the driving force, which in turn depends on the composition of the solutions bathing the membrane. The development of sodium inactivation is about 10 times slower than that for the sodium activation. In contrast, the later potassium conductance has only an activation phase (at this time scale), which develops about 10 times slower than the sodium activation phase and has a pronounced delay before its onset.

How can these voltage-dependent conductances account for the action potential? We note that sodium con-

tential. At a given voltage and time, the sodium current is thus proportional to the driving force on sodium, namely, the difference between the command potential E and the Nernst potential for sodium E_{Na}. If the Nernst potential is altered by partially substituting external sodium with an impermeant ion, the voltage dependence of the conductance remains the same even though the current magnitude, and possibly direction, may have changed. Note that the direction of the early current changes from inward to outward when the command voltage passes E_{Na} (as shown by traces B, C, and D in the left-hand panel of Fig. 8). The potential at which a current changes direction is called the *reversal potential.* At the

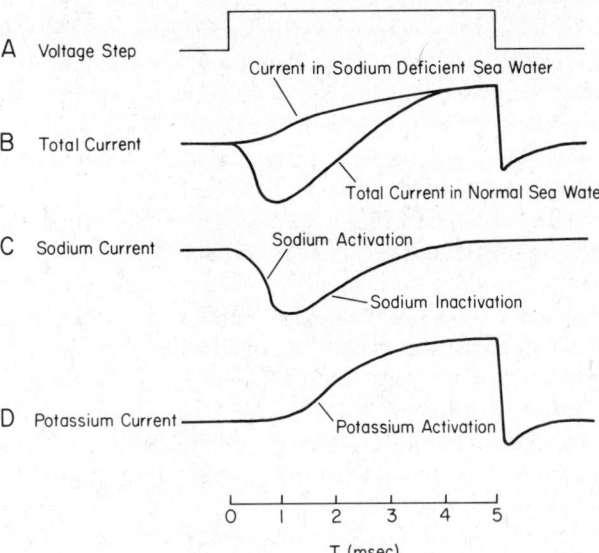

Figure 9 Current separation in a voltage clamped squid axon.

ductance is increased with depolarization. An increased sodium conductance, which implies an increased sodium permeability, would be expected to result in a further depolarization. Consider the constant field equation (Chapter 1); as the P_{Na} term becomes larger, the membrane potential moves toward E_{Na}. With increased depolarization more of the sodium system is activated, which in turn produces still more depolarization. This *positive feedback* system has often been called the *Hodgkin cycle*. The depolarization is however limited by E_{Na}. So far we have accounted for the depolarizing phase of the action potential. The action potential must now be repolarized. The squid axon system employs two mechanisms. First, the sodium activation that has produced the depolarization is followed by sodium inactivation. The inactivation would produce some repolarization, which in turn would decrease some of the activation—a sort of backward Hodgkin cycle. Recently such a system, that is, sodium activation and inactivation, has been shown to account completely for the action potential of peripheral myelinated nerve in rat. A second repolarization mechanism employs the potassium conductance system, that is, depolarization enhanced by the sodium system activates the slower potassium conductance. As the potassium permeability increases, the membrane potential tends toward E_{K}.

Unfortunately, for the sake of simplicity, in a number of different preparations other important voltage-dependent conductances occur. In some nerve cell bodies as well as smooth and cardiac muscle, at least part of the inward current, which produces depolarization, is carried by calcium ion. For cells that have large calcium currents recent evidence indicates, the increased *internal* calcium brought about by the inward calcium current activates a potassium conductance. The *calcium-activated potassium conductance* has been implicated in the process of adaptation. Muscle cells and some nerve cells have another potassium conductance, activated by membrane hyperpolarization rather than by depolarization. For this reason, it is called the *inward or anomalous rectifier* to distinguish it from the previously described potassium conductance activated by depolarization, which is sometimes referred to as the *delayed rectifier*. The consequences of the inwardly rectifying conductance are discussed with regard to cardiac muscle in Chapter 5.

EFFECT OF DIVALENT CATIONS

Alterations of external calcium and magnesium concentrations affect the threshold. Increases of divalent Ca ion concentrations increase the threshold potential, whereas decreases in concentration reduce threshold potential. In very low calcium concentrations some nerves actually fire spontaneously. The most popular explanation of this effect is the following. Recall that a membrane is composed of a lipid bilayer with polar phospholipid head groups oriented toward the aqueous phases. Many, if not most, of these head groups and membrane proteins bear *fixed negative charges* oriented toward aqueous phases surrounding the plasma membrane. These fixed charges produce a surface potential in series with the membrane potential measured by an intracellular electrode. This surface potential in the normal ionic environment extends for only tens of Ångstroms and is not measured by intracellular and extracellular electrodes, which are usually much further from membrane. Although the surface potential is not part of the resting potential measured with electrodes in the bulk aqueous phases, it is close enough to affect the voltage-dependent molecules responsible for conductance changes. Divalent cations can interact strongly with the fixed negative charges in such a way that the negative surface potential is neutralized. A neutralized external surface potential is equivalent to a hyperpolarization of the membrane potential, from the point of view of the membrane molecules—although not from our electrodes. In order to activate sufficient sodium current to generate an action potential, this additional hyperpolarization induced by surface potential neutralization by divalent cations must be overcome, that is, the threshold potential becomes more positive.

CONSEQUENCES OF ACTION POTENTIALS

The energy used to produce the nerve action potential comes from the energy stored in the electrochemical gradients for sodium and potassium. These gradients must be slightly dissipated every time an action potential occurs. In fact, with repetitive activation, the ion gradients would eventually be too small to produce all-or-none action potentials. Fortunately these gradients are maintained by the sodium–potassium pump, discussed in Chapter 1. Thus the sodium pump removes internal sodium accumulated during the depolarization phase and restores the potassium lost during the repolarization phase. The sodium–potassium pump, although restorative, is not responsible for the repolarization phase, a fact that is commonly misunderstood. For some cells the sodium–potassium pump generates a sizeable hyperpolarizing potential (Chapter 1). This potential would then act to reduce the firing frequency during a prolonged stimulus or to raise the current "threshold" for a second stimulus of short interval. Thus, the *electrogenic* sodium pump as well as a calcium-activated potassium conductance may contribute to adaptation.

Many nerve cells are enmeshed in a complex network of glial cells. The extracellular space is thus somewhat restricted to a thickness of 200–400 Å. Potassium ions move from the intracellular compartment into this *periaxonal* or *perisomal* space in response to depolarization. The glial cells act as a diffusion barrier for K ions to pass to the larger extracellular space. The result is that potassium accumulates in the restricted extracellular space causing the Nernst potential and the membrane potential during the repolarization phase to shift in a positive direction.

A MOLECULAR INTERPRETATION OF EXCITABILITY

How does the membrane conductance change as a result of an alteration of membrane potential? How does the membrane discriminate between sodium and potassium ions?

The macromolecules responsible for excitation, probably proteins, change their orientation in the membrane in response to voltage. A voltage-induced change in conformation, rotation, or translation within the membrane would be expected if the macromolecule behaved as a dipole or monopole in the membrane; a change in the potential gradient would act as a force on the particle, causing it to move. A moving dipole or monopole within the membrane field would constitute a current in the membrane. Recently a very fast current was discovered that has properties related to the sodium conductance system. These fast currents have been called *gating currents* for their presumed role in gating the sodium activation process. The gating results in a patent hole or pathway through the membrane known as a channel that ions may traverse. Once a channel is opened by such a voltage-dependent gating process, it may have one or more discrete states of conductance known as the *unit conductance*. Small currents flowing through channels have been detected and are referred to as single channel currents. These unit conductances have mean lifetimes that are voltage dependent. When the membrane potential is altered the opening and closing rates of these channels change in a probabilistic manner producing the macroscopic membrane conductance such as that shown in Figure 9.

The sodium and potassium conductance systems are capable of conducting different ions with varying degrees of *ion selectivity*. Thus the sodium channel discriminates for sodium over potassium in a ratio of about 10–1 and the potassium channel discriminates potassium over sodium in a ratio of about 100–1. Although these selectivities are not perfect, they are close enough to justify the labels "sodium channel" or "potassium channel." The selectivity is thought to result from the distribution of specific atoms that line the channel wall. For cation selectivity bearing partial negative charges, oxygen atoms from carboxyl, carbonyl, and ether groups have been proposed. Selectivity may result from both the *coulombic* (C) interaction of the ion with these negative charges and from size and shape limitations imposed by the geometry of the channel. Once within a channel an ion may confront several potential energy barriers separating binding sites in sequence. If the number of barriers is sufficiently great and their amplitude sufficiently small, ionic flow may be described adequately by the constant field equation, or by some modification of it. More recently it has been shown that ion conduction in Na and K channels may also be described by a theory that deals explicitly with ion transport across a few barriers in the channel. These are steps toward a molecular interpretation of ion transport in channels, but the adequacy of these interpretations requires more experimental information about channel structure.

PHARMACOLOGICAL MODIFICATION OF EXCITABILITY

The actions of drugs on nerve interest us not only for their clinical implications, but also for the information they provide on the mechanisms underlying excitability. For example, specific drugs are useful in separating the different ionic components previously discussed. Thus tetrodotoxin, a toxin contained in pufferfish, and saxitoxin, a toxin found in clams infested with certain protozoa, specifically block the sodium conductance system. The insecticide DDT and certain scorpion venoms block sodium inactivation, thereby producing action potentials of long duration. Certain local anesthetics appear to plug open sodium channels. Tetraethylammonium (TEA) and barium ions, when present in the cytoplasm, block the potassium conductance. Binding studies with tetrodotoxin and saxitoxin indicate that sodium channels are comparatively sparsely distributed in the membrane, about 300–500 channels/μm^2 in squid axon. Studies with TEA suggest a similar channel density for potassium channels.

PROPAGATION

Thus far we have treated excitation as a local event involving a small area of membrane under voltage clamp. But, as was pointed out earlier, the nervous system is often required to send signals over long distances, and transmission must be as rapid as possible. As we have seen, nervous tissue is not perfectly insulated from extracellular fluid, and the action potential must be locally amplified along the length of the cable to prevent the potential from dissipating.

When an experimenter passes inward or outward current across the membrane, there is a net current to ground; that is, the entire circuit includes the stimulating electrode, the membrane, and a path to ground, or to the other pole of the stimulator. When an action potential propagates along a nerve, the net current to ground is zero. The entire current path is confined to the membrane in a *local circuit,* as indicated in Figure 10. Figure 10b shows the local circuit in longitudinal section—local circuit currents do, of course, flow in three dimensions around the circumference of the excited section. Current flows into the axon at regions known as *sinks* and originates from regions known as *sources*. Obviously sink currents must equal source currents for the *total* membrane current to equal zero. However, the *current densities* of individual sources and sinks need not be equal, particularly when the sodium channels are localized in discrete areas of membrane. If we pass current locally to depolarize a region of axon mem-

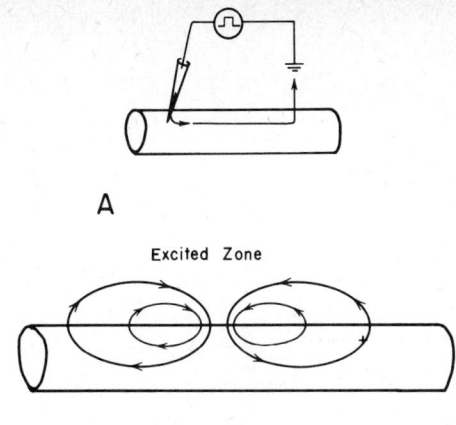

A

Excited Zone

B

Figure 10 Stimulating and local circuits.

brane the voltage dissipates exponentially along the length of the cable—as does the current density. If the voltage over a given region exceeds threshold, an action potential will fire. The voltage from the action potential will electrotonically spread down the axon as described in Eq. (7). Sodium current enters the fiber at the excited zone, and the local circuit path produces an outward current in adjacent current patches. The outward current depolarizes these adjacent patches, and when the potential reaches threshold, the adjacent patches fire an action potential. The action potential thus *propagates* by depolarizing adjacent patches in a continuous manner. The propagation velocity or *conduction velocity* clearly depends on λ, the length constant. When the length constant is long, distal sections of the axon are brought to threshold sooner than for a short length constant. The length constant itself varies inversely as the square root of internal resistance

$$\lambda = \left(\frac{r_m}{r_i} \right)^{1/2} \qquad \textbf{Eq. (8).}$$

The internal resistance per unit length of axon r_i increases as the cross-sectional area decreases:

$$r_i = \frac{R_i}{\pi(\text{radius})^2}$$

where R_i is the intracellular resistivity in Ωcm of axoplasm; eq. (1a). The membrane resistance r_m varies inversely with the fiber radius

$$r_m = \frac{R_m}{2\pi(\text{radius})}$$

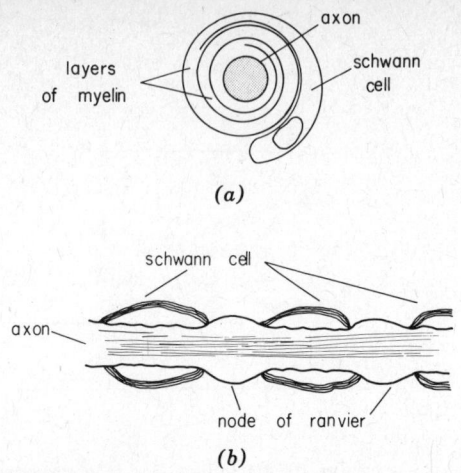

(a)

(b)

Figure 11 Structure of a myelinated nerve fiber. (*a*) Radial cross section; (*b*) longitudinal cross section.

where R_m is the specific resistance of a 1 cm^2 membrane section. Thus

$$\lambda = \left(\text{radius } R_m/2R_i \right)^{1/2}$$

The conduction velocity therefore *increases* as the fiber diameter increases essentially by increasing λ. For certain responses, a fast conduction velocity is essential. It is therefore not surprising that the escape responses of several invertebrates is mediated by axons of large diameter. The giant axons of squid may be more than 1 mm in diameter and have conduction velocities of about 25 m/s.

SALTATORY CONDUCTION

High speed propagation is also important for the axons of vertebrate nerves. In major nerves, such as the sciatic or optic nerves, the number of fast conducting fibers is great. If all the axons were about 1 mm in diameter, the diameter of the nerve would be absurdly large—a square optic nerve containing 1 million square-shaped axons of 0.5 mm sides would be 0.5 m on a side. Instead of possessing axons of large diameter, vertebrates have developed a unique strategy for increasing conduction velocity. Glial cells envelop the axon periodically (about every 2 mm) along the axonal length. The glial cells then wrap around the axon several times extruding the glial cytoplasm as indicated in Figure 11*a*. This process is known as *myelination*. Myelination is accomplished by glial cells called *Schwann cells* for peripheral nerves and *oligodendrocytes* for central nervous system cells. Between adjacent glial cells is a region relatively free of myelin known as the *node of Ranvier* as illustrated in Figure 11*b*. The axon membrane at the node is excitable. The layers of myelin increase the effective length constant by forcing current to flow to adjacent nodes (see Figure 12). The effective length constant is increased in two ways. First, the layers of myelin are composed of nonpolar fatty material and act as a good insulator. Current flow must pass through the myelin layers in series before reaching the external electrolytic solution; thus R_m under the myelin is effectively raised. Second, the layers of myelin act as capacitors in series and reduce the total capacitance of the cable; capacitance in series is given by

$$\left[\Sigma_1^n (\frac{1}{C_n}) \right]^{-1}$$

Thus both ionic and capacitative currents are impeded. More of the local current produced by an action potential must then flow to the next adjacent node of Ranvier. When this nodal membrane is brought to threshold, it fires an action potential. The action potential thus hops from node to node in a process known as *saltatory conduction*. The membrane underlying the myelinated portion, sometimes called the *internode,* appears to be unexcitable. In certain demyelinated conditions, such as those produced by diphtheria toxin, the resulting conduction velocity is greatly decreased.

EXTRACELLULAR RECORDING

Most of the potential related data you will encounter in the clinic will be recorded with extracellular electrodes. Obviously extracellular electrodes cannot record the resting potential of cells. However, excitable cells produce

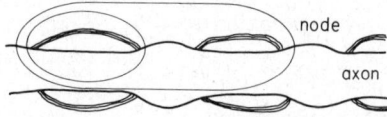

Figure 12 Currents for myelinated nodes.

currents that flow across the resistance of the extracellular medium. The potentials that result follow Ohm's law. To record extracellular potentials two electrodes must be used. One electrode may be placed near the tissue of interest and the other at a remote ground location. Alternatively both electrodes can be placed near the tissue. If the two electrodes are too close, the potentials at the two electrodes will be identical and so the potential difference will be zero. The form of the extracellular potentials depends greatly on the geometric arrangement of the two electrodes and the tissue. A detailed exposition of these considerations is far beyond the scope of this chapter. Instead we will examine how propagation affects the form of the extracellular potentials.

Current must flow in a circuit. As pointed out earlier, current flows from anode to cathode. Often a system is composed of many *sources* of current that flow into *sinks*. Figure 13 depicts a nonpropagating excitable cell with current flowing. The cell has one source that drives current into two lateral sinks. Lines of equal current density form complete loops. *Equipotential lines* (surfaces for three dimensions) intersect these current lines at right angles. The potential decreases with distance from the active surface. The actual potential recorded depends on which equipotential lines the two electrodes sample. As the electrode is moved away from the cell surface, the local current density decreases and the equipotential lines broaden and decrease in amplitude. Thus, if there are many oriented current generators, the electrode will sample a larger population when it is away from the surface.

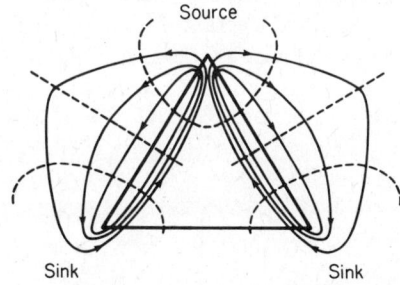

Figure 13 A hypothetical cell with a source and two sinks. The solid, closed curves are lines of equal current density. Note that the current density decrements with distance from the surface. The dashed lines are lines of equipotentiality and intersect the current lines at 90°.

Only when the electrode is close to the source or sink will individual generators be apparent. Such extracellular measurements are generally used to provide information on ensembles of cells acting in concert. Whole organs or layers of cells can be composed of oriented sources and sinks, dipoles, such that the whole unit exhibits orderly current flow. Such extracellular potentials are useful in locating structures involved in generating correlated responses. A disease process is often characterized by a particular arrangement of waveforms, the underlying origins of which are obscure.

Figure 14 illustrates a nerve with an action potential propagating from left to right. The electrodes are placed sufficiently far apart that potentials are recorded separately at each of them. When the inward current sink is directly below electrode X (Fig. 14*b*) a maximal potential is recorded at this electrode and no potential is noted on electrode Y. After the action potential has propagated to a point between the two electrodes, the small outward currents are about equal, and the difference in potential between electrodes X and Y is zero as noted in Figure 14*c*. The potential waveform then inverts—electrode Y is of opposite polarity to electrode X—as the propagating action potential passes beneath electrode Y, as indicated in Figure 14*d*.

Another method of recording extracellular potentials is shown in Figure 15. The two electrodes are shunted by a resistance and the remote electrode is grounded. Voltage is measured across the shunt resistance and is proportional to the local current at the probe. A propagating action potential is recorded as a triphasic potential that reflects the local current flow. Initially the membrane at electrode X acts as a current source for membrane to the left hand side of it (containing the action potential). As the propagating potential sweeps directly under, the membrane acts as a current sink; as the potential passes beyond electrode X, the membrane at X acts as a current source again, this time to the membrane to the right-hand side of electrode X.

An investigator usually encounters large bundles of axons that compose a whole nerve. Stimulation of such a nerve results in a complicated waveform known as a *compound action potential*. The complexity arises from a dispersion of conduction velocities and waveform shapes arising from the individual axons that vary in diameter

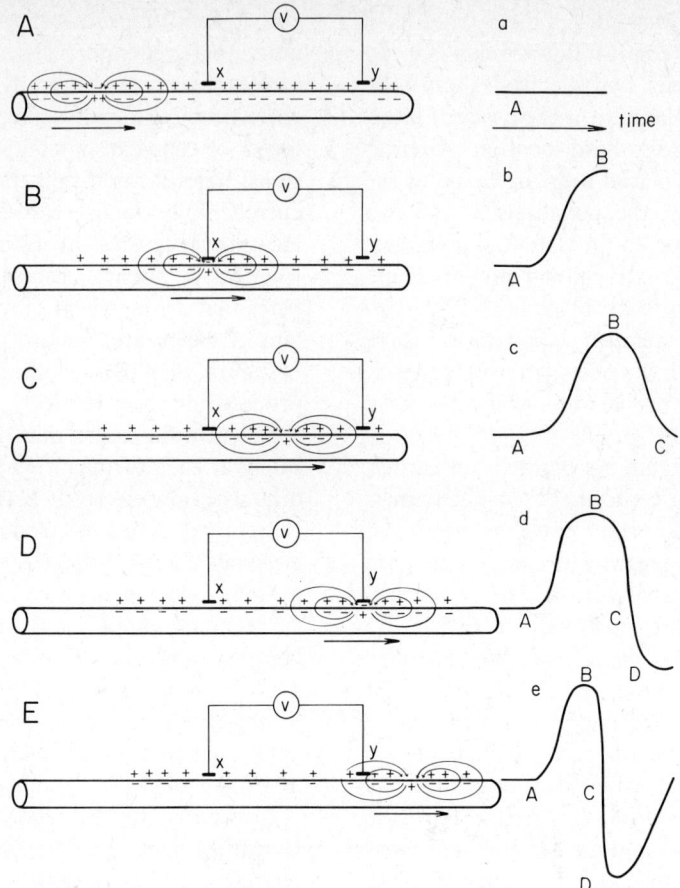

Figure 14 Extracellular recording of a propagating action potential.

and can be either myelinated or unmyelinated. Moreover, unlike the all-or-none action potential of a single axon, the compound action potential is proportional to stimulus parameters; as stimulus intensity or duration is increased, more fibers are brought to threshold and produce more propagating action potentials. The compound action potential is composed of several distinct conduction velocity groups. The significance of these groupings will be discussed in later chapters dealing with the central nervous system.

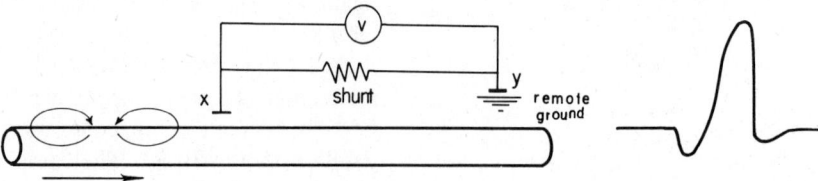

Figure 15 Extracellular recording of local current.

SELECTED BIBLIOGRAPHY

Aidley DH: *The Physiology of Excitable Cells.* New York, Cambridge University Press, 1971.

Cole KS: *Membranes, Ions and Impulses.* Berkeley, University of California Press, 1968.

Hodgkin AL: *The Conduction of the Nervous Impulse.* Springfield IL, Charles C Thomas, 1964.

Jack JJB, Noble D, Tsien RW: *Electric Current Flow in Excitable Cells.* Oxford, Clarendon Press, 1975.

Junge D: *Nerve and Muscle Excitation.* Sunderland, Sinaueur Associates, 1981.

SELF-STUDY QUESTIONS

1. Suppose a membrane had a resistance of 10,000 Ωcm^2 and a capacity of 1 $\mu F/cm^2$. What would be the time constant for a cell with an area of 1 cm^2? With an area of 2 cm^2?

2. Suppose λ, the length constant, is 3 mm. Suppose the voltage change at V_1 induced by current injection at the same spot is 100 mV. What is the voltage at V_2, which is 6 mm away?

3. How many sodium ions must enter an axon to produce a depolarization of 100 mV? Assume a cell with a capacitance of 1 $\mu F/cm^2$ and an area of 1 cm^2. (Note: 1 Faraday = 96,500 C/mol and 1 mol = 6 $\times 10^{23}$ particles.)

4. How is an action potential like a flushing toilet? Where does this simile break down?

5. What would presumably happen to the resting potential and action potential if we were to double the concentration of sodium ion outside (neglect osmotic water movements)? What would happen if we were to double external potassium?

6.

   A. Plot the early transient current and the steady state currents as a function of voltage from Figure 8.

   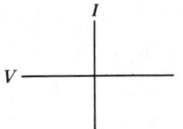

   B. For the early transient current a reversal potential is noted. What is the significance of this voltage?

   C. The slope of the graph you have constructed in 6A should give the conductance, $1/R$. What happens to this slope conductance as you progress to more depolarized potentials? What assumption regarding Ohm's law is being violated?

   D. The "real" conductance is given by the chord conductance

   $$\frac{I_{ion}}{V - V_{reversal}} = G_{ion}$$

   Plot the chord conductance versus voltage for the early transient current. Note that there are no negative conductances as there are with slope conductances.

7. What would happen to the sodium reversal potential if we were to double the external sodium?

8. How do the variables of sodium activation, sodium inactivation, and potassium activation account for the relative and absolute refractory periods?

9. What would happen if the sodium inactivation were removed? What if sodium activation had the same speed as sodium activation?

10. Certain amphiphilic drugs such as salicylate can partition into the membrane and leave a fixed negative charge on the surface of the bilayer. How might such an effect explain the pain-alleviating capabilities of salicylates (aspirin)?

11. In low calcium concentrations, nerves and muscles can fire spontaneously. Suggest a mechanism for this observation.

12. What would happen to action and resting potential at short and long times if ouabain were applied?

13.

    A. What would the current–voltage relationship (see question 6A) be after a nerve was treated with tetrodotoxin? With TEA?

    B. How might the action potential (Fig. 8) appear in a nerve treated with TEA? With TEA and scorpion venom?

14. What would happen to the internal resistance, length constant, and propagation velocity if we were to stick a low resistance wire inside the axon?

15. Are fiber diameter and internodal distance important in determining conduction velocity of myelinated nerves?

ANSWERS

1. $\tau = RC$ 1 cm: $10^4 \ \Omega \ cm^2 \cdot \dfrac{10^{-6} \ F}{cm^2} = 10^{-2}$

              2 cm: $2 \ cm^2 \cdot 10^4 \ \Omega cm^2 \cdot \dfrac{10^{-6}}{2 \ cm^2} \ F$

                                         $= 10^{-2}$

The time constant is insensitive to the membrane area.

2. For a spatially decaying voltage:

$$V_1 = V e^{-x/\lambda} = 100 \ mV \ e^{-6mm/3mm} = 13.53 \ mV$$

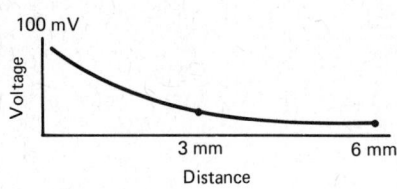

3. $C_T = \dfrac{capacitance}{unit \ area} \cdot area$ $C_T = \dfrac{1\mu F}{cm^2} \cdot cm^2 = 1 \ \mu F$

   $q = CV$ $q = 10^{-6} \ F \cdot 10^{-1} \ V = 10^{-7} \ C$

   $\dfrac{1 \ mol}{96,500 \ C} \cdot 10^{-7} \ C = 1.036269 \times 10^{-12} \ mol$

   $\dfrac{6 \times 10^{23} \ sodium \ ions}{mole} \cdot 1.036269 \times 10^{-12} \ mol$

                        $= 6.2176 \times 10^{11} \ sodium \ ions$

4. There are a number of similarities. The water pressure head is like the transmembrane voltage. The action potential would be equivalent to the all-or-none flushing. If you fail to rotate the flushing handle enough, you get only an abortive flush. Some toilets even have a strength–duration relationship. Refractory periods are also similar. As the water level falls at the pressure source, a floating device turns off the flow somewhat similarly to the sodium inactivation process. Differences include the fact that a toilet has only one current. Also flushing current is not composed of water pressure dependent unit conductances. The tetrodotoxin blockade analogy is left to your imagination.

5.   A. Na_o would affect the resting potential little, since $P_K >> P_{Na}$—refer to the constant field (Goldman) equation. At the peak of the action potential, the membrane potential approaches the Nernst potential for sodium. Thus the action potential overshoot would be increased by 17.5 mV.

$$E = 58 \log \dfrac{2[Na]_o}{[Na]_i} = 58 \log 2 + 58 \log \dfrac{[Na_o]}{[Na_i]}$$

  B. $2 \times K_o$ would depolarize the resting membrane potential.

$$V_m = 58 \log \dfrac{P_K[K]_o + P_{Na}[Na]_o}{P_K[K]_i + P_{Na}[Na]_i}$$

The amount of the depolarization clearly depends on the ratio $P_{Na}:P_K$. The action potential would be somewhat smaller because (1) the resting potential RP is somewhat depolarized, (2) the depolarized RP might cause some sodium inactivation and result in less sodium current available for depolarization, and (3) the repolarization driving force E_K would be smaller (i.e., more depolarized).

6. A.

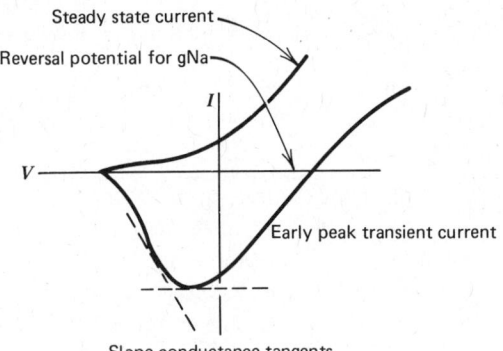

Steady state current
Reversal potential for gNa
I
V
Early peak transient current
Slope conductance tangents

B. $I_{Na} = g_{Na}(E - E_{Na})$. When $E = E_{Na}$ there is no driving force; thus the current equals zero. When E exceeds E_{Na}, the sodium current will actually be net outward.

C. For the early peak current I_{Na} the slope conductance is negative. At the maximum peak inward current, the slope conductance is zero. But clearly the membrane is conductive. The reason for this anomalous situation is that the assumption that the conductance is voltage independent, as for a fixed resistor, is violated; conductance is itself a function of voltage. If we measure the conductance at a particular potential instantaneously, however, it will be constant and the current–voltage relationship linear, as shown by the dashed lines. This conductance is called chord conductance.

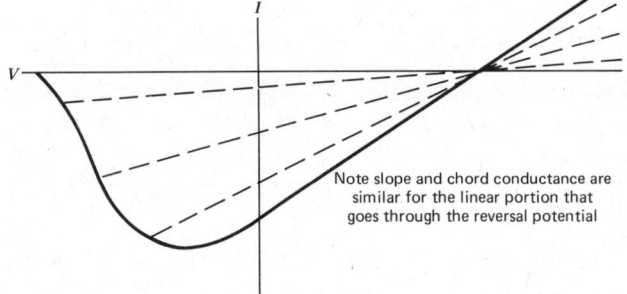

I
V
Note slope and chord conductance are similar for the linear portion that goes through the reversal potential

7. $2 \times$ Na would shift the reversal potential to more depolarized potentials (17.5 mV).

8. The relative refractory period would be characterized by a loss of available inward sodium current through sodium inactivation and an outward shunt of potassium current due to potassium activation. During the absolute refractory period these features would be so large that additional depolarization would be unable to enlist enough sodium current to yield a regenerative action potential.

9. If sodium inactivation were removed, the action potential duration would be prolonged, repolarization being dominated by potassium activation alone. If sodium inactivation and activation had the same time course and magnitude, no sodium current would occur at all in response to a voltage stimulus. The actual current would depend on the magnitude of the two processes, but they would clearly be working against each other, a poor and unlikely design.

10. As the negative charges insert into the outside face of the bilayer, the surface potential would increase. This increased negative surface potential would be equivalent to a tonic *de*polarization. Such a depolarization might result in sodium inactivation.

11. When calcium ion is removed, the magnitude of the surface negative potential increases. This increased external negativity would shift the $g_{Na}(V)$, where

$$g_{Na} = \frac{I_{Na}}{V - E_{Na}}$$

curve in a hyperpolarizing direction. If the resting potential RP is relatively unaffected by this procedure, more inward current would be available at rest. With sufficient inward current, threshold would be reached and the fiber would fire spontaneously.

12. At short times, ouabain would cause a small initial depolarization due to blockade of the electrogenic component of the sodium–potassium pump. At longer times, the gradients for sodium and potassium would run down. RP and overshoot would approach zero. Slowly the sodium system would inactivate, and excitability would disappear.

13.

   A.

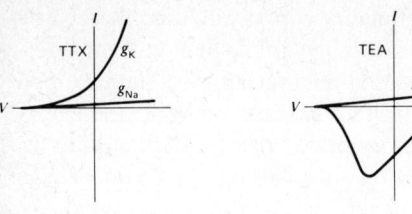

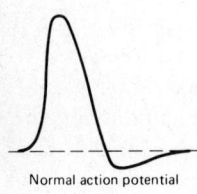
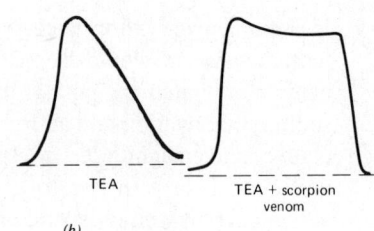

(a)

Normal action potential   TEA   TEA + scorpion
                                                venom

(b)

14. A low resistance wire placed inside would lower the internal resistance:

$$\lambda = \left(\frac{R_{\mathrm{m}}}{R_{\mathrm{i}}} \right)^{1/2}$$

If R_{i} is decreased λ would increase. If λ increases, so does the conduction velocity, since the depolarization can spread over a longer distance.

15. Fiber diameter is important since λ increases with fiber diameter:

$$\left(\frac{R_{\mathrm{m}}}{R_{\mathrm{i}}} \right)^{1/2}$$

R_{i} decreases as the square of the fiber radius, whereas R_{m} is given by $2\pi r$). A longer length constant means that the internodal distance can be increased, which in turn implies a faster conduction velocity. The internodal distance is important, because within the limit of zero distance, the conduction velocity would be the same as an unmyelinated fiber. When the internodal distance is too long, the decrementing voltage would be too great and conduction would fail. The nodes are spaced to make for a substantial safety factor.

Synaptic Transmission

BURGESS CHRISTENSEN

OBJECTIVES

1. Describe the ultrastructure of the synapse and the variety of synaptic relationships found in the central nervous system.

2. Describe the electrical and chemical events occurring at the postsynaptic membrane following release of neurotransmitter.

3. Describe the mechanisms underlying the release of transmitter from presynaptic elements.

4. Discuss diseases involving synaptic mechanisms.

The transfer of signals from one cell to another can be treated as an extension of the principles discussed for generation and conduction of the action potential. The fundamental problem to be considered is the transfer of the electrical impulse or action potential from one nerve cell to the next or from nerve cell to effector organ. Since the basic unit of information conducted along the cell axon is the action potential, it is reasonable to ask why this electrical event is not transmitted directly across an uninterrupted connection to the next neuron. In fact, this does occur in some situations across electrical synapses. However, one reason this might not be the best mechanism is the impedance mismatch between connected elements. Action currents generated by the action potential crossing electrical synapses into another low-impedance cell would not be of sufficient intensity to alter the membrane potential except by only a few millivolts. The mechanisms by which the transfer of the action potential occurs include a combination of electrical and chemical events. When the action potential arrives at the end of the axon or nerve terminal, the depolarization of the terminal membrane results in the liberation of a specific chemical called a neurotransmitter. This chemical produces a conductance change on the membrane of the next neuron in the pathway; this conductance change results in the continuation of the electrical events. Because of this intermediate step of chemically induced transmission, the synapse is a vulnerable site at which many therapeutic drugs as well as potent toxins can interfere with the process of signal transmission. As you now know, the change in membrane potential that follows a membrane permeability change will depend on the ion species that flow across the membrane. The direction of current and the resulting synaptic potentials depend on the particular ions, their concentrations on both sides of the membrane, and the value of the membrane potential of the postsynaptic cell at the time of transmitter action. We will examine each of these events in some detail.

ULTRASTRUCTURAL CONSIDERATIONS

It is particularly helpful in discussing the subject of synaptic transmission to become familiar first with the neuronal elements involved and with their structural arrangements within the nervous system. The term *synapse* comes from the Greek word meaning conjunction or connection. It is used in biology to mean the site at which two distinct and separate cellular elements form a specialized connection. This specialized connection permits the passage of electrical information from one excitable cell to the next. Each synapse has both a presynaptic and a postsynaptic element. The cellular elements that form a synapse can come from any part of the neuron. As you remember, a neuron consists of a cell body and an axon. Most neurons also have extensions of membrane from the cell body, called dendrites. The presynaptic element is that part of the neuron that will release the neurotransmitter following depolarization by the action potential. The postsynaptic element is that part of the neuron that will be acted on by the neurotransmitter. It should be pointed out, however, that any element can be both pre- and postsynaptic. Let us take the simplest case as an example. As shown in Figure 1 (axosomatic synapse), an axon terminal can form the presynaptic element and the cell body the postsynaptic element.

Once initiated at the cell body of the presynaptic neuron, the action potential travels down the axon, finally arriving at the terminal. Neurotransmitter is released and diffuses across a small intercellular space (0.05 μm) called the *synaptic cleft*. The transmitter then causes a membrane permeability change to occur on the membrane of the postsynaptic cell body. This type of synapse is called an axosomatic synapse. Synapses can be formed between axon terminals and dendrites (axodendritic), between axon terminals and other axons (axoaxonic), and among dendrites (dendrodendritic) (Fig. 1).

There are many possible combinations among these elements, for example, serial and reciprocal synapses, as shown in Figure 2. In a serial synapse, an axon terminal can be presynaptic to a postsynaptic dendrite, which then forms the presynaptic element to yet a third element, which can be another dendrite. In a reciprocal synapse, an axon terminal forms the presynaptic element to a dendrite, which then synapses back onto the axon terminal in a feedback relationship. These synaptic combinations are common in the central nervous system (CNS).

In summary, the synapse consists of a presynaptic element and a postsynaptic element, with a synaptic cleft interposed between. An important point to remember is

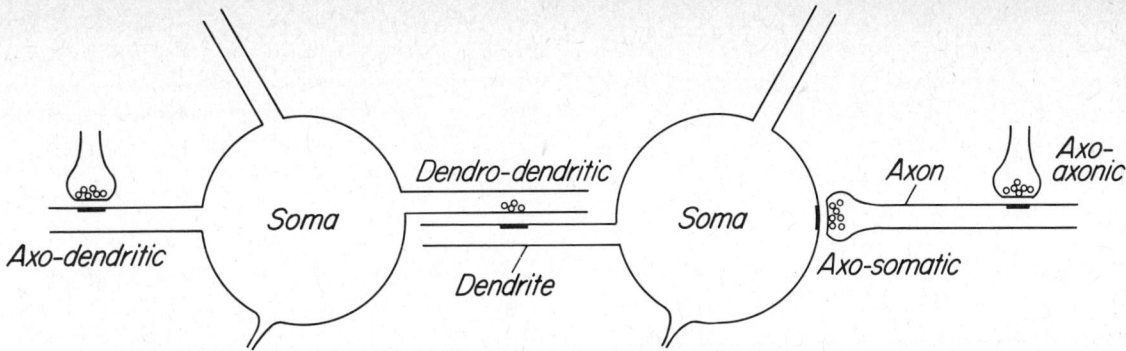

Figure 1 Schematic representation of the different types of synapses that can be formed by the different parts of a neuron. The spheres in the presynaptic element represent the synaptic vesicles that contain the transmitter. The dense bars in the postsynaptic element represent the site at which receptors exquisitely sensitive to the transmitter are concentrated.

that these elements are not continuous, but are separated by the synaptic cleft. One further ultrastructural characteristic that must be mentioned is the membrane compartments, which contain the neurotransmitter. These are represented by the usually round vesicles contained in presynaptic terminals. Each vesicle contains a large number of molecules (about 8,000 in the motor nerve) of neurotransmitter. The steps involved in packaging the transmitter within the vesicle remain unknown.

POSTSYNAPTIC EVENTS

To understand the mechanisms involved in the release of neurotransmitter and its subsequent action on the post-

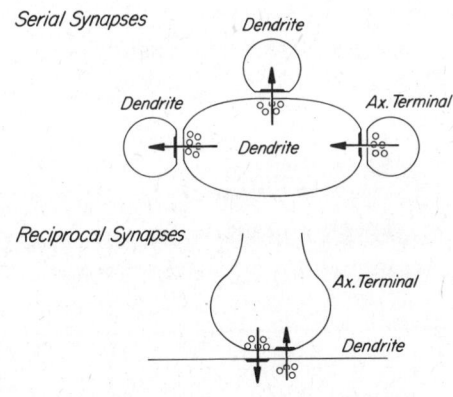

Figure 2 Two types of synaptic arrangements commonly found in the CNS. Top, Serial synapses; bottom, a reciprocal synapse.

synaptic element, we will consider the best understood example of a synapse—the connection between the motor nerve terminal and the muscle cell, the neuromuscular junction (NMJ). Although synapse is often used to describe this connection, the correct terminology is the motor endplate. In the vertebrate, the cells that give rise to the axons connecting to or innervating skeletal muscle cells are located in the spinal cord. These axons are heavily myelinated; when they reach the vicinity of the muscle, they branch. The myelin is lost as each branch terminates on a single muscle cell, leaving only a simple Schwann process covering the axon. The terminal at the mammalian NMJ looks like a bunch of grapes sitting on top of the muscle fiber, as shown in Figure 3. Under each grape-like ending, the muscle membrane is invaginated, as shown in Figure 3b–c. These invaginations are the sites embedded with receptors that bind the neurotransmitter (Fig. 3c).

The underlying mechanisms involved in synaptic transmission are essentially expressed as voltage changes across a membrane. The method of approach to the study of synaptic transmission has therefore been one of measuring the voltage across the membrane with a microelectrode, much as has been done in the study of the action potential. We will first concern ourselves with the electrical events that occur at the postsynaptic membrane, in this case the muscle fiber. The investigator most responsible for describing the events that occur during synaptic transmission is Dr. Bernard Katz, who won the Nobel prize in 1971 for his work in this area.

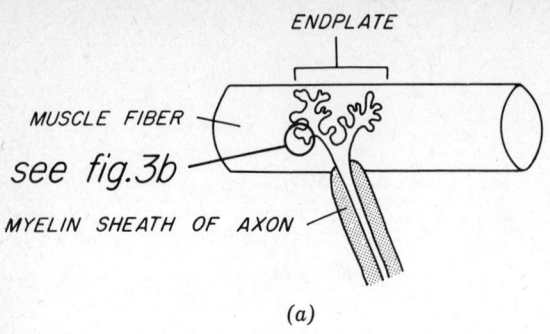

ENDPLATE

MUSCLE FIBER

see fig.3b

MYELIN SHEATH OF AXON

(a)

Figure 3 (*a*) A mammalian motor nerve ending on a muscle fiber. Note the absence of myelin over the nerve terminal. (*b*) Relationship of one bouton with the muscle cell membrane. Note the invaginations of the muscle cell where postsynaptic receptors are located. (*c*) Structure of one invagination and its proximity to the presynaptic release sites.

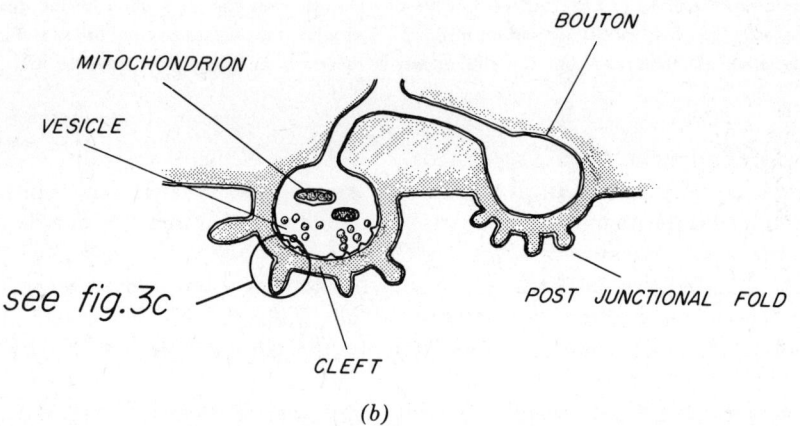

BOUTON

MITOCHONDRION

VESICLE

see fig.3c

POST JUNCTIONAL FOLD

CLEFT

(b)

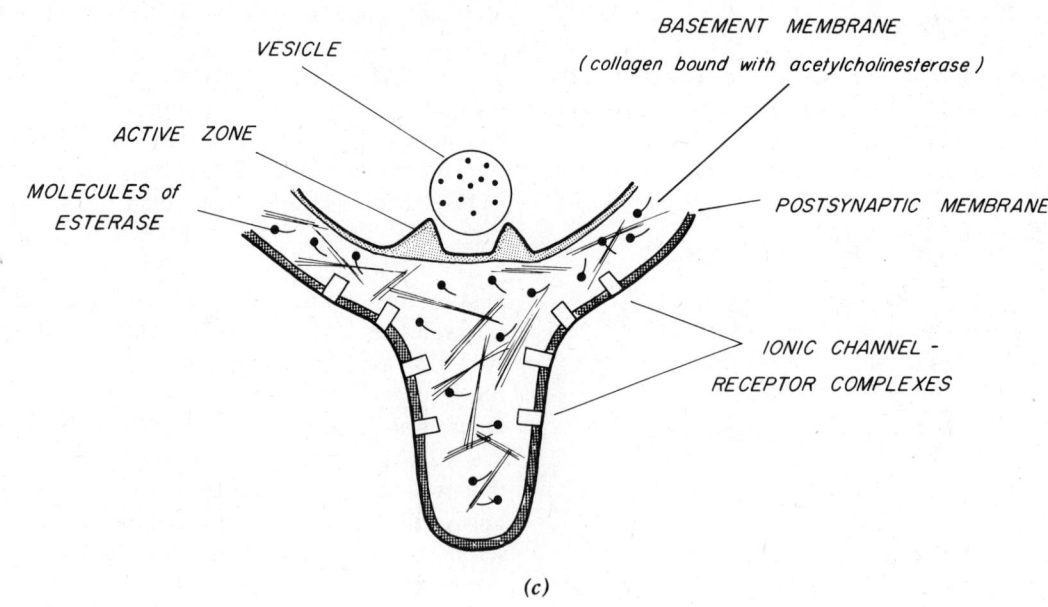

VESICLE

BASEMENT MEMBRANE
(collagen bound with acetylcholinesterase)

ACTIVE ZONE

MOLECULES of ESTERASE

POSTSYNAPTIC MEMBRANE

IONIC CHANNEL - RECEPTOR COMPLEXES

(c)

When a microelectrode is inserted into a muscle fiber in the endplate region, where the nerve contacts the muscle as shown in Figure 3a, the first change one records of course is the transmembrane potential or resting potential. For skeletal muscle this is usually about −90 mV. If one were to examine closely the electrical recording on, for example, an oscilloscope, one would notice very small rapid changes in membrane potential occurring randomly in time. These changes have a relatively constant amplitude of about 500 mV, with a duration of about 5 ms— all in the depolarizing direction (see Fig. 4c).

Now if one stimulates the motor nerve electrically so as to cause action potentials to travel down the axons, one will record after a short delay (why is there a delay?) a significant change in membrane potential, causing the muscle to contract (Fig. 4b). What you have observed is the fundamental process of synaptic transmission—the transfer of information from the presynaptic axon to the postsynaptic muscle fiber, resulting in a muscle contraction. Unfortunately, because of the size of the muscle action potential and the resulting muscle contraction, the microelectrode pops out of the muscle fiber, and you are unable to examine in detail the electrical events that result from the transmitter action.

Now let us review briefly what you have observed and what you know about action potential production and see if you cannot deduce some of the events. First, you know that to produce an action potential the membrane potential must cross some threshold level in order to activate the rapid sodium influx mechanism. To do this, the membrane potential must be moved toward zero volts, i.e., the process of depolarization. After you inserted the microelectrode into the muscle fiber, you first observed small depolarizations occurring randomly and spontaneously. When you stimulated the nerve, you observed a large depolarization and an action potential. If you use your imagination, you might surmise that in the absence of an active depolarization of the nerve terminal by an action potential, small amounts of transmitter spontaneously leak out of the nerve terminal. These small amounts of transmitter must be relatively constant, since you observed that the amplitude of the small depolarizations were constant in size. You then surmise that following the arrival of the action potential in the nerve terminal a large amount of transmitter is simultaneously released. This results in a large

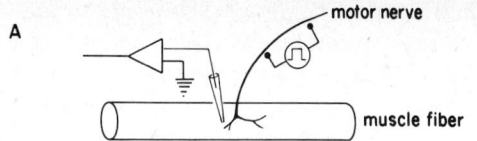

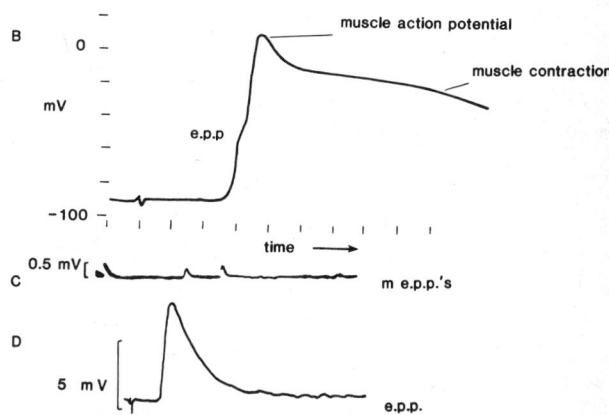

Figure 4 (a) A typical intracellular recording setup for recording synaptic and action potentials from the muscle fiber. An electrical stimulating device is attached to the nerve and is used to initiate action potentials in the axon. The triangle represents an amplifier having an input from the microelectrode. The output of the amplifier would be attached to a recording device, such as an oscilloscope. (b) The recording from the microelectrode following stimulation of the motor nerve. Note the small biphasic deflection of the trace just above the second time calibration. This represents the stimulus artifact. Near the fifth time calibration the potential starts to change rapidly from the resting value (−90 mV) toward zero. The first part of this change is due to the endplate potential (EPP) and is barely recognizable by the slight inflection on the rising phase. The muscle action potential then begins, followed by a muscle contraction. (c) Three spontaneously occurring miniature EPPs. Note the similar amplitude and time course. (d) A single EPP after blocking with curare. Again, note the stimulus artifact. The delay between stimulus artifact and the EPP constitutes conduction time for the nerve action potential plus synaptic delay.

depolarization of the muscle fiber membrane sufficient to cross the threshold and start a muscle action potential. This last step in your thinking process might represent a quantum leap. You set out to prove your theory, which is exactly what Katz did. The problem in examining the electrical events that occur following the substantial release of transmitter is the muscle action potential and subsequent muscle contraction which obscure these events. Your theory suggests that before the action potential is generated in the muscle, a depolarization occurs in the muscle membrane caused by the transmitter, and it is this depolarization that generates the action potential. These events are separate but occur so rapidly that you are unable to see them. The problem, then, is to suppress the muscle action potential. Now if your theory is correct, you can accomplish this in two ways. First, by reducing the amount of transmitter released from the nerve terminal, and second, by inhibiting its action on the muscle membrane. (Review Fig. 3 to see how a number of drugs can affect the two active sites at the synapse.) Both methods could reduce the amount of muscle membrane depolarization to below the action potential threshold. To achieve the first, you reduce the amount of calcium in the external solution that bathes the muscle. The effect of calcium and other divalent cations will be discussed under presynaptic mechanisms. To achieve the second, a poison called curare, which competes with the neurotransmitter for the postsynaptic membrane sites, is added to the bathing solution. Curare reduces the effect of the transmitter released from the motor nerve terminal. With either method, you record the depolarization following nerve stimulation depicted in Figure 4d. This depolarization is similar to that for the spontaneously occurring potentials, except that it is larger and lasts for a longer period of time. This potential is called the synaptic potential or, in the case of the endplate, the endplate potential (EPP). The small spontaneously occurring potentials are called miniature endplate potentials (MEPPs). The depolarizing excitatory synaptic potentials recorded from neurons are called excitatory postsynaptic potentials (EPSPs).

If the electrode is inserted some distance away from the motor nerve terminal, no EPP is recorded. As the electrode is moved closer to the endplate region, the EPP becomes larger, with a faster rise time and decay, until it reaches a maximum when the electrode is located right at the endplate (Fig. 5).

This is an extremely important property of synaptic transmission. The synaptic potential is different from the action potential because it is *not conducted in an all-or-none fashion*. Remember that no matter where the electrode position is located, the action potential would appear the same. However, the synaptic potential becomes

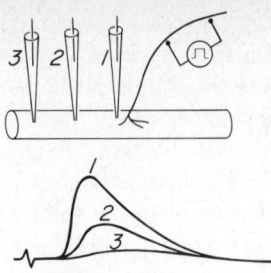

Figure 5 Recording setup similar to that in Figure 4, except the electrode is moved to three different sites. At each site the endplate potential (EPP) is recorded.

smaller the farther away from the site of its generation. In other words, the synaptic potential is attenuated. In the central nervous system (CNS) most synapses are made on the dendritic trees at some distance from the action potential initiation zone at the axon hillock. You can imagine that EPSPs generated on the distal portions of the dendrites can become completely attenuated by the time they reach the soma.

It is unfortunate that the neuromuscular junction is the example used as the basis for a description of synaptic transmission, since the most important functional aspect of synaptic transmission is not normally exhibited at the NMJ. This aspect is the integrative property of synaptic transmission. It has been clearly stated that the synapse is the site of information transfer from one excitable cell to the next. The bit of information is the electrical event or action potential. At the NMJ every action potential in the presynaptic axon normally results in an action potential in the postsynaptic muscle fiber and a muscle contraction. This occurs simply because a large amount of transmitter is released each time from the motor nerve terminal. This is fortunate in this case, since most mammalian skeletal muscle fibers are innervated by only one axon, and failure of this one-to-one relationship would result in loss of muscle activity. In the CNS the situation is quite different.

Each postsynaptic neuron can receive many hundreds of presynaptic inputs from almost as many different cells. If there were a similar one-to-one transmission between neurons as at the NMJ, the CNS would be in a constant state of electrical activity, and complete chaos would result. In contrast to the NMJ, a CNS neuron must be able to examine the activity of the many inputs and decide only when a certain number of inputs are active almost simultaneously to fire an action potential. This process is called integration, which is simply summation of membrane potential changes during a certain time period. The one property of the synaptic potential that enables this integrative process to occur is the ability of synaptic potentials to summate. From your study of action potential generation, you will recall that two action potentials cannot be generated *simultaneously* in the same axon. This is because the membrane becomes refractory following an action potential. Let us return for a moment to the NMJ and perform the following experiment. Suppose we have reduced the amount of transmitter released following a single electrical stimulus to the motor nerve so that the EPP is subthreshold. This is an experimental maneuver and does not normally occur. Now suppose we deliver two stimuli to the nerve with a short delay between each stimulus (Fig. 6a).

If the delay between each stimulus is subsequently decreased (Fig. 6b, c), we note that the second EPP begins to add or summate onto the tail of the first EPP. Finally, when the delay is sufficiently short, the addition of the two individual EPPs results in a voltage that crosses the threshold for action potential generation. Thus, the process of integration is artificially produced for this example. You can imagine how this process occurs in the CNS, in which the activation of a single input to a neuron normally does not release sufficient transmitter to produce a postsynaptic action potential. However, activity in several presynaptic inputs simultaneously (or almost so) will result in summation of the individual EPSPs, with the result that the membrane potential will cross threshold, producing an action potential. Think a minute about what would happen in the postsynaptic cell if the synaptic potential change were of long duration and of sufficient size to cross threshold.

We should re-emphasize the nature of the synaptic connections within the CNS. Before it makes a synapse, each presynaptic axon can divide several times, forming fine terminal branches, and thus several synaptic contacts with a postsynaptic cell. An action potential in the parent axon of the presynaptic cell will invade each terminal branch and depolarize each presynaptic element, releasing some transmitter that will act simultaneously on the postsynaptic cell. Furthermore, literally hundreds of axons can synapse on a single postsynaptic cell. This is called *convergence*. When activity arrives over several individual presynaptic elements and transmitter is released almost at the same time, the individual synaptic potentials sum to make a very large potential change. This type of summation is called *spatial* summation. If activity in the presynaptic elements is repetitive, the summation of the *individual* synaptic events is called *temporal* summation. The summing of the synaptic events is therefore an integrative process. When the potential change in the postsynaptic cell reaches threshold, an action potential is initiated.

POSTSYNAPTIC IONIC EVENTS

The next event we must consider is the way in which a chemical transmitter produces a change in membrane potential, as well as the subsequent fate of the transmitter. It is well accepted that the transmitter acting at the vertebrate skeletal NMJ is the chemical substance acetylcholine (ACh). Following its release from the presynaptic nerve terminal, ACh molecules diffuse across the synaptic cleft and attach to receptors in the postsynaptic muscle cell membrane. These receptors are specific proteins located in the membrane in great concentration just beneath the nerve terminal (Fig. 7a). A snake toxin, α-bungarotoxin, has been used to identify probable receptor sites at the neuromuscular junction. These experiments have indicated approximately 15,000 toxin binding sites per square micrometer in the postsynaptic membrane at the mouth of the junctional folds and about 10^7–10^8 sites

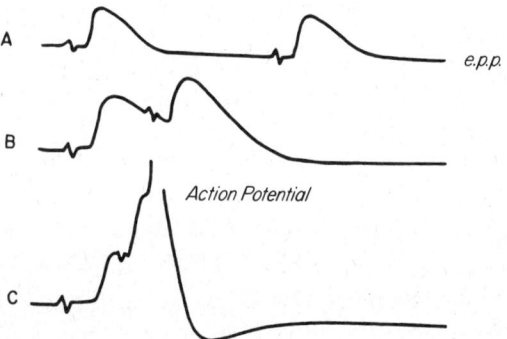

Figure 6 (*a*) Two EPPs following two electrical shocks applied to the motor nerve, several seconds apart. (*b*) The shocks are closer together, showing how one EPP begins to sum on the decaying limb of the first EPP. (*c*) Finally, when the shocks are close enough together, the second EPP crosses threshold to generate an action potential that is off the scale of this drawing.

in the whole NMJ. ACh has a specific affinity for binding to these receptors. Other chemicals can also bind to the receptors and thus block the receptor to the action of ACh. This is how curare and other so-called muscle relaxants work. They compete for the same ACh site and prohibit the binding of ACh. The poison α-bungarotoxin, isolated from krait snake venom, has a very strong affinity for the ACh receptor. In fact, it binds irreversibly with the receptor, and death results from asphyxiation, since the diaphragm is no longer stimulated into contraction. Once the ACh molecule has attached itself to the receptor, a physical conformational change presumably occurs in the receptor protein, or in nearby proteins, permitting a membrane channel to open (Fig. 7b). Evidence suggests that two ACh molecules must bind with the receptor for the change to occur. When the channel opens, ions can then flow through the membrane. As you know, a net inward flow of current results in depolarization. Conversely, a net outward flow of current will result in hyperpolarization. If a conductance change occurs, but without a net flow of current, the membrane potential will be clamped or held near its resting level. More about this important corollary later.

Obviously, if the transmitter were permitted to remain in the synaptic cleft, the muscle would always be depolarized and contraction would fail. The transmitter must somehow be removed from the extracellular space. At the NMJ, this occurs by a chemical reaction. The extracellular space contains an enzyme called acetylcholinesterase (AChE), which hydrolyzes ACh and leaves choline and acetic acid. Although inactive in changing the membrane conductance, choline is reused to make ACh. Active transport mechanisms in the presynaptic terminal take up choline from the synaptic cleft. Once inside the terminal, choline is used to manufacture ACh, which is repackaged into the synaptic vesicles. Enzymatic degradation of transmitter seems to be the exception, rather than the rule. At synapses where other transmitters are used, the mechanism that removes the transmitter from the extracellular space is active reuptake by the presynaptic terminal. This seems to be the most ubiquitous mechanism for transmitter removal in the CNS.

We must now consider what ion species are likely to flow across the membrane to produce appropriate changes in membrane potential to generate action potentials. As an

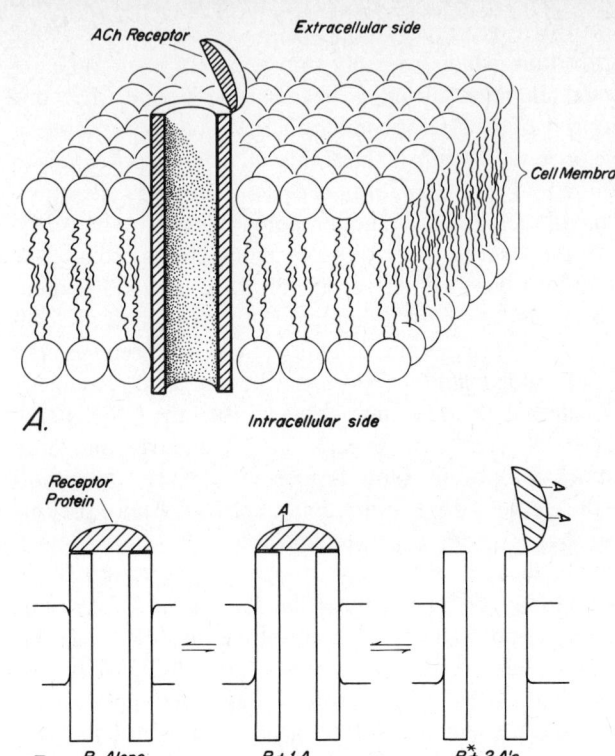

Figure 7 (a) Schematic diagram of synaptic membrane with a receptor protein channel embedded within the lipid matrix. (b) The process of a single synaptic membrane channel opening sequence. The opening can occur only after two acetylcholine (ACh) molecules have attached to specific binding sites on the receptor. R, Receptor; A, agonist; R*, open channel.

aid in intuiting this process, refer to Figure 9, which shows on a voltage scale the equilibrium potentials for the three important intracellular and extracellular ions—sodium, potassium, and chloride. It should be obvious that if the channel permits sodium ions to flow, a net inward current will result, and the membrane will depolarize toward E_{Na}. Conversely, if the channel permits only potassium to flow, a net outward current will result, and the membrane will hyperpolarize toward E_K, as indicated by the arrows in Figure 8.

Under such circumstances, what types of ions will flow

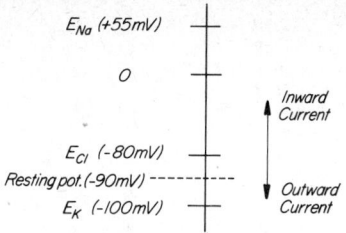

Figure 8 Relationship between the equilibrium potentials for sodium, potassium, and chloride. The resting potential of a muscle fiber of −90 mV is also indicated. Note that an inward current will produce a depolarization, and an outward current will produce a hyperpolarization.

across the membrane when ACh is applied to the muscle cell? The following thought experiments will provide the answer. Suppose that we could somehow alter the membrane potential artificially so as to make it equal to the sodium equilibrium potential and then we were to apply the transmitter ACh. If the effect of ACh were to open channels only to sodium, no net sodium current would flow across the membrane, since the membrane potential was already at the sodium equilibrium potential. The chemical gradient that would otherwise permit sodium to enter the cell is balanced by an electrical gradient to keep it out. Therefore, if the membrane is held artificially at the sodium equilibrium potential and the effect of ACh is to open channels only to sodium ions, no potential change—

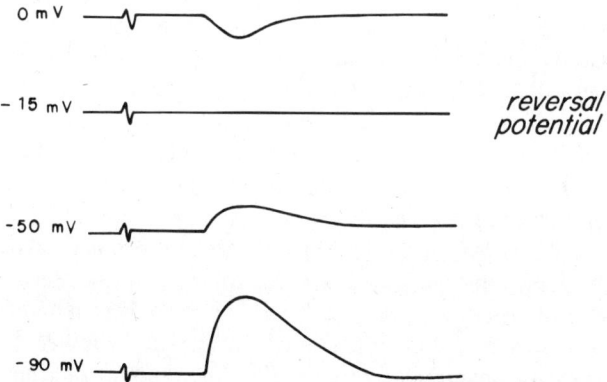

Figure 9 The endplate potential (EPP) recorded from a muscle fiber at four different membrane potentials indicated by each trace. Note the absence of the EPP at the reversal potential of −15 mV.

depolarizing or hyperpolarizing—will occur when ACh is applied to the muscle membrane. Now we need to do an actual experiment to determine which ions are able to flow when ACh is applied to the membrane. In designing the experiment we use the same reasoning we applied above. We want to determine the membrane potential at which the net membrane current flow will be zero when ACh is applied. If we are lucky, we will be able to deduce the ion species crossing the membrane. First, we alter the muscle membrane potential artificially and then apply ACh. The results are depicted in Figure 9. We find that when the membrane potential is about −15 mV, the addition of ACh does not change the membrane potential.

When the membrane potential is altered to a more positive potential, application of ACh results in a hyperpolarization of the membrane. The membrane potential at which ACh does not produce a net current is called the reversal potential for the transmitter. This is an important concept, because the experiment just described is the classical approach for determining the kind of transmitter liberated from CNS nerve terminals—that is, by comparing the reversal potential of the native transmitter with that of putative transmitters. All that we have learned from our experiment is that at least sodium crosses the membrane during ACh action. We know this simply because −15 mV is beyond both the equilibrium potentials for chloride and potassium. However, since the reversal potential is not at the sodium equilibrium potential, we can conclude that either potassium or chloride, or possibly both, cross the membrane as well. To determine which possibility is correct, we introduce a slight variation to our experiment. You remember that changing the concentration of ions on either side of the membrane will alter the equilibrium potential for that ion. Now, if this ion is permeant during the action of ACh, the reversal potential will also be changed in exactly the same direction. For example, if we increase the external potassium ion concentration, the potassium equilibrium potential will move toward zero. If ACh increases membrane permeability to potassium, the reversal potential for ACh will also move in the positive direction. When these experiments were carried out, it was determined that ACh increased membrane permeability both to sodium and to potassium, but not to chloride. The reversal potential for ACh of −15 mV represents a weighted average of the effects of the increase in membrane conductance to both potassium and sodium.

Since the reversal potential for ACh at the neuromuscular junction is about halfway between the sodium and potassium equilibrium potentials, the action of ACh is to increase the membrane permeability equally to sodium and potassium. It has been possible to measure the conductance experimentally through a single channel. At the NMJ the single channel conductance is 15 picosiemens.

So far we have discussed the action of a transmitter that results in the initiation of an action potential in the postsynaptic cell. In order for the proper routing of electrical information to occur in the CNS, such messages are sometimes inhibited. This process involves mechanisms that prevent the transfer of information from cell to cell. Why is such a mechanism necessary? Suppose you are walking on the beach and suddenly step on a piece of glass. A reflex action mediated by cutaneous receptors is activated, giving rise to painful sensations. The reflex action in this case is withdrawal of the injured limb and simultaneous contraction in the extensor muscles of the opposite limb. This action permits you to support yourself while hopping about. To ensure that nothing interferes with the activity in the motor neurons that activate the extensors of the leg, the flexor motor neurons are actively inhibited. The complex pathway that produces this withdrawal reflex is discussed under motor systems (Chapter 39). The mechanisms involve the production of inhibitory postsynaptic potentials (IPSPs) in flexor motor neurons. As you might imagine, in order to prevent the cell from firing an action potential, it will be necessary to prevent the membrane potential from crossing threshold. This can be accomplished in two ways. The first method would be to move the membrane potential further from threshold by hyperpolarizing. We could achieve this goal by applying a transmitter that opens channels to potassium ions. In this case, the membrane would act like a potassium electrode and move to the equilibrium potential for potassium. Another way to prevent the membrane potential from crossing threshold would be to create a leaky membrane. This method would be most effective if the site of inhibitory action were located between the site of excitatory action and the action potential initiation zone. The geometry of this relationship is given in Figure 10. Note that when the EPSP is conducted across the site of the inhibitory synapse, the leaky membrane produced by the action of

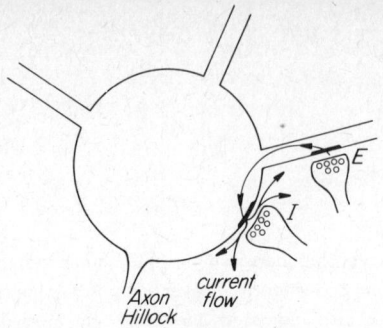

Figure 10 The geometric relationship between an excitatory (E) and inhibitory (I) synapse. The arrow indicates current flow for a synaptic potential generated at site E. Note that the current will tend to flow from the membrane at the I synapse when transmitter is released from the I terminal, thereby decreasing membrane resistance at this site and preventing this current from reaching the axon hillock region.

inhibitory transmitter would permit the excitatory synaptic current to shunt or short circuit (indicated by the arrow) through the membrane and into the extracellular space and therefore never reach the action potential initiation zone. Therefore, postsynaptic inhibition can occur by hyperpolarizing the cell, moving the membrane potential away from threshold or simply by creating a membrane short circuit (i.e., a patch of low-resistance membrane).

PRESYNAPTIC EVENTS

We shall now turn our attention to the mechanisms underlying the release of a chemical neurotransmitter. Unfortunately, this aspect of synaptic transmission has been much more difficult to study because presynaptic terminals are generally too small to be penetrated by microelectrodes. However, a few preparations exist, including the squid giant synapse and the synapses in lamprey eel spinal cord, in which the presynaptic axons are quite large.

We have already stated that it is necessary to depolarize the presynaptic terminal in order to effect transmitter release. This normally occurs by means of the action potential. However, it is important to determine whether the extent of depolarization can affect the amount of transmitter released. If it can, this might be an important mechanism used by CNS neurons to vary the amount of transmitter released from presynaptic elements. Such experiments were achieved in both squid and lamprey eel when the presynaptic action potential was blocked with tetrodotoxin and the presynaptic membrane artificially polarized by passing current from an intracellular electrode. As Figure 11 indicates, the amount of transmitter is directly proportional to the degree of depolarization of the presynaptic terminal. This result is important in explaining another type of inhibition that occurs in the CNS, that is, presynaptic inhibition. As the term suggests, the inhibition occurs by decreasing the amount of transmitter released from the presynaptic terminal. The synaptic relationships that produce this inhibition are depicted in Figure 12. Activity in terminal 1 releases transmitter, which depolarizes terminal 2. Because terminal 2 is depolarized, the action potential will be smaller and less transmitter will be released from terminal 2 when the action potential arrives. The net result will be a smaller EPSP in cell A.

When we first discussed the generation of the EPP at

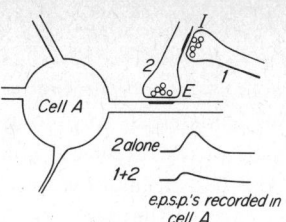

Figure 12 Presynaptic inhibition in the CNS. If the terminal marked 2 is activated alone, an EPSP (inset) can be recorded from the postsynaptic cell. If terminal 1 is activated just before terminal 2, terminal 2 will be depolarized by the action of the transmitter from terminal 1. The net effect is to reduce the size of the action potential in terminal 2 and thus reduce the amount of transmitter released by terminal 2. The overall effect is inhibition.

the neuromuscular junction we stated that the electrical stimulation of the motor nerve resulted, after a short delay, in an EPP in the muscle fiber. We must now consider what makes up the delay. Some of the delay results from the time it takes for the action potential to travel down the motor nerve. However, it is possible to measure the precise time in milliseconds that the action potential will arrive in the nerve terminal. A delay of 0.3–0.5 ms occurs before the EPP is produced. This is called the synaptic delay. This delay is composed of several elements, including the time for transmitter release, diffusion in the synaptic cleft, attachment to receptors, receptor change, and membrane permeability to increase. Most of the delay is taken up by the transmitter release mechanism, especially the influx of calcium ions across the presynaptic membrane.

ROLE OF CALCIUM

A clue that some intermediary event occurs between the depolarization and the liberation of transmitter is provided by the synaptic delay, which is markedly sensitive to temperature. For example, at the frog NMJ at room temperature at least 0.5 ms elapses between the depolarization of presynaptic endings and the first sign of postsynaptic depolarization. On the basis of electron micrographic measurements of the synaptic cleft, this is several times too slow to be accounted for by diffusion of transmitter across the cleft, which should require at most 50 μs. In fact, when ACh is applied to the synapse directly, postsynaptic depolarization occurs in less than 150 μs,

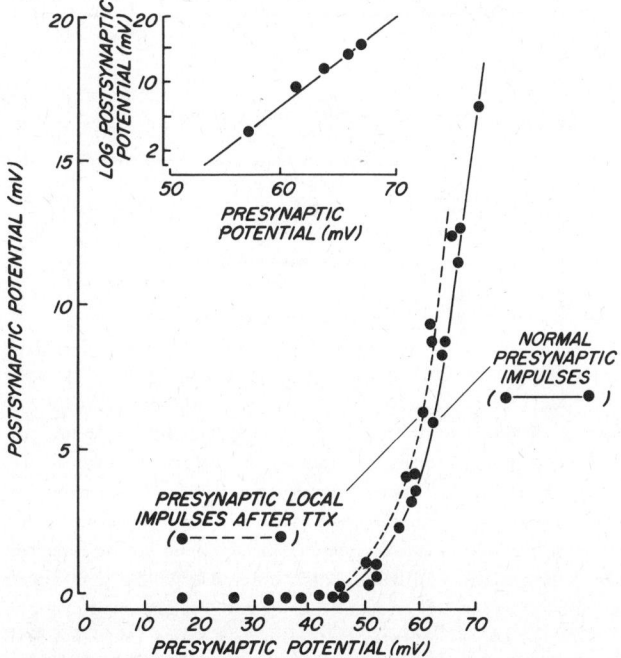

Figure 11 The relationship between the size of the presynaptic impulse and postsynaptic response (solid lines and inset). The dashed line shows the same relationship, even though all action potentials in the presynaptic terminal have been blocked with tetrodotoxin (TTX). (Modified from Katz B and Miledi R: A study of synaptic transmission in the absence of nerve impulses. *J Physiol* 192: 407–436, 1967.)

even though the application pipette is significantly farther from the subsynaptic membrane than the presynaptic terminal. Furthermore, if the preparation is cooled to 2°C, the synaptic delay is increased to as long as 7 ms, while the delay associated with directly applied ACh is unchanged. Two hypotheses, singly or in combination, can explain the sensitivity of the delay to temperature: (1) since diffusion is relatively unaffected by temperature, a metabolic process might interfere between depolarization and release; and (2) some substance might have to enter the presynaptic terminal before release can occur.

Calcium has long been known to be an essential link in the process of transmission. When its concentration in the extracellular fluid is decreased, release of ACh at the NMJ is reduced and eventually abolished. The importance of Ca^{2+} for release has been established at all chemical synapses tested, regardless of the nature of the transmitter.

Moreover, Ca^{2+} must not only be present for release to occur, but it must be physically present at the time of depolarization of the presynaptic terminal. This fact was shown by directly pipetting Ca^{2+} on to the synaptic terminal at different times before, during, and after the presynaptic depolarization (Fig. 13).

Subsequently, evidence was obtained to suggest that calcium itself enters the nerve terminal to facilitate release. Other experiments on nerve fibers have shown that membrane permeability to calcium (g_{Ca}) is increased with depolarization of the membrane and that some calcium enters with each action potential. In the presynaptic terminals of motor nerve fibers, calcium current has been shown to contribute to the impulse current. The concept of Ca^{2+} entering the presynaptic terminal was strengthened by experiments on squid synapse. The experiments were based on the hypothesis that if the presynaptic terminal were depolarized to the calcium reversal potential (E_{Ca}) or beyond, no calcium would enter during the pulse and no release would occur (just as internal positivity during the action potential tends to prevent Na^+ from running in). The inside of the presynaptic terminal was made strongly positive by the use of large currents and, as expected, release was blocked, even though the presynaptic terminal was depolarized. This test provides indirect evidence for the idea that a positively charged ion, Ca^{2+}, must enter for transmitter to be released.

A more direct test was the injection of Ca^{2+} into the presynaptic terminal, which caused release even in the absence of external Ca^{2+} and depolarization.

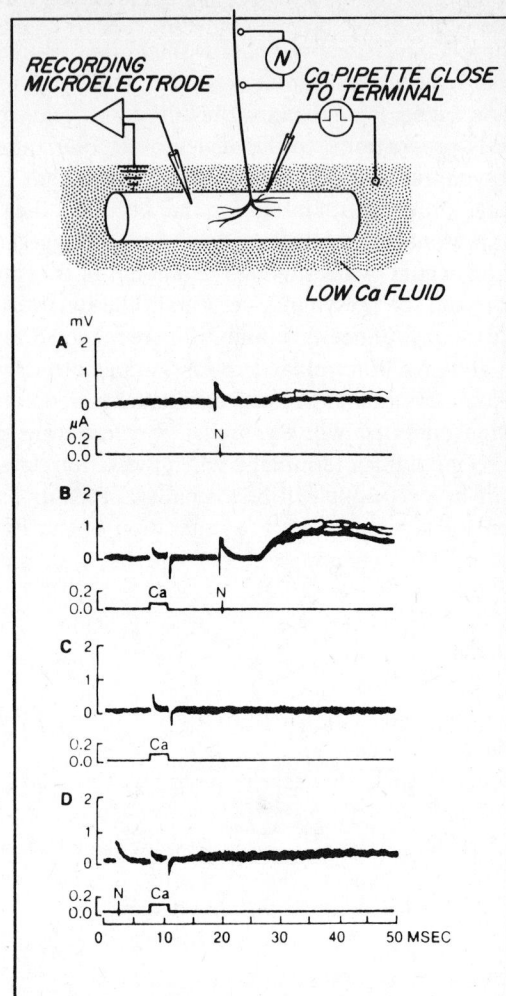

Figure 13 Effect of calcium on transmitter release. (*a*) In low-calcium medium, a motor nerve impulse causes little or no transmitter release at the skeletal neuromuscular synapse. (*b*) If the nerve impulse is preceded by application of calcium from a pipette to the terminal, transmitter is released. (*c*) Calcium alone has no effect on transmitter release. (*d*) If calcium is applied after the presynaptic impulse arrives in the terminal, but before the expected synaptic potential, it has no influence on transmitter release. (Modified from Katz B and Miledi R: Modification of transmitter release by electrical interference with motor nerve endings. *Proc Royal Soc Series B* 167: 1–22, 1967.)

So far, the general scheme can be depicted as follows:

Depolarization of presynaptic terminal → calcium entry → release of transmitter

QUANTAL RELEASE

Once the general framework for release has been established, it remains to be shown how the transmitter is secreted from the terminals. The basic clue to the mechanism of secretion was provided by two observations: (1) at rest, small spontaneous fluctuations of postsynaptic membrane potential—the *miniature synaptic potentials*—can be recorded; and (2) statistical variations of the same order of magnitude as the miniature synaptic potentials occurred when the nerve impulse liberated only small amounts of transmitter.

In a frog muscle fiber, the miniature potentials are less than 1 mV in size and occur with a mean frequency of about one per second. During their time course, they resemble little synaptic potentials. Their importance derives from the observation that they are the result of spontaneous release of ACh from the nerve endings and that these packets containing many molecules constitute the units out of which the normal synaptic potential is synthesized. Under normal conditions, the ACh contents of one presynaptic vesicle will produce one miniature potential. The potential represents a few thousand ACh molecules. During a normal synaptic event, many vesicles are released and the miniature potentials summate to produce the synaptic potential.

Under certain abnormal conditions, the packet size can be reduced. Regenerating nerve terminals release ACh in smaller quantal packages than under normal circumstances. Also, hemicholinium, a drug that interferes with ACh synthesis by preventing uptake of choline by the presynaptic terminal, produces lower quantal content. Finally, in some of the myoneural diseases, such as myasthenia gravis, it is important to determine whether individual quanta contain fewer ACh molecules than normal or whether the postsynaptic receptor system is impaired.

The various stages of synaptic transmission are summarized in Figure 14.

RECYCLING OF SYNAPTIC VESICLE MEMBRANE

Evidence has recently been reported for a recycling of synaptic vesicle membrane at the neuromuscular junction. Stimulation of the nerve causes a reduction in the population of synaptic vesicles, but an increase in the surface area

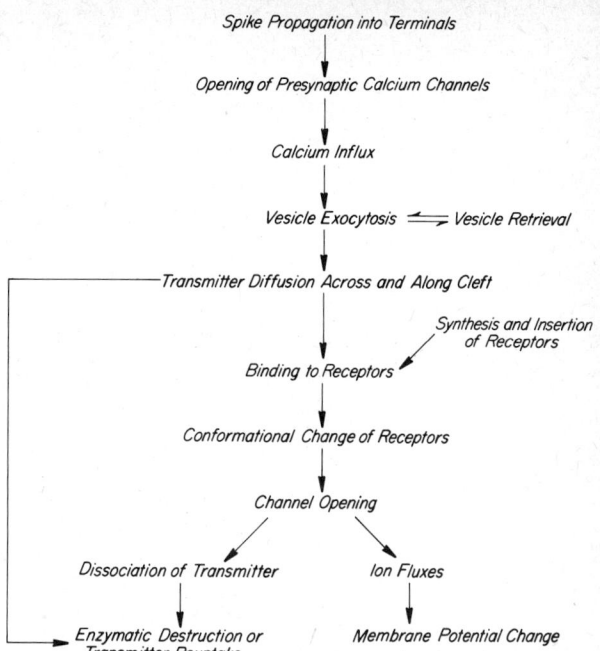

Figure 14 Mechanisms involved in synaptic transmission.

of the terminal. Presumably, the synaptic vesicles fuse with the surface membrane during the exocytotic process of transmitter release. After a period of stimulation, coated vesicles and membraneous cisternae appear in the cytoplasm of the ending. After a rest period, the cisternae are replaced by new synaptic vesicles. Meanwhile, the surface area of the terminal returns to normal. Horseradish peroxidase molecules, which serve as an electron-opaque marker when introduced into the extracellular space around the synaptic ending, appear in the coated vesicles and cisternae shortly after nerve stimulation. Later, the marker appears in the synaptic vesicles, suggesting that these form from the other organelles. Further stimulation causes the synaptic vesicles to release the peroxidase. Coated vesicles appear to be the initial intermediary in the recycling process. They probably form from membrane adjacent to the release sites, with several coalescing to form a cistern. Synaptic vesicles pinch off from the cisternae. In this fashion, membrane is conserved, reducing the need for resynthesis as well as for disposal. On a longer time scale, however, new synaptic vesicles presumably arrive at the synaptic ending after

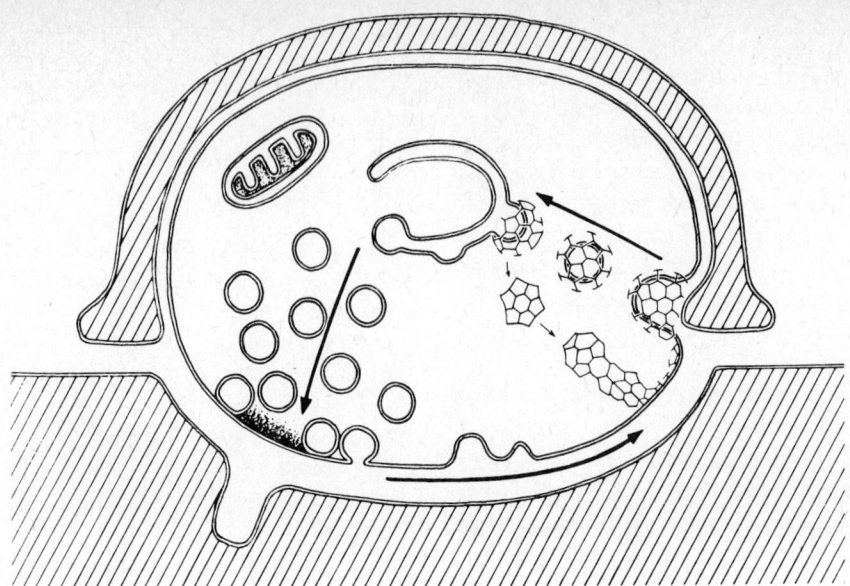

Figure 15 Diagrammatic summary of a hypothesis for synaptic vesicle membrane recycling at the frog neuromuscular junction. (1) Synaptic vesicles discharge their content of transmitter as they coalesce with the plasma membrane at specific regions adjacent to the muscle. (2) Equal amounts of membrane are then retrieved when coated vesicles pinch off from regions of the plasma membrane adjacent to the Schwann sheath. (3) The coated vesicles lose their coats and coalesce to form cisternae, which accumulate in regions of vesicle depletion and slowly divide to form new synaptic vesicles. (From Heuser JE and Reese TS: Evidence for recycling of synaptic vesicle membrane during transmitter release at the frog neuromuscular junction. *J Cell Biol* 57:315–344, 1973.)

axoplasmic transport from the neuronal soma, and old membrane is probably removed from the ending by retrograde axoplasmic transport, to be broken down in the soma by lysosomal action (Fig. 15).

SYNAPTIC PHARMACOLOGY

A whole variety of compounds have been suggested as transmitter substances at different locations of the nervous systems of various animals. Figure 16 presents the chemical structures of some of the established and proposed transmitter substances.

As already mentioned, the chemical synaptic trans-

mitter substance at the NMJ is ACh (Fig. 16). ACh is synthesized by an enzyme called choline acetyltransferase, which is stored in synaptic vesicles in the motor axon terminal to be released from the storage sites in response to the invasion of a nerve action potential into the presynaptic ending, as well as spontaneously. ACh diffuses across the synaptic cleft and reacts with acetylcholine receptors associated with the postsynaptic membrane. The reaction between ACh and the ACh receptors appears to result in conformational changes in the muscle membrane structure, which in turn cause an enhanced permeability of the membrane to sodium and potassium ions.

The action of ACh is stopped by several processes. Some of the ACh is probably taken up by the nerve ending by an active transport mechanism. Some might diffuse out of the synaptic cleft into the extracellular space. However, the most significant process at the NMJ is the hydrolysis of ACh by the enzyme acetylcholinesterase (AChE), present at the surface of the muscle membrane. The reaction products of the hydrolysis are acetate and choline, neither of which has any significant effect on ACh receptors.

Several drugs can mimic the action of ACh at the NMJ. One of these is nicotine, which in low doses has an excitatory action on skeletal muscle (like ACh, in high doses, it blocks neuromuscular transmission). Various other substances, including several choline esters, also have an excitatory or cholinomimetic action on the NMJ.

A number of important drugs have an influence on neuromuscular transmission. The sites of possible drug action include the synthesis and storage mechanism, the release mechanism, the ACh receptors, and AChE. For instance, hemicholinium blocks the synthesis of ACh, presumably by competing for active sites of choline uptake. Botulinus toxin, the agent that causes botulism from food contaminated by *Clostridium botulinum,* prevents the release of ACh from motor axon terminals. Curare, the poison used by South American Indians on the tips of their blowgun arrows, prevents the reaction of ACh with the ACh receptors on the muscle membrane by competing for the active sites. Finally, several compounds, including eserine and neostigmine, combine with AChE to prevent

it from hydrolyzing ACh. Such agents are called anticholinesterases and are widely used in insecticides.

The effect of these various drugs on neuromuscular transmission is generally predictable. Hemicholinium, botulinus toxin, and curare all block neuromuscular transmission, although by quite separate mechanisms. In addition, local anesthetics block nerve conduction, hence neuromuscular transmission (however, this might not be readily demonstrable, since there is generally a one-to-one relationship between motor axon discharge and activation of a motor unit contraction). The effect of anticholinesterases in enhancing transmission is best seen in junctions that are fatigued or partially blocked, for example, by curare.

Anticholinesterases have also been employed in investigations of the phenomenon of desensitization in neuromuscular transmission. In the presence of anticholinesterase, ACh can accumulate in excess of normal levels at the postsynaptic membrane, prolonging the reaction between ACh and the receptor molecules. The conformational change induced by this reaction (and the subsequent permeability increase) is believed to be transient, but an inactive coupling between ACh and receptor is thought to occur that actually interferes with the further synaptic

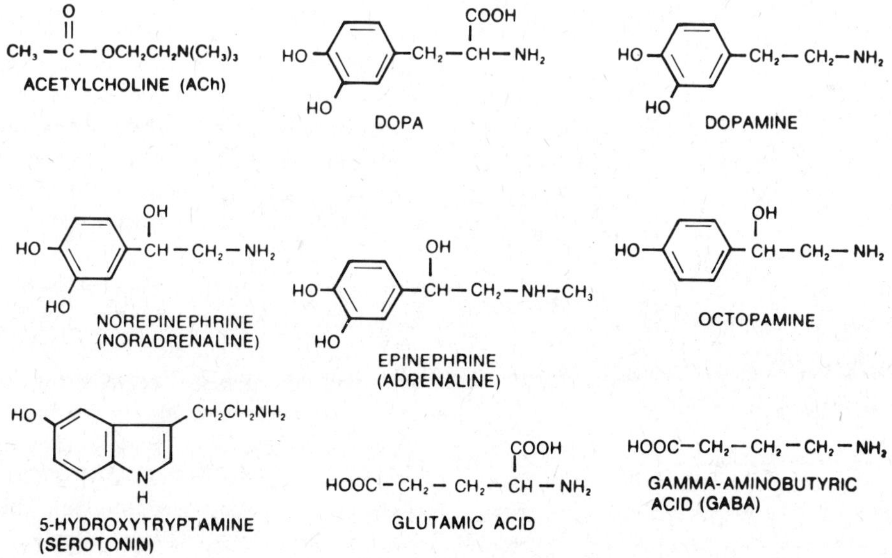

Figure 16 Chemical structures of transmitter substances. ACh, norepinephrine, epinephrine, and γ-aminobutyric acid (GABA) are established transmitters.

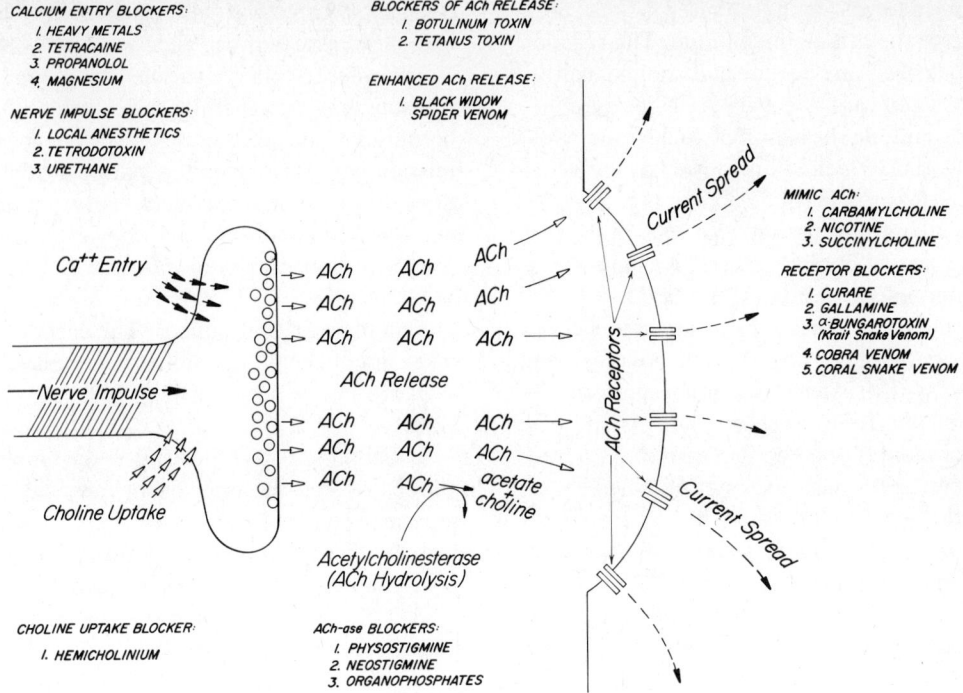

CALCIUM ENTRY BLOCKERS:
1. HEAVY METALS
2. TETRACAINE
3. PROPANOLOL
4. MAGNESIUM

NERVE IMPULSE BLOCKERS:
1. LOCAL ANESTHETICS
2. TETRODOTOXIN
3. URETHANE

BLOCKERS OF ACh RELEASE:
1. BOTULINUM TOXIN
2. TETANUS TOXIN

ENHANCED ACh RELEASE:
1. BLACK WIDOW
SPIDER VENOM

MIMIC ACh:
1. CARBAMYLCHOLINE
2. NICOTINE
3. SUCCINYLCHOLINE

RECEPTOR BLOCKERS:
1. CURARE
2. GALLAMINE
3. α-BUNGAROTOXIN
(Krait Snake Venom)
4. COBRA VENOM
5. CORAL SNAKE VENOM

CHOLINE UPTAKE BLOCKER:
1. HEMICHOLINIUM

ACh-ase BLOCKERS:
1. PHYSOSTIGMINE
2. NEOSTIGMINE
3. ORGANOPHOSPHATES

Figure 17 Neuropharmacological agents that affect synaptic transmission and their site or mode of action.

action of ACh. This process is called desensitization. Removal of the excess ACh usually leads to a return of the normal functional state.

Synaptic pharmacology can best be visualized by referring to a diagrammatic neuromuscular synapse with the actions of various agents on the different steps of synaptic transmission in the neuromuscular junction (Fig. 17).

ROLE OF SYNAPTIC TRANSMISSION IN CNS AND MUSCLE DISORDERS

Unfortunately, space does not permit going into great detail on the role that synapses play in disease. Obviously, since the synapse is the site of neural integration, this is one point at which alterations in the mechanisms of synap-

tic transmission or at which a variety of pharmacologically active agents could affect the transmission of information between excitable cells. One such example is Parkinson's disease, which is discussed under motor systems (Chapter 39). In this disease insufficient amounts of the neurotransmitter dopamine are released from presynaptic terminals involved in this motor pathway.

The neuromuscular disease myasthenia gravis was first described more than three centuries ago. However, it is only within the last 10 years that we have begun to understand the mechanism underlying this neuromuscular deficit. The main characteristic of the disease is easy fatigability, most commonly affecting the extrinsic muscles of the eyes. Other muscle groups are often involved, especially as the disease progresses. It was originally thought that the deficit was due to insufficient transmitter released from the presynaptic motor nerve terminal. The evidence that supported this hypothesis was obtained from intracellular recordings from the neuromuscular junctions of myasthenic patients. MEPPs were found to be substantially smaller and the EPP was also decreased in size. On the basis of experiments, it was concluded that the quantal content is decreased. However, an alternative explanation is a deficit in the postsynaptic membrane, for example, decreased receptor sensitivity or decreased numbers of receptors. It has more recently been shown that myasthenia gravis is the result of an autoimmune response in which an antibody to the ACh receptor is produced by the body. The antibody competes with ACh for the receptor much as curare does. It is also likely that receptor turnover is decreased, and there is a decrease in the number of ACh receptors in the postsynaptic muscle membrane.

Schizophrenia is a disease that is also believed to have part of its origin in defects at the synaptic junction. Catecholamines are a class of neurotransmitters that are actively removed from the synaptic cleft by active reuptake mechanisms. Amphetamines block this reuptake mechanism, and this drug can produce schizophrenic-like symptoms. It has been suggested that schizophrenics have a deficit in the reuptake mechanism in specific brain pathways, with the result that the transmitter remains in the synaptic cleft for a long period of time, overactivating these pathways.

Clearly, the synapse is a vulnerable point of contact between excitable cells. We have not discussed the role of plasticity within the CNS nor have we explored the possibility that alterations in synaptic interactions might play a role in learning and memory. However, the basic concepts presented here are important for the clinician's understanding of disorders involving the nervous system.

SELECTED BIBLIOGRAPHY

Katz B: *Nerve, Muscle and Synapse.* New York, McGraw-Hill, 1966.

Kuffler SW, Nicholls JG: *From Neuron to Brain.* Sunderland, Sinauer, 1976.

SELF-STUDY QUESTIONS

MULTIPLE CHOICE

Select the single best answer.

1. Synaptic potentials may
   A. summate.
   B. be refractory.
   C. not attenuate.
   D. be regenerative.
   E. none of the above.

2. After transmitter release, vesicle membrane is
   A. permanently lost.
   B. recycled.
   C. derived from neurofilaments.
   D. derived from glia.
   E. none of the above.

3. The decrease in the response of skeletal muscle in a patient with myasthenia gravis to a motor nerve action potential is a result of
   A. decrease in quantal content.
   B. increase in magnesium conductance in the presynaptic terminal.
   C. a decrease in available postsynaptic receptor sites.
   D. an increase in postsynaptic receptor turnover.
   E. a decrease in the miniature endplate potential size.

4. Postsynaptic inhibition can result from
   A. an increase in g_K.
   B. an increase in g_{Na}.
   C. a decrease in g_K.
   D. an increase in g_{Ca}.
   E. none of the above.

5. Calcium is an important cation involved in synaptic transmission because
   A. it increases postsynaptic receptor turnover.
   B. it increases the affinity of the neurotransmitter molecules for the postsynaptic receptor.
   C. it is directly involved in the process of transmitter release.
   D. it aids in the degradation of the neurotransmitter.
   E. it is released with the transmitter to activate the postsynaptic receptor site.

6. A synapse on a small dendrite located at one length constant from the soma will be more effective than one located at the same distance, but on a larger dendrite, because
   A. the same amount of current will produce a larger synaptic potential.
   B. the reversal potential will be further from resting.
   C. the action potential will be larger in the presynaptic axon.
   D. the resting sodium conductance will be larger.
   E. none of the above.

7. In the CNS most of the released neurotransmitter is removed by
   A. microglia.
   B. reuptake mechanisms.
   C. hydrolysis.
   D. the postsynaptic membrane.
   E. none of the above.

8. Synaptic delay at chemical synaptic junctions is primarily a function of
   A. the time used for the action potential to depolarize the terminal membrane.
   B. the time for neurotransmitter to diffuse across the synaptic cleft.
   C. the time for Ca^{2+} to enter the presynaptic terminal and effect fusion of synaptic vesicles with the axolemma.
   D. the time for vesicle membrane to be recycled and new vesicles to be manufactured.
   E. the turnover time for postsynaptic receptor proteins.

9. Presynaptic inhibition occurs because less transmitter is released from the presynaptic terminal. This is a result of
   A. an increase in potassium permeability at the inhibited terminal.
   B. failure of the action potential to invade the inhibited presynaptic terminal.
   C. a decrease in the amount of depolarization of the inhibited terminal by the action potential.
   D. blocking of the calcium channels of the inhibited terminal by neurotransmitter.
   E. desensitization of the receptor proteins of the inhibited presynaptic terminal.

10. Integration at electrical synaptic junctions is usually not as effective as at chemical synaptic junctions. This is a result of
    A. the refractoriness of the postsynaptic membrane at which the synaptic potential is generated.
    B. the slower action of chemical neurotransmitters.
    C. the slower conductance changes that occur at electrical synaptic junctions.
    D. the effect of neurotransmitters on the cell membrane time constant at electrical synaptic junctions.
    E. the long synaptic delay that occurs at electrical synaptic junctions.

11. A miniature endplate potential is the result of
    A. the release of a quantum of neurotransmitter.
    B. the action of a single molecule of neurotransmitter.
    C. increased calcium conductance at the presynaptic terminal.
    D. the release of several quanta of neurotransmitter following the invasion of the action potential in the presynaptic terminal.
    E. the action of neurotransmitter on a single postsynaptic receptor protein.

12. A common property of gallamine triethiodide (Flaxedil), and of α-bungarotoxin is
    A. their ability to influence transmitter release.
    B. their ability to block the presynaptic action potential.

C. their hydrolytic action on neurotransmitters.

D. their ability to block transmitter binding to post-synaptic receptors.

E. their ability to mimic the action of inhibitory neurotransmitters.

13. Release of neurotransmitter from a presynaptic terminal will occur if

A. the presynaptic terminal is depolarized.

B. calcium ions cross the presynaptic terminal membrane.

C. magnesium ions cross the pesynaptic terminal membrane.

D. vesicle membrane is recycled.

E. postsynaptic receptor sites are rapidly turned over.

14. The reversal potential for ACh action at the neuro-muscular junction is

A. the membrane potential at which calcium ions cross the membrane when ACh is applied.

B. the membrane potential at which no net current flows across the membrane when ACh is applied.

C. the membrane potential at which only sodium ions cross the membrane when ACh is applied.

D. the membrane potential, which is equal to threshold for the muscle action potential.

E. the membrane potential at which potassium and sodium ions are in equilibrium when ACh is applied.

15. When an inhibitory transmitter is applied to a postsynaptic neuron, the membrane potential changes from its resting level to a more hyperpolarized level. In this situation it is toward the chloride and potassium equilibrium potentials. The most likely action of the transmitter on the membrane permeability to sodium, potassium, and chloride is to

A. increase permeability to sodium ions.

B. decrease permeability to potassium ions.

C. increase permeability to chloride ions.

D. decrease permeability to all three ions.

E. have no change in permeability to any of the ions.

MULTIPLE CHOICE

Select the best answer(s). (In many instances, more than one answer is correct.)

16. Characteristics of the excitatory synaptic potential include

A. passive conduction.

B. summation.

C. increase in g_{Na}.

D. hyperpolarization.

17. Postsynaptic inhibition may involve

A. reduction in transmitter release.

B. increase in g_{Cl}.

C. blocking of postsynaptic receptors.

D. reduction in firing rate of postsynaptic cell.

18. Steps in transmitter release include

A. hyperpolarization of presynaptic terminal.

B. inhibition by magnesium of presynaptic receptor sites.

C. decrease in g_K at the presynaptic membrane.

D. calcium entry into presynaptic terminal.

19. The fate of transmitter following its release is

A. enzymatic breakdown.

B. diffusion out of the synaptic cleft.

C. reuptake into the presynaptic terminal.

D. transport out of the synaptic cleft by calcium.

20. The toxins curare and α-bugarotoxin affect neuro-muscular transmission by

A. blocking transmitter release.

B. blocking transmitter inactivation.

C. increasing transmitter release.

D. blocking the action of transmitter at postsynaptic receptor sites.

21. The reversal potential for a transmitter is

A. the membrane potential at which g_K is maximum.

B. the membrane potential at which no net current flows when transmitter is applied.

C. the membrane potential at which transmitter is most effective in depolarizing the membrane.

D. the membrane potential at which transmitter has no effect on membrane potential.

MATCHING

Select the single best answer.

A. Nerve impulse transmission
B. Synaptic transmission

22. Delay.
23. Bidirectional.
24. Refractory period.
25. Summation.
26. Inhibition.
27. Simultaneous increase in g_{Na} and g_K.

Select the best answer(s). (In many instances, more than one answer is correct.)

A. E_{Na}
B. E_K
C. E_{Cl}

28. Excitatory postsynaptic potential
29. Inhibitory postsynaptic potential

A. Temporal summation
B. Spatial summation

30. Input from one rapidly firing synapse triggers an action potential in the postsynaptic cell.
31. Input from multiple synapses of one presynaptic cell leads to an action potential in the postsynaptic neuron.
32. A synchronous input from multiple neurons at multiple synapses.

MULTIPLE CHOICE

Select the single best answer.

33. The most important aspect of postsynaptic activation is the necessity for
A. large value for E_{Na}.
B. large potential change.
C. large conductance change.

SEQUENCE

34. Place the following in the correct order.
A. Calcium influx into synaptic ending.
B. Reuptake of transmitter into synaptic ending.
C. Transmitter synthesis.
D. Presynaptic nerve impulse.
E. Postsynaptic potential.
F. Transmitter release.

COMPLETION

35. Transmitter release at the presynaptic terminal.
a. does not depend on sodium influx, because it is not blocked by_____.
b. does not depend on potassium efflux, because it is not blocked by_____.
c. depends on calcium influx, because it is blocked by_____.

MATCHES

Indicate correct (A) or incorrect (B) for each set.

36. Endplate potential–Acetylcholine.
37. Miniature endplate potential–Quantum.
38. Synaptic vesicle–Quantum.
39. Presynaptic hyperpolarization–Increased quantal release rate.
40. Endplate folds–Cholinesterase.

LISTING

41. List six candidate synaptic transmitters.

MATCHING

A. Increased neuromuscular transmission
B. Decreased neuromuscular transmission
C. Either, depending on concentration

42. Botulinum toxin
43. Acetylcholine
44. Decreased calcium

45. Decreased magnesium

46. Hemicholinium

47. Anticholinesterase

48. Nicotine

49. Local anesthetic

50. Curare

ANSWERS

1. A. An important property of the ESP is summation with other local passive membrane potentials.

2. B. Vesicle membrane is believed to be conserved by recycling.

3. C. Myasthenia gravis is an autoimmune disease in which the ACh receptors are blocked.

4. A. Only an increase in g_K will inhibit the cell. All others will result in depolarization.

5. C. Ca^{2+} influx during depolarization of the presynaptic terminal results in vesicle fusion.

6. A. The input resistance of the small dendrite is greater than a large dendrite. From Ohm's law, $E = IR$, the same current through a larger resistance will result in a larger voltage.

7. B. Reuptake is believed to be the method of transmitter removal at CNS synapses. Hydrolysis is important at the NMJ.

8. C. Turn-on of calcium current, of calcium influx, and of vesicle fusion appear to be the rate-binding steps.

9. C. Decreasing the size of the invading action potential reduces the amount of transmitter released by decreasing the number of voltage-dependent calcium channels that are opened. Blowdin theory does occur, which would totally block transmitter release.

10. B. Synaptic potentials produced by neurotransmitters last relatively longer than do synaptic potentials produced by direct transmission across electrical synapses. Therefore, they have a longer period of time with which to sum with other membrane potential charges.

11. A. This is believed to be physically equivalent to a single synaptic vesicle.

12. D. These compounds act at the postsynaptic receptor site.

13. B. Calcium influx is the necessary requirement. Depolarization in the absence of calcium will not result in transmitter release.

14. B. Although a conductance change will occur in the presence of ACh, no net current flows when at the reversal potential for ACh.

15. C. The membrane could also increase permeability to potassium. This is not, however, a choice.

16. A, B, C. A hypopolarization is synonymous with inhibition.

17. B, D. An increase in g_{Cl} will clamp the membrane potential at E_{Cl}. If the cell is active (i.e., generating action potentials), the rate of discharge will be reduced.

18. D. A necessary and sufficient requirement.

19. All are more or less important.

20. D. Blocks receptors. The toxin is irreversible.

21. B, D.

22. B. A delay in transmission across the synapse.

23. A. The action potential will travel in either direction along an axon. If initiated at the axon hillock, it will invade the soma and can do so in the dendrites in some instances. The regenerative response depends on the number of sodium channels in the membrane.

24. A.

25. B.

26. B.

27. B.

28. A and B, or just A. Can be transmitter dependent.

29. B and C, or just B, or just C.

30. A.

31. A and/or B.

32. B.

33. C. A large conductance change will permit larger current to flow, producing a larger voltage change (Ohm's law).

34. C, D, A, F, E, B.

35. a. *tetrodotoxin:* The terminal can be artificially depolarized by passing extrinsic current, as from a battery, to obtain transmitter release in the presence of a sodium channel blocker.

  b. *tetraethylammonium:* This drug prolongs the action potential by slowing down the repolarizing phase.

  c. *magnesium:* This ion competes for the calcium channels at the presynaptic membrane.

36. A.

37. A.

38. A.

39. B.

40. A.

41. Glycine, ACh, GABA, norepinephrine, glutamate, dopamine, serotonin.

42. B.

43. C. In small amounts, it increases the muscle depolarization with increasing concentration. However, as larger amounts are applied, desensitization of the receptors occurs.

44. B.

45. A.

46. B.

47. A.

48. C. Same reasoning as for ACh.

49. B.

50. B.

4

Skeletal Muscle

R. DAVID BAKER
LEE E. MOORE

OBJECTIVES

After completion of this chapter, you should be able to

1. Use the terminology.
2. Understand the microanatomy and micromechanics of skeletal muscle contraction.
3. Understand the process of excitation–contraction coupling.
4. Appreciate the role of Ca^{2+} in muscle contraction.
5. Describe the chemical composition and structure of the myofilaments.
6. Know the reaction pathway for hydrolysis of ATP by myosin ATPase and understand the role of actin as the activator of myosin ATPase.
7. Understand the cross-bridge theory of muscle contraction.
8. Understand the macromechanics of muscle contraction.
9. Describe the properties of the different fiber types.
10. Understand the concepts of the active state and series elasticity.
11. Understand the phenomenon of summation of contractions and how a tetanus can develop more tension than a twitch.
12. Know the length–tension relationship and how it might be explained by the sliding filament model and cross-bridge theory of muscle contraction.
13. Understand the concepts of preload and afterload.
14. Know the force–velocity relationship.
15. Understand the mechanical transient responses.
16. Understand the mechanisms for grading the strength of contractions.

MICROANATOMY OF CONTRACTION

Skeletal muscle is the largest tissue in the body, accounting for 40–45% of total body weight. A whole skeletal muscle contains long muscle fibers embedded in connective tissue. The fibers are arranged either longitudinally or obliquely with respect to the long axis of the muscle. Each fiber is a multinucleated cell, with the nuclei placed peripherally. Muscle fibers contain myofibrils, which are their contractile components. They also contain an extensive system of smooth-surfaced endoplasmic reticulum, the sarcoplasmic reticulum, which is involved in the excitation process, and many mitochondria that supply ATP for energizing contraction.

THE MYOFIBRILS

The manner in which myofibrils are arranged to form fibers as well as the way in which fibers are arranged to form whole muscle is illustrated in Figure 1. Electron micrographs of myofibrils are shown in Figure 2. Each myofibril extends the entire length of the fiber and has a characteristic banded or striated appearance. The names of the bands are given in Figures 1 and 2. The dark Z lines repeat at intervals of about 1.5–2.5 μm, depending on the degree of shortening or stretching of the myofibril. The region from one Z line to the next is called a sarcomere and can be regarded as the structural unit of muscular contraction.

The central 1.6 μm of each sarcomere is occupied by the A band, so named because it is *a*nisotropic with respect to polarized light. The H band is a region in the center of the A band, where there are no thin filaments. The M line is a dark stripe that bisects the H band. The light region between successive A bands is called the I band, which is *i*sotropic with respect to polarized light. Each I band is

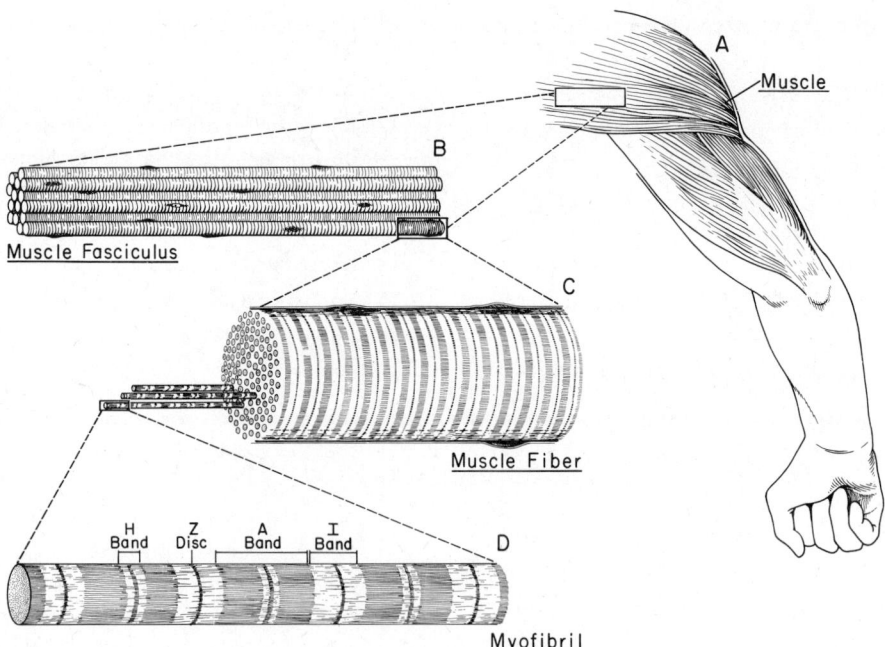

Figure 1 A diagram showing the manner in which myofibrils are arranged to form fibers and fibers are arranged to form whole muscle. (From Bloom W, Fawcett DW: *A Textbook of Histology,* ed 9. Philadelphia, Saunders, 1968.)

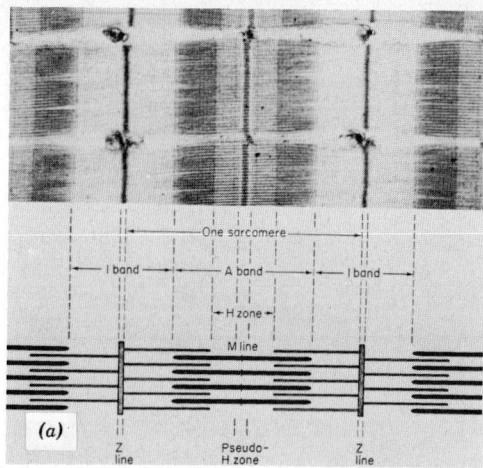

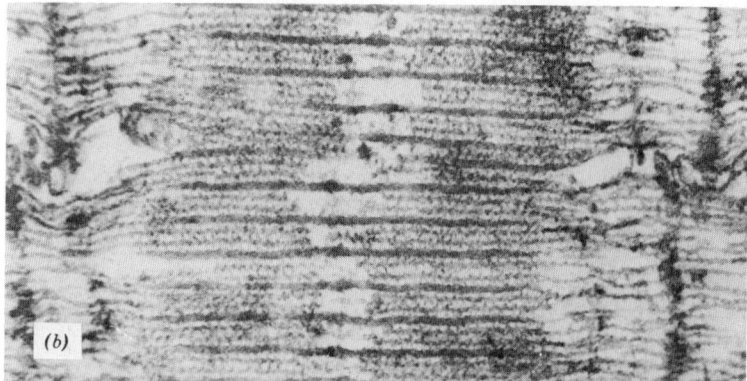

Figure 2 Electron micrographs of skeletal muscle sectioned longitudinally. (*a*) Longitudinal section of frog sartorius muscle together with a diagram showing the way in which the overlapping arrangement of thick and thin filaments gives rise to the A, I, and H bands. The A band is most dense in its lateral zones where the thick and thin filaments overlap. The central zone of the A band (the H zone) is less dense, since it contains thick filaments only. The I bands are less dense still because they contain only thin filaments. Sarcomere length here is about 2.5 μm. (*b*) Two adjacent myofibrils from rabbit psoas muscle. At this magnification the overlapping (or interdigitated) arrangement of the two types of myofilaments— thick and thin—can be clearly seen. (From Huxley HE: Molecular basis of contraction in cross-striated muscles, in Bourne GH (ed): *The Structure and Function of Muscle,* ed 2. New York, Academic Press, 1972.)

bissected by a Z line. The I bands increase in length during extension and disappear at very short muscle lengths.

Myofibrils are constructed of (1) Z lines; (2) thin myofilaments, which extend in both directions from each Z line to the beginning of the H band; and (3) thick myofilaments, which are located in the A bands. The arrangement of myofilaments to form a myofibril is shown diagrammatically in Figure 2a. Sarcoplasm occupies the space between myofilaments. The length of the A band (1.6 μm) is determined by the length of the thick filaments. The M line is produced by fine connections of unknown chemical composition that seem to join adjacent thick filaments at their centers. The regular overlapping arrangement of filaments produces the cross-striations that give striated muscle its name. The I band is the region in which there are no thick filaments. The A band, except for the H band, contains both thick and thin filaments. The striations of each myofibril lie approximately in register with those of adjacent myofibrils, giving the entire muscle fiber the same characteristic striated appearance as each of its component myofibrils.

If a myofibril is passively stretched, the two sets of filaments do not change length; they merely slide past each other so that the degree of overlap becomes less. The Z lines move further apart, the I bands and H bands both increase in length, as these are the two regions where there is no overlapping of thick and thin filaments. However, the length of the A band does not change, as it is determined by the length of the thick filaments. The positions of the filaments after a moderate degree of stretch are illustrated in Figure 3 (state 1). If this stretched myofibril is excited, forces suddenly develop between the thick and thin filaments that tend to make them slide along each other in the direction that reverses the effects of stretching. Unless the ends of the myofibril are mechanically fixed, the thin filaments are drawn toward the M line, thus reducing and finally obliterating the H band (Fig. 3, state 2). Since the thin filaments are attached to the Z lines, the Z lines come closer together, reducing the length of the I bands; each

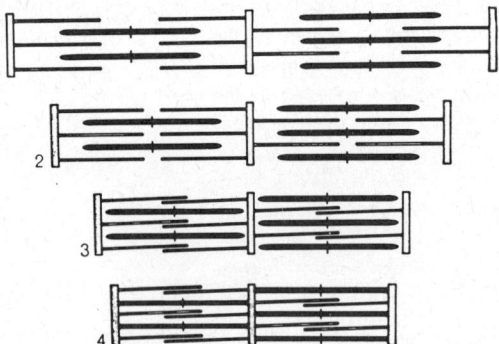

Figure 3 A diagram showing the manner in which the thick and thin filaments slide past each other during lengthening or shortening of a myofibril. Two sarcomeres are illustrated (see text for description). (From Huxley HE: Molecular basis of contraction in cross-striated muscles, in Bourne GH (ed): *The Structure and Function of Muscle*, ed 2. New York, Academic Press, 1972.)

sarcomere is shortened and, consequently, the entire myofibril is shortened.

As shortening proceeds further, the ends of the thin filaments overlap at the center of the A band, forming a dark region called a contraction band (Fig. 3, state 3). Shortening can proceed to such an extent that the thick filaments actually butt against the Z lines, thus obliterating the I bands (Fig. 3, state 4); and under some circumstances shortening can be so severe as to cause the ends of the thick filaments to crumple against each side of the Z lines.

As a myofibril shortens, the lateral separation between myofilaments increases and, consequently, myofibril thickness increases. Total myofibril volume remains constant. As all the myofibrils in a fiber modify their shape, becoming shorter and thicker, the shape of the entire fiber must change accordingly. If a large number of fibers change their shapes simultaneously, so does the whole muscle.

THE SARCOPLASMIC RETICULUM AND TRANSVERSE TUBULAR SYSTEM

Each myofibril is surrounded by a network of endoplasmic reticulum, which at regular intervals along the myofibril is confluent with flattened membranous sacs known as terminal cisternae. These two membranous components form the sarcoplasmic reticulum, which is illustrated in Figure 4. A single unit of sarcoplasmic reticulum consists of a continuous system of longitudinal reticulum encircling a myofibril together with the terminal cisternae at each end of this system.

At regular intervals along each muscle fiber, deep indentations of the plasma membrane form long tubules that pass transversely among the myofibrils between adjacent terminal cisternae (Fig. 4). These transverse tubules are open to the extracellular space; their membranes are continuous with the plasma membrane. They come into close contact with the terminal cisternae as they pass among the myofibrils. The structure formed by two adjacent terminal cisternae and a transverse tubule passing between them is called a triad. In some muscles the triads are at the level of the junctions between A and I bands, in other muscles the triads are at the level of the Z lines (as in Fig. 4).

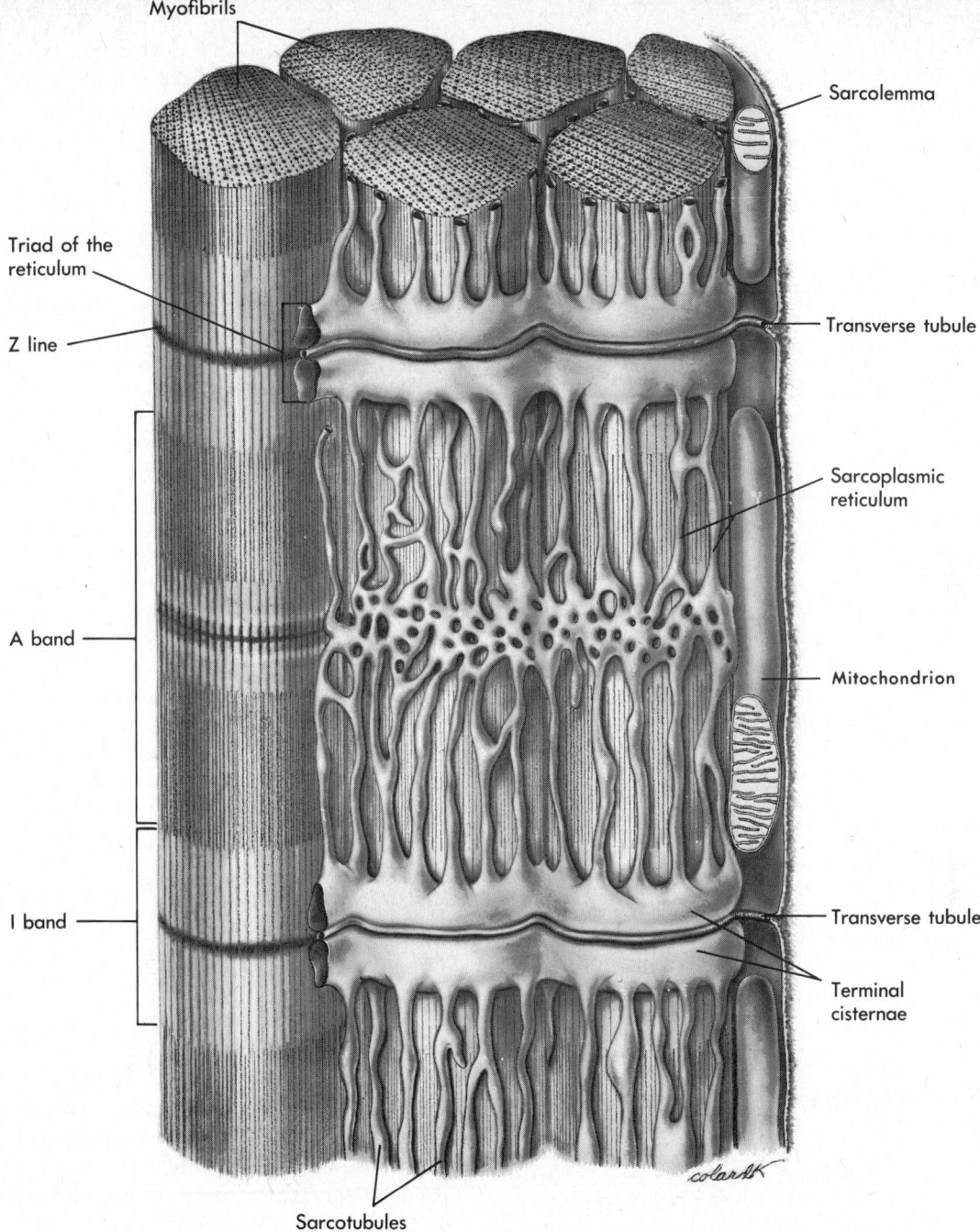

Myofibrils

Sarcolemma

Triad of the reticulum

Z line

Transverse tubule

A band

Sarcoplasmic reticulum

Mitochondrion

I band

Transverse tubule

Terminal cisternae

Sarcotubules

Figure 4 A drawing to illustrate the anatomical relationships among transverse tubules, sarcoplasmic reticulum, and myofibrils. (From Bloom W, Fawcett DW: *A Textbook of Histology,* ed 9. Philadelphia, Saunders, 1968.)

MITOCHONDRIA AND GLYCOGEN

Many mitochondria are located in the clefts between adjacent myofibrils where they can supply ATP directly to the contractile machinery. Granules of glycogen also accumulate between myofibrils and serve as one of the sources of chemical energy for production of ATP.

EXCITATION OF SKELETAL MUSCLE

Skeletal muscle fibers contract in response to action potentials traveling over the muscle plasma membranes. These muscle fiber action potentials originate at endplates in response to activity in nerve terminals. Skeletal muscle is neurogenic; in other words, its activity originates from nerve impulses. This requirement for neural input is an important characteristic that distinguishes skeletal muscle from cardiac muscle and visceral smooth muscle. The latter are myogenic (i.e., activity arises spontaneously within the muscle itself and does not require nerve impulses).

NEUROMUSCULAR TRANSMISSION

Transmission of activity across the neuromuscular junction is the first intramuscular process involved in muscle excitation. This process was discussed in Chapter 3.

THE MUSCLE ACTION POTENTIAL

The muscle action potential has the same basic characteristics as the action potential in an unmyelinated neuron and is propagated in the same manner. Its duration is a few milliseconds, and its conduction velocity is around 5 m/s in mammalian muscle. It spreads out in all directions from the endplate, circumferentially as well as longitudinally, and finally rings of activity pass to both ends of the fiber. As the impulse travels over the fiber, it initiates events within the fiber that lead to contraction of the myofibrils. Thus, a wave of contraction starts at the endplate and spreads to both ends of the fiber. The beginning of the contraction lags briefly behind the action potential. This delay between the action potential along the surface membrane and the contraction of the myofibrils is called the latent period.

ROLE OF THE TRANSVERSE TUBULES

Before the transverse tubules and sarcoplasmic reticulum were discovered and their function surmised, the coupling between excitation of the plasma membrane and contraction of myofibrils was quite mysterious. One of the main problems was in understanding how the process could occur so quickly. The latent period is only a few milliseconds, too brief to allow a chemical messenger enough time to diffuse from the plasma membrane to the more centrally placed myofibrils.

The function of the transverse tubules (T tubules) is to conduct an electrical signal from the plasma membrane to the interior of the fiber around each myofibril. Functionally, as well as anatomically, the T tubules can be considered an inward extension of the plasma membrane. This electrical signal along the T tubules is a continuation of the all-or-none muscle action potential; however, the conductive properties of the T-tubular membranes are quantitatively distinct from those of the plasma membrane. The densities of K^+ and Na^+ channels involved in excitation are less in the T system than in the surface membrane, and the Cl^- conductance is considerably lower, such that in the resting state the principal conductance is a K^+ conductance.

The necessity for the T system in excitation–contraction coupling is dramatically demonstrated by the finding that muscle fibers treated with hypertonic glycerol solutions for a brief period, and then returned to normal solutions, lose their ability to contract when electrically stimulated. This occurs despite the presence of a propagated action potential across the surface membrane. Electron microscopic evidence and measurements of the membrane capacitance indicate that the glycerol treatment disrupts the T system such that the surface membrane remains intact but isolated (broken off) from the T tubules.

ROLE OF CALCIUM IONS AND THE SARCOPLASMIC RETICULUM

The intracellular chemical messenger that directly triggers contraction of the myofibrils is Ca^{2+}. The concentration of free Ca^{2+} in the sarcoplasm of a resting muscle is very low (below 10^{-7} M), but large quantities of Ca^{2+} are stored within the terminal cisternae of the sarcoplasmic reticulum. An electrical signal in the transverse tubules causes a pulse of Ca^{2+} to escape from the terminal cisternae. This Ca^{2+} quickly diffuses among the myofilaments and triggers the chemical reactions that lead to development of longitudinal force between thick and thin filaments.

The link between excitation in the T tubules and the release of Ca^{2+} from the sarcoplasmic reticulum remains obscure and an area of intense research activity. The morphological link is the junction between T tubules and sarcoplasmic reticulum (SR), which consists of two distinct membranes, 120 Å apart; however, the SR membrane sends out projections or "feet" to within 50 Å of the T-tubular membrane. The projections are "connected" to the T membrane by an amorphous material of unknown composition. Recent experiments on "gating" currents have led to the speculation that there is a shift of certain charged structures within the SR membrane projections in response to changes in the potential across the T-tubular membrane. Presumably the movement of these mobile charged structures, in some unknown manner, triggers the release of Ca^{2+} from the SR.

The muscle action potential and the electrical signals in T tubules and SR last only a few milliseconds; Ca^{2+} then rapidly moves back into the sarcoplasmic reticulum, thus turning off the contractile process and allowing relaxation to ensue. This reuptake of Ca^{2+} by the SR is coupled to hydrolysis of ATP and takes place by active transport of Ca^{2+} across the SR membranes. Active reuptake of Ca^{2+} is accomplished by the longitudinal SR rather than the terminal cisternae. The Ca^{2+} then moves through the longitudinal SR to be stored in the terminal cisternae.

CHEMISTRY OF CONTRACTION

COMPOSITION OF MYOFILAMENTS

The thick filaments are primarily constructed from molecules of a protein called myosin. Each thick filament contains approximately 400 myosin molecules. An individual myosin molecule is diagrammed in Figure 5a; it consists of a globular head and a long tail. Myosin molecules combine in the manner shown in Figure 5b to form thick filaments. The long part of each tail, known as light meromyosin (LMM), joins side by side with the corresponding parts of other myosin molecules to form the "backbone" of the filament. The short part of the tail is called the S_2 segment, and the globular head is called the S_1 segment. S_1 and S_2 together are known as heavy meromyosin (HMM). The globular heads, S_1, project radially from the LMM backbone at longitudinal intervals of 143 Å. As shown in Figure 5c, these projections occur in a helical arrangement along the filament. Note from Figure 5b that myosin molecules are oriented in opposing directions in the two halves of the filament with the long tails always pointing toward the center, and that the central portion of the filament has no S_1 projections.

The thin filaments are made from three proteins: actin, tropomyosin, and troponin. A model of part of a thin filament is shown in Figure 6.

Globular subunits of actin (G actin) stick together in long fibers or chains called F actin. Two such F actin chains wind around each other in a double helix to form the core of each thin filament. Tropomyosin molecules are each about 400 Å long. They are bound to actin and arranged end to end along each chain of the actin double helix. Each tropomyosin molecule spans seven G actin subunits and is thought to be positioned just out of the groove formed by the double helix. Troponin is a complex of three subunits: TN-I, an inhibitory component that prevents actin–myosin interaction; TN-C, a Ca^{2+}-binding component; and TN-T, a tropomyosin-binding component. The troponin complex is attached to each end of the tropomyosin molecule; thus it is periodically spaced on the actin filament at 400 Å intervals.

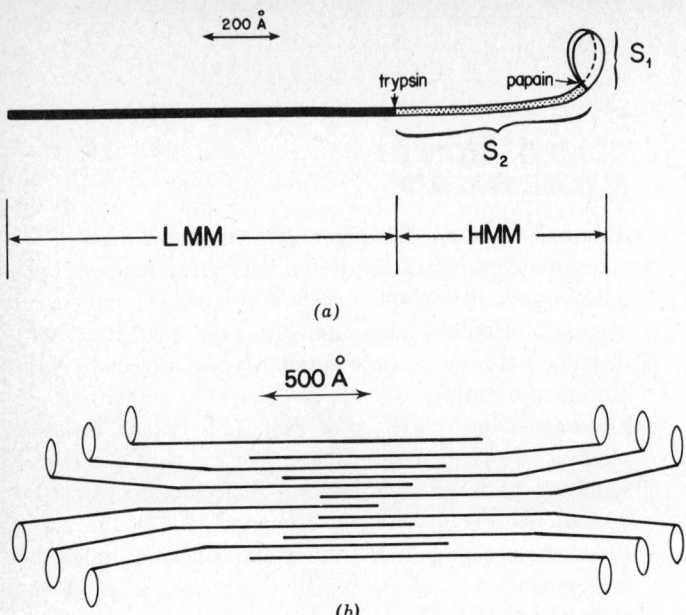

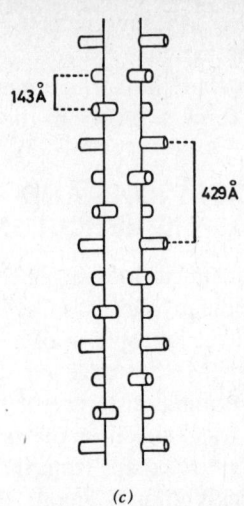

Figure 5 (*a*) Diagram of the myosin molecule. Total molecular weight is about 475,000. By brief trypsin digestion the molecule can be split into two parts called light meromyosin (LMM) and heavy meromyosin (HMM). If HMM is subsequently treated with papain, it splits into two subfractions, S_1 and S_2. LMM together with S_2 form the tail of the molecule (about 1,500 Å long), while S_1 forms the globular head of the molecule (about 100–150 Å long). S_1 is actually a double head, consisting of two apparently identical subunits. (*b*) A diagram showing the approximate way in which myosin molecules assemble to form thick filaments. This diagram depicts only the central portion of a thick filament where one-half the molecules point toward one end of the filament and one-half point toward the other end. Assembly of a complete filament would simply require more molecules to be inserted from each end oriented in the proper direction. The LMM parts of the molecules stick to each other and are responsible for self-assembly of filaments. The S_2 segments lie nearly flat along the surface of the LMM "backbone." The S_1 heads project from the surface at about 90°. The scale of this diagram is approximately correct in the longitudinal direction, but the model should be compressed in the vertical direction and then rolled up and probably twisted a little. The actual thickness of a thick filament is about 120 Å and a cross section of a thick filament would include about 20 molecules. (*c*) The helical distribution of S_1 projections along a thick filament. (From Huxley HE: The mechanism of muscular contraction. *Science* 164:1366, 1969.)

BINDING OF MYOSIN TO ACTIN

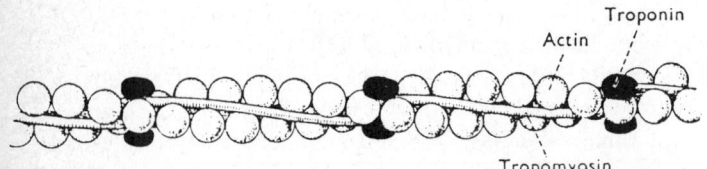

Figure 6 A model of thin filament structure. (From Ebashi S, Endo M, Ohtsuki I: Control of muscle contraction. *Q R Biophys* 2:351, 1969.)

The globular heads of HMM (i.e., the S_1 segments) have the property of binding to actin subunits. Each actin subunit can bind one S_1. Thus, thick and thin filaments can bind together via the S_1 projections, which are often called cross-bridges. However, in resting muscle, binding between S_1 and actin is blocked by tropomyosin, which physically covers the sites on actin that could otherwise bind to S_1.

BINDING OF Ca^{2+} TO TROPONIN —THE TRIGGER FOR INITIATING CONTRACTION

When Ca^{2+} is released from the terminal cisternae, some of it binds to troponin. X-ray diffraction data suggest that when this happens the tropomyosin molecules move into the center lines of the grooves formed by the double helix, thereby uncovering the sites on actin that can bind to S_1. Each troponin molecule controls the position of a tropomyosin molecule. In turn, the position of each tropomyosin molecule determines the ability of seven actin subunits to bind to the S_1 heads of HMM. When Ca^{2+} is not bound to troponin, tropomyosin physically covers the binding sites on actin; and when Ca^{2+} is bound to troponin, these sites are exposed.

HYDROLYSIS OF ATP BY MYOSIN

Another important property of the globular head of HMM is its ability to bind ATP and to catalyze the hydrolysis of ATP; in other words, S_1 is an ATPase. This enzymatic activity is greatly increased in the presence of thin filaments. The thin filament is a powerful activator of myosin-ATPase. But in resting muscle thin filaments cannot activate myosin-ATPase, because tropomyosin prevents binding between actin and S_1.

The reaction sequence for ATP hydrolysis is shown in Figure 7. When the free Ca^{2+} concentration is very low, as it is in resting muscle, actin cannot bind to S_1 and the only reaction sequence possible is the one shown within the dashed box. ATP binds to S_1 and is quickly split into its products, ADP and inorganic phosphate (P), but the products tend to remain attached to S_1 for a relatively long period of time (10–100 s). Consequently, the rate of myosin-catalyzed ATP hydrolysis in resting muscle is very low. After Ca^{2+} is released from the terminal cisternae and diffuses along the actin filaments, actin can combine with the [M · ADP · P] complex to form the [AM · ADP · P]$_I$ complex. This complex undergoes a conformational change to the [AM · ADP · P]$_{II}$ complex. In the crossbridge theory of muscle contraction, described below, it is this conformational change that is thought to generate active tension in the muscle—also believed to be the step at which the chemical energy originally residing in ATP is transformed into mechanical energy. Next, the products come off the actomyosin complex, leaving the actin tightly

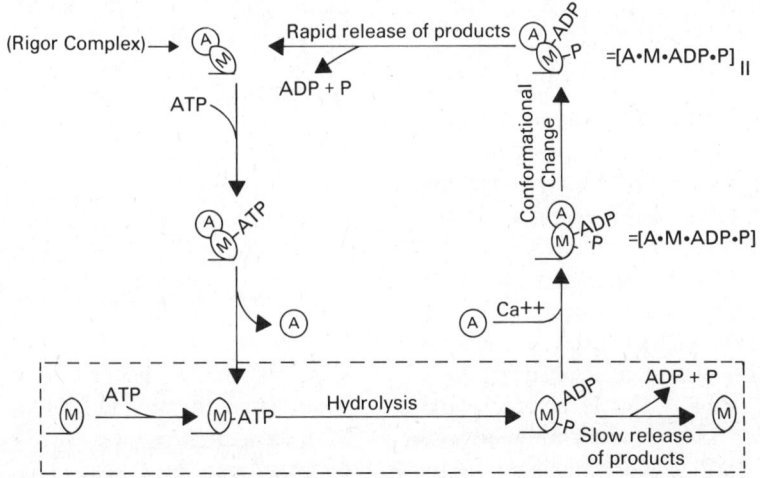

Figure 7 Reaction sequence for hydrolysis of ATP by both myosin and actomyosin. See text for description. A, An actin molecule; M, heavy meromyosin; $\frac{A}{M}$, the complex between actin and myosin (referred to as actomyosin); and P, inorganic phosphate.

bound to myosin. If another ATP molecule is available it then may attach to S_1. This attachment of ATP to S_1 weakens the bond between S_1 and actin and they separate; thus ATP is responsible for dissociating actin and myosin from each other. This new ATP is then split while on the catalytic site of S_1; if the free Ca^{2+} concentration remains elevated and the supply of ATP holds out, the cycle can be repeated over and over again. All the reactions in the cycle are very fast compared to the slow conversion of $[M \cdot ADP \cdot P]$ to $M + ADP + P$. Herein lies the activating effect of actin on myosin ATPase; it greatly accelerates release of the products.

RESYNTHESIS OF ATP

It should be clear from the above discussion that hydrolysis of ATP provides the energy required for muscular contraction. A muscle fiber contains only enough ATP to sustain vigorous activity for a fraction of a second. Clearly, a mechanism must be available during activity to replenish the available ATP at a rapid rate. This mechanism is the transfer of phosphate from creatine phosphate (CP) to ADP. Resting muscle has several times as much CP as it does ATP; it also contains an enzyme, creatine kinase, that catalyzes this phosphotransferase reaction. Owing to this reaction, the concentration of ATP in a muscle does not change detectably until fairly vigorous activity has occurred.

Nevertheless, the stores of ATP and CP could not be maintained for more than a few seconds during vigorous activity unless they were replenished by glycolysis and oxidative phosphorylation. Glycolysis is especially important during vigorous activity, because the Krebs cycle and oxidative phosphorylation cannot keep pace with the demands to supply more ATP (probably because O_2 cannot reach the mitochondria fast enough). Consequently, lactic acid accumulates in intensely active muscle, and much of it diffuses into the blood. Some of this lactic acid is oxidized by the heart and provides some of the energy required for the increased cardiac work during exercise. Some of it enters the liver parenchymal cells where it can mostly be converted back to glucose (at the expense of about one-sixth of it, which provides the energy for conversion to glucose by being oxidized). This glucose produced in the liver can then re-enter the circulation and ultimately provide substrate for more anaerobic glycolysis in muscle.

Following a bout of vigorous muscular exercise, oxygen continues to be used by the body in excess of basal requirements in order to burn off accumulated lactic acid and replenish CP and ATP supplies. This excess O_2 use is called the oxygen debt.

RIGOR MORTIS

All the skeletal muscles of the body stiffen 1–7 hr after death. This stiffening is called rigor mortis; it begins to disappear after a day or less as the muscle proteins undergo autolysis, caused by lysosomal enzymes. This muscle stiffening is associated with the loss of ATP from the muscles. You will remember that ATP is required to dissociate S_1 from actin (Fig. 7). In the absence of ATP, the bond between actin and S_1 is very strong, and the thick and thin filaments remain strongly attached to each other, presumably accounting for the rigor. In fact this bond between actin and S_1 is known as the rigor bond. In normal muscle ATP serves the extremely important function of weakening this bond, thereby allowing the thick and thin filaments to dissociate from each other and the resting muscle to be plastic rather than rigid.

THE CROSS-BRIDGE THEORY OF MUSCLE CONTRACTION

The cross-bridge theory of muscle contraction states that when a sarcomere is activated (by Ca^{2+} release from the SR), cross-bridges form between thick and thin filaments by attachment of S_1 heads to actin subunits. These cross-bridges generate the longitudinal sliding force between thick and thin filaments responsible for generating longitudinal tension in the whole muscle. The following observations, the first four of which have already been explained, form the basis for this theory:

1. The S_1 heads of myosin project radially from the thick filaments at 143 Å intervals; these projections are arranged helically along each thick filament (Fig. 5c).

2. Under certain circumstances the S_1 heads of myosin can bind to actin subunits, thereby forming cross-bridges between thick and thin filaments. The conditions necessary for formation of cross-bridges are (a) close proximity between S_1 and actin, and (b) accessibility of the binding sites on actin. Both of these conditions are expected to be met when a sarcomere, at normal length, is activated by Ca^{2+} release from the SR.

3. Under these same conditions the ATPase activity of S_1 is greatly increased by actin. This is because the hydrolytic products, ADP and P, are released from S_1 much faster when it is bound to actin than when it is not bound.

4. The $[AM \cdot ADP \cdot P]$ complex undergoes a conformational change prior to release of products.

5. Structures that look like cross-bridges have frequently been seen in electron micrographs of contracting striated muscle (e.g., Fig. 2b).

6. X-ray diffraction studies have been interpreted as indicating that in the activated state there is an asynchronous axial movement of S_1 projections.

7. The amount of active tension that a sarcomere is capable of developing seems to be a nearly linear function of the amount of overlap between thick and thin filaments. This important finding will be explained in detail later in the section on the "length–tension relationship." It leads to the conclusion that the greater the number of potential cross-bridges, the greater the active tension.

A model showing how cross-bridges might accomplish the development of tension is depicted in Figure 8. Look again at Figure 7, since it illustrates not only the cycle of actomyosin-catalyzed ATP hydrolysis, but the corresponding cycle of cross-bridge activity as well.

The mechanism depicted in Figure 8 is somewhat hypothetical, but it seems very appealing to most physiologists and currently enjoys wide acceptance. However, some investigators remain skeptical. For a different point of view, consult the article by Noble and Pollack listed in the Selected Bibliography.

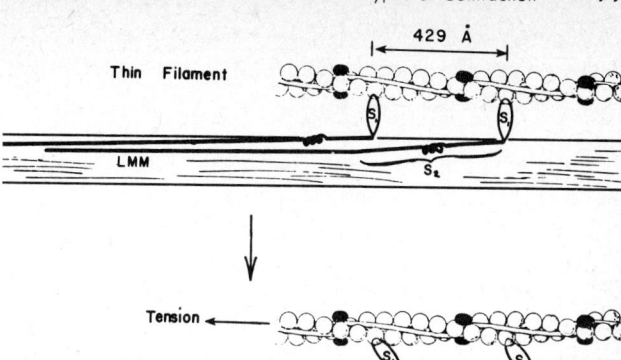

Figure 8 Model of a cross-bridge theory of muscle contraction. Portions of one thick and one thin filament are depicted. The S_1 projections from the thick filament that contact any one thin filament are spaced 429 Å apart (Figure 5c). In the upper drawing the S_1 projections have just formed cross-bridges by attaching to actin subunits, the binding sites on which have been made available by the Ca^{2+}–troponin–tropomyosin regulatory mechanism. When initially formed, the cross-bridges are approximately perpendicular to the filaments. The S_2 rods are depicted, by little springs, as having elastic properties; when they are stretched, tension is generated. Immediately after the cross-bridges are formed, according to this theory, they rotate to about 45° with respect to the filaments, as shown in the lower drawing. This rotation most likely results from a change in the manner of binding to actin and is the power stroke in the cross-bridge cycle. The rotation is allowed to take place because the S_2 rods are extensible and because the points of attachment between S_2 and S_1 and between S_2 and LMM are flexible. The power strokes generate tension in the S_2 rods, in turn generating the sliding force between thick and thin filaments. If the sarcomere is not fixed at a constant length, the filaments will slide past each other a distance of about 70 Å. The drawing of thin filaments is the same as that shown in Figure 6.

TYPES OF CONTRACTION

The characteristic response of muscle is the development of force directed longitudinally. If the opposing force is less than the force developed by the muscle, the muscle shortens and thus performs work. External work performed is equal to the opposing force multiplied by the distance moved. If the force is maintained constant throughout shortening, the contraction is called isotonic. It is common to refer erroneously to any shortening contraction as "isotonic," even if, as is usually the case, force is not maintained constant. The term "concentric contraction" is sometimes used to refer to a shortening contraction.

If the opposing force just balances the force developed by the muscle, then the muscle cannot shorten but remains at a constant length. This type of contraction is called an isometric contraction. No external work is performed by an isometric contraction.

If the opposing force is greater than the force exerted by the muscle, then the muscle is stretched; it may lengthen even though it is actively contracting. In this case work is done on the muscle by the opposing force. The term "contraction" does not necessarily refer to shortening; it does refer to the development of active tension. The term "eccentric contraction" is sometimes used to refer to a lengthening contraction.

Thus, there are three types of contraction: shortening (concentric), isometric, and lengthening (eccentric). The body's muscles perform all three types while carrying out their normal duties.

ELECTRICAL STIMULATION OF MUSCLE

For experimental purposes, muscle is often stimulated electrically. Electrical stimulation is also used clinically in cases of nerve injury.

THE STIMULUS–RESPONSE RELATIONSHIP AND THE ALL-OR-NONE LAW

If a single muscle fiber is isolated and laid across two electrodes from which a brief electrical current is passed to initiate an action potential, the fiber will contract if the electrical stimulus is strong enough. There is some critical stimulus strength for each muscle fiber, below which there is no action potential and therefore no contraction, and above which the fiber contracts. Changing the stimulus strength to values that are progressively greater than this critical value causes no progressive change in the strength

of contraction. The fiber either contracts, or it does not. Varying the strength of the stimulus above threshold does not influence the strength of the response—this is the "all-or-none law."

The myofibrils themselves do not follow an all-or-none law; their response is graded according to how much Ca^{2+} becomes attached to troponin, how much ATP is available, and probably other factors. But the muscle action potential does follow the all-or-none law and usually causes the same amount of Ca^{2+} release regardless of the strength of the stimulus. Consequently, the contractile response of the single fiber is all-or-none whenever excitation occurs via a single action potential.

If a whole muscle is stimulated electrically, the stimulus–response relationship is quite different from that for a single fiber. As shown in Figure 9, the response gradually increases over a considerable range of stimulus strengths. The first response (at just threshold stimulus strength) is caused by stimulation of the most sensitive fibers. As the strength of the stimulus increases, more and more fibers are stimulated, and the muscle contracts more forcefully, until eventually the stimulus is strong enough to excite all the fibers. A stimulus is called subminimal, minimal, or maximal according to whether it excites zero, one, or all the muscle fibers. A supramaximal stimulus is one that is more than strong enough to excite all the muscle fibers.

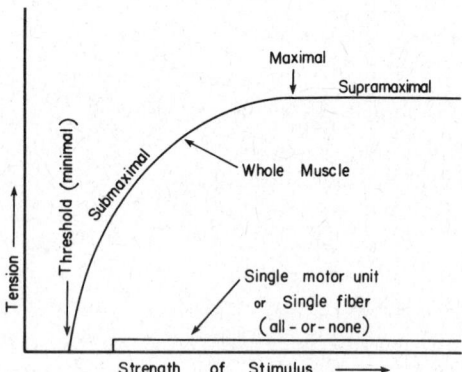

Figure 9 The stimulus–response relationship for a whole muscle or for a single motor unit.

MECHANICS OF CONTRACTION

THE TWITCH

If either a whole muscle or a muscle fiber is stimulated with a single stimulus (e.g., a brief electrical shock to its motor nerve), it responds with a single brief contraction–relaxation cycle called a twitch. The twitch is the basic unit of mechanical response of a muscle fiber. In a shortening contraction, the distance shortened can be plotted against time. In an isometric contraction the tension developed can be plotted against time. An isometric twitch is illustrated in Figure 10. The time from the beginning of the action potential to the first rise in tension is called the latent period. In mammalian muscle the latent period for an isometric twitch varies from about 1–10 ms. The latent period represents the time it takes for the electrical signal to reach the sarcoplasmic reticulum, for Ca^{2+} to be released and to diffuse to the troponin molecules on the thin filaments, and for the chemical and physical events involved in force generation between thick and thin filaments. The time from the beginning of the contraction to the peak of contraction is called the contraction period. The time from the peak of contraction to the full return to baseline is called the relaxation period.

It is of great importance to realize that the action potential in skeletal muscle is much briefer than the contraction period. The refractory period of the membrane hardly lasts beyond the beginning of contraction. Consequently, a second stimulus (e.g., a nerve impulse or an electrical shock) delivered during the contraction period is capable of generating a second action potential along the muscle fiber plasma membranes, and the myofibrils can respond to this second stimulus. This behavior of skeletal muscle contrasts with that of cardiac muscle. In cardiac muscle each action potential lasts the full duration of the contraction period. The importance of this to the functioning of the heart will be explained in Chapter 5.

FIBER TYPES

Individual skeletal muscle fibers are classified, according to the speed of their contraction period, into two general types: fast twitch fibers and slow twitch fibers.* The fast twitch fibers have contraction periods averaging about 35–40 ms, and the slow twitch fibers have contraction periods averaging about 75 ms. Various differences between these two major fiber types are summarized in Table 1. The fast twitch fibers are fast because of their relatively large amount of SR and high HMM-ATPase activity, but most tend to be susceptible to endurance fatigue because of their low rate of oxidative metabolism and low myoglobin content. The slow twitch fibers possess more endurance because of their high mitochondria and myoglobin content.

The muscles of the body all consist of mixtures of these two major fiber types. The proportion of slow twitch fibers ranges from about 10% to about 95% of the total. The higher the proportion of fast twitch fibers, the faster the whole muscle. Generally, muscles not required to contract quickly, but required to maintain activity over long periods of time, are composed of a high percentage of slow twitch fibers. The soleus muscle, for example, contains about 85% slow twitch fibers. Figure 11 shows the twitch contours for three different skeletal muscles.

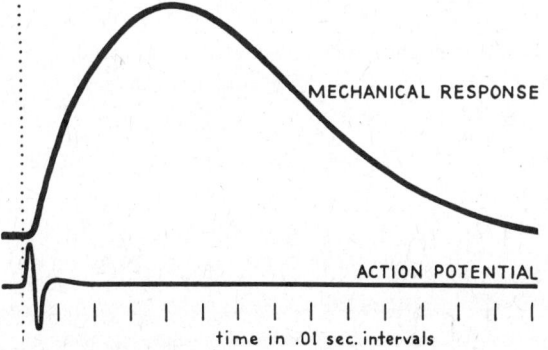

MECHANICAL RESPONSE

ACTION POTENTIAL

time in .01 sec. intervals

Figure 10 An isometric twitch and the action potential that initiated it. Upper trace, tension generated by the muscle. Lower trace, electrical potential difference between two electrodes placed on the surface of the muscle. (From Fulton JF (ed): *A Textbook of Physiology,* ed 16. Philadelphia, Saunders, 1949.)

---

*Another category of muscle fibers, called "slow fibers," has mechanical and electrical properties quite different from those of twitch fibers. These properties are not described here, as there is no solid evidence that slow fibers play any major role in mammalian muscle; they have been studied mainly in amphibian muscle.

Table 1 Comparison Between the Properties of Fast Twitch and Slow Twitch Skeletal Muscle Fibers

Property	Fast Twitch	Slow Twitch
Average contraction period	35–40 ms	75 ms
Average frequency of stimulation required for tetanus	50 Hz	20 Hz
Sarcoplasmic reticulum	High	Low
HMM-ATPase activity	High	Low
Rate of anaerobic glycolysis	High	Low
Endurance	Low[a]	High
Number of mitochondria and rate of oxidative metabolism	Low[a]	High
Myoglobin content	Low	High
Density of capillaries around fiber	Low	High
Color	White[a]	Red

[a]There are several subgroups of fast twitch fibers. Some have many mitochondria and rapid oxidative metabolism and appear redder and are more fatigue resistant than the usual fast twitch fibers.

THE ACTIVE STATE AND SERIES ELASTICITY

If a muscle is subjected to a sudden longitudinal stretch during various stages of the twitch, it can be observed that its stiffness increases and then returns to normal. In other words, it gets more difficult to stretch for a brief time. Of course, this response is not surprising, since interaction between thick and thin filaments would be expected to increase muscle stiffness. The interesting aspect of this observation is that the change in stiffness (or elasticity) is a considerably quicker event than the twitch as recorded by tension developed or distance moved. This change in

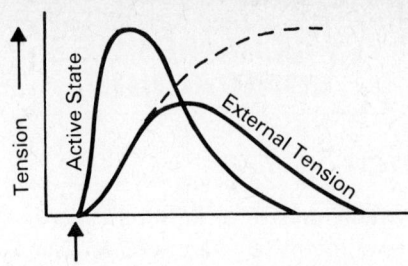

Figure 12 Comparison between intensity of the "active state" and external tension during an isometric twitch. The arrow indicates the time the stimulus was delivered. The curve labeled "active state" shows the tension the contractile components would be capable of developing if there were no series elastic components. The dashed line shows the external tension the muscle would generate if the active state were prolonged, as in a tetanus.

stiffness is believed to reflect the interaction between thick and thin filaments much more closely than does the actual twitch, which lags considerably behind. Figure 12 shows the time course of the intensity of the interaction between thick and thin filaments (as surmised from the change in stiffness); this is referred to as the intensity of the "active state." The intensity of the active state increases abruptly at the end of the latent period, reaching a maximum almost immediately; it is then maintained for several milliseconds, and then gradually disappears. The intensity of the active state begins to decline before the contraction period is over. To illustrate this important point a record of an isometric twitch is also shown in Figure 12. Recent studies of the free Ca^{2+} concentration in the sarcoplasm during isometric twitches indicate that the intensity of the active state follows a time course very similar to that of the free Ca^{2+} concentration.

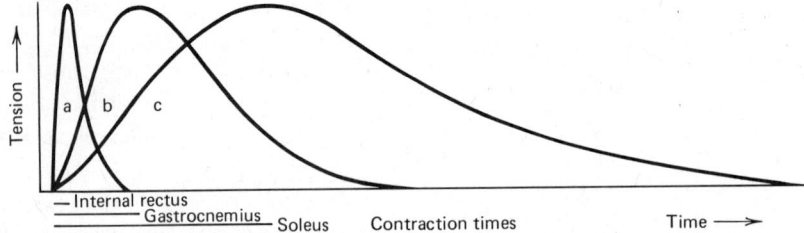

Figure 11 Isotonic twitches of (a) internal rectus of the eye, (b) gastrocnemius, and (c) soleus. The contraction period for the soleus is about 100 ms. (From Fulton JF (ed): *A Textbook of Physiology,* ed 16. Philadelphia, Saunders, 1949.)

The fact that the isometric twitch tension lags considerably behind the intensity of the active state led to the concept of passive elastic components operating in series with the contractile components. When the contractile components are activated, they develop tension rapidly and begin to stretch the "series elastic components." As the series elastic components are stretched, tension between the ends of the muscle *gradually* increases, even though the activity of the contractile elements *abruptly* reaches maximum intensity. Connective tissues, especially tendons, contribute to series elasticity in a whole muscle. But the presence of series elasticity can be demonstrated in single fibers having minimal connective tissue, and it is now widely believed that a major contribution to series elasticity is from the contractile components themselves. The elastic property of the contractile machinery was depicted in Figure 8 by little springs in each S_2 segment; however, the precise location of series elasticity is still uncertain.

SUMMATION OF CONTRACTIONS

If a muscle or a single fiber is stimulated with a supramaximal stimulus twice in succession, and if the second stimulus is delivered sometime during the first twitch, then the second contractile response summates upon the first response. In other words, two stimuli in rapid succession produce a greater amount of isometric tension (or of shortening) than is possible from a single maximal stimulus. If the second stimulus is delivered just before the peak of the first twitch, the second response may be almost as great as the first response and the total response is almost doubled. If the second stimulus is delivered at progres-

sively later times during the relaxation phase of the first twitch, the total summated response becomes less and less until finally there are only separate twitches with no summation when the second stimulus is given after the first twitch is over. Figure 13 illustrates summation of contractions.

The response to three successive stimuli can be even greater. There is, of course, some maximum summated response that is attained after several successive stimuli; the magnitude of this plateau depends on the frequency of stimulation. If frequency is fairly high, the summated response is a smoothly fused contraction termed a tetanus or tetanic contraction. At lower frequencies the contraction is jerky and only partly fused, and is termed a partial tetanus. A tetanic contraction may develop up to four times the tension of a single twitch. Complete tetanus and partial tetanus are illustrated in Figure 14.

The stimulation frequency required to produce a smooth tetanus varies with the speed of contraction of the muscle, which, in turn, depends on the proportion of fast twitch fibers it contains. A very fast muscle, such as the internal rectus of the eye, requires about 350 stimuli per second for complete tetanus. A slow muscle, such as the soleus, requires only 30 stimuli/s. The gastrocnemius is intermediate (about 100/s). Tetanic contractions do not often occur during normal activity except during very intense muscular exertion; the usual frequency of nerve impulses is not great enough. There are two pathophysiological conditions that apparently involve tetanic, or at least partial tetanic, contraction; one is a condition called "tetany" and is caused by hypocalcemia; the other is called "tetanus," referred to colloquially as lockjaw, and is caused by the toxin of a bacillus.

During a tetanic contraction, the action potentials that

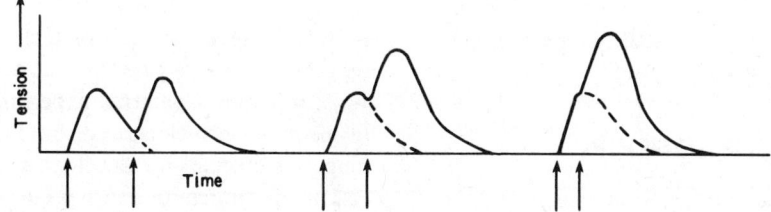

Figure 13 Summation of contractions. The arrows indicate the times the stimuli were delivered. The dotted lines show the tension that would have been recorded had the second stimulus of each pair not been delivered.

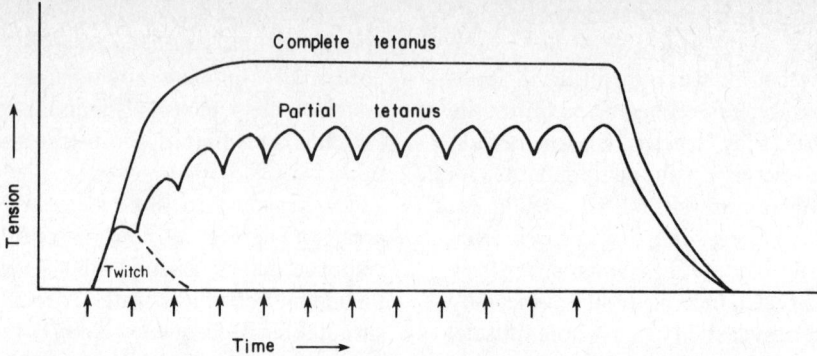

Figure 14 Complete tetanus and partial tetanus compared with a single twitch. The arrows indicate the stimuli that produced the partial tetanus. The stimulation frequency was about 2.4-fold greater than this for the complete tetanus.

trigger the contraction are not themselves summated. Although the contractions are fused and are indistinguishable from each other, there is a series of individual action potentials, one for each individual stimulation or augmentation of the muscle's active state.

TETANIC VERSUS TWITCH TENSION

The fact that the twitch tension is much less than tetanic tension shows that a single excitatory event does not elicit the full capability of a muscle fiber to develop tension. The series elastic component is thought to be mainly responsible for this. Because of series elasticity, the activity of the contractile elements is already declining by the time the total tension approaches a maximum. In other words, before tension has a chance to reach its full amplitude, the fiber begins to relax. In effect, the series elastic component damps the contractile response. But when the series elastic component is already stretched from the first response, a second burst of activity in the contractile elements can be more fully translated into muscle tension.

LENGTH–TENSION RELATIONSHIP

If a single skeletal muscle fiber is freed of connective tissue by careful dissection and is suspended in Ringer's solution, its "slack length" is such that each sarcomere is about 2.0 μm in length. At this length the H band is very

narrow, that is, the thin filaments almost reach to the center of the A band. This is the degree of filament overlap in a relaxed muscle fiber on which no stretch is being exerted. If this fiber is fixed at each end to rigid supports and is then stimulated electrically, it develops tension (isometrically) between its two ends. If one of the supports is a strain gauge, this tension can be measured; it is called *active tension* to distinguish it from the *passive tension* developed when an unstimulated fiber is stretched.

Now if the distance between the supports is slightly increased, thereby stretching the fiber, a small amount of *passive tension* can be measured. If the fiber is then electrically stimulated, active tension can be recorded superimposed on the passive tension. However, the active tension recorded from the stretched fiber is never as great as that from the unstretched fiber at its slack length.

On the other hand, if the distance between the supports is decreased so that the fiber is no longer held straight and it is then electrically stimulated, the fiber shortens until it is taut between the supports and then develops tension. However, again, the active tension developed at shorter lengths is never as great as that developed at slack length. It is apparent from this experiment, that the degree of overlap between thick and thin filaments present at slack length is optimal for developing tension.

The classic length–tension curve of single muscle fibers obtained by Ramsey and Street in 1940 is shown in Figure 15. The exact shape of the active tension curve is currently under dispute. However, one point is clear; when the muscle is stretched so much that there is no filament overlap, active tension cannot develop. Filament overlap is thus a requirement for the development of force. The amount of force at a given sarcomere length beyond l_o (the slack length) seems to be determined by the extent of thick and thin filament overlap, as well as by the degree of activation of contraction brought about by Ca^{2+} and by the concentration of ATP.

The principal difficulty in measuring accurate length–tension curves is the control of the uniformity of the sarcomere lengths along the entire fiber. Accurate control of sarcomere length was first attempted by A. F. Huxley and colleagues. They developed a servocontrol system for maintaining precisely a short segment of a single muscle fiber at various lengths. The experiments were done using the middle portion of each muscle fiber because the variation in sarcomere lengths in this region is quite small. Figures 16 and 17 summarize their results.

Figure 16 plots active tension against sarcomere length. The curve is composed of four segments having distinctly different slopes. These segments correspond to four different stages in the interaction between thick and thin filaments. These stages are diagrammed in Figure 17. At point A the sarcomere is stretched just enough so there is no overlap between thick and thin filaments; consequently, cross-bridges cannot be developed. Active ten-

sion progressively increases along segment A–B as overlap between thick and thin filaments increases. There is, in fact, a linear relationship between the number of potential cross-bridges and active tension. At point B the sarcomere is not stretched beyond the length at which a maximum number of cross-bridges can form. At point C filament overlap is greater than at point B, but there is no increase in the number of cross-bridges because the central portion of each thick filament has no HMM projections. Thus, along segment B–C the number of cross-bridges is maximal and remains constant, and along this whole segment maximum active tension can be developed. At lengths shorter than that corresponding to point C, forces develop that resist sliding of filaments. At point C the ends of opposing thin filaments start to slide past each other, creating increased internal resistance to filament sliding, and active tension declines along segment C–D. At point D the ends of the thick filaments make contact with Z lines, and at shorter lengths internal resistance is augmented as the thick filaments crumple on the Z lines.

There is also evidence that at lengths progressively below the slack length, there is a progressive decrease in the amount of Ca^{2+} released from the terminal cisternae during excitation. Currently it is not clear whether the mechanical factors discussed above or this decrease in Ca^{2+} release are more important in determining the shape

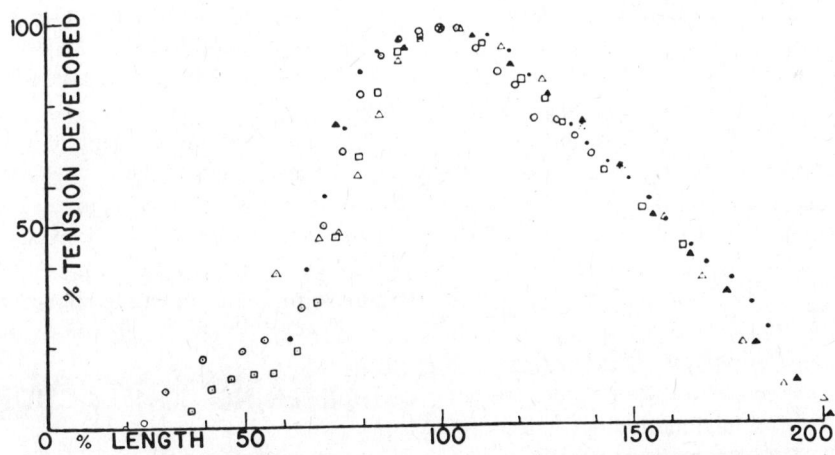

Figure 15 Length–tension diagram for single isolated frog muscle fibers. "Tension developed" is the active tension (i.e., total tension minus passive tension) and is expressed as percentage of maximum. Length is given as % of slack length. (From Ramsey RW, Street SF: *J Cell Comp Physiol* 15:11, 1940.)

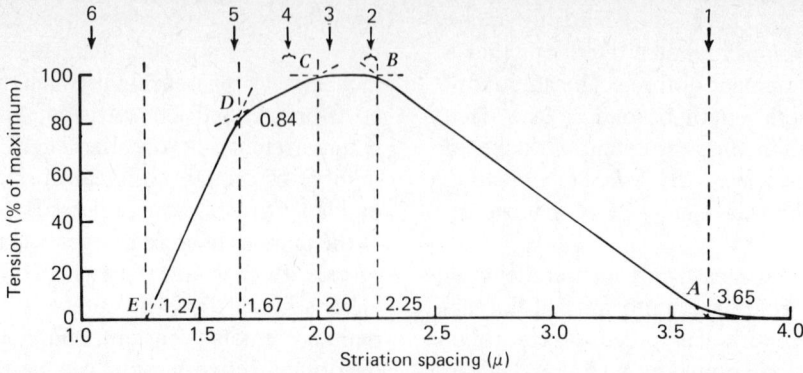

Figure 16 Active isometric tetanus tension as a function of sarcomere length. This length–tension diagram is probably somewhat more accurate than that presented by Ramsey and Street (cf. Fig. 15). Letters (A, B, C, D, and E) indicate the four nearly straight lines from which the curve can be considered to be constructed. The numbers (1–6) indicate the sarcomere lengths (i.e., striation spacings) at which certain critical stages in the process of filament sliding occur. These critical stages are illustrated in Figure 17. (From Gordon AM, Huxley, Julian FJ: The variation in isometric tension with sarcomere length in vertebrate muscle fibers. *J Physiol (Lond)* 197:709, 1968.)

of the length–tension relationship at short sarcomere lengths.

According to the cross-bridge theory, each cross-bridge is an independent force generator, and the total active tension is the sum of the minute tensions generated by all the cross-bridges. Thus, as mentioned earlier, the linear relationship between active tension and sarcomere length (beyond l_o) seen in Figure 16 is strong support for this theory. On the other hand, recent measurements by other investigators indicate that the falling phase of the length–tension curve beyond l_o may *not* be linear. These results call into question the independence of the force generators and preclude full endorsement of the cross-bridge theory.

IN SITU LENGTH OF SARCOMERES

We now must consider which part of the length–tension curve is involved during normal functioning of muscle fibers in situ. Happily, the length at which skeletal muscles are held by their origins and insertions usually maintains the fibers approximately at their optimal length. Consequently, this optimal length is often referred to as resting

length. The "resting length" of a whole muscle is defined as the length at which maximum isometric tetanic tension can be developed, and, as we have seen, this is the length at which the individual muscle fibers are at their slack length. It should be noted, however, that the whole muscle is not slack at its resting length. Even a resting muscle exerts some tension on its tendons, but this is due to passive stretch of connective tissue rather than of muscle fibers.

The range of muscle lengths encountered in normal bodily movements varies considerably among muscles, but is thought usually to involve only a fairly narrow portion of the length–tension curve. Most sarcomeres probably never get much longer than point B or shorter than point D on the curve shown in Figure 16.

SHORTENING CONTRACTIONS

Figure 18 illustrates the type of experimental arrangement often used to study shortening contractions. There are two types of loading: preloading and afterloading. The preload is the passive tension created in the muscle by stretching

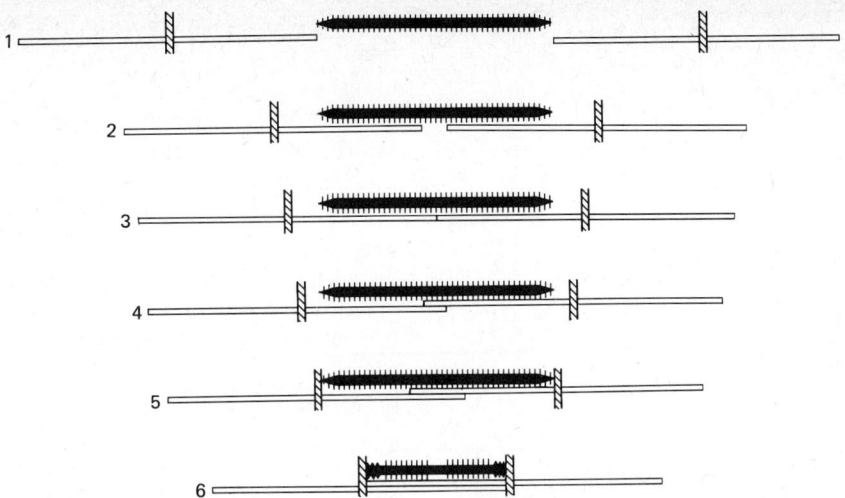

Figure 17 Critical stages in the increase of overlap between thick and thin filaments as a sarcomere shortens. The numbers correspond to those in Figure 16. 1 = no filament overlap and, therefore, no cross-bridges; 2 = overlap nearly to center of A band (maximum cross-bridges); 3 = overlap to center of A band (same number of cross-bridges as in 2), with thin filaments meeting at M line; 4 = thin filaments overlap each other; 5 = ends of thick filaments reach the Z lines; 6 = ends of thin filaments reach the Z lines, with thick filaments crumpled against Z lines. (From Gordon AM, Huxley AF, Julian FJ: The variation in isometric tension with sarcomere length in vertebrate muscle fibers. *J Physiol (Lond)* 197:709, 1968.)

it; in the example shown in Figure 18 the preload is produced by hanging an object *(W)* on one end of the muscle. If this object is not supported by the platform, the preload is simply equal to the weight of the object. If the object is supported by the platform, the preload is less than the weight of the object and is determined by the position of the platform. The afterload is always equal to the weight of the object to be lifted. Obviously, in order to lift the object (i.e., to shorten) the muscle must develop just enough tension to exceed the afterload. A so-called "afterloaded contraction" is one in which the afterload exceeds the preload; this would always be the case when the object is supported by the platform before the muscle is stimulated. Figure 19 shows the time course of afterloaded isotonic tetani and illustrates the effects of changing the afterload. In all cases just enough preload was used to stretch the muscle to its resting length. Increasing the afterload results in (1) an increase in the apparent latent period, (2) a decrease in the maximal shortening distance, and (3) a

decrease in the velocity of shortening. The apparent increase in the latent period with increasing afterload is caused by the fact that the development of active tension in a muscle is a gradual process, as shown in Figure 14. Consequently, as the afterload increases more time is required to develop enough active tension to start lifting the load.

The distance an afterloaded muscle can shorten can be predicted from its length–tension diagram (Figs. 15 and 16). It will always shorten to a length at which the maximum isometric tension it can develop is equal to the load being lifted. Consequently, the greater the load, the less shortening.

Figure 20 shows the effect of afterload on initial velocity of shortening. As the load increases, the velocity of shortening decreases, leading to a load-velocity curve that forms part of a rectangular hyperbola. The maximal velocity of shortening, V_{max}, occurs at zero load.

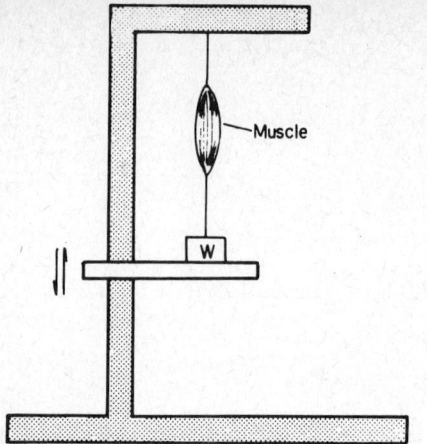

Figure 18 Arrangement for preloading and afterloading a muscle before an isotonic contraction. See text for explanation.

QUICK RELEASE EXPERIMENTS

A very informative experiment is to fix the object (W in Fig. 18) in position so that the muscle cannot shorten until an isometric tetanus is fully developed. Then the object is suddenly released and the velocity of shortening measured. For any given load it is found that the velocity of shortening is exactly the same as it was when the muscle was allowed to shorten without first developing full isometric tension. In other words, allowing more time for activation of the contractile machinery has no effect on the speed of shortening. This observation confirms and strengthens the notion that the intensity of the active state rises abruptly to a maximum immediately after the latent period.

The velocity of shortening after quickly releasing a tetanically stimulated muscle is not immediately constant. Instead there is a velocity transient, lasting about 15–20 ms in frog muscle, during which shortening is not linearly related to time. The upper part of Figure 21 shows that the velocity transient consists of three phases before the eventual nearly steady velocity of phase 4 is reached. The velocity of phase 4 was the measurement used to construct the classical load–velocity curve of Figure 20 and was the only determination of velocity available until recent advances in instrumentation allowed detailed study of the transients.

A similar experiment is to establish again an isometric tetanic contraction so that a maximum number of cross-bridges are attached. But then, while continuing to stimulate the muscle, the muscle length is abruptly reduced by externally moving the load a short distance toward the muscle. The tension in the muscle drops and then recovers as shown in the lower part of Figure 21. Again, there are four distinct phases in the tension response curve, tension being the response in this case rather than shortening. The phases of the velocity and tension transients can be interpreted (see below) in terms of the cross-bridge theory of contraction and provide information about the way cross-bridges develop force.

The study of transient responses to sudden perturbations is a standard approach to investigating kinetic mechanisms. In physiological systems one of the best known applications is the voltage clamp method, which measures the transient responses of the ionic conductances of excitable membranes. A sudden change in membrane potential leads to a kinetic response in the conductance due to the specific permeability changes, which, as you know, once started, are time dependent. The mechanical transients of muscle can be viewed in a manner similar to the membrane conductance transients. In one case, changes in load lead to transient changes in velocity; in the other case, changes in length lead to transient changes in tension. Once started, the muscle responses are both time dependent. A step change in length leading to tension transients is analogous to a voltage step followed by current transients.

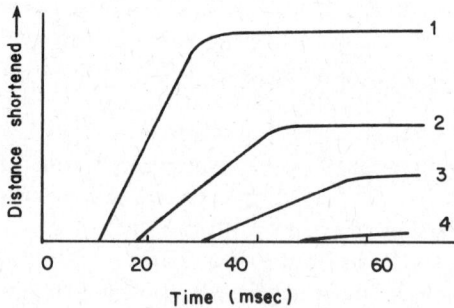

Figure 19 Afterloaded isotonic tetani. Initial length was equal to resting length in all cases. Curve 1 represents a lightly loaded contraction. For curves 2, 3 and 4 the afterload was two times, three times, and four times that of curve 1.

The interpretation of the tension transient is as follows. The first phase, seen as an abrupt drop in tension during the instant of muscle shortening (lower part of Fig. 21), is approximately proportional to the change in length; it is a consequence of relieving the strain on the passive elastic components of the muscle. When this experiment was performed on muscle fibers stretched beyond their slack length, it was found that the abrupt drop in tension decreased as the degree of thick and thin filament overlap decreased; the total stiffness of the elastic components was directly proportional to the number of cross-bridges that could be formed. This result implies that a major part of the series elasticity resides in the cross-bridges rather than in the thick or thin filaments or the Z lines.

Following this passive drop in tension attributable to cross-bridge elasticity, there is a very quick recovery of some of the tension (phase 2 in lower part of Fig. 21). In fact, if the original quick release does not exceed about 40 Å per half sarcomere, the recovery of tension during phase 2 is nearly complete. In terms of the cross-bridge theory, this quick recovery of tension in the fiber is attributable to movement of cross-bridges that had already been formed before the quick release took place, but were kept from moving (i.e., power stroking) by the tension in the S_2 segment (Fig. 8). When this tension is relieved by the quick release, the S_1 heads can rotate and tension is again generated.

The third phase in the tension transient is a plateau that may last 5–20 ms. During this time tension rises very

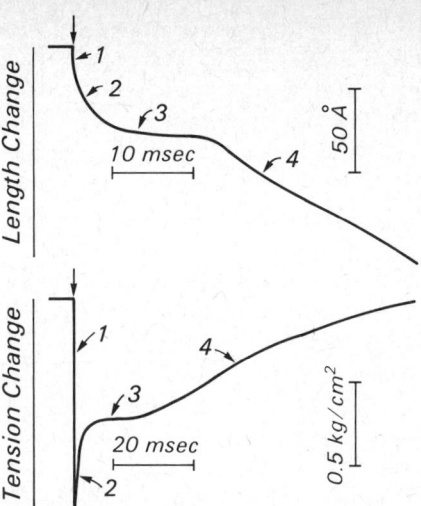

Figure 21 Mechanical transient responses of frog muscle fibers during tetanic contraction. The upper curve shows the time course of the change in length following a sudden decrease in the load; at the arrow the load was decreased by about 0.8 kg/cm². The lower curve shows the time course of the change in tension following a sudden decrease in length; at the arrow the stretch on the fiber was reduced by about 80 Å for each half sarcomere. Each response consists of four phases that are numbered accordingly. The length changes shown above are for each half sarcomere. (Redrawn from Huxley AF: *J Physiol (Lond)* 243:1, 1974.)

slowly, if at all, and may even decrease slightly. This phenomenon can be interpreted by the crossbridge theory. After the preformed cross-bridges go through their power strokes and accomplish the rapid rise in tension of phase 2, they detach approximately as rapidly as new cross-bridges are being formed. Because active tension depends on the total number of attached cross-bridges in each half sarcomere, as long as the rate of attachment equals the rate of detachment, no increase in tension is expected. Gradually the rate of attachment begins to exceed the rate of detachment, and phase 4, a slow recovery to the steady state isometric tension, takes place.

The four phases of the velocity transient can be similarly interpreted as follows. The first phase, a rapid decrease in muscle length simultaneous with the decrease in load, is simply due to reducing the stress on the passive elastic components (mostly residing in the cross-bridges). The second phase is the most interesting, a rapid shortening attributable to movement of previously formed cross-

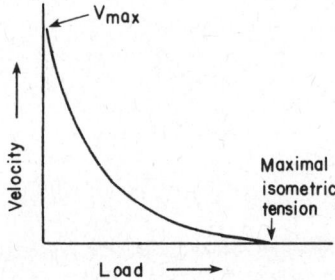

Figure 20 The load–velocity curve for a skeletal muscle preloaded to its resting length. V_{max} is the maximal velocity of shortening expected at zero load. This quantity cannot actually be measured and is obtained by extrapolation. The load that just prevents any shortening is equal to the maximal *isometric* tension.

bridges. The distance shortened per half sarcomere in phase 2 (about 50 Å in Fig. 21) is consistent with geometrical estimates of how far a single power stroke ought to move a thin filament with respect to a thick filament. Phases 3 and 4 of the velocity transient are again explained by simultaneous cross-bridge attachment and detachment.

These tension and velocity transients support the idea that the force-generating mechanism is part of a cyclic reaction sequence involving attachment and detachment of cross-bridges to actin, and that each cycle takes place within the time frame of a few milliseconds. Although this model is widely accepted, it should be emphasized that there is no absolute proof that the myosin projections attach to actin during contraction. One alternative would be an electrostatic interaction between the HMM projections and the thin filament without actual attachment. Whatever the correct mechanism may be, it is reasonably clear that the myosin-S_1 heads form a series of discrete sites that lead to the development of force between the filaments.

GRADATION OF MUSCULAR ACTIVITY

THE MOTOR UNIT

Each efferent nerve axon branches to supply more than one muscle fiber. The average number of muscle fibers in a given muscle supplied by each axon is called the innervation ratio. This ratio ranges from just a few muscle fibers per motor neuron in the extrinsic eye muscles to about 200–300 muscle fibers per motor neuron in large limb muscles. A smaller innervation ratio permits a greater delicacy in the gradation of contractile force.

A single motor neuron together with all the muscle cells it innervates is called a motor unit. The least contraction that a muscle normally produces is a twitch of all the fibers in a single motor unit. When one muscle fiber of a motor unit is excited, all other fibers of this same motor unit are also excited. The action potentials in all branches of an efferent axon are ordinarily more than adequate to excite all the associated muscle fibers.

SMOOTHNESS OF CONTRACTION—SUSTAINED CONTRACTIONS

During most muscular contractions not all the motor units are active. For simplicity let us assume that during a particular contraction only three motor units are active. In Figure 22 the tension produced by each of these motor units is shown during a succession of twitches. If these three motor units twitched simultaneously (or synchronously) as in Figure 22, the total contraction would be neither smooth nor sustained; it would just be a series of twitches referred to as clonus. Clonus occurs in certain pathological conditions of the central nervous system, but does not occur normally.

In normal movements the active motor units twitch *asynchronously* and at different frequencies, as illustrated in Figure 22, thus smoothing out the total contraction. If 100 motor units are visualized, instead of just the three shown in Figure 22, it is easy to see that asynchronous twitching would result in several motor units always being in some phase of a twitch and the total response being smooth and sustained rather than phasic.

The dampening effect of the series elastic component helps make muscle responses smooth by preventing each twitch from developing tension as fast as the contractile component is activated.

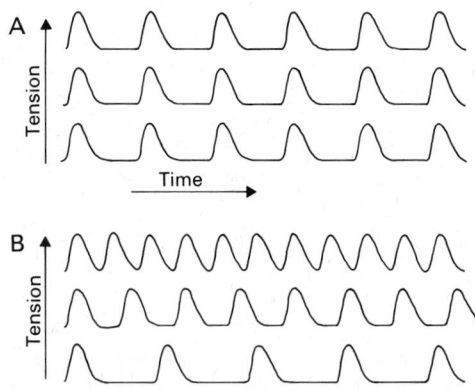

Figure 22 Trains of twitches in three motor units. (*a*) All three motor units twitch synchronously; the result is clonus. (*b*) Activity is asynchronous. The result is a smoother overall contraction.

MECHANISMS FOR GRADING THE STRENGTH OF CONTRACTION

Assuming the absence of summation of contraction, the strength of a sustained contraction is determined by the number of twitches in progress at any time. The number of twitches in progress can be increased either by increasing the number of active motor units or by increasing the twitch frequency of each active motor unit.

Activation of more motor units is called recruitment. As more motor units are recruited, the number of twitches in progress at any one time increases, with the result that the force of contraction increases. This increase obviously must be a stepwise process. Since there is a large number of motor units in a given muscle, the number of steps possible is large, and the increase in force with each new unit correspondingly small.

With increased frequency of nerve impulses over the motor axons, the motor unit twitches occur closer together, so that at any one time there are more twitches in progress. Thus, even in the absence of summation, increasing motor unit twitch frequency increases the total force of contraction.

As the frequency of nerve impulses increases still further, the individual responses start summating. Since the motor units do not all fire at the same frequency, some might develop partial tetani while others are still performing individual twitches. Stronger contractions would involve a greater proportion of motor units in partial tetanus.

As the frequency is increased still further, complete tetanus of some of the motor units occurs, although this is believed to be reserved for only the strongest contractions.

SKELETAL MUSCLE TONE

The maintenance of any bodily posture requires certain skeletal muscles to exert relatively small amounts of tension over relatively long periods of time. This sustained tension is called muscle tone, or postural tone. Muscle tone depends on low frequency nerve impulses to the muscle over a relatively few motor neurons. These impulses are delivered asynchronously over the individual motor neurons involved so there is a completely asynchronous contraction of motor units resulting in a steadily maintained active tension in the muscle. Activation of the motor neurons is brought about by facilitatory impulses from two general sources: the brain and the muscle spindles. The muscle spindles are stretch receptors located in the muscle itself and will be discussed in Chapter 39. If the sensory nerves from a muscle are cut, thereby eliminating sensory input from the muscle spindles, nearly all the postural tone in that muscle disappears.

Active tension in a normal muscle can be completely and promptly eliminated by voluntary relaxation. This has been demonstrated using the technique of electromyography (EMG). A resting muscle demonstrates complete electrical silence. A small amount of tension remains, which is sometimes referred to as tone, but this resting tension is entirely passive and is thought to be caused by stretch of the connective tissue surrounding the muscle fibers.

FATIGUE

The term *fatigue* is used ambiguously, but in exercise physiology it is usually taken to mean the inability to maintain a particular type of muscular exertion. Thus fatigue is distinguished from the feelings of tiredness, weariness, and exhaustion that may follow, or even preceed, muscular exertion. It is also distinguished from the termination of muscular exertion due to "giving up," or lack of motivation. Fatigue represents a real objective inability to continue at peak intensity or even to continue at all.

There are three types of fatigue: (1) central fatigue, (2) transmission fatigue, and (3) muscle fatigue. In central fatigue the central nervous system simply cannot supply the traffic of impulses over motor neurons necessary to maintain the particular exertion at peak or desired intensity. The mechanisms of central fatigue are not at all clear, but probably in endurance work are induced by elevated concentrations in the blood of certain metabolites (e.g., lactic acid) and/or reduced concentrations of other metabolites (e.g., glucose). There may also be intramuscular chemoreceptors that can signal the central nervous system of peripheral elevations and depletions in metabolites and thereby induce central fatigue. In short-term highly intense exertion, nerve impulses from stretch receptors in tendons are probably of more importance in limiting impulse traffic over motor neurons.

In transmission fatigue the motor nerve endings become depleted of acetylcholine and neuromuscular transmission fails. In muscle fatigue action potentials of normal amplitude and frequency continue to pass over the plasma membranes of the muscle fibers, but the fibers fail to contract as forcefully as before and might eventually fail to contract at all. The mechanism of muscle fatigue is not understood. The traditional view is that muscle fatigue is caused by depletion of energy stores, most critically ATP and CP, but recent evidence indicates that a failure of excitation–contraction coupling may be of prime importance.

Obviously, the mechanisms involved in fatigue are very poorly understood. To make matters worse, it is also not well understood which of these types of fatigue is most important in limiting each particular type of muscular exertion. Undoubtedly, the fatigue of the marathon runner is quite different from that of the weightlifter; perhaps future research will provide greater understanding.

SELECTED BIBLIOGRAPHY

Huxley AF: Muscular contraction. Review lecture. *J Physiol (Lond)* 243: 1–43, 1974.

Huxley HE: The mechanism of muscular contraction. *Science* 164:1356–1366, 1969.

Murray JM, Weber A: The cooperative action of muscle proteins. *Sci Am* 58, 1974.

Noble MIM, Pollack GH: Molecular mechanisms of contraction. *Circ Res* 40:333–342, 1977.

Tregear RT, Marston SB: The crossbridge theory. *Annu Rev Physiol* 41:723–736, 1979.

SELF-STUDY QUESTIONS

There may be more than one correct answer for each question.

1. If a relaxed skeletal muscle fiber is passively stretched,
   A. the A bands increase in length.
   B. the I bands increase in length.
   C. the H bands increase in length.
   D. the thin filaments increase in length.
   E. the thick filaments increase in length.

2. If a muscle fiber contracts and is allowed to shorten slightly, but not enough to form contraction bands,
   A. the A bands decrease in length.
   B. the I bands decrease in length.
   C. the H bands decrease in length.
   D. the thin filaments decrease in length.
   E. the thick filaments decrease in length.

3. If a muscle fiber contracts and is allowed to shorten just enough to start developing contraction bands around the M lines,
   A. the A bands decrease in length.
   B. the I bands decrease in length.
   C. the H bands disappear.
   D. the thin filaments disappear.
   E. the thick filaments move farther apart from each other.

4. In skeletal muscle, excitation–contraction coupling is mediated as follows:
   A. Action potential at plasma membrane directly induces activity of contractile elements.
   B. Action potential at plasma membrane induces action potential in transverse tubules, directly inducing activity of contractile elements.

C. Action potential at plasma membrane causes release of Ca^{2+} from plasma membrane, which directly induces activity of contractile elements.

D. Action potential at plasma membrane causes action potential in transverse tubules, which causes release of Ca^{2+} from sarcoplasmic reticulum, which in turn directly induces activity of contractile elements.

5. Net uptake of Ca^{2+} by the sarcoplasmic reticulum is associated with
   A. relaxation.
   B. ATP hydrolysis.
   C. contraction.
   D. action potentials.

6. Myosin ATPase
   A. is located in heavy meromyosin.
   B. is located in light meromyosin.
   C. is activated by actin.
   D. is inhibited when Ca^{2+} binds to troponin.

7. Ca^{2+} triggers contraction by
   A. binding to tropomyosin.
   B. binding to actin.
   C. binding to light meromyosin.
   D. binding to troponin.

8. In resting muscle, tropomyosin
   A. inhibits Ca^{2+} release from sarcoplasmic reticulum.
   B. prevents Ca^{2+} from binding to troponin.
   C. blocks the binding of heavy meromyosin globular heads to actin subunits.
   D. prevents the formation of cross-bridges.

9. The gradual (rather than abrupt) development of tension during the contraction period of an isometric twitch is explained by
   A. the fact that Ca^{2+} is released gradually from the sarcoplasmic reticulum.
   B. the fact that the activity of the contractile elements develops gradually.
   C. the gradual recruiting of more and more cross-bridges as tension increases.
   D. the presence of an elastic component in series with the contractile elements.

10. Single twitches of any particular muscle fiber (at resting length) are usually of about the same strength because
    A. force development by the contractile elements is an all-or-none phenomenon.
    B. each action potential causes about the same release of Ca^{2+} from the sarcoplasmic reticulum.
    C. the series elastic component determines the strength of contraction.
    D. all of the Ca^{2+} contained in the sarcoplasmic reticulum is released in response to a single action potential.

11. Tetanus
    A. results from smooth fusion of rapidly recurring twitches.
    B. produces more tension than a single twitch.
    C. probably does not often occur during normal activity.
    D. requires the same frequency of muscle action potentials for all types of skeletal muscle.

12. If active tension begins to develop at a sarcomere length about 15% less than resting length
    A. the maximum active isometric tension the sarcomere is capable of developing is less than it is at resting length.
    B. the actin filaments have started to overlap at the M line.
    C. the internal resistance against shortening has increased above that at resting length.
    D. the amount of Ca^{2+} released from the sarcoplasmic reticulum in response to each action potential is less than it is at resting length.

13. During normal functioning of skeletal muscles
    A. the H band never becomes as long as the A band (i.e., the two sets of filaments always overlap to some extent).
    B. the H band is seldom longer than the region along the center of the thick filaments having no heavy meromyosin projections.
    C. the two sets of filaments are usually in a position with respect to each other that a maximum number of cross-bridges could be formed.
    D. contraction bands seldom develop at the Z lines.

14. The smoothness of normal muscular contractions results from
    A. asynchronous activity of motor units.
    B. clonus.
    C. tetanus.
    D. recruitment.

15. Increased active tension developed by a whole muscle can be the result of
    A. recruitment of motor units.
    B. increased twitch frequency of active motor units.
    C. partial tetanus of some active motor units.
    D. stretching beyond the resting length.

ANSWERS

1. B and C are correct.
   A. The A bands remain the same length, since stretching the fiber does not stretch the thick filaments.
   B. The I bands increase in length because the degree of interdigitation between thick and thin filaments is reduced.
   C. The H bands increase in length as the tips of the thin filaments move farther apart.
   D. The thin filaments are not stretched.
   E. The thick filaments are not stretched.

2. B and C are correct.
   A. The A bands remain the same length until the fiber shortens beyond point 5 in Figure 16.
   B. Correct.
   C. Correct.
   D. The thin filaments probably never decrease in length (Fig. 17).
   E. Not until beyond point 5 in Figure 16.

3. B, C, and E are correct.
   A. The A bands do not decrease in length until contraction bands begin to develop around Z lines (i.e., beyond point 5 in Fig. 16).
   B. Correct.
   C. Correct.
   D. Never.
   E. Yes. The total volume of each fibril remains constant; therefore, as it shortens it gets thicker.

4. D is correct.
   A. The word "directly" makes this wrong.

B. This statement neglects role of SR and of Ca^{2+}.
C. This statement neglects role of T tubules and SR.
D. This is definitely the most accurate statement given.

5. A and B are correct. No explanation required. See discussion on page 75.

6. A and C are correct.
   A. Correct (see page 77).
   B. Wrong.
   C. Correct.
   D. When Ca^{2+} binds to troponin myosin ATPase can be *activated by actin*.

7. D is correct.
   A. Ca^{2+} binds to troponin, not to tropomyosin,
   B. nor to actin,
   C. nor to LMM.
   D. Correct.

8. C and D are correct.
   A. No reason to suspect this.
   B. Wrong.
   C. Correct (see page 76).
   D. Correct (see page 76).

9. D is correct.
   A. No. Ca^{2+} release is abrupt.
   B. No. The "active state" develops abruptly.
   C. No. They are activated essentially all at once, as compared with the time course of the twitch.
   D. Yes. This is the "series elastic component" (see pp. 82–83 and Fig. 8).

10. B is correct.

    A. Activity of the contractile elements themselves is not necessarily all-or-none; the force they generate would be graded *if* the amount of Ca^{2+} released from the SR were graded.

    B. True, because the action potential truly follows the all-or-none law. It is for this reason that each muscle fiber seems to obey the all-or-none law, as shown in Figure 9.

    C. This statement is partially true, but does not provide an explanation for the all-or-none behavior of muscle fibers.

    D. False. Only a small fraction is released with each action potential.

11. A, B and C are correct. No explanation required. See pages 83–84.

12. A, B. C and D are all correct. See pages 84–86.

13. A, B, C, and D are all correct. As stated on page 86, "most sarcomeres probably never get longer than point B or shorter than point D on the curve shown in Figure 16." If this is true, then Figures 16 and 17 clearly demonstrate that all four of the above statements are correct.

14. A is correct.

    A. Yes (page 90).

    B. No (page 90).

    C. Complete tetanus is a smooth contraction but probably seldom occurs and is not responsible for the smoothness of ordinary contractions.

    D. Recruitment of more fibers increases contraction strength.

15. A, B and C are correct. Straightforward (see page 91).

5

Heart Muscle

WAYNE R. GILES
ARTHUR M. BROWN

OBJECTIVES

After completion of this chapter, you should be able to

1. Understand electrical coupling between heart muscle cells: its anatomical basis and functional importance.

2. Describe the ionic basis of the action potential and the pacemaker potential in cardiac cells.

3. State the differences between the action potentials of skeletal and cardiac muscle.

4. Describe the relationship between intracellular recordings and extracellular electrograms in the heart.

5. Describe the pattern of impulse conduction in the heart, and approximate propagation velocities in the different anatomical locations.

6. Understand the main features of excitation–contraction coupling in the heart, and state how this process differs from E–C coupling in skeletal muscle.

7. Understand how the autonomic transmitters produce their characteristic inotropic and chronotropic effects.

8. Describe the length–tension relationship in cardiac muscle, and compare it with that in skeletal muscle.

9. Explain Starling's law of the heart and ventricular function curves and describe the relationship of these to the length–tension diagram.

The primary objectives of this chapter are to summarize the main electrophysiological and mechanical properties of cardiac muscle and to describe how these properties are integrated such that the heart can function as a pump. The characteristics of cardiac muscle and skeletal muscle will be compared and contrasted. An understanding of the physiology of skeletal muscle is an important prerequisite to this chapter.

Within the last 15 years very significant advances have been made in understanding the electrophysiology and pharmacology of cardiac muscle. It is therefore important to summarize some of these data and to attempt to integrate this material with the previously existing "laws" governing the electrical and mechanical activity of the heart. An understanding of these physiological principles is an essential background to one of your ultimate goals: protection and preservation of the myocardium.

ELECTRICAL COUPLING BETWEEN HEART CELLS

A characteristic feature of skeletal muscle is that a single fiber within a large muscle can be stimulated to contract without exciting any of the neighboring fibers. That is, the electrical impulses (action potentials) do not spread from fiber to fiber. As a consequence, the central nervous system may exert fine control over the strength and temporal pattern of contraction in skeletal muscle.

By contrast, in the heart or in smooth muscle, suprathreshold stimulation at one localized point can spread, or propagate, and trigger contraction throughout the entire organ. Before about 1950 this phenomenon was interpreted to mean that the heart was an *anatomic syncytium* —that the cytoplasm must be continuous throughout the tissue. However, electron microscopic studies have now clearly established that each heart cell is completely surrounded by a plasma membrane, or sarcolemma.

How can action potential travel from cell to cell? The answer is that specialized regions of contact between adjacent cells (nexuses or gap junctions) act as low resistance pathways for intercellular currents. Figure 1 illustrates a region of apposition between two cells: the *intercalated disc.* As viewed in electron micrographs, the intercalated disc runs between cells in a tortuous, steplike fashion, so that certain parts of it lie parallel to the long axis of the cell while others are perpendicular to this axis. In the "perpendicular" or transverse portions of the intercalated disc two types of desmosomes are often found: (1) the "spot" desmosome, and (2) the belt desmosome. At these sites, there is only a very small space between the two cell membranes (200–300 Ångström Units; 1 Å = 10^{-10} m).

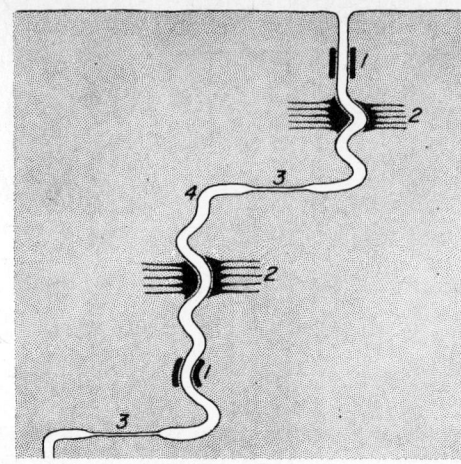

Figure 1 Drawing of the intercalated disc region between two mammalian cardiac cells. 1, Spot desmosome; 2, belt desmosome or fascia adherens; 3, nexus (gap junction); and 4, undifferentiated region of cell apposition. (Adapted from Urthaler F, Kawamura K, James TN: The anatomic basis for cardiac rhythm and conduction, pp. 831–873 in Andreoli TE, Hoffman JF, Fanestil DD (eds): *Physiology of Membrane Disorders.* New York, Plenum, 1978.)

Thin filaments, which contain the contractile protein actin, commonly insert into both types of desmosome. The functional role of the desmosome has not been definitely established, but it is thought to be important in force transmission between cardiac cells.

The aspect of the intercalated disc lying parallel to the long axis of the cell contains a third type of specialized structure: the gap junction, or nexus. At the gap junction the surface membranes of the two cells are very closely apposed (20 Å) and are bridged by tiny "stalks" (about 20 Å in diameter) arranged in a hexagonal pattern. The interior "tunnel" of these "stalks" provides a continuous pathway between adjacent cardiac cells. The diameter of this "tunnel" has been estimated to be 10 Å. Hence, metabolites as well as ions can move between cells. In mammalian cardiac muscle this gap junction, or nexus, occupies about 5% of the cell surface area, and is thought to be the site of intercellular current flow.

The problem of electrical communication between two adjacent heart cells and the basis for understanding propagation of electrical activity throughout the heart (about 10^{10} cells) both require a clear understanding of this concept of intercellular electrical communication. Consider first the case of a single column of cardiac cells, as shown in Figure 2a; and ask the question: If current is injected into a cell at one end of the bundle, how far along the bundle can it spread or produce a voltage change? The answer depends on the particular tissue under consideration, but it falls in the range of 600 μm (1 μm = 10^{-6} m) to 2 mm. Stated differently, the injected current produces a measurable voltage change at a distance of about three to nine cell lengths. This experimental result provides

strong evidence that current can flow from cell to cell. This current is thought to be carried by potassium ions moving across the gap junctions. For example, the pattern of current spread throughout a ventricular trabeculum may be as drawn in Figure 2b. Recently it has been shown that (1) anoxia, (2) mild anesthesia, and (3) certain commonly used therapeutic agents (e.g., the cardiac glycosides) can change impulse conduction in the heart by altering the effectiveness of this *intercellular electrical coupling*.

Propagation of the cardiac action potential is thought to occur by means of current flow from active (excited) regions to resting (quiescent) areas, as previously described for action potential propagation in nerve. This implies that during propagation the intracellular and extracellular fluid must form two "arms" of a closed electrical circuit. Therefore, if extracellular current flow can be interrupted; conduction should be abolished. An experiment that supports this "local circuit current" mode of impulse propagation in the heart is shown in Figure 3. A strand of cardiac muscle (e.g., an atrial trabeculum or a papillary muscle) has been removed from the heart and placed in a chamber where its central region can be superfused with a "cuff" of isotonic sucrose. Since this sucrose is essentially ion free, it forms a very high resistance in the extracellular "cuff" region. Hence, propagation *cannot* occur, and a monophasic action potential is recorded. Alternatively, when the sucrose is "shunted" by switching in a relatively small resistor across it, the familiar diphasic response is recorded.

Thus, in the heart as in nerve or skeletal muscle, impulse propagation along the *functional syncytium* occurs via local circulating currents. This concept is essential for understanding conduction or rhythm disturbances in the heart.

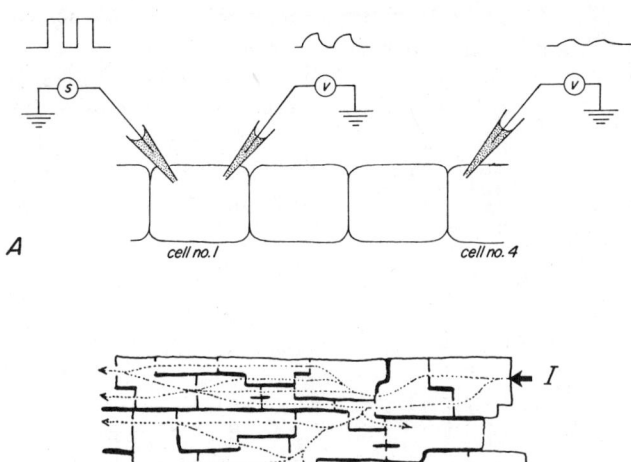

Figure 2 (*a*) Microelectrode method used to measure electrical coupling (intercellular communication) between cells of, for example, a mammalian Purkinje fiber. Pulses of constant current are injected into a cell (cell 1) and resulting *subthreshold* changes in voltage across the surface of cell 1 and adjacent cells (e.g., cell 4) are accurately measured and compared. Because a ΔV can still be measured some four cell lengths "away," low resistance intercellular pathways must extend a considerable distance down the fiber. The distance taken for initial ΔV to decay from 100 to 37% is termed the space constant (λ). (*b*) A drawing of the pathway(s) of current flow in a multicellular cardiac preparation. The arrow denotes the point of initial current injection. Thick black lines represent extracellular spaces. Note that the current spreads both longitudinally and laterally. However, nexuses are few in number between the "long" sides of adjacent cells. (Adapted from Sommer JR, Johnson EA: Ultrastructure of cardiac muscle, pp. 113–187 in Berne RM, Sperelakis N, Geiger SR (eds): *The Handbook of Physiology*, section 2, *The Cardiovascular System*, Volume 1, American Physiological Society, Bethesda, MD, 1979.)

ELECTRICAL ACTIVITY IN THE HEART

SITE OF INITIATION AND PATTERN OF CONDUCTION

In mammals the heartbeat is of myogenic origin; that is, the action potentials that produce the contractions are not

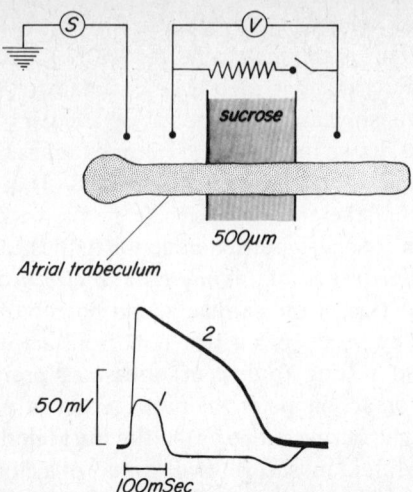

Figure 3 Experimental test of action potential conduction in cardiac muscle (frog atrial muscle). (*a*) Experimental arrangement. One end of muscle strip is isolated from the other by a high-resistance sucrose gap, across which the voltage can be recorded and shunt resistances placed. The preparation is excited by applying a brief pulse to the shielded electrode. (*b*) Records of action potentials. Curve 1 is obtained when the high extracellular resistance of the sucrose is shunted by a small resistor. Curve 2 shows the monophasic action potential, which can be recorded across the sucrose gap when the shunt is switched out. Width of sucrose gap is 500 microns. Conduction across the gap is indicated by diphasic wave when shunt resistance is in place. (Adapted from Barr LM, Dewey M, Berger W: Propagation of action potentials and the structure of the nexus in cardiac muscle. *J Gen Physiol* 48:797–823, 1965.)

elicited by excitatory nerve impulses. This spontaneous electrical activity normally originates in the sinoatrial (SA) node, which is made up of a collection of specialized cells lying near the junction of the superior vena cava and the right atrial appendage (Fig. 4). These pacemaker cells are smaller than the atrial cells and they contain many fewer myofibrils. Cells in the SA node characteristically exhibit a slow diastolic depolarization, or pacemaker potential, which has a steeper slope than subsidiary pacemaker activity elsewhere in the heart. This means that in the SA node this pacemaker depolarization reaches the threshold for initiation of an action potential more quickly than in the AV node. Hence, the regenerative action potential normally is initiated in the SA node.

The SA node action potential spreads into the crista terminalis and then excites the atrium. Intracellular records from atrial muscle fibers show an action potential that begins fairly abruptly from a more negative and steady resting potential. Normally no pacemaker potential is present. The conduction velocity of the atrial action potential is in the range of 0.5–1 meter per second (m/s). The longest conduction pathway over the atria is roughly 8 cm. Thus, an action potential beginning at the SA node can activate the entire atrium within some 40–80 ms. Since the duration of the atrial action potential is 100–300 ms, most atrial fibers are depolarized simultaneously and the atrial contraction is synchronous.

There is some evidence for the existence of specialized conduction pathways between the SA and AV nodes. This means that the impulse may quickly traverse rather than spread through the atrial musculature. Normally this does not happen, but it may occur in certain pathological states.

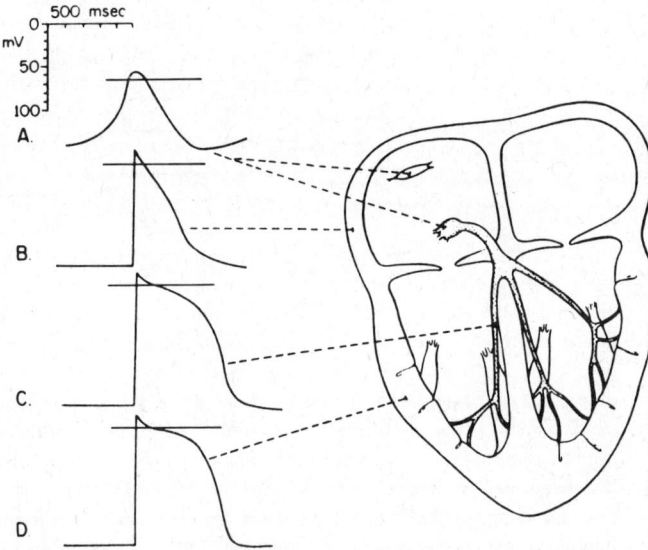

Figure 4 Schematic representation of the conduction system of the heart. A, Sinoatrial and atrioventricular node; B, atrium; C, Purkinje system; D, ventricle. (Modified from Scher AM, Spach MS: Cardiac depolarization and repolarization and the electrocardiogram, pp. 357–393 in Berne RM, Sperelakis N, Geiger SR, (eds): *Handbook of Physiology*, Section 2, *The Cardiovascular System*, Volume 1. American Physiological Society, Bethesda, Md. 1979.)

Upon reaching the AV node, conduction slows to about 0.2 m/sec, and the action potential becomes shorter in duration, smaller in amplitude, and exhibits a slower rate of rise. Fibers of the AV node also exhibit a very slow diastolic depolarization or pacemaker potential. Normally, however, this secondary pacemaker mechanism is suppressed by the faster "pacemaker activity" originating in the SA node. The slow conduction through the AV node permits atrial *repolarization* to occur simultaneously with ventricular *depolarization*. This has two important consequences. First, atrial contraction is completed before ventricular contraction begins. (This will be discussed further in Chapter 7.) Second, in surface records the electrical signs of atrial repolarization are obscured by ventricular depolarization.

Once the action potential has passed through the AV node, it reaches the bundle of His, moves into the Purkinje system, and is rapidly propagated throughout the ventricular muscle. The Purkinje system divides into two main bundles, the left and right. In some pathological states conduction may be blocked through just one of these branches. Conduction along the Purkinje fiber network is relatively fast (approximately 4 m/s); thus the entire ventricle is normally activated in a synchronous fashion.

Note the pattern of activation of the ventricular muscle. In the conducting system, the action potential sweeps downward from base to apex and then back in the opposite direction.

INTRACELLULAR ELECTRICAL ACTIVITY IN THE HEART

Figure 4 shows that the various anatomical regions of cardiac muscle exhibit distinct action potentials. To understand the genesis of these action potentials, one must think in terms of sodium and potassium currents, much like those in nerve and muscle. In addition, an inward-directed Ca^{2+} current is present in cardiac muscle. This Ca^{2+} current is functionally very important: (1) it is responsible for the long "plateau" of the cardiac action potential; (2) it triggers and partially controls the mechanical response; and (3) modulations of this Ca^{2+} current by the autonomic transmitters can produce marked changes in the strength of contraction of the heart.

The Resting Potential

The level of the resting potential in quiescent heart cells (atrium, ventricle, and Purkinje fibers) is mainly controlled by the high permeability of the sarcolemma to *potassium* ions. The values of $[K]_i = 155$ mM, and $[K]_o = 4$ mM typical of most mammalian cardiac cells, when substituted into the Nernst equation, give an E_K of about -97 mV. The resting potential (measured with glass micropipettes) is 10–15 mV positive to this; hence there must exist a *steady inward current*. This steady inward current is thought to be carried by *sodium ions* and will be referred to as the *sodium background current*. As shown below, this current also plays a very

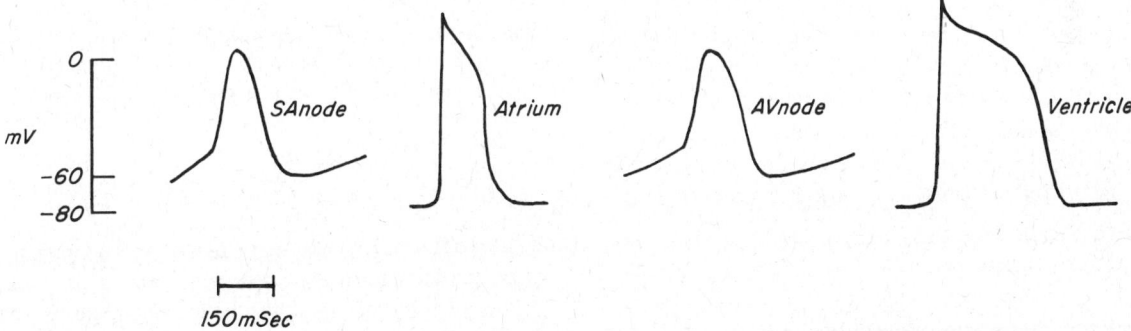

Figure 5 Drawings of action potentials recorded in various tissues (regions) of the mammalian heart. Note differences in (1) presence of pacemaker potential, (2) resting potential, and (3) duration of the action potential. These differences are functionally important and should be carefully considered in relationship to (1) ability of the tissue to generate spontaneous, rhythmic activity, (2) speed of conduction, and (3) ability to trigger and partly control the cardiac contraction.

important role in the generation of the pacemaker potential.

In summary, the resting potential of quiescent cardiac cells is set by the interaction of two *time-independent* ionic currents: (1) an outward K^+ current, and (2) a steady inward current, thought to be carried by Na^+. In contrast to skeletal muscle, the steady (background) Cl^- permeability is relatively small in cardiac muscle. As is the case in nerve and skeletal muscle, the Na^+ and K^+ concentration gradients are maintained by an ATP-requiring Na^+–K^+ "pump." If this pump operates in the *electrogenic* mode (e.g., exchanging $3Na^+$ for $2K^+$) it may also contribute to the observed level of the resting potential.

The Action Potential

Figure 6 illustrates an action potential that is representative of atrial or ventricular muscle. Below the action potential, the *time-dependent* transmembrane ionic currents underlying this response are shown. The sequence of ionic current changes underlying the action potential is as follows:

1. The stimulus depolarizes the membrane enough to exceed the threshold for the rapid inward sodium current, i_{Na}. This current activates very rapidly and drives the membrane potential toward its reversal potential (E_{Na}), resulting in the initial rapid depolarization of the action potential. Note, however, that this Na current is transient—it shuts off or *inactivates* very quickly.

2. The depolarization produced by the inward movement of Na^+ (i_{Na}) exceeds the threshold for activation of a *second inward current*, i_{si}, which is carried mainly by *calcium* ions. The turning on, or activation, of this current produces a relatively slow secondary depolarization. The slow time course of the turn-off of i_{si} is thought to be an important determinant of the plateau phase of the action potential. Note that the relatively long-lasting (50–150 ms) plateau depolarizes the cells to about 0 mV.

3. This depolarization activates an *outward potassium current*, i_K, which drives the membrane potential back toward E_K: repolarizing the cell. Repolarization therefore results from:

   a. activation of i_K
   b. reduction of i_{si}

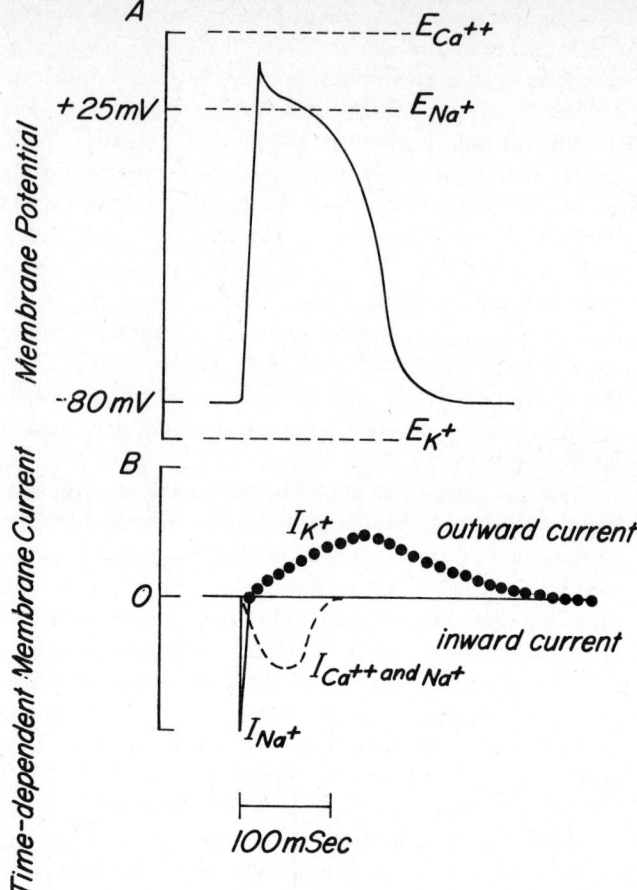

Figure 6 (*a*) Drawing of an action potential similar to those recorded from an atrial or a ventricular cell. The approximate Nernst potentials for the various permeable ions are indicated. (*b*) Illustration of three distinct *time-dependent* ionic currents, thought to generate these action potentials. Note the differences in (1) direction, (2) ionic nature, and (3) time course of these currents. The *time-independent* currents responsible for the resting potential have been omitted from this diagram.

Once the cell has almost fully repolarized, it is ready to produce another action potential—that is, it is no longer *refractory*. Recall what you have learned about the refractory period in nerve and skeletal muscle. But remember that the cardiac action potential is very long, so that the maximum discharge rate for the heart is much lower than in other excitable cells. Although the i_{Na} of the heart turns on (activates) and off (inactivates) with kinetics very similar to those of i_{Na} in nerve or skeletal muscle, it nevertheless requires an appreciable "rest" period before it can be activated a second time. This "slow" recovery of i_{Na} may be partly responsible for the long relative refractory period in the heart. A number of drugs that are used as antiarrhythmic agents (e.g., lidocaine or quinidine) may further prolong the i_{Na} recovery process.

The Pacemaker Potential

Perhaps the most striking feature of cardiac muscle is its ability to contract in a spontaneous, rhythmic fashion. The purpose of this section is to suggest an ionic mechanism for this important biological "oscillator." As shown in Figure 5, the membrane potential in the sinus node is not steady during diastole; instead, it slowly depolarizes. This slow depolarization is called the *pacemaker potential* —its slope is the most important determinant of heart rate.

For technical reasons, the most complete studies of the mechanism of the cardiac pacemaker potential have been done using Purkinje fibers. In this tissue, there is strong evidence that a *slowly decaying potassium current* controls the rate of the pacemaker depolarization. In attempting to understand this phenomenon, it may be useful to think back to your knowledge of nerve action potentials. In particular, reconsider the question: How is the afterpotential created in the squid axon? The pacemaker potential in the heart is somewhat similar to this afterpotential. In both cases, as the potassium current (which was activated by the initial rapid depolarization of the action potential) relaxes or decays, the membrane potential depolarizes because of the *steady* or *background inward current*. In the Purkinje fiber, this pacemaker depolarization exceeds the threshold for i_{Na}, an action potential is initiated, and another "heartbeat" ensues.

The ionic mechanism of the pacemaker potential in sinus tissue is not completely understood. There is, however, good evidence that a *decay* in *potassium current* is importantly involved.

Figure 7 illustrates an experiment demonstrating this point. Part A shows spontaneous pacemaker activity recorded from a frog sinus venosus preparation. At two points in time, the pacemaker cycle was interrupted, and the membrane potential was held or clamped at the maxi-

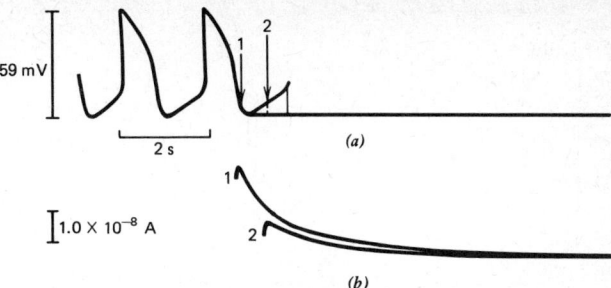

Figure 7 (*a*) Records of spontaneous pacemaker activity in frog sinus venosus. (*b*) Two superimposed records of slowly relaxing currents obtained when the membrane potential was rapidly held or "clamped" at the maximum diastolic potential. Note that a smaller relaxation of current is obtained as the pacemaker potential develops. (From Brown HF, Giles W, Noble SJ: Membrane currents underlying rhythmic activity in bullfrog sinus venosus. *J Physiol* 271: 783–816, 1977.)

mum diastolic potential. Note that just after the action potential had fully repolarized a large decay of current was recorded, whereas the next clamp test (occurring near the midpoint of the pacemaker potential) resulted in a much smaller relaxation of current. Additional experiments have shown that this relaxing current reverses (became a decaying inward current) near E_K.

A second physiologically important result has been obtained from voltage clamp analysis of primary pacemaker activity: no conventional (tetrodotoxin-sensitive) sodium inward current (i_{Na}) is present. This is not surprising if one recalls (1) that the maximum diastolic potential in primary pacemaker tissue is normally -65 ± 5 mV, and (2) that i_{Na} is very voltage dependent, being almost completely shut off or inactivated at these potentials (cf. Chapter 4). Therefore, primary pacemaker activity may be modelled using only two time-dependent ionic currents: (1) the slow inward (Ca^{2+}/Na^+) current, i_{si}, and (2) an outward potassium current. Figure 8 shows an intracellular record from the rabbit SA node, together with a drawing of a suggested time course of these two time-dependent transmembrane currents.

How do the autonomic transmitters alter heart rate? In sinus tissue the classical effect of acetylcholine is to increase a *steady, background potassium* current (this current was discussed under Resting Potential, above). Thus, the sinus cells hyperpolarize, (as shown in Figure 9a). In

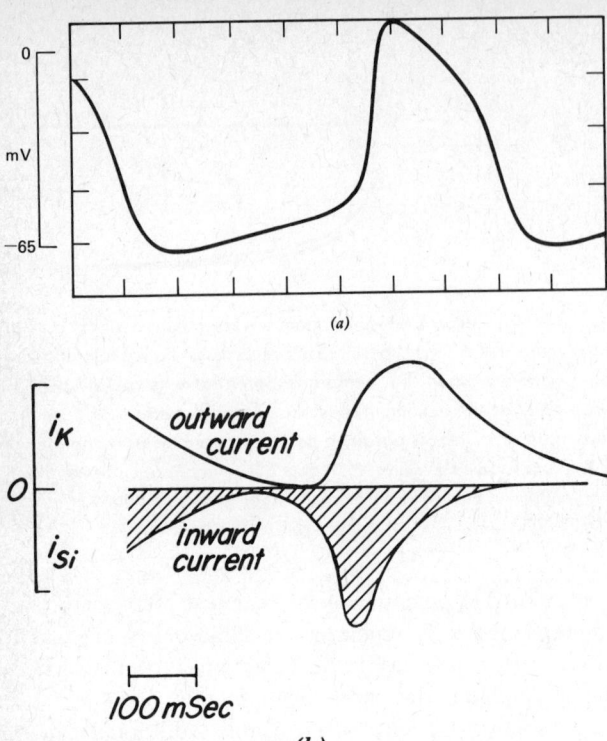

(a)

(b)

Figure 8 (a) Intracellular record of pacemaker activity in the rabbit SA node. (b) Hypothetical scheme of the two time-dependent ionic currents thought to underlie the pacemaker depolarization and the action potential. Remember that in addition to these time-dependent ionic currents a steady (background) inward current "holds" the maximum diastolic potential positive to E_K.

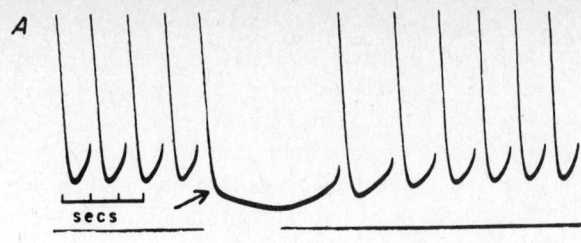

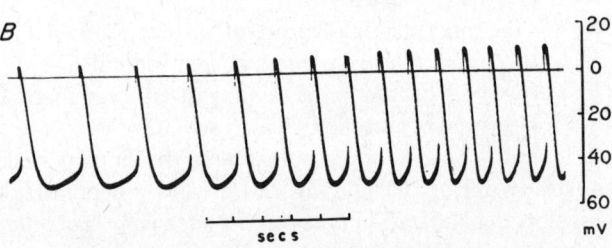

Figure 9 Electrophysiological effects of acetylcholine and adrenaline in pacemaker cells of the frog sinus venosus. The vagus nerve (A) or sympathetic trunk (B) were stimulated during the time period denoted by the interruptions of the horizontal line. (From Hutter OF, Trautwein W: Vagal and sympathetic effects on the pacemaker fibres in the sinus venosus of the heart. *J Gen Physiol,* 39, 715–733, 1956.)

addition, the slope of the pacemaker potential decreases. In combination, these two electrophysiological changes produce the *muscarinic negative chronotropic effect,* or decrease in heart rate.

About 10 years ago, it was shown that in Purkinje fibers epinephrine greatly speeds the rate of decay of the *pacemaker potassium current.* (How could this result in an increased heart rate?) By analogy, this mechanism is also thought to apply to sinus tissue, although it has not been demonstrated conclusively.

Epinephrine has recently been shown to significantly increase the slow inward current in the sinus node. In the scheme presented in Figure 8, an increase in i_{si} could (1)

increase the slope of the pacemaker potential, and (2) increase the peak height of the action potential (Fig. 9).

Research into the mechanism of generation of the primary pacemaker response (sinus node), and its alteration by the autonomic transmitters, is continuing. Some important changes in the above "story" are likely to emerge in the near future.

RELATIONSHIP BETWEEN INTRACELLULAR "POTENTIALS" AND EXTRACELLULAR "ELECTROGRAMS"

Medical practitioners are seldom able to record the *intracellular* electrical activity in the heart or other excitable tissues. Instead, they must rely on data obtained by extracellular electrodes. In the heart, this recording is named the electrocardiogram (ECG). This important clinical tool is described in detail in Chapter 7.

Figure 10a compares the electrical waveforms recorded from (1) an *intracellular* microelectrode in a ventricular cell with (2) an "active" extracellular electrode and an "indifferent", or reference, electrode (Fig. 10b). The extracellular electrode "sees" potentials that develop in the extracellular space from the local "circulating" ionic currents that are responsible for propagation. Note that the large, rapid current flow during depolarization produces a significant change of potential in the extracellular space, whereas the slower more gradual process of repolarization produces a smaller extracellular waveform. Moreover, when the tissue is quiescent or in the plateau phase of the action potential, little change in extracellular potential is recorded from healthy cardiac tissue.

Would an equivalent experiment in a skeletal muscle show a similar delay between the extracellular "wave" signaling depolarization, and the second "wave" indicating repolarization?

The ECG provides a clear signal of the excitation and repolarization of ventricular tissue. Since atrial depolari-

zations can also be detected, the ECG is a powerful clinical tool for the detection of abnormal conduction patterns, and/or arrhythmias. Further discussion of the pathophysiology of conduction is beyond the scope of this chapter, but is dealt with in Chapter 7; and is discussed in Urthaler et al. (1978), and Tsien and Siegelbaum (1978).

CONTRACTILE EVENTS IN CARDIAC MUSCLE

Many of the important features of excitation–contraction coupling in the heart are very similar to those in skeletal muscle. For that reason, only the processes that differ from skeletal muscle will be emphasized in this chapter.

In skeletal muscle the action potential sweeps along the surface of the fiber, invades the T system and *triggers* the release of Ca^{2+} from the intracellular storage site: the sarcoplasmic reticulum. As shown in Figure 11a, the action potential has nearly fully repolarized before any significant tension is recorded. In the case of cardiac muscle, however, the action potential is very long (200–500 ms),

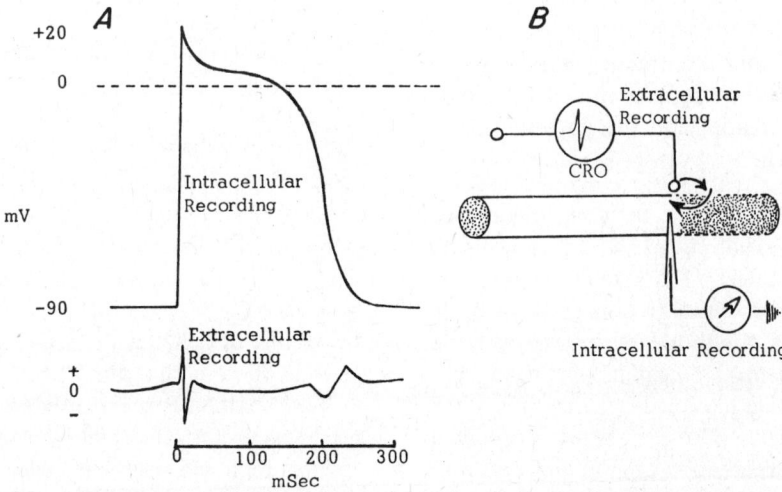

Figure 10 Comparison of intracellular and extracellular recordings from ventricular muscle. Since the extracellular record is very small (about 100–200 microvolts) the amplifiers used are designed so that they see only rapid *changes* in potential (see text for further explanation).

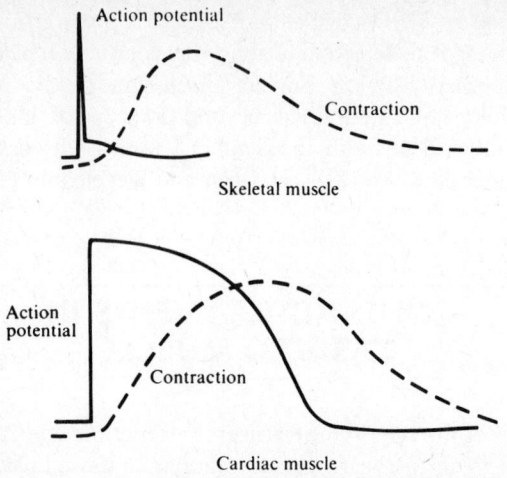

Figure 11 Comparison of the relative time courses of (1) the action potential, and (2) the resulting contraction in skeletal (top) and cardiac (bottom) muscle. The contraction is denoted as a dashed curve.

and therefore the mechanical event (contraction) and the action potential are of similar duration. This has two very important consequences:

1. In the heart the height and the duration of the action potential not only *trigger* but also *modulate* the strength of contraction.

2. In cardiac muscle a tetanic contraction normally cannot be obtained, since the length of the refractory period of the action potential and the duration of the contraction are similar.

An equally striking difference between mechanical events in skeletal and cardiac muscle was first observed about 100 years ago. Ringer (1883) found that unless Ca^{2+} is present in the extracellular (bathing) medium, the contraction of the heart rapidly decreases and may become undetectable. In contrast, a similar maneuver with a skeletal muscle shows that tension development is nearly independent of reductions in $[Ca^{2+}]_o$. It is now possible to give a straightforward explanation for this difference. In cardiac muscle there exists a Ca^{2+} conductance mechanism, the slow inward current i_{si}, which generates the plateau of the action potential and results in an increase in $[Ca^{2+}]_i$ with each action potential. In skeletal muscle,

the action potential is generated by the inward movement of Na^+ followed by an outward K^+ current. There is no significant transmembrane Ca^{2+} influx.

It is important to note, however, that in ventricular muscle the amount of Ca^{2+} that enters the cell via the conductance mechanism is *not* thought to be sufficient to fully activate the contractile proteins. Therefore, additional Ca^{2+} must be released from the sarcoplasmic reticulum. This could occur in two ways:

1. In a calcium-dependent fashion (i.e., Ca^{2+}-induced Ca^{2+} release).

2. In a voltage-dependent manner (i.e., depolarization of the cell triggers release).

The available evidence indicates that both the level of intracellular $[Ca^{2+}]$ and transmembrane voltage are important determinants of the amount of intracellular Ca^{2+} release.

REGULATION OF INTRACELLULAR CALCIUM LEVELS

In order for cardiac cells (and smooth muscle cells) to avoid being "loaded" by the Ca^{2+} that enters with each heartbeat, there must be an efficient method of Ca^{2+} extrusion. That is, the Ca^{2+} cannot simply be sequestered by the sarcoplasmic reticulum as it is in skeletal muscle; it must be pumped out or extruded against a large chemical ($[Ca^{2+}]_o = 2$ mM, $[Ca^{2+}]_i = 10^{-5}–10^{-7}$ M) and electrical ($E_m = -70$ to -85 mV) gradient. In principle, this could be accomplished by means of an ATP-requiring Ca^{2+} pump. Such a mechanism has been described in red blood cells and could operate in the heart. However, the majority of the Ca^{2+} extrusion in the heart is thought to occur by means of a novel "transport exchange" mechanism, such as the one diagrammed in Figure 12.

Note that ATP is not used as a *direct consequence* of Ca^{2+} extrusion. Instead, by some obscure mechanism, Ca^{2+} is removed from the cell in a way that depends on the size of the sodium gradient. ATP is, however, required to fuel the $Na^+–K^+$ pump, which establishes this Na^+ gradient.

As shown in Figure 12, this Na^+/Ca^{2+} exchange mech-

outside   inside

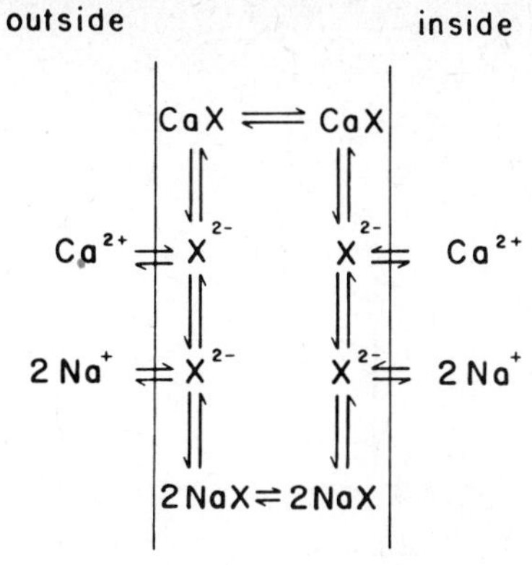

$$\frac{[Ca^{2+}]_i}{[Ca^{2+}]_o} = \frac{[Na^+]_i^2}{[Na^+]_o^2}$$

Figure 12 A hypothetical carrier scheme for Na–Ca exchange across a membrane; two Na ions and one Ca ion compete for a carrier (X^{2-}) at the inside and outside surfaces of the membrane. The carrier can move as 2NaX or as CaX across the membrane, but the unloaded carrier cannot move. In such a transport scheme, a Ca gradient can be established across the membrane as a consequence of an existing Na gradient. The respective distribution ratios at equilibrium are given by the equation below the scheme. Note that the carrier scheme illustrated here is an *electroneutral* one $(2Na^{++}$ for $1Ca^{++})$. Recent evidence indicates that 3 or 4 Na^+ ions enter for each Ca^{++} extruded. Thus this carrier exchange system may operate in an *electrogenic* mode. (From Reuter H: *Circ Res* 34:599, 1974.)

anism would operate in an *electroneutral* fashion; two Na^+ ions move in for every 1 Ca^{2+} ion being extruded. However, the possibility that this carrier could operate in a voltage-dependent fashion; or the exchange could be *electrogenic* must still be considered carefully.

CONTRACTILE PROTEINS IN CARDIAC MUSCLE

The contractile proteins in cardiac muscle, and the mechanism by which Ca^{2+} regulates their interaction and therefore controls tension development, are very similar to those in skeletal muscle (Chapter 6). As a reminder, a brief description is given in Figure 13 of the biochemical and mechanical events thought to be essential in cross-bridge formation and tension development.

1. The Nonactivated (Diastolic or Quiescent) State
Actin and myosin are separated and each myosin "head" has an ATP bound to it.

2. ATP Hydrolysis
ATP is cleaved or hydrolyzed, but although the H^+ is rapidly released, the $ADP \cdot P_i$ remains on the myosin, realigning itself (changing the angle) with respect to the actin. That is, the energy derived from the ATP splitting is transiently stored and the myosin head is activated or "cocked" for action.

3. Cross-Bridge Formation
When the sarcoplasmic $[Ca^{2+}]$ rises above 10^{-7} M and Ca^{2+} binds to troponin, the troponin–tropomyosin complex undergoes a change that permits the "active site" on the actin molecules to be exposed. The "activated" myosin head attaches to actin and the "cross-bridge" is formed.

4. Force Generation
Energy is used to produce a change in the actomyosin complex; the myosin head changes its position and in so doing moves the actin filament relative to the myosin. Hence the term *sliding filament*.

5. End-Product Release
The ADP and P_i are released from the "deactivated" myosin complex, but the myosin head remains in its "contracted" conformation.

6. ATP Binding
A "new" ATP binds to the myosin head, the cross-bridge is broken, and the biochemical cycle begins once again.

MYOCARDIAL ENERGETICS

Skeletal muscle relies heavily on anaerobic metabolism; during bursts of activity endogenous stores of high energy phosphate compounds are utilized rapidly. The energy pool in skeletal muscle (Chapter 4) is comprised of ATP and creatine phosphate. By contrast, cardiac muscle depends on aerobic metabolism and utilizes an exogenous supply of high energy compounds for its "fuel" requirements. In a normal heart about 70% of the energy require-

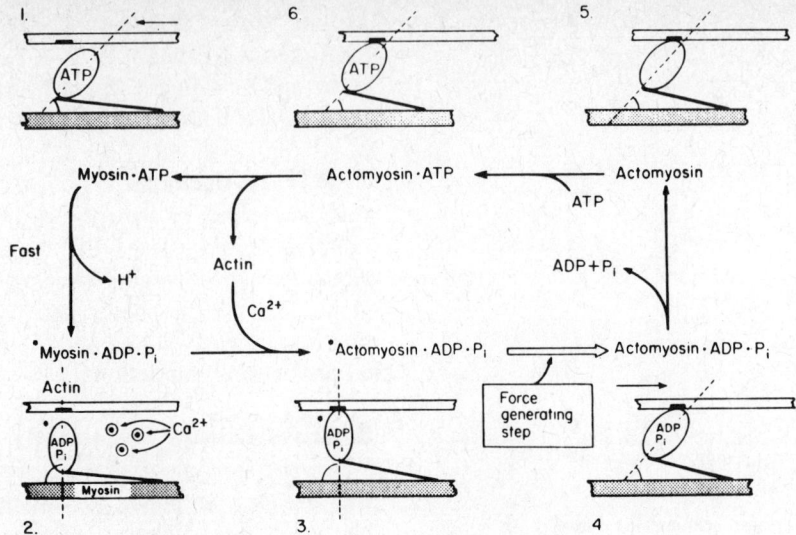

Figure 13 Six hypothetical stages in the biochemical and mechanical events that lead to the generation of force in skeletal and cardiac muscle. Asterisks (*) denote activated state. (Adapted from Merin RG, Pask HT: in Prys-Roberts C (ed): *The Circulation in Anesthesia.* Oxford, Blackwell Scientific, 1980.)

ments are derived from nonesterified fatty acids that enter the cells across the sarcolemma; the remaining 30% comes from metabolism of glucose.

CARDIAC MUSCLE MECHANICS

The available evidence strongly indicates that the sliding filament model of contraction (Chapter 4) adequately accounts for most aspects of cardiac muscle mechanics. Therefore, this section will aim to provide a brief review of important material from Chapter 4, and emphasize the ways in which the mechanics of cardiac and skeletal muscle may differ. Finally, an attempt will be made to account for some features of the mechanics in the intact mammalian ventricle.

Almost all the detailed information regarding the mechanics of cardiac muscle has been obtained in experiments using the isolated mammalian papillary muscle or ventricular trabeculum. These cardiac tissues are somewhat similar to a single skeletal muscle fiber, although it is important to remember that they are multicellular and that they cannot be tetanized.

In general the data from cardiac muscle has been analyzed in terms of the contractile element/series elasticity model presented in Chapter 4. Alpert (in Vassalle, 1976) has summarized the important, qualitative characteristics of this model. These are shown in Figure 14a, which is a diagramatic representation of this model. Figure 14b shows the changes in the components which are thought to occur during an *isometric* contraction in an isolated papillary muscle.

Small changes in heart rate may modulate the amplitude of the contraction; and some temporal summation (the so-called staircase effect) can be demonstrated. Nevertheless, due to the long duration of the action potential it is not possible, in normal physiological conditions, to obtain a *fused tetanic* mechanical response in cardiac muscle. For that reason, quantitative comparison of the length-tension relationships of cardiac and skeletal muscle should not be made. Figure 15 shows data obtained from an isolated rat ventricular trabeculum. From these measurements it is possible to relate muscle length (or sarcomere length) to active and passive tension, over a 20% change in muscle length, or a change in sarcomere length of 1.5–2.3 μm. Note that active force increases as sarcomere length changes from 1.5 to 2.2 μm. Further

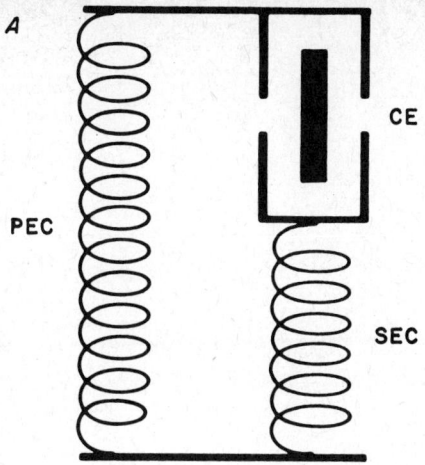

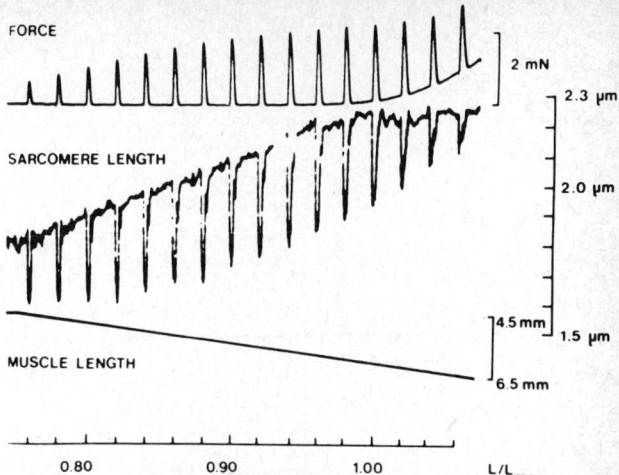

Figure 15 Summary of the changes in force developed by an isolated rat ventricular trabeculum (stimulated at 0.2 Hz) as a function of (1) sarcomere length, and (2) muscle length. Note that increases in *sarcomere length* are indicated by an *upward* displacement of the baseline, whereas increases in *muscle length* move the baseline *downward*. On the force record the *active force* is shown as transient spike-like events. The *passive force* can be measured from the baseline of the force trace. Note that as muscle length increases beyond L_{max}, passive force increases and active force decreases. For further explanation, see text. (Adapted from ter Keurs H, et al: *Circ Res* 46:703–714, 1980.)

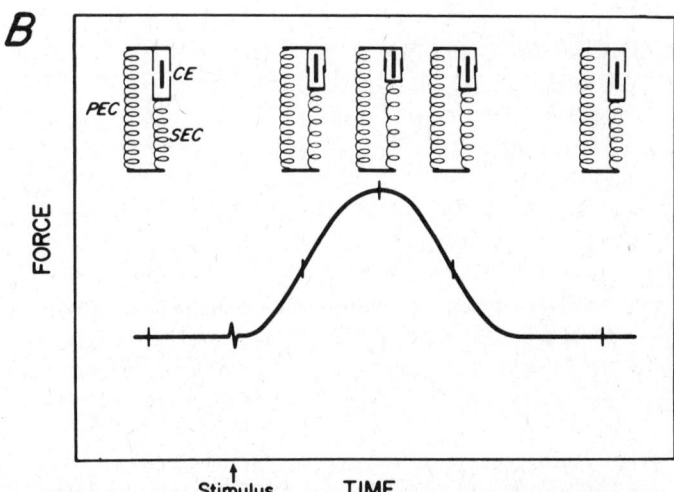

Figure 14 A physical model that accounts for the main features of the mechanical events in cardiac muscle. (*a*) CE, SEC, and PEC represent the contractile element, the series elastic components, and the parallel elastic components, respectively. (*b*) Drawing of the changes in these components thought to occur during an isometric contraction cycle. (Adapted from Alpert NR, Hamrell BB: Cardiac hypertrophy: a compensatory and anticompensatory response to stress, pp. 174–187, in Vassalle M (ed): *Cardiac Physiology for the Clinician.* New York, Academic Press, 1976.)

increases in sarcomere length, however, produce an *increase in total tension* (active plus passive) but a *decrease* in active tension. Quantitative comparisons of the descending limb (Chapter 4, Fig. 15) of the length–tension relationship in cardiac muscle and in skeletal muscle are not available. Thus, a major piece of evidence for (or against) the application of the crossbridge theory of muscle contraction in cardiac muscle is lacking.

Figure 16 compares the relationships between passive and active isometric tension in skeletal muscle and in cardiac muscle. Two apparent differences are worth noting:

1. At L_{max}, cardiac muscle exhibits significant resting (passive) tension whereas skeletal muscle does not.

2. Small changes in muscle length near L_{max} produce larger changes in active tension in cardiac muscle than in skeletal muscle.

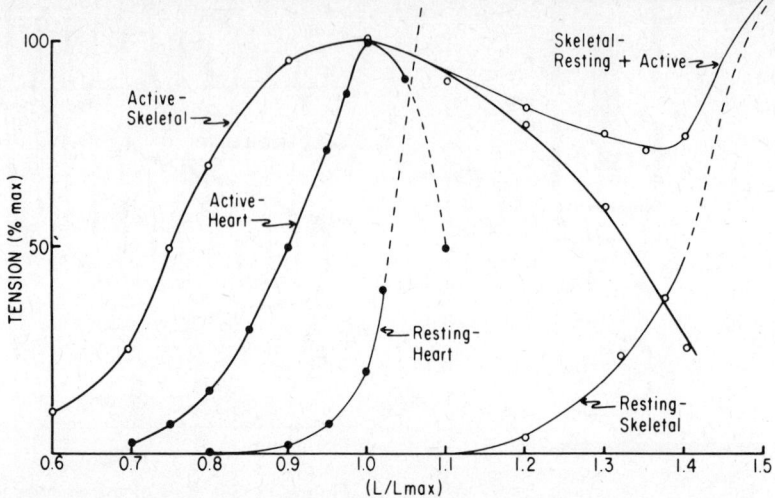

Figure 16 Relationship between muscle length and (1) resting or passive tension, and (2) actively developed tension in heart versus that in skeletal muscle. Length is expressed as a function of L_{max}; the length at which active isometric tension is maximum. Tension is normalized.

Obviously, each of these differences could be of considerable functional importance. At present, however, it is not possible to critically assess the apparent significance due to technical difficulties in the cardiac muscle experiments.

STARLING'S LAW OF THE HEART

The ability of the intact ventricle to alter its force on a beat-to-beat basis depending on its end-diastolic size is a very important principle of cardiovascular physiology. From the data shown in Figures 15 and 16, it is probable that the explanation for this "Frank-Starling phenomenon" is provided by the shape and position of the active and passive tension curves in relationship to L_{max}. Remember, however, that "muscle length" in the intact ventricle is an exceedingly difficult parameter to measure quantitatively. In addition, in most practical noninvasive situations "tension" can usually only be inferred from measurements of left ventricular pressure (Fig. 17). In Chapter 8 the problem of accounting for the complex relationship between tension development in the heart (or cardiac output) and end-diastolic volume (initial length) will be discussed in some detail.

INOTROPIC EFFECTS OF THE AUTONOMIC TRANSMITTERS IN THE HEART

The chronotropic effects (changes in heart rate) produced by the autonomic transmitters have been summarized previously. Although changes in heart rate may alter the mechanical response; it can be shown that when heart rate is held constant norepinephrine increases the strength of contraction (positive inotropic effect); whereas acetylcholine diminishes it (negative inotropic effect). Figure 18 shows the changes in the action potential and in tension development of a canine papillary muscle exposed to 10^{-6} M isoproterenol, a drug that mimics the β actions of the transmitter, norepinephrine. The plateau of the action potential becomes more convex in shape, and the tension increases (1) in magnitude, and (2) in rate of development and relaxation. Voltage clamp measurements have shown that many of these electrophysiological and mechanical changes can be explained by an increase in the inward current i_{si}. The increased rate of relaxation, however, appears to result from an augmented rate of Ca^{2+} sequestration by the sarcoplasmic reticulum. This phenomenon is

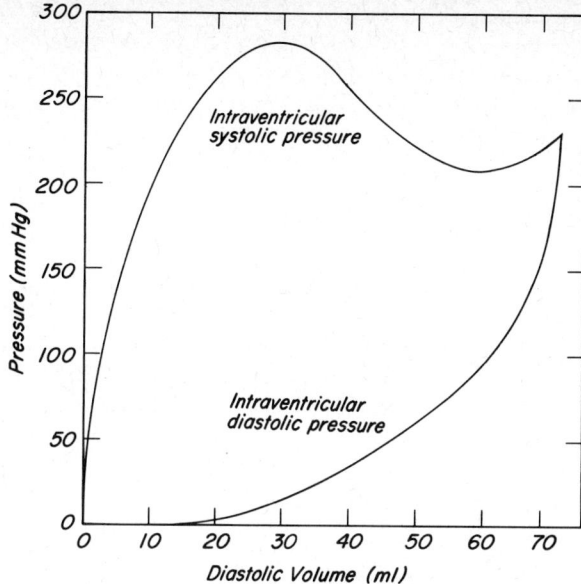

Figure 17 Relationship between intraventricular pressure and volume. The intraventricular diastolic volume partially controls the resting length of ventricular muscle (e.g., papillary muscle). Hence the intraventricular *diastolic pressure* provides an indication of the resting, or *passive tension*. Note that the intraventricular *systolic pressure curve* is similar in shape to the curve for active *isometric tension development* (see Fig. 16). However, the peak pressure (which is fairly flat over diastolic volumes of 20–40 ml) is actually measured when the ventricular is shortening under *isotonic* conditions (i.e., tension or pressure is constant).

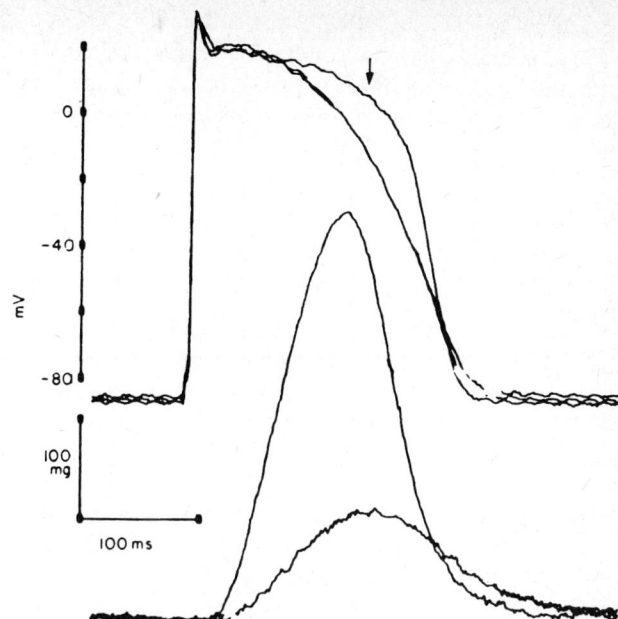

Figure 18 Effects of stimulation of β-adrenergic receptors on the action potential and active tension in canine papillary muscle. The traces marked by arrows were obtained after the agonist isoproterenol (10^{-6} M) had produced a maximal effect in a paced (0.5 Hz) trabeculum. For further description, see text. (From Nathan D, Beeler DW: Electrophysiologic correlates of the inotropic effects of isoproterenol in canine myocardium. *J Mole Cell Cardiology*, 7: 1–15, 1975.)

modulated by increased intracellular levels of cyclic AMP —the second messenger for many β actions of norepinephrine in the heart.

Acetylcholine does not exert a very powerful negative inotropic effect in the intact mammalian ventricle. The most likely reason for this is that the ventricle receives only very sparse vagal innervation. In isolated preparations of atrial or ventricular muscle, acetylcholine exerts essentially opposite effects to that illustrated in Figure 18; it shortens the action potential and depresses peak tension. The original explanation for this negative inotropic effect was based on the finding that acetylcholine increases the background K$^+$ permeability in *pacemaker* tissue. (For a summary of evidence, see Giles and Shibata, 1980.) Thus, it was reasoned that an increased P_K would shorten the action potential and that as a consequence of the reduced action potential duration, less tension would be developed. However, recent voltage clamp experiments have shown that acetylcholine strongly inhibits the slow inward current, i_{si}. Thus, less Ca^{2+} enters the cell per beat, and contractility is reduced.

SUMMARY

The important new concepts in this chapter should be considered in relationship to the important functions of the tissues in the heart: (1) rhythmicity, (2) contractility, and (3) conduction. In order to understand (1) and (2), it is essential to think in terms of the *slow inward current,* i_{si}. Understanding impulse conduction within the cardiac syncitium requires knowledge of the microanatomy and physiological properties of the *gap junction* or *nexus.*

SELECTED BIBLIOGRAPHY

Braunwald E, Ross J Jr, Sonnenblick EH: *Mechanisms of Contraction of the Normal and Failing Heart,* ed 2. Boston, Little, Brown, 1976.

Giles W, Shibata EF: Conductance changes produced by autonomic transmitters in cardiac pacemaker tissue: A brief review. *Fed Proc* 40: 2618–2624, 1981.

Jewell BR: The physiology of cardiac muscle contraction, in Dickinson CJ, Marks J (eds): *Developments in Cardiovascular Medicine.* Baltimore, University Press, 1979.

McNutt NS, Weinstein RS: Membrane ultrastructure at mammalian intercellular junctions. *Prog Biophys* 26:45, 1973.

Merin RG, Pask HT: The myocardial cell and its metabolism, in Prys-Roberts C (ed): *The Circulation in Anesthesia.* Oxford, Blackwell Scientific, 1980.

Noble D: *The Initiation of the Heartbeat.* Oxford, Clarendon Press, 1979.

Reuter H: Exchange of calcium ions in the mammalian myocardium. Mechanisms and physiological significance. *Circ Res* 34:599, 1974.

Tsien RW, Siegelbaum S: Excitable tissues: The heart, in Andreoli TE, Hoffman JF, Fanestil DD (eds): *Physiology of Membrane Disorders.* New York, Plenum, 1978.

Urthaler F, Kawamura K, James TN: The anatomic basis for cardiac rhythm and conduction, in Andreoli TE, Hoffman JF, Fanestil DD (eds): *Physiology of Membrane Disorders.* New York, Plenum, 1978.

Vassalle M: *Cardiac Physiology for the Clinician.* New York, Academic Press, 1976.

Berne RM, Sperelakis N, Geiger SR (eds): *The Handbook of Physiology,* section 2: *The Cardiovascular System,* vol 1: *The Heart.* Bethesda MD, American Physiological Society, 1979.

SELF-STUDY QUESTIONS

MULTIPLE CHOICE

Select the single best answer.

1. The longest delay in propagation of excitation in the heart occurs in the
   A. sinoatrial node.
   B. atrium.
   C. atrioventricular node.
   D. Purkinje system.
   E. ventricular myocardium.

2. The sinoatrial node is the normal pacemaker because it
   A. is most richly supplied with nerve endings.
   B. is triggered by excitatory nerve impulses.
   C. is the first portion of the heart to beat in the embryo.
   D. possesses the highest frequency of automatic discharge.
   E. is the only part of the heart capable of spontaneously generating action potentials.

3. The electrical events in a single myocardial muscle fiber, or trabeculum
   A. occur earlier for isotonic than for isometric changes.
   B. vary in amplitude directly with the strength of contraction.
   C. are the result of contraction.
   D. trigger mechanical events.
   E. occur simultaneous with mechanical events.

MATCHING

Ion	Charge	Concentration (mEq/liter)		Equilibrium Potential (mV)
		Inside Cell	Outside Cell	
A	+	155	4	−95
B	−	3.8	120	−90
C	+	12	145	+65
D	+	13×10^{-5}	4×10^{-5}	−32
E	−	8	27	−32

4. The ion most likely to be potassium.

5. The ion that would have equal tendency for efflux and influx if the cell's membrane potential was −90 mV.

6. The ion for which the concentration and electrical gradients are in the same direction at a membrane potential of −90 mV.

   A. Cardiac sympathetic nerve stimulation
   B. Cardiac parasympathetic nerve stimulation
   C. Both
   D. Neither

7. Increases the slope of pacemaker potential.

8. Modulates the size of the inward current i_{si}.

9. Extensively innervates ventricular myocardium.

   Action potential of:

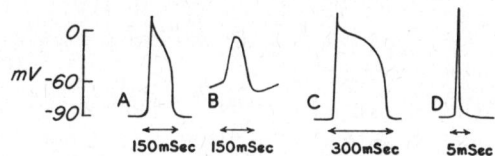

10. Skeletal muscle
11. Sinoatrial node
12. Purkinje fibers

MULTIPLE CHOICE

Select the correct answer(s). (In many instances, more than one answer is correct.)

13. In the action potential of a ventricular muscle cell
    A. membrane permeability to sodium is high during the initial rapid depolarization.
    B. the transmembrane potential is most negative during the plateau phase.
    C. membrane permeability to potassium is high during the process of repolarization.
    D. the duration of the plateau is increased in tachycardia.

14. The "prepotential" found in the sinoatrial node
    A. is characteristic of pacemaker cells.
    B. shows a decreased rate of change after acetylcholine application.
    C. is partly controlled by a decreasing i_K.
    D. is insensitive to catecholamines.

15. The site of electrical communication between cardiac cells
    A. is called the nexus or gap junction
    B. permits ionic current to flow through it.
    C. may become dysfunctional under such conditions as hypoxia or cardiac glycoside toxicity.
    D. is identical to a desmosome, or an intercalated disc.

16. The plateau phase of the cardiac action potential
    A. is generated by inward movement of Ca^{2+} and Na^+ via the so-called "slow inward current" i_{si}.
    B. is depicted by a large diphasic wave in an extracellular recording, such as the ECG.
    C. can be increased in height and duration by β-adrenergic drugs.
    D. is similar to that in skeletal muscle.

17. The Na^+/Ca^{2+} exchange mechanism in cardiac muscle:
    A. helps prevent the cardiac cells from becoming "Ca^{2+}" loaded.
    B. moves Ca^{2+} out of the cell against the existing electrochemical gradient.
    C. requires the existence of an electrochemical gradient for Na^+.
    D. is inhibited by norepinephrine and stimulated by acetylcholine.

18. The contractile activity in the heart
    A. is completely independent of heart rate.
    B. is triggered and partially controlled by the action potential.
    C. is not influenced by the autonomic nervous system.
    D. is thought to be generated by Ca^{2+}-dependent interaction of "contractile proteins" very similar to those in skeletal muscle.

ANSWERS

1. C. Propagation through the AV node is much slower than through the atrium, ventricles, or Purkinje system. The cells in the AV node have a small diameter (5 compared with 10 μm), gap junctions or nexuses occur less frequently, and the maximum diastolic potential is about -65 mV; hence i_{Na} is partially inactivated.

2. D. Virtually any part of the heart can exhibit "pacemaker" activity. These "subsidiary pacemakers," however, are normally suppressed or "overdriven" by the sinoatrial node. The major reason is simply that the SA node "paces" faster and therefore triggers or entrains, the rest of the heart. The cardiac pacemaker is myogenic, not neurogenic.

3. D. As in skeletal muscle, the action potential in the heart triggers the mechanical event. Since the cardiac action potential is relatively long, it can also modulate the strength of contraction.

4. A. Recall that in normal circumstances K^+ is *low outside* and *high inside* cells. The concentrations given in the Table are approximately correct for the mammalian myocardium.

5. B. Recall the definitions and significance of the Nernst potential.

6. C. The data given in *C* indicate that the (1) *concentration gradient* is directed inward. Since this ion is positively charged (a cation), if the membrane potential was set to -90 (inside negative), it would also tend to move inward down the *electrical gradient*.

7. A. The release of norepinephrine from adrenergic varicosities in the SA node produces an increase in the slope of the pacemaker potential; the parasympathetic transmitter, acetylcholine, has the opposite effect.

8. C. A fundamental but relatively new finding in cardiac electrophysiology is that the two autonomic transmitters have opposing actions on i_{si}. Norepinephrine increases it via a β action, and acetylcholine reduces i_{si} via a muscarinic action.

9. A. Only the *adrenergic* side of the autonomic nervous system *extensively* innervates the *ventricular myocardium*. Vagal effects in the ventricle do occur, but the innervation is relatively sparse.

10. (D); **11.** (B); **12.** (C).
    This is straightforward pattern recognition. The (1) relative durations, (2) presence or absence of a pacemaker potential, and (3) levels of the resting or diastolic potentials are the key pieces of information.

13. A, C. The ionic currents underlying the ventricular action potential are illustrated in Figure 6. From this it is apparent that the initial rapid depolarization is generated by i_{Na}, the plateau is mainly due to i_{si} and repolarization is triggered by i_K. The plateau is thus relatively *positive*. In tachycardia, the action potential shortens, and some of the *shortening* occurs in the plateau phase.

14. A, B, C. Please refer to Figures 8 and 9. In addition, recall that the sinoatrial node is exceedingly responsive to both autonomic transmitters; modulation of the slope of the pacemaker potential is a major determinant of the so-called chronotropic effects.

15. A, B, C. The anatomical site of electrical communication between mammalian cardiac cells is the nexus, or gap junction. The ultrastructure of a nexus and a desmosome differ significantly; moreover, the nexus is only one part of the intercalated disc. Recent data indicate that the conductance of the channels in the

nexus can be modulated by pH, toxic levels of drugs, and hypoxia.

16. A, C. Since the plateau phase of the action potential corresponds to a "steady voltage" (lack of a voltage gradient), only very small extracellular currents will flow during this part of the action potential. The plateau is generated by i_{si}, and catecholamine application increases i_{si}; hence the plateau will become more positive and the action potential duration *may* increase. Skeletal muscle action potentials have no plateau.

17. A, B, C. The Na^+/Ca^{2+} exchange mechanism is thought to play an important role in removing calcium from the cell against both electrical and concentration gradients. The energy required for this seems to be derived from the existing electrochemical gradient for Na^+. There is no evidence that the autonomic transmitters directly modulate it.

18. B, D. Recall that although cardiac muscle cannot normally be tetanized, the contractile activity is modulated by heart rate (positive and negative "staircase" effects) and by the autonomic nervous system (positive and negative inotropism). In cardiac muscle, the mechanism of the Ca^{2+}-dependent regulation of contractile protein interaction is the same as that in skeletal muscle—phosphorylation and dephosphorylation of one subunit of troponin.

6

Smooth Muscle

WAYNE R. GILES

OBJECTIVES

After completion of this chapter, you should be able to

1. Describe certain structural aspects of smooth muscle:
   a. Approximate cell size
   b. Presence and possible significance of cavaeolae and dense bodies in, or near, the sarcolemma
   c. Types and functions of intercellular connections
   d. Location and importance of the sarcoplasmic reticulum, and/or the transverse tubule (T) system

2. List some features of the contractile proteins in smooth muscle:
   a. Relative proportion of actin and myosin
   b. Arrangement of the myofilaments in relaxed versus contracted cells
   c. Mechanism of control of force generation, including, CA^{2+}-dependent phosphorylation and dephosphorylation of myosin light chains

3. Describe a number of interesting properties of smooth muscle mechanics:
   a. Existence of significant resting or "tonic" tension
   b. Distinction between tonic and phasic tension

   c. General information on
      1. Extent of shortening
      2. Maximal force generating capacity of smooth versus skeletal muscles
   d. Energy source(s) for tension generation, and the rate of ATP hydrolysis
   e. Response of smooth muscle to stretch

4. Compare excitation–contraction coupling in skeletal (fast twitch) and smooth muscle(s):
   a. Trigger for contraction
   b. Sources of Ca^{2+} for activation
   c. Control of intracellular Ca^{2+} levels

5. Briefly describe the electrophysiology of smooth muscle:
   a. Neurogenic versus myogenic tissues
   b. Ionic basis of the resting potential
   c. Distinction between, and the ionic basis of the "Slow" wave, and the "Spike"

6. Summarize the anatomy and physiology of autonomic neurotransmission in smooth muscle.

Detailed knowledge of the electrophysiology and mechanics of smooth muscle currently lags far behind that of skeletal muscle. There are two reasons for this:

1. Smooth muscles are extremely diverse in their physiology and pharmacology; as a result, general principles have been difficult to develop.
2. Various features of smooth muscles (e.g., small cell size and strong contractility) present major technical difficulties in electrophysiological experiments.

Nevertheless, knowledge of smooth muscle function is a very important part of muscle physiology. Understanding the physiology and pathophysiology of a number of organ systems—notably the digestive system, cardiovascular system, and reproductive tract—requires this information.

PHYSIOLOGICAL CLASSIFICATION OF SMOOTH MUSCLES

Nearly 40 years ago, it was suggested that smooth muscles could be divided into two distinct groups based on their physiological properties. *Multiunit smooth muscles* can be activated by motor nerves and usually are organized into motor units. Examples of this kind of smooth muscle are: pilomotor muscles, muscles of the nictitating membrane, cilary muscles, the iris, and certain blood vessels. *Unitary smooth muscles* are much more similar to cardiac muscle. Hence, they exhibit pacemaker activity and contract spontaneously; and they behave as a functional syncytium, although their electrical activity and mechanical responses are strongly modulated by the nervous system. Unitary smooth muscles, in general, respond to a stretch by developing tension, whereas multiunit tissues do not. Unitary smooth muscles include those in the hollow organs of the body, for example, the gut (visceral smooth muscle) and the genitourinary tract.

It is now apparent, however, that this kind of classification scheme is oversimplified. A number of well-characterized smooth muscles fall into an intermediate category. For example, smooth muscles in some arterioles and veins exhibit autorhythmicity, but they also contract in re-

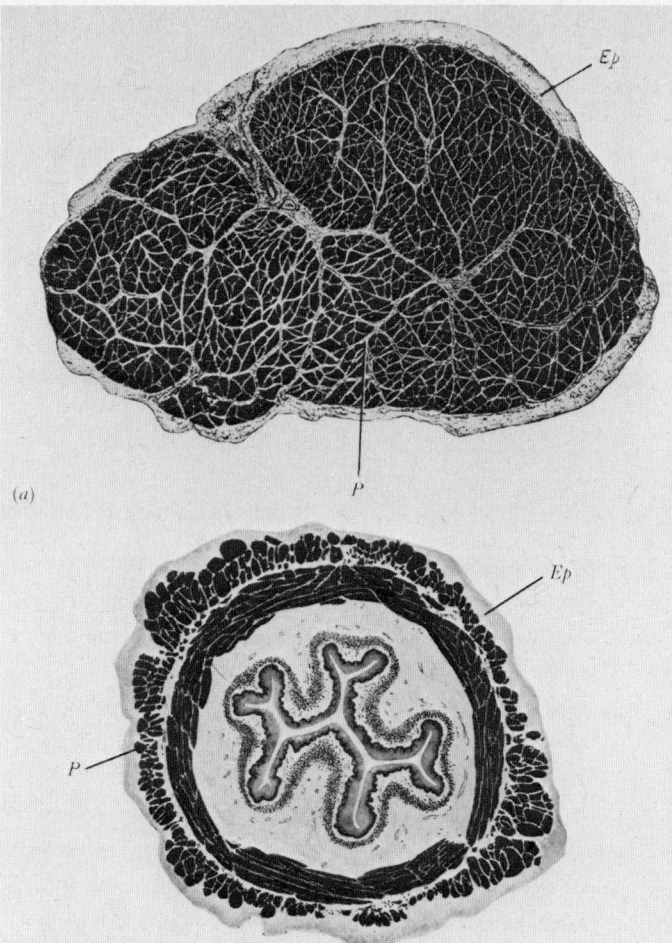

Figure 1 Comparison of the structure of skeletal muscle with that of smooth muscle. (*a*) Cross section through the human sartorius muscle (skeletal muscle). The epimysium (Ep) is the fibrous sheath which surrounds the muscle. It extends into the muscle forming the perimysium (P), which divides the muscle into bundles of striated muscle fibers. (*b*) Cross section through the human esophagus, showing the relationship between the smooth muscle and the other components that form the wall of this visceral organ. Note that the epimysium surrounding the whole organ also extends into the muscle to form the perimysium, which divides the muscle into bundles of smooth muscle fibers. Note also the outer longitudinal and inner circular muscle layers. (Adapted from Bennett MR: *Autonomic Neuromuscular Transmission.* New York, Cambridge University Press, 1972.)

sponse to nerve stimuli. A second example is the urinary bladder, which responds to distension by contracting, but also contracts when it receives a volley of nerve impulses.

STRUCTURE OF SMOOTH MUSCLE

Smooth muscle is usually arranged in sheets in the walls of hollow viscera and tubular structures. Well-known examples include the esophagus, stomach, and intestines; the uterus, the urinary bladder and ureters; the entire tracheal–bronchial tree; the ducts of the male reproductive system; and blood and lymph vessels of the circulatory system. In other organs, however, smooth muscle is found in small bundles and even in individual fibers, as in the pilomotor muscles of hair follicles, the muscle fibers of the capsule and trabeculae of the spleen, and the myoepithelal cells of mammary, sweat, and salivary glands.

Most hollow organs contain two coats or layers of smooth muscle: an outer coat oriented in the long axis of the organ, and an inner coat that is perpendicular to this axis, or circular (Fig. 1). Some organs have a third inner, longitudinal coat. It is important to note that different muscle layers from a single tissue often have widely differing anatomical, electrophysiological, and pharmacological properties. Therefore, an understanding of the physiology of these organs requires knowledge of the functions of each muscle layer, as well as the interactions between the layers.

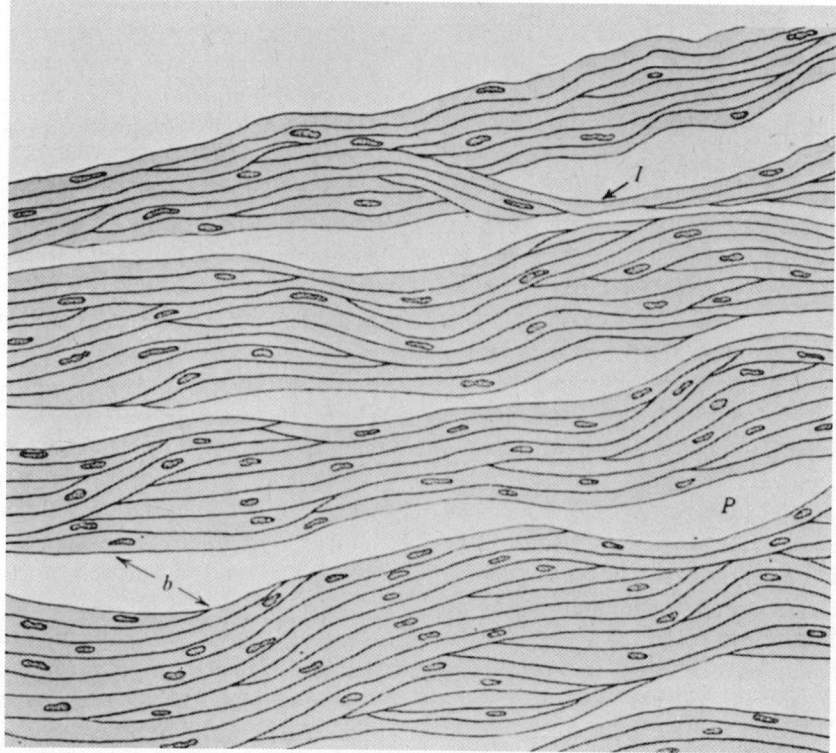

Figure 2 Structure of one muscle layer of a smooth muscle, for example, longitudinal muscle of the intestine. P, Perimysium; b, a larger bundle of muscle fibers; l, a small muscle bundle joining two larger ones. Note arrangement into branching bundles separated by intercellular spaces filled with perimysium. (From Bennett MR: *Autonomic Neuromuscular Transmission.* New York, Cambridge University Press, 1972.)

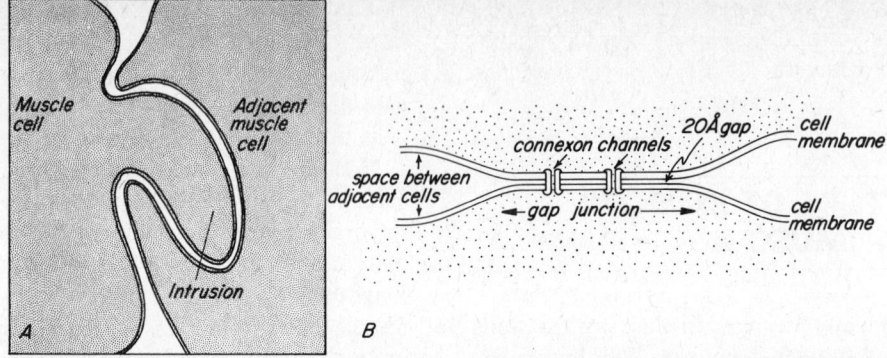

Figure 3 The right-hand side is a drawing of a gap junction or nexus between two smooth muscle cells. The left-hand side shows a different type of apposition between two cells: cell 1 protrudes into the end region of cell 2. Note, however, that the plasma membranes of the two cells are not as closely applied as in the nexus.

Within the muscle layers of visceral smooth muscles, cells are often grouped in the form of branching bundles surrounded by a matrix composed mainly of connective tissue (Fig. 2). Bundle cross sections show that the cells can be arranged in a polyhedral fashion. That is, each cell has approximately five "nearest neighbors." The gaps between cells are approximately 500–800 Ångstrom units (1 Å = 10^{-10} m). These gaps are partly filled with collagen fibrils. Detailed information regarding structure, chemical composition, and function(s) of this intercellular matrix is not currently available.

Smooth muscle cells are usually described as spindle-shaped structures having a single, centrally located nucleus and being 2–5 μm (1 μm is 10^{-6} m) in maximum diameter, and 50–100 μm in length. Recent ultrastructural evidence, however, indicates that a more accurate "average size" for a smooth muscle cell is: 5–10 μm in maximum diameter, and 200–300 μm in length. Extremely small cells are present in some blood vessels: in certain arterioles the cells are approximately 2 μm in diameter and 15–20 μm in length.

INTERCELLULAR JUNCTIONS

Early light microscopic studies of smooth muscles often indicated that there was protoplasmic continuity between adjacent cells, that is, that the cells were arranged as an anatomic syncytium. But more recent work, using electron microscopy, has shown that, although there exist areas of "close contact" between smooth muscle cells; each cell is completely surrounded by a sarcolemma.

In smooth muscle these regions of close contact or intercellular junctions comprise approximately 3–5% of the surface area of the cell. The conventional types of intercellular junction, the nexus or gap junction (Fig. 3), cannot always be demonstrated in smooth muscle. Other areas of "close apposition" can occur either on the lateral or on the end surfaces of the cells. These areas of "close contact" are thought to be the sites of electrical communication between adjacent smooth muscle cells, and thus responsible for the propagation of electrical activity through small muscle bundles. The evidence for this is very similar to that presented in Chapter 5 (see the description of electrical coupling between cardiac muscle cells).

In addition, desmosomes (Chapter 5) are common in a number of smooth muscle tissues. Desmosomes may be involved in the transmission of force between individual cells.

Smooth muscle cells are so named because they lack the "banded" or striated appearance characteristic of both skeletal and cardiac muscle. As described in detail later, this is because the myofilaments are not arranged in orderly repeating units, such as the sarcomeres of skeletal muscle.

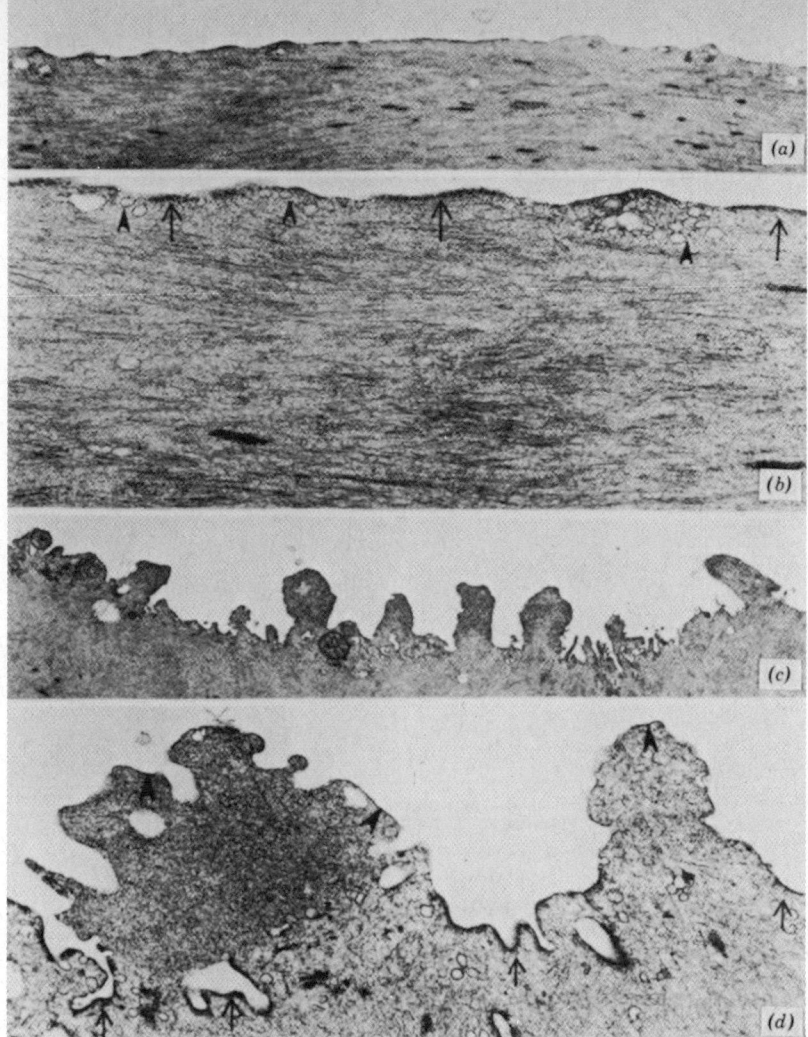

Figure 4 (*a*) Electron micrograph showing the smooth surface contour of a relaxed cell. (*b*) The plasma membrane of a relaxed cell consists of areas either subtended by amorphous dense material (arrows) or differentiated into micropinocytotic vesicles (darts). (*c*) Electron micrograph showing the rough surface contour of a contracted cell. (*d*) The plasma membrane of a contracted cell is subtended by amorphous dense material along the bases and "neck regions" of the surface evaginations (arrows). The membrane covering the evaginations proper (darts) is not subtended by these dense areas, and generally it does not have the micropinocytotic vesicles that characterize the plasma membrane of relaxed cells. (*a,c*) × 7,740. (*b,d*) × 21,635. (From Fay FS, Cooke PH, and Canaday PG, et al: Contractile properties of isolated smooth muscle cells. pp. 249–265 in: Bülbring E, Shuba MF (eds): *Physiology of Smooth Muscle.* New York, Raven, 1975.)

THE CELL SURFACE

In addition to these specialized "close contact" structures; microscopic studies of the surface of smooth muscles typically exhibit both *cavaeolae* and *dense bodies* (Fig. 4). All smooth muscle cells have numerous cavaeolae (inpocketings approximately 0.1 μm in diameter) on the sarcolemma. Although the function of these cavaeolae remains obscure, their presence increases the surface area of most smooth muscle cells by approximately 70%. Dense bands or dense bodies, which are darkly stained regions just beneath or near the sarcolemma have also been described. Thin filaments, thought to be contractile proteins, often insert into these dense bodies; thus the dense bodies are somewhat analogous to the Z lines of skeletal muscles.

CONTRACTILE APPARATUS

Actin filaments, or thin filaments, are readily identified in smooth muscles. In relaxed tissue, actin filaments lie parallel to the long axis of the cell. Evidence for the presence of myosin has been more equivocal, but recent work consistently shows that myosin is present in smooth muscle. It is important to note that although myosin can be identified in vertebrate smooth muscle, the myosin content is three to four times less than in vertebrate striated muscle. This difference may be the reason that myosin is often difficult to identify in electron micrographs of smooth muscle.

Ultrastructural evidence indicates that the myofilaments frequently attach to the sarcolemma at the location of the dense bodies. Thus, as noted previously, in smooth muscle there are no clearly defined sarcomeres as there are in striated muscle. When smooth muscle is activated, finger-like *evaginations* appear in the sarcolemma, and the cell tends to assume the shape shown in Figures 3 and 4.

Figure 5 diagrammatically illustrates the kind of forces that can be developed within a smooth muscle cell when it is activated. It is important to remember, though, that in intact smooth muscle the cells are mechanically interconnected and are supported by the intercellular matrix. It is therefore uncertain whether this "single cell" picture strictly applies to force transmission within the intact smooth muscle tissue.

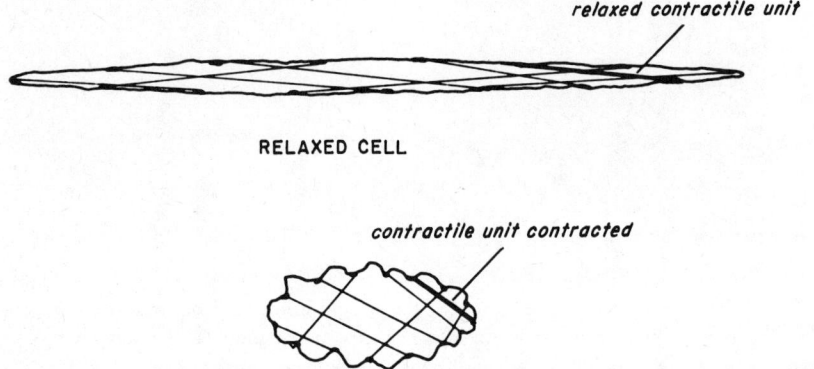

relaxed contractile unit

RELAXED CELL

contractile unit contracted

FULL CONTRACTION

Figure 5 Schematic representation of a smooth muscle cell showing how contractile units are attached to the cell surface. The dark areas along the cell membrane represent plasma membrane dense bodies and the lines connecting them represent the "contractile units." One of these lines has been widened to facilitate its identification in the two different states of contraction. Note that contractile units run for only relatively short lengths within the cell and that the angle between contractile units and the long axis of the cell is thought to increase during contraction. (Adapted from Fay FS, Cooke PH, and Canaday PG, et al: Contractile properties of isolated smooth muscle cells. pp. 249–265, in Bülbring E, Shuba MF (eds): *Physiology of Smooth Muscle.* New York, Raven, 1975.)

INTERCELLULAR ORGANELLES

The sarcoplasmic reticulum (SR) in smooth muscle cells is poorly developed (sparse) compared with that of striated muscle. Small sacs of sarcoplasmic reticulum are, however, usually found immediately beneath the sarcolemma, often near the cavaolae. Histochemical studies have shown that these organelles can accumulate divalent cations, but the precise role of the SR in the excitation–contraction coupling of smooth muscle remains obscure.

Mitochondria are present in smooth muscle cells. They make up about 7% of the cell volume in intestinal smooth muscle and are thought to be important in cellular metabolism. Mitochondria have been shown to be able to transport calcium, thus they *may* act as "sources" and "sinks" for intracellular calcium in smooth muscle.

EXCITATION–CONTRACTION COUPLING

REGULATION OF INTRACELLULAR CALCIUM LEVELS

The sequence of events underlying the initiation of smooth muscle contraction is still unknown. The rise in intracellular Ca^{2+} that triggers the contraction could be attributable to:

1. Ca^{2+} entry by means of a voltage-dependent inward current mechanism, similar to that in cardiac muscle.
2. Ca^{2+} released from the sarcoplasmic reticulum.
3. Ca^{2+} released from intracellular binding sites.

It has been shown that calcium influx into smooth muscle cells increases with mechanical activity. Moreover, in preparations from rat uterus and toad intestine, voltage clamp data indicate that a transient inward transmembrane ionic current, carried partly by calcium ions, is present.

Two striking features of the mechanical activity of visceral smooth muscle are:

1. The long latency (200–700 msec) between the electrical event and tension development.
2. The extremely slow time course of activation and relaxation.

Recently it has been possible to monitor the intracellular calcium levels in smooth muscle cells of the stomach. These results show that $[Ca^{2+}]_i$ rises coincident with or very shortly after the "spikes" of electrical activity. The long latency in the onset of contraction therefore *cannot* be explained by a delay in the increase of $[Ca^{2+}]_i$. Instead, it appears to be a property of the contractile proteins. The kinetics of the molecular interactions that generate force in smooth muscle seem to be very slow compared with those in skeletal or cardiac muscle.

The mechanism by which calcium regulates the interactions between actin and myosin in smooth muscle deserves some special attention, since it has recently been shown to be quite different than the regulatory scheme that governs the activation of skeletal and cardiac muscle. Tropomyosin is found in a wide variety of both vascular and visceral smooth muscles; and the proportion of actin to tropomyosin (about 50–1, by weight) is about the same in smooth and skeletal muscles. Largely on this basis, it was, until recently, thought that Ca^{2+} would exert its regulatory role in smooth muscle via the well-known troponin–tropomyosin system.

But there is now convincing evidence that in smooth muscle one component of *myosin* (a 20,000-dalton myosin light chain) is the site of Ca^{2+} regulation. This myosin light chain undergoes a Ca^{2+} dependent phosphorylation and dephosphorylation, and this appears to be the first step in the regulation of the contractile protein interaction. In a number of smooth muscles, combined biochemical and mechanical data have shown that:

1. The myosin light chain is the only contractile protein that is phosphorylated in a Ca^{2+}-dependent fashion.
2. The Ca^{2+} concentration dependence of this phosphorylation reaction corresponds closely to the Ca^{2+} dependence of tension development.

The mechanism underlying the control of intracellular calcium levels in smooth muscle is incompletely understood. Four possible modes of Ca^{2+} removal or sequestration are:

1. The sarcoplasmic reticulum has been shown to be able to accumulate divalent cations (e.g., Ca^{2+}).

2. In some vascular smooth muscles there is evidence for a Na^+–Ca^{2+} exchange mechanism (see Chapters 3 and 5).

3. Active (ATP requiring) extrusion, or pumping out, of Ca^{2+} has been demonstrated.

4. Intracellular binding sites for Ca^{2+} may quickly rebind Ca^{2+} after it has been released.

SMOOTH MUSCLE MECHANICS

Compared with striated muscle (Fig. 6) tension development in smooth muscle is extremely slow, partly because of the long latency in the onset of tension (see above). Interestingly, peak force development in skeletal, cardiac and smooth muscle is similar: approximately 3–5 g/mm². Other properties of smooth muscle, however, are quite different from those in "fast twitch" skeletal muscles:

1. During maximal contraction, smooth muscle cells can shorten to about 35% of their resting length. By comparison, skeletal muscle fibers shorten to only about 70% of their resting length during a tetanus.

2. ATP appears to be the major energy source for force development in smooth muscle. However, to maintain tension, taenia coli (visceral smooth muscle) uses high energy phosphate much more slowly (approximately 200 times *slower*) than striated muscle which is maintaining similar forces.

3. Normal electrophysiological activity in smooth muscles can elicit two distinct kinds of mechanical response: (a) a transient mechanical response, called phasic tension, and (b) a "tonic" or maintained tension. This "tonic" tension is not analogous to a partially fused or a fused tetanus in skeletal muscle, since it occurs in the absence of high-frequency electrical activity in the muscle.

Very few attempts to determine the force–length relationship for smooth muscle have been reported. One reason is that in the absence of well-defined sarcomeres "muscle length" is difficult, if not impossible, to measure accurately. Quantitative data describing the force–velocity relationship in smooth muscle is also lacking.

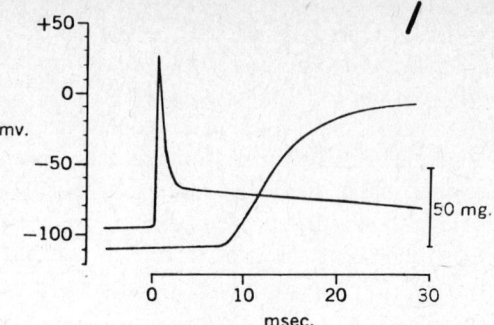

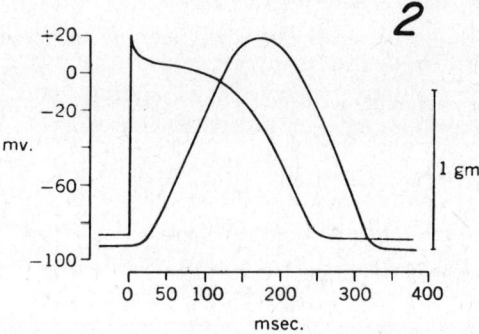

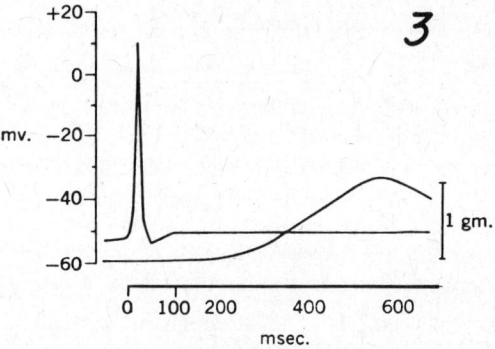

Figure 6 Diagrams showing relationship between transmembrane action potential and isometric tension. (1) Single muscle fiber from frog semitendinosus muscle. Temperature: 20°C. Excitation–contraction latency (time from onset of depolarization to initial development of tension) about 8 ms. (2) Small segment of ventricular muscle of cat. Temperature: 37°C. Excitation–contraction latency about 10–15 ms. (3) Small segment of uterine muscle of rat. Temperature: 37°C. Excitation–contraction latency about 200 ms. (From Marshall J: Vertebrate smooth muscle, pp. 121–151, in Mountcastle VB (ed): *Medical Physiology,* vol I. St. Louis, Mosby, 1974.)

ELECTROPHYSIOLOGY OF SMOOTH MUSCLE

RESTING POTENTIAL

The resting membrane potential of smooth muscle preparations is between -45 and -65 mV. Measurements of intracellular electrolyte concentrations indicate that in myometrium

$$\frac{[K]_o}{[K]_i} = \frac{1}{26}$$

giving a Nernst potential for potassium, E_K of about -83 mV. Hence, a substantial inward background (steady or time-independent) current must be present. The ionic basis of this current has not been established, although radioactive ion tracer experiments (^{24}Na and ^{36}Cl) indicate the permeability of quiescent smooth muscle to Na may be as much as *10* times that of skeletal muscle.

The presence of a substantial, steady influx of sodium into smooth muscle cells indicates that there must be an effective extrusion mechanism that can pump sodium ions out of the cell against the existing electrochemical gradient. A variety of smooth muscle tissues have been shown to possess an effective "sodium pump" (Na$^+$–K$^+$ ATPase). Moreover, in some preparations this pump has been shown to operate in the *electrogenic* mode. That is, because it transports unequal amounts of Na$^+$ and K$^+$ ions, *net* current is generated. As noted in Chapter 1, it is thought that *two potassium ions* are transported *into* the cell in exchange for *three Na$^+$ ions* moving *out* of the cell. Thus, a net outward, or repolarizing, *electrogenic current* is generated.

NEUROGENIC SMOOTH MUSCLE

Most *multiunit* smooth muscles are neurogenic; that is, they are quiescent unless activated by transmitter substances released from "motor" nerves. It is important to note, though, that all "motor" nerves to smooth muscles are part of the *autonomic nervous system*. Examples of neurogenic smooth muscles include ciliary muscle, muscles of the iris, pilomotor muscle, the trachealis muscle, and muscles of larger arteries. However, this kind of classification is not absolute; a growing number of tissues have been shown to be partly neurogenic but also capable of generating spontaneous electrical activity and mechanical responses. Well-known examples of this mixed behavior are the muscles in arterioles, venules, and veins.

MYOGENIC SMOOTH MUSCLE

Examples of smooth muscles in which mechanical activity is generated by electrical activity originating within the muscle itself include the muscle of the gastrointestinal tract, the uterus, and the ureters. Because of their locations, these *myogenic* muscles are often referred to as *visceral smooth muscle*. They are not excited by nerve impulses, but by spontaneous, rhythmic depolarization–repolarization cycles, often called *slow waves*. When a slow wave produces enough depolarization to reach threshold, an action potential is triggered, and a contraction ensues. Sometimes a whole volley of action potentials is produced by a single slow wave, and a stronger contraction is produced. However, not all slow waves are of sufficient amplitude to trigger action potentials. In smooth muscle of the small intestine, for example, rhythmic slow waves constantly occur (at a frequency of 12 per min in human duodenum), but on the average, only about 40–50% of these generate action potentials and contractions.

The ionic basis of the "slow wave" phenomenon has been studied extensively. Unfortunately, however, there is strong disagreement about fundamental aspects of its ionic basis. Figure 7 illustrates the components of electrical activity in one myogenic smooth muscle—the circular muscle of guinea pig stomach. There appear to be three components:

1. An initial depolarization that arises in the absence of any measurable change in membrane conductance.

2. A secondary voltage-dependent maintained conductance increase.

3. Spikes, or nerve-like action potentials, superimposed on the slow wave.

It appears that Na$^+$ or Ca^{2+} ions, or both, carry the inward currents that generate component 2 of the slow wave and underlie the spike(s). A potassium current has been identified in uterine smooth muscle; hence it is

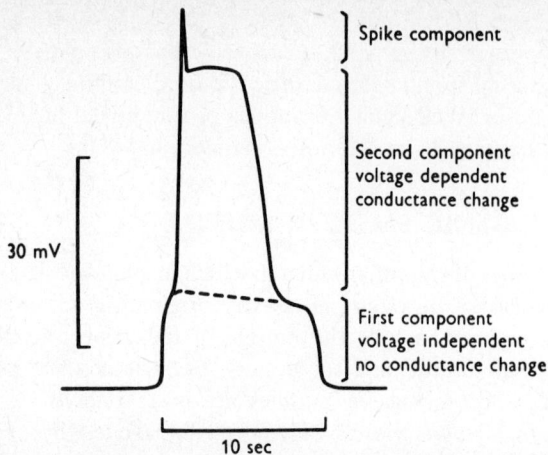

Figure 7 Schematic representation of the components of electrical activity in a myogenic smooth muscle (circular muscle layer of guinea pig stomach). Note the extremely long duration of first and second components of the slow wave, which can vary from 0.5 to 15 s. Although only 1 spike per slow wave is shown here, frequently trains of spikes (bursting) are triggered by the second component of the slow wave. (Adapted from Tomita T, Sakamoto Y: Electrical and mechanical activity in guinea pig stomach muscle, pp. 37–47, in Casteels R, Godfraind T, and Ruegg JC (eds): *Excitation-Contraction Coupling in Smooth Muscle.* New York, Elsevier/North Holland, 1977.)

hypothesized that repolarization in smooth muscle is triggered by activation of an outward flow of potassium ions. A major controversy is whether component 1 of the slow wave is generated by a conductance change, or whether modulation of the *electrogenic current* produced by the Na^+–K^+ pump is involved.

An important characteristic of myogenic smooth muscle is that the electrical activity (slow waves) is conducted from one cell to the next. As previously mentioned, this intercellular conduction is thought to take place via the *gap junctions, nexuses,* or other *"close contact"* areas. Intercellular conduction leads to synchronized activity in large areas of visceral smooth muscle, that is, many cells function as a single unit. Consequently, this type of smooth muscle is also referred to as *syncytial smooth muscle.*

Myogenic smooth muscle is often stimulated by stretch. For example, distension of the small intestine, urinary bladder, ureter, uterus, or small arterioles often elicit a contractile response. Longitudinal stretch of the smooth muscle fibers in these organs produces a depolarization of the membrane potential, which can reach the threshold for triggering action potentials, or it can increase the chances of spontaneous slow wave activity reaching the threshold for spike initiation. It is not known how stretch produces this depolarization. Muscles that are strictly neurogenic are not directly responsive to stretch.

AUTONOMIC NERVE–SMOOTH MUSCLE NEUROTRANSMISSION

As previously noted, most smooth muscles are composed of small bundles of cells, varying between 20 and 200 μm in diameter. Autonomic 'motor' nerves enter smooth muscles in relatively large aggregates (about 100 axons, each 20 μm in diameter) and then divide into smaller units (10–20 axons per unit, each about 10 μm in diameter) as they course deeper into the muscle. Finally, the axon units again divide (three to five axons, each about 2 μm in diameter) as they enter individual muscle bundles (see Figure 8).

The diameter of these "terminal" axons is not fixed; it increases and decreases each few microns as the axon passes into the muscle bundle. This produces a *beaded* appearance called the *axon varicosity* (Fig. 8b). These varicosities are thought to be the sites of release for the neurotransmitters. The distance between a varicosity and the nearest smooth muscle cell can be as large as 100–200 nm (1 nm is 10^{-9} m and is typical of blood vessels); or it can be as small as 20 nm—the *close-contact* varicosity.

The basis for understanding excitatory neurotransmission (cf. Chapter 7) has been provided by the work of Katz

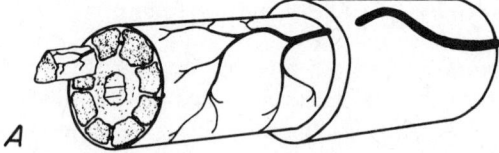

Figure 8 The pattern of autonomic (motor) innervation in smooth muscle. (a) The smooth muscle is shown as being divided into longitudinal and circular layers, each composed of muscle bundles. The nerve supply enters from the surface as a large bundle of axons. This then subdivides into smaller bundles that pass between the muscle bundles.

and his colleagues in their experiments at the frog neuro-muscular junction. No comparable analysis has been done at the autonomic nerve–smooth muscle junction, possibly because of the extreme technical difficulty of this task. However, excitatory junction potentials and miniature spontaneous junction potentials have been recorded in, for example, the mammalian hypogastric nerve–vas deferens preparation (Figure 9). Nerve stimulation at low frequencies (about 2 cycles/s, or 2 Hz) produces discrete junction

A

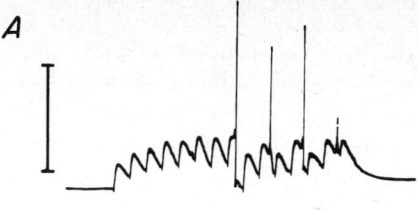

B

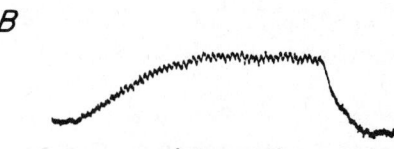

Figure 9 Membrane potential changes in smooth muscle produced by an excitatory transmitter. (*a*) Stimulation of the autonomic nerve trunk at a relatively low frequency (e.g., 2 Hz) produces discrete excitatory junction potentials (EJP) that can depolarize the membrane enough to produce action potentials. (*b*) Higher frequency stimulation (e.g., 10 Hz) "fuses" the EJP's and produces a maintained depolarization. Calibrations are (*a*) 40 mV and 300 msec; (*b*) 40 mV and 2 sec. (Adapted from Bennett MR: *Autonomic Neuromuscular Transmission.* New York, Cambridge University Press, 1972.)

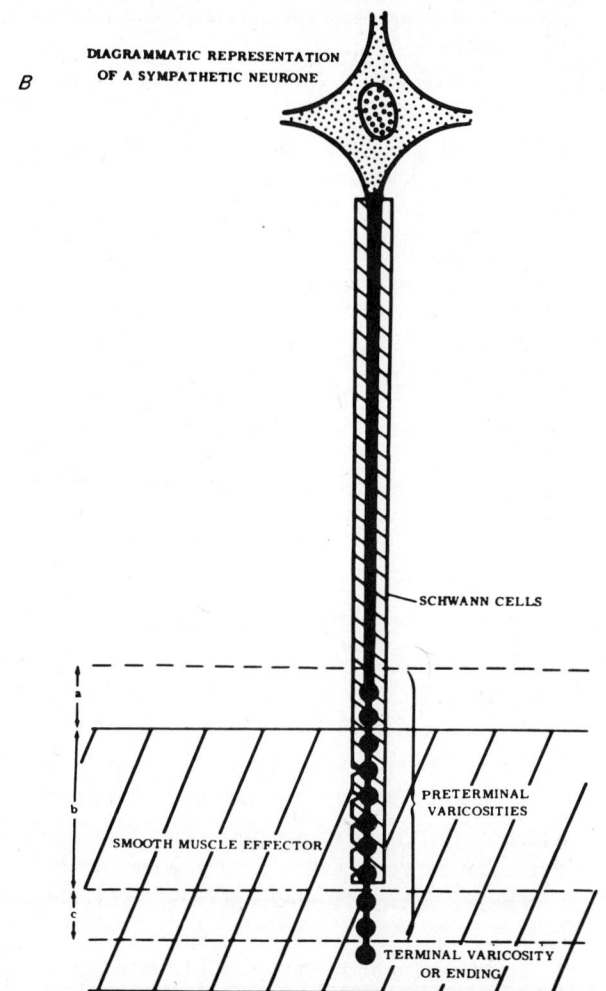

B

DIAGRAMMATIC REPRESENTATION
OF A SYMPATHETIC NEURONE

SCHWANN CELLS

PRETERMINAL
VARICOSITIES

SMOOTH MUSCLE EFFECTOR

TERMINAL VARICOSITY
OR ENDING

Figure 8 (con't) (*b*) Finally, as the small axons enter muscle bundles they assume a beaded appearance. These "beads" are called *varicosities* and are thought to be the site of transmitter release. (Adapted from Bennett MR: *Autonomic Neuromuscular Transmission.* New York, Cambridge University Press, 1972.)

potentials; higher frequency stimulation (approximately 10 Hz) can cause summation of the junction potentials, producing a maintained depolarization.

In contrast to the phenomenon at the neuromuscular junction, the junction potential in smooth muscle preparations can vary widely in *latency* and *duration.* This variation can be explained by recalling the differences in the anatomy of the two kinds of nerve-muscle junction. Thus, in striated muscle the transmitter is released from discrete, localized presynaptic regions that are in very close apposition to the post-synaptic receptors. As shown in Figure 9*b*, however, the varicosity type of junction can release transmitter either very close to the smooth muscle (effector) cell, or at a considerable distance from it. In the latter case, the transmitter substance must diffuse to its site of action—possibly giving rise to a considerable latency.

Two very important questions that remain are:

1. What substances act as inhibitory and excitatory transmitters in smooth muscles?

2. What is the mechanism of action of these transmitters?

Unfortunately, succinct answers to these questions do not exist. Thus, catecholamines and acetylcholine can act as *either* excitatory or inhibitory transmitters in various types of smooth muscle. In addition, at so-called *purinergic* synapses (examples are found in vertebrate gastrointestinal tract) ATP is thought to be the transmitter.

The postsynaptic conductance changes produced by each of these transmitters substances are also extremely diverse (see Bolton, 1979 for detailed information). Thus, no general hypothesis for transmitter action can be developed. A comprehensive textbook of pharmacology or the original literature, such as Bolton's review (see Selected Bibliography) must be referred to when questions arise.

A rather striking but obscure aspect of transmitter action on smooth muscle is the finding that catecholamines, and perhaps acetylcholine, can produce marked changes in mechanical activity, without changing the electrical activity. This phenomenon has been named *pharmacomechanical coupling.*

SELECTED BIBLIOGRAPHY

Adelstein RS (ed): *Phosphorylation of Muscle Contractile Proteins.* A symposiumin in *Fed Proc* 39:1544–1574, 1980.

Adelstein RS, Hathaway RD: Role of calcium and cyclic adenosine 3':5' monophosphate in regulating smooth muscle contraction. *Am J Cardiol* 44:783, 1979.

Bennett MR: *Autonomic Neuromuscular Transmission.* New York, Cambridge University Press, 1972.

Bolton TB: Mechanisms of action of transmitters and other substances on smooth muscle. *Physio Rev* 59:606, 1979.

Bülbring E, Bolton TB: Smooth muscle. *Br Med Bull* 35:209–305, 1979.

Bülbring E, Needham DM: A discussion of recent developments in vertebrate smooth muscle physiology. *Proc R Soc Lond* 265:1, 1973.

Bülbring E, Shuba MF (eds): *Physiology of Smooth Muscle.* New York, Raven, 1975.

Bülbring E, Brading A, Jones A, et al: *Smooth Muscle.* London, Edward Arnold, 1970.

Casteels, R, Godfraind T, Ruegg JC: *Excitation–Contraction Coupling in Smooth Muscle.* New York, Elsevier/North Holland, 1977.

Connor JA: On exploring the basis for slow wave potential oscillations in the mammalian stomach and intestine. *J Exp Biol* 81:153, 1979.

Daniel EE, Sarna S: The generation and conduction of activity in smooth muscle. *Annu Rev Pharmacol Toxicol* 18:145, 1978.

Marshall JM: Vertebrate smooth muscle, in Mountcastle VB (ed): *Medical Physiology,* vol I. St. Louis, Mosby, 1974.

SELF-STUDY QUESTIONS

MULTIPLE CHOICE

Select the best answer(s). (In many instances more than one answer is correct.)

1. In smooth muscle
   A. the excitation–contraction coupling mechanism is identical to that in fast twitch skeletal muscle.
   B. the sarcomeres range in length from 1.6 to 3.2 μm.
   C. desmosomes are thought to be the sites of intercellular electrical communication.
   D. there is a relatively long latency between the electrical activity (spikes) and the onset of contraction.

2. Which of the following are physiologically important differences between skeletal and smooth muscle?
   A. The presence and function of transverse tubule system.
   B. The presence and function of low resistance intercellular pathways.
   C. The Ca^{2+}-dependent phosphorylation and dephosphorylation of a "light chain" of myosin.
   D. The resting sarcomere length.

3. The autonomic nerves supplying smooth muscle
   A. release acetylcholine, which always inhibits the development of the slow wave.
   B. terminate in structures resembling strings of beads called varicosities.

C. release norepinephrine, which is thought to act preferentially at the nexuses or gap junctions.
D. may release transmitter agents that produce excitatory junction potentials.

ANSWERS

1. D.
   A. Wrong. Smooth muscle has no transverse tubule system and lacks a well-organized sarcoplasmic reticulum.
   B. Wrong. Smooth muscle has no sarcomeres.
   C. Wrong. The function of desmosomes remains obscure, but they are *not* thought to provide pathways for intercellular current flow.
   D. Correct. See Figure 5.
2. A–D. All are correct.
   A. Correct. Skeletal muscle has an extensive T system and its function in E–C coupling is very significant; smooth muscle lacks a T system.
   B. Correct. Smooth muscle cells function as an electrical syncitium, that is, they are joined by low resistance intercellular junctions. Skeletal muscle fibers, however, are not.
   C. Correct. In smooth muscle the major Ca^{2+}-dependent regulation of contraction is via phosphorylation and de-phosphorylation of the *myosin light chain(s);* in skeletal muscle one subunit of *troponin* is phosphorylated and dephosphorylated in a Ca^{2+}-dependent fashion.
   D. Correct. Smooth muscle does not contain discernible sarcomeres.
3. B, D.
   A. Wrong. The autonomic nerves release acetylcholine, norepinephrine, and possibly ATP; and the postsynaptic actions of each of these transmitters varies tremendously.
   B. True. This nerve ending morphology is in sharp contrast to the structure of the neuromuscular junction.
   C. Wrong. See answer 3A. No transmitter is thought to act only at the nexus.
   D. True. An example is provided in Figure 9.

7

Electrical Events of the Cardiac Cycle and the Electrocardiogram

DANIEL L. TRABER
DONALD W. STUBBS

OBJECTIVES

After completion of this chapter, you should be able to

1. Define the relationship between the size and orientation of a dipole and that of a vector.

2. Describe the size and orientation of a dipole that results from a fiber depolarizing in a volume conductor.

3. Describe the changes in galvanometer recordings that would result from changes in vector size and orientation.

4. Determine the direction and amplitude of a resultant vector by measuring galvanometer deflections from two lead axes of an equilateral triangle.

5. Identify the electrode placements for the recording of leads I, II, and III of the ECG.

6. Explain the manner in which the waves of depolarization spread across the atria and ventricles.

7. Delineate the various vectors that result from the waves of depolarization in the heart.

8. Identify the major waves of the ECG (i.e., P, Q, R, S, T).

9. Identify which areas of the myocardium depolarize with each of these waves.

10. Determine the direction and amplitude of deflection on each limb lead from a vector resulting from cardiac depolarization.

11. Recognize the unipolar lead connections and relate the galvanometer changes that would be seen when various vectors of depolarization occur.

12. Identify and explain the conduction and rhythm abnormalities responsible for the arrhythmias listed.

RECORDING FROM A VOLUME CONDUCTOR

From the initial development of the technique of electrocardiography by Einthoven, about 70 years ago, it has become established as a useful, nontraumatic, diagnostic tool for obtaining information about the heart in an intact individual. The *electrocardiograph* is used both clinically and experimentally to study the depolarization and repolarization of the heart. Chapter 5 discussed the depolarization and repolarization of cardiac cells from the point of view of intracellular electrodes. The electrocardiograph, however, records electrical changes in the heart as a whole from the point of view of *extracellular* electrodes. Not only are the electrodes extracellular, but they are placed at some distance from the heart, namely, at the body surface. Recording of electrical events in the heart from the body surface is possible because of the conducting properties of the extracellular fluid of the body. The body is said to act as a volume conductor, that is, potential differences and currents within the conducting body fluids have measurable effects throughout the body, for example, at the skin.

THE ELECTROCARDIOGRAPH

An electrocardiographic machine is an amplifier, a galvanometer, and a recorder (movements of the galvanometer pen are recorded on a moving strip of paper). The electrocardiograph records changes in electrical potential against time. The record produced by the electrocardiograph is called the *electrocardiogram,* or ECG. The positive and negative terminals of the galvanometer are connected to two wires called *electrodes.* When the distal ends of the electrodes are placed within an electrical field, the galvanometer measures differences in potential between the electrode ends at any given instant. The two electrodes within an electrical field constitute a *lead* and an imaginary line between them is called the *lead axis.* Consider the simple galvanometer illustrated in Figure 1a. If the positive electrode is connected to the positive pole of a battery and the negative electrode is connected to the negative pole, the galvanometer pen swings to the right.

For convenience in visualizing the swing of the pen of the simple galvanometer in comparison with that of a recording galvanometer, the simple galvanometer is turned on its back and sideways so that the pen swings upward instead of to the right (when similarly connected, a recording galvanometer pen swings upward and writes an upward-slanting line on the moving paper). If the battery is placed in a volume of conducting fluid, such as saline, and the electrodes are also placed in that fluid, each one opposite the poles of the battery (although not in direct contact), the galvanometer pen will again show an upward deflection (Fig. 1b). If the battery had been turned around 180° and then placed along the lead axis, the galvanometer pen would have swung in the opposite direction (Fig. 1d) because the polarity would have been reversed. If the battery were placed perpendicular to the lead axis (each pole equidistant from both electrodes), the galvanometer pen would remain at midposition (would register no potential difference between the two electrodes; see Fig. 1c).

The above sequence can be thought of in the following way. A positive and negative charge situated a short distance apart constitute what is known as a *dipole* (e.g., a battery is a dipole). When a dipole exists in a conducting medium, current flows between the two charges or poles and potential differences in the electrical field can be sensed by the electrodes of the electrocardiograph. In measuring such a potential difference the extent of excursion of the galvanometer pen depends on (1) the *strength* of the dipole (potential difference between the two charges of the dipole), and (2) the *directional orientation* of the dipole relative to the lead axis. The concept is useful because as the cardiac impulse propagates the heart may be thought of as a dipole in a volume conductor. This development is continued in the next section.

THE HEART AS A DIPOLE

As an action potential propagates across the heart, some of the cardiac cells will be depolarized (those cells behind the propagating wave of depolarization) while cells in front of the wave of depolarization are still in the polarized or "resting" state. An extracellular potential difference

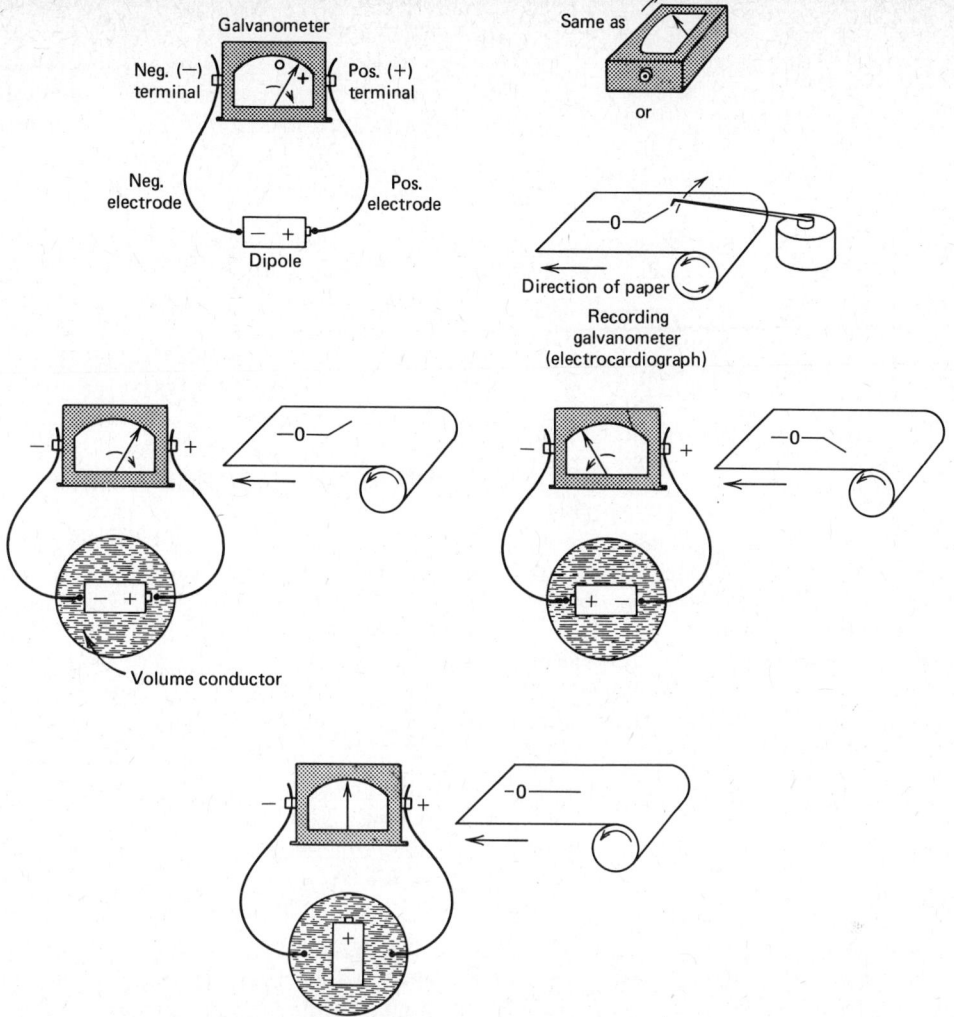

Figure 1 The use of a flashlight battery as a dipole to illustrate the influence of polarity on a galvanometer and in comparison with an electrocardiograph.

exists externally between these two areas. During propagation of the action potential current flows along the interior of the membrane from the active region to the resting region, then outward across the membrane of the resting region in front of the advancing potential and then back to the active region and inward across the membrane completing a circuit. This creates a potential field outside the fiber with the resting region positive relative to the active region. The depolarization process may be repre-

sented by a moving dipole with the positive charge in advance of the negative charge and is sometimes called an equivalent dipole (Fig. 2).

Once the atria or ventricles have become entirely depolarized, there is no further exterior potential difference, that is, all the cells have the same extracellular charge relative to each other. However, depolarization is soon followed by repolarization, and as soon as some of the cells become repolarized, there is once again an exter-

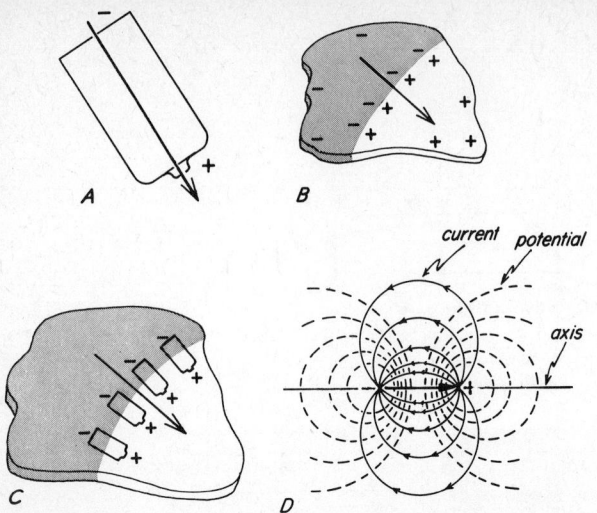

Figure 2 Vector representation of average potential difference (equivalent dipole) magnitude and direction. (a) A flashlight battery as a dipole. (b) A section of myocardium undergoing depolarization. (c) A section of myocardium undergoing depolarization in which charge differences are represented as miniature dipoles (drawn as small flashlight batteries). The average magnitude and direction (the resultant vector) are indicated by an arrow. By convention, the length of the arrow indicates magnitude, and the arrowhead points toward positivity. (d) Current flow between the charges of a dipole in a homogeneous volume conductor. Lines of constant potential cross the lines of current flow everywhere at right angles.

nal potential difference (between the depolarized cells and the newly repolarized cells). Now the outside of the repolarized cells is positive relative to the exterior of the cells still depolarized, that is, the direction of the dipole is reversed. As soon as repolarization is complete, there is once again no extracellular potential difference.

These dipoles have magnitude and direction, the magnitude being determined by the amount of current and the rate at which current flows between resting and depolarized regions. Hence, dipoles can be represented as vectors. Physiological quantities having magnitude alone are called scalars. Examples of scalars are temperature, pH, and density. Examples of vectors are force, and acceleration. The magnitude of a vector can be depicted graphically by the length of a line, and direction by an arrowhead at the tip of the vector.

The vector representing the cardiac dipole, since it exists in space, will have projections on the spatial coordinates representing the ECG leads. The projections of the vector onto these leads is shown in Figure 5b. As we shall see it is possible from these leads (which are the experimental data) to construct an imaginary spatial vector that would account for the voltage deflections seen with time on these leads.

RECORDINGS FROM MYOCARDIAL STRIPS

ATRIAL MYOCARDIAL STRIPS

Consider the events of depolarization and repolarization in a hypothetical strip of atrial myocardium within a volume of conducting medium. Initially the entire strip is in the resting state and the ECG shows a straight line along the midline of the paper (a baseline representing zero potential difference) as in Figure 3a. Assume that depolarization begins at the left and propagates toward the right. During the process of depolarization an upward deflection of the pen is recorded on the moving paper (Fig. 3b). As depolarization becomes complete there is no external potential difference, and the ECG record returns to the zero baseline (Fig. 3c). In *atrial* repolarization (as in many excitable tissues) the first area to have undergone depolarization is also the first to recover, that is, the first to begin repolarization. Therefore repolarization is shown as beginning at the left and moving toward the right but the equivalent dipole has a reversed polarity, and the recording pen swings downward (Fig. 3d). As soon as repolarization of the atrial strip is complete, the potential difference between all parts of the exterior is again zero, and the pen swings back to the zero baseline (Fig. 3e).

VENTRICULAR MYOCARDIAL STRIPS

The process of depolarization in ventricular strips is similar to that in atrial strips. However, in *ventricular* myocardium the last area to depolarize (the epicardium of the ventricular wall) is the first to repolarize. In other words, repolarization proceeds in a direction *opposite* to that of the preceding depolarization. Depolarization occurs from endocardium to epicardium, and repolarization occurs from epicardium to endocardium. Because of this difference in ventricular repolarization, the equivalent dipole is oriented in the same direction as in depolarization (unlike atria). This results in an upward deflection for repolarization (just as the deflection for depolarization is upward).

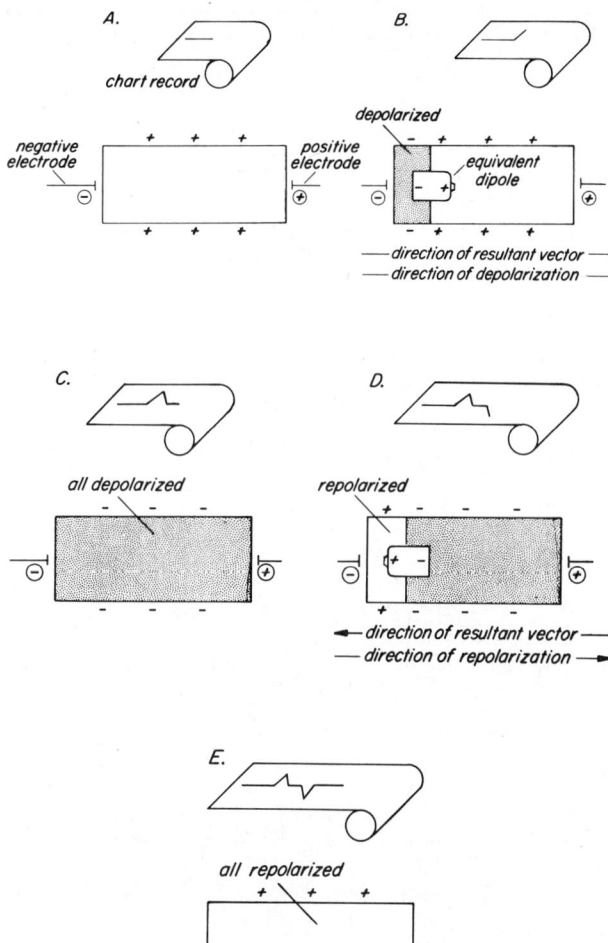

Figure 3 Atrial myocardial strips. (*a*) Resting state. (*b*) Depolarization begins. (*c*) Completion of depolarization. (*d*) Beginning of repolarization. (*e*) Completion of repolarization.

RECORDING OF ELECTRICAL EVENTS IN THE HEART FROM THE BODY SURFACE

In the above section, hypothetical strips of myocardium were aligned directly along the lead axis for convenience in visualizing events producing changes in the ECG without the necessity of regarding certain directional effects. However, in the heart as a whole the orientation of the equivalent dipole (and its resultant vector) undergoes changes in direction from moment to moment (particularly during ventricular depolarization). Because of changes in the direction of the propagating wavefront (the spreading wave of depolarization) and the effects of these changes in direction on the orientation of the equivalent dipole, the resultant vector is seldom aligned with any specific lead axis. This is the reason for using a system of lead axes. The manner in which the directional orientation of the equivalent dipole at any given instant affects the amplitude of the ECG is discussed below.

SIGNIFICANCE OF DIRECTIONAL ORIENTATION OF THE EQUIVALENT DIPOLE

The equivalent dipole of the heart may be thought of as existing in a three-dimensional space. However, for most purposes it is convenient to consider it as existing in two-dimensional space giving it a planar representation. A branch of electrocardiography does exist in which three-dimensional space is considered, but that is a specialized topic. A coordinate system based on the leads of the ECG system will then have different projections from the dipole on it depending on their orientation with respect to the dipole. Thus, if the lead is parallel to the dipole, it will have a maximum projection on it since all the lines of force generated between the ends of the dipole will cross it, whereas if the lead is perpendicular it will have a minimum projection since it will be crossed by few of the lines of force (Fig. 4). In practice, the equivalent dipole is not measured—rather, it is reconstructed from its representation (as voltage deflection) on the lead axes that are measured. This is considered in the next section.

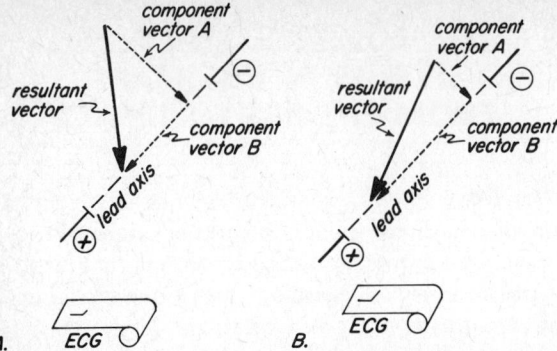

Figure 4 Effect of direction of equivalent dipole (resultant vector) on the recording from a specific lead axis. (*a*) The resultant vector of the equivalent dipole here has a small effect on the ECG record because of a small component vector B. (*b*) The equivalent dipole here is oriented more along the lead axis. Therefore, this dipole has a greater effect, because the component vector along the lead axis (component vector B) is larger. Note the greater excursion of the recording for (*b*).

EINTHOVEN'S HYPOTHESIS

The magnitude and direction of the resultant vector (of the equivalent dipole) cannot be determined from a recording from a single lead axis. Recall that the amplitude of the ECG recording depends on (1) the strength of the equivalent dipole, and (2) the directional orientation relative to a lead axis. For example, a given amplitude recorded from one lead axis might be due to a large resultant vector that is diagonal to the lead axis, or might be due to a smaller resultant vector that is more nearly aligned along the lead axis. In order to determine the actual magnitude and direction of the resultant vector, more than one lead axis recording is needed. Einthoven devised a scheme for geometrically determining the magnitude and direction of the resultant vector from three lead axes (each of which forms the side of an equilateral triangle).

The core of contemporary electrocardiographic theory is based on the hypothesis of Einthoven's triangle. This hypothesis includes the following assumptions: (1) the body acts as a volume conductor; (2) the electromotive forces created by charge differences in the heart at any given instant can be represented by a single equivalent dipole, or its resultant vector, located at the center of the triangle; and (3) the roots of the two arms and the left leg

form the three apices. The sides of the triangle are the axes of the three standard limb leads (leads I, II, and III) of the ECG. The triangle is a two-dimensional figure lying in the frontal plane of the body.

THE THREE STANDARD LIMB LEADS

Einthoven's triangle is an inverted triangle (Fig. 5), the base of which (now at the top of the inverted triangle) consists of a lead axis between a negative electrode at the right shoulder and a positive electrode at the left shoulder (lead I). A second lead axis makes up the right side (the

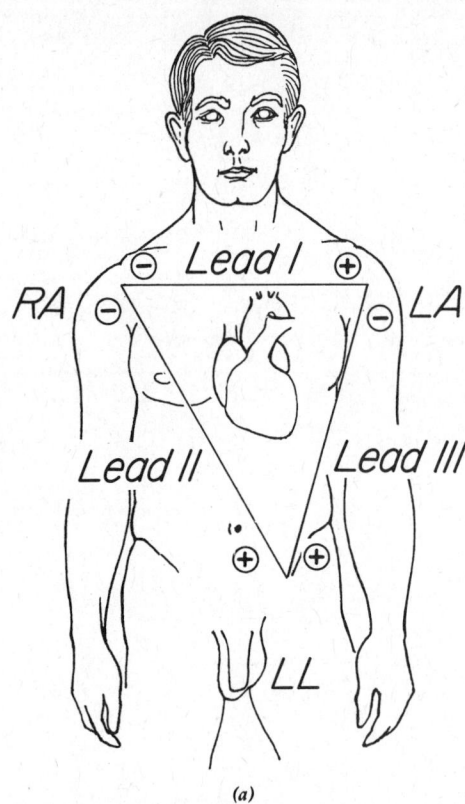

Figure 5 (*a*) Einthoven's triangle. Conventional arrangement of electrode polarities for recording the standard limb leads. Although the roots of the limbs are the apices of the triangle, electrodes are usually placed on wrists and ankle for convenience (essentially equivalent, since the limbs are volume conductors).

patient's right) of the triangle. This second lead axis consists of a negative electrode at the right shoulder and a positive electrode where the left thigh joins the torso, and is called lead II. The third lead, making up the remaining side of the triangle (the patient's left side), consists of a lead axis between a positive electrode at the base of the left thigh and a negative electrode at the left shoulder. This is lead III. These polarities for the three limb leads were chosen to give an upward (positive) deflection of the pen on the ECG record for the R wave in patients with an average, normal position of the heart and typical orientation of the R wave equivalent dipole vector for ventricular depolarization. These are called limb leads because the electrodes can just as easily be placed peripherally on the limbs (right wrist, left wrist, and left ankle). Essentially the same record is obtained, with the limbs, like the rest of the body, being volume conductors.

COMPONENTS OF THE ECG

ATRIAL COMPONENTS

The P wave represents depolarization of the atria (Fig. 6). Atrial depolarization begins at the SA node and spreads across the atria like a wave spreading outward from a splash in calm water. The wave of depolarization moves downward and to the left, since its site of origin, the SA node, is in the upper right portion of the atria. This wave spreads through the predominant cells of the atria, the contractile myocardial fibers, and reaches the most distant portions of the atria in about 0.08 sec. (80 msec.). Although part of the depolarization may be carried to the AV node along specialized conductile tracts, as shown in Figure 7, the total mass of these specialized fibers is small.

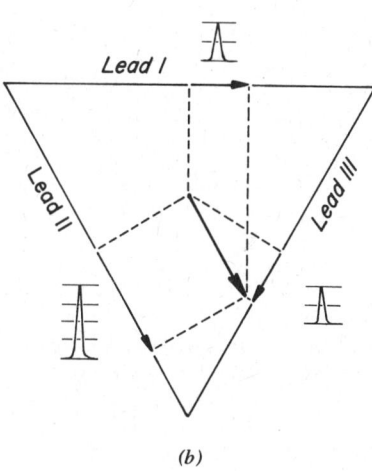

(b)

Figure 5 (con't) (b) Einthoven's triangle. If the recorded magnitude of a pen deflection (in mm) is measured for a given ECG wave from lead I and for the same wave from lead II or III, the resultant vector of the equivalent dipole can be determined. Count off from the midpoint of each lead the number of mm (magnitude of component vectors) and drop perpendiculars. A line from the center of the triangle to the point of intersection of the perpendiculars gives the resultant vector, also called an electrical axis.

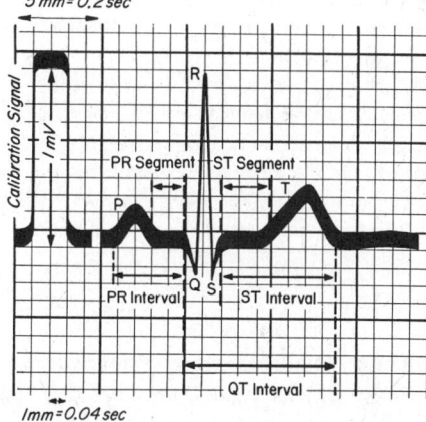

Figure 6 ECG conventions. By convention the sensitivity of the electrocardiograph is set to give a 1 cm deflection for a standard calibration signal of 1 mV. The paper speed is set at a standard rate of 25 mm/sec. The portions of the ECG in between waves are called *segments,* and certain distances from one wave to another are called *intervals.* The conventions for defining the waves of ventricular depolarizations are as follows: The first downward wave preceding an upward deflection is called a Q wave. The upward deflections are called R waves; if there is more than one, the second is designated R_1. The downward deflections following the R waves are called S waves; if there is more than one, then the second is designated S_1. In some instances there may be no upward deflection. In this case the large downward wave is referred to as a QRS complex.

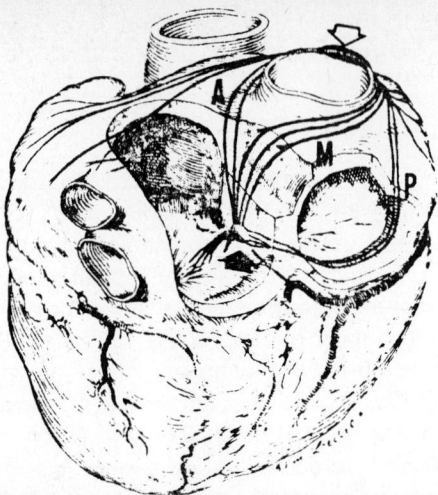

Figure 7 Location of three possible specialized internodal pathways in human heart. The open arrow indicates the position of the SA node; the closed arrow, the internal position of the AV node. A, the anterior internodal pathway (downward to AV node) as well as an interatrial division running to the left atrium (Bachmann's bundle). M, the middle internodal pathway, and P is the posterior internodal pathway. Although it is possible, and perhaps even likely, that these internodal pathways contribute in part to the conduction of impulses from the SA node to the AV node, this has not yet been established in normal humans. On the other hand, there is reasonable support for role in conducting impulses from the right atrium to the left atrium through Bachmann's bundle. (From James TN: in Hurst JW (ed): *The Heart.* New York, McGraw-Hill, 1978.)

Therefore, the resultant vector of atrial depolarization is determined primarily by the wave of excitation spreading through the typical contractile myocardial fibers. The resultant vector at its maximum has a magnitude of only 0.2 mV and points downward and to the left (Fig. 8a). This produces a small upward deflection, the P wave (Fig. 6) in the standard limb leads. The P wave is normally upright in leads I and II (generally tallest in lead II), but can be small (nearly flat) or even inverted in lead III. Since the P wave represents atrial depolarization, which requires approximately 0.08 seconds to spread throughout the atria, the duration of the P wave is also approximately 0.08 sec. (80 msec.). The first atrial fibers to have undergone depolarization are also the first to become repolarized. In other words, repolarization also appears to

spread outward from the region of the SA node. Actually, repolarization may not be propagated given the slow rate at which it occurs and may, in fact, take place independently in each cell after a certain duration of depolarization. However, because of the electrical coupling between cells it is possible to initiate repolarization in one cell and shorten the duration of the cellular action potential in an adjoining cell. The sequence of recovery results in an equivalent dipole the polarity of which is the opposite of that produced by depolarization. For the atria this is roughly upward toward the right clavicle and would be expected to produce a small downward deflection in the ECG. However, this is seldom seen, since the greater magnitude of ventricular depolarization occurring simultaneously completely obscures the much smaller effect of atrial repolarization.

About half way through the P wave (the average electrical deflection of all the atrial cells), the impulse propagating from the atria arrives at the AV node via the atrial contractile fibers for the most part and perhaps partly through the specialized internodal fibers. In the AV node the velocity of propagation is greatly slowed and remains slow as the impulse traverses the node and the cells around it. Once this barrier has been traversed, the impulse then rapidly passes through the bundle of His and its branches. The electrical changes produced by these depolarizations are very small and as a consequence do not produce a deflection on the conventional ECG leads. However, these waves of depolarization passing through the AV nodal area and Purkinje system can be recorded if electrodes, contained in a catheter, are passed into the right ventricle. The record is called a His bundle ECG.

VENTRICULAR COMPONENTS

Before correlating the sequential events of ventricular depolarization and repolarization with the lead vectors, it is useful to review the normal anatomical position of the heart, particularly that of the ventricles and the interventricular septum. The right ventricle is anterior to the left ventricle as shown in Figure 9. The apex of the heart is anterior and shifted toward the left. The degree to which the heart is shifted to the left varies in the population according to variations in general body build. For example, a short, fat person might have a heart lying almost

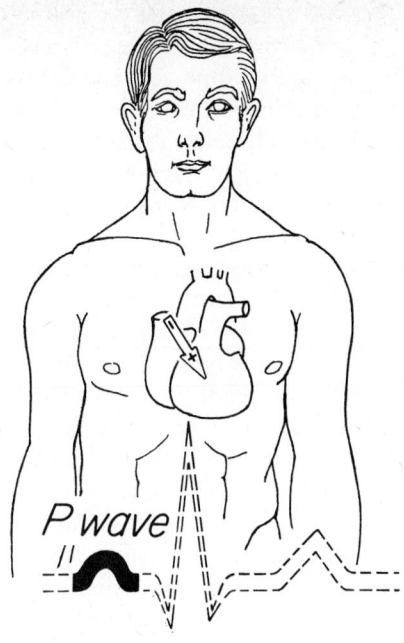

P wave

(a)

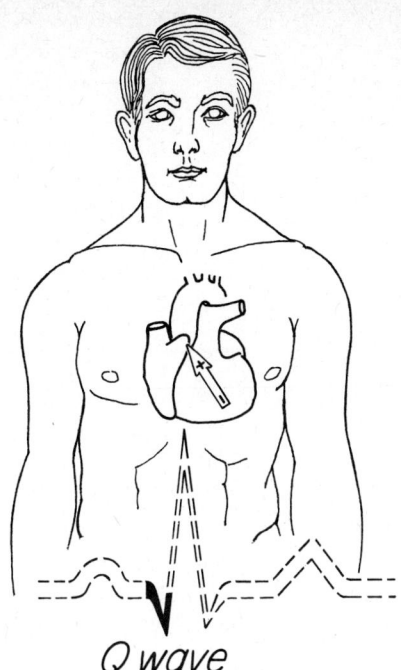

Q wave

(b)

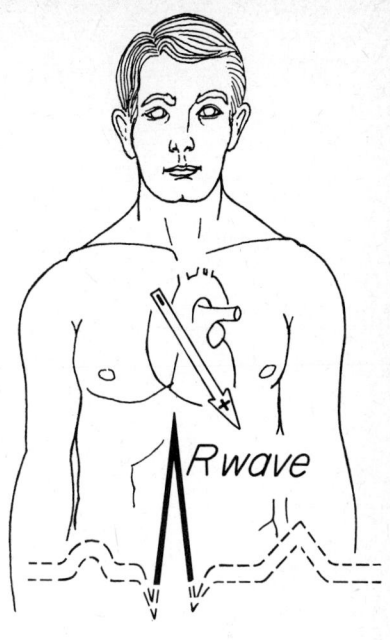

R wave

(c)

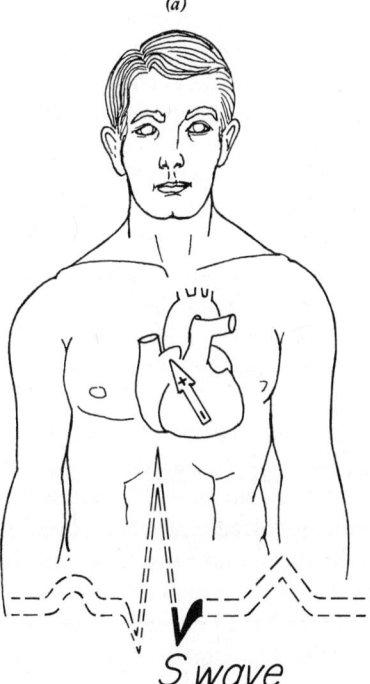

S wave

(d)

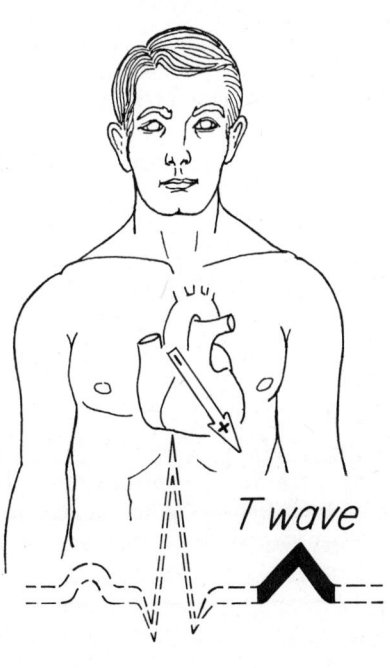

T wave

(e)

Figure 8
(a) Atrial depolarization.
(b) First stage of ventricular depolarization.
(c) Ventricular depolarization.
(d) Final stage of ventricular depolarization.
(e) Ventricular repolarization.

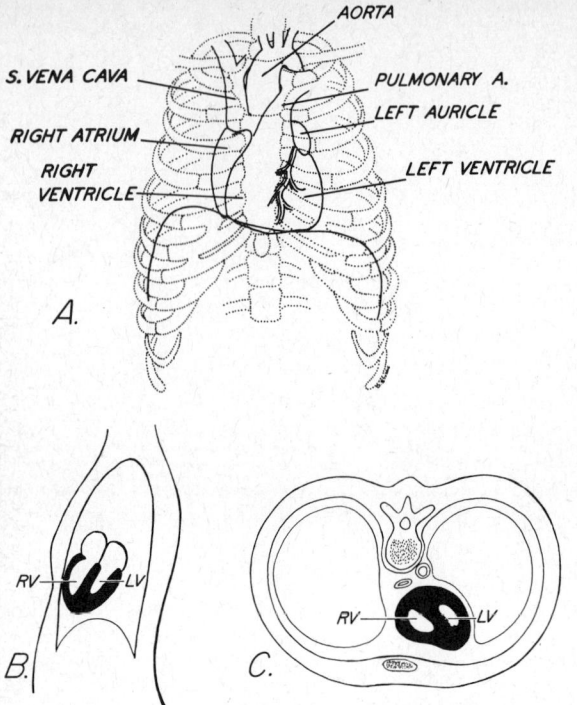

Figure 9 Anatomical position of the heart. (*a*) Position of the heart in the rib cage as seen from the front (looking toward the subject). (*b*) Lateral view of the heart as seen from the left side; shows the backward tilt of the interventricular septum. (*c*) Cross-sectional view of the heart as seen from above looking downward, showing the diagonal position of the interventricular septum.

completely on its side horizontally to the left, whereas a tall, lean person might have a nearly vertical heart.

Let us consider the case of a person of normal height and weight. The apex of the heart is angled to the left 40° from vertical. Excitation of the ventricles results from propagation of depolarization down the bundle of His, through its right and left bundle branches, and into the ventricular myocardium via terminal twigs of Purkinje fibers, which merge with working myocardial fibers. Since the initial bundle branches course along the walls of the interventricular septum, the septum is the first ventricular area to become depolarized. Furthermore, because the left bundle branch has numerous early divisions along the left septal wall, while the right bundle branch has a more sparse distribution to the right septal wall, depolarization

of the septum begins first in a sizable portion on the left. Considering the anatomical position of the heart, this initial septal depolarization spreading from the left toward midseptum can be visualized as producing an equivalent dipole the resultant vector of which points toward a point in space forward of the right clavicle. Figure 10*a* illustrates three-dimensionally the first area of the ventricle to be depolarized (first 5 milliseconds). The resultant vector is also illustrated three-dimensionally in Figure 10*a*. Since the septal tissue undergoing depolarization is relatively small compared with the remainder of the ventricular tissue, this vector is of correspondingly small magnitude, viz. about 0.2 mV. This resultant vector produces a small downward deflection, the *Q wave* (Fig. 8*b*), usually in at least two of the three standard limb leads. (Depending on the precise orientation of the vector, the Q wave is sometimes missing in one of the three limb leads.) The further course of depolarization invading greater amounts of ventricular mass results in both an increase in magnitude and a change in orientation of the equivalent dipole. The interventricular septum is first depolarized from the left and then from the right. Next, most of the right ventricle is depolarized, and a wave of depolarization spreads through the thick free wall of the left ventricle from within outward (from endocardium to epicardium). This excitatory activity is represented by a large resultant vector with an amplitude of up to 2 mV and pointing toward a region behind and just to the left of the left hip. Figure 10*b* illustrates this period of ventricular depolarization with its resultant vector and shows the portions of the ventricles depolarized after 40 msec. This results in a large upward deflection, the *R wave*, in all three standard limb leads (Fig. 8*c*). The R wave has the greatest magnitude in lead II, since this resultant vector is almost parallel to this lead axis.

The last regions to become depolarized lie within the central portion of the basal septum and the posterior-basal portion of the left ventricular wall near the left atrial junction. The mass of tissue involved in the last phase of depolarization of the ventricles is small, and the resultant vector is therefore of small magnitude: about 0.4 mV. This vector points to a region behind the head and slightly to the right (Fig. 10*c*). Accordingly, a small downward deflection, the *S wave*, may be recorded in all three standard limb leads. However, as was the case with the Q wave, this wave may be greatly diminished or absent in one of the

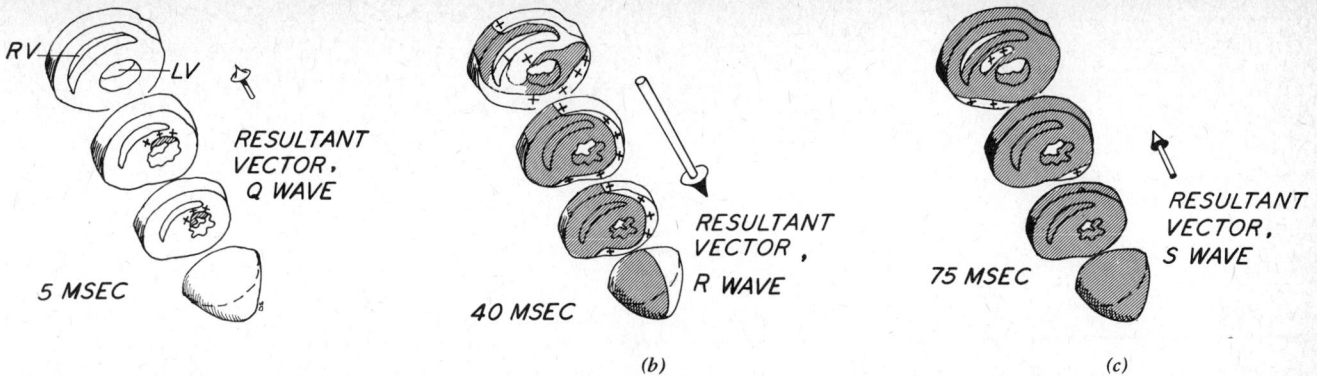

Figure 10 The spread of depolarization through the ventricles three-dimensionally (as extrapolations from dog heart experiments). A large number of extracellular recording electrodes can be implanted in dog hearts in a manner permitting recording of polarity throughout the entire thickness of the ventricular muscle (by inserting a large number of electrodes, the activity of as many as 900 different points can be recorded in a single experiment). The ventricle is shown as four separate sections arranged in the normal anatomical position of the heart (for orientation, note the thicker wall of the left ventricle). Depolarized areas are shaded, and areas yet to be depolarized are unshaded (and relatively positive to the depolarized areas).

leads, particularly lead I, depending on the precise orientation of the equivalent dipole (see Fig. 8d for two-dimensional representation). The *QRS complex,* comprised of Q, R, and S waves, represents ventricular depolarization and has a duration of about 0.08 sec. (80 msec.).

One might expect ventricular repolarization to follow the same sequence as depolarization. However, this is not the case, and the myocardial fibers depolarized last are the first to repolarize. The wave of depolarization in the left ventricle spreads from the endocardium outward to the epicardium, but the endocardial fibers have a delayed onset of repolarization. Several factors have been thought possibly responsible for this sequence of repolarization. According to one theory, the pressure differential within the ventricular wall favors early repolarization in the outer layers (lower pressure), and later repolarization near the endocardium (higher pressure).

Since flow of charge is in the opposite direction, the equivalent dipole has a resultant vector pointing in the same direction as that which previously existed during depolarization, viz. downward and to the left (Fig. 8e). Therefore the ECG wave representing ventricular repolarization is an upward deflection, the *T wave.* Repolarization spreads at a slower velocity than did depolarization (perhaps because it may not be propagated). As a result, the duration of repolarization and its T wave are pro-

longed; the T wave lasts about 0.16 sec. (160 msec., or about twice as long as the QRS complex).

PRACTICAL APPLICATIONS OF THE VECTOR MODEL

The specific direction of the vectors resulting from electrical activity in atria and ventricles (as an angle, in degrees from the horizontal) is called the *electrical axis* and can be determined at any point during the cardiac cycle in which an equivalent dipole exists (during depolarization or repolarization). This measurement is useful clinically, especially in the diagnosis of various cardiac hypertrophies. The electrical axis, for ventricular depolarization, is determined through the use of Einthoven's triangle (Fig. 5b). The height of the R wave in one of the standard limb leads is measured and marked along the lead axis as a component vector of so many mV with its direction given by the arrowhead. The same is done for an R wave in one of the other standard limb leads. Perpendicular lines are drawn from the arrowheads of each component vector until these perpendiculars intersect with each other. A line is drawn from this tip of the resultant vector to the center

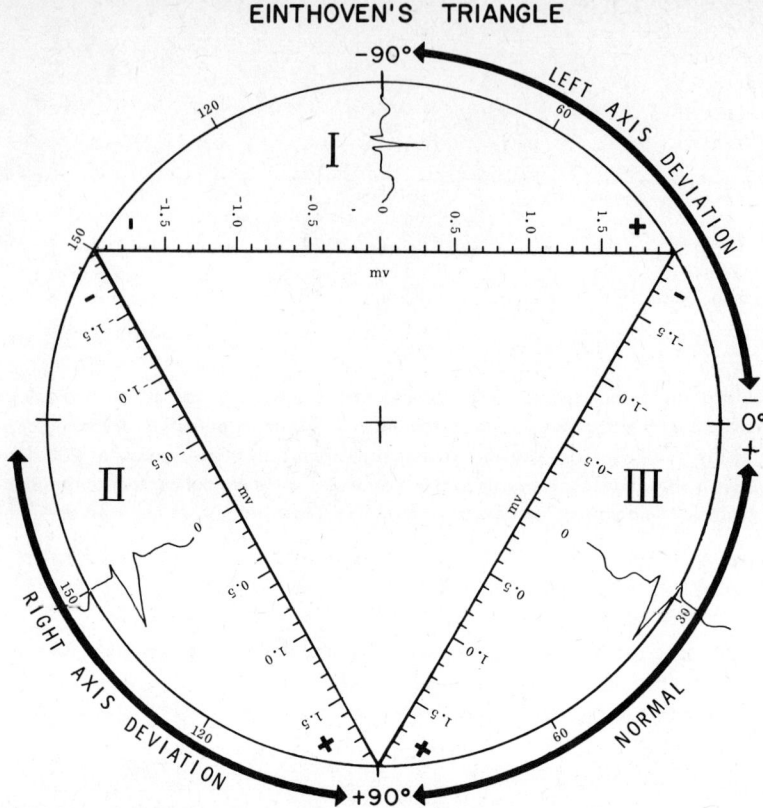

Figure 11 Electrical axis of ventricular depolarization. The specific direction of the mean vector for the general orientation of ventricular depolarization is noted in degrees according to its position on an imaginary circle superimposed over the patient's chest (frontal plane). Traditionally, a normal electrical axis is described as being between 0° and +90°. Deviations from normal are described as right axis deviation if the mean vector is from +90° to +180°, and left axis deviation if from 0° to −90°.

of the triangle to obtain the length or magnitude of the vector and its direction. The angle that the resultant vector makes with the horizontal can then be measured (Figure 11). The Lead I axis has an angle of zero degrees at its positive pole (Figure 5*b*) and -180° at its negative pole. A line drawn perpendicular to Lead I & intersecting it at zero amplitude has +90° above the horizontal and −90° below it.

The angle of the electrical axis for the R wave is considered normal if it is between 0° and +90°. A *left axis deviation* (LAD) would be from 0° to −90°. *Right axis deviation* (RAD) is from +90° to +180°.

Another aspect of the use of Einthoven's triangle is a relationship between the magnitudes of the three compo-

nent vectors of the three standard limb leads. This relationship, referred to as *Einthoven's law,* states that height of a wave in lead I plus that of the corresponding wave in lead III sum to give the height of that wave in lead III (i.e., I + III = II). As you will recall this is simply a restatement of the physical law that the sum of the voltages around a loop equals zero.

OTHER ECG LEAD SYSTEMS

In obtaining a standard ECG from a patient, the three standard limb leads (which are bipolar electrodes) are recorded, in addition to nine unipolar leads: three *augmented unipolar leads,* and six *unipolar chest leads.*

The unipolar chest lead system uses a probing electrode placed at various positions on the chest wall. The potentials recorded by this electrode are compared to zero potential obtained by connecting the three standard limb leads together (according to Kirchhoff's Voltage Law the sum of the potentials around a circuit equals zero, that is, $I + II + III = 0$). The exploring electrode is placed at six positions on the chest wall including its lateral surface, and some idea of the cardiac electrical vector in three-dimensional space can be obtained from this system.

Another lead system that has been devised in the augmented unipolar lead system. Originally, the potential difference between a standard limb lead was compared with zero potential, obtained by adding the three standard limb leads together. However, the deflections from this system were smaller than those obtained from the standard system; in order to increase their amplitude, the reference potential was changed from 0 V to a negative value. This was accomplished by disconnecting the input from the standard limb lead of interest from the zero volts reference lead. The reference potential is now less than zero volts. The deflection from a standard limb lead is now increased, hence the term augmented unipolar leads.

ARRHYTHMIAS

The sinoatrial node is the normal pacemaker of the heart because is has the greatest rate of rhythmicity. The SA node usually suppresses rhythmicity in other potential pacemakers (AV node, bundle of His, and terminal Purkinje fibers). However, if the SA node is depressed or if its conduction is blocked, one of these subsidiary pacemakers takes over and produces ectopic or escape rhythms. At times an irritable ectopic pacemaker may introduce individual ectopic beats, or even ectopic rhythms, despite normal SA nodal function.

Since the SA node is the normal pacemaker, the normal rhythm of the heart is called a sinus rhythm. If more rapid SA nodal activity produces a heart rate greater than 100

beats per minute, a condition of sinus tachycardia is said to exist. On the other hand, slower SA nodal activity producing a heart rate of less than 50 or 60 beats per minute constitutes sinus bradycardia. Both sinus tachycardia and sinus bradycardia have fairly normal-appearing ECGs, except that the rates are different and that the segment between the end of the T wave and the beginning of the next P wave is decreased in tachycardia and prolonged in bradycardia. A sinus bradycardia is not unusual in athletes.

Rhythmic changes in the heart rate associated with the respiratory cycle are commonly observed in young persons. This phenomenon of sinus arrhythmia is characterized by an increased rate during inspiration and decreased rate during expiration (Fig. 12). These changes are produced by changes in vagal activity associated with variations in activity of cardiac centers in the brain stem medulla.

ECTOPIC FOCI

Premature Beats

Premature beats (extrasystoles) are extraneous contractions produced by ectopic foci. The abnormal emission of

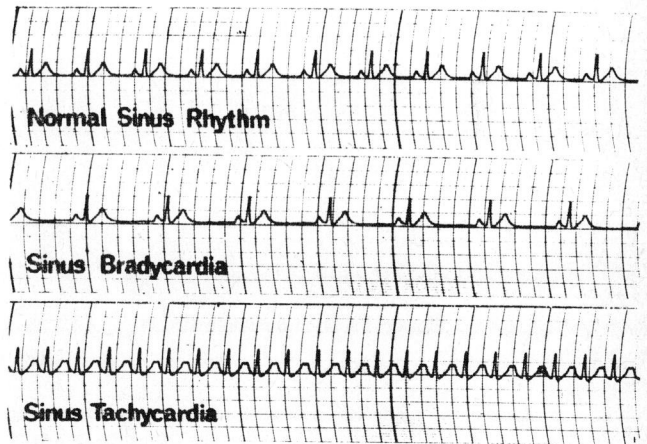

Figure 12 Changes in rate due to changes in sinoatrial nodal discharge rate.

an excitatory impulse from an ectopic pacemaker interposes an extra, early, contraction before the time at which the next normal contraction should occur. Premature beats may originate in either the atria or the ventricles, depending on the specific location of the ectopic focus (in the atria, AV node,* or in the ventricles). An ectopic focus in the AV node or in the bundle of His produces a premature atrial impulse followed by a ventricular impulse that has an essentially normal-appearing QRS complex (Fig. 13). Most ectopic foci are located below the bifurcation of the common bundle and produce premature ventricular impulses or contractions (called PVC's) whose QRS complexes are quite abnormal in appearance. Following a premature ventricular contraction, the next normal contraction does not occur at the expected moment. If the myocardium is in a refractory state (during the extrasystole) when the normal excitatory impulse from the S-A node has reached it, it is not excited and does not contract. This results in a *compensatory pause* before the next sinoatrial impulse reaches the myocardium in a nonrefractory state. The pause is usually not fully compensatory in the case of premature atrial impulses because the S-A node is invaded and reset to discharge earlier by the premature impulse. Although PVCs are occasionally the result of serious cardiac pathology, they are often found in apparently healthy persons and may disappear with decreased consumption of coffee or tobacco.

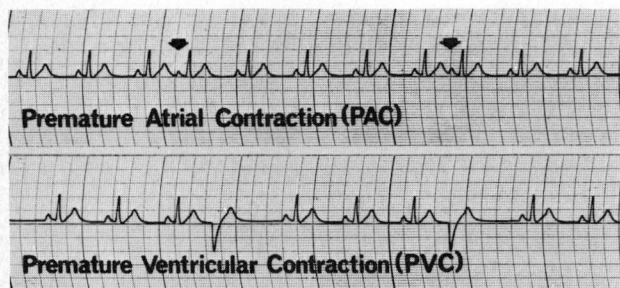

Premature Atrial Contraction (PAC)

Premature Ventricular Contraction (PVC)

Figure 13 Premature beats.

*For convenience, the conduction pathway or entry point between the atria and ventricle is referred to as the AV node. In reality this entry point includes, in addition to the anatomical AV node, the cells that surround it and should more precisely be referred to as the AV junctional tissue.

Atrial Tachycardia, Flutter, and Fibrillation

Much discussion has centered on the underlying mechanisms of atrial tachycardia, flutter, and fibrillation. Most authorities agree that *atrial tachycardia* is attributable to the rapid, repetitive firing of a single ectopic focus, that is, the equivalent of a run of consecutive atrial extrasystoles (100–200/min). There is less agreement as to how ectopic foci produce the more rapid beating of *atrial flutter* (250–350/min). Not all the atrial impulses cross the AV node; there is usually a 2–1 or 4–1 conduction ratio, since the AV node acts as a low pass filter and does not transmit high rates of depolarization. In other words, only one-half or one-fourth of the impulses reaching the AV node are transmitted into the ventricular conductile system. These supraventricular tachycardias are an imposition on the normal control mechanisms for heart rate; that is, the individual may not be able to increase heart rate on demand. This results in severe limitations to physical activity. The excessive rate that can result from this increased firing can lead to inadequate ventricular filling and a diminished cardiac output. The ECG pattern of an atrial flutter with a 4-1 conduction ratio is a "sawtooth" pattern of about four P waves preceding each QRS complex.

In *atrial fibrillation* there are numerous impulses spreading in all directions through the atria. The AV node receives repetitive atrial impulses with great irregularity, which, in turn, results in a rapid and irregular ventricular beating (Fig. 14). Not only is atrial pumping compromised, but the rapid irregular ventricular rate decreases cardiac efficiency. Atrial fibrillation is one of the most

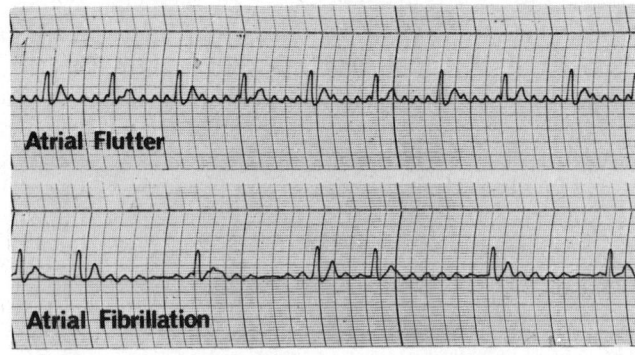

Atrial Flutter

Atrial Fibrillation

Figure 14 Atrial flutter and fibrillation.

common atrial arrhythmias and draws attention to possible heart disease (e.g., ischemic heart disease, thyrotoxic heart).

Ventricular Tachycardia, Flutter, and Fibrillation

Several ectopic beats of the ventricles occurring at a rate equivalent to more than 100 beats per minute constitute a run, or paroxysm, of *ventricular tachycardia*. A single ectopic PVC reaching the ventricles in a vulnerable state between refractory periods can initiate a repetitive response in the form of ventricular tachycardia, flutter, or fibrillation. Ventricular tachycardia is most often found in diseased hearts and only adds to the heart's difficulties, for example, it is often associated with a fall in blood pressure. Unlike supraventricular tachycardia (sinus or atrial tachycardia), ventricular tachycardia does not respond (by slowing) to vagal stimulation (by pressing with the thumbs over the carotid sinuses). A variant of ventricular tachycardia is *ventricular flutter*. Ventricular flutter is characterized by an unusually rapid rate and an ECG pattern of regular "zigzag" waves. It is often a transitional state predisposing to development of *ventricular fibrillation*. Ventricular fibrillation, which causes uncoordinated writhing and twitching of the ventricular myocardium, is almost invariably fatal (in the absence of life support capabilities). It is marked by low amplitude, irregular electrical activity (Fig. 15). The irregular electrical activity is accompanied by irregular mechanical activity that precludes any effective pumping action of the heart. The patient in ventricular fibrillation becomes pulseless, cyanotic, and has dilating pupils. Ventricular fibrillation can be caused by heart disease (e.g., myocardial infarct), anoxia, acidosis, certain anesthetics, or electrical shock.

ATRIOVENTRICULAR JUNCTIONAL CONDUCTION DEFECTS

The 2–1 or 4–1 conduction ratios between atrial and ventricular depolarization seen in atrial flutter are the result of a physiological block due to the refractoriness of the AV junctional tissue to high rates of impulses. Certain pathological conditions (e.g., myocarditis, scar compression, coronary ischemia) can cause hindrance or blocking of even normal rates of impulses arriving at the AV node, a condition described as heart block. This has classically been divided into three categories: first, second, and third degree.

In first-degree heart block there is one P wave for each QRS complex; however, the P–R interval is prolonged (greater than 0.20 sec). In second-degree heart block there is still some association between atrial and ventricular depolarization, but not every P wave is associated with a QRS complex. In third-degree heart block there is complete disassociation between atrial and ventricular depolarizations, also known as complete heart block (Fig. 16).

Conduction defects are also found in the ventricular

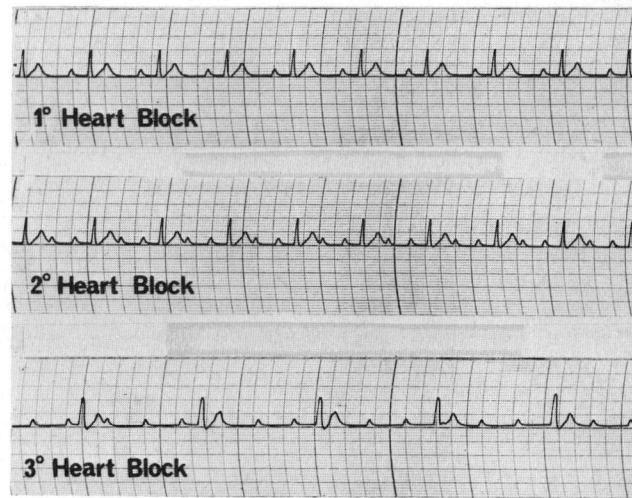

Figure 16 Heart block.

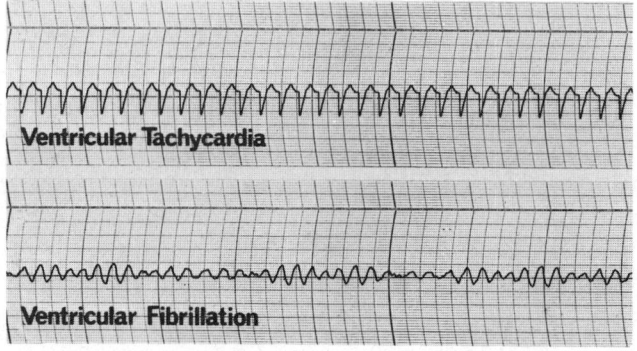

Figure 15 Ventricular tachycardia and fibrillation.

conductile system (right bundle branch block, left bundle branch block, and peri-infarction, or parietal block); however, these conditions and their underlying pathological causes will not be included here.

SELECTED BIBLIOGRAPHY

Abel FL, McCutcheon EP: *Cardiovascular Function: Principles and Functions.* Boston, Little, Brown, 1979.

Beckman VR: *Grant's Clinical Electrocardiography,* ed 2. New York, McGraw-Hill, 1970.

Dubin, D: *Rapid Interpretation of ECG's,* ed 3. Tampa, Cover, 1974.

Goldman MV: *Principles of Clinical Electrocardiography,* ed 10. Los Altos CA, Lange Medical Publications, 1979.

Lamb LE: *Electrocardiography and Vectorcardiography.* Philadelphia, Saunders, 1965.

Lipman BS, Massie E, Kleiger RE: *Clinical Scalar Electrocardiography,* ed. Chicago, Yearbook, 1972.

SELF-STUDY QUESTIONS

1. When a vector points toward the positive pole of a galvanometer an upward deflection will be seen. True or false?

2. The electrocardiogram records electrical changes in the heart as a whole from the point of view of extracellular electrodes. What property of the body allows these changes to be recorded at some distance from the heart?

3. A dipole has two major characteristics: strength and directional orientation. What is a dipole?

4. The ECG is a recording of transient extracellular potential differences that exist during the processes of depolarization and repolarization. In terms of extracellular recording, what potential differences exist across the membranes of resting, depolarized, and repolarized cells?

5. What differences exist in the depolarization–repolarization sequence of atrial and ventricular myocardial strips?

6. Which will produce the greatest deflection on a galvanometer: a vector parallel with the lead axis or one which is perpendicular to it?

7. Diagram the three standard limb leads of the ECG (Einthoven's triangle). Label the conventional arrangement of electrode polarities for recording each limb lead.

8. What is the sequence of depolarization of the heart?

9. Label the following diagram (limb lead I tracing). Discuss which stage of depolarization–repolarization each component represents, as well as the time interval of each component (assume a normal, resting individual with a heart rate of 72 beats per minute). Which wave represents atrial repolarization in the limb lead I tracing?

10. Which anatomical structures in the heart can function as ectopic foci?

11. Fill in the following table:

Condition	Heart Rate	Pacemaker	ECG Changes
Sinus rhythm			
Sinus tachycardia			
Sinus bradycardia			

12. Discuss sinus arrhythmia, its causes, and its mechanism of occurrence.

ANSWERS

1. True.

2. The body acts as a volume conductor; that is, electrical potential differences have measurable effects due to the electrolytic composition of the body fluids.

3. A dipole consists of a positive and a negative charge situated a short distance apart; this potential difference in an electric field is sensed by the electrodes of the ECG.

4. Resting or polarized cells have an exterior positivity relative to the interior, whereas depolarized cells have an exterior negativity relative to the inside. Repolarized cells show the same potential differences as resting cells.

5. In atrial repolarization the first area to have undergone depolarization is also the first to recover, that is, the first to begin repolarization. However, in ventricular myocardium the last area to depolarize is the first to repolarize. These facts lead to differing ECG records for atrial and ventricular myocardium: an atrial ECG tracing consists of one upward and one downward deflection, whereas the ventricular tracing consists of two upward deflections.

6. The vector that produces the greatest deflection on the ECG is that which is parallel. The perpendicular vector will produce no deflection on the ECG.

7. The diagram should be similar to that in Figure 6.

8. The wave of depolarization initiates in the sinoatrial node, excites the AV node, and is carried via the bundle of His to the terminal Purkinje fibers.

9. A. P wave   80 ms duration   Represents atrial depolarization

  B. P–R segment   80 ms duration

  C. QRS complex   80 ms duration   Represents ventricular depolarization

  D. S–T segment   120 ms duration

  E. T wave   160 ms duration   Represents ventricular repolarization

Atrial repolarization is not seen in the limb lead I tracing.

10. AV node, bundle of His, and terminal Purkinje fibers can function as ectopic foci.

11.

Condition	Heart Rate	Pacemaker	ECG Changes
Sinus rhythm	50–100	SA node	None Essentially
Sinus tachycardia	100	SA node	normal, except for decreased segment between T and P waves and ST segment
Sinus bradycardia	50	SA node	Essentially normal, except for increased segment between T and P waves and ST segment

12. The cardiac output varies rhythmically with the inspiratory–expiratory respiratory cycle. This, in turn, produces rhythmic variations in the systemic arterial blood pressure sensed by arterial baroreceptors, such as the carotid sinus, which monitor the arterial blood pressure. Changes in baroreceptor reflexive discharge in response to these changes in blood pressure produce changes in the autonomic influence on heart rate, as reflected by a waxing and waning of the rate with respiration.

8

The Nature
of the Cardiac Pump

DANIEL L. TRABER

OBJECTIVES

After completing this chapter, you should be able to

1. Define the functions of the cardiac valves and associated structures (e.g., chordae tendineae, papillary muscle).

2. Identify the various waves of the atrial pulse and the events that produce them.

3. Recognize the points on the ventricular pressure trace where diastole begins and ends.

4. Demonstrate the importance of the ventricular end diastolic pressure as a determinant of venous return and force of ventricular contraction.

5. Recognize the aortic pressure changes, as well as identify systolic and diastolic pressures and the incisura.

6. Identify the temporal relationship of the four heart sounds with the events of the cardiac cycle, as well as delineate the causes for these sounds.

7. Delineate the various segments of the ventricular volume curve and define the causes for these volume changes.

8. Identify the various events of the cardiac cycle on a pressure volume curve of the left ventricle.

9. Correlate the temporal relationships between the variables (aortic, ventricular, atrial pressure, ventricular volume, ECG, and heart sounds).

10. Define the differences between the cycle for the right heart and left heart.

11. Delineate the changes that would occur in the events of the cardiac cycle as a result of valvular defects.

12. Identify the differences between the components of right and left ventricular contraction.

13. Recognize the ability of each ventricle to pump against extreme increases in volume and pressure loads.

The goals of this chapter are to correlate the electrical and mechanical events of the cardiac cycle in order to make judgments about "pump" action through the use of noninvasive techniques.

VALVE ACTION AND OVERVIEW

One of the most significant features in the evolution of the cardiovascular system of tetrapod vertebrates is that blood pumped to the lungs by the heart returns to the heart to be pumped a second time, thereby receiving an additional impetus before distribution to the tissues. Throughout the phylogeny of higher vertebrates there has been a continuous improvement in the separation of blood pumped by the heart through the pulmonary circuit from that pumped through the systemic circuit. In warm-blooded animals the single four-chambered heart is actually two pumps in series but placed side by side within the same organ—a right pump consisting of the right atrium and right ventricle, and a left pump comprised of the left atrium and ventricle. Both sides share the same pacemaker and thus pump more or less simultaneously. Aside from certain differences in pressure, the sequence of events in the right side is essentially the same as that in the left.

That there is a unidirectional sequence of pumping from the right heart through the lungs to the left heart and through the various tissues before returning to the right heart was first clearly expressed by Sir William Harvey in his famous seventeenth century book, *De Motu Cordis*. The basis for the unidirectional movement of blood through the heart chambers lies in the action of the atrioventricular valves (tricuspid and mitral) and that of the semilunar valves (pulmonary and aortic). When the hydraulic pressure on the upstream side of the valve leaflets is higher than that on the downstream side, these valves are open; on the other hand, when the pressure differential is reversed, the valves are forced shut. The blood returning to the right heart through the inferior and superior venae cavae empties into the right atrium. In between ventricular beats the pressure in the ventricles is less than that in the atria and, as a result, the blood entering the atria flows freely into the ventricular chambers (Figure 1e). When the atria contract, an additional quantity of blood is forced into the ventricles (Fig. 1a). However, because a one-way valve does not exist between the right atrium and the vena cavae, some of the blood moved by atrial contraction flows

in a retrograde direction into the vena cavae (giving rise to an observable jugular "pulse").

Following atrial systole, the ventricles contract. As the atria relax and the ventricles begin their contraction, the pressure differential across the AV valves is reversed and they are forced shut (Fig. 1b). However, because of the much higher pressure on the arterial side of the semilunar valves, more time is required for a reversal of the pressure differential across these valves. For a brief period at the beginning of ventricular systole the ventricular chambers are closed off at both the inlets and outlets. However, as ventricular pressure continues to rise, the semilunar valves open, and the ventricles discharge their contents into the arterial outflow (Fig. 1c). As the ventricles begin to relax (past the peak of contraction) the ventricular pressure once again drops below that of the arterial outflow, and the semilunar valves snap shut again (Fig. 1d). With continued relaxation, the ventricular pressure falls until it is again below that of the atria. The AV valves open again, blood flows into the ventricles, and a new cycle is begun (Fig. 1e).

From this thumbnail sketch of the pumping action of the heart, it should become apparent that a dysfunctioning valve or a leaking chamber can seriously impair myocardial function. Such pathology is not uncommon. Its early diagnosis is essential to ensure that pharmacological or

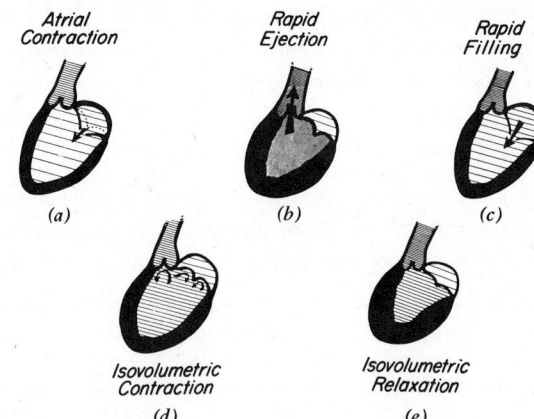

Figure 1 Model of a single side of the heart illustrating the valvular action that produces unidirectional flow. The distance between cross-hatched points indicates relative changes in pressure. When the lines are close together, the pressures are high. When they are apart, the pressures are low.

surgical intervention will be successful in restoring normal action before myocardial function progresses into a state of irreversible failure.

The position of the heart behind the protective armor of the thoracic cage helps guard against physical damage in accidents and other traumatic situations, but renders clinical evaluation of myocardial function more difficult. The average medical practitioner prefers to assess the cardiovascular status of a patient through the use of noninvasive techniques, such as the electrocardiogram (ECG) and auscultation. Before these techniques can be used intelligently, an understanding of their relationship to the different events of the cardiac cycle must be accomplished. In some cases the final diagnosis requires invasive procedures, such as cardiac catheterization, requiring that the pressures in the various atrial and ventricular chambers be measured. Therefore, the various pressure changes, electrical events, volume alterations, and sounds produced when the heart is pumping are of great importance to the student of medical physiology, for they provide the tools with which to diagnose and evaluate the state of the function of this vital organ.

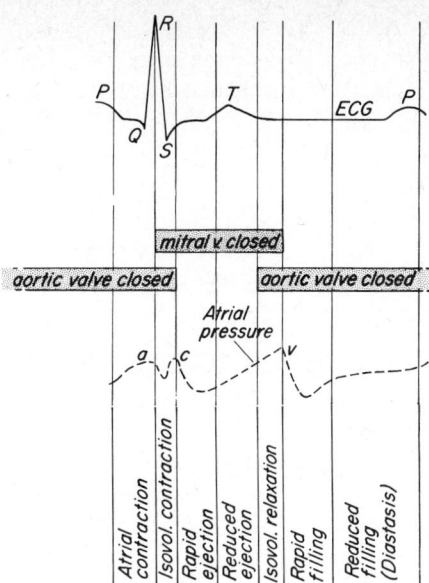

Figure 2 Atrial pressure trace. Shaded bars indicate period during which valves are closed. Note that during isovolumetric periods both the inflow and outflow valves for ventricle are closed.

LEFT ATRIAL PRESSURE

In a discussion of the events of the cardiac cycle it is traditional to consider the left side of the heart first, as the student is usually more familiar with the systemic pressures. The relationship of the pressure changes that occur in the various chambers of the left side of the heart to the ventricular volume, heart sounds, and the electrocardiogram is illustrated in Figure 6. This relationship will be presented step by step in Figures 2–5. About three-fourths of the way through the P (Fig. 2) wave of the ECG (the electrical event) the atria contract (the mechanical event). This contraction is always preceded by the P wave, since the atrium must depolarize before it can contract. The contraction of the atrium produces an increase in atrial pressure known as the a wave. As the atrium relaxes, the atrial pressure decreases, and with the onset of ventricular systole the pressure in the ventricle then becomes greater than that in the atrium. Thus the AV valve closes. At this

time another wave is usually observed. This is known as the c wave. The presence of this wave is variable; however, it is frequently seen and several explanations have been proposed. Each time the atrium contracts, it not only forces blood into the left ventricle but also into the pulmonary veins. Some cardiologists think that the c wave is produced by the return of this blood into the atrium. Another plausible explanation is that the wave was produced by the squirting of a small amount of blood from the ventricle into the atrium as the AV valve closes. The bulging of the mitral valve into the left atrial chamber might also produce this effect.

After the c wave occurs, the pressure of the atrium falls and then rises as blood continues to return from the veins, the mitral valve remaining closed. This rise is terminated at the end of ventricular systole as the initial valve opens and the accumulated left atrial blood rushes into the left ventricle. The wave that this produced in the atrial pressure curve is called the v wave. The pressure rapidly decreases as the effective volume available for atrial blood increases. Then, as the ventricle fills, ventricular pressure rises and atrial pressure increases similarly since the two chambers are connected via the open AV valves. This last pressure increase is terminated by the a wave of atrial contraction, signaling the beginning of a new cycle.

LEFT VENTRICULAR PRESSURE

The ventricular pressure follows that of the atrium quite closely during diastole when the ventricular musculature is relaxed since the AV valve is open (Fig. 3). It must be slightly lower since blood will enter from the atria (the open state of the AV valve is dependent on this slight pressure differential). The QRS complex then occurs as the ventricular fibers depolarize. About halfway through the QRS complex (the electrical event), the ventricle begins to contract (the mechanical event). The first fibers to contract are those of the papillary muscles. Their contraction pulls at the chordae tendineae, which in turn pull on the leaflets of the AV valve helping to approximate them. At about this time atrial contraction has been completed

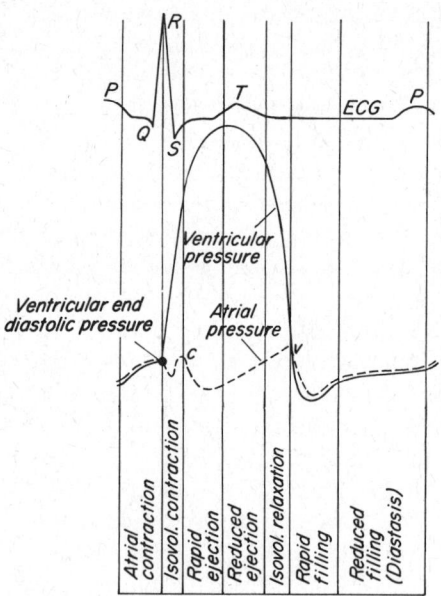

Figure 3 Ventricular pressure trace.

and as it relaxes, the atrial pressure decreases. At this time the atrial pressure is exceeded by that of the ventricle and the AV valve closes. As greater numbers of ventricular fibers depolarize, they contract, causing a steep rise in ventricular pressure. At this time both the semilunar and AV valves are closed, and the ventricle is thus contracting against a constant volume. Therefore this phase of the cycle is referred to as the period of isometric or *isovolumetric contraction*.

Although the volume of the ventricle does not change during this time, its geometrical configuration, or shape, does. In ventricular contraction the inner and outer longitudinal muscle fibers contract prior to that of the circular fibers. This pulls the base of the ventricle toward the apex and thereby shortens its length. Because the volume remains constant, the diameter of the ventricle must increase. This puts an additional stretch on the circular muscles, and they can thus contract with more force. To understand this process, review the length–tension diagram presented in Chapter 4.

During this period the ventricular pressure rapidly increases until it exceeds diastolic pressure in the aorta. The aortic valve then opens and blood rushes from the heart. As the blood is ejected, the ventricular muscle begins to shorten (isotonic contraction). The shortening muscle contracts less forcefully. This accounts for the slower rate of rise *(dp/dt)* of the ventricular pressure during this time. As the pressure reaches its maximum, the T wave occurs and the ventricle begins to repolarize and relax. As more and more fibers relax, the ventricular pressure begins to decrease until it falls below the aortic pressure; at this time the aortic valve closes. The pressure then rapidly falls until it is exceeded by the atrial pressure, and the mitral valve then opens. During this period, when both valves are closed, the volume of the ventricles is again constant. This time is referred to as the *isovolumetric relaxation* period. Following the opening of the AV valve, the pressure continues to fall. Some investigators have found the pressure in the ventricle at this time to be much more negative than the atrial pressure. This is attributed to the elastic recoil that results as the ventricle relaxes. If this is true, then the ventricle might "suck" some blood from the atrium. Once the ventricle begins to fill, the pressure within it begins to increase; this increased pressure is further augmented by atrial contraction.

The pressure recorded just before ventricular systole is known as the *end-diastolic pressure.* This is of some physiologic importance, as it determines the resting tension of the ventricular musculature and thus the subsequent force of its contraction (Chapter 9).

AORTIC PRESSURE

The aortic pressure is decreasing at the beginning of the cycle as the blood runs out of the great arteries (Fig. 4). Then the left ventricle contracts and after the isovolumetric period, the aortic valve opens, producing a rapid increase in aortic pressure. The aortic pressure seen at the time that the aortic valve opens is the lowest that pressure trace attains. This is the *diastolic blood pressure.* The maximum pressure attained during ventricular systole is the *systolic pressure.* Once the ventricle starts to relax, the amount of blood running out of the aorta into the body tissues exceeds the volume of blood entering the aorta from the ventricle. Thus the pressure decreases. This pressure decrease continues until the ventricular pressure falls below the aortic pressure and the aortic valve closes. Ventricular systolic ejection is a rather forceful event, and the blood is expelled from the ventricle with considerable momentum. This force is rapidly dissipated against the elastic fibers in the wall of the aorta (measurements indicate that the aortic diameter may increase by 30% with each stroke). As the ventricle relaxes, these elastic elements of the arteries recoil, hurling their contents against the aortic valves and slamming them shut. This produces a notch in the aortic trace called the *incisura.* It is thought to be produced by a rebound pulse reflecting from the closed aortic valve.

HEART SOUNDS

The heart sounds are produced by vibrations of the various parts of the heart and the blood they contain during the events of the cardiac cycle (Fig. 5). The most significant component of these audible sounds is considered the opening and closing of the heart valves and the accelera-

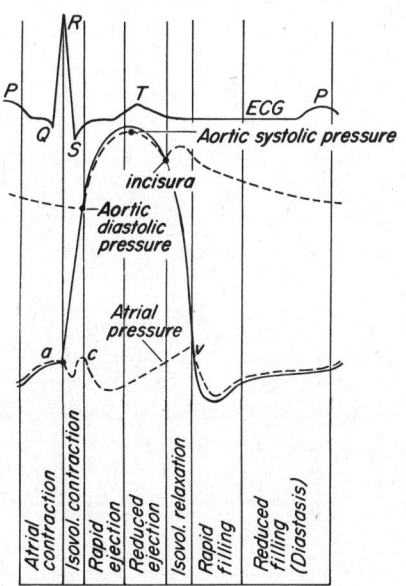

Figure 4 Aortic pressure trace.

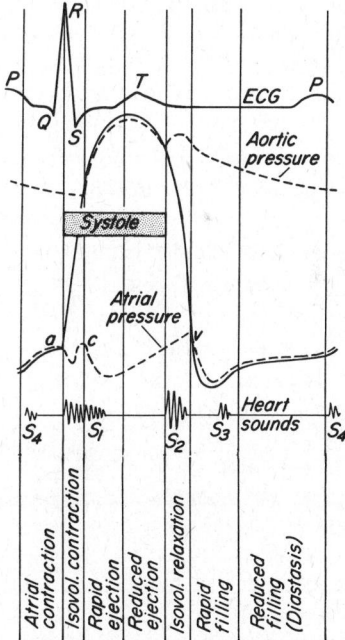

Figure 5 Heart sounds. Duration of ventricular systole noted by shaded bar.

tion of the blood. Normally, mammals have two heart sounds, which can be heard with a stethoscope, as well as two additional sounds, which can be recorded with appropriate electronic amplification.

The *first heart sound* (S₁) begins with the closure of the AV valves. The sound is produced by the closure of the AV valves and by ejection of blood with the opening of the semilunar valves. A high frequency component is associated with the acceleration of blood into the aorta and pulmonary artery (flow may be turbulent). The *second heart sound* (S₂) is a short-duration low-frequency thump that occurs almost simultaneously with the isovolumetric relaxation period. This sound is primarily produced by the abrupt closure of the semilunar valves and the sudden rebound of blood against the valves after they have shut. It is much more prominent in an amplified recording than with the stethoscope. The *third heart sound* (S₃) is prominent in children and adolescents and is believed to be caused by vibrations in the ventricular wall produced by rapid filling. It is usually seen as the atrial pressure begins to rise during the rapid filling period and is thought to be due to a vibration of the ventricular walls as they are stretched out by filling. The *fourth heart sound* (S₄) is associated with the vibrations set up in the ventricles by the acceleration of blood into them by atrial contraction.

Pathological states can be indicated within a heart and great vessels that produce abnormal sounds. These sounds are produced by turbulent blood flow and are called murmurs. Two types of valvular defects commonly produce these abnormal sounds: *stenosis* and *insufficiency*. Stenosis is a narrowing of the valvular orifice. Insufficiency exists when the valves cannot seal completely upon closure. Interventricular and interatrial septal defects can also produce murmurs. The location of these sounds in relationship to the other events of the cardiac cycle enables the physician to diagnose these maladies.

Streamline flow is silent,
Remember that, my boys,
But when the flow is turbulent
There's sure to be a noise.
So when your stethoscope picks up
A bruit, murmur, sigh,
Remember that it's turbulence
And you must figure why.

Alan Burton

VENTRICULAR VOLUME

The ventricular volume curve is used to divide the cardiac cycle into seven periods, as indicated in Figure 6: (1) atrial contraction, (2) isovolumetric contraction, (3) rapid ejection, (4) reduced ejection, (5) isovolumetric relaxation, (6) rapid filling, and (7) diastasis (reduced filling).

The duration of ventricular filling has been divided into three time periods: (1) rapid filling, (2) diastasis, and (3) atrial contraction (Fig. 6). During the normal cycle the filling period is roughly equally divided among the three (diastasis being somewhat longer than the other two, however). With increased heart rates the period of diastasis becomes disproportionally shortened. The period of rapid filling begins with the opening of the AV valves and terminates as the ventricular pressure begins to increase and level off. As the AV valve opens, blood pours into the

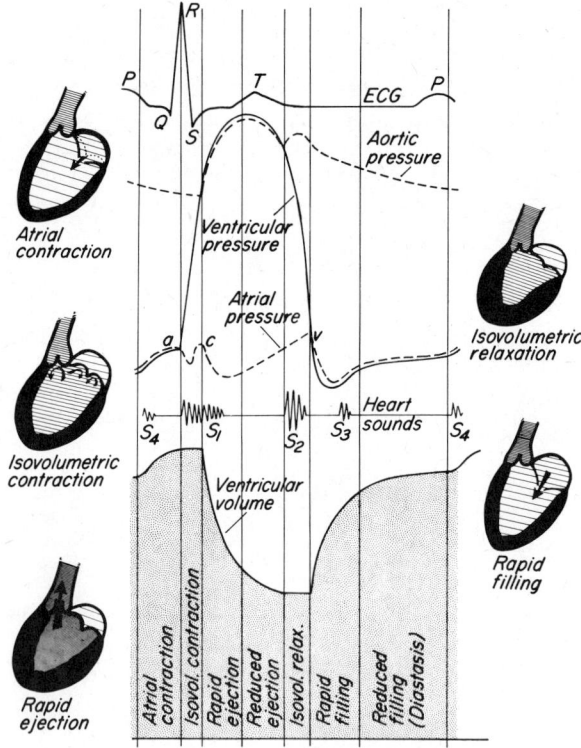

Figure 6 The cardiac cycle.

ventricle at a rapid rate. The initial valve is quite large and offers little resistance to flow. Approximately 70% of the ventricular filling is accomplished during this period. During the period of diastasis very little blood enters the ventricles. This period is terminated by atrial contraction, which contributes an additional 20–30% of the ventricular volume. This contribution is especially important during tachycardia, when the periods of diastasis and rapid filling are much reduced.

The time of ventricular systole is divided into three periods: the period of isovolumetric contraction, the period of rapid ejection, and the period of reduced ejection. These periods are labeled in Figures 2–6. This physiological definition of systole differs slightly from that of the clinician. For convenience, the cardiologist or internist usually defines ventricular systole as beginning with the first heart sound and ending with the second heart sound.

As mentioned previously, there is no change in ventricular volume during the period of isovolumetric relaxation. Most of the blood is ejected from the ventricles during the period of rapid ejection. The ventricle has been said to deliver its volume like a hammer hitting a piston rather than from a slow squeeze. This period is terminated just after the beginning of the T wave, which begins the period of reduced ejection. As one can observe from the figure, very little outflow occurs during this time. The ending of this period is signaled by the end of the T wave. During this time blood flow may be reversed and some blood may enter the ventricles as the semilunar valves close. Once the valves close, the period of isovolumetric relaxation, which has previously been described, is begun (Fig. 6).

VENTRICULAR PRESSURE–VOLUME CURVE

It is also common to express the relationship of ventricular volume and pressure in the form of a curve, two examples of which are depicted in Figure 7. The illustration on the left is from an individual at rest while that on the right is during work. Isovolumetric contraction begins at the lower right (A) and ends at the point at which the curve

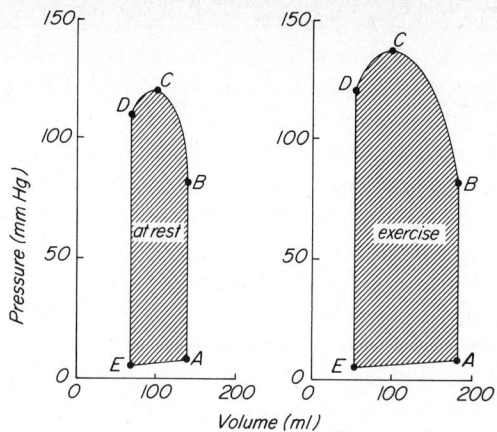

Figure 7 Pressure–volume diagrams for the left ventricle under basal conditions (rest) and during exercise. Normally under basal or resting conditions each ventricle contains about 70 ml/m² BSA of blood at the end of diastole. About 45 ml/m²BSA of this is expelled as the ventricle contracts. Letters (A–E) indicate different parts of the cardiac cycle as explained in text. BSA = body surface area.

begins to move toward the left hand ordinate. At this time rapid ejection begins (B). The curve moves to the left as the ventricular volume diminishes in a straight line. The point at which this begins to slow (C) signals the beginning of reduced ejection, a small increment on the curve, which ends at D. This then begins the period of isovolumetric relaxation, which ends at E with the opening of the mitral valve and the beginning of the filling period. The isovolumetric periods are easy to recognize since, as the name indicates, there are no volume changes. This figure is not time oriented. The period of reduced ejection appears short on the diagram, when in reality it accounts for some 60% of the ejection period during the resting state.

COMPARISON OF RIGHT AND LEFT SIDES OF HEART

The cycle for the right side of the heart is much the same as that on the left. The pulmonary and right ventricular pressures are much smaller than the left ventricular and aortic pressures, the normal systolic and diastolic pulmonary arterial pressures being 22/8 mm Hg, respectively. The periods of isometric contraction and relaxation are shorter than in the left heart since less pressure is necessary for opening the pulmonary valve.

As is apparent from their architecture, the manner in which the two ventricles pump is somewhat different. For ejection of the blood from the right ventricle, see Figure 8. (1) contraction of the spiral muscles draws the AV ring toward the apex of the heart. (2) The free wall of the right ventricle moves toward the convex surface of the interventricular septum as it contracts. (3) Contraction of the left ventricle puts additional traction on the wall of the right. The effect of the contraction is thus a bellows type of action. This is a very effective means of pumping large and varied volumes of blood against a low resistance, and the pulmonary circuit is a low resistance system. A sudden increase in pulmonary resistance, such as is seen with massive pulmonary embolism, will lead to death because the right ventricle cannot sustain the higher pressures

needed to provide adequate flow through the lungs. If, on the other hand, the pulmonary pressure rises gradually, the right ventricle will adapt itself for a chronic pressure load by assuming thicker walls and a more cylindrical shape, becoming similar to the left ventricle.

The left ventricle resembles a cylinder with a conoid segment at the apical end. The cylindrical portion is enclosed by a thick strong cuff of circular muscle. The conoid portion is mostly spiral muscle. When the left ventricle contracts, the circular muscles reduce the diameter of the cylinder. This action accounts for most of the power and volume of the ejection. It can easily be seen if the geometric equation for the volume of a cylinder is recalled:

$$V = \pi r^2 L$$

The volume changes with the square of the radius. Shortening of the chamber is brought about by contraction of the spiral muscles. This shortening of the longitudinal axis is very small and not very effective in reducing the volume of the chamber. It may be of some importance as far as the strength of contraction of the circular muscles are concerned.

The thick circular muscles are ideally situated for the development of high pressure pumping necessary for pushing blood into the high pressure systemic circulation. However, because of the toughness of its walls, the left ventricle is not as adaptable as the right ventricle in handling large volumes.

A. RIGHT VENTRICULAR EJECTION

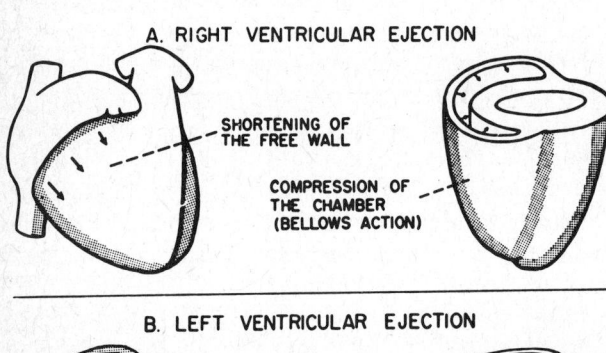

B. LEFT VENTRICULAR EJECTION

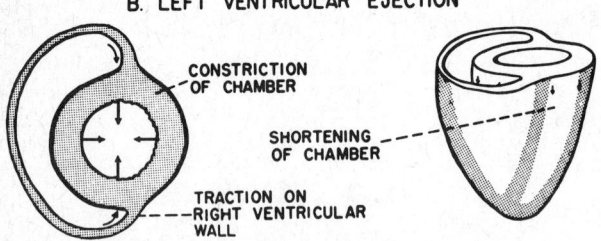

Figure 8 Components of ventricular contraction. From Rushmer RF: *Cardiovascular Dynamics*, ed. 4, Philadelphia, Saunders, 1976.

SELECTED BIBLIOGRAPHY

Berne RM, Levy MN: *Cardiovascular Physiology,* ed 4. St. Louis, Mosby, 1981.

Katz AM: *Physiology of the Heart.* New York, Raven, 1977.

Little RC: *Physiology of the Heart and Circulation.* Chicago, Year Book, 1977.

Rushmer RF: *Cardiovascular Dynamics,* ed 4. Philadelphia, Saunders, 1976.

SELF-STUDY QUESTIONS

1. At what time during the cardiac cycle are the left atrial and left ventricular pressures similar?

2. What factors contribute to the v wave? Under what pathophysiological condition will it be abnormally large?

3. How does the configuration of the ventricle change during isovolumetric contraction?

4. What is the ventricular end diastolic pressure? What is its physiologic importance?

5. Which are the diastolic and systolic arterial pressures? Which is the incisura of the aortic pressure?

6. What produces each of the four heart sounds?

7. At which points in the cardiac cycle would you find murmurs with stenosis or insufficiency with each of the four valves?

8. What changes are occurring in ECG; atrial, ventricular, and aortic pressures; and heart sounds during each of the seven periods of the cardiac cycle?

9. At what point in Figure 7 does the aortic valve open? At what point does it close?

10. Which ventricle (right or left) is more likely to fail with an increased volume load? Which with a pressure load?

ANSWERS

1. When the AV valves are open, that is, during diastole, the ventricular filling period (rapid filling, reduced filling, and atrial systole).

2. Blood returning to the left heart gets dammed up against the closed mitral valve, causing the atrial pressure to gradually increase. The valve then opens and this blood empties into the ventricle so the pressure falls. With mitral insufficiency (leaky mitral valve), blood will shoot from the ventricle into the atrium during ventricular systole accentuating the rising v wave.

3. A base to apex movement is produced because the spiral muscles contract before the circular. This increases ventricular diameter and stretches circular muscle.

4. The ventricular end-diastolic pressure is the pressure in the ventricle just before it contracts. It determines the amount of tension placed on the ventricular fibers

as they are activated. It is also a determinant of the venous return (Fig. 3).

5. The lowest pressure is the diastolic, the highest the systolic. The incisura is the notch on the descending limb of the aortic pressure pulse.

6. S_1 Closure of the AV valves, opening of semilunar valves, acceleration of blood into aorta.

   S_2 Closure of semilunar valves and rebound of blood against these.

   S_3 Vibrations on ventricular wall as they fill.

   S_4 Vibrations produced by atrial contraction.

7.

	Insufficiency	Stenosis
Aortic	Diastole	Systole
Mitral	Systole	Diastole
Tricuspid	Systole	Diastole
Pulmonary	Diastole	Systole

Insufficient valves make abnormal sounds when they are supposed to be sealed. Stenotic valves produce murmurs when they are open and have blood flowing through them.

8. See text and Figure 6.

9. The aortic valve opens at point B and closes at point D. The aortic valve is open during ventricular ejec- tion. During this time period the ventricular valve and pressure diminish.

10. The left ventricle fails with a volume load. Its thick walls are not very compliant and consequently do not stretch rapidly enough to accommodate the increased volume. The right ventricle is thin walled and compli- ant. This thin walled structure is not capable of devel- oping high pressures.

9

The Cardiac Output and Its Measurements

DANIEL L. TRABER

OBJECTIVES

After completion of this chapter, you should be able to

1. Recall approximate figures for normal stroke volume and heart rate.

2. Recall the relationship between heart rate, stroke volume, and cardiac output.

3. Define cardiac index and know the upper and lower normal limits for this variable.

4. Define stroke work.

5. Know the relationship between ventricular end-diastolic pressure and stroke work (the Starling work curve) and how it is affected by inotropic influences.

6. Recall the manner in which the sympathetic and parasympathetic systems control heart rate.

7. Calculate cardiac output given two variables of the Fick equation.

8. Calculate cardiac output given the proper variables of the Stewart-Hamilton equation.

The goals of this chapter are (1) to define cardiac output and discuss its relationship to the metabolic requirements of the body, (2) to examine the determinants of the stroke volume and heart rate, and (3) to detail the various methods currently being used in the clinical setting for the measurement of cardiac output.

The business of the heart is to pump blood, and the amount pumped by a single ventricle in 1 min is defined as the cardiac output. Cardiac output can also be defined as the product of the heart rate and stroke volume (the quantity of blood pumped with one systole). The normal stroke volume is about 75 ml, and the normal heart rate is 72 beats/min. Multiplying these two values gives a figure for the cardiac output of 5.4 L/min for a resting, healthy man weighing 70 kg.

The outputs of the right and left ventricles must be equal when the circulation is in a steady state. However, there can be considerable variation on a beat-to-beat basis, and the steady state cardiac output may differ greatly as we shall see on page 163.

THE CARDIAC INDEX

The cardiac output is markedly influenced by body size. Therefore, a means of comparing cardiac outputs among persons of different sizes is necessary. A good correlation is found between cardiac output and body surface area. Thus, it is common to see cardiac output expressed as L/min/m² of body surface area (BSA), the *cardiac index*. The normal surface area for a 70 kg male is 1.7 m². Therefore, given a normal cardiac output of 5.4 L/min and a surface area of 1.7 m², the average cardiac index in healthy resting males is about 3.2 L/min/m². A cardiac index below 2.5 or above 5 is definitely abnormal.

The correlation between cardiac output and body surface area is probably attributable to the fact that cardiac output is approximately proportional to the body's metabolic rate (see below), which in turn is approximately proportional to body surface area. In diseases that produce a considerable increase in the body's metabolic rate, such as thyrotoxicosis (elevated thyroid hormone), the cardiac output is also considerably increased. In hypothermia the metabolic rate is greatly diminished, resulting in a decrease in cardiac output. Cardiac output is increased in an attempt to meet metabolic requirements for adequate oxygen supply to the tissues under such conditions as (1) decreased oxygen in inspired air, (2) decreased ability of the blood to transport oxygen (as in anemia), or (3) decreased utilization of oxygen in cyanide toxicity.

RELATIONSHIP BETWEEN CARDIAC OUTPUT AND METABOLIC REQUIREMENTS

The apparently straightforward relationship between cardiac output and rate of metabolism is not really a simple one. Although the specific mechanisms controlling cardiac output are complex, they can be understood as adaptations for maintaining homeostasis, the tendency toward stability of the internal environment of the body. The homeostatic adaptations of both the heart and peripheral vascular system regulate the heart's pumping action in accordance with bodily requirements. For example, vigorous exercise is a physiological condition involving up to an 18-fold increase in the body's metabolic rate (based on oxygen consumption) in a well-trained athlete (Fig. 1). Since the ability of skeletal muscle to extract additional oxygen from each unit volume of blood can only increase by a factor of 3, a further supply of oxygen for exercising muscles must come from an increased supply of oxygenated blood (Fig. 1). A sixfold increase in cardiac output not only provides a greater flow of blood to the oxygen consuming tissues, but also circulates the blood through the lungs at a rate sufficient to maintain the normal degree of oxygenation of hemoglobin (97% saturated at an alveolar P_{O_2} of 100 mm Hg). Thus, the sixfold increase in cardiac output and the threefold increase in oxygen extraction from normally oxygenated blood support the

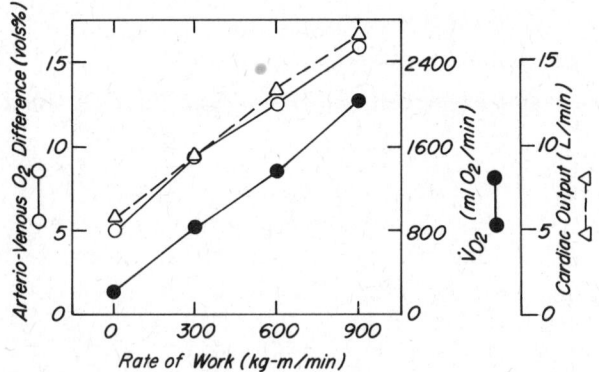

Figure 1 Effects of different rates of work (from different levels of exercise) on arteriovenous oxygen difference (o-o), oxygen consumption (•-•), and cardiac output (Δ-Δ). (From data of Carlsten A, Grimby C: *The Circulatory Response to Muscular Exercise in Man.* Springfield, Charles C Thomas, 1966.)

18-fold increase in oxygen demands brought about by severe exercise.

RELATIONSHIP OF STROKE WORK TO END-DIASTOLIC PRESSURE: STARLING'S "LAW OF THE HEART"

The control of cardiac output is vested in its two components: stroke volume and heart rate. In order to understand the regulation of stroke volume, it must be remembered (Chapter 5) that cardiac muscle contracts more forcefully when either its initial fiber length increases (within the normal range of fiber lengths) or its contractility increases. The first of these two mechanisms is used in two basic situations: (1) when the volume of blood coming back to the heart (the venous return) is increased, and (2) when the outflow resistance for expulsion (i.e., the arterial pressure) is increased. The second mechanism is described on page 162.

The effects of these two situations on the heart were studied by the great English physiologist, Ernest Henry Starling, just before World War I. In order to simplify the study of the myocardium, he isolated the heart and lungs from the rest of the circulation. Here the lungs are ventilated artificially. The left ventricle expells its contents into a catheter placed in the innominate artery. The remainder of the arterial tree is sealed shut with ligatures (the coronary circulation is left intact). Each stroke volume is expelled into an external circuit where the outflow resistance can be regulated by a screw clamp. This external circuit contains other functions normally performed by the body, such as maintenance of temperature and damping of pulsation. Ultimately the blood flows into a reservoir. This is connected by a tubing to a cannula inserted into the superior vena cava (all other veins being ligated). The blood flow from the reservoir back toward the heart is regulated by a screw clamp. The pressures and flows are monitored at several places in the system. Since the central nervous system is dead in this preparation—because it has no blood supply—there are no influences from autonomic nervous activity on the heart. Consequently, heart rate and contractility are considered constant unless some drug is injected into the system.

When Starling increased the venous return to this preparation, the stroke volume correspondingly increased. He found that if he suddenly increased the outflow resistance, the stroke volume suddenly dropped, gradually returning to normal over a period of a few beats. Simultaneous with these changes there was, in both situations, an increase in atrial pressure and ventricular volume. The ventricular volume augmentation would indicate that there was an increase in fiber length, which led Starling to formulate his law of the heart: "the energy of contraction is a function of the initial length of the muscle fibers." When he increased venous return the ventricles filled with more blood and the pressure within them increased. Consequently, the individual muscle fibers were stretched and could then contract with more force. When the outflow resistance was increased, the volume ejected by the next beat of the heart was reduced because the ventricle did not develop the additional force required to expel the previous stroke volume. The blood that was not expelled remained in the ventricle (increased ventricular systolic volume) increasing its diastolic pressure and placing additional stretch on the fibers. As a consequence of this, the next beat was more forceful. The process continued until a new balance between the inflow and outflow was obtained.

Starling's results can be expressed in terms of cardiac output, or stroke volume, or "stroke work." Left ventricular stroke work is the work performed on the blood by each stroke of the left ventricle. Work is defined as the product of the force acting on an object and the distance this object moves in the direction of the applied force. For pressure–volume work, force times distance converts to pressure times volume change. In the case of the heartbeat, left ventricular stroke work is equal to the area within the ventricular pressure–volume loop (Chapter 8, Fig. 7). However, stroke work is approximately equal to the product of mean arterial pressure and stroke volume. The latter way of estimating stroke work is usually used, since it is much easier than obtaining ventricular pressure–volume loops. A so-called "Starling work curve" is depicted in Figure 2. As the end-diastolic pressure increases, the stroke work increases. This relationship is nearly *linear* within the physiologic range of end-diastolic pressures. However, at extremes of pressure there is no longer a compensatory increase in stroke work. It has been determined that there is a linear relationship between fiber length and end-diastolic pressure, until the higher pressures are reached. Eventually the pericardial membrane

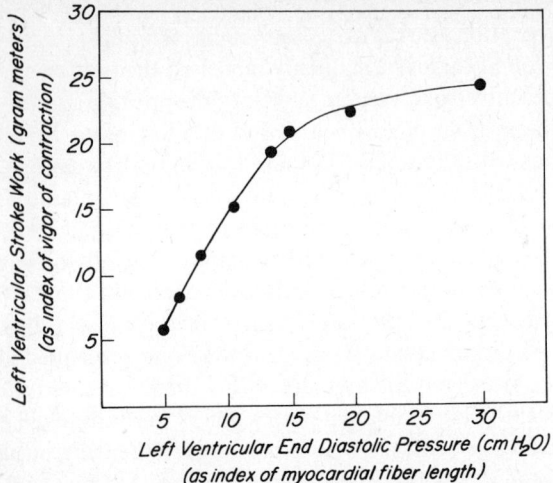

Figure 2 As the left ventricular end-diastolic pressure increases its walls are stretched. They therefore contract with more force thus a greater stroke work will result.

limits further expansion of the ventricle, and when this happens the relationship between end-diastolic pressure and fiber length (and consequently between end-diastolic pressure and stroke work) becomes nonlinear.

PATHOPHYSIOLOGY

The constraint of the pericardial membrane can reduce the stroke volume in certain instances. This condition is referred to as cardiac tamponade. The pericardium can become smaller as a result of scar formation or, more commonly, excessive fluid collects in the pericardial space. There is normally 5–10 ml of pericardial fluid in this space. However, if the pericardium becomes inflamed, fluid can be secreted and accumulated in excess. The fluid compresses the myocardium, limiting the amount of diastolic filling and, therefore, the stroke volume.

Normal hearts operate along the linear portion of the work curve. Failing hearts, on the other hand, tend to operate along the non-linear portion of the curve since they can pump out all of the venous return only if the end-diastolic pressure is increased above normal. This situation severely limits the degree to which any further compensation for increased venous return (as during exercise) can occur.

RELATIONSHIP OF STROKE WORK TO CONTRACTILITY

When Starling added *norepinephrine* to his system or stimulated the sympathetic nerves to the heart, he obtained a new work curve, which was shifted upward and to the left, as shown in Figure 3.

Thus at any given end-diastolic pressure (and ventricular volume), the ventricle produced a greater amount of stroke work. This effect is a function of the fact that norepinephrine increases the contractility of the myocardium (Chapter 5). This change in the work curve with changes in contractility is an important component of cardiac function. Under positive inotropic influence the stroke work may be maintained or increased at a shorter fiber length or lower left ventricular end-diastolic pressure. End-diastolic pressure changes, naturally, result in changes in the atrial pressure. In the case of the left atrium this has a dramatic influence on the pulmonary blood volume, and consequently, the venous return to the left heart. When the left atrial pressure is elevated, blood pools

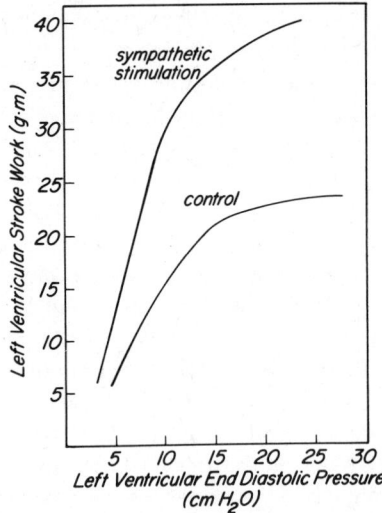

Figure 3 When the ventricles contract with varying left ventricular end-diastolic pressure, the force of contraction varies. This is affected when the sympathetic activity to the heart is increased such that at a given fiber length more work will be performed.

in the pulmonary circuit. As there are no valves in this vascular bed, elevations in pulmonary venous pressures result in elevations in microvascular pressure with the result that edema may occur. In addition, this pooling of blood elevates the work of breathing. Under normal states there is seldom an elevation in atrial pressure. With moderate exercise there is usually an increased sympathetic activity with its accompanying inotropic influences, resulting in the shifting of the work curve to the left and perhaps even a moderate fall in left atrial pressure. In normal situations the left atrial pressure remains relatively constant and consequently can be used as an index of the contractile state of the myocardium.

Unfortunately, left atrial pressure is probably one of the most difficult of the cardiac pressures to obtain. However, the absence of valves in the pulmonary circuit, as well as the lobular nature of the pulmonary vasculature, make it possible to obtain a reliable index of the left atrial pressure by measuring what is called the *pulmonary capillary wedge pressure.* This pressure is obtained by passing a catheter out into the pulmonary vessel. A continuous column of blood is produced from the top of the catheter to the left atrium. As the catheter occludes the vessel, blood flow through it ceases. Consequently, resistance in the column is zero and the tip of the catheter then measures the atrial pressure. To correct for gravitational influence on the column of blood, the transducer used for the pressure measurement is positioned and zeroed at the level of the left atrium. Situations can occur that make the analogy of left atrial pulmonary–capillary wedge pressure invalid. In situations in which the left atrial pressure is extremely low, the pulmonary veins might be occluded, thereby interrupting the column of fluid that must be obtained between the catheter tip and atrium. This column can also be interrupted with the use of positive-pressure ventilators that might place excessive pressure on the alveoli, occluding the pulmonary capillaries.

The measurement of pulmonary capillary wedge pressure has become routine in situations in which it is advisable to monitor left atrial pressure, that is, hypovolemia and left heart failure. The catheter placements has been simplified by the development of the Swan-Ganz flow-directed catheters. The placement of these catheters is described on page 164. To measure the pulmonary wedge pressure, the balloon on the tip of the catheter is inflated; the catheter is then advanced until the balloon wedges in the pulmonary vessels. Thus one can obtain the pulmonary wedge pressure with the balloon inflated. Deflation of balloon restores blood flow to the lobule and to the tip of the catheter, permitting measurement of both the pulmonary artery pressure and the wedge pressure.

REGULATION OF HEART RATE

The main regulators of heart rate are the parasympathetic and sympathetic nervous systems. Normally their actions are coordinated by the central nervous system (CNS). Thus when the heart rate is increased one will usually find that there has been both a decrease in parasympathetic and an increase in sympathetic activity. This type of mechanism in which there are both inhibitory and excitatory activities acting simultaneously is termed synergistic; it is found in many areas in which it is desirable to have control, for example, industrial temperature control and ship navigation, and indeed the heart rate is finely controlled. During the resting state there is considerable parasympathetic activity and much less sympathetic activity, and the resting heart rate may be somewhat slower than the spontaneous activity of denervated heart.

CONTROL OF CARDIAC OUTPUT

The normal heart pumps whatever blood comes to it from the great veins; it is not the rate-limiting step in the circulation. In other words, the heart normally does not set the pace for the circulation of blood. The cardiac output automatically adjusts to match any given venous return.

Mechanisms for accomplishing these adjustments have been discussed. The question that remains to be answered is, when does each become important? When one is at rest, the Starling, or stretch–tension, mechanism seems to predominate, since there is very little sympathetic activity.

In addition, this mechanism is of importance in maintaining the equalization of the outputs of the right and left ventricles. As one begins physical activity, there is an elevated sympathetic tone, the heart rate increases, and there is an augmentation of contractility. These mechanisms predominate, and the ventricular volume and end-diastolic pressure actually fall. With severe exercise, the venous return may become elevated to the extent that the Starling mechanism might again come into play.

MEASUREMENT OF CARDIAC OUTPUT

Until recently, measurement of cardiac output was more or less restricted to cardiac catheterization and laboratory studies. This was because the methodology was not only laborious and time consuming, but it involved the use of invasive techniques. The technological advances of the last few years have enabled us to make this measurement relatively rapidly. Although most of the techniques that will be described are still invasive, (i.e., arterial catheterization), they are becoming more and more commonplace in intensive care units and in the operating theaters of many hospitals.

THE FICK METHOD

The oldest and perhaps most reliable method for the determination of cardiac output is known as the *Fick method*. This method makes use of the principle that any oxygen taken into the body from the lungs must be combined with the venous (pulmonary arterial) blood in the alveoli. Once the timed amount of oxygen taken into the body is known, and the concentrations of oxygen in the venous and arterial blood determined, one can quantitate the volume of blood that would be necessary to carry the oxygen taken into the body from the lungs over a known time period.

To use an analogy, let us say a tanker arrives at the port of Galveston and unloads 6,000 tons of oil into railroad cars. Let us also say that each railroad car can hold 6 tons of oil. How many railroad cars would it take to unload the tanker?

$$6{,}000/6 = 1{,}000$$

Let us now examine the problem using a physiological example. Let us say that the subject has an oxygen consumption (the amount of oxygen taken up by the pulmonary blood per minute) of 300 ml/min. The quantity of oxygen in the venous blood is 14 ml/100 ml of blood. The quantity of oxygen in the arterial blood is 20 ml/100 ml of blood. Thus as the blood passes through the lungs it picks up 6 ml of oxygen for every 100 ml of blood that passes through the lungs. How many ml of blood would thus be required to take up the 300 ml/min of oxygen?

$$\frac{300}{6} \times 100 = 5{,}000 \text{ ml blood/min}$$

This principle can be expressed by the Fick equation:

Cardiac output =

$$\frac{\text{the oxygen absorbed by the lungs/min}}{\text{arteriovenous oxygen difference}}$$

The amount of oxygen absorbed by the lungs can be obtained through the comparison of inhaled and exhaled air. The concentration of oxygen in arterial blood can be determined from any systemic arterial sample because of the uniformity of systemic arterial blood.

The major drawback to determining cardiac output with the Fick method is that one must have a venous sample that is representative of the drainage from all areas of the body (a mixed venous sample). Because the various tissues extract differing amounts of oxygen from the blood, such a mixed venous sample can only be obtained from the pulmonary artery. Therefore, for many years use of the Fick method was limited to the cardiac catheterization laboratory, where a catheter could be introduced into the pulmonary artery under fluoroscopy. Modern developments in techniques now permit catheterization at the patient's bedside by means of the Swan-Ganz flow-directed right heart catheter. This instrument is inserted into a peripheral vein and passed to larger vessels, that is, the subclavian. The location of the tip is inferred from markings along the tube, which indicate the length of tubing in the vein. When the catheter tip has reached one of the larger veins, a balloon at the catheter tip is partially inflated with CO_2, and blood flow propels the balloon and catheter to the right atrium. The balloon is then fully inflated and advanced again so that blood flow propels it through the right ventricle and into the pulmonary artery.

Verification of the location of the catheter is accomplished by pressure measurements monitored through a lumen extending to the catheter tip, (e.g., about 22/8).

The Fick method has the advantage of providing additional information concerning the physiologic status of a patient over and above the cardiac output. The oxygen absorbed by the lungs and the arteriovenous oxygen difference are important pieces of data in and of themselves. For example, the oxygen consumption can give a great deal of information concerning the metabolic status of a patient. The arteriovenous oxygen difference also has value beyond just the determination of cardiac output. This difference tells the amount of oxygen extracted from the blood during its circulation. If this difference is high, that is, a large amount of oxygen has been extracted from the blood, it indicates that the tissues are inadequately perfused. In fact, some investigators look simply at the mixed venous oxygen saturation as their index of the circulatory status of a patient.

Although there are many advantages to the use of the Fick method, there are also some disadvantages, besides the need for a representative sample, described above. The measurements of O_2 consumption and AV oxygen difference are not easily determined. In addition, to be valid they must be taken simultaneously in a patient whose cardiovascular system is relatively stable during the time the data are obtained.

DYE DILUTION TECHNIQUE

Although the Fick principle of cardiac output determination is considered the gold standard for this measurement, the dye dilution method enjoys the greatest popularity at this time. This technique is based on the relationship

$$Concentration = quantity/volume$$

For example, if you want to make a 10% solution, you add 10 g per 100 ml of fluid (wt/vol). Rearranging the equation, it can be seen that volume then equals quantity divided by concentration. In other words, if 10 g of a substance is added to an unknown volume of water and the concentration is found to be 10 g/100 ml, we can determine that the volume of water must be 100 ml. This technique is used to measure the volumes of many fluid compartments.

In order to measure cardiac output, a dye is injected into an area of the cardiovascular system where it will be rapidly mixed with the blood flowing through the heart—usually into the right atrium. Blood is then withdrawn at a constant rate from a catheter placed in a systemic artery downstream. This blood is pulled through a densitometer so that a continuous measurement of the concentration of the dye can be obtained (Fig. 4). The volume that is determined is the volume of blood moving during the period of dye flow, which requires determination of the average concentration of dye. Figure 4 demonstrates two problems in determining the dye concentration in the blood that has diluted it. The first problem is that the dye does not suddenly appear in the arterial sample at full concentration and then suddenly disappear. Rather, its concentration gradually increases and then gradually decreases. Therefore, it is the *average concentration* over the entire period of dye flow that must be determined. The second problem is that the concentration of dye never returns fully to zero because some dye begins to recirculate before the first passage is completed. The latter problem is solved by assuming that the early part of the downslope of the curve is exponential and then extrapolating the curve to zero from this known segment.

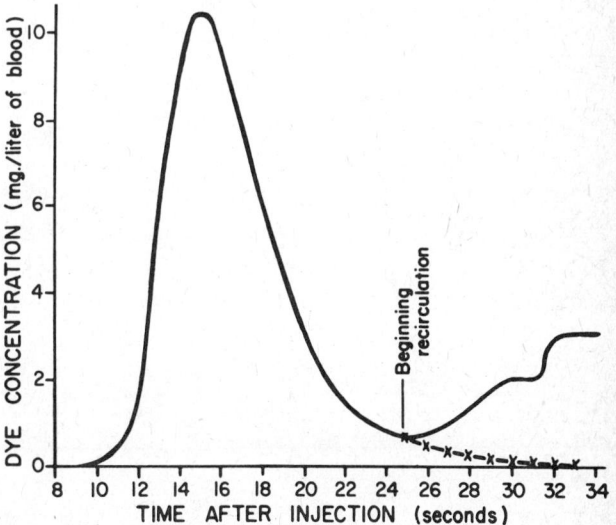

Figure 4. An idealized dye dilution curve. The curve has been extrapolated to zero in order that the area can be obtained. (From Guyton, Jones and Coleman, Cardiac Output and Its Measurement, in *Circulatory Physiology: Cardiac Output and Its Measurement.* WB Saunders Co., Philadelphia, London and Toronto, 1973.)

The reason the dye concentration is not constant during its passage by the sampling site is that the dye passes through the circulatory system in the shape of a parabolic bolus (Fig. 5). In order to obtain the average concentration of the dye in the bolus, the area of the curve is measured; the area is then divided by its duration (the time of dye appearance to the extrapolated point of disappearance). Once this is accomplished, the cardiac output can be determined by dividing the amount of dye injected by average concentration of the dye times the duration of the curve. The following equation, known as the *Stewart-Hamilton equation,* is customarily used:

$$\text{Cardiac output (ml/min)} =$$

$$\frac{\text{mg dye injected} \times 60}{\text{ave conc. of dye (mg/ml)} \times \text{duration of curve (sec)}}$$

The equation is multiplied by 60 to convert the answer to ml/min, rather than ml/s. That is, the volume determined was the volume flowing during the period of dye flow (duration of curve in seconds). Multiplying this volume by 60 s/duration of curve in seconds gives the volume per minute (vol/min). All the above calculations (e.g., extrapolations, area measurements) are usually accomplished by an on-line computer to make the cardiac output

information available immediately. The dye used in such determinations must stay within the vascular compartment and must not leak into the interstitial fluid.

THERMAL DILUTION

A modification of the dye dilution technique, which is gaining in popularity, is the thermal dilution method of cardiac output determination. The principles involved are essentially the same as those described above, except that instead of injecting dye and measuring its dilution, heated or cooled solutions are injected and the resulting downstream temperature changes recorded. This technique is facilitated by use of modified Swan-Ganz catheters. The flow-directed catheter used in this instance is modified so that, in addition to the usual lumen used for sampling and pressure measurements, there is a thermistor on the tip and another lumen and port for injection of solutions some 5 cm from the catheter tip. This device can be placed in the pulmonary artery as described previously. Warm or cool solutions may then be injected through the extra port. They will then be diluted by the cardiac output. Temperature changes are sensed by the thermistor and the electrical signal interpreted by an on-line computer with the determined cardiac output displayed instantaneously. The dye dilution and thermal dilution techniques have been compared with the Fick method under various experimental conditions and found to agree with it within 10%.

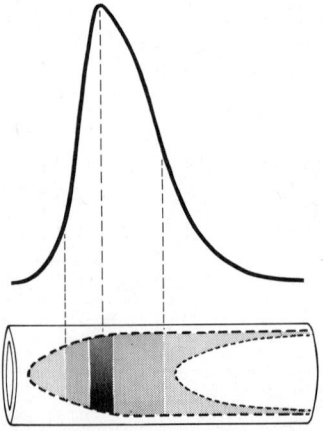

Figure 5 Relationship between the dye dilution curve and the blood concentration of dye. Since the dye flows through the circulatory system as a parabola rather than a cylinder, the concentration varies depending on which portion of the parabola you are sampling from.

SELECTED BIBLIOGRAPHY

Bloomfield, DA (ed): *Dye Curves: The Theory and Practice of Indicator Dilution.* Baltimore, University Park Press, 1974.

Guyton AC, Jones CE, Coleman TG: *Circulatory Physiology: Cardiac Output and Its Regulation,* ed 2. Philadelphia, Saunders, 1973.

Honig CR: *Modern Cardiovascular Physiology.* Boston, Little, Brown, 1981.

Richardson D: *Basic Circulatory Physiology.* Boston, Little, Brown, 1976.

Smith JJ, Kampine JP: *Circulatory Physiology—the Essentials.* Baltimore, Williams & Wilkins, 1979.

SELF-STUDY QUESTIONS

1. What change in stroke volume would occur if the heart rate were doubled and the cardiac output remained constant?

2. What transient variations between the stroke volume of the right and left ventricle would occur in the following situations:

   A. Going from the supine to the standing position.

   B. Rupture of the aortic valve.

3. The subject has a stroke volume of 100 ml, a heart rate of 70 beats/min, and a BSA of 1 m². Is his cardiac index within normal limits?

4. Would you expect the cardiac output to be increased, decreased, or unchanged in the following situations? Living at an altitude of 14,000 ft, anemia, hypothermia, exercise?

5. How can the denervated heart compensate for an increase in outflow resistance?

6. How will changes in inotropic activity affect the Starling work curve?

7. From what vessel can a venous blood sample be obtained that is representative of the drainage of the entire venous system?

8. If you were to design a cardiovascular catheter using gas-filled balloon, what gas would you fill it with to minimize aeroembolus in the event that the balloon were to fail?

9. The subject has a cardiac output of 6 L/min. His mixed venous O_2 concentration is 15 ml/100 ml. His arterial O_2 concentration is 20 ml/100 ml. What is his O_2 consumption?

10. The subject inhales room air and exhales into a Douglas bag (a special device for collecting gas samples) for 5 min. The concentration of O_2 in the bag is determined and found to be 160 ml/L. Room air has an O_2 content of 210 ml/L; the bag contains 30 L of gas. What is the O_2 consumption?

11. The subject has an oxygen consumption of 300 ml/min. His arterial O_2 is 20 ml/100 ml of blood, his mixed venous (pulmonary arterial) O_2 content is 5 ml/100 ml. BSA = 2 m². Is his cardiac index normal?

12. Why is it necessary to measure all the variables of the Fick equation simultaneously?

13. In determining the cardiac output by the dye dilution technique, dye is injected into the right atrium, and its average concentration and time of passage are measured in blood samples from a peripheral artery. What would be some guidelines to be used in selecting the dye to be used?

14. A sequence of dye dilution curves is obtained from a patient undergoing surgery. The amount of dye injected for each curve is constant, but the area under the curves becomes progressively larger. Is the patient's output going up or down? During this time the patient's blood pressure is 100 mm Hg and does not vary. Is his total peripheral resistance increasing or decreasing?

15. What is a Swan-Ganz flow-directed catheter? How can it be modified for cardiac output determinations?

ANSWERS

1. It would be halved.

2. A. Output from the right ventricle would fall transiently secondary to venous pooling in the lower extremities. The output of the left ventricle would increase transiently as blood drains from the lungs.

B. Output of the left ventricle would rise since a portion of the stroke volume would run back into the ventricle during diastole.

3. Cardiac index $= \dfrac{100 \times 70}{1} = 7$ L/min/m². Upper limit for normal is 5; therefore, patient's output is higher than normal.

4. Increased, increased, decreased, increased.

5. It gradually dilates, so that after a few beats it is operating far enough to the right on the Starling work curve that the ventricle can contract forcefully enough to expel a normal stroke volume against the increased resistance.

6. A positive inotropic influence will shift the curve upward and to the left. A negative inotropic influence will shift the curve downward and to the right.

7. The venous blood draining the various organs has varying O_2 contents (i.e., high from the kidney, low from skeletal muscle). These are mixed together in the right side of the heart. Thus the pulmonary artery is the most logical area for obtaining a mixed venous sample.

8. Carbon dioxide is some 30 times more soluble in blood than are the most commonly available gases (helium, nitrogen, oxygen, argon). Thus, if the catheter balloon fails, the CO_2 would dissolve in the blood minimizing the danger of embolus.

9. The Fick equation is $\dot{Q} = O_2$ consumption/A-Vo_2 difference. Therefore,

$$O_2 \text{ consumption} = \dot{Q} \times \text{A-}Vo_2 \text{ difference}$$

$$= 6000 \text{ ml/min} \times \frac{5 \text{ ml}}{100 \text{ ml}}$$
$$= 300 \text{ ml/min}$$

10. The O_2 consumption would be the difference in O_2 content between inhaled and exhaled gases (160 − 210) × 30 = 1,500 ml/5 min, or 300 ml/min.

11. Cardiac output is

$$\dot{Q} = \frac{O_2 \text{ consumption}}{\text{A-}Vo_2 \text{ difference}}$$

$$\dot{Q} = 300 \text{ ml/min} \times \frac{100 \text{ ml}}{15 \text{ ml}} = 2,000 \text{ ml/min}$$

Cardiac index is:

$$CI = \frac{\dot{Q}}{\text{m}^2 \text{ BSA}} = \frac{2,000 \text{ ml/min}}{2 \text{ m}^2}$$
$$= 1,000 \text{ ml/min/m}^2$$

These values are less than 2 L/min/m² and are thus abnormally low.

12. If the variables of the Fick equation are not measured simultaneously, they may change. If the O_2 consumption is measured at a different time than the A-Vo_2 difference, the $\dot{Q}$ value could be erroneous.

13. The dye must remain in the vascular system; otherwise it will be diluted by volumes other than the cardiac output (i.e., interstitial fluid), must be nontoxic, be measurable.

14. The patient's output is going down. The curve area is inversely related to the cardiac output. The peripheral resistance is increasing ($R = P/\dot{Q}$, and P is constant).

15. A Swan-Ganz flow-directed catheter (see page 164).

10

Hemodynamics

DONALD W. STUBBS

OBJECTIVES

After the completion of this chapter, you should be able to

1. Define and describe hydrostatic pressure in physical terms as force per unit area and, as used most often in practice, as the height of fluids of known density (mm Hg, or cm H_2O).

2. Relate Pascal's principle to clinical applications.

3. Understand the relationship between flow, pressure and resistance in rigid tubes (i.e., Poiseuille's law) and work problems dealing with these variables.

4. Describe the analogy between electrical currents defined by Ohm's law and fluid currents.

5. Understand and compare differences between series and parallel resistances.

6. Describe the relationship between average velocity, flow, and total cross-sectional area, as well as work problems involving these variables.

7. Understand relationship between changes in total cross-sectional area and changes in resistance resulting from branching of conduits, as well as describe the typical situation in bifurcation of vessels in the circulatory system.

8. Understand the relationship between kinetic and potential energy of flow in a tube or vessel (Bernoulli's principle).

9. Describe the variables that determine the Reynolds' number and explain the relationship between the critical Reynolds' number and development of turbulence.

10. Discuss the relevance of knowledge about streamline and turbulent flow to clinical and physiological aspects of circulatory function.

The general objectives of this chapter are to (1) describe and discuss fundamental principles and laws governing the flow of fluids in simple physical systems; (2) employ those basic principles from simple systems that permit approximations of the behavior of flowing blood in the more complex circulatory system; and (3) point out similarities and differences between fluid systems in rigid tubes and the flow of blood in distensible vessels.

The principal role of the cardiovascular system is to circulate blood. This circulation provides for distribution of essential substances (e.g., oxygen, glucose, amino acids, fatty acids, hormones, immunoproteins) to the various tissues and for the removal of wastes (e.g., carbon dioxide, ammonia, creatinine, urea) from the tissues. In considering the flow of blood through vessels of the circulatory system, we are naturally concerned with fundamental principles and laws governing flow. The simplest model for study is that of water under pressure flowing steadily through rigid cylindrical tubes. From such simple systems basic principles can be derived that can permit us to make approximations about the way in which blood flows in the circulatory system. A treatment of the flow of blood through the cardiovascular system in more precise physical and mathematical terms is beyond the scope of this chapter. The flow of blood in the arteries is not steady but pulsatile. The heart is an extremely complicated pump, the behavior of which is affected by a wide variety of physical and chemical factors. The blood vessels are multibranched distensible conduits of continuously varying dimensions. The blood itself is not a simple fluid like water, but is rather a suspension of blood cells in a colloidal solution of proteins. Nevertheless, much insight can be gained from an understanding of the more elementary principles of fluid mechanics as they apply to the simpler physical systems.

Hemodynamic principles have considerable relevance as applied in the elucidation of mechanisms of arterial disease (e.g., atherosclerosis) in clinical diagnoses (generation of acoustic energy manifest as murmurs) and cardiovascular reconstruction procedures (heart valve prostheses, vascular graft prostheses, and mechanical circulatory assist devices). Vascular surgery for correction of arterial lesions presents many problems of a hemodynamic nature. The entire step-by-step procedure, beginning with diagnosis, localization, and assessment of diseased arteries, and decisions regarding surgery, procedures, postoperative evaluation, and follow-up can all involve various considerations of hemodynamic principles at each stage.

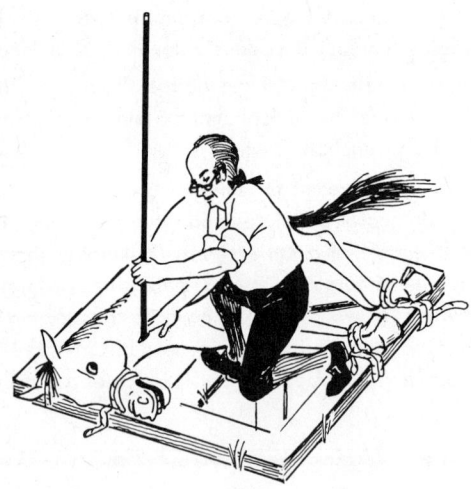

Figure 1 Stephen Hales performing a direct measurement of arterial blood pressure in 1731. He tied an old mare to a gate and cannulated the carotid artery. He connected the cannula to a glass tube and observed the rise of the blood to a height of 9 ft, 6 in. (290 cm of H_2O, or about 213 mm Hg). Such a high arterial pressure is not unusual, considering that the horse was unanesthetized.

HYDROSTATIC PRESSURE

When water occupies a beaker or other vessel, it exerts a pressure or force on every part of the area of the sides and bottom of the vessel. This pressure at any point is defined as *force per unit area* surrounding that point. The pressure exerted by an open, standing column of water depends on the weight of the column of water. Such a pressure of a column of water at rest is called a hydrostatic pressure. The pressure exerted by any column of fluid can be calculated from the height of the column and the density of the fluid:

$$P = \rho g h \qquad (1)$$

The specific units of pressure, of course, depend on the units used for the density ρ, acceleration due to gravity g, and height h. The most common system of units is that of the metric system.

A form of the metric system that has been popular with some physiologists and clinicians for many years is the centimeter-gram-second (cgs) system. The principal unit of force is the dyne (a force giving a mass of 1 gram an acceleration of 1 centimeter per second per second). The basic unit of pressure in the cgs system is dyn/cm². Thus, for Eq.(1) pressure P is dyn/cm² when the density ρ is in g/cm³, gravitational acceleration g is 980 cm/s/s, and the height h is in cm. However, because of the traditional and persistent use of mercury manometers to measure blood pressure (the most common pressure measured by clinicians), the practical unit used by most American clinicians is not dyn/cm², but millimeters of mercury (written as mm Hg or as torr). Since for any given liquid, such as mercury or water, the hydrostatic pressure is proportional to the height of a fluid column,* comparing or expressing pressures simply in terms of column height of a specific fluid is not unreasonable. Whereas those interested primarily in the cardiovascular system use mm Hg, other workers, such as respiratory physiologists, who often deal with much lower pressures (e.g., intrapleural pressure), commonly use cm H_2O. It is sometimes necessary to be able to convert from one to the other. Numerically this is simple enough: multiply cm H_2O times 0.74 to get mm Hg; multiply mm Hg times 1.36 to get cm H_2O.

More recently, most physical scientists and many European clinicians have adopted a new modification of the metric system: the Système International d'Unites (SI units). SI units are a version of the meter–kilogram–second–ampere (mksa) system. The basic SI unit of force is the newton, the force giving a mass of 1 kilogram an acceleration of 1 meter per second per second. The basic SI unit of pressure is the pascal, a force of 1 newton acting over an area of 1 square meter. The clinical derivative used to express blood pressures is the kilopascal (kPa). Mathematically, 1 kPa is equal to 10,000 dyn/cm². Since 1 mm Hg is equal to 1,332.8 dyn/cm², and 1 kPa is equal to 10,000 dyn/cm², 1 mm Hg is the same as 0.13328 kPa. Thus, to convert mm Hg into kPa, multiply by 0.13328. A convenient way to do this in your head is to take mm Hg and move the decimal one place to the left (as if converting mm to cm) and then add one-third. For example, what is 120 mm Hg in kPa? To find out, change 120 to 12 and add one-third of 12. That is, 120 mm Hg equals 16 kPa. To convert from kPa to mm Hg multiply by 7.5.

*For any given liquid, the density ρ and the acceleration due to gravity g are constants; the pressure therefore varies only with the height and is thus directly proportional to that height.

Since the vaguely familiar units of dyn/cm² have not been particularly popular with American clinicians, it is perhaps understandable that the more unfamiliar Pa and kPa units have gained little acceptance. However, the use of kPa along with many other SI units is quite common in Europe. A number of European medical journals currently use SI units. Because it is possible that the SI units will prevail as international standard units, it is important that American clinicians and medical students acquire some familiarity with the SI system.

TRANSMISSION OF PRESSURES— PASCAL'S PRINCIPLE

The hydrostatic pressure of a liquid in an *open* vessel caused by its weight is equal to the density times the depth at the point being considered and acts equally in all directions *at that point.* However, whenever an additional pressure is exerted on a liquid in a *closed* container, that pressure is transmitted undiminished to *all parts* of the liquid. This principle was demonstrated by the French mathematician and religious philosopher, Blaise Pascal (1623–1662), when he burst a strong wine cask (fortunately full only of water) with a small amount of water poured into a tall pipe of small diameter connected to the cask.

CLINICAL APPLICATIONS OF PASCAL'S PRINCIPLE

Chronically ill patients who must remain in bed for long periods of time have a tendency to develop decubitus ulcers (bed sores) when confined on an ordinary mattress. The use of a water mattress has been found to help prevent the formation of decubitus ulcers. This is a direct application of Pascal's principle. The ulcers tend to develop on the elbows, buttocks, and other areas of contact that have the largest pressures exerted on them by an ordinary mattress. Since the water mattress constitutes a closed fluid system, the pressure is evenly distributed within it. When the patient lies on the water mattress, the same pressure is evenly exerted against every part of the body that is in contact with the mattress. To obtain the full advantage, the entire body must be supported by the enclosed fluid. The mattress must be sufficiently filled with liquid so that no part of the body touches the bottom layer of the mattress.

Pascal's law applies to any enclosed fluid. There are several examples of enclosed or nearly enclosed fluids in the body. One interesting case is that of the cerebrospinal fluid (CSF). This fluid circulates around the spinal cord and up into the subarachnoid space around the lower part of the brain. Although it is not, strictly speaking, a perfectly closed liquid system, the CSF acts as if it were over short periods of time. An increase in pressure at any point in the fluid will increase the pressure in all parts of the fluid. Thus, a rapidly growing brain tumor protruding into the space normally occupied by the fluid can cause a measurable increase in pressure throughout the fluid. A measurement of the pressure of the fluid at any convenient location can detect the increase in pressure. The measurement of the fluid pressure is usually made by means of a spinal tap. A water manometer is normally used for this measurement, and the pressure is then recorded in cm H_2O.

The Queckenstedt test for obstructions in the CSF is based on Pascal's principle. If the flow of blood from the venous sinuses into the internal jugular veins is shut off temporarily by squeezing the jugular veins, the cerebral venous pressure rises, in turn causing a rise in intracranial pressure, including the cerebrospinal fluid pressure. If there is no obstruction in the cerebrospinal fluid, this pressure increase will be transmitted to all parts of the fluid and will be manifested by a rise in the hydrostatic pressure in the spinal tap manometer. If the manometer pressure is not altered by the application of pressure on the jugular veins, an obstruction is indicated.

HYDRAULIC PRESSURE

Fluids can be under pressures other than those caused by a column of the fluid acted on by gravity (hydrostatic pressure). Hydraulic pressure can be exerted on fluids by pumps. When a pump circulates fluid in a closed system, it raises the pressure of the fluid to overcome the system resistance in order to produce flow from an area of high pressure at the pump to one of lower pressure through the system (down a pressure differential). Figure 2 illustrates the pumping action of the left ventricle as an analogy to that of a force pump.

When the body is lying in the horizontal position, the long columns of blood (e.g., from the heart to the feet) are all at, or very near, the same level (i.e., no significant hydrostatic pressure due to height of a column of blood). The pressure gradient causing flow through the circulatory system is that produced by the hydraulic pumping action of the heart (this is true for either a standing or supine subject). In other words, the only significant pressure of blood in a reclining subject is that due to the hydraulic pumping action of the heart (necessary for causing flow through the circulatory system). For example, the mean systemic arterial pressure in any artery of a horizontal individual is about 90 mm Hg (12 kPa).

If a person stands up after lying in a horizontal position, the arteries in the feet will then have two kinds of pressure: (1) the same hydraulic pressure imparted to the blood by the heart, as described above, plus (2) the true hydrostatic

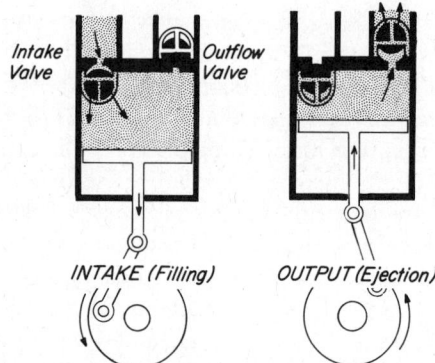

Figure 2 The left ventricle as a force pump with one-way ball valves. The intake valve (analogous to the mitral atrioventricular valve) permits inflow into the pump during the intake stroke (filling of the ventricle), while the outflow valve (aortic valve) is shut. During the output stroke (contraction of ventricle) the intake valve is closed and the output valve opens. The analogy here illustrates valve action and the development of hydraulic pressure, but it is not accurate in terms of the filling of the heart. The force pump above exerts considerable suction on the down stroke whereas in the heart only a little suction is produced during ventricular diastole. Ball valves similar to those illustrated have actually been implanted during cardiac surgery to replace damaged heart valves.

pressure of the long column of blood from approximately the level of the heart to the feet. For a person of average height, the total pressure of the blood in an artery on the dorsum of the foot would be about 180 mm Hg (24 kPa). This is the sum of the hydraulic pressure of about 90 mm Hg (12 kPa) and the true hydrostatic pressure of a column of blood about 120 cm from the heart to the top of the foot; that is, an additional 90 mm Hg (12 kPa).

RELATIONSHIP BETWEEN PRESSURE DIFFERENCE AND FLOW

Flow, symbolized by $\dot{Q}$, has the dimensions of quantity or volume per unit time. Flow is sometimes more specifically designated as *volume flow* (or rate of flow) to distinguish it from flow velocity. In this chapter flow or $\dot{Q}$ will always mean volume flow and will be expressed in units of cm³/sec (ml/s) or L/min. *One of the two principal determinants of $\dot{Q}$ is pressure difference* (ΔP). Consider the flow through a tube connecting the open-topped reservoirs A and B in Figure 3. If reservoir A is filled with water to a certain height h_1, and reservoir B is empty (contains only air), the fluid would flow from A into B, which begins to fill. After a time the fluid levels in the two vessels would become equal and flow would stop. At any point before

this, the difference in heights of the two columns represents a pressure difference and therefore contains the potential energy that drives the flow of fluid. When this energy has been dissipated (as manifest by loss of any pressure difference), no further flow can occur. *A pressure difference is essential to flow.*

Figure 3 shows that initially the maximum pressure difference is entirely due to the original height of fluid in reservoir A (h_1). This is responsible for the initial inflow pressure P_{in} for the connecting tube. Because reservoir B is initially empty, it has no hydrostatic pressure, only atmospheric pressure, which is zero or reference pressure. This zero pressure is the initial outflow pressure P_{out}. Let us say that as a result of this initial pressure difference ($\Delta P = P_{in} - P_{out}$), the initial flow ($\dot{Q}$) through the connecting tube is 4 ml/s.

If reservoir A had been filled to a greater initial height h_2, which was exactly twice as high as h_1, and reservoir B was empty as before (zero P_{out}), the initial flow through the connecting tube would have been twice as great (i.e., $\dot{Q}$ would have been 8 ml/s). The flow would have been double that shown in Figure 3 because the pressure difference ΔP would have been doubled (Fig. 4). *Flow is proportional to the pressure difference.*

To illustrate the point that it is the pressure *difference* ΔP that is of principal importance, rather than the absolute values of P_{in} and P_{out}, consider what would happen if reservoir B had a fluid level of height h_1 and the fluid level in A was of height h_2, that is, twice as high as h_1. The difference between the two would be h_1 and, therefore ΔP the same as in Figure 3. Thus the initial flow would be also the same, that is, 4 ml/s (Fig. 5).

If reservoir A were filled further to a height h_3, three

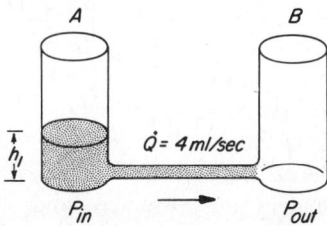

Figure 3 Relationship Between $\dot{Q}$ and ΔP. Reservoir A is filled with water to height h_1, providing the inflow pressure P_{in} for the connecting tube between the two reservoirs. Reservoir B has no water and the outflow pressure P_{out} for the connecting tube is simply the atmospheric pressure. Under these conditions the initial ΔP ($P_{in} - P_{out}$) causes an initial flow of 4 ml/sec from reservoir A to reservoir B.

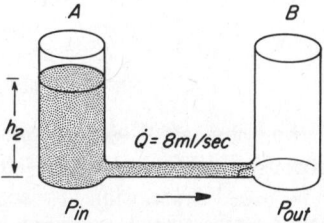

Figure 4 Relationship between $\dot{Q}$ and ΔP. Reservoir A has been filled with water to a height h_2, which is twice that of h_1 in Figure 3. The initial ΔP is also twice that in part 1 and, consequently, the initial flow is also doubled, that is, 8 ml/s.

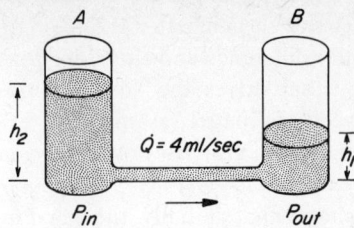

Figure 5 Relationship Between $\dot{Q}$ and ΔP. Although the individual heights of water in reservoirs A and B are different from those given in Figure 3, the initial ΔP is the same; therefore, the initial flow is the same (4 ml/s). It is the ΔP that is of principal importance, rather than the individual absolute values of P_{in} and P_{out}.

times as high as the level h_1 in reservoir B, the difference would be h_2, and the ΔP would be double that shown in Figure 5. The initial $\dot{Q}$ would then also be doubled from 4 to 8 ml/s. Again, it should be apparent that $\dot{Q}$ *is directly proportional to* ΔP. This is the basis for the statement that ΔP is one of the principal determinants of the flow $\dot{Q}$. The other principal determinant is resistance, which is discussed in the next section.

RELATIONSHIP BETWEEN RESISTANCE AND FLOW

The second principal determinant of flow is resistance. Resistance can be defined as hindrance to flow; that is, factors tending to impede or retard the free flow of fluids through vessels. Some of these factors contributing to resistance are easily recognized intuitively.

VISCOSITY

Consider first the influence of *viscosity*, η. Viscosity was defined in rigorously physical terms by Newton, but for now consider his word description of viscosity as a kind of internal friction of liquids causing "a defect of slipperiness." Liquids of high viscosity, such as molasses, have such a high internal friction that they do not flow readily. Liquids of low viscosity, such as water, flow more freely.

There is no uniform relationship between density and viscosity.

DIMENSIONS OF TUBE

The dimensions of the tube or conduit through which flow occurs are another important aspect of resistance. The resistance to flow through a long tube of given radius is greater than that of a short tube of the same radius. That is, flow is inversely proportional to length. For example, a reservoir containing fluid to a certain height is connected to an outlet tube of length l_1, and the outflow is measured as 10 ml/s. If an additional section of tubing is added, making the outflow tube twice as long ($l_2 = 2 \times l_1$), the flow will be found to have been reduced by one-half to 5 ml/s. Flowing within the confines of tubing impedes the volume moved per unit time because of the influence of the tubing on the internal friction (viscosity) of the moving fluid. The longer the tube, the greater the total frictional resistance and thus, the less the rate of flow $(\dot{Q})$. That flow is inversely proportional to length might not be so surprising, but what might be surprising to you is the profound influence of the tube radius. Of course, you would guess that resistance to flow is greater in a narrow tube than in a wide one, but you probably would not have guessed that flow is proportional to the radius raised to the fourth power (r^4). Other things being equal (i.e., a constant ΔP, length, and viscosity), only a mere doubling of tube radius causes a $16 \times$ increase in flow. If the radius were tripled the flow would increase $81 \times$. If the radius were decreased to one-half its former value, flow would decrease to one-sixteenth its former rate. To determine the change in flow, take the relative value by which radius has been changed and raise it to the fourth power (e.g., if radius is decreased by one-half, raise one-half to the fourth power, viz. one-sixteenth. If the radius is increased by 2, raise 2 to the fourth power, viz. 16). It should be obvious from these figures that relatively small differences in radius can produce large changes in resistance—and, thereby large changes in flow.

The factors involved in resistance to flow through cylindrical tubes were carefully studied by a French physician named Jean Leonard Marie Poiseuille. Poiseuille was primarily interested in the determinants of blood flow, but substituted simpler liquids for blood in his measurements of flow through glass tubes. The influences of pressure difference, viscosity, length, and radius were measured with impressive precision, permitting the empirical derivation of the fundamental principle called *Poiseuille's law.* In effect, Poiseuille determined the specific components of resistance R, and their mathematical relationship to each other:

$$R = \frac{8l\eta}{\pi r^4} \qquad (2)$$

Resistance is directly proportional to length l and viscosity η, and inversely proportional to the fourth power of the radius (r^4); $8/\pi$ can be regarded as a proportionality constant. Describing flow $\dot{Q}$ in terms of its two principal determinants, the pressure difference and resistance, *Poiseuille's law* is written as

$$\dot{Q} = \frac{\Delta P \pi r^4}{8l\eta} \qquad (3)$$

which is equivalent to

$$\dot{Q} = \frac{\Delta P}{R} \quad \text{where} \quad \frac{1}{R} = \frac{\pi r^4}{8l\eta} \qquad (4)$$

or, rearranging

$$R = \frac{\Delta P}{\dot{Q}} \qquad (5)$$

That is, R can be more simply described in terms of the pressure difference and flow. Instead of attempting to measure precisely the specific variables of resistance to flow in an organism's circulatory system (e.g., the lengths of all the individual vessels, the blood viscosity, and the vessels' various radii), the physiologist usually chooses to describe the resistance simply in terms of the ratio of pressure difference over the flow, Eq. (6). To simplify matters, one can divide ΔP in mm Hg by the flow in volume per unit time and simply call the units of peripheral resistance "peripheral resistance units" (PRUs). As originally described, ΔP in mm Hg was divided by the flow in cm³/min. Many physiologists prefer to use cm³/s. Some clinicians use L/min for flow. As you can see, there must be at least three kinds of PRU, depending on the particular units of volume or time used for flow (cm³/min, cm³/s, or L/min). Whenever describing resistance in terms of PRU, one must state specifically which units of flow one is using. For the remainder of

this chapter we will define our PRU as ΔP in mm Hg divided by flow as cm³/s:

$$\text{PRU} = \frac{\Delta P \text{ in mm Hg}}{\dot{Q} \text{ in cm}^3/\text{s}} \qquad (6)$$

For example, renal resistance is calculated as 4.5 PRU from a ΔP of 90 mm Hg and a flow of 20 cm³/s (90/20 = 4.5).

The value of calculating resistance in terms of ΔP divided by $\dot{Q}$ is not so much in the quantification of the resistance per se, but in what may be inferred about changes in the circulation that are responsible for changes in resistance. As pointed out above, resistance is directly proportional to length and to viscosity, while inversely proportional to the fourth power of vessel radius, Eq. (2). Since vessel length in vivo usually does not change much, one can assume that changes in resistance (as measured by changes in $\Delta P/\dot{Q}$) are due to changes either in viscosity or in radius. Although viscosity can change (decreased in anemia, increased in polycythemia), it normally does not change or influence the resistance of a given vascular bed. Thus, if viscosity can be shown to be constant, it can be concluded that changes in resistance are due to changes in radius (vasoconstriction or vasodilation).

HYDRAULIC ANALOGY TO OHM'S LAW

Defining resistance to flow in terms of pressure difference divided by flow is analogous to defining electrical resistance in terms of voltage drop divided by current flow. Indeed, the analogy of hydraulic flow of liquids to the flow of electrical current is a particularly good one. Figure 6 shows the principal elements of a simple circuit: E (electromotive force) represents a voltage difference (e.g., from a battery) and is analogous to a pressure difference in a hydraulic system. E could just as easily be represented by ΔV, but is traditionally written as E. The current I that flows is analogous to fluid flow $\dot{Q}$, and of course both systems have resistance R. *Ohm's law* states that the current flowing through a resistor is a function of the ratio of the voltage/resistance $(I = E/R)$. The hydraulic equiva-

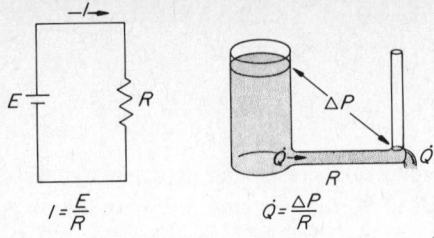

Figure 6 Simple electrical circuit to illustrate Ohm's law. For comparison, the analogous hydraulic circuit is also illustrated.

lent of Ohm's law is $\dot{Q} = \Delta P/R$. In both cases flow is inversely proportional to the resistance.

RESISTANCES IN SERIES

When a second resistor is added in series with the first, as in Figure 7, the resulting circuit has some interesting properties that are pertinent to our understanding of the circulation. First, the total resistance in the circuit is simply equal to the sum of the individual series resistances $(R_T = R_1 + R_2 + R_3, \ldots, R_n)$. Second, a portion of the source voltage is dropped across each resistor. The magnitude of this drop is proportional to the resistance. That is, the higher the resistance of a particular element, the greater will be the voltage drop across that element. This series resistance circuit is also called a voltage divider circuit in that the source voltage is divided into two voltages: E_1 and E_2. The magnitude of E_1 is equal to the total current times resistance R_1, and E_2 is equal to the total current times the resistance R_2. This relationship also applies to the pressure flow in vessels. As blood flows from

one end of a vessel to another, a pressure drop from one end of the vessel to the other will occur. The magnitude of this drop depends on the resistance through vessel and the flow rate. If the two vessels are in series, a portion of the total pressure drop occurs across each of them (Fig. 7).

$$E_T = IR_T \qquad\qquad \Delta P_T = \dot{Q}R_T$$
$$E_T = I(R_1 + R_2) \qquad \Delta P_T = \dot{Q}R_1 + \dot{Q}R_2$$
$$E_T = IR_1 + IR_2 = E_1 + E_2 \qquad \Delta P_T = \Delta P_1 + \Delta P_2$$

In other words, just as the total voltage drop in an electrical series resistance circuit is proportionally divided across each element according to that element's magnitude of resistance, so too is the total pressure difference (ΔP_T) divided across each resistance in series in a hydraulic system in proportion to each element's magnitude of resistance. Also, in a series electrical circuit, the total flow of electrical current through any resistance in the circuit is the same as that through any of the others *(Kirchhoff's current law)*. Similarly, because liquids are essentially incompressible, the total flow through one resistance in the series is the same as that through the others (the flow is subject neither to disruption nor compression), that is, $\dot{Q}$ is constant through the entire system for the hydraulic system in Figure 7.

For series resistance circuits—either electrical or hydraulic—the total resistance is the sum of the individual resistances.

$$R_T = R_1 + R_2 + R_3, \ldots R_n \qquad (7)$$

Figure 8 shows the circulation as a simple series resistance circuit. The seven series resistances represent the large arteries, the small arteries, arterioles, capillaries, venules, small veins, and large veins. As blood flows

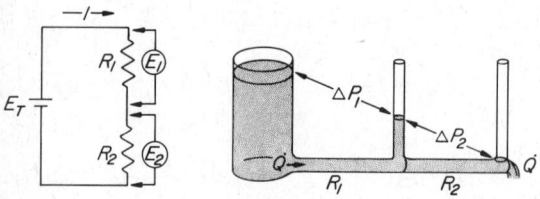

Figure 7 Electrical series resistance curcuit with analogous hydraulic circuit.

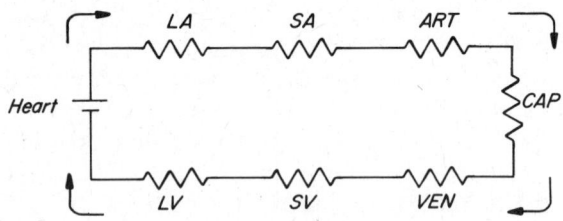

Figure 8 Circulatory system as a series resistance circuit. LA, large arteries; SA, small arteries; ART, arterioles; CAP, capillaries; VEN, venules; SV, small veins; LV, large veins.

through these channels, a pressure drop occurs across each type of vessel. It will be of interest later to quantitate these pressure drops and thus identify the high resistance and low resistance elements in the vascular tree.

This series circuit model is too simple for much of the circulatory system, for, as we know, there are a number of parallel channels from the arterial side of the circulation to the venous side (e.g., through the kidney, through the gut, through the muscles). Therefore, we must understand how current flow in a parallel resistance circuit, and how current flow through individual elements of such a circuit, can be altered.

PARALLEL RESISTANCES

Two electrical resistances in parallel are illustrated in Figure 9. Each resistor in a parallel circuit has the same voltage applied across it. All the electrical appliances in household circuits are connected in parallel so that each receives the same voltage. The total amount of current that flows is the sum of the individual currents. In the circulatory system numerous vascular beds (resistances) are in parallel with each other. As mentioned above, at the same time that blood is flowing through the kidneys it is also flowing through skeletal muscles, both resistances being in parallel. Both the renal vascular bed and the skeletal muscle vascular bed experience essentially the same large artery high pressure on the one side and essentially the same low pressure in the large veins of the other. It can then be appreciated that the amount of current or flow going through the resistors will vary inversely with the resistance. All the elements will have flow across them

all the time as long as their resistance is not infinite; however, the low resistance elements will experience greater flow than will the higher resistance elements. These flows can be calculated from Ohm's law. Such a parallel resistance circuit is also called a current divider circuit because in such a circuit the total current (flow) is divided among the individual resistors in inverse proportion to their resistances.

That the total flow is divided among the individual resistors is the same as saying that the total flow is the sum of the individual flows in parallel:

$$\dot{Q}_T = \dot{Q}_1 + \dot{Q}_2 \tag{8}$$

Although the flow is divided, ΔP across each resistance is the same. Thus, since $\dot{Q} = \Delta P/R$:

$$\Delta P/R_T = \Delta P/R_1 + \Delta P/R_2 \tag{9}$$

and since ΔP is the same for all resistors,

$$1/R_T = 1/R_1 + 1/R_2 \tag{10}$$

In a parallel resistance circuit (Fig. 9), the total resistance of the circuit is *less than* the resistance of any one element alone (true for both electrical and hydraulic systems). Intuitively one should see that the total flow from several holes (of same size) in a bucket will be greater than the flow through any one hole alone. That is, the total resistance of all the holes together is less than the individual resistance of each. What might not be readily apparent is that this is the same thing as saying the reciprocal of the total resistance is equal to the sum of the reciprocals of the individual resistances. One can either derive this mathematically, as above, or simply memorize:

$$1/R_T = 1/R_1 + 1/R_2 + \ldots + R_n \tag{11}$$

Figure 10 shows major circulatory beds in the body as variable parallel resistors. It can be appreciated from this diagram how blood might be shunted from one area to another when needed. For example, suppose that for some reason the kidneys needed more blood flow than they were getting. It would only be necessary for the resistance of the kidneys to drop. Once this were to occur, an increased portion of the total blood flow would then go through the kidneys, decreasing the total peripheral resistance. Routinely, other vascular beds might have to increase their resistance slightly so that the total peripheral resistance

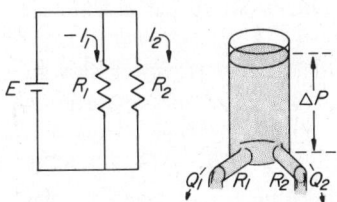

Figure 9 Electrical parallel resistance circuit with hydraulic analog.

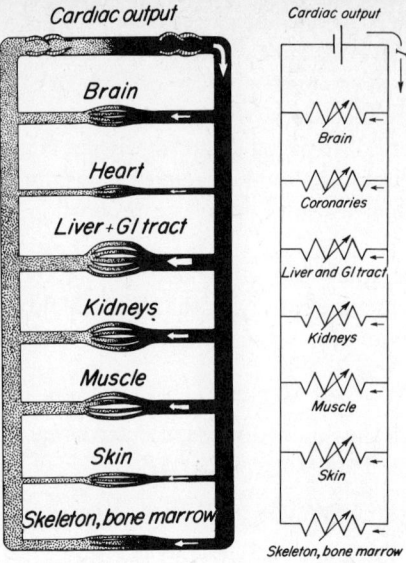

Figure 10 Circulatory system shown as electrical circuit with variable parallel resistances. Hydraulic analog is included for comparison.

Table 1 Blood Flow and Resistance in Various Vascular Beds in the Average Man at Rest[a]

Organ	ml/100 g/min	ml/min	% Cardiac Output	PRU[b]
Brain	55	770	14	7.0
Heart	70	200	4	27.0
Liver and GI tract	100	1,500	27	3.6
Kidneys	400	1,200	21	4.5
Skeletal muscle	5	1,300	23	4.2
Skin	10	200	4	27.0
Other	3	430	7	12.6
Totals		5,600/min	100%	1.0

[a]Total resistance is less than any one individual organ bed. This is because many of the circulatory beds are in parallel with each other.
[b]Peripheral resistance units calculated as mm Hg/cm^3 per second.

would not decrease and the main pressure head in the system would not drop. Maintenance of an adequate arterial pressure is necessary to assure normal brain and heart function. When the kidney is satisfied, its resistance could then increase, allowing blood to be shunted elsewhere. This kind of dynamic change in resistance in various vascular beds would permit second-to-second changes in flow as need would arise.

It would be interesting to see, under normal conditions, which vascular beds had the lowest resistance, and consequently the greatest flow. Table 1 shows these figures in different ways. It can be seen that the liver, GI tract, kidneys, and muscles command the largest percentage of the total cardiac output. These figures can be somewhat deceiving, however, for it seems intuitively obvious that a very large tissue should have a larger total blood flow than a very small tissue. Therefore, it is of interest to look at the figures for flow per minute per unit weight of tissue. When these figures are examined it can be seen that gram for gram the kidneys, the liver, the heart, and the brain have very high rates of blood flow. Two organs not included in Table 1, but that deserve mention, are the carotid bodies and the thyroid gland. Both have very high flow rates when expressed as ml/min/100 g. Indeed, the carotid bodies have the very highest relative flow rate (2,000 ml/min/100 g), and the thyroid gland has the second highest relative flow rate (600 ml/min/100 g).

VELOCITY

Velocity, sometimes designated more specifically as linear velocity, is the rate of movement with respect to time. It has the dimensions of distance per unit time (e.g., cm/s). If no turbulence is present, the smooth flow of liquid in a cylindrical tube occurs in concentric layers or laminae, and is called laminar or streamline flow. The fluid layer immediately adjacent to the vessel wall does not flow at all (has zero velocity). The next layer outward moves slowly past the stationary one. It moves slowly because of the frictional drag (viscosity) of the stationary layer on the moving layer. The next layer outward from the slowly moving one moves a little faster, and so on, with increasing velocities until maximal velocity is reached at the very center of the vessel (Fig. 11).

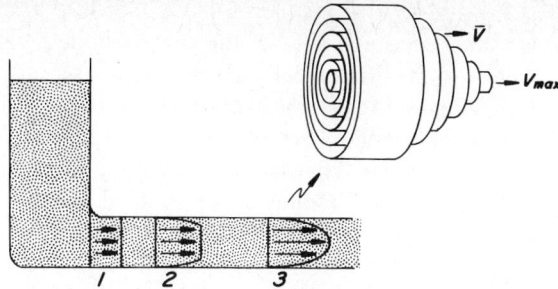

Figure 11 Laminar flow in a cylindrical tube. At inlet 1, the velocities are equal in all concentric laminae. Near the inlet at location 2, the velocity profile is flat for the laminae near the center of the tube, but a velocity gradient is established near the wall. When flow becomes fully developed, at location 3, the velocity profile is parabolic.

This kind of smooth flow, called laminar or streamline flow, is the usual type of flow found in most parts of the circulatory system. Indeed, the direct relationship between ΔP and $\dot{Q}$ described in Poiseuille's law applies only if there is laminar flow. Under conditions of turbulent flow (discussed below) the flow rate does not increase linearly with increases in ΔP. To describe the velocity of the flow of fluid in a vessel with laminar flow one often uses the average velocity $(\bar{v})$, which is one-half the maximal velocity ($\bar{v} = \frac{1}{2} v_{max}$).

RELATIONSHIP BETWEEN VELOCITY AND CHANGE IN CROSS-SECTIONAL AREA

As a liquid flows through a tube of varying radius (Fig. 12) or through a vascular system of vessels dividing and redividing into vessels of differing radii, *the average velocity* (v̄) *at any point varies inversely with the total cross-sectional area* (A) *at that point.* For a single tube of varying radius, the cross-sectional area is simply πr^2. For a vascular system of differing vessel size, the total cross-sectional area for any given category of vessels is the sum of the individual cross-sectional areas for each of the vessels of that type.

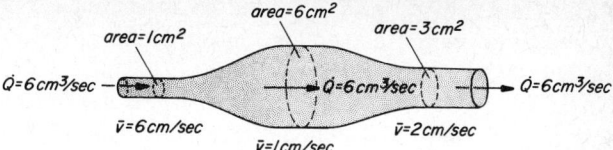

Figure 12 Relationships among average velocity, flow, and total cross-sectional area. Because of continuity of flow, $\dot{Q}$ is the same through all parts of the system (6 cm³/s). As the liquid flows through each part of the tube, the average velocity $\bar{v}$ varies inversely with the total cross-sectional area A of that part.

A common misconception among students as to why blood velocity slows down in going from the heart to the peripheral circulation is to attribute it to frictional resistance (viscosity).* Perhaps this stems from our early experience with frictional devices for slowing vehicular velocity (e.g., bicycle brakes, foot-dragging). Although frictional resistance may influence the rate of flow for the entire system (the influence of viscosity on flow), it does not increase it at one point or decrease it at another; the flow rate $\dot{Q}$ into a system of rigid tubing is the same as the volume flow through the system and the flow rate out of the system (hydraulic analog of Kirchhoff's law). This is the principle of continuity of flow and is often described by a formula known as the *continuity equation:*

$$\bar{v}_1 A_1 = \bar{v}_2 A_2 \qquad (12)$$

Equation (12) states that the product of average velocity times the total cross-sectional area is a constant at all points. If the flow is known, it can be included with velocity and area:

$$\bar{v} = \dot{Q}/A \qquad (13)$$

Thus, for a given flow (which is the same throughout the system) the only thing that determines the velocity is the total cross-sectional area.

This interrelationship among average velocity, flow, and total cross-sectional area is illustrated with a single tube of varying radius in Figure 12. The wide midsection has a total cross-sectional area six times that of the first section, and consequently an average velocity one-sixth as

---

*The principal frictional consequence of viscosity is the gradual loss of pressure with length or distance.

great. The third section has a total cross-sectional area one-half that of the midsection, and, therefore, has an average velocity twice as great as that of the midsection. That for a given flow the average velocity is dependent only on the total cross-sectional area is as true for the single tube of varying radius (Fig. 12) as for a vascular system that divides into a number of smaller vessels in parallel with each other (Fig. 13).

Thus it can be seen that the basis for the slowed velocity of blood in going from the heart to the peripheral circulation must be one of increased total cross-sectional area. As the small vessels of the peripheral circulation rejoin each other to form the larger veins for return of the blood to the heart, the total cross-sectional area again becomes smaller, and the average velocity again increases. As the cardiac output flows from the heart through the low or minimal cross-sectional area of the aorta (about 3 cm²), and on to the maximal total cross-sectional area of the venules (about 4,000 cm²), the average velocity slows from a maximum of about 35 cm/s (aorta) to a minimum of 0.026 cm/s (venules). That is, there is a little more than a 1,000-fold increase in total cross-sectional area along with a little more than a 1,000-fold decrease in average velocity. As the blood continues to flow from the venules back to the heart, the total cross-sectional area decreases from the maximum of about 4,000 cm² down to about 8 cm², for the two vena cavae together at their entrance into the right atrium. Along with this 500-fold decrease in total cross-sectional area is a 500-fold increase in average velocity (to about 13 cm/s).

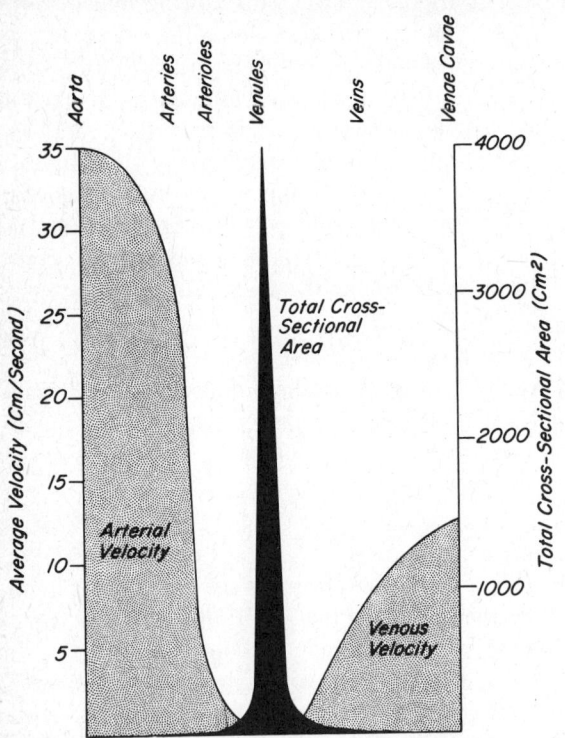

Figure 13 Total cross-sectional areas and average blood velocities for the different categories of vessels in human systemic circulation. Note that the minimal blood velocity occurs in the vessels with the maximal total cross-sectional area (the venules). The formula for the inverse relationship between blood velocity and total cross-sectional area is given in Eq. **(13)**.

EFFECT OF BRANCHING ON RESISTANCE AND PRESSURE GRADIENT

When a vessel bifurcates into two daughter vessels, the total cross-sectional area usually increases. However, the total resistance per unit length of the two daughter vessels (two resistances in parallel) may be either higher or lower than the parent vessel, depending on the extent to which the cross-sectional area has increased, assuming daughter branches to be of same length as parent branch.

As a simple example, consider the change in total cross-sectional area when a hypothetical main vessel of 5 mm radius divides into two daughter vessels, each having the same radius as that of the parent vessel. In this example the cross-sectional area will have increased by +100% (will have doubled). Since each of the daughter vessels has the same radius and, therefore, the same resistance per unit length as the parent vessel and are in parallel, the total resistance per unit length of the daughter vessels together is one-half that in the parent vessel (is halved). This is an example of a large increase in total cross-sectional area with a decrease in total resistance.

Blood vessels in actual vascular beds seldom bifurcate into daughter branches of the same radius as the parent and, therefore, have smaller increases in total cross-sectional area than the example above. A more physiological example would be for a main vessel of 5 mm radius to divide into two daughter vessels of 4 mm radius each. The total cross-sectional area of the daughter vessels together is greater than that of the parent vessel by only +28%. In this example the total resistance per unit length will have *increased* (by +22%). This is quite similar to the changes found in arterial bifurcations (+26% increase in total cross-sectional area). *A rule of thumb is that resistance per unit length increases if the total cross-sectional area increases by less than +41% and decreases if the total cross-sectional area is more than +41%.*

As the arteries branch into smaller arteries there is a slight increase in total cross-sectional area and a slight *increase* in total resistance per unit length with each bifurcation. The average velocity slows because of this increase in total cross-sectional area. There is some decrease in pressure due to frictional losses even in vessels that do not branch. However, with the increase in resistance per unit length that accompanies arterial branching there is an even greater drop in pressure. The main drop in pressure occurs as the blood moves through the arterioles (the arterioles are the principal resistance vessels in the circulation).

BERNOULLI'S PRINCIPLE

Poiseuille's law, discussed in a previous section, is concerned with only part of the energy in a moving fluid; namely, the part associated with the pressure difference (a potential energy difference that initiates the flow). A full accounting of the energy in a flowing fluid must include the kinetic energy component in addition to that of potential energy as pressure. Up to this point, we have been considering the driving force for flow as being only the pressure difference between two points (ΔP). Actually, this concept requires some modification. The true driving force for flow is not just a difference of pressure alone, but of a difference in the "total fluid energy" (kinetic plus potential energies) between any two points. For example, consider the flowing jet of water from the nozzle of a garden hose. The water continues to move forward after

it leaves the nozzle even though there is no pressure difference (no ΔP) between the nozzle end and the most distant forward point at which the water strikes the ground. However, although there is no potential energy of pressure difference between the end of the nozzle and the flowing stream, there is a kinetic energy component that causes the stream to flow forward from the nozzle.

Fluid moving within a tube or vessel also has both kinds of energy. The potential energy is represented by the pressure. As a viscous fluid moves down its pressure gradient, some of the potential energy is lost as heat because of friction (the principal cause of pressure loss with length as fluid flows through a tube). Furthermore, in initiating flow, part of the original potential energy of pressure is converted into the kinetic energy of movement. Bernoulli developed a principle to describe the relationship between the potential and kinetic energies in a flowing system. Although the theorem on which this principle was based involved the theoretical behavior of hypothetical "frictionless" or nonviscous fluids, the conclusions that followed are applicable to the flow of real fluids (those that have viscosity).

In general, Bernoulli's principle is nothing more than a restatement of the first law of thermodynamics (conservation of energy) as applied to the flow of fluid in a tube. Anything increasing one form of energy at any one point in a flowing system must be accompanied by an equal decrease in another form of energy at that point. For example, as blood flows through a stenosed or narrowed segment of a vessel, the velocity increases (review the section on the relationship between velocity and change in cross-sectional area). This increase in velocity causes the kinetic energy of the blood moving through the narrowed segment to increase. The resultant increase in kinetic energy must be at the expense of the potential energy and is manifested as a decrease in blood pressure within the narrowed segment.

A more significant consideration of Bernoulli's principle in relation to hemodynamics is that of directly measuring blood pressure by means of a cannula or catheter. If a saline-filled catheter is threaded into a peripheral artery and pushed centrally into the aorta, aortic blood pressure can be measured by means of an appropriate pressure transducer. If the catheter attached to the pressure transducer at one end is simply left open at the other end facing the oncoming bloodstream within the blood vessel, blood will flow around the fluid-filled catheter but will not flow into it. There is a small area of stagnation at the open end of the catheter. Since there is no flow at this stagnation point, there is no kinetic energy component, and the total fluid energy is entirely expressed as potential energy of pressure (measuring what some physiologists refer to as "end pressure"). This is not the true pressure of the flowing blood. On the other hand, if the catheter tip has a lateral opening at a right angle to the flow, the pressure measured ("lateral pressure"), which is not as great, is considered the true blood pressure that normally exists in the aorta. In a normal individual at rest, this difference in aortic pressures is negligible (about 1 mm Hg). However, during heavy exercise when the velocity of flow in the aorta can increase severalfold, the difference between the artifactual end pressure and the true pressure could be as much as 20–40 mm Hg.

TURBULENT FLOW

Poiseulle himself observed that the relationship between ΔP and $\dot{Q}$ ceased to be linear at high rates of flow and rightly attributed this to a breakdown of laminar flow into turbulent flow. However, it was Osborne Reynolds (1883) who provided the first serious analysis of the problem. Using a long rigid cylindrical tube, he showed that at low velocities the motion of the flowing fluid was smooth and regular (laminar flow). Any disturbance introduced into this stable flow (e.g., by irregularities of the tube wall) was soon damped out and did not persist long. At higher velocities the liquid became more sensitive to disturbances. At a critical point a stage was reached in which stable flow could no longer be maintained since the slightest disturbance would cause the motion of the fluid to become wildly irregular so that it contained vortices and eddy currents having largely random movements. This transition from laminar to turbulent flow comes about because at the critical point small disturbances are amplified and quickly spread throughout the flow. "Big whirls have little whirls which feed on their velocity. Little whirls have smaller whirls and so on to vorticity" (L. F. Richardson).

Reynolds found the critical point to be dependent on the mean velocity $\bar{v}$, the tube radius r, the fluid density ρ, and its viscosity η. This was expressed in the form of an equation for a dimensionless quantity known as the Reynolds' number Re:

$$Re = \frac{\bar{v}r\rho}{\eta} \tag{14}$$

When $\bar{v}$ is measured in cm/s, r in cm, and viscosity in poise, turbulence usually appears when Re exceeds 1,000, the critical Reynolds' number (Re_c).*

For the flow of blood (having a given density and viscosity), the significant physiological variables in this Reynolds equation are $\bar{v}$ and r. For two different-size vessels having the same flow velocities, the tendency toward turbulence will be greater in the larger vessel. For two vessels of the same radius, but having different velocities flow, the tendency toward turbulence will be greater in the one with the higher velocity. Thus far, the comparison of two different vessels equal in one variable and different in the other, provides a straightforward way to visualize the effects of a change in one variable alone. In both comparisons the flows are also different between the two tubes. A possible source of confusion can arise from comparisons in conditions in which the flow is regulated to a constant value. For example, consider the influence of a decrease in radius in the narrowing of a valvular orifice in the heart (a stenotic valve). If the degree of narrowing is not severe, the heart will increase the ΔP sufficiently to maintain a normal cardiac output through the valve ($\dot{Q}$ will remain normal). If one were to ignore the effect that this ΔP would have on the $\bar{v}$, one might look at Eq. (14) and assume that a decrease in r in vivo would cause a decrease in Re. Such an assumption would be incorrect because r is not the only variable changed in this instance. Because of the increased ΔP, in order to force a normal $\dot{Q}$ through the restrictive valve opening at a high velocity, there would be a disproportionately greater increase in $\bar{v}$ as compared with the decrease in r. The net effect would be an increase in Re and an increased tendency for turbulence. Indeed, such stenotic valves often have an audible murmur, which is thought to be a classic manifestation of turbulence.

---

*Note that if the diameter of the tube is used in the formula instead of the radius, as is sometimes the case, Re_c has a value of 2,000.

The best way to demonstrate the onset of turbulence is to find the discontinuity in the influence of ΔP on $\dot{Q}$. With stable laminar flow the ΔP per unit length (pressure gradient) varies linearly with the flow; however, when turbulence develops, there is less increase in $\dot{Q}$ for each additional increase in ΔP. Beyond Re_c a much greater pressure gradient is required to maintain a level of flow than would be the case if the flow were laminar (Fig. 14). A greater pressure gradient is required to supply extra energy to offset the energy wasted by increased random movements of vortices and eddy currents in turbulent flow.

The relationship between Re and turbulence for steady flow through straight tubing in vivo is fairly clear. A relevant question is, what happens with blood flow in vivo? For vessels with steady flow the relationship between ΔP and $\dot{Q}$ is essentially the same as it is in vitro with one notable exception: that geometric considerations (e.g., curving and branching) appear to favor lower Re_c values in vivo than the Re_c of 1,000 in vitro. Since the circulatory system is an extensive distributory system, bifurcations and junctions are a prominent feature. In a bifurcation of an artery into two daughter branches, the total cross-sectional area of the daughter branches together increases, whereas the radius and velocity in each daughter branch is decreased. Thus, each time an artery branches, the blood flows from an area of higher Re to one of lower Re, a situation that is supposedly unfavorable to the development of turbulence. However, the point or apex of the bifurcation constitutes a discontinuity akin to a physical obstacle in the flow, and the potential for development of

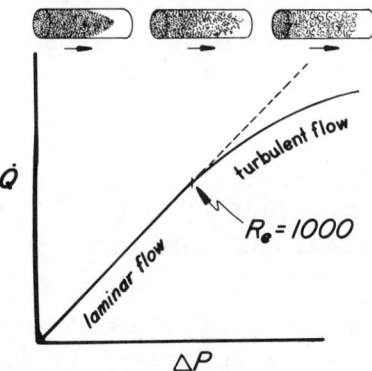

Figure 14 Relationship between pressure head (ΔP) and flow ($\dot{Q}$) in a rigid tube during streamline flow and its transition to turbulent flow.

a disturbance or turbulence is actually increased. The greater the angle of branching, the greater the likelihood of turbulence despite a lowered Re number. Another complication in vivo is that of pulsatile flow. Although flow is relatively steady in capillaries and small veins, it is pulsatile in arteries and in the larger central veins. The larger veins have fluctuations in pressure resulting from the cycle changes in intrathoracic venous pressure with inspiration and expiration. Close to the heart, venous flow becomes quite pulsatile (e.g., the jugular pulse). The problem with pulsatile flow is that the concept of a Re_c distinguishing laminar from turbulent flow is based on steady flow conditions. Reynolds' equation (14) cannot be applied with accuracy to conditions of pulsatile flow. Therefore, whereas flow in capillaries and small veins is both steady and of low velocity (low Re values) and, therefore, laminar in nature, the question of what kind of flow occurs in larger pulsatile vessels has been somewhat controversial until recently.

Recent studies have measured instantaneous velocities (discrete moment-to-moment changes in velocities) by means of hot film anemometry, in which the amount of electrical current necessary to maintain a constant temperature in a heated metallic film on a flow probe is proportional to the velocity of the blood flowing past it. Once this current has been quantified, one can determine the velocity. Any disturbances show up as rapid oscillations in velocities. Furthermore, minute sound vibrations, not audible at the chest wall, can be detected by means of a catheter tip micromanometer. Such methods have given further evidence of turbulence in vivo, but only in very limited areas of the circulatory system. True turbulence appears to be found only in the vicinity of the heart valves in normal human subjects at rest (Fig. 15).

Clinical interest in disturbance of flow arises from two principle sources: (1) diagnostic interest in audible murmurs from patholigical heart valves, and (2) etiological interest in possible participation of disturbed flow patterns in the pathogenesis of various focal lesions in the aorta and arteries.

MURMURS

The existence of audible sound (the heart sounds and murmurs, when present) indicates that acoustic energy is being dissipated over and above other losses. Among pio-

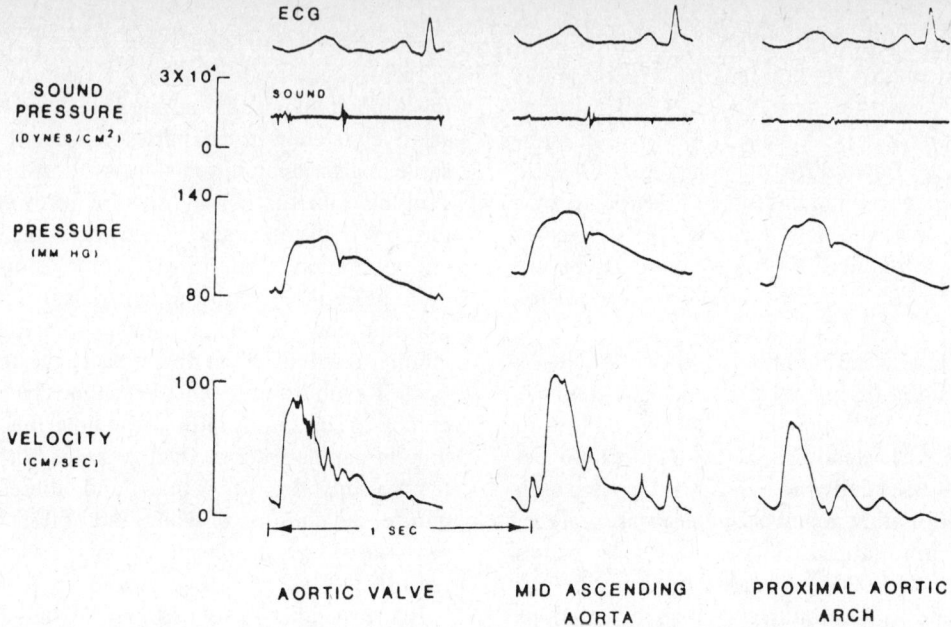

Figure 15 Comparison of simultaneous events of ECG, sound pressure vibrations, arterial blood pressure, and blood velocity in a normal human subject at the level of the aortic valve, midascending aorta, and proximal aorta. At the aortic valve, low amplitude sound vibrations (picked up by the use of a catheter tip micromanometer but inaudible at chest wall) occurred simultaneously with peak disturbances (turbulence) of flow. No significant turbulence or sound (other than normal second heart sound) were recorded in the midascending aorta and proximal aortic arch. (From Sabbah HI, Stein PD: *Circ Res* 38:513, 1976, by permission of the American Heart Association, Inc.)

neers of auscultation attempting to explain origins of this acoustic energy were Corrigan (1829), who stated that murmurs were caused by "an alteration in the movement of blood, instead of its equable progressive motion en masse," and Rouanet (1844), who stated that the basis of murmurs was to be found in the formation of turbulence *(tourbillons).* Conditions that favor the production of murmurs are those of greatly increased velocity of flow. The long-held idea that turbulence produces murmurs now seems well established. Murmurs appear to occur especially easily where blood is rapidly pushed through a narrow segment into a wider one. Aortic or mitral valve stenoses, coarctation of the aorta, and patent ductus arteriosus, are all associated with murmurs. The Korotkow sounds of sphygmomanometry may also have their origin in high velocity flow.

In severe anemia functional cardiac murmurs—murmurs not caused by structural abnormalities—are frequently detectable. The physical basis for such murmurs is commonly thought to reside in (1) the reduced viscosity of blood caused by the low red blood cell content, and (2) the high velocity of flow associated with marked augmentation of cardiac output that usually occurs in anemic patients.

POSSIBLE RELATIONSHIP BETWEEN DISTURBED FLOW AND ARTERIAL DISEASE

Interest in the nature of disturbed flow patterns in arteries continues to increase because of possible participation in the pathogenesis of certain localized arterial disease processes. Arterial disease is more common at bifurcations and branch orifices; it is at these locations that the velocity profile is no longer perfectly parabaloid. However, although the flow is not typical of ideal laminar flow, neither is it turbulent (does not have randomized scattering of eddy currents in all directions). The arterial wall at such nonideal locations might be more susceptible to disease than others. Model studies with glass tubing and animal studies suggest that changes in hemodynamic flow patterns could produce changes in wall structure and function that might lead to atheromatous plaque formation. However, a line of evidence from early wall changes to chronic atheromatous lesion production remains to be established.

SUMMARY

Contraction of the ventricle supplies hydraulic pressure for flow of blood through the resistance of the circulatory system. The blood moves down its pressure gradient from the arterial side through the capillaries and veins to return to the heart (pressure near zero at the right atrium). The rate of flow ($\dot{Q}$) through the circulatory system, that is, the cardiac output, is proportional to this pressure difference ΔP and inversely proportional to the resistance R. As the blood flows, it gradually loses pressure due to frictional losses of energy as it moves through the resistance. The velocity of flow $\bar{v}$ also decreases as the blood moves from the arterial side into the capillaries. Unlike the diminution in pressure, however, this decrease in velocity is not the result of frictional losses, but is simply the result of an increase in total cross-sectional area. As the blood moves into the veins the total cross-sectional area decreases and velocity rises again. Pressure continues to fall as the blood moves through the veins.

Many of the various vascular beds are resistances in parallel with each other (with the exception of notable series resistance systems such as the hypophyseal and hepatic portal systems, and the relationship between the glomerulus and peritubular capillaries in the renal circulation). The distribution of the cardiac output to the different vascular beds is determined by their relative resistances (primarily a function of arteriolar radius). Many vascular beds regulate their flows to meet their needs. However, the brain and heart have first priority and reflex mechanisms to maintain an adequate mean arterial pressure will cause increases in the resistance of numerous vascular beds to assure this. This increase in total peripheral resistance is also primarily a function of arteriolar radius (in this case a widespread vasocontriction of arterioles in many beds).

As blood flows through the circulatory system, most of the energy in the system is potential energy (as pressure), and the remainer is kinetic energy (energy of motion). Normally the amount of total energy as kinetic energy is small in most vessels, but it can increase significantly with high flow rates, as in exercise. When the velocity of flow is fairly high, the development of turbulence is possible. This is particularly true for the pulsatile flow of blood in the immediate proximal aorta. Turbulent flow can give rise to audible murmurs of diagnostic significance. Disturbed flow that is not turbulent might be implicated in the pathogenesis of focal arterial lesions. Consideration of the various principles governing hemodynamics is not only helpful in understanding normal cardiovascular function, but is also of importance clinically in the diagnosis and correction of cardiovascular flow problems.

SELECTED BIBLIOGRAPHY

Berne RM, Levy MN: *Cardiovascular Physiology,* ed. 4. St. Louis, Mosby, 1981.

Caro CG, Pedley TJ, Schroter RC, et al: *The Mechanics of the Circulation.* New York, Oxford University Press, 1978.

Hwang NH C, Normann NA: *Cardiovascular Flow Dynamics and Measurements.* Baltimore, University Park Press, 1977.

McDonald D: *Blood Flow In Arteries,* ed 2. Baltimore, William & Wilkins, 1974.

Stehbens WE: *Hemodynamics and the Blood Vessel Wall.* Springfield, Charles C Thomas, 1979.

SELF-STUDY QUESTIONS

1. What is the pressure, in dyn/cm², of a column of blood of 100 cm height? (Assume the density of blood to be essentially the same as that of water.)

   A. 9.8
   B. 98
   C. 980
   D. 9,800
   E. 98,000

2. Express the above pressure in mm Hg (torr).

   A. 0.74
   B. 7.35
   C. 73.5
   D. 735
   E. 7,350

3. Express the above pressure in kPa.

   A. 0.1
   B. 1.0
   C. 9.8
   D. 98
   E. 980

   Two organ vascular beds, C and D, are arranged in parallel, as shown in the drawing below. The arterial inflow pressure P_a and the venous outflow pressure P_v are the same for both organs. These pressures and the blood flows, $\dot{Q}_C$ and $\dot{Q}_D$, are given below:

   $P_a = 100$ mm Hg
   $P_v = 0$ mm Hg
   $\dot{Q}_C = 20$ ml/s
   $\dot{Q}_D = 30$ ml/s

4. The total resistance to flow R_t offered by both organ vascular beds together, in peripheral resistance units (PRU) in mm Hg/ml/s is

   A. 80
   B. 40

   C. 20
   D. 2.0
   E. 0.2

5. When the vessels in organ vascular bed D are constricted to one-third their former radius without a change in P_a or P_v, the mean blood flow through organ vascular bed D,

   A. is unchanged.
   B. is diminished by one-third.
   C. is tripled.
   D. is diminished to one-ninth its former value.
   E. is diminished to one-eighty-first its former value.

6. In the system illustrated below, both the radius and the length of the tube are doubled. According to Poiseuille's law, the resistance R will

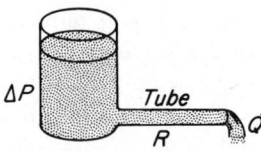

   A. remain unchanged.
   B. double.
   C. increase to a level 16 times its previous value.
   D. decrease to a level one-sixteenth its previous value.
   E. decrease to a level one-eighth its previous value.

7. Suppose the system illustrated in question 6 has a tube of 1 cm radius and a flow of 102.4 ml/s. Assuming a constant pressure head (e.g., if the reservoir is replenished to maintain the same height) and a constant viscosity, what would the new flow rate be (in ml/s) if the tube radius were decreased to 0.25 cm?

   A. It would be essentially unchanged.
   B. 25.6
   C. 2.56
   D. 0.40
   E. 0.10

8. Note the following values:

Cardiac output = 100 ml/s; mean systemic arterial pressure = 100 mm Hg; right atrial pressure = O mm Hg; mean pulmonary arterial pressure = 14.5 mm Hg and left atrial pressure = 2 mm Hg.

By calculating the total peripheral resistances of the systemic and pulmonary circuits, one can conclude that the resistance of the pulmonary circuit is

A. essentially the same as that of the systemic circuit.

B. 12.5 times higher.

C. 8 times higher.

D. 12.5 times lower.

E. 8 times lower.

9. Two vessels in series have a total ΔP of 100 mm Hg across them (together). The second one has four times the resistance of the first. What are the individual ΔP values (in mm)?

A. 100 for each.

B. 80 for each.

C. 20 for each.

D. 80 for the first and 20 for the second.

E. 20 for the first and 80 for the second.

Two organ vascular beds are connected in parallel as illustrated below. The total flow ($\dot{Q}_t$) is maintained at a constant 100 ml/s.

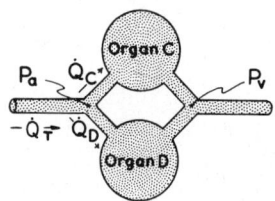

10. If organs C and D have equal resistances, what is the flow (in ml/s) through each?

A. 100

B. 80

C. 50

D. 25

E. 10

11. Which of the following vessels (with collective total cross-sectional area in cm²) would you expect to have the most rapid blood velocity?

A. Small arteries (80 cm²)

B. Arterioles (100 cm²)

C. Venules (3,000 cm²)

D. Small veins (800 cm²)

E. Vena cavae (8 cm²)

12. If a vessel has a cross-sectional area of 1 cm² and a flow of 10 cm³/s, what is the average velocity of blood (in cm/s) in the vessel?

A. 100

B. 10

C. 1

D. 0.1

E. 0.01

13. Suppose the vessel described in question 12 carrying 10 cm³/s, develops an aneurysm with a total cross-sectional area of 2 cm². What would be the average velocity (in cm/s) in the aneurysm?

A. Absolute stagnation (no velocity)

B. Higher velocity than in question number 12

C. 50

D. 5

E. 0.5

14. If a small person has an aorta with a cross-sectional area of 2 cm² and an average aortic blood velocity of 30 cm/s, what would you expect the venular blood velocity to be (in cm/s) if the total cross-sectional venular area were 3,000 cm²?

A. 50

B. 2.0

C. 0.5

D. 0.02

E. 0.005

15. Only a small segment of a blood vessel has undergone vasoconstriction (see illustration below).

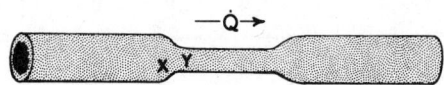

In comparing points X and Y, one would conclude that at point Y there is

A. increased velocity and increased pressure.

B. increased velocity and decreased pressure.

C. decreased velocity and decreased pressure.

D. decreased velocity and increased pressure.

E. no change in either velocity or pressure.

16. A patient has a cardiac output of 9 L/min (150 cm³/s) and an aortic radius of 1 cm. Assume the blood viscosity to be 0.04 poise and the density to be 1. The Calculated Reynolds' number in the aorta is

A. nearly 800.

B. just over 900.

C. nearly 1,200.

D. over 2,000.

E. unusually high Re and not found in vivo.

17. The correctly calculated Re above is

A. much too low for turbulence.

B. just below the critical Re for turbulence.

C. the critical Re.

D. greater than the critical Re.

E. unusually high and not found in vivo.

18. An adult has a constriction of the aorta just below the insertion of the ductus arteriosus (a postductal coarctation of the aorta). In the case illustrated here, the constricted segment has a radius of 0.1 cm, the flow through this segment is 50 cm³/s, and viscosity and density are as in question 16. The Re in the segment is

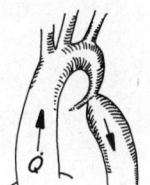

A. below the critical Re.

B. identical to the critical Re.

C. nearly 2,000.

D. nearly 4,000.

E. more than 8,000.

ANSWERS

1. E. Substitute the following into Eq. (1): density = 1 g/cm³, acceleration of gravity = 980 cm/s², and height = 100 cm. $P = 1 \times 980 \times 100 = 98,000$ dyn/cm².

2. C. One way to do this is to substitute 98,000 dyn/cm² in Eq. (1). Use 13.6 g/cm³ for density of mercury and 980 cm/s² for acceleration of gravity. Solve for h (be careful to convert the answer in cm Hg to mm Hg). Another way to work this is to convert 100 cm H₂O to mm Hg by multiplying by 0.74, for a close approximation.

3. C. 100 cm of blood (essentially the same as 100 cm H₂O) is equal to 73.5 mm Hg or equal to 9.8 kPa. To convert 73.5 mm Hg to kilopascals multiply by 0.13328, to get an answer of 9.8 kPa. Another way to work this is to change 73.5 to 7.35 and add one-third: $7.35 + 2.45 = 9.8$ kPa.

4. D. Recall that with parallel resistances the total flow $(\dot{Q}_t)$ is the sum of the individual flows $(\dot{Q}_c + \dot{Q}_d)$ and

that the ΔP is the same for each. Thus $R_t = \Delta P / \dot{Q}_t$, or $100/50 = 2$ PRU. A longer method would be to calculate the individual resistances and substitute into $1/R_t = 1/R_c + 1/R_d$. (Hint: Be sure to convert $1/R_c$ and $1/R_d$ to decimal fractions before adding them together.)

5. E. If everything else is unchanged, the change in flow is easily calculated from the relative change in radius raised to the fourth power: $(\frac{1}{3})^4 = \frac{1}{81}$.

6. E. $R = \dfrac{8l\eta}{\pi r^4}$. Although resistance would tend to double from a doubling of length, it also tends to decrease to one-sixteenth by doubling the radius. The net change is a combination of the two with the l term in the numerator and the r^4 term in the denominator $(\frac{2}{16} = \frac{1}{8})$.

7. D. Everything remaining unchanged, the relative change in $\dot{Q}$ is equal to the relative change in radius raised to the fourth power $(\frac{1}{4}) = \frac{1}{256}$. 102.4 ml/s $\times \frac{1}{256} = 0.4$ ml/s.

8. E. The resistance of the systemic circuit in PRU values is the ΔP divided by the cardiac output. The resistance of the pulmonary circuit is the ΔP for that circuit divided by the cardiac output (the same cardiac output that goes through the systemic circuit goes through the pulmonary circuit). The systemic resistance is 1 PRU and the pulmonary resistance is 0.125 PRU; that is, 8 times less.

9. E. ΔP is directly proportional to R for a given $\dot{Q}$. $\Delta P = \dot{Q}R$.

10. C. Since the resistances are equal, the flow is equally divided.

11. E. Of the vessels listed, the vena cavae have the smallest total cross-sectional area (A) and would, therefore, have the highest velocity $(\bar{v} = \dot{Q}/A)$.

12. B. $\bar{v} = \dot{Q}/A$, again. (10 cm/s = 10 cm³/s divided by 1 cm²).

13. D. $\bar{v} = \dot{Q}/A$, again. (5 cm/s = 10 cm³/s divided by 2 cm²).

14. D. First, determine the cardiac output $(\dot{Q})$ from a rearrangement of Eq. (13) as $\dot{Q} = \bar{v}A$. $\dot{Q} = 30$ cm/s $\times 2$ cm² = 60 cm³/s. Second, apply Eq. (13) as $\bar{v} = \dot{Q}/A$ for the venules, and substitute the $\dot{Q}$ just deter-

mined: $\bar{v} = 60$ cm/s divided by 3,000 cm² = 0.02 cm/s through the venules.

15. B. As the kinetic component increases (due to increased velocity), the potential component of total hydrodynamic energy decreases (decreased pressure).

16. C. Using Eq. (13), calculate $\bar{v}$ from $\bar{v} = \dot{Q}/A$. $\bar{v} = 47.7$ cm/s. Substitute $\bar{v}$, r, ρ, and η into Eq. (14) to calculate Re of 1,194.

$$\bar{v} = \frac{150 \text{ cm}^3/\text{s}}{\pi \times 1^2} = 47.7 \text{ cm/s}$$

$$Re = \frac{47.7 \times 1 \times 1}{0.04} = 1,194$$

17. D. Re_c is 1,000. Since the calculated Re in this problem is 1,194, some turbulence would be expected on the basis of Reynolds' number alone.

18. D. Using Eq. (13), calculate $\bar{v}$ from $\bar{v} = \dot{Q}/A$. $\bar{v} = 1592$ cm/s. Substitute $\bar{v}$, r (0.1 cm in this problem), ρ, and η in Eq. (14) to calculate a very high Re of 3979.

$$\bar{v} = \frac{50 \text{ cm}^3/\text{s}}{\pi \times 0.1^2} = 1,592 \text{ cm/s}$$

$$Re = \frac{1592 \times 0.1 \times 1}{0.04} = 3,979$$

11

Systemic Arteries and Veins

R. DAVID BAKER

OBJECTIVES

After completion of this chapter, you should be able to

1. Know pressure–volume (compliance) curves for the arterial system.

2. Understand how the four independently adjustable parameters influence mean arterial volume and pressure.

3. Know the determinants of pulse amplitude.

4. Understand the effects of exercise, aging, and chronic hypertension on arterial pressures (mean, systolic, and diastolic).

5. Appreciate the importance of the arterial *Windkessel* effect.

6. Understand the factors that influence arterial pulse wave velocity and contour.

7. Understand the venous pressure gradient and be familiar with venous pressure–volume curves.

8. Understand the various ways by which the venous system can serve as an adjustable blood reservoir.

This chapter discusses the physical properties of systemic arteries and veins that contribute to the hydraulic behavior of the cardiovascular system. The physical factors that determine mean arterial pressure and the amplitude of the pressure pulse are explained, as are the factors that influence the shape and velocity of the pulse wave as it travels along the arterial tree. The reservoir function of the venous system is discussed. This chapter introduces the concept of four major independently adjustable parameters of the cardiovascular system: cardiac effectiveness, total peripheral resistance, total blood volume, and venous capacity. Adjustments in these values are responsible for changes in the hydraulic behavior of the cardiovascular system.

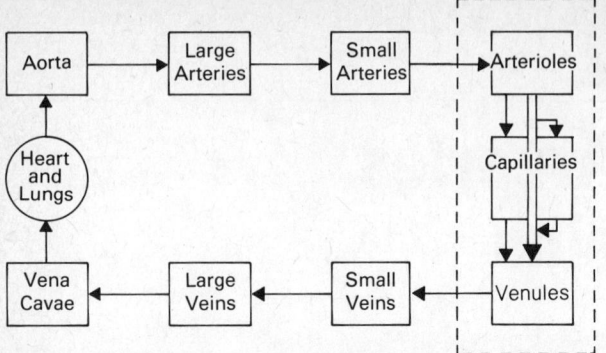

Figure 1 Block diagram of the circulatory system. The components within the dashed box are called the microcirculation because a microscope is needed to see them. The large arrow from arterioles to venules represents the throughfare channel or metarteriole found in certain vascular beds.

The systemic circulatory system consists of all the vascular plumbing from aorta to vena cavae. Figure 1 names the components of the system; to complete the circuit, the pump oxygenator (heart and lungs) is included. As shown in Figure 2, the total cross-sectional area of this system progressively increases to the level of the venules and then decreases again to the vena cavae. The distribution of blood volume in the major parts of the systemic circula-

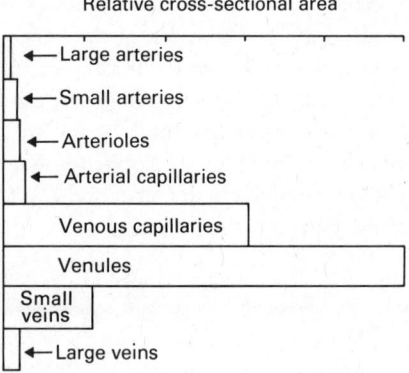

Relative cross-sectional area

Figure 2 Total cross-sectional areas of various parts of the systemic vascular system. In some vascular beds, such as the one studied by Wiedeman, there are two kinds of capillaries in series with each other. The "arterial capillaries" do not have a particularly broad total cross section, but they branch into a much broader system of "venous capillaries." (From the data of Wiedeman MP: *Circ Res* 12:375, 1963.)

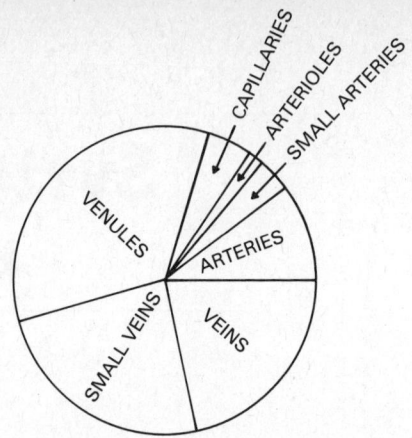

Figure 3 Distribution of blood volume in various parts of the systemic vascular system (% of total volume). (From Wiedeman MP: *Circ Res* 12:375, 1963.)

tion is illustrated in Figure 3. Note that despite their large total cross-sectional area, capillaries contain relatively little blood; this is because they are so short. Also note that most of the systemic blood is in veins and venules. The pressures in various parts of the system are illustrated in Figure 4. Most of the pressure drop from arterial to venous sides of the circulation occurs in the terminal arteries and arterioles, especially the latter. The relative resistances to flow through the various components are shown in Figure 5. The arterioles have the highest resistance. The diameter, and therefore the resistance through the arterioles, is adjustable (by smooth muscle contraction) and can be adjusted differently and independently in the various organs of the body. It is largely the arterioles, then, that are responsible for regulating the resistance to flow through each vascular bed. The physiology of the microcirculation will be discussed in Chapter 17. Here we treat the arterial and venous systems.

THE SYSTEMIC ARTERIAL SYSTEM

DETERMINANTS OF MEAN ARTERIAL PRESSURE

The pressure in any segment of an artery depends on the volume of blood distending it and its viscoelastic proper-

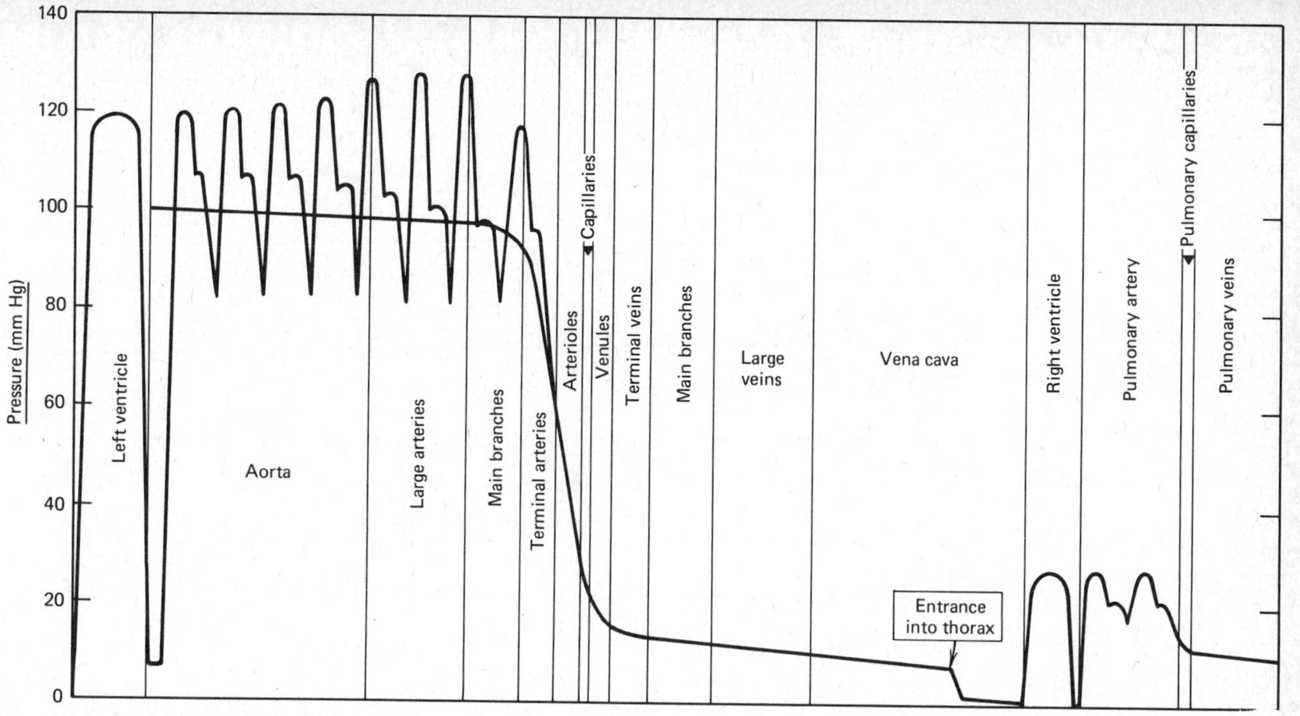

Figure 4 Pressures throughout the cardiovascular system. Relative length; arteriole; capillaries, and venules are shorter than indicated. (From Selkurt EE, et al: *Physiology,* ed 3. Boston, Little, Brown, 1971.)

ties. If the volume of a healthy aorta is measured as a function of the pressure applied to it, a pressure–volume curve like that shown in Figure 6 is obtained. The greater the volume, the greater the pressure. The slope *(dV/dP)* at any point along such a pressure–volume *(P–V)* curve is called the compliance. Arterial compliance is a measure of how easily an artery can be expanded from any given volume to a slightly larger volume.

Mean arterial pressure (MAP) is normally about 90 mm Hg. This pressure would require a mean volume of blood in the aorta of 225 ml (Fig. 6, point M). If mean arterial volume (and consequently mean aortic volume) increases or decreases, MAP changes accordingly along the *P–V* curve. For example, if mean aortic volume rises to 300 ml, MAP will rise to 125 mm Hg; if aortic volume drops to 150 ml, MAP will fall to 50 mm Hg. Four independently adjustable factors cause changes in mean arterial volume and, therefore, in MAP: (1) the total blood volume contained in the entire cardiovascular system, (2) the capacity

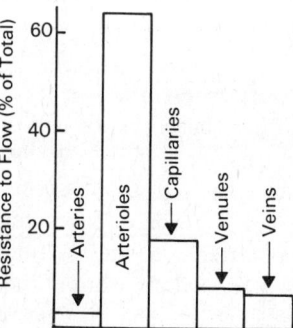

Figure 5 Relative resistances to blood flow in various components of the systemic vascular system.

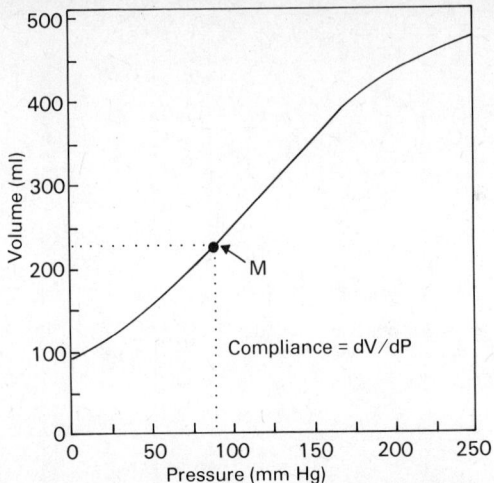

Figure 6 Pressure–volume diagram for the human aorta. M, mean pressure and volume. (Based on data of Hallock P, Benson IC: *J Clin Invest* 16:595, 1937.)

of the venous system, (3) the total peripheral resistance, and (4) the effectiveness of the heart as a pump.

Blood Volume

If there is an increase in total blood volume, the excess distributes throughout the cardiovascular system. Some of this excess is pumped into the systemic arteries and, unless compensatory changes take place, MAP will rise.

Venous Capacity

If veins constrict because of contraction of smooth muscle in their walls, or if veins are compressed by pressing on them, some blood tends to be redistributed from the veins to the arteries; consequently, unless compensatory changes take place, mean arterial volume and MAP rise.

Peripheral Resistance

Changes in total peripheral resistance lead to redistribution of blood between the systemic arteries and the rest of the cardiovascular system. If total peripheral resistance is suddenly decreased it becomes easier for blood to flow from the systemic arteries to the systemic veins, and for a few heartbeats blood will flow out of the arterial system faster than it is being pumped in. Consequently, during these few heartbeats, the mean arterial blood volume and

pressure decrease. A sudden increase in peripheral resistance has the opposite effect, that is, a redistribution of blood from the venous side to the arterial side with a consequent rise in MAP.

Effectiveness of the Pump

The ability of the heart to pump blood can be changed by altering contractility or heart rate. If contractility or heart rate or both increase, then, for a few heartbeats, cardiac output will exceed the flow rate through the microcirculation, and mean arterial blood volume and pressure will increase.

FUNDAMENTAL EQUATION FOR MAP

Total peripheral resistance is defined as the pressure difference that drives flow around the systemic circuit divided by the total flow rate. In other words

$$R_{TP} = \frac{MAP - RAP}{CO} \tag{1}$$

where R_{TP} is the total peripheral resistance, RAP is right atrial pressure, and CO is cardiac output. This definition of resistance is the same as that given in Chapter 10, Eq. (5), except that it specifically applies to the entire systemic circuit. Since right atrial pressure is ordinarily close to 0 mm Hg, it can usually be ignored, and Eq. (1) becomes

$$R_{TP} = \frac{MAP}{CO} \tag{2}$$

or

$$MAP = CO \times R_{TP} \tag{3}$$

Equation (3) is extremely important. It tells us that MAP is always simply the product of cardiac output and total peripheral resistance. If you are wondering what happened to the other factors (blood volume, venous capacity, and cardiac effectiveness), they have not been lost. These factors, as well as R_{TP}, all influence cardiac output in ways that will be discussed in Chapter 12, and their effects on MAP can all be understood in terms of their effects on cardiac output. Consequently, Eq. **(3)** is valid and extremely useful even though it does not explicitly define the effects of blood volume, venous capacity, and cardiac effectiveness. Equation **(3)** tells us that MAP increases whenever cardiac output increases unless there is a concomitant decrease in peripheral resistance. It makes no difference whether the cardiac output is changed by a change in stroke volume, or in heart rate, or in both. The effects on MAP are the same.

ARTERIAL PRESSURE PULSE

Since the heart is a discontinuous pump, blood enters the aorta in spurts. Each spurt immediately increases the blood pressure just beyond the aortic valve and consequently causes distension of that part of the aorta. This pulse of increased pressure and distension rapidly moves as a wave along the arteries and can be palpated over any large superficial artery. The general shape of the pressure pulse from the aorta is shown in Figure 7. The maximum pressure attained is called the systolic pressure, the minimum is the diastolic pressure, and the difference between these is called the pulse pressure. The mean pressure can be obtained by dividing the area under a pulse curve by the duration of the pulse. The mean pressure in the brachial artery at a heart rate of about 60–80/min is approximately equal to the diastolic pressure plus one-third the pulse pressure. The steep ascending limb of the pulse is caused by rapid ventricular ejection and the more gradual descending limb by escape of blood through the microcirculation to the veins. A perturbation along the descending limb is always found in the aorta, the indented portion of which is named the dicrotic notch (or incisura). The dicrotic notch is usually followed by a wave called the dicrotic wave. The time interval from the dicrotic notch to the next ascending limb is known as the diastolic run-off period.

AMPLITUDE OF THE ARTERIAL PRESSURE PULSE

The amplitude of the arterial pulse (i.e., the pulse pressure) is determined by the difference between the volume of blood ejected from the left ventricle during the rapid ejection period and the volume of blood leaving the arterial system via the microcirculation during this same period. The greater this pulsatile expansion of the arterial system during each rapid ejection period, the greater the amplitude of the pulse. The pulse pressure is also influenced by the compliance of the arterial system. For any given volume expansion during the rapid ejection period, the less the arterial compliance, the greater the pulse amplitude.

The volume of arterial expansion during the rapid ejection period is influenced by several factors, but most importantly by stroke volume. When stroke volume increases, pulse pressure increases. The volume of arterial expansion during rapid ejection is also influenced to some extent by the duration of the rapid ejection period. When myocardial contractility is increased, rapid ejection may occur in a shorter time, allowing less time for peripheral run-off, and the pulse pressure will then increase. Another factor is the arterial diastolic pressure. The arterial diastolic pressure constitutes an afterload to the left ventricle. If this afterload is increased the left ventricle ejects its contents more slowly, thereby allowing more time for peripheral run-off during rapid ejection, and the pulse pressure decreases. Figure 8 may be helpful in visualizing how the degree of volume expansion of the arteries during the rapid ejection period (given by the vertical arrows) determines the pulse pressure (horizontal arrows).

SYSTOLIC AND DIASTOLIC PRESSURES

The systolic and diastolic pressures are determined by the mean pressure and by the amplitude of the pulse (i.e., the pulse pressure). For any given mean pressure the greater the amplitude of the pulse, the higher the systolic pressure and the lower the diastolic pressure. This can be readily

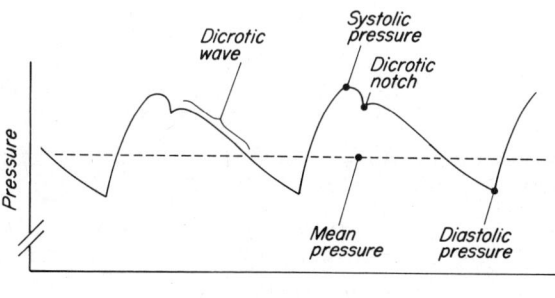

Figure 7 The aortic pressure pulse.

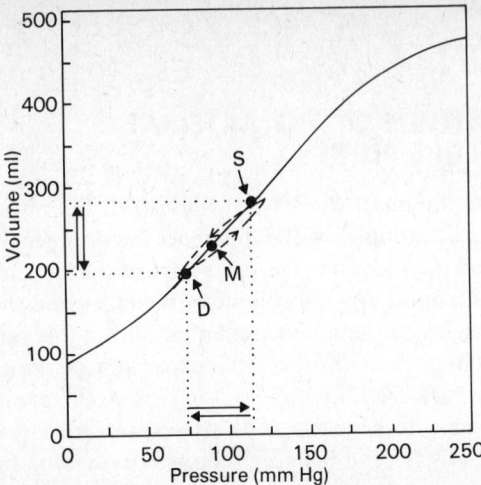

Figure 8 Pressure–volume diagram for the human aorta, showing the effect on pressure of the phasic changes in volume occurring during a cardiac cycle. M, Mean pressure and volume: D, diastolic pressure and volume; S, systolic pressure and volume.

visualized on Figure 8. For example, if stroke volume is increased, but MAP is experimentally held constant, systolic pressure rises and diastolic pressure falls. If stroke volume is increased and MAP is, as a result, allowed to increase, systolic pressure rises considerably more than does diastolic pressure.

EFFECTS OF EXERCISE ON ARTERIAL PRESSURES

The cardiovascular events occurring during vigorous exercise include (1) reduction in total peripheral resistance (due to vasodilation of arterioles in skeletal muscle), and (2) increased cardiac output (mediated by increased stroke volume and increased heart rate). These events have opposing effects on MAP; see Eq. (3). However, cardiac output increases more than peripheral resistance decreases, with the result that MAP increases. The increased stroke volume results in an increased pulse pressure. A briefer rapid ejection period contributes to this increase in pulse pressure. Thus, we have in exercise an increased MAP and an increased pulse pressure. Therefore, systolic pressure goes up markedly, but the diastolic pressure goes up only moderately. In fact, in many forms of exercise

diastolic pressure does not rise at all. It may be helpful to draw for yourself on Figure 8 the changes expected during exercise.

EFFECTS OF AGING ON ARTERIAL PRESSURES

Figure 9 shows average values for systolic and diastolic pressures of normal men and women of various ages. The following observations can be made: (1) systolic, diastolic, and mean pressures all gradually increase with age, apparently a normal consequence of aging.*; (2) systolic pressure increases with age more than does diastolic pressure, so that pulse pressure gradually increases; and (3) the

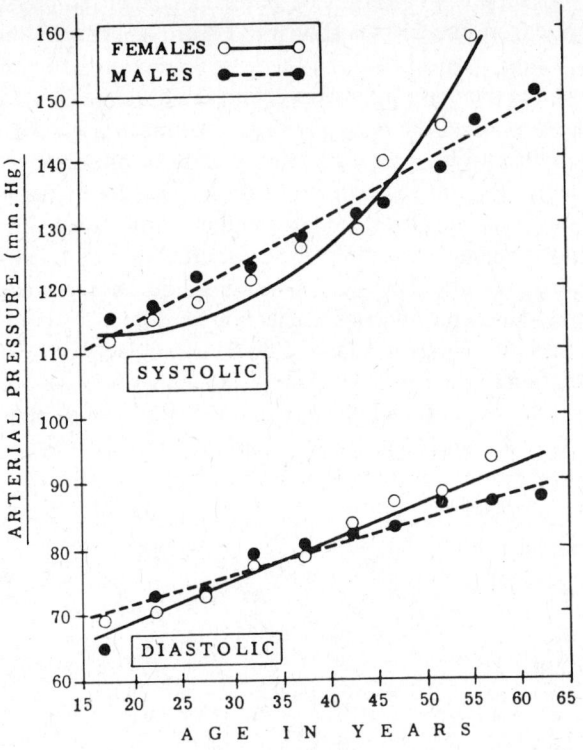

Figure 9 Normal arterial pressures for men and women as a function of age. (From Morris JN: *Mod Concepts Cardiovasc Dis* 30:635, 1961.)

---

*Of course, there are those who argue that, although growing older is normal, aging is not.

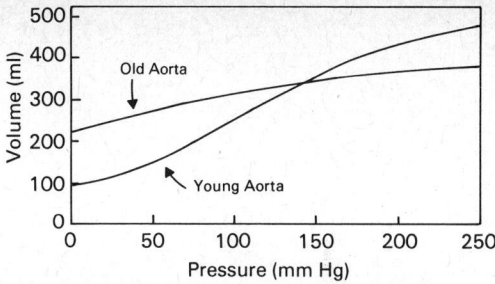

Figure 10 Pressure–volume curves for young (about 23 years) and old (about 75 years) human aortas. (Based on data of Hallock P, Benson IC: *J Clin Invest* 16:595, 1937.)

systolic pressure rises faster with aging in postmenopausal women than in men of the same ages.

The gradual increase in MAP and, therefore, in systolic and diastolic pressures is caused by a gradual increase in total peripheral resistance with aging, rather than by an increase in cardiac output. In fact, cardiac output tends to decrease somewhat with aging. The increased peripheral resistance is not well understood but probably is caused by arteriosclerotic changes in the precapillary resistance vessels. The increased pulse pressure results from the tendency during aging (especially at rather advanced ages) for the shape of the P–V curve to change. Figure 10 compares a P–V curve for a young aorta to a P–V curve for an old aorta. The two major differences are (1) the unstressed volume (i.e., the volume at zero pressure) is greater for the old aorta, and (2) the compliance throughout the physiological range of pressures is much less for the old aorta. As a result of decreased arterial compliance, any given volume expansion of the old arterial system during the rapid ejection period causes a pulse amplitude that is greater than it would be for the young arterial system. Reduced compliance in old arteries is the result of arteriosclerotic changes that seem to be a "normal" accompaniment of aging.

ARTERIAL PRESSURES IN CHRONIC HYPERTENSION

Patients with chronic hypertension generally have an abnormally high total peripheral resistance and also often have arteries that are less compliant than normal. In such a situation, assuming that stroke volume and heart rate are normal, MAP will be elevated to the same degree that peripheral resistance is elevated [Eq. (3)]. The decrease in compliance results in a greatly increased pulse pressure. Since the oscillation around the abnormally high MAP is larger than normal, the systolic pressure is elevated much more than is the mean, but the diastolic pressure, although higher than normal, is not elevated as much as the mean. Consequently, patients with chronic hypertension usually have greatly elevated systolic pressures, but may have only moderately elevated diastolic pressures.

IMPORTANCE OF ARTERIAL COMPLIANCE

If the arteries were rigid tubes instead of compliant tubes and could not expand at all during systole, systolic pressure would be tremendously high and diastolic pressure would be zero. Furthermore, a volume of blood equal to the entire stroke volume would be delivered into the microcirculation during systole and none during diastole. Flow through the capillaries, as well as through the arteries, would be extremely jerky. With compliant vessels only a portion of the stroke volume (about one-third at resting heart rates) moves through the microcirculation during systole. The remainder (about two-thirds) distends the arteries during systole and then is gradually pushed through the microcirculation during diastole because of the elastic recoil of the arterial walls. In this behavior the arteries act like a *Windkessel* (air kettle), as pointed out by Ernst Weber in 1834. In those days, the fire engines were rigged with a large kettle of air interposed between the water pump and the hose nozzle. The pumps were manually operated and intermittent. During the "systolic" phase of the pump the air was compressed (since air is highly compressible and the nozzle resistance was fairly high). Then during "diastole" the air pressure forced a continuing stream of water through the nozzle. If the pump was cycled fast enough, a high pressure could be maintained in the system and flow at the nozzle was fairly steady. The arterial system operates this same way except that it uses the distensible arterial walls to store part of the energy of each pump stroke, rather than compressible air.

TRANSMISSION OF THE PULSE WAVE IN SYSTEMIC ARTERIES

The arterial pulse wave can be pictured as a ripple traveling along the following stream of blood. Its velocity depends only slightly on the velocity of blood flow. If a rope is fixed at one end and the other end flipped, a transverse wave of displacement travels toward the fixed end. The pulse wave is to the arterial system what this displacement wave is to the rope. Arterial pulse waves can also be likened to sound waves traveling in air. Sound waves are much faster than the wind, and their velocity is relatively independent of the velocity of convective air movement (wind), just as the velocity of pulse waves is much faster than, and relatively independent of, the velocity of the blood.

The arterial pulse wave travels at a velocity of about 3 m/s in the ascending aorta*, increases to about 10 m/s in large arterial branches and up to 25–30 m/s in small arteries.† This progressive increase in velocity of the pulse wave is caused by decreasing distensibility of the arterial walls and increasing ratio of wall thickness to diameter as the arteries are progressed.

Since velocity of the pulse wave depends on distensibility, and distensibility in turn depends on the degree to which an artery is already distended by pressure, velocity becomes greater as the diastolic pressure in the artery goes up (Fig. 11). Since distensibility of arteries decreases with age, the pulse wave velocity increases with age (Fig. 11). In patients with arteriosclerosis, the arterial wall becomes much less distensible than normal; the diastolic pressure is usually somewhat elevated as well. Consequently, the pulse wave has a much faster velocity than it does in a normal person of the same age. Therefore, measurement of pulse wave velocities can be used clinically for assessing the condition of the large arteries.

One should not get the impression that only a short stretch of the arterial system experiences the pulse at any instant, with all the rest of the system uninvolved. Actually the propagation velocity of the pulse is so fast relative

---

*In contrast, the mean velocity of blood flow in the aorta is less than 0.5 m/s.

†Thus the velocity of the arterial pulse is roughly the same as the velocity of nerve impulses in small myelinated axons.

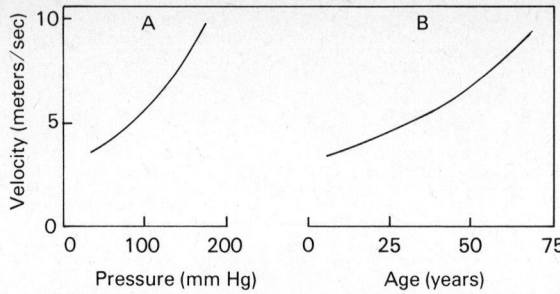

Figure 11 Effect of diastolic pressure and age on pulse wave velocity in the aorta. Curve A is from experiments on dogs in which the diastolic pressure was experimentally varied over a wide range. Curve B is from data on more than 500 humans of various ages. (A: From Dow P, Hamilton WF: "An experimental study of the velocity of the pulse wave propagated through the aorta." *Am J Physiol.* 125:60, 1939. B: From Hallock P: "Arterial elasticity in man in relation to age as evaluated by the pulse wave velocity method." *Arch Intern Med* 54:770, 1934.)

to its duration that each wave is spread over the entire systemic arterial tree during most of its period. In fact, the smallest most distal arteries have already begun their pulsation by the time the incisura is developing at the root of the aorta.

ORIGIN OF THE INCISURA AND DICROTIC WAVE

At the root of the aorta, the incisura (or dicrotic notch) is coincident with closure of the aortic valve; at more distal sites the entire wave, including the incisura, is delayed by the propagation time. The incisura results from the fact that at the moment of valve closure the column of blood in the ascending aorta tends to continue moving forward by inertia, even though no more blood is being expelled by the heart. This creates, just beyond the valve, a sharp reduction in pressure that lasts only an instant, however, because very promptly the flow of blood in the ascending aorta reverses its direction in response to the reversed pressure gradient. There is then a brief moment of retrograde flow in the ascending aorta that sharply elevates the pressure at the root of the aorta thus completing the incisura. The incisura is then propagated distally as a transitory negative deflection superimposed on the primary wave. The retrograde flow in the ascending aorta usually causes a rebound pressure at the root of the aorta that is momentarily higher than it would be otherwise. This positive pressure wave is reflected off the closed aortic valve and propagates distally as the dicrotic wave.

CHANGES IN SHAPE OF THE ARTERIAL PULSE WAVE DURING TRANSMISSION

As the pulse wave travels away from the heart its shape changes. This is shown in Figure 12. The major changes are (1) the sharp incisura characteristic of ascending aorta is gradually smoothed out and finally disappears entirely in the abdominal aorta; (2) the ascending limb becomes steeper and reaches a higher systolic pressure; (3) a more definite dicrotic wave appears; and (4) the diastolic pressure decreases. The increasing systolic and decreasing diastolic pressures lead to a greatly increasing pulse pressure; in fact, the pulse pressure in the femoral artery may be almost twice that at the root of the aorta.

The most important reasons for these changes are the following:

1. The viscoelastic properties of the large arteries are such that very rapid pressure fluctuations are not transmitted well, even though slower fluctuations are. The arterial system behaves like a high frequency filter and the rapid fluctuations are damped out. This is what happens to the incisura.

2. Wherever there are branches in the arterial system a fraction of each pulse wave is reflected backward toward the heart. The major sites for wave reflection are the precapillary resistance vessels (terminal arteries, arterioles, and precapillary sphincters), which, when constricted, allow almost no pulse to get through to the capillaries. The reflected waves are large when these vessels are constricted and small when they are dilated. Reflected waves traveling back toward the heart interact with the oncoming primary wave, thereby distorting it. Reflection of the systolic portion of the pulse wave from the precapillary resistance barrier probably influences the shape of the dicrotic wave, and may contribute to the peaking of the primary wave; the exact contribution is uncertain.

3. The drop in diastolic pressure (and also mean pressure)

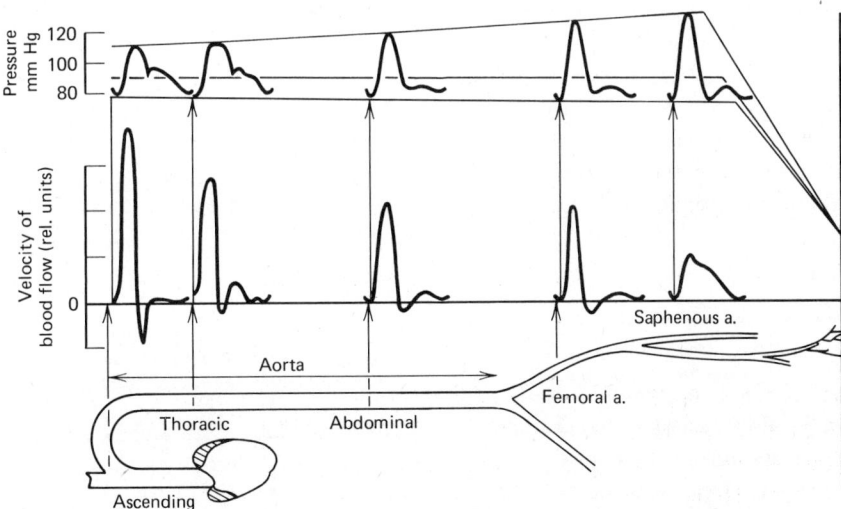

Figure 12 (Top) The arterial pulse at different locations in the arterial tree. (Middle) Relative rates of blood flow that take place simultaneous with the above pressures. It is interesting to note that coincident with the dicrotic notch there is actually a brief retrograde flow in the aorta. (From McDonald DA: *Blood Flow in Arteries.* Baltimore, Williams & Wilkins, 1974.)

along the large arteries is determined mainly by the rate of blood flow. Because blood is viscous, pressure energy is converted into heat during flow, and the pressure drops. There is no satisfactory explanation for why the peak systolic pressure increases along the large arteries, although the "resonant properties" of the system are sometimes rather vaguely invoked as the probable explanation.

ATTENUATION OF ARTERIAL PULSE IN MICROCIRCULATION

Although the arterial pulse wave gradually increases in amplitude as it passes along the large arteries, it is abruptly attenuated in the precapillary resistance vessels, and is very small in the capillaries. A pulse of only 1 mm Hg or so normally reaches the systemic capillaries (this is somewhat larger in pulmonary capillaries).

This attenuation of the pulse is usually attributed to damping in the precapillary resistance vessels. The geometry and visco-elastic properties of these vessels are such that pressure fluctuations at the frequency of the heart beat are not transmitted well, instead the pulse energy is dissipated as heat. At least two other factors besides damping contribute to the abrupt attenuation of the pulse in the microcirculation. The first of these is reflection off the precapillary resistance barrier. Obviously, that portion of the pulse energy that is reflected back toward the heart is not available for transmission into the microcirculation. The other factor is simply a dilution of the pulse energy as the pulse wave enters levels of the microcirculation having far greater total cross-sectional areas than present in the larger arteries.

BLOOD FLOW IN THE LARGE ARTERIES

Blood flow in the large arteries is markedly pulsatile. In fact, as shown in Figure 12, flow becomes retrograde (i.e., back toward the heart) briefly during each cardiac cycle, especially in the upper aorta. Here, to some extent, the blood sloshes back and forth (mostly forth) with each cycle. Even as far away from the heart as the saphenous artery, flow is markedly pulsatile, dropping to zero during diastole. It is not until fairly small arteries are reached that flow becomes continuous, and not until well into the microcirculation that flow becomes essentially smooth.

Because of the very rapid changes in blood velocity in the large arteries, flow here is not entirely laminar, but rather is disturbed by brief bursts of somewhat disorderly systolic flow. This has been discussed in Chapter 10.

RELATIONSHIP BETWEEN ARTERIAL PRESSURE AND FLOW THROUGH THE MICROCIRCULATION

One might expect the flow rate through the microcirculation from arteries to veins to be directly proportional to the mean arterial pressure and a plot of flow versus MAP to be a straight line that would go through the zero–zero origin. Such is not the case. Figure 13 shows a pressure–flow relationship for a "typical" vascular bed. Wide variations from this curve are seen in different vascular beds, but this curve shows most of the more interesting features that are often encountered:

1. If MAP is very small, there is no flow at all. The driving pressure has to reach some critical value before flow begins. This critical value is often called the critical closing pressure. This term may be a misnomer because it probably represents not the pressure required to open closed arterioles as formerly believed, but rather the pressure required to distort blood cells enough so that they can be forced through the microcirculation.

2. There is a range of pressures above the critical closing pressure in which flow rate increases out of proportion

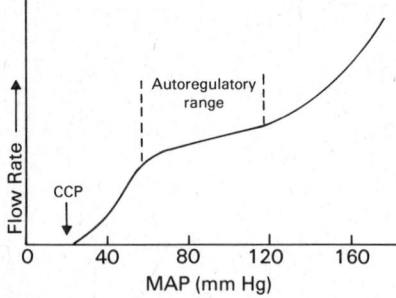

Figure 13 Effect of mean arterial pressure on flow rate through a "typical" vascular bed.

to increases in MAP. Within this range, the vessels of the microcirculation expand to some extent with each increase in MAP because the transmural (or distending) pressure increases. When the resistance vessels expand in diameter, the resistance to flow decreases approximately according to Poiseuille's law:

3. At higher values of MAP, within the physiological range, the curve tends to plateau. Increases in driving pressure within this range cause less than expected increases in flow. The degree to which flow is stable over this pressure range is quite variable from one vascular bed to another. In some organs, such as the kidneys, flow rate is nearly independent of driving pressure over a considerable range of pressures. This phenomenon is called autoregulation; its explanation and significance will be presented in Chapter 17.

THE SYSTEMIC VENOUS SYSTEM

VENOUS PRESSURE GRADIENT

The discussion in this section will be simplified by ignoring the effects of gravity on venous pressures. This approach is justifiable when referring to pressures in a reclining subject in whom all parts of the vascular system are at approximately the same height. In Chapter 12 we shall discuss the effects of postural changes on arterial and venous pressures and flows.

To visualize the venous pressure gradient it might be helpful to refer back to Figure 4. In a reclining individual venous pressure is greatest in the smallest peripheral veins and least in the right atrium (assuming blood is circulating through the system). The average pressure in the right atrium is close to zero; that is, close to atmospheric pressure. With a normal circulatory flow rate of about 5–6 L/min, the pressure in the smallest veins is about 15 mm Hg. Most of the pressure drop from smallest veins to right atrium occurs either within the peripheral venules and small veins themselves, or at certain obstructions along the course of larger veins. For example, large veins in the abdominal cavity are often partially collapsed because in-

tra-abdominal pressure sometimes rises above intravenous pressure and because various organs sometimes impinge on the veins. At regions of compression or collapse, resistance to flow is increased and the pressure drop through these regions may be a few mm Hg. However, large veins, when they are open, offer almost no resistance to flow and the pressure drop through them can hardly be detected. This is generally true of intrathoracic veins, and it is found that venous pressures anywhere within the thorax are equal to within a fraction of a mm Hg.

VENOUS PRESSURE–VOLUME RELATIONSHIP

The average pressure in the venous system is determined by the volume of blood in it and the accommodative properties of the system. A P–V curve for a large vein is shown in Figure 14; a typical arterial P–V curve is shown for comparison. These curves illustrate four extremely important features of veins.

1. If the pressure inside a vein falls just below the pressure outside of it (i.e., if the transmural pressure becomes just slightly negative), essentially no blood remains in the vein—it is all sucked out. This result stems from

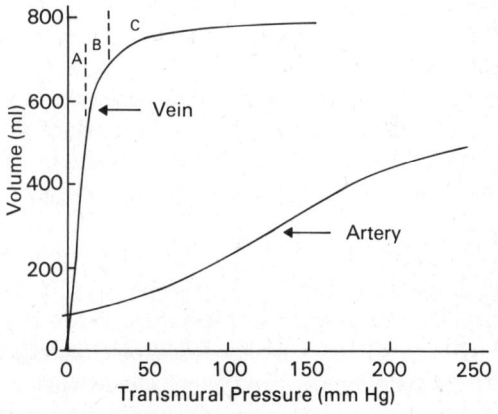

Figure 14 Comparison between the approximate "accommodative" properties of veins and arteries of comparable size. A, partial collapse range with high venous compliance; B, range in which venous compliance is greater than arterial compliance, C, range in which venous compliance is less than arterial compliance.

the fact that veins, unlike arteries, readily collapse. Collapse from a circular cross section to an elliptic cross section begins, according to Figure 14, at a transmural pressure of about 6–7 mm Hg. As the transmural pressure is reduced further, collapse proceeds to flatter and flatter ellipses until finally, at just below zero transmural pressure, the vein is entirely depleted of contents.

2. Throughout this range of transmural pressures (roughly 0–7 mm Hg) veins can accommodate large increases in blood volume with very small increases in pressure. In other words, over this pressure range, the compliance of veins is very large compared with that of arteries. This is because when veins are in a state of complete or partial collapse, expanding them by adding more blood does not require stretching of their walls, but merely requires establishing a more circular cross section. Thus to change the average pressure in the venous system by just a few mm Hg requires adding or subtracting quite a lot of blood. At low venous pressure, the veins are roughly 15–20 times more compliant than are arteries at normal arterial pressures.

3. The total volume of blood that the veins can hold before their walls begin to stretch is very large. This accounts for the fact that even though the average pressure in the veins is less than 12 mm Hg, at least 75% of the systemic blood volume is in the systemic veins (Fig. 3).

4. Further expansion of a fully rounded vein becomes progressively more difficult as the pressure increases. Just beyond full rounding the compliance of a vein is greater than that of an artery of comparable size, but at venous transmural pressures in the approximate range of 25–50 mm Hg, venous compliance becomes less than arterial compliance. This results from the fact that at these pressures the distensibility of the walls of veins becomes considerably less than that of comparable arteries. At even higher transmural pressures the veins become almost completely nondistensible. This low compliance of veins at relatively high pressures is extremely important in helping to prevent pooling of blood in the legs during standing, at which time venous hydrostatic pressures in the legs and feet may be quite high.

EFFECT OF SYMPATHETIC NERVE ACTIVITY ON VENOUS CAPACITY

The peripheral veins are richly endowed with smooth muscle that can adjust their capacity. The venous smooth muscle is under the control of the sympathetic nervous system. Figure 15 shows the effect of increased sympathetic activity on the P–V curve for the venous system. Increased sympathetic activity, acting via norepinephrine and α-receptors, leads to constriction of the venous system so that for any pressure (within a wide range), the system contains less blood.

RESERVOIR FUNCTION OF THE VENOUS SYSTEM

Since the volume of blood contained in the veins is so large and so variable, the venous system is often thought of as a blood reservoir. Figure 16 shows the results of an experiment that demonstrates how blood is shifted out of veins as a result of sympathetic nerve stimulation. In this experiment the hindquarters of an anesthetized cat are enclosed in a box. The box is filled with water, a watertight seal being made around the waist at the opening where the cat is inserted. The pressure of the water inside the box is then recorded. Changes in pressure inside the box, called a plethysmograph, reflect changes in the volume of the hindquarters. Blood flow rate from the hindquarters is measured at the level of the inferior vena cava. The arterial pressure in the hindquarters is held constant in this experiment by adjusting a screw clamp on the aorta. When the

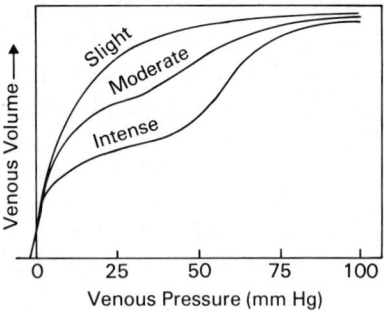

Figure 15 Effects of sympathetic nerve activity on the accommodative properties of veins.

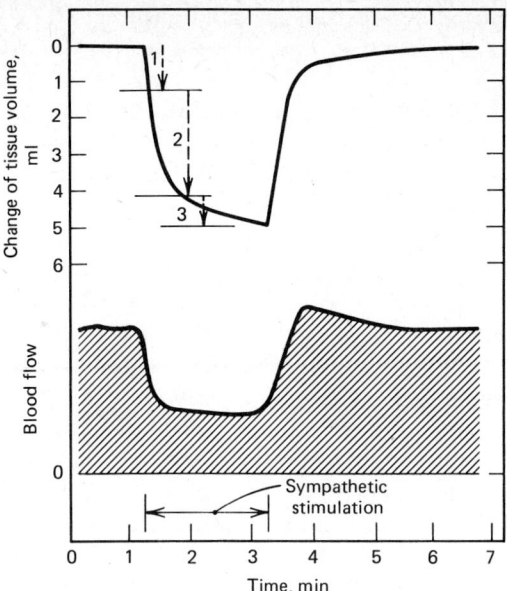

Figure 16 Effect of sympathetic nerve stimulation (2 impulses/sec) on blood flow and tissue volume in the hindquarters of the cat. Steps 1, 2, and 3 correspond to the three mechanisms of blood "mobilization" referred to in the text. (Slightly modified from a drawing in Berne RM, and Levy MN: *Cardiovascular Physiology,* ed. 4. St Louis, Mosby, 1981, based on original data of Mellander S: "Comparative studies of the adrenergic neuro-hormonal control of resistance and capacitance blood vessels in the cat." *Acta Physiol Scand* 50(suppl 176):1–86, 1960.)

sympathetic nerves to the hindquarters are stimulated at a frequency of 2 impulses/sec, flow rate decreases and remains low, until stimulation is stopped. This effect is mainly due to constriction of precapillary resistance vessels, principally the arterioles. Sympathetic stimulation also results in a decrease in *volume* of the hindquarters. There are three components to this decrease in volume. The first, and quickest, component results from the fact that when the arteriolar resistance increases, the flow rate into the small veins momentarily becomes less than the flow rate out (toward the heart), and the volume goes down, with the walls of the veins just passively recoiling or collapsing, or both, to contain the reduced volume. The second, and largest, component results from sympathetic excitation of vascular smooth muscle in the walls of the

veins themselves. Venoconstriction ensues, and blood is squeezed out of the hindquarters toward the heart. The third, and slowest, component of volume decrease results from the fact that when the arterioles constrict, the pressure in the capillaries goes down. When this happens, the balance of pressures normally maintained between capillary blood and tissue spaces is upset, and a slow net movement of extravascular fluid into the capillary blood takes place. This fluid becomes part of the blood and flows out of the hindquarters. This mobilization of extravascular fluid augments the reservoir function of the veins themselves. At somewhat higher frequencies of sympathetic nerve stimulation, a maximum of about one-third of the total blood volume in the hindquarters can be shifted out of the veins.

In certain special circumstances shifting of blood from the veins to other parts of the circulation is particularly important. Two of these circumstances will be mentioned here. The first is exercise. During vigorous exercise the arterioles in skeletal muscle dilate and there is a considerable drop in total peripheral resistance and a tremendous increase in blood flow through the muscles. This in itself tends to increase the volume of blood in the veins, and if the capacity of the venous system were not diminished during vigorous exercise the venous volume would rise. The increased venous blood volume would be at the expense of arterial blood volume and consequently MAP would decrease. This would be self-defeating, since the decrease in MAP would tend to decrease blood flow through the active muscles. Fortunately, during vigorous exercise there is increased sympathetic activity to veins and they constrict. This ability of the veins to constrict during exercise is absolutely essential to attain the increases in MAP and in circulatory flow rate that are observed during exercise. It is greatly aided, however, by compression of the veins by contracting skeletal muscle; this is an auxiliary mechanism for shifting blood to the arterial side during exercise, which will be further discussed in the next chapter.

The second special circumstance to be mentioned here is hemorrhage. Following severe hemorrhage (say 25% of the total blood volume) there is a great deal of trouble keeping the mean arterial pressure high enough to keep blood flowing through vital organs fast enough. Arterioles in many organs constrict and this helps to support MAP;

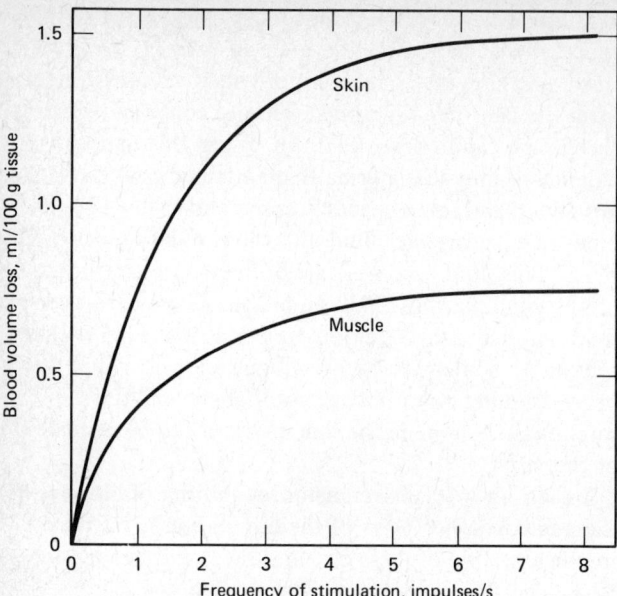

Figure 17 Comparison of the blood volume loss from skin and muscle with sympathetic nerve stimulation. (From Berne RM, and Levy MN, *Cardiovascular Physiology*, ed. 4. St. Louis, Mosby, 1981, based on data of Melander S: "Comparative studies of the adrenergic neuro-hormonal control of resistance and capacitance blood vessels in the cat." *Acta Physiol Scand* 50(suppl 176):1–86, 1960.)

side can be regarded as an autotransfusion. The reflex pathways involved in this response will be discussed in Chapter 13.

Not all organs participate equally as blood donors. The most important ones are the skin, the liver, the spleen, the splanchnic veins, and the lungs. Figure 17 shows that the skin can deliver much more blood per unit weight than can skeletal muscle in response to sympathetic stimulation. The differences are a function of (1) how much venous blood is contained in the organ in the first place, (2) how sensitive its venous smooth muscle is to sympathetic impulses, (3) how much smooth muscle its veins have, and (4) how much basal level of tone its smooth muscle has. For example, skeletal muscle, which is a poor venous reservoir, has rather poorly innervated veins that are, however, well constricted without sympathetic activity. Both of these features help explain why the amount of blood shifted out of skeletal muscle by sympathetic stimulation is not very great.

SELECTED BIBLIOGRAPHY

1. Berne RM, Levy MN: *Cardiovascular Physiology,* ed. 4. St. Louis, Mosby 1981.
2. Folkow B, Neil E: *Circulation.* New York, Oxford University Press, 1971.
3. Little RC: *Physiology of the Heart and Circulation.* Chicago, Year Book, 1977.
4. Rushmer RF: *Structure and Function of the Cardiovascular System,* ed 2. Philadelphia, Saunders, 1976.

but more to the present point, veins constrict, thereby shifting a considerable fraction of the surviving blood to the arterial side. This supports MAP and thereby helps keep the circulation going at a rate consistent with life. This shifting of blood from the venous side to the arterial

SELF-STUDY QUESTIONS

MULTIPLE CHOICE

Select the correct answer(s). (In many instances, more than one answer is correct.)

1. The greatest total cross-sectional area in the systemic vascular system is at the level of the

A. arterioles.

B. capillaries.

C. venules.

D. small veins.

E. large veins.

2. The greatest blood volume is contained in the
    A. arteries.
    B. capillaries.
    C. venules.
    D. large veins.

3. The greatest total resistance to blood flow occurs at the level of the
    A. arteries.
    B. arterioles.
    C. capillaries
    D. venules.

4. Mean arterial volume and, therefore, MAP tend to increase as
    A. cardiac output increases.
    B. total peripheral resistance increases.
    C. total blood volume increases.
    D. veins constrict.

5. The total volume of blood in the systemic veins tends to be increased by
    A. an increase in cardiac contractility and heart rate.
    B. an increase in total peripheral resistance.
    C. a decrease in contraction of venous smooth muscle.
    D. a decrease in total blood volume.

6. With a heart rate of 70 beats/min, a systolic arterial pressure of 120 mm Hg, and a pulse pressure of 45 mm Hg, the mean arterial pressure would be about
    A. 85 mm Hg.
    B. 90 mm Hg.
    C. 100 mm Hg.
    D. 360 mm Hg.

7. If heart rate is doubled while stroke volume is halved, and there is no change in peripheral resistance, the MAP will
    A. increase.
    B. decrease.
    C. not change.

8. A resting person with a MAP of 90 mm Hg begins to exercise. His heart rate and stroke volume both double, while his total peripheral resistance reduces to one-third its value at rest. At this time his MAP is

A. 90 mm Hg.
B. 120 mm Hg.
C. 180 mm Hg.
D. 360 mm Hg.

9. The difference between left ventricular ejection and peripheral runoff during the rapid ejection period is increased by
    A. increased stroke volume.
    B. increased myocardial contractility, with no change in stroke volume.
    C. increased MAP.

10. Assuming no change in MAP or heart rate, a decrease in arterial compliance would cause
    A. increased systolic pressure.
    B. decreased diastolic pressure.
    C. decreased pulse pressure.
    D. reduction in the Windkessel function of the arteries.
    E. smoother (i.e., more steady) flow through the microcirculation.

11. Suppose the parasympathetic fibers in the left vagus nerve of an anesthetized animal are stimulated electrically so that the heart rate is reduced to one-fifth its control value. If, as a result of this procedure, MAP is reduced to four-fifths its control value, and peripheral resistance is reflexly increased to four-thirds its control value, what is the new stroke volume?
    A. Same as control
    B. $\frac{1}{2}$ control
    C. 2 × control
    D. 3 × control
    E. 4 × control

12. Aging is often associated with
    A. increased systolic pressure.
    B. decreased diastolic pressure.
    C. increased pulse pressure.
    D. increased total peripheral resistance.
    E. increased arterial compliance.
    F. increased pulse wave velocity.
    G. increased cardiac output.

13. Vigorous exercise is associated with
    A. increased MAP.
    B. increased pulse pressure.
    C. markedly increased diastolic pressure.
    D. increased total peripheral resistance.
    E. increased heart rate.

14. The velocity of the arterial pulse wave
    A. tends to increase with age.
    B. increases as the diastolic pressure increases.
    C. increases as the distance from the heart along the arteries increases.
    D. is about the same as the velocity of blood flow in the arteries.

15. As the arterial pulse wave travels away from the heart along the large arteries
    A. the systolic pressure increases.
    B. the pulse pressure increases.
    C. the diastolic pressure increases.
    D. the incisura disappears.
    E. the dicrotic wave disappears.
    F. the ascending limb becomes less steep, that is, the pulse develops less abruptly.

16. Let us say that a sensitive sphygmograph (a device for recording the mechanical movements of a superficial artery) is strapped on the wrist over the radial artery. The device registers pulsations with twice the frequency of the heartbeat, the latter being determined by electrocardiography. Is anything necessarily wrong?

17. Not much pulsation can normally be detected in capillaries because
    A. capillaries are rigid tubes.
    B. capillaries are very compliant tubes.
    C. it is impossible to measure capillary pressures.
    D. much of the pulse energy is lost by damping in small arteries and arterioles.
    E. much of the pulse energy is reflected from the precapillary resistance vessels back toward the heart.
    F. the total cross-sectional area of the capillaries is larger than that at any generation of arterial branching.

TRUE–FALSE

18. In a reclining person, the average pressure in the right atrium nearly equals (to within less than 1 mm Hg) the average pressure in the inferior vena cava. True or false?

19. At a transmural pressure of 90 mm Hg the compliance of veins is greater than that of equivalent-size arteries. True or false?

20. At -2 mm Hg transmural pressure, the volume of blood in a vein is
    A. less than
    B. greater than
    C. equal to
    the volume of blood in an artery of equivalent size.

MULTIPLE CHOICE

Select the correct answer(s). (In many instances, more than one answer is correct.)

21. Increased sympathetic nerve activity causes
    A. increased contraction of venous smooth muscle mediated by α-receptors.
    B. increased capacity of the venous system at any transmural pressure.
    C. increased average pressure in the venous system at any given total venous volume.
    D. increased compliance of the venous system at high transmural pressures.

22. The decreased volume of the hindquarters of a cat during sympathetic nerve stimulation is partially caused by
    A. arteriolar constriction.
    B. venous constriction.
    C. passive elastic recoil of the walls of veins.
    D. increased flow of lymph from the hindquarters.

23. Upon sympathetic stimulation, skeletal muscle loses ("mobilizes") a smaller fraction of its total venous blood volume than does skin because

A. the veins in skeletal muscle have little or no smooth muscle.

B. the veins in skeletal muscle are relatively poorly innervated.

C. the smooth muscle in veins of skeletal muscle has a relatively high level of basal tone.

D. the veins in skeletal muscle have no valves.

ANSWERS

1. C is correct (Fig. 2).

2. C is correct (Fig. 3).

3. B is correct (Fig. 5).

4. A, B, C, and D are all correct. See pages 193, 194.

5. C is correct. All the other changes have the opposite effect.

6. B is correct. At this heart rate mean pressure approximately equals diastolic pressure plus one-third of the pulse pressure. Diastolic = 75 mm Hg (i.e., 120 − 45); Mean = $75 + \frac{1}{3} 45 = 90$ mm Hg.

7. C is correct, since there is no change in either cardiac output or peripheral resistance [Eq. (3)].

8. B is correct, since cardiac output is increased four-fold, whereas peripheral resistance is cut to one-third. $4/3 \times 90 = 120$.

9. A and B are correct.

   A. Obvious.

   B. The stroke volume will be delivered in a shorter time, thereby allowing less time for peripheral runoff during rapid ejection.

   C. An increased afterload lengthens the rapid ejection period, allowing more time for peripheral runoff.

10. A, B, and D are correct.

    A and B are true and C is false because the pulse amplitude would be increased.

    D is true because with decreased compliance the arteries distend less during systole. E is wrong; just the opposite occurs.

11. D is correct.
    $$MAP = CO \times R_{TP} = SV \times HR \times R_{TP}$$
    $$SV = \frac{MAP}{(HR \times R_{TP})}$$

12. A, C, D, and F are correct. See pages 196, 197.

13. A, B, and E are correct. See page 196.

14. A, B, and C are correct. See page 198.

15. A, B, and D are correct. See page 199.

16. No. The main wave (sometimes called the "tidal wave") and the dicrotic wave can readily be detected as a double pulse over a peripheral artery.

17. D, E, and F are correct. See page 200.
    A and B are obviously wrong.
    C. It is difficult but not impossible to measure pressure in some capillaries.

18. True, because there is practically no resistance to flow in the large intrathoracic veins. See page 201.

19. False (Fig. 14).

20. A is correct. It is much more difficult to collapse arteries by reducing the transmural pressure than it is to collapse veins.

21. A, C, and D are correct.

    A. See Figure 15.

    B. Just the opposite.

    C. See Figure 15.

    D. It may seem surprising that at high transmural pressure smooth muscle contraction actually increases the compliance, but this is apparent in Figure 15.

22. A, B, and C are correct.

    A. Phase 1 in Figure 16.

B. Phase 2 in Figure 16.

C. When the arterioles constrict the pressure in the venules and small veins decreases, permitting the walls of these veins to recoil passively so that they contain less blood.

D. Lymph flow tends to *decrease,* since the net filtration pressure across the capillary and venule walls decreases.

23. B and C are correct. See page 204.

12

The Control
of Circulatory Flow Rate

R. DAVID BAKER, PhD

OBJECTIVES

After completion of this chapter, you should be able to

1. Discuss the major determinants of cardiac performance and understand the cardiac performance curves.

2. Understand the characteristic curves for the systemic system and how the adjustable parameters of the system influence these.

3. Understand how the preload cardiac performance curve interacts with the systemic characteristic curve to establish an operating point, as well as how changes in the adjustable parameters influence this operating point.

4. Understand why a decrease in peripheral resistance has a

great effect on cardiac output and a lesser effect on MAP, whereas an increase in peripheral resistance has a small effect on cardiac output and a relatively large effect on MAP.

5. Describe the auxiliary mechanisms for driving blood around the circuit: cardiac suction, the respiratory pump, and the skeletal muscle pump.

6. Understand the hydraulic problems imposed on the system when a person assumes an upright posture, and the compensatory mechanisms that ordinarily solve these problems.

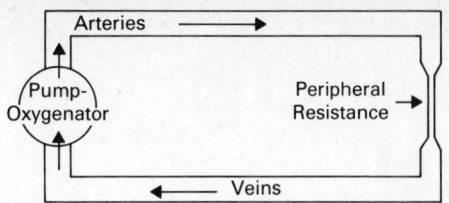

Figure 1 Simplified model of the circulatory system.

The simplified model of the circulatory system shown in Figure 1 is intended as a guide in studying this chapter. In this model the circulatory system is divided into four basic components: (1) the pump oxygenator (heart and pulmonary circulation), (2) the arteries, (3) the peripheral resistance, and (4) the veins. Most of the resistance to flow is conceptually lumped into one component: the "peripheral resistance." The veins, it should be noted, are distensible, collapsible, compressible, and contractable. Therefore, their volume is quite variable.

As blood passively circulates through the systemic circuit a slight pressure drop occurs through the arteries, a large pressure drop through the microcirculation, and another small drop through the veins; the pump oxygenator then actively boosts the blood into the arteries back up to arterial pressure. During steady flow, which is the usual condition of the circulatory system except for brief transients during development of new steady states, the flow rate is the same in all series components of the system. In other words, during steady flow the cardiac output, venous return, microcirculatory flow rate, and arterial flow rate are all quantitatively the same, and in this discussion we will often use the more general term, "circulatory flow rate." The circulatory flow rate is determined by the characteristics of the pump oxygenator and by the characteristics of the systemic system.

CARDIAC PERFORMANCE

There are four major determinants of cardiac performance: (1) preload, (2) afterload, (3) contractility, and (4) heart rate.

The preload is most conveniently represented by the right atrial pressure (RAP); this is a measure of end-diastolic expansion of the right ventricle. During most of ventricular diastole, ventricular and atrial pressures are essentially the same since the atrioventricular (AV) valves are open. The afterload is most conveniently represented by the mean arterial pressure (MAP); this is a measure of the load against which the pump oxygenator must force blood. Contractility has been defined in an earlier chapter. Changes in either contractility or heart rate are said to change the "effectiveness" of the heart as a pump.

PERFORMANCE CHARACTERISTICS OF THE PUMP OXYGENATOR

The performance characteristics of any pump are generally represented by performance curves, sometimes called characteristic curves. Of the several types of performance curves used in hydraulics two are most useful for describing the performance of the heart. The first of these shows the effect of changing the pressure head against which pumping occurs on the rate of pumping, the other shows the effect of changing the pressure at the inflow side of the pump on pump output. We will refer to these two types of curves as afterload performance curves and preload performance curves respectively. Figure 2a depicts afterload performance curves for the heart. Cardiac output (CO) is plotted as a function of MAP. The preload (i.e., RAP) is held constant at a normal value (i.e., about 0 mm Hg). Sympathetic and parasympathetic nervous activity are also held constant, and there is no change in heart rate. It is apparent that CO remains nearly constant over a wide range of MAPs. Not until MAP rises to roughly 200 mm Hg or more is CO severely depressed. This is an important characteristic of the heart; it is able to increase its work output automatically to meet demands over the entire range of MAPs normally encountered.

Figure 2b shows preload performance curves for the heart. CO is plotted as a function of RAP. In this case the afterload (i.e., MAP) is held constant at a normal value and there are no changes in contractility or heart rate. This type of curve is commonly called the Frank-Starling relationship. It demonstrates that CO is strongly influenced by RAP from about −2 to +4 mm Hg. Normal RAP is about 0 mm Hg. The mechanism of the Frank-Starling relationship was discussed in Chapter 9.

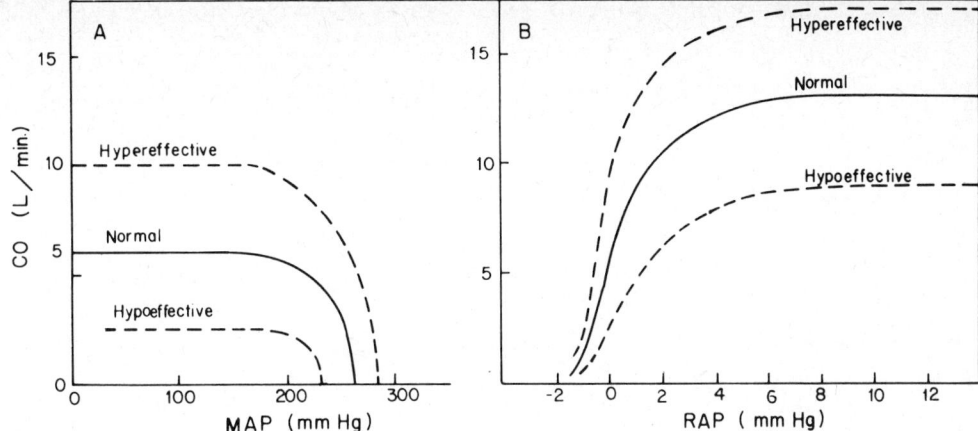

Figure 2 Cardiac performance curves (*a*) Effect of afterload (i.e., MAP) on cardiac output (CO). The preload (i.e., RAP) is experimentally kept at a constant normal value, whereas the afterload is varied. The solid line is for a normal heart with a "normal" degree of sympathetic and parasympathetic tone. The dashed lines are for hearts made hypereffective or hypoeffective by increasing either sympathetic activity or parasympathetic activity, respectively. (*b*) The effect of preload on cardiac output. The afterload is experimentally kept at a constant normal value, while RAP is varied.

Figure 3 shows a series of afterload performance curves, each corresponding to a different preload. Also shown is a series of preload performance curves, each corresponding to a different afterload. Although this figure may at first seem difficult to comprehend, it is useful in visualizing how CO responds to simultaneous changes in MAP and RAP.

There are two reasons for the near absence of an influence of MAP on CO until high MAPs are reached: First, for a few beats following a moderate rise in aortic pressure, the left ventricle puts out less blood per beat than it had previously. However, this moderate rise in aortic pressure is, for the most part, not transmitted to the pulmonary artery and so the *right* ventricle does not become overloaded and its output does not change. Consequently, for a few beats following the rise in aortic pressure, the right ventricle puts out a little more blood than the left ventricle; this slightly increases the left ventricular filling pressure, and the left ventricle increases its output by the Frank-Starling mechanism to match that of the right ventricle. Since this readjustment is accomplished with almost no increase in RAP, the cardiac performance curves are hardly influenced. Second, the contractility of the left ventricle automatically increases in response to an increase in

afterload. This increase in contractility requires several beats to become fully developed. The mechanism of this effect is not understood but it does not require maintained stretch of the myocardial fibers. Since this automatic response of the heart does not depend on a change in fiber length it is called homeometric (same length) autoregulation. Since the Frank-Starling mechanism does depend on changes in fiber length, it is sometimes called heterometric autoregulation.

EFFECTS OF CHANGING CONTRACTILITY AND HEART RATE ON THE PERFORMANCE CURVES

The dashed lines in Figure 2 depict performance curves for hearts made "hypereffective" or "hypoeffective" by increasing or decreasing contractility or heart rate, or both. Increasing contractility or heart rate makes the heart hypereffective compared with normal, and the performance curves are elevated. Decreasing contractility or heart rate has the opposite effect. Table 1 lists the influences that commonly lead to hypereffective or hypoeffective cardiac performance.

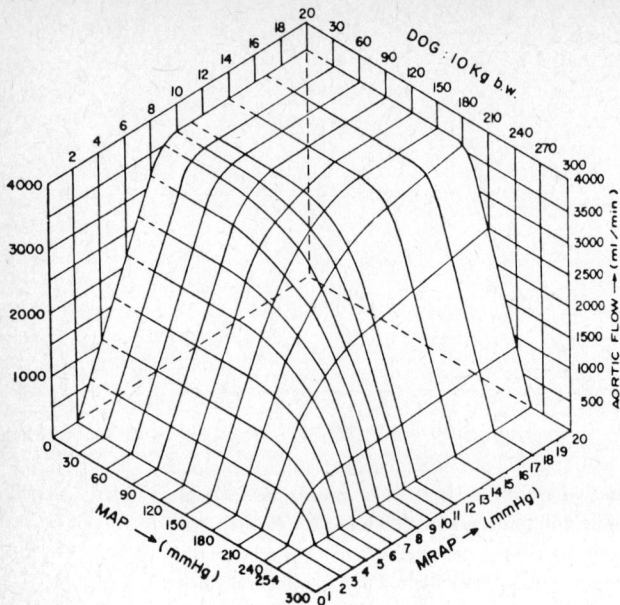

Figure 3 The simultaneous effects of MAP and RAP on CO. MRAP is the mean right atrial pressure. The data are from experiments on anesthetized dogs with autonomic reflexes blocked. The dogs had open chests; this increased the "normal" or control MRAP to about 4 mm Hg. (From C. W. Herndon and K. Sagawa, *Am J Physiol* 217:65, 1969.)

FACTORS THAT ALTER THE PRELOAD AND AFTERLOAD

Four independently adjustable parameters of the cardiovascular system determine preload (RAP) and afterload (MAP): (1) total blood volume, (2) venous capacity, (3) total peripheral resistance, and (4) cardiac effectiveness (including both contractility and heart rate). The ways by which these parameters determine MAP were discussed in Chapter 11. RAP is determined by the same parameters of the system. If blood volume increases or venous capacity decreases, the whole system becomes more tightly filled and the pressures increase everywhere, including that in the right atrium. If peripheral resistance decreases, blood passively shifts from arteries to veins; venous pressures (including RAP) go up and MAP goes down. If cardiac effectiveness goes up, blood is actively shifted from veins to arteries; MAP goes up and RAP goes down.

The adjustable parameters of the systemic system:

Table 1 Some Factors That Can Make the Heart Hypereffective or Hypoeffective

Hypereffective	*Hypoeffective*
Increased sympathetic activity (or increased circulating catecholamines)	Decreased sympathetic activity
Decreased parasympathetic activity	Increased parasympathetic activity
Myocardial hypertrophy (as in long-distance runners)	Valvular heart disease Myocarditis Toxic effects of various drugs Arrhythmias

blood volume, venous capacity, and peripheral resistance, will be further discussed below.

CHARACTERIZATION OF THE SYSTEMIC SYSTEM

CHARACTERISTIC CURVES FOR THE SYSTEM

In the previous section we examined the influence of inflow and outflow pressures on the performance of the pump. When a pump is connected to a system of tubes, the pressures within the tubes are influenced by the output of the pump. If, experimentally, we vary CO independently and determine the effect on MAP the characteristic curve shown in Figure 4*a* is obtained. The position of this curve is determined by the characteristics of the systemic system. The experiment must be done in such a way as to preclude any significant changes in peripheral resistance, venous capacity, and blood volume. In other words, if the system parameters are held constant while varying CO, we obtain Figure 4*a*. As CO increases, MAP increases along a nearly straight line. The slope of this line is related to the peripheral resistance; as peripheral resistance increases the slope decreases. At zero cardiac output, the characteristic curve intercepts the MAP axis at a pressure of about 7 mm Hg. This is the pressure that would be measured if the heart were suddenly stopped and the blood allowed to flow passively from arterial to venous

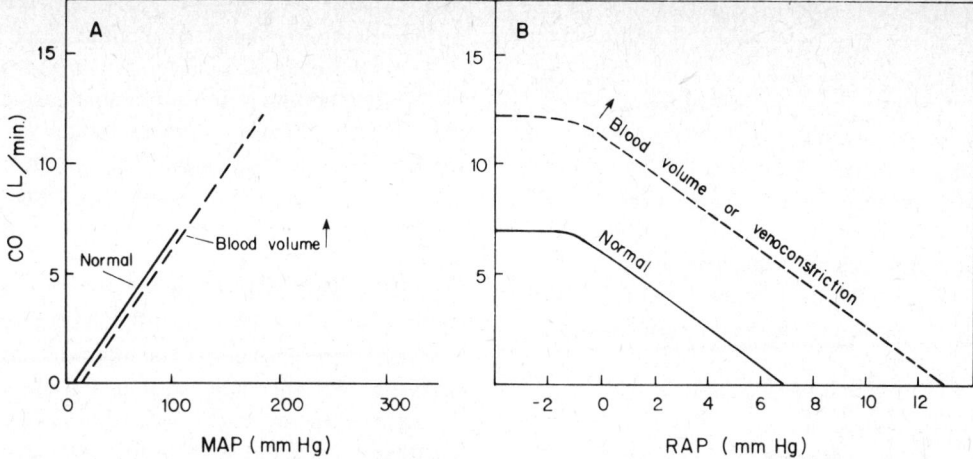

Figure 4 Characteristic curves for the systemic system. (*a*) The effect of CO on MAP. Peripheral resistance, blood volume, and venous capacity are kept constant experimentally while CO is varied. The intercept on the MAP axis is the general static blood pressure. The slope of the line is roughly inversely proportional to peripheral resistance. (*b*) The effect of CO on RAP. Again, the systemic system parameters are kept constant while CO is varied. The intercept on the RAP axis is the general static blood pressure. The steepness of the line is roughly inversely proportional to peripheral resistance. The dashed line shows the effect of increasing the blood volume by about 15% or of increasing the degree of venoconstriction.

sides until the pressure was the same throughout the entire cardiovascular system. This pressure, termed the "general static blood pressure," is a measure of how tightly the circulatory system is filled with blood when there is no flow.*

---

*A note on terminology: This quantity was originally called the "mean systemic pressure" by Weber. Starling also used this term. In 1903, Bolton used the term "hydrostatic mean pressure." In 1940, Starr coined the term preferred by the present author, "general static blood pressure." More recently, Guyton has called it the "mean circulatory filling pressure," the "mean circulatory pressure," and the "circulatory filling pressure." Unfortunately, there has been a tendency to confuse the static blood pressure with the average pressure of all the blood in the vascular system *while blood is flowing;* but it is a simple matter to show that these quantities would be the same only if the compliances of all parts of the vascular system were constant and identical. Since this is obviously not the case, the static blood pressure should not be identified with the mean pressure in the system while blood is flowing. This was pointed out as early as 1900 by Hill. Normally the mean pressure in the systemic circulatory system, while blood is flowing, is expected to be roughly three times greater than the static pressure. Since such terms as mean *circulatory* pressure can contribute to misinterpretation, they are not favored by the present author.

The characteristic curve showing the effect of cardiac output on RAP is shown in Figure 4*b*. This curve has three important features. First, at zero cardiac output, RAP equals about 7 mm Hg; this is the general static blood pressure. Second, as the pump output is increased, blood is actively shifted from veins to arteries and RAP decreases. Over a wide range of pumping rates, RAP is roughly inversely proportional to CO. Again, the slope of this curve is related to peripheral resistance; as peripheral resistance decreases, the curve becomes steeper.

The third important feature of the curve shown in Figure 4*b* is the plateau at the left end of the curve. This plateau is caused by venous collapse. When the pumping rate is increased enough to drop RAP a little below 0 mm Hg the veins just outside the thorax start to collapse, and when RAP is about −2 to −4 mm Hg, these veins completely collapse. The output of the pump is limited by venous collapse. Beyond this point, no matter how hard the pump might try it cannot increase its output. The pressure in the intrathoracic veins can drop a few mm Hg below atmospheric pressure before they begin to collapse, since intrathoracic pressure is normally subatmospheric. This negative intrathoracic pressure helps to hold the central veins open. However, because there is hardly any pressure gradient along the large intrathoracic veins, when RAP drops below 0 mm Hg so does the pressure in the veins just outside the thorax and they collapse. When the veins collapse, the resistance to flow through them grows considerably and limits the rate at which the pump receives blood. Consequently, if the heart is already pumping well enough that RAP is just below 0 mm Hg, then increases in contractility or in heart rate cannot speed up the circulation of blood.

EFFECTS OF VARYING THE SYSTEM PARAMETERS

Three independently adjustable parameters of the systemic system influence the characteristic curves of Figure 4: (1) total blood volume, (2) venous capacity, and (3) total peripheral resistance.

Blood Volume

If total blood volume is increased abruptly (by intravenous infusion) or gradually (by renal retention of fluid) the entire cardiovascular system becomes more tightly filled with blood, and the general static blood pressure increases. The characteristic curves of Figure 4 both move toward the right, which means that for any given cardiac output MAP and RAP are both elevated. This effect is illustrated by the dashed line in Figure 4*b*. The dashed characteristic curve in Figure 4*b* (CO $\rightleftharpoons$ RAP) also shows that increasing the blood volume increases the maximum attainable cardiac output, that is, the cardiac output that can be achieved before the veins collapse. If blood volume is decreased, for example, as a result of hemorrhage, the opposite trends result. These effects on general static blood pressure and on RAP can be quite large. An increase in blood volume of 15% can double the general static blood pressure.

Venous Capacity

As discussed in Chapter 9, the capacity of the venous system is large and quite variable. It can be adjusted by changes in the degree of contraction and compression of the veins. A decrease in venous capacity has similar effects on the characteristics of the systemic circuit as an increase in blood volume: the general static blood pressure is increased, the characteristic curves are moved toward the right, and the maximum attainable cardiac output is increased. An increase in venous capacity has the opposite effects. Vigorous sympathetic stimulation of the venous system can more than double the general static blood pressure.

Peripheral Resistance

If total peripheral resistance is reduced by dilation of arterioles, it becomes easier for blood to flow from arteries to veins, and the volume of blood on the venous side increases at the expense of the volume in the arteries. The pressure drop from arteries to veins decreases and the average pressure in the veins, including the right atrium, increases, while that in the arteries decreases. In terms of the characteristic curves for the systemic system, the CO $\rightleftharpoons$ MAP curve is rotated counterclockwise, while the CO $\rightleftharpoons$ RAP curve is rotated clockwise; in other words, they both get steeper. For any given cardiac output, MAP will be lower than normal and RAP higher. A change in peripheral resistance has no appreciable effect on the static blood pressure. This is because the arterioles, which provide most of the peripheral resistance, have a very small total capacity; therefore, changing their capacity does not change the tightness with which the whole circulatory system is filled under static conditions.

THE COMBINED CIRCULATORY SYSTEM

We have seen how MAP and RAP influence cardiac output and then how cardiac output influences MAP and RAP. We have also examined the influences of the four independently adjustable parameters of the cardiovascular system: cardiac effectiveness, peripheral resistance, venous capacity, and blood volume. But so far we have ignored the fact that there are mutual simultaneous interactions between CO and MAP and between CO and RAP; at the same time that CO is influencing MAP and RAP, MAP and RAP are influencing CO. The curves in

Figures 2 and 3 were obtained by experimentally establishing a series of fixed preloads and afterloads and determining the effects on CO; CO was not allowed to influence preload and afterload. On the other hand, the curves of Figure 4 were obtained by establishing a series of fixed cardiac outputs and determining the effects on MAP and RAP; the latter were not allowed to influence CO. Various clever and elaborate experimental arrangements were necessary to make the appropriate measurements, such as Starling's heart-lung preparation and Guyton's right ventricular bypass preparation. But in the real world of the cardiovascular system, interactions take place that must be taken into account. In order to analyze the system as it actually functions we use a graphical analysis borrowed from hydraulics and introduced into cardiovascular physiology by Guyton.

THE OPERATING POINT

When Figure 2 is superimposed on Figure 4 we obtain, simultaneously, the effects of CO on MAP and RAP and the effects of MAP and RAP on CO. These combination graphs are shown in Figure 5. In both cases the curves cross each other at a point known as the operating point.

The operating point defines the only set of values (CO and MAP in one case; CO and RAP in the other) at which the system can operate in the steady state. In other words, given the position of the curves in Figure 5 there is only one possible stable set of values for CO, MAP, and RAP; these values can be read on the graphs from the positions of the operating points. All other combinations of values for CO, MAP, and RAP are impossible because in the steady state there can be only one CO, only one MAP, and only one RAP.

In the following sections we will drop the CO ⇌ MAP curves and discuss only the CO ⇌ RAP curves. The reason is that over normal pressure ranges the cardiac performance curves are influenced far more by RAP than by MAP. As a result the afterload cardiac performance curve (Figure 2a) is not only raised and lowered by changes in cardiac effectiveness, but also by changes in RAP and indirectly by changes in blood volume, venous capacity, and peripheral resistance. On the other hand, the preload cardiac performance curve (Fig. 2b) is not influenced very much by changes in MAP until very high MAPs are reached; we can consider it to be specifically controlled by cardiac effectiveness (contractility and heart rate).

A very important fact is made readily apparent by the

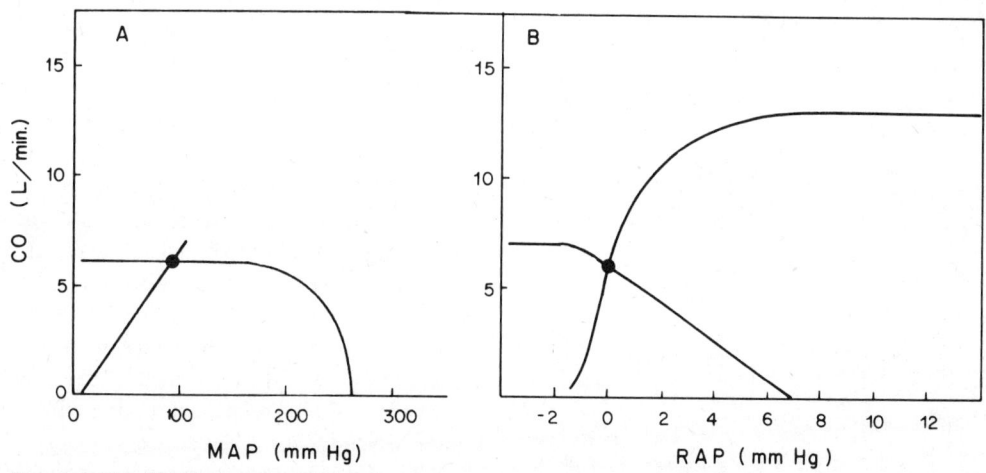

Figure 5 Combined cardiac and systemic characteristic curves. In both cases the curves cross each other at a point known as the operating point. The operating points define the only set of values for CO, MAP, and RAP at which the system is stable for any given values of the adjustable parameters of the system. The positions of the curves, and therefore the operating points, change when the adjustable parameters of the system are varied.

CO ⇌ RAP diagram in Figure 5b; that is, the normal heart has the ability to pump blood at a far greater rate than RAP ordinarily permits it, and this is true even without making it hypereffective. As we have seen, the position of the systemic characteristic curve is determined by the peripheral resistance, blood volume, and venous capacity. Consequently, these systemic factors determine where along the cardiac performance curve the operating point will lie. The heart is at the mercy of these systemic factors.

EFFECTS OF THE ADJUSTABLE PARAMETERS ON THE OPERATING POINT

Cardiac Effectiveness

Figure 6 is a CO ⇌ RAP diagram showing preload cardiac performance curves for a heart operating in normal, hypereffective, and hypoeffective modes. The normal systemic performance curve intersects these three curves at different operating points. If the heart becomes hypoeffective and we assume that no compensatory changes take place in either peripheral resistance, blood volume, or venous capacity, we can predict from the CO ⇌ RAP diagram that when a new steady state is reached, cardiac output will be lower than normal and RAP higher than

normal. For a step-by-step analysis assume that there is a sudden decrease in cardiac effectiveness so that cardiac output immediately drops along the vertical arrow to 2.5 L/min. Then over the next few heart beats RAP will gradually rise because of redistribution of blood from arteries to veins, and cardiac output will rise along the new cardiac performance curve until the new stable operating point is reached. This gradual return toward a more normal cardiac output depends upon the Frank-Starling relationship and helps protect against the effects of unwanted cardiac hypoeffectiveness.

If the heart becomes hypereffective, and we assume again that no compensatory changes take place that would alter the systemic characteristic curve, the new cardiac output will be higher than normal and RAP lower than normal. The increase in cardiac output that can be brought about by making the heart hypereffective is limited by collapse of the large veins entering the thorax. If cardiac effectiveness were increased even more than that shown in Figure 6, cardiac output would not increase much more.

Peripheral Resistance

The CO ⇌ RAP diagram shown in Figure 7 depicts a normal cardiac performance curve and three different systemic characteristic curves. The middle curve represents a normal or control situation, the upper curve a situation in which total peripheral resistance is reduced to about one-half normal by dilation of precapillary resistance vessels, and the lower curve a situation in which total peripheral resistance is increased to about twice normal. With decreased peripheral resistance, the new operating point indicates that in the steady state CO and RAP are both increased above control. With increased peripheral resistance, CO and RAP are both decreased below normal. The effect of a large reduction in peripheral resistance is much greater than the effect of a reciprocal increase in peripheral resistance. This is because of the shape of the venous pressure–volume relationship and will be explained in a later section.

Changes in peripheral resistance result in changes in the maximal attainable cardiac output; that is, the plateau caused by venous collapse is elevated as peripheral resistance is decreased. This is an extremely important effect. Remember, with a "normal" systemic characteristic

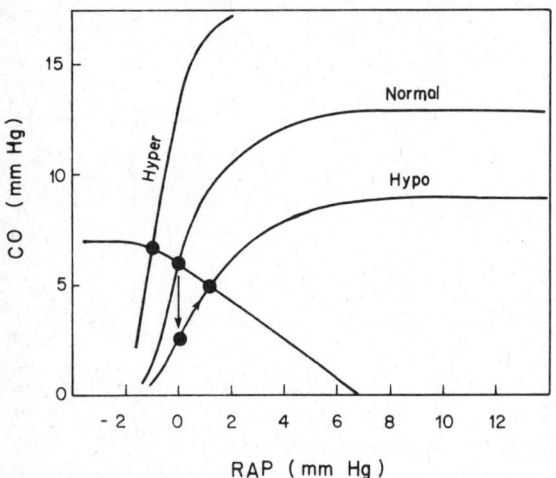

Figure 6 A CO ⇌ RAP diagram showing the effect of changing cardiac effectiveness on the operating point.

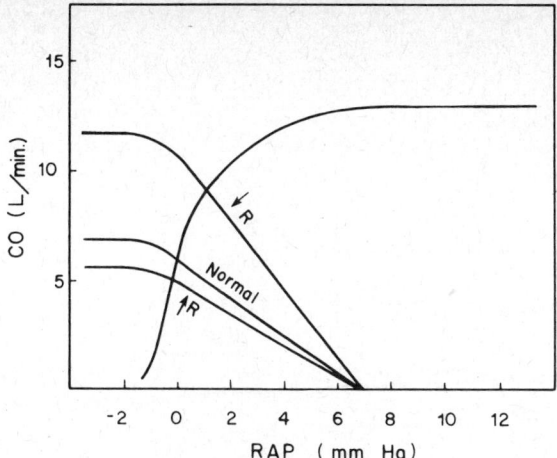

Figure 7 A CO ⇌ RAP diagram showing the effect of changing total peripheral resistance on the operating point. The curve labeled ↓ R represents a total peripheral resistance of about one-half normal, whereas the curve labeled ↑ R represents a total peripheral resistance of about twice normal.

curve, increasing the effectiveness of the heart could increase cardiac output by only a very limited amount. But if an increase in cardiac effectiveness is combined with a decrease in peripheral resistance the cardiac output can be greatly increased. In effect, the decreased peripheral resistance permits the heart to go ahead and do its thing.

Blood Volume

The CO ⇌ RAP diagram in Figure 8 shows systemic characteristic curves for hypervolemic and hypovolemic states, which might be brought about by intravenous infusion of blood or by hemorrhage, respectively. In hypervolemia, assuming that no compensatory changes take place, the new operating point is above and to the right of the normal operating point so that CO and RAP are both increased. The opposite occurs with hypovolemia.

Venous Capacity

The effects of changing the degree of venoconstriction or compression are qualitatively similar to the effects of changing blood volume. Increased venoconstriction shifts the systemic characteristic curve to the right, increases the general static blood pressure, and increases the maximum attainable cardiac output. If no other changes take place,

the new operating point sets at an increased CO and an increased RAP. Decreased venoconstriction does the opposite. If increased venoconstriction is accompanied by hypereffectiveness of the heart, the curves cross each other at a considerably higher than normal CO. This is another case in which the systemic factors can help allow a hypereffective heart to increase its output.

INCREASED CARDIAC OUTPUT DURING EXERCISE

Three primary cardiovascular changes occur during exercise: (1) increased effectiveness of the heart as a pump, (2) decreased total peripheral resistance, and (3) venoconstriction and venocompression (the latter being the result of skeletal muscle contractions). All these changes aid in increasing cardiac output. The appropriate CO ⇌ RAP diagram is shown in Figure 9. The heart becomes hypereffective as a result of sympathetic stimulation. The systemic characteristic curve is shifted to the right because of venoconstriction (which is also the result of sympathetic stimulation) and because of venocompression. The systemic characteristic curve is also rotated clockwise because of decreased total peripheral resistance. This decreased total peripheral resistance is the result of arteriolar vasodilation in the exercising muscles, the

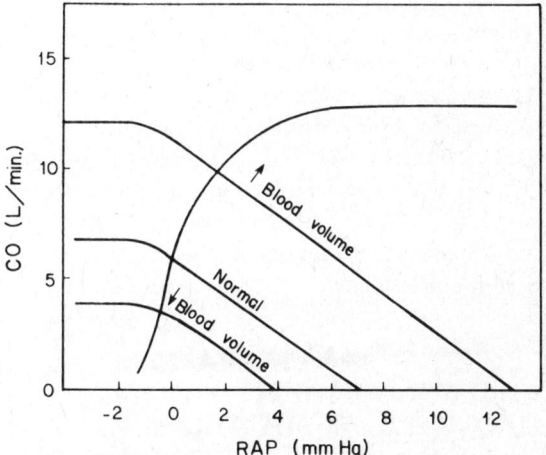

Figure 8 A CO ⇌ RAP diagram showing the effect of changing blood volume on the operating point. The effect of venoconstriction is similar to that of increased blood volume.

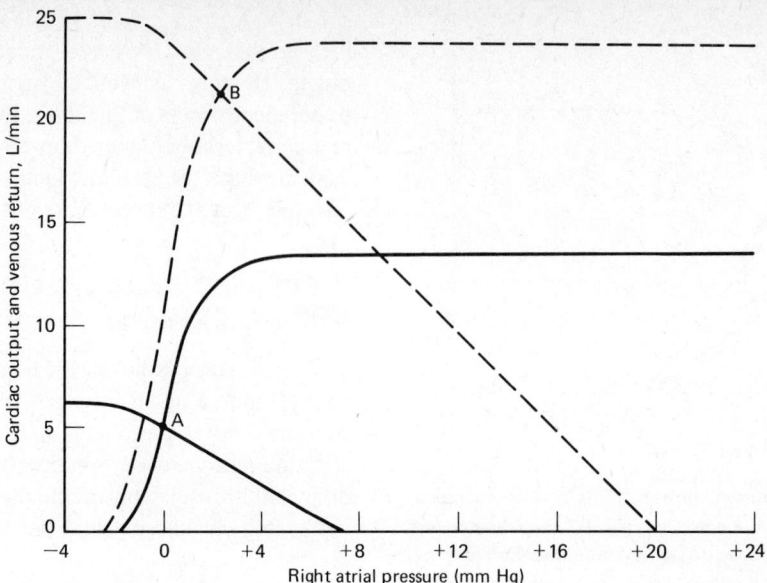

Figure 9 Graphical analysis of the changes in cardiac output and right atrial pressure with the onset of strenuous exercise. (From Guyton AC: *Textbook of Medical Physiology*, ed 6. Philadelphia, Saunders, 1981.)

mechanism of which will be discussed in Chapter 16. The new operating point in this example is 4.5 times above normal cardiac output; it can go as much as six times above normal in well-trained athletes. One of the most interesting things to note is that the real effectiveness of the sympathetically stimulated heart can be brought into play if, and only if, the changes in peripheral resistance and venous capacity occur.

The mean arterial pressure increases during exercise. Remember, $MAP = CO \times R_{TP}$. Therefore, during exercise cardiac output must increase more than peripheral resistance decreases. It is the increase in venoconstriction and venocompression that accomplishes this by shifting blood to the arterial side.

USE OF CO ⇌ RAP DIAGRAMS IN ANALYZING OTHER CARDIOVASCULAR SITUATIONS

CO ⇌ RAP diagrams are extremely useful for analyzing the responses in a number of other cardiovascular situations, such as heart failure, hemorrhagic shock, and car-

diac tamponade. Discussion of these situations is beyond the scope of this chapter, but it is strongly recommended that interested students consult the book by Guyton, Jones, and Coleman listed in the Selected Bibliography.

ROLE OF PERIPHERAL RESISTANCE IN CONTROLLING CIRCULATORY FLOW RATE AND MEAN ARTERIAL PRESSURE

As total peripheral resistance decreases, the rate of blood flow through the microcirculation increases and, therefore, the flow rate back to the heart increases. The flow rate back to the heart is called the venous return. As venous return increases RAP increases slightly, thereby inducing the Frank-Starling response of the heart. Because of this response, the normal heart has the capability of pumping blood at higher rates than are ordinarily required by venous return. Consequently, when total peripheral resistance decreases, cardiac output increases, and very quickly a new steady state is established in which the flow rate around the whole circulation is increased. Figure 10 shows the results of an experiment that illustrates this effect. In this experiment an anesthetised dog is fitted with a large tube running from the abdominal aorta to the right atrium. When this tube is opened by releasing a clamp, blood can flow directly from the abdominal aorta to the right atrium through the tube, called an arteriovenous fistula. In this way the total peripheral resistance can be suddenly and drastically reduced. The dog is given a spinal anesthetic to eliminate autonomic reflexes. When the fistula is opened cardiac output promptly rises to a new steady value, which is maintained until the fistula is closed again. The changes in cardiac output are very rapid following opening or closing of the fistula, and new steady values are reached within just a few heart beats. This experiment demonstrates the extreme importance of the total peripheral resistance in controlling cardiac output. The response of the heart can be entirely via the Frank-Starling mechanism. The normal heart is poised to accept and pump onward far more blood per minute (or per beat)

than ordinarily flows back to it through the veins (at least while the individual is resting or engaged in quiet activity). The heart has this reserve capacity even without any increase in its effectiveness that might be brought about by sympathetic activity. It is because of the reserve capacity of the heart, mediated by the Frank-Starling mechanism, that the total peripheral resistance can be such an important determinant of cardiac output.

Note in Figure 10 that the increase in cardiac output when opening the fistula was not proportionately as great as the decrease in peripheral resistance. In this particular experiment the peripheral resistance was nearly cut in half; however, the cardiac output did not nearly double. This accounts for the fact that MAP dropped appreciably. The reason for this behavior of the system has nothing to do with any limitation in the pumping ability of the heart unless very extreme decreases in total peripheral resistance are accomplished. Over the normal range of peripheral resistance, the failure of cardiac output to increase to the same degree that peripheral resistance decreases is explained by the fact that when peripheral resistance is suddenly decreased a redistribution of the total blood volume quickly occurs, with a greater percentage of it now in the systemic veins and in the pump oxygenator and less in the systemic arteries. As a result, MAP decreases, in turn preventing the flow through the microcirculation plus fistula, and subsequently back to the heart, from rising to the same degree that total peripheral resistance falls. This redistribution of blood is entirely dependent on compliant systemic veins and cardiopulmonary system. *If they were rigid tubes, MAP would not be influenced by changes in total peripheral resistance, since then there could be no redistribution of blood.* The change in cardiac output would be the reciprocal of the change in total peripheral resistance, and there would be no change in MAP. This principle is important because it applies to normal situations in which peripheral resistance is changed by changing arteriolar tone, and not just to the situation of an artificial arterio-venous fistula. The ability of the precapillary resistance vessels to influence mean arterial pressure is entirely dependent on the accommodative properties of the venous and cardiopulmonary systems, especially the former. *The more compliant the venous system, the more effect changes in peripheral resistance will have on MAP, and the less effect they will have on circulatory flow rate.*

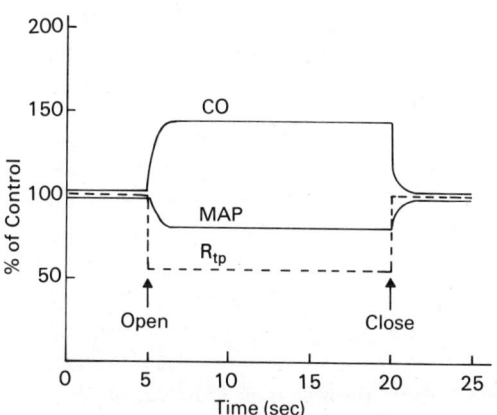

Figure 10 Effects of opening and closing an arteriovenous fistula on total peripheral resistance, cardiac output, and MAP. (Adapted from the data of Guyton AC, Sagawa K: "Compensations of cardiac output and other circulatory functions in areflex dogs with large A-V fistulas." *Am J Physiol* 200:1157, 1961.)

Peripheral resistance can also be experimentally increased. This has been done in dogs by injecting plastic microspheres into the aorta. The microspheres flow with the blood through the large arteries but get stuck in the microcirculation, thereby increasing peripheral resistance. As a result, the circulatory flow rate is reduced. However, the reduction in circulatory flow rate that results from increasing the peripheral resistance above normal is much less than the increase in circulatory flow rate that results from reducing the peripheral resistance below normal. This result might seem surprising. It is illustrated in Figure 11. Note that as the peripheral resistance increases, its effect on circulatory flow rate becomes less and less until at very high peripheral resistances there is almost no effect. This result can be explained by the compliance properties of the venous system. When the peripheral resistance is high the average venous pressure tends to be low, and in the low range of venous pressures the venous system is very compliant (Chapter 11, Fig. 14). Consequently, a large increase in peripheral resistance induces a substantial redistribution of blood from the venous system to the arterial system and MAP goes way up. In fact, the percentage increase in MAP is nearly as great as the percentage increase in peripheral resistance and, consequently, the flow rate does not go down very much. On the other hand, a large fall in peripheral resistance leads to an increase in average venous pressure, driving the venous system up into its low compliance range. There will still be some redistribution of blood (from arteries to veins in this case), but not nearly as much redistribution as when peripheral resistance increases above normal.

Thus, in the low range of peripheral resistances, expected during vigorous exercise, changes in peripheral resistance cause relatively large changes in circulatory flow rate, but small changes in MAP. This is probably of considerable advantage during exercise. On the other hand, in the high range of peripheral resistances, changes in peripheral resistance cause relatively large changes in MAP and small changes in circulatory flow rate. This may be of advantage in combating the effects of moderate hemorrhage.

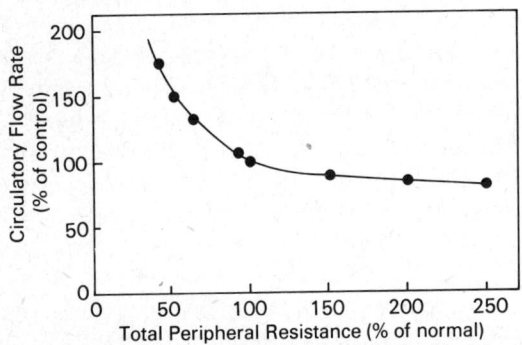

Figure 11 Effects of changing the total peripheral resistance on circulatory flow rate in anesthetized dogs. Cardiovascular reflexes were eliminated by spinal anesthesia. Increases in peripheral resistance above normal were accomplished by arterial injection of microspheres. Decreases in peripheral resistance below normal were accomplished by opening an AV fistula. (Drawn from the data of Guyton AC, Abernathy B, Langston JB, et al: Relative importance of venous and arterial resistances in controlling venous return and cardiac output, *Am J Physiol* 196:1008, 1959; and from the data of Guyton AC, Sagawa K: Compensations of cardiac output and other circulatory functions in areflex dogs with large A-V fistulas, *Am J Physiol* 200:1157, 1961.)

AUXILIARY MECHANISMS FOR INCREASING CIRCULATORY FLOW RATE AND KEEPING THE PRESSURE LOW IN CAPILLARIES AND VEINS

In preparation for discussing auxiliary mechanisms I shall try to make it explicitly clear what the *major* mechanism is for getting blood around the circulation. Of course, the prime mover is the heart, and it acts mainly by *pushing* blood around the circulation rather than by pulling it. In other words, the heart is mainly a pressure pump, rather than a suction pump. This pushing force that the heart imparts to the blood is known as the *vis à tergo* (force from behind). Even though it is reduced considerably in the precapillary resistance vessels, the *vis à tergo* is still the major cause of blood flow through the capillaries and veins, and all the way back to the heart. Let there be no doubt that the *vis à tergo*, provided by the heart, is the major force of blood circulation.

SUCKING ACTION OF THE HEART

Erasistratus suggested in about 290 BC that the heart is both a pressure pump and a suction pump. Galen (201 AD) agreed; Harvey (1628) did not. Following Harvey, the issue was debated for centuries and was finally settled by Brecher in 1956. The heart is both a pressure pump and a suction pump. The sucking force of the heart is called the *vis à fronte,* (force from in front). There are two kinds of cardiac suction: diastolic suction and systolic suction. Diastolic suction works like this. During systole, elastic elements in the myocardial wall are compressed to some extent and, as diastole begins, they recoil; this elastic recoil helps to draw blood into the ventricles during diastole.

Systolic suction is probably more important. As the ventricles contract during systole, the base of the heart (i.e., the AV junction) is pulled toward the apex*. By this action, the atria are greatly expanded during ventricular systole (Fig. 12) and, therefore, they suck blood from the vena cavae and pulmonary veins. Then, when the AV valves open, more blood is available in the atria to flow into the ventricles than would otherwise be the case, and diastolic filling is augmented.

The importance of *vis à fronte* is hard to evaluate. No one thinks it has anything like the importance of *vis à tergo*. Nevertheless, it seems certain that if the heart did not suck, right atrial pressure would have to be considerably higher than it normally is in order to accomplish the diastolic filling required for normal cardiac output. A higher than normal RAP is probably, in itself, no problem. But RAP can be higher than normal only if venous pressure everywhere, including the venules, is higher than normal. If pressure in the venules is elevated, then pressure in the capillaries must also be elevated. The capillaries and venules are the "exchange vessels" of the circulatory system; in other words, it is across these vessels that

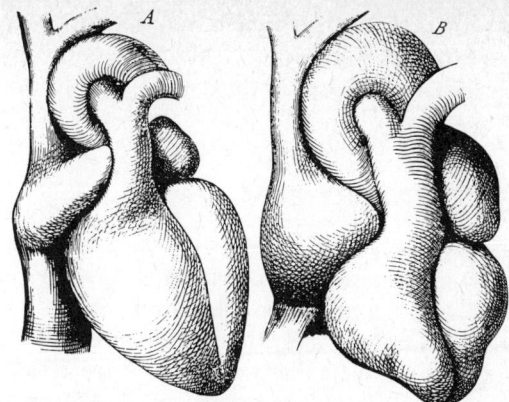

Figure 12 Illustration of the expansion of atria and venae cavae by the descent of the atrioventricular junction during ventricular systole. It can be seen how ventricular contraction might draw blood into the atria by expanding them. (*a*) Anterior view of the exposed heart and great vessels during ventricular diastole. (*b*) The same view during ventricular systole. (Drawing by Rollet 1880: *Hermann's Handbook of Physiology,* after an original by Henke, 1872.)

water and solutes move between blood and tissue spaces in accordance with their pressure and concentration gradients. If capillary and venular pressures are chronically higher than normal, a tendency to form excess volumes of fluid in the interstitial tissue spaces will be present. These excess volumes of fluid are called edema. I view the *vis à fronte* as primarily a way of helping to keep venous and capillary pressures as low as possible, thereby preventing edema.

NEGATIVE INTRATHORACIC PRESSURE

By virtue of elastic recoil, the lungs exert a continuous inward pull on the chest wall, and the chest wall exerts a continuous outward pull on the lungs. These elastic forces result in a negative pressure in the intrathoracic space, which oscillates between about −3 to −7 mm Hg and averages about −4 mm Hg. This negative pressure around the heart helps it to open up and accept the incoming blood during diastole. Diastolic filling of the right ventricle is dependent on the difference between RAP and intrathoracic pressure. If intrathoracic pressure were elevated,

---

*"One can see how during ventricular systole the atrioventricular junction approaches the apex, which remains almost stationary, whereas during diastole, the atrioventricular junction returns to its previous position. This descent and ascent of the atrioventricular junction is the most conspicuous movement which can be perceived in the acting heart" (Rollet, 1880).

RAP would have to be elevated as well in order to maintain a normal cardiac output. Thus the negative intrathoracic pressure aids in keeping venous and capillary pressures low.

THE RESPIRATORY PUMP

During quiet breathing intrathoracic pressure drops to about −7 mm Hg during inspiration and rises to about −3 mm Hg during expiration. Therefore, with each inspiration, the central veins expand and blood is sucked into them from the extrathoracic veins, with the result that venous return is momentarily increased. Increased intraabdominal pressure during inspiration contributes to this increase in venous return. Of course, the right ventricle is also influenced by intrathoracic pressure and fills with more blood during inspiration. It automatically pumps this extra blood (Frank-Starling mechanism). With expiration the reverse of this process occurs and venous return decreases. The increase in venous return caused by inspiration is greater than the decrease in venous return caused by expiration. The reason this is so is not entirely clear, but is probably at least partly attributable to the valves in the splanchnic veins, which retard retrograde flow. Regardless of the mechanism, it is generally accepted that respiration provides an auxiliary pump that helps move blood through the veins back to the heart. Again, the real significance of the respiratory pump is that it permits the heart to maintain a normal output without the necessity of high peripheral venous and capillary pressures.

VENOUS COMPRESSION DURING EXERCISE AND THE SKELETAL MUSCLE PUMP

When a skeletal muscle contracts it compresses the intramuscular veins thereby forcing blood from these veins back toward the heart. The blood is squeezed toward the heart by this action rather than back through the microcirculation because it takes the path of least resistance. When the skeletal muscle relaxes, its intramuscular veins refill with blood from the microcirculation. The venous valves prevent refilling from the more central veins. Rhythmic contraction-relaxation cycles keep pumping blood onward toward the heart. During strenuous running this action of the leg and thigh muscles keeps the average pressure in the small intramuscle veins less than that in the femoral veins. Thus, the muscle pump actually performs work on the blood and can be regarded as a booster pump in series with the heart. It has been estimated that during strenuous running the skeletal muscle pump in the legs and thighs provides almost one-third of the total power required to keep the blood circulating, and the heart provides the remaining two-thirds.

In addition to the effectiveness of this mechanism as a booster pump, it also helps reduce total venous volume. If venous volume is reduced by this action, arterial volume and pressure must be increased. In other words, the skeletal muscle pump helps to shift blood from venous to arterial sides of the circulation, just as does venoconstriction. The increased MAP resulting from this shift increases the flow rate through the microcirculation and, therefore, in the steady state, increases the cardiac output. Many investigators refer to rhythmic venocompression, and also to venoconstriction, as mechanisms for increasing the volume of "circulating" blood. This is misleading, since the blood involved was already circulating. The important thing that is actually accomplished is a shifting of volume from the venous to the arterial sides.

EFFECTS OF POSTURAL CHANGES ON CIRCULATORY FLOW RATE

Until now we have ignored the influence of gravity on the circulatory system, as well as those hydrostatic pressures that are related to depth in a column of blood. When a person is lying horizontally, depth differences in the circulatory system are relatively unimportant, but when a person stands upright such differences are very significant.

When a completely relaxed person is suddenly tilted from a horizontal posture to an upright posture blood shifts toward the feet. Arterial and venous pressures in the upper body decrease, while they increase in the lower body. There is a point located a few centimeters below the diaphragm at which the venous pressure is the same in the vertical posture as it is in the horizontal posture. This is sometimes called the hydrostatic indifference point (HIP). For the sake of convenience we will use this term, although it is actually a misnomer. In the upright posture the venous and arterial systems have the hydrostatic properties of long vertical columns of blood, and pressure increases with depth below the HIP and decreases with height above the HIP. These changes in hydrostatic pressure are proportional to distance from the HIP. Thus the increase in pressure in the foot vessels for a man of normal height is about 120 cm of blood, which is about 90 mm Hg. The venous valves do nothing to impede this hydrostatic effect, except perhaps momentarily, because as long as blood is flowing around the circuit the valves are open. This hydrostatic effect is added to the usual hydraulic pressures in the arteries and veins of the feet and so the total pressures become about 180 mm Hg and 105 mm Hg respectively. This, in itself, does nothing to impede blood flow through the feet or legs since the arterial and venous pressures are both increased equally. Furthermore, it is no more difficult for blood to flow from the heart to the feet than it was in the horizontal posture, even though now the total pressure in the arteries of the feet exceeds that in the root of the aorta. This might seem paradoxical since we have learned that flow rate is proportional to pressure difference, but this is true only if the effects of gravity can be ignored. Actually, as explained in Chapter 10, the driving force for flow is the difference in total fluid energy. Total fluid energy is the sum of the energies due to pressure, height, and velocity. Pressure energy increases with depth, but this increase is offset precisely by the decrease in potential energy of position (height). The end result is that hydrostatic pressures in the arterial system, brought about by taking an upright posture, have no *direct* influence on flow through the arteries. The same is true on the venous side; the high hydrostatic pressure in the feet while upright does not aid venous return, nor does the fact that the blood must flow upward retard venous return.

Nevertheless, shifting to an upright posture has a very serious effect on the cardiovascular system. Remember that *the HIP in the upright posture is below the heart* and that venous pressure decreases as the height above the HIP increases. Therefore, when a person rises to stand, the right atrial pressure decreases and, consequently, cardiac output decreases. Unless compensatory mechanisms promptly come into play, the cardiac output decreases by roughly 2 L/min, and the subject faints. The CO ⇌ RAP diagram in Figure 13 will be helpful for visualizing this effect; it shows a normal operating point at N for a reclining subject. Upon standing, the systemic characteristic curve shifts to the left, thus lowering the cardiac output and right atrial pressure. The change in the systemic characteristic curve is qualitatively similar to the change that would result from a reduction of blood volume. In fact, the translocation of blood from the upper parts of the body to the legs is often called "venous pooling." This blood is still circulating, but as far as circulatory flow rate is concerned, it is as though this blood were temporarily pooled somewhere. In the absence of compensatory adjustments, roughly 500 ml of blood are pooled in the legs upon standing. Far more than this would be pooled if the venous system were as compliant at high pressures as it is at low pressures. Fortunately, venous pooling is limited by the veins becoming almost completely nondistensible in the high pressure range.

In addition to problems resulting from reduced cardiac output upon standing, the elevated pressures in the microcirculation of the feet and lower legs cause increased filtration of fluid across the endothelium of capillaries and venules into the interstitial spaces. If this situation prevails for any length of time, edema of the feet and lower legs might occur.

Of course, we do not faint every time we stand up. There *are* compensatory mechanisms. The most immediate compensatory mechanism is tensing of the muscles of

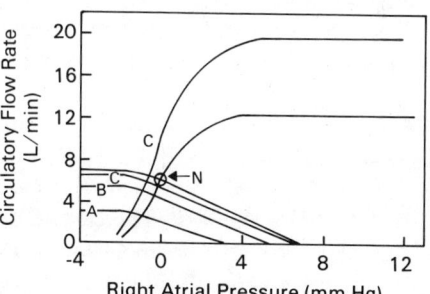

Figure 13 A CO ⇌ RAP diagram showing the effects of tilting to an upright posture. Curve A is without any compensatory mechanisms; curve B is after muscle tensing; curves C are after reflex adjustments.

the legs and abdomen. This ordinarily occurs along with standing and causes venous compression in the legs and abdomen. Consequently, the venous capacity and venous pooling in these regions are reduced. The important effect is elevation of the HIP upward toward the right atrium so that right atrial pressure is not reduced very much. Consequently, cardiac output is not reduced very much. In addition to muscle tensing, phasic contraction–relaxation cycles, as in walking, cause muscle pumping of blood toward the heart as described earlier. This process reduces average intramuscle venous and capillary pressures and helps to prevent edema.

In addition to contractions of skeletal muscle, certain autonomic reflexes also assist in adjusting the cardiovascular system to changes in posture—in particular the baroreceptor reflexes. These reflexes will be described in detail in Chapter 13. All we need to point out here is that if the arterial pressure is reduced because of decreased cardiac output during standing, a generalized increase occurs reflexly in sympathetic nervous activity. This results in (1) increased effectiveness of the heart as a pump, (2) increased venoconstriction, and (3) increased arteriolar constriction in many vascular beds. The first two of these effects result in an increase in cardiac output, as can be visualized in Figure 13. The third effect, increased arteriolar constriction, tends to reduce cardiac output slightly by increasing total peripheral resistance, but it also tends to increase MAP, or rather keep MAP from falling very much. This effect on MAP is advantageous because even if cardiac output decreases somewhat upon standing, if MAP is kept from dropping too low, the brain will continue to be perfused adequately with blood. These reflex readjustments may require a few seconds to develop; consequently, it is important that muscle tensing occur immediately and automatically on standing.

Some of the receptors involved in the baroreceptor reflexes are located in the carotid sinuses. Remember that in the upright position arterial pressure decreases in proportion to height above the HIP. The carotid sinuses are roughly 40 cm above the HIP, so upon standing the MAP in the carotid sinuses drops from about 95 mm Hg to about 65 mm Hg. This decrease would reflexly induce the sympathetic activity described above even if the cardiac output did not go down. The carotid sinus reflex helps maintain the circulatory flow rate and MAP (at heart level) during periods of quiet standing and sitting at values not much below those that exist during reclining.

When an upright posture is taken, pressures in the cerebral blood vessels fall by roughly 40 mm Hg. This drop has no effect on blood flow through the brain, since arterial and venous pressures are affected equally. Flow continues to be driven by the difference between inflow and outflow pressures (at any particular level of the system), as it would be through any other *siphon*. Note that a 40 mm Hg drop in pressure results in cerebral venous pressures that are considerably subatmospheric. The cerebral veins do not collapse since they are held open by surrounding tissues and ultimately by the rigid cranium. The superficial veins in the neck collapse during standing, since there is nothing to hold them open, and venous return from the head occurs mainly through deeper cervical veins.

If MAP decreases very much, cerebral perfusion may become inadequate, and the individual may "feel faint." Fainting can often be avoided by taking a horizontal posture or just by lowering the head with respect to the heart. This maneuver is effective not because it makes it easier for blood to flow to and through the brain, as is commonly supposed, but by raising right atrial pressure and, consequently, cardiac output. The most effective position should be the one at which right atrial pressure is maximal, and this occurs at a slightly head down position.

SELECTED BIBLIOGRAPHY

Berne RM, Levy MN: *Cardiovascular Physiology,* ed. 4. St Louis, Mosby, 1981.

Brecher GA: *Venous Return.* New York, Grune & Stratton, 1956.

Folkow B, Neil E: *Circulation.* New York, Oxford University Press, 1971.

Guyton AC, Jones CE, Coleman TG: *Circulatory Physiology: Cardiac Output and Its Regulation.* Philadelphia, Saunders, 1973.

Little RC: *Physiology of the Heart and Circulation.* Chicago, Year Book 1977.

Rushmer RF: *Structure and Function of the Cardiovascular System,* ed 2. Philadelphia, Saunders, 1976.

SELF-STUDY QUESTIONS

1. List the four major determinants of cardiac performance.

2. List the four independently adjustable parameters (variables) of the cardiovascular system that are of major importance in determining circulatory flow rate (i.e., cardiac output, venous return) and MAP.

3. List the influences on the heart that can make it hypereffective. List those that can make it hypoeffective.

MULTIPLE CHOICE

Select the correct answer(s). (In many instances, more than one answer is correct.)

4. In the absence of reflexes, cardiac output is not greatly influenced by MAP until very high MAPs are reached. Major explanations for this characteristic of the heart include
   A. left ventricular heterometric autoregulation.
   B. left ventricular homeometric autoregulation.
   C. right ventricular heterometric autoregulation.
   D. right ventricular homeometric autoregulation.
   E. increased RAP.

5. The general static blood pressure is increased by
   A. a reduction of venous capacity.
   B. an increase in blood volume.
   C. an increase in arteriolar constriction.
   D. a reduction of cardiac output.

6. Under certain circumstances venous collapse limits the degree to which increased cardiac effectiveness can increase circulatory flow rate. This maximal flow rate is
   A. increased by reduced peripheral resistance.
   B. increased by increased blood volume.
   C. increased by reduced venous capacity.
   D. determined by collapse of large intrathoracic veins.

7. The increased cardiac output associated with exercise is partly the result of
   A. increased ventricular contractility.
   B. increased heart rate.
   C. increased peripheral resistance.
   D. increased venous capacity.
   E. increased blood volume.

8. The Frank-Starling mechanism helps keep the circulatory flow rate from dropping too low during periods of reduced cardiac effectiveness.
   A. True.
   B. False.

9. With a decrease in cardiac effectiveness, RAP
   A. increases.
   B. decreases.
   C. does not necessarily change.

10. A large decrease in total peripheral resistance shifts the operating point on the CO $\rightleftharpoons$ RAP diagram more than does a reciprocal increase in total peripheral resistance.
    A. True.
    B. False.

11. A sudden and drastic reduction of total peripheral resistance below normal leads to
    A. a large increase in cardiac output.
    B. a large decrease in MAP.
    C. both.
    D. neither.

12. A sudden and drastic increase in total peripheral resistance above normal leads to
    A. a slight decrease in cardiac output.
    B. a large increase in MAP.
    C. both.
    D. neither.

13. The principal reason that increasing the total periph-

eral resistance above normal causes a large increase in MAP is that increasing the peripheral resistance induces a redistribution of blood from veins to arteries.

A. True.

B. False.

14. The reasons that decreasing the total peripheral resistance below normal causes a large increase in cardiac output are

A. the force of ventricular contraction increases as end-diastolic fiber length increases.

B. the compliance of veins diminishes as venous pressure increases above about 7 mm Hg.

C. flow rate through the microcirculation tends to increase as the resistance to flow decreases.

D. dilation of arterioles leads to mobilization of fluid from interstitial spaces.

15. During ventricular systole

A. the base of the heart moves downward toward the apex.

B. the pressures in the right and left atria decrease.

C. blood is sucked into the right and left atria.

16. If it were not for *vis à fronte*

A. normal cardiac outputs could not be achieved.

B. abnormally high capillary pressures would probably exist.

C. edema might be a much more frequent occurrence.

17. The "respiratory pump" contributes to the maintenance of high capillary pressures.

A. True.

B. False.

18. The "skeletal muscle pump" relieves the heart of an appreciable fraction of the total power output required for circulating blood during running.

A. True.

B. False.

19. Immediately after shifting from a reclining to an upright posture while intentionally keeping skeletal muscle as relaxed as possible,

A. cardiac output goes down.

B. MAP (at heart level) goes down.

C. right atrial pressure goes down.

D. pressure in the capillaries of the feet goes up.

20. The HIP in the vertical head-up posture is

A. located below the heart.

B. elevated by muscle tensing.

C. lowered by the carotid sinus reflex.

21. List four mechanisms that help prevent excessive "venous pooling" during standing.

ANSWERS

1. See page 210.

2. See page 212.

3. See Table 1.

4. A and B are correct (see page 211). C, D, and E are not correct because the RAP and right ventricular end diastolic pressure are not elevated appreciably.

5. A and B are correct (see page 214). C is wrong because only a small volume of blood is contained in the arterioles (page 214). D is wrong because by definition the general static blood pressure exists only when the heart is not pumping.

6. A, B, and C are correct (see CO ⇌ RAP diagrams). D is wrong because the intrathoracic veins are held open by the negative intrathoracic pressure—it is the large veins just outside the thorax that are likely to collapse.

7. A and B are correct.

C. *Reduced* peripheral resistance contributes to increased CO.

D. *Reduced* venous capacity contributes to increased CO.

E. The blood volume does not increase during exercise—if anything, it decreases.

8. A. True (Fig. 6).

9. A is correct (Fig. 6).

10. A. True (Fig. 7).

11. A is correct. The decrease in MAP is much less than the increase in CO (Fig. 10).

12. C is correct. With only a small decrease in CO (Fig. 11) and a large increase in peripheral resistance, there will be a large increase in MAP.

13. A. True. This is because a small reduction in venous pressures leads to a very large decrease in venous volume (Chapter 11, Fig. 14).

14. A, B, and C are correct.

    A. The Frank-Starling response (heterometric autoregulation) allows the heart to keep up with increased venous return.

    B. True (Chapter 11, Fig. 14). Therefore, the venous volume need not increase very much to produce a substantial increase in RAP.

    C. Obviously true.

    D. Just the opposite.

15. A, B, and C are correct (see page 221).

16. B and C are correct (see pages 221–222).

17. B. False. Just the opposite (see page 222).

18. A. True (see page 222).

19. A, B, C, and D are all correct (see pages 222–223).

20. A and B are correct (see pages 222–224). C is wrong. The carotid sinus reflex is one of the mechanisms for elevating the HIP.

21. 1. The substantial decrease in venous compliance as the venous pressure increases.

    2. Venous compression resulting from muscle tensing.

    3. The skeletal muscle pump.

    4. Venoconstriction as part of the baroreceptor reflexes.

13

Reflex Control of the Circulation

ARTHUR M. BROWN

OBJECTIVES

After completion of this chapter, you should be able to

1. Know how an arterial baroreceptor works.
2. Understand the reflex effects of arterial baroreceptor stimulation on heart rate, blood pressure, and peripheral resistance.
3. Express the range of pressures over which baroreceptor control occurs.
4. Describe the reflex effects of stimulating chemoreceptors and other circulatory mechanoreceptors.
5. Describe integrated responses such as the changes that result from standing up, or hemorrhage, or exercise.

CIRCULATORY REFLEX PATHWAYS

The nervous control of the circulation is reflex in nature and is carried out at subcortical levels. It is possible, however, to initiate circulatory responses from conscious experience, viz. the emotional faint, blushing, and it might even be possible to condition these responses. The two major variables that are controlled reflexly are blood pressure and blood volume. Both must be maintained within fairly narrow limits. Thus, too high a blood pressure can cause rupture of blood vessels, strokes, and other problems, whereas too low a blood pressure can cause fainting and renal malfunction. Likewise, blood volume changes can alter pressure and flow. As you have learned, the pressure in any of the components of the vascular system is the consequence of the volume of blood in that segment and the distensibility of the wall of that segment. In addition, volume transported per unit time is flow and is therefore a function of the available blood volume. The basic ingredients for reflex control of the circulation are the same as those for reflex control of somatic muscles and other organ systems. Three components must be identified: an afferent or input component, a central processing component, and an efferent or output component. For regulation, the arrangement provides negative feedback in much the same way as a thermostat regulates room temperature (Fig. 1). An increase in the parameter of interest, say arterial blood pressure, excites baroreceptors; this excitation in turn causes medullary neurons to reduce efferent sympathetic discharge and increase cardiac vagal efferent discharge. These effects result in cardiac slowing, decreased cardiac output, and reduced peripheral resistance thereby restoring blood pressure. Receptors sensitive to blood pressure and blood volume (mechanoreceptors) or to chemical composition (chemoreceptors) are located in the heart and blood vessels. Pressure receptors are located in the large arteries, where small changes in volume produce large changes in pressure. Some pressoreceptors are also located in the ventricles. Volume receptors are located in low pressure regions such as the atria, where small changes in pressure produce large changes in volume. In either case, the high and low pressure mechanoreceptors are stimulated by stretch and therefore

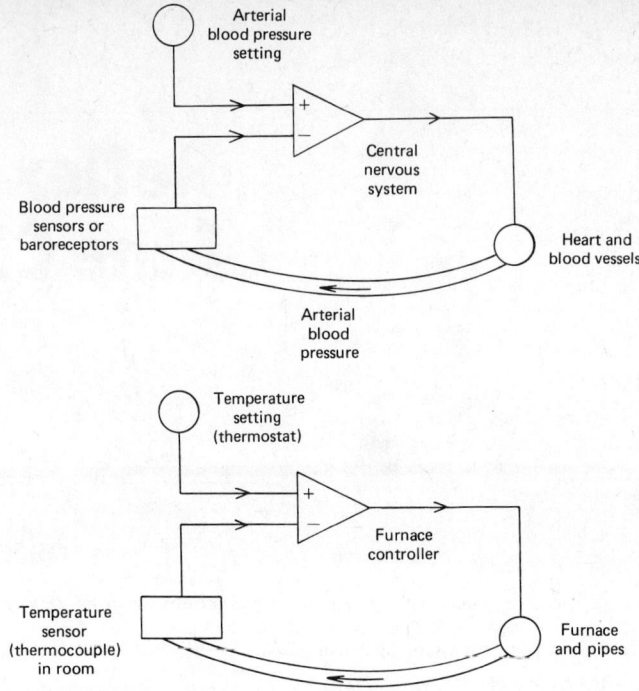

Figure 1 Comparison of baroreceptor regulation of blood pressure with thermocouple regulation of room temperature. Minus input of the sensors indicates the negative feedback effect on the controlling system and plus input indicates the reference value for the system.

respond indirectly to intravascular pressure or volume. The chemoreceptors are sensitive to changes in partial pressure of arterial oxygen (Pao$_2$), and to lesser extent pH and Paco$_2$. Both types of receptors are connected by fibers running in the IXth and Xth cranial nerves to an integrative area in the brain, the medullary reticular formation. Another set of receptors, including mechanoreceptors and possibly chemoreceptors and located mainly in the heart, have axons that run to the spinal cord frequently with sympathetic nerves. The cell bodies are in the dorsal root ganglia, and this spinal input is sometimes referred to as the afferent sympathetic input. For both inputs—spinal and medullary—the central receptors are connected via polysynaptic pathways to effector nerves that are distributed to the heart and to the smooth muscle of the resistance and capacitance vessels of the body. The efferent pathways are mainly (1) in the vagus nerve (parasympathetic), which has its central cell bodies in the nucleus ambiguus and dorsal motor nucleus; and (2) in the sympathetic nervous system, which flows out of the spinal cord between T1-L2. The cell bodies of sympathetic preganglionic neurons are in the intermediolateral horns of the gray matter; the nerve fibers run in the white rami to the prevertebral and paravertebral ganglia from which the postganglionic nerves arise. Parasympathetic preganglionic nerves that affect local flow run to the salivary glands, the cerebral and coronary vessels, and the penis.

At this point it is useful to review certain anatomical and functional aspects of the autonomic nervous system (ANS). The autonomic nervous system consists of two outflows from the central nervous system (CNS) to target organs (heart and blood vessels): namely, the sympathetic nervous system (SNS) and the parasympathetic nervous system (PNS). The emergent axons have synapses in peripheral ganglia; these axons and the associated cell bodies are therefore called preganglionic neurons. For the sympathetic nervous system, the peripheral ganglia are lateral (paravertebral), such as the stellate ganglion, or collateral (prevertebral), such as the coeliac ganglion; the postsynaptic neurons are called postganglionic neurons. The axons of the postganglionic neurons then have synapses on the target organ cells, which for the circulation are the smooth muscle cells of arteries, arterioles, venules, and veins, as well as cardiac muscle cells. Another target cell indirectly involved in neurocirculatory control is the adrenal medullary cell, which secretes norepinephrine and epinephrine (mainly the latter) into the bloodstream.

The second component of the autonomic nervous system, the parasympathetic nervous system, contributes to local flow regulation, but the relationship of this aspect to overall neurocirculatory control is poorly understood. The Xth cranial or vagus nerve, however, is importantly involved in control of heart rate. The preganglionic cells are in or near the nucleus ambiguus and dorsal motor nucleus, and the ganglion cells are located in the heart. The postganglionic axons therefore travel only a short distance to make synaptic contact with the cardiac muscle cells. The functional significance of these differences between the sympathetic and parasympathetic divisions is unknown. Synaptic action in the autonomic nervous system is key to an understanding of this system. The receptors, along with their agonists and antagonists, have been described in the section on the autonomic nervous system.

An outline of the most important receptor sites, afferent pathways, CNS integrative areas, and efferent pathways is shown in Table 1.

Table 1 Outline of Reflex Pathways

A. Afferent Sites and Pathways

1. Cardiovascular afferents: Fibers run mainly in the glossopharyngeal and vagus nerves. Cell bodies are in the petrosal and nodose ganglia, respectively. Some afferents run with the cardiac sympathetic nerves toward the spinal cord. Their cell bodies lie in the dorsal nerve root ganglia; these afferents probably carry the impulses that signal cardiac pain. The principal locations of the receptors are as follows:

   a. Mechanoreceptors   Nerve
      (1) Carotid sinus   IX
      (2) Aortic arch   X
      (3) Atria   X
      (4) Ventricles   X
      (5) Main pulmonary artery and branches   X
      (6) Coronary arteries   X
   b. Chemoreceptors
      (1) Carotid body   IX
      (2) Aortic body   X
   c. Mechano- and chemo-   Visceral
      receptors with afferent   sympathetic
      cardiac sympathetic nerves   nerves

2. Noncardiovascular afferent sites: Fibers run in somatic afferent nerves from skin and muscle and in the visceral afferent nerves from nonvascular areas of the gastrointestinal tract, lungs, and so on.

B. Efferent Pathways and Effector Sites

1. Heart
   a. Xth cranial nerve (vagus): Cell bodies lie in the nucleus ambiguus and dorsal motor nucleus. Their axons have cholinergic synapse in heart with postganglionic cells. Cholinergic action is exerted mainly on nicotinic receptors. Acetylcholine is also released by postganglionic

Table 1 (*continued*)

fibers, but here the synaptic action is on muscarinic receptors. Vagal excitation decreases SA and AV pacemaker discharge, decreases atrial contractility, and decreases AV conduction velocity. Vagal fibers also run to the coronary arteries and release ACh, which has a vasodilator action.

  b. Cardiac sympathetic nerves: Fibers of cell bodies of the intermediolateral horn cells at T1-L2 make synaptic connections at the paravertebral or prevertebral ganglia. The transmitter at the ganglia is acetylcholine, and the action is mainly nicotinic.

  The transmitter released from the postganglionic ending is norepinephrine, which excites α and β-receptors. Sympathetic discharge causes increased heart rate (chronotropic action); increased contractility (inotropic action) of both atria and ventricles; and increased conduction velocity. These actions take place on β-receptors. The sympathetics also innervate the coronary arteries via α-receptors (constrictor) and β-receptors (dilators).

2. Resistance and capacitance vessels: Arterioles, venules, and veins

  a. Sympathetic vasoconstrictor fibers. Preganglionics T1-L2. Ganglia are para- and pre-vertebral.
  Transmitter is norepinephrine, which causes constriction of arterioles, venules, and veins. α- and β-receptors are activated, and either or both may have constrictor actions.

  b. Sympathetic vasodilator fibers arising from thoracic and upper lumbar segments.
  Transmitter is acetylcholine, which dilates arterioles of skeletal muscle in proximal parts of extremities. Significance is questionable in primates and humans.

  c. Parasympathetic vasodilator fibers. Prevertebral neurons in certain cranial nerves III, VII, IX, and S_2-S_4 segments of spinal cord.
  Transmitter is generally acetylcholine. Dilates arterioles of salivary glands, cerebral and coronary arteries, and penis, although intermediary transmitters such as bradykinin can mediate flow changes in the salivary glands.

C. Central Nervous System Centers

1. Spinal centers: In chronic spinal animals, the spinal cord is capable of integrating afferent information from sympathetic afferents and spinal afferents to regulate efferent outflow to heart and vessels.

2. Medullary centers. An area in rostral medulla and lateral reticular formation, the pressor area, causes an increase in mean arterial pressure (MAP) when stimulated. An area in caudal medulla and medial reticular formation, the depressor area, produces decreased MAP. The medulla is probably a major integration area receiving afferent input (carotid and aortic baroreceptor afferents) and influences from central areas above medulla, as well as from spinal areas. An area in the medulla near the nucleus tratus solitarius, when lesioned, can produce fulminant hypertension in some species, such as the rat. The synaptic transmitters involved in central nervous system control of the circulation are not established definitely but norepinephrine is probably one. Thus, a drug called clonidine has strong hypotensive actions due to its α_2-agonistic action on CNS neurons.

3. Supramedullary centers

  a. Hypothalamus is involved in temperature control, defense or rage reaction, and other functions that have strong autonomic nervous system components.

  b. Cortex involvement is apparent from faint initiated by sight of blood or the blush from embarrassment.

  c. Cerebellum contains the fastigial nucleus, which may mediate certain pressor efferents.

CAROTID SINUS REFLEX

The most widely studied neurocirculatory reflex is the carotid sinus reflex. In the late 19th century, it was shown that a decrease in carotid artery pressure caused an increase in aortic blood pressure and tachycardia. This effect was attributed to the cerebral ischemia consequent to lowering carotid arterial pressure. This was the prevailing view until the 1920s despite some clever experiments by two Italian researchers named Pagano and Siciliano, who worked independently of each other in Naples and Palermo. Because their findings were published in an Italian journal they were largely unknown to other physiologists. These investigators showed that the fall in carotid arterial pressure was effective in causing tachycardia when it occurred in the common carotid artery, but that it was ineffective when it occurred in either the external or internal carotid arteries distal to the carotid bifurcation. This proved quite clearly that cerebral ischemia was not the important factor. Moreover, they were aware of the structural differentiation of the carotid bifurcation: the wall is thinner, dilated, and eccentric and has more elastic tissue and less muscle, as well as many nerve endings. These workers concluded that the effect they had obtained was initiated in the carotid bifurcation. There were many other leads that might have confirmed these findings, but they were largely ignored; the earlier view prevailed until a clinician named Hering discovered that merely stroking the area of the neck over the carotid bifurcation caused some sensitive people to faint. In his laboratory he then showed in animals that distension of the carotid sinus caused bradycardia and hypotension, whereas occlusion of the common carotid artery elicited the opposite results. He showed the response was mediated by the carotid sinus nerve because it was prevented by cutting this nerve.

It was not long before physiologists began working away on this lead; subsequently many of the characteristics of the carotid baroreceptors or mechanoreceptors were elucidated. For example, if the pressure in the carotid sinus (or aortic arch) is suddenly increased above threshold, the afferent nerve fibers from these receptors almost as suddenly start firing impulses. Figure 2*a* shows the time course of the discharge frequency for various step increases in pressure. It can be seen that for each pressure jump shown, the response (in terms of impulses/sec) rises to a peak in about 0.25 S and then gradually fades to a

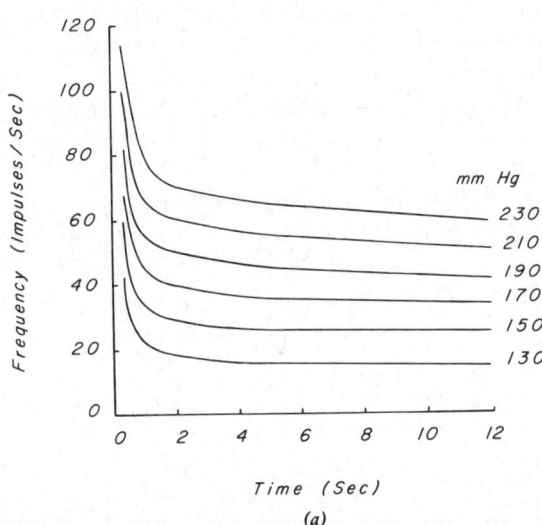

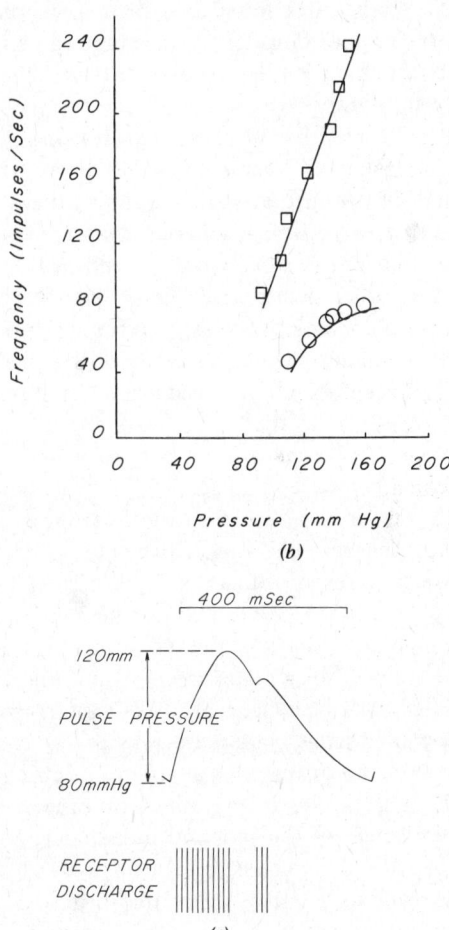

Figure 2 (*a*) Responses of baroreceptors in the aortic arch to sudden step increases in pressure. At zero time the pressure jumped from 0 to 130, 150, 170, 190, 210, or 230 mm Hg. The response has two phases: transient and steady state; the latter being characteristic of a slowly adapting baroreceptor. (*b*) Response of the baroreceptor shown in (*a*) as a function of pressure. The squares represent transient responses and the circles represent steady-state responses. (*c*) Action potentials (receptor discharge) in an afferent nerve fiber of a baroreceptor in the carotid sinus or aortic arch during the arterial pulse. Most action potentials occur during the ascending limb and a few at the beginning of the dicrotic wave. Note that for any given instantaneous pressure the receptors are much more responsive when the pressure is rapidly rising than when it is falling.

steady level of impulse frequency. We see from Figure 2a that (1) the transient peak response depends on the magnitude of the pressure jump; (2) the receptors gradually adapt during the first few seconds of a maintained pressure increase, but for the pressure jumps shown do not completely lose their responsiveness; and (3) the final "steady-state" response also depends on the magnitude of the pressure jump.

Figure 2b shows receptor response (either at the transient peak or the steady state) plotted against pressure. For this particular receptor there was a threshold of about 90 mm Hg for a transient response (not shown) and about 130 mm Hg for a steady-state response. The transient peak response is a linear function of pressure and the steady-state response shows saturation. If a large number of receptors is studied it is found that their thresholds vary over a wide pressure range but the average value is about 80 mm Hg for transient responses and about 105 mm Hg for steady-state responses.

It was of interest that when the carotid sinus was encased in a plaster of Paris cast so that its volume was independent of pressure over a wide range, there was no change in afferent neural discharge. This is strong evidence that the carotid sinus receptors respond to *stretch* or distortion rather than pressure itself. Hence, the term baroreceptor is not as accurate as the term mechanoreceptor, but it is commonly used. In any event, the mechanism whereby the receptors are stretched usually is the pressure within the vessel walls, that is, pressure changes volume.

The receptors are sensitive not only to the mean pressure to which they are exposed, but also to the rate at which the pressure rises, that is, the faster the rise, the greater the discharge. This is why they show a peak transient discharge in response to a step input of pressure (Fig. 2a). Since the pressure pulse to which these receptors are normally exposed is phasic, if the systolic upstroke quickens and the pulse pressure increases, the number of receptors activated and their discharge will be increased. The relationship of discharge to pulse pressure is shown in Figure 2c. Note that except for a burst of impulses at the time of the dicrotic notch, they show no other response during the decline of the pressure pulse. This is partly because the rate of change of pressure is now negative.

The receptors have either myelinated or unmyelinated axons that run to the central nervous system. The un-

myelinateds predominate. The myelinated group has lower thresholds, is more sensitive to changes in pressure, and has higher discharge rates, but both groups respond to steady pressures and changes in pressure. The functional differences between the two sets of receptors is unknown at present.

The receptor membrane consists of small areas of the nerve terminations, which are bare (of Schwann cells), and are linked by amorphous material to the connective tissue of the vessel walls in which they lie. Pressure produces wall stress through the Laplace relationship

$$T = \bar{P} \times r/h$$

where $\bar{P}$ is mean distending pressure, r is radius of the vessel over which $\bar{P}$ acts, h is wall thickness, and T is tangential wall stress. The stress then deforms or strains the material of the vessel wall (collagen, elastin, and smooth muscle) and produces strain of the receptor membrane. This in turn causes depolarization of the ending called the generator potential, and current (generator current) flows outward across the spike-initiating zone, an area of low threshold for action potential initiation located near the receptor. For small displacements the current is a linear function of the potential, and the spike discharge is in turn a linear function of the generation current. A linearized mechanotransduction model is shown in Figure 3.

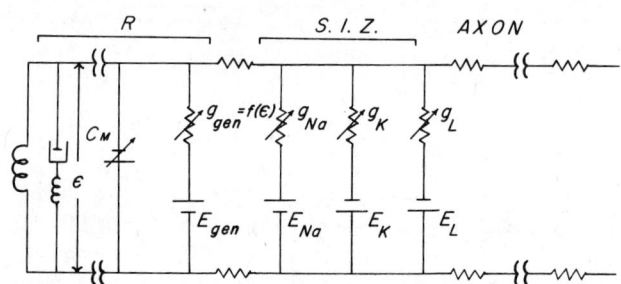

Figure 3 Electromechanical analog of an arterial baroreceptor. R is the receptor potential and includes the spring and dashpot on the extreme left-hand side; E is mechanical strain produced by pressure acting on the vessel wall. Interrupted lines indicate our lack of understanding of the mechanisms of mechanoelectrical transduction. C is membrane capacitance, and g_{gen} and E_{gen} are generator conductance and generator battery (or electrochemical potential), respectively. G_{Na}, G_K, and G_L are Na, K and leakage conductances, respectively. SIZ refers to spike-initiating zone.

The precise mechanism of mechanotransduction is unknown, hence the broken lines in the figure.

The possibility of efferent control of baroreceptor discharge by sympathetic innervation has frequently been raised using as a parallel example the muscle spindle. As you will recall, the intrafusal muscle fibers of the spindles have a gamma fiber efferent supply that can modify spindle discharge despite no change in the length of extrafusal skeletal muscle fibers. It is known that norepinephrine applied to the wall of the carotid sinus will increase the afferent discharge elicited by a given pressure pulse. It has also been suggested that changes in sympathetic drive to the carotid sinus wall can therefore modify the output of the sinus. Whether these effects are due to direct action on the receptors or to indirect action secondary to a change in smooth muscle tension is unknown. The functional significance of these observations is also unknown.

One of the most interesting properties of baroreceptors is that they are reset in hypertension of all sorts. The threshold pressure at which they begin to discharge is increased, and their sensitivity to suprathreshold pressures is reduced. These lesions are of fundamental significance in the maintenance of hypertension, since the central nervous system receives improper information concerning the blood pressure; that is, the information indicates that the pressure is lower than it actually is. The mechanism of resetting is not known with certainty. In some cases, especially in long-standing hypertension, vessel wall distensibility is reduced. This may be the main cause of resetting. In other cases, particularly early stages, the overall wall properties are unchanged, and changes may occur in the structures coupling the receptors to the vessel wall or in the receptors themselves.

Studies on carotid sinus reflexes made it apparent that the efferent limbs were vagal and sympathetic. If the vagi were cut, most but not all of the bradycardia in response to a rise in carotid sinus pressure was prevented. On the other hand, the rise in blood pressure provoked by a fall of pressure in the sinus was mediated by the sympathetics. It was also shown that after those parts of the medullary

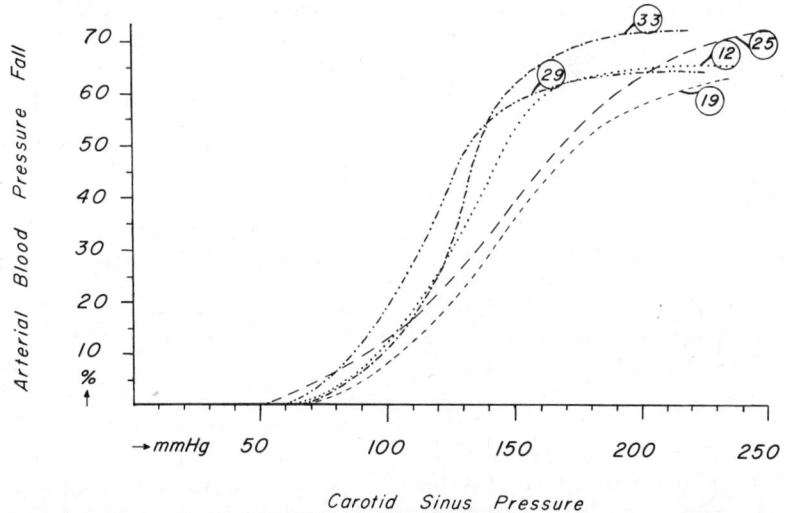

Figure 4 Regulatory importance of the carotid sinus as shown by Koch (1933). See text for explanation. If a change in body position or other factor causes the pressure in the carotid sinus to increase from 100 to 150 mm Hg, the circulatory reflexes will bring about a systemic pressure drop of about 40%. Percent changes in heart rate can be substituted on the ordinate of this graph. Hence, an increase in carotid sinus pressure from 100 to 150 mm Hg will cause a 40% decrease in heart rate. It is the increased discharge of the carotid sinus baroreceptors that produces these inhibitory effects.

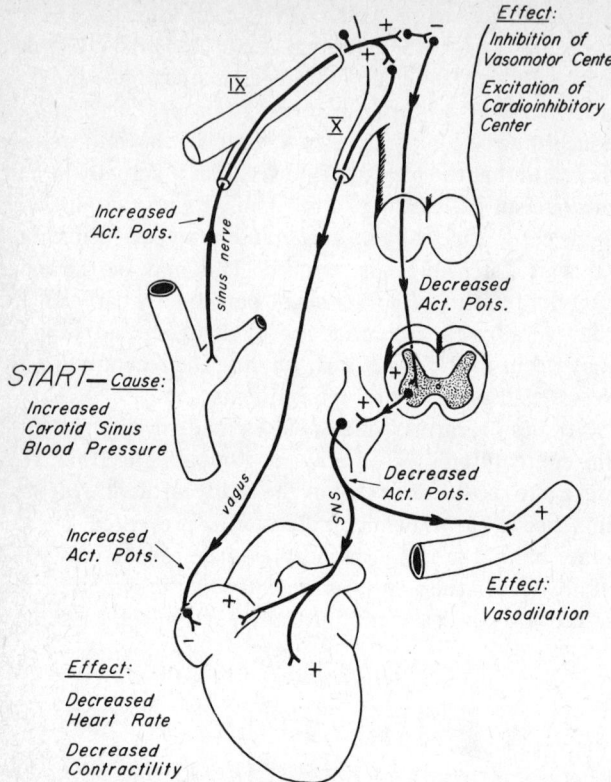

Effect:
Inhibition of Vasomotor Center

Excitation of Cardioinhibitory Center

IX

X

Increased Act. Pots.

sinus nerve

Decreased Act. Pots.

START—*Cause:*

Increased Carotid Sinus Blood Pressure

vagus

SNS

Decreased Act. Pots.

Increased Act. Pots.

Effect:
Vasodilation

Effect:
Decreased Heart Rate

Decreased Contractility

Figure 5 Carotid sinus reflex. The cause of the reflex (label on the left) is increased carotid sinus blood pressure. This causes increase in action potentials of the sinus nerve (a branch of the glossopharyngeal, IX). The sinus nerve neurons excite neurons in the medulla. A plus (+) sign indicates action on postsynaptic neuron is excitatory; a minus (−) sign indicates inhibition. The cardioinhibitory center is excited, and the rate of firing of the vagus (X) to the heart increases with inhibition of heart rate (−). The sinus nerve neurons also excite inhibitory neurons (depressor area of medulla), which inhibit (−) influence of vasomotor center (pressor area of medulla) on preganglionic fibers of the sympathetic nervous system. The normal action of the sympathetic nervous system (SNS) is excitation of heart rate and contractility (+) and stimulation of vasoconstriction (+). Along with decreased firing of these sympathetic nervous system fibers, there is decreased heart rate, decreased contractility, and decreased peripheral resistance (vasodilation). Venodilation and decreased venous return also occur. These effects will tend to reduce the blood pressure back to normal.

reticular formation known as the vasomotor and cardiac centers were destroyed, the reflex could no longer be obtained. The central synaptic mechanisms are also unknown except for the fact that the pathways are polysynaptic. The efferent sympathetic mechanisms have been described in the section on the Autonomic Nervous System.

It was Koch in 1933 who first demonstrated the regulatory importance of the carotid sinus reflex (Fig. 4). The graph shows the S-shaped relationship between blood pressure and carotid sinus pressure. Heart rate may be substituted for blood pressure. The graph indicates the negative feedback feature of the control, that is, as sinus pressure increases, blood pressure decreases. Figure 4 can be interpreted on the basis of known features of receptor discharge. This implies that little modification occurs at subsequent stages. The flat portion in the initial stage is due to the fact that at low sinus pressures relatively few receptors are discharging. The curve begins to run upward at a pressure that defines the threshold of the reflex and becomes steepest over that range that produces the greatest changes in total receptor discharge. When all the receptors are maximally discharging, the curve flattens again. The graph shows that it is in the range of physiological pressures that the greatest effects of changes of sinus pressure are produced. That is, the S-shaped curve is steepest between 100 and 150 mm Hg, so that pressure increases over this range produce the largest reflex fall in systemic arterial blood pressure. The ratio of changes in arterial pressure to changes in sinus pressure is the open-loop gain A of the system. As noted, the gain is maximal at physiological pressures where the best control is required.

The carotid sinus reflex represented in Figure 4 is an experimental convenience. In the intact animal or human, the carotid sinus arterial pressure will increase only if the aortic or arterial blood pressure increases as well. The control is now a closed loop function, since arterial and carotid blood pressures are identical, and the loop gain G is given by

$$A/(1 + \beta A)$$

where β is a factor relating pressure to baroreceptor discharge. The greater A is, the closer to a maximum value of 1 the gain becomes. For the carotid sinus reflex, $A = 2$.

However, other inputs, such as aortic and ventricular mechanoreceptors, will be stimulated as well making the overall control much greater. The structural basis for the reflex is shown in Figure 5. Another observation of interest is that if the mean pressure is kept constant, but the pulse pressure decreased, then the afferent activity falls, inhibition of the medullary vasomotor center and cardiac centers is less, and the heart rate and blood pressure will rise. This answered a well-known clinical phenomenon; in mild hemorrhage of say less than 15% of the blood volume, the heart rate increases when the mean blood pressure remains unchanged.

OTHER NEUROCIRCULATORY REFLEXES

AORTIC ARCH AND SUBCLAVIAN BARORECEPTOR REFLEXES

Mechano- or baroreceptors similar to those in the carotid sinus are also located in the aortic arch, and the right subclavian artery and baroreceptor reflexes similar to those elicited from the carotid sinus may be initiated from these areas. Differences between carotid sinus and aortic arch baroreflexes have been described, but there is some uncertainty concerning their magnitude or even their occurrence.

CARDIAC AND CARDIOPULMONARY REFLEXES

We have recently become more aware of the importance of the heart as a reflexogenic organ. Much of the lethal pathophysiology associated with heart disease, such as sudden death during acute myocardial infarction, may be attributable to reflexes originating from cardiac receptors or reflexes mediated by the efferent cardiac sympathetic nerves. The heart is unique as a cardiovascular reflexogenic area since it has major inputs to the CNS via two pathways, afferent vagal and afferent cardiac sympathetic fibers. The receptors are located in every cardiac chamber and in subendocardial, myocardial and epicardial locations. The methods used to produce cardiac reflexes often affect discharge from receptors in the lungs so that the broader term, cardiopulmonary reflexes, is often used. The pulmonary component of these reflexes is initiated by receptors in the airways and pulmonary blood vessels described in more detail in the respiratory section of this book (Chapters 18–25).

With the exception of the Bainbridge reflex, to be discussed below, cardiac reflexes mediated by afferent vagal fibers are depressor type; they are similar to reflexes elicited from systemic arterial baroreceptors (carotid sinus, aortic arch). In fact, one of the first neurocirculatory reflexes, the Bezold reflex (1867), was elicited by chemical excitation (using veratridine) of cardiac receptors. One difference from arterial baroreflexes, however, is that cardiac reflexes appear to have a more significant effect on the kidneys than do carotid sinus and aortic arch reflexes.

The Bainbridge reflex was described in 1915. Bainbridge found that enhanced venous return caused tachycardia, which was abolished by vagotomy; he ascribed the effects to excitation of afferent vagal fibers. Why would anyone care about the Bainbridge reflex? There are at least two reasons. First, one approach to understanding the function of the nervous system, namely that of Sherrington, is that all integrated neural responses are built from reflex components and this was the first reflex associated with the nervous control of the circulation. A second reason is that the reflex was thought to account for the tachycardia of muscular exercise. However, the Bainbridge reflex has become very controversial, mainly because reproducing it requires careful attention to initial conditions. If the initial heart rate is fast, enhanced venous return can, in fact, cause bradycardia. Thus at the moment, the role of the Bainbridge reflex in cardiovascular responses is uncertain. Excitation of atrial mechanoreceptors of which there are two types: (A), which discharge during atrial systole, and (B), which discharge during atrial diastole, causes a tachycardia mediated exclusively by the cardiac sympathetics. The functional significance of this atrial reflex produced by local stimuli using atrial balloons is, at present, uncertain, and it may not be directly related to the reflex described by Bainbridge, since enhanced venous return might excite many cardiac receptors in addition to those located in the atria. The reflex is associated with a diuresis. The mechanism of the diuresis is unknown; it may involve either a decrease in ADH (the antidiuretic hormone) or an increase in an as yet unidentified diuretic hormone.

Atrial receptors are reset in chronic heart failure in the same way as arterial baroreceptors are reset in hypertension. The fluid retention associated with congestive heart failure may be due to this resetting. These receptors may also be involved in the various renin profiles associated with hypertension and discussed in a subsequent section.

In contrast to depressor vagal reflexes, cardiac reflexes mediated by afferent cardiac sympathetic fibers are mainly pressor. However, at low frequencies of stimulation depressor reflexes may result. The importance of the pressor reflexes is that the efferent cardiac and peripheral sympathetic fibers are excited causing tachycardia, increased cardiac output and increased peripheral resistance. Some of the receptors in this pathway are excited during myocardial ischemia and probably activate efferent cardiac sympathetic fibers. It has been known for many years that the incidence of lethal arrhythmias, namely ventricular fibrillation, during acute myocardial infarction is significantly reduced by ablation of cardiac sympathetics. In fact, the current long-term treatment for patients at risk of sudden cardiac death is use of the β-blocker propranolol. It appears likely that one mechanism underlying the enhanced cardiac sympathetic efferent discharge during coronary occlusion is excitation of afferent cardiac sympathetic receptors.

REFLEXES FROM PULMONARY BARORECEPTORS

Baroreceptors are located in the large pulmonary arteries, but their reflex effects are uncertain. Other receptors in the pulmonary parenchyma will be discussed in the chapter on respiration.

CHEMORECEPTORS

The carotid bodies lying next to the carotid bifurcation and the aortic bodies lying along the arch of the aorta receive the termination of sensory fibers, which respond principally to a reduction of oxygen tension. A fall in oxygen tension, particularly below 50 mm Hg, causes an increased frequency of discharge from these receptors. The response to hypoxia is increased when the tension of CO_2 is elevated and when pH falls. The sensory input from these peripheral chemoreceptors plays its most important role in the reflex regulation of respiration (Chapter 25), but this input also has significant effects on the circulation. Hypoxia, or stimulation of the chemoreceptor fibers, leads to an increase in vasomotor tone and thereby brings about a pressor reflex response.

Severe hypoxia of the central nervous system, particularly the brain stem, also increases blood pressure. A diminution in blood flow through the brain produced by an increase in intracranial pressure, for example, causes a compensatory rise in blood pressure to overcome the increased resistance. This response is sometimes referred to as the Cushing reflex or reaction.

NONCARDIOVASCULAR RECEPTORS

Stimulation of somatic afferents from the skin or muscle produces reflex changes in blood pressure and heart rate as does activation of many visceral afferents. Their role in the control of circulation is not a direct one but because their input converges on neuronal pools shared by afferents from cardiovascular reflexogenic areas, they are able to influence the neurocirculatory control system.

EFFERENT LIMBS OF NEUROCIRCULATORY REFLEXES

VAGAL INNERVATION OF THE HEART

Circulatory responses to carotid baroreceptor stimulation, as well as those of other mechanoreceptors and of chemoreceptors, have been described. This section describes how activity of the sympathetic and vagal efferent pathways produces these responses.

The vertebrate heart receives an important motor innervation from both the right and left vagus nerves. These nerves contain large numbers of axons whose cell bodies lie in the nucleus ambiguous and dorsal motor nucleus within the medulla. The vagal fibers are preganglionic; they terminate on nerve cells lying within the heart. The postganglionic neurons, in turn, innervate the atria and nodal tissues. Except for the conduction system, they do not appear to innervate the ventricles to any great extent. In addition to the efferent fibers, the cardiac vagus also contains a variety of sensory neurons from the heart.

Stimulation of the peripheral ends of the vagi results in a slowing of the heart. The amount of slowing depends on the number of motor fibers stimulated and the frequency of stimulation. (Sufficiently strong stimulation of the right vagus can cause temporary cardiac arrest. However, even though stimulation is continued, the heart "escapes" vagal influence and begins to beat again, although at a slower than normal rate.) The vagal innervation of the atrium in the region of the SA node is predominantly from the right vagus. Hence, the right vagus often displays a greater effect on heart rate than the left vagus. A reduction in heart rate is accompanied by greater diastolic filling and hence greater stroke output.

The left atrium and AV node receive a proportionately larger innervation from the left vagus. Stimulation of the left vagus therefore tends to have a greater effect on the AV nodal region than does stimulation of the right vagus. Stimulation of the left vagus causes a slowing of AV nodal conduction and, if sufficiently intense, nodal block. In this situation the SA node may continue to generate impulses; this is evident from the rhythmic occurrence of P waves on the electrocardiogram. Slowing of conduction through the nodal tissue may be seen as a prolongation of the interval between the P wave and the onset of ventricular excitation, the QRS complex. At critical levels of inhibition, every other impulse generated by the SA node may be conducted to the ventricles resulting in a two to one heart block. Still more intense inhibition blocks nodal conduction completely. Stimulation of the vagi is also known to reduce the force of atrial contraction.

The vagal motor terminations in the heart are known to release acetylcholine (ACh) on stimulation. This, in fact, is where release of a chemical transmitter was first shown, and the demonstration was the first evidence of chemical mediation of a nervous response. During the 1920s, Otto Loewi bathed two isolated beating frog hearts in Ringer's solution in such a way that the solution bathing the first heart subsequently bathed the second heart. Stimulation of the vagi to the first heart caused a slowing of both hearts. The substance released behaved pharmacologically like a choline ester.

The effect of ACh on pacemaker and conducting tissue is to increase the membrane permeability to K ions. This causes hyperpolarization and reduction in the rate of diastolic depolarization. The reduction in membrane resistance lowers the length constant of electrotonic spread and slows conduction velocity as a result. ACh may also reduce Ca currents in pacemaker cells.

There is normally a tonic background activity in vagal efferent fibers, an action that can be demonstrated by blocking or cutting the vagi or by the administration of atropine, a drug that prevents the action of acetylcholine on receptor sites in the effector organ membrane.

SYMPATHETIC INNERVATION OF THE HEART

The sympathetic innervation of the heart originates from the upper thoracic segments of the intermediolateral cell column in the spinal cord. Preganglionic fibers synapse in the paravertebral and stellate ganglia, and postganglionic fibers innervate both atria and ventricles. The effect of sympathetic activity on pacemaker cells is to speed up the rate of diastolic depolarization, thereby increasing heart rate. The mechanism underlying this change in diastolic depolarization is not thoroughly understood. It is related to a shift in the voltage dependence of the hyperpolarizing pacemaker K^+ conductance in such a way that diastolic depolarization is enhanced. It is important to note that all potential pacemaker tissue in the heart is influenced in this manner by epinephrine and norepinephrine, and the pacemaker activity in some areas may be so increased as to cause the production of ectopic beats. The second major effect of the sympathetics on cardiac muscle is an increase in the force of contraction, when the muscle is operating at the same length and activated by an action potential of the same duration. The possible mechanisms involved in this effect have been discussed in Chapter 5. It appears that increased membrane Ca current is involved. This action is referred to as the positive inotropic effect of the sympathetics.

CONTROL OF PERIPHERAL CIRCULATION

Sympathetic fibers innervate arteriolar smooth muscle and cause changes in arteriolar diameter, which in turn regulate the peripheral resistance. The effect of sympathetic fibers on arterioles may differ from one vascular bed to another. On stimulation of the peripheral sympathetic constrictor fibers, the overall effect on vasomotor tone in a skinned perfused hind limb is one of vasoconstriction. Such a preparation perfused at a constant pressure shows a conspicuous decrease in flow on sympathetic stimulation. Accompanying this is a reduction in volume of the tissue, indicating that the so-called capacitance vessels have been emptied. However, the sympathetic supply to the hind limb contains vasoconstrictor and vasodilator fibers that activate α- and β-receptors, respectively. Ordinarily the α-effects predominate, but during α-blockade a vasodilator response is elicited by stimulation of the sympathetic nerves.

Whereas circulating norepinephrine appears to have, principally, a vasoconstrictor effect and causes pressor responses, epinephrine in small doses produces a fall in blood pressure consequent to peripheral vasodilation. The vasodilation is the result of the effect of epinephrine on β-adrenergic receptors, which have higher sensitivity to epinephrine than do α-receptors. Larger doses of epinephrine bring about pressor responses through action on the α-receptors, which are generally more numerous. The sympathetic outflow to blood vessels in various tissues and organs can also show selective reflex activation. That is, the system can respond in a selective manner. In exercise, for example, a larger fraction of cardiac output may be shifted to muscle than is the case at rest. One also has selective changes in arteriolar resistance in the skin associated with temperature regulation.

The sympathetic efferent fibers producing vasoconstriction have their preganglionic outflow along most of the length of the intermediate lateral cell column of the spinal cord. The preganglionic fibers synapse with postganglionic neurons in the paravertebral or more peripheral sympathetic ganglia. This outflow maintains the overall

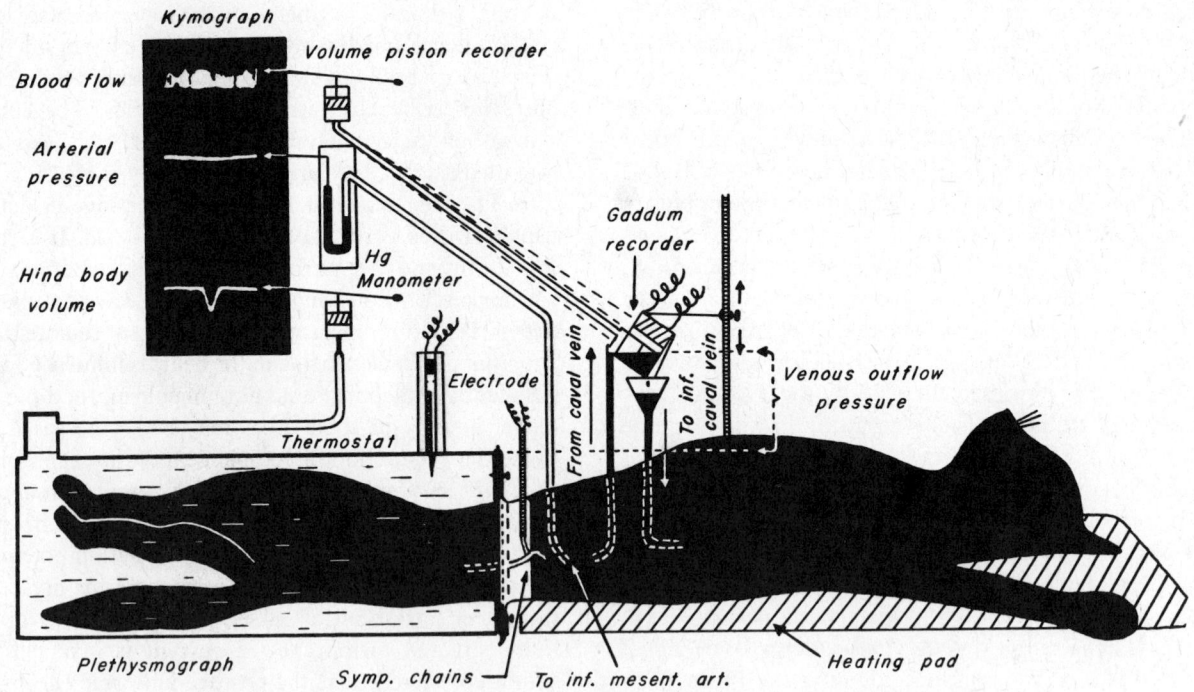

Figure 6 Experimental arrangement for the study of peripheral neurocirculatory control.

systemic peripheral resistance at a level sufficient to keep the blood pressure within a certain range. The activity of these sympathetic vasoconstrictor neurons is maintained by a tonic excitatory activity descending from the brain stem to sympathetic neurons in the spinal cord. This is made quite evident when the cord is transected above the T1 level. Blood pressure falls abruptly as a consequence of the loss of vasoconstrictor tone.

We now consider in more detail the efferent neural control of the resistance and capacitance vessels. Whereas efferent cardiac control was mediated by both the vagus and sympathetics, peripheral vascular neural control is mediated by the sympathetic outflow alone (except for some local beds, such as salivary glands and sexual organs, in which the parasympathetics are involved).

The peripheral vascular bed that has been most studied is skeletal muscle. The procedure is to leave the sympathetic nerves and the aorta and inferior vena cava intact and to isolate the rest of the hind limbs from the body. The hind limb is then skinned, and the circulation to the paws is cut off by a ligature. What are then left are the muscles of both hind limbs, the arterial and venous supply of these muscles, and the sympathetic nerves to the muscle blood vessels. If the preganglionic sympathetics are left intact,

and the ventral nerve roots of the lower cord are cut below L2, any effect of motoneurons to skeletal muscle can be avoided as well. We can measure the blood flow coming out of the hind limbs and we can keep the pressure head perfusing the hind limbs constant. We can also measure the change in volume of this bed by placing the hind limbs into a plethysmograph (Fig. 6).

When the sympathetics to the hind limb are stimulated, the blood flow falls and, since the systemic pressure in this experiment remains constant, what has happened to the resistance in this bed? The volume of the hind limb also falls precipitously, due to an emptying of the capacitance vessels (Chapter 11, Fig. 16). The emptying is the result partly of decreased inflow from the arterioles and mainly because of increased venomotor tone. There is, however, a slower subsequent fall in tissue volume, which is caused by the absorption of extravascular fluids (see arrow no. 3, Fig. 16, Chapter 11). This absorption is greater when sympathetic drive increases. Why do you suppose this slow absorption occurs? Why does it increase at greater frequencies of nervous discharge? Examination of Figure 7 shows the percentage of maximal response of the resistance and capacitance vessels when the sympathetics are stimulated at different frequencies. Clearly, the slow ab-

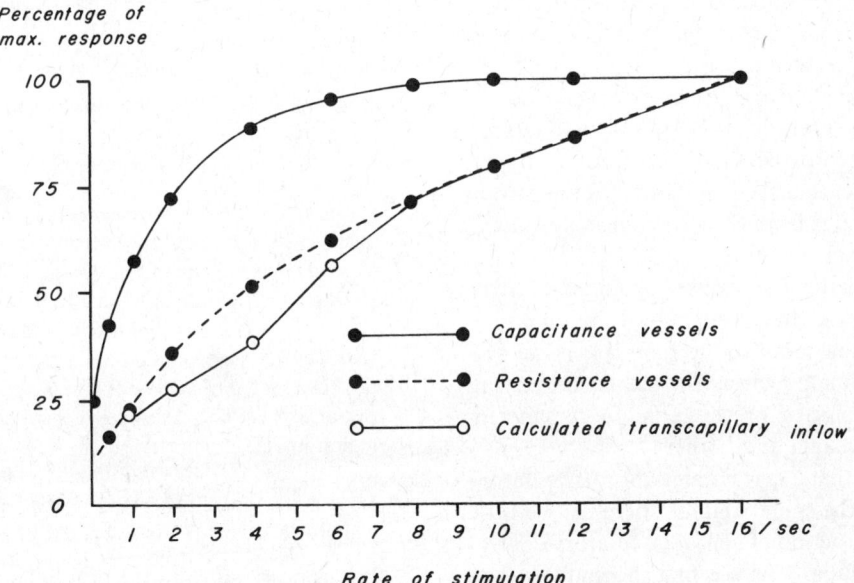

Figure 7 Relationship between rates of sympathetic stimulation and response.

sorption occurs because at all frequencies the pressure drop across the arterioles is greater than the pressure increase in the venous system, causing the capillary hydrostatic pressure to drop and fluid to be resorbed from the extravascular extracellular space. The resorption increases at higher frequencies because the effect of stimulation on resistance vessels is more nearly linear over the physiological frequency range. Thus, the answer to our questions is that the resistance goes up.

ROLE OF THE CENTRAL NERVOUS SYSTEM IN NEUROCIRCULATORY CONTROL

Whereas a considerable amount is known of the afferent and efferent limbs of circulatory reflexes, very little is known of the part played by the central nervous system in integrating these responses. It is certainly true that nervous adjustments are possible in the spinal animal, more so in the chronic spinal animal after the effects of spinal shock have worn off. For example, during asphyxia, very large increases in pressure can occur. Whether these reflexes are initiated from some receptor site outside the spinal cord or whether they are the direct result of asphyxia on the preganglionic sympathetic neurons has not been determined. In humans with high spinal lesions, bladder distension can provoke the most enormous hypertension, so great in fact that the stimulation of the carotid baroreceptors that results from the hypertension produces vasodilation of the neck and face.

Not only can sympathetic responses be initiated in the spinal animal, but they show considerable differentiation. Thus, stimulation of one input to the central nervous system of the spinal animal, such as the sciatic nerve, can excite one postganglionic or preganglionic nerve and inhibit or produce no change in another.

Although it is true that a certain amount of circulatory control is possible in the spinal animal, there can be little doubt that most of the action of neural circulatory control takes place in the medulla. This was first shown in the late nineteenth century by Dittmar and Ovsannikow, working in the laboratory of the most famous circulatory physiologist of that time, Karl Ludwig. These workers stimulated the central end of the cut sciatic nerve at frequencies that elicited a rise in blood pressure and tachycardia and then proceeded to cut the brain transversely from the cortex caudally. Section had no effect when the medulla was connected to the spinal cord, but the response to sciatic stimulation became very much weaker when the medulla was separated from the cord. In another series of experiments they repeated these transverse sections and measured the blood pressure; there was no change until the pons was reached, and then the pressure fell progressively with each more caudally placed section. At a point 4–5 mm above the calamus scriptorius, the pressure fell to the low level obtained by section of the cervical cord. Subsequently, direct stimulation of various parts of the medulla accompanied by measurements of arterial blood pressure and heart rate, followed by histological studies have shown that the pressor area is in the rostral and lateral reticular areas of the medulla, whereas the depressor area is more caudal and medial (Fig. 8). Moreover, it appears that the pressor area or center is tonically active and independent of the afferent input from circulatory receptors, whereas the depressor area is modified or controlled by input from circulatory receptors.

It is appropriate at this point to define what is meant by a central nervous system center, since we will be referring to such areas as the pressor center, the depressor center, the vasomotor center, the cardiac acceleratory center, and the cardioinhibitory center. A description given by Bard many years ago still seems applicable:

In the study of the central nervous system, it has sometimes been possible to show very clearly that a complex motor act depends upon the functional integrity of a sharply bounded portion of the brain, a central nervous system center. In such instances, it is commonly found that although destruction of the essential region permanently abolishes all possibility of evoking the total response, some item of the response is still to be elicited. Further, with the essential region intact, it is invariably discovered that other parts of the brain exert modifying influences either directly upon it or at some point along the chains of neurones through which it controls the peripheral motor pathways.

Let us now return to the pressor (increased arterial BP) and depressor (decreased arterial BP) responses. As we

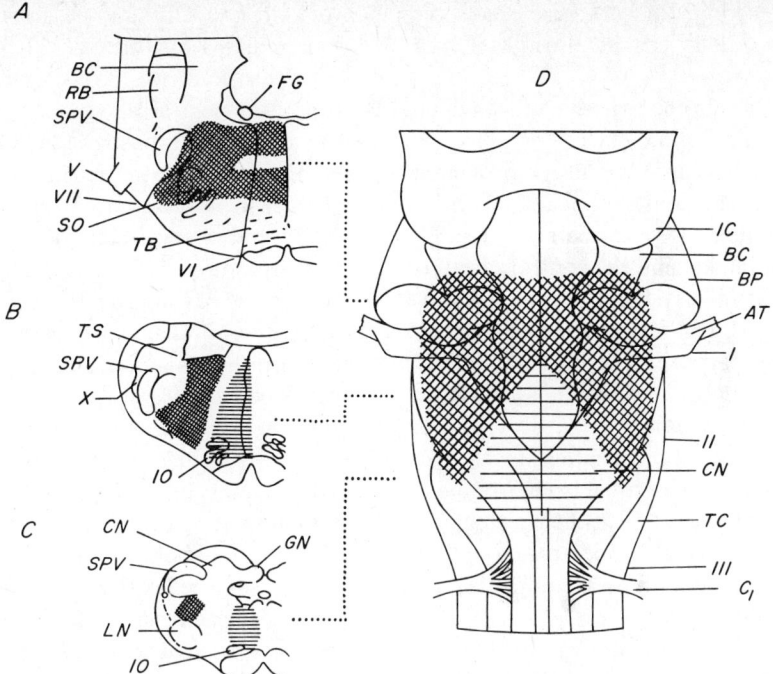

Figure 8 Localization of pressor (cross-hatching) and depressor (horizontal ruling) centers in the brain stem of the cat. (a–c) Cross sections through medulla at levels indicated by guidelines to D. (d) Semidiagrammatic projection of pressor and depressor regions onto the dorsal surface of the brain stem viewed with the cerebellar peduncles cut across and the cerebellum removed. AT, auditory tubercle; BC, brachium conjunctiva; BP, brachium pontis; C_1, first cervical nerve; CN, cuneate nucleus; FG, facial genu; GN, gracile nucleus; IC, inferior colliculus; IO, inferior olivary nucleus; LN, lateral reticular nucleus; RB, restiform body; SO, superior olivary nucleus; SPV, spinal trigeminal tract; TB, trapezoid body; TC, tuberculum cinerum; TS, tractus solitarius. V, VI, VII: corresponding cranial nerves; I, II, III: levels of transection. (From Alexander : 1946.)

have pointed out, when sections are made progressively caudally along the medulla, the blood pressure falls, and the fall becomes maximal at a point some 4–5 mm above the calamus scriptorius (level I). But an interesting finding was that with sections even more caudal (level II), the pressure actually began to rise. When the cervical spinal cord was reached (level III), pressure fell to the low levels seen in spinal shock. Thus, a depressor area (level II) had been established, and it was possible to elicit depressor responses from this area by electrical stimulation. Moreover, it was clearly shown that activity of the depressor area depended on the afferent input from the barorecep-

tors for its effect. Another question that was asked was whether the effects elicited by stimulation of the depressor area depended on inhibition of already present vasoconstrictor drive or whether it depended on the activation of specifically vasodilator pathways. It turns out that the effects are due mainly to suppression of vasoconstrictor drive from the pressor areas. In summary, the baroreceptor input acts upon the neurons of the depressor area so as to cause inhibition of the neurons of the pressor area.

With our definition of a central nervous system center in mind, we can consider the effects of more rostral structures on the circulation. It is important to point out that with only two exceptions so far known, these effects are mediated through the medullary centers already referred to.

The next important area is in the hypothalamus, which is something more than an autonomic center. That is, something more than the autonomic system is represented at this level. Thus, we have such different neural activities as the expression of emotion, body temperature regulation, water balance, food intake, and the control of pituitary hormones. The circulatory system is particularly involved in emotional behavior and temperature regulation. Stimulation of the dorsal part of the hypothalamus between the optic chiasm and the anterior commissure in the unanesthetized goat has been shown to cause vigorous panting and cutaneous vasodilation identical to the response obtained on exposure of the animal to environmental heat. In animals exposed to cold, both the panting and the vasodilatation caused the rectal temperature to fall to 30°C. This cutaneous vasodilation was accompanied by renal vasoconstriction.

There is another hypothalamic area, activation of which leads to an integration of cardiovascular and other visceral changes with a specific pattern of overt behavior called the rage reaction. In conscious cats threshold stimulation of the hypothalamus in the region of the fornix caused the cat to prick up its ears, dilate its pupils, and increase the blood flow to its skeletal muscles, the latter mediated by a sympathetic cholinergic pathway that runs from the cerebral cortex to the spinal cord without passing through or making synaptic contact with the medullary centers. Stronger stimulation provoked piloerection, unsheathing of the claws, hissing, and spitting, as well as running movements that might end in flight or attack. There was also cardioacceleration, positive inotropism, and vasoconstriction in skin and intestine. Hess used the term *Abwehrreaktion* (defense reaction) to describe the integrated response. The circulatory components of the response are, by the way, very similar to the changes that occur during muscular exercise.

There have been many studies that have dispelled any doubts that marked visceral changes, particularly cardiovascular ones, can be evoked by stimulation of several cortical areas and that removal of certain cortical areas, particularly the somatomotor and somatosensory areas may give rise to such changes. But with few exceptions, one being the sympathetic cholinergic vasodilator pathway, the exact functional significance of these observations remains unclear.

To this point we have considered the hierarchical nature of the central nervous system control of circulation. The central nervous system centers can also generate reflex response patterns that are different depending on whether the input comes from carotid sinus or aortic arch baroreceptors or cardiac mechanoreceptors or peripheral chemoreceptors. Whereas increased input from arterial and cardiac baroreceptors, except for certain atrial receptors, results in depressor reflexes, the input from cardiac receptors has a much greater inhibitory effect on renal sympathetic outflow than on splanchnic or skeletal muscle outflow. The emphasis is opposite when the carotid sinus pressure is increased above normal. However, the various patterns may be species dependent, and it is probably premature to extrapolate the details of experimental results from animals to man. Moreover, distension of the atria causes a reflex tachycardia and a diuresis. The influence of cardiac receptors on renal function may have very important clinical consequences. Not only does excitation of atrial receptors produce a diuresis, but the sympathetics regulate renin production and thereby the renin–angiotensin–aldosterone system. Since the circulatory receptors affect renal sympathetic drive, they probably exert important effects on the control of blood volume and body electrolytes.

INTEGRATED RESPONSES OF THE CIRCULATION

The CNS is the integrator or processor of the information it receives from cardiovascular reflexogenic areas, and from the foregoing account most of the integration is done in the medulla. What are the principles by which integration is carried out? At present, the concepts of nervous system function introduced by Sherrington seem to apply here, namely that inputs project onto pools of processing neurons, which they may share. If the inputs are supramaximal to common neurons, occlusion will result, and the sum of the reflex effects will be less than the individual reflexes. If the inputs are submaximal, facilitation may occur, and the summed response may actually be greater than the individual responses. The problem as to how many neurons are shared, if any, and the extent to which they receive inputs from more than one source is an anatomical one; it is a direct reflection of the specificity of neural pathways.

From our present knowledge it appears clear that afferent inputs from the circulation project to common pools of neurons. In addition, the projections showing specificity for certain inputs have stronger projections to specific target organs.

With these principles in mind we can now consider some integrated responses.

HEMORRHAGE

When less than 15% of the blood volume is lost, the mean arterial blood pressure is unchanged, but heart rate increases reflexly. We have already discussed the reflex mechanism responsible for this effect.

For more severe hemorrhage, the mean arterial blood pressure drops as well. Refer to Figure 4 for the sequence that ensues. A patterned response emerges, the essential feature of which is that flow to the heart and brain is maintained at the expense of flow to other organs, such as liver and kidney.

EXERCISE

In anticipation of exercise, a sequence of circulatory changes similar to those associated with the defense reaction occurs. During exercise, cardiac output can increase from three to five times, mostly as a result of increased heart rate. Most of the increased flow is directed toward the skeletal muscle vascular beds. Hence, skeletal muscle flow can increase as much as 15-fold; coronary flow trebles, and renal and gut flows are reduced. Mean arterial pressure is also increased. This matter is discussed more thoroughly in Chapters 16 and 26.

FAINTING

Fainting or syncope is due to a sudden fall in arterial blood pressure so that cerebral flow falls below levels necessary to maintain consciousness. It is self-limiting since the supine posture of the faint increases cerebral flow and results in consciousness. A common cause of the faint is a sudden drop in heart rate due to increased cardiac vagal drive. This is associated with a reduction in peripheral resistance and cutaneous vasoconstriction and is sometimes called vasovagal syncope. It may be emotionally induced or some exteroceptive input such as an unpleasant sight may be involved.

EMOTION

Although the emotional faint is well known, emotional upsets usually turn on the sympathetic outflow to the cardiovascular system. In this situation the affective component may differ from that associated with an emotional faint. Heart rate and blood pressure are increased, both of which are restrained by the ensuing increased afferent drive from neurocirculatory reflexogenic areas. The time constant for the turn-off of this increased heart rate and arterial blood pressure may be on the order of minutes, so that sustained upsets may be the cause of hypertension.

REFLEX RESPONSES DURING HEART ATTACK

Acute coronary occlusions are usually accompanied by severe precordial pain. The afferent pathway is mainly via afferent cardiac sympathetic nerves that run from the heart to the spinal cord. The exact nature of the stimulus to the nerve endings may be chemical, mechanical or both. The impulses are transmitted to the cortex via spinothalamic pathways, and the pain is referred to the chest wall. The mechanisms of referred pain is beyond the scope of this chapter.

The disorder may be accompanied by tachycardia or bradycardia. The former response is medicated by arterial baroreceptors in response to the fall in stroke volume associated with decreased contractility. The latter response probably reflects a large excitatory input from cardiac afferents if the heart size increases locally or globally. Hence, the heart rate response in coronary occlusions may be quite variable and by itself will not be a useful prognostic sign.

The incidence of lethal arrhythmias accompanying acute coronary occlusion is markedly reduced by sympathectomy in experimental animals and by the use of the β-blocker propranolol. In these conditions, the sympathetic drive to the heart may be increased or the ischemic myocardium may be unduly sensitive to the sympathetic transmitter. Increased sympathetic drive, when it occurs, may be initiated from arterial baroreceptors (a fall in activity) or afferent cardiac sympathetic fibers (an increase in activity).

HYPERTENSION

This is often defined as a maintained diastolic pressure of 90 mm Hg or more. The systolic pressure is usually 140 mm Hg or more. The most common form is essential hypertension, which means that we do not know the cause. However, we do know the following:

1. The baroreceptors are reset to operate at the new higher level.
2. The vascular smooth muscle is more reactive to sympathetic drive or norepinephrine.
3. The levels of the renin-angiotensin system can be normal, low, or high. The low renin pattern may be associated with some increase in central blood volume or may result from inhibition of sympathetic release of renin. The inhibition may be initiated from distended cardiac receptors.

4. The activity of the sympathetic nervous system may be higher than normal even taking into account baroreceptor resetting.

There are two reasons for this emphasis in hypertension. First, hypertension is the leading cause of disability due to disease in the United States today. Second, it is a disease in which the control mechanisms of the body are deranged and therefore provides an opportunity to understand more fully these mechanisms.

POSTURAL ADJUSTMENTS

Postural adjustment involves an integrated response of the circulation. It is discussed in detail in Chapter 12.

SELECTED BIBLIOGRAPHY

Berne, M., Levy MN: *Cardiovascular Physiology.* St Louis, Mosby, 1972.

Folkow, B, Neil, E: *Circulation.* New York, Oxford University Press, 1971.

SELF-STUDY QUESTIONS

MATCHING

Questions 1–3.

A. *Increases total peripheral resistance*
B. *Increases arteriovenous oxygen difference*
C. *Both*
D. *Neither*

1. Exercise
2. Hemorrhage
3. Baroceptor stimulation

Questions 4 and 5.

A. *Increased cardiac output*
B. *Increased right atrial pressure*
C. *Both*
D. *Neither*

4. Hemorrhage
5. Efferent cardiac vagal activity increased

Questions 6–8.

A. *Cortical control of cardiovascular response*
B. *Hypothalamic control of cardiovascular response*
C. *Medullary control of cardiovascular response*
D. *Spinal control of cardiovascular response*

6. Primary center affected by baroreceptor activity
7. Primary center responsible for basal vasoconstrictor activity

8. Primary center responding to anxiety

MULTIPLE CHOICE

Select the correct answer(s). (In many instances, more than one answer is correct.)

9. Increased pressure in the carotid sinus causes
   A. increased frequency of discharge in afferent fibers from the sinus.
   B. reflex inhibition of medullary vasopressor centers.
   C. reflex slowing of heart rate.
   D. reflex inhibition of medullary respiratory centers.

10. A sympathetically induced increase in heart rate will be accompanied by
    A. a decrease in the duration of the S–T segment of the electrocardiogram.
    B. an increase in the rate of ventricular tension development.
    C. an increase in rate of diastolic depolarization of fibers in sinoatrial node.
    D. a decrease in magnitude of ventricular action potential.

11. A rise in arterial pressure induces increased frequency of
    A. impulses in the lumbar sympathetic adrenergic nerve fibers.
    B. afferent impulses in the carotid sinus nerve.
    C. impulses in the cardioaccelerator nerve fibers.
    D. efferent impulses in the cardiac fibers of the vagus nerve.

12. Fainting of neurogenic origin is characterized by
    A. an increase in parasympathetic activity to the heart.
    B. a decrease in cardiac output.
    C. a decrease in total peripheral resistance.
    D. an increase in neurogenic vasoconstrictor activity to skeletal muscle vasculature.

13. The central nervous system is capable of altering peripheral resistance in the vascular system in the
    A. spinal cord.
    B. medulla.

C. pons.
D. hypothalamus.

14. Central nervous system regulation of cardiovascular system
    A. is not significantly impaired when the brain stem is severed above the level of the pons.
    B. results in bradycardia when afferent baroceptor impulses reach the medullary centers.
    C. results in decreased peripheral resistance when baroceptor impulses reach the medullary centers.
    D. results in cessation of blood flow when all centers are removed.

15. Increased sinus nerve activity results in
    A. a decrease in the slope of prepotential in sinoatrial nodal cells.
    B. a marked hyperpolarization of vascular smooth muscle cells in cerebral arterioles.
    C. hyperpolarization of intestinal vascular smooth muscle.
    D. excitation of the medullary vasopressor centers.

Select the single best answer.

16. The nerve leading from the carotid sinus shows an increased number of impulses per unit time when
    A. mean systemic arterial pressure increases.
    B. mean systemic arterial pressure decreases.
    C. arterial P_{O_2} increases.
    D. arterial pH increases.
    E. venous P_{CO_2} increases.

17. Circulation is influenced *least* by neurogenic vasomotor activity in the
    A. kidney.
    B. brain.
    C. skin.
    D. spleen.
    E. intestine.

18. Strong stimulation of aortic and carotid chemoreceptors would be caused by
    A. severe anemia.
    B. carbon monoxide poisoning.

C. congestive heart failure.

D. low P_{O_2}.

E. low oxygen content.

Again, select correct answer(s)

19. The relationship between carotid sinus pressure and systemic arterial pressure is known as the *Blutdruchcharacteristik*. It demonstrates that

    A. greatest control by the carotid sinus reflex is exerted over the physiological range of arterial blood pressure.

    B. control is poor at very high and very low pressures.

    C. the *Blutdruchcharacteristik* is steepest in the physiological range.

    D. positive feedback is the outcome of the carotid sinus reflex.

20. In response to a step increase in blood pressure

    A. baroreceptor discharge is decreased.

    B. baroreceptor discharge rises steadily to a new level.

    C. baroreceptor discharge is unchanged.

    D. baroreceptor discharge is transiently increased.

21. Cardiac reflexes are

    A. mainly depressor when mediated by vagal afferents.

    B. characterized by a dual input to the central nervous system.

    C. mainly pressor when mediated by sympathetic afferents.

    D. mediated by mechanoreceptors exclusively.

22. The Bainbridge reflex

    A. may be involved in bradycardia of muscular exercise.

    B. may be involved in the tachycardia of the muscular exercise.

    C. is mediated by increased efferent cardiac vagal discharge.

    D. is mediated by increased efferent cardiac sympathetic discharge.

23. Noradrenergic transmission involves

    A. α-receptor activation.

    B. β-receptor inactivation.

    C. vasoconstriction.

    D. direct depression of SA nodal cells.

24. Administration of atropine

    A. produces bradycardia.

    B. increases heart rate.

    C. dilates some skeletal muscle arterioles.

    D. constricts some skeletal muscle arterioles.

25. The β-blocker propranolol would be expected to

    A. increase heart rate.

    B. decrease myocardial contractility.

    C. speed a–v conduction.

    D. cause bradycardia.

26. Vasovagal syncope (fainting) in a previously healthy person is accompanied by

    A. bradycardia.

    B. cutaneous vasoconstriction.

    C. decreased peripheral resistance.

    D. increased central venous pressure.

ANSWERS

1. B. During exercise, the increase in cardiac output (CO) is proportionally less than the increase in O_2 consumption. Nevertheless, large increases in CO do occur with very little change in arterial blood pressure, and peripheral resistance actually falls.

2. C. In hemorrhage, CO falls, although O_2 consumption is unchanged. The fall in BP reduces baroreceptor discharge, which leads to increased TPR.

3. D. Lying down might transiently increase carotid

sinus baroreceptor input and, if anything, will reduce TPR. It will have little or no effect on CO or O_2 consumption.

4. D.

5. B. A reduction in heart rate is produced by efferent cardiac vagal stimulation. The stroke volume is increased because the prolonged diastolic filling period increases right atrial pressure.

6. C

7. C

8. B

9. A, B, C, D. All are correct.

10. A, B, C. The ventricular action potential may be enhanced, if anything, by sympathetic stimulation.

11. B, D. Negative feedback is the neural mechanism here.

12. A, B, C.

13. A, B, C, D. All are correct.

14. A, B, C.

15. A, C. The cerebral arterioles are not significantly influenced by baroreceptor input.

16. A

17. B. This is as good a time as any to emphasize that the cerebral and coronary circulation are more controlled by local factors than neural factors.

18. D

19. A, B, C.

20. D

21. A, B, C.

22. B, D.

23. A, C.

24. B, D.

25. B, D.

26. A, B, C.

14

Coronary Circulation

DONALD W. STUBBS

OBJECTIVES

After completion of this chapter, you should be able to

1. Describe the estimated proportions of total coronary blood flow through the major coronary arteries and veins.

2. Compare skeletal muscle with cardiac muscle in terms of muscle fiber size, capillary density, and diffusion distances as bases for differences in exchange rates between capillaries and tissues.

3. Discuss the physical, metabolic, and neural factors that affect coronary blood flow.

4. Compare flow rates between systole and diastole for left coronary artery, right coronary artery, and coronary sinus.

5. Explain the basis for the statement that the primary means of increasing the supply of oxygen to the heart is by increasing the supply of blood.

6. Describe the individual effects of increased sympathetic nervous activity to the heart and the overall net effect on coronary blood flow.

7. Discuss coronary blood flow reserve in terms of coronary resistances.

8. Describe the three principal determinants of myocardial oxygen consumption.

9. Relate the influence of these determinants on cardiac efficiency.

10. Compare the energy costs of "pressure work" with "volume work."

As a result of the influence of medical knowledge on Western civilization, the average life expectancy today is estimated to be approximately twice that of primitive peoples. It is unlikely that medicine will ever again duplicate the feat of doubling our average life-span, unless it can solve the problem of gradual deterioration of the cardiovascular system with age. Of primary concern is the role of coronary arterial inadequacy in ischemic heart disease, which is the most common serious health problem today. More than 600,000 patients die each year from ischemic heart disease in this country alone. Since almost everyone undergoes some impairment of the coronary circulation with advanced age, information about this special circulation continues to be of considerable relevance to the physician.

ANATOMY

In an average 70-kg (154-lb) man, the heart weighs about 300 g, of which half constitutes the massive thick-walled left ventricle (i.e., the weight of the left ventricle in grams is approximately equal to a person's weight in pounds).

THE CORONARY ARTERIES

The coronary arterial tree can be divided functionally into two categories: (1) the large coronary arteries and their tributaries, which lie on the epicardial surface of the myocardial wall; and (2) small arteries descending into the myocardium, where they and their arterioles lie in close proximity to the contracting myocardial fibers. The large arteries are primarily conducting conduits and have little resistance to flow unless they become occluded by atherosclerosis.

The Large Coronary Arteries

The two main coronary arteries (left and right) that supply the myocardium arise from pouchlike dilations (sinuses of Valsalva) at the root of the aorta. The left coronary arises from the left aortic sinus. As it passes to the left and downward, it shortly divides into two main branches, the anterior descending artery and the circumflex artery. The anterior descending artery is the most conspicuous vessel on the surface of the heart. It is all too often of consider-

able concern to pathologists. The extensive distribution of the anterior descending branch of the left coronary artery makes occlusion of it or of its tributaries particularly dangerous. The left coronary artery and its branches supply most of the massive left ventricle and the left atrium. In human physiology, the left coronary artery is estimated to supply about 60% of the total coronary arterial flow. The right coronary artery arises from the right aortic sinus. The right coronary and its branches supply the right atrium; a portion of the posterior left ventricle, in most people; and the right ventricle. The right coronary artery is estimated to carry about 40% of the total coronary arterial flow in human circulation.

Anatomists often refer to the anatomic configuration of the coronary arteries on the posterior surface of the heart as "right predominant" or "left predominant." This topographical designation has often been misconstrued to signify whether the right or left coronary artery supplies the predominant blood flow to the entire heart. Whereas it is true that the right coronary artery supplies a part of the left ventricle in most people—so-called "right predominant" hearts—*most* of the massive left ventricle and the left atrium are supplied blood by the left coronary artery even in "right predominant" hearts (i.e., left coronary flow is always functionally predominant) (Fig. 1).

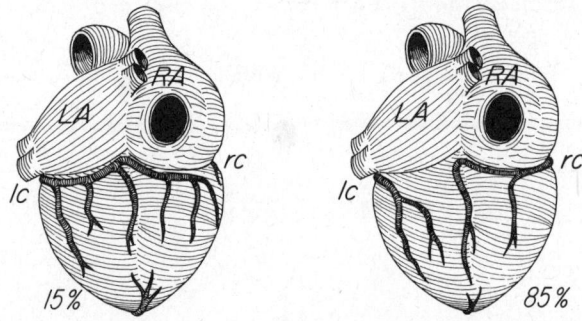

Figure 1 Left and right anatomical predominance, the two anatomical configurations of coronary artery supply to the posterior (diaphragmatic) surface of the human heart. (Left) Left predominance (15% of the population), in which the left circumflex artery (lc) crosses the crux (an intersection of the circumferential coronary sulcus with the posterior longitudinal sulcus). (Right) Right predominance, in which the right coronary artery (rc) crosses the crux to supply this area (85% of the population).

The Small Coronary Arteries

Epicardial, or surface, branches of the major coronary arteries send smaller tributaries into the wall of the heart, branching at right angles to the surface vessels and descending vertically through the wall. These small intramyocardial arteries are of two types: (1) subepicardial vessels, which soon branch into a network supplying the subepicardial (outer) part of the wall; and (2) subendocardial vessels, which are longer and penetrate almost the entire thickness of the wall before branching into a network just beneath the subendocardial (inner) surface of the wall. Changes in the radii of these two types of intramyocardial vessels and their arterioles are largely responsible for regulating the flow of blood through the coronary circulation (i.e., they are the principal resistance vessels of the heart).

Small Collateral Arteries

Not all the small intramyocardial arteries end directly in arterioles and capillary networks (i.e., not all are "end arteries"). Some of these small arteries connect with other small arteries, which are subbranches of the same parent vessel. Others connect with subbranches of a different parent artery (e.g., interconnections between small tributaries at the end of the left anterior descending artery near the apex of the heart and those of the right posterior descending artery). At birth, these interconnecting collaterals are quite small in diameter and wall thickness (i.e., little more than that of arterioles), but with the development of gradual atherosclerotic occlusion in adults, particularly the elderly, these collaterals can grow and become transformed into larger intramyocardial arteries that can carry significant amounts of blood to areas that would otherwise become ischemic. Because gradual occlusion permits time for this transformation to occur, chronic development of occlusion is usually less serious than abrupt or acute occlusion of a main coronary artery.

THE CORONARY CAPILLARIES

The sole purpose of all the coronary vessels mentioned thus far is to make possible the function of the coronary capillaries. The function of these coronary capillaries is to permit exchange of metabolic substrates (glucose, unesterified fatty acids, O_2) and waste products (CO_2, inosine)

between the myocardium and the blood. This is accomplished by two processes: (1) *convection,* or bulk flow, that is, the flow of blood linearly through the capillaries down a pressure gradient; and (2) *diffusion,* the lateral interchange of molecules between blood in the capillaries and cardiac interstitial fluid across the endothelium. Although the ultrastructure and permeability of the individual coronary capillaries are essentially the same as those of skeletal muscle capillaries, the rate of exchange of substances in the coronary capillaries is much greater than in skeletal muscle, for each gram of tissue—largely the result of a greater density of capillaries in cardiac muscle. Although both kinds of striated muscle have a 1–1 ratio between muscle fibers and capillaries, the smaller diameter of myocardial fibers permits a closer packing of fibers and capillaries in the myocardium. The greater density of fibers and capillaries per unit area permits diffusion between blood and muscle fibers to take place over shorter distances (Fig. 2).

Several lines of evidence indicate that the rate of exchange of molecules between capillary blood and myocardium is not maximal in the resting state. The diffusion or exchange rate is inversely proportional to capillary resistance, as regulated by precapillary sphincters. In the resting person, not all precapillary sphincters are open, and consequently not all coronary capillaries have blood flowing through them—about one-half are open. With increased metabolic demand (e.g., with exercise) or with ischemia, all capillaries can be recruited to supply blood to the myocardium. Regulation of the number of open capillaries can be almost as important as regulation of small artery and arteriole resistance in maintaining a convection of blood adequate to supply myocardial needs.

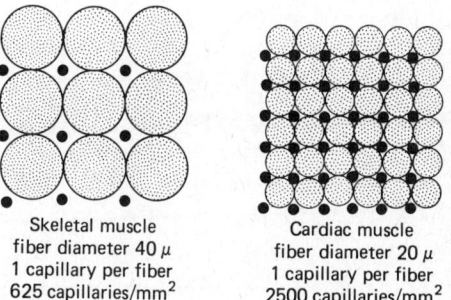

Skeletal muscle
fiber diameter 40 μ
1 capillary per fiber
625 capillaries/mm^2

Cardiac muscle
fiber diameter 20 μ
1 capillary per fiber
2500 capillaries/mm^2

Figure 2 Comparison of fiber and capillary densities in skeletal and cardiac muscle.

THE CORONARY VEINS

The great majority of the cardiac veins drain into a single large vein, just slightly smaller than a person's little finger, located in the atrioventricular sulcus on the posterior surface of the heart. This large vein, the *coronary sinus,* which empties into the right atrium, receives about 60% of the total venous outflow from the coronary circulation. This outflow is derived primarily from the highly vascular left ventricle. The remainder of the venous outflow returns by two routes: (1) about 35% from the right ventricle to the right atrium by the *anterior cardiac veins;* and (2) the remaining 5% from different parts of the myocardial wall, particularly the atrial walls, directly into the underlying heart chambers by a number of very short, minute veins, the *thebesian veins* (venae cordis minimae).

PHYSIOLOGY OF CORONARY BLOOD FLOW

Although the heart constitutes only a minute fraction of body weight (0.4%), it receives a disproportionally large share of cardiac output (4%). The total blood flow to the heart is estimated to be almost 200 ml/min. For technical reasons, it is difficult to obtain direct measurements of flow in individual coronary arteries in humans. A recent indirect method still in the developmental stages involves measurement of isotope washout from myocardium after injection of the isotope into coronary arteries. Xenon-133 can be selectively injected into either the right or the left coronary artery by catheter. Isotope washout, which is a function of myocardial blood flow, is determined externally by means of a multicrystal scintillation camera. Results from this method show a left coronary artery flow of approximately 65 ml/100 g myocardium/min and a right coronary artery flow of about 50 ml/100 g myocardium/min in patients with normal coronary arteries. Factors that affect this flow are physical, metabolic, and neurohumoral.

PHYSICAL FACTORS

The moment-to-moment fluctuations in coronary flow are phasic. The periodic changes in flow in each cardiac cycle of systole–diastole are the result of phasic changes in two physical or mechanical factors affecting flow: (1) the phasic changes in aortic pressure (systolic and diastolic pressures) as changes in the inflow or perfusing head of pressure, and (2) the phasic variations in resistance to flow resulting from compression of intramyocardial arteries during systolic contraction of the ventricles. This latter factor is of greater significance within the left ventricle, since the greater muscularity of the left ventricular wall has a more profound influence during contraction (systole). As a result of this muscular squeezing or strangulation, not as much blood can get through the tributaries of the left coronary during systole—only about 25% of total left coronary arterial flow occurs during systole. The greater flow of left coronary arterial blood occurs during diastole when the ventricular myocardium is relaxed— about 75% of total left coronary flow—even though the arterial pressure is lower (Fig. 3). Muscular strangulation of intramyocardial veins pushes the venous blood into the larger veins (e.g., coronary sinus), so that venous flow is higher during systole than in diastole. During ventricular systole, the intramyocardial tissue pressure is highest within the inner half of the myocardial wall (subendocardium). Because of the high tissue pressure developed by the muscular *left* ventricle, it is in the subendocardial layer of the *left* ventricle that the maximal intramyocardial tissue pressure is developed. During systolic contraction, the intramyocardial pressure in this layer exceeds the systolic arterial blood pressure in the intramyocardial vessels. The subendocardial intramyocardial arteries extending into this layer are the most susceptible to external compression by the high tissue pressure. This compression, up to and including temporary occlusion, is responsible for the greatly diminished flow in the left coronary artery during ventricular systole. This compression of the intramyocardial vessels could also explain why the subendocardial layer has a lower Po_2, is more susceptible to ischemic damage, and is the most common site of myocardial infarction.

After considering the physical factors influencing flow, you are apt to appreciate the constraining effects that

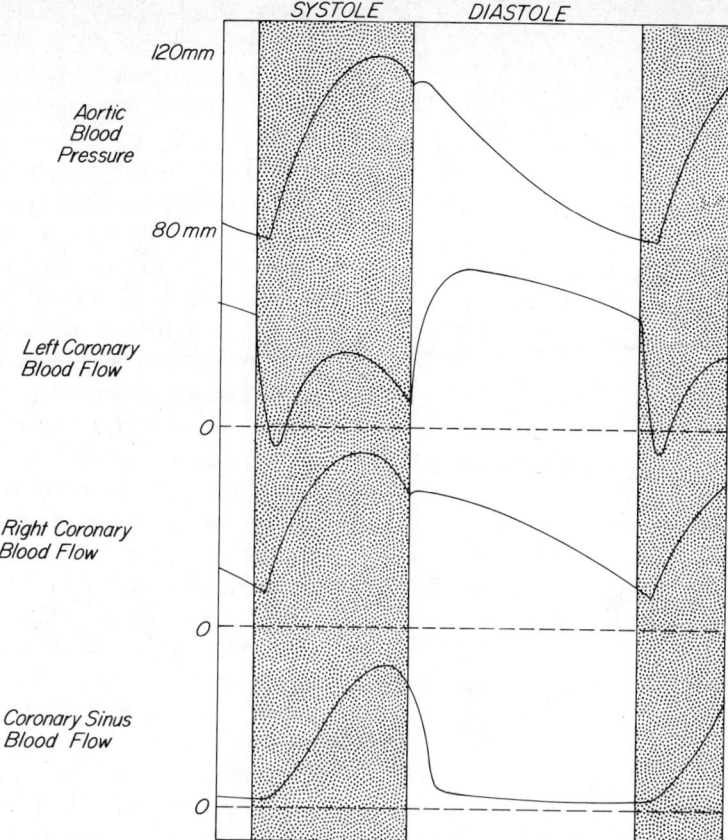

Figure 3 Approximate relationship between aortic blood pressure and phasic blood supply to the left and right ventricles and coronary sinus throughout the cardiac cycle. Note the far stronger mechanical interference with the blood supply of the left ventricle during systole.

various diseases and conditions tend to have on the hydro-dynamic perfusion of the coronary circulation. Either a diminution of the coronary arterial pressure (e.g., in aortic stenosis) or an increase in the venous pressure (e.g., in congestive heart failure) decreases the perfusion pressure (i.e., the ΔP, or arterial pressure minus venous pressure) and tends to impede flow through the coronary circula-tion. An accelerated heart rate is accompanied by a greater shortening of the duration of diastole than of sys-tole, thereby reducing the period of time during which the greater proportion of coronary flow occurs, that is, dias-tole. Fortunately, compensatory changes in resistance to flow brought about by the intrinsic chemical regulation of flow (see next section, Metabolic Factors) can effectively counter most of these physical impediments in a normal heart. For example, increased heart rates up to 200 beats/min are actually accompanied by an increase in coronary flow, rather than by a decrease (Fig. 4).

METABOLIC FACTORS

The overall flow of blood through the coronary circulation is almost exclusively regulated by local intrinsic control of vascular resistance in response to the metabolic require-

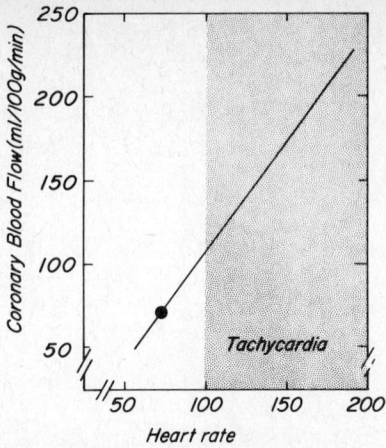

Figure 4 Coronary blood flow in relationship to heart rate.

ments of myocardial tissue. Whenever the nutritive needs of heart muscle are increased (e.g., as the result of increased rate or vigor of contraction), the rate of flow through the myocardium simultaneously increases. By contrast, diminution of activity is accompanied by a decreased flow. This intrinsic regulatory control of blood flow in heart muscle in response to metabolic requirements is very similar to that in a number of tissues.

The most important metabolic substrate of the heart is oxygen. Even under quiescent or basal conditions, the heart extracts a nearly maximal amount of oxygen from each milliliter of arterial blood flowing through the coronary circulation. The arterial blood supplied to the heart, as well as other organs, has a partial pressure of oxygen (Po_2) of 90 mm Hg, which gives about a 97% saturation of hemoglobin with oxygen. This amount of oxygen reversibly bound to hemoglobin plus a small amount of dissolved oxygen in plasma makes up a total oxygen content in arterial blood of about 20 ml O_2/100 ml blood (20 vol%). As the blood passes through the coronary circulation, the heart extracts nearly three-fourths of the oxygen brought to it, leaving the venous blood of the coronary sinus with a content of only 6 ml O_2/100 ml blood (6 vol%) at a Po_2 of 18 mm Hg, which gives a 30% saturation of hemoglobin with oxygen (Fig. 5)—the lowest Po_2 of venous blood in the body at rest.* Compare this percentage with the average extraction of oxygen by all tissues combined: about one-fourth extraction leaving mixed venous blood with a Po_2 of 40 mm Hg, 75% saturation of hemoglobin, and an oxygen content of 15 vol%.

Even during heavy exercise with a fivefold increase in cardiac output there is only a slight decline in the coronary venous oxygen tension. Therefore, since the relative amount of oxygen extracted from each milliliter of blood is nearly maximal, the only way to meet requirements for an increase in the absolute amount of oxygen is by an increase in the total milliliters of blood delivered. That is, the total quantity of oxygen delivered is mainly governed by the rate of coronary blood flow. Fortunately, the control of coronary flow is determined by oxygen requirements. The most urgent stimulus for increased coronary flow is myocardial hypoxia.

The overall metabolic activity of the myocardium, as reflected by its oxygen consumption, correlates closely with the coronary flow rate under numerous physiological conditions, from rest to severe exercise. That there is close coupling between metabolic activity (or oxygen consumption) and coronary flow is well established. However, there is no unanimity of opinion as to the specific mechanism by which increased oxygen requirements elicit coronary vasodilation. A number of the chemical factors or metabolites suggested as controlling vascular resistance in skeletal muscle have been proposed as operating in coronary circulation as well (e.g., local increases in H^+, adenosine, Pco_2, K^+, or regional osmolarity, or a decrease in Po_2). Although there is some evidence that coronary arteriolar smooth muscle might be directly sensitive to decreased Po_2, the bulk of evidence favors the idea of vasodilation in response to increased oxygen requirements being brought about by the local release of metabolites such as K^+ and adenosine. Whatever the mechanism, the principal regulator of coronary blood flow is the oxygen requirement of the myocardium.

NEUROHUMORAL FACTORS

The resistance vessels of the coronary circulation are innervated by both divisions of the autonomic nervous sys-

---

*The only venous Po_2 that is ever lower than 18 mm Hg is that produced by vigorously exercising skeletal muscle.

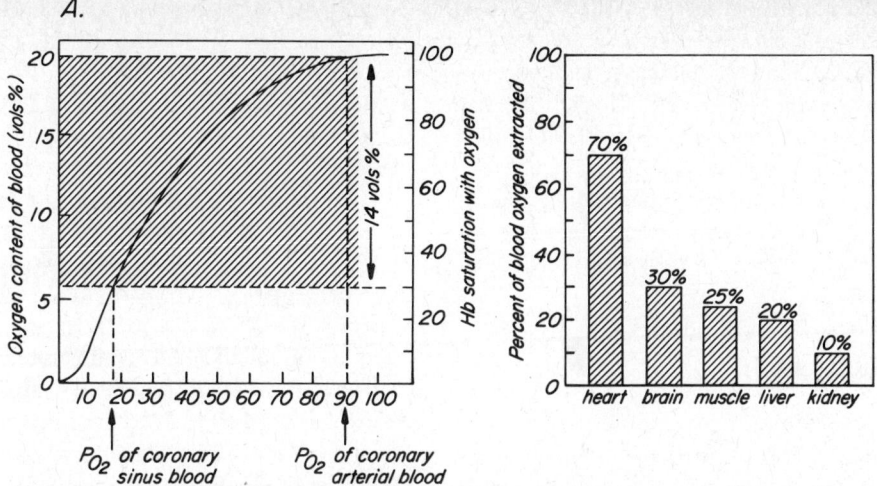

Figure 5 Oxygen extraction from blood by the heart. (*a*) The difference in oxygen content (ml O$_2$/100 ml blood, more commonly called vol%) is shown between arterial and venous blood in the coronary circulation. The myocardium normally extracts 14 ml O$_2$ from each 100 ml/blood (14 vol%) flowing through the coronary circulation. Removal of 14 vol% from the 20 vol% normally present in arterial blood is a 70% extraction. (*b*) Compares oxygen extraction by the heart with that of other tissues in an individual at rest.

tem. It is possible to bring about a change in resistance to flow by the action of the locally released neurotransmitter agents (e.g., norepinephine from sympathetic postganglionic fibers and acetylcholine from parasympathetic postganglionic fibers), or by the action of circulating catecholamines (mainly epinephrine) of adrenal origin.

Recent evidence has indicated the presence of parasympathetic cholinergic fibers from the vagus that can produce vasodilation of the coronary resistance vessels (e.g., in response to stimulation of carotid chemoreceptors by nicotine) in dogs. However, the importance of these vasodilator fibers and their possible role in normal function in humans remain to be established.

The functional influence of endogenous catecholamines (norepinephrine and epinephrine) appears well established: the net or overall effect of an increase in either locally released norepinephrine or circulating epinephrine is a decrease in coronary vascular resistance and an increase in blood flow. This effect is an indirect result of the positive chronotropic (increased heart rate) and positive inotropic (increased contractility) effects of these catecholamines wherein myocardial activity is increased, which produces vasodilation via the metabolically influenced intrinsic regulatory mechanism (see Metabolic Factors above). This indirect influence of the catecholamines completely overwhelms any other actions they have on coronary resistance vessels.

Other indirect actions of endogenous catecholamines include the mechanical effects of increases in heart rate and contractility, which, in the absence of the intrinsic regulatory system, tend to increase resistance in the following ways: (1) an increase in heart rate increases the proportion of time spent in systole when intramyocardial compression of resistance vessels is maximal, and (2) an increase in contractility increases the force of contraction for a given fiber length, thereby further augmenting this intramyocardial compression, or strangulation, of intramyocardial vessels during systole (see Physical Factors above). In addition to these various indirect actions, either norepinephrine or epinephrine directly stimulates α-receptors on the coronary resistance vessels. In the absence of the metabolically influenced intrinsic regulatory system, this direct effect tends to increase resistance by producing vasoconstriction. However, because of the marked predominance of the indirect effects of catecholamines on coronary vascular resistance as mediated by the increased metabolic activity of the myocardial muscle fibers, increased sympathetic activity results in a *net decrease in the resistance,* hence an increase in the flow of blood to cardiac muscle.

BLOOD FLOW RESERVE

The resistance vessels of the coronary circulation are similar in many respects to those of skeletal muscle. In both cases there is a pronounced basal tone of vascular smooth muscle, which is relaxed (vasodilation) primarily under the influence of metabolites according to the metabolic rate of striated muscle fibers. Both have a large "blood flow reserve," that is, both can undergo considerable vasodilation to match flow to striated muscle requirements. Since coronary vascular tone is so readily affected by the local chemical changes, induced by even small changes in myocardial metabolism, a primary reduction of coronary blood supply (e.g., narrowing of the lumen due to atherosclerosis), at a constant level of myocardial metabolism, is followed by a compensatory vascular dilation downstream. Recall that with series resistances, the total resistance is the sum of the individual resistances. In this case, the first resistance in the series is the partial occlusion of a coronary artery by atherosclerosis. The second resistance in the series is the principal resistance vessels (i.e., small arteries and arterioles). An increase in the first resistance can be offset by a decrease in the second. Thus, if the atherosclerotic narrowing of the lumen increases resistance there, a vasodilation of the principal resistance vessels downstream can reduce their resistance and maintain a normal total resistance, hence a normal flow. However, this uses up part of the "blood flow reserve" or potential for vasodilation even in resting conditions. It correspondingly limits the degree to which further vasodilation can occur in response to a greater level of myocardial metabolism, such as that caused by exercise or by emotional excitation. Thus a patient with partial occlusion of a coronary artery can compensate flow at rest adequately, yet experience symptoms of myocardial ischemia on excitement or exertion, the principal symptom of which is pain or angina.

MYOCARDIAL OXYGEN CONSUMPTION AND CARDIAC EFFICIENCY

The mammalian myocardium is so highly dependent on aerobic metabolism that it is unable to maintain normal contractility during even brief periods of lowered oxygen tension. When completely deprived of oxygen, it ceases to contract at all after a few minutes. In this respect cardiac muscle is unlike skeletal muscle; the myocardium cannot derive very much energy from anaerobic glycolysis, nor can it sustain a significant oxygen debt. Like the brain, the heart has an acute obligatory requirement for aerobic metabolism. The continuous delivery of oxygen by the coronary circulation is therefore of critical importance in maintaining normal cardiac function. The rate of energy use by the heart can be closely estimated by measuring the rate of oxygen consumption, determined by measuring coronary flow and arteriovenous differences in the amount of O_2/ml blood. Each milliliter of oxygen consumed is equivalent to the release of about 2 kg-meters of energy from oxidative metabolism of substrates—about the same as the energy spent in raising a large textbook to a height of 3 ft.

The three principal determinants of myocardial oxygen consumption, $M\dot{v}o_2$, are (1) myocardial wall *tension,* as related to arterial pressure and ventricular radius in the law of Laplace; (2) state of *contractility;* and (3) *heart rate* (Fig. 6).

CARDIAC EFFICIENCY

Work has to do with the application of force through distance (e.g., the forced hydraulic movement of fluid by a pump). In the heart the force exerted by the myocardial wall on the blood produces blood pressure. This force, acting through the distance of wall movement as the left ventricle becomes smaller during the ejection phase of

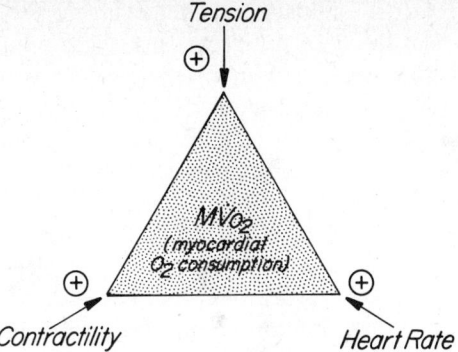

Figure 6 The three principal determinants of myocardial oxygen consumption.

systole, displaces a volume of blood into the systemic circulation. In terms of mean arterial pressure and stroke volume, the stroke work of the heart is easily calculated.

Efficiency, which is simply the ratio of work performed to energy used, is a measure of the economy of converting energy to work. Most of the energy used by a pump is lost as heat; only a fraction is responsible for the mechanical work of pumping. For most pumps, energy use is directly related to work performance at a relatively fixed level of efficiency. However, this is not true for the heart. The efficiency of the heart is not constant; it changes, within limits, with such variables as heart size, type of work performed ("pressure work" versus "volume work"), heart rate, and state of contractility. The two most important determinants of energy costs for performing work are (1) myocardial fiber tension development, and (2) velocity of shortening, as a measure of contractility:

$$\% \text{ Cardiac efficiency} = \frac{\text{cardiac work}}{K(\text{M}\dot{\text{v}}\text{o}_2)}$$

where $\text{M}\dot{\text{v}}\text{o}_2$ is myocardial oxygen consumption (in ml/min) and K is the energy equivalent of 1 ml O_2.

Under basal conditions, that is, cardiac output of 5.6 L/min, mean arterial pressure of 100 mm Hg, and cardiac oxygen consumption of 30 ml/min, the efficiency of the heart is nearly 15%—about the same as the efficiency of your car's gasoline engine.

Effect of Heart Size on Efficiency

As mentioned above, one of the most important determinants of oxygen consumption, or energy costs, by the

heart is the tension developed by the myocardial fibers in contraction. This tension (as a circumferential wall tension) acts on the incompressible blood by imparting a pressure to it that is responsible for moving that blood when the appropriate heart valves are open. Whereas the tension in the myocardial wall itself is not easily measured, the pressure produced is. Except for one problem, pressure would provide an excellent measure of wall tension, and therefore of energy use or oxygen consumption. However, the problem is this: since the tension producing a certain pressure in a heart of a given size is not the same as the tension producing that pressure in a heart of different size, intraventricular pressure, by itself, is not a good measure of wall tension. An equation that relates wall tension, intraventricular pressure, and size of the ventricle can be derived from the law of Laplace as applied to a sphere with walls of finite thickness:

$$T = \frac{P \cdot r}{2d}$$

where T is wall tension, P is intraventricular pressure, r is the radius of ventricle; and d is the wall thickness.

If the heart does "pressure work," that is, maintains the same cardiac output against a higher arterial pressure, as in hypertension, there is an increase in tension that is proportional to the increase in pressure. Since the $\text{M}\dot{\text{v}}\text{o}_2$ is determined to a large extent by the tension, there is also a proportional increase in oxygen consumption (and energy utilization) and, therefore, the ratio of work performed/energy used remains the same (the efficiency is not changed). However, if the heart does "volume work," that is, increases cardiac output without increasing arterial pressure, the increase in tension is relatively small and, therefore, the increase in $\text{M}\dot{\text{v}}\text{o}_2$ is also relatively small. In this case, the ratio of work performed/energy used becomes greater because of a smaller increase in the denominator, and the efficiency is actually increased (Fig. 7).

The Laplace equation provides a ready explanation for these differences in tension requirements, in addition to demonstrating that "volume work" is cheap in terms of energy costs. In "pressure work" (e.g., an attempt to move the same cardiac output against a higher peripheral resistance), the increase in pressure results from, and is directly proportional to, an increase in tension. Since myocardial tension is an important determinant of pressure, the

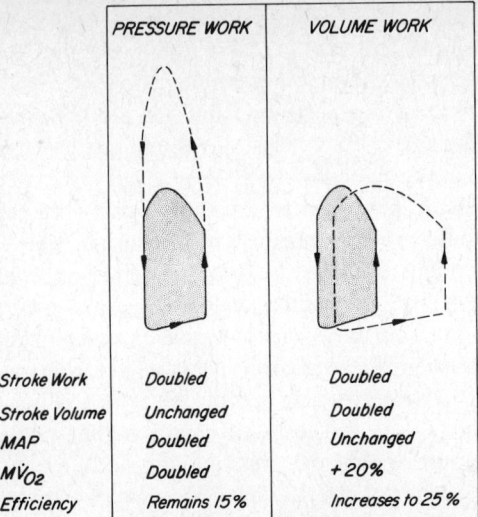

	PRESSURE WORK	VOLUME WORK
Stroke Work	Doubled	Doubled
Stroke Volume	Unchanged	Doubled
MAP	Doubled	Unchanged
$M\dot{v}_{O_2}$	Doubled	+20%
Efficiency	Remains 15%	Increases to 25%

Figure 7. Examples of two ways in which stroke work can be doubled: as increased pressure work and as increased volume work. In each case the normal stroke work is indicated by the stippled area and the increase by dashed lines. For convenience stroke work is estimated by the product of stroke volume times mean arterial pressure (MAP). For a doubling of stroke work as pressure work, the stroke volume is illustrated as remaining unchanged while the MAP is doubled (example on the left). For a doubling of stroke work as volume work, the pressure developed is shown as being unchanged while the stroke volume is doubled (example on the right). Although in both examples the stroke work is doubled, the effect on myocardial oxygen consumption ($M\dot{v}_{O_2}$) is quite different. A 100% increase in pressure work is accompanied by a 100% increase in $M\dot{v}_{O_2}$. Since the increase in $M\dot{v}_{O_2}$ is proportional to the increase in work, the efficiency (work/$M\dot{v}_{O_2}$) remains unchanged. However, a 100% increase in volume work is accompanied by only a 20% increase in $M\dot{v}_{O_2}$, and therefore an increase in efficiency.

$M\dot{v}_{O_2}$ also increases in proportion to the increase in pressure. Now, on the other hand, an increase in volume output is not directly proportional to an increase in tension. A slight increase in tension is sufficient to produce a large increase in volume. Since the volume of a sphere is a function of r^3 (volume of sphere = $\frac{4}{3}\pi r^3$), a small change in radius r brings about a large change in volume. It can be seen from the Laplace equation that a small change in radius will involve an equally small change in tension, and therefore a small change in $M\dot{v}_{O^2}$.

There are considerable differences in tension requirements—and therefore differences in oxygen requirements—for a given work load in "pressure work" versus "volume work," as described by the Laplace relationship. Likewise, the efficiency and chronic adaptation of the heart with excessive work loads are different for pressure overload as opposed to volume overload. Increased left ventricular systolic pressure (e.g., hypertension, aortic stenosis) results in a less than optimal efficiency and a hypertrophy of the myocardial wall. The level of hypertrophy in aortic stenosis can be marked, with a maximally hypertrophied heart weighing as much as 1 kg). In aortic insufficiency (regurgitation), the left ventricle is faced with a volume overload. However, the degree of hypertrophy associated with volume overload is not as great, and the efficiency is better. These differences might explain the greater incidence of coronary insufficiency and anginal pain in aortic stenosis, as compared with aortic insufficiency.

The Laplace equation can also explain the decreased efficiency of the dilated heart, which has a larger radius and requires a greater tension for a given pressure than does a normal heart. The dilated heart also has a thinner wall, further increasing tension requirements and decreasing efficiency. The hypertrophied heart also has a greater radius than does the normal heart, but the greater thickness of the wall, expressed in the denominator in the Laplace equation, counters the influence of the radius to some extent.

Effect of Heart Rate on Efficiency

Since work is force acting through distance, no work is performed if nothing is moved. From a physical point of view, two evenly matched tug-of-war teams straining at a rope can develop considerable tension and consume great quantities of oxygen, yet they will perform no work on the rope if it is not moved. Similarly, during the isovolumetric phase of systole, tension develops with increased consumption of oxygen, but blood does not move from the ventricle, hence there is no external work; however, internal work is performed through shortening contractile elements within the myocardial fibers. Therefore, a part of each systole consumes energy without performing the external work of pumping blood. It follows that, other things being equal, the greater the number of systoles per minute, the lower the efficiency in relationship to external work. For example, simply increasing the heart rate by means of an electrical pacemaker has a slight effect on blood pressure and cardiac output, in the absence of any increase in venous return or inotropic influences, stroke volume will decrease as heart rate increases; however, oxygen consumption increases without increasing external work. Thus increased heart rate will produce decreased cardiac efficiency.

As mentioned above, patients with coronary insufficiency display a compensatory dilation beyond the partial occlusion, which covers their *resting* metabolic demands. Sudden increases in heart metabolism then lead to serious decreases in the vital ratio of coronary flow to myocardial metabolism. An increase in heart rate caused by emotional excitement can decrease efficiency sufficiently to precipitate an anginal attack or, in serious cases, hypoxia of the myocardium of such intensity that ventricular fibrillation and sudden death ensue. John Hunter, "the father of surgery," who suffered from severe coronary insufficiency, said, "My life is at the mercy of any rascal who chooses to annoy me." He died in the midst of a heated argument (i.e., he was dead right).

Effect of Contractility on Efficiency

The force of contraction of myocardial fibers can be increased either by increasing the initial length (preload) before contraction (the Frank-Starling relationship), or by increasing the contractility with inotropic agents. Both incur increased oxygen requirements; however, in the case of increased contractility, the oxygen requirements are greater than might be expected on the basis of tension alone. In other words, for a given tension, the $M\dot{v}O_2$ in a heart with increased contractility will be higher than in a heart that has less contractility. Indeed, as a result of shortening the duration of tension development and decreasing the radius of the heart, the tension might actually fall below normal after the administration of positive inotropic agents, while the oxygen consumption is increased above normal. Thus whereas the ability to do work is increased with increased contractility, the energy costs are disproportionately greater and the efficiency is decreased.

Clinicians should be aware of the energy costs of positive inotropic agents and weigh this along with the possible beneficial effects such agents might have.

Consider the use of digitalis as a positive inotropic agent. The treatment of heart failure patients with digitalis glycosides has several influences on cardiac efficiency. Digitalis will improve efficiency in most patients with a dilated failing heart. Although the increased contractility tends to decrease efficiency, the heart size eventually decreases under the inotropic influence of digitalis, so that ultimately less tension is required (Laplace relationship) and, therefore, less oxygen is needed for tension development. The improved efficiency resulting from decreased size and a slower heart rate, another effect of digitalis, usually offsets the increased energy costs of increased contractility. Although most patients with failure do not have angina, this net increase in efficiency with digitalis therapy might explain the relief of anginal pain in those few who do. However, in a patient without cardiac enlargement, digitalization can either produce angina, or intensify existing angina, by increasing contractility and $M\dot{v}O_2$ without substantially decreasing heart size and wall tension, as there are no offsetting effects to counter the increased costs of greater contractility.

CORONARY ARTERY DISEASE

As mentioned at the beginning of this chapter, ischemic heart disease caused by coronary artery insufficiency is the most serious health problem today, that is, it is the major cause of death in our society. The principal cause of coronary artery insufficiency is atherosclerosis. Coronary atherosclerosis is a pathological condition of the coronary arteries, but not of the arterioles, characterized by abnormal accumulation of lipids and fibrous tissue in the arterial wall with various degrees of obstruction to blood flow. Lesions of coronary atherosclerosis are more frequent and more severe in older persons, occur more often in men than in premenopausal women, and occur generally more often in more affluent societies. The United States is second only to Finland in the death rate from coronary atherosclerosis. Attempts to determine factors responsible for the higher incidence in persons from the more affected countries of Western civilization have shown a positive correlation between mortality from coronary atherosclerosis and the level of personal income, standard of living, economic development, and dietary caloric intake, particularly with diets high in total fat, saturated fat, and cholesterol. Additional epidemiological studies have demonstrated an association of a number of risk factors. Major risk factors amenable to modification by concerned persons and their physicians are (1) elevated serum lipids, (2) habitually high caloric intake, (3) obesity, (4) smoking, and (5) hypertension. Those risk factors about which little or nothing can be done are (1) age, (2) male sex, and (3) a familial history of premature coronary atherosclerosis.

Major coronary atherosclerosis mostly involves the proximal 4 cm of either the right coronary artery or the left coronary artery and its two principal branches, the left anterior descending artery and the left circumflex artery. Mild obstruction in large coronary arteries can be countered or compensated for by vasodilation in the small arteries and arterioles farther downstream. Whereas this compensatory action might prevent ischemia of the myocardium in a person at rest, it uses up coronary blood flow reserve (see section on Blood Flow Reserve above). Exertion or emotional excitement creates a problem in such patients, because the small resistance vessels, which are already maximally dilated, cannot compensate further, thereby preventing increased myocardial requirements for blood flow from being met. The resultant myocardial ischemia gives rise to a characteristic type of chest pain known as *angina pectoris* (literally, "strangling in the chest"). In most patients with angina of mild obstruction, the pain is of a few minutes' duration, subsiding when the person stops and rests.* On the other hand, severe to complete occlusion, usually from thrombus formation, can produce an area of myocardial damage and necrosis called a *myocardial infarct,* commonly referred to as an MI by clinicians and a "heart attack" by the average person.

Anginal pain can begin with a choking sensation and a feeling of a crushing weight in the chest. Substernal pain is most common, but the sensation can also radiate up into the neck, toward the back, or into the left arm, particularly down the inner aspect of the arm and into the ring and little fingers. The Roman philosopher, Seneca, who apparently had anginal attacks, wrote, "The attack is very short and like a storm. It usually ends within an hour. I have undergone all bodily infirmities, but none appear to be more grievous."*

An important source of relief, in addition to resting, is the use of vasodilatory nitrites, particularly nitroglycerin, the mainstay of anginal treatment for the past century. The nitrites cause vasodilation both in the arteries and in the veins through a generalized direct relaxation of vascular smooth muscle. This action results in venous pooling due to the venodilation, a reduction in total peripheral resistance due to arteriolar dilation, and a decrease in both ventricular filling pressure and volume. As a result of these effects, cardiac output is reduced and the systemic arterial blood pressure is decreased, greatly decreasing the work load of the heart and reducing myocardial oxygen requirements. Despite evidence of some coronary vasodilation, there is no reason to believe that diseased coronary vessels are dilated in preference to other nonoccluded vessels. Thus the principal means by which nitroglycerin relieves anginal pain appears to be by decreasing the work load of the heart.

Pain of myocardial ischemia lasting longer than 15–30 min in spite of rest is more likely to be associated with the death of myocardial cells (i.e., infarction). It is similar to anginal pain, except that it is usually of greater intensity and longer duration and might not be associated with exertion. The pain of myocardial infarction can be slightly diminished, but not relieved completely, by nitroglycerin. Sixty percent of victims dying from myocardial infarction expire within the first 24 h after onset of symptoms: of these, most die suddenly, from fibrillation, before seeking help or reaching the hospital. After admission to a coronary care unit (CCU), arrhythmias can be controlled with currently available therapy. Today the principal cause of death in the CCU is not fibrillation, but rather insufficient myocardial function due to cardiac failure ("pump fail-

---

*Anginal pain not brought on by exertion (e.g., while at rest or sleeping) occurs in some persons and is thought to be the result of neurally mediated arterial spasm.

*DeBakey M and Gotto A: *The Living Heart.* New York, David McKay, 1977, p. 122.

ure" as cardiogenic shock and/or congestive heart failure), or to a blowout or rupture of the damaged myocardial wall.

SUMMARY

Two main coronary arteries comprise the arterial supply to the coronary circulation. The massive left ventricle, which constitutes about half the entire weight of the heart, is mainly supplied by divisions of the left coronary artery. Because of the considerable intramyocardial pressures developed by contraction of the left ventricle, systolic flow in the left coronary arterial system is much less than diastolic flow. By contrast, right coronary flow and venous flow (e.g., anterior cardiac veins, coronary sinus) are greater during systole than during diastole. Coronary blood flow is tightly coupled to the metabolic requirements of the myocardial cells for oxygen. Since the extraction of oxygen from coronary blood is normally near maximal, the sole means of increasing the supply of oxygen is by increasing the flow of blood. Agents that increase myocardial activity and oxygen requirements, such as increased heart rate, increased contractility, and increased wall tension, all result in increased coronary blood flow. The coupling of the blood supply to metabolic oxygen requirement is probably mediated by metabolic agents, such as adenosine or K^+.

In the normal person, differences in the efficiency of cardiac work are not important. However, if there is any compromise in the coronary arterial blood supply (e.g., coronary artery disease), factors increasing low efficiency work of the heart can be of critical significance. Pressure work is less efficient than volume work. Cardiac efficiency is decreased by an increase in size (dilation) and by an increase in heart rate. Whenever the metabolic requirements exceed the ability of the coronary circulation to supply oxygen, the myocardium becomes ischemic. Temporary ischemia of exertion, which is symptomatic of coronary artery disease, results in a characteristic chest pain called angina pectoris. Severe ischemia (e.g., from arterial occlusion) can produce damage and death of myocardial cells—a myocardial infarct. The pain of myocardial infarction is qualitatively similar to that of angina, but it is of greater severity and duration.

SELECTED BIBLIOGRAPHY

Braunwald E: Chapters 235, 236, 244, 245, in Isselbacher KJ, Adams RD, et al (eds): *Harrison's Principles of Internal Medicine,* ed 9. New York, McGraw-Hill, 1980.

Folkow B, Neil E: Coronary circulation, in *Circulation.* New York, Oxford University Press, 1971.

James TN: Anatomy of the coronary arteries and veins, in Hurst JW (ed): *The Heart,* ed 4. New York, McGraw-Hill, 1978.

Marchetti G, Taccardi B (eds): *International Symposium of the Coronary Circulation and Energetics of the Myocardium.* Basel, Karger, 1967.

Morris JJ, Peter RH: Coronary circulation, in Conn HL, Horwitz O (eds): *Cardiac and Vascular Diseases,* vol I. Philadelphia, Lea & Febiger, 1971.

SELF-STUDY QUESTIONS

MATCHING

A. *Left coronary artery*
B. *Right coronary artery*
C. *Coronary sinus*
D. *Anterior cardiac veins*
E. *Thebesian veins*

1. Empties no more than 5% of total coronary blood flow directly into luminal chambers, particularly the atria.

2. Carries most of the total arterial supply to the heart in most people.

3. Channel of venous drainage derived mainly from the left ventricle.

4. Estimated to carry an average of 40% of total arterial supply to heart.

MATCHING

A. *Cardiac muscle only*
B. *Skeletal muscle only*
C. *Both*
D. *Neither*

5. One capillary per muscle fiber.

6. 2,500 capillaries per mm² cross-sectional area.

7. Striated muscle fibers.

MULTIPLE CHOICE

Select the single best answer.

8. The highest (peak) left coronary arterial blood flow occurs during
   A. ventricular rapid ejection.
   B. ventricular reduced ejection.
   C. ventricular filling.
   D. the same period as peak aortic blood pressure.
   E. isovolumetric contraction.

MATCHING

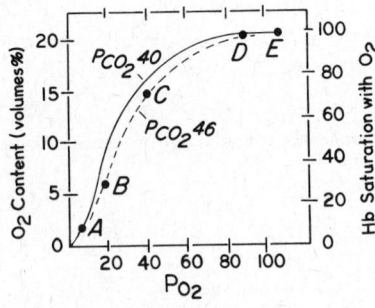

9. Point representing normal coronary arterial blood.

10. Point representing normal coronary sinus blood.

11. Point representing normal mixed venous blood (e.g., from the pulmonary artery).

MULTIPLE CHOICE

Select the single best answer.

12. A potent coronary vasodilator is
    A. adenosine.
    B. hypocarbia (hypocapnia, $\downarrow P_{CO_2}$).
    C. direct action of norepinephrine on coronary vascular α-receptors.
    D. low coronary arterial H^+ concentration.
    E. high coronary arterial pH.

13. Myocardial oxygen consumption ($M\dot{v}_{O_2}$) is increased by an increase in each of the following except
    A. sympathetic activity to the heart.
    B. parasympathetic activity to the heart.
    C. heart size.
    D. heart rate by artificial pacer.
    E. systemic arterial blood pressure.

14. Normally, the heart has a stroke work load of about 0.11 kg-m per stroke. If this stroke work is doubled by a doubling of the stroke volume (e.g., by increased venous return) while the mean systemic arterial pressure and the heart rate remain unchanged,
    A. the efficiency of the heart will increase.
    B. the efficiency of the heart will decrease.
    C. the efficiency of the heart will remain about the same.
    D. the myocardial oxygen consumption ($M\dot{v}_{O_2}$) will decrease.
    E. the tension developed by the left ventricular myocardium will double.

MULTIPLE CHOICE

Select the correct answer(s). (In many instances, more than one answer is correct.)

15. A patient with a dilated failing heart and occasional anginal pain is given an effective therapeutic dose of digitalis each day. This should cause
    A. a decrease in heart size.
    B. a decrease in myocardial oxygen consumption ($M\dot{v}_{O_2}$).

C. an increase in contractility.

D. a decrease in efficiency and increased incidence of anginal pain.

16. A patient with partial occlusion of the coronary arteries complains of anginal pain with exertion. Which of the following is/are likely to precipitate an episode of angina?

A. Excitement with increased heart rate.

B. Administration of a pharmacological agent that activitates peripheral vascular β-receptors.

C. Administration of a pharmacological agent that activates peripheral vascular α-receptors.

D. Increased vagal activity to the heart.

ANSWERS

1. E. Five percent or less of the coronary venous blood is carried by a number of very small veins, the thebesian veins (also called venae cordis minimae), which arise in the wall of the heart (particularly the atria) and empty directly into the luminal chambers.

2. A. Because the great majority of people have right coronary arteries that cross the crux of the posterior surface of the heart (the crux is a cross-shaped intersection of the coronary sulcus with the posterior longitudinal sulcus) and are called "right predominant" by anatomists, many investigators have mistakenly assumed that the predominant blood supply is also from the right (i.e., the right coronary artery). Not so. Even in people with anatomical "right predominance," the major supply of arterial blood (about 60%) is via the left coronary artery.

3. C. Most of the large veins of the heart, which are from left ventricle primarily, open into the coronary sinus —a vein in spite of its name. Veins not emptying into the coronary sinus are the anterior cardiac veins, which drain part of the right ventricle, and the small thebesian veins.

4. B. The right coronary primarily supplies the less massive right ventricle and only a part of the posterior left ventricle. Since the right ventricle contains less myocardium and the right coronary has a lesser flow rate per gram, as shown in animal studies, it follows that the right coronary artery carries less blood than the left coronary. The right coronary carries an estimated 40% of the total arterial flow to the heart.

5. C. Both skeletal muscle and cardiac muscle are shown to have one capillary per muscle fiber in cross section (Fig. 2).

6. A. The density of capillaries per unit cross-sectional area is greater in cardiac muscle than in skeletal muscle.

7. C. Both are striated.

8. C. The highest left coronary flow occurs in ventricular diastole (e.g., during ventricular filling). All the other choices occur during ventricular systole when flow is much less.

9. D. The arterial blood to the heart and other organs normally has a Po_2 of 90 mm Hg, 97% saturation of hemoglobin with O_2, and an oxygen content of about 20 vol% (20 ml O_2/100 ml blood).

10. B. The coronary sinus venous blood, primarily derived from the venous drainage of the left ventricle, has the lowest Po_2 value of any venous blood in the body, except for skeletal muscle under conditions of vigorous exercise. The normal coronary sinus blood has a Po_2 value of 18 mm Hg, a 30% saturation of hemoglobin with O_2, and an oxygen content of only 6 vol% (6 ml O_2 per 100 ml blood).

11. C. Normally, the other tissues extract much less oxygen from each milliliter of blood than does the heart. Venous blood from all the tissues (mixed venous blood) obtained from the pulmonary artery shows a Po_2 value of 40 mm Hg, a 75% saturation of hemoglobin with O_2, and an oxygen content of 15 vol% (15 ml O_2/100 ml blood).

12. A. Of the choices, only adenosine is a coronary dilator. Hypocarbia is a vasoconstrictor in the cerebral circulation (see Chapter 15). The *direct* action of norepinephrine on coronary vascular α-receptors is to produce vasoconstriction, although the *net* effect is usually vasodilation from increased oxygen consumption. Low H^+, or high pH, in coronary arterial blood produces no significant change in coronary resistance. If it had effects like those produced on skeletal muscle vasculature, however, it would probably produce a slight vasoconstriction.

13. B. An increase in sympathetic firing, heart size (by Laplace relationship), heart rate, or systemic arterial blood pressure would increase $M\dot{v}O_2$ by the heart. Only a decrease in heart rate by increased vagal firing would decrease myocardial $\dot{v}O_2$.

14. A. This is an example of volume work. The total work can be greatly increased by a small increase in average ventricular radius. Since a small increase in radius involves only a small increase in tension, and since tension is an important determinant of $M\dot{v}O_2$, the myocardial oxygen consumption will increase only slightly. A large increase in work with a small increase in oxygen consumption constitutes an increase in efficiency (Fig. 7).

15. A, B, C. Digitalis has a beneficial effect on the dilated failing heart. Although it increases contractility (which should in itself increase $M\dot{v}O_2$), the decrease in size of the dilated heart (and decrease in tension) more than makes up for the effects of increased contractility, so that the $M\dot{v}O_2$ decreases. This should reduce the incidence of anginal episodes.

16. A, C. An increase in heart rate decreases efficiency and can precipitate angina in a patient with coronary insufficiency. An agent that acts via peripheral vascular β-receptors produces vasodilation, which should be beneficial by lowering arterial pressure, thereby lowering heart work and $M\dot{v}O_2$. An agent acting via peripheral vascular alpha receptors has the opposite effect and could precipitate anginal episodes. Increased vagal activity would decrease heart rate and improve efficiency (reduce likelihood of angina).

15

Cerebral Circulation

DONALD W. STUBBS

OBJECTIVES

After completion of this chapter, you should be able to

1. Compare heart and brain in terms of (a) percentages of cardiac output received by each, and (b) their relative blood flow rates (ml blood/100 g tissue per minute).

2. Describe the Monro–Kellie doctrine and how it was originally applied to explain constancy of blood flow in the brain.

3. Describe in what way the resistance vessels of the cerebral circulation differ from the general characteristics for resistance vessels in most other vascular beds.

4. Describe what is meant by the blood brain barrier (BBB) and give two possible explanations for it.

5. Describe the factors affecting cerebral perfusion pressure

ΔP, and cerebrovascular resistance R, as well as how these influence cerebral blood flow (CBF, or $\dot{Q}$).

6. Give a theory for the mechanism of autoregulation.

7. Explain the possible role of arterial P_{CO_2} (Pa_{CO_2}) in the intrinsic control of cerebrovascular resistance in terms of the metabolic theory.

8. Describe the role of extravascular H^+ in mediating the effects of CO_2 on CBF.

9. Compare capabilities of changes in Pa_{CO_2} with those of Pa_{O_2} in affecting CBF.

10. Describe the relationship between regional neuronal activity and regional blood flow in the cerebral cortex.

The distinctive characteristic of the mammalian brain is its considerable development of a greatly expanded mid-dorsal roof of the cerebral hemispheres, the general somatic cortex. This development has reached its peak in humans and, because of our cerebral superiority and our advanced ability to think, sets us apart from other animals. Thus many people regard the cerebral circulation as the most important of the special circulations.

Human cerebral blood flow (CBF) is 14% of the cardiac output at rest, whereas that in mammals with less extensive cerebral development constitutes a smaller proportion of the cardiac output. The overall flow rate in human circulation averages 55 ml blood/100 g brain/min. There are regional differences, with flow being higher in gray matter than in white matter, and higher in the cerebral hemispheres than in the spinal cord. Typical regional cerebral blood flows (rCBF) in humans are cortical gray matter, 70–80 ml/100 g/min; and hemisphere white matter, 18–20 ml/100 g/min.

FUNCTIONAL ANATOMY

A unique anatomical feature of the brain and its circulation is that they are enclosed within a rigid container, the cranium. Although blood flows into and out of this sytem, in a steady state in which inflow equals outflow, the total volume of blood, cerebrospinal fluid (CSF), and brain tissue—all virtually incompressible fluids—must be a constant. The brain normally has a relatively constant overall blood flow (55 ml/100 g/min). This constancy was recognized or guessed at by Alexander Monro (1783). The unyielding nature of the skull led Monro to suppose that expansion of cranial contents was not possible and that expansion of cerebral vessels (vasodilation) was unlikely. In 1824 Kellie supported this idea, and the theory came to be known as the Monro–Kellie doctrine. It was reasoned that any tendency for increased arteriolar flow by expansion of the arteriolar volume (vasodilation) would be met by an equal and opposite compression due to increased intracranial pressure. However, this notion is incorrect. A minor increase in the radii of the resistance vessels of the cerebral circulation (arteries and arterioles),

which would have a substantial effect on inflow, can easily be compensated for by a small decrease in the voluminous venous sinuses that collect the cerebral venous outflow. Since these venous sinuses are large vessels with very low resistance, a small decrease in their radii has a negligible effect on their resistance. The small increase in venous resistance does not appreciably offset the large decrease in arterial and arteriolar resistance from vasodilation; therefore, CBF increases. Although the Monro–Kellie doctrine is partly correct, that is, the *total* volume of the cranial contents cannot change—it is not a valid explanation of the relative constancy of CBF (i.e., under the conditions discussed below, the CBF *can* change as a result of changes in the radii of the resistance vessels).

THE LARGE CEREBRAL ARTERIES

In human circulation, the major source of arterial blood for the brain is from the internal carotid arteries, in contrast to many mammals in which the vertebral arteries are more prominent. Thus the carotid sinus baroreceptors in humans, in effect, monitor the pressure head to the brain, and the carotid sinus reflex assures an adequate arterial pressure to maintain CBF. The internal carotid arteries and the basilar artery, derived from the vertebral arteries, unite in the circle of Willis at the base of the brain. Branches from the circle of Willis (to most of the cerebral cortex and upper brain stem) and from the basilar artery (to medulla, pons, cerebellum, and occipital cortex) constitute the arterial supply to the brain.

Although the circle of Willis (Fig. 1) appears to provide a functional collateralization of blood from one side to the other, the situation is somewhat variable. Normally, there is no mixing of the blood from one side to the other via the circle of Willis, as demonstrated by angiography, which shows that a radiographically opaque dye injected into one carotid artery usually does not appear on the contralateral side. However, in about three out of four persons, compression of the carotid artery on one side, which lowers the pressure on that side, does permit blood from the other side to cross over under these circumstances. This is particularly true in younger persons and explains why a totally occluded internal carotid artery is occasionally found in a symptomless patient. On the other

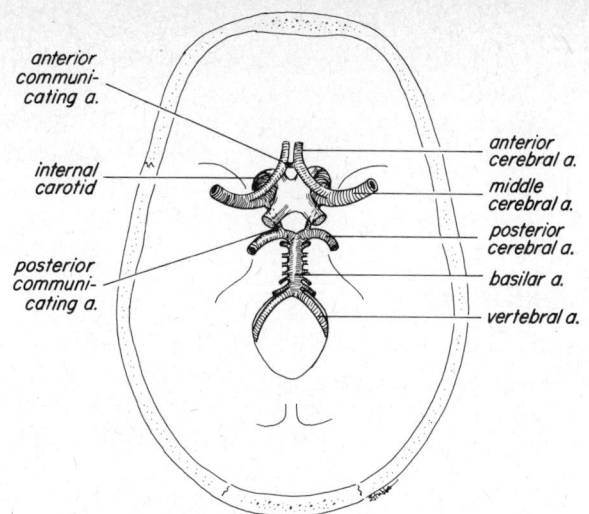

anterior communicating a.

internal carotid

posterior communicating a.

anterior cerebral a.

middle cerebral a.

posterior cerebral a.

basilar a.

vertebral a.

Figure 1 Relationship of the arterial supply, including the circle of Willis, to the base of the skull.

hand, the connecting vessels (e.g., the anterior and posterior communicating arteries) might be too small in other people to permit adequate flow to the opposite side if the internal carotid artery on one side is suddenly occluded. This is more likely to be the case in older persons with atherosclerotic narrowing of arterial conduits. It seems clear that the ancient Greeks were aware of some of the effects of carotid occlusion, for they named these arteries the carotids (from *karoun,* meaning to stupefy); it was known that partial occlusion of both carotid arteries simultaneously by judicious external pressure produces sufficient stupefaction or anesthesia for minor skin surgery.

SMALL ARTERIES

The large arteries arising from the circle of Willis curve around the cerebral hemispheres and divide into smaller branches. These small branches are of two types: (1) parenchymal arteries, which arise immediately from the large arteries and penetrate deeply into the substance of the brain; and (2) pial arteries, the dividing branches of which lie on the surface of the brain, between the dura and pia mater. Unlike the parenchymal arteries, the pial arteries undergo considerable branching while on the surface,

thereby becoming fairly small before entering the substance of the brain. Both parenchymal and pial arteries terminate in arterioles. In most vascular beds the arterioles are the principal resistance vessels. However, in the brain the small arteries preceding the arterioles carry a substantial portion of the overall resistance. By the time blood reaches arterioles, which are only 25 μ in diameter in the pial artery system, at least half the arterial pressure originally present in the aorta has been lost.

CEREBRAL CAPILLARIES

Most cerebral capillaries do not have fenestrations, and the plasma membrane of adjacent endothelial cells is partially fused at tight junctions. These anatomical features partly explain the existence of a blood brain barrier (BBB). The possible existence of such a barrier is a concept originally derived from early pharmacological studies by investigators who injected dyes into the circulatory system and who examined the brain for their presence or pharmacological effects, or both. In the early part of this century, Goldman established that the brain failed to stain with the dye trypan blue when injected intravenously, but did stain when injected into the CSF. There appeared to be a barrier between the brain and the blood. The principal site of the BBB is the primary interface between the brain and blood: the cerebral capillaries. Another barrier site is at the choroid plexus—a barrier between the blood and the CSF. These barrier sites are characterized by cells having tight junctions that restrict intercellular diffusion from one side of a cell layer to the other: (1) the endothelial cells of the cerebral capillaries, and (2) the epithelial cells of the choroid plexus. Closely connected as they are by the tight junctions between them (Fig. 2), these cells act like a continuous layer through which solute exchange takes place mainly by the transcellular route (through the plasma membrane on one side, the cytoplasm, and out through the plasma membrane on the other side). Thus because large intercellular pores or channels are absent, permeability approximates that of the cells' plasma membranes. Lipophilic solutes (e.g., O_2 and CO_2) penetrate plasma membranes easily and therefore equilibrate rapidly between blood and brain. Water is also in osmotic equilibrium between brain and blood, although it does not penetrate as rapidly as lipophilic substances. A number of

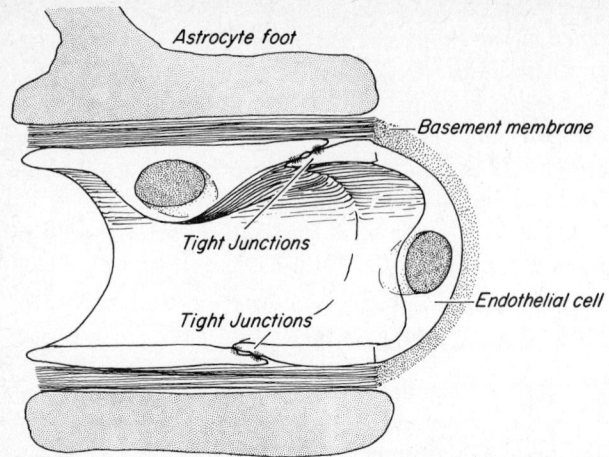

Figure 2 Typical cerebral capillary. Endothelial cells of most cerebral capillaries are connected by continuous belts of tight junctions that restrict intercellular diffusion.

hydrophilic substances (acidic dyes, H⁺) enter very slowly or not at all. Thus the absence of fenestrations and the presence of tight junctions between endothelial cells of most cerebral capillaries provide a partial explanation for an exclusionary interface separating brain from blood.

An exclusionary barrier that operated merely on the basis of plasma membrane lipid solubility would constitute a functional nuisance where essential hydrophilic metabolites are concerned. The cellular membranes have therefore been adapted to permit carrier-mediated transport of desirable hydrolphilic metabolites (e.g., glucose, amino acids). These transport systems not only take up solutes into the brain from the blood, but can also secrete substances from the brain into the blood, such as prostaglandins, I⁻, para-aminohippurate, and iodopyracet (Diodrast). Other substances, such as HCO_3^-, can move in either direction according to need, in this case permitting control of CSF pH independently of plasma pH. The arrangement of specific transport systems for absorption and secretion bears resemblance to a similar situation in renal tubule cells. Thus it is now evident that the blood brain barrier is actually a regulatory interface rather than simply a physical barrier.

The density of capillaries (number per unit cross-sectional area) varies somewhat, being higher in gray matter than in white, and averages around 1000 capillaries/mm² in gray matter, which is more than that in skeletal muscle and less than that in cardiac muscle. The greater density of capillaries in gray matter plus the higher metabolic rate of neurons explains the greater regional blood flow in gray matter than in white matter.

CEREBRAL VEINS

The venous drainage of the brain differs from that of most other organs and tissues in that (1) the veins contain no valves; (2) their walls, having little musculature, are very thin; and (3) their distribution often does not correspond to that of the arteries, and they usually do not accompany them. Superficial and deep veins empty into dural venous sinuses that lie between layers of the dura mater. These dural sinuses also lack valves. The principal drainage of the venous sinus blood is via the internal jugular veins. However, some venous blood drains by way of the vertebral plexus of veins. Just what proportion of the cerebral venous outflow is normally carried by the vertebral plexus is unknown. It is known, however, that the vertebral plexus is able to carry the entire cerebral venous outflow if both internal jugular veins are ligated.

INNERVATION OF THE CEREBRAL VASCULATURE

The presence of nerve fibers in the larger cerebral arteries has been known since 1664, when first described by Thomas Willis. Despite discrepancies between subsequent studies using light microscopy, electron microscopy, and histochemical methods, the principal weight of evidence supports the presence of adrenergic vasoconstrictor innervation of some of the cerebral arteries. These adrenergic fibers are abundantly distributed to the large arteries forming and leaving the circle of Willis, moderately distributed to pial arteries, and sparsely supplied to all branches penetrating into the substance of the brain.

Functionally, the significance of vasomotor control of CBF is considered by many cardiovascular physiologists to be of little importance. Regional sympathetic blockage does not significantly increase cerebral flow by decreasing cerebral resistance. Furthermore, even supramaximal electrical stimulation of the cervical sympathetic fibers to cerebral vessels decreases CBF only a little by increasing resistance 20–30% above normal. This is a rather unimpressive increase in resistance compared with the ability of sympathetic stimulation to increase resistance of skeletal muscles vasculature by 600%. In animals (e.g., dogs) a greater decrease in CBF has been obtained by supramaximal electrical stimulation of the stellate ganglion; however, it is doubtful that this degree of vasoconstriction could ever be obtained reflexively.

From a teleological point of view—one regarding possible adaptive advantage or purposeful design—it has been argued convincingly that this relative lack of significant adrenergic vasoconstriction in the cerebral vasculature is desirable. Many other vascular beds (e.g., splanchnic, renal, skeletal muscle, cutaneous) participate in reflex control of the mean systemic arterial blood pressure (e.g., the baroreceptor reflex). Hypotensive influences, such as changing from a reclining to a standing position and loss of blood volume, are opposed, in part, by a generalized increased vasoconstriction, which helps maintain a normal level of arterial blood pressure by increasing the total peripheral resistance. This maintenance of adequate arterial blood pressure is important for the normal function of vital organs, such as the heart and brain. For example, CBF begins to reach ischemic levels when the mean arterial blood pressure becomes less than half-normal. If the cerebral arteries and arterioles themselves were to participate in such general reflex vasoconstriction, the advantage of sustaining the arterial blood pressure to maintain CBF would be offset by increased resistance to flow through the constricted cerebral vessels. Indeed, the opposite is often seen: in severe hemorrhagic blood loss, during which an intense peripheral vasoconstriction occurs, the cerebral resistance vessels *dilate,* thanks to their intrinsic mechanisms for regulating CBF (described below). These intrinsic mechanisms are not dependent on autonomic innervation.

The pial arteries are supplied with parasympathetic cholinergic fibers emerging from the facial nerve. However, the functional role of these vasodilatory fibers remains to be determined.

FACTORS REGULATING CEREBRAL BLOOD FLOW

As in all vascular beds, blood flow in the cerebrovascular bed depends on both the perfusion pressure ΔP across the bed and the resistance: $Q = \Delta P/R$.

PERFUSION PRESSURE (ΔP)

Cerebral perfusion pressure ΔP (or pressure difference) is the inflow pressure (mean arterial blood pressure of the internal carotid artery) minus the outflow pressure (mean venous blood pressure of the internal jugular vein). Since the venous outflow pressure is normally quite low (< 10 mm Hg), it can usually be ignored, and the mean arterial blood pressure can be taken to be essentially the same as the perfusion pressure for a rough approximation.

Although the cerebral venous pressure is normally low enough to be considered negligible in estimating the cerebral perfusion pressure, it can become substantially elevated in some conditions. A large increase in cerebral venous pressure (the outflow pressure) in the absence of any change in the mean arterial pressure (the inflow pressure), decreases the cerebral perfusion pressure ΔP. For a given resistance, a decrease in the perfusion pressure causes a decrease in flow. Thus for a given mean arterial pressure and a given resistance, an increase in cerebral venous pressure might be expected to cause a decrease in CBF. One way in which venous pressure is abnormally elevated is by an increase in intracranial pressure, measured in the CSF space. Thus an increase in intracranial pressure (e.g., from a subdural hematoma or following rapid growth of a space-occupying tumor) would be expected to decrease the CBF for the reasons given above (i.e., a decrease in perfusion pressure). Fortunately, increases in CSF do not decrease CBF unless extremely high (Fig. 3).

The CBF is prevented from falling to ischemic levels, until the CSF pressure becomes quite high, by two active mechanisms: (1) a compensatory decrease in cerebrovascular resistance to counter the drop in perfusion pressure, and (2) a rise in the mean arterial pressure (the inflow pressure) to diminish the fall in perfusion pressure. Thus

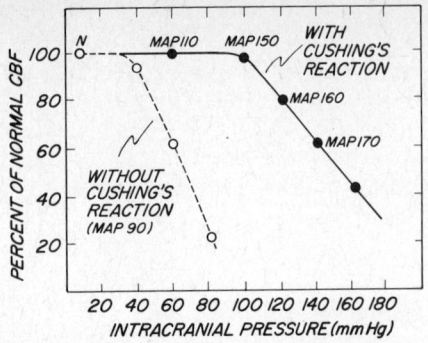

Figure 3. Effect of increasing intracranial pressure (CSF pressure) on cerebral blood flow. In this hypothetical example an increase of CSF pressure to 30 mm Hg decreases the perfusion pressure of the brain to 60 mm Hg if the mean arterial pressure (MAP) is 90 mm Hg. However, normal cerebral blood flow can be maintained if, for example, the cerebral vascular resistance decreases from 7 PRUs (mm Hg/ml per second) to 4.7 PRUs. To maintain a normal CBF in the presence of further increases in CSF pressure would require additional compensatory mechanisms. The Cushing's reaction is such a compensatory mechanism and can maintain CBF, up to a point, by increasing the MAP. At very high CSF pressures, not even the Cushing's reaction can maintain CBF and coma results. If there were no Cushing's reaction to increase MAP (the cerebral vascular inflow pressure), CBF would be decreased to intolerable levels at much lower CSF pressures. The hypothetical curves illustrated are similar to those obtained in experimental animals with artificial CSF infusion at increased pressures. (See Häggendal E et al: *Acta Physiol Scand* 79:262, 1970.)

as CSF pressure increases, there is an active compensatory decrease in the resistance of the small arteries and arterioles, so that CBF remains normal despite the initial decrease in perfusion pressure. However, as CSF pressure increases further and begins to approach the normal level of systemic arterial pressure, the decrease in perfusion pressure begins to exceed the ability of the resistance vessels to compensate and some ischemia results. This initiates the second mechanism, that is, whenever the medullary vasomotor center becomes ischemic (either from severe hypotension or, as in this case, from greatly increased CSF pressure), its activity begins to increase considerably and raises the total peripheral resistance in the body. The greater the fall in cerebral perfusion pressure, the greater the rise in mean arterial pressure. This re-

sponse of the vasomotor center to ischemia has the ability to raise the mean arterial pressure to the maximal level, which can be sustained by the heart. By increasing cerebral inflow pressure, this elevated mean arterial pressure helps improve the cerebral perfusion pressure. Although this response was originally called the Cushing reflex,* it is probably not a true reflex and should be called either the *Cushing reaction* or the cerebral ischemic response. A very high arterial blood pressure following head injury or a cerebrovascular accident usually indicates intracerebral hemorrhage, as well as operation of the Cushing reaction due to increased CSF pressure. The Cushing reaction is usually a triad consisting of (1) elevated systemic arterial blood pressure, (2) bradycardia, and (3) slowed respiration.

A decrease in mean arterial blood pressure, for a given cerebrovascular resistance, should also decrease cerebral perfusion pressure. Fortunately, the cerebrovascular resistance is not fixed and can decrease to some extent in order to offset or counter the drop in arterial blood pressure. This autoregulation of blood flow has a lower limit beyond which further decreases in arterial pressure result in substantial decreases in cerebral blood flow. In normotensive persons autoregulation falls off markedly at about 60 mm Hg of arterial blood pressure. At lower arterial blood pressures the cerebral blood flow is greatly diminished, to the detriment of neuronal function. Unconsciousness occurs at about 30 ml/100 g/min. Measurable electrical activity of neurons is absent at a flow of about 15 ml/100 g/min. If the cerebral blood flow is allowed to diminish further to about 8 ml/100 g/min, there is a massive release of intracellular K^+, which is indicative of cell damage. If this low level of blood flow is maintained for more than a half-hour, cortical electrical activity cannot be restored by resumption of a normal cerebral blood flow. The K^+ release threshold is an irreversible "lethal ischemic threshold" below which cell death results after several minutes.

In addition to the lower limit of autoregulation (60 mm Hg), there is also an upper limit, that is, if arterial blood pressure is acutely increased to above 140 mm Hg in

---

*Named for Harvey Cushing, a Boston surgeon who first analyzed it quantitatively.

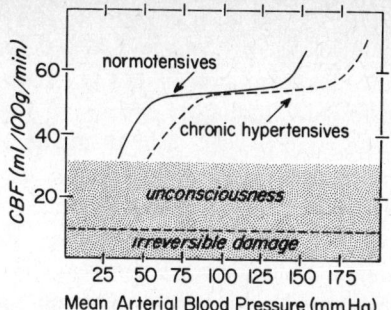

Figure 4 Relation between mean arterial blood pressure and cerebral blood flow. (After Lassen NA: "Brain" in Johnson PC (ed): *Peripheral Circulation, John Wiley & Sons, Inc. N.Y.,* 1978.)

otherwise normotensive persons, there is a marked increase in cerebral blood flow. Above this limit, the increase in perfusion pressure can no longer be adequately countered by an increase in cerebrovascular resistance; the pressure "breaks through" the vasoconstrictor response, permitting increased flow. Excessively high pressures above the upper autoregulatory limit are associated with a patchy, multifocal disruption of the BBB with edema formation. Thus cerebrovascular autoregulation in response to changes in arterial blood pressure has both upper and lower limits. In patients with chronic hypertension the upper and lower limits are both elevated, that is, the autoregulatory curve is shifted to the right (Fig. 4). The cerebral resistance vessels become adapted after long-term exposure to a chronically elevated arterial blood pressure. Although the pressures at which the upper and lower autoregulatory limits exist are higher in chronic hypertensives, the level of cerebral blood flow maintained between these limits by autoregulation is the same as in normotensive persons (55 ml/100 g/min).

CEREBROVASCULAR RESISTANCE

The decrease in cerebral blood flow that results from a very high cerebral venous pressure, due to a very high cerebrospinal fluid pressure (as seen in Fig. 3), or from a very low arterial blood pressure (as seen in Fig. 4), as well as the increase in cerebral blood flow resulting from a very high arterial blood pressure (also seen in Fig. 4), all illustrate the point that extremely low or high perfusion pres-

sures can greatly alter cerebral blood flow. However, another important factor regulates cerebral blood flow: cerebrovascular resistance. Indeed, change in resistance to flow is responsible for the ability of the cerebral circulation to autoregulate. Since $\dot{Q} = \Delta P/R$ and cerebral blood flow ($\dot{Q}$) remains fairly constant through autoregulation in spite of substantial changes in perfusion pressure ΔP (within the limits illustrated in Fig. 4), it follows that there must be compensatory changes in cerebrovascular resistance R in the face of a change in arterial blood pressure. Similarly, a compensatory change in resistance must occur in order to maintain a normal cerebral blood flow despite a diminution in perfusion pressure due to increased cerebrospinal fluid pressure. As can be seen in Figures 3 and 4, there are limits as to how much a change in resistance can compensate for a change in perfusion pressure; within these limits, however, the cerebrovascular resistance vessels have an effective and vital control over cerebral blood flow.

EFFECTS OF CO$_2$, O$_2$, and H$^+$ ON CEREBRAL BLOOD FLOW

That increased arterial P_{CO_2} (Pa_{CO_2}) causes dilation of cerebral resistance vessels and an increase in CBF has been known for some time. Increases in the caliber of pial arteries viewed through a cranial window in asphyxiating animals were noted by several workers during the late nineteenth century. With the introduction of accurate quantitative methods for measuring CBF, a series of studies in numerous species, including humans, confirmed the earlier qualitative observations, namely that CBF varies with Pa_{CO_2}. An increase in Pa_{CO_2} dilates cerebral arteries and arterioles, thereby decreasing resistance and increasing flow. A decrease in Pa_{CO_2} (e.g., from hyperventilation) results in the opposite changes. The relationship between CBF and Pa_{CO_2} is sigmoid (Fig. 5).

Changes in arterial P_{O_2} (Pa_{O_2}) have effects qualitatively opposite in direction to those of Pa_{CO_2}. When air mixtures containing only half the normal oxygen content are breathed, cerebrovascular resistance is reduced and CBF

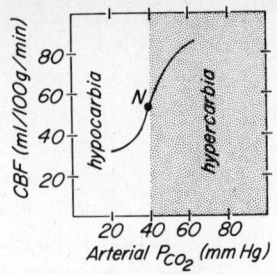

Figure 5 Relation between cerebral blood flow (CBF) and arterial P_{CO_2} in human circulation. N, Normal point for arterial P_{CO_2} (40 mm Hg) and CBF (55 ml/100 g/min). An arterial P_{CO_2} above normal (> 40 mm Hg) is called hypercarbia (shaded area) and is usually associated with an increase in CBF. An arterial P_{CO_2} below normal (< 40 mm Hg) is called hypocarbia and is usually associated with a decrease in CBF.

is increased. However, a low Pa_{O_2} (hypoxemia) is not quantitatively as potent a stimulus for increasing CBF as is an increase in Pa_{CO_2} (hypercarbia) on a mm-per-mm Hg basis. An increase in Pa_{O_2} produced by breathing oxygen-rich mixtures causes little or no cerebral vasoconstriction and virtually no reduction in flow rate (Fig. 6).

It can be concluded that P_{CO_2} is the prime chemical regulator of CBF, just as oxygen availability determines coronary vascular resistance. However, it has become increasingly clear that it is not the CO_2 molecule per se that

acts on the cerebral resistance vasculature, but rather the local smooth muscle H^+ concentration that is in large part determined by the formation of H_2CO_3 from CO_2 and H_2O. The profound vasodilatory effect of increased Pa_{CO_2} results from the rapid passage of lipid-soluble CO_2 across the BBB into extravascular interstitial fluid spaces of the brain, where it ultimately increases the intracellular H^+ concentration, from H_2CO_3 formation, in arterial and arteriolar smooth muscle fibers. This increase in intracellular H^+ causes vasodilation. It must be emphasized that it is the extravascular H^+ concentration that affects cerebrovascular tone, not the H^+ of the blood. Blood H^+ has no direct access to the vascular smooth muscle in brain because it cannot cross the endothelium of cerebral vessels. An increase in blood H^+ by the presence of non-volatile acid—acid other than H_2CO_3 (e.g., lactic acid)—does not directly affect CBF. Indeed, it can indirectly affect CBF in the opposite direction by causing hyperventilation, which lowers Pa_{CO_2} and decreases vascular smooth muscle H^+ concentration. Any change in arterial pH not associated with a change in Pa_{CO_2} will not affect CBF.

Figure 7 schematically represents the currently accepted mode of action of CO_2 on the vascular smooth

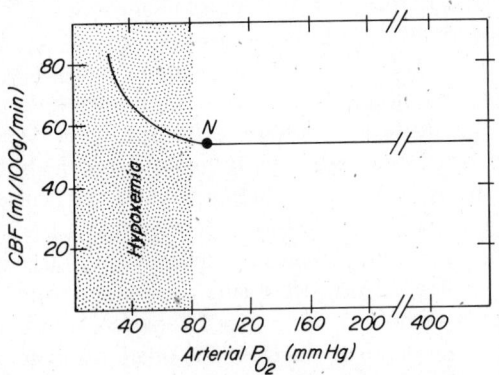

Figure 6. The relation between cerebral blood flow (CBF) and arterial P_{O_2}. The arterial P_{CO_2} is assumed to be constant as the P_{O_2} is varied. Point N represents the normal CBF at the normal arterial P_{O_2} of 90 mm Hg. Extrapolated to human values from animal experiments. (See James IM et al: *Circ Res* 25:77, 1969; Koguire K et al: *J Appl Physiol* 29:223, 1970.)

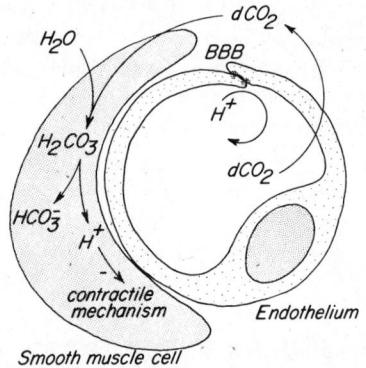

Figure 7 Vasodilatory effect of CO_2 on cerebrovascular smooth muscle by means of extravascular H^+. Dissolved CO_2 (dCO_2) readily crosses the endothelium into the interstitial fluid space of the central nervous system. H_2CO_3 can form at that point, or more likely dCO_2 enters smooth muscle fibers and forms H_2CO_3 intracellularly. In some manner an increase in H^+ inhibits the contractile mechanism and causes vasodilation. H^+ in blood can cross neither cerebral vessel endothelium membranes nor the tight junctions between endothelial cell-to-cell connections, the site of the BBB.

muscle of cerebral resistance vessels by means of H^+ concentration.

REGIONAL DIFFERENCES OF BLOOD FLOW IN CORTICAL GRAY MATTER

Technological advances have made measurements of circumscribed or regional blood flows in different areas of the cerebral cortex in human subjects a practicality. A radioactive isotope of xenon (^{133}Xe) is either injected as a solution into a carotid artery or inhaled as a gas. The arrival and washout of radioactivity from different cortical regions is followed by a gamma ray camera. This consists of numerous externally placed scintillation detectors, each of which is focused to scan approximately 1 cm^2 of brain surface. The information received from each individual area is processed by a computer, and all are displayed as an assembly of individual pixels, or picture images, on a color television monitor, the different magnitudes of flow being coded as different colors.

Methods that permit estimation of regional metabolic rates (e.g., measurement of ^{14}C-labeled 2-deoxyglucose uptake, or clearance of ^{15}C-labeled hemoglobin) have shown that regional differences of blood flow reflect regional differences of neural metabolic activity. Just as in many other organs (e.g., the heart), blood flow is coupled to metabolic requirements. Voluntary unilateral limb movements enhance the blood flow in the appropriate contralateral rolandic region. An increase in the blood flow of distinct areas can also be observed during increased sensory input from different receptors. For example, as illustrated in Figure 8, the normal person at rest and awake, but with eyes closed, has a typical "hyperfrontal" resting pattern of cortical blood flow; however, upon listening to spoken words, the frontal flow decreases while the temporal lobe flow increases, particularly in the area corresponding to Wernicke's area, which mediates understanding of speech. On the other hand, the solving of mathematical problems causes a widespread increase of blood flow over the entire cerebral hemisphere, particularly in the premotor and frontal areas.

The coupling between neuronal activity of cortical gray

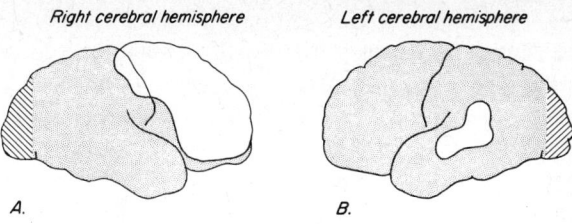

□ rCBF Above Average
▨ rCBF Average or Below
▧ not accessible for measurement

Figure 8. Differences in regional cerebral blood flow (rCBF) in the cortex of humans. A gamma-ray camera consisting of a large number of individual scintillation detectors can detect regional differences in cortical blood flow as differences in the rate of radioisotope washout. (a) The pattern of rCBF in the right cerebral hemisphere of an individual in a quiet room, at rest with eyes closed and awake. The rCBF of the frontal lobe is greater than that for the rest of the cortex. The unstippled area represents a rCBF greater than average (about 75 ml/100 g/min). This is described as a "hyperfrontal" pattern. (b) The pattern of rCBF in the left cerebral hemisphere of a person at rest with eyes closed, but listening to spoken words. This pattern shows increased rCBF in the auditory cortex including Wernicke's area (mediates the understanding of spoken language).

matter and regional blood flow is explainable on the basis of a dependency of blood flow rate on the metabolic rate, a metabolically influenced intrinsic regulatory mechanism. Although the specific mediator of this coupling remains to be established, changes in perivascular H^+, in response to changes in neuronal CO_2 production, seems the likely stimulus.

CEREBROVASCULAR ACCIDENTS

A cerebrovascular accident (CVA), one of the first clinical states to be described by an acronym and also known as apoplexy, or stroke,* can be defined as a sudden interrup-

---

*The word stroke conveys the meaning of an abrupt, unexpected neurological deficit frequently causing the victim to fall as if struck down by surprise.

tion of blood flow to a region of the brain. A small, brief interruption produces a transient ischemic attack (TIA), or little stroke, and causes only a temporary loss of normal function, usually without permanent damage. More prolonged ischemia causes death of neural tissue—a cerebral infarct—and is always associated with a degree of permanent damage to the brain. CVAs can result from either occlusion due to vascular embolism or thrombosis on the one hand, or from hemorrhage on the other (i.e., the opposite ends of the hemostatic spectrum). Most CVAs (about 50%) are the result of a local thrombosis, usually in an atherosclerotic artery. About 30% of all CVAs are the result of occlusion from an embolism formed elsewhere (e.g., a thromboembolism from the heart) that circulates to the brain. CVAs can also result from intracerebral hemorrhage (about 20% of all CVAs). Such hemorrhages cause damage not only from interruption of the blood supply, but also from the pressure of an enlarging hematoma. Hemorrhage can be the result of damage to the arterial wall, which can be another consequence of atherosclerosis, or from hypertension, or from rupture of a malformation, such as an aneurysm.

Loss of function as a result of a CVA can take many forms, depending on the specific site of the lesion and the extent of the damage. The degree of infarction can be so severe that death ensures within hours. For example, massive hemorrhage can destroy areas on both sides of the brain. On the other hand, injury might be so slight that signs and symptoms are transient and go unrecognized. Between these extremes certain features are commonly found: (1) unconsciousness (from minutes to days), (2) loss of motor funciton on the contralateral side, (3) difficulty of speech (aphasia is most likely to be present if the dominant hemisphere is involved; i.e., the left hemisphere in 97% of the population), (4) difficulty in swallowing, (5) visual disturbances, and (6) other sensory dysfunction (e.g., paresthesias, numbness, loss of proprioception). Cerebrovascular diseases remain the third leading cause of death in Western civilization, a situation unlikely to change as long as the population continues to increase in mean age.

SUMMARY

Even though the total volume of the brain, CSF, and blood within the rigid cranium is constant, it is possible for the resistance vessels of the cerebral circulation to alter the rate of blood flow to the brain (vasomotor control). Although cerebral vessels are innervated by both divisions of the autonomic nervous system, autonomic neurally mediated vasomotor control is considered to be of little importance. The principal means of vasomotor control is by a metabolically influenced intrinsic regulatory system. A unique feature of the cerebral capillaries is the existence of a blood brain barrier between the blood and brain that selectively determines what substances can exchange between them.

Overall blood flow through the cerebral circulation depends on the perfusion pressure ΔP and the resistance R. Compensatory mechanisms, within limits, prevent large changes in the cerebral blood flow in spite of changes in ΔP. Thus the cerebral blood flow tends to remain normal despite changes in systemic arterial blood pressure (by autoregulation). A decrease in perfusion pressure caused by an increase in venous outflow pressure (e.g., by increased CSF pressure) is compensated for, within limits, by two mechanisms: (1) a decrease in the resistance, by the metabolically influenced intrinsic regulatory system; and (2) the Cushing reaction, which raises systemic arterial blood pressure. The partial pressure of carbon dioxide in arterial blood (Pa_{CO_2}) has a profound influence on cerebral blood flow by its effect on extravascular H^+. Regional differences in cortical blood flow are related to regional differences in neuronal activity, an effect that is probably mediated as well by carbon dioxide via extravascular H^+—an increase in regional neuronal CO_2 production decreases extravascular pH, resulting in vasodilation in that region.

Vascular disease of the nervous system is the most common cause of neurological disorders. Interruption of the blood supply to part of the brain, either from local thrombosis, embolism, or hemorrhage, results in a CVA (apoplexy or stroke).

SELECTED BIBLIOGRAPHY

D'Alecy LG: in Ruch TC, Patten HD. (eds): *Physiology and Biophysics,* vol II. Philadelphia, Saunders, 1974.

Folkow B, Neil E: *Circulation.* New York, Oxford University Press, 1971.

Lassen NA: in Johnson PC (ed): *Peripheral Circulation.* New York, Wiley, 1978.

Purves MJ: *The Physiology of the Cerebral Circulation.* New York, Cambridge University Press, 1972.

SELF-STUDY QUESTIONS

GREATER THAN/LESS THAN

If item in column A is greater than in column B, mark your answer A. If item in column B is greater, mark your answer B. If both are essentially equal, mark your answer C.

Column A	Column B
1. Percentage of cardiac output going to brain in dog.	Percentage of cardiac output going to brain in humans.
2. Regional cerebral blood flow (rCBF) in cerebral cortical gray matter.	rCBF in cerebral hemisphere white matter.
3. Adrenergic innervation of large arteries of brain.	Adrenergic innervation of parenchymal arteries of brain.
4. Ability of trypan blue dye in blood to cross blood brain barrier (BBB).	Ability of blood gases (CO_2, O_2) to cross BBB.
5. Average density of capillaries (per mm^2 cross-sectional area) in cerebral gray matter.	Average density of capillaries in myocardium.
6. Increase in resistance to flow with supramaximal electrical stimulation of adrenergic fibers to resistance vessels of brain.	Increase in resistance to flow with supramaximal electrical stimulation of adrenergic fibers to resistance vessels of skeletal muscle.

CASE HISTORY FOR QUESTIONS 7–11

The school nurse at a local junior high school was presented with a boy, aged 13, of normal appearance, who was reported to have fainted while playing with his classmates during the noon hour. The nurse found nothing obviously wrong, but sent the student home with a note. His mother was alarmed and took him to their family physician. The physician gave him a complete physical examination but also found no basis for the fainting episode. After close interrogation—both the authoritative physician and the stern parent having pressured him into telling the truth—the boy revealed the following history. One of his friends had shown him a new game in the schoolyard. It involved breathing rapidly for 50 breaths while in a squatting position with both hands locked together. Upon completion of the fiftieth breath, the subject suddenly stands while holding his breath and bearing down as he attempts to pull his hands apart while gripping them together as hard as possible (exerts powerful isometric tension). The kindly physician writes "schoolboy's fainting lark" in the student's medical records, assures the mother that nothing is physically wrong with her son, and advises the boy to seek other friends.

Multiple Choice

Select the single best answer.

7. What effect does hyperventilation have on the arterial blood gases?

A. Increases Pa_{CO_2} and decreases Pa_{O_2}.

B. Increases both Pa_{CO_2} and Pa_{O_2}.

C. Decreases both Pa_{CO_2} and Pa_{O_2}.

D. Decreases Pa_{CO_2} and increases Pa_{O_2}.

E. Does not change blood gases.

8. What effect does hyperventilation have on the cerebral circulation?

A. An intense vasodilation that greatly increases CSF pressure, thereby decreasing CBF.

B. A profound vasodilation in the cerebral circulation and other vascular beds decreases the total peripheral resistance and produces a substantial hypotension.

C. An intense cerebral vasoconstriction shuts off all supply of blood to the brain.

D. A cerebral vasoconstriction occurs and decreases CBF, but usually not to the point of cerebral ischemia.

E. None.

9. What effect does suddenly standing up have on venous return to the heart and cardiac output?

A. It momentarily increases venous return and cardiac output.

B. It momentarily decreases venous return and cardiac output.

C. It increases venous return and decreases cardiac output.

D. It decreases venous return and increases cardiac output.

E. It has no effect on either.

10. Straining against a closed glottis is called the Valsalva maneuver. This produces a brief rise in arterial pressure as blood is squeezed out of the thorax, followed by a sustained decrease in cardiac output because of decreased venous return to the heart from the damming effect of increased thoracic pressure on the great veins. What will happen to the arterial pressure after the brief increase?

A. It will remain elevated because arterial blood cannot escape from the arteries into the veins.

B. It will remain elevated as long as the intrathoracic pressure pushes against the thoracic aorta.

C. It will return to its normal level.

D. It will fall because of decreased cardiac output.

E. It will fall to zero because of the cardiac arrest that invariably follows this dangerous maneuver.

11. Which of the following explanations or combinations explain the syncope?

A. Toxic hypercarbia (hypercapnia), increased CSF pressure, and increased arterial blood pressure causing myogenic vasoconstriction with cessation of CBF.

B. Hypoxemia, hypercarbia (hypercapnia), and intense hypertension to the point of cerebral apoplexy.

C. Autosuggestion reinforced by group hysteria.

D. Adolescent malingering, to avoid class attendance.

E. Hypocarbia (hypocapnia), cerebral vasoconstriction, decreased venous return to the heart, decreased cardiac output, and cerebral ischemia (cerebral hypoxia).

Multiple Choice

Select the single best answer.

12. In a normal person, which of the following will produce the greatest change in CBF?

A. An increase of 20 nEq H^+/L blood (decrease in arterial pH from 7.4 to about 7.2) in the absence of a change in alveolar ventilation.

B. A 20 mm Hg drop in mean arterial blood pressure.

C. A 20 mm Hg rise in CSF pressure.

D. A 20 mm increase in Pa_{O_2}.

E. A 20 mm Hg decrease in Pa_{CO_2}.

13. A sudden increase in CSF pressure (e.g., the formation of a subdural hematoma) can decrease CBF. This initial decrease in blood flow *primarily* comes about because the increased CSF pressure

A. presses on all intracranial blood vessels thereby decreasing their luminal radii and increasing cerebrovascular resistance.

B. presses on the thin-walled veins and sinuses, which flatten, thereby increasing cerebral venous resistance to flow.

C. presses on the thin-walled veins and sinuses which raises cerebral venous pressure thereby decreasing cerebral perfusion pressure ΔP.

D. triggers the Cushing reaction, which raises systemic arterial blood pressure, which, in turn, causes myogenic vasoconstriction of cerebral arteries and arterioles.

E. triggers the Cushing reaction, which causes an intense generalized vasoconstriction in numerous vascular beds, including the cerebral circulation.

14. The blood-borne factor most effective in regulating CBF is

A. Pa_{CO_2}.

B. Pa_{O_2}.

C. arterial pH.

D. lactic acid.

E. pyruvic acid.

15. An acute increase in arterial pH without a change in alveolar ventilation (e.g., after ingesting a large amount of sodium bicarbonate, but before any compensatory change in respiratory ventilation) will cause

A. no effect on cerebral resistance to blood flow.

B. cerebral vasodilation.

C. cerebral vasoconstriction.

D. increased danger of cerebrovascular accident (stroke).

E. symptoms of incipient cerebral ischemia (blurred vision, dizziness).

16. An adult male typist has received ^{133}Xe, and his pattern of cortical rCBF has been scanned by a gamma ray camera. After the typist was asked to type his name, which of the following would you expect to find as he begins typing?

A. Generalized activation of rCBF to the frontal cortical areas.

B. Only activation of rCBF to the hand–finger area of the primary (premotor) cortex.

C. Activation of rCBF to hand–finger areas of both primary and supplementary motor areas of the cortex.

D. Activation of rCBF to hand–finger areas of both primary and supplementary areas of motor cortex plus related regions of the somatosensory cortex.

E. No change in any of the cortical regions, since all gray matter has the same flow rate per unit weight.

ANSWERS

1. B. The proportion of cardiac output going to the brain in mammals with less extensive cerebral development than in humans is smaller (e.g., in the dog only 5% of the cardiac output goes to the brain).

2. A. Regional CBF is usually about four times higher in gray matter than in white matter.

3. A. The large arteries forming and leaving the circle of Willis have abundant adrenergic innervation. Smaller arteries on the surface of the brain have only moderate adrenergic innervation. Small arteries that penetrate the brain substance (parenchyma), such as the parenchymal arteries, have very sparse innervation.

4. B. The blood brain barrier (BBB) prevents entry of

various dyes, such as trypan blue, from blood into brain. CO_2 and O_2, being lipid soluble, easily cross cell membranes and readily penetrate the BBB.

5. B. The average density of capillaries in brain gray matter is about $1000/mm^2$, whereas that in myocardium is about $2,500/mm^2$.

6. B. Supramaximal electrical stimulation of adrenergic fibers to cerebral resistance vessels causes only a 20–30% increase in resistance. The resistance in muscle vasculature can be increased by 600–700%.

7. D. Hyperventilation lowers alveolar P_{CO_2} and raises alveolar P_{O_2}, resulting in decreased Pa_{CO_2} and increased Pa_{O_2}.

8. D. Decreased Pa_{CO_2} from hyperventilation causes cerebral vasoconstriction and a decrease in CBF. Although the Pa_{O_2} is above normal, this decrease in CBF can decrease the supply of O_2 to the brain because an increase in Pa_{O_2} above normal causes very little effect on the oxygen content of the blood (hemoglobin is already 97% saturated with oxygen at normal Pa_{O_2}). Thus hyperventilation has a negligible effect on the oxygen *content* of the blood, but it can have a significant effect on the supply of blood to the brain. With voluntary hyperventilation CBF decreases, usually not to the point of unconsciousness. The usual symptoms produced by voluntary hyperventilation are lightheadedness and tingling paresthesias of fingers, toes, and/or lips.

9. B. Suddenly standing up after squatting for awhile causes a momentary decrease in venous return and cardiac output. The increase in hydrostatic pressure to the leg veins results in their distention and venous pooling of blood. The baroceptor reflex is usually able to cope with this postural influence, which, by itself, seldom produces fainting.

10. D. The arterial pressure falls because of decreased cardiac output. The Valsalva maneuver, by itself, usually does not decrease arterial blood pressure to dangerous levels. We use the Valsalva maneuver several times a day in performing numerous daily tasks (e.g., lifting a heavy object, straining with defecation).

11. E. The syncope of the "schoolboy's fainting lark" results from a combination of events: decreased CBF from hypocarbia (hypocapnia) and decreased mean arterial blood pressure. The mean arterial blood pressure falls to a low level because of a combination of influences acting to decrease venous return: (1) sudden standing after squatting awhile, and (2) the Valsalva maneuver.

12. E is correct. A is false because arterial pH does not directly influence CBF. It can have an indirect effect if it produces a change in alveolar ventilation and thereby a change in Pa_{CO_2}, in which case it is the CO_2 affecting CBF, and not the blood H^+. An increase in blood H^+ usually does increase alveolar ventilation and lowers Pa_{CO_2}, which helps return the pH back toward normal. B is false. A drop in mean arterial blood pressure of only 20 mm Hg below normal does not change CBF, since the cerebral circulation has excellent autoregulation. C is false. A 20 mm Hg rise in CSF pressure has about the same influence on cerebral perfusion pressure as a 20 mm Hg fall in mean arterial pressure, but neither causes decreased CBF because of cerebral autoregulatory ability—the ability of resistance vessels to compensate for changes in perfusion pressure within this range. D is false. A 20 mm Hg rise in Pa_{O_2} produces no appreciable decrease in CBF (Fig. 6). However (E) a 20 mm Hg fall in Pa_{CO_2} significantly decreases CBF (by about 22 ml less than normal/100 g/min). The intrinsic control of CBF cannot counter this decrease because the intrinsic mechanism is medicated by changes in CO_2 in the cerebral circulation. CO_2 is probably the principal vasodilator metabolite that regulates CBF. When the supply of CO_2 in the arterial blood coming to the brain is decreased, CBF decreases just as it would if increased blood flow had washed out locally produced CO_2.

13. C is correct. A is false. Although increased CSF pressure does increase cerebrovascular resistance somewhat, it decreases CBF primarily because it decreases cerebral perfusion pressure. B is false. The large veins do become slightly flattened, but because they are fairly large to begin with, moderate diminution of radius does not substantially increase the overall resistance of the cerebral circulation. D is false. Increased CSF pressure can trigger the Cushing reaction, which would return flow back toward normal, but not by producing myogenic contraction of the cerebral vessels. E is false. The Cushing reaction does increase vasoconstriction in many vascular beds in order to increase mean arterial pressure by increasing total peripheral resistance, but fortunately the cerebral circulation does not participate in this vasoconstriction—this would be self-defeating.

14. A. Pa_{CO_2} is the principal blood-borne factor regulating CBF. Pa_{O_2} has only a minor effect. Blood-borne nonvolatile acids (acids other than H_2CO_3), such as lactic or pyruvic acids and the H^+ that makes them acidic do not directly affect CBF.

15. A. Changes in blood H^+ not accompanied by changes

in ventilation and Pa_{CO_2} have no effect on CBF. These H^+ concentration changes also have no direct influences on cerebrovascular accidents, nor do they produce symptoms of cerebral ischemia.

16. D. The hand–finger portions of both the primary and supplementary motor areas would be expected to show increased activity, as evidenced by increased rCBF to these areas. Feedback information from proprioceptive sensors concerning position and movement of the fingers would also activate corresponding somatosensory areas. It has been suggested that the planning and programming of movements involves the supplementary motor area, execution of the movements involves the primary motor area, and the feedback information necessary for control involves the somatosensory areas.

16

Skeletal Muscle Circulation

DONALD W. STUBBS

OBJECTIVES

After completion of this chapter, you should be able to

1. Describe the changes in blood flow and oxygen extraction in active skeletal muscle, which support a 50-fold increase in aerobic muscle metabolism during maximal steady-state exercise.

2. Relate the functional differences between white and red skeletal muscle to their biochemical differences.

3. Explain what is meant by a state of inherent basal tone in the smooth muscle of arterioles and precapillary sphincters in skeletal muscle.

4. Define the extrinsic and intrinsic mechanisms that regulate blood flow in skeletal muscle vasculature.

5. Elucidate the specific mechanisms by which the extrinsic and intrinsic regulatory systems operate.

6. Describe the circumstances under which the intrinsic mechanism predominates and those under which the extrinsic mechanism predominates.

Skeletal muscle, comprising about 40% of the body weight (28 kg in a 70-kg adult), is by far the most abundant tissue of the body. Under basal conditions—when the person is comfortably at rest and awake—about one-fourth of the total oxygen consumption of the body is used by resting muscle (about 60 ml/min). However, with maximal effort during severe steady-state exercise (steady-state exercise is a level of exercise in which oxygen can be supplied to muscles at a rate sufficient to meet aerobic energy requirements) an average person consuming a maximum of 3.5 liters oxygen/min has 90% of total oxygen consumption being used by muscle (about 3.2 L/min). Thus, the metabolic activity of skeletal muscle varies over a wide range, about 50-fold, between rest and maximal steady-state exercise.

At rest, venous blood from skeletal muscle contains about 15 ml O_2/100 ml blood (15 vol%), which represents a 25% extraction of the oxygen content originally present in the arterial blood. With maximal steady-state exercise, the venous effluent from active muscle has only about 2 ml O_2/100 ml blood (2 vol%), which represents a 90% extraction of oxygen content. However, this 3.6-fold increase in oxygen extraction only partly supports the 50-fold increase in oxygen requirements by muscle in maximal steady-state exercise. By far the greatest increase in availability of oxygen to muscle is that provided by increased blood flow. The flow of blood to skeletal muscle can increase about 14-fold from 1.4 L/min at rest to 20 L/min with maximal exercise in the average person. Thus, the range of blood flow through the vascular bed of muscle is quite large because the range of aerobic metabolism in muscle is quite large—greater than in other tissues.

FUNCTIONAL ANATOMY

Arterioles divide off from the network of anastomosing arteries supplying skeletal muscle. These arterioles branch off with considerable regularity, about every millimeter or so, and at right angles to the longitudinal direction of the muscle fibers. Capillaries from the arterioles run parallel with the muscle fibers, with which they have a ratio of approximately 1–1, as in myocardium. All veins from skeletal muscle, including the smallest tributaries, have valves that permit flow only in the direction of the heart.

DIFFERENCES BETWEEN WHITE AND RED MUSCLE FIBERS

Meat eaters have long been aware of differences in the color of various muscles. The traditional question asked of each guest at Thanksgiving dinner is, "White or dark meat?" Hunters have noted the leg and back muscles of European rabbits are pale or white, whereas those of hares are red. These differences in muscle color, attributable to the presence or absence of myoglobin, are associated with differences in function. The white fibers of pale muscle are capable of rapid and/or intense contractile activity, but such activity cannot be sustained for a long period of time. The red fibers of dark muscle are capable of less rapid or less intense activity, but can maintain their activity over long periods of time. (See Figure 1). In humans most skeletal muscles have mixtures of white fibers (the majority) and red fibers (the minority). The percentage of white fibers is slightly higher in females than in males. Although white fibers are capable of greater strength than are red fibers, females are not as physically strong as males because of the smaller bulk of female musculature. Muscle biopsies from the gastrocnemius muscle in well-trained runners show a higher than normal proportion of white fibers in sprinters, but a higher than normal proportion of red fibers in endurance (marathon) runners.

BIOCHEMICAL DIFFERENCES BETWEEN WHITE AND RED MUSCLE FIBERS

White muscle fibers are often called fast-twitch, glycolytic muscle fibers. These fibers, which are specialized for brief but intense activity, require periods of rest to restore their stores of glycogen and phosphocreatine. During bursts of activity, the rate of expenditure of total energy can temporarily greatly *exceed* that of energy produced aerobically from oxidative metabolism. Under these circumstances,

Figure 1. Correlation between ecological adaptation and physiological function of skeletal muscle fiber types in European rabbits and hares. When necessary to flee for life from a predator (such as a fox), the rabbit and hare depend upon different physiological strategies. The rabbit, which has primarily pale, fast-twitch, glycolytic skeletal muscle fibers, tries a fast sprint to a nearby burrow or hiding place, while the hare, which has a greater proportion of red, slow-twitch, oxidative skeletal muscle fibers, depends upon his endurance and attempts to outlast the pursuing predator (along with evasive zig-zagging).

energy is derived primarily from the breakdown of phosphocreatine and ATP from anaerobic glycolysis. Since the bursts of activity do not depend on the highly efficient mechanisms of oxidative metabolism within mitochondria, the number of mitochondria in white fibers is few and the volume that would otherwise be occupied by mitochondria is occupied by contractile proteins, adding to the strength of these muscle fibers. During intervening periods of rest (e.g., the rabbit in its burrow after successfully outsprinting a fox), any accumulated lactate from anaerobic glycolysis is oxidized and ATP and phosphocreatine stores replaced. Because of the increased requirements of oxygen in the postexercise period, over and above the baseline level of oxygen consumption in the resting pre-exercise muscle, which is needed for restoration of

high energy phosphates and for lactate oxidation, the muscle receives a supply of blood greater than the resting level of blood flow that existed before the bout of exercise. The increased flow of blood is said to be repaying an *oxygen debt.* This oxygen debt is incurred during the period of anaerobic energy use and is repaid during the postexercise period of rest. Being free of the immediate need to balance the rate of aerobic energy supply to total energy use at the time of contraction, white fibers have a high rate of adenosine triphosphate (ATP) hydrolysis, permitting a high velocity of shortening, hence the high rapidity of the twitch.

Red muscle fibers are often called slow-twitch, oxidative muscle fibers. In these fibers, which are specialized for sustained activity without rest periods, it is essential that the blood-dependent supply of oxygen for oxidative metabolism match the rate of energy use during contraction. Because of this dependence on oxidative metabolism, red fibers have an abundance of myoglobin to facilitate transfer of oxygen from oxyhemoglobin of blood to the muscle mitochondria. ATP is produced in large quantities in red fibers by the highly efficient pathways of oxidative metabolism of glucose and fatty acids to CO_2 and H_2O. However, the apparent luxury of a high rate of ATP production is not without cost. The price that red fibers must pay is threefold: (1) these red fibers are dependent on an uninterrupted supply of oxygen from blood, (2) they must contain large numbers of bulky mitochondria that occupy space, and (3) the rate of ATP hydrolysis must not be excessive (i.e., is lower than that in white fibers to help maintain the balance between energy utilization and oxidative energy production). The net results are a lower velocity of shortening and a reduced force of contraction.

The differences between white and red fibers were concisely summarized by Mommaerts, who said that white fibers operate on a "twitch now, pay later" plan, whereas red fibers are strictly "pay as you go."

PHYSIOLOGICAL DIFFERENCES BETWEEN WHITE AND RED MUSCLE FIBERS

Because red muscle fibers have a greater dependence on their blood supply, it should not be surprising to find that red fibers receive a greater flow of blood than do white

Table 1 White Versus Red Skeletal Muscle Fibers

Parameter	White Fiber	Red Fiber
Descriptive name	Fast-twitch, glycolytic	Slow-twitch, oxidative
Distribution in human muscle	Majority of fibers	Minority of fibers
Contractile properties	Greater strength, less endurance	Greater endurance, less strength
Oxygen debt	Can sustain a considerable O_2 debt, less dependent on blood supply	Cannot sustain much of an O_2 debt, more dependent on blood supply
Resting blood flow	3 ml/100 g/min	20 ml/100 g/min
Maximal blood flow	60 ml/100 g/min	150 ml/100 g/min
Blood flow with maximal vasoconstriction	15% of resting flow (666% increase in resistance)	60% of resting flow (66% increase in resistance)

fibers. The resting level of blood flow to red fibers is about 20 ml blood/100 g fibers/min, whereas that to white fibers is a mere 3 ml/100 g/min. With severe exercise, the flow of blood to red fibers can increase up to a maximum of about 150/ ml/100 g/min, whereas the flow to white fibers reaches a maximum of only 60 ml/100 g/min. That is, the intrinsic mechanism of regulating blood flow provides active red fibers with a much greater supply of blood than that for active white fibers. This greater dependency of red fibers on their blood supply is also reflected by the lesser extent to which the extrinsic neural mechanism can decrease blood flow to nonexercising muscle. The flow to inactive red fibers can only be decreased to 60% of the normal resting level, whereas that to inactive white fibers can be decreased to a mere 15% of normal. Table 1 summarizes the differences between these two types of skeletal muscle fibers.

PHYSIOLOGICAL ADJUSTMENTS IN BLOOD FLOW TO SKELETAL MUSCLE

The vascular bed of skeletal muscle exhibits a *dual control* of blood flow to a greater extent than do other vascular beds. Flow can be regulated by either extrinsic or intrinsic mechanisms. Extrinsic control is only significant in nonexercising muscle. Extrinsic control is that exerted by vasomotor fibers of the sympathetic nervous system. Because of the considerable dimensions of the vasculature of the most abundant tissue of the body, skeletal muscle vasculature in nonexercising muscle can play a significant role in vasomotor reflexes concerned with the overall control of cardiovascular function (see Chapter 13, Reflex Control of the Circulation). For example, extrinsic control can be the predominant influence of blood flow to nonexercising muscles when it is necessary for increased sympathetic adrenergic activity to increase the total peripheral resistance. On the other hand, when muscle becomes active, local intrinsic control predominates. This local regulation in exercising muscle is attributed to release of vasodilator metabolites, which are produced in proportion to the level of metabolic activity. The 14-fold increase in blood flow through the skeletal muscles with maximal exercise, discussed above, is entirely the result of this intrinsic mechanism, which operates independently of neural control.

BASAL TONE

The vascular bed of skeletal muscle has a considerable capacity for increased blood flow when maximally dilated, that is, when resistance is lowest. If the resistance to flow in the resting subject were diminished to this same extent, the mean arterial pressure would be drastically decreased and cause syncope, from inadequate cerebral perfusion pressure. It is therefore essential that resistance be maintained, that is, the skeletal muscle vascular bed must have a substantial vasomotor tone under basal conditions—basal tone. Only part of this basal tone is the result of sympathetic adrenergic activity; the principal factor is the strong inherent tone of the vascular smooth muscle of arterioles and precapillary sphincters. Interruption or blockade of the sympathetic fibers to skeletal muscle resistance vessels only doubles flow, that is, decreases resistance by one-half. On the other hand, arterial injection of vasodilator substances (ATP, ACh) can markedly reduce basal tone and increase flow severalfold. Thus, basal tone represents the normal background state of vascular resistance around which vasodilation and vasocontriction can occur. The vascular resistance under any given set of circumstances is the result of the extent to which vasomotor effects are added to or substracted from this basal tone.

CHANGES IN MUSCLE BLOOD FLOW WITH EXERCISE

When a muscle is active, the blood flow through that muscle rapidly increases. A single brief contraction is almost immediately (within a second) followed by an increase in blood flow. With repetitive or rhythmic contractions, the changes in flow are also rhythmic (Fig. 2). Whereas blood flow is less during a contraction than between contractions, it is usually greater than the flow preceding exercise.

ROLE OF THE INTRINSIC MECHANISM IN ACTIVE MUSCLE

The most important mechanism for increased blood flow in exercising muscle is the intrinsic metabolically linked

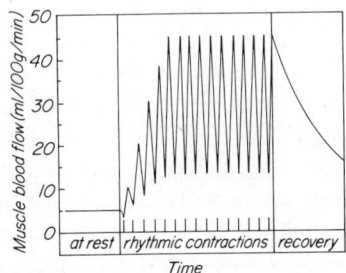

Figure 2 Diagrammatic representation of changes in blood flow in the calf muscles of the human leg during strong rhythmic contractions.

mechanism, that is, local vasodilation of arterioles and precapillary sphincters in response to increased metabolic activity. In resting muscle, only 10–20% of the total number of capillaries are thought to be open to flow. With increased metabolic activity from contraction, the number of capillaries open is increased (by opening of more precapillary sphincters), and the resistance of flow in arterioles is decreased. That metabolically linked vasoactive chemicals might cause local regulation of blood flow in exercising muscles was first suggested by Walter Holbrook Gaskell in 1877, in the first quantitative measurement of flow changes consequent to muscular activity (active hyperemia). The essential validity of this idea has been confirmed and enhanced by numerous subsequent physiological investigations and appears to be applicable to intrinsic regulation of blood flow in a number of tissues. In skeletal muscle, decreases in the local P_{O_2} and increases in the concentrations of vasodilator metabolites resulting from increased metabolic activity with contractions are thought to interact directly with the smooth muscle of arterioles and precapillary sphincters to cause vasodilation, in turn increasing blood flow to a level appropriate to the metabolic needs of the active muscle. Convincing evidence for this comes from experiments in which a first set of limb muscles has its venous blood diverted into the arterial supply of a second set of muscles from a different limb. When the first group of muscles is stimulated to contract, vasodilation is elicited in the second set of muscles, even though they are inactive.

No single chemical factor is likely to be solely responsible for metabolically linked vasodilation in active muscle. Increments in blood flow to active muscle are paralleled by decreases in (1) venous P_{O_2} and (2) venous pH, and by increases in (1) venous K^+ concentrations; (2) venous adenosine, inosine, and adenine nucleotides (ATP, ADP, and AMP); (3) venous P_{CO_2}; and (4) venous osmolarity. Individually, none produces large enough effects in maximal physiological concentrations to be solely responsible for all the vasodilation that occurs. It seems likely that it is the additive or synergistic effects of several or all of these factors that produces the total level of vasodilation that results from interaction with the basal tone. For example, it is known that decreased P_{O_2} potentiates the vasodilator response to K^+ and that these effects are further potentiated by increases in local osmolarity (Fig. 3).

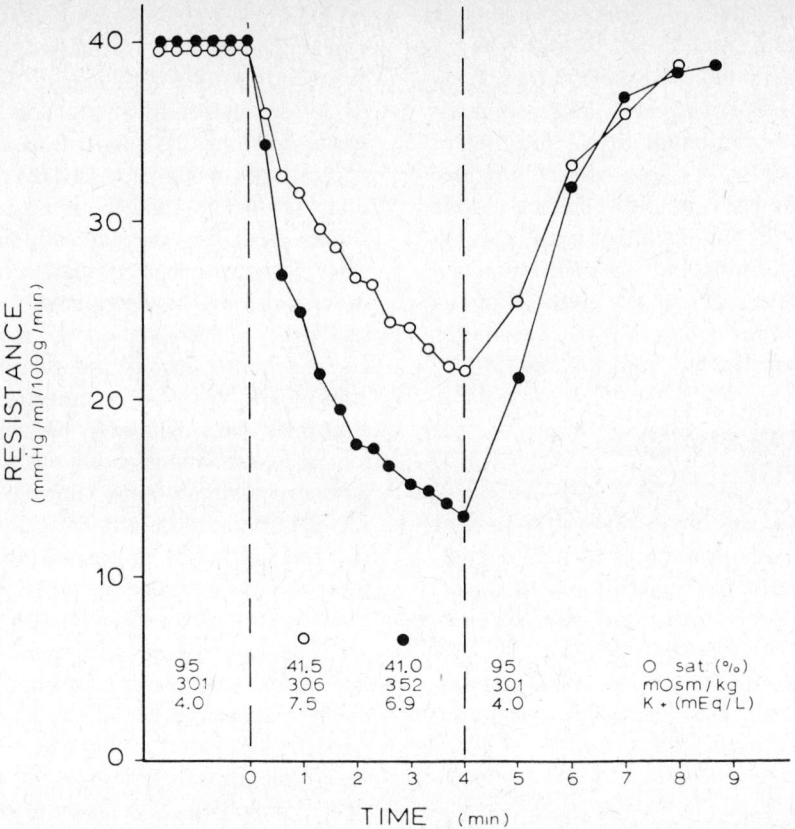

Figure 3 Effect of combined hypoxia, hyperkalemia, and hyperosmolarity (filled circles) and of only hypoxia with hyperkalemia (open circles) on muscle vascular resistance. (From Skinner NS Jr, Costin JS: "Interaction of vasoactive substances in exercise hyperemia: O_2, K^+, and osmolality," *Am J Physiol* 219:1386, 1970.)

ROLE OF THE EXTRINSIC MECHANISM IN ACTIVE MUSCLE

SYMPATHETIC CHOLINERGIC VASODILATOR FIBERS

With initiation of voluntary muscular activity in some experimental animals, the cerebral motor cortex simultaneously activates a neural pathway (with a relay in the hypothalamus) that causes increased discharge of sympathetic cholinergic vasodilator fibers to skeletal muscle.

Whereas such extrinsic vasodilation at the onset of muscular activity might augment local intrinsic vasodilation, it is not thought to play a very important role. In fact, no discernable differences in vasodilation are found in comparing normally innervated with sympathetically denervated active muscles. In unanesthetized dogs the vascular resistance of skeletal muscle in a sympathectomized leg decreases with muscular activity to the same extent as that in a normal contralateral leg at all levels of exercise (i.e., from mild to severe). That hyperemia of muscular activity does not depend on sympathetic cholinergic vasodilation is also true of humans. Patients with sympathectomized limbs have no difficulties with any form of normal exercise. Barcroft (1963) reported an interesting example of a policeman who had to undergo a lumbar sympathectomy. Before the operation, the policeman was able to run a distance of 380 yards in an average of 63 sec. Ninety-nine days after the sympathectomy he ran the same distance in an average of 61 sec, clearly not a significant difference. There is considerable doubt as to whether a sympathetic cholinergic system to skeletal muscles is even functional in humans and other primates. In any case, there is no doubt that for active hyperemia of exercise, the intrinsic mechanism is alone sufficient.

SYMPATHETIC ADRENERGIC VASOCONSTRICTOR FIBERS

Conceivably, a degree of vasodilation in actively contracting muscles might be brought about by a diminution of sympathetic adrenergic vasoconstrictor discharge—such diminution is the principal means of neural vasodilation in most vascular beds. However, as discussed above, the resistance vessels of skeletal muscle have a pronounced inherent myogenic basal tone independent of innervation, so that decreased sympathetic adrenergic discharge to muscle vasculature has only a moderate effect. Furthermore, with exercise there is a generalized increase in sympathetic adrenergic vasoconstrictor activity to both resting and contracting muscles, and to numerous internal organs as well. The powerful local intrinsic mechanism for vasodilation easily overcomes this vasoconstrictor influence in the active muscle. The result is a greatly diminished flow in the inactive muscles and an increased flow in the active muscles. In other words, there is a redistribution of blood to those muscles that have the need for increased circulation.

HUMORAL ADRENERGIC VASODILATION

The general vasoconstrictor effect of adrenergic nerve fiber activity is the result of the action of norepinephrine on α-vascular receptors. However, muscle vasculature also has many β-receptors. The effect or result of β-receptor activity is vasodilation. Norepinephrine, either from injection or from endogenous release from adrenergic neurons, stimulates both α- and β-receptors, but to a different degree. The principal effect is on the α-receptors, hence the vasoconstriction. Norepinephrine stimulation of β-receptors does occur, but only to a minor extent—not enough to oppose significantly the vasoconstrictor effect mediated by the α-receptors. Therefore, neural adrenergic stimulation of skeletal muscle vascular β-receptors by norepinephrine does not appear to be of any obvious functional significance. However, norepinephrine is not the only adrenergic agent capable of acting on the receptors of skeletal muscle vasculature. A blood-borne adrenergic agent is epinephrine, which is secreted by the adrenal medulla. The action of epinephrine on the β-receptors of skeletal muscle vasculature is quantitatively different from that of norepinephrine. Epinephrine also significantly stimulates both α- and β-receptors. However, stimulation of β-receptors by epinephrine is not to a minor extent, as is the case with norepinephrine. Low concentrations of epinephrine, such as might result from secretion by the adrenal medulla, administered either intravenously or intra-arterially, produce a transient vasodilation of skeletal muscle resistance vessels, thereby increasing the flow of blood through muscle (Fig. 4). This transient increase in flow is attributed to effective stimulation of the β-receptors, which can be said to have a lower threshold for epinephrine than for norepinephrine. This humorally mediated adrenergic vasodilation in skeletal muscles can be of significance during the immediate anticipation or onset of exercise, particularly where emotion is involved (e.g.,

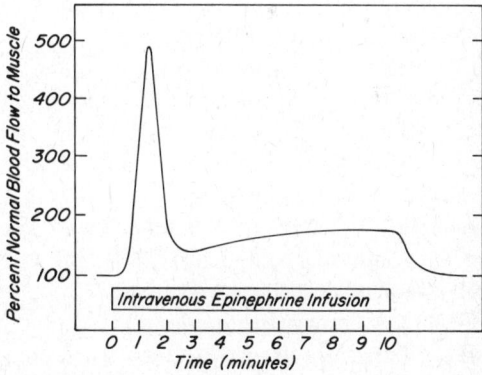

Figure 4. Epinephrine-induced vasodilation of skeletal muscle resistance vessels. Intravenous infusion of epinephrine produces a transient increase in blood flow to skeletal muscle due to the action of epinephrine on arteriolar beta receptors.

excitement of competition, fear of excessive body contact, anger at the jeers and slurs from opponents). Emotional stress at the beginning of exercise increases secretion of epinephrine from the adrenal medulla. "Like greyhounds straining in the slips" (Shakespeare, *Henry V*), the command "ready" to a sprinter at the starting blocks gets the circulation going, that is, increases cardiac output and increases flow to muscles. Although epinephrine secretion continues in heavy exercise, it is doubtful that it continues to be of further importance to muscle vasculature. Even if the initial vasodilation were not transient, it would still be small in comparison with the overwhelming active hyperemia of muscular exercise due to the local intrinsic mechanism.

ATHEROSCLEROTIC ISCHEMIA OF SKELETAL MUSCLE

The same pathological processes that lead to atherosclerotic changes in coronary and cerebral arteries can also affect arteries supplying skeletal muscles, particularly the larger arteries supplying leg muscles. However, because of the rich collateral circulation in most skeletal muscles, infarction of muscle tissue occurs relatively infrequently— necrotic infarction occurs only if the collateral circulation is also compromised. Gradual narrowing of the arterial lumen results in a chronic development of muscle ischemia, particularly evident during muscular activity, when there is a greater demand for circulation. Ischemic symptoms will not occur if the capacity for increased circulation to muscle provides for at least a 10-fold increase in blood flow. The characteristic symptom of exercise ischemia is *intermittent claudication* ("recurrent limping"): a cramp-like pain along with fatigue, occurring in one or several muscle groups, invariably associated with increased muscle activity (e.g., walking). The pain usually goes away within a minute or two after resting, but returns with resumption of muscular activity, hence the term intermittent. A unique feature of intermittent claudication is the presence of a latent period before the onset of pain; the patient can walk a certain distance before pain and

fatigue develop. The shorter the walking distance before these symptoms appear the more severe the occlusive disease.

If the terminal abdominal aorta is occluded, the pain of intermittent claudication is bilateral and can occur in buttocks, thighs, and calves. If the occlusion is in the femoral artery or its branches, pain and fatigue are more localized. With few exceptions, pain begins in the most distal muscle mass, most frequently in the calf.

Oxygen deprivation, as a result of the inadequate blood flow, is an important factor in the pain of intermittent claudication. However, it is probably not the direct result of a decrease in P_{O_2} per se, but is indirectly attributable to ischemic production of an as yet unidentified pain factor. The normal removal of this pain factor appears to depend on oxidative processes.

Acute occlusion of the arterial supply to skeletal muscle can result from either local thrombosis or embolism. Sudden occlusion results in ischemic muscular contracture within a few hours. The most serious aspect of acute occlusion is not contracture, however, but rather derives from the potential for development of gangrene and the subsequent need for amputation of the affected limb.

SUMMARY

Skeletal muscle is the most abundant tissue of the body. It receives a substantial proportion of the total cardiac output—about one-fourth, even in a resting subject. With exercise, the blood flow to active muscles increases, as a result of both an increase in cardiac output and the proportion of cardiac output going to muscle. This increased blood flow along with an increase in oxygen extraction is necessary to supply the O_2 requirements of oxidative metabolism. Human muscles contain both white and red muscle fibers. Most are white fibers, which are capable of both rapid and intense contraction. Less abundant are red fibers, which are neither as fast nor as strong, but are capable of longer endurance. The white fibers are capable of operating for a time at a level of energy consumption in excess of that supplied by oxidative metabolism, resulting in the production of an oxygen debt, and are therefore not as dependent on their blood supply. Red fibers are unable to operate for long at a level of energy consumption in excess of that supplied by oxidative metabolism (i.e., cannot sustain much of an oxygen debt) and are therefore more dependent on their blood supply. Because of their greater dependence on the blood supply, red fibers receive a greater blood flow per unit weight than do white fibers.

Although varous neurohumoral mechanisms can alter the flow of blood in nonactive muscles (e.g., vasoconstriction from adrenergic sympathetic stimulation, and vasodilation from either cholinergic sympathetic stimulation or increased epinephrine release from the adrenal medulla), the principal mechanism for regulating blood flow in active muscle is by the metabolically influenced intrinsic mechanism. The mediators of this intrinsic control, which are probably several in number, can act together synergistically to produce local changes in O_2, CO_2, H^+, K^+, adenosine, and osmolarity.

Chronic reduction in the supply of blood to skeletal muscle, most often to leg muscles, particularly the inability to increase blood flow adequately during exercise, results in an ischemic pain called intermittent claudication. Intermittent claudication usually occurs after a latent period following the onset of muscle activity (e.g., after walking a certain distance).

SELECTED BIBLIOGRAPHY

Barcroft H: Circulation in skeletal muscle, in *Handbook of Physiology.* Section 2: *Circulation,* vol 2. Baltimore, Williams & Wilkins, 1963, pp 1353–1385.

Folkow B, Neil E: *Muscle Circulation in Circulation.* New York, Oxford University Press, 1971.

Katz AM: *Physiology of the Heart.* New York, Raven Press, 1977.

Rowell LR: Circulation of skeletal muscle, in Ruch TC, Patton HD (eds): *Physiology and Biophysics,* vol II. Philadelphia, Saunders, 1974.

SELF-STUDY QUESTIONS

Matching

Questions 1–6

A. *Increases with increasing levels of exercise*
B. *Decreases with increasing levels of exercise*
C. *Does not change with increasing levels of exercise*

1. Cardiac output.

2. Blood flow (ml/min) to heart (coronary circulation).

3. Blood flow (ml/min) to the brain (cerebral circulation).

4. Blood flow (ml/min) to viscera (splanchnic circulation).

5. Blood flow (ml/min) to active skeletal muscles.

6. Milliliters of oxygen extracted from each 100 ml of blood passing through active skeletal muscle.

Questions 7–12

A. *White skeletal muscle fiber*
B. *Red skeletal muscle fiber*
C. *Both*
D. *Neither*

7. Best adapted to short bursts of rapid or intense activity, or both.

8. Best adapted to maintain sustained activity.

9. Use(s) ATP as immediate source of energy for contraction.

10. Resting blood flow of 3 ml/100 g/min.

11. Extrinsic vasodilation by increased activity of parasympathetic cholinergic nerve fibers.

12. Leg muscles of humans.

Questions 13–15

A. *Increased activity of sympathetic adrenergic fibers*
B. *Decreased activity of sympathetic adrenergic fibers*
C. *Increased activity of sympathetic cholinergic fibers*
D. *Decreased activity of sympathetic cholinergic fibers*
E. *Increased local release of vasodilator metabolites.*

13. Predominant influence on blood flow in nonactive

skeletal muscle of an individual with hemorrhagic blood loss.

14. Predominant influence on blood flow in nonactive skeletal muscle of an individual whose carotid sinus reflex has been stimulated by a rise in mean arterial blood pressure.

15. Predominant influence on blood flow in contracting muscles.

ANSWERS

1. A. Cardiac output increases as the level of muscular activity increases.

2. A. The increased work of the heart in increasing the cardiac output with increasing levels of exercise results in increased coronary flow.

3. C. Blood flow to the brain remains relatively constant at most levels of exercise.

4. B. Blood flow to the viscera decreases with increased levels of exercise.

5. A. The largest increase in blood flow with increased levels of exercise is that to the active skeletal muscles (from about 5 ml/100 g/min at rest to nearly 75 ml/100 g/min with maximal steady-state exercise).

6. A. In addition to increased blood flow through active skeletal muscle, there is increased oxygen extraction —from 25% at rest to about 90% with severe exercise.

7. A. White muscle fibers are best adapted to short bursts of rapid or intense activity, or both, but soon require a period of rest to restore glycogen stores and repay oxygen debt.

8. B. Red muscle fibers usually function at a level of energy consumption equal to aerobic energy production and therefore can sustain this level for fairly long periods of time.

9. C. ATP is the immediate source of energy for all muscle fibers.

10. A. The low flow rate of only 3 ml/100 g/min is characteristic of resting white muscle fibers.

11. D. There are no known parasympathetic cholinergic fibers to the vasculature of skeletal muscle.

12. C. Our leg muscles, like most of our muscles, contain mixtures of both white and red skeletal muscle fibers. For the body as a whole, the proportion is estimated to be 85% white and 15% red.

13. A. The carotid sinus reflex is decreased in activity with hemorrhagic blood loss. This results in an increase in sympathetic adrenergic activity to many vascular beds including skeletal muscle. The increase in total peripheral resistance helps maintain the mean arterial blood pressure.

14. B. A rise in mean arterial blood pressure stimulates the carotid sinus reflex, resulting in inhibited sympathetic adrenergic activity to a number of vascular beds, including skeletal muscle. This decreases the total peripheral resistance and helps lower the blood pressure.

15. E. The predominant influence on blood flow in active skeletal muscle is that of the local intrinsic mechanism, which depends on release of vasodilator metabolites or decreased P_{O_2}, or both.

17

Microcirculation: Capillary Dynamics, Lymph Circulation, and Local Vascular Regulation

M. MASON GUEST

OBJECTIVES

After completion of this chapter, you should be able to

1. Recognize the structural and functional relationships within the microcirculation.

2. Describe the rheological properties of blood and their relationships to the microvasculature.

3. Identify the forces and factors involved in the exchange of substances between blood and the interstitial fluid— the Starling hypothesis.

4. Describe the alterations resulting in edema formation, and identify the responsible mechanisms.

5. Describe how and explain why the composition of lymph differs from the composition of blood.

6. Relate the structural and functional aspects of the lymphatic system.

7. Evaluate the physiological role of the local vascular control mechanisms and the relationships of local vascular control to the more generalized neural and endocrine control of blood flow and blood pressure.

This segment of our study of the physiology of the heart and circulation is concerned with the functional features of the microcirculation and the lymphatic system.

The microcirculation consists of the blood flowing in the terminal arterioles, metarterioles, capillaries, and venules. The lymphatic system, composed of closed-ended endothelial tubes and larger collecting vessels with elastic and smooth muscle investments, is functionally interrelated with the microcirculation.

The *raison d'être* of a circulation is to maintain homeostasis, that is, a relatively constant internal environment containing adequate substrate for the survival and well-being of the cells composing the organism. The exchange between the blood and the interstitial fluid, which constitutes the immediate environment of the cells, takes place from and to capillaries and venules. Other parts of the circulation, including the heart, arteries, and veins, function solely to deliver blood to the capillaries and venules at an adequate rate and sufficient pressure to ensure that the critically essential exchange of materials across the blood vessel wall occurs without major interruption.

The lymphatic system functions in part as an adjunct to the microcirculation. The quantity of fluid from capillaries and venules that enters the interstitial space exceeds the amount that returns to the capillaries and venules from the interstitial space. Hence, more fluid leaves the microcirculation than is returned directly to the microcirculation. The lymph capillaries take up the excess interstitial fluid, and the lymph vessels empty the fluid into veins, primarily central veins in the neck region. In addition, lymph vessels return large colloid molecules (i.e., protein), as well as formed elements (erythrocytes and leukocytes) to the circulatory system.

APPROXIMATE DIMENSIONS OF THE MICROCIRCULATORY COMPONENTS*

Do not memorize the numbers in Table 1. However, you should have a concept of the approximate size of the microcirculatory structures and of the order of magnitude of the rheologic quantities. The following dimensions are for a medium-size dog, with a 1-cm-diameter aorta.

*Based on data from Mall (1888) and Weideman (1963); modified from Schmid-Schonbein (1976) (see Selected Bibliography).

Table 1 Approximate Dimensions of Microcirculatory Components

	Small Arteries	Arterioles	Capillaries	Venous Capillaries[a]	Venules
Diameter (lumen) (μm)	19	7	3.7	7.3	21
Tissues in wall	Endothelium, smooth muscle, collagen, elastin	Endothelium, smooth muscle, collagen, elastin	Endothelium only	Endothelium only	Endothelium, infrequent smooth muscle, collagen
Length (mm)	3.5	0.9	0.2	0.2	1.0
Total cross-sectional area (cm²)	110	150	180	2150	3700
Percent of total blood volume	2.7	1.0	0.3	3.6	25.6
Mean pressure (mm Hg)	76	56	25	4.5	4.1
Approximate velocity in active vessels (mm/se)	5	4	3	0.3	0.2
Diameter (equatorial) of human or canine red blood cell = 7.5μm.					

[a] Also termed "postcapillary venules." Arterial capillaries, usually called simply "capillaries," branch to form the venous capillaries. There are about three venous capillaries for each arterial capillary. Each venous capillary is somewhat larger in diameter than an arterial capillary. Often when the term "capillary" is used, it refers to both types.

DIAGRAMMATIC SUMMARY OF CERTAIN IMPORTANT FEATURES OF THE CIRCULATORY SYSTEM

For purposes of orientation, three diagrams taken from various sources are presented. The first, Figure 1, is a schematic diagram of the parallel and series arrangement of the vessels comprising the circulatory system.

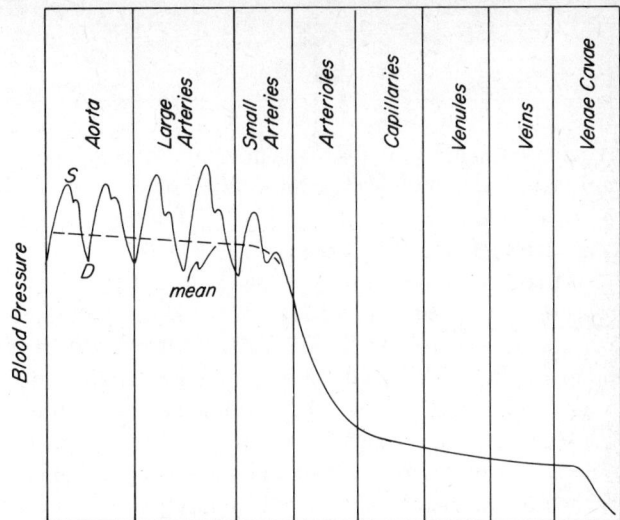

Figure 2 The most rapid fall in pressure, associated with loss of the pressure pulse, occurs in the arterioles.

The second schematic diagram, Figure 2, shows the blood pressure in the various parts of the systemic circulation.

The third schematic diagram, Figure 3, depicts the effect of vasomotor tone on blood flow.

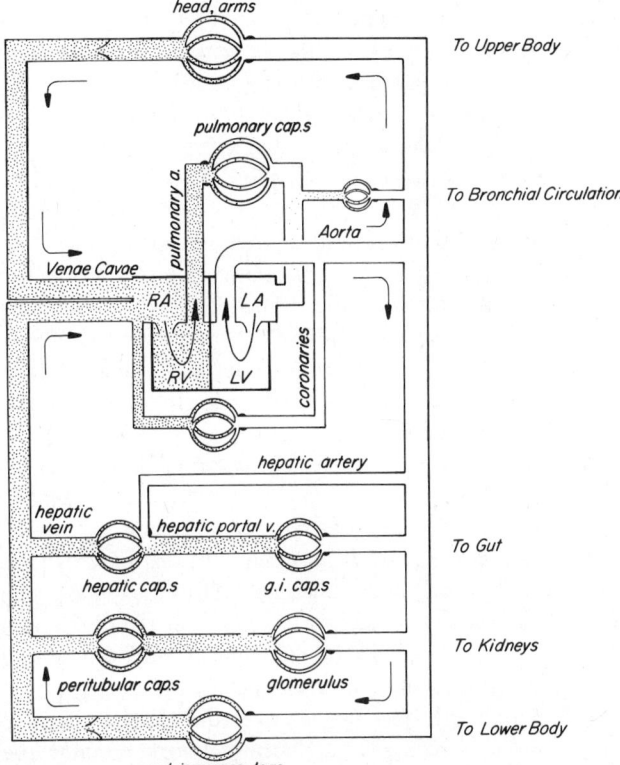

Figure 1 A simplified schematic representation of the circulatory system showing series and parallel pathways between the aortic outflow (from the left ventricle) and the venous return (through the venae cavae to the right atrium). Deoxygenated blood (venous blood in systemic circulation and arterial blood in pulmonary circulation) is indicated by the stippled area. The variable resistances of the arterioles (the principal resistance vessels) are illustrated as black semicircular thickenings proximal to the capillary beds.

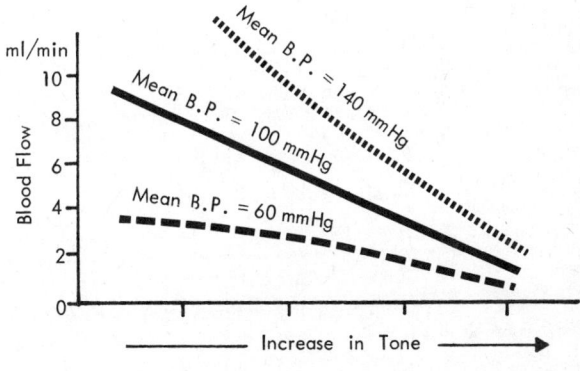

Figure 3. Effect of vasomotor tone on blood flow. At a given pressure, flow of blood through organs and tissues is dependent primarily on the tone of the smooth muscles in its blood vessels. The tone is modified by neural influences, blood-borne substances, and local influences (see Fig. 9).

THE MICROCIRCULATORY UNIT

A microcirculatory unit is made up of an arteriole, metarteriole(s), capillaries, and venules. Figure 4 depicts a representative unit. The precise arrangement of the microvessels varies in different tissues.

Metarterioles differ from arterioles in that the smooth muscle investment in metarterioles is discontinuous, whereas it is continuous in arterioles. In many tissues, as indicated in Figure 4, the metarteriole serves as a thoroughfare channel between the arteriole and the venule.

Precapillary sphincters are a feature of the microvasculature in most tissues. These sphincters are functionally related to the metarterioles; the capillary per se is a tube formed by interdigitating endothelial cells and contains no smooth muscle investment. Therefore, it cannot contract actively.

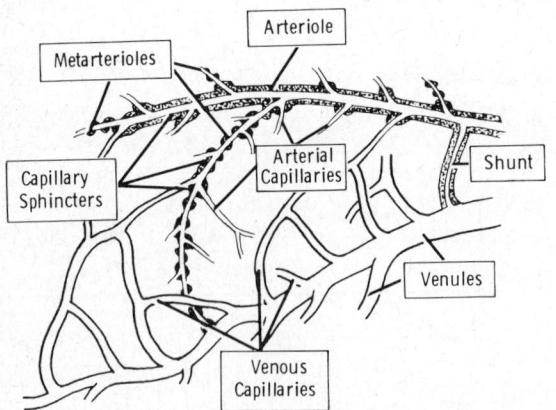

Figure 4 Schematic drawing of microcirculation. Circular structures on arteriole, metarteriole, and venule represent smooth muscle fibers ng solid lines represent sympathetic nerve fibers.

CONTROL OF FLOW AND PRESSURE IN MICROCIRCULATION

Capillaries, venous capillaries, and most venules are passive conduits with respect to the movement of blood from arteries to veins. These microvessels are endothelial tubes, devoid of smooth muscle. The only possible ways in which the size of the lumens of these tiny tubes could be altered would be by swelling or shrinking of the endothelial cells or by marked changes in the transmural pressures. From direct observation and cinephotographic recording of the microcirculation, it appears that the luminal diameters of capillaries and venous capillaries remain relatively constant during normal physiological conditions.

Assuming an adequate cardiac output, since capillaries cannot actively change their diameters, the quantity of blood flowing through a given capillary–venous capillary network is dependent on the relative amount of shortening of smooth muscle in arterioles, metarterioles, and precapillary sphincters preceding the network and in veins into which the network empties. The degree of shortening (contraction) of smooth muscle is controlled by the sympathetic outflow from the central nervous system (CNS), hormones such as the catecholamines, and local vascular influences (see below).

You will recall that Poiseuille's formulation for quantifying flow through tubes can be stated as $\dot{Q} = \Delta P/R$, where $\dot{Q}$ is the volume of flow per unit time, ΔP is the pressure difference, and R represents the resistances in the tube or system of tubes. When applied to a capillary network of a designated size, we can consider the resistances to be constant, since (1) the diameter of individual capillaries does not change appreciably, (2) the length of the capillary network does not change, and (3) the viscosity of the blood flowing through the network remains the same during the period under consideration. Hence the flow in the network $\dot{Q}$ varies directly with changes in the difference between the inflow and outflow pressures ΔP. A large ΔP (e.g., dilated arterioles, metarterioles, precapillary sphincters, and veins) results in a high volume of flow through the capillary network, whereas constriction either on the arterial side or of the veins, or both, will lead to a reduced ΔP and reduced capillary flow.

NEUROGENIC CONTROL

Sympathetic efferent nerve fibers, which have their preganglionic outflow along most of the length of the intermediate lateral cell column of the spinal cord (T1–L2), are controlled by hypothalamic and medullary centers and innervate smooth muscle cells in the walls of blood vessels. Those nerve endings that release norepinephrine cause contraction, and therefore vasoconstriction, when the receptors are α-adrenergic receptors. Some smooth muscle cells in blood vessels of skeletal muscles have β-adrenergic receptors; with appropriate stimulation of these β-receptors, the smooth muscles relax (vasodilation). Sympathetic cholinergic fibers also innervate blood vessels of skeletal muscle in carnivores. When these sympathetic fibers release acetylcholine (ACh) from their efferent terminations, the smooth muscles in the walls of the innervated vessels relax (vasodilation). Parasympathetic cholinergic vasodilator nerves supply salivary and some gastrointestinal glands, the coronary and cerebral circulations, and the external genitalia. Note that stimulation of the parasympathetic fibers to the heart and brain has only a slight—possibly insignificant—effect on blood flow. These parasympathetic nerves have their outflows from the CNS via cranial nerves VII, IX, and X and via the parasympathetic sacral nerves.

HORMONES

The catecholamines (epinephrine and norepinephrine) when released from the adrenal medulla in sufficient quantities, cause contraction of smooth muscle fibers in the walls of most blood vessels. Secretion of the adrenergic hormones usually occurs when the organism is under stress.

LOCAL CONTROL

Blood flow in many tissues is modulated by local factors that vary with the metabolic activity in the tissue. Local control usually takes precedence over neural and hormonal control. The term *autoregulation of blood flow* is sometimes used synonymously with local control of blood flow. However, autoregulation in the strict sense constitutes maintenance of essentially normal blood flow, despite changes in pressure in the artery supplying the organ or tissue.

VISCOSITY

As a variable characteristic of blood, viscosity has a major influence on flow in the microcirculation. You will recall that resistance to flow in tubes (Poiseuille formulation)

$$R = \frac{8l\eta}{\pi r^4}$$

is a function of the length of the tube, the viscosity, and the fourth power of the radius. The length of the tube and the radius are the tube factors. The viscosity, on the other hand, is a property of the fluid flowing through the tube.

In several diseases, but espcially those in which the hematocrit is high (polycythemias), viscosity can be a critical factor in the survival of the patient. A hypodynamic heart is frequently unable to maintain an adequate cardiac output, partly because of the increased viscosity of the blood that results from an abnormally high hematocrit value. A vicious circle develops (positive feedback) in which renal hypoxia with the release of erythropoietin creates a further increase in red blood cell (RBC) production.

The circulation of blood at a relatively low pressure and with the expenditure of only a small amount of energy is a remarkable engineering achievement, especially in view of the fact that a large fraction of the blood during each complete circuit passes through capillaries smaller in diameter than the erythrocytes. A precisely regulated flow of blood at a small P is possible only because the circulating fluid is not very viscous and because the erythrocytes are very flexible.

Viscosity was described by Sir Isaac Newton as the lack of slipperiness between adjacent layers of a fluid. The unit of viscosity is the poise, named in honor of Poiseuille. However, in recording viscosities of water and blood, the unit used is the centipoise (cP). One centipoise equals 0.01

poise. At the velocities of flow that normally occur in large arteries, filled with blood of a normal hematocrit, the absolute viscosity is 3–4 cP. The viscosity of plasma is about 1.5 cP and is independent of velocity of flow (see below).

Water and most other homogeneous fluids, including plasma, are newtonian, that is, they have a viscosity independent of shear rate. (Shear rate or rate of shear in a flowing fluid is the velocity gradient between adjacent layers or laminae of the fluid.) Whole blood, like many other heterogeneous fluids, is nonnewtonian (anomalous viscosity); its viscosity increases as the shear rate decreases. The greater the number of formed particles in blood (especially erythrocytes), the greater the increase in viscosity; this effect is most marked at low shear rates. However, in small vessels, that is, the microcirculation, the viscosity of blood at low shear rates appears to be less than would be predicted from measurements of viscosity done in a viscometer. This reduction in viscosity within microvessels when compared with the viscosity of blood measured at various shear rates in vitro is known as the sigma or Fahraeus-Lindqvist effect.

One explanation for the difference between the viscosity of blood as measured in a viscometer and its viscosity in microvessels is that the movement of blood cells and plasma occurs as a unit (slug flow) in the center of small blood vessels, while the shearing forces (friction between laminae) occur almost entirely in the plasma annuli between the central "slug" and the vessel wall. Since plasma is a newtonian fluid and its internal friction is relatively low, the viscosity in microvessels is lower than that measured in a viscometer, in which friction occurs throughout the heterogeneous medium (whole blood).

Measurements in a viscometer are more predictive of the internal frictional forces during flow in larger vessels, where friction between blood elements occurs essentially throughout the cross-sectional area of the blood within the vessels. However, in the large vessels shear rates are relatively high, and therefore viscosity is not augmented as much by an increase in hematocrit as it would be at low shear rates.

The internal (intracellular) viscosity of the formed elements has an effect on flow in the microcirculation. This is particularly so in capillaries with internal diameters less than the equatorial diameter of red blood cells. To traverse a capillary, a red blood cell must be flexible. The flexibility (deformability of an RBC) is a function of its low internal viscosity; this, in turn, is a function of the fluid nature of its contents and the fact that the surface area of the cell membrane is large with respect to the volume of the contents of the cell. When an RBC is released from the bone marrow, its nucleus is extruded. Loss of the nucleus results in reduction in the volume contained within the cell membrane; furthermore, the nucleus is much more rigid than the semi-fluid cytoplasm.

The mature erythrocyte assumes the convex disk shape, partly because its membrane is too large for its contents. The disk shape is easily altered by rheological forces because of the large membrane surface relative to volume and because of the low internal viscosity of the erythrocyte.

In fact, the convex disk shape is altered when the RBC enters a true capillary. To squeeze through a true capillary, having an internal diameter of about 3.5 μm, the RBC with its disk diameter of about 7.5 μm, must convert to some other shape. High-speed photography of the microcirculation demonstrates that erythrocytes convert into hollow paraboloids (parachute shape) as they traverse small diameter microvessels (Fig. 5).

Change in shape permitting entry into a true capillary is not possible if the cell is rigid (having an inflexible membrane, a high internal viscosity, or a content volume that maximally fills the space within the limiting membrane). Hence, rigid cells cannot traverse true capillaries. Rigid erythrocytes are characteristic of certain kinds of anemia (e.g., some hemolytic anemias and sickle cell anemia). Formation of rigid cells also follows bites from some venomous snakes. The rigidity of the abnormal RBC—especially in sickle cell crises, following venomous snake bites, and in some anemias—blocks the transit of erythrocytes through true capillaries, causing tissues to become hypoxic.

Note that leukocytes, white blood cells (WBC), are much more rigid than are normal mature erythrocytes. Their contents completely fill their membranes, ensuring the spherical shape. White blood cells are unable to traverse true capillaries because of their size and rigidity. Their transit to the venous side of the circulation can occur only through thoroughfare channels or shunts.

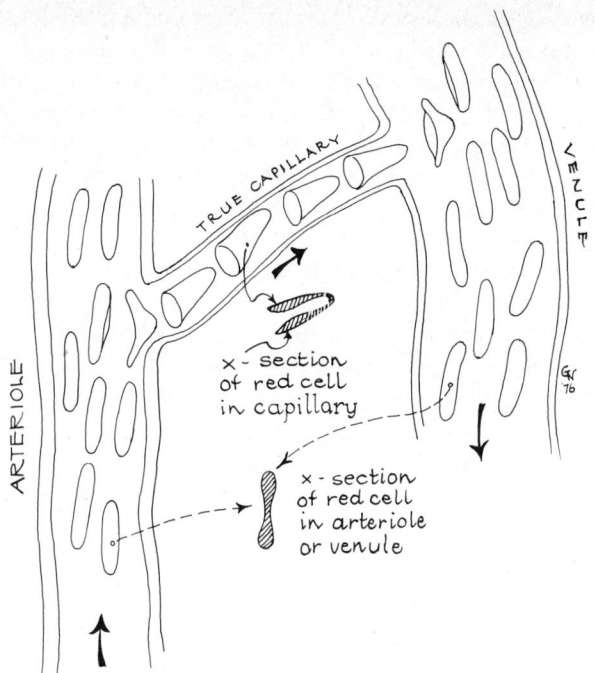

Figure 5 Diagram showing change in the shape of erythrocytes from biconcave disks to hollow paraboloids (parachute shape) as they enter an arterial or true capillary.

STRUCTURE OF CAPILLARY WALLS

The vessels through which exchange of substances between blood and interstitial fluid takes place, the arterial capillaries, venous capillaries, venules, and sinusoids, vary in structural characteristics from organ to organ (Fig. 6). For example, liver and spleen sinusoids have regions of discontinuity between endothelial cells; in striated muscle and CNS, the endothelial cells of capillaries are thin; in the dermis, thicker endothelial cells are found; and in kidney glomeruli, intestinal villi, many of the glands of internal secretion, and other organs, fenestrated capillaries have been described. Fenestrated capillaries have thin endothelial cells (200–400 Å) and a continuous basement membrane. The adjacent endothelial cells are tightly connected, and the cells are pierced with fenestrae or

windows, 0.1 μm or less in diameter, either open or closed by a diaphragm. Some capillaries have a continuous basement membrane; around other capillaries the basement membrane is discontinuous or absent.

Capillaries in skeletal muscle and the brain are composed of low (thin), continuous endothelial cells. The junctions between the endothelial cells in capillaries of skeletal muscles are not as tight as in brain capillaries. Both morphological and kinetic studies indicate the presence of interendothelial apertures or pores in capillaries of skeletal muscle having narrowest diameters of about 40 Å.

Each tissue has unique requirements for the delivery of nutrients and the removal of metabolic products. The vari-

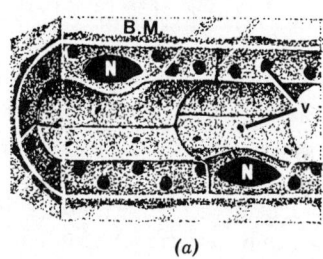

(a)

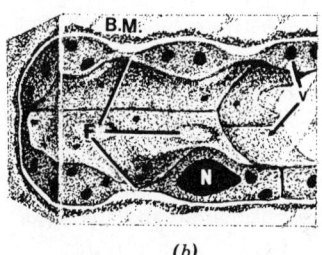

(b)

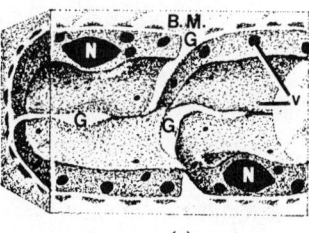

(c)

Figure 6 Schematic illustration of the different types of capillaries. (From Majno, 1965.)

ously structured walls of capillaries and venules in different tissues appear to be related to specific tissue requirements. A single mode of transfer between the lumen of all microvessels and surrounding interstitial spaces is unlikely.

The tighter junctions between endothelial cells in capillaries and venules of the cerebral circulation might explain in part the inability of certain substances to cross the blood–brain barrier.

EXCHANGE BETWEEN BLOOD AND INTERSTITIAL FLUID

Transfer of substances across the capillary wall occurs by diffusion, filtration, and possibly vesicular transport (pinocytosis). The precise routes of transport across capillary endothelium have not been unequivocally established.

Although the exact path taken by a low molecular weight substance in penetrating the capillary wall is uncertain, it is clear that the structure and composition of the capillary wall are the primary determinants of which substances exchange between blood and interstitial fluid and how rapidly the exchange takes place. The characteristic features of capillaries that make them better candidates for the site of exchange than other blood vessels include the following: (1) the capillary wall is as thin as a single endothelial cell, hence the relatively short diffusion or filtration path; (2) the luminal diameter of a true capillary is usually less than that of an erythrocyte, hence the small distance between any point within the capillary lumen and the intimal surface; (3) for a unit of length the ratio of intimal surface area to the enclosed volume is greater than in other vessels; and (4) the velocity of blood flow in capillaries is low, permitting time for exchange.

Diffusion of water and water soluble substances occurs primarily via pores or slits, either through the endothelial cells or between them. It is powered by heat; due to thermal energy all molecules and ions in the body fluids, including water and dissolved substances, are in constant motion. Hence, if a passageway is available—either a pore or a membrane in which the substance is soluble— particles will randomly move through. Water and its dissolved particles, if small enough to traverse the passageways (pores), move back and forth equally in both directions assuming that concentrations are the same on the opposite sides of the partial barrier. Most of the transfer across the capillary wall is by diffusion; this lateral diffusion amounts to about 40 times the amount of material moved linearly along the capillary; however, because it occurs nearly equally in both directions, the net loss of fluid by diffusion from the capillary is minimal. Only if a concentration gradient is present will there be a net movement in one or the other direction due to diffusion. Substances used by the tissue cells, such as glucose, occur in higher concentration within the capillary. A gradient is established for substances utilized by tissue cells or produced by tissue cells, and net movement by diffusion occurs down the concentration gradient.

Lipids and lipid soluble substances diffuse through the membranes of the endothelial cells. Pores are not required for lipid diffusion because the membranes are primarily composed of lipids. Lipid soluble substances include oxygen and carbon dioxide. In the case of these substances, concentration gradients are present, since actively metabolizing cells use oxygen and give off carbon dioxide. Therefore, more oxygen molecules diffuse out of a capillary than into it, and more carbon dioxide molecules diffuse into a capillary than out of it.

FICK'S LAW OF DIFFUSION

Fick's law, (Chapter 1, Movements of Molecules and Ions Across Cell Membranes) also applies to diffusion across the capillary wall. In summary,

$$\frac{ds}{dt} = -DA \frac{dc}{dx}$$

where ds is the amount of a substance moving across the membrane or capillary wall per unit of time dt and is equal to the product of a diffusion coefficient D in cm²/sec, the cross-sectional area A in cm², and the concentration gradient dc of the substance across the thickness of the wall dx. Note that the Fick formula is for *net flux,* which depends on a concentration gradient. It does not give information about the total number of molecules moving through the membrane or across the capillary wall in both directions per unit of time. Since the net movement of a substance (net flux) is into the compartment in which its concentration is lower (downhill along a concentration gradient), the sign is negative.

Superimposed on the characteristics of the capillary membrane are properties of the exchangeable substances. The properties include (1) electrical and chemical potential differences across the wall; (2) selective non-diffusibility of some substances (colloidal osmotic pressure), thereby influencing solvent and solute movement (Gibbs-Donnan effect); (3) solubility in the capillary wall; (4) physical ability to pass through capillary pores; and (5) transport through the cells.

FILTRATION

The capillary acts like a filter or sieve. Its pores are usually too small to permit most of the protein molecules to pass through. (However, the pores of venous capillaries and venules are large enough to permit the escape of small amounts of protein.) Because, under normal conditions, only small amounts of protein escape from the circulating blood, the concentration of protein in plasma is higher than concentration of protein in the interstitial fluid. Resulting from the difference in protein concentration, and hence the number of colloid particles, *colloid osmotic pressure* (COP) is greater in plasma than in interstitial fluid. Fundamentally, it is because of the greater concentration of colloid particles in the capillary that water is in higher concentration outside the capillary and therefore it tends to move down its concentration gradient into the capillary. On the other hand, the capillary blood pressure, frequently termed the capillary hydrostatic pressure,* is greater than the hydrostatic pressure in the interstitial

---

*Strictly speaking, hydrostatic pressure is the pressure exerted by a standing column of fluid (Chapter 10). Pressure within a blood vessel is the sum of true hydrostatic pressure (from blood column above the level of measurement) plus hydraulic pressure imparted by ventricular systole. Since blood pressure has been traditionally measured with mercury manometers, it is common to find blood pressure expressed as "hydrostatic pressure," even though only part of the pressure measured is a true hydrostatic pressure. In this chapter the term capillary blood pressure is used, but keep in mind that this refers to the same pressure within capillaries that many investigators term "capillary hydrostatic pressure."

fluid. Therefore, if the difference between the capillary blood pressure and the interstitial hydrostatic pressure is greater than the COP difference between the two compartments, fluid will filter out of the capillary. Thus, the effective net filtration pressure that forces fluid through pores or slits in the capillary wall is the algebraic sum of forces (pressures) acting on the opposite sides of the capillary wall. Contrariwise, if the colloid osmotic pressure difference is greater than the blood and the interstitial hydrostatic pressure difference, fluid will enter the capillary from the interstitial space (resorption). (See examples under Starling Hypothesis.)

THE GIBBS–DONNAN EQUILIBRIUM

The distribution of fluids and dissolved or suspended substances between the blood and interstitial fluid is determined primarily, as has been discussed, by the balance between the forces (pressures) across the wall of the capillary. The major factor in producing the osmotic pressure is the relative impermeability of the capillary wall to the plasma proteins. The concentration of plasma proteins is about 0.9 mmoles/kg H_2O; this concentration of proteins produces an osmotic pressure of approximately 17 mm Hg. However, because of the Gibbs–Donnan equilibrium, the concentration of diffusible ions in plasma water exceeds that in the interstitial water by about 0.5 mOsm/kg H_2O. This excess of ions in plasma contributes an additional osmotic pressure of 10 mm Hg, bringing the effective osmotic pressure of plasma to 27 mm Hg.

In addition to the greater osmotic pressure in plasma that results from the Gibbs–Donnan effect, a trans-capillary wall difference in electrical potential is produced. Because of nondiffusible ions (proteins) in the plasma and the resulting Gibbs–Donnan effect, the concentration of cations is higher in plasma than in the interstitial fluid. The Gibbs–Donnan concentration ratio for Na^+ is

$$\frac{155 \text{ mEq/kg } H_2O \text{ (interstitial fluid)}}{164 \text{ mEq/kg } H_2O \text{ (plasma)}} = 0.945$$

The resulting potential difference across the capillary wall at 37°C is

$$61 \log_{10} 0.945 = 1.5 \text{ mV (plasma negative to interstitial fluid)}$$

SOLVENT DRAG

The transport of a solute through a membrane, if solute and solvent follow the same pathway, is influenced by the direction and rate of movement of the solvent. The effect of solvent movement on the solute is called *solvent drag,* which permits a solute to move across a membrane against its electrochemical concentration gradient.

THE STARLING HYPOTHESIS

Mechanisms and forces other than diffusion responsible for movement of fluid out of (filtration) and into capillaries (resorption) were first described by the English physiologist Ernest Henry Starling. According to his hypothesis, two kinds of forces or pressures algebraically combine for a net force that, depending on its direction and magnitude, causes either filtration of water and solutes out of the capillary, or resorption into the capillary. Water, electrolytes, and other dissolved (or suspended) substances move out of the capillary if the net outward force (pressure) is greater than the net inward force (pressure) (Fig. 7). On the other hand, if the net inward force is greater than the net outward force, the fluid movement is into the capillary.

Note that when Starling enunciated his hypothesis, it was assumed that the exchange occurred only through pores in the capillary wall. More recent evidence indicates that the largest pores are in the walls of venules.

The small amount of protein that can enter the interstitial space (primarily via pores in venous capillaries and in venules) is continuously removed by the lymphatic system. Lymphatics also remove water, electrolytes, and other solutes from the interstitial space. In tissues with a functioning lymphatic system, the sum of the net driving forces tending to cause fluid to leave the capillaries is slightly greater than the sum of the net driving forces tending to cause fluid to enter the capillaries. The excess filtered fluid enters the lymph capillaries and is returned to the blood via lymphatic channels.

Major emphasis has been placed by Guyton and his co-workers on a hypothesized negative interstitial pressure; it could be that some tissues have a negative interstitial pressure. Guyton's hypothesis is supported by experiments in which pressure is measured in perforated

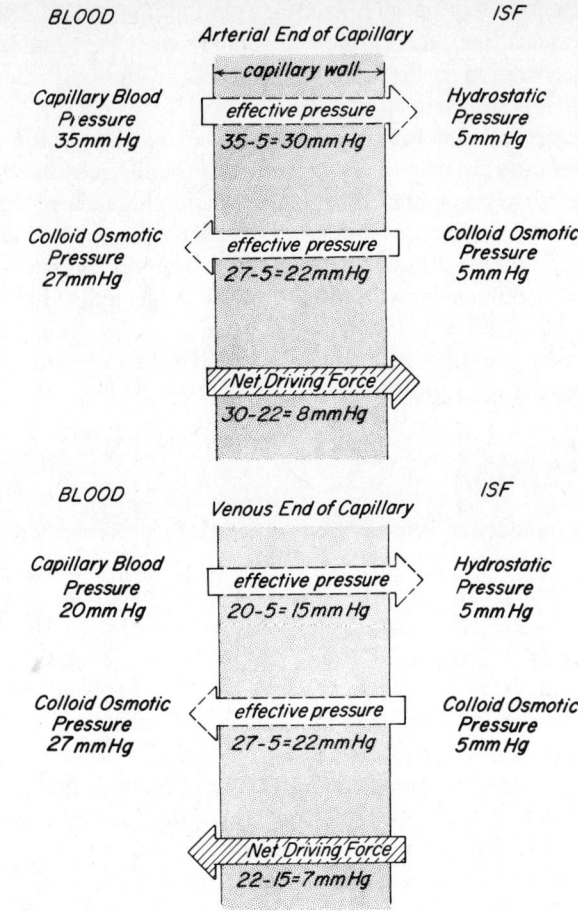

Figure 7 Example of forces acting in one capillary at one instant in time (positive tissue pressure).

capsules implanted in tissue and by several other kinds of evidence. Some investigators claim that Guyton's measurements are artifactual. Nevertheless, regardless of the validity of the negative tissue pressure concept, it can be fitted into the Starling hypothesis.

In Guyton's balance sheet with a negative tissue pressure for the Starling relationship, the capillary blood pressure is considered to be 25 mm Hg and the tissue hydrostatic pressure − 7 mm Hg. If the plasma COP is 27 mm Hg and the interstitial COP is 4.5 mm Hg, the net driving force out of the capillary is (25 + 7) − (28 − 4.5) = 9.5 mm Hg.

Guyton suggests that a negative tissue pressure is an important functional entity that (1) helps hold cells of tissues in close proximity because of a partial vacuum, and (2) keeps the amount of interstitial fluid relatively small (calling this the "dry" state of the interstitial spaces). Because cells are close together and the fluid between them is small, the diffusion distance for nutrients is short.

Capillary blood pressures, interstitial hydrostatic pressures, and interstitial protein concentrations vary in different tissues and during altered metabolic states of tissues. For example, partly because of local control, capillary pressure and capillary flow increase in actively metabolizing tissue; interstitial hydrostatic pressure is higher in encapsulated organs than in nonencapsulated organs; and interstitial protein concentration is low in skeletal muscle but high in the intestine. The exchange between plasma and interstitial fluid is a dynamic, constantly changing process that depends to some extent on pressures and flow in the local capillary network and on the level of metabolic activity of adjacent tissue cells. Hence, the numerical values that are used can only illustrate the status of one capillary in one tissue at a single instant in time. If you understand the concept and you know the ranges of capillary blood pressures, interstitial pressures, and colloid osmotic pressures, you can set up your own illustrative example of the Starling relationship for a capillary under the conditions you have chosen.

EDEMA

Edema is the abnormal accumulation of fluid in tissues. It can occur in cells (cellular edema) or in the interstitial space (interstitial edema). The Starling hypothesis is a fundamental concept that explains the mechanisms responsible for interstitial edema. If the amount of fluid filtered exceeds the amount resorbed plus the amount returned to the circulation via the lymphatics, edema occurs. The unbalanced forces that result in edema involve capillary blood pressure, interstitial hydrostatic pressure, and osmotic pressures in capillaries, venules, and the adjacent interstitial spaces.

Conditions that result in filtration greatly exceeding resorption include lowered plasma protein concentration,

an abnormal increase in interstitial protein concentration, or a marked rise in capillary blood pressure. Plasma protein concentration falls when protein is deficient in the diet (nutritional edema) or when excessive protein is lost via the kidneys (nephrosis). Interstitial protein concentrations rise if capillaries, and especially venules, are damaged (injury can increase the size of the pores).

An abnormal rise in capillary blood pressure results from a rise in venous pressure. Venous pressure is increased in dependent parts of the body, increasing as a result of venous thrombosis and increasing during cardiac decompensation, that is, cardiac failure. Capillary and venule blood pressures also rise if excessive dilation of small arteries and arterioles occurs (allergic conditions and angioneurotic edema) or in renal failure (anuria or oliguria).

Although the precise sequence of events leading to interstitial edema in patients suffering from cardiac failure is currently controversial, responsible mechanisms clearly involve imbalances in the forces delineated in the Starling hypothesis, that is, blood pressure, hydrostatic pressure, and osmotic pressure.

Two theories are extant that attempt to explain mechanisms responsible for edema in patients with cardiac failure. The earlier theory, called the *backward failure hypothesis,* postulates an increase in venous pressure, thereby augmenting capillary blood pressure and promoting excessive filtration relative to resorption. The increase in venous pressure is said to result from the inability of the failing heart (right ventricle) to accept a normal amount of blood from the great veins because the ventricular chamber at the completion of systole retains a larger fraction of its end-diastolic volume.

Many investigators have questioned the validity of the backward failure hypothesis. They have claimed that an increase in venous pressure is a consequence of an increase in plasma volume, rather than an inability of the failing heart to accept its quota of blood from the great veins. In 1913 MacKenzie proposed the *forward failure hypothesis,* in which the signs and symptoms of congestive failure were postulated to result from a reduced cardiac output with decreased blood flow to tissues. Although the initial explanations for the loss of fluid from the microcirculation were made on the basis of damage to capillaries caused by ischemia, a renal mechanism was later proposed.

The essence of the current concept is that the decreased cardiac output in congestive failure brings about a reduction in glomerular filtration, with little or no change in tubular function, leading to retention of sodium and water. The Starling equilibrium is thereby upset because plasma proteins are diluted by the retained fluid with a consequent decrease in the capillary protein osmotic pressure. As the fluid volume of the body increases, the blood volume increases as well, leading to augmented venous and capillary blood pressures. More fluid enters the interstitial space, and interstitial edema results.

Inadequate perfusion of the kidney also results in the production and release by the kidney of the proteolytic enzyme *renin.* Renin splits an α_2-globulin, angiotensinogen, to form a decapeptide called angiotensin I; this is converted to an octapeptide, angiotensin II, by "converting enzyme" in the lungs. Angiotensin II acts on the adrenal cortex, causing an increase in the rate of secretion of aldosterone. Aldosterone brings about an increase in the resorption of sodium by the renal tubules. Increased sodium in body fluids fosters the retention of water in the extracellular compartment. Excess plasma water, hypervolemia, can lead to increased capillary and venous blood pressures with consequent movement of fluids into the interstitial spaces. Current evidence indicates that a high level of aldosterone is not a primary cause of interstitial edema, but that it can augment edema produced through other mechanisms. These other mechanisms are discussed in Chapter 33.

Blockage of lymphatic channels causes edema (lymph edema). Since more fluid is ordinarily filtered than is reabsorbed into microvessels, absence of lymph flow results in the accumulation of fluid in the interstitial space. Furthermore, if the lymphatics are blocked, protein will accumulate in the interstitial space, thereby reducing the effective osmotic pressure, the force tending to move water into the microvessels.

LYMPHATICS

The following information is generally accepted, representing direct observation and reasonably sound experimental evidence.

Distribution

Although lymphatic vessels are present in most tissues, they have not been found in cartilage, bone marrow, the CNS, epithelia, and fetal parts of the placenta.

Architectural Arrangement

Lymphatic vessels are a branching system and are closed at their distal ends (lymph capillaries).

Connections with Blood Circulatory System

Lymphatic vessels make numerous connections with the venous system. However, the major points of drainage into veins are in the neck region via the thoracic duct emptying into the left innominate vein, and on the right side via the right lymphatic duct emptying into the right subclavian vein.

Structure of Lymphatics

Lymph capillaries are endothelial tubes closed at their distal ends. The endothelium is thinner than that of most blood capillaries. Medium-size lymphatic vessels have muscle fibers, and the larger lymphatics have a connective tissue sheath containing scattered elastic and muscle fibers. Valves are present in the larger vessels and are usually either unicuspid or bicuspid.

Functional Aspects of the Lymphatic System

The lymphatic system is a transport system that removes excess fluid and protein from the tissue spaces. It is a pathway through which lipids enter the blood via the lacteals of the intestine. Lipids and proteins that leave the plasma are returned via the lymphatic system. Proteins, such as fibrinogen, which are synthesized in the liver, are delivered into the circulation partly via lymphatic channels. The lymph nodes are the site of production of the small lymphocytes involved in cellular immunity; these nodes also filter out and trap bacteria.

Composition of Lymph

Lymph varies greatly in composition, depending on the organ from which it is drained and on the physiological or pathological state of the organ from which it is drained.

Lymph contains water, salts, and small organic molecules in about the same concentrations as in plasma or interstitial fluid. It usually contains slightly more protein than interstitial fluid. Red blood cells are absent, except when microvessels are damaged.

Factors Influencing Lymph Flow

A pressure differential is necessary to permit lymph to move from its capillaries via its lymphatic channels and ducts into the venous system. The lymphatic system is a low pressure system, and actual pressures are difficult to measure. Lymph flow is aided by skeletal muscle contraction and relaxation ("milking action"), changes in pressure within the abdominal and thoracic cavities with respiratory movements, and by a peristaltic-like contraction and relaxation of lymphatic vessels. The valves in the larger lymphatic vessels determine the direction of flow. Perhaps a venturi-like action helps deliver lymph into veins (Bernoulli's principle).

Lymph flow is increased when lymph is being produced at a more rapid rate. Exercise increases lymph production as do other increases in metabolic activities of tissues. Substances that cause increased lymph flow when injected into the blood are called lymphagogues. Lymphagogues, as pointed out by Heidenhain about 100 years ago, act in either of two ways. Some lymphagogues, such as histamine, peptone, and the enzyme hyaluronidase, increase microvessel permeability, whereas other lymphyagogues act by increasing the plasma volume without increasing its colloid osmotic pressure. Among the second class of lymphagogues are hypertonic solutions of sodium chloride or glucose. Injected intravenously, these solutions promptly leave the capillary and, by osmotic withdrawal of water from parenchymal cells, cause increased lymph flow.

LOCAL VASCULAR REGULATION

Although the term autoregulation was originally used to specify maintenance of relatively constant flow in an organ during changes in arterial pressure, it is now commonly used to designate all mechanisms involved in local vascular regulation. In this chapter the term "autoregulation" is used interchangeably with local vascular regulation.

LONG TERM LOCAL BLOOD FLOW ALTERATIONS

In addition to acute changes in blood flow, adjustments are made over longer time periods. Before discussing rapid alterations in blood flow, a few comments about the slower, long term adjustments are appropriate.

Long term adjustments primarily involve increases or decreases in vascularization. During the growth and repair of tissues, vascularization usually keeps abreast of the increase in mass of the other tissue elements. However, even without a major increase in mass of an organ or tissue, the vascularization of the tissue can be augmented. The number of capillaries in a given tissue is continuously being adjusted to the metabolic needs of the tissue.

Although the precise stimulus that sets in operation the outgrowth of additional capillaries is unknown, a deficiency in the delivery of oxygen to the tissue is a likely candidate. Vascularization of many tissues increases if the tissue is subjected over an appreciable period of time to low oxygen tensions. Thus vascularization tends to be enhanced in mammals living for months to years at high altitudes. It also is increased in patients with coarctation of the aorta.

The enhancement or development of a collateral circulation is a special case of long term blood flow regulation. Fortunately, when a major vessel is gradually occluded, collateral blood vessels usually hypertrophy. Blood flow in collateral vessels can supply the needs of an organ even though its normal blood supply has been severely compromised.

ACUTE BLOOD FLOW REGULATION

The volume of blood circulating to and through a tissue depends on (1) the pressure in the artery that supplies the tissue, (2) the pressure in the vein that drains the tissue, and (3) the resistance to flow in the network of vessels between the artery and vein. Resistance to flow in a vessel is dependent on its cross-sectional area, its length, and the viscosity of the flowing fluid

$$R = \frac{8l\eta}{\pi r^4}$$

To decrease flow, reduction in the radius of the vessel is the most effective adjustment ($\dot{Q} \approx r^4$), and this is the adjustment that can be rapidly made. Smooth muscle in walls of arteries, arterioles, metarterioles, precapillary sphincters, and veins contracts when norepinephrine is released from sympathetic nerve endings (vasoconstriction). Norepinephrine brings about vasoconstriction when it acts on α-receptors in the membranes of the smooth muscle cells. Contrariwise, blood flow is increased by an increase in the radius of a vessel (vasodilation). Reduction in sympathetic tone usually results in vasodilation.

A chemical agent that causes smooth muscles of blood vessels to relax is acetylcholine (ACh). Cholinergic nerves to blood vessels are limited to the skeletal muscles, the salivary and some gastrointestinal glands, the external genitalia, the heart, and the brain.

Vasoconstriction mediated via efferent nerves from the CNS is essentially a generalized phenomenon. Precise control of gradation in the degree of vasoconstriction in the various tissues results from locally controlled modifications of resistance to blood flow. These local adjustments are made by means of non-neural mechanisms acting directly on smooth muscles in the walls of blood vessels.

When the metabolic activity in a specific organ or tissue is increased, more blood flows through that tissue. Neural signals via efferent sympathetic nerves result in enough vasoconstriction in relatively inactive tissues to ensure that sufficient blood is provided to supply the active tissue. Local control in an organ usually takes precedence over the CNS signal for vasoconstriction. Consequently, vasodilation occurs in a metabolically active organ and its blood flow increases, while blood flow in the less metabolically active organs decreases. Thus a redistribution of blood flow results. Refer to Figure 8 in which flow in resting and in maximally active organs is compared.

The ultimate details of the mechanisms by which local adjustment in blood flow is brought about are not yet completely understood. In fact, there could be various mechanisms acting in the different kinds of tissues throughout the body.

It must be emphasized that the adjustments involving local regulation, or for that matter regulation via the nervous system, occur mainly in vessels that precede the capillaries, that is, small arteries, arterioles, metarterioles, and precapillary sphincters. The flow and pressure being regulated are capillary flow and capillary pressure, but the diameter of the capillaries appears to be relatively constant. The alteration in flow in an individual capillary is primarily a function of the diameter of the precapillary sphincter; the diameter of the true capillary undergoes little or no change.

Since some degree of contraction is usually present in smooth muscles of blood vessels as a result of baseline sympathetic activity, and this sympathetic activity is augmented whenever demands are made for increased blood flow, the local control of blood flow could be mediated by the extent of release of some substance that either blocks the sympathetic vasoconstrictor effect or directly causes relaxation of smooth muscles in blood vessels even while they are receiving sympathetic vasoconstrictor impulses from the CNS. Keep in mind, however, that vasodilation could also be brought about by the removal of some substance needed to maintain contraction of smooth muscles in blood vessels.

No single specific agent acting directly or indirectly to relax the smooth muscle of blood vessels in the different organs has been identified. However, it appears that the relaxation of the smooth muscles of the resistance vessels and the resultant increase in the diameter of these vessels are mediated either by a deficit in some nutrient substance, such as oxygen or glucose, and/or by the excess of some product of metabolism, such as adenosine, potassium, carbon dioxide, and acids, and/or by an increase in osmotic pressure attributable to a breakdown of large into smaller molecules.

A generally accepted mechanism for vasodilation in many tissues involves an oxygen deficit in the active tissue. The deficit occurs because blood flow is inadequate to supply the oxygen needed by that tissue to maintain its metabolic activity (e.g., its metabolic activity might have increased, or blood flow might have decreased because of a decrease in ΔP). The smooth muscles of arterioles, metarterioles, and precapillary sphincters compete with other metabolically active cells in a tissue; therefore, they might not be able to obtain sufficient oxygen for their metabolic needs. The direct effect of the oxygen deficit or the release

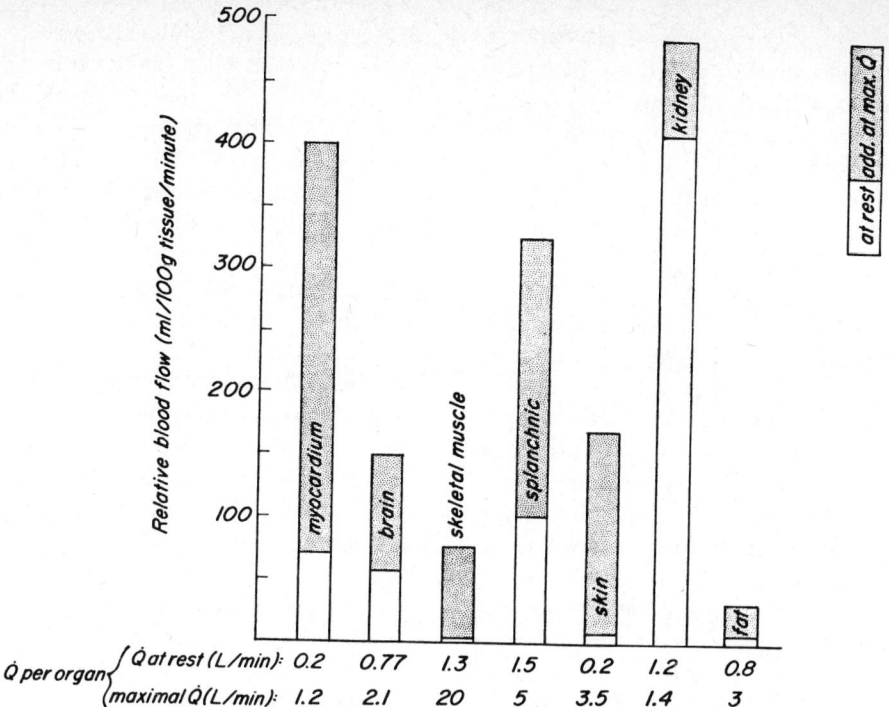

Figure 8 Regional blood flows at "rest" and at maximal dilation, adapted to fit approximately the situation in 70 kg man. The flow figures are given both for 100 g of tissue and for the entire organs. (Modified from Mellander and Johansson, 1968.)

of a metabolite from the anaerobic tissues causes the smooth muscles of the resistance vessels to relax. With relaxation (dilation), resistance to flow decreases and blood flow and pressure within the capillaries increase. When blood flow is greater than that required to provide sufficient oxygen for the needs of the tissue and the smooth muscle of the resistance vessels, the smooth muscles, probably because of sympathetic adrenergic nerve impulses (basal vasomotor tone), tend to contract, reducing blood flow to the tissue. The local regulation that occurs in coronary vessels is believed to be primarily dependent on the level of the oxygen tension in the myocardial tissue.

The local regulatory adjustment is rarely precise, and an overshoot in adjustment is frequently observed, possibly the reason that intermittent contraction and relaxation of smooth muscle in metarterioles, and precapillary sphincters *(vasomotion)*, occurs in some tissues. In other words, a kind of "hunting reaction" occurs.

The carbon dioxide tension in some tissues modifies the tone of their vascular smooth muscle. With an increase in metabolism and inadequate blood flow to carry away the excess CO_2, smooth muscles of resistance vessels relax and blood flow increases. Carbon dioxide tension is reported to be the primary factor modifying blood flow in the brain.

Hydrogen ion concentration has been implicated as an important factor in the control of the tone in vascular smooth muscle. However, hydrogen ion concentration is partly dependent on both the P_{O_2} and P_{CO_2}, making it difficult to identify the controlling stimulus. For example, when oxygen lack occurs, lactic and other metabolic acids are produced by anaerobic metabolism. A rise in P_{CO_2} due to increased metabolism and inadequate blood flow results in a local accumulation of carbonic acid

$$H_2O + CO_2 \rightleftharpoons H_2CO_3$$

Other substances that might be involved in autoregulation are potassium (ratio of extracellular to intracellular K^+), adenosine, and lactate. Oxygen lack, or possibly carbon dioxide or hydrogen ion excess, might act indirectly in causing relaxation of the smooth muscle in resistance vessels through a secondary release of potassium, adenosine or lactate. For example, an increase in hydrogen ion concentration in extracellular fluid results in a potassium shift from cells; an oxygen deficit is responsible for the production and release of lactic acid and for the release of adenosine and adenosine phosphate compounds. All these substances have been shown to be capable of causing relaxation of vascular smooth muscle.

In addition to the lack of specific nutrients or an excess of a metabolite or metabolites, the local osmolarity may influence the tone of vascular smooth muscle. During active metabolism, large complex molecules are broken down to a larger number of smaller molecules. The increase in the number of molecules increases the osmotic pressure. The local increase in osmotic pressure causes a decrease in tone of the smooth muscles and a consequent increase in the diameter of resistance vessels. Also, the diameter of a vessel may be increased by the shrinking of endothelial and other cells in the vessel wall and, of course, when endothelial cells swell the lumenal diameter of a vessel is decreased.

It was recently suggested that certain prostaglandins are responsible for the local control of blood flow. Some prostaglandins are vasodilators, others vasoconstrictors. They are synthesized from free fatty acids, among them arachidonic acid. Prostaglandins are short-lived in the body and hence could act locally and acutely to modulate peripheral resistance.

In addition to local effects that adjust blood flow in a tissue to approximate the metabolic needs of the tissue, at least two physical mechanisms tend to stabilize local blood flow. These stabilizing mechanisms adjust the blood flow to a tissue when the pressure within the artery supplying blood to the tissue is significantly increased or decreased. The first of these mechanisms is the myogenic reflex of Bayliss. (This is not a true reflex, as it does not involve the nervous system.) When pressure within arteries or arterioles rises, the smooth muscle in their walls is stretched, and the muscle responds by increasing its tone. This local muscular response helps prevent an excessively large flow in a tissue when the pressure in its supplying artery suddenly rises.

A second physical means by which flow in a tissue is kept relatively constant has been described by Rodbard. If pressure within capillaries of a tissue rises, more fluid is forced out of the capillaries (and venules). This fluid increases the pressure in the interstitial space and acts to decrease the diameter of blood vessels, which traverse the region in which the augmented interstitial pressure occurs. Most of this effect is on thin walled vessels, such as capillaries and venules.

Figure 9 depicts some of the control mechanisms, including the myogenic reflex.

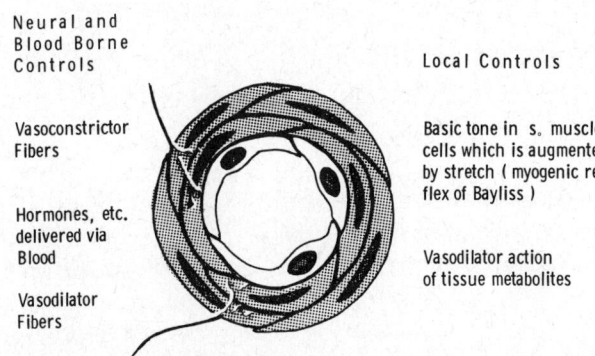

Neural and Blood Borne Controls

Vasoconstrictor Fibers

Hormones, etc. delivered via Blood

Vasodilator Fibers

Local Controls

Basic tone in s. muscle cells which is augmented by stretch (myogenic reflex of Bayliss)

Vasodilator action of tissue metabolites

Figure 9 Basic machinery of local vascular control: myogenic "pacemaker" activity, reinforced by the continuous stretch offered by the blood pressure, constitutes, via cell-to-cell excitation spread, the basal vascular tone, being steadily counteracted by the continuously produced tissue metabolites. Extrinsic excitatory and inhibitory factors, in the form of nervous and blood-borne influences modulate, but rarely dominate, the local control system. (After Folkow and Neil, 1971.)

REACTIVE HYPEREMIA

An extreme vasodilation, called reactive hyperemia, occurs in some tissues after ischemia. This response is an exaggerated autoregulation. If the blood supply to a limb is blocked (a tourniquet can be used) for a period of 3–10 min, and then the block is removed, blood flow to the limb increases within 10 sec to about four times normal and subsequently gradually returns to the preischemic flow level. If the period of occlusion has not been too long, the oxygen debt acquired during the ischemic period is almost completely paid off.

Heart muscle, like skeletal muscle, undergoes reactive hyperemia after a brief period in which the coronary circulation has been blocked. The skin also shows reactive hyperemia. However, reactive hyperemia does not occur in most of the other organs. In fact, in some tissues vasoconstriction occurs after an ischemic period. This vasoconstrictive response is especially characteristic of the kidney and the liver.

SELECTED BIBLIOGRAPHY

Berne RM, Levy MN: *Cardiovascular Physiology,* ed 3. St Louis, Mosby, 1977.

Burton AC: *Physiology and Biophysics of the Circulation.* Chicago, Year Book, 1965.

Folkow B, Neil E: *Circulation.* London, Oxford University Press, 1971.

Guest MM: Circulatory effects of blood clotting fibrinolysis and related hemostatic processes, Hamilton WF, Dow P (eds): *Handbook of Physiology, Circulation, vol. 3.* Washington DC, American Physiology Society, 1965.

Schmid-Schonbein H: Microrheology of erythrocytes, blood viscosity, and the distribution of blood flow in the microcirculation, in Guyton AC Cowley AW Jr (eds): *Cardiovascular Physiology, vol 2.* Baltimore, University Park Press, 1976.

Wells R (ed): *The Microcirculation in Clinical Medicine.* New York, Academic Press, 1973.

SELF-STUDY QUESTIONS

MATCHING

In each of the following indicate the vessel(s) by selecting the appropriate letter(s).

A. *Small arteries*
B. *Arterioles*
C. *Capillaries (arterial)*
D. *Venous capillaries*
E. *Venules*

Of the choices given, which vessels have

1. greatest internal diameter?
2. greatest wall thickness?
3. smallest internal diameter?
4. greatest total number in body?
5. greatest total cross-sectional area?
6. lowest mean pressure?
7. greatest mean velocity of erythrocytes?
8. highest ratio of surface area per unit of length to volume within the cylindrical space?
9. greatest increase in relative viscosity of blood between their proximal and distal ends? (Before answering, review sections on the Starling *Hypothesis*, p. 302, and viscosity, p. 303.)
10. highest ratio (per unit length) of endothelial cells in wall to cells within the lumen of the vessel?

MULTIPLE CHOICE

Select the **single best** answer.

11. In arterioles
    A. the greatest ΔP occurs.
    B. systolic and diastolic pressures are essentially equal.

C. mean pressure is less than in the aorta.

D. blood velocity is greater than in capillaries.

12. Viscosity

A. of plasma is newtonian, of whole blood is anomalous.

B. of whole blood increases as shear rate decreases.

C. increases at different shear rates as hematocrit increases.

D. of whole blood is independent of size of vessel through which it flows.

13. A hollow paraboloid-like shape of RBC is

A. observed in true capillaries.

B. observed in large veins.

C. dependent on the interior viscosity of RBCs and their volumes relative to the area of their cell membranes.

D. an interesting phenomenon but it has no physiologic importance.

14. Capillaries

A. in intestinal villi have fenestrated endothelium.

B. in the brain have thin, nonfenestrated endothelium with a continuous basement membrane and tight junctions.

C. in kidney glomeruli have fenestrated endothelium.

D. in the heart have discontinuous endothelium.

15. Movement across capillary walls of

A. glucose occurs mainly by pinocytosis.

B. oxygen occurs by diffusion.

C. water occurs mainly by filtration–resorption.

D. protein is limited by pore size.

16. With respect to transfer across membranes and capillary walls,

A. the Fick diffusion equation permits calculation of the total number of molecules of a substance moving across a capillary in both directions per unit of time.

B. the Gibbs–Donnan equilibrium amplifies the osmotic effects of nondiffusible ions and molecules.

C. in most tissues the total number of molecules filtered exceeds the total number of molecules moving across the capillary wall in both directions by diffusion.

D. the concentration of sodium in plasma is greater than in interstitial fluid.

17. According to the Starling hypothesis,

A. capillary blood pressure minus tissue hydrostatic pressure is usually greater at the arteriolar end of the capillary than the plasma colloid osmotic pressure minus the tissue colloid osmotic pressure.

B. capillary blood pressure minus tissue hydrostatic pressure is usually less at the venous end of the capillary than the plasma colloid osmotic pressure minus the interstitial colloid osmotic pressure.

C. more fluid leaves the capillary at its arteriolar end than enters the capillary at its venous end.

D. some of the water, some of the dissolved substances, and all or most of the colloid, are removed from the interstitial space by lymph capillaries.

18. Conditions that could be responsible for tissue edema in the feet and ankles are

A. low protein diet.

B. renal disease.

C. lymphatic or venous obstruction.

D. either right or left heart failure.

19. Conditions that could lead to pulmonary edema are

A. septic shock.

B. inhaling a toxic gas.

C. chronic left heart failure.

D. pulmonary thrombosis or embolism.

20. For lymph to flow, a pressure differential is necessary. This may be contributed to or maintained by

A. skeletal muscle contractions.

B. respiratory movements.

C. smooth muscle contractions.

D. lymphatic valves.

21. Lymph normally contains

A. albumin.

B. fibrinogen.

C. factor VIII.

D. erythrocytes.

Indicate the one *incorrect* statement.

22. A. Local control of coronary blood flow is highly dependent on the oxygen tension in the myocardial tissue.

B. Local control of cerebral blood flow is highly dependent on the carbon dioxide tension in arterial blood.

C. The diameter of a specific capillary remains relatively unchanged, even though the metabolic activity of the tissue in which it is located varies greatly.

D. The smooth muscles of all blood vessels are innervated by either cholinergic sympathetic or parasympathetic fibers.

23. A. At altitudes above 3,600 m (12,000 ft), one would expect the coronary blood vessels to be more dilated than at sea level, assuming same level of physical activity.

B. Bayliss stated that the smooth muscle of blood vessels tends to increase its tone when stretched.

C. Both hypoxia and hypercapnea can cause a drop in the pH of tissues, and it is possible that the direct stimulus for coronary dilation and for cerebral dilation is actually an increase in H^+ ions within the smooth muscle cells.

D. Reactive hyperemia occurs in all tissues of the body.

ANSWERS

1. E. According to Table 1, on p. 294, venules have a larger internal diameter than small arteries. Keep in mind though, that the numbers given are averages; many venules have a smaller internal diameter than some small arteries. Venules vary in diameter from slightly larger than capillaries to about 40 μm.

2. A. Small arteries have an investment of smooth muscles in their walls. Arterioles also have smooth muscle but not as much as small arteries.

3. C.

4. D. See footnote, Table 1.

5. E. However, note that the total cross-sectional area of the venous capillaries is also very large.

6. E. Pressure continues to drop as the distance from the heart increases. Energy is lost because of friction.

7. A. Small arteries have the smallest total cross sectional area of the vessels listed. Since this total cross-sectional area carries the entire cardiac output, velocity must be greater than in vessels with a larger total cross-sectional area.

8. C. Surface area equals circumference ($2\pi r$) times a unit of length; hence it varies directly with the radius.

Volume equals the cross-sectional area πr^2 times the unit of length of the same cylinder; hence it varies with the square of the radius. The ratio of surface area to the contained volume of a cylinder having a length (or height) of one unit is

$$\frac{2\pi r l}{\pi r^2 l} = \frac{2r}{r^2} = \frac{2}{r}$$

This indicates that the smaller the radius, the greater the ratio of surface area to volume. Arterial (true) capillaries have the smallest radius, and therefore the largest ratio of surface area to volume of contained blood. This is important physiologically because the large surface relative to volume is necessary for rapid exchange of oxygen and carbon dioxide between blood and interstitial fluid.

9. E. Currently available information indicates that venules are more leaky (have larger pores) than venous capillaries. Consequently, more fluid and electrolytes leave the lumen of the venule, bringing about a higher concentration of blood cells (higher hematocrit). As the hematocrit increases, relative viscosity increases.

10. C. Capillary walls contain only endothelial cells. Arterial (true) capillaries are very small in diameter (or radius); therefore, the ratio of endothelial cells to blood cells in lumen must be greater than in vessels with larger diameters (see answer to question 8). Furthermore, white blood cells do not transit true capillaries, and plasma skimming tends to reduce the number of red blood cells in true capillaries.

11. All are correct.
  A. True. See Figure 2, p. 295.
  B. True. See Figure 2, p. 295.
  C. True. See Figure 2, p. 295.
  D. True. See Table 1, p. 294.

12. A, B, C are correct.
  A. True. See under Viscosity, p. 297–298.
  B. True. See under Viscosity, p. 297–298.
  C. True. See under Viscosity, p. 297–298.
     Viscosity increases as the number of RBC in blood increases. However, the hematocrit effect is greatest at low shear rates.
  D. False. See under Viscosity, p. 297–298.
     In microvessels, the sigma or Fahraeus–Lindqvist effect results in a lower viscosity than would be observed in viscometers.

13. A, C.
  A. True. See p. 299.
  B. False. Because of the large cross-sectional areas of large veins and the relatively slow flow in these vessels, the rheological forces generated are insufficient to cause change of shape of RBC.
  C. True. See p. 298.
  D. False. It is an interesting phenomenon, but it also has physiological importance: First, it permits RBCs to transit through true capillaries where they can give up oxygen in close proximity to tissue cells; second, the change in shape brings a larger amount of surface area close to the capillary wall.

14. A,B,C. See p. 299.
  D. False. Heart capillaries have low continuous endothelium. Discontinuous endothelium in capillaries and sinusoids is found in liver and spleen.

15. B, D. Almost all the oxygen is transported by diffusion. Pore size limits the free movement of proteins across the capillary wall.
  A. False. Most of the glucose moves across the capillary wall by diffusion.
  C. False. Most of the water movement is by diffusion, not by filtration–reabsorption.

16. B, D. This phenomenon is discussed on p. 301.
  A. False. The Fick equation gives only the net transfer.
  C. False. Most of the transfer occurs as a result of diffusion.

17. All are correct. They are simple reiterations of the features of the Starling hypothesis.

18. All are correct. The explanations are given on pp. 303–304.

19. All are correct. Although the pulmonary circulation is a low pressure system, and normally the colloid osmotic pressure of the plasma greatly exceeds the capillary blood pressure keeping the lungs "dry," several abnormal conditions can lead to pulmonary edema. Among these are septic shock in which a bacterial toxin, such as endotoxin, damages the pulmonary endothelium and can be responsible for platelet or white blood cell "thrombi," or both. A toxic gas such as smoke, ammonia, or chlorine damages the pulmonary epithelium; left heart failure results in a marked increase in the pulmonary capillary pressure; and thrombi or emboli block pulmonary vessels, leading to increased pressure in the remaining nonblocked vessels.

20. All are correct. See p. 305 for a discussion of the factors that assist in lymph flow.

21. A, B, C. As pointed out on p. 304, lymph contains the same substances found in plasma, including the protein clotting factors. However, the concentration of proteins is less in lymph than in plasma.
  D. False. In the normal person under normal conditions, red blood cells enter lymphatic vessels only when the microvessels have been damaged.

22. D. This statement is *incorrect*. Most blood vessels are innervated only by sympathetic adrenergic fibers.
  A, B. These statements are correct, as discussed on p. 307.

C. This statement is correct. Capillaries do not contain smooth muscle, and they appear to have a relatively constant diameter, even though the pressure within them varies somewhat.

23. D. This statement is incorrect, since some tissues vasoconstrict rather than vasodilate following ischemia.

A. This statement is correct, because the arterial O_2 tension is reduced above 12,000 ft and hypoxia causes coronary vasodilation.

B. This statement is correct; it is the basic tenet of the myogenic reflex.

C. This statement is correct; it is a restatement of the explanation given on p. 308.

18

The Volume and Elastic Properties of the Lung

JOHN E. REMMERS

OBJECTIVES

After completion of this chapter, you should be able to

1. Appreciate the elastic recoil properties of the excised lung and understand the meaning of lung compliance.

2. Describe the role of alveolar surface forces in generating lung recoil.

3. Apply the concept of elastic recoil of the excised lung to the lung in situ.

4. Know how lung compliance is measured in humans.

5. Appreciate the elastic recoil properties of the chest wall.

6. Describe the actions of the respiratory muscles and the resultant motions of the rib cage and abdomen during breathing.

7. Know the meaning of the terms residual volume, functional residual capacity, vital capacity, and total lung capacity.

8. Appreciate the forces that determine the functional residual capacity.

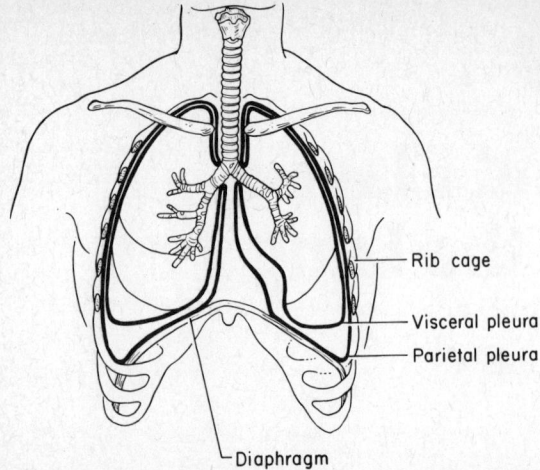

Figure 1 The respiratory system, illustrating the chest wall (rib cage and diaphragm) and the lungs, each with its pleural lining. An intrapleural space is shown for convenience, but this is not normally present. Intrapleural pressure is measured between these surfaces and gives the pressure acting on the surface of the lung. The lung is inflated and deflated as a result of increases and decreases in intrapleural pressure. These changes in intrapleural pressure are produced by contraction of the respiratory muscles.

The respiratory system is composed of two parts, the lungs and the chest wall. The chest wall also has two parts, the diaphragm and the rib cage. Figure 1 shows the anatomic relationships between these structures. The lungs and the chest wall are each lined by the *parietal* and *visceral pleurae*. These two smooth, slippery pleural surfaces are opposed, so that the lung readily conforms to the internal surface of the chest wall. During breathing, the chest wall enlarges and contracts in size, and the lung is able to follow these motions easily.

The diaphragm resembles a dome. When it contracts, it flattens and descends. This exerts a pistonlike action on the lungs and draws inspired air down the tracheobronchial tree and into the alveoli. Contraction of the inspiratory intercostal muscles moves the rib cage up and out. This action also stretches the lungs and draws gas into the alveoli.

As inspired air flows down the tracheobronchial tree, it moves through the lobar and segmental bronchi, through bronchioles and respiratory bronchioles, and into alveoli (Fig. 2). In transit from mouth to alveolus, a gas molecule passes 23 branch points. These airways serve only to conduct gas between the mouth and the alveolus. Transfer of O_2 to blood and CO_2 from blood occurs only in the alveoli. Since the conducting airways do not participate in gas exchange, they are termed *dead space*.

An important feature of the airways is the cartilaginous support that holds the airway open. Figure 2 shows that as the airways become smaller, their cartilaginous support lessens. However, these airways (segmental bronchi and smaller) are embedded in the lung parenchyma. This means that they receive support from the surrounding lung tissue that tends to pull them open. This is particularly true for the bronchioles and respiratory bronchioles (Fig. 3), which depend entirely on the outward pull of the lung parenchyma to maintain a patent lumen.

DISTENSIBILITY OF THE LUNG

The most obvious property of the lung is its distensibility. When a relatively low pressure is applied to the trachea, the lung inflates. When the lung inflates, *all* parts of the

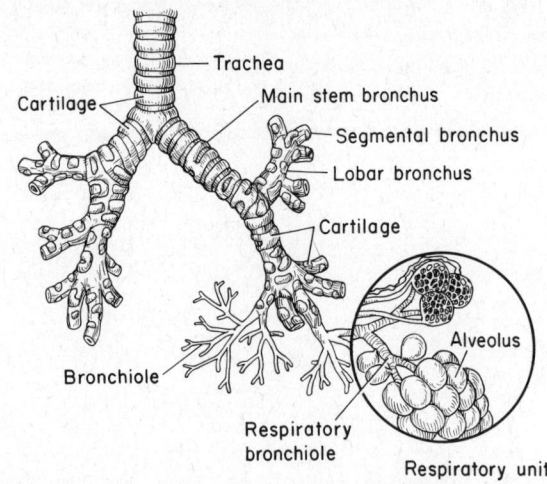

Figure 2 The tracheobronchial tree is depicted. Bronchioles are devoid of cartilaginous support and are held open by the outward pull of the lung parenchyma. A higher magnification drawing of the respiratory unit is shown in the lower right inset.

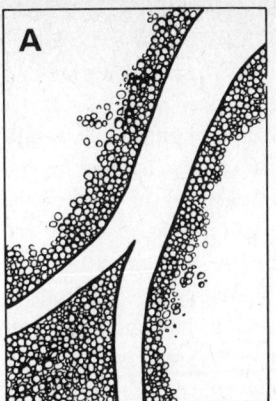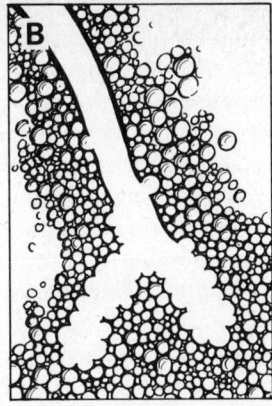

Figure 3 Drawing of a section of freeze dried lung showing (*a*) a small branching bronchiole, cut longitudinally; (*b*) a respiratory bronchiole, cut longitudinally, with alveolar ducts arising from the wall. Note that these small airways lack cartilage, and hence, depend on the outward pull of the lung parenchyma to remain open.

lung enlarge, the alveoli increase in volume, and the airways expand. The increase in size of the airways results from an increased outward pull exerted on the airways by the lung parenchyma as it is stretched. This action is clearly outlined in Figure 4, which shows the lungs fixed at various degrees of expansion. At the *low lung volume* (Fig. 4*a*), the alveoli are small and convoluted and the small bronchioles are thick walled and kinked. At the *high volume* (Fig. 4*b*), the bronchioles are widely patent, and the alveoli are 200–300 μm in diameter. This distending characteristic of the lung stands as its first important mechanical feature.

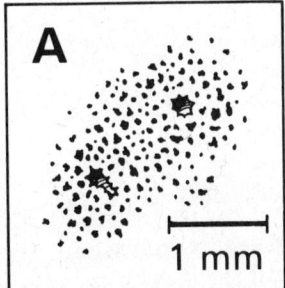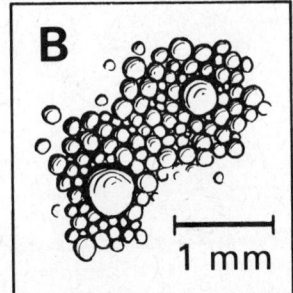

Figure 4 Drawing of sections from freeze-dried lung inflated to two different volumes: (*a*) low volume, and (*b*) high volume. Terminal bronchioles are widely patent at high lung volume and collapsed at low lung volume.

ELASTICITY OF THE LUNG

A second and closely related feature of the lung is its elasticity. An elastic structure is one that, when deformed, tends to return to its original shape and size. For example, when a rubber band is stretched, it exerts force because of its elasticity (Fig. 5). This is a restoring force; it tends to return the rubber band to its original size and shape. The excised lung is like a balloon; it displays three-dimensional elasticity. When it is inflated through the trachea, it is stretched in all directions. Once the lung is inflated, it can be held in this inflated or distended state only if a pressure is maintained at the trachea. If the pressure applied to the trachea is removed, the lung collapses to its original size and shape. To characterize the elastic properties of a rubber band, we relate the length to the restoring or elastic recoil force. To characterize the elasticity of a balloon or

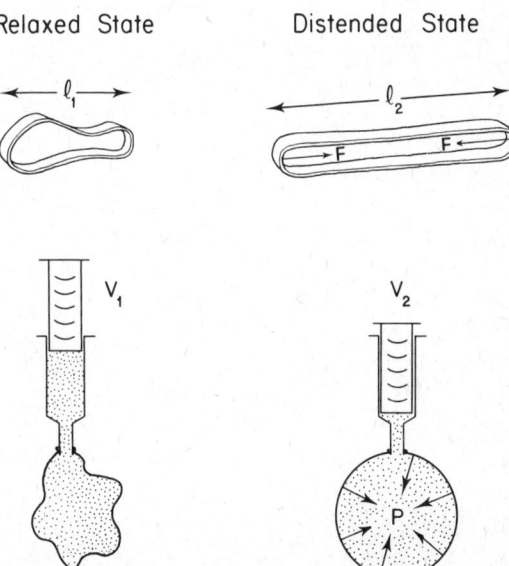

Figure 5 The elastic recoil of two structures is illustrated. The rubber band is stretched from length l_1 to length l_2; in its distended state it generates an elastic recoil force F. The balloon is inflated from volume V_1 to volume V_2; in its distended state it generates an elastic recoil pressure P. Inflating the balloon is exactly analogous to stretching the rubber band; recoil pressure is comparable to recoil force and volume is comparable to length.

a lung, we relate the lung volume to the restoring or elastic recoil pressure. Elastic recoil pressure of the excised lung is equal to the difference in pressure between the inside of the lung and the lung surface required to keep the lung inflated to a particular volume (Fig. 5).

COMPLIANCE OF THE LUNG

DEFINITION OF COMPLIANCE

An experiment in which the lung is inflated is depicted in Figure 6. The lung is considered to be a balloon. The lung is connected to a large syringe, and pressure is measured in the tube connecting the two *when there is no flow.* In condition 1 the lung is held at a low lung volume V_1, and the static airway pressure is measured P_1. Because there is no flow, pressure is the same throughout the inside of the lung, and tracheal pressure equals alveolar pressure (P_A). Distending pressure is the pressure that must be applied to the trachea to offset precisely the elastic recoil pressure developed by the lung. It can be measured as the difference in pressure between that inside the lung (alv. pressure) and that outside the lung, barometric pressure (P_B). The lung is then inflated by the syringe. After flow has stopped, condition 2 is achieved; the lung is then held at this larger lung volume V_2, and a larger static airway pressure P_2 is noted. The values for pressure and volume

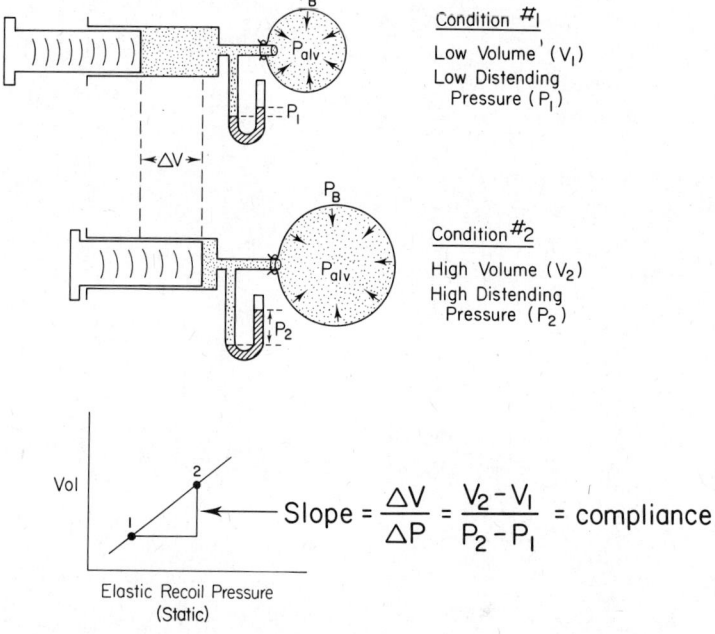

Figure 6 A schematic representation of measurement of static compliance of the lung. Condition 1: Volume of the lung is low, and a low distending pressure is measured. Condition 2: The lung has been inflated and a condition of no flow exists. The lung volume is higher and the distending pressure has increased. Volume is plotted against static distending pressure at the bottom for Conditions 1 and 2, and the slope of the line connecting the two measures the compliance of the lung in this range of lung volume.

for conditions 1 and 2 are plotted on a graph of distending pressure (alv. pressure − PB) versus lung volume and connected by a straight line. The slope $\Delta V / \Delta P$ reflects the intrinsic distensibility of the lung and is referred to as compliance. Compliance equals the change in volume produced by a unit change in distending pressure.

The relationship between lung volume and elastic recoil pressure is comparable to the relationship between length l and force F of a rubber band, as shown in Figure 5. In such a one-dimensional system, the ratio of change in length Δl to change in recoil force ΔF, that is, $\Delta l / \Delta F$, is the equivalent expression for compliance of a spring. The greater the spring's compliance, the larger will be the change in length per unit change in applied force. Similarly, for the three-dimensional analog (the lung), the measured compliance gives the change in volume that will result when there is a unit change in distending pressure. A high compliance means that there will be a large change in volume per unit change in applied pressure.

Commensurate with its primacy as a physical property of the lung, changes in pulmonary compliance are fundamental manifestations of disease. Some diseases decrease compliance and some increase it—in either case functional impairment results. Pulmonary scarring, infiltration with cells, or excess water stiffen the lung (decrease its compliance), whereas pathological processes that destroy the pulmonary parenchyma impair the recoil property and increase pulmonary compliance, as in pulmonary emphysema.

In summary, the fundamental feature of the static lung is its compliance. Changes in pulmonary compliance are often the key alteration in lung disease.

STATIC PRESSURE–VOLUME CURVE

Having mastered the concept of pulmonary compliance, one can examine the static pressure–volume relationship for the lung in detail, as shown in Figure 7. The lung is suspended in a jar, and the trachea is connected to a volume measuring device (a water-bell spirometer). The gas within the jar, outside the lung, is partially withdrawn by the syringe. When this gas is removed, the pressure around the lung decreases and this causes the lung to inflate (the pressure inside the lung is greater than the pressure outside it). The gas is withdrawn in steps, that is, after a volume is withdrawn, time is allowed for flow to occur, and then measurements of pressure and volume are made in static condition. In this case, as the lung is inflated, a negative (subatmospheric) pressure is recorded. This negativity should not be confusing; a particular elastic recoil tendency of the lung generates a particular trans-

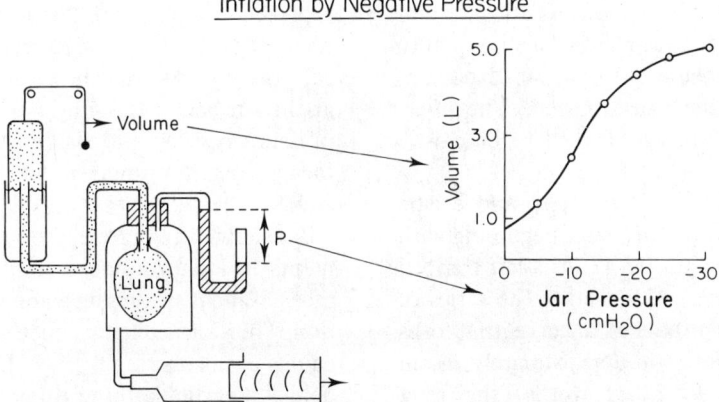

Figure 7 The measurement of the static compliance of an excised lung is illustrated. The lung is suspended within a jar, and it is inflated in steps by withdrawing gas from the jar using the syringe. The pressure in the jar is measured when the lung volume is constant (i.e., no flow condition). This elastic recoil pressure is plotted against lung volume.

pulmonary pressure, and the absolute pressure inside or outside the lung is not important. What is important is that the pressure is greater inside the alveoli than outside the lung. This is always true. In other words, elastic recoil pressure can be thought of as alveolar transmural pressure, that is, the difference between alveolar pressure and the pressure on the surface of the lung. Because of the elasticity of the lung, the pressure in the alveoli will always be greater than the pressure at the lung surface. Elastic recoil pressure is defined as the absolute value of the pressure difference between alveolar pressure and pressure on the pleural surface.

At any particular lung volume, the elasticity of the lung generates a particular alveolar transmural pressure or elastic recoil pressure, regardless of the absolute pressure inside or outside the alveoli. In this experiment (Fig. 7), the negative pressure around the lung corresponds to the positive pressure inside the lung in the previous experiment (Fig. 6). The transmural pressure across the alveolar wall will be the same if the volume is the same. The elastic properties of the alveoli generate restoring forces, and if the alveolar volume is to be held constant, these elastic forces must be offset by an externally applied pressure across the alveolar wall. The pressure must be greater inside the alveolus than at the pleural surface. In the experiment shown in Figure 7, the pressure in the alveolus is equal to atmospheric, and the pressure at the pleural surface is below atmospheric. In the experiment shown in Figure 6, the pressure inside the alveolus is above atmospheric and the pressure at the pleural surface is equal to atmospheric. In both cases, the pressure in the alveolus is greater than the pressure at the pleural surface. The difference between the two pressures is defined to be elastic recoil pressure.

In Figure 7, the lung is inflated in steps, and elastic recoil pressure is measured *after each step,* beginning with the minimal lung volume (about 0.5 L air is left trapped in the excised, collapsed lung). Connecting the series of static values (open circles) produces a curve that rises steeply and then begins to flatten at approximately 20 cm H_2O distending pressure. In other words, the lung becomes less compliant at high lung volumes. Distending pressure is given negative values to indicate that the pressure acting on the pleural surface of the lung is less than atmospheric. In the intact person, the pressure between

the parietal pleura and the visceral pleura is called the intrapleural (IP) pressure. Intrapleural pressure is the pressure acting on the surface of the lung. It is equivalent to extramural alveolar pressure. Normally, IP pressure is less than atmospheric pressure and is therefore given negative values. If the elastic recoil forces of the lung increase, the value of IP pressure will become lower, that is, there will be a greater difference between atmospheric and intrapleural pressures.

MEASUREMENT OF COMPLIANCE IN HUMANS

The pressure–volume characteristics of the lung in a normal person can be measured easily, as shown in Figure 8. We measure changes in lung volume at the mouth with a commonly available volume-measuring device called a spirometer. To measure the pressure acting on the pleural surface of the lung, we must measure the pressure between the visceral and parietal pleurae, the so-called IP pressure. Direct measurement of IP pressure is not ordinarily possible in humans, but a close approximation can be obtained by measuring the pressure in the esophagus. Because the esophagus is usually flaccid, we can measure the pressure inside the thoracic segment of the esophagus; this pressure represents IP pressure. For such a measurement, intraesophageal pressure can be assumed to be equal to intrapleural pressure. As was shown in Figure 7, the IP pressure at the surface of the lung must be less than atmospheric pressure because the elastic recoil tendency of the lung means that the lung tends to pull away from the chest wall.

As shown in Figure 8, the subject makes stepwise increments in lung volume, pausing after each step to allow measurement of pressure and volume in the static situation. The glottis remains open so that alveolar pressure P_A equals atmospheric pressure. In this situation, the lung is held at a particular lung volume by the subatmospheric IP pressure acting on its surface. That is, P_A is greater than IP pressure, and this pressure difference across the alveolar wall offsets the elastic recoil or collapsing tendency of the lung and maintains the lung in a static situation. Thus,

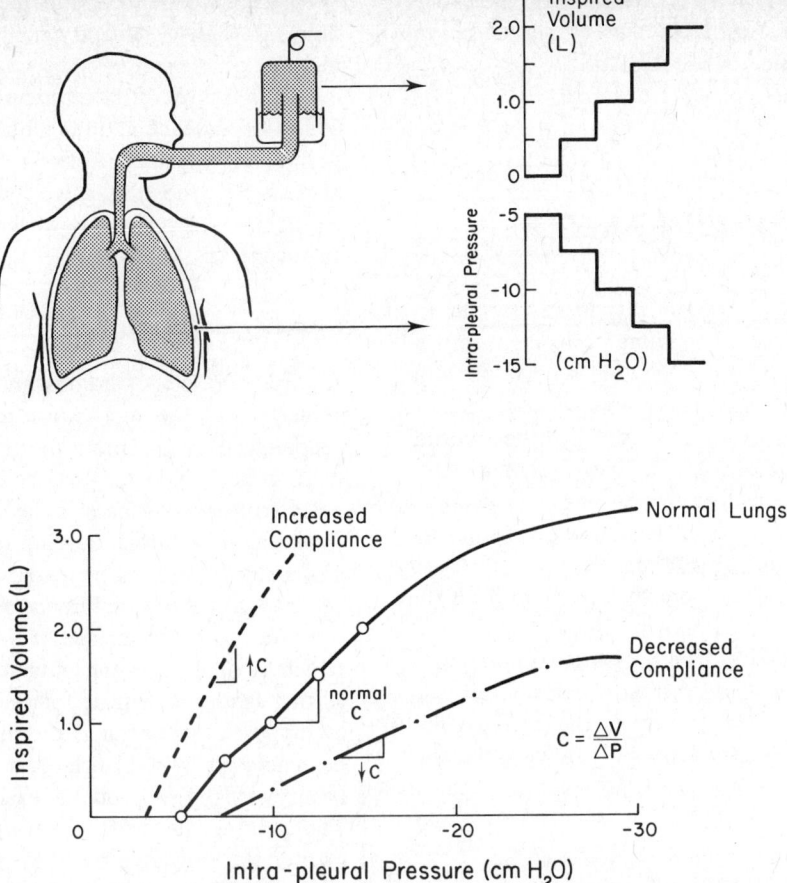

Figure 8 The measurement of static lung compliance in a normal subject is illustrated. Intrapleural pressure is measured from measurement of intraesophageal pressure. The lungs are voluntarily inflated by the subject in a stepwise fashion. Pressure and volume are measured after each increment in lung volume. The bottom panel shows a resultant plot of IP pressure versus lung volume for normal lungs. The P–V relationship for a patient with increased compliance and patient with decreased compliance are shown.

this experiment is exactly analogous to the measurement of static compliance in the excised lung (Fig. 7), except that the subatmospheric pressure at the pleural surface is created by the inspiratory muscles rather than by the syringe.

The resulting plot of volume versus distending pressure (in this case, IP pressure) is shown in the lower panel of Figure 8. Once again, note that the curve is linear (constant compliance) over a range, and then at high lung volumes the compliance decreases as the limits of lung elasticity are reached. Results obtained in pulmonary disease causing increased or decreased compliance are also shown.

Note that we shall describe measurements of IP pressure that have negative values. This simply means that IP pressure is less than zero (atmospheric pressure). Elastic recoil pressure causes IP pressure to be less than atmospheric pressure and therefore causes negative or subatmospheric values of IP pressure. If the elastic recoil force of the lung increases, a lower IP pressure is measured, that is, a more negative value of IP pressure.

SOURCE OF LUNG RECOIL

What structures in the alveoli are responsible for the retractile or recoil tendency of the lung? This is an important question, since alterations in lung recoil and lung compliance are common features of lung disease. The answer to this question lies in the fact that the internal surface of the alveoli is covered with fluid. This means that the internal surface of the alveoli is a sharply curved air–fluid interface. A simple experiment illustrates the importance of this curved liquid–air surface. We take an excised lung and remove all the gas from the alveoli and fill the alveoli with saline, thereby removing the air–fluid interface at the alveolar surface. The pressure–volume curve for the gas filled lung is compared with that for the fluid filled lung, as shown in Figure 9. Note the obvious shift in the elastic recoil curve produced by removing the air–

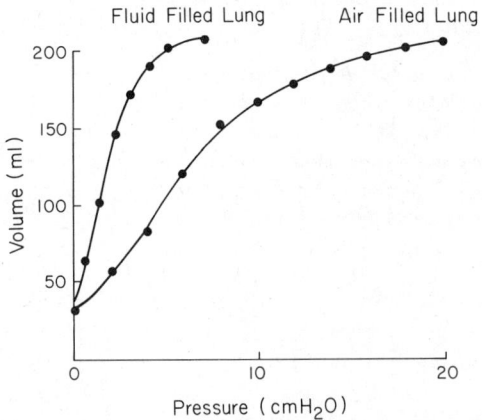

Figure 9 Effect of loss of air–fluid interface. The static pressure–volume relationship for excised cat lung is shown. The right-hand curve is generated by filling the degassed lung with air and measuring pressure and volume under static conditions at each point. The left-hand curve is generated by filling the degassed lung with saline and measuring static pressure–volume values.

fluid interface. Specifically, the fluid filled lung (no air–fluid interface) is much steeper. That is, the fluid filled lung is much more compliant. This surprising result means that most of the elastic property of the lung derives from the presence of an air-filled interface, since elimination of this interface abolishes most of the recoil pressure generated by the lung. If we consider the forces generated by the alveolar surface, however, this result does not seem so surprising.

SURFACE TENSION

All free surfaces of liquids and solids tend to contract. Molecules at the surface are exposed to an unbalanced preponderance of intermolecular attractive forces that tend to pull them toward the interior of the substance, away from the surface. If the surface is prevented from shrinking, it will then develop tension within itself. If the surface is curved, as in a soap bubble, this tension will generate a pressure difference across the wall, greater on the concave, inner surface. If the surface encloses a space which is opened to the outside through a duct, the resulting tendency will be for the contents of the space to empty itself out through the duct, collapsing the chamber enclosed by the surface, as shown in Figure 10. In other words, a soap bubble exerts elastic recoil pressure. The greater the surface tension of the soap film, the greater the elastic recoil pressure of the bubble. The importance of the air–liquid interface within the lung is therefore understandable. The pulmonary airways and alveoli, being lined by a layer of moisture at the air-tissue interface, tend to collapse even in the absence of elasticity in the tissue itself. In fact, the surface tension of the alveolar lining layer accounts for most of the tendency of the lungs to collapse. The elastic properties of the lung tissue comprise a minor part of the total elastic recoil pressure of the lung.

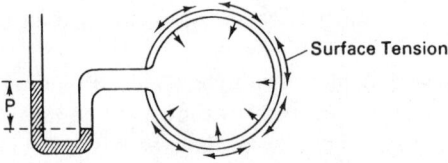

Figure 10 Pressure generated in a soap bubble as a result of the tension of the spherical surface of the bubble.

LUNG SURFACTANT

The internal surfaces of the alveoli are not lined with water. Rather, the internal alveolar surface is covered with a solution with very peculiar surface properties. An extract from the alveolar surface reveals a solution, called lung surfactant, with a very low surface tension. The alveolar surface tension is much less than it would be if the alveoli were lined with water. This property results from the presence of a phospholipid containing dipalmitoyl lecithin, known to be a surface active agent. In addition, the alveolar surface lining material (lung surfactant) has the special property of changing surface tension as it is expanded or contracted. Surface tension of lung surfactant increases if the surface is stretched and decreases if it is compressed. This feature of lung surfactant also tends to promote alveolar stability and increases the compliance of the lung. In other words, a lung with surfactant on its alveolar surface has a greater compliance than one without surfactant.

In summary, the alveoli are lined with lung surfactant, a phospholipid produced by the type II cells of the lung. This agent reduces the surface tension and promotes alveolar stability by allowing surface tension to increase and decrease as the alveoli expand and contract. Lack of surfactant occurs in some newborns (a disease called respiratory distress syndrome or hyaline membrane disease). The disease is characterized by decreased pulmonary compliance and alveolar instability. This increased recoil tendency and lack of stability lead to collapse of alveoli (atelectasis).

ELASTIC PROPERTIES OF THE CHEST WALL

The chest wall (rib cage and diaphragm) is also a distensible, three-dimensional structure that, when displaced from its resting position, tends to recoil. Accordingly, we can speak of the elastic recoil pressure of the chest wall and compliance of the chest wall. Consider a closed eviscerated thorax (Figure 11b) with a chest tube connected to a large syringe. As was done for the lung, the chest wall can be inflated and deflated in stepwise fashion, and static intrathoracic pressure and static volume can be measured after each step. Static pressure and static volume for the human chest wall are plotted in Figure 11b. A similar static pressure–volume relationship derived for the excised lung as described previously is shown in Figure 11a. Although the curve resembles that for the excised lung (Fig. 11a), there is one striking difference, namely, when the transmural pressure is zero, the chest wall contains a substantial volume (7 L in this example). At smaller volumes, it generates a negative (subatmospheric) intrathoracic pressure; at larger volumes, it exerts a positive pressure within the thorax.

As is the case for lung compliance, abnormalities of chest wall distensibility occur (e.g., fibrothorax or kyphoscoliosis). These result in a decrease in compliance and functional abnormalities.

RESTING POSITION OF THE RESPIRATORY SYSTEM

We now have enough knowledge to predict the resting position of the respiratory system, that is, we can predict the volume after we place the lungs back inside the thorax and remove the air in the pleural space. As shown in Figure 12, when the respiratory muscles are completely relaxed and the glottis is open, the alveolar pressure is atmospheric. The lung volume is at its normal end-expiratory volume, which is referred to as functional residual capacity (FRC). However, the lungs generate an elastic recoil pressure; therefore, the intrapleural pressure must be less than atmospheric (-5 cm H_2O). This means that a 5 cm H_2O transmural pressure exists across the chest wall. In other words, at the end of a normal expiration the elastic recoil pressure of the lungs is exactly offset by an equal and opposite recoil pressure across the chest wall. As depicted in Figure 12, the lungs tend to collapse inward and the chest wall tends to spring outward. This can also be seen by comparing Figure 11a with Figure 11b. There is only one volume where the elastic recoil pressure developed by the lungs is matched by an equal and opposite elastic recoil pressure devel-

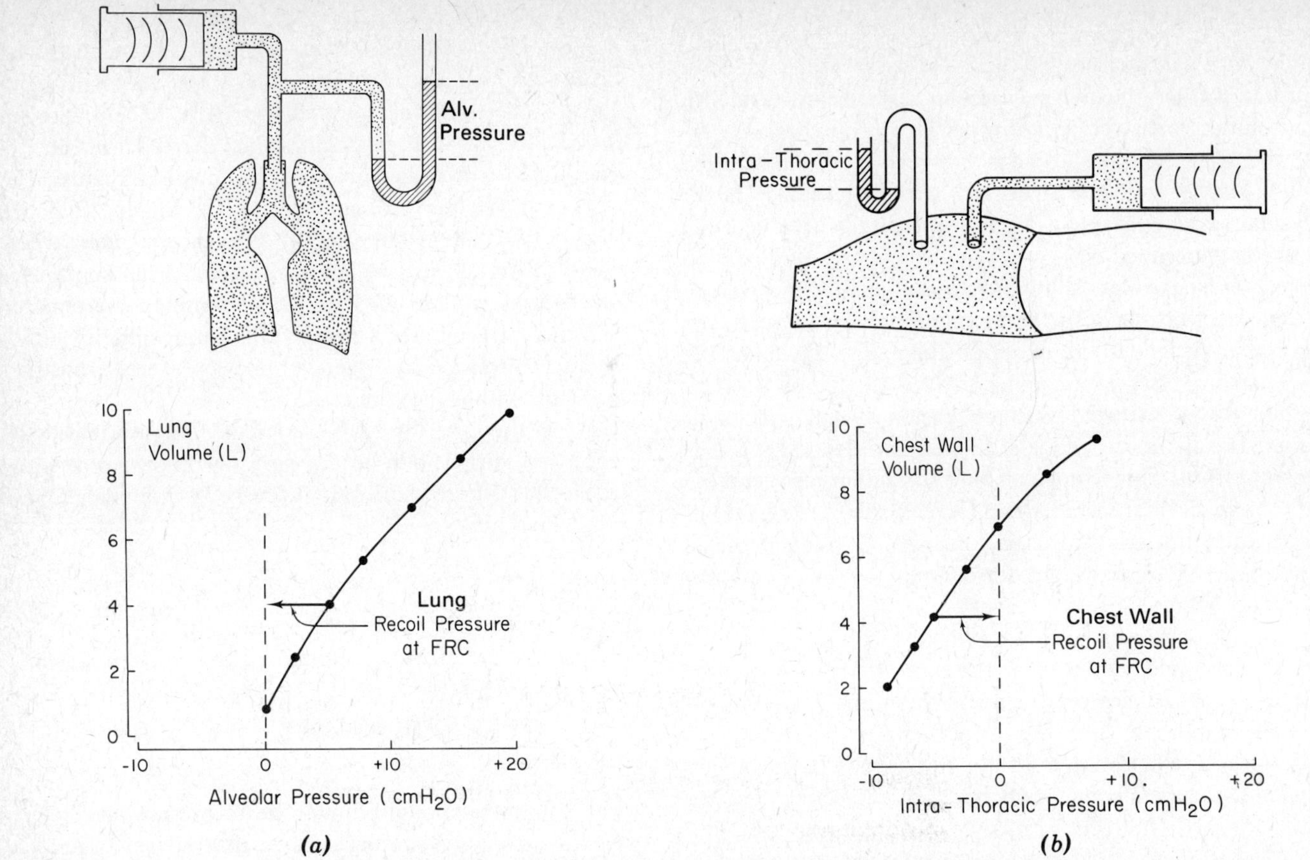

Alv. Pressure

Intra–Thoracic Pressure

Lung Volume (L)

Lung Recoil Pressure at FRC

Alveolar Pressure (cmH$_2$O)

(a)

Chest Wall Volume (L)

Chest Wall Recoil Pressure at FRC

Intra-Thoracic Pressure (cmH$_2$O)

(b)

Figure 11 (*a*) Inflation of excised lungs. Measurement of the static pressure–volume characteristics of the lung. The resultant curve (below) shows the relationship between static distending pressure (elastic recoil pressure) and lung volume. Note that at a lung volume of 4 L, an alveolar pressure of +5 cm H$_2$O is required to keep the lungs distended statically. (*b*) Distension of the eviscerated thorax. Measurement of the static pressure–volume characteristics of the eviscerated thorax. The resultant static relationship between thoracic volume and intrathoracic pressure describes the static recoil properties of the chest wall (rib cage and diaphragm). Note that at zero transmural pressure, the thorax contains approximately 7 L. When the volume is less than this, the chest wall tends to spring outward, with an intrathoracic pressure of −5 cm H$_2$O. Therefore, at this volume the inward recoil of the lungs exactly matches the outward recoil of the chest wall—the relaxation volume of the respiratory system (FRC).

oped by the chest wall (lung volume = 4 L, horizontal arrows in Fig. 11*a, b*).

This lung volume is the equilibrium volume, the FRC of the respiratory system. If either the chest wall or the lungs were punctured, air would leak between the pleural surfaces, creating a pneumothorax. If this continued, one might have no transmural pressure across the chest wall, and the lungs would be completely collapsed (Fig. 12). In this case, the chest wall volume would be 7 L, and the lung volume would equal the minimal air volume (about 0.5 L trapped gas).

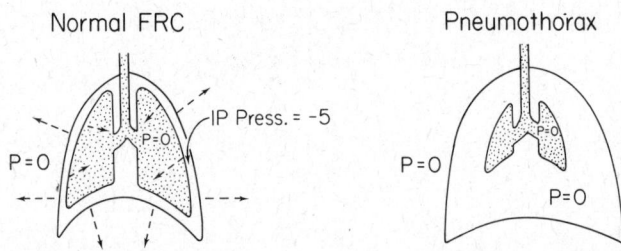

Figure 12 Left, The forces acting on the lung and chest wall at the end of a normal expiration (FRC). Right, the changes in lung and rib cage volume caused by pneumothorax.

MOVEMENTS OF THE RESPIRATORY SYSTEM

ACTIONS OF THE RESPIRATORY MUSCLES

The Diaphragm

Contraction of the principal respiratory muscle moves the dome of the diaphragm down and lifts the ribs, assisting the action of the inspiratory intercostals (Fig. 1). As inspiration proceeds, the effectiveness of the diaphragm diminishes, owing to two consequences of an increase in lung volume: (1) the length of diaphragmatic fibers decreases as the lung volume increases, and (2) the diaphragm becomes flatter at higher lung volumes. The first change means that for the same number of action potentials in the diaphragm, less force is generated. This is simply a manifestation of the force–length relationship of muscle. The second change means that the "mechanical advantage" of the diaphragm is compromised, that is, the same force develops less transdiaphragmatic pressure because of the increased radius of curvature (Laplace law).

Both factors compromise the efficiency of conversion of muscle action potentials into inspiratory pressure. Both play an important role in lung diseases associated with an increased compliance and an increased lung volume (obstructive lung disease). In these situations, the diaphragm is flat, so that fiber length is reduced and the radius of curvature is large.

Abdominal Muscles

Contraction of the abdominal muscles increases IP pressure and causes exhalation. During quiet breathing, the abdominal muscles are relaxed, and expiratory flow occurs because of the passive collapse of the respiratory system. However, during periods of increased breathing, expiratory airflow will increase because of contraction of the abdominal and internal intercostal muscles during expiration.

MAXIMUM VOLUME EXCURSION

Figure 13 displays a spirogram (a record of lung volume) recorded during spontaneous breathing and during maxi-

$$TLC = RV + VC$$

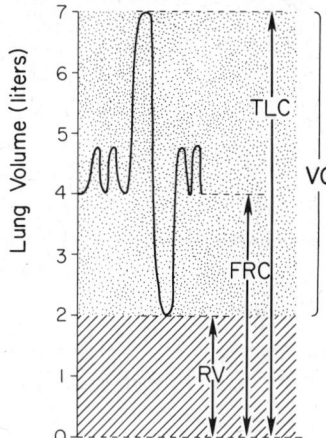

Figure 13 Normal forced inspiratory and expiratory spirogram. The maximum range of volume excursion is the vital capacity (VC). The volume remaining in the lung at the end of a normal expiration is the functional residual capacity (FRC). The volume remaining in the lung at the end of a maximum expiratory effort is the residual volume (RV). Total lung capacity (TLC) is comprised of the vital capacity and the residual volume.

mal inspiratory and expiratory efforts. The volume excursion with each inspiration and each expiration is termed the tidal volume (VT). The maximum possible volume excursion (i.e., the volume expired during a maximal expiration begun from the largest possible lung volume) is termed the vital capacity (VC). The residual volume (RV) is the volume remaining in the lung following the maximal expiratory effort. The largest possible lung volume (RV + VC) is referred to as the total lung capacity (TLC).

SELECTED BIBLIOGRAPHY

Bouhuys A: *Breathing.* New York, Grune & Stratton, 1974, p 121.

Comroe JH: *Physiology of Respiration,* ed 2. Chicago, Year Book Medical Publishers, 1974, p 94.

Goerke J: Lung surfactant. *Biochim Biophys Acta* 344:241, 1974.

Murray JF: *The Normal Lung.* Philadelphia, Saunders, 1976, p 77.

Mines AH: *Respiratory Physiology.* New York, Raven Press, 1981, pp 1, 17.

SELF-STUDY QUESTIONS

MULTIPLE CHOICE

Select the single best answer.

1. Which of the following measures the elastic recoil tendency of the lung?
   A. Alveolar pressure.
   B. Airway pressure.
   C. The difference between the alveolar pressure and the pressure on the pleural surface of the lung.
   D. Intrapleural pressure.
   E. Lung interstitial pressure.

2. An excised lung is inflated by a syringe. One liter is injected into the lung. The intratracheal pressure before injection was 5 cm H_2O. After injection, the pressure was 10 cm H_2O. What is the compliance of the lung?
   A. 10 cm H_2O/L.
   B. 5 cm H_2O/L.
   C. 1 L/cm H_2O.
   D. 0.2 L/cm H_2O.
   E. Cannot be determined from the data given.

Select the correct answer(s). (In many instances, more than one answer is correct.)

3. A lung with a patent trachea can be inflated by
   A. increasing the pressure acting on the pleural surface (intratracheal pressure = atmospheric pressure).
   B. decreasing the pressure acting on the pleural surface (intratracheal pressure = atmospheric pressure).
   C. decreasing the pressure in the alveoli (pleural surface pressure = atmospheric pressure).
   D. increasing the pressure in the alveoli (pleural surface pressure = atmospheric pressure).

4. When a lung is inflated,
   A. elastic recoil tendency increases.
   B. all parts of the lung increase in size.
   C. the outward pull (radial) on small airways increases.
   D. the relation between pressure and volume is linear over the entire range of lung volumes.

5. Alveolar surfactant differs from saline in which of the following ways?
   A. Alveolar surfactant provides a surface tension greater than water.
   B. Alveolar surfactant displays a surface tension that varies with area.
   C. Alveolar surfactant forms bubbles that are unstable.
   D. Alveolar surfactant contains a phospholipid.

6. The elastic recoil pressure of the lungs
   A. is always equal to alveolar pressure.
   B. changes with lung volume.
   C. is analogous to the compliance of a spring.
   D. is attributable, in large measure, to the presence of an air–fluid interface in the alveoli.

7. Which of the following influence the compliance of the lungs?
   A. Surfactant.
   B. Contraction of airway smooth muscle.
   C. Scar tissue in the lungs (pulmonary fibrosis).
   D. Hyperbaric conditions for 1 h.

8. The pressure–volume relationship for the lungs measured under static conditions with the glottis open
   A. provides the basis for estimating the compliance of the lung.
   B. is not feasible in humans.
   C. is not influenced by airway resistance.
   D. is linear over the entire lung volume range.

9. Movement of air into the lungs is associated with
   A. upward movement of the ribs.
   B. descent of the diaphragm.
   C. rotation of the ribs.
   D. increase in elastic recoil tendency of the lung.

10. The static elastic recoil pressure of the lung (alveolar pressure–intrapleural pressure)
    A. can be measured only in excised lungs.
    B. increases when lung volume increases.
    C. disappears when airflow occurs.
    D. is used in calculating lung compliance.

11. Two liters of gas are introduced into the intrapleural space of a paralyzed man (refer to data shown in Fig. 11). Which of the following consequences can be anticipated?
    A. The lung volume decreases by 2 Ls.
    B. The recoil pressure of the lungs cannot equal recoil pressure of the chest wall.

C. The chest wall volume will increase by 2 L.
D. Intrapleural pressure becomes less negative.

12. Which of the following measurements is *not* obtainable with simple spirometry?
    A. Tidal volume.
    B. Lung compliance.
    C. Vital capacity.
    D. Residual volume.

Select the single best answer.

13. In which condition listed below would FRC be equal to residual volume?
    A. Snorkle breathing at the surface of the water.
    B. Bilateral phrenic nerve transection.
    C. A patient with restrictive lung disease.
    D. A patient with spinal cord transection at C6.
    E. A patient with obstructive lung disease.

ANSWERS

1. C. Recoil tendency is measured by the difference between the pressure in the alveoli and the pressure on the pleural surface; this transmural pressure across the alveolar wall is the pressure that is distending the lung under static conditions. If the pressure acting on the surface of the lung is zero, alveolar or airway pressure measures recoil tendency. If the alveolar pressure is zero, the intrapleural pressure measures elastic recoil tendency.

2. D. Compliance is given by the ratio $\Delta V/\Delta P$, where ΔV is the change in lung volume and ΔP is the change in pressure distending the lung under conditions of no airflow. Accordingly, in this excised lung, the distending pressure rises from 5 cm H_2O to 10 cm H_2O, so that $\Delta P = 5$ cm H_2O. This change in distending pressure inflated the lung 1 L, so that $\Delta V = 1$ L. Hence, the compliance

$$\frac{\Delta V}{\Delta P} = \frac{1}{5} = 0.2 \text{ L/cm } H_2O$$

3. B, D. A lung can be inflated by making the alveolar pressure greater than the pressure at the pleural surface. This can be accomplished either by increasing the alveolar pressure or by decreasing the pressure at the pleural surface.

4. A, B, C. Inflation of a lung increases the stretch of the lung and, hence, it tends to recoil with greater force. This elastic recoil tendency is measured as the pressure difference between the alveolus and the pleural surface. Hence, as volume increases, recoil pressure increases. As mentioned, all parts of the lung expand and the airways caliber increases because of the radially directed forces of the lung parenchyma acting to pull the airway open. The curve for the static $\Delta P/\Delta V$ relationship of the lung plateaus at high lung volumes.

5. B, D. Without the surfactant phospholipid, lung compliance decreases and alveoli collapse, as outlined earlier. An extract of surface lining material

forms tiny bubbles that are extraordinarily stable, because (1) surface tension is less than that of water, and (2) the surface tension of surfactant varies with surface area.

6. **B, D.** Recoil pressure is equivalent to transmural alveolar pressure, not alveolar pressure. It is analogous to the force, not compliance, of a spring and is relatively low because pulmonary surfactant lowers surface tension below that of water. As lung volume increases, the alveolar surface is expanded or stretched, and this causes increased recoil pressure. Without an air–fluid interface, however, the recoil tendency is very low.

7. **A, C.** Surfactant is the major determinant of recoil pressure and how this pressure changes as volume increases. The alveolar surface area increases when the lung is inflated, thereby increasing the alveolar surface tension and increasing the static recoil pressure. Recoil pressure and compliance depend on transmural alveolar pressure, not on absolute pressure. Therefore, unless hyperbaric exposure were complicated by oxygen toxicity, hyperbaric conditions should not influence pulmonary compliance. Contraction of airway smooth muscle increases airway resistance but does not influence static recoil pressure or compliance. However, extensive deposition of collagen in the alveolar wall (fibrosis) greatly restricts the expansion of the lung (restrictive disease) because the compliance is decreased.

8. **A, C.** Pulmonary compliance is commonly measured in diagnosis of lung disease by plotting the relation between lung volume and intrapleural pressure (assumed to be equal to esophageal pressure). Since the measurement is made under conditions of no flow (static) conditions, the airway resistance is irrelevant.

9. **All are correct.** During inspiration, the rib cage increases in volume because the ribs move up and out in a rotating motion. The diaphragm descends and causes the abdomen to protrude. Whenever the lung volume increases, the elastic recoil forces increase.

10. **B, D.** Compliance can be estimated in humans using esophogeal pressure and measuring this pressure under static conditions with the glottis open. The pressure difference (alveolar pressure–IP pressure) is the transmural pressure of the alveolus, which increases as lung volume increases, whether airflow is present or not.

11. **D.** If the chest wall were fixed, the lung volume would decrease 2 L, and if the lungs were fixed, the chest wall would increase by 2 L. Neither occurs, since both structures move. However, after the new volumes of chest wall and lungs are established, their recoil tendencies will be less but will be equal, each tending to recoil in the opposite direction. Intrapleural pressure will be less negative because the lung volume will be lower.

12. **B, D.** Compliance requires measurement of IP pressure (usually approximated by esophageal pressure), and residual volume must be measured indirectly (e.g., by an indicator dilution method).

13. **D.** Snorkle breathing will reduce the volume that can be inspired above FRC because of the hydrostatic pressure acting on the chest wall, but it will not reduce FRC to RV. Similarly, restrictive disease (decreased lung compliance) will reduce FRC, but it will not make FRC equal RV. Loss of diaphragm contraction decreases the volume that can be inspired but does not shift FRC. Obstructive disease (increased compliance) increases RV and FRC, but the two do not become equal. Section of the cord at C6 leaves the phrenic innervation intact but causes paralysis below that point (quadriplegia). This means that the expiratory muscles are paralyzed. Accordingly, the patient cannot expire below FRC. In this case, FRC is normal, but RV is equal to FRC.

19

Airflow in the Lungs

JOHN E. REMMERS

OBJECTIVES

After completion of this chapter, you should be able to

1. Know what is meant by airway resistance and to appreciate the driving pressure that produces flow through the airways.

2. Understand the two components of intrapleural pressure: elastic recoil pressure and flow pressure.

3. Know the relationship between lung volume and airway resistance.

4. Understand how a forced expiratory maneuver produces collapse of airways and limitation of flow.

THE DYNAMIC STATE

The respiratory system is in constant motion. Pauses in the respiratory rhythm are abnormal, except for brief pauses during sleep. The static mechanics of the respiratory system described in Chapter 18 are only part of the story of pulmonary mechanics—superimposed on the static pressure–volume relationship are pressures generated when gas is caused to flow in the airways. These dynamic pressures add to or subtract from the elastic recoil pressure of the lung, as can be seen in the following example.

Let us consider in detail the events that occur during a single breath. Just before inspiration begins, the respiratory system is at its equilibrium position of functional residual capacity (FRC), and intrapleural (IP) pressure is about -5 cm H_2O, as described in Chapter 18, Figure 11. The inspiratory muscles then develop force, tending to enlarge the chest wall. This is transmitted to the intrapleural space, thereby further reducing IP pressure (i.e., IP pressure becomes more negative) (Fig. 1). Because the lungs are still at FRC, recoil pressure has not changed and, therefore, alv. pressure is reduced by an equal amount. This means that the pressure at the mouth is greater than in the alveolus, and inspiratory flow begins.

Pressures during Breathing

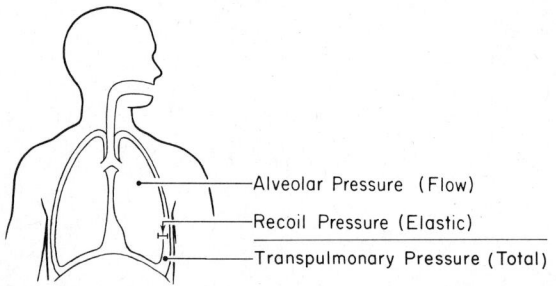

Alveolar Pressure (Flow)
Recoil Pressure (Elastic)
Transpulmonary Pressure (Total)

Figure 1 The pressures developed during flow. The difference between atmospheric and alveolar pressures gives the pressure necessary to overcome resistance to airflow. The difference between alveolar pressure and intrapleural (IP) pressures gives the pressure necessary to overcome the elastic properties of the lung. The total pressure across the lung (transpulmonary pressure) is equal to the sum of the flow-related pressure and the elastic recoil pressure. This total pressure must be equal and opposite to the IP pressure.

As inspiration progresses, lung volume increases; therefore, elastic recoil pressure increases as well (i.e., as lung volume increases, the absolute value of the difference between alveolar pressure and IP pressure increases). However, to maintain flow, alveolar pressure must be maintained less than mouth pressure. That is, total pressure difference from mouth to intrapleural space (transpulmonary pressure) has two components: (1) the pressure necessary to overcome the elastic forces of the lung, and (2) the pressure necessary to produce flow through the airways. Intrapleural pressure is equal to the transpulmonary pressure, but has the opposite sign during inspiration.

During inspiration, (as shown in Fig. 1), the pressure at the surface of the lung, IP pressure, must perform two operations: (1) produce airflow down the airways, and (2) overcome the elastic forces of the lung. To produce flow, alveolar pressure must be less than atmospheric pressure. To offset the elastic properties of the lung, IP pressure must be less than alveolar pressure. We shall repeatedly use the concept that these two pressures summate. A person wishing to inspire must use the inspiratory muscles to generate a more subatmospheric pressure in the IP space. This pressure will do two types of work: (1) it will stretch the lungs and thereby do elastic work; and (2) it will cause flow down the airway and thereby do frictional work against the resistance of the airway. We shall refer to the latter as the flow-resistive component of IP pressure and to the former as elastic recoil component of IP pressure. Intrapleural pressure represents the total pressure drop from mouth to the lung surface (i.e., it is transpulmonary pressure). This total always equals the sum of two components, namely, the elastic recoil component and the flow-resistive component.

The behavior of the two components of intrapleural pressure is illustrated in Figure 2, which shows how the two components of IP pressure change from the beginning of inspiration until a 1 L tidal volume has been inspired. Changes at the beginning, midpoint, and end of a normal inspiration are shown in Figure 2.

Condition A gives the pressures at the end of a normal expiration (just before inspiration begins). There is no airflow. Therefore, alveolar pressure equals atmospheric. Intrapleural pressure is less than atmospheric by the amount dictated by the elastic recoil properties of the lung (alveolar pressure = 0; elastic recoil pressure = 5 cm

A.   NO AIRFLOW: Lung Volume = FRC

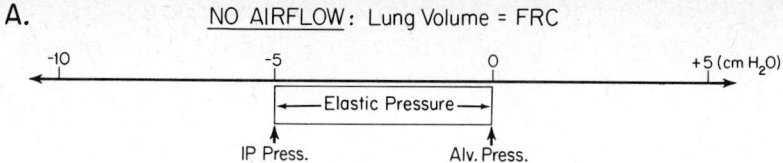

B.   INSPIRATORY AIRFLOW : Lung Volume = FRC + ½V_T

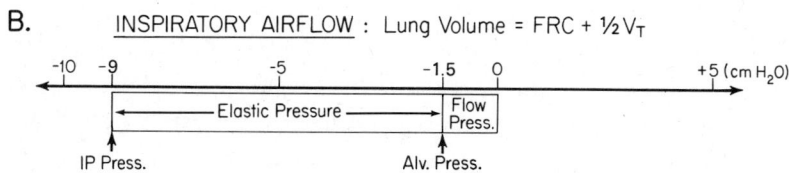

C.   NO AIRFLOW: Lung Volume = FRC + V_T

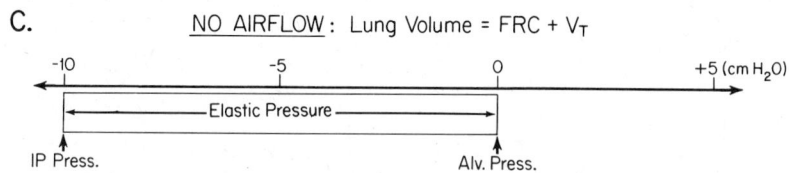

Figure 2 Changes in the elastic and flow component of intrapleural pressure during inspiration of a 1 L tidal volume.

H_2O; transpulmonary pressure = 5 cm H_2O). Since IP pressure is equal to transpulmonary pressure with the sign reversed, IP pressure = −5 cm H_2O.

Condition B depicts what happens midway through inspiration, when air is actually flowing down the airways and into the alveoli. In order to generate this airflow, alveolar pressure must be less than mouth (atmospheric) pressure; in this example it is shown as −1.5 cm H_2O (i.e., flow pressure = 1.5 cm H_2O). In addition, the lung volume has increased to 0.5 L above FRC, so that the pressure necessary to offset the elastic recoil forces of the lung is greater. (In this example, elastic recoil pressure is 7.5 cm H_2O.) Transpulmonary pressure represents the sum of these two pressures, namely 9 cm H_2O; therefore, IP pressure is −9 cm H_2O.

Condition C shows the pressures at the end of inspiration. A 1 L tidal volume has been inspired, and airflow has stopped. Alveolar pressure equals atmospheric pressure again, because there is no airflow. However, the lung volume is greater than in condition A, hence a greater pressure is required to offset the elastic recoil forces of the lung (elastic recoil pressure = 10 cm H_2O). Therefore, transpulmonary pressure = alveolar pressure + elastic recoil

pressure = 0 + 10 = 10 cm H_2O; IP pressure = −10 cm H_2O.

These same events can be represented graphically, as shown in Figure 3. For a normal size tidal volume, the compliance of the lung is constant; therefore, as lung volume increases, the pressure necessary to offset elastic forces of the lung increases linearly, as shown in Figure 3a. This static pressure–volume curve is measured when there is no flow by stepwise inflation of the lungs, as described in Chapter 18. In this case, the horizotal distance between the y axis and the static curve represents the IP pressure required to offset the recoil pressure of the lungs. This pressure must be exerted on the lungs whenever the lung is being inflated. The pressure equals the static elastic pressure, and pressure required to produce flow will be added to this pressure.

The pressure driving flow down the airways is the difference between atmospheric and alveolar pressure. For a constant airway resistance, flow will increase as driving pressure increases. If alveolar pressure is very near atmospheric, the driving pressure for flow will be very small and airflow will be low, as depicted in Figure 3b for a slow inspiration. The breath begins at FRC with IP pressure

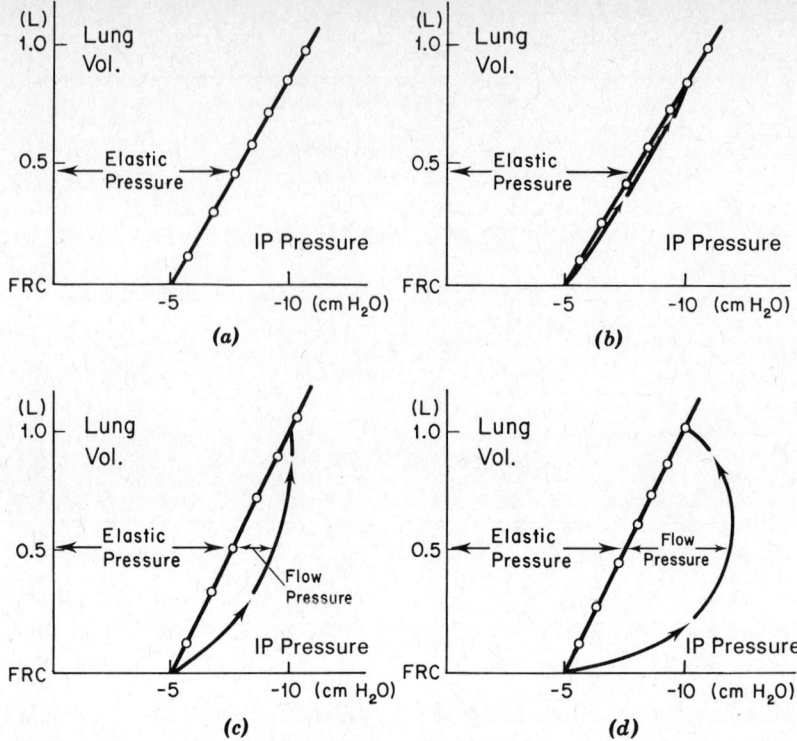

Figure 3 The relationship between IP pressure and lung volume for a normal person is shown. (*a*) Static pressure–volume relationship (same as Chapter 18, Fig. 8). (*b*) The subject makes a very slow inspiration, so that alveolar pressure is only slightly less than atmospheric pressure. Hence, the IP pressure is almost entirely the result of elastic recoil of the lung. (*c*) The subject inspires normally; IP pressure has two components: elastic recoil and flow pressure. (*d*) The subject inspires rapidly and IP pressure has a large component-related airflow. The component of IP pressure related to elastic recoil forces of the lung does not depend on the magnitude of airflow, but on lung volume alone. The component of IP pressure related to airflow resistance varies directly with the rate of airflow and the airway resistance. The measured IP pressure is the sum of these two components.

equal to the elastic recoil pressure. If inspiratory airflow is very small, alveolar pressure is near zero, and the breath is described by the solid line moving up just to the right of the static *P–V* line. The driving pressure required to produce a very low flow is extremely small, so that virtually all the IP pressure is used to overcome elastic forces —"flow pressure" is small, and virtually all the pressure applied to the lung is used to offset elastic recoil pressure of the lung.

With more rapid inspirations, inspiratory airflow is

greater. To produce a higher airflow, the driving pressure must be greater. That is, alveolar pressure must be below atmospheric, as shown for a normal inspiration in Figure 3*c*. This means that the breath moves up along a line that lies to the right of the static P–V line, because IP pressure is more subatmospheric than is necessary to offset elastic forces. At the end of inspiration, airflow is zero once again, hence IP pressure is again equal to the elastic recoil pressure. During the inspiration, however, IP pressure was the sum of the flow-resistive component and the elastic recoil component. The pathway followed during this normal inspiration corresponds to the situations depicted in Figure 2.

With a very rapid inspiration, a large driving pressure for airflow (atmospheric pressure minus alveolar pressure) is generated, producing a high inspiratory flow rate, as shown in Figure 3a. Accordingly, alveolar pressure must be lower (more subatmospheric) than before. To accomplish this, IP pressure must be lower than before, and the breath follows a path that deviates considerably from the elastic recoil curve. The solid curve gives IP pressure (total pressure). The distance between the *y* axis and the static P–V curve indicates pressure required to offset elastic recoil of the lung. The horizontal distance between this static line and the dynamically observed curve reflects that part of IP pressure used to overcome the flow resistance of the airways. Again, during inspiration, IP pressure is the sum of the flow-resistive component and the elastic component.

SPONTANEOUS BREATHING

With the aid of a spirometer and a device for measuring the rate of airflow, we can observe the dynamic aspects of spontaneous breathing in a normal person and relate flow and volume to the pressure in the alveolus and in the IP space, as shown in Figure 4. Recall that the difference between these two pressures is the pressure required to offset the elastic recoil force of the lung. This transmural pressure of the alveolus is linearly related to lung volume (constant lung compliance) over the tidal range. Figure 4 shows the time course of lung volume, airflow, recoil pressure of the lung, and alveolar pressure for a single respiratory cycle in a normal person.

Inspiration

The recordings in Figure 4 begin at time zero, just as expiration has ended, and the spirometer volume (top trace) at end-expiration is arbitrarily referred to as zero. A 1 L tidal volume is inspired over the ensuing 2 seconds beginning from FRC. As lung volume increases from FRC to FRC + 1 L, the elastic recoil pressure goes from 5 cm H_2O to 10 cm H_2O (lung compliance is 1 L/5 cm H_2O = 0.2 L/cm H_2O). Note that elastic recoil pressure changes linearly with the change in lung volume. As volume in-

creases, the pressure across the alveolar wall required to offset elastic forces increases progressively also. That is, the difference between alveolar and IP pressure increases progressively with lung volume. In contrast, alveolar pressure (fourth trace from top) changes in a fashion that parallels the rate of airflow. As alveolar pressure falls below atmospheric pressure, flow increases, because the difference between atmospheric pressure and alveolar pressure represents the driving pressure for flow in the airways. Inspiratory airflow rises to a maximum value and then declines during the subsequent portion of inspiration; alveolar pressure displays a similar pattern, decreasing to − 2 cm H_2O and increasing thereafter. Intrapleural pressure measures the total pressure drop across the lung, and therefore must be equal to the sum of elastic and alveolar pressures. The bottom trace gives summation continuously during inspiration. Intrapleural pressure becomes progressively more negative during inspiration and passes through a maximum of − 10.5 cm H_2O shortly before the end of inspiration.

Expiration

During expiration (Fig. 3), elastic recoil pressure goes from − 10 cm H_2O to − 5 cm H_2O as the tidal volume is expired and lung volume decreases back to FRC. Alveolar pressure becomes positive during this period, thereby providing the driving pressure that causes gas to flow out of the airways. Total pressure across the lung (IP pressure) is less negative than recoil pressure because alveolar pressure is positive during expiration. For instance, at 3 s, alveolar pressure = + 1, elastic pressure = −7. The total IP pressure = + 1 + (−7) = −6 cm H_2O.

THE P–V LOOP

The manifestation of inspiratory and expiratory flow can also be seen on a continuous plot of lung volume against IP pressure shown in Figure 5. Inspiration begins at the dot on the abscissa (− 5 cm H_2O), which gives the elastic recoil pressure of the lung at FRC. The dashed line gives the static recoil line for the lung. During inspiration, the dynamic *P–V* relationship lies to the right of this line— at any lung volume, total pressure is more negative than recoil pressure, since additional pressure is required to generate flow through the airways. At the end of inspira-

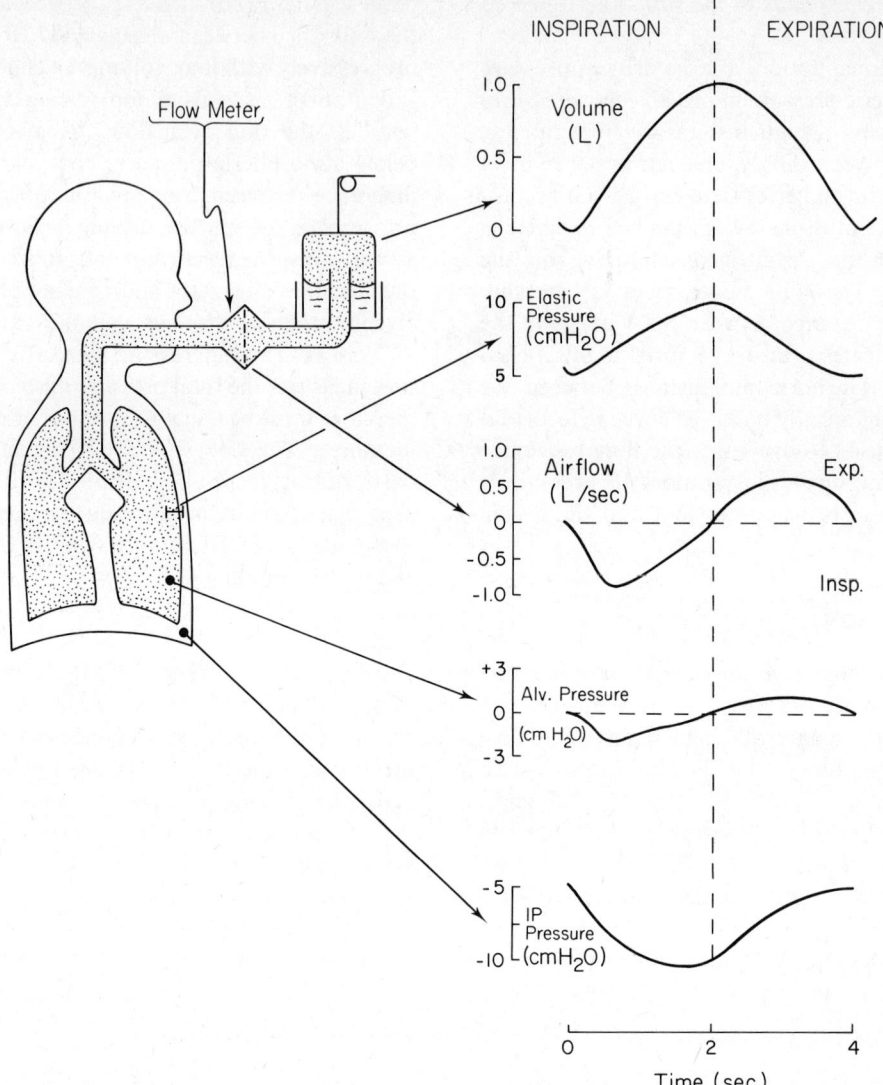

Figure 4 Volume, flow, and pressure during normal quiet breathing are recorded for a single respiratory cycle. During inspiration, the lung volume increases progressively and elastic recoil pressure of the lung changes as expected for a constant lung compliance. Also during inspiration, airflow occurs, hence alveolar pressure must be subatmospheric. The total pressure (IP pressure) is the sum of recoil and alveolar pressure, as shown at the bottom. During expiration, alveolar pressure must be greater than atmospheric, hence the total pressure is less than recoil pressure.

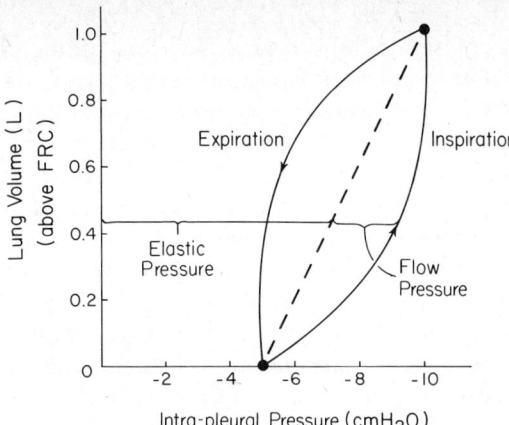

Figure 5 Relationship between instantaneous intrapleural pressure and lung volume, shown in the records of Figure 3. Inspiration begins with lung volume at FRC and IP pressure equal to -5. During inspiration, the IP pressure lies to the right of the elastic recoil line (dashed line). The difference between the two equals alveolar pressure. During expiration, IP pressure is less than recoil pressure because of pressures developed in generating flow through the airway resistance.

tion, when there is no airflow, the observed trace lies on the static recoil line (upper dot). During expiration, the situation is the reverse of that during inspiration: for any volume, IP pressure is less negative than recoil pressure because alveolar pressure is positive, and therefore the sum is less negative than if IP pressure were equal to zero. Accordingly, the P–V relationship during expiration lies to the left of the static recoil line and merges with this line as the lung volume returns to FRC and expiratory flow ceases.

In summary, spontaneous breathing forms a P–V loop with the inspiratory and expiratory limbs being displaced from the static recoil line. The width of the loop depends on the flow rates and the airway resistance.

AIRWAY RESISTANCE

We have seen that the airways present resistance to gas flow during breathing and that this resistance to flow is manifest by an IP pressure that is different from that expected from the static recoil properties of the lung. Airway resistance can be calculated by dividing the rate of airflow into the driving pressure (atmospheric–alveolar pressure difference):

$$\text{Airway resistance} = \frac{\text{alveolar pressure}}{\text{airflow}}$$

We now ask, What factors influence airway resistance? This is a deserving question in view of the central role of airway obstruction in many respiratory diseases. Like many conduits in the body, the size of the bronchi and bronchioles depends on both active factors (bronchomotor tone) and transmural pressures of the airway. We will now consider each of these.

Bronchomotor Tone

Airway diameter varies greatly with variations in the force generated by intramural smooth muscle. Whereas a systematic treatment of the factors controlling this force is beyond the scope of this presentation, it is important to point out that various humoral agents (e.g., histamine) and neural factors (sympathetic and parasympathetic component of the autonomic nervous system) greatly influence bronchomotor tone, hence airway caliber and airway resistance.

Radial Traction and Lung Volume

The pulmonary parenchyma surrounding a collapsible airway pulls outward on the lung, and this distending action tending to dilate the airway is termed radial traction. This outward pull on the small airways of the lung is depicted in Figure 6, in which the lung parenchyma is represented by springs exerting radial forces on the collapsible airway. As mentioned previously, and as shown in

Radial Traction

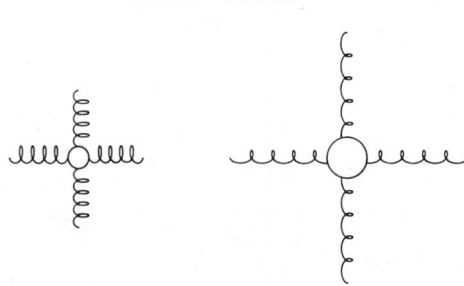

Low Lung Volume   High Lung Volume

Figure 6 The outward pull of the lung parenchyma and a small collapsible airway (radial traction) is represented by radially positioned springs. When lung volume increases, the springs are stretched and the airway is dilated by increased radial traction.

Chapter 18, Figure 4, all parts of the lung enlarge as lung volume increases; this action of an increase in lung volume to increase airway caliber was emphasized. As lung volume decreases from near its maximum value (transpulmonary = 30 cm H_2O and lung volume = 7 L) to its minimal volume, average airway diameter decreases 35%. The change in caliber is particularly significant when changes in volume near FRC or below are considered. The consequence of this influence of lung volume on airway size can be seen in Figure 7. As lung volume decreases 50% from its largest value, airway resistance increases somewhat. Below this value, resistance rises dramatically, reflecting the great decrease in airway diameter that occurs at low lung volumes.

Transmural Airway Pressure

Transmural pressures of the airways can be estimated by calculating the difference between intra-airway pressure and intrapleural pressure. As shown in Figure 8, intra-airway pressure usually exceeds IP pressure, tending to hold the airways open. Estimates for the transmural airway pressure at FRC, during inspiration, or at the end of inspiration (FRC + V_T) are shown in this figure. Intrapleural pressure is usually 5–8 cm H_2O *less than* airway pressure, indicating that the airways are normally distended by transmural pressure. However, during expiration, the transmural airway pressure depends on the way in which expiration is achieved. For example, if expiration is a passive act, IP pressure remains substantially negative and acts to keep the airways distended. However, this situation is altered whenever large expiratory efforts are made. In such situations, IP pressure is *greater than* airway pressure, as shown in Figure *8d*. Alveolar pressure also rises during a forced expiration (IP pressure—alveolar pressure is still dictated by recoil pressure), but because high flows are generated during such efforts, the pressure within certain airways is appreciably less than alveolar and IP pressures. This transmural pressure has the net effect of narrowing of the airways during large expiratory efforts. Such decrease in airway caliber caused by large expiratory efforts and high expiratory flow rates is called *dynamic airway collapse.*

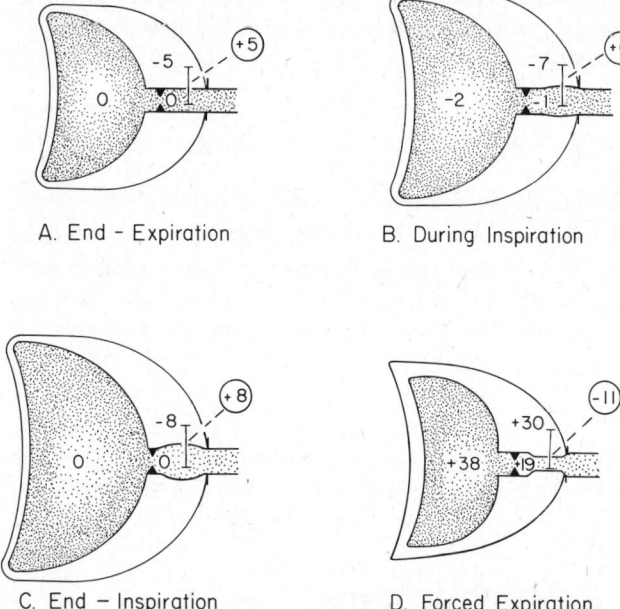

A. End - Expiration   B. During Inspiration

C. End – Inspiration   D. Forced Expiration

Figure 8 (*a–c*) Airway, alveolar, and IP pressures at end-expiration, during inspiration, and at the end of the normal inspiration. In all three, IP pressure is less than airway pressure, so that the transmural pressure acts to distend the airways. (*d*) Pressures existing during forced expiration. Recoil pressure of the lung is the same as in (*c*), but because of flow through the airway resistance, the IP pressure is greater than airway pressure. Consequently, the transmural pressure of the airways tends to collapse them.

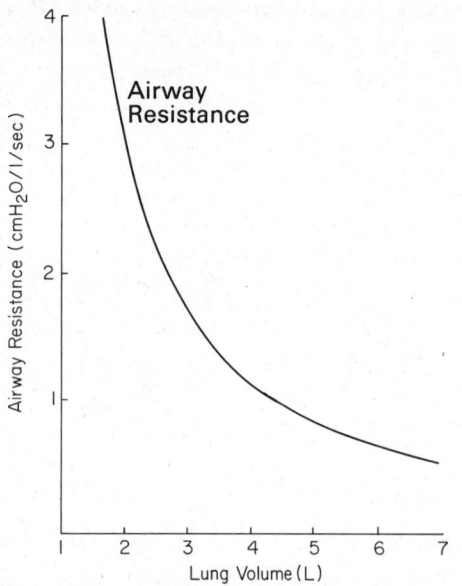

Figure 7 The normal relationship between airway resistance and lung volume is shown. The decrease in airway diameter at low lung volume causes a pronounced increase in airway resistance.

An important ultimate consequence of dynamic collapse is the appearance of expiratory flow limitation. This arises from the interrelationship between transmural airway pressure and resistance. A large increase in IP pressure during a forced expiratory maneuver causes a great increase in alveolar pressure, which tends to increase expiratory flow rate. However, any increase in expiratory flow rate leads to more negative transmural airway pressure and, hence, more "dynamic collapse." In a normal person, this collapse occurs at the level of the segmental bronchi. This narrowing causes higher resistance, thereby minimizing the increase in flow. The interplay of these actions is such that progressively greater efforts are accompanied by progressively smaller increases in flow rate. Ultimately, expiratory airflow rate becomes virtually fixed. In other words, a self-limiting process normally develops during severe expiratory efforts whereby a further increase in IP pressure causes no further increase in flow rate because an offsetting rise in airway resistance occurs owing to a greater collapsing action of transmural pressure. This situation normally occurs when substantial expiratory efforts are made at lung volume equal to 50% of VC or below. Once IP pressure is high enough to produce a large amount of dynamic collapse, flow rate is effectively independent of the degree of effort. This situation arises at higher lung volumes and less extreme pressures in the presence of obstructive disease of the lungs, because not as great an increase in IP pressure is required to produce a great deal of dynamic collapse of the airways.

SELECTED BIBLIOGRAPHY

Bouhuys A: *Breathing*. New York, Grune & Stratton, 1974, pp 145, 174.

Comroe JH: *Physiology of Respiration,* ed 2. Chicago, Year Book Medical Publishers, 1974, p 94.

Murray JF: *The Normal Lung*. Philadelphia, Saunders, 1976, p 77.

SELF-STUDY QUESTIONS

Multiple Choice

Select the correct answer(s). (In many instances, more than one answer is correct.)

1. Contraction of the diaphragm causes

    A. intrapleural pressure to fall (to become more subatmospheric).

    B. alveolar pressure to rise.

    C. an increase in anterioposterior dimension of the abdomen.

    D. a decrease in recoil tendency of the lung.

2. During an inspiration (while airflow is occurring) in a quietly breathing normal subject,

    A. the abdomen decreases in anteriorposterior dimension.

    B. the intrapleural pressure is less (more subatmospheric) than during expiration.

    C. the airways constrict.

    D. the pressure in the alveoli is less than at the beginning or end of inspiration.

3. Inspiratory airflow occurs whenever

    A. alveolar pressure equals intrapleural pressure.

    B. alveolar pressure is less than atmospheric pressure (assuming the airways are patent).

    C. intrapleural pressure is less than atmospheric pressure.

    D. the total pressure difference between mouth and intrapleural space exceeds the elastic recoil pressure of the lung.

4. At the end of a normal inspiration (just before expiration begins, no airflow),

A. alveolar pressure is greater than intrapleural pressure.

B. alveolar pressure is less than atmospheric pressure.

C. intrapleural pressure is less than atmospheric pressure.

D. intrapleural pressure is greater than elastic recoil pressure of the lung.

5. Airway resistance is increased

A. by "radial traction" (outward pull of lung parenchyma).

B. during a forced expiratory maneuver.

C. by a decrease in bronchomotor tone.

D. by a decrease in lung volume.

6. At the end of a normal expiration (no airflow), intrapleural pressure is

A. less than atmospheric pressure.

B. a measure of the elastic recoil pressure of the lung.

C. less than alveolar pressure.

D. a measure of the elastic recoil pressure of the chest wall.

7. During a normal inspiration (i.e., while airflow is present), intrapleural pressure is

A. less than atmospheric.

B. equal to the recoil pressure of the lung.

C. less than alveolar pressure.

D. equal to the recoil pressure of the thorax.

8. During a normal expiration (i.e., while airflow is present), intrapleural pressure is

A. greater than atmospheric pressure.

B. greater than alveolar pressure.

C. linearly related to lung volume.

D. equal to the sum of lung recoil pressure plus alveolar pressure.

9. Which of the following influence the rate of pulmonary airflow during normal quiet breathing?

A. Airway resistance.

B. Intrapleural pressure

C. Alveolar pressure

D. Dynamic airway collapse

ANSWERS

1. A, C. Force generated by the diaphragm decreases intrapleural pressure because it tends to expand the chest wall (tends to make the diaphragm descend). This, in turn, expands the lung and decreases alveolar pressure. As the lung expands, the recoil tendency of the lung increases and the abdomen increases in size.

2. B, D. The intrapleural pressure decreases during inspiration, and the alveolar pressure must decrease below atmospheric pressure or else inspiratory flow would not occur. The airways get larger during inspiration and smaller during expiration.

3. B, D. During inspiration, alveolar pressure decreases below atmospheric pressure and this *causes* inspiratory flow. Intrapleural pressure is always less than alveolar pressure; this is simply a result of the elastic recoil of the lung. Inspiratory flow will not occur unless the total pressure drop from mouth to pleural space is

greater than that required to offset the elastic recoil tendency of the lung. When IP pressure is exactly equal to elastic recoil pressure, there is a balance of equal and opposite forces, and the respiratory system is static.

4. A, C. At the end of inspiration, flow has ceased and the alveolar pressure is exactly equal to atmospheric pressure. However, IP pressure is less than alveolar (= atmospheric) pressure because of the elastic recoil of the lung. Because there is no flow, the pressure at the pleural surface of the lung (IP pressure) exactly equals the elastic recoil pressure at this time.

5. B, D. "Radial traction" refers to the outward pull of the pulmonary parenchyma on the bronchioles—it acts to decrease airway resistance. Dynamic collapse of airways occurs during a forced expiratory maneuver. Accordingly, resistance is increased by a forced expiratory maneuver but is decreased by radial traction. A de-

crease in lung volume relaxes the radial pull exerted by the lung parenchyma on the airways. Bronchoconstriction results from an increase in contraction of bronchial smooth muscle and causes an increase in airway resistance.

6. All are correct. At FRC, with no airflow, the intrapleural pressure results only from the elastic recoil pressure of the lungs and of the chest wall. These two structures recoil in opposite directions, and the IP pressure is a result of the equal elastic recoil tendency of both structures. Since there is no airflow, alveolar pressure equals atmospheric pressure, and IP pressure is less than both.

7. A, C. Because there is inspiratory airflow, alveolar pressure must be less than atmospheric pressure. Intrapleural pressure is less than alveolar pressure because of lung recoil. However, the *difference,* IP pressure minus alveolar pressure, is a measure of the recoil pressure in the presence of inspiratory flow.

8. D. When expiratory flow is present, alveolar pressure is greater than atmospheric pressure. Intrapleural pressure is less than alveolar pressure because of the recoil tendency of the lung. Intrapleural pressure is not linearly related to lung volume; the difference, alveolar pressure minus IP pressure, is linearly related to lung volume. Why? Only during active expiration or during passive expiration from a large lung volume will IP pressure exceed atmospheric pressure. The IP pressure will always equal recoil pressure plus alveolar pressure, since the difference between the pressure inside the alveolus and outside the lung is solely attributable to the tendency of the alveolar wall to contract, that is, to elastic recoil of the lung.

9. A, B, C. If IP pressure changes because of contraction of the respiratory muscles, alveolar pressure will change. Dynamic collapse occurs only during substantial expiratory efforts, and therefore is not a factor during normal quiet breathing. Airway resistance will determine the airflow rate that will be produced by any particular driving pressure (i.e., for any particular atmospheric minus alveolar pressure difference). For any particular airway resistance, flow will be proportional to the driving pressure, that is, it will be directly related to alveolar pressure. These relationships are described by the following relationship (an expression of Ohm's law):

$$\text{Airflow} = \frac{\text{alveolar pressure}}{\text{airway resistance}}$$

20

Circulatory Transport of Respiratory Gases

JOHN E. REMMERS

OBJECTIVES

After completion of this chapter, you should be able to

1. Understand how CO_2 and O_2 are reversibly bound and stored in blood, and have facility with a graphic description of this process (i.e., the CO_2 and O_2 dissociation curves for whole blood).

2. Describe the relationship between gas pressure and content and explain its molecular basis.

3. Identify factors influencing the above relationship, describe qualitatively the direction of dissociation curve shifts, and explain the molecular basis of this action.

TRANSPORT OF O_2 BY THE BLOOD

Oxygen enters the blood in the pulmonary capillaries by diffusing down its partial pressure gradient from alveolar air. It is taken up by the blood and carried to the tissues in two forms: physically dissolved O_2 and oxygen in combination with hemoglobin.

PHYSICALLY DISSOLVED O_2

Physically dissolved oxygen refers to free O_2 molecules in physical solution in the water content of blood. It is interesting that oxygen, the energetically essential molecule, is only slightly soluble in blood and other body fluids. At body temperature, blood equilibrated with a normal arterial Po_2 of 90 mm Hg contains only 0.3 vol% O_2 (0.3 ml O_2/100 ml blood) in physical solution. This represents less than 2% of the total amount of oxygen carried by arterial blood. Since most of the water in the blood is extracellular, most of the physically dissolved O_2 resides in plasma. If blood contained only dissolved O_2, a man at rest, consuming O_2 at a rate of 250 ml/min, would require a cardiac output of 80 L/min, that is, about 16 times the normal resting value.

Fortunately, O_2 transport does not solely depend on the ability of blood to carry O_2 in physical solution. More than 98% of the total amount of oxygen carried by arterial blood is carried in *reversible* chemical combination with hemoglobin, as described below.

O_2 COMBINED WITH HEMOGLOBIN

An important property of hemoglobin (Hb) is the ability of the ferrous ion in the heme portions of the hemoglobin molecule to combine with O_2 reversibly. This combination does not involve oxidation of Fe^{2+} to Fe^{3+} (the iron remains in the ferrous state); therefore, the formation of oxygenated hemoglobin (oxyhemoglobin, or HbO_2) is a process of oxygenation, and not of oxidation. The release of oxygen from oxyhemoglobin forms deoxygenated hemoglobin (deoxyhemoglobin, or Hb), a process called deoxygenation, not reduction. Each molecule of Hb con-

tains four iron-containing heme groups, each of which is capable of binding one molecule of O_2. The oxygenation reactions are more accurately expressed as

$$Hb + O_2 \rightleftharpoons HbO_2 \qquad (1)$$
$$HbO_2 + O_2 \rightleftharpoons Hb(O_2)_2 \qquad (2)$$
$$Hb(O_2)_2 + O_2 \rightleftharpoons Hb(O_2)_3 \qquad (3)$$
$$Hb(O_2)_3 + O_2 \rightleftharpoons Hb(O_2)_4 \qquad (4)$$

The average number of oxygenated heme groups per Hb molecule, and thus the amount of O_2 bound per unit volume of blood, depends on the concentration of dissolved oxygen, in turn determined by the partial pressure of oxygen in the solution. This general idea can be summarized as follows:

$$Po_2 \rightleftharpoons \text{dissolved } O_2 + Hb \rightleftharpoons HbO_2$$

For example, if one begins with a completely deoxygenated Hb solution and exposes it to room air, O_2 molecules will diffuse into the aqueous phase, where they will be present as physically dissolved oxygen. By mass action, this shifts reaction (1) to the right and forms HbO_2 and distrubs the equilibrium of the other three reactions. As more O_2 diffuses into the solution, the concentration of dissolved O_2 continues to rise and the amount of HbO_2, $Hb(O_2)_2$, $Hb(O_2)_3$, and $Hb(O_2)_4$ increases progressively. Ultimately, after the concentration of dissolved O_2 has reached 0.3 ml/100 ml (equivalent to a Po_2 of 90 mm Hg), virtually all the Hb will be in the form of $Hb(O_2)_4$, that is, all the binding sites for O_2 will be filled. This state is referred to as completely saturated hemoglobin, and the total amount of oxygen bound to hemoglobin at 100% saturation is called the oxygen capacity of Hb. Each gram of hemoglobin, when fully oxygenated, binds 1.34 ml O_2. Since 100 ml blood normally contains 15 g Hb, we can calculate that 100 ml completely oxygenated blood normally contains 20 ml O_2 combined with Hb (1.34 ml $O_2 \times$ 15 g Hb/100 ml blood). That is, the oxygen capacity of Hb is 20 ml/100 ml blood, or 20 vol%.

It is important to note that the total amount of oxygen contained in 100 ml blood is referred to as oxygen content, the units of which are referred to as vol%. Under normal circumstances, virtually all the oxygen in blood is present in the form of oxyhemoglobin, that is, HbO_2, $Hb(O_2)_2$, $Hb(O_2)_3$, and $Hb(O_2)_4$. Oxygen concentration usually refers only to the small amount of oxygen present as physically dissolved O_2.

Each of the intermediate oxygenation reactions (1)–(4) has a different equilibrium constant, and each constant determines the "affinity" of Hb for oxygen. Oxygenation of each heme group of the Hb molecule increases the affinity of the remaining heme groups for O_2. This effect is brought about by subtle changes in the geometry of the molecule that accompany the oxygenation of each heme group. To begin with, deoxyhemoglobin has a relatively low affinity for O_2, but oxygenation of one of the heme groups greatly increases the O_2 affinity of the remaining heme groups. A free molecule of dissolved O_2 is 70 times as likely to combine with a Hb molecule that already has three of its hemes occupied by O_2, as with a completely deoxygenated hemoglobin molecule.

The results of these rather complicated relationships are expressed simply and usefully in the oxyhemoglobin dissociation curve for whole blood (Fig. 1, top). This curve relates the oxygen content of whole blood (plotted as the ordinate), to the Po_2 with which the blood is equilibrated (the abscissa). The ordinate gives the amount of oxygen loaded on Hb, and the abscissa gives the partial pressure of oxygen required to produce and maintain that loading. The unusual shape of the curve reflects the effects of partial oxygenation on the O_2 affinity of remaining binding sites. the ordinate scale at the right in Figure 1 shows "percentage saturation," an alternative notion that is useful for some purposes. The percentage saturation of Hb in a solution of blood sample is simply the percentage of the O_2 binding capacity of the Hb that is actually occupied by O_2.

We can now understand why the oxyhemoglobin dissociation curve has this sigmoid shape. The principal determinant of how much O_2 will be reversibly bound to Hb is the Po_2 of the gas with which the blood is equilibrated. Loading the first O_2 molecule onto a Hb molecule requires a relatively large Po_2. Once this initial step in oxygenation has been accomplished, subsequent O_2 molecules are loaded onto Hb rather easily. Consequently, the slope of the oxyhemoglobin dissociation curve is less steep near the origin of the graph than it is at somewhat higher Po_2 (e.g., $Po_2 = 25$ mm Hg, 50% saturation). Part A of the bottom panel depicts the equilibrium situation at 50% O_2 saturation (point A in the top panel). At 50% saturation, half of all oxygen binding sites on the Hb molecule are occupied by O_2 molecules, that is, each Hb molecule holds an average of two O_2 molecules. These bound O_2 molecules

are in reversible equilibrium with free dissolved O_2 molecules in the gas phase. An increase in the Po_2 of the gas phase causes an increase in the concentration of free dissolved O_2, which in turn increases the amount of O_2 bound to Hb. Ultimately, all available O_2 binding sites will be filled with O_2, and the curve asymptotically will approach the O_2 content value corresponding to the oxygen capacity of the blood. This situation is depicted in B of the lower panel in which virtually all the O_2 binding sites are filled and the blood is 99–100% saturated. Further increases in Po_2 now cause no further loading of O_2 onto Hb molecules. Rather, additional O_2 can be present only in the form of additional physically dissolved O_2, so that a further increase in Po_2 causes only a small increase in O_2 content of the blood.

The complete O_2 dissociation curve of whole blood, shown in Figure 2, combines the oxyhemoglobin dissociation curve shown in Figure 1 and the straight, nearly horizontal dissociation line for O_2 in physical solution. The total O_2 content of blood at a particular Po_2 is equal to the sum of the O_2 bound to Hb and the physically dissolved O_2. Since at normal Po_2 values the amount of dissolved O_2 is tiny in comparison with that bound to Hb, physically dissolved O_2 can usually be neglected.

The following discussion covers a number of factors affecting the ability of Hb to combine with O_2. It should be emphasized, however, that although these factors are of some physiological significance, the most important and essential factor determining the loading or unloading of O_2 is the local Po_2 in the immediate vicinity of the red blood cells (RBCs). For example, in the alveolar capillaries the Po_2 is relatively high, and the Hb becomes completely saturated as HbO_2, loading up O_2. When this HbO_2 reaches the tissue where the Po_2 is lower, it cannot hold that much O_2 and releases some O_2, thereby becoming less saturated, unloading some of its O_2.

FACTORS THAT MODIFY THE AFFINITY OF HEMOGLOBIN FOR OXYGEN

The shape and position of the dissociation curve are subject to modification by a number of factors. This section describes modifying influences that are of physiological significance.

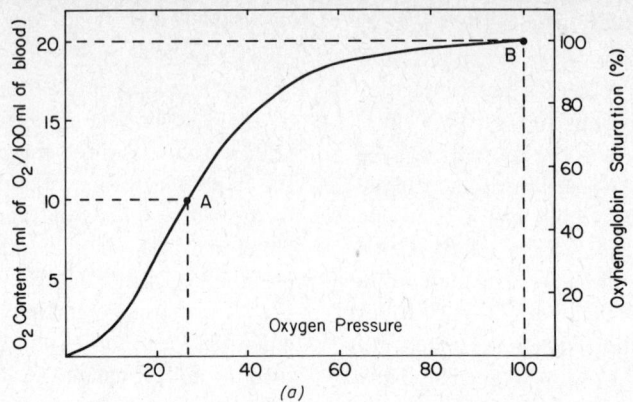

A.

$P_{O_2} = 25\,mm\,Hg$

$O_2\,Sat = 50\%$

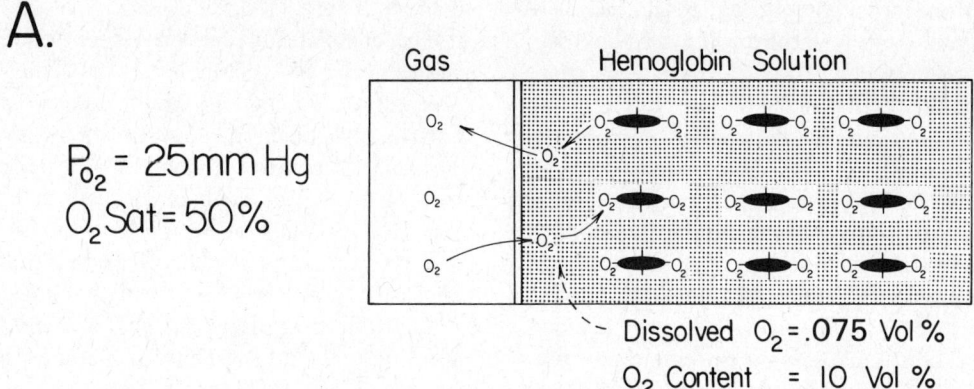

Dissolved $O_2 = .075$ Vol %

O_2 Content $= 10$ Vol %

B.

$P_{O_2} = 100\,mm\,Hg$

$O_2\,Sat = 100\%$

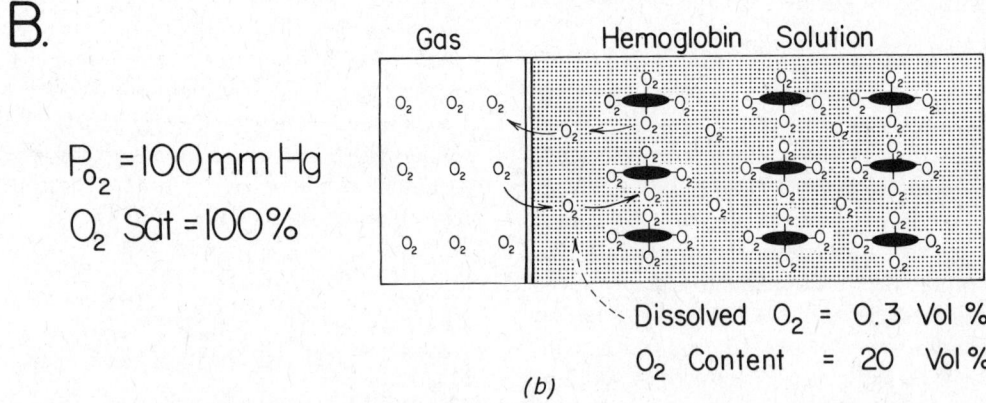

(b)

Figure 1 (Top) Oxyhemoglobin dissociation curve of normal whole blood. (Bottom) A and B indicate 50% and 100% oxyhemoglobin saturation. The Hb–O₂ relationship for these two examples is depicted.

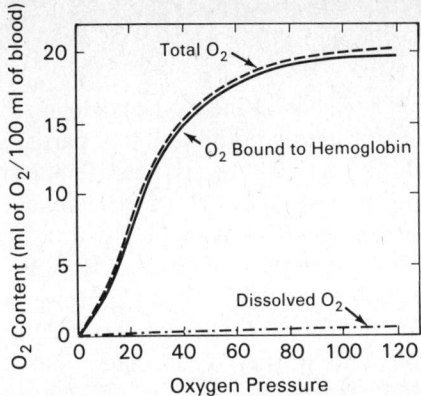

Figure 2 Complete O_2 dissociation curve of whole blood, showing dissolved and chemical bound components of the O_2 content.

Acidity

Figure 3 shows the sensitivity of the dissociation curve to blood pH. Increasing acidity shifts the curve to the right, particularly in its middle region. This phenomenon, called the *Bohr effect* after its discoverer, is of considerable importance, particularly in tissue gas exchange, as described in Chapter 21.

Modern evidence indicates that the Bohr effect results

from changes in conformation that the Hb molecule undergoes when it combines with O_2. Oxygenation alters the environment of several ionizable groups of Hb such that they tend to lose protons, that is, they become stronger, more dissociated acids. This effect can be expressed simply, although without proper attention to stoichiometry, as follows:

$$Hb\text{-}H + O_2 \rightleftharpoons HbO_2 + H^+ \qquad (5)$$

On the other hand, the addition of H^+ ions from some external source forces the equilibrium to the left by mass action, reducing the extent of oxygenation of the Hb molecule. This corresponds to the effect of increasing acidity on the dissociation curve shown in Figure 3—at a constant P_{O_2} value, acidification causes the release of O_2 and a decrease in the percentage saturation of the Hb.

A shift in the oxyhemoglobin curve to the right means that for the same O_2 content of blood, the P_{O_2} will be higher. That is, to load the same amount of O_2 onto the Hb molecule, one must supply a higher driving pressure or P_{O_2}. Accordingly, it is reasonable to conclude that the affinity of Hb for O_2 is decreased by a rightward shift in the O_2–Hb curve. The converse is equally valid, that is, a leftward shift means that the same O_2 content is achieved with a lower P_{O_2} value, implying that the affinity of Hb for O_2 is increased.

Organic Phosphates

2,3-Diphosphoglycerate (2,3-DPG), a phosphorylated by-product of glycolysis, is present in RBCs in a molar concentration roughly equal to that of Hb. In recent years this substance and, to a lesser extent, other organic phosphate compounds have been found to reduce the O_2 affinity of Hb, that is, to move the dissociation curve to the right. Unlike the blood pH, which can undergo rapid changes, and which in fact differs between arterial and venous blood, the concentration of 2,3-DPG is relatively stable, changing only slowly in response to various stimuli over periods of hours or days. 2,3-DPG thus provides a mechanism for long-term regulation of O_2 affinity, which complements the moment-to-moment control provided by the Bohr effect.

The mechanism of the effect of 2,3-DPG on O_2 affinity is analogous to that outlined above for the Bohr effect. Deoxyhemoglobin binds 2,3-DPG rather strongly in a 1:1

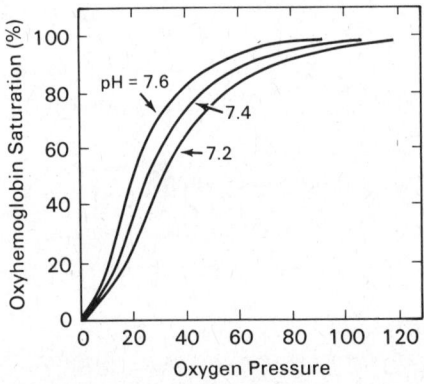

Figure 3 Oxyhemoglobin dissociation curves, showing the effect of varying pH (Bohr effect). An increase in H^+ (↓pH) shifts the curve to the right. A decrease in H^+ (↑pH) shifts the curve to the left. A shift in the curve has minor effects on loading (higher P_{O_2} values), but has large effects on unloading for a given low P_{O_2} value (e.g., $P_{O_2} = 40$ mm Hg).

molar ratio, whereas oxygenated Hb has little or no affinity for organic phosphates. During oxygenation, 2,3-DPG is displaced from its binding site by changes in the structural geometry of the Hb molecule. Hence, the following expressions can be written, in analogy with that for the Bohr effect:

$$Hb\text{-}DPG + O_2 \rightleftharpoons HbO_2 + DPG \qquad (6)$$

Increases in the RBC concentration of 2,3-DPG force this equilibrium to the left, reducing the proportion of Hb that is oxygenated. Stated differently, increasing the concentration of 2,3-DPG shifts the dissociation curve to the right and decreases the affinity of Hb for O_2.

The concentration of 2,3-DPG in RBCs is altered in a variety of circumstances, all of which will not be listed here. High-altitude populations serve as an example. The RBC 2,3-DPG concentration is increased in such people, and their oxyhemoglobin dissociation curves are shifted to the right. This change is advantageous for life in hypoxic environments, because it facilitates unloading of O_2 at the tissues.

Temperature
Figure 4 shows the dissociation curve that is shifted to the right with increasing blood temperature. During exercise, the blood temperature rises and the curve shifts to the right. This action means a decrease in affinity of Hb for O_2 and facilitates the unloading of O_2 from blood to the exer-

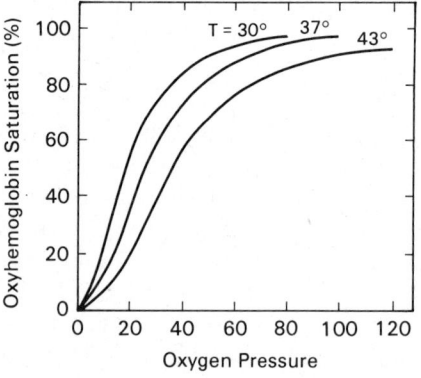

Figure 4 Oxyhemoglobin dissociation curves, showing the effect of varying temperature. By comparing this figure with that of Figure 2, it can be seen that both hydrogen ion and heat shift the curve to the right, (as does 2, 3-diphosphoglycerate).

cising muscles, where the P_{O_2} value is very low. The usefulness of this shift during exercise and fever will be described later.

Note that the shift in the oxyhemoglobin dissociation curve occurs primarily in the region of partial saturation of Hb with O_2 (i.e., 10–90% oxyhemoglobin saturation). This is true for the effect of H^+, 2,3-DPG, and temperature. As a consequence, the shift has little significance when the P_{O_2} is greater than 80 mm Hg, that is, in the lungs, where the P_{O_2} is 90–100 mm Hg. However, the shift is important in the tissues, where the P_{O_2} is 30–40 mm Hg. Clearly, the shifts in the O_2 dissociation curve of blood have little influence on the "loading" of oxygen into blood in lung, but have substantial influence on the "unloading" of oxygen in the tissues, where the P_{O_2} is relatively low.

TRANSPORT OF CO₂ BY THE BLOOD

Like oxygen, carbon dioxide is carried by the blood mostly in reversible chemical combination. The problem is somewhat more complicated than that of O_2 transport, however, because CO_2 enters into several different types of chemical unions in the blood. In addition, differences between the two compartments of blood—cells and plasma—with regard to protein (buffer) concentration and carbonic anhydrase activity are important in determining the rates and equilibria of the various chemical reactions.

TRANSPORT MECHANISMS

CO_2 is present in both RBCs and plasma in three forms (dissolved CO_2, carbamino CO_2, and bicarbonate ion).

Dissolved CO₂
The solubility of CO_2 in water is 20 times that of O_2, so blood contains an appreciable amount of CO_2 in physical solution (2.7 vol% at the normal arterial P_{CO_2} of 40 mm Hg). However, extensive exchange of dissolved CO_2 in the tissues or lungs would require large differences in P_{CO_2} levels between arterial and venous blood, which in turn would cause severe abnormalities of acid-base balance. If

the blood carried all of the CO_2 in the physically dissolved form only, the difference between the amounts of dissolved CO_2 in venous and arterial blood would be so great (and therefore the difference in amounts of H_2CO_3) that the pH difference would overwhelm the ability of the blood to maintain the acid–base balance. Fortunately, dissolved CO_2 actually accounts for only a small fraction of the CO_2 exchanged in the lungs and tissues.

Carbamino CO₂

CO_2 reacts with free amino groups of proteins according to the following rapidly reversible reaction:

$$R\text{-}N^{H}_{H} + CO_2 \rightleftharpoons R\text{-}N^{H}_{COO-} + H^+ \qquad (7)$$

The extent to which this reaction flows to the right, providing a carrier mechanism for CO_2, depends on the availability of free NH_2 groups and of mechanisms for buffering the H^+ ions produced by the reactions. Hemoglobin has many free amino groups and is an excellent buffer, so the high Hb concentration in RBCs provides them with the ability to carry considerable CO_2 as carbamino compounds. Plasma, on the other hand, contains relatively little protein, and thus binds only trace amounts of CO_2 in this form.

Bicarbonate

The largest fraction of CO_2, both in plasma and in RBCs, is in the form of HCO_3^- ions, in equilibrium with dissolved CO_2 according to the hydration and ionization reactions described in Figure 5.

Reaction (1) is very slow in the plasma, which does not contain carbonic anhydrase (CA). RBCs contain enough of the enzyme to speed the reaction rate many thousand times, permitting equilibrium to be reached within a frac-

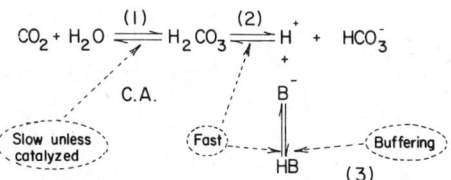

Figure 5 Hydration and buffering of CO_2. C.A., Carbonic anhydrase. B⁻, Buffer groups on the Hb molecule.

tion of a second following a change in P_{CO_2}. Therefore, reaction (1) occurs primarily inside the RBC. Reaction (2), the dissociation of carbonic acid, and reaction (3), H^+ buffering by buffer groups on the Hb molecule, proceed very rapidly.

DISTRIBUTION OF CO₂ ADDED TO BLOOD

Figure 6 displays the CO_2 movements and reactions in blood in metabolizing tissue. Metabolically produced CO_2 molecules move by diffusion down a partial pressure gradient from tissue cells into capillary blood. When CO_2 enters the capillary blood plasma, it is distributed within the blood in several ways:

1. *Plasma CO_2.* A small fraction—roughly 10% of the total CO_2 entering the blood from the tissues—remains in the plasma as physically dissolved CO_2, as carbamino CO_2 bound to plasma proteins, and as bicarbonate formed by the slow uncatalyzed hydration of CO_2. This last reaction does not even approach equilibration during the brief time available for gas exchange in the tissue capillaries.

2. *Red Blood Cell CO_2.* Most of the added CO_2 diffuses into the RBCs, where a small fraction remains in physical solution, and a considerably large proportion is bound to Hb as carbamino CO_2. The largest fraction—about 60% of the total CO_2 entering the blood—is rapidly hydrated to carbonic acid through the action of carbonic anhydrase. The carbonic acid is largely dissociated to form HCO_3^- and H^+ ions because virtually all the H^+ produced combines with buffer groups on the Hb molecule [reaction (3)], and much of the HCO_3^- formed by this reaction diffuses out of the RBCs into the plasma. Other anions—predominantly Cl^-—move into the RBCs in exchange for the bicarbonate ion moving out into the plasma, thereby preserving electrical neutrality within both compartments. This exchange of Cl^- for HCO_3^- is called the "chloride shift." The H^+ ions produced within the RBCs by HCO_3^- production and the formation of carbamino complexes are buffered by Hb. In essence, the products of these reactions are being removed so that relatively large amounts of CO_2 are taken up through these path-

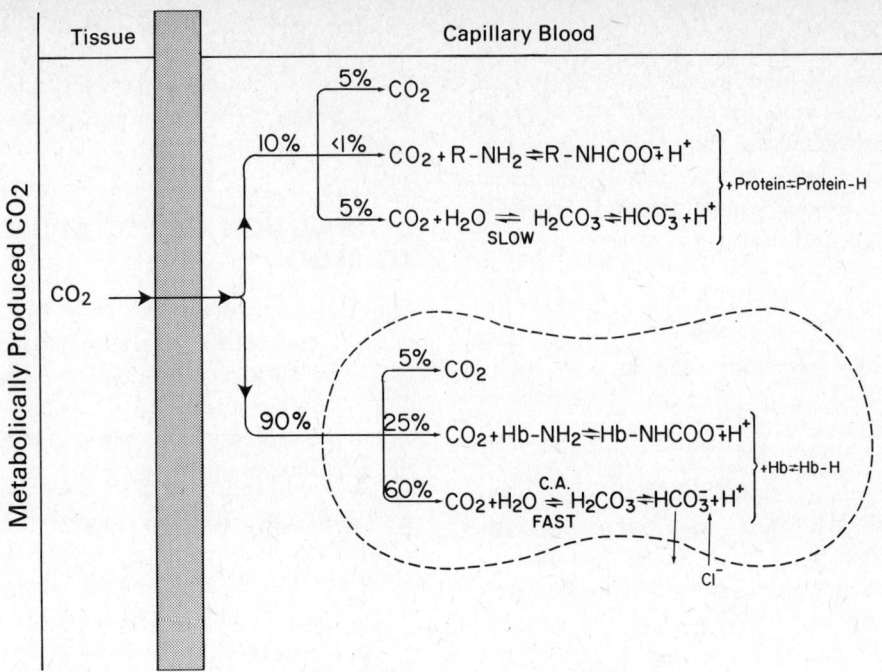

Figure 6 Schematic diagram showing disposition of added CO_2 within the blood in a tissue capillary. The numbers with percentage (%) signs indicate the approximate proportions of added CO_2 handled by the various pathways.

ways with little increase in the P_{CO_2} value and with only a slight increase in free H^+.

CO_2 DISSOCIATION CURVE

As was the case for the reactions of O_2 in blood, the reactions of CO_2 in blood are complex. Once again, we can circumvent these problems by using a dissociation curve for CO_2 just as we employed an oxygen dissociation curve for whole blood. The CO_2 dissociation curve is a graph of the relationship between the total amount of CO_2 present in all forms (HCO_3^-, dissolved CO_2, carbamino CO_2, and carbonic acid) and the partial pressure of CO_2. Figure 7 shows a CO_2 dissociation curve for whole blood. Note that over the physiological range this curve is nearly linear. Furthermore, note that the CO_2 dissociation curve is much steeper than is the oxyhemoglobin dissociation curve in the physiological range. This means that the increase in CO_2 content caused by a 1 mm increment in P_{CO_2}

is greater than the increase in O_2 content brought about by 1 mm increment in P_{O_2}. Conversely, a 1 vol% change in content is associated with a larger change in P_{O_2} than in P_{CO_2}. The linearity and steepness of the CO_2 dissociation curve will have important implications for gas exchange in the lungs.

As shown in Figure 7, the position of the whole blood CO_2 dissociation curve depends on the extent of oxygenation of the hemoglobin molecule. An increase in the O_2 saturation of hemoglobin shifts the curve downward and to the right. Stated differently, oxygenation of blood reduces the total amount of CO_2 present at any particular P_{CO_2}. This phenomenon is known as the *Haldane effect*. This effect of oxygenation of hemoglobin on the CO_2 dissociation curve is shown in detail in Figure 8. Normally, arterial blood is nearly 100% saturated with O_2, so that the arterial values for CO_2 content and pressure lie on the bottom curve. Venous blood is 70–75% saturated, so that the venous values for CO_2 content and pressure lie on the middle curve in Figure 8.

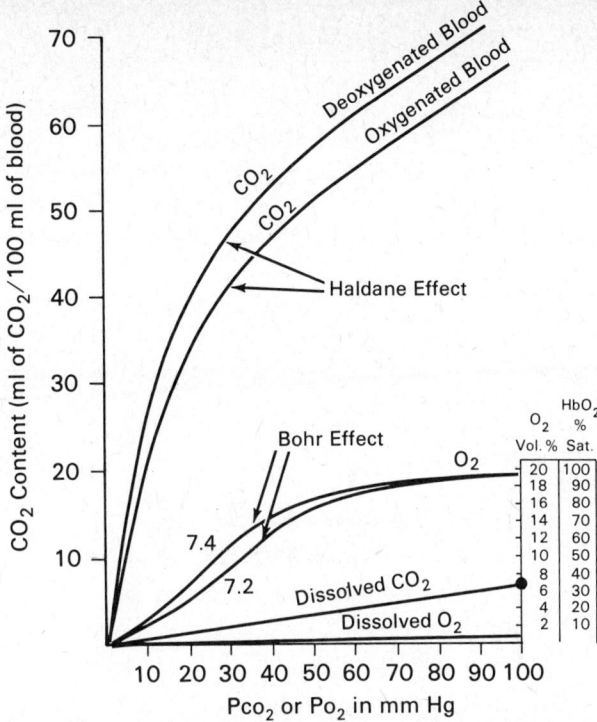

Figure 7 The entire CO$_2$ dissociation curve for whole blood is compared with the O$_2$ dissociation curve for whole blood. Note that the CO$_2$ curve is much steeper and nearly linear in the physiological range.

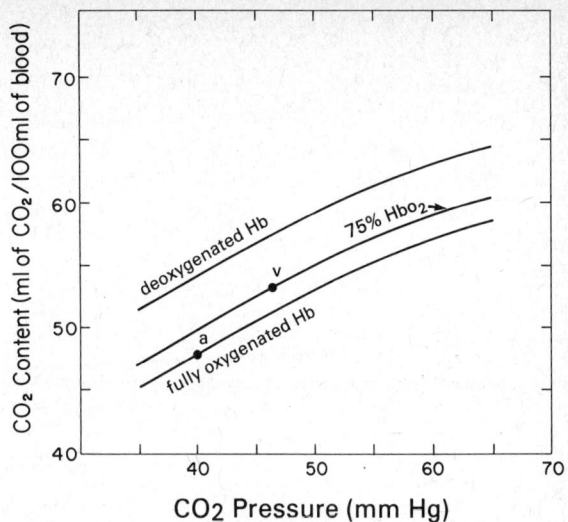

Figure 8 An expanded section of the CO$_2$ dissociation curve in the physiological range. The arterial point (a) lies on the 100% HbO$_2$ curve and the venous point (v) lies on the 75% HbO$_2$ curve.

The principal chemical mechanism responsible for the Haldane effect is closely linked to that described for the Bohr effect (see the section on O$_2$ transport). Recall that oxygenation of Hb releases H$^+$ ions into solution (i.e., oxyhemoglobin is a stronger acid than is deoxyhemoglobin). Both chemical reactions that bind CO$_2$ in the blood also release H$^+$ ions. Since the H$^+$ ions involved in all three reactions constitute a single pool, the three equilibria are made mutually interdependent by a common ion effect, which can be shown as follows:

Common Pool

$$CO_2 + H_2O \rightleftharpoons H_2CO_3 \rightleftharpoons HCO_3^- \; + \; H^+$$

$$H^+ \; + \; HbO_2 \rightleftharpoons Hb\text{-}H \; + \; O_2$$

$$CO_2 + R\text{-}NH_2 \rightleftharpoons R\text{-}NHCOO^- \; + \; H^+$$

Oxygenation of Hb increases the H$^+$ and forces the equilibria involving HCO$_3^-$ (Fig. 5) and carbamino compounds [reaction (7)] to the left, reducing the amount of chemically bound CO$_2$ (Haldane effect). Conversely, increasing the H$^+$ either by raising the P$_{CO_2}$, and thus forming more HCO$_3^-$ and carbamino compounds, or by adding a mineral acid (not shown) drives the equilibrium involving Hb and O$_2$ to the right, reducing the degree of oxygenation of the Hb molecule (Bohr effect). This reciprocal relationship between O$_2$ and CO$_2$ transport increases the efficiency of gas exchange in the lungs and tissues, as described in Chapter 21.

Bohr Effect

The increase in CO$_2$ at the peripheral *tissue* level favors enhanced unloading of O$_2$ from HbO$_2$.

Haldane Effect

The increase in HbO$_2$ formation in pulmonary capillaries favors enhanced formation and unloading of CO$_2$ from HCO$_3^-$ and carbamino compounds. Similarly, in the tissues, deoxygenation of Hb favors loading of CO_2 into blood.

Whereas these two effects increase the efficiency of gas exchange, they are only helpful and not essential to the normal resting person. During muscular exercise or in disease states in which the circulation is impaired, these two effects greatly facilitate O_2 delivery and CO_2 removal in the tissues.

SELECTED BIBLIOGRAPHY

Finch CA, Lenfant C: Oxygen transport in man. *N Engl J Med* 286:407, 1972.

Mines AH: *Respiratory Physiology.* New York, Raven Press, 1981, pp 61, 73.

Murray JF: *The Normal Lung.* Philadelphia, Saunders, 1976, pp 151, 171.

SELF-STUDY QUESTIONS

MULTIPLE CHOICE

Choose the single best answer.

1. The oxygen dissociation curve of normal human blood
   A. shows the amount of O_2 contained per 100 ml blood (O_2 content) when the blood is equilibrated with any given partial pressure of O_2 (Po_2).
   B. shows that raising the Po_2 from 100 to 200 mm Hg doubles the O_2 content.
   C. is a plot of oxygen capacity versus Po_2.
   D. is a straight line, the slope of which depends on the solubility of O_2 in blood plasma.
   E. shows that the affinity of Hb to O_2 is constant.

2. The percentage saturation of hemoglobin refers to
   A. the fraction of the oxygen capacity made up by physically dissolved O_2.
   B. the fraction of the total Hb in the deoxygenated form.
   C. the fraction of the total O_2 binding capacity of Hb occupied by O_2.
   D. the fraction of total Hb in the form of $Hb(O_2)_4$.
   E. is calculated by multiplying the Hb concentration by 1.34.

3. When breathing air at sea level, physically dissolved O_2
   A. refers to all oxygen physically bound in blood.

   B. is so small as to be negligible in most calculations.
   C. contributes 1.34 ml O_2/g Hb.
   D. contributes substantially to the oxygen capacity of blood.
   E. exists only in the plasma.

4. An arterial blood sample with a Hb concentration of 15 g/100 ml is found to have a Po_2 value of 40 mm Hg and is 75% saturated with O_2. The oxygen content of this blood is
   A. $40 \times 1.34 = 5.36$ vol%.
   B. $75 \times 1.34 = 10.0$ vol%.
   C. $15 \times 1.34 = 20.1$ vol%.
   D. $0.75(15 \times 1.34) = 15.0$ vol%.
   E. not calculable without knowing the shape and position of the oxyhemoglobin dissociation curve for this sample of blood.

5. A sample of blood is obtained from the pulmonary artery (a "mixed venous" sample) and found to have a Po_2 value of 25 mm Hg and an O_2 content of 5 vol%. The measured Hb concentration was 7.5 g/100 ml. The percentage O_2 saturation of this blood is
   A. not calculable without knowledge of the O_2 dissociation curve.
   B. 50%.
   C. 25%.
   D. 75%.
   E. 100%.

6. The Bohr effect
   A. influences "loading" of O_2 in the lungs more than "unloading" of O_2 at the tissues.
   B. has nothing to do with O_2 delivery to the tissue.
   C. aids CO_2 transport by blood.
   D. facilitates "unloading" of O_2 in the tissues.
   E. is the primary determinant of O_2 movement in the body.

7. All the following shift the HbO_2 dissociation curve to the right *except*
   A. decreasing the temperature.
   B. 2,3-DPG.
   C. protons.
   D. increasing the P_{CO_2} value.
   E. a decrease in pH.

8. Carbon dioxide
   A. facilitates O_2 binding by Hb.
   B. is transported in arterial blood principally as molecular CO_2 in physical solution.
   C. when added to blood, causes almost no change in the concentration of bicarbonate ion (HCO_3^-).
   D. enters capillary blood by passive diffusion from the tissues.
   E. is transported principally as carbamino compound.

9. When CO_2 is added to blood, the buffering of H^+ produced within the RBCs causes
   A. most of the CO_2 to be converted to bicarbonate ion.
   B. dissolved CO_2 to increase.
   C. a limitation in the conversion of CO_2 to HCO_3^-.
   D. CO_2 to be converted to HCO_3^- primarily in the plasma.
   E. pH to stay constant.

10. The CO_2 dissociation curve for whole blood
    A. is independent of hemoglobin concentration.
    B. gives the relationship between $HbCO_2$ and P_{CO_2}.
    C. has a sigmoid shape.
    D. describes the relationship between all forms of CO_2 in blood and P_{CO_2}.
    E. is flatter (less steep) than the oxyhemoglobin dissociation curve in the physiological range.

11. When CO_2 is added to blood
    A. the concentration of H^+ and HCO_3^- rises by equal amounts.
    B. bicarbonate ions are formed very slowly.
    C. the affinity of Hb for O_2 increases.
    D. bicarbonate ions diffuse out of the RBCs.
    E. chloride ions diffuse out of the RBCs.

12. Oxygenation of Hb
    A. decreases the affinity of Hb for O_2.
    B. is enhanced by 2,3-DPG.
    C. increases the dissociation of acid groups on the Hb molecule.
    D. decreases when the P_{O_2} is raised above 100 mm Hg.
    E. converts Fe^{2+} to Fe^{3+}.

ANSWERS

1. A. The volume of O_2/100 ml blood is the sum of oxygen bound to Hb and dissolved O_2 and is referred to as the oxygen content of whole blood. This value is plotted on the ordinate, and the partial pressure of O_2 is plotted on the abscissa. B is incorrect because at $P_{O_2} = 100$ mm Hg, Hb is 100% saturated with oxygen, so that further increments in P_{O_2} increase O_2 content only by increasing the concentration of dissolved O_2. The O_2 affinity of Hb changes when the degree of oxygenation of the hemoglobin molecule changes, and this property contributes to the alinearity of the dissociation curve.

2. C. The oxygen saturation of Hb can be calculated as the ratio of O_2 content to O_2 capacity of Hb. Hence, it refers to the relative amount of O_2 capacity actually occupied by O_2 bound to Hb, that is, to the fraction of total Hb in the oxygenated (not the deoxygenated) form.

3. B. Oxygen in physical solution constitutes only 2% of the total oxygen capacity present in normal blood at oxygen pressure of 90 mm Hg, and solubility is independent of Hb concentration. O_2 molecules dissolve in all aqueous elements of blood, both outside and inside the RBC.

4. D. One must first calculate the oxygen capacity of the blood, and then use the information that the blood is 75% saturated, to calculate the oxygen content. In this case, 75% of the total O_2 binding capacity of the blood is being used to hold oxygen. Therefore, calculate the O_2 capacity ($15 \times 1.34 = 20.1$ vol%) and multiply by the % O_2 sat/100 ($20.1 \times 0.75 = 15.0$ vol%).

5. B. Again, calculate the oxygen capacity of the blood sample ($1.34 \times 7.5 = 10.0$ vol%). The observed O_2 content, 5 vol%, is half the calculated oxygen capacity. Therefore, the blood is 50% saturated with oxygen. If the patient were not anemic, that is, if Hb = 15 gm%, the correct answer would be 25% (oxygen capacity = $1.34 \times 15 = 20$ vol%, 5/20 = 25%). Neither this question nor the preceding one requires knowledge of the O_2 dissociation curve. This curve describes the pressure that will accompany any particular O_2 saturation or oxygen content. These questions, however, ask only that you convert O_2 content to percentage saturation and vice versa. All you need to know is the Hb concentration and the factor 1.34 to make this conversion. The P_{O_2} is irrelevant.

6. D. The action of the Bohr effect is minimal at high values of P_{O_2} because the HbO_2 curves for different values of pH converge at the upper end. The curves are clearly separated at lower values of P_{O_2}, that is, in the range of capillary blood P_{O_2} (the average P_{O_2} of blood in the capillary of a metabolizing tissue is approximately 50 mm Hg). Accordingly, the rightward displacement in HbO_2 curve resulting from the decrease in pH as blood circulates through the tissues decreases the affinity of Hb for O_2 and thereby facilitates unloading of O_2 from blood. However, in the lung, the P_{O_2} value is high and the action of the Bohr effect is negligible.

7. A. 2,3-DPG and protons (H^+) produce a rightward shift. An increase in the P_{CO_2} value causes an increase in H^+ concentration. However, a decrease in temperature produces a leftward shift.

8. D. Addition of CO_2 to blood diminishes the tendency of Hb to bind O_2 (Bohr effect), because it decreases the pH levels. Most CO_2 transport occurs in the form of bicarbonate ion. This ion is formed in significant amounts, when CO_2 is added to blood, only because the H^+ generated by dissociation of H_2CO_3 is buffered by Hb and other proteins. Disappearance of the products (i.e., buffering of H^+ and diffusion of HCO_3^- out of the RBCs) "pulls" the reaction in that direction.

9. A. (See answer to 8.) Dissolved CO_2 increases because its concentration is proportional to the pressure of CO_2. Buffering promotes the conversion of CO_2 to HCO_3^-, but this occurs primarily within the RBC, where the principal buffer, Hb, is located. If a limited amount of CO_2 were available, this would decrease the dissolved CO_2, not increase it. Although buffering minimizes pH change, some decrease does occur, because the pH level cannot stay constant.

10. D. The ordinate of the CO_2 dissociation curve represents all forms in which CO_2 is reversibly bound in blood. Hemoglobin influences the curve because it combines with CO_2, and buffers H^+ formed when CO_2 is hydrated. This linear curve is steeper than the oxyhemoglobin dissociation curve.

11. D. Addition of CO_2 to blood generates HCO_3^- inside the RBC, which diffuses out while Cl^- reciprocally diffuses into the RBC. Bicarbonate ion is formed rapidly because of the presence of carbonic anhydrase. The buffering action of Hb limits the increase in $[H^+]$, and HCO_3^- must rise to a considerably greater degree before the reaction reaches equilibrium.

12. C. Additions of O_2 to the Hb molecule increase its affinity for oxygen and cause it to release protons (i.e., oxyHb is a stronger acid than is deoxyHb). Oxidizing agents (e.g., nitrates) convert Fe^{2+} to Fe^{3+}; that is, they produce methemoglobin, a form of Hb that cannot function in delivering O_2 to the tissues. 2,3-DPG facilitates O_2 unloading, not O_2 loading.

21

Tissue Gas Exchange

JOHN E. REMMERS

OBJECTIVES

After completion of this chapter, you should be able to

1. Have a working knowledge of the factors that determine whether the metabolic demands of O_2 supply and CO_2 removal are met.

2. Appreciate the significance of venous P_{O_2} and P_{CO_2} as indices of tissue gas pressure, as well as predict the dependence of these variables on blood flow, metabolic rate, and arterial blood gas composition.

3. Describe the influence of shifts in the position of the dissociation curves (Bohr and Haldane effects, temperature, 2,3-DPG) on venous and tissue gas pressures.

DEFINITIONS AND CONVENTIONS

This chapter uses a special set of abbreviations in referring to respiratory gases in blood:

Term	Definition
art P_{CO_2} or art P_{O_2}*	Partial pressure or tension of CO_2 or O_2 in a sample of arterial blood
ven P_{CO_2} or ven P_{CO_2}:†	Partial pressure or tension of CO_2 or O_2 in a sample of venous blood

Note that the P_{CO_2} or P_{O_2} value of a blood sample obtained from the pulmonary artery, which reflects a mean venous value for the entire body, is referred to as a P_{CO_2} or P_{O_2} of *mixed venous blood* and is abbreviated $\overline{ven}$ P_{CO_2} or $\overline{ven}$ P_{O_2}. It can also be abbreviated as $P\overline{v}_{CO_2}$ or $P\overline{v}_{O_2}$).

GENERAL CONSIDERATIONS

In order to survive, a cell must have its energy requirements satisfied. In the cells of most tissues, this requires the constant influx of oxygen and outflux of carbon dioxide. Gas exchange and transport by the respiratory and cardiovascular systems function to this end; the remarkable physicochemical properties of blood and the reactions of blood with O_2, CO_2, and H^+ enable the heart and lungs to accomplish their task. As was pointed out in Chapter 20, without hemoglobin (Hb) in the circulating blood, the respiratory and cardiovascular systems could not supply O_2 and remove CO_2 at rates satisfactory for a person at rest.

Neither arterial nor venous *contents* of O_2 or CO_2 di-

---

*Also abbreviated Pa_{CO_2} and Pa_{O_2}.

†Also abbreviated Pv_{CO_2} and Pv_{O_2}.

rectly influence the cell; however, they are of considerable indirect influence. Rather, the *pressures* of O_2 and CO_2 in the metabolizing tissue are critical factors that directly influence the cell. This must be so because O_2 and CO_2 move across the capillary wall by passive diffusion, and the rate of passive diffusion is determined by the partial pressure gradient. Accordingly, the partial pressures of the gases, both in the capillary and in the cell, are the critical factors in determining the rate of gas diffusion into or out of the cell. Furthermore, there are certain critical limits. For instance, if the P_{O_2} value in the tissue decreases to a very low level, the rate of aerobic metabolism will decrease, resulting in possible cell injury. Tissue P_{O_2} and P_{CO_2} values are difficult to measure, but the venous P_{O_2} and P_{CO_2} provide good indices of the average tissue gas pressures. In fact, under most circumstances, venous P_{O_2} and P_{CO_2} are found to approximate closely the *average* tissue levels of P_{O_2} and P_{CO_2} in an organ.

When the rate of oxygen transport to cells is impaired, the venous P_{O_2} can be abnormally low, indicating inadequate tissue P_{O_2}, that is, insufficient oxygen to maintain metabolic requirements. On the other hand, one may have a modest reduction in O_2 content of blood with little impairment in the O_2 supply to the metabolizing cells. For instance, anemia (decreased Hb concentration) reduces both arterial and venous O_2 contents. However, anemia also decreases blood viscosity and produces cardiovascular compensations that can increase cardiac output. The net effect is that the venous P_{O_2} value decreases only slightly, and the metabolizing cells are well oxygenated in a person at rest. We know that venous P_{O_2} and P_{CO_2} provide useful information about oxygen availability and use in tissues; we shall now explore the factors that determine the magnitudes of these tensions, taking three approaches: (1) an intuitive analysis, (2) use of CO_2 and O_2 dissociation curves for whole blood, and (3) consideration of the dynamic events occurring in an idealized capillary. These three approaches are complementary and all aim at helping you understand one central theme; namely, the action of factors that determine the venous and tissue values of P_{O_2} and P_{CO_2}.

AN INTUITIVE ANALYSIS OF BLOOD FLOW–METABOLISM RELATIONSHIPS: APPROACH 1

Intuitively, one can see that the partial pressures of O_2 and CO_2 in a tissue will be influenced by two factors: the rate of blood flow to the tissue ($\dot{Q}$) and the rate of aerobic metabolism of the tissue ($\dot{V}O_2$). Consider what would happen if the flow of blood to a tissue were to increase, but the rate of oxygen consumption were to remain constant. Less oxygen would be extracted from each milliliter of blood that passed through the tissue (i.e., venous O_2 content would be higher than previously). Conversely, if blood flow decreases, but O_2 consumption rate remains constant, the extraction of oxygen from each milliliter of blood circulating through the tissue must increase (i.e., venous O_2 content would be lower than before). Overall, we can state that either an increase in blood flow or a decrease in O_2 consumption rate will produce less oxygen extraction and, conversely, a decrease in blood flow or an increase in oxygen consumption rate will cause an increase in oxygen extraction.

Oxygen extraction refers to the amount of O_2 removed from arterial blood and therefore refers to the difference in O_2 content between arterial and venous blood, that is, arterial minus venous O_2 content difference. An increase in O_2 extraction signifies an increase in (art − ven) O_2 content difference, and vice versa. Similar considerations apply to carbon dioxide transport. An increase in blood flow or a decrease in CO_2 production rate will result in a decrease in (ven − art) CO_2 content difference. Conversely, a decrease in perfusion rate or an increase in CO_2 production rate leads to an increase in (ven − art) CO_2 content difference. Whenever the (art − ven) O_2 content difference increases or the (ven − art) CO_2 content difference increases, we can expect a change in venous gas composition so long as arterial values remain constant. Specifically, the venous O_2 content will fall as more oxygen is extracted from each milliliter of blood; the venous CO_2 content will rise because more CO_2 is added to each milli-

liter of blood. This, in turn, means that venous and tissue PO_2 values will be lower, and venous and tissue PCO_2 values will be higher than normal.

A third variable to be considered is the effect of changes in the content of CO_2 and O_2 in the arterial blood entering the tissue capillaries. Arterial oxygen content and arterial carbon dioxide content indicate how much O_2 and CO_2 is carried to the tissue by each milliliter of arterial blood. If blood flow and metabolism are constant, O_2 extraction will be constant and venous gas content will depend only on arterial gas content. The greater the art O_2 content, the higher will be the ven O_2 content, all other things being equal.

AN ANALYSIS USING CO_2 AND O_2 DISSOCIATION CURVES FOR BLOOD: APPROACH 2

The intuitive notions just advanced can be placed in a quantitative framework, using the Fick principle. We shall examine the situation for O_2, but a completely analogous exercise for CO_2 could be carried out as well. The Fick principle was derived in Chapter 9. We shall review it here by defining the rate of O_2 delivery to the tissues and the rate at which O_2 is returned to the lungs from the tissues as follows:

$$O_2 \text{ delivery by arterial blood} = \dot{Q} \text{ (art } O_2 \text{ content)}$$

$$O_2 \text{ return by mixed venous blood} = \dot{Q} \text{ (}\overline{\text{ven}} \text{ } O_2 \text{ content)}$$

Considering the entire body, we can state that the body's oxygen consumption rate ($\dot{V}O_2$) is equal to the difference between the oxygen delivery rate and the oxygen removal rate. In this case, $\dot{Q}$ is equal to the cardiac output, and the venous oxygen content refers to the O_2 content of mixed venous blood ($\overline{\text{ven}}$) O_2 content. Therefore, the O_2 consumption rate of the body ($\dot{V}O_2$) equals the difference between rate of O_2 delivery and rate of O_2 removal for the whole body.

$$\dot{V}O_2 = O_2 \text{ delivery} - O_2 \text{ return}$$

$$\dot{V}O_2 = \dot{Q} \text{ (art } O_2 \text{ content)} - \dot{Q} \text{ (}\overline{ven} \text{ } O_2 \text{ content)}$$

$$\dot{V}O_2 = \dot{Q} \text{ (art } O_2 \text{ content} - \overline{ven} \text{ } O_2 \text{ content)}$$

$$\text{art } O_2 \text{ content} - \overline{ven} \text{ } O_2 \text{ content} = \dot{V}O_2/\dot{Q} \qquad (1)$$

In common parlance, we say that the "A-V" difference for oxygen is equal to the ratio of O_2 consumption rate to cardiac output. Normal values for an adult would be $\dot{V}O_2 = 0.25$ L/min and $\dot{Q} = 5$ L/min. The ratio $\dot{V}O_2/\dot{Q}$ is 0.25/5, or 0.05 ml O_2/ml blood. However, we usually express O_2 content in units of vol% (ml O_2/100 ml blood), so that the normal (a-v) O_2 content difference is equal to 5 vol%. We will use the Fick principle [Eq. (1)] together with the oxygen dissociation curve to predict the mixed venous PO_2 in a normal subject and in situations in which cardiac output or O_2 consumption rate change.

NORMAL RESTING SUBJECT

Step 1. The lungs and the respiratory control system are functioning normally. This means that the arterial PO_2 value is 90 mm Hg. Using the oxyhemoglobin dissociation curve for normal blood (Fig. 1), you can see that PO_2 of 90 mm Hg corresponds to an art O_2 content of 20 vol% on the y axis.

Step 2. Since the $\dot{V}O_2$ and $\dot{Q}$ are normal, the (art $- \overline{ven}$) O_2 content difference must be normal, that is, 5 vol%. Therefore, one moves five units down the y axis to obtain a mixed venous value of 15 vol% ($\overline{ven}$ O_2 content $=$ 15 vol%).

Step 3. The mixed venous PO_2 can now be predicted; it is the value corresponding to $\overline{ven}$ O_2 content of 15 vol%; that is, mixed venous $PO_2 = 35$ mm Hg. Here is another way to look at it: When the $\overline{ven}$ PO_2 has fallen 35 mm Hg, the percentage saturation of Hb is decreased (and therefore the amount of O_2 bound with Hb is decreased) and blood contains only 15 vol% of O_2. In this manner, 5 vol% has been unloaded.

It is important to note that in step 3 we have ignored the Bohr effect. To predict accurately the mixed venous PO_2 that corresponds to a venous O_2 content of 15 vol%, we should have used a curve for a slightly lower pH, since venous blood is more acid than arterial owing to the increase in CO_2 that occurs as blood passes metabolizing tissue. This will become clearer in approach 3.

PATIENT WITH A REDUCED CARDIAC OUTPUT (FOR EXAMPLE, HEART FAILURE)

Assuming that the lungs are functioning normally, the arterial PO_2 value will again be 90 mm Hg and the art O_2 20 vol%. However, the $\dot{V}O_2/\dot{Q}$ ratio will be greater than normal (Fig. 2, right-hand panel). For instance, if the cardiac output is half-normal, the ratio $\dot{V}O_2/\dot{Q}$ will be twice normal, and the difference (art $- \overline{ven}$) O_2 content will be twice normal, or 10 vol%. Consequently, the mixed venous will be 10 vol%; from the O_2 dissociation curve, one can predict that mixed venous PO_2 will be 25 mm Hg, again ignoring the Bohr effect. Therefore, because cardiac output is decreased by a factor of two, oxygen extraction will be increased by twofold. The $\overline{ven}$ O_2 saturation will be less than normal because more O_2 has been unloaded from each milliliter of blood. Hence, $\overline{ven}$ PO_2 will be less than normal. This is an abnormally low value of $\overline{ven}$ PO_2 and indicates that some cells of the body are inadequately oxygenated and might not have an oxygen supply commensurate with their energetic needs.

PATIENT WITH AN INCREASED CARDIAC OUTPUT

If the cardiac output were twice normal, the condition depicted in the left-hand panel of Figure 2 would apply. The ratio $\dot{V}O_2/\dot{Q}$ would be half-normal, the difference (art $- \overline{ven}$) O_2 content, would be half that of normal (2.5 vol%), and $\overline{ven}$ O_2 content would be 17.5 vol%. In this case, the O_2 dissociation curve predicts that $\overline{ven}$ PO_2 will be 55 mm Hg. That is, with a twofold increase in cardiac output, oxygen extraction is one-half that of normal. Less O_2 is unloaded from each milliliter of blood, with the result that $\overline{ven}$ PO_2 is greater than normal.

Just as blood flow relative to metabolism sets the arteriovenous content difference for O_2, it also determines the carbon dioxide content difference between arterial and

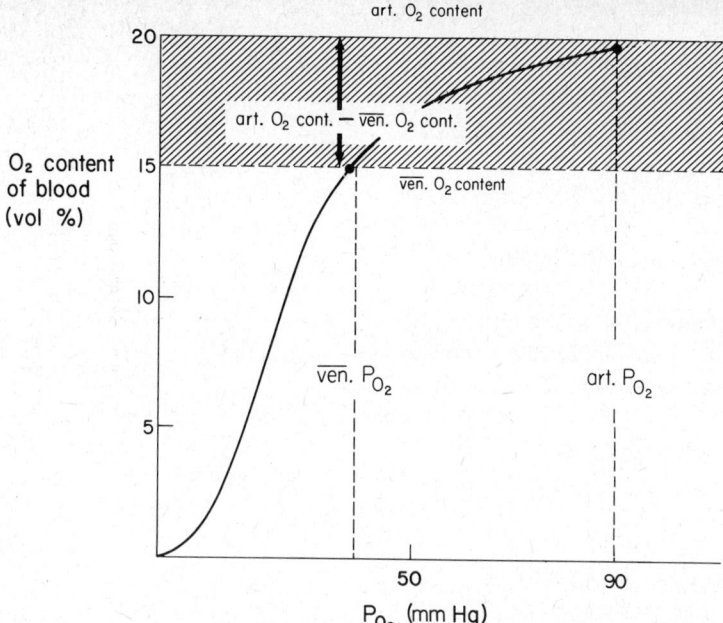

Figure 1 Oxyhemoglobin dissociation curve for whole blood. Normal arterial and mixed venous O_2 contents are indicated by the horizontal lines. The vertical distance between the lines, the arterial—mixed venous O_2 content difference, is equal to the ratio of oxygen consumption ($\dot{V}O_2$) to the cardiac output ($\dot{Q}$).

Alterations In Blood Flow – Metabolism Relationship

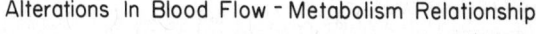

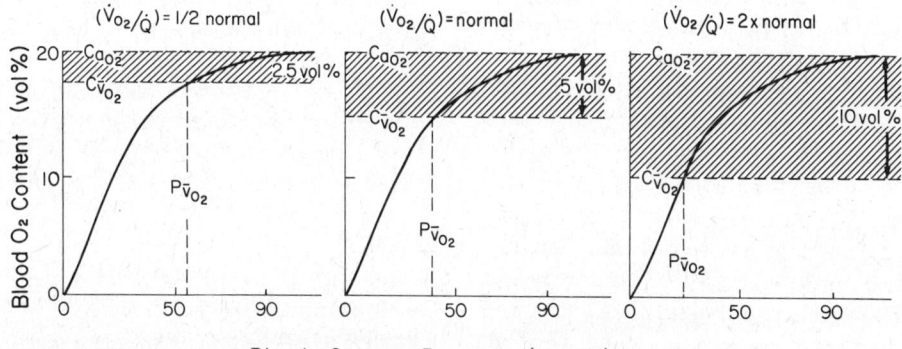

Blood Oxygen Pressure (mm Hg)

Figure 2 Three conditions with different metabolism/cardiac output relationships are compared. Left, decreased $\dot{V}O_2/\dot{Q}$. Right, increased $\dot{V}O_2/\dot{Q}$. The (art — ven) O_2 difference (height of the hatched area) changes as the ratio $\dot{V}O_2/\dot{Q}$ changes. Arterial PO_2 is assumed to be control at 90 mm Hg. These examples show that mixed venous PO_2 increases as the (art — ven) O_2 difference decreases, and vice versa.

venous blood. In the first example, where (art − $\overline{\text{ven}}$) O_2 content is twice normal ($\overline{\text{ven}}$ − art) CO_2 content will be twice normal. The changes in mixed venous P_{CO_2} are somewhat smaller than those for P_{O_2} values because of the steeper slope of the CO_2 dissociation curve (compare steepness of the two curves in Chapter 20, Fig. 7).

To simplify the discussion of factors determining venous gas pressures, we have neglected the effect of a change in venous P_{CO_2} and pH values on the affinity of Hb for O_2 or the effect of a change in percentage O_2 saturation on CO_2 storage by blood. Although the Bohr and Haldane effects have not been included, their influence is considered in the next section of this chapter.

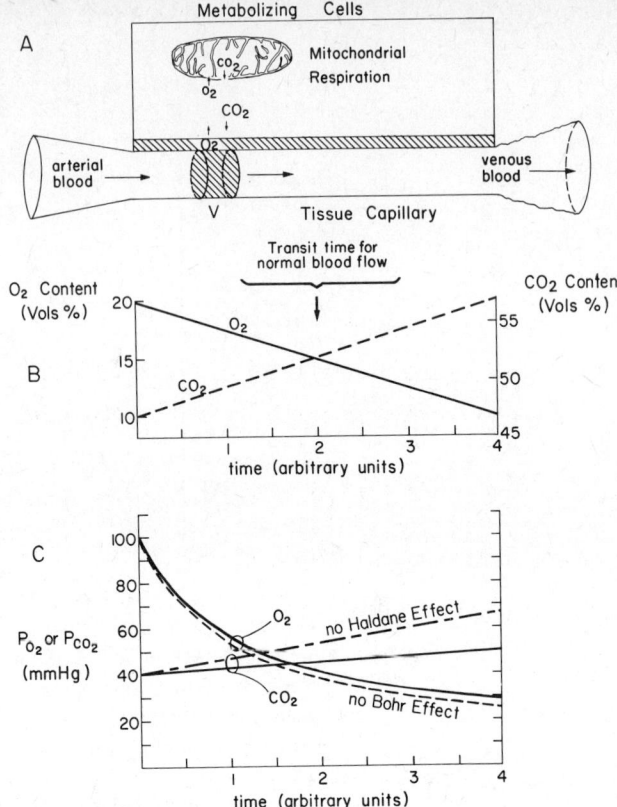

Figure 3 (a) Simplified version of CO_2 and O_2 dynamics in a small volume of blood V moving through a tissue capillary. (b) O_2 content and CO_2 content of V change linearly with time and reach the normal resting values of ven CO_2 content and ven O_2 content at time = 2. With greater transit time (i.e., lesser rate of blood flow) gas contents at the end of the capillary will be lower for O_2 and higher for CO_2. (c) P_{O_2} and P_{CO_2} values, which correspond to the gas contents in (b) are plotted against time. Solid curves indicate the actual time course of P_{O_2} and P_{CO_2} in V when the Bohr and Haldane effects are included. Double-dashed curves indicate the time course of P_{CO_2} without the Haldane effect. Dashed curves indicate the time course of P_{O_2} without the Bohr effect.

DESCRIPTION OF DYNAMIC EVENTS IN TISSUE CAPILLARIES: APPROACH 3

Consider the events that occur when a small volume of arterial blood enters a hypothetical capillary of a metabolizing tissue (Fig. 3a). Oxygen leaves the blood and CO_2 enters it. This transport results from diffusion of these gases across the capillary endothelium down their partial pressure gradients; the cells surrounding the capillary have a lower P_{O_2} and a higher P_{CO_2} than that of capillary blood.

As a first approximation, the *content* of oxygen and the *content* of carbon dioxide of the blood packet change with time at constant rates (Fig. 3b), which depend on the $\dot{V}_{O_2}$ and $\dot{V}_{CO_2}$ of the surrounding tissue. As the packet of blood proceeds through the capillary, O_2 content and CO_2 content continue to change at nearly constant rates because the metabolic rate throughout the tissue is nearly uniform. As a result, each curve in Figure 3b is linear, and the slope depends on the metabolic rate, which we shall assume is constant.

The values to which O_2 content falls and CO_2 content rises, before the packet exists from the capillary, depend on the time required to traverse the capillary. Transit time is inversely proportional to the rate of blood flow; at low flow rates and long transit times, O_2 content will decrease

to a greater extent, and CO_2 content will increase to greater extent before the packet leaves the tissue. At high flow rates, the reverse will be true. This is simply a manifestation of the Fick principle.

The partial pressures of CO_2 and O_2 in this small volume of blood can be predicted at every instant, using the blood dissociation curves for CO_2 and O_2 (Fig. 4). One simply determines the P_{O_2} and P_{CO_2} values corresponding to the content of O_2 and CO_2 at each instant in time in Figure 3b. This calculation has been done for the example shown; Figure 3c gives resulting curves that describe the time course of Po_2 and Pco_2 in the small volume of blood as it travels through the capillary. At every point along the capillary, the surrounding tissue has a Po_2 value that is lower and a Pco_2 higher than the corresponding values for the blood. Likewise, the average tissue Po_2 value is lower than the average capillary blood Po_2 value; the average tissue Pco_2 value is higher than the average capillary blood Pco_2 value. However, average tissue pressures for CO_2 and O_2 happen to be nearly equal to the end-capillary or venous pressures for these gases under a variety of circumstances.

Certain features of tissue gas exchange can be noted in Figures 3b, c. First, O_2 content and CO_2 content of the blood packet change at equal rates. This occurs because we have, for convenience, assumed that RQ = 1. RQ is the ratio of CO_2 production to O_2 consumption ($\dot{V}co_2/\dot{V}o_2$). In contrast, Po_2 and Pco_2 values change at drastically different rates owing to the different slopes of the CO_2 and O_2 dissociation curves.

Second, because of the S shape of the O_2 dissociation curve, Po_2 initially falls rapidly and then more slowly, whereas the Pco_2 value rises steadily and slowly throughout the capillary. At the arterial end of the capillary the Hb is fully saturated with O_2. Because the O_2 dissociation curve is very flat in this region, removal of 1 vol% O_2 greatly decreases the Po_2 (90–70 mm Hg). Near the venous end of the capillary, the Hb is about 75% saturated and the O_2 dissociation curve is very steep. Removal of 1 vol% decreases Po_2 from 50 to 40 mm Hg.

Third, the effect of changing the rate of blood flow on the composition of end-capillary or venous blood can be easily appreciated. At *normal blood flow* (art − ven), O_2 content difference and (ven − art) CO_2 content difference equal 5 vol%. This corresponds to a transit time of 2 and means that ven Po_2 and ven Pco_2 are 40 and 46 mm Hg, respectively. If blood flow is half-normal (transit time = 4), the (art − ven) O_2 and (ven − art) CO_2 content differences are 10 vol%. In this case, ven Po_2 falls to 33 mm Hg and ven Pco_2 rises to 50 mm Hg. If blood flow is twice normal (transit time = 1, art − ven content difference = 2.5 vol%), ven Po_2 rises to 58 and ven Pco_2 falls to 43. The venous pressures given here for half-normal and twice-normal blood flow include the Bohr and

Haldane effects, and are therefore slightly different from the values given in Figure 2.

SIGNIFICANCE OF CO_2–O_2 EXCHANGE

It is obvious that oxygen loss and CO_2 accumulation in blood flowing through a capillary occur simultaneously. This process of trading O_2 for CO_2 is called *gas exchange*. It is itself of importance because removal of O_2 from blood is aided by the addition of CO_2, and the addition of CO_2 to blood is aided by the removal of O_2. This situation is the consequence of the interaction between the O_2 molecule and the H^+ ion on the Hb molecule, as described in Chapter 18.

BOHR AND HALDANE EFFECTS

As blood traverses a capillary of a metabolizing tissue, its Pco_2 value and $[H^+]$ increase. This action, in turn, decreases the binding of O_2 by Hb, so that for any particular O_2 content, the Po_2 value is higher than if the $[H^+]$ had not changed (Bohr effect). On the other hand, as oxygen is removed from Hb, some free $[H^+]$ is bound to Hb. This minimizes the rise in $[H^+]$ and means that, for a particular Pco_2 value, more CO_2 is stored as HCO_3^- and carbamino compound (Haldane effect).

The net effect of these dissociation curve shifts on gas exchange is to lessen the changes in Pco_2 and Po_2 of blood traversing the capillary, so that the Po_2 value falls less and the Pco_2 value rises less. This point is illustrated by the dashed lines in Figure 3c, in which the Pco_2 and Po_2 values of blood traversing the capillary are calculated without the Bohr and Haldane effects. Without the Bohr effect, the Po_2 value of capillary flow is less; without the Haldane effect, the Pco_2 level of capillary blood is greater. The influence of the Bohr and Haldane effects is small when transit time is low (high $\dot{Q}$), but increases steadily the longer the blood remains in the capillary. For this reason, the Bohr and Haldane effects are of great importance in diseases in which the blood flow to an organ or to the whole body is compromised.

Another way of viewing the effects of dissociation shifts on gas exchange is to plot the values for arterial and mixed venous blood on the CO_2 and O_2 dissociation curves shown in Figure 4. Note that in Figure 4 (top) because the

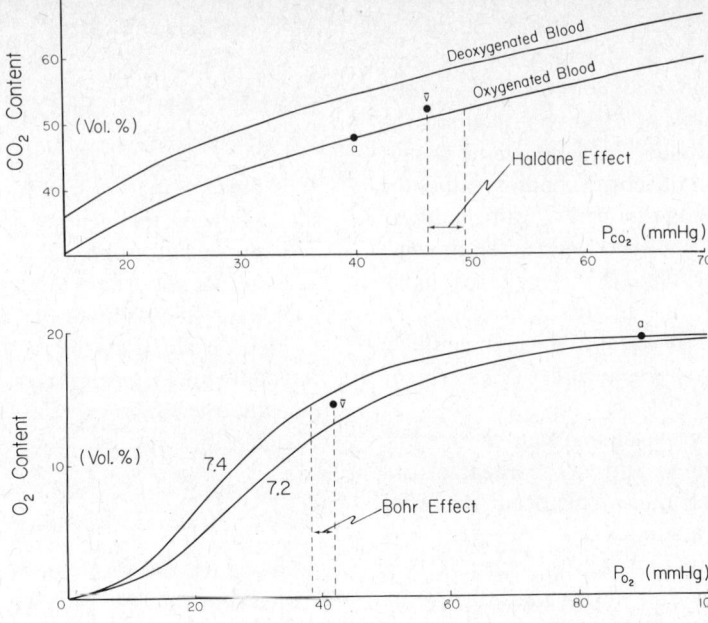

Figure 4 Whole blood dissociation curves for CO_2 (top) and O_2 (bottom) showing the Haldane and Bohr effects, respectively. a = normal arterial point; $\bar{v}$ = normal mixed venous point. Because of the Haldane effect, $\overline{ven}$ P_{CO_2} is 3 mm Hg lower than if blood were to remain completely oxygenated. The Bohr effect causes $\overline{ven}$ P_{O_2} to be 4 mm Hg higher than if venous blood pH were 7.4.

venous point for CO_2 lies to the left of the CO_2 dissociation curve for oxygenated blood, P_{CO_2} is about 3 mm Hg lower than it would have been had the blood remained completely oxygenated. Similarly, in Figure 4 (bottom) note that mixed venous P_{O_2} lies to the right of the oxyhemoglobin dissociation curve of pH 7.4, owing to the relative acidity of venous blood. This means that because of the Bohr effect, the mixed venous P_{O_2} is 4 mm Hg higher than it would otherwise be. The magnitude of these actions by the Bohr and Haldane effects is greatly increased during muscular exercise or in disease states in which the cardiac output or regional perfusion is compromised. A similar shift of the oxyhemoglobin dissociation curve results from the increase in temperature of an exercising muscle. This effect probably also becomes significant both in heavy exercise and in fever.

It is important to note that the Bohr and Haldane effects *facilitate* unloading of O_2 and loading of CO_2 at the tissues, they do not cause it. The movement of O_2 out of capillary blood occurs because the P_{O_2} value of cells surrounding the capillary is less than the P_{O_2} of the blood inside the capillary. Similarly, the "loading" of CO_2 into capillary blood results because tissue P_{CO_2} is greater than capillary blood P_{CO_2}. Tissue gas exchange would occur without the Bohr and Haldane effects, but these effects aid gas transfer and maintain a higher P_{O_2} and lower P_{CO_2} value in the tissue.

SELECTED BIBLIOGRAPHY

Comroe JH: *Physiology of Respiration,* ed 2. Chicago, Year Book Medical Publishers, 1974.

Mines AH: *Respiratory Physiology.* New York, Raven Press, 1981.

West JB: *Respiratory Physiology,* ed 2. Baltimore, Williams & Wilkins, 1979.

SELF-STUDY QUESTIONS

MULTIPLE CHOICE

Choose the single best answer.

1. The P_{O_2} and P_{CO_2} values of venous blood leaving an organ
   A. reflect the average tissue P_{O_2} and P_{CO_2} of that organ.
   B. are relatively insensitive to changes in blood flow rate.
   C. approximate the average capillary P_{O_2} and P_{CO_2} in most cases.
   D. are not influenced by changes in art O_2 content and art CO_2 content.
   E. are uninfluenced by changes in Hb concentration.

2. When you are given the value for the oxygen content of arterial blood, you can calculate the content of oxygen in the mixed venous blood if you know the
   A. cardiac output.
   B. magnitude of the Bohr effect.
   C. metabolic rate.
   D. cardiac output relative to the metabolic rate ($\dot{Q}/\dot{V}_{O_2}$).
   E. blood oxygen capacity.

3. A patient in heart failure (reduced cardiac output) can be expected to have
   A. a greater than normal venous P_{O_2}.
   B. an abnormally large arteriovenous content difference for oxygen.
   C. a greater than normal arterial P_{O_2}.
   D. a greater than normal hemoglobin concentration.
   E. anemia.

4. A patient with severe anemia (Hb concentration equals 7 gm%) can be expected to have

A. an oxyhemoglobin dissociation curve shifted to the left.
B. a decreased arteriovenous O_2 content difference.
C. an increased arteriovenous CO_2 content difference.
D. a decreased cardiac output.
E. a mixed venous P_{O_2} greater than 40 mm Hg.

5. As blood passes from the arterial to the venous end of a capillary of a metabolizing tissue,
   A. the decrease in the P_{O_2} of the blood is lessened by the simultaneous entry of CO_2 into the blood.
   B. the decrease in P_{O_2} equals the increase in P_{CO_2}.
   C. the P_{O_2} of the blood decreases at a nearly constant rate.
   D. the P_{CO_2} of the blood rises but the hydrogen ion concentration remains the same owing to the buffering action of Hb.
   E. the P_{CO_2} rises more than the P_{O_2} decreases.

6. The Bohr effect (i.e., the effect of pH change on the dissociation of HbO_2)
   A. acts to decrease tissue P_{O_2}.
   B. acts to increase tissue P_{O_2}.
   C. has no effect under normal conditions.
   D. is mediated by 2,3-DPG.
   E. retards oxygen unloading from blood in the tissues.

7. The Haldane effect (i.e., the effect of a change in O_2 saturation of Hb on the CO_2 dissociation curve of blood)
   A. acts to decrease tissue P_{O_2}.
   B. acts to increase tissue P_{CO_2}.
   C. helps maintain a normal tissue P_{O_2}.
   D. aids CO_2 loading into blood from the tissue.
   E. means that less CO_2 is stored in blood as it becomes deoxygenated.

ANSWERS

1. A. The partial pressure of CO_2 and O_2 in blood at the venous end of the capillary of a metabolizing tissue is always different from the average value of P_{O_2} and P_{CO_2} in the capillary blood. For O_2, the partial pressure is highest at the arterial end and lowest at the venous end of the capillary; for CO_2, the reverse is true. Therefore, the average value must lie between the arterial and venous values. The mean capillary value must always be greater than the average tissue value for O_2 and less than the average tissue value for CO_2, because the gases are moving down their partial pressure gradient by passive diffusion. It is an empirical observation that venous partial pressures of CO_2 and O_2 nearly equal the average *tissue* partial pressures for CO_2 and O_2. Changes in blood flow, arterial content of CO_2 and O_2, and hemoglobin concentration will all likely change the venous values.

2. D. The ratio of O_2 consumption rate to cardiac output is the sole determinant of the $(art - \overline{ven})\ O_2$ content difference. One must know both $\dot{V}_{O_2}$ and $\dot{Q}$ or their ratio to calculate the $(art - \overline{ven})\ O_2$ content difference.

3. B. The only safe prediction that can be made is that the $(art - \overline{ven})\ O_2$ difference will be increased, that is, it can be assumed that the oxygen consumption rate is near normal. The mixed venous P_{O_2} is likely to be less than normal, because the $(art - \overline{ven})\ O_2$ difference must be increased in order to achieve extraction of the needed amount of O_2.

4. B. In severe anemia the cardiac output is increased, decreasing the $(art - \overline{ven})\ O_2$ difference. The Fick principle states that $(art - \overline{ven})\ O_2 = \dot{V}_{O_2}/\dot{Q}$. If $\dot{Q}$ is increased $(art - \overline{ven})\ O_2$ will be decreased. This change tends to maintain the mixed venous P_{O_2} value, but with anemia of this degree, the cardiac compensation is inadequate and the mixed venous P_{O_2} is likely to be less than normal. The O_2 dissociation curve is usually shifted to the right in anemia, a factor that promotes O_2 unloading and also tends to maintain a normal mixed venous P_{O_2}.

5. A. The Bohr effect causes the P_{O_2} value to fall less than would otherwise be the case if the P_{CO_2} and pH values were to remain constant. The P_{O_2} value of blood flowing through the capillary of a metabolizing tissue follows an alinear time curve because of the shape of the oxyhemoglobin dissociation curve. The O_2 and CO_2 contents change at similar rates, but their pressures change at different rates because of the different slopes and shapes of the two dissociation curves. Similarly, the P_{O_2} value falls more than the P_{CO_2} value rises because the dissociation curve for CO_2 is steeper than that for O_2. Buffering minimizes the rise in $[H^+]$ as CO_2 enters blood, but the $[H^+]$ concentration rises, nonetheless.

6. B. The entry of CO_2 into capillary blood increases the P_{O_2} of blood throughout the capillary, thereby promoting a higher tissue P_{O_2} than if there were no Bohr effect. The effect, mediated by $[H^+]$, promotes unloading of O_2 from Hb.

7. D. The unloading of O_2 from Hb causes the Hb molecule to take up H^+. This shifts the CO_2 hydration reaction to the right ($CO_2 + H_2O \rightarrow H_2CO_3 \rightarrow H^+ + HCO_3^-$), thereby converting CO_2 to HCO_3^-. The same change can be seen in the upward shift of the CO_2 dissociation curve when O_2 is removed from hemoglobin (i.e., $\downarrow O_2$ sat). This means that for any particular venoarterial content difference, venous and tissue P_{CO_2} will be less than if there were no Haldane effect. Either approach leads to the conclusion that deoxygenation of Hb promotes loading of CO_2 into blood from the tissue.

22

Alveolar Ventilation

JOHN E. REMMERS

OBJECTIVES

After completing this chapter, you should be able to

1. Appreciate the dynamics of CO_2 and O_2 movement across the alveolar–capillary membrane that lead to diffusional equilibrium between the gas and blood in an alveolus.

2. Understand the difference between total pulmonary ventilation and alveolar ventilation in the idealized, "perfect" lung and to calculate alveolar ventilation from total pulmonary ventilation, dead space volume, and respiratory frequency.

3. Comprehend how the movement of inspired gas into the alveoli dilutes the metabolically produced CO_2 and replenishes the metabolically consumed O_2.

4. Anticipate the effects of alveolar hypoventilation and alveolar hyperventilation and calculate the alveolar P_{CO_2}, given alveolar ventilation and CO_2 production rate.

Blood leaves the right ventricle, flows through the pulmonary artery, passes into the pulmonary arterioles, and reaches the pulmonary capillaries (pulmonary perfusion). Air is inhaled into the bronchi, moves through the respiratory bronchioles, and ultimately arrives in the alveoli (pulmonary ventilation). Teleologically, the "objective" is to create a situation in which oxygen will leave the air and enter the blood, and carbon dioxide will leave the blood and enter the air. To accomplish this, the inhaled gas and the blood are brought into intimate contact within the alveolus (Fig. 1). Oxygen molecules in the alveolar gas are traded for carbon dioxide molecules in the blood, a process called gas exchange. The movement of O_2 and CO_2 molecules results entirely from their passive diffusion across the alveolar–capillary membrane.

FUNCTIONAL ANATOMY OF THE ALVEOLUS

For our purposes, the alveolus consists only of gas and blood: the gas is contained in the alveolar space and the blood lies in capillaries in the alveolar wall. The capillary network is so extensive that virtually the entire wall is occupied by capillary blood.

The alveolar wall is best thought of as a sandwich: a sheet of blood lying between two alveolar–capillary membranes. The lung is an example of "sheet" flow; the blood enters the alveolar capillaries and spreads out to form a thin film covering virtually the entire wall of the alveolus. The alveolus appears to be designed to serve one principal function—to bring the alveolar gas into intimate contact with the pulmonary capillary blood.

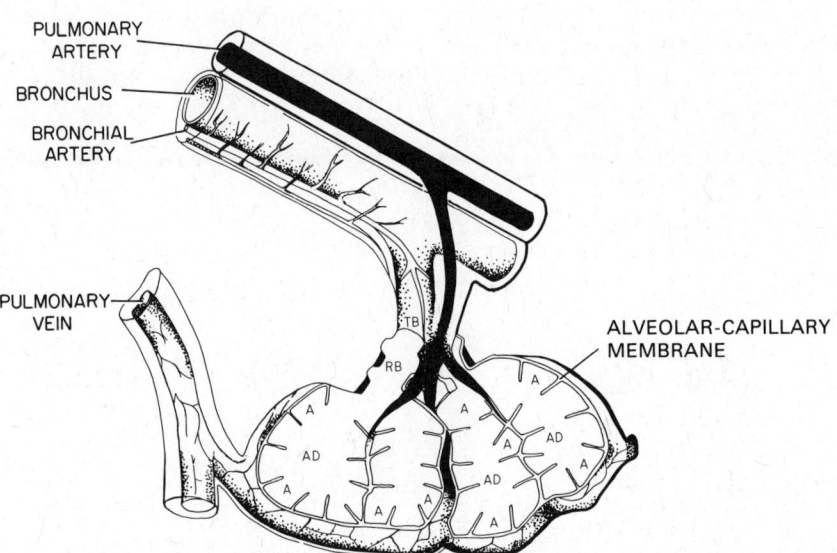

Figure 1 Microscopic anatomy of an alveolus. Blood flows from the pulmonary artery, through arterioles and then into pulmonary capillaries in the alveolar wall. Pulmonary capillary blood drains into the pulmonary vein. Air flows down the bronchus during inspiration, enters terminal bronchioles (TB), respiratory bronchioles (RB), and alveolar ducts (AD), and ultimately reaches the alveoli (A).

DIFFUSION OF CO$_2$ AND O$_2$ ACROSS THE ALVEOLAR–CAPILLARY MEMBRANE

All gas movement within the alveolus occurs as a result of passive diffusion, that is, gas moves from one region to another only when the partial pressure of the gas is greater in one region than another. Because the lung is partitioned into so many small alveolar compartments, its internal surface area is enormous (about the size of a tennis court). Furthermore, as shown in Figure 2, the membrane separating alveolar gas from the capillary blood (the "alveolar–capillary membrane") is extremely thin (0.3 μm). Recall that the resistance to diffusion is directly proportional to distance and inversely proportional to area. Hence, we can conclude that spreading the pulmonary blood over a large area and separating it from alveolar gas by a thin membrane creates conditions *extremely* favorable for diffusion of O$_2$ from the alveolar gas into the film of blood

and for diffusion of CO$_2$ from the blood film into the alveolar gas.

Figure 2 shows the total diffusion pathway for an O$_2$ molecule moving from the alveolar gas. First, it dissolves in the surfactant lining, then it crosses the "alveolar–capillary membrane" (epithelium + endothelium), enters the plasma, and diffuses into the red blood cell (RBC), where it binds to hemoglobin (Hb). Because the total distance is very small, oxygen moves from the alveolar gas and into the red cell so readily that the red cell need reside in the pulmonary capillary for only a fraction of a second in order to take up its full quota of oxygen. The same applies for CO$_2$; as the red cell enters the alveolar capillary it begins to unload its "excess" CO$_2$ very quickly. In other words, after a red cell has been exposed to alveolar gas for only a fraction of a second, it has taken on enough O$_2$ so that its Po$_2$ is equal to the alveolar Po$_2$. Similarly, in a fraction of a second the blood has given up enough CO$_2$ that its Pco$_2$ is the same as the Pco$_2$ of the alveolar gas. Then, of course, no more net movements of O$_2$ and CO$_2$ occur because the blood and the gas are in diffusional equilibrium.

The important point is that the process of diffusion is completed long before the cell reaches the end of the capillary. Therefore, you can safely assume that blood leaving an alveolus has partial pressures of oxygen and carbon dioxide equal to those prevailing in the alveolar gas.*

GRAPHIC PRESENTATION OF ALVEOLAR–CAPILLARY DIFFUSION

Taking an approach similar to that used in Chapter 21, we can follow the changes in the pressures of oxygen and carbon dioxide in a small volume of blood as it traverses a pulmonary capillary, going from the arterial to the venous ends of the capillary (Fig. 3). Although the move-

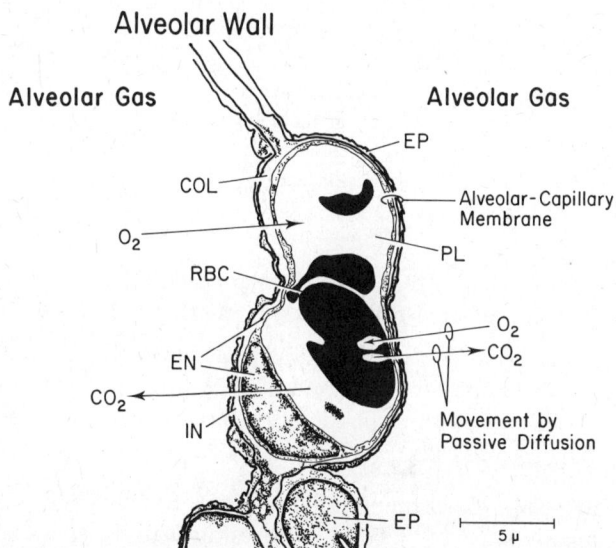

Figure 2 An electron micrograph of the alveolar wall containing a capillary. Note the thin alveolar–capillary membrane. EP, Epithelium; COL, Colagen; PL, plasma; RBC, red blood cell; EN, endothelium; IN, interstitium.

---

*The pressures of oxygen and carbon dioxide might not be the same in all alveoli. The fundamental fact that we will call upon frequently is that in any particular alveolus, the pressure of O$_2$ in the gas of that alveolus is equal to the pressure of O$_2$ in the blood leaving that alveolus, and for CO$_2$ the pressure in the alveolus equals that in the blood leaving the alveolus.

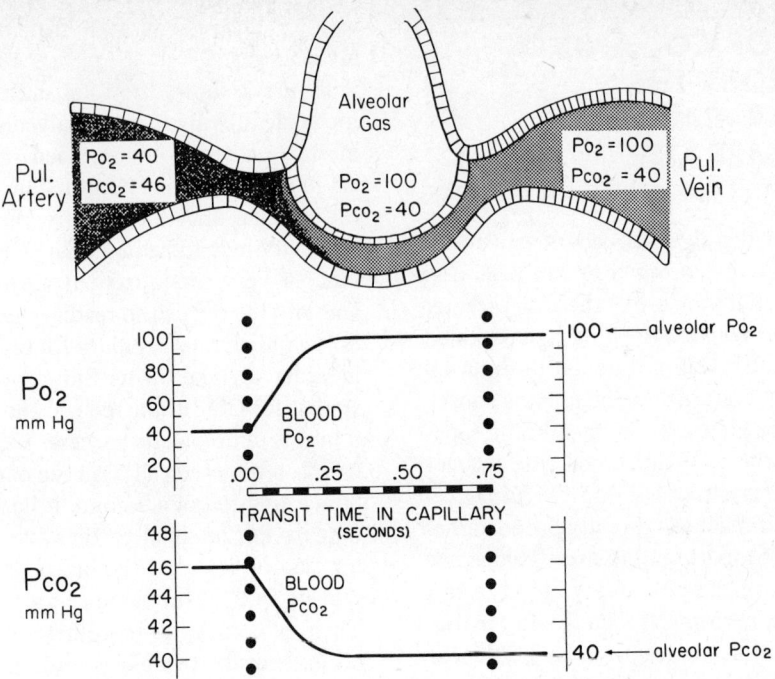

Figure 3 (Top) An idealized alveolus and capillary. (Bottom) The partial pressures of O_2 and CO_2 in the pulmonary artery (left) and pulmonary vein (right). Curves show the PCO_2 and PO_2 at points in the pulmonary capillary in between. The blood enters the pulmonary capillary with the composition of mixed venous blood ($PO_2 = 40$, $PCO_2 = 46$) and leaves as arterialized blood having the same PO_2 and PCO_2 as the alveolus ($PO_2 = 100$, PO_2 $PCO_2 = 40$). After the blood enters the capillary, its PO_2 rises rapidly and PCO_2 falls rapidly and reaches the alveolar values for these gases in 0.3 s. Thereafter, no net movement of CO_2 or O_2 occurs, because capillary blood and alveolar gas are in diffusional equilibrium. Therefore, the resistance to diffusion between the blood and gas phases is very low.

ments of CO_2 and O_2 are the reverse of those described for blood flowing through the capillary of a metabolizing tissue, there is really only one important difference, that is, because the gases diffuse so readily across the alveolar–capillary membrane, the PO_2 and PCO_2 values change rapidly at the beginning of the capillary, and diffusional equilibrium is reached early in the course through the capillary. Thereafter, the PCO_2 and PO_2 values of the blood do not change as it continues through the capillary. This period can be considered a safety factor.

Looking at the representation of an alveolus shown in Figure 3, we begin the analysis by making the following assumptions:

1. The PO_2 of the gas in the alveolar space is 100 mm Hg and the PCO_2 of this gas is 40 mm Hg.

2. The partial pressures of these gases in the alveolar air are held constant at these values by a normal level of ventilation with inspired air (inspired $PO_2 = 150$ mm Hg; inspired $PCO_2 = 0$ mm Hg).

We can now examine the changes in PO_2 and PCO_2 in a small volume of blood as it moves through the alveolar capillary. Just before it enters the capillary, the blood has the composition of "mixed venous" blood. For a normal man at rest, mixed venous PO_2 and PCO_2 are 40 and 46 mm Hg, respectively. Once in the capillary, the blood is surrounded by gas with a P_{O_2} value of 100 mm Hg and a P_{CO_2} value of 40 mm Hg. Consequently, O_2 molecules diffuse across the alveolar–capillary membrane into the blood, and CO_2 molecules move out of the blood and cross the alveolar–capillary membrane to reach alveolar gas. As this occurs, the content of O_2 in the blood rises and the content of CO_2 in the blood falls. As the small volume of blood continues to move along the capillary, the process continues. The rapid movement of CO_2 and O_2 molecules across the alveolar–capillary membrane causes the oxygen content and carbon dioxide content of the blood to change quickly. Consequently, the P_{O_2} rises quickly and the P_{CO_2} falls quickly. Before the blood has traversed half the length of the capillary, it has gained enough O_2 and lost enough CO_2 that the pressures of these gases in the blood are equal to the pressures in the overlying alveolar space. At this point, net movement of the gas molecules ceases because diffusional equilibrium has been achieved. In other words, within about 0.3 s, the P_{O_2} has risen to the alveolar value of 100 mm Hg and the P_{CO_2} has fallen to the alveolar value of 40 mm Hg. From this point on no net flux of gas occurs, and the blood exits from the pulmonary capillary after approximately 0.75 has elapsed. With increases in cardiac output, transmit time decreases, but there is still adequate time for diffusion to proceed to completion.

THE "PERFECT" LUNG

Turning to a consideration of the whole person, we must deal with 300 million alveoli connected to a common airway, a complex problem, to say the least. To simplify the problem we shall, for the moment, assume that all alveoli are identical, that is, we shall consider a perfect lung that is completely homogeneous. In essence, then, we have one gigantic alveolus communicating with the outside world via the tracheobronchial tree. The tracheobronchial tree provides no gas exchange with the blood. Because this part of the respiratory system does not participate in gas exchange, it is termed the anatomic dead space, and its volume is given the symbol V_D. The blood leaving the gigantic alveolus has the same P_{O_2} and P_{CO_2} as the alveo-

lar gas. This means that in the "perfect" situation, the values of alveolar P_{CO_2} and P_{O_2} (alv P_{CO_2} and alv P_{O_2}) are equal to the values of arterial P_{CO_2} and P_{O_2} (art P_{CO_2} and art P_{O_2}).

TOTAL VENTILATION VERSUS ALVEOLAR VENTILATION

The respiratory cycle of a gigantic alveolus is depicted in Figure 4. At the end of expiration (just before inspiration begins), the dead space contains gas of alveolar composition (i.e., $P_{O_2} = 100$, $P_{CO_2} = 40$), as depicted in Figure 4a. During the initial phase of inspiration, inspired air

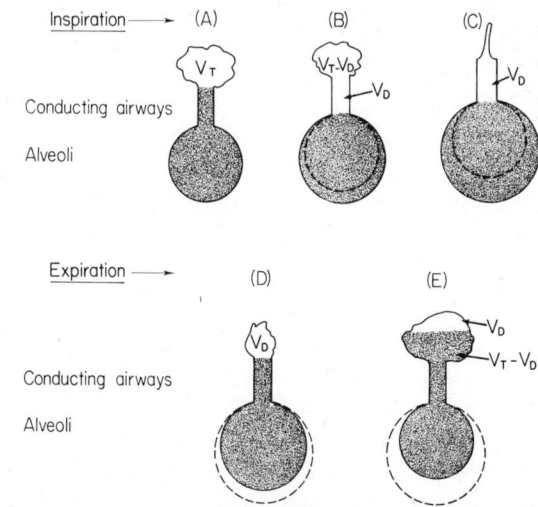

Figure 4 The respiratory cycle of a "perfect" lung. The spherical balloon represents all alveoli, and the tube represents the conducting airways (dead space). The floppy bag connected to the top of each airway measures volume changes during the respiratory cycle. Shaded areas represent CO_2 molecules. During inspiration (A), CO_2-free air is first brought into the dead space (B) and then into the alveoli (C). At the end of inspiration, V_D is filled with CO_2-free air (C). During expiration, the dead space air exits first (D), followed by the alveolar component of the tidal volume (V_T–V_D) (E). The mixed expired gas (the gas in the bag at the end of expiration) has a lower CO_2 concentration than alveolar gas because it is a mixture of both dead space gas and CO_2-containing alveolar gas.

flows into the conducting airway and the dead space gas moves into the alveolar or gas exchanging space (Fig. 4b). At the end of inspiration, the dead space contains gas of inspired composition (Fig. 4c). During expiration, the first gas that appears comes from the conducting airway. This dead space gas has the composition of inspired gas since no CO_2 has been added and no O_2 has been removed. Assuming that the inspired gas was room air, this first portion of the expired tidal volume contains essentially no CO_2 (Fig. 4d). The rest of the expired gas comes from the alveoli and contains a high CO_2 and lower O_2 concentration. At the end of expiration, the bag contains a mixture of dead space gas and gas derived from the alveoli. In other words, the tidal volume (VT) can be divided into two parts: the dead space volume, VD (CO_2-free gas), and the volume derived from the alveoli, VA (CO_2-rich gas). This can be stated as follows:

$$VT = VD + VA$$

Multiplying both sides of the equation by the respiratory frequency f,

$$f(VT) = f(VD) + f(VA)$$

or

$$\dot{V}E = (\dot{V}A) + f(VD)$$

Where $\dot{V}E$ is the total volume of gas expired per minute and is termed the minute volume or total pulmonary ventilation. $\dot{V}A$ is the total volume derived from the alveoli per minute and is called the alveolar ventilation. Alveolar ventilation represents effective ventilation, that is, that part of the total ventilation that is functionally useful in gas exchange. Total pulmonary ventilation represents the sum of alveolar ventilation and dead space ventilation (f · $\dot{V}D$). The latter is referred to as wasted ventilation— alveolar ventilation is the difference between total minute volume and wasted ventilation.

$$\dot{V}A = \dot{V}E - f(VD) \qquad (1)$$

This formulation leads to a useful insight—for the same $\dot{V}E$, one can increase $\dot{V}A$ by decreasing dead space ventilation. This can be accomplished either by decreasing VD (e.g., tracheostomy) or by decreasing respiratory frequency (i.e., breathing with a lower frequency and a larger VT).

COMPOSITION OF ALVEOLAR GAS

The alveoli essentially serve as mixing chambers in which inspired air, rich in oxygen and free of carbon dioxide, is equilibrated with blood returning from the tissues. Normally, the rate of flow of air into the alveoli (alveolar ventilation) is strictly maintained at 17–20 times the rate of flow of CO_2 from the tissues to the blood. Accordingly, each milliliter of CO_2 introduced into the alveoli is diluted by 17–20 ml CO_2 free air, as shown in Figure 5.

The fractional concentration of CO_2 in the alveoli therefore ranges from $\frac{1}{20}$ to $\frac{1}{17}$, or 5–6%. This value corresponds to a PCO_2 of 36–43 mm Hg. Similar considerations apply to the concentration of oxygen in the alveoli. One ml O_2 is extracted from every 14–15 ml air inspired. This means that the concentration of oxygen in alveolar gas decreases from 21 to 14% less than it is in inspired air. That is, alveolar O_2 concentration equals 21% − 7%, or 14%. Accordingly, alveolar PO_2 is approximately 100 mm Hg.

We can calculate the volume of CO_2 expired in a single breath as follows:

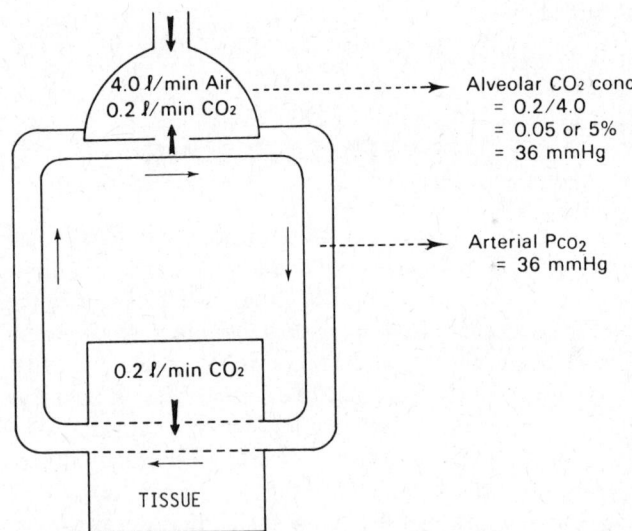

Figure 5 The alveolar concentration of CO_2 is calculated. In a minute, 0.2 L CO_2 is added to 4 L air, so that the final concentration is 5% and the alveolar PCO_2 (PA_{CO_2}) = 36 mm Hg.

$$\left(\begin{array}{c} \text{Vol } CO_2 \\ \text{expired} \end{array} \right) = V_A \left(\begin{array}{c} CO_2 \text{ concentration} \\ \text{in alveolar gas} \end{array} \right) x$$

fractional concentration of CO_2 in
alveolar gas

Multiplying both sides of the equation by frequency, we have,

$$f \left(\begin{array}{c} \text{Vol } CO_2 \\ \text{expired} \end{array} \right) = f \cdot V_A \left(\begin{array}{c} CO_2 \text{ concentration} \\ \text{in alveolar gas} \end{array} \right) x$$

fractional concentration of CO_2 in
alveolar gas

This equation can be restated, replacing the words with symbols:

$$\dot{V}_{CO_2} = \dot{V}_A \times (\text{alv } F_{CO_2}) \qquad (2)$$

Where $\dot{V}_{CO_2}$ is the rate of CO_2 production, $\dot{V}_A$ is the rate of alveolar ventilation, and alv F_{CO_2} is the fractional concentration of CO_2 in the alveolar space.

Since the fractional concentration of gas is equal to the partial pressure of the gas divided by the barometric pressure P_B minus the pressure of water vapor (47 mm Hg at 37° C), we can rewrite the equation above as follows:

$$\dot{V}_{CO_2} = \frac{\dot{V}_A \ (\text{alv } P_{CO_2})}{P_B - 47} \qquad (3)$$

Let $P_B - 47 =$ a constant K; thus this relationship can be arranged to yield

$$\text{alv } P_{CO_2} = \frac{\dot{V}_{CO_2} \ K}{\dot{V}_A}$$

or

$$\text{alv } P_{CO_2} = K \ \frac{CO_2 \text{ production}}{\text{alv vent}} \qquad (4)$$

This concise statement is extremely useful in clinical situations. It says that alveolar P_{CO_2} is directly proportional to the rate of CO_2 production and is inversely proportional to the rate of alveolar ventilation, which is believable. When the rate of alveolar ventilation increases, more inspired (CO_2-free) gas is brought into the alveolus

per unit time; consequently, the concentration (and pressure) of CO_2 will be less, and vice versa. If the alveolar ventilation is greater than normal (alveolar hyperventilation), the alveolar P_{CO_2} will be less than normal—the person "blows off" CO_2. If the alveolar ventilation rate is less than normal (alveolar hypoventilation), the alveolar P_{CO_2} will be greater than normal—the person will retain CO_2.

A similar relationship applies to alveolar P_{O_2}.

$$\text{insp } P_{O_2} - \text{alv } P_{O_2} = \frac{\dot{V}_{O_2}}{\dot{V}_A} (P_B - 47)$$

or

$$(\text{insp} - \text{alv}) \ P_{O_2} = K \ \frac{O_2 \text{ consumption}}{\text{alv vent}} \qquad (5)$$

This relationship states that the value of alveolar P_{O_2} will approach that of the inspired P_{O_2} if $\dot{V}_A$ increases and, conversely, that the difference between inspired and alveolar P_{O_2} increases if $\dot{V}_A$ decreases.

ALVEOLAR HYPERVENTILATION AND ALVEOLAR HYPOVENTILATION

Although the level of alveolar ventilation is tightly controlled normally, disease of lungs or central nervous system commonly causes alveolar ventilation to be less than or greater than normal. Such cases are referred to as alveolar hypoventilation or alveolar hyperventilation, respectively. Equations (4) and (5) predict how the P_{O_2} and P_{CO_2} of alveolar gas will change when alveolar ventilation increases above or decreases below normal. This relationship is depicted in Figure 6a as a gravimetric balance in which metabolic rate and alveolar ventilation are being compared. If the rate of alveolar ventilation increases in comparison with metabolic rate (alveolar hyperventilation), the balance tips and the scale indicates a drop in alveolar P_{CO_2} and a rise in alveolar P_{O_2}. Conversely, a decrease in alveolar ventilation in comparison with meta-

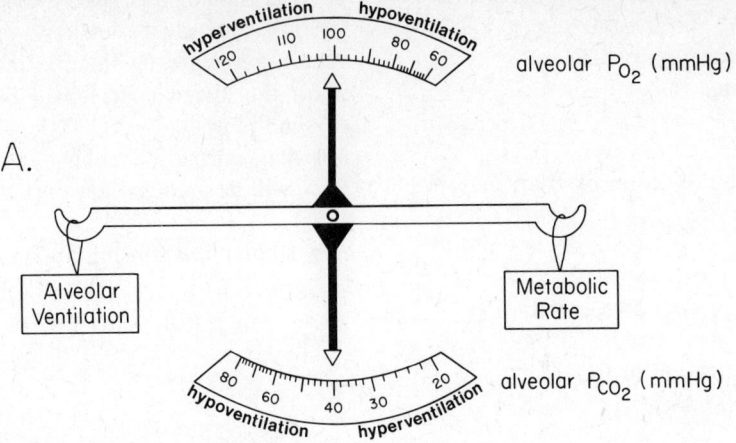

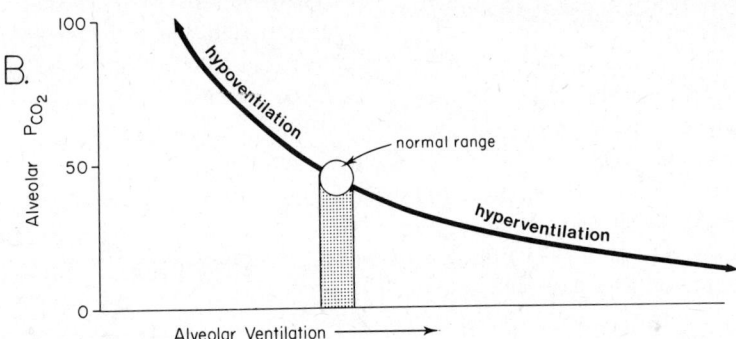

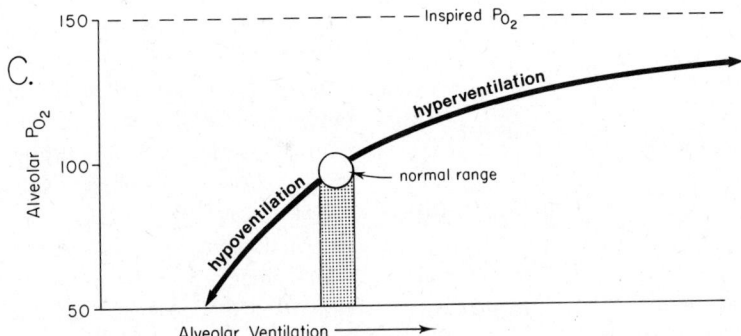

Figure 6 (A) Schematic representation of effects of changes on V_A and metabolic rate on alveolar P_{CO_2} and alv P_{O_2}. (B) Dependence of alveolar P_{CO_2} on V_A; (C) dependence of alveolar P_{O_2} on V_A.

bolic rate (alveolar hypoventilation) causes the scale to tip in the opposite direction, and the scale registers a rise in alveolar P_{CO_2} and a drop in alveolar P_{O_2}.

These relationships are shown graphically in Figure 6 *b,c.* Alveolar hypoventilation (left of normal range) causes P_{O_2} to fall and P_{CO_2} to rise. Alveolar hyperventilation (right of normal range) causes P_{O_2} and P_{CO_2} of alveolar gas to move toward the inspired values ($P_{O_2} = 150$ and $P_{CO_2} = 0$.)

THE "IMPERFECT" NORMAL LUNG

We have considered the exchange of CO_2 and O_2 molecules in a single alveolus and found that this elementary unit operates "perfectly," that is, there is no diffusion limitation at the alveolar level. However, the normal lung is not comprised of 300 million identical alveolar units. Even though each individual alveolus operates "perfectly," the lung as a whole does not because of the functional heterogeneity among alveoli in different parts of the lung. In fact, the normal lung is not a "perfect" lung. A manifestation of "imperfection" is that not all alveoli in the same lung have the same P_{O_2} or the same P_{CO_2}. Even though all receive the same inspired gas and the same mixed venous blood, the P_{O_2} and P_{CO_2} in the gas within each alveolus differ substantially from one alveolus to the next. This situation exists because all regions of the lung do not receive the same quantity of inspired air and the same quantity of mixed venous blood, per unit item. We shall call the quantity of inspired air delivered to a particular region of the lung per unit time regional ventilation (reg $\dot{V}A$), and the quantity of mixed venous blood delivered to a region of the lung regional perfusion (reg $\dot{Q}$). Actually, the P_{CO_2} and P_{O_2} in the gas of an alveolus or a region of the lung is determined by the regional ventilation in relationship to the regional perfusion, or by the ratio of regional ventilation to regional perfusion (i.e., reg $\dot{V}A$/reg $\dot{Q}$). It is fair to say that the *functional state* of an alveolus or a group of alveoli in a region of the lung is adequately characterized by a single variable, the ratio between its ventilation and its perfusion, or its $\dot{V}A/\dot{Q}$ ratio. Let us examine why the regional $\dot{V}A/\dot{Q}$ ratio plays such a pivotal role.

DEPENDENCE OF REGIONAL P_{CO_2} AND P_{O_2} ON REGIONAL $\dot{V}A/\dot{Q}$ RATIO

The question is: What determines the P_{O_2} and P_{CO_2} of a particular alveolus? The P_{O_2} of an individual alveolus will be the outcome of two opposing actions: (1) the action of oxygen-rich, inspired air entering the alveolus to raise the P_{O_2} value; and (2) the action of oxygen-poor blood entering the alveolar capillary to lower the P_{O_2} value. The P_{O_2} of the gas in an alveolus and its end-capillary blood represent the net result of these two opposing actions. If the rate of delivery of blood to the capillary is low, the resultant P_{O_2} value will be high (Fig. 7*b*). On the other hand, if the rate of delivery of inspired gas to the alveolus is low and the rate of blood flow through the capillary is high, the P_{O_2} value will, as a consequence, be low (Fig. 7*a*). In essence, it is the ventilation rate relative to the perfusion rate that dictates the P_{O_2} in the gas of an alveolus and in the blood leaving it.

Similarly, the P_{CO_2} in an alveolus is dictated by the balance between ventilation rate and perfusion rate in that alveolus. If ventilation with CO_2-free inspired air is high and the perfusion by CO_2-rich mixed venous blood is low, then P_{CO_2} of that alveolus will be low (Figure 7*b*). Conversely, if the rate of gas flow is low and the rate of blood flow is high, the resultant P_{CO_2} will be high (Figure 7*a*). In summary, the P_{CO_2} and P_{O_2} in a single alveolus, or in a group of alveoli in a small region of the lung, reflects the balance between ventilation rate and perfusion rate of that alveolus or region of the lung. Stated differently: The local P_{O_2} and P_{CO_2} in the lung depend *exclusively* on the local alveolar ventilation (reg $\dot{V}A$) in relationship to the local perfusion (reg $\dot{Q}$), that is, the local ventilation–perfusion ratio ($\dot{V}A/\dot{Q}$).

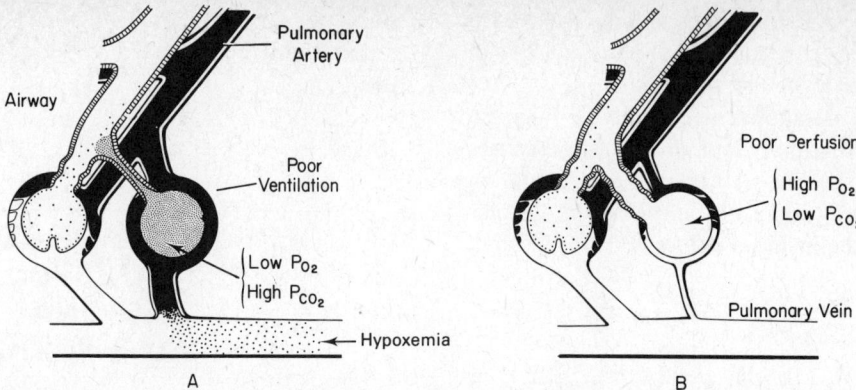

Figure 7 The effect of alteration of the ventilation–perfusion relationship in a region of the lung. (*a*) A region with a low $\dot{V}_A/\dot{Q}$ ratio that has a lower P_{O_2} and a higher P_{CO_2} than the rest of the lung. (*b*) a region with a high $\dot{V}_A/\dot{Q}$ ratio that has a higher P_{O_2} and a lower P_{CO_2} than the rest of the lung. The effects of these differing $\dot{V}_A/\dot{Q}$ ratios on the blood leaving the lung are also shown. (*a*) A large quantity of blood with a low P_{O_2} leaves the region with a low $\dot{V}_A/\dot{Q}$ ratio, reducing the P_{O_2} of the pulmonary vein blood below normal (hypoxemia). (*b*) The region with a high $\dot{V}_A/\dot{Q}$ ratio creates an inefficiency in gas exchange, but does not cause hypoxemia in the pulmonary venous blood (nor does it result in much increase in P_{O_2}, since its volume of blood is small).

CONSEQUENCES OF DIFFERENCES IN REGIONAL LUNG FUNCTION

Consider the instructive, albeit improbable, clinical case of a medical student who had been driving for 24 h to reach the Colorado slopes for Spring skiing. Taking a break, our unlikely hero steps from the car and begins eating a peach. The movement of the student's legs dislodges a clot that had formed in his leg veins and causes a large pulmonary embolus that completely obstructs the pulmonary artery of the right lung. He gasps and aspirates the peach pit, which lodges in his left main stem bronchus. The outcome is fatal, and the pathophysiology represents an extreme example of "ventilation–perfusion mismatching," as depicted in Figure 8. The patient's total pulmonary blood flow is normal and so is his total minute ventilation. Pulmonary gas exchange ceases, however, because the ventilation and perfusion are going to different lungs. The lesson is a simple one: aspirate your peach pit to the "right" lung.

The above case demonstrates that the efficiency of gas exchange in the lungs does not depend on the *total* pulmonary blood flow and *total* ventilation. Rather, it is the regional ventilation in relationship to the regional blood flow that is crucial. If poorly ventilated regions are poorly

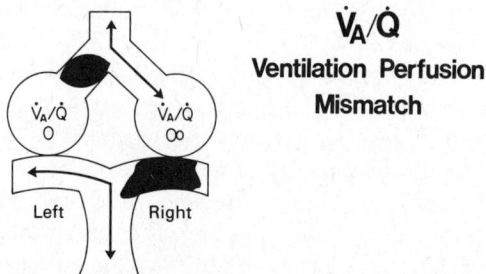

Figure 8 A "clinical" case illustrating the potentially devastating effects caused by regional differences in $\dot{V}_A/\dot{Q}$ ratio. For the right lung, $\dot{V}_A/\dot{Q}$ = infinity (no perfusion). For the left lung, $\dot{V}_A/\dot{Q}$ = zero (no ventilation). Overall minute volume and cardiac output are not grossly different from normal.

perfused while well ventilated regions are well perfused, the lung is functionally homogeneous, and it operates efficiently. If it contains regions that are excessively ventilated in relationship to perfusion (high $\dot{V}A/\dot{Q}$ regions), ventilation is wasted; if other regions receive excessive perfusion in relationship to ventilation (low $\dot{V}A/\dot{Q}$ regions), perfusion is wasted. In other words, so-called "ventilation–perfusion mismatching" is inherently inefficient. Normally, the lungs operate with a mild degree of inefficiency of this sort, owing to the effects of gravity on regional ventilation and regional perfusion (Chapter 23), that is, a slight degree of ventilation–perfusion mismatching occurs in the normal lung. This mechanism is frequently at play in disease states, however, and will be considered in detail in Chapter 24.

SELECTED BIBLIOGRAPHY

Bates DV: Measurement of regional ventialtion and blood flow distribution, in Fenn WO, Rahn H (eds): *Handbook of Physiology,* section 3: *Respiration,* vol II. Washington DC, American Physiological Society, 1964, p 1425.

Bouhuys A: *Breathing.* New York, Grune & Stratton, 1974, pp 59, 79.

Rahn H, Farhi LE: Ventilation, perfusion, and gas exchange—the $\dot{V}A/\dot{Q}$ concept, in Fenn WO, Rahn H (eds): *Handbook of Physiology,* section 3: *Respiration,* vol I. Washington DC, American Physiological Society, 1964, p 735.

West JB: *Ventilation/Blood Flow and Gas Exchange,* ed 3. Oxford, Blackwell Scientific Publishers, 1977.

SELF-STUDY QUESTIONS

MULTIPLE CHOICE

Choose the single best answer.

1. All the following describe gas exchange in a normal lung, *except*
   A. the P_{O_2} in the gas of an alveolus equals the P_{O_2} of blood leaving that alveolus.
   B. the P_{CO_2} in the gas of an alveolus equals the P_{CO_2} of blood leaving that alveolus.
   C. CO_2 and O_2 molecules traverse the alveolar–capillary membrane by passive diffusion.
   D. alveolar gas and pulmonary capillary blood almost always reach diffusional equilibrium.
   E. the transit time of blood in the pulmonary capillary normally limits the amount of O_2 taken up from alveolar gas.

2. The P_{O_2} value of blood leaving an alveolar capillary depends on

   A. the resistance to diffusion across the alveolar–capillary membrane.
   B. the hemoglobin concentration.
   C. the fractional concentration of oxygen in the gas of that alveolus.
   D. the metabolic rate of the lung.
   E. the pulmonary capillary transit time.

3. The content of oxygen in blood leaving an alveolus normally depends on all the following, *except*
   A. the hemoglobin concentration.
   B. the position of the oxyhemoglobin dissociation curve.
   C. the fractional concentration of oxygen in the gas of that alveolus.
   D. the P_{O_2} of the gas in that alveolus.
   E. the oxygen capacity of blood.

4. Alveolar hyperventilation will be present whenever one observes

A. reduced alveolar P_{O_2}.

B. reduced alveolar P_{CO_2}.

C. reduced metabolic rate.

D. a state of anxiety.

E. muscular exercise.

5. Alveolar hypoventilation will

A. increase alveolar P_{O_2}.

B. decrease alveolar P_{O_2}.

C. increase $\dot{V}_{O_2}$.

D. decrease alveolar P_{CO_2}.

E. cause cerebral vasoconstriction.

MULTIPLE CHOICE

Select the correct answer(s). (In many instances more than one answer is correct.)

6. A patient breathing room air at sea level is observed to have a respiratory frequency of 10 breaths/min. Tidal volume is measured and found to be 0.350 L. Assuming the dead space volume to be 150 ml and the metabolic rate for CO_2 equal to 200 ml/min, which of the following is true?

A. Total pulmonary ventilation = 2 L/min.

B. Alveolar ventilation = 2 L/min.

C. Alveolar P_{O_2} = 90 mm Hg.

D. Alveolar P_{CO_2} = 72 mm Hg.

7. A person at rest and breathing room air voluntarily begins to hyperventilate. If the alveolar ventilation is maintained at a value twice the normal value, and if the metabolic rate does not change, which of the following will result?

A. The plasma pH will decrease.

B. The alveolar P_{O_2} will double.

C. The cerebral vascular resistance will decrease.

D. The alveolar P_{CO_2} will be half the normal value.

ANSWERS

1. E. You can answer the question in simply knowing that there is so much time available for O_2 and CO_2 to diffuse across the alveolar–capillary membrane that this diffusion represents no impediment to gas exchange in the lung. Only during heavy muscular exercise at high altitude (low inspired P_{O_2}) would this be incorrect.

2. C. As mentioned above, diffusional resistance of the alveolar–capillary membrane is negligible. Normally, the time required for blood to traverse the pulmonary capillary is twice that required to complete diffusion across this membrane. The metabolic rate of the lung is of no quantitative importance. The hemoglobin concentration determines only the O_2 content, not the P_{O_2} of blood leaving the alveolar capillary. The P_{O_2} value of end-capillary blood is almost always equal to the P_{O_2} of alveolar gas, because there is always ample time for alveolar capillary blood and alveolar gas to reach diffu-

sional equilibrium. The P_{O_2} value of the alveolar gas is equal to the fractional O_2 concentration times barometric pressure minus P_{H_2O}.

3. B. Hemoglobin determines the oxygen capacity of blood; thus it sets the amount of O_2 that will be present in blood (O_2 content) for any value of P_{O_2}. The concentration of O_2 in alveolar gas dictates the partial pressure of oxygen in that gas, which in turn determines the P_{O_2} of end-capillary blood. The position of the O_2 dissociation curve is of little importance because, regardless of shifts to the right ($\uparrow H^+$, $\uparrow$temp, $\uparrow$2,3-DPG) or to the left, the curves converge at the upper end. That is, at a P_{O_2} value of 100 mm Hg, blood is virtually 100% saturated, regardless of the shape or position at lower values of P_{O_2}.

4. B. A decreased alveolar P_{CO_2} (40 mm Hg) must be present if the patient has alveolar hyperventilation. The term indicates nothing regarding either the metabolic

rate or the underlying cause of the abnormality, although anxiety is a common cause of alveolar hyperventilation.

5. **B.** Alveolar hypoventilation simply means that the effective ventilation is subnormal in relationship to metabolic demands. Consequently, the concentration of O_2 in the alveolus will necessarily be less than normal. Because of the increased arterial CO_2, cerebral vasodilation usually accompanies alveolar hypoventilation.

6. **B, D.** Total pulmonary ventilation, given by the product $V_T \times f$, is 3.5 L/min. Dead space ventilation, $V_D \times f$, is 1.5 L/min. Alveolar ventilation, the difference, is 2.0 L/min. Calculate alveolar CO_2 concentra-

tion from the ratio $\dot{V}CO_2/\dot{V}_A$. That is, if 0.2 L CO_2 is added to 2 L air per minute, the final concentration will be 10%. Multiplying by $(P_B - 47)$ yields the correct value for alveolar PCO_2. Alveolar PO_2 can be calculated, assuming RQ = 1, as $150 - \dot{V}O_2/\dot{V}_A (P_B - 47) = 150 - 72 = 78$ mm Hg.

7. During alveolar hyperventilation, the alveolar PCO_2 falls and the alveolar PO_2 rises. If $\dot{V}_A$ doubles, PCO_2 and PO_2 in the alveolus will move halfway toward their inspired values. This results in a $PCO_2 = 20$ and a $PO_2 = 125$. Note that the alveolar PO_2 does not double. The decrease in alveolar PCO_2 will decrease arteriolar. PCO_2, causing an increase in pH and cerebral vasoconstriction.

23

The Pulmonary Circulation

JOHN E. REMMERS

OBJECTIVES

After completion of this chapter, you should be able to:

1. Appreciate the special features of the pulmonary circulation that allow it to maintain relatively low pressures in spite of large increases in cardiac output.

2. Understand how gravity, alveolar hypoxia, and increased alveolar pressure influence pulmonary blood flow.

3. Describe the effects of gravity on ventilation, perfusion, $\dot{V}A/\dot{Q}$, PCO_2, and PO_2 at the base and apex of the lung.

The pulmonary circulation is not a regional circulation in the sense that the coronary or renal circulation are usually referred to as "regional circulations." The difference is that the lungs receive the entire cardiac output, whereas the other organs receive only a fraction of the cardiac output. The pulmonary circulation is analogous to the entire systemic circulation; it has, in fact, been called the "lesser" circulation. A change in pulmonary vascular resistance has the same implication for the right ventricle as a change in total peripheral resistance has for the left ventricle. To a large extent, the behavior of the pulmonary circulation becomes understandable if you bear in mind that the lungs are perfused for one reason, to bring alveolar air and blood into intimate contact so that gas exchange can occur. To achieve this goal, the entire cardiac output passes through the pulmonary circulation.

PRESSURES WITHIN PULMONARY CIRCUIT

Typical pressure pulses in the right ventricle and pulmonary artery are shown in Figure 1. The relatively low

values of the pressures at these two sites contrast with the much higher pressures observed in the aorta. The right ventricle ejects its stroke volume in 0.3 s, producing a systolic pressure of about 22 mm Hg in humans. During diastole the right ventricular pressure falls to 1.0 mm Hg, whereas pulmonary artery diastolic pressure is approximately 8 mm Hg. The estimated pressures in smaller pulmonary arteries, pulmonary capillaries, veins, and left atrium are shown in Figure 2.

The mean pressure in the left atrium is approximately 4 mm Hg. This pressure can be measured with reasonable accuracy by the so-called "wedge" pressure, the pressure measured through a catheter inserted into the pulmonary artery that has been wedged into a pulmonary arteriole. As depicted in Figure 3, the catheter occludes flow in that branch of the pulmonary vascular bed. In this situation, the pressure recorded provides an index of pulmonary venous, hence, left atrial pressure.

The pressures in the pulmonary circulation are remarkably low. This makes sense because higher pressures would cause fluid movement across the alveolar capillary membrane and would result in fluid filling the alveolar space—you would drown in your own juices. It is also noteworthy that the walls of the pulmonary arteries contain relatively little smooth muscle. In contrast to the systemic circulation, which regulates the supply of blood to various organs (including those which may be far above the level of the heart), the pulmonary circulation is rarely

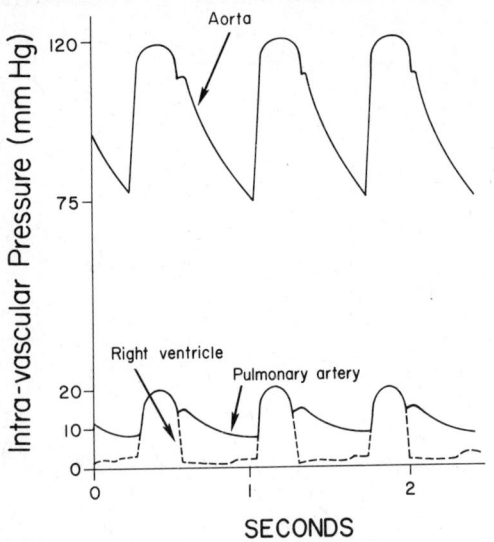

Figure 1 Pressure waves recorded in the aorta (top) and in the pulmonary artery and right ventricle (bottom).

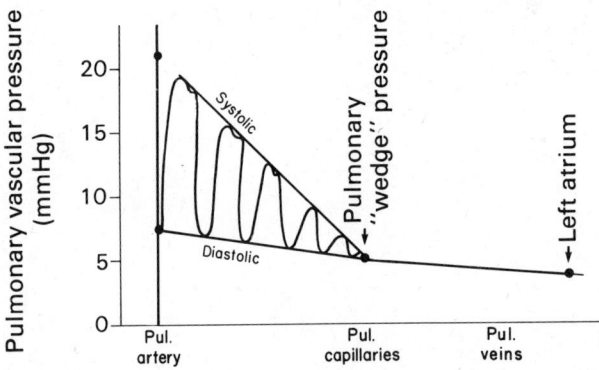

Figure 2 A composite drawing of pressures recorded and estimated at various points in the pulmonary circulation. "Wedge" pressure gives pressure recorded with catheter positioned as shown in Figure 3 and provides an estimation of left atrial pressure.

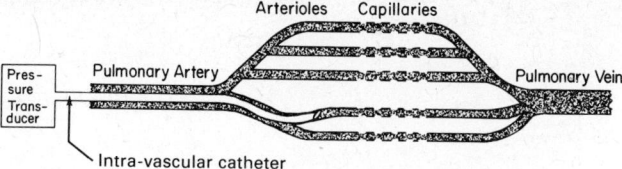

Figure 3 The position of a catheter for measuring pulmonary "wedge" pressure. The catheter is inserted through a peripheral vein, through the right heart, and into a pulmonary arteriole. It is wedged into the arteriole, thereby occluding flow in that part of the vascular bed. The pressure, therefore, measures pulmonary venous pressure.

concerned with directing blood from one region to another (an exception is localized alveolar hypoxia, see below).

PULMONARY VASCULAR RESISTANCE

The pulmonary circulation must accept the whole cardiac output and, as pointed out above, its arterial and capillary pressures must be kept very low in order to avoid pulmonary edema. This means that the pulmonary vascular resistance must be extraordinarily small. An equally striking feature of this circulation is its great capacity to produce additional decreases in resistance. As shown in Figure 4, increasing pulmonary artery pressure above normal causes a substantial decrease in pulmonary vascular resistance, and the reverse is also true.

This decrease in pulmonary vascular resistance with increasing flow means that large increases in cardiac output (as occur during muscular exercise) can be well accommodated by the pulmonary vascular bed without great increases in pulmonary artery pressure. In other words, the pulmonary vascular resistance adjusts to the cardiac output. The resistance is relatively low at rest, and it decreases greatly when the cardiac output rises during muscular exercise. Pulmonary artery pressure rises when cardiac output increases, but much less than it would if the pulmonary vascular resistance had remained constant.

The pulmonary resistance is virtually uninfluenced by the autonomic nervous system and, hence, we can safely

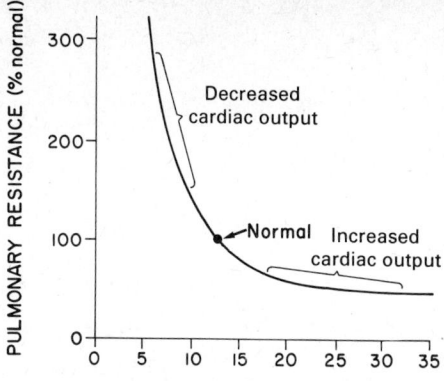

Figure 4 The effect of changing pulmonary artery pressure on pulmonary resistance. Note that resistance decreases greatly when pulmonary artery pressure increases.

attribute these large variations in resistance to local factors intrinsic to the pulmonary bed. This behavior of the pulmonary vasculature in response to increases in pulmonary artery pressure is primarily attributable to a change in overall size of the pulmonary vascular bed, that is, to opening or recruitment of previously closed capillaries with increasing pressure (as shown in Fig. 5, bottom left). In essence, the available vascular bed of the pulmonary circuit seems to adjust its size to th eperfusion pressure. Another factor involved in this adjustment of the pulmonary vascular resistance is distention of already open vessels (Fig. 5, bottom right). Both factors (recruitment and distention) allow the pulmonary vasculature to decrease its resistance when pulmonary artery pressure rises so that large pulmonary artery and pulmonary capillary pressures are avoided.

It is tempting to ascribe a "purpose" to this striking behavior of the pulmonary vasculature. A major threat to the lungs and a major complication in cardiopulmonary disease is pulmonary edema. In left heart failure, left atrial pressure increases and this "congests" the pulmonary vasculature ("congestive failure"). This condition is associated with an elevated pulmonary venous and pulmonary capillary pressure. This increase in hydrostatic pressure in the capillaries causes extravasation of water into the interstitial space and into the alveolar space (pulmonary edema). Because of the lack of supporting tissue around the pulmonary capillaries and the disastrous

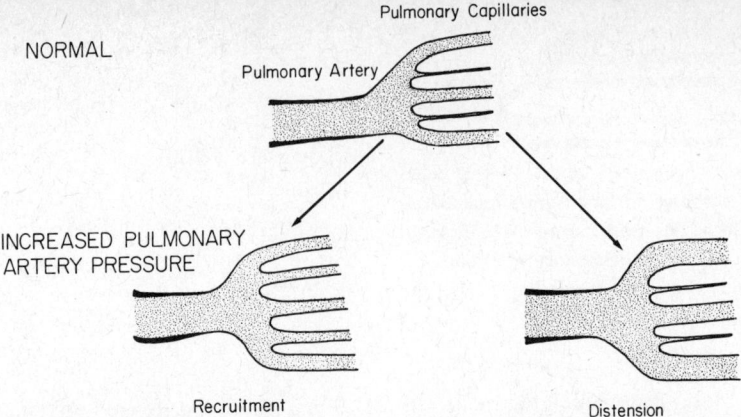

Figure 5 Two factors causing pulmonary vascular resistance to decrease as a consequence of an increase in pulmonary artery pressure are depicted. The normal pulmonary vascular bed has a large number of closed capillaries, which open (they are recruited) when pulmonary artery pressure rises (bottom left). In addition, capillaries that are already open normally increase in size (they are distended) when faced with an increase in pulmonary artery pressure (bottom right).

consequences of fluid in the alveoli, it is not surprising that, in the "wisdom of the body," precautions have been taken to prevent increases in pulmonary capillary pressure when the cardiac output increases. The increase in size of the pulmonary vascular bed, caused by the opening of unperfused regions, is the principal means whereby large increases in cardiac output can be achieved without pulmonary edema.

EFFECT OF GRAVITY ON PULMONARY VASCULAR RESISTANCE AND BLOOD FLOW

In a man standing upright, gravity influences the hydrostatic pressure in pulmonary capillaries. The pressure inside arteries at the *top of the lung* (apex) is approximately equal to pulmonary artery pressure minus 10 cm H_2O (the distance from the heart to the apex); the pressure inside the arteries at the *bottom of the lung* is roughly equal to pulmonary artery pressure plus 10 cm H_2O. This means that the hydrostatic pressure within the pulmonary capillaries is very low at the top of the lung and rather high at the bottom. One important consequence is that the basilar capillaries are most prone to extravasate water (pulmonary edema generally occurs first at the base of the lung). Another is that the capillaries at the base are distended, whereas those at the apex are rather collapsed. Accordingly, the *resistance* to blood flow is low at the base and high at the apex. This difference in capillary size caused by gravity can be seen in Figure 6, in which sections of a dog lung are shown. The lung was fixed when the animal was held in the vertical position. The net effect of gravity on the alveolar capillaries can be readily appreciated here. Compare the size of apical capillaries and basilar capillaries in this vertical lung. Notice that the capillaries of apical alveoli are collapsed (Fig. 6a), whereas those around basilar alveoli are distended with blood (Fig. 6b). This results from the hydrostatic pressure inside the capillaries being much greater at the bottom of the lung than at the top. Consequently, in the upright posture the resistance to blood flow is less at the base of the lung than at the top, and accordingly pulmonary blood flow is directed preferentially to the base.

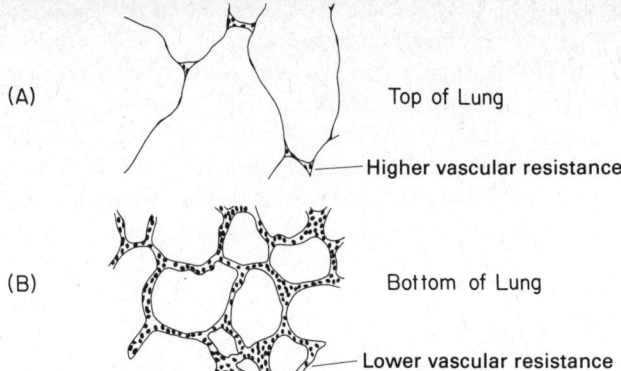

(A)   Top of Lung

—— Higher vascular resistance

(B)   Bottom of Lung

—— Lower vascular resistance

Figure 6 Drawings from sections of a dog lung, fixed with the animal in the upright position demonstrating the effect of gravity on regional alveolar size and capillary filling. *(a)* Section from the apex of the lung showing large alveoli and little blood in the capillaries. *(b)* Section from the base of the lung showing small alveoli and capillaries engorged with blood.

EFFECT OF GRAVITY ON THE DISTRIBUTION OF ALVEOLAR VENTILATION

Because the lung itself has mass, it also is influenced by gravity. As a result, in the upright posture the intrapleural pressure is more negative at the top of the lung (apex) than at the bottom of the lung (base). This means that normally the alveoli at the apex (Fig. 6*a*) are more completely distended than those at the base (Fig. 6*b*). In a sense, the basal alveoli are rather slack or incompletely stretched, whereas apical alveoli are stretched taut. During inspiration the alveoli at the bottom expand readily (their compliance is high). Consequently, inspiratory flow is directed preferentially to the alveoli in the base rather than to those in the apex. In other words, regional ventilation is greater at the base than at the apex, tending to compensate for the vertical inequality in blood flow.

EFFECT OF GRAVITY ON REGIONAL V̇A/Q̇ RATIOS

The final point is that the effect of gravity to cause over-perfusion of dependent alveoli is greater than the effect of gravity to increase the ventilation of these alveoli. The net effect is that the base of the lung is poorly ventilated in relationship to its perfusion, whereas the apex of the lung receives excessive ventilation in relation to its perfusion. Accordingly, apical alveoli have a high P_{O_2} and low P_{CO_2}, while basilar alveoli have a low P_{O_2} and a high P_{CO_2}. This effect of gravity on the normal lung is a major reason why the normal lung is not a "perfect" lung.

HYPOXIC VASOCONSTRICTION

The foregoing emphasizes that the passive mechanics of the pulmonary vasculature are dominant in determining the distribution of blood flow within the lung. In contrast, a remarkably *active response* occurs when the P_{O_2} of alveolar gas is reduced below normal. Although the steps responsible for this effect are not understood, hypoxic vasoconstriction occurs in the excised lung and, therefore, the autonomic nervous system is not involved. *Low alveolar* P_{O_2} stimulates the local pulmonary arterioles, causing contraction of smooth muscle in the walls of small arterioles, causing contraction of smooth muscle in the walls of small arterioles, thereby leading to an increase in pulmonary vascular resistance.

If the entire lung becomes hypoxic, as would occur if a 10% O_2–90% N_2 mixture were breathed, pressure in the pulmonary artery rises dramatically. This is an invariable and pronounced response in normal humans; it is responsible for the pulmonary hypertension seen in disease states associated with alveolar hypoxia. Since this hypoxic vasoconstriction of pulmonary arterioles results from a direct action of oxygen lack in the alveoli, it occurs in a localized part of the lung whenever a local region becomes hypoxic. For instance, if the flow of inspired gas to one lobe of the

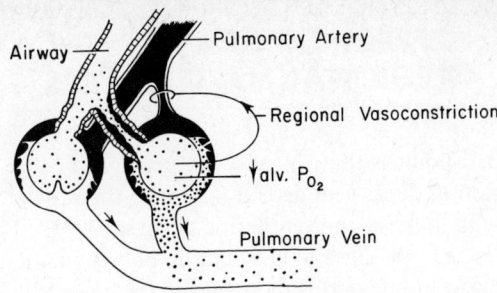

Figure 7 Two alveolar regions are depicted. The alveolar region on the left is well ventilated, but that on the right is poorly ventilated. Because of the low $\dot{V}_A/\dot{Q}$ ratio for the region on the right, its alveolar PO_2 (PA_{O_2}) is abnormally low. This low PA_{O_2} causes vasoconstriction in the artery supplying the right-hand region, tending to return the $\dot{V}_A/\dot{Q}$ ratio toward normal.

lung is impaired, the PO_2 of that lobe will decrease below normal. This is depicted in Figure 7, which shows two alveoli. One alveolus (right) is poorly ventilated in relation to its perfusion, that is, its regional $\dot{V}_A/\dot{Q}$ ratio is low. Accordingly, as the alveolar PO_2 of the region decreases, arterioles supplying blood to the lobe are affected by the local hypoxia and constrict, thereby decreasing blood flow to that region. This is a useful response; the decreased perfusion reestablishes a more normal ventilation/perfu-

sion ratio in the lobe so that the alveolar and end-capillary PO_2 of the region are more normal. In other words, regional alveolar hypoxia induces compensatory reactions in the local pulmonary arterioles, which tends to raise the regional $\dot{V}_A/\dot{Q}$ ratio and, thereby, increase regional PA_{O_2}. However, these reactions are generally less than completely effective—gravity and disease processes still cause $\dot{V}_A/\dot{Q}$ mismatching, even though regional alveolar hypoxia causes regional vasoconstriction.

SELECTED BIBLIOGRAPHY

Barer G: The physiology of the pulmonary circulation and methods of study. *Pharmacol Ther* 2:247, 1976.

Harris P, Heath D: *The Human Pulmonary Circulation,* ed 2. Edinburgh, Churchill Livingstone, 1977.

Murray JF: *The Normal Lung.* Philadelphia, WB Saunders, 1976, p 113. 5.

Racz GB: Pulmonary blood flow in normal and abnormal states. *Surg Clin North Am* 54:967, 1974.

West JB: Blood flow to the lung and gas exchange. *Anesthesiology* 41:124, 1974.

SELF-STUDY QUESTIONS

MULTIPLE CHOICE

Select the single best answer.

1. All the following characterize the pulmonary circulation, *except* one.
   A. It is a low pressure circuit.
   B. It has a great capacity to accommodate increases in flow.
   C. It is subject to pronounced regulation by the autonomic nervous system.
   D. Its arteries are relatively deficient in smooth muscle.
   E. It transmits all the cardiac output.

2. Pulmonary vascular resistance is
   A. relatively low and constant.
   B. relatively low and variable.
   C. uninfluenced by emboli.
   D. relatively uninfluenced by changes in pulmonary vascular pressure.
   E. greatly influenced by the automatic nervous system.

Select the correct answer(s). (In many instances, more than one answer is correct.)

3. Which of the following describe(s) the behavior of the pulmonary circulation during muscular exercise?

A. Mean pulmonary artery pressure increases (fractionally) less than cardiac output increases (fractionally).

B. Hypoxia induces pulmonary vasodilatation.

C. The capillary bed is recruited.

D. End-capillary P_{O_2} decreases.

4. Pulmonary "wedge" pressure

A. is measured with a catheter wedged into a pulmonary vein.

B. indicates right ventricular filling pressure.

C. approximates mean pulmonary artery pressure.

D. approximates pulmonary venous pressure.

Select the single best answer.

5. When a person is in the upright position, the alveolar ventilation at the base of the lung is

A. smaller than that at the apex.

B. nearly equal to that at the apex.

C. exactly equal to that at the apex.

D. larger than that at the apex.

E. the major factor determining gas exchange at the base.

6. When a person is in the upright position, the ventilation–perfusion ratio ($\dot{V}A/\dot{Q}$) at the base of the lung is

A. smaller than that at the apex.

B. nearly equal to that at the apex.

C. exactly equal to that at the apex.

D. larger than that at the apex.

E. bears no consistent relation to that at the apex.

Select the correct answer(s). (In many cases, more than one answer is correct.)

7. In comparing the apex of the lung with its base, in an upright (standing) individual, the apex of the lung has lower

A. blood flow ($\dot{Q}$).

B. ventilation ($\dot{V}A$).

C. alveolar P_{CO_2}.

D. alveolar P_{O_2}.

8. Hypoxic vasoconstriction in the pulmonary circulation

A. is caused by a low partial pressure of oxygen in mixed venous blood (decreased $P\bar{v}_{O_2}$).

B. is the result of vagal reflex activity.

C. does not occur if there is interruption of the sympathetic innervation to the lungs.

D. is caused by a low partial pressure of oxygen in the alveolus.

9. Regional inequalities of $\dot{V}A/\dot{Q}$ ratios

A. are not influenced by local mechanical factors.

B. are minimized by hypoxic vasoconstriction of regional pulmonary arteries.

C. are minimized by the autonomic compensatory responses of the lung.

D. reduce the overall efficiency of gas exchange in the lung.

ANSWERS

1. C. (See answer to 2 for explanation.)
2. B. Both questions 1 and 2 require that you recall that the pulmonary vasculature is a low-resistance, low-pressure circuit that accepts the entire cardiac output. Moreover, because of recruitment and distention of capillaries, the pulmonary vasculature has a great capacity to accommodate an increase in cardiac output. In a sense, the pulmonary vasculature can be thought of as a relatively passive, compliant vascular tree that is relatively deficient in smooth muscle and lacking in prominent autonomic control. Slight increases in pressure cause closed capillaries to open, thereby providing parallel channels for the flow (recruitment). Also, increases in pressure also distend arterioles and capillaries, causing less resistance to flow. Together, these two changes decrease the resistance of the pulmonary cir-

cuit when the cardiac output increases. Embolization greatly increases the pulmonary vascular resistance, of course.

3. **A, C.** Pulmonary vascular resistance decreases during muscular exercise. This means that the fractional rise in cardiac output must exceed the fractional increase in mean pulmonary artery pressure. Opening of closed capillary beds (recruitment) is the principal reason for this change in resistance. End-capillary P_{O_2} remains at normal values during muscular exercise. Alveolar hypoxia, if it were to occur, would cause pulmonary arteriolar constriction.

4. **D.** "Wedge" pressure, measured through a pulmonary artery catheter wedged into one of the peripheral radicals of the arterial tree, reflects pulmonary venous pressure. This is because the flow past the catheter is occluded, hence the pressure in the pulmonary arteries exerts no direct influence on the pressure measured by the catheter, as depicted in Figure 3.

5. **D.** In the normal standing person, vertical differences result from the effects of gravity. Because of the weight of the lung, intrapleural pressure is greater at the base than at the apex. Consequently, the apical alveoli are larger and less compliant than units at the base. It follows, therefore, that alveolar ventilation of apical alveoli will be less than that of basilar alveoli.

6. **A.** Although the base–apex differences in gas flow and blood flow are qualitatively similar, the difference in vascular resistance is much greater than the difference in compliance. Quantitatively, effects of gravity on perfusion exceed those on alveolar ventilation. The effects on regional $\dot{V}_A/\dot{Q}$ are summarized as follows:

| Regional | Alveolar and end-capillary pressure | |
$\dot{V}_A/\dot{Q}$	P_{O_2}	P_{CO_2}
Apex > Base	Apex > Base	Base > Apex

7. **A, B, C.** Blood flow is relatively low at the apex because alveolar pressure exceeds the hydrostatic pressure in the pulmonary capillaries in this region if the individual is standing. Regional ventilation is relatively low (at the apex) because the alveoli are relatively expanded and incompliant. However, the *magnitude of reduction in blood flow* (flow is virtually absent at the top of the lung) exceeds the magnitude of the decrease in regional $\dot{V}_A$. As a result the ratio $\dot{V}_A/\dot{Q}$ for the top of the lung exceeds that for the bottom. Regional P_{CO_2} is low, and regional P_{O_2} is high at the apex compared with that at the base.

8. **D.** Hypoxic vasoconstriction is triggered by a low P_{O_2} value in the alveolus, *not* by a low P_{O_2} in the pulmonary artery. The low alveolar gas P_{O_2} in a region of the lung induces pulmonary artery constriction in that region by an unknown mechanism, and it does *not* require intervention by the autonomic nervous system.

9. **B, D.** Regional differences normally arise because of differences in local mechanical factors (e.g., regional compliance and regional vascular resistance). Hypoxic vasoconstriction produces a higher vascular resistance in overperfused, underventilated alveoli, thereby tending to increase the $\dot{V}_A/\dot{Q}$ ratio of these regions. Autonomic factors play no role. Whenever regional differences occur in $\dot{V}_A/\dot{Q}$, there are regional differences in alveolar P_{O_2} and P_{CO_2} values, and gas exchange efficiency will be less than in the "perfect," homogeneous lung.

24

Oxygen and Carbon Dioxide Pressures in Arterial Blood

JOHN E. REMMERS

OBJECTIVES

After completion of this chapter, you should be able to

1. Describe how arterial and mixed venous P_{O_2} and P_{CO_2} change when a "perfect" lung is hyperventilated or hypoventilated.

2. Understand why intrapulmonary shunt or $\dot{V}_A/\dot{Q}$ mismatch cause arterial P_{O_2} to be lower than alveolar P_{O_2}.

3. Understand why a true shunt or $\dot{V}_A/\dot{Q}$ mismatch causes only a minor arterial–alveolar P_{CO_2} difference.

4. Appreciate how alveolar hypoventilation and ventilation/perfusion mismatch can combine to cause arterial hypoxemia.

FACTORS INFLUENCING ARTERIAL P_{O_2} AND P_{CO_2}

In Chapter 22 we examined the factors that determine the average concentrations and pressures of CO_2 and O_2 in alveolar gas. In the "perfect" or homogeneous lung, arterial P_{O_2} and P_{CO_2} will be equal to alveolar P_{O_2} and P_{CO_2}. This is because in the homogeneous lung the P_{O_2} and P_{CO_2} are the same in all alveoli. Accordingly, the blood leaving all alveoli has the same P_{CO_2} and P_{O_2}. The resulting pulmonary venous and systemic arterial blood must have the same composition as alveolar end-capillary blood. The situation for the "imperfect" lung is more complex.

Because the $\dot{V}_A/\dot{Q}$ ratio is not the same in all regions, the alveolar and end-capillary P_{O_2} and P_{CO_2} differ from one alveolus to the next. Figure 1 shows an example that we will refer to repeatedly in this chapter. The lung is represented by two alveoli. We shall refer to these as alveolar compartment 1 and alveolar compartment 2. The gas in alveolar compartment 1 will have a different concentration of CO_2 and O_2 than alveolar compartment 2 if the ventilation/perfusion ratio of the two regions differ. Consequently, the blood leaving each compartment will not have the same P_{O_2} or the same P_{CO_2} value. However, the blood from these two sources flows into the pulmonary vein, where they mix and ultimately appear in the systemic circulation as arterial blood. Similarly, the gas ex-

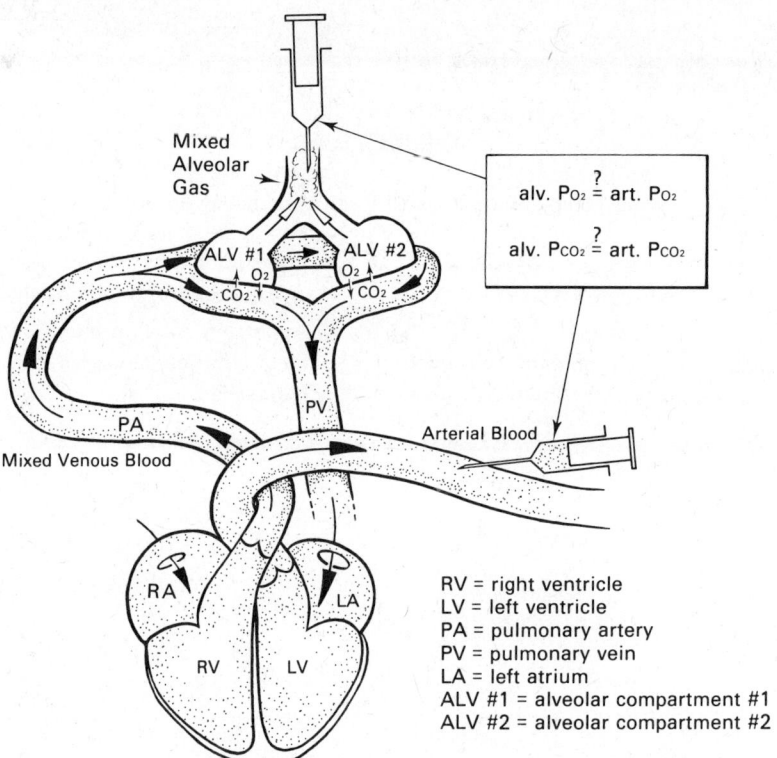

RV = right ventricle
LV = left ventricle
PA = pulmonary artery
PV = pulmonary vein
LA = left atrium
ALV #1 = alveolar compartment #1
ALV #2 = alveolar compartment #2

Figure 1 The central issue in this chapter is to determine what factors cause arterial P_{O_2} (Pa_{O_2}) to be different from alveolar P_{O_2} (PA_{O_2}). A lung is depicted with two sets of alveoli: alveolar compartment 1 and alveolar compartment 2. Alveolar gas is sampled as it is expired from both alveolar compartments. Arterial blood represents the mixture of blood derived from both alveolar compartments.

pired from the two alveolar compartments mixes as it flows out through the airways; ultimately the mixture appears at the mouth, where it can be sampled. This gas is referred to as mixed alveolar gas. We now ask, "How do the pressures of CO_2 and O_2 in alveolar gas compare with those in arterial blood?" We shall use this simple example to answer this question and to provide a basic understanding of how $\dot{V}A/\dot{Q}$ mismatching influences arterial blood gases.

> *A note on relevance:* Po_2 and Pco_2 of arterial blood is commonly measured in patients with known or suspected lung disease because the values *directly* reflect the functional state of the lungs. The job of the respiratory system is to maintain a normal Po_2 and Pco_2 in arterial blood. It is the job of the circulatory system to transport O_2 to the cells and remove metabolically produced CO_2. But, remember that arterial Po_2 and Pco_2 reflect the function of the respiratory system. Accordingly, interpretation of these two measurements are central to laboratory diagnosis of respiratory disease.

The common cause for differences between arterial and alveolar Pco_2 and arterial and alveolar Po_2 is mismatching of regional ventilation in relation to regional perfusion, that is, regional differences in the ratio, $\dot{V}A/\dot{Q}$. In order to examine how the alveolar–arterial differences arise and to appreciate their magnitude, we shall examine in detail the behavior of a two-compartment lung. That is, we shall study the consequences of there being two regions of the lung with different values for $\dot{V}A/\dot{Q}$. Let us first review the changes in Pco_2 and Po_2 caused by changes in alveolar ventilation when the lung is uniform, that is, where $\dot{V}A/\dot{Q}$ is the same for each compartment.

THE HOMOGENEOUS LUNG

NORMAL ALVEOLAR VENTILATION

The normally ventilated "perfect" or homogeneous lung with two compartments is shown in Figure 2. Each compartment receives 50% of the ventilation, and each receives 50% of the cardiac output. Alveolar Po_2 and Pco_2 are the same in each compartment (104 and 41 mm Hg),

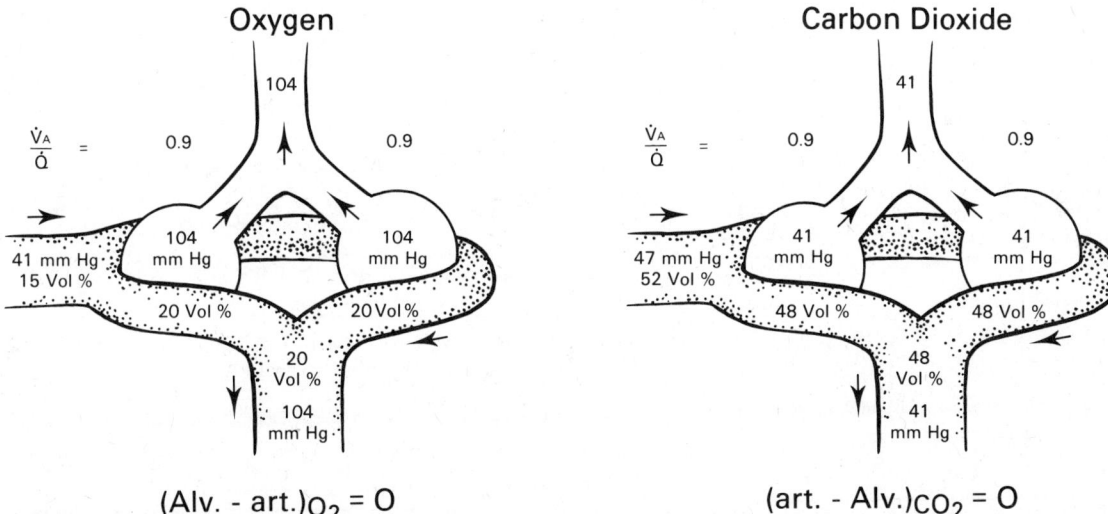

$$(\text{Alv.} - \text{art.})_{O_2} = O \qquad\qquad (\text{art.} - \text{Alv.})_{CO_2} = O$$

Figure 2 A two compartment example of the "perfect" or homogeneous lung is shown. Both compartments have the same $\dot{V}A/\dot{Q}$ ratio and the same Po_2 and Pco_2. Since the lung receives a normal level of alveolar ventilation, alveolar Po_2 and Pco_2 are normal.

and the blood leaving each has the same partial pressures and contains 20 vol% O_2 and 48 vol% CO_2. Of course, mixed alveolar expirate and arterial blood also have the same partial pressures of each gas.

GENERALIZED ALVEOLAR HYPOVENTILATION

Figure 3 shows that the ventilation of each compartment is reduced by one-half, so that the ratio $\dot{V}_A/\dot{Q}$ is reduced from a normal of 0.9 to 0.45 *in each compartment.* In other words, generalized alveolar hypoventilation is present, but there is no regional mismatching of ventilation and perfusion. The lung is homogeneous. In each compartment, the CO_2 pressure rises to 82 mm Hg, and the O_2 pressure falls to 57 mm Hg. The blood flowing from each has the same composition, containing 16.5 vol% O_2 and 64 vol% CO_2. Consequently, the arterial blood is hypoxic ($Pa_{O_2} = 57$ mm Hg) and hypercapnic ($Pa_{CO_2} = 82$ mm Hg). Note that similar changes have occurred in the mixed venous blood, but the decrease in venous P_{O_2} (Pv_{O_2}) is small because of the Bohr effect. This results from the addition of CO_2 to blood in the tissues, making the blood

more acid, and, therefore, decreasing the affinity of hemoglobin for O_2. No alveolar–arterial pressure differences exist for CO_2 and O_2, because both alveoli are ventilated at the same, albeit reduced, rate.

GENERALIZED ALVEOLAR HYPERVENTILATION

Figure 4 shows a situation in which both alveoli are overventilated to an equal extent, such that each receives twice normal ventilation. The $\dot{V}_A/\dot{Q}$ ratio is 1.8 in each compartment. In both compartments, the Po_2 level is elevated and the Pco_2 reduced; since these changes are of equal magnitude in each compartment, the mixed alveolar gas and the arterial blood have values equal to those in each compartment and thus to each other. Once again, there is no alveolar–arterial difference for CO_2 and O_2. The arterial values deviate from normal by the same amount that the alveolar values deviate from their normal values. The example also illustrates that even though Po_2 is above normal, the O_2 content of end-capillary and arterial blood is increased only slightly. This contrasts with the situation for CO_2; the abnormally low arterial Pco_2 level is as-

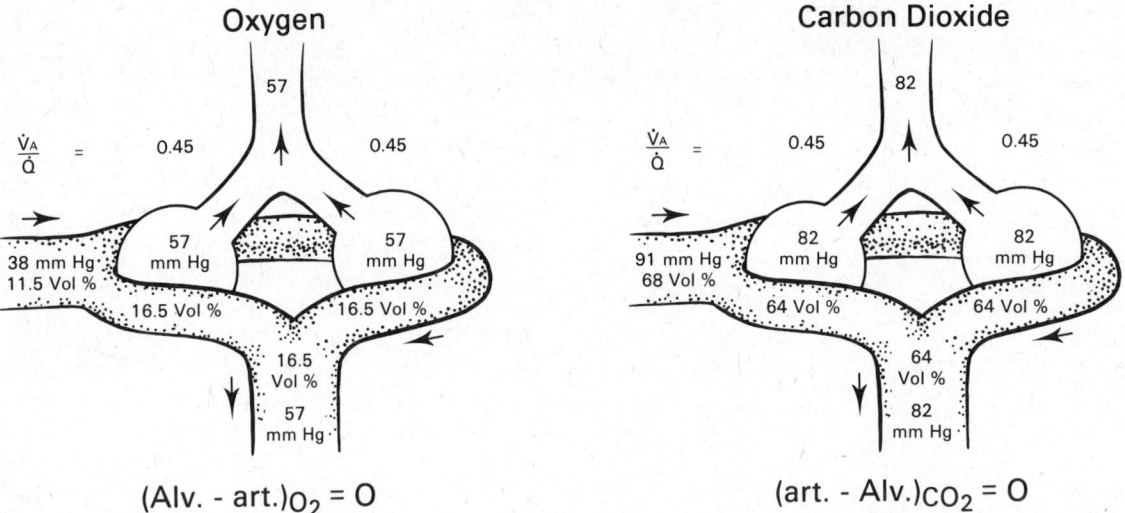

$$(\text{Alv.} - \text{art.})_{O_2} = 0$$
$$(\text{art.} - \text{Alv.})_{CO_2} = 0$$

Figure 3 A two-compartment example of a "perfect" or homogeneous lung inadequately ventilated in relationship to the metabolic rate (alveolar hypoventilation). The Po_2 and Pco_2 values are equal in both compartments, and the arterial blood has a high Pco_2 and low Po_2. This abnormality directly reflects the high Pa_{CO_2} and low Pa_{O_2} levels, that is, there is no A–a Po_2 and no a–A Pco_2 difference, since the lung is homogeneous.

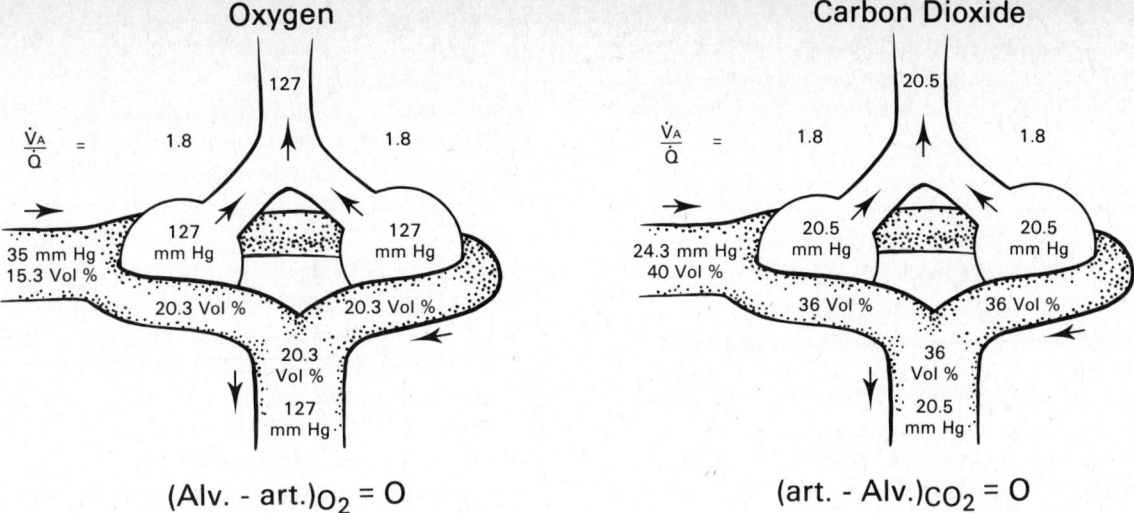

Figure 4 A two-compartment example of a "perfect" or homogeneous lung that is overventilated in relationship to the metabolic rate (alveolar hyperventilation). $P_{A_{O_2}}$ and $P_{A_{CO_2}}$ are equal in both compartments, but $P_{A_{O_2}}$ is increased and $P_{A_{CO_2}}$ decreased. The arterial blood gases show the same changes.

sociated with a substantial reduction in CO_2 content of arterial and mixed venous blood. Because this decrease in mixed venous P_{CO_2} increases its pH value, there is a larger Bohr effect. Consequently, the affinity of hemoglobin for oxygen increases and the P_{O_2} of mixed venous blood is somewhat less than normal.

FACTORS DETERMINING ARTERIAL PO₂ AND PCO₂

We have reviewed factors that determine alveolar gas P_{O_2} ($P_{A_{O_2}}$) and P_{CO_2} ($P_{A_{CO_2}}$). Clearly, these partial pressures are a major determinant of arterial P_{O_2} (Pa_{O_2}) and P_{CO_2} (Pa_{CO_2}). However, other factors may be acting within the lung to make arterial blood gas pressures different from the mixed $P_{A_{CO_2}}$ and $P_{A_{O_2}}$. These factors, which operate within the lung, impair the ability of the lung to "arterialize" the mixed venous blood. Remember that increased diffusion resistant is *not* one of these factors. There is conclusive evidence that in almost all situations, end-capillary blood has the same P_{O_2} and P_{CO_2} as that of re-

gional alveolar gas. Therefore, we shall assume that *there is never a diffusion impairment in the lung*. This assumption is valid, *even in most disease states*. This means that diffusion resistance of the alveolar–capillary membrane is almost never great enough to cause arterial blood gas pressures to differ from mixed alveolar gases.

There are two reasons why Pa_{O_2} and Pa_{CO_2} may not equal $P_{A_{O_2}}$ and $P_{A_{CO_2}}$, namely: (1) "true" shunt, and (2) ventilation/perfusion mismatch. If either of these events occurs in a lung, one can expect to see a difference between arterial and alveolar P_{O_2} and P_{CO_2}. Ventilation/perfusion mismatch means that one region of the lung has a different $\dot{V}_A/\dot{Q}$ ratio than another. As pointed out in Chapter 22, this means that regional $P_{A_{O_2}}$ and $P_{A_{CO_2}}$ are not the same throughout the lung. A shunt can arise either from an anastomosis between the pulmonary artery and pulmonary vein or because a region of the lung is totally unventilated (atelectasis). Both true shunt and ventilation/perfusion mismatch are present in normal persons, hence the normal lung is not a "perfect" lung. The "perfect" or homogeneous lung represents the condition of most efficient gas exchange. The imperfections, introduced by either $\dot{V}_A/\dot{Q}$ mismatch or shunt, decrease the overall efficiency of O_2 and CO_2 transfer.

The resultant alveolar–arterial pressure differences reflect the magnitude of this impairment. Another feature of the resultant alveolar–arterial discrepancy is that the difference is always greater for oxygen than for carbon dioxide.

EFFECT OF SHUNT ON PULMONARY GAS EXCHANGE

Let us examine the consequence of introducing a "true" shunt into a perfect lung. A shunt means that some portion of the cardiac output is shunted past the lungs, that is, blood passes from the pulmonary artery to the pulmonary vein without contacting ventilated alveoli, as depicted in Figure 5. The pulmonary vein receives blood from two sources: (1) completely oxygenated blood comes from the "perfect" lung (Po_2 = 100 mm Hg; Pco_2 = 40 mm Hg), and (2) incompletely saturated blood comes from the shunt having a composition of mixed venous blood (Po_2 = 40 mm Hg; Pco_2 = 46 mm Hg). Blood from both sources mixes in the pulmonary vein ultimately appearing in the systemic artery as poorly oxygenated blood containing an excess of CO_2.

The deficit in O_2 and the excess in CO_2 caused by a shunt can be calculated using the O_2 and CO_2 dissociation curves. A simple example is shown in Figure 5, in which half the cardiac output passes through a perfect lung and half traverses a "true" shunt. Fifty percent of the arterial blood is derived from the normal lung having a Pco_2 of 100 mm Hg (O_2 content, 20 vol%). The other half comes from shunt flow having Po_2 equal to 40 mm Hg (O_2 content, 15 vol%). Clearly, the resultant oxygen content of the combination will lie midway between the O_2 content of each of the sources (17.5 vol%), since equal volumes of blood from the two sources are mixed together per unit of time. However, the Po_2 value of the mixture is *not* equal to the average of the Po_2 of the two sources of blood. Arterial Po_2 is equal to the pressure of the O_2 associated with arterial O_2 content of 17.5 vol% or 55 mm Hg. Since the alveolar Po_2 in the perfect lung is 100 mm Hg, one can readily see that a "true" shunt produces arterial hypoxemia and results in a substantial difference between mixed alveolar Po_2 and arterial Po_2. In this case, the alveolar–

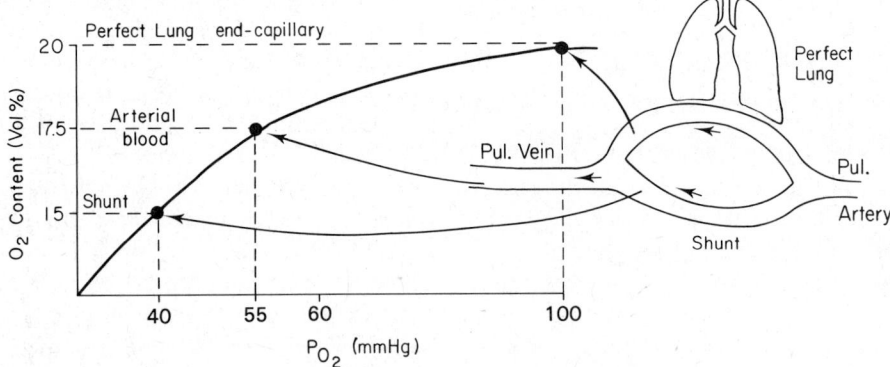

Arterial Hypoxemia Resulting from a 50% Shunt

50% C.O. comes from normal lung (O_2 content = 20 vol.%)
50% C.O. comes from shunt   (O_2 content = 15 vol.%)
Mixed arterial blood O_2 content = 17.5 vol.%, art. Po_2 = 55 mm Hg

Figure 5 The effect of a shunt on arterial Po_2 is calculated. Fifty percent of the cardiac output (CO) passes through the shunt and mixes with 50% of the CO that passes through normal lung. The arterial O_2 content lies midway between the O_2 content of shunt blood and blood from normal lung. The arterial Po_2, however, is closer to shunt Po_2 because of the shape of the O_2 dissociation curve.

Near Normal Art. P_{CO_2} Despite a 50% Shunt

    50% C.O. comes from normal lung (CO_2 content = 48 Vol%)
    50% C.O. comes from shunt (CO_2 content = 52 Vol%)
    Mixed art. blood CO_2 content = 50 Vol%; art. P_{CO_2} = 43 mm Hg

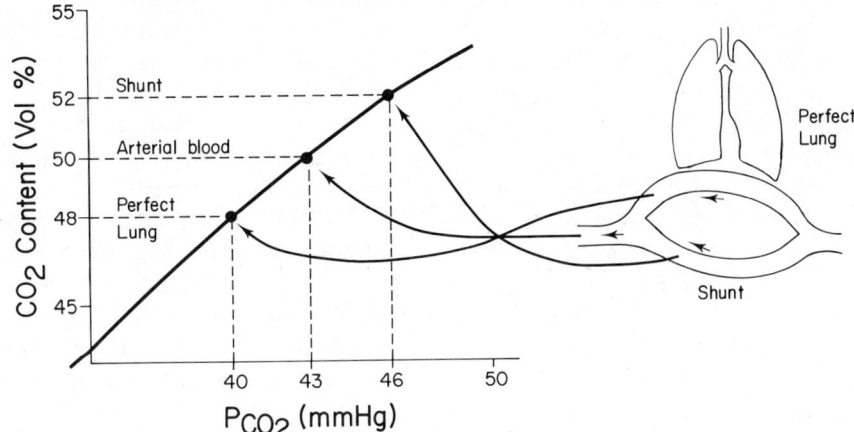

Figure 6 The effect of a 50% shunt on arterial P_{CO_2} is calculated as in Figure 5. The arterial CO_2 content lies midway between the CO_2 content of shunt blood and blood from normal lung. The arterial P_{CO_2} also lies midway between the shunt blood P_{O_2} and P_{O_2} of blood from normal lung. This is because of the straight CO_2 dissociation curve.

arterial P_{O_2} difference or $(A - a)$ P_{O_2} equals 45 mm Hg. This calculation is somewhat unrealistic because it assumes that the composition of mixed venous blood is normal. Actually, there will be a decrease in venous P_{O_2}.

The situation is qualitatively the same for CO_2, but the arterial P_{CO_2} changes are quantitatively less significant than those for arterial P_{O_2}. As shown in Figure 6, the blood flowing through the shunt contains 52 vol% CO_2 at a pressure of 46 mm Hg. Blood from the "perfect" lung contains 48 vol% at a CO_2 partial pressure of 40 mm Hg. The CO_2 content of mixed arterial blood, therefore, has a CO_2 content midway between the two, or 50 vol%. Since the CO_2 dissociation curve is virtually linear within this range, the P_{CO_2} of the mixture also lies midway between that of the two and equals 43 mm Hg. In other words, because of the steepness and linearity of the CO_2 dissociation curve, the introduction of a very large shunt has a minor effect on arterial P_{CO_2}. This simple calculation reveals that the arterial–alveolar P_{CO_2} difference or $(a - A)$ P_{CO_2} is only 3 mm Hg, or about $\frac{1}{15}$ of the $(A - a)$ P_{O_2}. These same considerations apply when alveolar–arterial differences result from $\dot{V}A/\dot{Q}$ mismatch.

The lesson from these simple calculations is that factors that impair gas exchange within the lung ($\dot{V}A/\dot{Q}$ mismatch or "true" shunt) cause a prominent alveolar–arterial P_{O_2} difference and very little arterial–alveolar P_{CO_2} difference, so that arterial hypoxemia rather than hypercapnia results. By contrast, pure alveolar hypoventilation produces no alveolar–arterial differences and causes a decrease in arterial P_{O_2} and an increase in arterial P_{CO_2} of comparable magnitudes.

EFFECT OF VENTILATION/PERFUSION MISMATCH ON GAS EXCHANGE

To illustrate the effects of regional differences in alveolar ventilation and perfusion, we shall examine three situations: (1) normal degree of $\dot{V}A/\dot{Q}$ inequality; (2) narrowed airway; and (3) airway closure. The magnitude of alveolar–arterial P_{O_2} differences, $(A - a)$ P_{O_2}, can be compared with the value of arterial–alveolar P_{CO_2} differences, (a — A) P_{CO_2}, in each situation. Begin by noting the normally ventilated, "perfect" lung in Figure 2. Note that alveolar P_{O_2} and P_{CO_2} are 104 and 41 mm Hg, respectively. Each compartent receives 50% of the alveolar ventilation, and each receives 50% of the cardiac output. The important feature is that the ratio $\dot{V}_A/\dot{Q}$ is the same (0.9) in both compartments. Consequently, the partial pressures of CO_2 and O_2 are the same in both compartments. Because there is no alveolar–end-capillary pressure difference, the blood leaving both compartments also has P_{O_2} and P_{CO_2} values of 104 and 41 mm Hg, respectively. Mixing blood from these two sources, therefore, causes no change in composition so that the arterial blood gas pressures equal those of the mixed alveolar gas. In other words, the differences (A — a) P_{O_2} and (a — A) P_{CO_2} are equal to zero.

THE NORMAL LUNG

Figure 7 represents the amount of ventilation/perfusion inequality present in a normal person in the upright position. In this example, compartment 1 receives one-third of the total alveolar ventilation, while the remaining two-thirds of total $\dot{V}_A$ is delivered to compartment 2. For convenience, we have not changed the fraction of the cardiac output delivered to each compartment. The crucial feature is that we create a situation in which the value of $\dot{V}_A/\dot{Q}$ is not the same in each compartment, that is, compartment 1, $\dot{V}_A/\dot{Q}$ = 0.6; compartment 2, $\dot{V}_A/\dot{Q}$ = 1.2. Total alveolar ventilation is maintained constant, so that mixed alveolar P_{CO_2} and P_{O_2} are the same as in the "perfect" lung. However, even though the total alveolar ventilation remains constant, it is distributed unequally to the two compartments; as a result, compartment 1 is under-ventilated and compartment 2 is overventilated. Accordingly, P_{O_2} is depressed in compartment 1 (86 mm Hg) and elevated in compartment 2 (112 mm Hg).

Note that the effects on O_2 content of end-capillary blood are not parallel to these P_{O_2} changes; O_2 content is depressed in the blood leaving compartment 1 (19.6 vol%) but is *not elevated* in compartment 2 (20 vol%). This is because hemoglobin is virtually 100% saturated at the normal alveolar P_{O_2} value of 100 mm Hg, so that increases in P_{O_2} above the normal to 112 mm Hg only increases dissolved O_2, and this is a negligible amount of oxygen. Mixed arterial blood has an O_2 content of 19.8 vol%, corresponding to an arterial P_{O_2} of 96 mm Hg. Overall, therefore, the maldistribution of alveolar ventilation re-

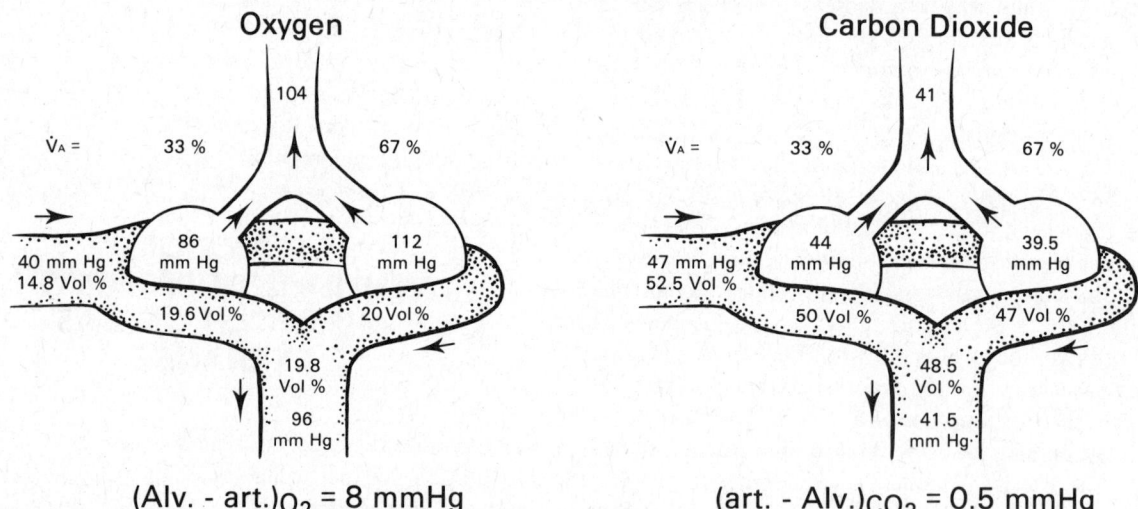

Non-Homogeneous Lung (Regional Inequalities)

$(Alv. - art.)_{O_2}$ = 8 mmHg $(art. - Alv.)_{CO_2}$ = 0.5 mmHg

Figure 7 This example illustrates how a small amount of $\dot{V}_A/\dot{Q}$ mismatching, comparable to that normally present, causes an (A–a) P_{O_2} difference, but no appreciable (a–A) P_{CO_2} difference.

sults in an 8 mm Hg decrease in Pa_{O_2}, or an $(A - a) Po_2$ difference of 8 mm Hg.

The changes in alveolar Pco_2 of the two compartments are less striking. Compartment 1 alveolar Pco_2 = 44 mm Hg, and compartment 2 alveolar Pco_2 = 39 mm Hg. The end-capillary content of CO_2 is 50 and 47 vol% in compartments 1 and 2, respectively. Mixing equal portions of blood from these two sources results in a mixed arterial CO_2 content of 48.5 vol%, corresponding to an arterial Pco_2 of 41.5 mm Hg. This last value is 0.5 mm Hg higher than alveolar Pco_2, so that we can conclude that the normal degree of $\dot{V}A/\dot{Q}$ mismatching brings about an $(a - A) Pco_2$ difference 0.5 mm Hg. This difference is not large enough to be measurable in practice.

THE LUNG WITH A NARROWED AIRWAY

Lung disease frequently causes a profound disturbance in the regional distribution of ventilation in relationship to perfusion. Figure 8 illustrates a condition caused by narrowing the airway to one alveolar compartment. The ventilation to compartment 1 is one-fifth that of compartment 2. For purposes of illustration, let us assume that each compartment receives half the cardiac output. Since flow is the same in both compartments, the $\dot{V}A/\dot{Q}$ ratio is quite low in compartment 1 ($\dot{V}A/\dot{Q}$ = 0.3) and rather high in compartment 2 ($\dot{V}A/\dot{Q}$ = 1.5). Consequently, the Po_2 values of the two compartments are quite different (compartment 1 Po_2 = 55; compartment 2 Po_2 = 113 mm Hg). However, the action of the hemoglobin dissociation curve is the same as before. That is, the reduced Po_2 in compartment 1 reduces O_2 content of end-capillary blood (O_2 content = 17.5 vol%), whereas the supranormal Po_2 in compartment 2 does not appreciably increase the O_2 content of the blood leaving that compartment. The result is that when blood from each compartment is mixed, the O_2 content of mixed arterial blood is 1.2 vol% lower than normal and, as a result, its Po_2 is reduced (arterial Po_2 = 66 mm Hg). This represents moderately severe arterial hypoxemia, and one calculates an $(A - a) Po_2$ difference of 38 mm Hg.

In Figure 8, compartments 1 and 2 have substantially different value of Pco_2 (49 and 39 mm Hg), but this difference is still much less than that for O_2. The difference in CO_2 content of end-capillary blood is very large. However, because the CO_2 dissociation curve is linear, the content of CO_2 is elevated in compartment 1 and depressed in

Non-Homogeneous Lung (Regional Inequalities)

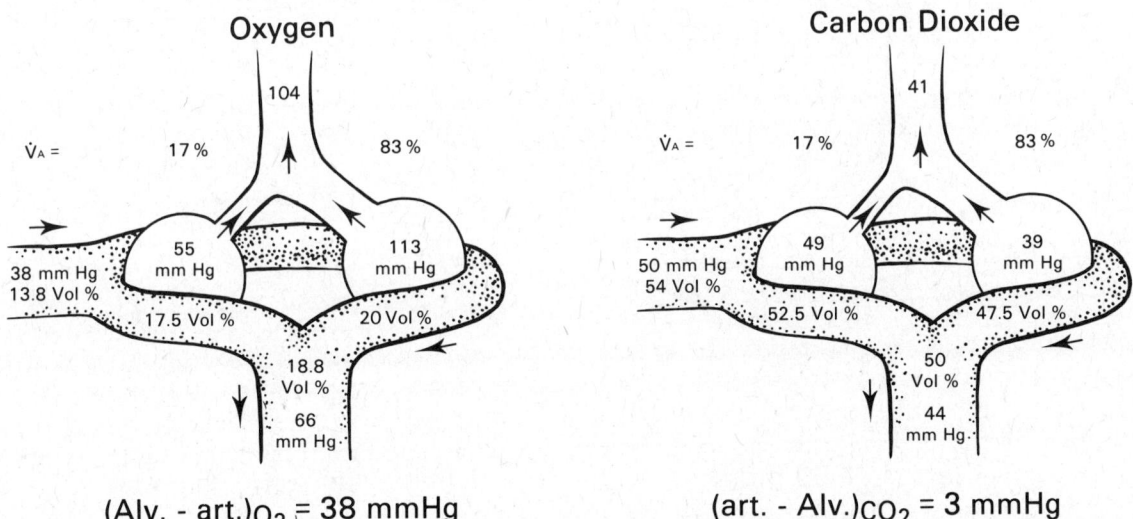

(Alv. - art.)$_{O_2}$ = 38 mmHg   (art. - Alv.)$_{CO_2}$ = 3 mmHg

Figure 8 The airway to compartment 1 is narrowed and its $\dot{V}A$ reduced, causing a substantial increase in $(A–a) Po_2$ but little change in $(a–A) Pco_2$.

compartment 2, so that these two changes tend to offset each other. The final mixture of the two (arterial blood) has a CO_2 content 2 vol% above normal. The important fact is that this causes arterial Pco_2 to be *only* 3 mm Hg above normal. This is because the CO_2 dissociation curve for blood is extremely steep. In other words, a 2 vol% rise in arterial CO_2 content causes only a 3 mm Hg rise in arterial Pco_2. Because of this, the arterial–alveolar Pco_2 difference is only 3 mm Hg. Note that the $(a - A) Pco_2$ difference is one-tenth the $(A - a) Po_2$ difference.

AIRWAY CLOSURE

Figure 9 shows the final and most extreme example of $\dot{V}A/\dot{Q}$ mismatch, the situation in which compartment 1 receives no alveolar ventilation and compartment 2 receives all the alveolar ventilation. If no ventilation goes to compartment 1, it will simply have the same Po_2 and Pco_2 as the blood perfusing it, that is, its Po_2 and Pco_2 will be equal to mixed venous Po_2 and Pco_2. In other words, it functions as a shunt whereby half the cardiac output flows from the pulmonary artery to the pulmonary vein without changing its composition. In this realistic example, the

mixed venous O_2 content is 10 vol% in the steady state. Accordingly, arterial blood will consist of a mixture of 50% mixed venous blood and 50% of blood derived from compartment 2 having a normal alveolar and end-capillary composition (i.e., 104 mm Hg and 20 vol%). The O_2 content of this mixed arterial blood is 15 vol%, and the Po_2 value is 41 mm Hg. The $(A - a) Po_2 = 63$ mm Hg, an extremely high value.

The situation for CO_2 shown in Figure 9 is similar, but the disturbances in Pco_2 are of much smaller magnitude. Arterial blood is made up of a mixture of end-capillary blood from compartment 2 with a normal CO_2 content (48 vol%) added to blood with a mixed venous CO_2 content (56 vol%). The resulting mixed arterial blood has a CO_2 content of 52 vol% and a Pco_2 value of 47 mm Hg. This value is 6 mm above normal, that is, $(a - A) Pco_2 = 6$ mm Hg. Again, the alveolar–arterial pressure differences for CO_2 are one-tenth the difference for O_2.

The values of $(A - a) Po_2$ difference and $(a - A) Pco_2$ difference are compared in Figure 10. The value of each is plotted against the $\dot{V}A/\dot{Q}$ ratio of alveolar compartment 1. Beginning with the ideal situation—the "perfect" or homogeneous lung—the $\dot{V}A/\dot{Q}$ ratio is 0.9, and the $(A -$

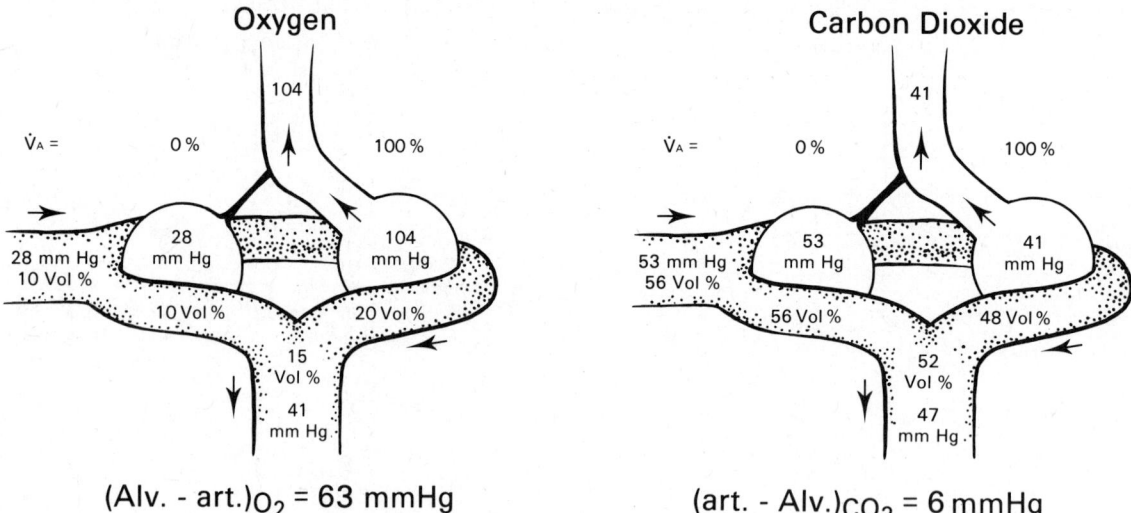

Non-Homogeneous Lung (Regional Inequalities)

(Alv. - art.)$_{O_2}$ = 63 mmHg   (art. - Alv.)$_{CO_2}$ = 6 mmHg

Figure 9 The airway to alveolar compartment 1 is completely closed, constituting severe lung disease. The result is a very large (A–a) Po_2 difference and a relatively small (a–A) Pco_2 difference.

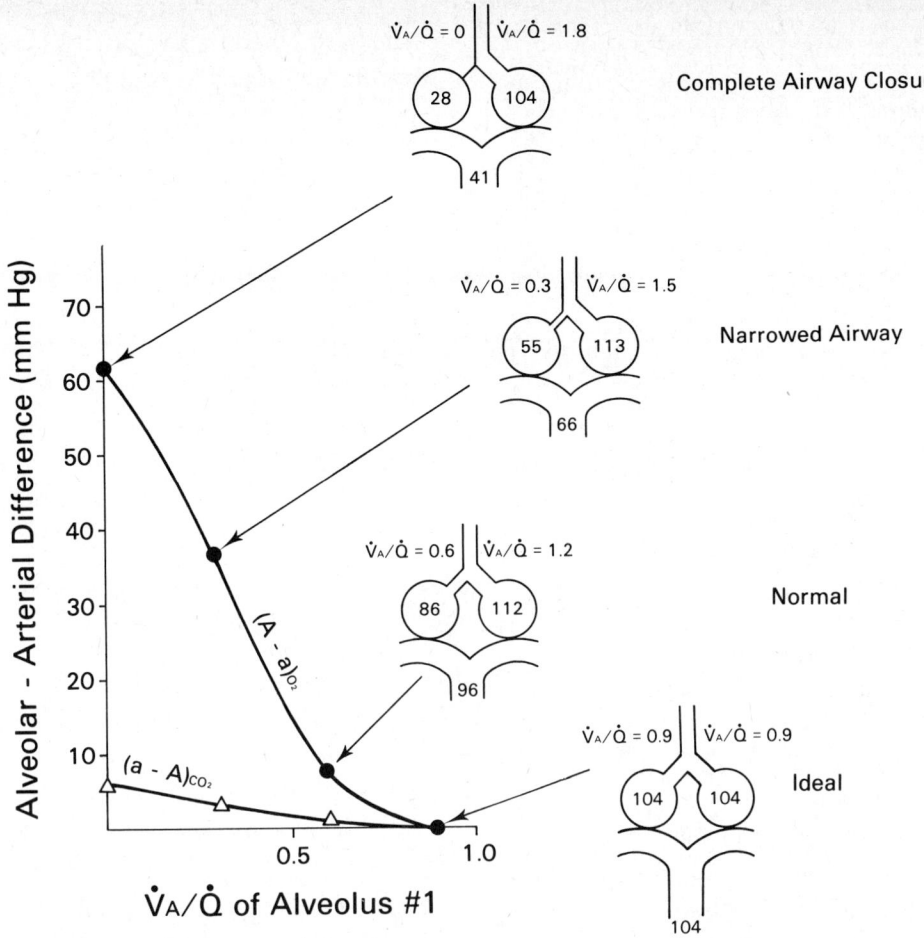

Figure 10 The change in (A–a) P_{O_2} and (a–A) P_{CO_2} is shown for various degrees of airway narrowing in the bronchus leading to compartment 1. The values in the alveoli and in the pulmonary vein give the P_{O_2} values for each case.

a) P_{O_2} and (a − A) P_{CO_2} differences are equal to zero. As the $\dot{V}_A/\dot{Q}$ ratio of alveolar compartment 1 is decreased, the (A − a) P_{O_2} difference increases progressively toward a value of 63 mm Hg when complete airway closure is present (i.e., $\dot{V}_A/\dot{Q}$ ratio of compartment 1 = 0).

The (a − A) P_{CO_2} difference rises slowly and is always small in comparison with the (A − a) P_{O_2} difference. This striking disparity between the two is commonly seen in a variety of lung diseases in which the basic abnormality is either $\dot{V}_A/\dot{Q}$ mismatching or "true" shunt. The (A − a) P_{O_2} difference is a sensitive index of gas exchange impair-

ment, e.g., regional airway narrowing, but the (a − A) P_{CO_2} difference is so small that it can be ignored in most cases. In fact, as a first approximation, arterial blood gas measurements are interpreted by assuming that the (a − A) P_{CO_2} difference is equal to zero. In that case, arterial P_{CO_2} equals alveolar P_{CO_2}. Therefore, Pa_{CO_2}, like PA_{CO_2}, is a direct reflection of the level of alveolar ventilation in relationship to metabolic rate. This is not the case for arterial P_{O_2}, because most lung diseases bring about a large difference between PA_{O_2} and Pa_{O_2}.

SELECTED BIBLIOGRAPHY

Lenfant C: Measurements of ventilation/perfusion distribution with alveolar–arterial differences. *J Appl Physiol* 18:1090, 1963.

Mines AH: *Respiratory Physiology.* New York, Raven Press, 1981.

West JB: *Respiratory Physiology,* ed 2. Chicago, Year Book, 1974.

West JB: State of the art: Ventilation–perfusion relationships. *Annu Rev Respir Dis* 116:919, 1977.

SELF-STUDY QUESTIONS

Multiple Choice

Select the correct answer(s). (In many instances, more than one answer is correct.)

1. Which of the following tend to produce arterial hypoxemia?
   A. Muscular exercise.
   B. Ventilation/perfusion mismatching.
   C. Increased cardiac output.
   D. Alveolar hypoventilation.

2. Alveolar hypoventilation tends to
   A. increase arterial P_{CO_2}.
   B. increase arterial hydrogen ion concentration.
   C. decrease mixed venous P_{O_2}.
   D. increase pulmonary vascular resistance.

3. Alveolar hyperventilation causes
   A. increased arterial P_{CO_2}.
   B. decreased mixed venous hydrogen ion concentration.
   C. increased pulmonary artery pressure.
   D. decreased mixed venous P_{O_2}.

4. The lung of a normal man
   A. produces no alveolar–arterial P_{O_2} difference.
   B. is essentially homogeneous.
   C. has a constant vascular resistance.
   D. displays measurable differences in regional $\dot{V}_A/\dot{Q}$ ratio.

5. Which of the following produce an alveolar–arterial P_{O_2} difference?
   A. Pulmonary artery-to-pulmonary vein shunt.
   B. Atelectasis (collapse of alveoli).
   C. Ventilation/perfusion mismatching.
   D. Effect of gravity on the lung.

6. A pulmonary shunt or regional ventilation/perfusion mismatching produces relatively small arterial–alveolar P_{CO_2} differences because
   A. CO_2 is highly diffusible.
   B. the CO_2 dissociation curve for blood is nearly linear.
   C. membrane diffusion resistances are rare.
   D. the CO_2 dissociation curve for blood is very steep in the physiological range.

7. Which of the following applies/apply to a region of the lung in which the airway is totally occluded?
   A. $\dot{V}_A/\dot{Q} = 0$.
   B. Alveolar P_{CO_2} and P_{O_2} equal mixed venous P_{O_2} and P_{CO_2}.
   C. Blood flow to the region is probably less than if the airway were patent.
   D. There will be a large alveolar–end-capillary P_{O_2} difference for this region.

Select the single best answer:

8. Intrapulmonary shunt or regional ventilation/perfusion mismatch can cause
   A. alveolar–arterial P_{O_2} difference approximately equal to arterial–alveolar P_{CO_2} difference.

B. alveolar–arterial P_{O_2} difference ten times larger than the arterial–alveolar P_{CO_2} difference.

C. diffusion impairments.

D. CO_2 retention.

E. arterial–alveolar P_{CO_2} differences that exceed the alveolar–arterial P_{O_2} difference.

9. What will be the arterial O_2 content of a patient in whom 20% of the cardiac output passes through a shunt? (Assume a mixed venous O_2 content of 15 vol% and that the lungs are functioning normally.)

A. 17 vol%.

B. 19 vol%.

C. 17.5 vol%.

D. 18 vol%.

E. Cannot be calculated from the data provided.

10. What will be the alveolar–arterial P_{O_2} difference in the patient described in question 9? (Use O_2 dissociation curve shown in Figure 5.)

A. 10 mm Hg.

B. 5 mm Hg.

C. 25 mm Hg.

D. 40 mm Hg.

E. 60 mm Hg.

ANSWERS

1. **B, D.** Alveolar hypoventilation and regional mismatching of ventilation to perfusion are the two most common causes of decreased arterial P_{O_2}. The former lowers alveolar P_{O_2}, whereas the latter increases the difference between PA_{O_2} and Pa_{O_2}. Muscular exercise does not decrease PA_{O_2} or Pa_{O_2}, and an increase in CO would have no predictable effect on PA_{O_2} or Pa_{O_2}.

2. **All are correct.** An increase in Pa_{CO_2} and PA_{CO_2} is the hallmark of an abnormally low $\dot{V}A$, which produces an increase in $[H^+]$ by shifting the CO_2 hydration reaction to the right:

$$CO_2 + H_2O \rightleftharpoons H_2CO_3 \rightleftharpoons H^+ + HCO_3^-$$

The decrease in arterial O_2 content will result in an abnormally low O_2 content. This means that $P\bar{v}_{O_2}$ will also be subnormal. The decrease in PA_{O_2} will result in increased pulmonary vascular resistance and pulmonary artery pressure.

3. **B, D.** Increased Pa_{CO_2} and elevated pulmonary artery pressure are features of alveolar hypoventilation. $P\bar{v}_{CO_2}$ is decreased in alveolar hyperventilation because the CO_2 dissociation curve for whole blood is linear. (When Pa_{CO_2} decreases, $P\bar{v}_{CO_2}$ will decrease if the mean venous–arterial CO_2 content difference is constant.) The low $P\bar{v}_{CO_2}$ will be associated with a low venous $[H^+]$. This alkalinity of venous blood shifts the

oxyhemoglobin dissociation curve leftward. Since arterial O_2 content is near the normal value in alveolar hyperventilation (why?), $\bar{v}O_2$ content will be virtually the same unless cardiac output changes. However, for the same $\bar{v}O_2$ content, the Bohr effect will produce a lower $P\bar{v}_{O_2}$.

4. **D.** The normal, but slightly imperfect, lung produces a 10 mm Hg (A − a) P_{CO_2} difference because the lung is not normally homogeneous, that is, regional differences in $\dot{V}A/\dot{Q}$ are readily measurable.

5. **All are correct.** A and B create a situation wherein blood passes from the pulmonary artery to the pulmonary vein without contacting ventilated alveoli. This decreases arterial blood P_{O_2} and creates an alveolar–arterial P_{O_2} difference. Gravity produces regional $\dot{V}A/\dot{Q}$ mismatching, a situation that always causes an (A − a) P_{O_2} difference.

6. **B, D.** Diffusibility or diffusion resistance is not a factor. However, the steepness and linearity of the CO_2 dissociation curve cause arterial P_{CO_2} to nearly equal alveolar P_{CO_2}. This is not the case for O_2, because the O_2 dissociation curve is flat and nonlinear in the physiological range.

7. **A, B, C.** If the airway is occluded, regional $\dot{V}A = 0$, and the blood perfusing the region leaves with the composition of mixed venous blood. This means that

alveolar Po_2 is low, and this local hypoxia causes vasoconstriction in the artery supplying the region.

8. B. Either situation causes $(A - a)$ Po_2 to exceed the $(a - A)$ Pco_2 difference by a factor of 10. In severe $\dot{V}A/\dot{Q}$ mismatching, the $(a - A)$ Pco_2 difference can reach 6 mm Hg. However, the rise in Pa_{CO_2} and fall in Pa_{O_2} bring about a powerful ventilatory response, so that Pa_{CO_2} is usually observed to be normal.

9. B. Make the calculation shown in Figure 5. The systemic artery receives 20% of its blood flow from the shunt (O_2 content = 15 vol%) and 80% from the normal lung. The O_2 content of the mixture lies 80% of the way between 20 and 15 vol% (i.e., 19 vol%).

10. C. Alveolar Po_2 is 100 mm Hg, and arterial Po_2 can be estimated to be 75 mm Hg.

25

Control of Breathing

JOHN E. REMMERS

OBJECTIVES

After completion of this chapter, you should be able to:

1. Understand how the respiratory control system maintains normal arterial blood gas composition.

2. Describe the feedback loops that operate, as well as identify the components of these loops.

3. Outline the central neural structures responsible for rhythmic breathing.

4. Describe the location, function, and special features of the respiratory chemoreceptors.

5. Explain how readjustments in plasma and cerebrospinal fluid bicarbonate concentration influence the responses to chronic hypercapnia.

The regulation of bodily processes is a central theme in medical physiology. A century ago, Claude Bernard called attention to the constancy of the *milieu intérieur* and developed the concept of automatic regulation of the body. Since then, understanding of control systems has developed greatly, and control system analysis has been applied throughout the field of physiology. But the essence of regulation in the body is still what Walter Cannon called "homeostasis" of factors crucial to normal function.

The respiratory system is a complex, multicomponent control system with several feedback loops. Despite its complexities, however, the system is rather easily characterized:

1. It has a motor component consisting of the bellows-like motion of the thorax.
2. This motion causes pulmonary ventilation and results in the elimination of CO_2 from the supply of O_2 to the blood passing through the pulmonary capillaries.
3. The brain continuously monitors the motor output so that *alveolar ventilation will exactly match metabolic rate;* as a consequence, arterial PCO_2 and PO_2 are held constant.

In other words, the system strives to maintain homeostasis with respect to arterial PO_2, PCO_2, and pH. The most remarkable feature of the respiratory control system is the precision with which $\dot{V}A$ is adjusted to match $\dot{V}O_2$—even with enormous increases in metabolic rate during muscular exercise, $\dot{V}A$ increases in proportion to $\dot{V}O_2$. As a result of this precise control of alveolar ventilation, arterial PO_2, arterial PCO_2, and arterial pH are held constant at normal values. It is important to realize that the arterial composition is held constant, and not the mixed venous blood gas composition. During muscular exercise, $\overline{ven}$. PO_2 falls, $\overline{ven}$. PCO_2 rises and $\overline{ven}$. pH falls. The respiratory control system regulates the arterial blood gas and pH values. This section describes the various regulatory processes that act in concert to produce *respiratory homeostasis,* that is, constancy of arterial PCO_2, PO_2, and pH. The automatic process of breathing lends itself well to control system analysis, since its checks and balances can be tested and described fairly completely.

REGULATION OF BREATHING THROUGH FEEDBACK

Physiological regulation (i.e., maintenance of homeostasis) is usually achieved by feedback, a condition in which the response of a system to a stimulus in turn influences the stimulus. In negative feedback, deviation of the stimulus from some "normal" value alters the response such that the original change in stimulus is reversed, thereby restoring the value of the stimulus toward the "normal" value and promoting stability.

For the respiratory control system, Pa_{CO_2} and Pa_{O_2} act as stimuli, and the volume of air breathed per minute ($\dot{V}E$) is the response or output. Negative feedback is operative inasmuch as the response acts to regulate or maintain the stimuli at their "normal" values. Either an *increase* in Pa_{CO_2} or a *decrease* in Pa_{O_2} evokes an increase in $\dot{V}E$, which returns the level of the stimulus toward its normal value. Hence, Pa_{CO_2} and Pa_{O_2} are the controlled variables; homeostasis in the respiratory system implies the constancy of arterial blood gases. This principle, termed chemical feedback, is represented diagrammatically in Figure 1.

Also shown in Figure 1 is a second feedback loop that operates in addition to the first. This second loop, termed volume feedback, controls the volume of the lungs and ensures that breathing takes the form of regular inspiratory and expiratory movements.

These two feedback loops, together with their respective controlled variables, are listed below:

Feedback Loop	*Controlled Variables*
1. Chemical feedback	Arterial PO_2 and PCO_2 (Pa_{O_2}, Pa_{CO_2})
2. Volume feedback	Lung volume

The two feedback loops serve entirely different ends, that is, they (1) function to maintain respiratory gas homeostasis, and (2) operate to ensure that the motor act of breathing is executed by the respiratory muscles in a coordinated fashion that results in regular increases and decreases in lung volume.

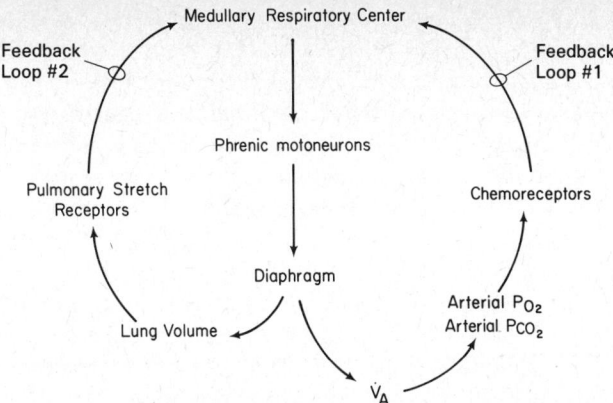

Figure 1 A representation of the components of volume feedback (left) and chemical feedback (right).

ORGANIZATION OF THE NEURAL COMPONENTS OF RESPIRATION

The centers responsible for the motor act of breathing and their organization can best be described in relationship to the *medullary respiratory center.* This center is the focal point of the entire respiratory control system (Fig. 2). Most neuronal impulses that descend from the brain to reach the muscles of respiration emanate from the medullary respiratory center. This center is also the final integrating station for neural activity from higher centers and from a variety of receptors. Its neurons project down the spinal cord to reach the motor neurons of the respiratory muscles. In other words, the medullary respiratory center constitutes the final integrating station, and its neurons directly control the activity of diaphragmatic and intercostal motoneurons.

Two other centers located in the pons, the apneustic center, and the pneumotaxic center have important interconnections with the medullary respiratory center (Fig. 3). These pontine centers profoundly influence the form of the final rhythmic neural output from the medulla, as described below.

Although the medullary respiratory center constitutes a major integrating station for a variety of afferent influences, it has another fuction of equal importance, that of generating intermittent discharge. To emphasize the obvious, breathing is nothing more than the to-and-fro movement of gas into and out of the lungs; it requires *intermittent* contraction of inspiratory muscles. Therefore, it requires efferent inspiratory discharge, sustained until sufficient inspired gas enters the lungs and then ceasing long enough for the lungs and thorax to collapse back to functional residual capacity. The medullary respiratory center is the structure responsible for *initiating each breath* in this cyclic process. The medullary respiratory center is the pacemaker for the respiratory system—if it is destroyed, breathing ceases.

The neurons comprising the medullary respiratory center are diffusely interconnected and are capable of generating, by themselves, periodic inspiratory discharge. In

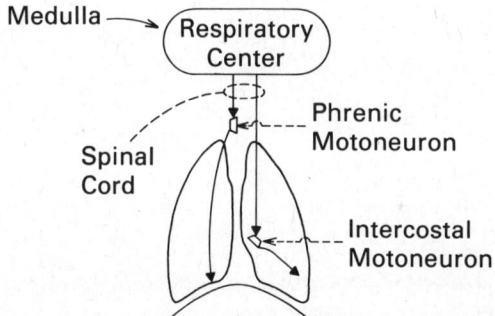

Figure 2 Relationship between respiratory center discharge and efferent impulses to the muscles of respiration. Motor neurons of the respiratory muscles depend on descending impulses from the respiratory center for rhythmic activation.

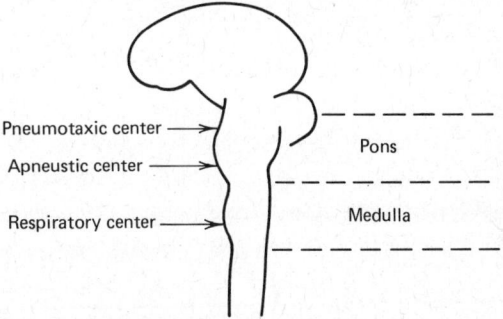

Figure 3 The anatomical location of the three centers in the brain stem that control respiration.

other words, after the brain is transected between the pons and medulla, rhythmic breathing persists (Fig. 4a). Note, however, that the inspiratory movements in this situation are feeble and the respiratory pattern irregular.

By contrast, when the pons is connected to the medulla a normal breathing pattern is observed (Fig. 4c). Therefore, we conclude that while the medullary respiratory center is the pacemaker, the pons exerts a strong action on the medullary neurons to convert the irregular, weak neural respiratory rhythm with a neural output of normal magnitude.

ACTION OF PONTINE CENTERS

The action of the pons on the medullary respiratory center is complex and, in fact, is the result of the action of two separate centers, the *pneumotaxic center* in the rostral

pons and the *apneustic center* in the caudal pons. *To appreciate the action of these centers, the vagus nerves must be cut bilaterally,* since the vagal afferent impulses to the medullary respiratory center also promote a regular respiratory rhythm. That is, one must perform a bilateral vagotomy to unmask the action of the pontine centers.

The key observation can be made after transecting the pons between the pneumotaxic and apneustic centers in the vagotomized animal. When this is done, a very peculiar type of breathing, termed apneustic breathing, occurs. Apneustic breathing consists of very deep prolonged inspirations separated by short expirations (Fig. 4b). In other words, when *only* the apneustic center acts on the medulla, the medullary respiratory neurons are greatly activated, disrupting the respiratory rhythm. An ineffective respiratory pattern results, consisting of prolonged inspiratory efforts. (Fig. 4c). From this we deduce that the rostral pons contains a second center that modulates the excitatory action of the apneustic center. This rostral pontine center, called the pneumotaxic center, converts the ineffective inspiratory spasms of apneustic breathing into

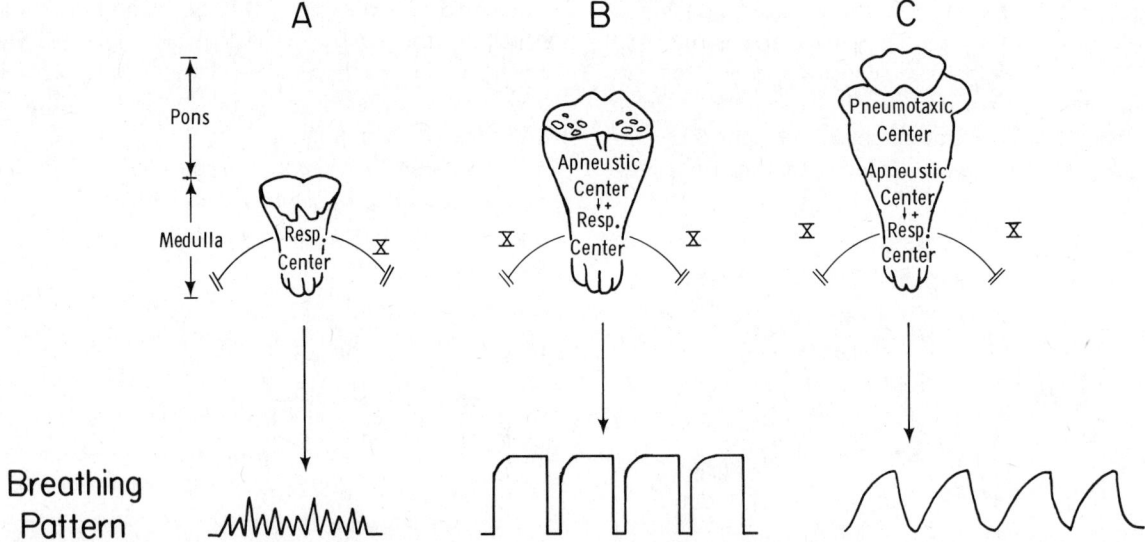

Figure 4 Effects of brain stem section on the respiratory pattern. The vagus nerve X has been transected bilaterally. *(a)* Respiratory center alone generates an irregular, ineffective pattern. *(b)* Respiratory center plus apneustic center produces prolonged inspirations (opneustic breathing). *(c)* Respiratory center plus both pontine centers result in a normal breathing pattern. To produce the abnormal respiratory pattern shown in *(b)*, volume feedback must be eliminated, that is, the vagus nerves must be sectioned bilaterally, and the pneumotaxic center must be detached from the caudal pons.

regular to-and-fro breathing. Once again, it should be emphasized that apneustic breathing occurs only when both the pneumotaxic center and the vagus nerves have been destroyed. Either of these is sufficient to convert the apneustic pattern into an efficient, to-and-fro breathing pattern.

Little is known about the mode of action of the pneumotaxic center, that is, how it exerts its modulating influence on apneustic breathing. However, the action of the vagus nerves is reasonably well understood. The vagus nerves constitute the afferent pathway, which feeds back neural information relating to the volume of the lungs to the medullary respiratory center (volume feedback, loop 2, in Fig. 1).

VOLUME FEEDBACK

Volume feedback refers to afferent or ascending information derived from receptors in the lung that travels up the vagus nerve to the medullary respiratory center. Volume feedback serves to switch off inspiration by removing inhibition when lung volume has reached the "desired" level and to switch on inspiration again when lung volume has returned to functional residual capacity (FRC). In essence, therefore, volume feedback is analogous to the pneumotaxic center inasmuch as it controls the inspiratory activity and promotes regular breathing. Volume feedback is composed of two components:

1. Pulmonary stretch receptors "measure" the volume of the lungs and send afferent impulses up the vagus nerves when the lungs are inflated.
2. These afferent impulses produce inhibition of inspiratory activity of the medullary respiratory neurons.

The inhibition of inspiratory activity by lung inflation results from the inhibitory action of afferent impulses from pulmonary stretch receptors (traveling in the vagus nerves) and is termed the inflation reflex or the *Hering–Breuer reflex*. In spontaneous breathing, the volume of the lungs increases during inspiration, and the neural impulses from stretch receptors increase steadily, as well. When the volume of the lungs has reached a critical level,

the neural traffic traveling up the vagus nerves is sufficient to arrest the activity of inspiratory neurons in the medullary respiratory center. At this point inspiration ceases and expiration begins. When the lung volume returns to FRC, the inspiratory neurons are no longer inhibited by neuronal traffic traveling up the vagus, and inspiration begins again.

SYNTHESIS OF THE MOTOR ACT OF BREATHING

The preceding is a presentation of the classical view of the role of the nervous system in the control of breathing. Recent research has provided information that makes this classical schema somewhat less mysterious and, it is hoped, less mystifying to the student. Three salient findings help elucidate the function of these centers and reflexes:

1. The medullary respiratory center plus regions of the caudal pons generate inspiratory activity, which causes the phrenic motoneurons to begin to fire and steadily increase the frequency of their discharge.
2. The intensity of phrenic efferent activity continues to increase linearly until it is switched off. If it is not switched off, it will ultimately reach a maximal value and continue for a prolonged period at that "saturation" level. When this happens we call the breathing pattern apneustic breathing.
3. Inspiration can be switched off either by volume feedback transmitted up the vagus (Hering-Breuer reflex), or by a center in the rostral pons (the pneumotaxic center).

CHEMICAL FEEDBACK

Chemical feedback enables the brain to find out whether the pulmonary gas exchange is normal and, if it is not, to make the appropriate adjustments in alveolar ventilation.

As pointed out in Chapter 22, arterial P_{CO_2} and P_{O_2} information is specifically related to whether the lung is "doing its job," that is, whether alveolar ventilation is appropriate for the metabolic rate.

We can say, therefore, that Pa_{CO_2} and Pa_{O_2} are the *controlled variables* and that "respiratory homeostasis" is achieved whenever the values of Pa_{CO_2} and Pa_{O_2} are normal. What this means for the respiratory control system is that deviations of these variables from their normal or "desired" values will initiate a compensatory ventilatory response that will tend to restore their values to normal. However, detailed examination of chemical feedback will reveal that Pa_{CO_2} per se is not ultimately the stimulus that actually initiates the respiratory response. In this sense, Pa_{CO_2} is not the specific stimulus in chemical feedback, since changes in the pressure of CO_2 molecules at specialized receptors (chemoreceptors) do not directly influence their activity. Rather, it is the action of the hydrogen ion in the vicinity of the receptor that is important. Changes in Pa_{CO_2} are sensed *indirectly* by the receptor. Pa_{CO_2} is a controlled variable, but it is not the specific stimulus in regulation of breathing through chemical feedback. As you will see, this distinction has important implications in various clinical and environmental abnormalities.

SITE OF ACTION OF CHEMICAL STIMULI

As shown in Figure 5, chemical feedback arises from specialized neural receptors, called chemoreceptors, that sense the value of Pa_{O_2} and also sense Pa_{CO_2}, or something related to this value. These receptors transmit this information to the medullary respiratory center. Two types of chemoreceptors initiate chemical feedback: peripheral and central chemoreceptors. Two sets of *peripheral chemoreceptors* (aortic and carotid bodies) lie outside the central nervous system (CNS), and one *central chemoreceptor* lies inside the CNS. We will first consider, in a general sense, where the individual chemical stimuli act:

Hypoxia

A decrease in alveolar P_{O_2} below normal and the resulting decrease in arterial P_{O_2} stimulates breathing by activating the peripheral chemoreceptors *only*. When an anesthetized animal becomes hypoxic, total pulmonary ventilation increases, and this ventilatory response (an increase of $\dot{V}_E$) to hypoxia can be eliminated by cutting the nerves from the peripheral chemoreceptors. In fact, after denervating the peripheral chemoreceptors, hypoxia decreases the total volume of gas breathed per minute due to a direct depressant action of oxygen lack on the respiratory neurons of the pons and medulla. Accordingly, we conclude that hypoxia does not stimulate the central chemoreceptors.

Hypercapnia

An increase in alveolar P_{CO_2} and the resulting increase in arterial P_{CO_2} above normal exerts a *dual action;* it activates both the peripheral and the central chemoreceptors. If an anesthetized animal breathes a CO_2-enriched gas mixture, pulmonary ventilation increases. This ventilatory response to the CO_2 stimulus is decreased when the nerves from the peripheral chemoreceptors are cut. However, after denervation of the peripheral chemoreceptors, the animal still displays an unequivocal ventilatory response to hypercapnia, manifesting the action of the central chemoreceptor. It is estimated that about half the total increase in $\dot{V}_E$ that accompanies the CO_2 stimulus results from the action of CO_2 on the peripheral chemoreceptors.

THE CAROTID BODIES

The carotid and aortic bodies are the principal peripheral chemoreceptors in mammals; the carotid bodies can be considered prototypes. Each carotid body is about 1 mm across and is situated at the junction of the external and internal carotid arteries. The chemoreceptor cells within the glomus are exposed to arterial blood that perfuses the organ through a richly sinusoidal vascular system. Impulses from these receptor cells travel to the CNS over the IXth cranial nerve; those from the aortic body travel in afferents in the cranial nerve X. That the peripheral chemoreceptors sense some aspect of both arterial P_{CO_2} and arterial P_{O_2} can be demonstrated by recording afferent neuronal traffic in the nerve leading from the carotid body while changing Pa_{O_2} and Pa_{O_2} to varying degrees. Either a decrease in Pa_{O_2} or an increase in Pa_{CO_2} stimulates the receptor cells. If both occur simultaneously, the activation of receptor cells is greater than would be expected on a simple summation basis. Therefore, each carotid body can

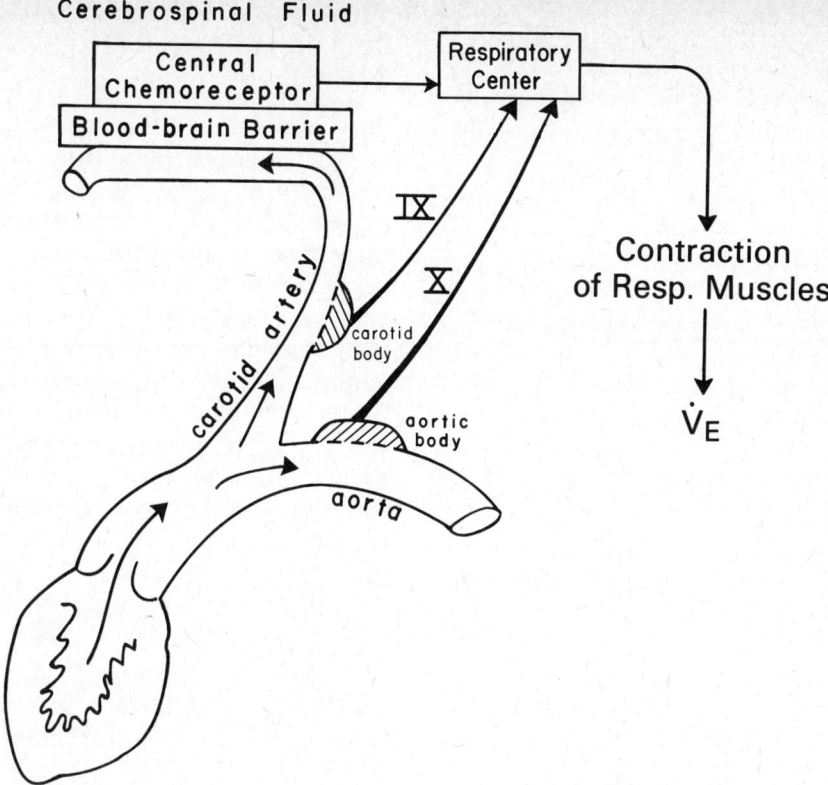

Figure 5 The location of central and peripheral chemoreceptors. The central chemoreceptor lies within the brain. It is separated from blood by the blood–brain barrier and is bathed by cerebrospinal fluid (CSF). The two sets of peripheral chemoreceptors are the carotid and aortic bodies. They are placed in intimate contact with arterial blood and transmit neural information to the respiratory center over cranial nerves IX and X.

be thought of as being particularly sensitive to simultaneous changes in Pa_{CO_2} and Pa_{O_2}. Its sensitivity to the CO_2 stimulus is heightened by the hypoxic stimulus, and vice versa, its sensitivity to the hypoxic stimulus is increased by coexisting hypercapnia. The peripheral chemoreceptors appear to be "hypoventilation" receptors and particularly well suited to detect the immediate effects of a decrease in alveolar ventilation, which produces a confirmed change in the hypoxic and hypercapnic stimuli, that is, an increase in Pa_{CO_2} and a decrease in Pa_{O_2}.

The carotid body actually senses the P_{O_2} value, and *not the O_2 content* of arterial blood. When an animal is poisoned with carbon monoxide (CO), its hemoglobin (Hb)

is largely tied up as COHb, and the oxygen capacity is reduced. Consequently, the content of O_2 in arterial blood is greatly reduced in spite of a normal arterial P_{O_2} value. However, CO poisoning does not activate the carotid body cells and does not stimulate breathing. We therefore conclude that the carotid body senses Pa_{O_2} directly. In contrast, the carotid body does *not* directly sense Pa_{CO_2}. Rather, *it senses arterial H^+ concentration.* Experiments in which arterial P_{CO_2} and arterial $[H^+]$ were changed separately, demonstrate that arterial P_{CO_2} *has no unique ability to stimulate the chemoreceptors;* increases in $[H^+]$ caused by increasing Pa_{CO_2} or by acidifying blood with nonvolatile acid both increase carotid body discharge.

CENTRAL CHEMORECEPTORS

One important consequence of the fact that the central chemoreceptors lie within the brain substance in the medulla is that they are bathed by a special extracellular fluid, the cerebrospinal fluid (CSF) (Fig. 5). This means that the receptors do not lie in intimate contact with blood. Rather, they are separated from blood by the blood–brain barrier, which is relatively impermeable to ions, but which permits CO_2 to diffuse readily into the vicinity of the chemoreceptors cells, as shown in Figure 6.

Although most ions are unable to diffuse freely across the blood-brain barrier, some ions can be actively transported into or out of the CSF, or both. This is particularly important with regard to the bicarbonate ion, since $[HCO_3^-]$ of the CSF is automatically readjusted in certain situations, with important implications for the control of breathing. As is the case for the carotid body, the central chemoreceptor does not sense CO_2 molecules directly. Rather, it senses $[H^+]$ of the extracellular fluid, which is in turn dependent on Pco_2 and the $[HCO_3^-]$ of CSF.

STIMULUS–RESPONSE CURVES OF THE RESPIRATORY CONTROL SYSTEM

What is the overall behavior of the respiratory control system? To answer this question, two stimuli need to be considered, Pa_{O_2} and Pa_{CO_2}. The ventilatory response to each stimulus can be described by a stimulus–response curve for each stimulus obtained by increasing the stimulus and artificially *maintaining* it at some level while the response is measured.

Arterial Pco_2 can be increased to a value greater than normal and artificially maintained there by the addition of CO_2 to the inspired air. This experiment permits us to note the ventilatory response to graded increases in the CO_2 stimulus, as plotted in Figure 7. Note that in producing values of Pa_{CO_2} progressively greater than the "normal" value, Pa_{O_2} was held at its "normal" value, and a linear relationship between $\dot{V}_E$ and Pa_{CO_2}

The stimulus–response curve for PO_2 is obtained in exactly the same way (Fig. 8). Pa_{CO_2} is held constant at its "normal" value, whereas Pa_{O_2} is reduced by decreasing inspired PO_2 (a decrease in PO_2 represents an increase in hypoxic stimulus). The resulting curvilinear relationship

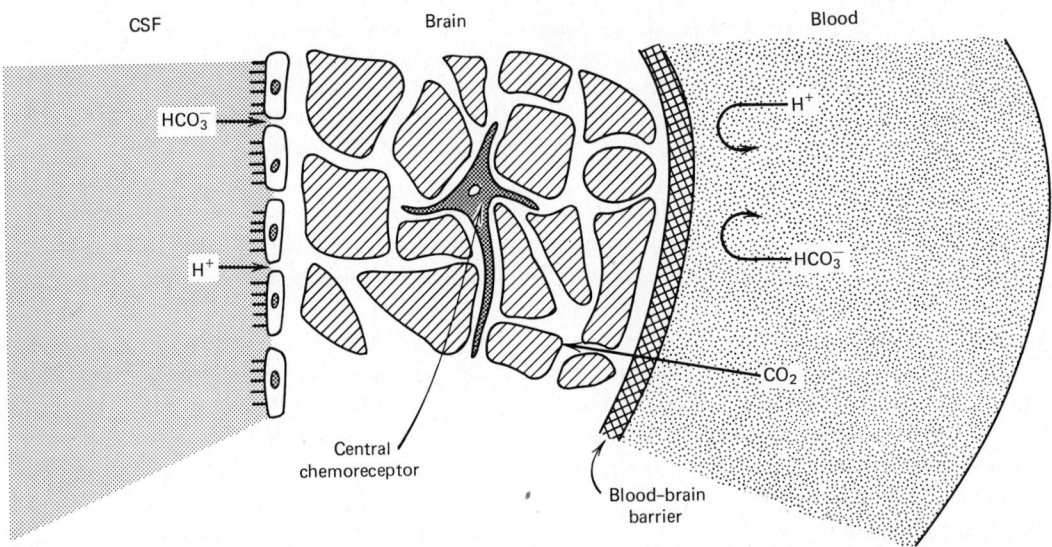

Figure 6 Schematic drawing of a central chemoreceptor neuron positioned between CSF and cerebral capillary blood. The blood-brain barrier prevents H^+ and HCO_3^- from reaching it from blood. However, CO_2 molecules can readily diffuse from blood to the chemoreceptor.

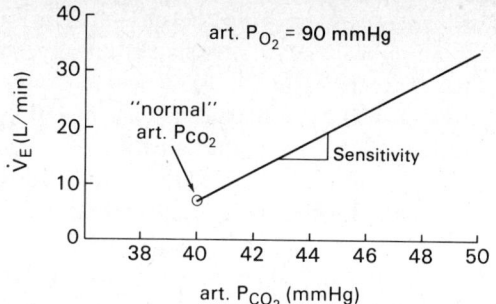

Figure 7 A stimulus–response curve of the respiratory control system for the CO_2 stimulus. Pa_{CO_2} is increased above the "normal" value, whereas Pa_{O_2} is held at the "normal" value.

indicates that there is little response until Pa_{O_2} falls below 70 mm Hg. With further increases in stimulus, the changes in response become progressively greater, that is, the slope of the curve increases.

Sensitivity

The slope of a stimulus–response curve indicates the sensitivity of a system to a particular stimulus; the slope equals the change in output for a unit change in stimulus (Fig. 7). For CO_2, the sensitivity when Pa_{O_2} equals the normal value (90 mm Hg) is simply the slope of the curve in

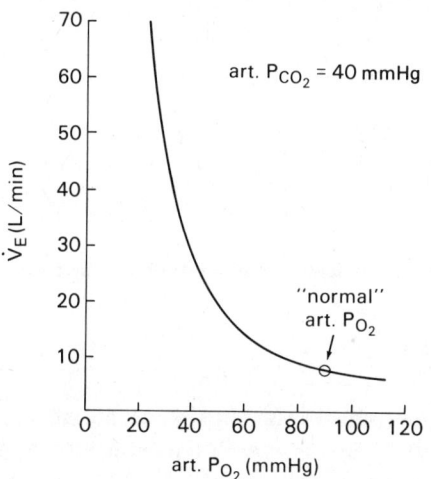

Figure 8 A stimulus–response curve for the respiratory control system for the hypoxic stimulus. Pa_{O_2} is decreased below its "normal" (an increase in the hypoxic stimulus to breathing), whereas Pa_{CO_2} is held at the "normal" value.

Figure 7. For O_2, slope varies with stimulus (Fig. 8); hence, there is no unique value for sensitivity to the hypoxic stimulus. An increase in one stimulus increases the sensitivity of the control system to the other stimulus.

RESPONSE TO CHRONIC HYPERCAPNIA: ROLE OF THE CEREBROSPINAL FLUID

When hypercapnia persists for a period of days, active transport mechanisms intervene to increase $[HCO_3^-]$ and restore a near-normal $[H^+]$. This occurs first in the brain, where an elevated P_{CO_2} value causes active transport mechanisms to increase the $[HCO_3^-]$ in CSF. $[H^+]$ is returned to normal despite the elevated P_{CO_2}, and the net effect is that the central chemoreceptor is no longer activated. Similarly, the kidney responds to an abnormally high P_{CO_2} value, and in a period of 1–2 weeks the plasma $[HCO_3^-]$ will have increased sufficiently to return the blood $[H^+]$ to normal, thereby eliminating the excessive stimulation of the carotid and aortic bodies. Overall, the ventilatory stimulation resulting from acute hypercapnia slowly declines as the $[HCO_3^-]$ of CSF and blood increases over a period of 1 day to 2 weeks. Once the $[HCO_3^-]$ has been readjusted so that the $[H^+]$ is normal at an elevated P_{CO_2}, *the receptors have been reset.* In other words, just as one resets a thermostat controlling the temperature of a house, the chemoreceptors (both central and peripheral) have been reset. This means that the chemoreceptors will act to control Pa_{CO_2} at the elevated level, because that is the P_{CO_2} value that provides chemoreceptors with a normal $[H^+]$ in their environment.

A note of uncertainty: There is evidence to suggest that the central chemoreceptor is not completely isolated from ions in blood, as depicted in Figure 6. Changes in $[H^+]$ or $[HCO_3^-]$ of blood also influence the $[H^+]$ in the vicinity of a central chemoreceptor. However, it is certain that the H^+ concentration of CSF exerts a major influence on the activity of this receptor.

ACUTE AND CHRONIC HYPOXIA

Earlier we considered the ventilatory response to hypoxia at a normal Pa_{CO_2}. Let us now look at the common clinical situation of hypoxia and its analog, ascent to high altitude. A normal person ascending to high altitude develops what might be called a "hypoxic drive to breathe." Acute stimulation of the peripheral chemoreceptors by the reduced P_{O_2} of arterial blood causes $\dot{V}_E$ to increase. The condition of alveolar hyperventilation results in a reduced P_{CO_2} value. This hypocapnia causes a decrease in $[H^+]$ in the blood and CSF, which is sensed by both the central and peripheral chemoreceptors. This central and peripheral alkalinity to some extent suppresses the ventilatory response to hypoxia because the central and peripheral chemoreceptors are inhibited by alkaline fluid:

↓arterial P_{CO_2} causes ↓arterial $[H^+]$ and ↓CSF $[H^+]$

The events are summarized below:

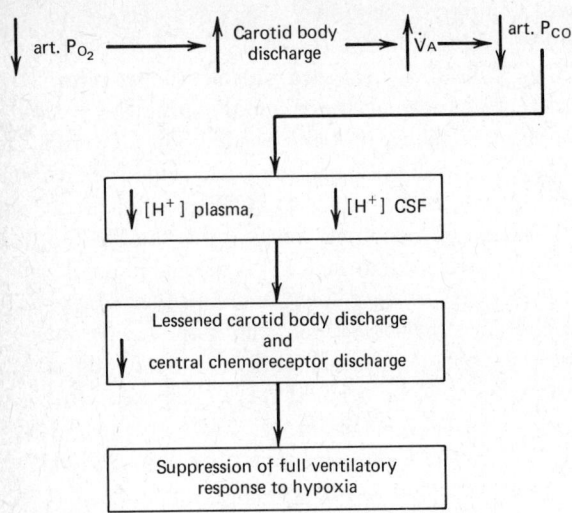

As one continues to stay on the mountain or, in the clinical situation, as hypoxia persists chronically, the active transport processes in the brain act to correct the alkaline state of the CSF. This means that the $[HCO_3^-]$ of the CSF is decreased, so that after a period of 1–2 days a normal $[H^+]$ prevails in the CSF. Consequently, the activity of the central chemoreceptor returns to normal, and ventilation increases further. Similarly, the kidneys respond to blood alkalinity by decreasing plasma $[HCO_3^-]$. This process is slower, but after 1–2 weeks at high altitude, the $[H^+]$ of plasma will be normal. Once again, this means that the suppression of the hypoxic response has been further diminished, and $\dot{V}_E$ increases still further. Only after restoration of a normal $[H^+]$ in plasma as well as CSF will the *full* response to the hypoxic stimulus be apparent.

SELECTED BIBLIOGRAPHY

Bouhuys A: *Breathing.* New York, Grune & Stratton, 1974, p 205.

Murray JF: *The Normal Lung.* Philadelphia, WB Saunders, 1976, p 223.

Widdicombe JC: Nervous receptors in the respiratory tract and lungs, in Hornbein TF (ed): *Regulation of Breathing,* Part I, New York, Dekker, 1981, p 429.

Younes MK, Remmers JE: Control of tidal volume and respiratory frequency, in Hornbein TF (ed): *Regulation of Breathing,* Part I, New York, Dekker, 1981, p 621.

SELF-STUDY QUESTIONS

MULTIPLE CHOICE

Select the correct answer(s). (In many instances, more than one answer is correct.)

1. Which of the following describe(s) the respiratory control system?
   A. Either an increase in arterial P_{CO_2} or a decrease in arterial P_{O_2} stimulate a ventilatory response.
   B. The control system maintains normal arterial and mixed venous P_{O_2}, P_{CO_2}, and pH at rest and during muscular exercise.
   C. The control system has at least two feedback loops.
   D. The control system acts to maintain a constant alveolar ventilation.

2. Which of the following ensure(s) constancy of arterial P_{CO_2}?

   A. A constant alveolar ventilation.

   B. A constant expired ventilation.

   C. A constant ratio of alveolar ventilation to cardiac output ($\dot{V}_A/\dot{Q}$).

   D. A control system that causes appropriate changes in alveolar ventilation whenever there are changes in metabolic rate.

3. In the respiratory control system, chemical feedback tends to produce constancy of arterial P_{O_2} and arterial P_{CO_2} because

   A. deviation of these variables from normal causes a compensatory ventilatory response.

   B. these variables are the controlled variables in the negative feedback loop.

   C. an increase in alveolar P_{CO_2} causes an increase in arterial P_{CO_2}, which stimulates ventilation.

   D. a decrease in alveolar P_{O_2} causes a decrease in arterial P_{O_2}, which decreases ventilation.

4. The apneustic center

   A. influences the behavior of medullary respiratory center neurons.

   B. is essential for a respiratory rhythm.

   C. can cause large, prolonged inspiration under certain circumstances.

   D. causes rapid shallow breathing in the vagotomized animal after midpontine transection.

5. The medullary respiratory center

   A. activates phrenic motoneurons.

   B. receives afferent messages from chemoreceptors.

   C. is the source of the respiratory rhythm.

   D. receives afferent input from pulmonary mechanoreceptors.

6. Which of the following contain(s) a respiratory center?

   A. Rostral pons.

   B. Caudal pons.

   C. Medulla.

   D. Cerebellum.

7. Pulmonary stretch receptors

   A. sense changes in lung volume.

   B. augment inspiratory discharge.

   C. send fibers to the medulla that inhibit inspiration.

   D. send fibers to the medulla that inhibit expiration.

8. The Hering–Breuer reflex

   A. involves chemoreceptors in the lung parenchyma.

   B. originates in mechanoreceptors in the lung.

   C. is mediated by chemical feedback.

   D. promotes rhythmic breathing.

9. The carotid and aortic bodies are stimulated by

   A. low arterial P_{O_2}.

   B. molecular CO_2.

   C. low pH.

   D. low arterial O_2 content, but a constant arterial P_{O_2}.

10. The respiratory stimulation during hypoxia results from

    A. cerebral hypoxia.

    B. stimulation of central and peripheral chemoreceptors.

    C. stimulation of the peripheral chemoreceptors by the direct action of a decreased arterial O_2 content.

    D. stimulation of the carotid and aortic bodies by a decreased arterial P_{O_2}.

11. The carotid and aortic bodies are stimulated by which of the following?

    A. arterial P_{O_2} = 40 mm Hg.

    B. arterial P_{CO_2} = 20 mm Hg.

    C. arterial pH = 7.20.

    D. arterial P_{O_2} = 700 mm Hg.

MATCHING

   A. *Carotid body*
   B. *Central chemoreceptor*
   C. *Both*
   D. *Neither*

12. Hypoxic ventilatory response.

13. Hypercapnic ventilatory response.

14. Stimulated by elevated blood pressure

MULTIPLE CHOICE

Select the single best answer.

15. A normal person breathes a CO_2-enriched gas mixture (normal inspired O_2 concentration). Which of the following is present after 2 weeks?
   A. Increased ventilatory drive because of the action of CO_2 on the central chemoreceptors.
   B. Increased ventilatory drive resulting from an action of CO_2 on the peripheral chemoreceptors.
   C. Increased ventilatory drive resulting from an action of H^+ on the central and peripheral chemoreceptors.
   D. Near-normal ventilatory drive with normal H^+ concentration in the CSF and plasma.
   E. Near-normal ventilatory drive in spite of elevated H^+ concentration at chemoreceptors.

16. A normal person breathes a CO_2-enriched gas mixture (normal inspired O_2 concentration). Which of the following is present after 2 days?
   A. $\dot{V}E$ comparable to that observed 10 min after onset of hypercapnia.
   B. $\dot{V}E$ greater than normal but less than that observed 10 min after onset of hypercapnia.
   C. Less than normal ventilation because of HCO_3^- readjustments in the CSF.
   D. Normal ventilation because of HCO_3^- readjustments in CSF and plasma.
   E. Elevated $\dot{V}E$ because of acid CSF and normal plasma pH.

17. A normal person suddenly becomes hypoxic. His pulmonary ventilation will
   A. decrease acutely, owing to the depressant effects of hypoxia.
   B. decrease acutely, owing to the inhibitory effects of hypocapnia.
   C. increase, but not to the level that will be achieved after 2 weeks of sustained hypoxia when acid–base readjustments have been completed.
   D. increase and then slowly decrease as acid–base adjustments occur.
   E. increase somewhat immediately and then increase more as the P_{CO_2} returns to normal.

ANSWERS

1. A,C. The system does not behave in a way that holds alveolar ventilation constant. Rather, it continuously adjusts $\dot{V}A$ to match changes in $\dot{V}_{CO_2}$. $P\bar{v}_{CO_2}$ and $P\bar{v}_{O_2}$ are not controlled—only the arterial values are controlled because they are sensed by the chemoreceptors, and changes in their values elicit a ventilatory response. $P\bar{v}_{CO_2}$ and $P\bar{v}_{O_2}$ are only indirectly influenced by changes in Pa_{O_2} and Pa_{CO_2}. The (a-$\bar{v}$) O_2 content difference is not constant; it increases in muscular exercise because oxygen consumption ($\dot{V}_{O_2}$) increases fractionally more than cardiac output (CO). (a-$\bar{v}$) $O_2 = \dot{V}_{O_2}/CO$. The respiratory system does not respond to these changes in $P\bar{v}_{O_2}$ and $P\bar{v}_{CO_2}$. Its job is simply to keep the arterial hemoglobin fully saturated (i.e., $Pa_{O_2} = 90$ mm Hg) and to keep Pa_{CO_2} and arterial pH normal.

2. D. Constancy of $\dot{V}A$, $\dot{V}E$, or $\dot{V}A/CO$ does not ensure a constant Pa_{CO_2} value. Only a moment-to-moment adjustment of $\dot{V}A$ to meet metabolic demands will enable Pa_{O_2} and Pa_{CO_2} to remain constant. Making these adjustments is the role of the respiratory control system.

3. A,B,C. If Pa_{O_2} were to produce $\downarrow\dot{V}E$, this would be a positive, not a negative, feedback. The correctness of A,B, and C can be seen in the answer to question 1.

4. A,C. Although the apneustic center is not an essential part of the neuronal mechanism that generates the respiratory rhythm (i.e., it is not a pacemaker), the center has a profound influence on the activity of the medullary respiratory center. This center, located in the caudal pons, causes the medullary respiratory center to generate powerful inspiratory activity. If

these inspirations are not checked by the action of the pneumotaxic center in the rostal pons or by the action of volume feedback (via the vagus nerve), prolonged inspirations will result.

5. All are correct. This question reviews all four major features of the medullary respiratory center, which is really "center" of action in breathing. These four features are as follows: (1) Neurons of the medullary respiratory center project monosynaptically to phrenic and intercostal motor neurons; (2) the center is the source of the respiratory rhythm; (3) the center receives chemical feedback (input from chemoreceptors); and (4) the center receives volume feedback (input from pulmonary stretch receptors).

6. A,B,C. A normal pattern of breathing requires the participation of three centers (pneumotaxic, apneustic, and medullary respiratory center). The cerebellum is not essential and contains no known "center" related to breathing.

7. A,C. These receptors sense lung volume and are the source of "volume" feedback, which acts in parallel with the pneumotaxic mechanism of the rostral pons. Both act to inhibit inspiratory activity and to prevent the development of a prolonged, apneustic inspiration. In contrast, the action of pulmonary stretch receptors on expiration is to excite expiratory neurons and prolong the expiratory phase.

8. B, D. The Hering–Breuer reflex is equivalent to "volume" feedback. It originates in volume receptors in the lung. It promotes rhythmic breathing because without its action and without the action of the pneumotaxic center, apneustic breathing results. Neither pulmonary chemoreception nor chemical feedback is involved in this reflex.

9. A,C. The carotid and aortic bodies respond when arterial P_{O_2} or arterial $[H^+]$ change. If arterial P_{CO_2} or arterial O_2 content are changed, but arterial P_{O_2} and arterial $[H^+]$ are held constant, no stimulation of the peripheral chemoreceptor occurs. A patient with anemia has a reduced arterial O_2 content, but a normal arterial P_{O_2} value—if the lungs are normal. Anemic patients do not hyperventilate at rest. During exercise, however, they breath excessively (alveolar hyperventilation). This occurs because arterial $[H^+]$

increases greatly, owing to metabolically produced lactic acid. The latter is a consequence of anaerobic metabolism in the exercising muscles when there is inadequate O_2 delivery to the tissues by the blood.

10. D. (See the explanation to question 9.) Cerebral hypoxia, if severe, exerts a direct action on respiratory neurons, depressing their activity.

11. A,C. Hypoxemia ($\downarrow P_{O_2}$) and arterial acidity stimulate the receptor.

12. A. The carotid body, but not the central chemoreceptor, senses a decrease in P_{O_2} and initiates a ventilatory response.

13. C. Both receptors sense the increase in H^+ concentration caused by an increase in P_{CO_2}.

14. D. Neither receptor is influenced by increase in arterial blood pressure.

15. D. The immediate effect of hypercapnia is an increased arterial $[H^+]$ and CSF $[H^+]$, because the reaction below is shifted to the right:

$$\uparrow CO_2 + H_2O \rightarrow H_2CO_3 \rightarrow \downarrow H^+ + HCO_3^-$$

This peripheral and central acidity stimulates breathing. However, after 2 weeks of hypercapnia, both CSF and plasma have a normal pH, owing to the effect of the kidney- and brain-elevated $[HCO_3^-]$ in each. Accordingly, there is no greater than normal stimulation of central or peripheral chemoreceptors. In this sense, an increased Pa_{CO_2} value is an evanescent stimulus. Even though the Pa_{CO_2} remains elevated, after 2 weeks the $[H^+]$ is back to normal and the stimulus is removed.

16. B. After 2 days of hypercapnia, CSF acid–base balance has been restored because of increased $[HCO_3^-]$ in CSF, but renal compensation is not complete and, hence, plasma pH is still low. Accordingly, $\dot{V}_E$ is greater than normal but less than would be observed acutely, without CSF compensation.

17. C. The immediate response to hypoxia exposure is to *increase* $\dot{V}_E$ as a result of stimulation of the peripheral chemoreceptors by a low Pa_{O_2}. This ventilatory response means alveolar hyperventilation and, hence, hypocapnia and an alkaline state throughout the body:

$$\uparrow \dot{V}E \rightarrow \uparrow \dot{V}A/\dot{V}CO_2 \rightarrow \downarrow PCO_2 \rightarrow \downarrow H^+$$

The resulting alkalinity at central and peripheral chemoreceptors means that these receptors will not be as stimulated by the hypoxia as they would be if they were not alkaline. As CSF and renal compensation occurs, the $[H^+]$ returns toward normal at these chemoreceptors and $\dot{V}E$ increases progressively. Of course, as this occurs, the PCO_2 decreases more (it does *not* return toward normal). However, the pH returns toward normal as CSF and plasma $[HCO_3^-]$ decreases.

26

The Physiology of Exercise

DIANA L. KUNZE

OBJECTIVES

After completion of this chapter, you should be able to:

1. Explain why oxygen consumption often can be used as a measure of the energy cost of dynamic exercise.

2. Understand how the heart is able to increase its cardiac output from a resting value of 5 L/min to as much as 25–30 L/min.

3. Describe how the respiratory system adjusts in order to increase oxygen delivery to the circulation.

4. Describe the local and neural adjustments of arterioles and veins that occur in the renal, splanchnic, skeletal, and cutaneous vasculature to allow most of the increased cardiac output to flow through the active skeletal muscle.

5. Define oxygen debt.

6. Explain what ensures an increased venous return during exercise.

7. Define maximal oxygen consumption.

8. Explain the differences between the physiological adjustments to dynamic and to static exercise.

The study of exercise provides an excellent opportunity to appreciate the ability of the circulatory, respiratory, and nervous systems to interact for the purposes of both supplying the demand of contracting muscles for oxygen and other substrates, and of relieving the muscles of accumulating metabolites.

We will consider forms of exercise from mild to severe, as well as the factors that determine maximum ability to exercise. It is important to realize that you spend most of your working hours engaged in some form of "exercise" —even walking to class or sitting in a chair while studying requires more energy than your basal metabolic rate provides. The circulatory, respiratory, and nervous systems must adjust to provide the muscles with the needed oxygen.

Exercise is generally classified as dynamic (rhythmic or intermittent contraction and relaxation) or static (sustained contraction), although most exercises contain both static and dynamic components. Both types of exercise are discussed in this chapter, although the emphasis is on dynamic exercise.

ENERGY COST OF DYNAMIC EXERCISE

The energy cost of an exercise is used to quantitate the severity of the exercise.

Skeletal muscles contract using the chemical energy stored as adenosine triphosphate (ATP), converting it to mechanical energy and heat. The chemical energy (ATP) must be provided continuously from the metabolism of the substrates, carbohydrates, and fat—protein is seldom used in muscle metabolism. The production of ATP is one step in a long series of reactions arising with the substrates and terminating in CO_2 and H_2O if metabolism is aerobic and lactic acid if it is anaerobic. Anaerobic metabolism (glucose → lactic acid) produces much less ATP per mole of glucose and, as will be discussed shortly, is mainly used in severe dynamic or isometric exercise in which O_2 supply can not meet demand. In general, the energy cost of exercise is met primarily by aerobic metabolism.

QUANTITATING DYNAMIC EXERCISE

In order to assess the response of a person's circulatory and respiratory systems to exercise, which is what we are leading up to, we must have some way of quantitating the exercise. This is commonly done by measuring either (1) the oxygen consumption during exercise, or (2) the rate at which work is performed during the exercise.

Calculation of Energy Cost of Exercise Using the Oxygen Consumption

A given exercise requires a specific amount of energy. When using aerobic metabolism during submaximal exercise, the cost of the exercise can be calculated by measuring the amount of oxygen used during the exercise. Why is this so? Consider the following equation for the aerobic metabolism of glucose:

$$C_6H_{12}O_6 + 6O_2 \rightarrow 6CO_2 + 6H_2O + \text{heat.}$$

You see that the amounts of oxygen, glucose, and CO_2 are fixed, so that 1 mole of glucose is oxidized by 6 moles of O_2 producing 6 moles of CO_2. The energy released from that reaction is 686 kcal. If any one of the components is known, the others can be calculated. Thus, if the amount of O_2 or glucose used or the amount of CO_2 produced in excercise is known, the energy cost of that exercise can be calculated. The easiest of these three factors to measure, hence the most commonly used, is the *oxygen consumption*.

Example
If 0.250 liter O_2/min is used to maintain basal metabolic rate, what is the energy cost when glucose is the substrate? 6 moles O_2 release 686 kcal. Since 1 mole O_2 = 22.4 liters O_2 (STP), 134.4 liters of O_2 release 686 kcal, or *0.250 liter of O_2 releases 1.28 kcal.*

The same type of calculation can be made for fats. Thus, it can be shown that 1 liter of O_2-oxidizing palmitate gives 4.69 kcal, whereas 1 liter of O_2-oxidizing glucose gives 5.05 kcal.

Exercise involves metabolism of a mixture of fat and carbohydrate. Carbohydrate metabolism supplies 25% of total energy at rest, with the proportion increasing with the intensity of exercise to 75–80% at near-maximal oxygen consumption. As the duration of exercise is lengthened, fat metabolism provides more of the total energy cost. For ease of calculation, an approximate value of *5.00 kcal/liter of O$_2$* will be used for mixed substrates. Therefore, a person consuming 2.5 liters O$_2$/min during exercise is doing so at an energy cost of

$$(5.0 \text{ kcal/liter O}_2) \ (2.5 \text{ liters of O}_2/\text{min}) = 12.50 \text{ kcal/min.}$$

At rest the average person (70 kg) maintains basic function at an energy cost of approximately 1.3 kcal/min. Values of 4 kcal/min in moderate work, 4–8 kcal/min in heavy work, and 10–25 kcal/min near-maximal exercise are to be expected for the non athletic person. At maximum exercise, some athletes show an energy cost increase of up to 50–65 kcal/min. Exercise of this severity is maintained for only a short period. Table 1 gives you an idea of the kilocalorie consumption of various types of exercise. From the kilocalorie values you can calculate approximately how much glucose or palmitate must be metabolized per minute to provide adequate energy to maintain the level of exercise.

Table 2 shows what stores of energy are available in the muscle. As you can see, the total ATP and phosphocreatine (CP) stores available amount to a mere 4.6 kcal. *This would supply only 1 minute of moderate exercise and only a few seconds of maximal exercise.* Therefore, glycogen and fat oxidation must supply the ATP to replenish the stores being depleted by increased muscle activity. In addition to the muscle stores of glycogen and fat, glucose and free fatty acids from other parts of the body are taken up by the muscle for use as substrate. In the healthy person, except in long periods of submaximal exercise, or during fasting, substrate availability is not a limiting factor to the continuation of exercise.

Calculation of the Work Rate of Exercise and its Relationship to Energy Consumption

You will encounter another method of quantitating intensity of exercise that will be covered briefly. The energy cost of exercise includes the transformation of chemical energy to both heat energy and mechanical energy. The percent-

Table 1 Energy Required by a 70-kg Person for Various Physical Activities

Activity	Total (kcal/min)[a]
Sleeping	1.21
Lying quietly	1.30
Sitting	1.67
Mental work seated	1.75
Standing	1.83
Driving a car	2.33
Office work	2.42
Light housework	2.50
Horizontal walking, 2 mph	2.83
Walking up stairs, 1 mph	3.00
Riding a bicycle, 5.5 mph	3.17
Walking down stairs, 2 mph	3.33
Bowling	3.58
Billiards	3.92
Dancing, moderate	4.17
Volleyball, 6-man (noncompetitive)	4.25
Horizontal walking, 3.5 mph	4.83
Gardening	4.92
Dancing, vigorous	5.67
Table tennis	5.75
Horizontal walking, 3 mph, carrying 19-kg (43-lb) load	5.83
Walking up 3% grade, 3.5 mph	6.17
Basketball	6.58
Swimming breaststroke, 1 mph	6.83
Bicycle riding, rapid	6.92
Swimming crawl stroke, 1 mph	7.00
Walking up 8.6% grade, 2.4 mph	7.17
Walking up 8.6% grade, 3.5 mph	9.33
Walking up 10% grade, 3.5 mph	9.67
Horizontal running, 5.7 mph	12.00
Walking up 14.4% grade, 3.5 mph	12.33
Wrestling	13.17
Horizontal running, 7 mph	14.5
Marathon running	16.5
Horizontal running, 11.4 mph	21.67
Horizontal running, 13.2 mph	38.83
Horizontal running, 14.8 mph	48.00
Horizontal running, 15.8 mph	65.17

Source: From Morehouse L, Miller A: *Physiology of Exercise.* St. Louis, CV Mosby, 1976.

[a]The oxygen consumptions from which the caloric values in this table have been calculated are only approximations. These values include resting metabolic needs. Variables that must be considered in any interpretation of this table are size, body type, and age of the subjects; differences among persons of the same build, physical fitness, and skill in the particular activity; and nutritional condition and environmental conditions—whether they help or hinder the person.

Table 2 Available in Muscle[a]

	Energy mole^{-1} (kcal)	Concentration (mmol kg^{-1} wet muscle)	Total Energy in Humans (body weight 75 kg, muscle weight 20 kg) (kcal)
ATP	10	5	1
Phosphocreatine (CP)	10.5	17	3.6
Glycogen	700	80	1,100
Fat	2400	Variable	75,000

Source: From Astrand, PO, Rodahl K: *Textbook of Work Physiology.* New York, McGraw-Hill, 1977.
[a]The figures are very approximate. The glycogen concentration can be anything from almost zero up to 250 mmol/kg, and certainly the fat content is subject to large variations. It is assumed that only part of the muscle mass is activated.

age of the total amount of energy converted to mechanical energy determines the efficiency of the exercise.

$$\text{Efficiency} = \frac{\text{work done}}{\text{total energy cost}}$$

Work is defined as the amount of energy transformed from chemical energy to mechanical energy and is the product of force times distance moved. For example, certain types of swimming are only 8% efficient, whereas most human exercise performed is 20–25% efficient. For a given exercise, as the intensity of the exercise is increased, the amount of chemical energy transformed to mechanical energy/time is proportional to the total energy cost/time. This is another way of saying that work is proportional to total energy cost/time. In this situation, rate of work can be used as an indication of the energy cost. It is very difficult to measure the work involved in most forms of human exercise. In the clinical situation, you will encounter three types of ergometer:* steps, the treadmill, and the bicycle. The rate of work of walking on the treadmill up various inclines at a constant speed has been calculated and shown to be proportional to oxygen consumption. The rate of work of pedaling the bicycle at a given rate against various loads is also proportional to oxygen consumption, as is the work of climbing steps. Therefore, without directly measuring oxygen consumption, you can use any ergometer to present a patient with various intensities of exercise in order to test the ability of the circulatory, respiratory, and nervous systems to meet the challenge of increased oxygen demand. This chapter discusses cardiovascular and respiratory changes in relationship to

*An ergometer is a machine used to measure rate of work.

the oxygen consumption/time, although we could also discuss these adjustments in relationship to the rate of work performed (watts).

OXYGEN CONSUMPTION DURING DYNAMIC EXERCISE

How much oxygen must be supplied to oxidize the carbohydrates and fats? The resting energy need of 1.3 kcal/min is met by the consumption of about 250 ml oxygen (since 1 liter oxygen yields an average of 5.0 kcal). At maximal exercise when energy used is 10–25 kcal/min, oxygen uptake is 2.0–5.0 L/min—in fact, some athletes are able to consume more than 7 liters O_2/min. The body must be able to meet a wide range of oxygen requirements, from 250 ml/min at rest to 2,000–5,000 ml/min in maximal exercise. Each individual has an absolute maximal O_2 consumption ($\dot{V}O_{2max}$) for strenuous exercise involving large muscle groups. This is an exercise that can be sustained at a given intensity for 2–5 min, but leads to exhaustion in 5–10 min. The oxygen consumption cannot be increased beyond this level. The maximum oxygen uptake is remarkably reproducible for a given subject from day to day and from trial to trial. Its value is a function of sex (being greater in male than female), age (highest in early twenties), and physical condition (which can be improved by training as will be discussed later). Maximum oxygen uptake by tissues appears to be limited by the ability of the circulatory system to deliver oxygen. This chapter designates those exercises that require up to 25% of a person's maximum oxygen consumption as mild exercise. Moderate exercise will encompass a range of 25–60% of $\dot{V}_{O_{2max}}$ and heavy exercise as 60–100% of $\dot{V}_{O_{2max}}$.

When a specific exercise is initiated, the energy require-

ment, and therefore the oxygen requirement, goes immediately to its steady-state value (Fig. 1a). However, the circulatory and respiratory systems are not yet delivering the amount of O_2 necessary to meet the energy demand by aerobic metabolism. The first part of the exercise (1–2

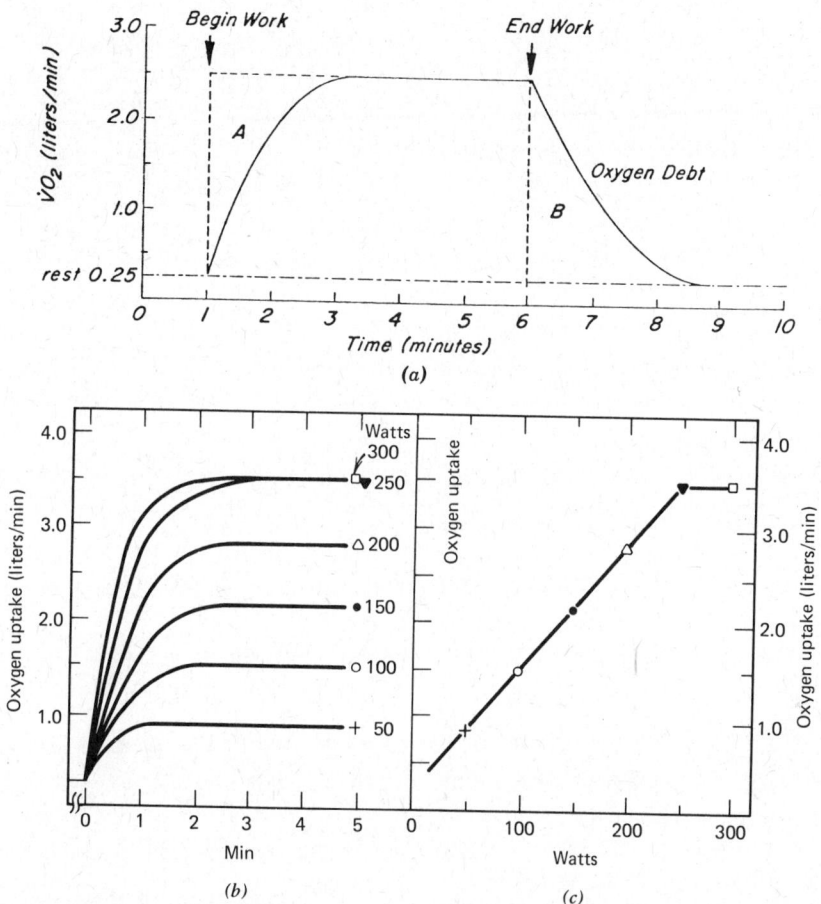

Figure 1 *(a)* As exercise begins at a steady rate, the oxygen requirement is 2.5 L/min. The circulatory and respiratory systems are not immediately prepared to deliver adequate oxygen. Depletion of ATP and CP stores, as well as myoglobin oxygen stores, makes up the difference (area A) until oxygen delivery meets the requirements of the tissue in approximately 1.5–2.0 mins. During near-maximal exercise, lactic acid production also contributes to energy requirements initially until oxygen delivery meets oxygen demand. At the end of the work period, oxygen consumption remains elevated until the oxygen debt is paid back (area B). *(b)* When the work rate of a given type of exercise is increased, the oxygen uptake is proportional to the rate of work of that exercise. *(c)* The oxygen consumption for each rate of work in *(b)* is plotted. (Redrawn from Astrand PO, Rodahl K: *Textbook of Work Physiology.* New York, McGraw-Hill 1977.)

min) must then be partly anaerobic. At this time, oxygen delivered from the blood and oxygen stored in the myoglobin is used to oxidize substrate to produce ATP. Since there is not enough oxygen delivered yet to maintain the ATP and CP levels, a portion of the ATP and CP stores is decreased. If the exercise is near-maximal for the subject, new ATP is produced as well from the breakdown of muscle glycogen and blood glucose to lactic acid. As you can see in Figure 1a, within 2 min the oxygen supplied to the tissues increases to match the O_2 needed to supply the energy for ATP production at the level of exercise. As discussed earlier in this chapter, the O_2 uptake at this time is linearly related to the work rate as long as exercise is submaximal (Fig. 1b, c).

When exercise is terminated, oxygen consumption does not drop immediately to zero, but remains elevated, declining toward pre-exercise levels over a time period that depends on the severity of the exercise. The amount of oxygen above resting level consumed during this postexercise period is called the *oxygen debt*. The O_2 debt consists of repayment of O_2 (1) to replenish the ATP and CP stores, (2) for the myoglobin stores, and (3) to remove any lactate produced during exercise either to metabolize it to $CO_2 + H_2O$ or to reconvert it to glycogen in the liver. At submaximal exercise this debt is primarily incurred during the first 2 min of exercise before O_2 delivery meets demand. Oxygen debt has been divided into lactic debt and alactic debt. The alactic debt includes factors (1) and (2) and is paid back within a few minutes (half-time, 0.5 minutes) after exercise has been terminated, whereas the lactic debt repayment proceeds more slowly (with a half-time of 15 minutes). In mild and moderate exercise, any small amount of lactic debt incurred during the first few minutes of exercise appears to be paid back as oxygen consumption reaches a steady state, so that blood lactate levels are not elevated during the exercise. In heavy exercise, blood lactate levels are elevated in proportion to the intensity of exercise. At maximal exercise, lactate levels continue to rise throughout the exercise.

The subject tested on the bicycle ergometer shown in Figure 1 reached a maximum aerobic metabolism at 250 watts of work when his oxygen uptake was 3.5 L/min. When he increased his work load to 300 watts, his oxygen consumption did not increase further; therefore, he must have been using anaerobic metabolism to accomplish this work.

REGULATION OF OXYGEN DELIVERY TO THE TISSUES DURING DYNAMIC EXERCISE

How is O_2 supplied to the tissues? There are two potential mechanisms for increasing the supply of O_2 to the tissues. The first is an increased delivery, the second is an increased extraction at the level of the metabolizing tissues. Increased delivery depends on an increase in cardiac output, a preferential distribution to exercising muscle, and an increase in ventilation to oxygenate the increased circulatory flow.

VENTILATORY RESPONSE TO DYNAMIC EXERCISE

Pulmonary ventilation increases as the demand for oxygen uptake by the tissues increases (Fig. 2). The increase is

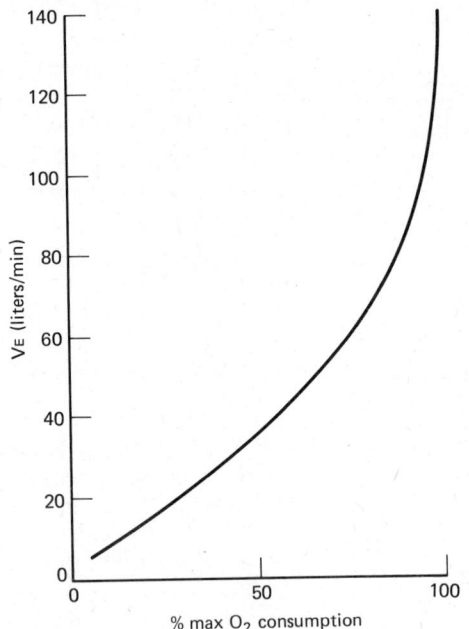

Figure 2 Pulmonary ventilation increases in proportion to oxygen uptake through 50–70% of maximum oxygen uptake. The ventilation increase is greater than the increase in oxygen uptake (hyperventilation).

fairly linear at first and then rises sharply at higher work loads. At the point of the sharp increase, ventilation exceeds demand. At lower intensities of exercise the response is primarily an increase in tidal volume in which both inspiratory and expiratory reserve volume are used. At higher intensities of exercise, both tidal volume and rate contribute to the increased ventilation. At higher ventilation rates, as much as 10–20% of the increased oxygen uptake goes to supply the contracting respiratory muscles. One of the mysteries of exercise is the mechanism initiating the increased ventilation. The usual mechanisms controlling ventilation, such as peripheral and central chemoreceptors, seem not to be the initiating stimuli, since arterial Po_2, pH, and Pco_2 are not changed when ventilation is already increased (Fig. 3). Recent evidence suggests that mechanoreceptors or chemoreceptors lying in exercising muscle might initiate the increased ventilation, but the evidence is far from conclusive. A neural component must be involved, because ventilatory flow rate can increase during the first breath of the period in which the exercise is initiated, or even before exercise commences.

Does delivery of oxygen by the lungs limit the amount of exercise that a person is capable of? Probably not, because pulmonary ventilation can be increased even above that at maximal oxygen consumption, but the O_2 consumption does not increase. Diffusion of O_2 from the lungs to the capillaries is not seriously impaired by the fast transit time of the blood through the lungs that results from an increased circulatory flow rate.

CIRCULATORY RESPONSE TO DYNAMIC EXERCISE

A number of circulatory changes occur in exercise, all of which are geared toward increased cardiac output directed to the exercising muscles. During exercise, the heart becomes a more effective pump. Increased cardiac output, as you recall, depends on increased stroke volume or heart rate, or both ($CO = SV \times HR$). Stroke volume increases because of an increased contractility of the cardiac muscle. This is the result of increased sympathetic nerve activity to the heart muscle. The diastolic volume before contraction is unchanged or only slightly increased during exercise. The volume in the heart at the end of contraction is less during exercise than it is during rest (i.e., more volume ejected). Stroke volume continues to

increase for exercise requiring up to 40–50% maximal oxygen consumption (Fig. 4). Beyond this level, stroke volume generally does not increase much. The other contribution to the cardiac output is that made by an increase in heart rate (Fig. 4). Heart rate increases because of an increase in sympathetic activity and a decrease in the vagal activity to the Sinoatrial (SA) node. The amount of increase in heart rate is usually dependent in an approximately linear fashion on the O_2 uptake. For that reason heart rate is sometimes used as a measure of intensity of exercise when oxygen uptake cannot be measured. However, heart rate can be varied by a number of factors and is not always a reliable indicator. Unlike stroke volume, heart rate continues to increase to maximal oxygen consumption. Together, increases in the stroke volume and heart rate increase cardiac output (Fig. 4).

Increased cardiac output is dependent on increased venous flow—the heart cannot put out what it does not receive. Venous flow is aided initially by increased activity of sympathetic nerves to the venous capacity vessels. This activity causes constriction of the veins returning more blood to the heart. Venous flow is also aided by the pumping action of the contracting exercising muscles (this includes thoracic and abdominal muscles involved in respiration). If contractions are too strong, however, venous flow can be impeded as veins are collapsed by the force of the contractions. A decrease in peripheral resistance also aids venous flow, since what is pumped out returns more easily to the right heart (Fig. 4).

The decrease in peripheral resistance that occurs in exercise ensures two points: (1) mean arterial blood pressure will not rise very much in spite of increased cardiac output, and (2) an adequate portion of the increased flow will reach the exercising tissue. How does the decrease in resistance occur? When exercise is initiated, sympathetic nerve activity increases to the splanchnic, renal, cutaneous, and nonexercising skeletal muscle arterioles, thereby decreasing flow to these organs and diverting it to lower resistance beds.

The arterioles of the skin are vasoconstricted until body temperature begins to rise 5–10 min after exercise begins, at which time they begin to dilate. The sympathetic noradrenergic vasoconstrictor fibers to the exercising skeletal muscle also increase their discharge; however, most of the effect they might have in increasing resistance of the muscle vessels is overridden by the vasodilating effect of the

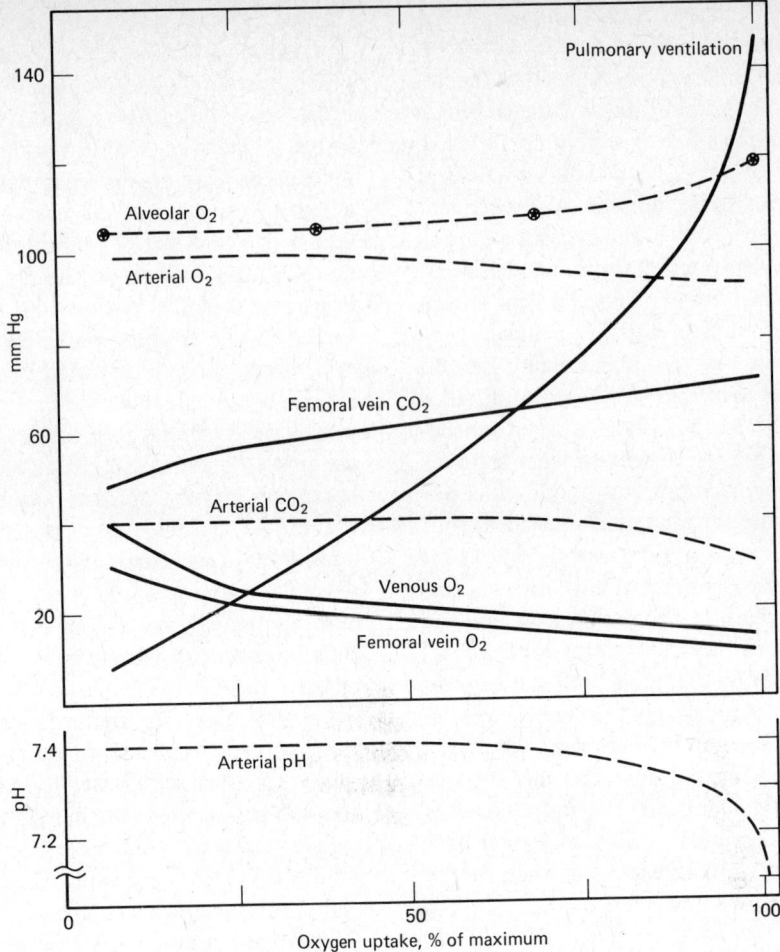

Figure 3 The following mechanisms are demonstrated: (1) Arterial O_2 arterial CO_2, and arterial pH have not changed when ventilation has already increased; (2) as ventilation increases out of proportion to oxygen uptake, arterial CO_2 falls (hyperventilation); (3) alveolar-arterial oxygen difference increases (the cause is uncertain—possibilities include an increased venous arterial admixture because of increased shunting or inadequate equilibration in hypoventilated areas of the lung); and (4) arterial pH decreases as lactic acid is formed at oxygen uptake greater than 50-60% of maximum. (From Astrand PO, Rodahl K: *Textbook of Work Physiology.* New York, McGraw-Hill, 1977.)

accumulating products of metabolism in these vessels (adenosine, H^+, CO_2, and K^+). The sympathetic cholinergic vasodilator fibers, which supply vessels of skeletal muscle, are also active, at least initially. The significance of their contribution is unknown, since blockage of their action does not reduce flow during exercise. They may be active, at the beginning of exercise, to increase flow before

local vasodilator substances are released. The net result of the increase in arteriolar resistance of the nonexercising muscles, the splanchnic and renal beds, and the decrease in resistance of the exercising skeletal muscle beds is a decrease in total peripheral resistance.

Mean arterial pressure is slightly increased as a result of circulatory adjustments, which have opposite effects.

The increase in cardiac effectiveness and venoconstriction would tend to shift blood to the arterial side, increasing pressure, but the decrease in peripheral resistance shifts most of the blood to the venous side. In exercise involving a large percentage of the skeletal muscle, such as leg exercise, mean arterial pressure rises only slightly. Pulse pressure, which is determined primarily by the stroke volume, is increased. The increase in pulse pressure superimposed on a slightly increasing mean pressure means that diastolic is nearly constant (Fig. 4), whereas systolic is increased. Plasma norepinephrine levels increase three- to fourfold in an approximately linear relationship, with exercise requiring up to 70% maximal oxygen consumption. Above this point the levels rise rapidly. This primarily reflects the release of norepinephrine from sympathetic nerve endings.

Another mystery of exercise is the control of the mechanism that adjusts output of the sympathetic and parasympathetic nerves to the heart and vessels. It is this adjustment, together with the local dilating effects of metabolites in the exercising skeletal beds, that ensures an appropriate increase in cardiac output to match the oxygen requirement. It is likely that this adjustment involves the cerebral cortex, the medulla, and the hypothalamus, but the inputs initiating the response are unknown. It is clearly possible for circulatory adjustments (like the respiratory adjustments) to be made in anticipation of exercise (i.e., an increase in heart rate before a race commences). Therefore, some of the adjustments are likely to be cortical.

Afferent nerve fibers from the muscles as well as collaterals from the motor cortex have been suggested as regulating the sympathetic and parasympathetic outflow in exercise.

OXYGEN EXTRACTION DURING EXERCISE

The arterial–venous oxygen difference increases with intensity of work. The oxygen gradient between the capillaries and the metabolizing tissue is increased as a result of the decreased tissue P_{O_2}. This results in an increased extraction of oxygen from the blood. Remembering the oxygen saturation curves from Chapter 20, you will recall that a decreased pH, increased P_{CO_2}, or an increased temperature of the blood will cause the curve to shift to the right. Figure 3 demonstrated such changes in exercise. This means that at the tissues the oxygen will be released more readily from the hemoglobin.

EXERCISE AND BODY TEMPERATURE

Heat is formed as a product of metabolism. For most exercise this amounts to 75–80% of the energy cost. Ini-

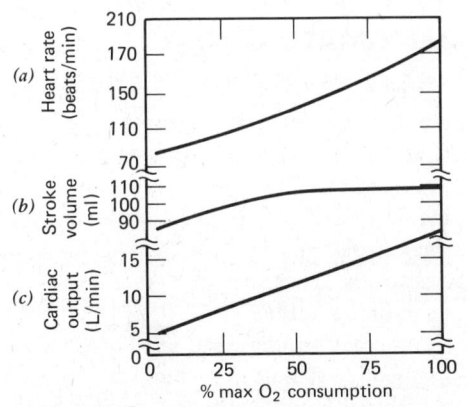 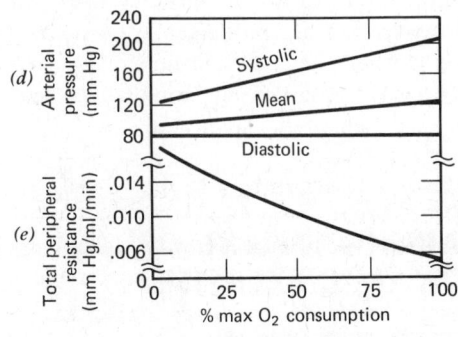

Figure 4 (a) Heart rate increases in an approximately linear relation with increasing O_2 consumption. (b) Stroke volume increases as exercise increases to about 40% of maximum and then remains constant. (c) Cardiac output (SV × HR) relationship to oxygen uptake. (d) While pulse pressure rises, mean arterial pressure rises only slightly as O_2 consumption increases. (e) Total peripheral resistance falls with increasing O_2 consumption.

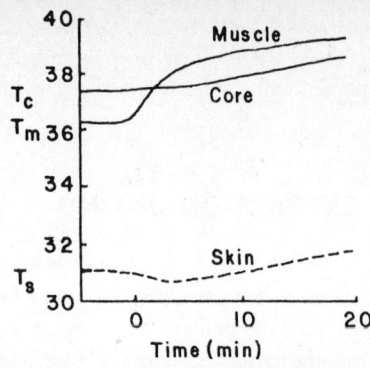

Figure 5 Time course of muscle, core, and skin temperatures during the first 20 min of exercise. (From Nadel ER: *Problems with Temperature Regulation During Exercise.* New York, Academic Press, 1977.)

tially muscle temperature rises, followed by a slower increase in core temperature as the rest of the body acts as a sink for the heat produced in the exercising muscle (Fig. 5). Without an adequate means of disposing of the heat, the body temperature would soon rise to values incompatible with life. Regulatory mechanisms are thus brought into play to dissipate the heat. The vessels of the skin that are initially constricted begin to vasodilate after 4–5 min of exercise when the body temperature begins to rise. This increases flow to the skin, warming the skin and increasing heat loss through radiation and convection. Heat is also lost to the environment through the evaporation of sweat. The proportion of heat lost by evaporation as compared with that by convection and radiation is dependent on the ambient temperature. More heat is lost by sweating, and less by convection and radiation as external temperature increases. The temperature to which the body rises is a function of O_2 uptake. During prolonged exercise, the temperature rises during the first half-hour and then becomes fairly stable. Temperature regulation will be covered in detail in a later chapter (Chapter 36).

TRAINING EFFECTS OF DYNAMIC EXERCISE

The effects of training are studied either by comparing physically trained with untrained persons or by submitting untrained subjects to a training program. Changes occur in the circulatory, respiratory, metabolic, and neural adjustments to exercise when a training program is followed. For research purposes, training programs are usually carried out on the bicycle ergometer or treadmill, for which conditions can be closely controlled and physiological variables easily measured.

SUBMAXIMAL EXERCISE

When the cardiovascular parameters are compared for a specific exercise before and after training, the following results generally are obtained. Cardiac output is either unchanged or slightly reduced; however, the components of cardiac output are altered. Heart rate is reduced and stroke volume increased. Myocardial contractility is increased, accounting for the increased stroke volume. Heart rate is reduced, probably through an adaptation of the central nervous system, which increases vagal activity even at rest. Mean arterial blood pressure and total peripheral resistance are not changed. The arteriovenous oxygen difference of the exercising muscle is increased, supplying the tissue with adequate O_2 in spite of decreased flow. There are now a great many biochemical studies that indicate increased ability for aerobic metabolism through increased levels of many mitochondrial enzymes. The skeletal muscle also develops an increased ability to metabolize fats.

Another effect of training is increase in heart size. Wrestlers and weight lifters increase ventricular wall thickness, while endurance trained athletes increase ventricular end-diastolic volume, often without effecting a change in wall thickness.

MAXIMAL EXERCISE

A major effect of a training program involving large muscle groups is an increase in maximal oxygen consumption (Fig. 6). The increase in maximal oxygen consumption is provided by an increase in maximal cardiac output and an increase in the maximal arteriovenous O_2 difference. Maximal stroke volume is increased, while heart rate (which has started from a lower resting level) reaches its maximal pretraining value.

Whereas an increased challenge to the circulatory and metabolic systems is prerequisite to the training changes, the actual stimuli initiating these changes are unknown. From a series of complicated studies involving the testing of trained and untrained muscles of the same subject, it is clear that the adaptions are both central (in the nervous

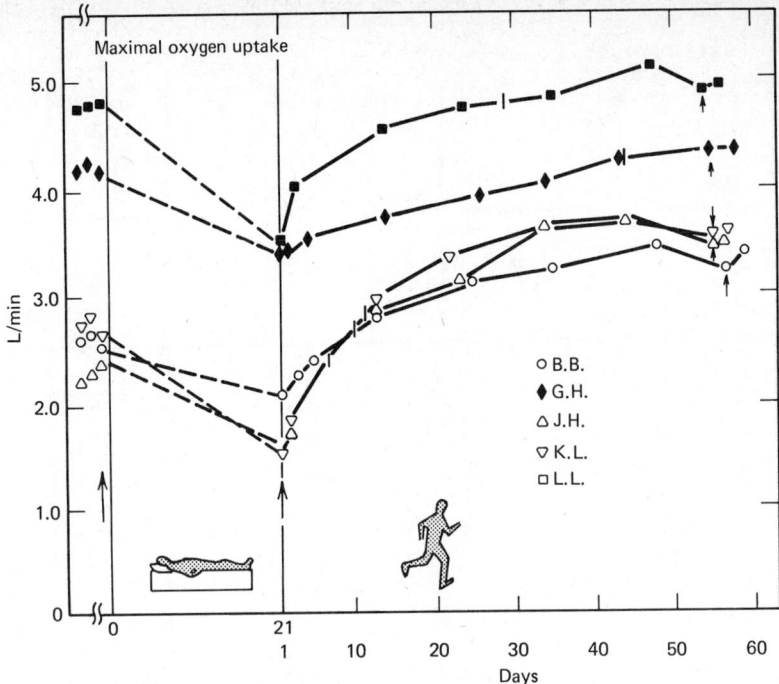

Figure 6 Changes in maximal oxygen uptake for five subjects measured on a tread-mill before and after bed rest at various intervals during training. (From Saline B, Blomqvist B, Mitchell JH, Johnson RL, Wildenthal K, Chapman CB, et al: *Response to Submaximal and Maximal Exercise after Bedrest and Training,* in *Circulation* 38 (suppl 7), 1968.)

system) and peripheral (at the heart and skeletal muscle). For further information, consult the references listed in the Selected Bibliography.

TIME FOR TRAINING EFFECTS

The duration and intensity of a training program influence the adaptations reported above. Hundreds of studies have been conducted. Results vary widely. However, in general there appear to be no training effects when heart rates of fewer than 130–150 beats/min are achieved during the training sessions. Three to four sessions a week for 30–45 min each at 60% of maximal oxygen consumption will regularly improve cardiovascular and respiratory performance, although some studies show improvement with less exercise.

STATIC (ISOMETRIC) EXERCISE

Thus far we have considered the response to exercise when the muscle contractions are dynamic or rhythmic. The cardiovascular responses to static (or isometric) exercise in which muscle contractions are maintained, such as weight lifting, differ in several ways. Much of static exercise uses anaerobic metabolism, hence oxygen consumption cannot be used to quantitate the intensity of static exercise. Static exercise is usually quantitated by expressing a given sustained contraction of a muscle or set of muscles as a percentage of the maximum tension that can be developed when the person uses those particular muscles, that is, % of maximum voluntary contraction. Isom-

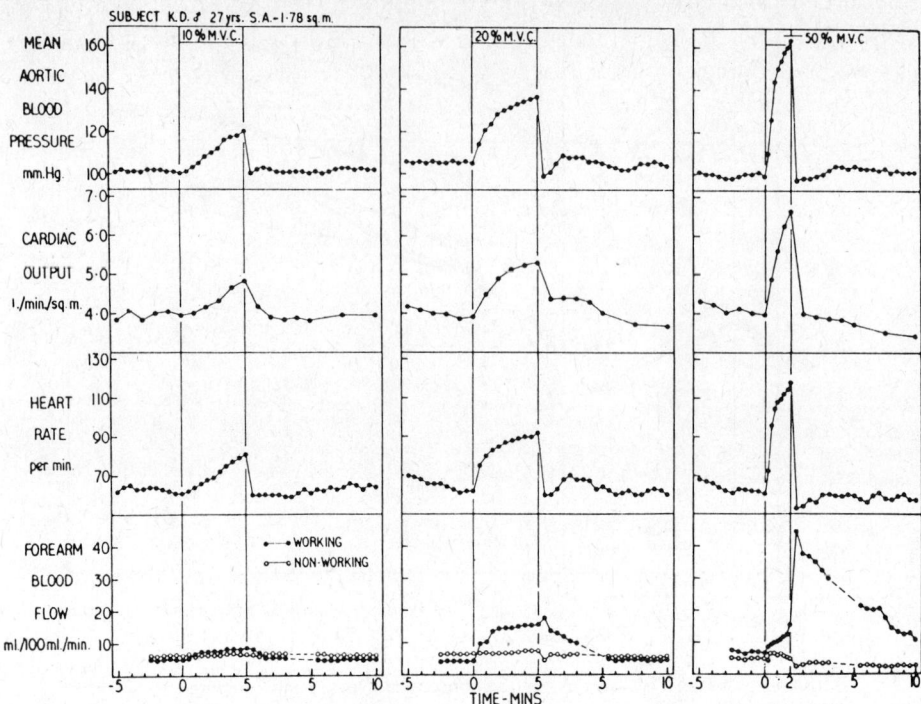

Figure 7 Circulating responses to static exercise in a subject in response to sustained hand grip.

etric contractions impede blood flow through the muscle. When a static exercise is approximately less than 20% of the maximum voluntary contraction, the impedance to flow is minimal, so that the flow matches the metabolic requirements of the tissue. Circulatory changes are minimal, with only slight increases in heart rate, arterial pressure, and cardiac output. Although there are variations from muscle to muscle, generally if the contractions are greater than 20% of the maximum voluntary contraction, flow is insufficient to meet metabolic needs. The circulatory responses are increased heart rate, increased systolic, diastolic, and mean pressure, as well as increased cardiac output, which continue to rise until work is terminated by fatigue (Fig. 7). Vasoconstriction occurs in the nonworking muscles and splanchnic beds, as it does in dynamic exercise. Mechanical occlusion of the working muscles, if

sufficient, will increase resistance of the working muscles in spite of local mechanisms causing dilation of these muscles.

SELECTED BIBLIOGRAPHY

Astrand PO, Rodahl K: *Textbook of Work Physiology.* New York, McGraw-Hill, 1977.

Clausen JP: Effect of physical training on cardiovascular adjustments to exercise in man. *Physiol Rev* 57:779–815, 1977.

Margaria R: *Biomechanics and Energetics of Muscular Exercise.* Oxford Clarendon Press, 1976.

Rowell LB: Human Cardiovascular Adjustments to Exercise and Thermal Stress. *Physiol Rev* 54:75–159, 1974.

SELF-STUDY QUESTIONS

1. What is the energy cost of walking on level terrain 1 mile in 30 min (use Table 1)?
2. What is the additional oxygen consumption above basal rate to perform the exercise described in question 1?
3. What would you predict would be the blood pressure response to arm exercise alone as compared with leg exercise alone?
4. How does maintaining a low blood pressure help reduce O_2 needed by heart muscle?

MULTIPLE CHOICE

Select the correct answer(s). (In many instances, more than one answer is correct.)

5. Maximal oxygen consumption
   A. is the same for everyone.
   B. is expected to decrease for a patient confined to bed.
   C. increases with age after 30 years.

D. can be achieved with strenuous and exhausting arm exercises.

6. Which characteristics are expected by an untrained subject as the result of a training program?
   A. Increased maximal oxygen consumption.
   B. Increased maximal cardiac output.
   C. Increased maximal heart rate.
   D. Increased maximal stroke volume.

7. Respiratory response to exercise is characterized by
   A. hyperventilation at low intensities of exercise (20% of maximal oxygen consumption).
   B. an increased tidal volume with decreased frequency.
   C. inability to deliver sufficient oxygen to the blood at high intensities of exercise.
   D. the fact that it is produced by effects of reduced arterial P_{O_2} value on carotid chemoreceptors at low intensities of exercise.

ANSWERS

1. At 1 mile in 30 min, the rate is 2 mph. At a rate of 2 mph, the cost is 2.83 kcal/min. Therefore, 2.83 kcal $\times$ 30 min = 84.9 kcal consumed in 30 min.

2. First, assuming oxidation of a mixed carbohydrate and fat substrate, 5.00 kcal is provided by 1 liter O_2.

$$\frac{5.00 \text{ kcal}}{1 \text{ liter } O_2} = \frac{84.90 \text{ kcal}}{x \text{ liters } O_2}$$
$$(5.00 \text{ kcal}) (x) = 84.90$$
$$x = 16.98 \text{ liters } O_2$$

Next, 0.250 L/min is the basal metabolic rate. Therefore, 0.25/min $\times$ 30 min = 7.50 liters O_2 as the basal O_2 requirements; 9.48 liters represents the additional O_2 requirement.

3. Since the vascular beds of the nonexercising muscle vasoconstrict while the vascular beds of exercising muscle vasodilate, you should expect the blood pressure during arm exercise at a given work level to be higher than that during leg exercise. During arm exercise, the larger vascular bed of the leg muscle vasoconstricts.

4. The oxygen consumption of the heart is dependent on heart rate and the development of tension. (See Chapter 12 for review.) The heart must develop more tension to open the valves against a higher pressure.

5. B.

6. A, B, D.

7. None.

27

Gastrointestinal Physiology: The Control Mechanisms

R. DAVID BAKER

OBJECTIVES

After completion of this chapter, you should be able to:

1. Understand the forms and general features of gastric and small intestinal smooth muscle electrical activity.

2. Describe the general organization of the gastroenteric nervous system.

3. Describe the pathways by which parasympathetic and sympathetic input to the gastrointestinal tract influence smooth muscle activity.

4. Know the types of reflex pathways available for regulating gastrointestinal motility and secretion.

5. Know the names, sites of secretion, stimuli for secretion, and physiologically important effects of the five established gastrointestinal hormones.

6. Describe the hormonally mediated processes (listed in Table 2) for which the hormones have not yet been identified, and list the names of these "phantom" hormones.

Gastrointestinal (GI) physiology includes the functions of the pharynx, esophagus, stomach, small intestine, large intestine, salivary glands, and gall bladder, as well as those functions of the liver and pancreas that relate to digestive processes. This list includes some of the most complicated organs of the body. Five different categories of events must be considered: (1) control mechanisms, (2) motility, (3) secretion, (4) digestion, and (5) absorption.

Chapters 27–31 assume an understanding of the gross anatomy and microanatomy of the GI system, as well as a general knowledge of membrane transport mechanisms, smooth muscle physiology, and the autonomic nervous system. This chapter introduces the three types of control mechanisms: myogenic, neural, and hormonal.

MYOGENIC CONTROL

Visceral smooth muscle is myogenic, that is, it initiates its own excitation, usually rhythmically. Neural activity often modifies the excitability (responsiveness) of visceral smooth muscle, but the primary exciting event is generated by the muscle itself, rather than by neurons. In this respect, visceral smooth muscle is like cardiac muscle and unlike skeletal muscle. Visceral smooth muscle cells are electrically coupled via low-resistance junctions (gap junctions and probably also other types of junctions); therefore, electrical activity in one cell can be transmitted to adjacent cells. Also in this respect visceral smooth muscle is like cardiac muscle. The excitatory event in visceral smooth muscle is the pacesetter potential, also called the slow wave and sometimes the control potential. The pacesetter potential is a membrane depolarization of about 10–30 mV, followed by repolarization. Gastric and small intestinal pacesetter potentials are spontaneous, almost perfectly rhythmic, and omnipresent. Their constancy of rhythm far exceeds that of cardiac pacemaking. However, gastric and small intestinal pacesetter potentials do not always result in action potentials and contractions. In this respect visceral smooth muscle is unlike cardiac muscle.

A brief discription of gastric and small intestinal myoelectric activity follows. The ways in which this activity times and coordinates contractions in the stomach and small intestine are discussed in Chapter 28.

ELECTRICAL ACTIVITY OF GASTRIC SMOOTH MUSCLE

If a smooth muscle cell in the distal half of the stomach is impaled with a microelectrode, spontaneous electrical complexes, repeating at a frequency of three per minute, can be recorded. Figure 1a depicts a typical electrical complex—a relatively rapid depolarization and partial repolarization phase is followed by a prolonged "plateau" phase and then a complete repolarization. With respect to shape, this gastric electrical complex somewhat resembles the cardiac electrical complex. The resemblance is more than superficial, since the initial depolarization is mainly due to an inward Na^+ current, and the plateau is mainly attributable to an inward Ca^{2+} current. The inward Ca^{2+} current during the plateau phase triggers a weak contraction. The plateau phase is quite variable in amplitude and duration. Sometimes it does not appear at all, leaving only the true pacesetter potential (Fig. 1b), in which case there is no contraction. The strength of contraction increases as the plateau phase (and therefore Ca^{2+} influx) increases in amplitude and duration.

Sometimes a burst of spike potentials is superimposed on the plateau phase, as shown in Figure 1c. These spikes are Ca^{2+} action potentials; when they occur the contraction is stronger. Ca^{2+} spikes superimposed on the plateau have no counterpart in cardiac muscle.

ELECTRICAL ACTIVITY OF SMALL INTESTINAL SMOOTH MUSCLE

The pacesetter potentials (slow waves) in small intestinal smooth muscle are monophasic and do not generate pla-

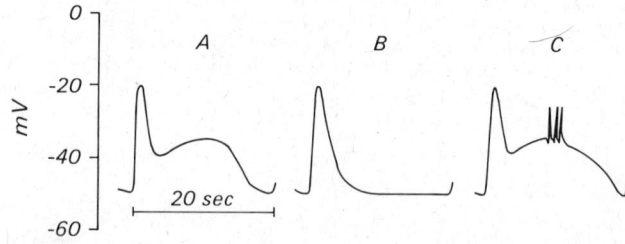

Figure 1 Smooth muscle electrical activity in the distal stomach. *(a)* Pacesetter potential followed by a "plateau" phase. *(b)* Pacesetter potential with no plateau. *(c)* Pacesetter potential with plateau and superimposed spikes.

teaus. Their ionic mechanism is uncertain. One point of view holds that intestinal slow waves are caused by rhythmic variation in activity of the electrogenic Na^+–K^+ pump. Other investigators believe that oscillation of membrane permeability t. Na^+ and Cl^- is more important. In the duodenum and upper jejunum, slow waves occur at a frequency of 12 per minute (depicted in Fig. 5 of Chapter 28). Intestinal slow waves do not themselves initiate contractions, but they sometimes trigger a burst of spikes. These spikes are Ca^{2+} action potentials; they trigger contractions.

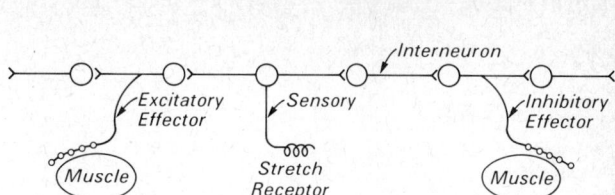

Figure 2 The basic elements of the gastroenteric nervous system. The particular arrangement shown here would mediate the effect of stretch on smooth muscle excitability; muscle orad to the stimulus would be excited while muscle aborad to the stimulus would be inhibited. The black dots on the effector nerve terminals represent the so-called varicosities from which the neurotransmitters are released.

NEURAL CONTROL

THE GASTROENTERIC NERVOUS SYSTEM

The GI tract has its own nervous system. It is called the gastroenteric nervous system or intrinsic nervous system and has also, more flamboyantly, been dubbed the "visceral brain." The unique element of the gastroenteric nervous system is the intrinsic neuron—a neuron having its cell body within the wall of the GI tract. Intrinsic neurons include (1) receptor neurons, (2) effector neurons, and (3) interneurons. The cell bodies are clustered into ganglia, most of which lie either between the longitudinal and circular muscle layers, or in the submucosa. Bundles of neuronal processes run between the ganglia. The ganglia and neuronal processes between the muscle coats are called the myenteric (Auerbach's) plexus; those in the submucosa are called the submucosal (Meissner's) plexus. Postganglionic sympathetic fibers and preganglionic parasympathetic fibers contribute to these plexuses. Intrinsic receptor neurons can be either mechano- (stretch) receptors or chemoreceptors. Intrinsic effector neurons can be either excitatory or inhibitory and innervate either smooth muscle cells or secretory cells. The basic elements of the gastroenteric nervous system are diagrammed in Figure 2.

The neurotransmitter released by intrinsic excitatory neurons is acetylcholine. The neurotransmitter(s) released by intrinsic inhibitory neurons is (are) not firmly established, but two current candidates are ATP and a small polypeptide called vasoactive intestinal peptide (VIP). Because of the probable involvement of ATP, the intrinsic

inhibitory neurons are sometimes called purinergic neurons. There is a long list of polypeptides that are probably used as transmitters by intrinsic sensory neurons and interneurons. This list includes VIP, substance P, enkephalin, and some of the GI hormones. The amine, serotonin, might be a neurotransmitter as well. It is of great interest that most of these materials are also found in the central brain, where they probably function as neurotransmitters.

THE EXTRINSIC NERVES

Extrinsic nerves run between the GI system and the central nervous system (CNS), or between the GI system and the prevertebral autonomic ganglia. They include (1) parasympathetic preganglionic fibers found in the vagus nerves and pelvic nerves, (2) postganglionic sympathetic fibers from the prevertebral ganglia (celiac, superior mesenteric, and inferior mesenteric), and (3) sensory fibers from the GI system that accompany either the parasympathetic or sympathetic nerves.

Parasympathetic Input

Preganglionic parasympathetic fibers enter the organs of the GI system and make synapses with intrinsic neurons. These intrinsic neurons can be excitatory neurons, inhibitory neurons, or interneurons. The neurotransmitter released by preganglionic parasympathetics is acetylcholine. Parasympathetic input is diagrammed in Figure 3. It should be appreciated that the intrinsic effector neurons are not simply postganglionic parasympathetic neurons;

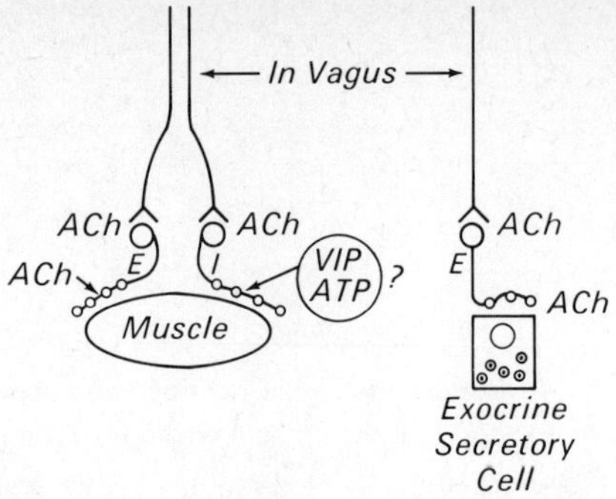

Figure 3 Parasympathetic input to smooth muscle and exocrine secretory cells. E indicates an excitatory intrinsic neuron; I indicates an inhibitory intrinsic neuron.

they are also influenced by intrinsic sensory neurons and interneurons and by postganglionic sympathetic neurons (see below). The effect of parasympathetic activity is modulated by activity from these other inputs. In other words, the intrinsic effector neurons (excitatory and inhibitory) do not merely relay information from preganglionic parasympathetic fibers to muscle and secretory cells, but rather integrate information from a variety of sources.

Sympathetic Input

The prevertebral autonomic ganglia are connected to the CNS by the splanchnic nerves and white rami. Sympathetic fibers from the prevertebral ganglia run along with mesenteric blood vessels; they enter the GI organs and can influence smooth muscle cells either directly or indirectly. Direct pathways are shown in Figure 4. Most of the sympathetic nerve terminals release norepinephrine, which inhibits the muscularis externa of the GI tract via both α- and β-receptors on the smooth muscle cells. Some of the sympathetic nerve terminals release acetylcholine, which excites the smooth muscle via muscarinic receptors; this seems to be of importance mainly in the gastrocolic reflex, described in Chapter 28. At GI sphincters, especially the ileocecal sphincter, direct sympathetic activity releases norepinephrine which, acting via α-receptors, causes the sphincter to constrict. Vascular smooth muscle also constricts in response to direct sympathetic activity, and this again involves the action of norepinephrine on α-receptors.

The indirect pathways to the muscularis externa involve intrinsic effector neurons, as shown in Figure 5. This representation, although somewhat speculative, shows that sympathetic activity can excite intrinsic inhibitory neurons via α-receptors and inhibit intrinsic excitatory neurons via β-receptors. Both effects inhibit smooth muscle cells of the muscularis externa.

Strong sympathetic activity to secretory organs of the GI system generally inhibits secretion. An exception is the salivary glands, which secrete in response to either parasympathetic activity or sympathetic activity. Inhibition of secretion (gastric, pancreatic, and biliary) by sympathetic activity is probably caused primarily by vasoconstriction, with consequent reduced blood flow, rather than by a direct effect on the secretory cells.

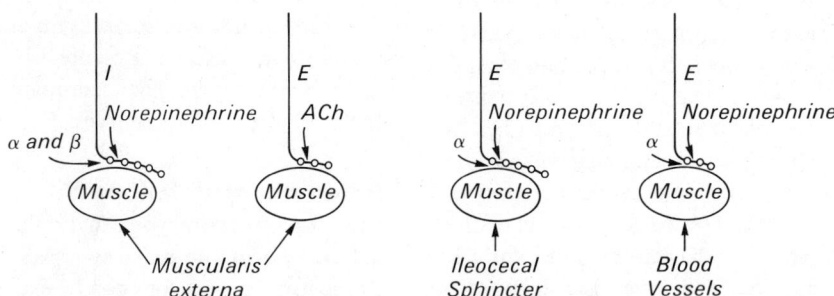

Figure 4 Direct sympathetic influences on GI smooth muscle. The sympathetic fibers depicted here are postganglionic, with cell bodies in the prevertebral autonomic ganglia (celiac, superior mesenteric, and inferior mesenteric). The fibers run along with mesenteric blood vessels.

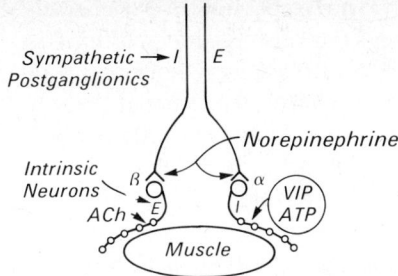

Figure 5 Indirect sympathetic influences on GI smooth muscle.

Sensory Output

Sensory information from GI mechano- and chemoreceptors travels centrally in two different types of neurons. One of these is the ordinary sensory neuron, with cell body located in a dorsal root ganglion or nodose ganglion and with processes entering the CNS. The other is an intrinsic neuron, with cell body in an intrinsic ganglion and with processes again going to the CNS. The latter type of sensory neuron can also terminate in one of the prevertebral autonomic ganglia, rather than in the CNS; in fact, the prevertebral ganglia are probably the usual relay stations for information carried by this type of intrinsic sensory neuron.

REFLEX PATHWAYS

With the wiring arrangements described above, three different types of reflex pathways can be used: (1) intrinsic, (2) prevertebral, and (3) central. An intrinsic reflex is one in which only elements of the intrinsic nervous system are directly involved. A prevertebral reflex is one in which sensory fibers make synaptic contact with adrenergic sympathetic neurons in the prevertebral ganglia. A central reflex is one having synaptic connections in the CNS. In most cases central GI reflexes are vagovagal reflexes (i.e., both afferent and efferent limbs run in the vagus nerves).

Many of the adrenergic neurons with cell bodies in the prevertebral ganglia receive input from a variety of sources, including (1) preganglionic sympathetic fibers from the splanchic nerves, (2) sensory fibers from the GI system, and (3) fibers that connect the prevertebral ganglia to each other. Therefore, the adrenergic neurons in the prevertebral ganglia are able to integrate information from various sources and should be considered something more than merely postganglionic sympathetics.

HORMONAL CONTROL

Hormones are chemical messengers secreted into blood by endocrine cells in response to specific stimuli, are carried by blood to their sites of action, and then have regulatory effects on specific target cells. A GI hormone is one secreted by cells in the GI tract. There are five established GI hormones: secretin, gastrin, cholecystokinin (CCK), gastric inhibitory peptide (GIP), and motilin—all of which are small polypeptides. Their chemical structures are known. Additional GI hormones almost certainly exist and await identification.

Most other hormones are secreted by cell types clustered into distinct endocrine glands or islets. The endocrine cells of the GI tract, however, are distributed individually among the other epithelial cells of the GI mucosa. There are at least five types of GI endocrine cells, corresponding to the five hormones listed above. Each of these types of GI endocrine cell secretes only one of the hormones. All these cell types belong to a cell line that originates embryologically in the neuroectoderm and is responsible for secreting many peptide hormones from a variety of endocrine organs in addition to the GI mucosa. The cells belonging to this cell line are known as APUD cells (which stands for their biochemical attributes of *a*mine *p*recursor *u*ptake and *d*ecarboxylation). Endocrine cells of the stomach, intestine, and pancreas are often called GEP (gastroenteropancreatic) cells. The GEP endocrine system includes not only the five cell types mentioned above, but also insulin cells, glucagon cells, and at least seven other endocrine-like cell types, all of which synthesize and secrete peptide messengers.

The endocrine cells responsible for secreting GI hormones are, as mentioned above, dispersed among the other epithelial cells of the GI mucosa. They secrete their hormones into the bloodstream and, accordingly, their secretory surfaces are on the blood side of the epithelium. However, most of these cells also make contact with the

LUMEN

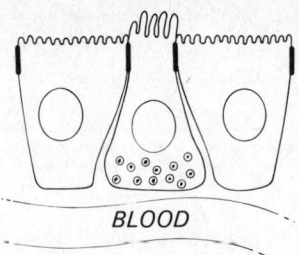

BLOOD

Figure 6 A GEP endocrine cell nestled among ordinary epithelial cells. The endocrine cell has long microvilli that probably act as sensory receptors. It also has basal granules in which its hormone is stored. When stimulated, these granules are secreted by exocytosis toward the blood vessel.

lumen of the GI tract, as depicted in Figure 6. This arrangement provides the opportunity for GI endocrine cells to "taste" the luminal contents and then relay information about these contents, via the secreted hormones, to the target cells. Some GI endocrine cells, particularly those that secrete gastrin, are also stimulated by acetylcholine released from intrinsic cholinergic neurons.

In 1902 Bayliss and Starling found that shortly after a 0.4% HC1 solution was placed in the upper small intestine of dogs, the pancreas began secreting its juice at a greatly increased rate, *even after all conceivable neural connections between the two organs had been cut.* HCl injected directly into the bloodstream had no such effect; therefore, these workers concluded that luminal HC1 induced release of some chemical messenger from the intestine into blood that, after arriving at the pancreas, caused it to secrete. They then extracted duodenal mucosa with an HC1 solution and found that this extract contained some ingredient that, when injected intravenously, caused the pancreas to secrete. They called this ingredient secretin and assumed it to be the same as the chemical messenger responsible for inducing secretion after acid was placed in the lumen. The work of Bayliss and Starling is often proclaimed as the first discovery of a hormone. The discovery of other hormones, especially testosterone, might compete with this claim, but there is no doubt that the basic concept of blood-borne chemical messengers was crystallized and firmly established by the work of Bayliss and Starling

on secretin. Their work successfully challenged the Pavlovian concept that prevailed at the time (i.e., that neuronal reflexes controlled everything in the GI system) and introduced the concept of hormonal "reflexes".*

In similar fashion, Edkins in 1905 established the existence of a blood-borne chemical messenger involved in stimulating HC1 secretion by the stomach; he called it gastrin. In 1928, Ivy and Oldberg discovered the existence of a chemical messenger that signaled the gallbladder to contract when fat was placed in the small intestine; they called their hormone cholecystokinin. Much later, during the 1960s, these three hormones—secretin, gastrin, and CCK—were finally isolated in pure form and their chemical structures determined. More recently, two other hormones, GIP and motilin, have been added to the list. We will now proceed to a brief discussion of the five established GI hormones. Additional information about their physiology will be found in subsequent chapters.

THE ESTABLISHED GI HORMONES

Secretin

Secretin is a polypeptide amide having 27 amino acid units. It is synthesized in, and secreted by, a type of GEP cell located in the duodenal epithelium. The stimulus for secretion of secretin is acid in the duodenum. Duodenal pH must reach a value of 4.5 or lower to stimulate secretin cells. The pH in the upper duodenum does sometimes drop below 4.5, and secretin is secreted. There are also secretin cells in the jejunum, but since jejunal pH is always well above 4.5 (generally about 7.0–7.5), jejunal secretin would seem to have no importance (?).

The principal target organ for secretin is the pancreas. Secretin stimulates the pancreatic ducts to secrete a $NaHCO_3$ solution. This solution flows into the duodenum and buffers the HC1, which originally stimulated the secretin cells. Thus secretin is part of a negative feedback mechanism that keeps duodenal pH from dropping too low. Secretin also potentiates the effect of CCK on enzyme secretion by pancreatic acinar cells (see below). Furthermore, secretin stimulates growth of the pancreas via a

---

*The 1902 paper of Bayliss and Starling also included the first, but by no means the last, secretin versus secretion typographical error.

trophic effect. In addition to these effects on the pancreas, secretin stimulates intrahepatic bile ducts to secrete a $NaHCO_3$ solution. Biliary HCO_3^- aids pancreatic HCO_3^- in regulating intestinal pH.

Ever since its discovery in 1902, it has been assumed that secretin is secreted mainly during digestion of a meal and therefore that the principal flow of bicarbonate and water from the pancreas occurs during digestion. Apparently this is not necessarily true. Recent experiments on dogs show that only a small amount of secretin is secreted during digestion of a meal (Purina dog chow), and that consequently only a small volume of pancreatic juice is secreted. This juice is rich in enzymes, but not in bicarbonate. The apparent reason for this result is that beyond the first part of the duodenum the pH value seldom drops below 4.5 during digestion. In these same experiments, it was discovered that roughly 6–10 h after ingestion of a meal, the pH did frequently drop below 4.5 over a large portion of the duodenum and that large amounts of secretin were secreted, leading to a large flow of pancreatic juice containing high levels of bicarbonate. Perhaps there is a greater need for intraduodenal buffering by pancreatic bicarbonate, and therefore secretion of secretin, during interdigestive periods than during digestion. The degree to which this is true for other types of meals having less buffering capacity than Purina dog chow remains to be determined.

Gastrin

Gastrin is a polypeptide amide that is present in several molecular sizes. The two most common gastrins have 17- and 34-amino acid units, respectively. Accordingly, they are called G17 and G34, or sometimes little gastrin and big gastrin. The chemistry of gastrin and of the other GI hormones can be found in several of the sources listed in the Selected Bibliography at the end of this chapter.

Gastrin is synthesized, stored, and secreted by so-called G cells located in the pyloric glands in the antral region of the stomach. A relatively small number of G cells are also present in the small intestine, but the importance of intestinal gastrin has not been established. The principal stimulus for gastrin secretion is acetylcholine released from intrinsic cholinergic neurons at the secretory surface of the G cells. Various reflexes, described in Chapter 29, are involved in activating these intrinsic neurons. G cells can also be stimulated by partially digested proteins in the lumen of the stomach. G cells are inhibited by acid in the gastric antrum; in fact, if antral pH drops below 1.5, no gastrin can be secreted, even when the intrinsic cholinergic neurons are activated.

Gastrin has two principal functions: (1) to stimulate the oxyntic (acid forming) cells of the stomach to secrete HCl solution into the lumen, and (2) to promote growth of the gastric and intestinal mucosae.

All the actions of gastrin can be elicited by the C-terminal tetrapeptide fragment ($Try-Met-Asp-Phe-NH_2$) of the total molecule. A pentapeptide containing this C-terminal tetrapeptide is commercially available; it is called pentagastrin and can be used clinically to test for gastric secretory ability.

Cholecystokinin

CCK is a polypeptide amide having 33 amino acid residues. Part of its structure is similar to that of gastrin and, when injected in large amounts, the two hormones have many of the same pharmacological effects. However, their physiological effects (i.e., those produced by the small amounts of hormone normally secreted) are quite different.

CCK is secreted by GEP cells located throughout the small intestine. CCK cells secrete in response to long-chain fatty acids and certain amino acids in the intestinal lumen.

CCK has several physiologically important effects. First, CCK induces contractions of the gallbladder, which cause ejaculation of stored bile into the duodenum; CCK was named on this basis. Second, CCK signals the pancreatic acinar cells to secrete digestive enzymes. Third, CCK potentiates secretin-induced $NaHCO_3$ secretion by pancreatic and biliary ducts. Fourth, CCK reduces the rate at which the stomach empties its contents into the duodenum. Fifth, CCK stimulates growth of the pancreas. All these actions of CCK promote digestion of nutrients as follows: (1) bile salts in bile are necessary for lipid digestion and absorption; (2) pancreatic enzymes are necessary for digestion of carbohydrates, fats, and proteins; (3) pancreatic and biliary bicarbonate are necessary for maintaining a near-neutral intestinal pH at which the pancreatic enzymes can function; (4) regulated gastric emptying is necessary so that the digestive and absorptive processes are not suddenly overwhelmed; and (5) the pancreas must not be permitted to atrophy.

One further historical note is important. In 1943 Harper and Raper discovered that food in the small intestine signaled the pancreas to secrete enzymes via a bloodborne chemical messenger. Appropriately, they named their hormone pancreozymin. Unfortunately for their terminology, pancreozymin and cholecystokinin later turned out to be exactly the same peptide, and the historically earlier term (cholecystokinin) has taken charge.

Gastric Inhibitory Peptide

Gastric inhibitory peptide (GIP) is a peptide having 43 amino acids. It has a structure similar to that of secretin. GIP is secreted by cells located in the duodenal and jejunal epithelium in response to the presence of fatty acids and glucose in the lumen.

GIP has an important effect on the insulin-secreting β cells of pancreatic islets. Insulin is a peptide hormone that promotes glucose and amino acid uptake by muscle and adipose cells. The β cells of pancreatic islets secrete insulin in response to the elevated blood glucose concentration that occurs following carbohydrate digestion and absorption. The effect of insulin keeps blood glucose concentration from rising too high. GIP potentiates the effect of blood glucose on insulin secretion. In other words, in the presence of GIP, glucose stimulates the β cells more strongly than it does in the absence of GIP. GIP cells in the intestine sense glucose as it is being absorbed; they signal β-cells in the pancreas hormonally to respond strongly to blood glucose in a way that encourages the body to handle the extra glucose.

GIP also inhibits both gastric motility and gastric HCl secretion of the stomach. These are the effects for which gastric inhibitory peptide was originally named. Recently, however, it has been shown that in order for GIP to inhibit the stomach, a blood concentration several times higher than normally attained must be produced. Thus gastric inhibition apparently is not a normal function of GIP, and the hormone seems to be misnamed. (It has been pointed out, however, that the same initials, GIP, can stand for *g*lucose-dependent *i*nsulinotropic *p*eptide.)

Motilin

Motilin is a polypeptide having 22 amino acids. It is secreted by cells in the duodenal and jejunal mucosa. The stimulus for secretion is probably acetylcholine released by intrinsic cholinergic neurons, although this is not certain. Motilin is secreted in rhythmic cycles every 1.5–2 hours during interdigestive (i.e., fasting) periods. Motilin is involved in actuating interdigestive migrating motor complexes in the stomach and small intestine, described in Chapter 28.

Motilin is the newest of the established GI hormones. In fact, some GI physiologists might still be reluctant to accept motilin as a true hormone.

Summary

The established GI hormones and their physiologically important effects are summarized in Table 1.

OTHER EFFECTS OF THE ESTABLISHED GI HORMONES

Injected in large amounts, the established GI hormones have many effects in addition to those mentioned above; these are called their pharmacological effects. One of the great problems in GI endocrinology is trying to determine which effects have physiological significance under normal circumstances and which occur only in the presence of abnormally large amounts of the hormone.

Gastrin, for example, in addition to stimulating gastric acid secretion and GI mucosal growth, is also capable of stimulating pancreatic enzyme and HCO_3^- secretion, biliary HCO_3^- secretion, intestinal secretions, gallbladder contractions, gastric and intestinal motility, tone of the lower esophageal sphincter, and insulin secretion from pancreatic islets. At pharmacological doses secretin can inhibit GI motility, inhibit gastric HCl secretion, stimulate gastric pepsinogen secretion, and stimulate insulin secretion. The other GI hormones also have many effects currently thought not to occur to a significant degree under ordinary circumstances. In-depth discussions of these pharmacological effects can be found in more comprehensive sources.

GI HORMONES OF THE FUTURE

Certain GI events are known to involve hormones, and yet the peptide messengers involved have not been chemically identified. The more prominent of these events and the names of the proposed hormones are listed in Table 2.

Table 1 Summary of the Established GI Hormones

Hormone	Stimuli for Secretion	Physiological Actions
Secretion	Duodenal pH below 4.5.	Stimulates $NaHCO_3$ and water secretion by pancreatic and biliary duct epithelium. Potentiates effect of CCK on pancreatic acinar cells. Trophic effect on pancreas.
Gastrin	Acetylcholine released from intrinsic neurons in wall of antrum in response to various reflexes. Peptides and amino acids in lumen of antrum.	Stimulates secretion of isotonic HCl solution by oxyntic (parietal) cells of stomach. Trophic effect on GI mucosa.
CCK	Long-chain fatty acids and certain amino acids in lumen of small intestine.	Stimulates contractions of gallbladder. Stimulates secretion of enzymes by pancreas. Potentiates effect of secretin on pancreatic and biliary ducts. Slows gastric emptying. Trophic effect on pancreas.
GIP	Glucose and fatty acids in lumen of duodenum and jejunum.	Potentiates the effect of blood glucose in stimulating insulin secretion by β cells of pancreatic islets.
Motilin	Spontaneous rhythmic cycles of secretion with period of about 1.5–2 h. Probably the immediate stimulus is acetylcholine.	Initiates interdigestive migrating motor complexes in stomach and duodenum.

Perhaps a peptide messenger will eventually be identified for each of these control processes, and the phantom hormones will be added to the list of established hormones. On the other hand, some of these processes could prove to be mediated by one or more of the established GI hormones or candidate hormones (see below), and the names of these phantom hormones would be dropped—this is what happened to pancreozymin.

In addition to the established GI hormones, many other biologically active peptides have been found in the GI tract.*Some of these peptides (e.g., chymodenin, neurotensin, somatostatin, and glucagon) could eventually prove to be GI hormones; meantime they can be regarded as candidate hormones. Some of these peptides (e.g., somatostatin and VIP) might be paracrines. A paracrine is like a hormone, except that instead of being delivered to distant targets via blood, it is delivered to nearby targets by diffusion through interstitial spaces. Some of these peptides (e.g., VIP, substance P, and enkephalin) are almost certainly neurotransmitters.

Another candidate hormone, a 36 amino acid peptide

called pancreatic polypeptide (PP), deserves special mention. PP is secreted into blood by cells in the pancreas rather than by cells in the GI tract proper; nevertheless, it is usually listed as one of the candidate GI hormones. PP is secreted during digestion, especially if the meal is

Table 2 A Partial List of Hormonally Mediated Processes for Which the Hormones Have Not Yet Been Identified

Hormonally Mediated Process	Name of Phantom Hormone
Fat in the small intestine inhibits gastric acid secretion	Enterogastrone
Acid in the duodenal bulb (first part of the duodenum) inhibits gastric acid secretion	Bulbogastrone
Partially digested protein in the duodenum stimulates gastric acid secretion	Entero-oxyntin (intestinal-phase hormone)
Chyme in the small intestine stimulates secretion of intestinal juice	Enterocrinin
Chyme in the small intestine stimulates secretion of mucus from duodenal (Brunner's) glands	Duocrinin
Chyme in the small intestine stimulates contractions of intestinal villi	Villikinin

---

*A concise list of the peptides of the GI tract and pancreas can be found in a paper by Grossman (in Zimmermann et al: Peptides of the Brain and Gut, *Fed Proc* 38: 2286, 1979.).

high in protein content. The signal from duodenum to pancreas involves poorly understood cholinergic reflex pathways and perhaps also a hormone. PP inhibits pancreatic secretion, both enzymes and bicarbonate, and might have a significant role in modulating the pancreatic responses to CCK and secretin. The usefulness of this effect remains to be established.

Some of the peptides found in the GI tract and pancreas are also found in the brain. These include gastrin, CCK, motilin, VIP, somatostatin, substance P, neurotensin, and enkephalin. The subject of brain and GEP peptides is of tremendous current interest and is revolutionizing our understanding of the control systems of the body—neural, endocrine, and paracrine. Further discussion would take us beyond the scope of this chapter. (Recommended sources are Zimmermann et al, Glass, and Miyoshi; see Selected Bibliography.)

SELECTED BIBLIOGRAPHY FOR GASTROINTESTINAL PHYSIOLOGY*

MAJOR TREATISES

Code CF (ed): *Handbook of Physiology, section 6: Alimentary Canal.* Baltimore, Williams & Wilkins, 1968, 6 vols.

Jacobson ED, Shanbour LL (eds): *International Review of Physiology—Gastrointestinal Physiology.* Baltimore, University Park Press, 1974, vol I.

*This general list applies to Chapters 27–31.

Crane RK (ed): *International Review of Physiology—Gastrointestinal Physiology.* Baltimore, University Park Press, 1977, vol II; 1979, vol III.

Johnson LR (ed): *Physiology of the Gastrointestinal Tract.* New York, Raven Press, 1981, 2 vols.

TEXTBOOKS

Brooks FP (ed): *Gastrointestinal Pathophysiology.* New York, Oxford University Press, 1974.

Duthie HL, Wormsley KG: *Scientific Basis of Gastroenterology.* New York, Churchill Livingston, 1979.

Davenport HW: *Physiology of the Digestive Tract,* ed 5. Chicago, Year Book, 1982.

Davenport HW: *A Digest of Digestion,* ed 2. Chicago, Year Book, 1978.

Johnson LR (ed): *Gastrointestinal Physiology,* ed 2. St Louis, CV Mosby, 1981.

Sircus W, Smith AN: *Scientific Foundations of Gastroenterology.* Philadelphia, WB Saunders, 1980.

SELECTED BIBLIOGRAPHY FOR CONTROL MECHANISMS

Gabella G: Innervation of the Gastrointestinal Tract. *Int Rev Cytol* 59:129, 1979.

Glass GBJ (ed): *Gastrointestinal Hormones.* New York, Raven Press, 1980.

Miyoshi A (ed): *Gut Peptides.* New York, Elsevier—North Holland, 1979.

Zimmermann EG, Grossman MI, Walsh TH, Stewart TM (Chairmen of Conference): Peptides of the brain and gut. *Fed Proc* 38:2286, 1979.

SELF-STUDY QUESTIONS

MULTIPLE CHOICE

Select the best answer(s). (In many cases, more than one answer is correct.)

1. The smooth muscle electrical complex in the distal stomach
   A. repeats at a frequency of 3 per minute.
   B. always includes a plateau phase.
   C. sometimes, but not always, includes bursts of Ca^{2+} spikes.
   D. always induces a contraction.

2. Small intestinal slow waves
   A. generally do not generate plateaus.
   B. occur at a frequency of 12 per minute in the duodenum.
   C. are actually Ca^{2+} action potentials that initiate contractions.
   D. sometimes trigger Ca^{2+} spikes that initiate contractions.

3. Candidates for the neurotransmitters relased by intrinsic inhibitory neurons include
   A. acetylcholine.
   B. VIP.
   C. norepinephrine.
   D. ATP.

4. In order to block the inhibitory effect of norepineph-rine on small intestinal smooth muscle, both α-blocking and β-blocking drugs must be used simultaneously. Explain why this is so.

5. What are the three different types of reflex pathways available for regulating GI motility and secretion?

6. Name the hormones that have primary responsibility for mediating the following effects.
   A. pH value of 4.5 or lower in duodenum (secretion of pancreatic bicarbonate).
   B. Sugar in duodenum (potentiation of the effect of blood glucose on insulin secretion by pancreatic islets.)
   C. Parasympathetic nerve activity to gastric antrum (secretion of HC1 by oxyntic glands of stomach).
   D. Fat in duodenum (contractions of gallbladder).

7. In animals provided nutrition entirely by intravenous "hyperalimentation" and given no food by mouth for a few weeks, their GI mucosa is shown to atrophy. Explain why.

8. What is a paracrine.

TRUE OR FALSE

9. The adrenergic neurons of the prevertebral ganglia are influenced only by nerve fibers in the splanchnic nerves.

10. The major action of secretin might occur during interdigestive periods.

ANSWERS

1. A, C. Spikes and even the plateau phase are sometimes absent; therefore, contractions are sometimes absent.

2. A, B, D. Intestinal slow waves do not require Ca^{2+} and do not themselves directly initiate contractions.

3. B, D. These are sometimes called noradrenergic inhibitory fibers.

4. Activation of either α- or β-receptors on small intestinal smooth muscle apparently causes inhibition. In addition, activation of intrinsic inhibitory fibers via

α-receptors and inhibition of intrinsic excitatory fibers via β-receptors both cause inhibition. Therefore, neither an α-blocker nor a β-blocker will by itself completely prevent the effect of adrenergic nerve activity on the small intestine.

5. Intrinsic, prevertebral, and central.

6. A, Secretin. B, GIP. C, Gastrin. D, CCK. You should be able to perform this same exercise for all the other hormonally mediated events listed in Tables 1 and 2.

7. If no food is ingested, gastrin secretion is greatly diminished, and its trophic effect on the GI mucosa is reduced accordingly. See p. 433.

8. See definition on p. 435.

9. False. They are also influenced by certain intrinsic sensory neurons and by fibers from other prevertebral ganglia.

10. True. See p. 433.

28

Movements of the Digestive Tract

R. DAVID BAKER

OBJECTIVES

After completion of this chapter, you should be able to:

1. Understand the physiology of swallowing.

2. Explain the events that take place preceding and during vomiting.

3. Describe the movements of the proximal and distal stomach and how the functions of these regions differ from each other.

4. Understand the myoelectrical activity of the distal stomach.

5. Describe the factors that influence the rate of gastric emptying.

6. Describe the types of contractions and special patterns of motility in the small intestine.

7. Understand the myoelectrical activity of the small intestine.

8. Contrast the interdigestive pattern of gastrointestinal motility against the digestive pattern.

9. Describe the interdigestive migrating motor complex, explain its function, and have some notions as to how it is initiated, and how its migration is controlled.

10. Describe the movements of the large intestine, the gastrocolic reflexes, and the defecation reflex.

Ingested materials are chewed if necessary, swallowed, and passed down the esophagus to be stored in the proximal part of the stomach. Food is then gradually pressed toward the distal stomach, where peristaltic contractions mix it with gastric juice and triturate it to a semiliquid mush called chyme. Chyme is gradually squeezed and squirted through the pylorus into the duodenal bulb. Eventually all the meal is chymified and pumped into the duodenum. Contractions of the duodenum and jejunum spread chyme along the upper small intestine and mix it with pancreatic juice and bile. Chyme sloshes back and forth in the upper small intestine while digestion and absorption take place. Most absorption of nutrients occurs there. Gradually the remains are delivered into the lower ileum, where additional absorptive processes occur. The ileocecal sphincter retards progress down the tract; chyme can dwell for some time in the lower ileum while water and other materials are absorbed. Chyme is gradually propelled into the cecum. Most of the residue from a meal enters the large intestine about 3–6 h after ingestion. Progress through the colon is usually slow, providing ample time for absorption of water and consolidation of fecal material; occasionally, however, colonic propulsion occurs rapidly. In either case, feces accumulate in the rectum and sigmoid colon. When these structures are distended, the defecation reflex is initiated and the journey is likely at an end.

Along the tract various kinds of contractions occur; the most prominent are circumferential constrictions that *do not progress,* and circumferential constrictions that *do progress.* The latter are called peristaltic contractions. These and other types of gastrointestinal (GI) contractions are described below. They serve two principal functions—mixing and propulsion. Propulsion of GI contents along the tract is promoted by pressure gradients and by direct mechanical squeezing and pushing.

SWALLOWING

DESCRIPTION OF THE SWALLOWING PROCESS

Swallowing takes place in three stages: buccal, pharyngeal, and esophageal. The buccal stage is voluntary in that it can be started or stopped at will. The pharyngeal and esophageal stages are completely out of voluntary control.

After food has been suitably crushed and mixed with saliva during the chewing process, the slimy mass is formed into a bolus and squeezed toward the pharynx by elevating the tongue against the hard and soft palate (Fig. 1), constituting the buccal stage of swallowing.

Contact of food with the upper pharynx stimulates sensory receptors in the pharyngeal mucosa that initiate the swallowing reflex. During the pharyngeal stage, a number of highly coordinated events occur. The soft palate is elevated to seal off the nasopharynx from the oropharynx. The larynx is elevated, the true and false vocal cords are adducted, and the epiglottis swings backward and downward over the raised larynx. Thus the glottis is closed. The mouth cavity is shut off by the position of the tongue against the palate and by contraction of the palatoglossi muscles. Thus all exits from the pharynx are closed. The pharyngeal constrictors (superior, middle, and inferior) then contract, forcing the bolus of food toward the esophagus. At the same time, the upper esophageal sphincter relaxes, permitting food to enter the esophagus. Contraction of the pharyngeal muscles proceeds through the pharynx as a peristaltic wave. The pharyngeal stage of deglutition (swallowing) is very rapid; the bolus passes through the pharynx in less than a second. The pharyngeal stage of swallowing can be visualized in Figure 1, although the various events are not actually depicted.

The esophagus is a muscular tube with a muscular sphincter at each end. The upper esophageal sphincter (UES) consists mainly of a thickened portion of the inferior pharyngeal constrictor called the cricopharyngeus muscle. Between swallows this muscle is constricted, and the sphincter is closed. The UES prevents ventilation of the esophagus during breathing. The lower esophageal sphincter (LES) cannot be identified anatomically as anything but the lower end of the esophagus, that is, there is no thickening of the muscle. Nevertheless, the circular muscle there remains tightly constricted between swallows, thereby helping to prevent regurgitation of gastric contents into the esophagus. The body of the esophagus between the two sphincters remains relaxed except during a swallow.

The muscle of the pharynx, UES, and upper third of the esophagus is striated. The muscle of the lower third of the esophagus and LES is smooth muscle. In the middle third

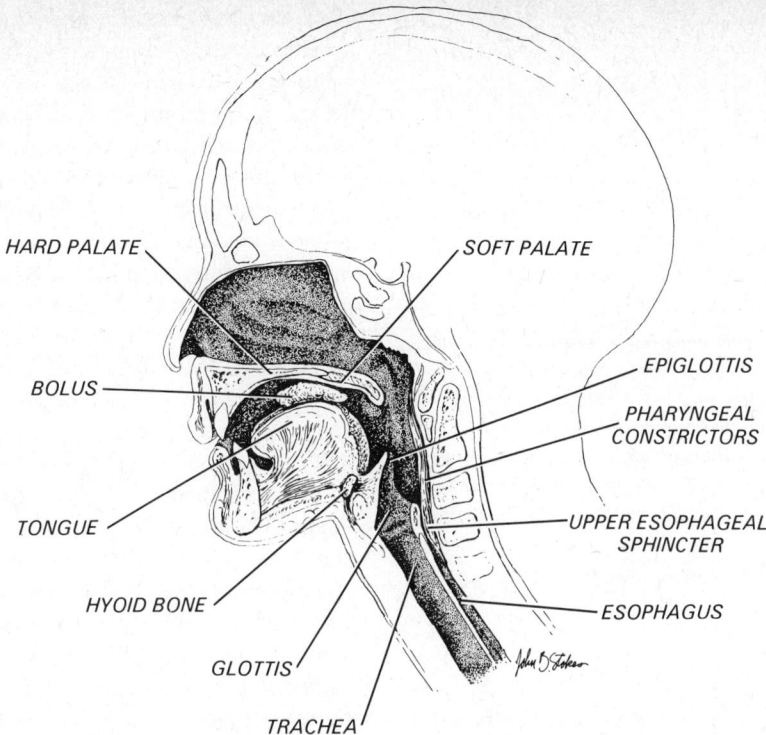

HARD PALATE

SOFT PALATE

BOLUS

EPIGLOTTIS

PHARYNGEAL
CONSTRICTORS

TONGUE

UPPER ESOPHAGEAL
SPHINCTER

HYOID BONE

ESOPHAGUS

GLOTTIS

TRACHEA

Figure 1 Sagittal diagram of the buccal and pharyngeal structures involved in swallowing.

there are both striated and smooth fibers, with the proportion of the latter gradually increasing distally.

The peristaltic contraction of the pharyngeal wall continues through the UES into the esophagus. Once in the esophagus it slows greatly and takes roughly 10 s to reach the stomach. Shortly after the peristaltic wave starts down the esophagus, the LES relaxes. A semisolid bolus of food proceeds along the esophagus with the peristaltic wave and takes about 10s to reach the stomach. By contrast, in an upright person a liquid bolus very rapidly runs down the esophagus ahead of the peristaltic wave. The strength of the peristaltic contraction depends on the size of the bolus. Sometimes the peristaltic wave is not successful in emptying the esophagus completely, and the lower part of the esophagus remains somewhat distended. The distention initiates a new peristaltic wave that begins at the upper end of the esophagus and travels all the way down. This process is known as secondary esophageal peristalsis.

The neural mechanisms that control esophageal peris-

talsis somehow permit only one peristaltic contraction to pass at any one time. If a second swallow is initiated while an esophageal peristaltic wave is in progress, the first wave dies out as the second wave begins. Consequently, during repetitive swallowing, as in chug-a-lugging a beverage, a complete peristaltic wave passing the entire length of the esophagus is associated only with the final swallow.

SWALLOWING PRESSURES

Figure 2 depicts the pressures developed during a normal swallow. Rhythmic respiratory pressure changes would normally be superimposed, but have been omitted in Figure 2. The resting pressure in the thoracic esophagus is equal to intrathoracic pressure and therefore is usually subatmospheric. The "resting" pressures within the UES and LES are above atmospheric because of tonic contraction of these muscles. Examination of swallowing pressures is useful clinically in the diagnosis of dysphagias.

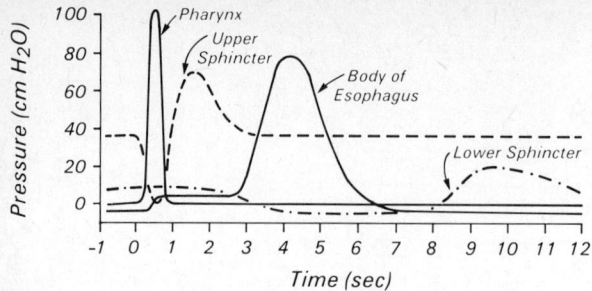

Figure 2 Swallowing pressures. The swallow is initiated at zero time. Pharyngeal pressure increases sharply at the same time that pressure in the UES decreases. When the UES opens and the bolus is propelled into the esophagus, there is a small, sustained rise in pressure along the entire body of the esophagus. About 2 s after the UES relaxes, the LES relaxes and the pressure within it decreases. Progression of the peristaltic contraction is represented by the succession of major pressure increases along the tract. The position of the esophageal pressure peak along the time axis depends on the location of the pressure-sensing device in the esophagus.

THE SWALLOWING REFLEX

The neural pathways and neurochemical transmitters involved in the swallowing reflex are not completely understood. The block diagram in Figure 3 depicts the minimal circuitry required. Touch receptors in the pharynx send

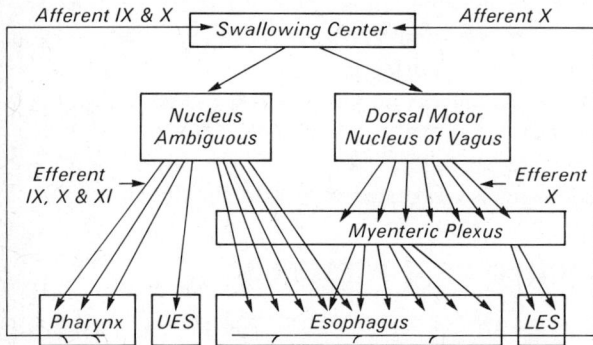

Figure 3 Block diagram of the pathways involved in the swallowing reflex. Efferents from the nucleus ambiguous innervate striated muscle fibers directly. Efferents from the dorsal motor nucleus of the vagus innervate intrinsic neurons. Information fed into the intrinsic nervous system is processed in unknown ways and is then manifested as activity in intrinsic excitatory and inhibitory neurons that innervate smooth muscle.

impulses along afferent fibers in cranial nerves IX and X to a swallowing center in the lateral reticular formation. This center controls the activity of certain efferent neurons in the nucleus ambiguous; these neurons innervate the striated muscle of the pharynx, UES, and upper part of the esophagus via cranial nerves IX, X, and XI. The peripheral neurotransmitter is acetylcholine. The fibers from the nucleus ambiguous are activated sequentially, so that the resulting contraction begins at the upper pharyngeal constrictor and passes aborally through the pharynx, UES, and upper part of the esophagus in peristaltic fashion. Relaxation of the UES just preceding its contraction is presumably caused by a well-timed reduction of tonic activity in the appropriate nerve fibers.

The swallowing center also controls the activity of certain efferent neurons in the dorsal motor nucleus of the vagus. These neurons are parasympathetic preganglionics that travel in the vagus nerve and innervate intrinsic neurons lying between the longitudinal and circular muscle layers of the esophagus. The intrinsic neurons innervate smooth muscle fibers in the lower two-thirds of the esophagus and also communicate with each other. The transmitter used by excitatory intrinsic neurons is acetylcholine; transmitters used by inhibitory intrinsic neurons and interneurons are unknown. Presumably, the fibers from the dorsal motor nucleus are activated sequentially as were those from the nucleus ambiguous, thereby producing a wave of activity in excitatory intrinsic neurons, causing a continuation of the peristaltic contraction. However, this mechanism of peristalsis in the lower esophagus is not certain.

Another theory is that the inhibitory intrinsic neurons in the lower esophagus are activated all at once and that they inhibit the smooth muscle. When these inhibitory intrinsic neurons stop their activity, the smooth muscle contracts as sort of a rebound phenomenon. Furthermore, the latent period preceding this rebound contraction increases as the lower esophagus is descended. Consequently, according to this theory, the rebound contraction begins at about the middle of the esophagus and progresses to and through the LES. The relaxation of the LES preceding its contraction is not well understood, but probably involves intrinsic inhibitory neurons.

Stretch receptors in the wall of the esophagus, when stimulated by a bolus of food, send impulses to the swallowing center via afferent fibers in the vagus nerves. These stretch receptors, especially those located in the lower esophagus, are responsible for initiating secondary peristalsis. Similar receptors in the upper esophagus are responsible for reflexly increasing the strength of the peristaltic contraction as the size of the bolus of food increases.

The esophagus also possesses stretch receptors that feed information directly into the intrinsic nervous system. This intrinsic pathway (not shown in Fig. 3) is thought to assist the extrinsic pathway in regulating the strength and timing of esophageal peristalsis, especially in the smooth muscle portion of the esophagus. If the vagus nerves are cut, all esophageal peristalsis stops. The striated muscle portion remains permanently paralyzed. But after several days the smooth muscle portion usually regains weak peristalsis in response to distention as a result of coordinated activity in the intrinsic nervous system.

DYSPHAGIA

Dysphagia is difficulty in swallowing. It can result from breakdowns in any of the stages of swallowing. One condition causing dysphagia is achalasia; in this disease contractions of the lower esophagus become very weak or absent altogether and do not progres peristaltically. Furthermore, the LES is hypertonic and fails to relax properly. Consequently, food is not properly propelled into the stomach, accumulating in the esophagus. Achalasia is caused by degeneration of the myenteric plexus and degeneration of certain efferent fibers in the vagus nerves.

VOMITING

Vomiting (emesis) is a highly coordinated reflex act involving paired vomiting centers in the lateral reticular formation of the medulla. Afferent impulses can arise from many sites, but most frequently are initiated by irritation of the oropharynx or the GI mucosa. Another common site of origin is the semicircular canals. Certain emetic drugs, such as apomorphine, act directly on a center in the floor of the fourth ventricle, called the chemoreceptor trigger zone, which in turn stimulates the vomiting centers. Other emetics cause vomiting by irritating the GI mucosa. Vomiting is usually preceded by nausea and excessive salivation. The saliva, mixed with air, can accumulate at the lower end of the esophagus, causing some distention. Signs of widespread sympathetic activity (e.g., tachypnea, tachycardia, mydriasis, sweating, and pallor) generally accompany vomiting. Parasympathetic inhibitory fibers cause the LES to relax.

The act of vomiting is initiated by a deep inspiration during which the diaphragm descends on the stomach. The glottis is then closed and active expiratory efforts take place that include very strong retching contractions of the abdominal muscles. These contractions greatly increase the pressure on the gastric and esophageal contents, which are consequently propelled out the mouth. During vomiting, the fundus and cardiac portion of the stomach, together with the esophagus and LES, are relaxed. However, the pyloric sphincter is constricted, and the antrum develops deep constrictions, thus impeding movement of material between stomach and duodenum. In severe vomiting, the pyloric sphincter may relax, permitting duodenal contents to be vomited. Reverse peristaltic contractions in the upper small intestine have been observed, though their importance is not known. It should be noted that propulsion of gastric contents toward the mouth is not driven by gastric contractions—the stomach serves a relatively passive role.

GASTRIC MOTILITY

Diagrams of the stomach are presented in Figure 4. The anatomical demarcation between body and antrum is useful for discussions of secretion since, as we shall see in Chapter 29, the secretions of the body and fundus are quite different from those of the antrum. For purposes of discussing motility, however, the stomach has two regions, proximal and distal. The distal stomach includes the distal part of the body as well as the antrum and pyloric sphincter. The dominant forms of motility and the purposes of these two regions are quite different. The proximal stomach is a receiving bin and storage vat; vigorous motility is minimal, and ingested food is stored un-

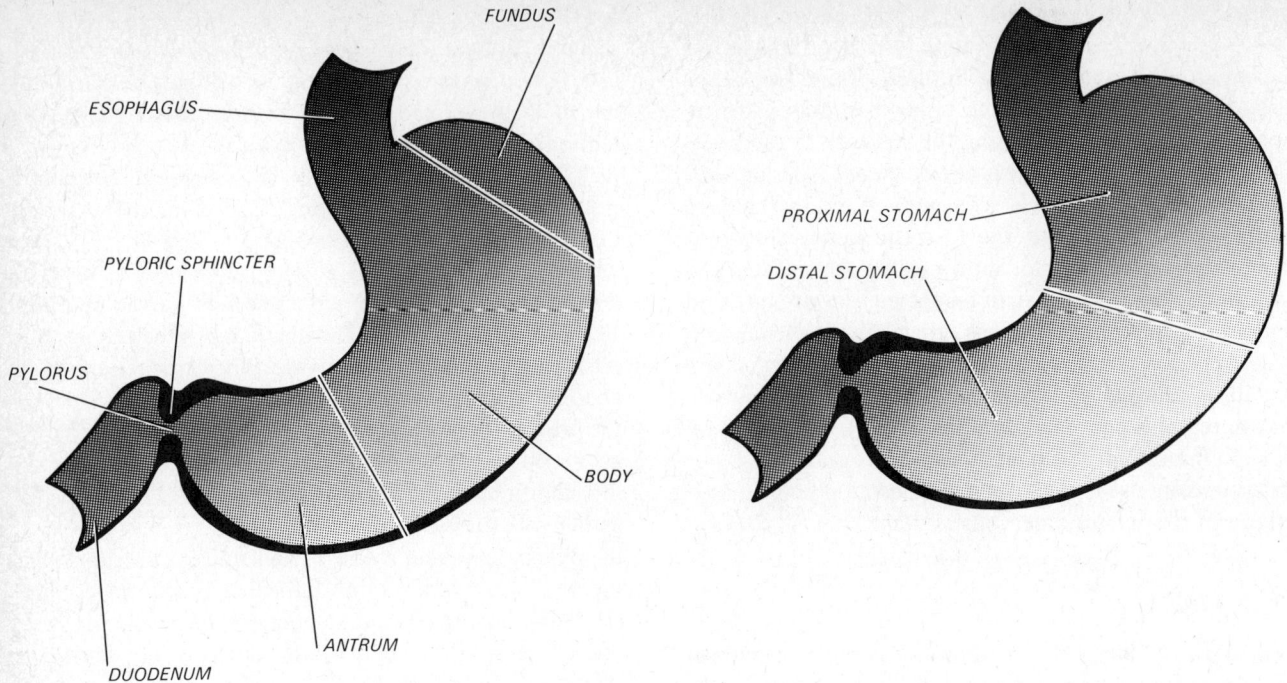

Figure 4 Diagrams of the stomach. Left, major anatomical divisions. Right, functional divisions with respect to motility.

mixed while it gradually enters the distal stomach. The distal stomach is a mixing chamber and pump; its strong peristaltic contractions mix food with gastric juice and churn the mixture into chyme, which is then gradually pumped into the duodenal bulb.

MOVEMENTS OF THE PROXIMAL STOMACH

The dominant form of motility in the proximal stomach is gentle and is called the tone contraction. A tone contraction is a slow increase in tension over a wide area of the gastric wall, which results in decreased capacity within that area and an increase in intragastric pressure. Tone contractions usually last about 1 min, but sometimes as long as 6 min; they recur at regular or irregular intervals. There can also be a background level of tonic activity in the smooth muscle on which tone contractions are superimposed. The electrical activity that induces and controls tone contractions has not been elucidated. Tone contractions in the proximal stomach help deliver ingested

material into the distal stomach and help regulate intragastric pressure.

MOVEMENTS OF THE DISTAL STOMACH

The dominant form of motility in the distal stomach is peristalsis. If an electrode is placed anywhere along the surface of the distal stomach rhythmic electrical activity can be recorded at a frequency of 3 cycles/min. The potentials that can be detected with an intracellular microelectrode were shown in Chapter 27; they consist of slow waves and spikes. Slow waves are omnipresent. Spikes are not always present, but when they are, a relatively strong contraction results. Gastric slow waves tend to have plateaus. The plateaus can be quite variable in both amplitude and duration. Slow waves with relatively large plateaus initiate weak contractions even in the absence of spikes; however, strong contractions occur only if spikes are present. All parts of the distal stomach are capable of generating slow waves. Furthermore, the electrical activity in one cell is coupled to the electrical activity in adjacent cells, so that the cells with the highest frequency of spontaneous activity establish the frequency for the entire distal stomach. The cells with the highest spontaneous frequency (i.e., the pacemakers) are located about midway down the body of the stomach along the greater curvature. Their frequency is three per minute in the human stomach. The rest of the distal stomach is driven at exactly the same frequency. Although the frequency of driven cells is the same as that of pacemaker cells, they are out of phase. Driven cells lag behind pacemaker cells, and the closer a cell is located to the pylorus, the greater the lag. Thus, a wave of depolarization passes from the body of the stomach along the antrum and into the pyloric sphincter.

When a slow wave having spikes passes along the distal stomach, a peristaltic contraction follows at the same velocity as the slow wave. The width of the contracting band is usually 1–2 cm. The strongest peristaltic contractions form indentations into the antrum that nearly occlude its lumen. As each electrical complex spreads along the antrum toward the pylorus, it speeds up. In fact, it often gets so fast in the terminal antrum and pyloric sphincter that the resulting contraction is nearly simultaneous throughout this region and is called the terminal antral contraction or antral systole. When slow waves that do not have spikes pass along the distal stomach, either a very weak peristaltic wave occurs or there is no contraction at all.

The pylorus is open except when a peristaltic wave reaches it. Therefore, for a few seconds during the early stages of each antral contraction, chyme is squirted through the pylorus into the duodenal bulb. However, even when the pylorus is open, its diameter is small; it thereby acts as a sieve to retain all but finely divided material. Propulsion of material through the pylorus is quite brief during a terminal antral contraction, since the sphincter closes before maximum pressure is reached in the antrum. With each terminal antral contraction, most of the antral contents are forced back into the proximal antrum through the aperture formed by the peristaltic wave. This retrograde movement of material is called retropulsion. The aperture through which antral contents are forcefully squirted is small during strong peristalsis. Consequently, during retropulsion, antral contents are vigorously ground and emulsified with gastric juice. This process triturates ingested material into chyme.

GASTRIC MOTILITY DURING INTERDIGESTIVE PERIODS

During interdigestive periods, when the upper GI tract is nearly empty, there is always a region along the stomach and small intestine that is undergoing vigorous contractions. A more general discussion of the interdigestive pattern of motility will be given later; for now, we need only point out that the distal stomach periodically becomes extremely active, with repetitively spiking (three per minute) electrical complexes and vigorous peristaltic waves. Between these periods of frantic activity the distal stomach is at first almost completely quiet (mechanically) and then gradually returns to full activity. These cycles repeat every 1.5–2 h, with the period of most intense activity lasting about 12 min. Recent evidence indicates that motilin is involved in initiating these moments of intense activity during interdigestive periods.

EFFECTS OF FOOD INGESTION

When the LES relaxes during swallowing, the proximal stomach also relaxes. Its background level of muscular tension diminishes, and tone contractions disappear. Also, if peristalsis was proceeding in the distal stomach, it stops. These events are called receptive relaxation. During repetitive swallowing, the LES and stomach relax and stay relaxed until after the final swallow. Receptive relaxation of the stomach lasts for 30 s or more; it is caused by activity in parasympathetic inhibitory fibers. Receptive relaxation of the stomach helps prevent regurgitation into the esophagus when the LES relaxes.

As food enters the stomach and its volume increases, the intragastric pressure increases very little. The stomach can be filled nearly to the bursting point with an increase in intragastric pressure of only a few millimeters of mercury. This property of the stomach is explained by the Laplace equation. If the stomach is assumed to be spherical, the Laplace equation tells us that

$$P_t = \frac{2T}{r}$$

where P_t is the transmural pressure, T is wall tension, and r is radius. If it is assumed that T increases as r increases, according to Hooke's law, one should expect that P_t

would be fairly constant over a wide range of r. We must point out, however, that some investigators believe there is actually a relaxation of the smooth muscle in the proximal stomach as it expands, which they refer to as gastric accommodation.

GASTRIC MOTILITY DURING DIGESTIVE PERIODS

If a meal is ingested suddenly (e.g., a hungry dog gobbling down a bowl of food), the stomach remains quiet for roughly 30 s to several minutes after completion of the meal, after which gastric contractions characteristic of the digestive period begin. If a large meal is ingested leisurely (e.g., by a civilized person), it is not known when the digestive phase of motility begins. In any event, at some point during or after a meal, tone contractions in the proximal stomach return, and peristaltic contractions in the distal stomach begin. The return of motility is in response to distention of the stomach and stimulation of stretch receptors. These initiate local intrinsic reflexes and vagovagal extrinsic reflexes, resulting in reduced activity in intrinsic inhibitory neurons and increased activity in intrinsic excitatory neurons. Gastrin and cholecystekinin (CCK) might also be involved in promoting increased excitability in the distal stomach, but this is uncertain.

The digestive period of gastric motility is characterized by peristaltic waves in the distal stomach, which continue to occur at the three-per-minute rhythm throughout the digestive period. Gone are the cycles of activity and quiescence characteristic of the interdigestive period; these are replaced by nearly continuous activity that gradually becomes stronger and more regular during the first hour or two, and then gradually subsides. After roughly 3–6 h, when the stomach is about empty, the interdigestive cycles of activity return. Peristalsis during digestive periods is seldom as powerful as it is during interdigestive periods.

GASTRIC EMPTYING

Gastric contents move through the pylorus into the duodenum at a rate proportional to the pressure difference between these two organs and inversely proportional to the resistance to flow through the pylorus. The resistance for large chunks is very high, and they are seldom emptied

from the stomach. The resistance for ingested liquids is very low right from the start, and they are emptied rapidly. In fact, if water is drunk on top of a solid meal, it tends to flow along the surface of the gastric mucosa around the food in the body and antrum, and rapidly on through the pylous, being driven principally by the intragastric pressure established by tone contractions of the proximal stomach. Solid foods empty more slowly, since they must first be chymified and also because the viscosity of chyme is much greater than that of water. Eventually, however, solid foods are chymified and driven through the pylorus by the pressure established by tone contractions and by antral peristalsis.

One of the principal functions of the stomach is to control the rate at which materials enter the duodenum. If gastric emptying were to occur too fast, the ability of the duodenal mucosa to cope with such physicochemical problems as acidity and hypertonicity might be exceeded, causing damage to the mucosa. Excessive distention of the duodenum can also be a problem. A person who has undergone a gastrectomy usually must eat very small meals; otherwise, too much material is dumped into the duodenum all at once, resulting in many unpleasant symptoms (e.g., pain, nausea, vomiting, faintness), collectively known as the "dumping syndrome."

The rate of gastric emptying is governed to some extent by the physical consistency of the food and its volume. It is also governed by some very important but, unfortunately, poorly understood negative feedback mechanisms that originate in the duodenum. These feedback influences from the duodenum involve both nervous and hormonal mechanisms. The nervous mechanism is called the enterogastric reflex. Little is known about it, except that it apparently uses all three types of reflex pathways: intrinsic, prevertebral, and central nervous system (CNS). Of these, the prevertebral pathway seems the most important. The hormonal mechanism uses CCK. The receptors in the duodenum that initiate negative feedback are of at least three types, responding to H^+, hypertonicity, and long-chain fatty acids, respectively. Fat in the duodenum is the most powerful inhibitor of gastric emptying; it probably acts mainly via CCK, but the enterogastric reflex might also be involved. The inhibitory effects of duodenal H^+ and hypertonicity are probably mediated mainly by enterogastric reflexes, but CCK might also be involved.

Feedback inhibition of gastric emptying can take place in four ways: (1) the proximal stomach can relax, thereby reducing intragastric pressure; (2) peristalsis in the distal stomach can get weaker, thereby reducing the rate of chymification and the vigor of the antral pump; (3) the pyloric sphincter and terminal antrum can constrict, thereby increasing the resistance to flow from stomach to duodenum; and (4) duodenal motility can increase, thereby increasing intraduodenal pressure. Unfortunately, it is not yet possible to state which of these mechanisms is most important in any particular circumstance. Distention of the duodenum with chyme also inhibits gastric emptying, but whether this effect involves reflex and hormonal pathways, or is simply a mechanical effect of the increased intraduodenal pressure is not known.

MOTILITY OF THE SMALL INTESTINE

Gastrointestinal motility exhibits two striking features: rhythmicity and polarity. The rhythmicity of contractions is myogenic and depends on rhythmically recurring electrical slow waves. The polarity is such that chyme, located anywhere in the gut, has a much greater tendency to move in the aboral direction than in the oral direction. Polarity is also principally myogenic and depends on the direction in which slow waves are usually propagated.

Smooth muscle all along the small intestine is capable of generating rhythmic slow waves. Each region, when separated from other regions, has a unique slow wave frequency that diminishes as the gut is descended. But normally the cells are electrically coupled so that higher frequency regions tend to bring more distal regions up to their pace. The pacemaker for the duodenum and upper jejunum is located in the proximal duodenum; it controls the pace of this entire stretch of gut at a frequency of about 12 waves/min. More distal regions of gut cannot maintain this duodenal pace and consequently keep slower rhythms. Along the gut there are stretches of inconstant length, having progressively slower pacesetter rhythms, until finally in the lower ileum the frequency is about eight per minute. Over these stretches of gut the slow waves usually travel aborally at a velocity of roughly 1–4 cm/s. Occasionally, they travel in the oral direction. Intestinal slow waves do not themselves initiate contractions, but if membrane depolarization reaches threshold, a burst of action potentials is triggered. These spike bursts initiate contractions with strength determined mainly by the number of spikes per burst. The relationship among slow waves, spikes, and contractions is illustrated in Figure 5. Contraction occurs simultaneously in both muscle layers, but the circular muscle usually dominates, and the result is a circumferential constriction that partially or completely closes the lumen of the gut. Each constriction lasts a few seconds.

Sometimes the smooth muscle along a stretch of gut is

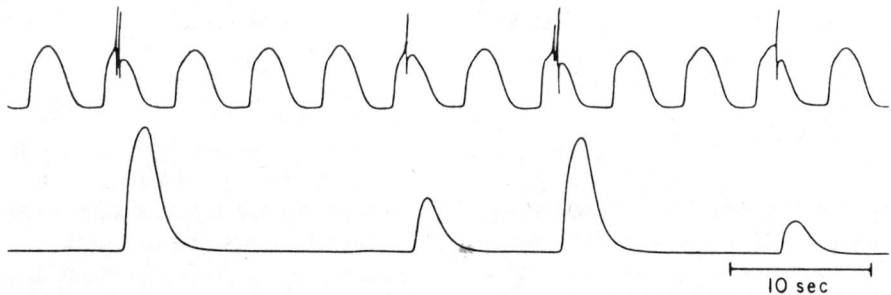

10 sec

Figure 5 Relationship among slow waves, spikes, and contractions in cat jejunum. Top, electrical activity recorded from a "pressure electrode," which picks up activity from a fairly large group of cells simultaneously. Bottom, mechanical activity of approximately this same group of cells recorded from a force transducer. Vertical calibrations are not important here. (From Bortoff A: Myogenic control of intestinal motility. *Physiol Rev* 56:418–434, 1976.)

not uniformly excitable, and any given slow wave might not initiate spiking and contraction continuously along its entire course. Instead, the contraction might skip along the gut from one responsive site to the next. The constriction at each site is about 1–2 cm wide. Sometimes a given site remains responsive during a succession of slow waves, constricting rhythmically at the slow wave frequency. At other times the responsive sites change from one slow wave to the next resulting in rather randomly spaced constrictions.

Often, however, a slow wave is successful in inducing a continuous constriction along a stretch of gut, which results in a fast peristaltic wave, traveling at the velocity of the slow wave. Fast peristaltic waves usually progress aborally about 5–40 cm, urging chyme ahead. Occasionally, fast peristaltic waves move in the oral direction for several centimeters; this event is known as reverse peristalsis. Sometimes weak peristaltic ripples travel alternately up and down the gut for short distances. This pattern of motility has been called a pendular movement.*

Occasionally a set of strong simultaneous constrictions occurs evenly spaced over a fairly long stretch of gut, dividing the contained chyme into a series of short sausage-like segments. As these sites relax a new set of constrictions appears, each one approximately halfway between the previous ones, and the chyme is redivided into new segments. This process is rhythmically repeated at a frequency, in humans, of about 7 cycles per minute, sometimes for hundreds of cycles, dividing and redividing the column of chyme. This extraordinary process was discovered in 1902 by Cannon, who named it, appropriately, rhythmic segmentation.† Figure 6 presents a diagram of the process. No one yet has satisfactorily explained how the frequency and velocity of slow waves can account for the temporal and spatial synchronization of rhythmic segmentation.

---

*The term pendular movement originally meant any rhythmic contraction of the gut (Carl Ludwig, 1861) but has since been applied ambiguously to a variety of vermiform intestinal contortions including the to-and-fro peristaltic ripple described here. The term should probably be abandoned.

†More recent workers have often applied the term rhythmic segmentation to *all* patterns of rhythmic constriction (other than peristalsis), rather than just to this highly coordinated segmenting pattern discovered by Cannon. This practice has led to considerable confusion. Here we shall reserve the term for its original intent.

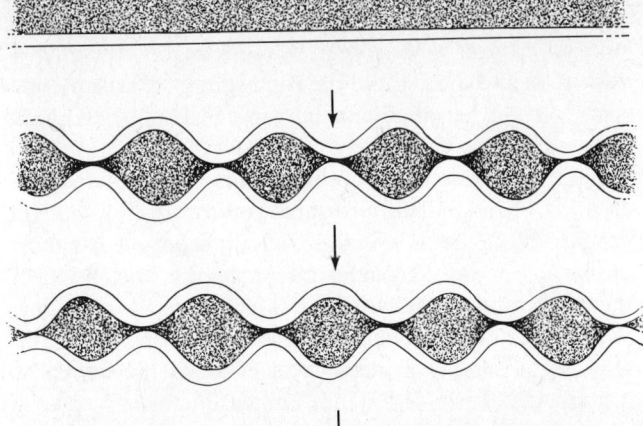

Figure 6 Diagram of rhythmic segmentation. Shaded areas represent chyme.

Another well-coordinated pattern of rhythmic activity is a discontinuous succession of stationary constrictions, each one located a little farther along the gut than its predecessor. This is a common pattern of motility in the upper small intestine, probably much more common than true rhythmic segmentation, and is thought to help propel chyme aborally. The manner in which slow waves coordinate these sequential rhythmic constrictions is not understood.

Thus, we see that rhythmic constrictions can be stationary or progressive and can sometimes occur with amazing spatial and temporal synchronization. These constrictions spread and propel chyme along the small intestine, mix chyme with pancreatic juice and bile, keep absorbable materials available for absorption, and rhythmically plunge the absorptive epithelium into the luminal mass of chyme.

The small intestine is capable of another type of contraction, the "slowly advancing peristaltic wave." This contraction appears as a strong, broad band of constriction that very slowly progresses aborally at a velocity of about 2 cm/min. Investigators of the nineteenth and early twentieth centuries thought this was the major form of propulsive motility. In 1899 Bayliss and Starling discovered that when a bolus of cotton was inserted into dog small intestine, the smooth muscle for several centimeters above the stimulus was promptly excited, and the smooth muscle for an even longer distance below the stimulus was inhibited. They dubbed this response the "law of the intestine." In 1912, Cannon called it the myenteric reflex. At first, in the region above the bolus, rhythmic constrictions at the slow wave frequency increase in regularity and vigor; then, if the stimulus is strong, these contractions fuse into a spasm that completely closes the lumen. The bolus is squeezed aborally into the relaxed region below. Irritation and distention at progressively aboral sites cause a wave of excitation and constriction to move aborally, driving the bolus ahead. Progress is very slow, the velocity being only about 2 cm/min. These waves progress for distances up to 20 cm or so and then die out. It is reasonable to suppose that slow waves continue to be involved in stimulating the smooth muscle and that they find the region above the bolus highly excitable and the region below it very inexcitable, a situation established by the intrinsic nervous system. The factors controlling velocity of this slow peristaltic wave are not understood but could involve the rate at which proximal excitation can overcome the previous inhibition at any site along the myenteric plexus. With modern methods of investigation, which are able to record intestinal movements under more normal circumstances than could the earlier investigators, these strong, slowly progressing peristaltic waves are seldom observed. Perhaps they are reserved for emergencies, when the gut is in danger of obstruction by clumps of accumulated material.

INTESTINAL MOTILITY DURING INTERDIGESTIVE PERIODS

During interdigestive periods, when the tract is relatively empty of chyme, mechanical activity tends to be localized. There is always some region in which intense activity is taking place—in which virtually every slow wave induces spike bursts throughout the entire length of that region, resulting in rhythmic peristaltic contractions at the slow wave frequency. This band of intense activity begins simultaneously in the distal stomach and duodenum, and then slowly migrates along the tract all the way to the terminal ileum. The band of most intense activity is preceded by a somewhat longer band of less intense and less regular activity and is followed by a long band of relative inactivity. In the region of inactivity, electrical slow waves continue but seldom initiate spikes and contractions. This migrating complex of electrical and mechanical activity has various names*; we will use "migrating motor complex" and abbreviate it MMC. Just as one MMC arrives in the lower ileum, another begins in the distal stomach and duodenum, after which it migrates to the terminal ileum. The entire migration takes about 2 h. The MMC cycles repeat unchanged for as long as fasting continues. The length of the band of most intense activity is roughly 40 cm in the duodenum, but it diminishes as the gut is descended to about 10 cm in the lower ileum. The duration of most intense activity is about 12 min in the stomach and 5 min in the small intestine.

The MMC is the "interdigestive housekeeper" of the stomach and small intestine. It cleans the tract of any residues left from the previous meal and keeps interdigestive secretions and sloughed epithelium moving along. Were it not for MMCs, interdigestive secretions would stagnate in the small intestine and bacteria would multiply. There are patients in whom MMCs are deranged or absent—these people suffer from overgrowth of bacteria in the small intestine.

The mechanisms responsible for turning on each MMC at just the right time in the stomach and duodenum and for making it migrate down the tract are not well understood and are currently under intensive investigation. MMCs appear to be initiated in the stomach and duodenum by motilin. During interdigestive periods, the concentration of motilin in blood changes cyclically with a period equal to the MMC period (1.5–2 h). Each time the motilin concentration begins to rise, contractions begin in the stomach and duodenum; gradually increasing in regularity and vigor until, just when the motilin concentration is at its peak, the period of most intense contractions occurs. Then, as the motilin concentration subsides, so do the contractions. This relationship is illustrated in Figure 7. Secretion of motilin from the duodenal mucosa seems to occur in response to cholinergic intrinsic excitatory neurons.

Extrinsic denervation does not influence the generation

---

*Interdigestive myoelectric complex, interdigestive migrating complex, migrating motor complex, migrating myoelectric complex, and migrating motility complex.

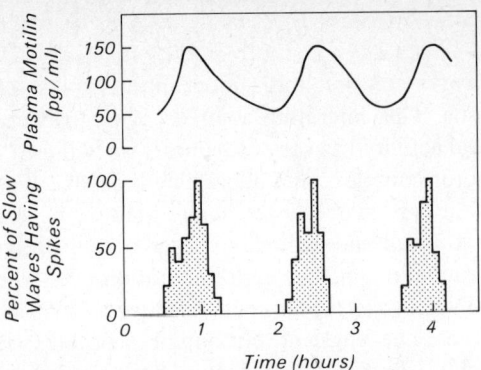

Figure 7 Relationship between plasma immunoreactive motilin concentration and contractions in dog duodenum following an 18-h fast. Contractile activity was recorded as the percentage of slow waves having spike bursts during each 5-min period. (Based roughly on the data of Lee KY, Kim MS, Chey WY: Effects of a meal and gut hormones on plasma motilin and duodenal motility in dog, in *Am J Physiol* 238:G280–G283, 1980.)

of MMCs, so apparently the intrinsic nervous system itself can time the secretion of motilin. The intrinsic nervous system also has the ability to control the migration process to the terminal ileum without the help of motilin, the extrinsic nerves, or the CNS. It appears, however, that the CNS, via sequential firing in parasympathetic neurons, can also take care of the migrating process. This functional redundancy is indicated by the fact that migration of motor complexes usually, but not always, continues unabated across transections that destroy continuity of the intrinsic nervous system.

EFFECT OF FEEDING AND THE DIGESTIVE PATTERN OF MOTILITY

When feeding begins, all interdigestive housekeeping immediately stops, and the stomach and small intestine become mechanically quiet along their entire length. The promptness of this response indicates that it must be a neural reflex, but the pathway(s) is (are) not understood. Inhibition of MMCs sometimes occurs even before the first bite of food is taken, showing that sight and smell or even anticipation can initiate the reflex. The efferent limb of the reflex probably involves increased activity in parasympathetic inhibitory neurons. The tract may remain quiescent for just a few seconds or for as long as 2 min. Then a pattern of motility characteristic of digestive periods begins. The digestive pattern of motility is quite different from the interdigestive pattern. Activity is no longer localized to a particular region of gut. Instead, the entire small intestine becomes active with more or less randomly spaced rhythmic constrictions, both stationary and progressive. These constrictions are seldom as strong and seldom progress as far as those that occur during interdigestive periods, but they continue along the entire small intestine for the duration of the digestive period. Occasionally, bouts of true rhythmic segmentation take place here and there, as well as episodes of sequential rhythmic constrictions.

During the most intense period of each interdigestive MMC, every slow wave induces spikes and a continuous contraction along an entire stretch of gut. During the digestive period, the responses are much less continuous, that is, they tend to be either stationary constrictions or relatively short peristaltic waves. If a recording is made early during the digestive period from a single site in the duodenum, it is found that about 60% of the slow waves produce a contraction at that site. In the upper jejunum the response rate is somewhat higher. As digestion and absorption proceed, the response rates diminish until finally, when the small intestine is about empty, the response rate might be only about 20–30%. Then a new MMC suddenly begins, and the digestive pattern stops. The first MMC usually starts in the jejunum rather than in the stomach and does not depend on motilin.

The mechanisms involved in initiating and controlling the digestive pattern of motility are not understood. There is evidence for both neural and hormonal involvement. The neural mechanism probably involves parasympathetic excitatory fibers; currently the most likely candidate for the hormonal mechanism is CCK.

THE ILEOCECAL JUNCTION

Protruding flaps composed of mucosa, submucosa, and circular muscle form the ileocecal valve. This valve resists rather large pressures in the cecum and prevents flow of cecal contents into the ileum. The wall of the terminal ileum has a thickened muscular coat called the ileocecal sphincter, which is normally slightly contracted and slows the passage of chyme into the cecum. Sphincter tone is increased by impulses over the sympathetic nerves. This response is mediated by α-receptors, either on the smooth muscle cells themselves or perhaps on excitatory intrinsic neurons. Sphincter tone is also increased when the pressure in the cecum rises or when there is chemical irritation of the cecum. The sphincter relaxes as chyme is propelled into the terminal ileum; however, it still offers significant resistance to the aboral movement of chyme. Surgical removal of the ileocecal junction causes chyme to be emptied from the small intestine so rapidly that malabsorption can occur.

MOTILITY OF THE LARGE INTESTINE

The large intestine consists of the cecum, the colon (ascending, transverse, descending, and sigmoid portions), the rectum, and the anal canal. In the large intestine, chyme is converted to feces by absorption of water and electrolytes and by the action of bacteria. The feces are stored for a time and are then eliminated. There are two functionally important anatomical features of the cecum and colon: the taenia coli and the sacculations, or haustra. The haustra, formed by deep indentations of the wall into the lumen, give the cecum and colon a beaded appearance. These indentations are puckers caused by the shortness of the taenia coli; they disappear if the taenia are cut. Within each indentation, the circular muscle is thicker than elsewhere and, when it contracts, the depth of the indentations increases. The haustra are not permanent structures; they subside or even disappear when the muscles relax.

By far the most common form of motility in the colon is haustration. The interhaustral folds constrict, sometimes to the point of nearly occluding the lumen. The haustral wall also contracts, causing increased pressure in the haustral lumen. Each contraction lasts anywhere from 12 to 60 s. Several such contractions sometimes succeed each other in rhythmic sequence. Their frequency is quite variable, but averages about two per minute. These constrictions are not propulsive; in fact, their main function might be to increase the resistance to flow along the colon. This type of motility is usually exaggerated during constipation and suppressed during diarrhea. Haustration contributes to the mixing of colonic contents.

A very dramatic form of motility in the colon is the mass movement—a massive propulsion of a long column of fecal material analward through the colon. Mass movements occur at infrequent intervals, usually following a meal. As seen radiologically, the haustra disappear over a long segment of colon; that segment then narrows, forcing a column of fecal material into the next segment of colon, which has also lost its haustra. A mass movement takes only a few seconds and is followed by a prompt return of haustra and a gradual return to normal diameter of the segment. A mass movement, or a succession of mass movements, propels fecal material into the sigmoid colon and rectum. Mass movements themselves produce no sensation, but when a mass movement has carried considerable material into the rectum, distention of the rectum can bring the matter to consciousness.

Peristalsis occurs only rarely in the human colon, although it might occasionally occur in the descending colon, assisting the movement of feces into the rectum. The taenia coli sometimes gradually increase their tone over a period of several hours, causing a long segment of large intestine to shorten. Slow changes in tone of the circular muscle lasting up to 15 min also occur. Tone changes might be important in propulsion if the resistance to flow is low.

During fasting, the large intestine is quiescent most of the time. During and following eating, there is greatly increased segmenting activity of the haustra, and often a mass movement. These responses to feeding are called the gastrocolic segmenting reflex and the gastrocolic propulsive reflex, respectively. The afferent limb of the gastrocolic reflexes is in the vagus nerves from the stomach and duodenum. The efferent limb is in sympathetic nerves to the colon and apparently involves cholinergic sympathetic fibers. Which of these reflexes occurs seems to depend on how full the cecum and colon are; if they are already distended with feces, the gastrocolic propulsive reflex is likely to occur; otherwise, the gastrocolic segmenting reflex is more likely. Some investigators believe that these colonic responses to feeding are partially mediated hormonally, the most likely candidate being CCK.

DEFECATION

About 1 L of chyme per day passes the ileocecal junction into the large intestine. By colonic absorption of water, this volume is reduced to about 150 ml feces per day. The feces consist of food residues (mainly indigestible material, such as cellulose), the unabsorbed remains of intestinal secretions, desquamated epithelial cells, and bacteria.

Defecation is a combination of an involuntary reflex act and a voluntary act. The reflex portion of defecation is initiated by distention of the rectum. When a sufficient volume of feces has been moved into the rectum, stretch receptors are excited that reflexly induce (1) relaxation of the haustral indentations in the descending colon; (2) a mass constriction and shortening of the entire descending colon, sigmoid colon, and rectum; and (3) relaxation of the internal anal sphincter. This reflex is carried over the intrinsic nervous system and also over the extrinsic nerves. Afferents from the rectum carry impulses to the spinal cord. There is a spinal defecation center in the second, third, and fourth sacral segments. Efferent impulses pass from the spinal cord to the distal part of the colon, the rectum, and the internal anal sphincter via the pelvic nerves. Defecation then occurs when the external anal sphincter is voluntarily relaxed and the abdominal muscles and diaphragm are contracted. If it is not convenient to defecate at this time, the act can be prevented by voluntary contraction of the external anal sphincter. The rectum adjusts its capacity by diminished muscular tone and the reflex may stop. Even if the spinal reflex is blocked (e.g., by injury to the spinal cord or the pelvic nerves), defecation can still take place as long as the intrinsic plexuses are intact.

EXTRINSIC GASTROINTESTINAL MOTOR REFLEXES

Although the GI tract enjoys a considerable degree of autonomy with its myogenic, intrinsic neural, and hormonal control mechanisms, a number of important reflexes operate via extrinsic neural pathways. In some cases these pathways seem to provide for strengthening the intrinsic responses or for functional redundancy. An example already given is the reflex increase in excitability of the distal stomach when it is distended with food. This response can be mediated by the intrinsic nervous system, but it is augmented by a vagovagal reflex. Similar examples are the enterogastric reflex and the defecation reflex. Initiation and migration of MMCs seem to involve redundant intrinsic and extrinsic neural pathways, whereas conversion to and maintenance of the digestive pattern of motility may require both extrinsic neural and hormonal pathways. It has also been suggested that CCK augments colonic motility during gastrocolic reflexes.

However, other extrinsic reflexes seem to stand alone. One of these is elicited by moderate distention at any site along the tract. This procedure, in addition to sometimes inducing the myenteric reflex, also sometimes results in inhibition well above the stimulus and excitation well below the stimulus. These responses are mediated by a vagovagal reflex called the enteroenteric reflex. It has been suggested that this reflex is a traffic control mechanism. If too much material begins to accumulate at some site, progress of chyme is halted above and accelerated below the site of congestion.

Very strong distention of a segment of intestine, enough to be painful, causes reflex inhibition along the entire small intestine. This reflex is known as the intestinointestinal inhibitory reflex. The efferent impulses travel over inhibitory sympathetic fibers, and the afferent impulses travel over sensory fibers that run in the same nerve trunks as the sympathetic fibers. The major relays for this reflex are probably in the prevertebral sympathetic ganglia. The distention resulting from intestinal obstruction elicits this reflex.

Trauma and irritation of abdominal contents, including the peritoneum, induce reflex inhibition in the small and large intestines. The inhibitory impulses are carried over sympathetic nerves to the gut. In addition, the adrenal medullae are stimulated to secrete catecholamines, which further inhibit the gut. Profound intestinal relaxation from this cause is frequently observed during and after abdominal surgery and can lead to functional obstruction of the gut. The condition is known as adynamic ileus. Since intestinal contents are not propelled, they accumulate and cause distention. Unless the distention is relieved (by aspiration), the patient might die. The most important cause of adynamic ileus is peritonitis, but a large variety of abdominal and extra-abdominal injuries, infections, and diseases can also produce this condition.

GAS IN THE ALIMENTARY TRACT

The student should be armed with the proper vocabulary.

Flatus: Gas in the stomach and intestine.

Flatulence: Distention by an abnormal volume of flatus.

Eructation: Belching.

Borborygmus: The noise produced by movement of gas within the GI tract.

Crepitus: A noisy discharge of gas from the rectum (note, however, that this term is rarely used).

About 50 cc of gas is normally found in the fundus of the stomach, originating from swallowed air contained in frothy saliva or fluffy foods. Some of the swallowed air is eructed, some is passed into the small intestine, and some remains in the stomach. Borborygmi can often be detected by stethoscope at a frequency of three per minute as gas is pumped through the pylorus. Not much gas is ordinarily present in the small intestine, because it passes through very rapidly. The volume of gas entering the small intestine from the stomach is augmented by formation of CO_2 in the duodenum (from gastric H^+ and pancreatic and biliary HCO_3^-). As gas passes through the small intestine, it produces borborygmi at a frequency of 8–12 per minute. A so-called silent abdomen is a sign of adynamic ileus and often occurs for 1–3 days following abdominal operations and during peritonitis. About 100 cc of gas is normally contained in the large intestine. About half this volume is formed in the colon by bacterial action and includes methane, hydrogen, and hydrogen sulfide. Colonic gas is therefore combustible. Intermittently, colonic flatus is expelled through the anus. The volume of flatus passed from the colon of the normal male medical student has been measured and found to average 408 cc/day. After a 2-week caloric intake consisting of 25–50% beans, average output rose to 4,872 cc/day.

SELECTED BIBLIOGRAPHY

Christensen J: *Gastrointestinal Motility.* New York, Raven Press, 1980. (Note: This book records the Seventh International Symposium on Gastrointestinal Motility. Other volumes in this series are recommended as well.)

Friedman MHF: *Functions of the Stomach and Intestine.* Baltimore, University Park Press, 1975.

SELF-STUDY QUESTIONS

MULTIPLE CHOICE

Select the best answer(s). (In many instances, more than one answer is correct.)

1. The UES normally
   A. remains tonically contracted when swallowing is not in progress.
   B. relaxes during the pharyngeal stage of swallowing.
   C. produces a pressure above atmospheric within the sphincter region while swallowing is not in progress.
   D. is composed principally of the cricopharyngeus muscle.
   E. prevents ventilation of the thoracic esophagus during breathing.
2. The LES
   A. remains tonically contracted when swallowing is not in progress.

B. relaxes during the pharyngeal stage of swallowing.

C. produces a pressure above atmospheric while swallowing is not in progress.

D. fails to relax properly in patients with achalasia.

3. During the swallowing reflex

A. sequential activity in nerve fibers from the nucleus ambiguous results in peristalsis in the upper part of the esophagus.

B. the pressure within the LES decreases while the peristaltic contraction is still located mainly in the upper esophagus.

C. there is reduced activity in efferent nerve fibers from the dorsal motor nucleus of the vagus.

D. intrinsic inhibitory neurons are definitely not activated.

4. Important mechanical features of vomiting include

A. contraction of the diaphragm.

B. reverse peristalsis in the stomach.

C. contractions of abdominal muscles.

D. maintenance of an open glottis.

5. Prominent features of motility in the proximal stomach include

A. vigorous peristalsis.

B. tone contractions.

C. retropulsion and chymification.

D. haustration.

6. Peristalsis is a prominent form of motility in the

A. esophagus.

B. proximal stomach.

C. distal stomach.

D. small intestine.

E. colon.

7. Fast peristaltic waves in the small intestine

A. usually travel in the oral direction.

B. travel at the same velocity as slow waves.

C. depend on sequential firing of parasympathetic preganglionic neurons.

D. travel at a velocity of about 2 cm/min.

8. The "slowly advancing peristaltic wave"

A. depends on a vagovagal reflex.

B. depends on an intrinsic reflex, called the myenteric reflex.

C. is normally the most important form of propulsive motility in the small intestine.

D. is not related to myogenic myoelectrical activity (slow waves).

9. Generation of MMCs in the stomach and duodenum

A. is usually related to secretion of motilin by the duodenal mucosal.

B. probably depends on activity in cholinergic intrinsic excitatory neurons.

C. always precedes MMCs in the jejunum.

10. The intestinointestinal inhibitory reflex

A. is elicited by strong distention of the small intestine.

B. is part of the myenteric reflex.

C. is mediated by an increase in efferent impulses traveling over sympathetic nerve fibers.

D. is involved in normal digestion.

11. Functional redundancy involving simultaneous extrinsic and intrinsic neural pathways is illustrated by

A. the enterogastric reflex.

B. the intestinointestinal inhibitory reflex.

C. the myenteric reflex.

D. the defecation reflex.

E. the enteroenteric reflex.

TRUE OR FALSE

12. The frequency of slow waves in the duodenum is about four times that in the distal stomach.

13. Receptive relaxation of the stomach is mediated by sympathetic adrenergic fibers.

14. True rhythmic segmentation occurs at the same frequency as slow waves.

15. At the peak of each MMC anywhere along the small intestine, fast peristaltic waves occur at the slow wave frequency.

16. The efferent limb of the gastrocolic reflexes uses parasympathetic cholinergic fibers.

17. List four stimuli, acting in the duodenum, that result in inhibition of gastric emptying.

ANSWERS

1. All are correct.
2. A, C, D. B is incorrect, since the LES does not relax until after the esophageal stage of swallowing begins.
3. A and B are correct. There is *increased* activity in fibers from the dorsal motor nucleus of the vagus. One theory holds that this activity excites intrinsic inhibitory fibers and that a rebound contraction progresses along the smooth muscle portion of the esophagus as a peristaltic wave. Almost certainly, intrinsic inhibitory fibers are involved in relaxation of the LES.
4. A, C. Reverse peristalsis in the stomach is thought not to be important. The glottis is closed during the actual retching movements.
5. Vigorous peristalsis, retropulsion, and chymification are features of the distal stomach. Haustration is a property of the large intestine.
6. A, C, D. In the proximal stomach, the prominent form of motility is the tone contraction. In the colon, haustration and mass movements dominate the scene.
7. B. Fast peristaltic waves in the small intestine usually travel aborally (anally). They are not synchronized by activity in extrinsic neurons, but rather by myogenic mechanisms; 2 cm/min is the velocity of slowly progressing peristaltic waves; fast peristaltic waves travel about 30–120 times this velocity.
8. B. Extrinsic nerves are not involved. Currently, the slowly advancing peristaltic wave is thought not to be especially important under normal circumstances.
9. A, B. Current evidence makes it appear likely that secretion of motilin is governed by activity in intrinsic excitatory neurons. Sometimes MMCs, especially the first one or two at the beginning of an interdigestive period, originate in the jejunum rather than in the stomach and duodenum; these do not seem to be related to secretion of motilin.
10. A, C.
11. A, D.
12. True. Twelve per minute in duodenum, three per minute in the distal stomach. Incidentally, this same ratio seems to hold in other animals, even though the frequencies might be different. For example, in dogs gastric frequency is slightly under five per minute, whereas duodenal frequency is 19 per min.
13. False. It is mediated by preganglionic parasympathetic cholinergic fibers acting on intrinsic inhibitory neurons, the mediator of which is unknown.
14. False. This seems to be a common misconception found in some textbooks, no doubt the result of confusing true rhythmic segmentation with ordinary rhythmic constriction. In humans the frequency is about seven per minute.
15. True.
16. False. It uses *sympathetic* cholinergic fibers.
17. H^+, long-chain fatty acids, hyperosmolarity, distention.

29

Secretions of the Digestive System

R. DAVID BAKER

OBJECTIVES

After completion of this chapter, you should be able to:

1. Know the terminology.
2. Know the events.
3. Know the usefulness (functions) of the various secretions.
4. Know the control mechanisms.
5. Understand, to the very limited extent presented here, the mechanisms of secretion of digestive juices.

SECRETION OF SALIVA

SITES OF SECRETION

The salivary glands contain two types of cells: serous and mucous cells. The secretory cells of the parotid gland are entirely of the serous type, those of the submaxillary gland are of both serous and mucous types, and those of the sublingual gland are predominantly of the mucous type. The serous cells secrete a thin, watery fluid containing salivary amylase (serous means thin or watery). The mucous cells secrete a thick, highly viscous fluid rich in mucins.

VOLUME AND COMPOSITION

In the absence of reflex stimuli, the total secretion of saliva in humans is only about 0.25 ml/min, which would amount to a daily output of about 360 ml. But the daily output is actually about 1.0–1.5 L, which indicates that a considerable increase in secretion occurs when food is ingested. A maximally active parotid or submaxillary gland can secrete its own weight of saliva within 20 min. The pH value of mixed saliva is about 6.5 and is maintained close to this level by the buffering action of bicarbonate, phosphate, and proteins. In addition to these anions, saliva also contains Cl^-. Cations include Na^+ and K^+.

Organic substances include the mucins (a variety of mucopolysaccharides and glycoproteins), salivary amylase, and another enzyme called lysozyme. Lysozyme has a bactericidal action—it destroys certain microorganisms. Human saliva is hypotonic with respect to blood plasma.

FUNCTIONS OF SALIVA

Digestion of Carbohydrate

Salivary amylase continues to digest starch and glycogen after food enters the stomach, even though it would be inactivated by a low pH. This digestion is made possible by the poor mixing between gastric juice and swallowed food in the body and fundus of the stomach.

Aid in Formation of Bolus During Swallowing

Bolus formation requires that the food material be wet. The bolus must also be given a lubricant coating of saliva before it can be swallowed.

Solvent Action for Taste

Solid substances must be dissolved before they can stimulate the taste buds.

Cleansing and Bactericidal Action

A constant flow of saliva is necessary to keep the mouth comparatively free of food residues, shed epithelium, and other substances. In this way, saliva indirectly inhibits growth of bacteria by removing material that would otherwise serve as culture media. In addition, by virtue of its lysozyme content, saliva is directly bactericidal.

Moistening and Lubricating Action

The moistening and lubricating action of saliva on the surface of the buccal cavity is essential to normal speech.

Excretion

Heavy metals, such as mercury and lead, are excreted in the saliva in rather large amounts after entry into the body. Iodide is actively transported into saliva by the tubular epithelium. In chronic nephritis, saliva contains a high concentration of urea. Several types of virulent microorganisms (e.g., the viruses of rabies and poliomyelitis) are excreted in saliva.

Aid in Regulation of Water Balance

Any condition tending toward general body dehydration, such as excess sweating or diuresis or insufficient water intake, results in a markedly decreased output of saliva, with a resulting dry mouth. A dry mouth arouses the sensation of thirst.

Buffering Action

Saliva has a considerable buffer capacity for acids and tends to keep the mouth contents at a surprisingly uniform pH. This buffering action of saliva within the bolus of food after it is swallowed may be important in prolonging the action of salivary amylase for some time within the stomach.

REGULATION OF SECRETION

Nervous

Parasympathetic fibers from the inferior and superior salivatory nuclei run in the chorda tympani branch of the facial nerve (VII) and the tympanic branch of the glossopharyngeal (IX). The salivary glands are also supplied by fibers from the sympathetic system via the superior cervical ganglion. Stimulation of the chorda tympani nerve produces an abundant secretion from the submaxillary gland, which is relatively thin and watery, and also causes a vasodilation within the gland. Stimulation of the sympathetic nerves to the submaxillary produces a scanty secretion, which is thick and contains much mucus, and also causes vasoconstriction. Both types of nerve supply are secretory in nature, but they produce quite different volumes and types of secretion. The parotid glands secrete only in response to parasympathetic impulses—not to sympathetic impulses.

The unconditioned stimuli for causing reflex secretion of saliva are touch and pressure acting on receptors in the oral mucosa and chemical stimuli acting on taste buds on the tongue. We are all conditioned to salivate in response to the sight and smell of appetizing food. These visual and olfactory stimuli are called conditioned stimuli. We can even salivate merely by thinking about good food.

Hormonal

No hormonal regulation has been demonstrated for the salivary glands, except that aldosterone promotes lower sodium concentration and higher K^+ concentration in saliva.

MECHANISM OF SECRETION

It has been known for more than 130 years that salivary secretion must be an active process (in contrast to the passive ultrafiltration in the renal glomeruli), because saliva can be secreted against a considerable hydrostatic pressure difference. Pressures as high as 300–400 mm Hg can be developed in an occluded submaxillary duct when the chorda tympani are stimulated. The acinar cells produce the primary secretion, which has a composition similar to that of plasma, except for higher K^+ and HCO_3^- concentrations. It is possible that K^+ is actively transported into the lumen of the alveoli, with HCO_3^- following by electrical coupling and water following by osmotic coupling. This primary secretion is hyperosmotic with respect to plasma. The mucous cells contain mucinogen storage granules that are released into the alveolus by exocytosis.

As the primary secretion passes through the inter-calated ducts, its inorganic components equilibrate with blood plasma. Farther along the system of ducts, probably especially in the striated ducts, the composition of the secretion is modified by active transport processes, losing a large fraction of its Na^+ and Cl^- and gaining some K^+ and HCO_3^-. The ducts are relatively impermeable to water; consequently, removal of NaCl makes the saliva hypotonic.

GASTRIC SECRETION

SITES OF SECRETION

The oxyntic glands, located in the body and fundus, contain chief cells, which secrete pepsinogen; oxyntic cells, which secrete an isotonic solution of HCl (about 165 mM), and intrinsic factor; and mucous neck cells, which secrete a soluble mucus. The pyloric glands, located in the antrum, contain cells that secrete mucus, cells that secrete gastrin into the blood, and a relatively small number of chief cells that secrete pepsinogen. The cardiac glands contain mainly mucus-secreting cells, but also a few oxyntic cells. The surface epithelial cells secrete an insoluble mucus.

VOLUME, ACIDITY, AND OSMOLARITY

The daily volume of gastric juice is about 2.5 L. Gastric secretions are isosmotic with respect to plasma. Most of the water is thought to come from the oxyntic cells. The pH of oxyntic cell secretion is about 0.78 (165 mEq of H^+/L). This acid solution is diluted and buffered by the alkaline secretions from other cell types and by ingested food. The pH of gastric contents is about 1–2 during fasting. After ingestion of a meal, the pH in the body and fundus rises to that of the meal mixed with saliva; in the antrum the pH value rises to roughly 3 or 4.

To test gastric secretory function clinically, gastric juice is collected through a tube and titrated with NaOH. The number of milliequivalents per liter neutralized in getting to pH 7 measures the concentration of acid, and this, multiplied by volume secreted per unit time, measures the rate of acid secretion. The number of milliequivalents of acid secreted per hour is a valuable index of parietal cell function. For reference purposes, the upper limit of normal basal acid output (BAO) is 6 mEq/h in men and 4 mEq/h in women; the upper limit of normal maximal acid output (MAO) after injection of histamine or gastrin is 40 mEq/h in men and 30 mEq/hr in women.

FUNCTIONS OF GASTRIC JUICE

Digestion

The hydrolysis of protein is believed to be the only major digestive function of gastric juice. Pepsinogen, an inactive precursor secreted from chief cells, is converted into the active proteolytic enzyme, pepsin, by the action of H^+ and by pepsin itself.

Bactericidal Action

Gastric juice has an important bactericidal action on the stomach contents due to its high acidity. Consequently, under normal conditions, the duodenal contents are nearly sterile. After gastric resection, or in a pathological condition called achlorhydria in which there is no acid secretion, the duodenum is sometimes invaded with microorganisms.

Supply of Intrinsic Factor

The oxyntic cells secrete a mucoprotein, the presence of which is essential for intestinal absorption of vitamin B_{12}. This substance is known as intrinsic factor (IF). If IF is not secreted, pernicious anemia eventually develops due to failure to absorb enough vitamin B_{12}.

Solvation of Calcium and Iron Salts

Insoluble calcium and iron salts of the diet are solubilized by gastric acid and thereby put into an absorbable form. They must be dissolved before they can be absorbed. Iron deficiency anemia can develop after total gastrectomy.

REGULATION OF SECRETION

Excitation

In humans slow gastric secretion occurs even during fasting, but secretion is greatly increased during feeding. The excitatory influences during feeding and digestion are divided into three phases: cephalic, gastric, and intestinal. These mechanisms exert constant stimulatory influences on gastric secretion from the time food enters the mouth (or even earlier), until it leaves the duodenum.

Cephalic Phase

The presence of appetizing food in the mouth excites taste buds that reflexly produce gastric secretion. The food must be appetizing and capable of stimulating taste buds. Food that is disliked or that is inert as far as taste is concerned will not call forth this secretory response. With normal conditioning, the smell, sight, or even thought of food can also cause the reflex, as can the sounds associated with receiving good food. The efferent limb of the reflex is carried by vagal parasympathetic fibers. Preganglionic parasympathetic fibers in the vagus enter the stomach wall and synapse with intrinsic excitatory neurons. In the body and fundus, the intrinsic excitatory neurons to the oxyntic glands directly excite the oxyntic cells and the chief cells, and both HCl and pepsinogen are secreted. In the antrum, the intrinsic excitatory neurons excite gastrin cells in the pyloric glands, and they secrete gastrin into the blood. When gastrin reaches the oxyntic glands via the vascular system, it causes the oxyntic cells to secrete HCl. There is apparently no direct effect of gastrin on the chief cells. The cephalic phase can be completely abolished by cutting the vagus nerves. A diagram of the cephalic phase of gastric secretion is given in Figure 1.

The cephalic phase of gastric secretion can be induced by injecting insulin. Insulin produces hypoglycemia, which stimulates hunger centers in the lateral hypothalamic nuclei; these in turn stimulate the vagal centers in the medulla, which stimulate gastric secretion.

Gastric Phase

Distention of the body and fundus with food stimulates stretch receptors, which activate excitatory intrinsic neurons; these in turn stimulate the oxyntic and chief cells, the neurotransmitter being acetylcholine. This event is sometimes called an oxynto-oxyntic reflex, since both sensory and effector limbs of the reflex are confined to the oxyntic (acid-producing) region of the stomach. From these same stretch receptors, impulses pass up the vagus nerves to the medulla; these impulses induce activity in parasympathic neurons that pass down the vagus to ex-

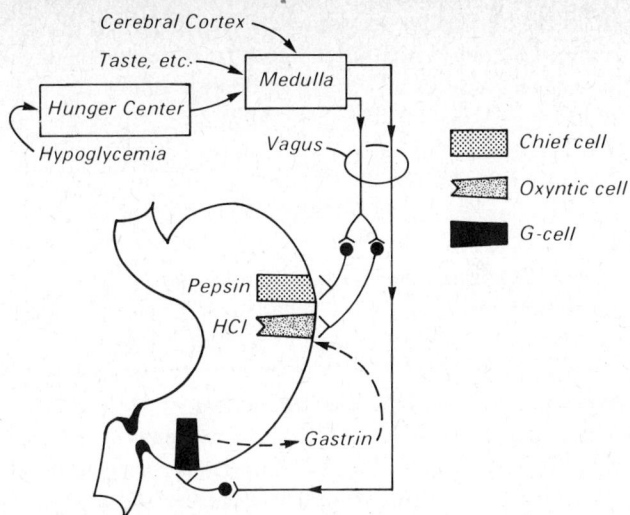

Figure 1 Cephalic phase of gastric secretion. The postganglionic neurons are in the intrinsic plexus. Stippled area, chief cell. Shaded area, oxyntic cell. Filled area, G cell.

citatory intrinsic neurons, which then stimulate the oxyntic and chief cells. Since both limbs of this reflex use the vagus nerves, it is called a vagovagal reflex. Also from these same stretch receptors neural activity spreads through the intrinsic nervous system to the pyloric gland area of the stomach (i.e., the antrum). This activity results in excitation of gastrin cells, which secrete gastrin into the blood. The neurotransmitter responsible for stimulating the gastrin cells is, once again, acetylcholine. Circulating gastrin then stimulates the oxyntic cells. This reflex through the intrinsic nervous system is called an oxyntopyloric reflex. Thus we see that stretch of the body and fundus causes acid and pepsinogen secretion by three different reflex pathways: oxynto-oxyntic, vagovagal, and oxyntopyloric.

Distention of the antrum with food also initiates reflexes that stimulate the oxyntic and chief cells. Again, three different reflex pathways are used. One is a local pyloropyloric reflex, which induces secretion of gastrin. Another is a long vagovagal reflex, which also induces secretion of gastrin. The mediator at the gastrin cells is acetylcholine for both reflexes. The third reflex from the antrum is a pyloro-oxyntic reflex, which travels through the intrinsic nervous system of the stomach and ends by stimulating the oxyntic and chief cells with acetylcholine.

Note the considerable redundancy of excitatory neural control mechanisms. We have listed six reflexes associated with the gastric phase of excitation; three of these directly stimulate the oxyntic and chief cells with acetylcholine, and three indirectly stimulate the oxyntic cells via the hormone gastrin. No single one of these reflexes can properly stimulate gastric secretion on its own—all apparently work in concert. Part of the reason for this is that neither acetylcholine nor gastrin can exert full effect on the oxyntic cells unless the other agent is simultaneously present. In other words, they potentiate each other. Potentiation of the effect of gastrin by acetylcholine is especially pronounced. In fact, gastrin has very little effect on oxyntic cells in the absence of acetylcholine. For this reason, selective vagotomy of the oxyntic region of the stomach greatly reduces the amount of acid secreted in response to a meal, even though gastrin continues to be secreted.

The reflexes involved in the gastric phase of gastric secretion are partially diagrammed in Figure 2. To avoid an excessive number of lines, the oxyntopyloric and pyloro-oxyntic reflexes have been omitted. You might wish to fill these in.

In addition to distention, acid secretion can be induced by certain chemicals found in chyme. The most natural of these is an ill-defined variety of polypeptides produced by

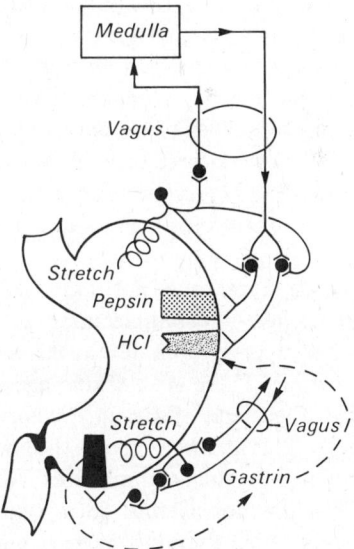

Figure 2 Gastric phase of gastric secretion, partially diagrammed.

the action of pepsin on ingested protein. Also, meat extracts or liver extracts containing peptides and amino acids stimulate acid secretion. Caffeine also has this effect. Ethanol stimulates acid secretion in some species, but probably not in humans. Acid secretion in response to these agents is thought to involve direct stimulation of the oxyntic cells and of the gastrin cells by these chemicals, rather than reflexes.

Intestinal Phase

Under certain experimental conditions, it can be demonstrated that partially digested protein in the duodenum results in increased acid secretion by oxyntic cells. This is the intestinal phase of gastric secretion. The intestinal phase is hormonally mediated. Intestinal gastrin is probably involved, but this is not certain. Another hormone might also be involved; it is sometimes called entero-oxyntin, and sometimes intestinal phase hormone, but it has not yet been chemically identified. The intestinal phase is thought to be relatively unimportant.

Role of Histamine

Another potent stimulus to the oxyntic cells is histamine. It has been suggested that histamine is the final common mediator for the stimulatory effects of both acetylcholine and gastrin. According to this hypothesis, acetylcholine (released from intrinsic excitatory neurons) and gastrin (which diffuses from the blood) both induce a local release of histamine from the gastric epithelium, which in turn binds to histamine receptors on the oxyntic cell membrane. There are two types of histamine receptors in the human body; these are referred to as H_1- and H_2-receptors. The oxyntic cells have H_2-receptors. When these receptors are activated, the oxyntic cells are stimulated to secrete HCl. Ordinary antihistamine drugs do not block these H_2-receptors and therefore do not inhibit HCl secretion. However, a new class of histamine antagonists that act at H_2-receptors has recently been developed. An example is the drug cimetidine. These H_2-blocking agents have been found to inhibit the effects of gastrin and acetylcholine on the oxyntic cells. This finding supports the above hypothesis that histamine is the final common mediator. However, an alternative suggestion is that the oxyntic cells have separate receptors for histamine, gastrin, and acetylcholine and that they secrete maximally only when all types

of receptor sites are occupied; the H_2-blockers prevent histamine from occupying its receptor site, and the oxyntic cell response to all stimuli is diminished.

Inhibition

Acidification of Antrum

Secretion of gastrin is inhibited by a low pH value. If antral pH drops below about 1.5, no more gastrin is secreted. Consequently, HCl is secreted less rapidly and antral pH increases again, a negative feedback mechanism that protects against too rapid secretion of acid.

Negative Feedback from the Duodenum

If the pH of duodenal contents drops below a value of about 3 or 4, gastric acid secretion is strongly inhibited. The effect is especially potent in the duodenal bulb, where the pH often drops to very low values during gastric emptying. The presence of fat in the duodenum also inhibits secretion of gastric acid and pepsinogen. To be effective, these lipids must be solubilized in biliary micelles (see Chapter 31). In addition, if duodenal contents become markedly hypertonic, gastric secretion is inhibited. Note that these are the same three duodenal stimuli (low pH, fats, and hypertonicity) that result in inhibition of gastric emptying (Chapter 28).

These inhibitory effects on gastric secretion, like those on gastric emptying, are mediated by both neural and hormonal mechanisms. The neural mechanism is sometimes called the enterogastric reflex, although historically this term applied only to the effect on motility. As pointed out in Chapter 28, the enterogastric reflex uses both extrinsic and intrinsic pathways. The hormonal mechanism is probably more important, but unfortunately the hormones involved have not yet been identified.

Although not chemically identified, the hormone mediating the effect of fat in the duodenum has a time-honored name: enterogastrone. For a time it seemed likely that enterogastrone is really GIP, but recent radioimmunoassay studies seem to rule out this contention. It is possible that enterogastrone is really cholecystokinin (CCK); however, it is also possible that a separate hormone, distinct from those that have already been chemically characterized, remains to be identified. The hormone mediating the effect of acid in the duodenum has also eluded identification. Secretin has been a candidate, although not a very successful one. The phantom hormone mediating this response has been tentatively called bulbogastrone.

The significance of gastric inhibitory influences, both motor and secretory, arising from the duodenum should be pondered. Antral chyme is quite acidic and proteolytic, and is also generally not isotonic. One of the main functions of the stomach is to deliver this potentially harmful material into the duodenum at a rate slow enough that neutralization and isotonicity can be promptly attained in the duodenum. Failure of the feedback mechanisms would be expected to lead to duodenal ulceration, because with a low duodenal pH, pepsin could then digest the duodenal mucosa.

SUMMARY OF EXCITATORY AND INHIBITORY INFLUENCES ON GASTRIC ACID SECRETION

Figure 3 summarizes the various agents that stimulate and inhibit the oxyntic cells. It is important to note that the neurons that stimulate the oxyntic cells, the chief cells, and the gastrin cells are all cholinergic. Clinically, gastric secretion can be intentionally reduced by administration of atropine or other cholinergic blocking agents.

MECHANISM OF SECRETION

Pepsinogen is synthesized in the chief cells and stored in the form of zymogen granules. Upon stimulation of the cells, these granules are secreted by exocytosis. H^+ and Cl^- are both actively transported across the membranes bordering the canaliculi of the oxyntic cells. The concentration of H^+ in the lumen of the canaliculus is about 165 mM, that in the cytoplasm of the oxyntic cell is roughly a million-fold less; thus, H^+ is transported up a very steep gradient. The source of H^+ for secretion is H_2CO_3. H_2CO_3 is formed by hydration of CO_2, a reaction catalyzed by carbonic anhydrase (CA). For each H^+ released from dissociation of H_2CO_3 and secreted, a HCO_3^- is produced and diffuses into the blood in exchange for Cl^-. These reactions are diagrammed in Figure 4. During a period of rapid HCl secretion, so much excess HCO_3^- enters the blood from oxyntic cells that the blood pH, and consequently the urine pH, rises. The transitory rise in urine pH following a meal is called the alkaline tide.

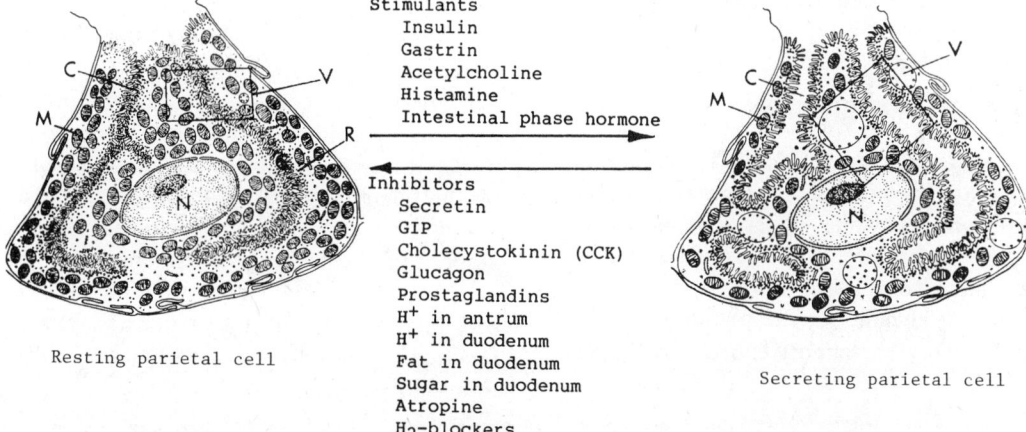

```
                    Stimulants
                      Insulin
                      Gastrin
                      Acetylcholine
                      Histamine
                      Intestinal phase hormone


                    Inhibitors
                      Secretin
                      GIP
                      Cholecystokinin (CCK)
                      Glucagon
                      Prostaglandins
                      H⁺ in antrum
                      H⁺ in duodenum
                      Fat in duodenum
                      Sugar in duodenum
                      Atropine
                      H₂-blockers
```

Resting parietal cell   Secreting parietal cell

Figure 3 The agents listed above the arrows stimulate the oxyntic (parietal) cells directly or indirectly, and those listed below the arrows inhibit the oxyntic cells directly or indirectly. Inclusion of secretion, GIP, CCK, and glucagon in the list of inhibitors is not meant to imply that inhibiting gastric secretion is one of their important functions; currently this seems doubtful. N, Nucleus; R, smooth endoplasmic reticulum; M, mitochondria; C, intracellular canaliculi; V, vacuole-containing body. (Diagrams of resting and active oxyntic cells are from Davenport HW: *Physiology of the Digestive Tract,* ed 3., Year Book Medical Publisher, Chicago, 1971.)

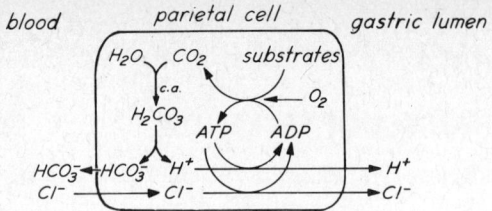

Figure 4 HCl secretion by oxyntic cell.

SECRETION OF BILE

SITE OF SECRETION

Bile is secreted continuously by the parenchymal cells of the liver into the bile canaliculi. This primary secretion is modified as it passes through the ducts. During interdigestive periods, the gallbladder is relaxed and the resistance at the terminal end of the common bile duct is high; therefore, bile, which is secreted at about 10 cm H_2O pressure, collects in the gallbladder without flowing into the duodenum. In the gallbladder, bile is concentrated five- to 10-fold by active absorption of NaCl and passive absorption of water. The gallbladder also adds a mucous secretion to the bile. Gallbladder bile along with fresh hepatic bile enters the duodenum when food arrives in the intestine.

VOLUME AND COMPOSITION

About 1.0 L of bile is secreted per day. The major constituents are bile acids, bile pigments, cholesterol, lecithin, mucins, inorganic salts, and water. Bile is iso-osmotic with respect to plasma. The primary bile acids of human bile are cholic and chenodeoxycholic; they are conjugated with either taurine or glycine. At the pH value of bile, which is about 7.4, they are almost entirely in the salt form.

FUNCTIONS OF BILE

Digestion and Absorption of Lipids

Bile salts are essential for normal intestinal absorption of fats, phospholipids, cholesterol, and the lipid vitamins (A, D, E, and K).

Excretion of Bile Pigments

The final breakdown product of hemoglobin is bilirubin. In the parenchymal cells bilirubin is conjugated with glucuronic acid and is then actively transported into the bile canaliculi.

Other Excretory Functions

Numerous substances, especially those foreign or toxic to the body, are detoxified by the liver, usually by conjugation with glucuronic acid, and excreted in the bile.

Neutralization and Buffering of Gastric Acid

The HCO_3^- in bile helps neutralize H^+ from the stomach and buffer chyme in the duodenum to a pH of around 6.0–6.5. However, pancreatic juice is probably of greater quantitative significance in this regard (see below).

BILIARY MICELLES

Amphiphilic molecules in aqueous solution often form aggregates known as micelles. In bile, the bile salts and lecithin are both amphiphiles and form mixed micelles containing both types of molecules. Most of the bile salt and lecithin in bile are in micellar form—only small amounts remain as free monomers. If this were not the case, the osmolarity of bile would be far higher than that of plasma.

Cholesterol is completely insoluble in water, yet it is dissolved in gallbladder bile at a concentration of about 0.9%. It can be dissolved in bile because it is solubilized by the biliary micelles. Normally these micelles contain nearly their maximum capacity of cholesterol. If the amount of cholesterol in bile increases, or anything happens to reduce the concentration of micelles, cholesterol precipitates and can form gallstones. The structure and functions of biliary micelles are discussed more thoroughly in Chapter 31.

ENTEROHEPATIC CIRCULATION OF BILE SALTS

Of the total bile salt secreted by the liver and entering the duodenum, more than 95% is reabsorbed by the small intestine, mostly by the lower ileum. Reabsorbed bile salt is carried back to the liver in portal blood. Hepatic parenchymal cells extract the bile salt from the blood and resecrete it (by active transport) into the canaliculi. This cycle is called the enterohepatic circulation. A single bile salt molecule may circulate enterohepatically several times before finally reaching the colon. The small amount of bile salt lost in the feces is exactly balanced by synthesis in the liver. During digestion of a single meal a bile salt molecule may circulate three or four times. Further information about the enterohepatic circulation of bile salts is given in Chapter 31.

REGULATION OF SECRETION

Nervous

Parasympathetic impulses increase bile secretion by the liver, while sympathetic impulses decrease bile secretion (probably by causing vasoconstriction). A cephalic phase of bile secretion exists and is mediated by the vagus nerves.

Hormonal

Secretin increases the HCO_3^- concentration of bile and therefore the pH; it has no effect on the amount of bile salt secreted. To a lesser extent, CCK also has this effect. Any substance that induces increased bile flow into the duodenum is called a cholegogue. A cholegogue that acts by causing the liver to secrete at a faster rate is called a choleretic. The most powerful and important choleretic is bile salt itself. When absorbed bile salts are resecreted by the liver, water must also be secreted by osmotic coupling, causing the volume of secretion to increase. This is probably the major mechanism causing increased bile secretion following ingestion of a meal—bile salt which had been stored in the gallbladder is reabsorbed by the ileum and causes increased bile secretion by the liver.

REGULATION OF BILE ENTRY INTO THE DUODENUM

Nervous

Nervous influences exist but are probably not of much importance.

Hormonal

Chyme in the duodenum causes contraction of the gallbladder and relaxation of the sphincter of Oddi, causing bile to flow into the duodenum. Newly secreted hepatic bile then continues to flow into the duodenum during digestion without being concentrated in the gallbladder. These effects are mediated by CCK. Fat in the small intestine is the most potent stimulus for CCK secretion and, therefore, for contraction of the gallbladder and relaxation of the sphincter of Oddi.

MECHANISM OF BILE SECRETION

Secretion of bile involves active transport processes. This is made obvious by the facts that (1) bile can be secreted against a higher pressure than the blood pressure in the liver, and (2) certain materials such as bile salts and bile pigments are far more concentrated in bile than in blood. The parenchymal cells produce a primary secretion that includes all the bile salts, lecithin, bile pigments, and cholesterol. This primary secretion is then modified as it passes through the ducts, principally by active secretion of Na^+ and HCO_3^-. Secretin stimulates this active secretion of HCO_3^- by the biliary ducts.

SECRETION OF PANCREATIC JUICE

SITES OF SECRETION

Two quite different juices are secreted by the pancreas and are mixed together in the ducts. One is secreted from the acinar cells; it is low in volume but has a high concentration of enzymes. The other is secreted from the duct cells (mainly the intralobular ducts) and is essentially a solution of $NaHCO_3$ that is iso-osmotic with plasma. The latter secretion provides most of the volume of pancreatic juice.

VOLUME AND COMPOSITION

A small amount of pancreatic juice is secreted continuously, but the secretory rate is increased during digestion of a meal. The total daily output in humans is roughly 0.5 L/day. Pancreatic juice is iso-osmotic with plasma; it has an alkaline pH. As the rate of secretion rises, so does the HCO_3^- concentration and, consequently, the pH to as high as 8.2.

Besides water and inorganic electrolytes (mainly $NaHCO_3$), the important constituents of pancreatic juice are the enzymes and mucins. The enzymes include pancreatic amylase, pancreatic lipase, cholesterol esterase, lecithinase A_2, ribonuclease, deoxyribonuclease, and five precursors of proteolytic enzymes—trypsinogen, chymotrypsinogen, proelastase, procarboxypeptidase A, and procarboxypeptidase B. In the small intestine, trypsinogen is converted to trypsin by the action of an intestinal enzyme called enterokinase and by the autocatalytic action of trypsin itself. The other proteolytic enzyme precursors are converted to their active forms by the action of trypsin.

FUNCTIONS OF PANCREATIC JUICE

Digestion
Hydrolysis of polysaccharides, lipids, polypeptides, and polynucleotides depends mainly on pancreatic enzymes. This subject is discussed in Chapter 31.

Neutralization and Buffering of Gastric HCl
This function is performed in conjunction with hepatic and intestinal secretions.

REGULATION OF SECRETION

Nervous
Impulses over parasympathetic nerves to the pancreas (vagus) cause secretion of a small volume of thick juice rich in mucus and enzymes. Distention of the body of the stomach induces a vagovagal reflex that causes secretion of an enzyme-rich pancreatic juice. This is called the gastropancreatic reflex.

Hormonal
Hormonal regulation of pancreatic secretion is far more important than neural regulation. Cholecystokinin is secreted by endocrine cells in the duodenal and jejunual epithelium in response to amino acids and fats. Cholecystokinin stimulates the acinar cells of the pancreas to secrete enzymes. Secretin, released in response to a low duodenal pH, stimulates the pancreatic ducts to secrete a $NaHCO_3$ solution. Cholecystokinin by itself has little or no effect on $NaHCO_3$ secretion, and secretin by itself has hardly any effect on enzyme secretion. However, in the presence of CCK, secretin becomes much more effective in promoting $NaHCO_3$ secretion; likewise, in the presence of secretin, CCK becomes much more effective in promoting enzyme secretion. In other words, CCK and secretin potentiate each other. The mechanism for this potentiation is unknown.

PHASES OF PANCREATIC SECRETION

As with gastric secretion, there are three stimulatory phases of pancreatic secretion: cephalic, gastric, and intestinal. The cephalic phase operates via parasympathetics in the vagus nerves. The gastric phase operates by the gastropancreatic reflex. The intestinal phase, by far the most important, operates by the secretion of CCK and secretin, and also perhaps, according to recent evidence, by a vagovagal enteropancreatic reflex.

MECHANISM OF SECRETION

The enzymes are released from the acinar cells by exocytosis of zymogen storage granules. Na^+ and HCO_3^- are both actively transported by the duct cells, with H_2O following by osmotic coupling. As the juice passes through the main pancreatic duct, HCO_3^- is reabsorbed in exchange for Cl^-. Consequently, final pancreatic juice contains a higher Cl^- concentration and a lower HCO_3^- concentration than does the primary secretion. When flow rate is increased, there is less time for this exchange, and HCO_3^- concentration rises, thereby elevating the pH.

SECRETIONS OF THE SMALL INTESTINE

MUCUS FROM BRUNNER'S GLANDS

The duodenal glands (Brunner's glands) secrete a small volume of a very viscous solution of mucoprotein. Mucus from Brunner's glands coats the duodenal surface and helps protect it. Regulation of this secretion is not understood, although it is known that acetylcholine, gastrin, CCK, and secretin are all capable of stimulating Brunner's glands.

MUCUS FROM GOBLET CELLS

Goblet cells in the crypts of Lieberkuhn and on the villi secrete mucus in response to mechanical stimulation of the mucosa. The apparent function of this secretion is to provide lubrication to aid the movement of chyme. Mucus from goblet cells may also help keep the epithelial surface moist, since mucus tends to retain water.

INTESTINAL JUICE

It is usually said that each day the intestinal epithelium secretes about 2–3 liters of a serous fluid called intestinal juice. There is no justification for this quantitative estimate, and we are really quite ignorant of the normal rate of secretion. The reason for this ignorance arises from the fact that the net rate of fluid movement into or out of the intestinal lumen represents a balance between absorption and secretion, and it has not been possible to measure the rate of secretion selectively. In some circumstances secretion exceeds absorption, and intestinal juice (also called succus entericus) accumulates in the lumen. Normally, however, absorption exceeds or just matches secretion, and no succus entericus accumulates.

Intestinal juice is iso-osmotic with respect to plasma; it has a considerably higher HCO_3^- concentration and a lower Cl^- concentration than that of plasma and contains almost no protein. It is usually assumed that intestinal juice is secreted from the crypts, but participation of the intervillous epithelium and even the villous epithelium cannot be ruled out. As this secretion flows over the absorptive cells on the villi it is reabsorbed. The functions of succus entericus are not well understood, although it probably contributes to the slightly alkaline pH of intestinal contents and could be important in maintaining the fluidity of chyme.

Succus entericus is secreted in response to mechanical irritation of the mucosa and distention of the gut. These responses are mediated by the intrinsic nervous system. Impulses over parasympathetic nerves augment secretion. Secretion of intestinal juice can be increased by gastrin, CCK, GIP, and VIP, but the physiological importance of these responses remains to be established.

Although the normal role of intestinal secretion is not understood, it is clearly a most important and dramatic concomitant of various bacterial infections of the intestine. The diarrhea associated with these infections is mainly the result of increased intestinal secretion. Bacteria, such as pathogenic *Escherichia coli, Shigella dysenteriae,* and staphylococci release polypeptide enterotoxins that activate the secretory mechanism. The most dramatic example is *Vibrio cholerae.* Patients with cholera secrete a huge volume of intestinal juice that completely overwhelms the reabsorptive capacity of the gut and leads to vomiting and watery diarrhea. The enterotoxin from the cholera vibrio is called cholera toxin. It acts on the apical membranes of intestinal epithelial cells to increase the activity of adenyl cyclase. Consequently, intracellular cyclic adenosine monophosphate (cAMP) levels go up. Cyclic AMP is the intracellular messenger that leads to increased secretion. It is highly probable that the other enterotoxins and the GI hormones also promote secretion via cyclic AMP.

In addition to stimulating secretion of intestinal juice from the crypts, cyclic AMP markedly inhibits absorption of NaCl and water by the villi. This effect contributes to the water and electrolyte loss associated with cholera and probably other enteric infections. It has not been established which effect—stimulation of secretion or inhibition of absorption—is the more important.

EPITHELIAL CELL EXTRUSION

Epithelial cells continually migrate up the sides of villi. New cells are produced by mitosis in the crypts and gradually crowd the older cells toward the tip of each villus. At the tip, the old epithelial cells are extruded into the intestinal lumen. Complete turnover of epithelial cells takes about 2 or 3 days. These cells degenerate in the lumen and release their contents.

ORIGIN OF THE ENZYMES FOUND IN INTESTINAL JUICE

Intestinal juice, as ordinarily collected, is a mixture of all the above secretions. It contains many enzymes, including maltase, sucrase, lactase, isomaltase, enterokinase, and oligopeptidases. Most of these enzymes are believed to come from epithelial cells after extrusion from the villi. These enzymes are thought to function primarily within intact cells on the villi, or attached to their brush borders, and it is only fortuitous that they are found free in the lumen.

COLONIC SECRETION

There are two secretions from the colon: (1) mucus is secreted from goblet cells, and (2) a serous fluid is secreted from other cells. The rate of secretion is normally very low. This juice contains fairly high concentrations of K^+ and HCO_3^-, and has a pH value of 8.0–8.4. Secretion is stimulated by impulses over parasympathetic nerves and by local mechanical or chemical irritation of the mucosa. The normal stimulus is fecal material. Mucus helps bind the fecal mass and lubricates its surface and the mucosal surface. The bicarbonate buffers against H^+ produced by bacterial fermentation in the colon. In a condition known as mucous colitis, the colon secretes so much mucus that mucus can become a large fraction of total fecal output. Hypokalemia and acidosis can result from diarrhea because of the loss of colonic K^+ and HCO_3^-, respectively.

SELECTED BIBLIOGRAPHY

Binder, HJ (ed): *Mechanisms of Intestinal Secretion.* New York, Alan R Liss, 1979.

Schneyer EH, Emmelin N: Salivary secretion, in Jacobson ED, Shanbour LL (eds): *Gastrointestinal Physiology.* Baltimore, University Park Press, 1974.

SELF-STUDY QUESTIONS

MULTIPLE CHOICE

Select the best answer(s). (In many instances, more than one answer is correct.)

1. Salivary secretion is normally stimulated by
   A. gastrin.
   B. sympathetic nerve impulses.
   C. CCK.
   D. parasympathetic nerve impulses.

2. The oxyntic (parietal) cells secrete
   A. HCl.
   B. intrinsic factor.
   C. pepsinogen.
   D. gastrin.
   E. mucus.

3. The gastric phase of gastric secretion is mediated

   A. entirely by gastrin.
   B. entirely by vagovagal reflexes.
   C. entirely by intrinsic reflexes
   D. entirely by the direct action of chemicals such as peptides on the oxyntic cells.
   E. by all of the above acting in concert.

4. Gastric acid secretion can be inhibited by
   A. acidification of the antrum.
   B. acidification of the duodenum.
   C. fats in the antrum.
   D. fats in the duodenum.

5. Enterogastrone
   A. inhibits acid secretion by the stomach.
   B. is probably the same as GIP.
   C. might be the same as CCK.
   D. might be the same as secretin.

6. The most important choleretic stimulus to the liver during digestion is
   A. secretin.
   B. cholecystokinin.
   C. bile salts.
   D. parasympathetic nerve activity.

7. Cyclic AMP
   A. inhibits intestinal absorption of NaCl and water.
   B. stimulates secretion of intestinal juice.
   C. concentration in intestinal epithelium is elevated in cholera patients.

TRUE OR FALSE

8. The cephalic phase of gastric secretion can be abolished by vagotomy.

9. The pyloro-oxyntic and oxyntopyloric reflexes can both be abolished by vagotomy.

10. Gastrin is much more effective in stimulating oxyntic cells when intrinsic excitatory neurons are active than when they are inactive.

11. The gastric secretory response to injection of insulin can be markedly inhibited by a previous injection of cimetidine.

12. Injection of secretin reduces the concentration of bile salt in bile.

13. The effect of secretin on pancreatic ducts is potentiated by CCK.

14. As the rate of pancreatic secretion increases in response to secretin, the bicarbonate concentration and pH decrease.

15. Sham feeding (in which the food runs out through an esophageal fistula and never reaches the stomach) is expected to increase pancreatic secretion by a hormonally mediated mechanism.

ANSWERS

1. B, D. Both divisions of the ANS are stimulatory. There is no hormonal regulation of any importance.

2. A, B. Pepsinogen comes from chief (zymogen) cells, gastrin from G cells, and mucus from mucous neck cells, surface epithelial cells, pyloric gland cells, and cardiac gland cells.

3. E.

4. A, B, D. There is no evidence for an inhibitory effect of fat in the antrum, as long as it stays in the antrum.

5. A, C. GIP is probably not entergastrone. CCK possibly is enterogastrone, although this is far from certain. Secretin is not secreted in response to fat in the duodenum and therefore cannot be enterogastrone.

6. C. Secretin is also a choleretic, but its effect is not as important as that of reabsorbed bile salts.

7. All are correct.

8. True.

9. False. These reflexes travel entirely in the intrinsic nervous system.

10. True. Intrinsic excitatory neurons release acetylcholine, which potentiates the action of gastrin.

11. True. Insulin lowers blood sugar concentration, stimulating hunger centers and resulting in increased parasympathetic activity to the stomach. Consequently, acetylcholine is released at oxyntic cells. Cimetidine is an H_2-blocker. If histamine cannot attach to its receptor, acetylcholine is relatively ineffective in stimulating the oxyntic cells.

12. True. Secretin increases the volume of bile secreted per unit time, but has no effect on secretion of bile salts; therefore, their concentration decreases.

13. True. Therefore, in the presence of secretin, CCK increases enzyme secretion from the acini and also increases $NaHCO_3$ and water secretion from the ducts. The increased flow of $NaHCO_3$ solution helps wash the enzymes through the ducts.

14. False. They *increase* because there is less time for $HCO_3^- - Cl^-$ exchange across the duct epithelium at higher flow rates.

15. True. Sham feeding initiates the cephalic phase of gastric secretion. When gastric HCl reaches the duodenum, especially when undiluted and unbuffered with food, duodenal pH drops and secretin is secreted.

30

Intestinal Absorption of Water and Electrolytes

R. DAVID BAKER

OBJECTIVES

After completion of this chapter, you should be able to:

1. Have a semiquantitative appreciation for the daily loads on the small intestine of Na^+, K^+, Cl^-, and H_2O and understand to what degrees these loads are derived from exogenous or endogenous sources.

2. Understand the mechanisms and pathways for *passive* movements of Na^+, K^+, Cl^-, and H_2O across the intestinal epithelium.

3. Understand the pathway and mechanism for active absorption of Na^+.

4. Explain how Na^+ can be actively absorbed, while K^+ is not actively secreted, even though both are actively transported in opposite directions by a Na^+-K^+ pump in the basolateral membranes.

5. Explain how diffusion potentials and electrogenic Na^+ pumping both contribute to the electrical potential profile across the epithelium.

6. Understand the genesis of the transepithelial potential difference and its importance in influencing Na^+, K^+, and Cl^- movements.

7. Understand both the active and the passive features of

Cl^- absorption and be able to explain how adenosine triphosphate is indirectly involved in active Cl^- absorption.

8. Describe the generally accepted mechanism for absorption of water and NaCl as an isotonic solution, as well as the way in which H_2O can be transported across the intestinal epithelium against its concentration difference, even though there is no mediated transport of H_2O itself.

9. State the four pathways for Na^+ entry across the apical membranes of small intestinal epithelial cells.

10. Calculate the minimal K^+ concentration in small intestinal contents if you know the transepithelial potential and the effective K^+ concentration in blood plasma.

11. Understand the mechanisms for absorption of HCO_3^-.

12. Relate the principal ways in which the colonic epithelium differs from the small intestinal epithelium with respect to Na^+, K^+, Cl^-, and HCO_3^- transport.

13. Understand the mechanism by which the colon passively secretes K^+.

Each day large loads of water, Na⁺, and Cl⁻ enter the small intestine—roughly 6–9 L of water and 1 mole NaCl. Of these quantities, only about one-third is ingested, the remainder comes from the various gastrointestinal (GI) secretions.* The small intestine absorbs all but about 10% of these loads; the rest enters the large intestine. The large intestine absorbs all but about 10% of the water and 2.5% of the Na⁺ and Cl⁻ that it receives.

Smaller amounts of K⁺ (mostly from diet) and HCO_3^- (mostly from secretions) are also presented to, and absorbed by, the small intestine; the large intestine secretes these ions.

Intestinal absorption can be defined as the net transfer of materials across the absorptive epithelium from the intestinal lumen to blood or lymph. The pathway for absorption can be transcellular or paracellular. A model absorptive epithelium is depicted in Figure 1. The apical end of each cell is encircled by a tight junction that binds that cell to its neighbors. Tight junctions prevent direct movement of large molecules, such as sugars and amino acids, into the intercellular spaces. Thus, during absorption these materials must traverse two plasma membranes, the apical membrane and the basolateral membrane. This is the transcellular path. However, water molecules and small ions, such as Na⁺, K⁺, and Cl⁻, can readily pass between adjacent cells down their electrochemical potential gradients, and thus follow a paracellular route.

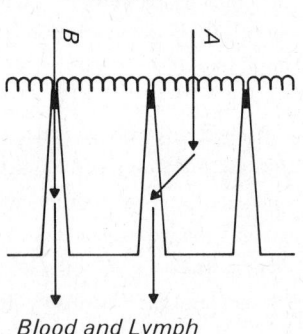

Lumen

Blood and Lymph

Figure 1 A model absorptive epithelium showing transcellular (a) and paracellular (b) pathways. The tight junctions are represented by black areas between cells at their apical ends.

*For a diagram depicting the water loads at various levels of the GI tract, see Chapter 33, Figure 6.

PASSIVE MOVEMENTS ACROSS INTESTINAL EPITHELIUM

The passive, unmediated pathways for movement of small ions and water are illustrated in Figure 2. The apical and basolateral membranes are permeable to H_2O but are much less permeable to Na⁺, K⁺, and Cl⁻. The tight junctions are rather permeable to all these materials. If a concentration difference for Na⁺, K⁺, or Cl⁻ is set up across the small intestinal epithelium, or if a transepithelial electrical potential difference is set up, most of the resulting transepithelial ion movements take place via the intercellular spaces and tight junctions, rather than through the cells. Therefore, the tight junctions must be extremely permeable to small ions, in view of the fact that their total cross-sectional area is only a small fraction of the total epithelial surface. If an osmotic pressure difference is set up across the epithelium, some of the resulting passive flow of water is transcellular and some of it is paracellular.

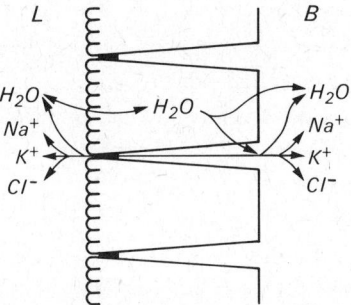

Figure 2 The major pathways taken by small ions and water in response to concentration gradients, electrical potential gradients, or osmotic gradients. Small ions move mainly through paracellular shunts. Water moves through cells and paracellular shunts.

OSMOTIC EQUILIBRATION IN THE DUODENUM

Meals are ordinarily not isotonic with respect to blood. Gastric juice is isotonic; therefore, chyme in the distal stomach tends to be closer to isotonicity than is the original meal. Nevertheless, when chyme or ingested liquids enter the duodenum they still are not usually isotonic. One of the main functions of the duodenum is to bring its contents to isotonicity. If hypertonic material enters the duodenum, water enters from the blood across the leaky duodenal epithelium and osmotic equilibrium is achieved quickly. Diffusion of Na^+ and Cl^- out of the chyme into the blood aids in achieving this osmotic balance. Rapid movements of water and ions in the opposite directions occur when hypotonic chyme enters the duodenum. If a hypertonic meal is dumped too rapidly into the duodenum, massive flux of water from blood to lumen can markedly reduce plasma volume and cause intestinal distention. The set of symptoms produced by these changes is called the dumping syndrome and is occasionally observed in postgastrectomy patients.

After duodenal osmotic equilibration, water and electrolytes are gradually absorbed all along the small and large intestines by a mechanism to be described below. Absorption of isotonic chyme obviously must involve at least one active transport process. This was proved in 1869, when Voit and Bauer found that an animal's own blood serum, placed in the small intestine, was absorbed.* It will be shown that the "vital" basis for Na^+, Cl^-, and water absorption from isotonic intestinal contents is the active extrusion of Na^+ from the absorptive cells across their basolateral membranes. The discussion will emphasize the small intestine. A briefer presentation of transport in the large intestine will be given later.

---

*In spite of this early demonstration that intestinal absorption could not be explained entirely by passive processes, many investigators, until as recently as the 1920s, could not accept the concept of "vital cell activity," and an historically important controversy raged.

THE SODIUM PUMP

In common with all other cells, the intestinal epithelial cells actively pump Na^+ out and K^+ in, thereby providing the control of intracellular cation concentration and cell volume necessary for life. What distinguishes epithelial cells from other cells is that the Na^+, K^+-ATPase that does the pumping is selectively located in the basolateral membrane and not in the apical membrane. This extremely important feature of the intestinal epithelium is indicated in Figure 3. Na^+ is pumped only toward the interstitial spaces, and not toward the intestinal lumen.

Another asymmetrical feature of intestinal absorptive cells is important for Na^+ transport, that is, the basolateral membrane has almost no passive permeability for Na, so Na cannot leak back into the cell after it is pumped. On the other hand, there are several mechanisms that allow downhill entry of Na^+ into the cell across the apical membrane. These entry mechanisms will be discussed below and are lumped together as one arrow in Figure 3.

Intracellular Na^+ concentration is kept low (about 15–20 mEq/L) by the basolateral Na^+ pump. Consequently, Na^+ can trickle downhill into the cell via its apical entry paths. In the steady state, the rate of entry is exactly matched by the rate of extrusion, producing a net lumen-to-interstitial transfer of Na^+. The basolateral membrane is far more permeable to K^+ than is the apical

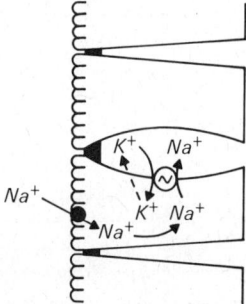

Figure 3 The basolateral Na^+ pump. The $\sim$ sign signifies that the pump is energized by hydrolysis of ATP. Na^+ is pumped out and K^+ is pumped in. The dotted arrow $--\rightarrow$ indicates that K^+ passively leaks back out as fast as it is pumped in. Various Na^+ entry mechanisms are lumped as one arrow across the apical membrane.

membrane. Consequently, in the steady state, K^+ leaks back into the interstitial space as rapidly as it is pumped into the cell, and there is no active movement of K^+ across the epithelium toward the lumen.

The Na^+ pump at the basolateral membrane is electrogenic, that is, it pumps Na^+ out faster than it pumps K^+ in. Consequently, the Na^+ pump generates an electromotive force across the basolateral membrane, oriented so that the inside is negative with respect to the outside of the cell.

THE ELECTRICAL POTENTIAL DIFFERENCES

The cytoplasm of intestinal absorptive cells is a potential "well"; its electrical potential is about 40 mV less than that of the mucosal solution and about 44 mV less than that of the serosal solution. This electrical potential profile is illustrated in Figure 4. The basolateral membrane is polarized about 4 mV more than is the apical membrane; consequently, there is a transepithelial potential difference of about 4 mV, the interstitial spaces and blood being

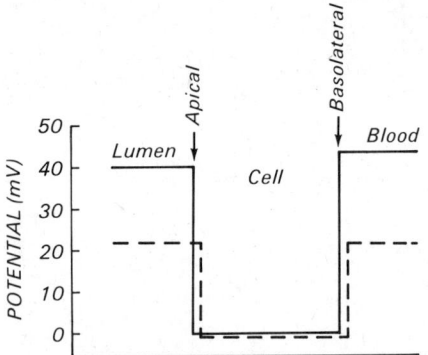

Figure 4 The electrical potential profile across the small intestinal epithelium. Solid line is with electrogenic Na^+ pump operating. Dashed line is with the pump not operating, but nevertheless with normal ion concentration gradients. Here, the intracellular potential is the reference potential; that is, it is arbitrarily called zero and the potentials in the lumen and blood are compared to it. The potential profile with the Na^+ pump inhibited is drawn slightly to the right of and below the control profile, so that it can be seen.

positive with respect to the lumen. This transepithelial potential difference is of great importance in influencing passive Na^+, K^+, and Cl^- movements across the epithelium.

Two mechanisms are involved in generating these potential differences: (1) passive diffusion of ions down their concentration gradients through selectively permeable membranes (this is the diffusion potential ordinarily handled by the Goldman equation), and (2) electrogenic pumping of Na^+ by the Na^+ pump located in the basolateral membrane. If the second mechanism is experimentally halted by inhibiting the Na^+ pump with ouabain or by cooling the tissue to stop metabolism, the potential profile changes to the one shown by the dashed line in Figure 4. Both membranes are depolarized to some extent, and the transepithelial potential difference is abolished. If the pump is allowed to start again, both membranes repolarize and the transepithelial potential difference is reestablished. We see that the transepithelial potential difference is entirely due to the electrogenic Na^+ pump. We also see that the apical and basolateral membrane potentials are determined by the diffusion potentials and, in addition, by the sodium pump.

Now it is easy to understand how turning on the pump polarizes the basolateral membrane since, after all, Na^+ is positively charged. But since there is no Na^+ pump in the apical membrane, how is it that turning on the basolateral Na^+ pump polarizes the apical membrane? The answer is that the low electrical resistance through the tight junctions causes the apical membranes to be electrically coupled to the basolateral membranes. Any change in basolateral membrane potential results in a change in apical membrane potential. The change in apical membrane potential is not as great as the change in basolateral membrane potential because of a small voltage drop through the tight junctions. The transepithelial potential difference is really this voltage drop through the tight junctions.

CHLORIDE TRANSPORT

Of course Na^+ cannot move across the epithelium all by itself; this would violate the basic physical requirement for macroscopic electroneutrality. Small charge separations can occur across the very thin hydrocarbon layer of the plasma membrane, but in any macroscopic region of the system, such as the lateral intercellular space, there must always be equal concentrations of cations and anions. Therefore, as Na^+ is transported across the epithelium, either an anion must move along with it at an equal rate, or another cation must move in the opposite direction. In the small intestine this electrical coupling is mainly with chloride; Cl^- moves downhill into the intercellular spaces as Na^+ is pumped in. There are two routes whereby Cl^- can enter the intercellular spaces; these are shown in Figure 5. The simplest of these Cl^- pathways is through the tight junctions. Small intestinal tight junctions are only about one-third as permeable to Cl^- as they are to Na^+ and K^+; nevertheless, Cl^- can diffuse through them fairly readily in response to the transepithelial potential difference.

The second Cl^- pathway is through the cells. The apical membrane contains a type of carrier that can transfer

Na^+ and Cl^- ions together. Both the cation and the anion must be bound to the carrier simultaneously in order for the transfer to take place. This transport system can operate in either direction, but normally, since Na^+ is much more concentrated outside than inside, the inward direction predominates, with a net movement of NaCl into the cell. This is one of the Na^+ entry mechanisms mentioned above, which provides a steady supply of Na^+ to the basolateral pump. Just as quickly as net Cl^- entry occurs by this mechanism, there is a net escape of Cl^- across the basolateral membrane into the intercellular space. This basolateral Cl^- pathway is mediated by a carrier or special channel, but very little is known about it.

The neutral NaCl entry mechanism in the apical membrane deserves further comment. It can be classed as one of the sodium cotransport systems discussed in Chapter 1. The downhill movement of Na^+ drives an uphill movement of Cl^- into the cell. Evidence for the uphill movement of Cl^- is provided by the equilibrium potential for Cl^- across this membrane, which is about -24 mV (when luminal Cl^- concentration is 145 mM), while the actual membrane potential is about -40 mV. Analysis using the Nernst equation shows that the inside Cl^- concentration is nearly twice what it would be if Cl^- were passively distributed. This elevated concentration of Cl^- within the cell drives its net escape from the cell across the basolateral membrane. Thus, Cl^- is actively transported across the cell.

We see that Cl^- can be absorbed from isotonic chyme by two mechanisms: (1) passive paracellular electrodiffusion in response to the transepithelial potential difference established by Na^+ pumping, and (2) active transport, driven by carrier-coupled downhill movement of Na^+ across the apical membrane.

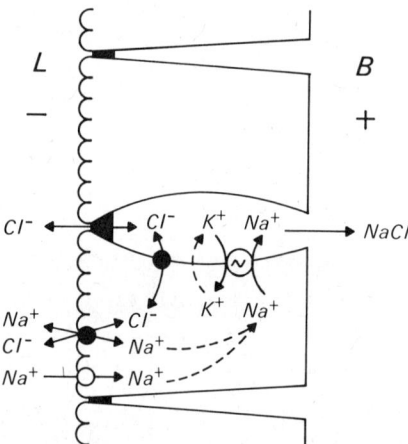

Figure 5 Diagram of sodium chloride transport by small intestinal epithelium. The neutral NaCl influx mechanism in the apical membrane is indicated by a solid black circle; all other Na^+ influx mechanisms are lumped as an open circle. In the steady state, for every Na^+ that enters the cell by these other mechanisms, a Cl^- has to cross the tight junction (assuming the situation is not complicated by movement of other ions such as bicarbonate). Although not indicated in this diagram, HCO_3^- can substitute for Cl^- on either black carrier.

ABSORPTION OF WATER FROM ISOTONIC INTESTINAL CONTENTS

Na^+ and Cl^- transport into the lateral intercellular spaces increases the osmolarity there, drawing water along by osmosis; this process is depicted in Figure 6.

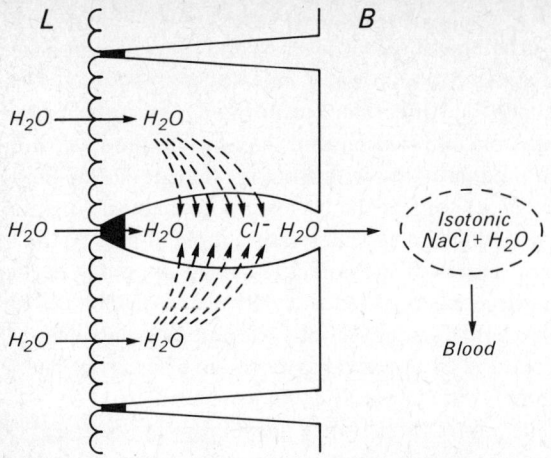

Figure 6 Isotonic absorption of water by the small intestinal epithelium.

All the water that enters the lateral intercellular spaces by osmosis comes from the lumen of the intestine, rather than from the blood side of the epithelium. This is because the reflection coefficient to NaCl is essentially zero at the basal end of the lateral intercellular space where there is only a very permeable basement membrane; at that end of the lateral space there is no effective osmotic pressure difference. However, at the apical end of the lateral space the reflection coefficient to NaCl is relatively high because of the porous properties of the tight junctions. Furthermore, the reflection coefficient for NaCl at the lateral plasma membranes is essentially 1.0. Therefore, when the osmolarity within the lateral spaces increases, water is drawn in selectively across the cells and across the tight junctions. As water enters the lateral intercellular spaces they swell, and hydrostatic pressure within them increases. Resistance to hydrodynamic flow is high at the apical end of the lateral spaces, but low at the basal end. Consequently, the solution formed in the lateral spaces flows preferentially toward the blood.

The ileal epithelium is a better water absorber than are the duodenal and jejunal epithelia. In fact, the ileal epithelium, unlike the duodenal and jejunal epithelia, can transport water uphill. This ability can be demonstrated by placing a 350 mOsM NaCl solution in the ileum. One would expect water to flow from blood to lumen by osmosis until the luminal solution was diluted to isotonicity;

indeed, this is what happens if the experiment is done in the duodenum or jejunum. But the ileum absorbs this solution, and H_2O moves from lumen to blood against its concentration difference. In the ileum, to produce net osmotic flow of water from blood to lumen, a NaCl solution having an osmolarity greater than 400 mOsM must be used. Now, it should be emphasized that this transport of water against an osmotic gradient is not believed to represent primary active transport of water molecules; it is accomplished by passive osmotic coupling of water flow with Na^+ and Cl^- transport. In the ileum, a relatively high osmolarity can be achieved and maintained in the lateral intercellular spaces, evidently at least 400 mOsM. Apparently, in the duodenum and jejunum, the osmolarity that can be achieved in the lateral spaces, and therefore the osmotic force driving water absorption, is less than in the ileum.

It is important to recognize that the uphill movements of both Cl^- and H_2O that can be accomplished by the ileum are both energized *indirectly* by ATP hydrolysis and the Na^+ pump.

Absorption of water against its osmotic pressure difference is important in the lower ileum, as well as in the colon, because in these regions nonabsorbable materials, principally dietary fiber and also secreted mucus, become increasingly concentrated as water is absorbed and cause ileal and colonic contents to be somewhat hypertonic.

SODIUM ENTRY MECHANISMS

Figure 7 depicts the various Na^+ entry mechanisms across the apical membrane. All are downhill; none of these mechanisms is energized except by the electrochemical potential gradient for Na^+ across the apical membrane.

The first pathway represents simple movement of Na^+ through Na^+ channels. The second pathway represents cotransport of Na^+ with organic molecules. Although shown in Figure 7 as a single pathway, there are actually several such cotransport mechanisms; they are discussed in Chapter 31. The downhill movement of Na^+ on these carriers drives the uphill movement of the

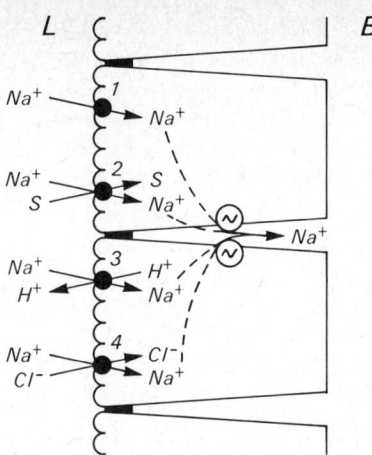

Figure 7 Pathways for Na$^+$ entry across the apical membrane of small intestinal epithelium. All these pathways are thought to be reversible, but they normally operate in the direction shown as long as Na$^+$ is continuously pumped out of the cell across the basolateral membrane. S, actively absorbed organic solutes, such as sugars and amino acids.

organic molecules. The third pathway is a Na$^+$–H$^+$ exchange diffusion mechanism. As Na$^+$ enters on this carrier, H$^+$ leaves the cell. Na$^+$–H$^+$ exchange across the apical membrane is inhibited by carbonic anhydrase inhibitors, such as acetazolamide. The fourth Na$^+$ entry pathway is the NaCl cotransport mechanism. This mechanism is also inhibited by carbonic anhydrase inhibitors. Apparently, carbonic anhydrase is involved somehow in Na$^+$–H$^+$ exchange and in Na$^+$–Cl$^-$ cotransport, but the details of this involvement are not understood. Na$^+$–Cl$^-$ cotransport is also inhibited by cyclic AMP, an effect that could be of great clinical importance in cholera and other enteric infections, as discussed in Chapter 29.

POTASSIUM TRANSPORT

About 60 mEq of K$^+$ enters the small intestine each day, mostly from food. Potassium absorption is strictly passive; in the upper small intestine it is absorbed downhill through the paracellular shunts, driven by its concentra-

tion difference against an electrical potential difference. If the transepithelial potential difference is 4 mV and the K$^+$ concentration in plasma is 5 mEq/L, the Nernst equation predicts that the K$^+$ concentration in the lumen cannot get below 5.8 mEq/L. For this reason, about 6 mEq/day of K$^+$ escapes absorption in the small intestine and enters the large intestine.

BICARBONATE TRANSPORT

Bicarbonate is secreted into the small intestine by the liver and pancreas. Some may also be ingested. Most bicarbonate from these sources reacts with H$^+$ from the stomach to form H_2CO_3, which dissociates into H_2O plus CO_2. The CO_2 thus produced, being fairly lipid soluble, readily diffuses down its partial pressure gradient across the epithelium, equilibrating with the CO_2 in venous blood. The Na$^+$–H$^+$ exchange mechanism in the apical membrane can also provide H$^+$ to neutralize HCO_3^- in the lumen. The CO_2 thus produced diffuses into the cells and, in the presence of carbonic anhydrase, is converted to H$^+$ and HCO_3^-. The H$^+$ can then be resecreted in exchange for Na$^+$, and the resulting intracellular $NaHCO_3$ can cross the basolateral membrane, the Na$^+$ going via the Na$^+$ pump and the HCO_3^-, presumably, by the same carrier (or channel) that Cl$^-$ moves on. This set of events is illustrated in Figure 8. In a third mechanism of HCO_3^- absorption, bicarbonate can compete with chloride for a place on the neutral Na$^+$–anion cotransport mechanism. This third pathway is illustrated in Figure 8 as well.

In the duodenum and jejunum, bicarbonate is absorbed by the above mechanisms. By the time chyme reaches the lower ileum, however, most of the bicarbonate secreted in pancreatic juice and bile has already been absorbed, and bicarbonate secreted by the crypts usually results in net bicarbonate movement *into* the lumen. Bicarbonate secretion buffers H$^+$ produced by bacterial metabolism in the lower ileum.

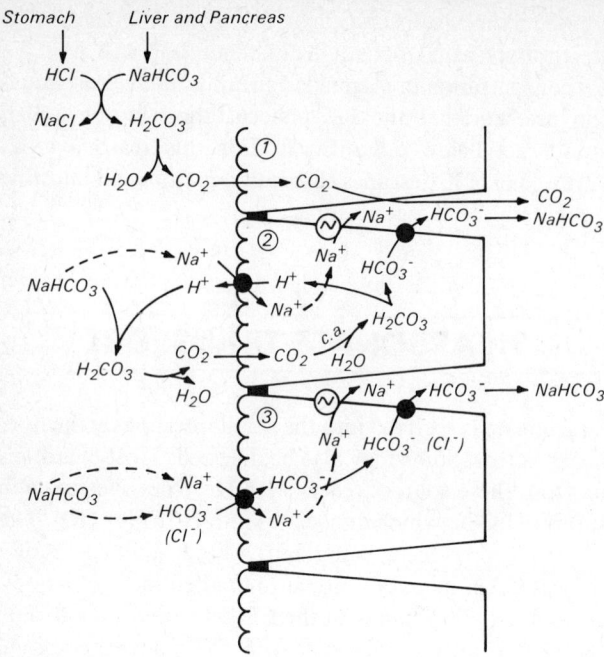

Figure 8 Bicarbonate absorption by small intestinal epithelium. Three separate mechanisms are shown in the cells numbered 1, 2, and 3.

WATER AND ELECTROLYTE TRANSPORT BY THE LARGE INTESTINE

The colonic epithelium actively absorbs Na^+ and Cl^-, and water follows osmotically, as in the small intestine. The colon normally secretes HCO_3^- at approximately the same rate that it absorbs Cl^-. The mechanism of this Cl^-–HCO_3^- exchange in the large intestine is not understood. Bicarbonate secretion is important for buffering H^+ produced by bacterial metabolism. Cyclic AMP induces active Cl^- secretion by the colonic epithelium and abolishes Cl^-–HCO_3^- exchange, but the mechanism of this effect is not understood. There are no cotransport systems for Na^+ and organic molecules in the colon, and there is probably no Na^+–H^+ exchange.

It was pointed out earlier that the colon secretes K^+. This movement of K^+ is passive; there is no active transport of K^+ across the epithelium. The Na^+–K^+ pump is unproductive in moving K^+ across the epithelium because of the relatively low apical permeability with respect to basolateral permeability to K^+. The secretion of K^+ is driven entirely by the transepithelial electrical potential difference, which is considerably higher in colon (10–20 mV) than in small intestine (4 mV). This electrodiffusion takes place entirely through the paracellular path. As a consequence of this passive secretion of K^+, fecal water has a higher K^+ concentration than blood. If the transepithelial potential difference is 20 mV and the K^+ concentration in blood plasma is 5 mEq/L, then, by the Nernst equation, the concentration in fecal water can be as much as 10.5 mEq/L. If the volume of fecal water is increased above normal, as in various diarrheas, excess K^+ may be eliminated from the body and hypokalemia may result. This is, in fact, a serious consideration in diarrheas.

ABSORPTION OF IRON

Of the 15–20 mg iron normally ingested per day, only about 0.5–1.5 mg is absorbed, mostly in the duodenum by an active transport process. The stomach plays an important role: gastric juice dissolves various iron salts and reduces Fe^{3+} to Fe^{2+}, thereby converting iron to an absorbable form (the apical membrane is much less permeable to Fe^{3+} than to Fe^{2+}). After gastrectomy, iron deficiency anemia frequently occurs.

Fe^{2+} is actively transported from lumen to blood by the epithelial cells of duodenum and jejunum. The rate of Fe^{2+} transport depends on Fe^{2+} concentration in chyme and upon how much iron is in the body. Following hemorrhage or during iron deficiency anemia, Fe^{2+} transport is augmented by an unknown mechanism; and following hyperingestion of iron, Fe^{2+} transport is diminished. This effect on iron absorption is of prime importance in maintaining normal iron balance. Body levels of iron are not controlled by renal excretion, only by intestinal absorption. This intestinal regulation of iron balance has misled many authors into stating that iron is not absorbed at all, except when needed by the body. This statement leads to the dangerous misconception that excessive amounts of iron can be ingested with impunity. Actually, a small fraction of ingested iron is always absorbed and, if the dose is high, even a small fraction can be lethal. Chronic iron overloading is not common, but it is sometimes observed in people who use iron pots for cooking or who drink huge quantities of certain wines. Acute iron poisoning is sometimes encountered in children who accidentally ingest iron pills.

One of the major storage sites for iron in the body is the intestinal epithelium. Iron is stored mainly in the form of ferritin (a complex of iron plus a protein called apoferritin). Iron is released from ferritin and transported into blood whenever increased iron is needed for hematopoiesis, thereby depleting the iron stores. On the other hand, if less iron is needed by bone marrow for hematopoiesis than is available, iron moves from blood into storage sites and is again converted to ferritin. Intestinal epithelial cells, containing ferritin, continually slough from the tips of villi. Almost none of the iron contained in ferritin is reabsorbed; nearly all is excreted in feces. Loss of iron from sloughed epithelial cells is the major way of eliminating excess iron from the body, especially in males. Urinary excretion is relatively minor. Since less iron is stored as ferritin during periods of iron deficiency than is stored normally, less will be eliminated in feces and iron will tend to be conserved. Since more iron is stored as ferritin during periods of iron overloading, more will be eliminated in feces and the overload will tend to be corrected.

Thus the intestinal mucosa regulates iron balance in two ways: (1) by adjusting the rate of active absorption to meet demands, and (2) by changes in the amount of iron lost by epithelial cell sloughing.

ABSORPTION OF CALCIUM

Ca^{2+} is absorbed by an active transport process located in the duodenum and upper jejunum. The rate at which Ca^{2+} is absorbed by this system depends on the serum Ca^{2+} concentration; consequently, intestinal transport is of considerable importance in the control of body Ca^{2+} levels. If serum Ca^{2+} concentration drops below normal, the parathyroid glands secrete parathyroid hormone. This hormone does not directly influence intestinal Ca^{2+} transport, but does increase the rate of synthesis of 1,25-dihydroxy vitamin D_3 by the kidney. This 1,25-$(OH)_2$-D_3 is secreted into the blood by the kidney; its effects on both bone and intestine tend to result in a return to normal serum Ca^{2+} concentration. The effect on the intestine is to increase the rate of active Ca^{2+} transport. This effect involves the stimulation, by 1,25-$(OH)_2$-D_3, of the synthesis of Ca^{2+}-binding protein, which is located in the brush borders and is essential for Ca^{2+} transport.

Ca^{2+} absorption is defective in vitamin D deficiency. Vitamin D_3, which is derived from diet or from the reaction of ultraviolet radiation with 7-dehydrocholesterol in the skin, is necessary for the synthesis of 25-OH-D_3 by liver, which in turn is necessary for the synthesis of 1,25-$(OH)_2$-D_3 by kidney. The latter compound is the active form of the vitamin and should properly be regarded as a *hormone,* since it is synthesized and secreted into the blood by a specific organ (the kidney) in response to a specific stimulus (parathyroid hormone), and acts on specific target organs (intestine and bone). The action on bone is to mobilize Ca^{2+}, an effect that is also of great importance in regulating serum Ca^{2+} concentration (see Chapter 47).

SELECTED BIBLIOGRAPHY

Kramer M, Lauterbach F: *Intestinal Permeation.* Excerpta Medica, 1977.

Robinson JWL: *Intestinal Ion Transport.* Baltimore, University Park Press, 1976.

SELF-STUDY QUESTIONS

TRUE OR FALSE

1. Most of the Na^+, Cl^-, and water absorbed by the small intestine is derived directly from oral intake.

2. The tight junctions in the small intestine are somewhat cation selective.

3. In the small intestinal epithelium, intracellular Cl^- concentration is generally greater than 145 mEq/L.

4. Reabsorption of HCO_3^- derived from pancreatic juice and bile is augmented by H^+ from gastric juice.

5. Oral ingestion of large amounts of iron is not dangerous, since the intestinal epithelium will not absorb iron unless it is needed by the body.

MULTIPLE CHOICE:

Select the correct answer(s). (In many instances, more than one answer is correct.)

6. The amount of water absorbed per day is greatest in the
   A. stomach.
   B. small intestine.
   C. large intestine.

7. In response to concentration and electrical potential differences, Na^+ and K^+ move passively across intestinal epithelium mainly via the
   A. paracellular pathway.
   B. transcellular pathway.

8. In response to an osmotic pressure difference, water moves across intestinal epithelium mainly via the
   A. paracellular pathway.
   B. transcellular pathway.

9. If a hypertonic solution is ingested,
   A. it will mostly be absorbed in the duodenum as a hypertonic solution.
   B. it will be diluted in the duodenum by osmosis.
   C. it will be absorbed eventually as an isotonic solution.
   D. it will still be hypertonic when it reaches the ileum.

10. The Na^+ pump is located in
    A. the apical cell membranes.
    B. the basolateral cell membranes.
    C. the tight junctions.
    D. the basement membrane.

11. Downhill Na^+ entry into intestinal epithelial cells takes place predominantly across the
    A. apical cell membranes.
    B. basolateral cell membranes.

12. K^+ is not ordinarily actively secreted by intestinal epithelium because
    A. it is not actively transported in exchange for Na^+ by the basolateral Na^+ pump.
    B. the basolateral membranes are virtually impermeable to K^+.
    C. it is actively transported into the cells across the apical membranes just as fast as it is across the basolateral membranes.
    D. the apical membranes are much less permeable to K^+ than are the basolateral membranes.

13. The normal transepithelial electrical potential difference in small intestine of about 4 mV
    A. is entirely dependent on electrogenic Na^+ pumping by the basolateral membranes.
    B. provides one of the major driving forces for K^+ absorption.
    C. provides one of the major driving forces for Cl^- absorption.
    D. provides a driving force for Na^+ leakage from lateral intercellular spaces back into the lumen.

14. The apical membrane potential (normally about -40 mV)
    A. is entirely the result of passive ionic diffusion across the apical membrane.
    B. is strongly influenced by the electrogenic Na^+ pump in the basolateral membranes.
    C. provides an important driving force for downhill movement of Na^+ into the cells.
    D. can be promptly eliminated by stopping the Na^+ pump.

15. Influx of Cl^- across the apical membrane of small intestinal absorptive cells
   A. requires the simultaneous influx of Na^+.
   B. can be uphill (i.e., against its electrochemical potential difference).
   C. is thought to be directly energized by hydrolysis of ATP.
   D. hyperpolarizes the apical membrane directly.

16. The ileum can transport water uphill because it
   A. can develop relatively high osmolarities in the lateral intercellular spaces.
   B. performs a great deal of pinocytosis.
   C. contains an H_2O carrier in the basolateral membranes, which can transfer energy directly from ATP to H_2O translocation.

17. Let us say that the lumen of the jejunum contains an isotonic Na_2SO_4 solution. You should expect that unidirectional influx of Na^+ into jejunal epithelial cells across their apical membranes would be increased by adding to the solution in the lumen
   A. glucose.
   B. Cl^-.
   C. HCO_3^-.
   D. K^+.
   E. H^+.

18. The coupled electrically neutral influx of Na^+ and Cl^- across apical membranes of small intestinal epithelium
   A. can be inhibited by drugs that inhibit carbonic anhydrase.
   B. can be inhibited by cyclic AMP.

C. can be inhibited by cholera enterotoxin.
   D. can apparently involve transfer of energy from the electrochemical potential gradient for Na^+ to the electrochemical potential gradient for Cl^-.

19. In the colon, Cl^-–HCO_3^- exchange across the epithelium
   A. is abolished by cyclic AMP.
   B. contributes significantly to the transepithelial potential.
   C. is important for buffering H^+ in the lumen of the colon.

20. Factors that contribute significantly to secretion of K^+ by colonic epithelium include
   A. active transport of K^+ into the cells across their basolateral membranes.
   B. the relatively high permeability of the tight junctions to K^+.
   C. the relatively high transepithelial potential.

21. The total amount of iron in the body is significantly regulated by
   A. adjustments in the rate at which iron is absorbed by the intestinal epithelium
   B. changes in the amount of iron stored in sloughing intestinal epithelial cells.
   C. adjustments in the rate of renal excretion of iron.

22. Which of the following has the most direct effect on Ca^{2+} transport across the small intestine?
   A. Serum Ca^{2+} concentration.
   B. Parathyroid hormone.
   C. $25\text{-}OH\text{-}D_3$.
   D. $1,25\text{-}(OH)_2\text{-}D_3$.

ANSWERS

1. False. About two-thirds is derived from secretions.

2. True. They are roughly three times more permeable to Na^+ and K^+ than to Cl^-.

3. False. If Cl^- were passively distributed its intracellular concentration would be about 32 mEq/L (by the

Nernst equation, assuming luminal Cl^- is 145 mEq/L, E_m is 40 mV and T 310°K). It was pointed out that intracellular Cl^- can be about twice this equilibrium value because of Na–Cl cotransport.

4. True. $H^+ + HCO_3^- \rightarrow H_2CO_3 \rightarrow H_2O + CO_2$. The

CO_2 and H_2O thus produced are both readily absorbed passively.

5. False. A small fraction of any dose will be absorbed.

6. B. Water is not absorbed in the stomach. The large intestine is a good water absorber, but most of it is normally absorbed in the small intestine before it has a chance to reach the large intestine.

7. A. The tight junctions are vastly more permeable to Na^+ and K^+ than are the cell membranes.

8. B. The tight junctions are permeable to water, but so are the cell membranes, and the latter present a far greater surface area.

9. B, C.

10. B.

11. A. The basolateral membranes have extremely low passive permeability to Na^+. Downhill entry across the apical membranes keeps the basolateral pump supplied with Na^+.

12. D. K^+ is pumped in by the basolateral pump, but it leaks right back out across the basolateral membrane —not appreciably across the apical membrane.

13. A,C,D. The transepithelial potential tends to drive K^+ secretion, not absorption. A small fraction of the Na^+ pumped into the lateral spaces cycles back into the lumen via the tight junctions, driven by the transepithelial potential.

14. B,C. The apical potential consists of diffusion potentials and the basolateral pump potential, attenuated by the IR drop across the tight junctions. If the baso-

lateral pump is stopped about half the apical membrane potential promptly disappears—the remainder dies out gradually as the ion concentration gradients dissipate.

15. A,B. This is the NaCl neutral cotransport system. No other influx mechanism for Cl^- in the small intestine has been discovered.

16. A. Pinocytosis occurs very slowly. There are probably no H_2O carriers in mammals.

17. A,B,C. The Na^+–glucose cotransport system is discussed in Chapter 31. HCO_3^- is effective because it buffers H^+ in the unstirred layer, thereby keeping the Na^+–H^+ exchange mechanism going.

18. All are correct. The role of carbonic anhydrase in this transport process is not understood. Cholera toxin stimulates adenyl cyclase (in the apical membranes) to generate cyclic AMP. Cyclic AMP inhibits the NaCl influx mechanism. This is one of the main problems in cholera. Most investigators believe that cyclic AMP also stimulates fluid secretion by the crypts, although this is not certain.

19. A,C.

20. B,C. Some investigators believe that tight junctions in the colon are especially permeable to K^+, more than to Na^+, thus contributing to rapid equilibration of K^+ between blood and lumen.

21. A,B. The kidneys do not regulate iron balance—this is a function of the small intestine.

22. D.

31

Digestion and Absorption of Nutrients

R. DAVID BAKER

OBJECTIVES

After completion of this chapter, you should be able to:

1. Describe the hydrolytic processes involved in the digestion of carbohydrates, both in the lumen and on the epithelial surface, as well as name the enzymes responsible.

2. Understand the mechanism of glucose and galactose absorption, and describe the role of Na^+.

3. Understand the concept of "disaccharidase-related" transport of sugars.

4. Describe the hydrolytic processes involved in the digestion of proteins and smaller peptides, and explain how the proteolytic enzyme precursors are activated.

5. Appreciate the variety of processes involved in amino acid absorption.

6. Understand the physical and chemical processes involved in the digestion of lipids.

7. Explain the role of biliary micelles in the digestion and absorption of lipids.

8. Describe the intracellular transformations undergone by lipids during absorption.

9. Understand the subject of bile salt absorption.

10. List the essentials of absorption of vitamins B_{12} and C.

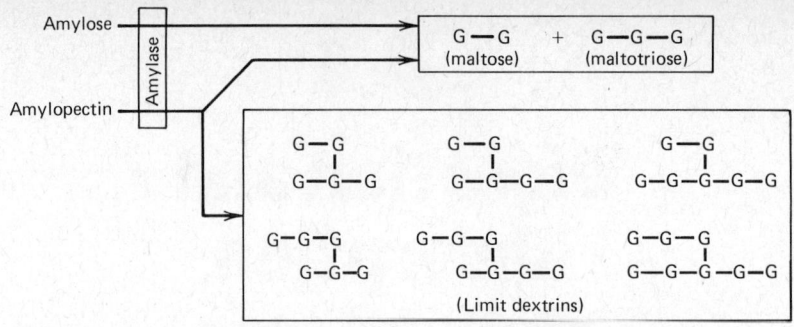

Figure 1 Digestion of starches by salivary and pancreatic amylase. G signifies a glucose molecule. Horizontal bonds between Gs are $\alpha - 1,4$ glucosidic bonds, verticals are $\alpha - 1,6$ bonds.

Carbohydrates, proteins, and fats—quantitatively, these are the major nutrients. Most are ingested as large polymers or insoluble lipids that cannot be absorbed as such. The polymers are hydrolyzed to smaller molecules; the fats are partially hydrolyzed and the hydrophobic products solubilized by detergents. These processes constitute digestion. The products of digestion are absorbed by the small intestinal epithelium. Virtually all the digestible carbohydrates, proteins, and fats we eat are absorbed. The normal intestine has considerable reserve capacity; huge amounts must be ingested before there is appreciable loss. Intestinal absorption does not regulate caloric input to metabolism; *ingestion* of food does.

CARBOHYDRATES

The principal nutritional carbohydrate is starch, of which there are two types: amylose and amylopectin. Amylose is a straight-chain polymer of glucose molecules* connected by α-1,4-glucosidic bonds. Amylopectin is a branched polymer of glucose in which the branches are formed by α-1,6-glucosidic bonds. Glycogen, another branched-chain glucose polymer, is ingested in much smaller amounts, from meat and liver. Glycogen is like amylopectin, except the branching is denser. The disaccharides,

*We always refer to the D, rather than the L, configuration of the monosaccharides.

sucrose and lactose, are common dietary constituents. Small amounts of the monosaccharides, glucose and fructose, are also eaten.

Cellulose is an abundant dietary glucose polymer, but the linkages are β-1,4-glucosidic bonds. We have no enzyme for catalyzing hydrolysis of this bond; therefore, these glucose molecules remain locked within the polymer, unable to be absorbed. The same is true of other less abundant polysaccharides, such as pectins and pentosans. Nondigestible polysaccharide fibers are important in the intestinal tract, acting osmotically to retain fluid in the lumen and, with the resulting bulk, promoting propulsive motility.

DIGESTION BY SALIVARY AND PANCREATIC AMYLASE

Salivary and pancreatic amylases appear to be identical. They attack the interior α-1,4-glucosidic bonds of starches and glycogen randomly. The main products of amylose digestion are maltose and maltotriose; only very small amounts of free glucose are produced. The products of amylopectin digestion are the same, but also include various small oligosaccharides called limit dextrins. The limit dextrins have an α-1,6-glucosidic bond that cannot be cleaved by salivary and pancreatic amylase. The digestion of starches is diagrammed in Figure 1.

Salivary amylase is inactivated at a low pH value and cannot function after food is mixed with gastric juice. However, it does function in the mouth and esophagus, and sometimes for a considerable time in the proximal stomach before mixing occurs. Gastric HCl can hydrolyze starch, but only at elevated temperatures; at body temperature this process is negligible. Most starch is digested in the upper small intestine by pancreatic amylase.

DIGESTION BY EPITHELIAL DISACCHARIDASES

The products of amylase action are maltose, maltotriose, and the limit dextrins. These, together with dietary sucrose and lactose, are hydrolyzed by a group of disaccharidases located in the apical membranes of absorptive cells. The active sites of these enzymes are accessible to the solution in the lumen. The most important of the membrane disaccharidases are maltase, isomaltase, sucrase, and lactase. Digestion by these enzymes is diagrammed in Figure 2. Maltose and maltotriose are converted to free glucose by maltase. Sucrose is hydrolyzed to glucose and fructose by sucrase. Lactose is hydrolyzed to glucose and galactose by lactase. The α-1,6 bonds of limit dextrins are cleaved by isomaltase to form maltose, maltotriose, and the four- and five-unit unbranched oligosaccharides, all of which are converted to glucose by maltase.

Since terminal carbohydrate digestion occurs on the surface of absorptive cells, glucose, galactose, and fructose are supplied directly to the monosaccharide absorptive machinery. About 80% of the monosaccharide presented for absorption is glucose, with 5% galactose and 15% fructose.

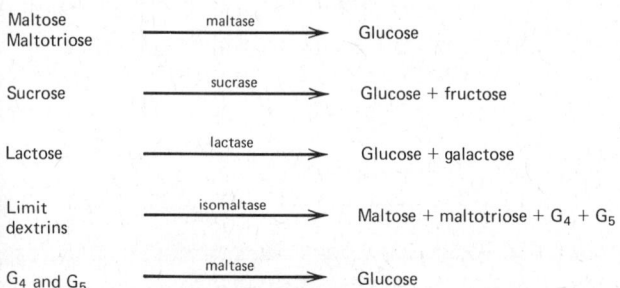

Figure 2 Digestion by membrane (brush-border) disaccharidases. G_4 and G_5 signify unbranched glucose oligosaccharides consisting of four glucose and five glucose molecules, respectively, joined by $\alpha-1,4$ bonds. The final products of these reactions are glucose, galactose, and fructose.

ABSORPTION OF GLUCOSE AND GALACTOSE

Hexoses are too large to penetrate tight junctions at rapid rates, and their absorption is almost entirely transcellular. They are also too large for membrane pores and too lipid insoluble for diffusion through the hydrocarbon bilayer of membranes. They must use carriers in both the apical and basolateral membranes of the absorptive epithelium. The carrier for glucose and galactose in the basolateral membrane accomplishes only facilitated diffusion, much like the sugar carriers in most nonepithelial cells; there is no uphill transport, no requirement for metabolic energy, and no Na^+ dependence. The carrier in the apical membrane is more interesting—it is Na^+-dependent and can accomplish uphill transport of glucose and galactose from the lumen into the epithelial cells. These sugars then accumulate in the cells at fairly high concentrations and move downhill across the basolateral membranes by facilitated diffusion; they then diffuse into capillaries. By means of the apical transport system, the epithelium can extract glucose and galactose from the lumen until there is virtually none left. During interdigestive periods the apical glucose–galactose transport system prevents escape of glucose from blood to lumen.

In addition to uphill transport, the glucose–galactose transport system displays all the other typical features of active transport systems discussed in Chapter 1. For example, specificity is illustrated by the fact that mannose cannot participate in this system, although it differs from glucose only in the configuration around C-2. Saturation kinetics and competitive inhibition can also be demonstrated, as can dependence on metabolism.

There is abundant evidence that apical glucose–galactose transport involves cotransport with Na^+; in fact, the concept of cotransport was first conceived for this transport process by R. K. Crane (1960). (Cotransport with Na^+ was discussed in Chapter 1.) The carrier in the apical membrane has two binding sites, one for Na^+ and the other for glucose or galactose. When both sites are occupied simultaneously, the carrier is somehow much more effective than when only one site is occupied. Consequently, luminal Na^+ is required for glucose and galactose influx, and these sugars stimulate Na^+ influx, that is, there is mutual facilitation. This is one of the Na^+ entry mech-

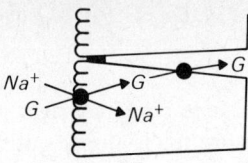

Figure 3 Na⁺–sugar cotransport. G, glucose or galactose.

anisms mentioned in Chapter 30 that helps make Na^+ available to the basolateral Na^+ pump. The pathway for Na^+-coupled sugar transport is illustrated in Figure 3.

The basolateral Na^+ pump keeps Na^+ concentration much lower within epithelial cells than in the lumen. Thus Na^+ has a strong tendency to move into the cells via its various apical entry paths. Glucose moves along with Na^+ into cells via the cotransport carrier, and this process can continue even after the concentration of glucose becomes higher in the cells than in the lumen. The downhill movement of Na^+ can drive an uphill movement of sugar. The electrical potential difference across the apical membranes can contribute an additional force for Na^+ influx and, consequently, for sugar influx.

The notion that the electrochemical potential gradient for Na^+ across the apical membrane can, by means of cotransport, provide energy for uphill transport of sugars, is known as the Na^+ gradient hypothesis. This hypothesis is universally accepted. However, the *degree* to which the Na^+ gradient provides the energy for sugar transport is not yet established. Many investigators believe that the Na^+ gradient provides *all* the energy required for sugar transport and that other sources of energy, such as adenosine triphosphate (ATP), are not directly used. This contention is supported by the fact that when the basolateral Na^+ pump is stopped with ouabain or other cardiac glycosides, sugar transport stops as well. On the other hand, it is possible to observe uphill sugar transport under conditions in which there is no stimulation of Na^+ transport. This observation would seem to indicate the presence of another energy source, but nothing is known about its nature.

ABSORPTION OF FRUCTOSE

Movement of fructose into the absorptive cells is mediated by a carrier separate from that for glucose and galactose. There is little need for this carrier to pump uphill, since fructose concentration in blood is quite low; nevertheless, a modest degree of uphill transport and Na^+ dependence has been demonstrated. The carrier for mediating fructose movement across the basolateral membrane could be the same as that for glucose and galactose, but this is not certain. In some animals much of the fructose that enters absorptive cells is converted to glucose. Conversion to glucose aids fructose absorption by maintaining a more favorable concentration gradient for movement into the cells. In humans, only a minor amount of fructose is converted to glucose in the intestine.

DISACCHARIDASE-RELATED TRANSPORT

When the Na^+-dependent glucose–galactose transport system is operating at maximal capacity (i.e., is saturated with a high concentration of glucose in the lumen), the rate of glucose transport across the epithelium can be increased still further by the addition of maltose, sucrose, or lactose. In an attempt to explain this observation, it has been proposed that the disaccharidases in the apical membrane sometimes release their products from the inner surface of the membrane *into the cells,* rather than always back to the outer surface, thus bypassing the Na^+-dependent monosaccharide transport mechanism. This interpretation is supported by the fact that in some experimental animals sugar transport does not require Na^+ if the monosaccharides are generated from disaccharides by membrane disaccharidases.

Disaccharidase-related sugar transport seems promising as a possible mechanism for generating high concentrations of glucose, galactose, and fructose within the absorptive cells without using any metabolic energy. However, disaccharidase-related transport has not yet been demonstrated in human intestine, and we must reserve judgment about its importance.

PROTEINS

Nearly all dietary amino acids* are ingested as proteins. Ingested protein accounts for roughly 50–70% of the pro-

---

*We refer exclusively to the L-, rather than the D-, amino acids.

tein that enters the small intestine—the rest comes from secretions and sloughed epithelium. Digestion and absorption of the latter materials, as well as dietary protein, are important for maintaining nitrogen balance.

DIGESTION BY PEPSIN

Protein digestion begins in the stomach by action of the proteolytic enzyme pepsin. The chief cells of the gastric glands, and to a lesser extent the pyloric glands, secrete pepsinogen, the inactive precursor of pepsin. When pepsinogen mixes with HCl from the oxyntic cells, pepsin is formed. Pepsin hydrolyzes *interior* peptide bonds of proteins, that is, it is an *endo*peptidase. Pepsin has two pH optima, one around pH 2.0 and the other at roughly 3.5. If pH gets much over 4.0, pepsin is inactivated.

Pepsinogen is changed to pepsin by hydrolytic detachment of about one-sixth its original length. The remaining five-sixths (about 35,000 daltons) is pepsin. This activating hydrolytic process is accomplished first by HCl. The pepsin initially formed then acts autocatalytically on pepsinogen to form more pepsin, which in turn causes more activation. This positive feedback results in an explosive generation of pepsin, and all the pepsinogen is quickly consumed.

Actually, there is not just one single pepsin, but rather a group of pepsins, at least four in number—perhaps as many as seven. The various pepsins have different specificities toward interior peptide bonds, the overall result of which is a fairly random attack. Pepsins do not carry protein digestion very far. Polypeptides of various lengths are generated, and little free amino acid is produced. Even if pepsins are absent, as after gastrectomy, digestion of proteins by pancreatic proteases is usually adequate.

DIGESTION BY PANCREATIC PROTEASES

Five protease precursors are secreted by the acinar cells of the pancreas: trypsinogen, chymotrypsinogen, proelastase, procarboxypeptidase A, and procarboxypeptidase B. These precursors are converted to the active proteolytic enzymes, trypsin, chymotrypsin, elastase, carboxypeptidase A, and carboxypeptidase B after arriving in the duodenum. Activation is initiated by an enzyme, enterokinase, which is located on the surface of duodenal epithelial cells. Enterokinase splits a hexapeptide from the N-terminal end of trypsinogen; the remainder is trypsin. Then trypsin itself converts more trypsinogen to trypsin autocatalytically. Activation of the other pancreatic proteases is catalyzed by trypsin. Activation of pancreatic proteases in the duodenum is diagrammed in Figure 4.

Trypsin, chymotrypsin, and elastase are endopeptidases, that is, they attack interior peptide bonds. Trypsin preferentially attacks peptide bonds formed with a basic amino acid. Chymotrypsin attacks peptide bonds formed with an aromatic amino acid. Elastase cleaves bonds between aliphatic amino acids. The concerted action of these three enzymes produces peptides of various lengths. Trypsin generates peptides having basic amino acids at their C-terminal ends, chymotrypsin generates peptides having neutral aromatic C-terminal units, and elastase generates peptides having neutral aliphatic C-terminal units.

The carboxypeptidases are exopeptidases, that is, they attack carboxyterminal peptide bonds. Carboxypeptidase A lops off aliphatic and aromatic neutral amino acids, while carboxypeptidase B liberates basic amino acids. Digestion by pancreatic proteases is diagrammed in Figure 5. The end result is neutral amino acids, basic amino acids, and a large variety of dipeptides, tripeptides, and oligopeptides having up to six amino acid units. More than 60% of the amino acids in the lumen of the small intestine during digestion are in the form of small peptides, the rest are monomers.

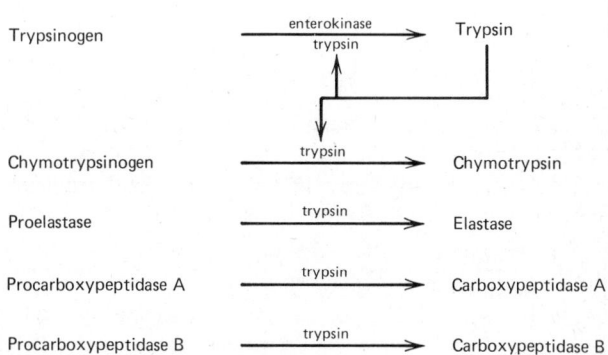

Figure 4 Activation of pancreatic proteolytic proenzymes, initiated by enterokinase and pursued by trypsin.

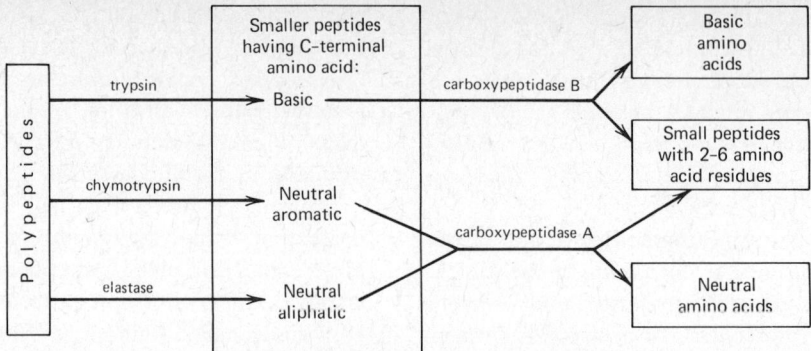

Figure 5 Digestion by pancreatic proteases.

DIGESTION BY MEMBRANE OLIGOPEPTIDASES

A group of oligopeptidases are located in the apical membranes of absorptive cells. Their active sites are accessible from the lumen. These oligopeptidases reduce the length of the small peptides produced by pancreatic proteases, generating amino acids, dipeptides, and tripeptides on the surface of absorptive cells. Membrane oligopeptidases are aminopeptidases, that is, they hydrolyze peptide bonds from the amino-terminal end of the chain, rather than from the carboxy-terminal end. The number and substrate specificities of these enzymes are not certain.

ABSORPTION OF AMINO ACIDS

Amino acids are absorbed by processes very much like that for glucose, with Na^+-dependent carriers capable of pumping uphill in the apical membrane and Na^+-independent facilitated diffusion processes in the basolateral membrane. The Na^+ gradient hypothesis is widely accepted for amino acid transport, but, again, the *degree* to which amino acid transport is energized by the Na^+ gradient rather than by other means is not known.

There are at least four different amino acid transport systems in apical membranes: two for neutral amino acids, one for basic amino acids, and one for acidic amino acids. Both neutral amino acid transport systems, which we shall designate N_1 and N_2, are used by most neutral amino acids, but each system shows preference for particular structural features. Neutral amino acids having large hy-

drophobic side chains, such as methionine, leucine, and phenylalanine, have greater affinity for N_1 than do amino acids with short side groups, such as glycine. Glycine and the imino acids (proline and hydroxyproline) are absorbed better by N_2 than by N_1. The pump for the basic amino acids, lysine and arginine, also handles cystine, a neutral amino acid. This is because cystine, like the basic amino acids, has two amino groups. This structural property confers upon cystine much more affinity for the basic system than for the N systems.

ABSORPTION OF DIPEPTIDES

Dipeptides are also actively transported into absorptive cells across their apical membranes. Whether there is just one or a whole group of dipeptide transport systems is uncertain. Tripeptides might also be transported by this system, but evidence for this is not conclusive in humans. Na^+ dependence has been reported, but the evidence for this dependence in humans is conflicting. It is likely that as much as 50% of the amino acids absorbed enter the epithelial cells as dipeptides or tripeptides. Thus the importance of this system is obvious despite the present obscurity of its details.

After entering the epithelial cells, di- and tripeptides are hydrolyzed to free amino acids by peptidases located in the cytoplasm. Nearly all the amino acid that diffuses into the vascular capillaries is in the monomere form, except for dipeptides containing the imino acids proline and hydroxyproline; appreciable amounts of the imino acids enter blood as dipeptides.

ABSORPTION OF INTACT PROTEINS

Intact proteins can be absorbed by pinocytosis, but only in very small, nutritionally unimportant, amounts. Pinocytotic vesicles develop between the microvilli of absorptive cells; they pass through the cytoplasm and extrude their contents by exocytosis at the basolateral membranes. Although the proteins absorbed by this process are too large to enter blood capillaries, lymph capillaries are more permeable. Consequently, the small amounts of protein absorbed by pinocytosis are carried from the intestine in lymph, rather than in blood. A good illustration of this point is the finding that diversion of intestinal lymph from rats (by cannulation of the thoracic duct) protects them from the toxic effects of ingested botulinus toxin. Food allergies are probably caused by absorption of intact proteins by pinocytosis in antigenically significant quantities.

Absorption of proteins by pinocytosis has special importance for newborn animals of certain species. For horses and pigs, absorption of antibodies from colostrum is the only way in which passive immunity can be transmitted from mother to offspring. In such animals, intestinal pinocytosis is exceptionally well developed for the first few days or weeks of life. In many mammals, prenatal transfer of antibodies occurs across the placenta, conferring passive immunity. In humans placental transfer is well developed, whereas neonatal intestinal transfer is poorly developed.

It was recently discovered that the ileum is capable of absorbing intact pancreatic enzymes. The absorbed enzymes return to the pancreas, where they can be resecreted, thus forming an enteropancreatic circulation of enzymes. A large fraction of secreted pancreatic enzymes is digested in the gut; however, the surviving fraction can be recycled by this ileal transport system, thereby reducing to some extent the necessity for the pancreas to synthesize new enzymes. The mechanism of transport of intact enzyme is not understood, but there is some evidence that it is a Na^+-coupled process. Other proteins are not transported by this system.

LIPIDS

Triglyceride fats are the most abundant dietary lipids; phospholipids (mainly lecithin), cholesterol, and the lipid-soluble vitamins (A, D, E, and K) are also ingested. In the stomach, fats tend to separate out into oily globs distinct from the aqueous and particulate phases of chyme. Large, oily globs are reduced to smaller globules by churnings of the distal stomach; these smaller globules are pumped into the duodenum as a coarse emulsion—just how coarse is not known. The main problem with absorbing lipids is not at the cell membrane level, as it is with hydrophilic nutrients. Lipid molecules readily penetrate cell membranes if given a chance. The problem is that since they are nearly insoluble in water, they cannot diffuse to the absorptive epithelium at a significant rate; they must first be solubilized by detergent. Bile salt is the main detergent of the intestinal tract. Mixed micelles consisting of bile salt, lecithin, and cholesterol enter the duodenum in bile. These micelles solubilize lipid molecules and carry them to the epithelium. The micelles cannot solubilize triglycerides very well; this problem is overcome by lipases, which hydrolyze triglycerides to monoglycerides, and fatty acids, which the micelles can solubilize quite nicely.

DIGESTION BY LINGUAL LIPASE

There are serous glands associated with the circumvallate papillae in the base of the tongue that secrete a lipase. This lingual lipase hydrolyzes one ester bond of triglycerides, producing diglycerides and fatty acids. Lingual lipase can function within a broad acidic pH range of about 2–6; therefore, it digests fats during their stay in the stomach. The degree to which triglycerides are hydrolyzed by lingual lipase in the stomach is not known; it is probably only slight for most fats. Milk fats, however, might be especially susceptible to this enzyme, and an appreciable percentage of their hydrolysis might occur in the stomach.

EMULSIFICATION IN THE DUODENUM

As stated earlier, bile salts are detergents. Detergency consists of two different processes: emulsification and solubilization. Emulsification will be considered here.

All detergents are amphiphiles, that is, they have both hydrophilic and lipophilic regions. Bile salts are rather flat steroids having two or three hydroxyl groups all on one surface, making that surface hydrophilic; the other side is lipophilic. Because of this dual affinity, bile salts tend to accumulate at the interface between fat globules and the surrounding aqueous phase, with their lipophilic sides dissolved in fat and their hydrophilic sides dissolved in water. Interfacial accumulation of detergent reduces the surface tension of fat globules, enabling them to be broken into smaller globules. Duodenal motility provides the required mechanical agitation and leads to emulsification of fat globules into very small oily particles having diameters of roughly 1–5 μm. Since their surfaces are covered with hydrophilic hydroxyl groups, the globules tend not to coalesce. The negatively charged side chains of conjugated bile salts also promote emulsion stability. Dietary phospholipids and the products of lingual lipase, being amphiphiles themselves, also accumulate at the oil–water interface and contribute to lowering of surface tension and emulsification. Emulsification in the duodenum greatly increases the area of fatty surface on which pancreatic lipase can operate.

DIGESTION AT THE OIL–WATER INTERFACE BY PANCREATIC ESTERASES; THE ROLE OF COLIPASE

Three esterases in pancreatic juice catalyze hydrolysis of lipids, namely, pancreatic lipase, phospholipase A_2, and cholesterol esterase. Pancreatic lipase is secreted in active form, with no intraluminal activation required. It splits off a fatty acid from triglycerides and another from diglycerides, leaving 2-monoglycerides and long-chain fatty acids. Pancreatic lipase performs on the surface of fatty globules. Curiously, pancreatic lipase cannot get at this surface when bile salt and phospholipid molecules cover it. This dilemma is overcome by pancreatic phospholipase and by a small protein, also secreted by the pancreas, called colipase. Pancreatic phospholipase, which is secreted as a proenzyme and activated by trypsin, hydrolyzes an ester bond of lecithin (at the 2 position), producing fatty acids and lysolecithins. These products leave the surface to be incorporated into biliary micelles. Coli-

pase attaches to the bile salt coat of fatty emulsion globules. Pancreatic lipase binds to colipase and, having become bound, gains access to the surface triglycerides, converting them to 2-monoglycerides and fatty acids.

SOLUBILIZATION BY BILIARY MICELLES

The liver secretes bile salts, lecithin, and cholesterol. Lecithin is not soluble in water, and it has a tendency to form extensive bilayers. Cholesterol molecules, also not soluble in water, interdigitate among the lecithin molecules in these bilayers. Bile salts fragment the bilayers into tiny bilayer disks rimmed with bile salt molecules. The resulting aggregates of bile salt, lecithin, and cholesterol are called mixed micelles. Biliary micelles have a diameter of roughly 30–100 Å. Lipophilic on the inside and hydrophilic on the outside, they are small enough and externally hydrophilic enough to be completely soluble in water.

The products of pancreatic lipase and phospholipase—fatty acids, monoglycerides, and lysolecithin—also interdigitate among lecithin molecules in biliary micelles and become solubilized in chyme. As pancreatic lipase and phospholipase generate fatty acids, monoglycerides, and lysolecithin at the surface of fatty globules, these products are carried off by biliary micelles into the aqueous medium. By mass action, removal of products promotes rapid hydrolysis at the surface of fatty globules.

Dietary cholesterol and cholesterol esters also shift from fatty emulsion droplets to biliary micelles. The same is true of lipid vitamins and of certain unhydrolyzed phospholipids, such as sphingomyelin.

It is important to note that the liver secretes about three to five times the amounts of cholesterol and lecithin ordinarily ingested. One day's production of bile includes as much cholesterol and lecithin as found in five or six eggs.

DIGESTION OF MICELLAR LIPID

The role of pancreatic lipase is completed once its products enter the micelles. Nevertheless, although the evidence is conflicting, pancreatic phospholipase probably continues to act on micellar lecithin, converting most dietary and biliary lecithin to fatty acids and lysolecithin. Cholesterol esterase acts mainly on micellar cholesterol esters, rather than at the surface of fat globules; its products are cholesterol and fatty acids. For some mysterious reason, all cholesterol must be converted to the unesterified form before it can enter the absorptive cells.

DIFFUSION TO THE EPITHELIAL SURFACE

Between the intestinal villi lies a relatively stagnant unstirred region. The villi are also covered with an unstirred aqueous layer that extends into the lumen for a surprising distance, roughly 100–400 μm. The thickness of this unstirred layer depends on the vigor of intestinal motility, but it is probably never reduced below about 100 μm. Anything that is absorbed must first diffuse across the unstirred layer. The problem is not severe with hydrophilic nutrients, since they readily dissolve in the lumen, thereby creating favorable concentration gradients for diffusion across the unstirred layer.* For lipids, diffusion across the unstirred layer constitutes a major problem, since they cannot dissolve at high enough concentrations in the mainstream of intestinal contents to diffuse at significant rates across the unstirred layer. This problem is surmounted by the process of solubilization in biliary micelles. The micelles are soluble in water and, with their cargo of solubilized fatty acid, monoglyceride, lysolecithin, other phospholipids, cholesterol, and lipid vitamins, they diffuse to the surface of absorptive cells. Thus, the biliary micelles, besides functioning as a sink for removing the products of lipid digestion from the surface of fatty globules, are also carriers that facilitate the diffusion of lipids across the unstirred aqueous layer. They are analogous to membrane carriers that facilitate the diffusion of hydrophilic materials across lipid layers.

ENTRY INTO EPITHELIAL CELLS

Fatty acids, monoglycerides, and so forth, are not absolutely insoluble in water, so wherever the micelles go, there is a very low monomolecular, as opposed to micellar,

---

*Nevertheless, during the terminal stages of carbohydrate and protein absorption, when the concentrations of absorbable products become very low in the mainstream of intestinal contents, diffusion across the unstirred layer limits the rate of the entire absorption process.

concentration of these materials in the surrounding water. At the epithelial surface, dissolved lipid molecules readily partition into the apical membranes. As this happens, more molecules come out of the micelles into the surrounding water, moving on across apical membranes, all by mass action and passive diffusion. The bile salts remain in the lumen, mostly in micellar form; they are absorbed later. Some of the cholesterol also remains in the lumen; only about 40% is absorbed. Within the cells, fatty acids and presumably other lipids bind to specific cytoplasmic carrier proteins; little is known about these carrier proteins, except that they serve the important function of carrying lipid molecules from apical membranes to endoplasmic reticulum.

INTRACELLULAR METABOLISM AND TRANSPORT TO LYMPH

When fatty acids and monoglyerides reach the smooth-surfaced endoplasmic reticulum in the supranuclear region of absorptive cells, they are converted back to triglycerides by at least two different synthetic processes. The most important of these is direct reesterification of absorbed monoglycerides by absorbed fatty acids; the fatty acids must first be activated by formation of fatty acid–coenzyme A (CoA) complexes, a process catalyzed by fatty acid thiokinase. A portion of the lysolecithin absorbed is reacylated to lecithin; the rest is further hydrolyzed intracellularly to fatty acids, glycerol, phosphate, and choline. Roughly one-half of absorbed cholesterol is reesterified in the cells.

The triglyceride synthesized in the cells accumulates as minute droplets within the endoplasmic reticulum and Golgi apparatus. These droplets grow in size as more triglyceride is synthesized; they are then adorned with a hydrophilic coating of lecithin and β-lipoprotein, also synthesized by the endoplasmic reticulum. Some absorbed cholesterol is dissolved within these droplets. The resulting complex of triglyceride, cholesterol, lecithin, and β-lipoprotein is called a chylomicron. The diameter of chylomicrons ranges from about 750 Å to 10,000 Å. Their composition is roughly 85% triglyceride, 10% phospholipid, 3% cholesterol and cholesterol esters, and 2% protein. Chylomicrons are extruded from the absorptive cells across the lateral membranes by exocytosis. They then move across the epithelial basement membranes into lymph. The intercellular junctions of lymph capillaries are permeable enough to permit entry of chylomicrons. Blood capillaries are not this permeable; consequently, most absorbed fat leaves the intestine in lymph, rather than in blood.

Another lipoprotein particle generated in the intestinal epithelium is the very low density lipoprotein (VLDL). VLDL particles are smaller than chylomicrons (about 300–500 Å). VLDLs have higher proportions of cholesterol, phospholipid, and protein, but less triglyceride. VLDL particles are produced in the epithelium even when no dietary fat is being absorbed. Their full significance is not understood, but they might serve as an important vehicle for transporting cholesterol from the intestine.

Absorption of short-chain fatty acids, such as acetic, propionic, butyric, and hexanoic, which are derived from butter and bacterial metabolism, is altogether different from that of long-chain fatty acids. Short-chain fatty acids are sufficiently water soluble not to require solubilization in micelles, but they are sufficiently hydrocarbon soluble to diffuse readily across the absorptive cells into portal blood. The short-chain fatty acids (C-2 to C-6) are not synthesized to triglycerides in the absorptive cells and do not enter chylomicrons.

Medium-chain fatty acids (C-8 to C-10), derived from so-called medium-chain triglycerides, are also readily absorbed without solubilization by bile salt micelles. They are also not extensively incorporated into triglycerides and chylomicrons by the absorptive cells, and they do diffuse into portal blood. Medium-chain triglycerides are used clinically as a caloric substitute for long-chain fats in cases of bile salt deficiency and in the rare disease, abetalipoproteinemia, characterized by defective formation of chylomicrons.

ABSORPTION OF BILE SALTS

The body normally contains about 3–4 g conjugated bile salts, nearly all of which are restricted to the enterohepatic circulation (liver, gallbladder, bile ducts, small and large intestine, and portal vein). This bile salt pool circulates enterohepatically about six to eight times per day. Thus, about 24 g bile salt enters the duodenum each day. Less than 1 g/day escapes intestinal absorption and is excreted in feces. The 23 g or more that is absorbed is returned to the liver in portal blood. The liver efficiently clears portal blood of bile salt and resecretes it toward the duodenum. Bile salt lost in feces is made up by new synthesis in the liver from cholesterol, but resynthesis requirements are less than 3% of the bile salts used in the gut. The enterohepatic circulation is a remarkable conservation mechanism.

The total concentration of conjugated bile salts in the jejunum during digestion of a meal is about 4–6 mM, primarily in the form of micelles. Concentration of bile salts in mesenteric blood is almost zero, so there is always a large concentration difference favoring absorption by passive diffusion. However, at the pH level of intestinal contents, conjugated bile salts are mainly ionized and the absorptive cells are nearly impermeable to the ionic form. Only a small amount of conjugated bile salt is normally absorbed in the duodenum and jejunum by passive diffusion. Consequently, the concentration of conjugated bile salt remains high in the duodenum and jejunum, where fat digestion and absorption take place, providing a clear physiological advantage.

Bile salts are mainly absorbed in the lower ileum, via two important mechanisms. First, various bacteria that normally live in the lower ileum enzymatically deconjugate bile salts. Thus, a considerable amount of unconjugated bile salt is produced in the ileum. Unconjugated bile salts—mainly cholate and chenodeoxycholate—have much higher pK values (about 6) than do the corresponding conjugated bile salts. Consequently, an appreciable amount of bile salt is undissociated (i.e., is in the bile acid form). The absorptive cells are faily permeable to undissociated bile acids, and considerable amounts are absorbed by passive nonionic diffusion. Second and most important, absorptive cells in the lower ileum possess an active transport system for bile salts; this transport system retrieves a large percentage of ileal bile salt and is an extremely important link in the recycling process. The ileal bile salt transport system is Na$^+$ dependent. Some bile salt escapes active transport in the ileum and is deconjugated by colonic bacteria. An appreciable amount of deconjugated bile acid is absorbed in the colon by nonionic diffusion.

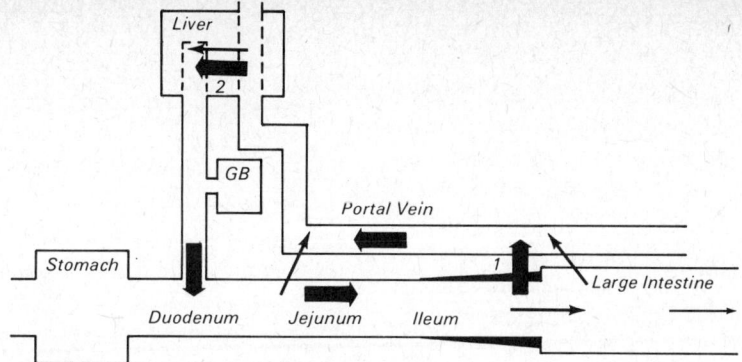

Figure 6 The enterohepatic circulation of bile salts. The active transport systems in the lower ileum and hepatic parenchymal cells have been labeled 1 and 2, respectively.

Figure 6 depicts the pathways taken by bile salts during their enterohepatic circulation. The active transport mechanisms—one across the ileal epithelium and another across the hepatic parenchymal cells—keep bile salts where they belong (i.e., in the enterohepatic circulation) and, except for small amounts, out of places where they do not belong (i.e., systemic blood and colon). A number of important clinical problems arise when one or the other of these transport systems becomes defective and bile salts are no longer kept in their proper place. These matters are beyond the scope of this chapter, but for your own edification you might try to figure out why resection of the lower ileum can lead to steatorrhea, diarrhea, weight loss, reduced blood cholesterol, and gallstones.

WATER-SOLUBLE VITAMINS

VITAMIN B$_{12}$

Vitamin B$_{12}$ is required for normal maturation of red blood cells during hematopoiesis. Failure of the small intestine to absorb vitamin B$_{12}$ results in pernicious anemia. The most common cause of failure to absorb enough vitamin B$_{12}$ is lack of a mucoprotein secreted by gastric oxyntic cells, called intrinsic factor. Ingested vitamin B$_{12}$ is loosely bound to ingested proteins. In the acid environment of the stomach, B$_{12}$ is released from ingested proteins and becomes bound to intrinsic factor. In the absence of gastric acid, a condition known as achlorhydria, binding to intrinsic factor does not occur, and pernicious anemia can result. Vitamin B$_{12}$ is not absorbed very well in the duodenum and jejunum. In the ileum the B$_{12}$–intrinsic factor complex binds to specific receptor sites in the glycocalyx of absorptive cells. These binding sites are not present in the upper small intestine. After binding, B$_{12}$ is released from intrinsic factor and transported into portal blood. The mechanism of transport is not understood.

L-ASCORBIC ACID (VITAMIN C)

Ascorbic acid is absorbed mainly by an active transport system closely resembling those for monosaccharides, amino acids, and bile salts. The uphill step occurs at the apical membranes and apparently involves cotransport with Na$^+$. This transport system is preferentially located in the ileum. It is present in primates and guinea pigs, species that require ascorbic acid in the diet. It is absent in other mammals in which adequate endogenous ascorbic acid can be synthesized.

Ascorbic acid can also be absorbed passively all along the small intestine, most likely by the paracellular route. This mechanism probably becomes important only after very large amounts of the vitamin are ingested.

OTHER WATER-SOLUBLE VITAMINS

Other water-soluble vitamins have not been adequately studied. However, there is evidence for active absorption of biotin, nicotinic acid, thiamine, and folic acid by Na^+-dependent processes. There is also evidence for carrier-mediated transport, not necessarily uphill, of choline and riboflavin. No convincing evidence for carrier-mediated transport of other vitamins has been published, possibly in part because of inadequate investigation.

SELECTED BIBLIOGRAPHY

Crane RK: Digestion and absorption: Water-soluble organics. *Int Rev. Physiol 12: 325, 1977.*

Gray G, Cooper H:. Protein digestion and absorption. *Gastroenterology* 61:535, 1971.

Rommel K, Goebell H, Böhmer R: *Lipid Absorption: Biochemical and Clinical Aspects.* Baltimore, University Park Press, 1976.

Rose RC: Water-soluble vitamin absorption in the intestine. *Annu Rev Physiol* 42:157, 1980.

Silk DBA, Dawson AM: Intestinal absorption of carbohydrate and protein in man. *Int Rev Physiol 19:151, 1979.*

Simmonds WJ: Absorption of lipids, in Jacobson ED, Shanbour LL (eds): *Gastrointestinal Physiology.* Baltimore, University Park Press, 1974, p 343.

Sleisenger MH, Kim YS: Protein digestion and absorption. *N Engl J Med* 300:659, 1979.

Taylor WH: Biochemistry of pepsins, in C. F. Code (ed): *Handbook of Physiology. section 6: Alimentary Canal, vol V;* (section ed). Baltimore, Williams & Wilkins, 1968, p 2567.

SELF-STUDY QUESTIONS

TRUE OR FALSE

1. Pancreatic amylase cannot hydrolyze the α-1,6-glucosidic bonds of amylopectin and glycogen.

2. Chymotrypsinogen is activated directly by enterokinase.

3. Elastase is a pancreatic endopeptidase that has major specificity for peptide bonds where there is a neutral aliphatic amino acid.

4. All the aminopeptidases are located on the outer surface of absorptive cells.

5. You should expect that bile salt absorption in the lower ileum could be drastically inhibited by ouabain.

6. You should expect that bile salt absorption in the colon could be drastically reduced by ouabain.

MULTIPLE CHOICE

Select the best answer(s). (In many instances, more than one answer is correct.)

7. Most of the free glucose liberated by hydrolysis during digestion of a potato results from the action of
   A. salivary and pancreatic amylase.
   B. enzymes located at the apical surface of absorptive cells.
   C. sucrase.
   D. maltase.

8. The rate of glucose absorption by small intestinal epithelium studied in vitro
   A. can be decreased by eliminating Na^+ from the lumen.
   B. can be decreased by adding galactose to the lumen.
   C. can be decreased by adding fructose to the lumen.
   D. is a linear function of glucose concentration in the lumen.

9. You should expect that galactose in the lumen of the small intestine

A. increases the transepithelial electrical potential difference.

B. depolarizes the apical membrane.

10. Which of the following is not absorbed in human small intestine by a Na^+-dependent active transport process?

A. Methionine.

B. Taurocholate.

C. Vitamin C.

D. Vitamin D.

E. Galactose.

11. You should expect that L-methionine absorption by the small intestine can be inhibited by

A. ouabain.

B. L-leucine.

C. cyanide.

12. A fairly common genetic defect that leads to failure of the small intestine (and of the renal tubules) to absorb cystine at normal rates is called cystinurea. You should expect a patient with cystinurea also to have defective absorption of

A. lysine.

B. phenylalanine.

C. glycine.

D. aspartic acid.

13. Colipase is secreted by

A. lingual glands.

B. stomach.

C. liver.

D. pancreas.

E. small intestine.

14. Important functions of biliary micelles (mixed micelles) include

A. serving as a "sink" into which the products of pancreatic lipase and phospholipase enter.

B. solubilization of biliary cholesterol.

C. carrying fatty acids and monoglycerides through the unstirred layer covering the absorptive epithelium.

D. carrying lipids through the apical cell membrane of absorptive cells.

15. Intestinal absorption of bile salts

A. takes place mainly in the lower ileum.

B. is aided by bacterial deconjugation.

C. is needed to prevent steatorrhea.

D. normally results in recycling, enterohepatically, at least 95% of the bile salt secreted by the liver.

16. Fatty acids and monoglycerides enter the intestinal epithelial cells across their apical membranes mainly

A. in the lower ileum.

B. by pinocytosis.

C. while solubilized by bile salt–lecithin micelles.

D. by monomolecular diffusion.

17. Medium-chain fatty acids (C-8 and C-10)

A. do not require bile salts for absorption.

B. are actively transported by small intestinal epithelium.

C. leave the intestine, following absorption, mainly via lymph.

D. can be well absorbed by people with abetalipoproteinemia.

18. In certain clinical conditions, in which large numbers of bacteria infect the upper small intestine, you should expect that

A. greater than normal amounts of bile salt are absorbed by the upper small intestine.

B. steatorrhea might be present.

C. the total amount of bile salt in the enterohepatic circuit is greatly reduced.

19. Intestinal absorption of vitamin B_{12}

A. takes place mainly in the ileum.

B. is greatly augmented by intrinsic factor.

C. is greatly reduced in achlorhydria.

ANSWERS

1. True. It hydrolyzes only α-1,4 glucosidic bonds.

2. False. Enterokinase initiates activation of trypsinogen. Trypsin activates all the other pancreatic peptidases, as well as pancreatic phospholipase A_2.

3. True. You should be able to answer similar statements about the other peptidases.

4. False. Some are located in the cytoplasm.

5. True. Active bile salt absorption is Na^+ dependent.

6. False. There is no active bile salt absorption in the colon.

7. B, D. Little or no free glucose is released by amylase, since it acts only on interior bonds. Most glucose is released by maltase, most of which is located on the apical surface of absorptive cells. The main carbohydrate in a potato is starch; sucrase has no action on either starch or its hydrolysis products.

8. A, B. Galactose competes with glucose; fructose does not. The rate of glucose absorption is a saturable function of glucose concentration. In vitro studies are specifically referred to here since, surprisingly, Na^+ dependence has not been clearly demonstrated under in vivo conditions.

9. A, B. Since Na^+ enters the cells along with galactose, the apical membrane depolarizes. Increased Na^+ entry across the apical membrane leads to increased electrogenic Na^+ extrusion across the basolateral membrane, thereby increasing the transepithelial potential.

10. D.

11. All are correct. Ouabain inhibits all Na^+ cotransport systems, presumably indirectly by interfering with the Na^+–K^+ pump. Leucine competes with methionine for the N_1 transport system. Cyanide inhibits oxidative metabolism, thus inhibiting the Na^+ pump.

12. A. Lysine is a basic amino acid; it is transported by the same system that transports cystine. The others are not. In this disease, adequate amounts of the basic amino acids and cystine can be absorbed as di- and tripeptides, even though the free amino acids cannot be absorbed.

13. D.

14. A, B, C. The micelles do not carry lipids across the cell membranes, only through the aqueous medium over the surface of the epithelium.

15. All are correct. Deconjugation in the ileum and colon aids absorption, by elevating the pK of the bile salts and thereby generating some undissociated bile acid. However, active transport in the lower ileum is by far the most important means of reabsorption.

16. D. The ileum can absorb lipids, but ordinarily, the upper intestine absorbs most ingested lipid, since it has more opportunity to do so. Pinocytosis is a discarded theory of fat absorption.

17. A, D. Consequently, medium-chain fats are sometimes useful for providing calories in patients with bile salt deficiency or with abetalipoproteinemia.

18. A, B. Deconjugation in the upper small intestine leads to increased absorption by nonionic diffusion and a short circuiting of the enterohepatic circulation. The unconjugated bile salts are relatively poor micelle formers and also tend to precipitate. They can also damage the intestinal epithelium to some extent. For all these reasons, steatorrhea is likely.

19. All are correct. The advantage of preferential ileal absorption of vitamin B_{12} is not clear. Achlorhydria (absence of gastric acid) reduces binding to intrinsic factor (IF). In addition, most patients with achlorhydria also have deficiency of intrinsic factor, since both HC1 and IF are secreted by the oxyntic cells.

32

Renal Physiology

DOUGLAS C. EATON
JOHN M. RUSSELL

OBJECTIVES

After completion of this chapter, you should be able to:

1. Understand the important relationship between renal anatomy and histology and renal function.

2. Know the factors affecting the glomerular filtration rate.

3. Know the discrete processes responsible for the tubular handling of Na^+, K^+, Cl^-, H^+, HCO_3^-, glucose, amino acids, and water and how their interrelationships can influence body fluid and urine composition.

4. Understand the countercurrent mechanisms, both the multiplier and the exchanger, as well as how they operate and maintain a high medullary osmotic pressure.

5. Understand the importance of the high medullary interstitial osmolarity for the maintenance of body fluid composition.

6. Understand the means by which the kidney is believed to operate in maintaining a constant extracellular fluid environment for the rest of the body.

7. Understand the concept of renal clearance for both solutes and water.

The primary function of the kidneys is to maintain the constancy of the fluid environment that bathes the cells of the body. This role involves the excretion of unwanted substances, for example, the end products of metabolism and excess salt, as well as the retention of necessary substances. The kidneys accomplish this task by first filtering the blood free of cells and of large protein molecules and then selectively removing from the resultant filtrate those substances required by the body in the amounts necessary for normal bodily functions. In addition to filtering out unwanted substances, the kidney is also able to secrete some of these substances directly into the filtrate. The remarkable thing about this organ is that it is able to perform these functions over a wide range of blood pressures, fluid volumes, fluid osmolarities, and salt loads, enabling individual cells of the body that have ionic and metabolic requirements within only narrow tolerance ranges to function normally.

ANATOMY OF THE KIDNEY

GROSS MORPHOLOGY

The kidneys are paired organs lying behind the peritoneum on either side of the vertebral column. In the average human adult, they are about 11 cm long, 5–7.5 cm wide, and 2.5 cm thick. Together they weigh approximately 300g, or only about 0.4% of the total body weight. Located on the concave or medial surface of each kidney is a slit called the hilus through which pass the renal artery and vein, the lymphatics, nerves, and the renal pelvis. Each kidney is completely covered by a loosely adherent, tough, fibrous capsule.

A kidney bisected along its long, narrow axis, shows a characteristic appearance, as in Figure 1. On the surface, two distinct regions are detected: a pale inner region, the medulla, and a darker outer region, the cortex. The

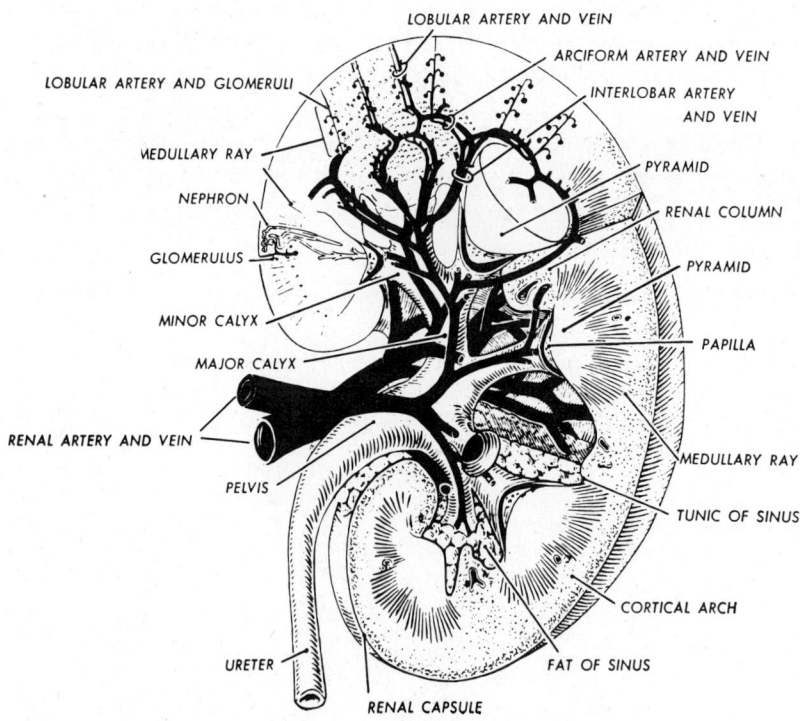

Figure 1 Gross structure of the kidney, sagittal section. (After Smith H: *Principles of Renal Physiology.* New York, Oxford University Press, 1956.)

medulla in the human kidney is divided into eight to 18 striated conical masses called the renal pyramids. The base of each pyramid is at the corticomedullary boundary, and the apex forms a papilla that extends into the renal pelvis. There are 10 to 25 small openings in each papilla, which represent the distal ends of the collecting ducts of Bellini. The cortex in the human kidney is about 1 cm thick, forming a cap over the pyramids and extending between them to form renal columns.

The renal pelvis has two and sometimes three out-pouchings, the major calices. From each major calix, several minor calices extend toward the papilla and drain the urine formed by each pyramidal unit. The walls of the calices and the pelvis contain smooth muscle, which contracts rhythmically to propel urine into the ureter.

MICROANATOMY

The functional unit of the kidney is the nephron. Each human kidney contains about 1.3×10^6 nephrons. The essential components of a nephron are displayed in Figure 2. The filtration takes place across the glomerular capillaries into *Bowman's capsule,* which together with the glomerular capillaries, form a distinct part of the nephron called the *glomerulus.* Bowman's capsule then narrows to form the renal tubule, which can be subdivided into three main regions: the proximal convoluted tubule, a loop of Henle, and a distal convoluted tubule that empties into a collecting duct.

The human kidney possesses two populations of nephrons: (1) those situated entirely within the cortex and outer medulla, called *cortical* nephrons, and (2) those having glomeruli that originate near the corticomedullary junction and processes extending from the inner cortex into the inner medulla, called the juxtamedullary nephrons. The cortical nephrons constitute about 85% of all the nephrons in the human body and have short loops of Henle with very short or nonexistent thin-limb segments. By contrast, juxtamedullary nephrons have well-developed descending and ascending thin segments in their loops of Henle (Fig. 2). This is believed to have considerable functional significance, as will be seen.

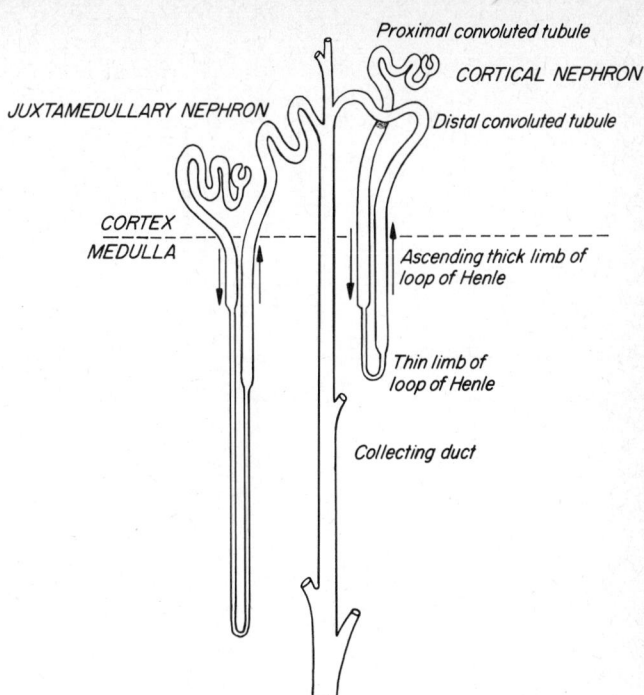

Figure 2 Schematic diagram showing the two general types of renal tubules with their characteristic subdivisions.

BLOOD SUPPLY TO THE NEPHRONS

The gross blood supply to the kidney is illustrated in Figure 3. The renal artery enters the hilus and divides into a series of branches that pass between the calices and penetrate the parenchyma between the calices. These are called the interlobar arteries.

At the junction between the cortex and medulla, the interlobar arteries bend over the bases of the pyramids to form a series of incomplete arches called the arciform arteries. Arising at right angles to the arciform arteries are the interlobular arteries, which run radially toward the cortical periphery. On their way through the cortex they give rise to short, lateral branches called the afferent arterioles, each of which enters a Bowman's capsule. As the afferent arteriole enters Bowman's capsule, it widens and then subdivides into four to eight trunks to form the

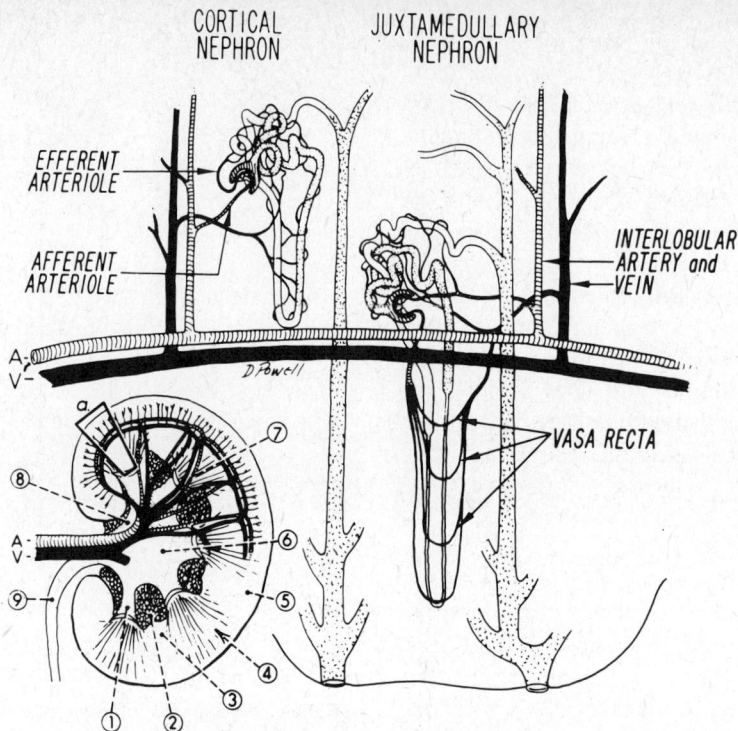

CORTICAL
NEPHRON

JUXTAMEDULLARY
NEPHRON

EFFERENT
ARTERIOLE

AFFERENT
ARTERIOLE

INTERLOBULAR
ARTERY and
VEIN

D.Powell

VASA RECTA

Figure 3 The sagittal surface of a bisected kidney is illustrated diagrammatically (lower left). 1, Minor calix; 2, fat in sinus; 3, renal column; 4, medullary ray; 5, cortex; 6, pelvis; 7, interlobar artery; 8, major calix; and 9, ureter. A, Renal artery; V, renal vein. Insert *a* from the upper pole of the kidney is enlarged in the upper portion of the diagram to illustrate the relationships among the two types of nephrons and their vasculature. A′, Arciform artery; V′, arciform vein. (From Brenner BM, Rector FC (eds): *The Kidney,* vol 1. ed. 2, Philadelphia, WB Saunders, 1981.)

glomerular capillaries. Each of these trunks further subdivides to form capillary loops, each subdivision thereby constituting a separate glomerular lobule. Thus, there are between 20 to 40 such capillary loops associated with the four to eight entering afferent trunks within each glomerulus. The capillaries within any loop are joined by numerous anastomoses. These loops finally recombine to form the efferent arteriole, which then exits the Bowman's capsule. The efferent arterioles soon subdivide again to become the peritubular capillaries. These peritubular capillaries finally join an interlobular vein.

Juxtamedullary nephrons are accompanied by a specialization of the peritubular capillaries called the vasa recta

(Fig. 3). These nephrons serve an important role in the urine-concentrating function of the kidney, as will be seen. The interlobular veins are collected into interlobar veins that then form the renal vein, which exits the kidney via the hilus.

JUXTAGLOMERULAR APPARATUS

At the point where the afferent arteriole enters and the efferent arteriole leaves Bowman's capsule, these two arterioles, especially the former, come into close contact with the straight portion of the distal tubule, just before

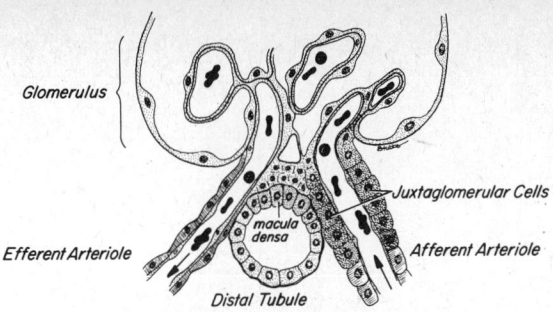

Figure 4 The juxtaglomerular apparatus.

it begins to convolute (Fig. 4). Several interesting cellular specializations occur in this region. There are three types of cell specialization:

1. *Granular cells:* The granular cells exhibit both endocrine and smooth muscle features. They are found in both the efferent and afferent arterioles, but occur much more commonly in the afferent arteriole. The granular cells contain the enzyme renin, a substance involved in the production of angiotensin I.

2. *Extraglomerular mesangium:* Located in the hilus between the afferent and efferent arterioles, these cells make gap junction connections among themselves, as well as with other vascular component cells and the cells of the glomerular tuft. This type of connection suggests a functional coupling of these cells; however, the nature of this coupling is unknown.

3. *Macula densa cells:* A third type of specialized cell arises from the portion of the distal convoluted tubule that is in contact with the vascular component of the juxtaglomerular apparatus. These *epithelial* cells are characterized by closely packed nuclei and by some unique biochemical specializations. Among the latter is a complete absence of Na, K-ATPase, which you will recall is believed to be the biochemical expression of the sodium pump (Chapter 1). In view of the cellular specializations of this region, it is no wonder that the juxtaglomerular apparatus is believed to be of crucial importance in regulating effective renal blood for optimal renal function.

GLOMERULAR FILTRATION

The final product of the kidney is urine. But, behind the formation of this inelegant fluid lie a number of elegant processes and interrelationships of processes. Every minute, approximately 600 ml of plasma flows through the kidneys of an adult human. Through the process of ultrafiltration, about 125 ml of this is filtered out from the blood vessels into the renal tubules, for a filtration fraction that is normally about 0.21. By definition, ultrafiltration not only retains in the blood the particular matter, that is, the blood cells, but also colloidal materials, such as proteins and lipoproteins, which are characterized by large molecular radii. The anatomical basis of this ultrafiltration is the glomerulus, a specialized structure consisting of three parts (Fig. 5):

1. *Basement membrane:* About 3,000 Å thick in the adult.

2. *Fenestrae:* Numerous holes or pores through the capillary endothelial cells. In humans, the diameter is about 700 Å.

3. *Podocytes:* Specialized epithelial cells called visceral epithelial cells, which abut against the basement membrane from the Bowman's capsule side in such a way as to leave slit or rectangular pores with dimensions of about 40 × 140 Å.

The precise anatomical basis of ultrafiltration is unclear. According to a tentative view, the three layers described above represent three filters in series. The endothelial layer is regarded as a coarse filter primarily concerned with keeping circulating blood cells and fat globules away from the basement membrane. The basement membrane is also a relatively coarse filter that permits penetration of molecules with radii of up to 60 Å. The epithelial slit pores appear to provide the most physically restrictive filter. The presence in the glomerular membranes of fixed negative charges that result from glycoproteins and glycolipids appears to be important in restricting the movement of negatively charged proteins across the glomerular capillary wall.

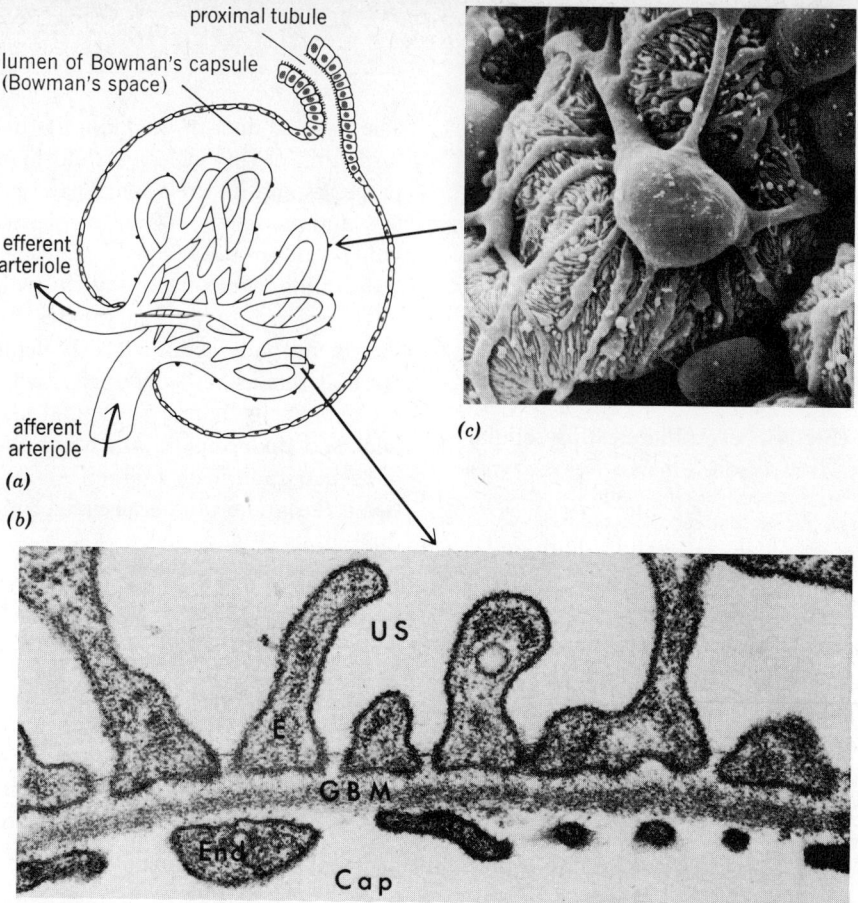

Figure 5 (*a*) Schematic diagram of the microanatomy of the glomerulus. (*b*) The three layers that comprise the glomerular ultrafiltration membrane. (*c*) Scanning electron micrograph of podocytes covering glomerular capillary loops (viewed from inside Bowman's space). (From Vander AJ: *Renal Physiology,* ed 2. 1980)

DETERMINANTS OF GLOMERULAR FILTRATION RATE

The initial event in the process of urine formation involves the production of a nearly ideal, that is, nearly protein free, ultrafiltrate of plasma. The rate of glomerular ultrafiltration is governed by the same driving forces governing fluid movements across other capillary membranes, namely, the imbalance between transcapillary hydraulic and oncotic pressures (Chapter 1). Starling's hypothesis states that the filtration pressure for any capillary is the algebraic sum of the opposing hydrostatic and oncotic pressures acting across the capillary. This law can also be applied to ultrafiltration across the glomerular capillaries:

$$P_{uf} = (P_{GC} + \pi_{BC}) - (P_{BC} + \pi_{GC}) \quad \textbf{(1)}$$

forces inducing   forces opposing
filtration   filtration

where P_{uf} is ultrafiltration pressure, P_{GC} is glomerular capillary hydrostatic pressure, π_{BC} is oncotic pressure in Bowman's capsule, P_{BC} is hydrostatic pressure in Bowman's capsule, and π_{GC} is oncotic pressure in glomerular capillaries.

Because there is virtually no protein in the ultrafiltrate, π_{BC} can be taken as zero. Thus,

$$P_{uf} = P_{GC} - P_{BC} - \pi_{GC}$$

Estimates of the P_{GC} have been difficult to obtain, and reliable measurements exist only for certain animal species. Figure 6 shows the relative pressures in various segments of the renal circulation for two such animal species. These values can be extrapolated to humans to permit calculations of P_{uf}, as shown in Table 1.

Table 1 shows that the net filtration pressure falls from 12 to 0 mm Hg as the plasma traverses the length of the glomerular capillaries. This drop is attributable to two factors: (1) a slight decrease in hydrostatic pressure that results from the resistance to flow offered by the glomerular capillaries, and (2) a much greater increase in the oncotic pressure of the capillary fluid. The oncotic pressure increase results from the concentration of protein within the glomerular capillary plasma caused by filtering

Table 1 Estimated Forces Involved in Glomerular Filtration in Humans[a]

Forces	Afferent End of Glomerular Capillary (mm Hg)	Efferent End of Glomerular Capillary (mm Hg)
Favoring filtration		
P_{GC}	47	45
π_{BC}	0	0
Opposing filtration		
P_{BC}	10	10
π_{GC}	25	35
Net filtration pressure		
P_{uf}	12	0

[a]Adapted from Vander AJ: *Renal Physiology*, ed 2. New York, McGraw-Hill, 1980.

an essentially protein-free fluid into the renal tubules. Thus, 20% of the water is filtered out, so that the original amount of protein is now present in a smaller volume of plasma. That a 40% increase in oncotic pressure can result from a 20% decrease in volume is due to the nonlinear nature of the relationship between oncotic pressure and protein concentration, a result of the Donnan effect on diffusible ions (Chapter 1).

Thus, the rise in oncotic pressure completely offsets the hydrostatic pressure, causing the net filtration pressure to fall to zero. The point along the glomerular capillary at which P_{uf} becomes zero is unknown. Current evidence suggests that this happens well before the blood has traversed the entire length of glomerular capillary (see below). We can only guess at the value for the mean net filtration pressure. It obviously lies between 12 and 0 mm Hg, probably around 5–6 mm Hg. We are therefore faced with the fact that a very small pressure difference can result in formation of a protein-free plasma filtrate at the rate of about 125 ml/min. This remarkable fact can be explained by two differences between glomerular capillaries and extrarenal capillaries, such as those of skeletal muscle: (1) the glomerular capillaries occupy a large surface area, about 1 sq m (11 sq ft) in the adult human, relative to the volume of tissue; and (2) glomerular capillaries are 100 times more permeable to water and small solutes.

The fact that $P_{uf} = 0$ before blood reaches the efferent end of the glomerular capillary means that the overall

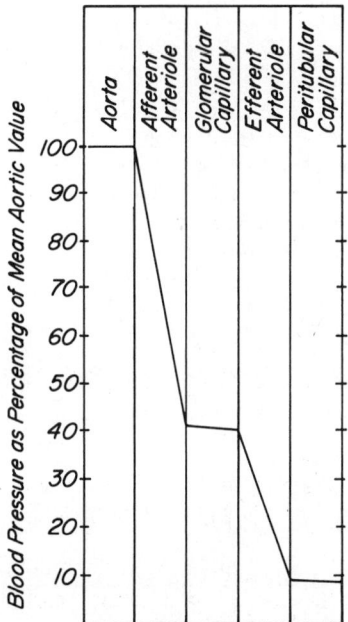

Figure 6 Profiles of mean hydraulic pressure along the renal microcirculation expressed as percentage of the mean aortic pressure. (After Brenner BM, Deen WM, Robertson CR: Glomerular filtration, in Brenner BM, Rector FC (eds): *The Kidney*. Philadelphia, WB Saunders, 1976.)

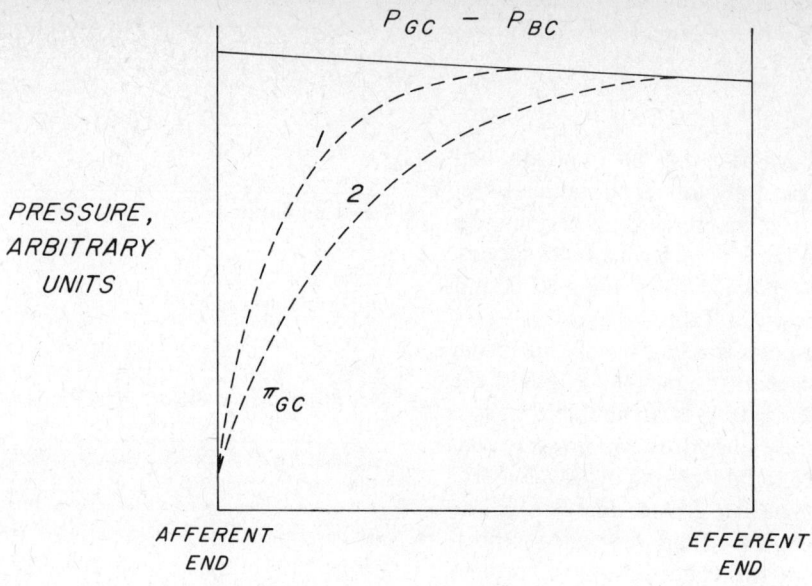

$$P_{GC} - P_{BC}$$

PRESSURE,
ARBITRARY
UNITS

π_{GC}

AFFERENT
END

EFFERENT
END

DISTANCE   ALONG   GLOMERULAR   CAPILLARY

Figure 7 Diagram illustrating the relationship between the forces favoring net ultrafiltration and those opposing it as blood moves through the glomerular capillary. The distance between the curve labeled $P_{GC}-P_{BC}$ and the curves labeled π_{GC} (1 and 2) represents the net filtration pressure (P_{uf}) at each point along the glomerular capillary. Curve 1 represents the situation at a given, arbitrary blood flow rate, while curve 2 represents the situation at an increased flow rate. The total area between the line labeled $P_{GC}-P_{BC}$ and the curves labelled π_{GC} (1 and 2) is proportional to the mean net ultrafiltration pressure. Since this area is greater at higher glomerular capillary blood flow rates, the glomerular filtration rate is likely to be higher at higher blood flow rates. (Modified from Deen WM, et al: *J Clin Invest* 52:1500, 1973.)

system is in equilibrium, that is, hydrostatic pressures and oncotic pressures precisely balance one another. Figure 7 shows that it also implies that under normal flow conditions (e.g., curve 1), there is a certain amount of "spare" capillary available for filtration. An increase in blood flow rate is believed to change the π_{GC} relationship to something like curve 2. Note that with curve 2 the area between the lines labeled $P_{GC}-P_{BC}$ and π_{GC} has increased. Also notice that this has occurred in the absence of any pressure change measurable at either end of the glomerular capillary. The increased area is a measure of the increase in the mean net filtration pressure. It is well known that the glomerular filtration rate (GFR) is directly proportional

to changes in glomerular blood flow rate, although the precise mechanism is not known.

AUTOREGULATION OF GLOMERULAR FILTRATION RATE

The kidney, as well as other organs characterized by high rates of blood flow per gram of tissue weight (heart, brain, and liver), possesses the ability to maintain a relatively constant blood flow in spite of variations in mean arterial pressure for mean systemic arterial pressures greater than 100 mm Hg. Since the changes in renovascular resistance that accompany these variations in mean arterial pressure are demonstrable in innervated, denervated, and isolated kidneys, this phenomenon, termed autoregulation of renal blood flow, has generally been assumed to be mediated by events intrinsic to the kidney. It has also been shown that the GFR remains relatively constant in spite of changes in mean arterial (systemic) pressures above 100 mm Hg. This seems to be because the ultrafiltration pressure remains constant despite changes in mean arterial pressure and because the blood flow through the glomerular capillaries is maintained (Fig. 7).

The explanation for this remarkable constancy of P_{uf} is that the resistances to blood flow in the afferent and efferent arterioles changes in such a way as to maintain a constant P_{GC}. With decreases in mean arterial pressure, P_{GC} is kept nearly constant by a simultaneous decrease in afferent arteriolar resistance and an increase in efferent arteriolar resistance.

It was recently shown that each individual nephron can regulate the rate at which its glomerulus filters plasma. This tubuloglomerular feedback mechanism is generally believed to be mediated through the juxtaglomerular apparatus. The juxtaglomerular apparatus senses some property of the flow of the filtrate through the distal tubule. Thus, when the GFR is high, the flow of filtrate through the distal tubule increases, resulting in a reduction of the GFR. This negative feedback system involves changes in the resistance of the afferent or the efferent arteriole(s), or both. Smooth muscle relaxants can be shown to block the autoregulatory process completely. The signal that causes the change in arteriolar resistance is still a matter of serious contention among investigators in this area.

Although the kidney possesses this intrinsic mechanism for regulating its blood flow and GFR, it is important to realize that the kidney is also subject to overriding extrinsic factors, such as hormonal and neural influences, that can and do affect renal plasma flow and GFR. It would therefore be a mistake to consider renal blood flow and GFR to be fixed constant values at all times. In fact, changes in these parameters can be used to make physiological adjustments of body fluid volume and composition.

TUBULAR REABSORPTION

The process of glomerular filtration poses a rather large problem for the body's ability to economize water and electrolytes. Column 1 in Table 2 expresses the enormous amounts of fluid and solutes filtered out of the plasma every day. Not only are these amounts large, but they greatly exceed total body stores, since total body water is 42 L and total body exchangeable sodium is only 2,400 mEq. There can be little doubt, then, of the necessity and importance of tubular reabsorption.

Renal tubular reabsorptive processes use all the mechanisms of transmembrane transport discussed in Chapters

Table 2 Estimates of the Normal Amounts of Water and Some Representative Solutes Handled Daily by the Human Kidney

Substance	Amount Filtered[a] (mEq/day)[b]	−	Amount Excreted (mEq/day)[b]	=	Amount Reabsorbed (mEq/day)[b]
Water	169.3 L/day		1.5 L/day		167.8 L/day
Na	24,565		75–100		24,465
K	694		22.5		671.5
Ca	485		10.0		475
Cl	19,588		90.0		19,498
HCO$_3$	4,960.5		1.5		4,959
PO$_4$	389		72		317
Glucose	1,080		0		1,080
Urea	810		380		430

[a]These values are corrected for a plasma water content of 94% and the Donnan distribution across the glomerular membrane.
[b]Except for water, obviously (L/day).

1 and 30. Specific mechanisms will be mentioned in conjunction with the discussion of the particular substances being reabsorbed. The various anatomical regions of the nephron have rather sharply different roles in the overall reabsorptive process. The proximal tubules reabsorb by far the largest amount of material, usually around two-thirds (64–87%) of the fluid that is filtered. Quantitatively, the loop of Henle contributes very little directly to the amount of fluid and electrolyte that is reabsorbed. However, as we shall see in later sections, this portion of the nephron performs a crucial role in overall renal function. The amount of fluid and solute absorbed by the distal tubule and collecting duct will vary, depending on the hydration state of the individual. It is this portion of the nephron that acts as the control site in the final determination of urine volume and composition.

PROXIMAL TUBULAR REABSORPTION OF WATER AND SOLUTES

Na, Cl, AND WATER

Water is currently believed to be reabsorbed by a mechanism identical to that for the intestine (Chapter 30, Fig. 6). This isotonic absorption of water depends on the presence of an active sodium transport process located in the basolateral membrane of the proximal tubular cell. Thus, sodium is actively transported into the lateral intercellular space; chloride (or bicarbonate) follows to preserve electroneutrality. The concentration of salt in the small space rises, creating a local osmotic gradient that draws water into the space, water being highly permeant in the proximal tubule. The volume of the lateral space is increased, creating a slight increase in hydrostatic pressure. Since the resistance to flow is less at the basolateral border of the epithelial cells, the water and solutes move out into the interstitial space and are then drawn into the peritubular capillary. The peritubular capillary plasma, you will recall, has a relatively large oncotic pressure due to the concentration of plasma proteins caused by the glomerular filtration process. This oncotic pressure "pulls" the water and solutes from the interstitial space into the capillaries, completing the reabsorption process.

Net water absorption will in turn cause substances like urea to be more concentrated in the tubule, resulting in their passive reabsorption by diffusion down their concentration gradients. The degree of reabsorption of these substances will depend on the permeability of the tubular epithelium to them. For example, the epithelium is less permeable to urea than to water; consequently, urea is not as rapidly reabsorbed as is water. Inulin, a large polysaccharide, cannot permeate the membrane and is thus not reabsorbed at all.

Many substances besides Na^+, like glucose and amino acids (see below), are actively reabsorbed by the proximal tubules; they tend to facilitate water reabsorption, hence the passive reabsorption of other substances to which the tubular epithelium is permeable.

The Na^+ pump in the proximal tubule cannot produce a net movement of Na^+ from tubule lumen to capillary when luminal Na^+ concentration falls below a certain critical value. Therefore, there is some maximal concentration gradient between lumen and blood above which the Na^+ pump is ineffective in causing net reabsorption of Na^+, and therefore of water. Thus we refer to the Na^+ reabsorption mechanism as a gradient-limited system. This gradient limitation is probably attributable mainly to the fact that passive backflux of Na^+ through the paracellular shunts increases as the Na^+ gradient increases, until it finally matches the rate of Na^+ transport by the pump. Mammalian proximal tubules are capable of generating no more than about a 30–50 mEq/L difference in Na^+ concentration between peritubular blood and tubular lumen.

One can calculate that nearly 6% of the resting oxygen consumption of the body is involved in supplying energy for the active sodium reabsorption by the kidney; that is, nearly 4 kcal/h is required for Na^+ reabsorption (the basal metabolic rate is around 70 kcal/h).

POTASSIUM

Little information is currently available as to the precise mechanism of K^+ reabosrption by the proximal tubule. It could involve both active and passive processes at the apical membrane, as well as passive mechanisms across the so-called "tight junctions" between epithelial cells. Some K^+ might exchange for Na^+ due to active Na^+–K^+ exchange transport at the basolateral membrane. This exchange could be important in regulating intracellular concentrations of Na^+ and K^+, as well as cellular volume.

BICARBONATE

Approximately 90% of the filtered bicarbonate is reabsorbed in the proximal tubules. The remaining bicarbonate is normally reabsorbed in the distal tubules. In alkalosis some bicarbonate is not reabsorbed, helping to correct the alkalosis. The mechanism of bicarbonate reabsorption is apparently about the same all along the nephron and is also the same as one of the mechanisms for bicarbonate reabsorption by the intestine (Chapter 30).

You should recall the Henderson-Hasselbalch equation, which, for the HCO_3^-–H_2CO_3 buffer pair, looks like this:

$$pH = 6.1 + \log \frac{[HCO_3^-]}{[H_2CO_3]} \qquad (2)$$

For a normal plasma pH of 7.4, HCO_3^-/H_2CO_3 must be 20, since log 20 = 1.3. The concentration of carbonic acid in arterial blood plasma is controlled by alveolar ventilation (Chapter 22). The concentration of bicarbonate is controlled by the kidneys. Clearly, the kidneys have an extremely important involvement in regulating the pH of body fluids. This topic will be more thoroughly discussed in Chapter 34, but two points regarding bicarbonate should be made here.

First, at the normal HCO_3^- concentration in arterial blood plasma of 24 mEq/L, the tubular reabsorption mechanism is poised just on the verge of failing to reabsorb all that is filtered. If the plasma concentration rises above normal, bicarbonate is spilled into the urine and the plasma concentration tends to return to normal.

Second, we must ask whether the kidneys can do anything to elevate plasma bicarbonate concentration. Obviously they cannot reabsorb HCO_3^- from the glomerular filtrate at a rate greater than normal, since all HCO_3^- is essentially reabsorbed normally. What the kidneys can do is to generate new bicarbonate. They do this by precisely the same mechanism as that shown as mechanism 2 Chapter 30, Figure 8, but with one important difference. There, it is shown that for every HCO_3^- generated within the cell and diffusing into the blood, a HCO_3^- is consumed within the tubular lumen. In effect, HCO_3^- ions are transferred from lumen to blood, although those that enter the blood are not the same ions that were filtered in the

glomerulus. But what happens after all the bicarbonate has been removed from the lumen (usually somewhere in the distal tubule)? Now, at this time and place, if bicarbonate is generated intracellularly by the carbonic anhydrase reaction using CO_2 from the blood, it will diffuse into the blood without an equivalent consumption of bicarbonate in the lumen. In this way, the kidneys generate new bicarbonate and can elevate the plasma bicarbonate concentration. This process takes place in the distal tubules and collecting ducts, and is of paramount importance to overall body pH regulation (Chapter 34).

GLUCOSE

Under normal conditions, all the glucose filtered in the glomeruli is reabsorbed in the proximal tubules by active transport. This transport system is Na^+ dependent and is thought to be similar to the glucose transport system in the small intestine. It is diagrammed in Figure 8. Movement of glucose into the lateral intercellular spaces along with Na^+ and Cl^- increases the osmolarity there, drawing water in across the cells and across the tight junctions. Thus, glucose reabsorption stimulates water reabsorption. Conversely, glucose excretion will facilitate water loss (e.g., diuresis of diabetes mellitus).

Given enough time, this transport system can keep pumping glucose out of the tubules until the luminal concentration has reached virtually zero. In other words, there is no gradient limitation for the glucose transport

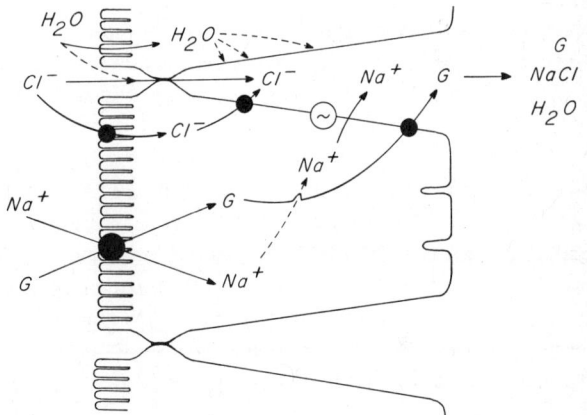

Figure 8 Reabsorption of glucose (G) by proximal tubular epithelium.

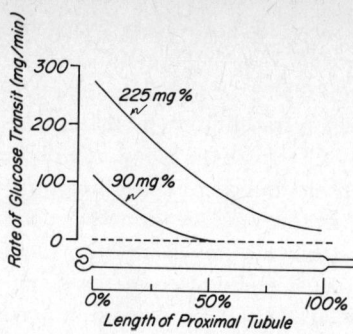

Figure 9 Transit rate of glucose as a function of position along the proximal tubules and as a function of the glucose load presented by the plasma. See text for description.

system—at least not with normal plasma concentrations. This amazing and important characteristic of the glucose transport system, also possessed by the equivalent system in the small intestine, is not understood.

Normally, there is enough time during passage of filtrate through the tubules for glucose to be almost completely reabsorbed. Figure 9 plots the total amount of glucose passing through the lumen of all proximal tubules per minute as a function of position along the tubules.

With a normal human plasma glucose concentration of 90 mg% and a normal GFR of 125 ml/min, the glomeruli will present the proximal tubules with 112 mg/min of glucose. Note that glucose transit is reduced to zero (because of complete reabsorption) well before the end of the proximal tubule, when plasma glucose level is normal (90 mg%).

Now let us examine the consequences of increasing the concentration of glucose in the blood plasma and, therefore, in the filtrate (i.e., hyperglycemia). Figure 9 also shows the transit–distance curve expected if the plasma glucose concentration is increased to 225 mg% with no change in GFR. We now see that not quite all the glucose is reabsorbed. This unreabsorbed glucose will now be flushed on through the renal tubules to be excreted in the urine, since no glucose reabsorption takes place anywhere else in the kidney. It turns out that glucose begins to show up in the urine when the plasma glucose concentration exceeds 180 mg%, a level termed the renal threshold for glucose.

This value for renal threshold is high enough so that under normal circumstances no glucose will be spilled into the urine, even after a meal fairly high in carbohydrate content has been consumed. Thus the kidney conserves blood glucose.

It should be noted that essentially no glucose passively diffuses from blood to tubular fluid in the distal portions of the nephron—the epithelium is impermeable to glucose. If it were not for this important, but frequently neglected, aspect of renal function, the kidneys would be poor glucose conservers.

Figure 10 shows the relationship between the concentration of glucose in the glomerular filtrate (or blood plasma) and the rates of (1) filtration, (2) reabsorption by proximal tubules, and (3) excretion in the urine, for a constant GFR. Note that for any given plasma concentration, the sum of the rates of reabsorption and excretion must equal the rate of filtration. As the plasma concentration is elevated above the renal threshold (180 mg%), the rate of reabsorption can no longer keep up with the rate of filtration, and some glucose is excreted. Gradually, the

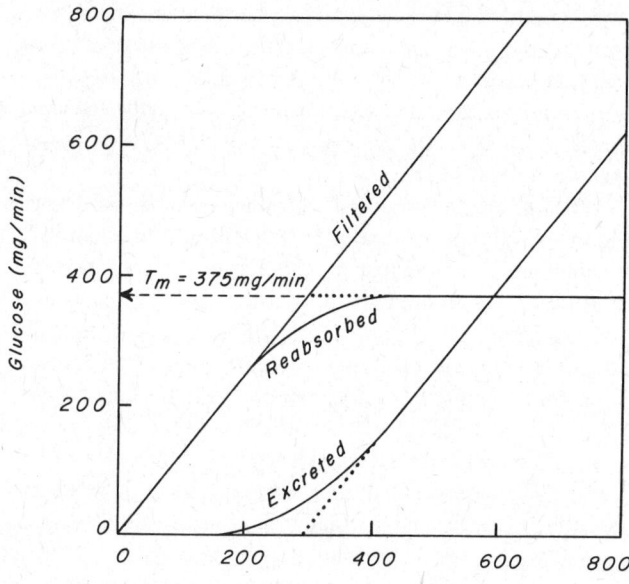

Figure 10 Rates of glucose filtration, reabsorption, and excretion as functions of the plasma glucose concentration. GFR is assumed to be constant at 125 ml/min. (Adapted from Pitts RF: *Physiology of the Kidney and Body Fluids*, ed 3. Chicago, Year Book, 1974.)

rate of reabsorption reaches a maximum value called the transport maximum, T_m. The T_m for glucose is normally about 375 mg/min and is reached at a plasma concentration of roughly 400 mg%. As the plasma concentration of glucose is further elevated above the latter value, no additional increase in reabsorption rate is possible, since by now the transport system is completely saturated, even at the ends of the proximal tubules. Therefore, above a plasma concentration of about 400 mg%, any increase in filtration rate is matched exactly by an increase in excretion rate, and these two lines run parallel to each other, the vertical distance between them being equal to the T_m.

The rounding off of the reabsorption curve, as opposed to the cornering depicted by the dotted curve in Figure 10, is referred to as splay. You should be able to appreciate that splay is caused, at least partly, by the gradual, rather than abrupt, saturation of the transport process as the glucose concentration increases. However, there is another contribution to splay, that is, the nephrons are not all identical to each other. The length of the proximal tubules varies around some mean, as does the rate of glomerular filtration. Therefore, the ratio of T_m to filtered load varies somewhat among nephrons. It is easy to see how this circumstance would increase the degree of splay.

Certain substances other than glucose also show fairly well-defined T_m's; these include amino acids, phosphate, and sulfate. Like glucose, these substances are all actively reabsorbed in the proximal tubules and cannot passively diffuse downhill across the epithelium. The tight junctions are not very permeable to these substances. Thus, for each of these materials, raising its concentration in the tubules above that required to saturate its active transport system results in no further increase in the rate of reabsorption; also, a T_m can be observed. Other actively reabsorbed materials, particularly Na^+ and K^+, have no T_m's, since raising their concentrations above levels required to saturate transcellular pathways simply results in greater rates of passive diffusion through paracellular pathways. The principal reason that active reabsorption of Na^+ and K^+ is gradient limited is that these ions can easily back-diffuse across the tight junctions. We now see that this is also the principal reason that their reabsorption is not T_m limited.

AMINO ACIDS

Normally, the proximal tubules completely reabsorb all the filtered L-amino acids presented to them. At least four separate transport systems are involved: two for the neutral L-amino acids and imino acids, one for basic L-amino acids (which also transports L-cystine), and one for the acidic L-amino acids. All these transport systems are Na^+ dependent and can pump the amino acid uphill across the luminal membranes of the epithelial cells. It is generally accepted that the energy for this process is derived directly from the electrochemical potential gradient for Na^+ across the luminal membrane, and therefore, indirectly from ATP hydrolysis by the Na^+ pump at the basolateral membrane (Chapters 1 and 30).

As for glucose, there is no reserve capacity to reabsorb amino acids in more distal portions of the nephron. Also, like glucose, there is essentially no passive downhill reabsorption. Thus, amino acid reabsorption is T_m limited. However, the T_m's may be quite different for different amino acids.

The amino acids transported by a common system compete for that system. Thus, if there is a large excess of some amino acid that happens to have a high affinity for its transport system, it displaces other amino acids from this transport system. These amino acids are then incompletely reabsorbed. The presence of free amino acids in the urine is called aminoaciduria. The reality and importance of this effect are illustrated by certain genetic diseases of amino acid metabolism.

The most common disease of amino acid metabolism is cystinuria. In cystinuria the tubular transport mechanism for the basic amino acids and cystine is absent or defective both in the proximal tubules and in the small intestine. Consequently, aminoaciduria of cystine, arginine, lysine, and ornithine results, and these amino acids are lost in the feces and urine.

PROTEIN

Although filtered in such small quantities that there is effectively no colloid osmotic pressure in the glomerular fluid, there is a filtration of perhaps 3.0–4.0 g/day of proteins from the plasma in the normal subject. This protein is reabsorbed (by pinocytosis) mainly in the proximal tubules. This mechanism is easily saturated, so that any large increase in filtered protein resulting from increased glomerular permeability will result in the excretion of large quantities of protein.

CALCIUM

Normally, 98% of the filtered calcium is reabsorbed by an active transport system located in the cells of the proximal convoluted tubules, distal convoluted tubules, and collecting ducts. Calcium transport, like sodium transport, occurs primarily in the proximal convoluted tubules, while transport by cells of the distal convoluted tubules and collecting ducts constitutes approximately 30% of the filtered load.

Calcium reabsorption by this system is increased by parathyroid hormone, although administration of the hormone generally produces an increase in urinary calcium loss. This hormone elevates blood calcium levels by enhancing the dissolution of bone salts and by augmenting intestinal absorption of Ca^{2+}. The resulting increase in blood calcium concentration raises the filtered calcium load by an amount that is more than sufficient to overcome the effects of increased tubular reabsorption.

PHOSPHATE

Phosphate reabsorption is located in the proximal convoluted tubule. It is a T_m-limited system, differing from the glucose or amino acid transport systems in that the T_m is below, not substantially above, the normal filtered load. Therefore phosphate reabsorption is seldom complete. Elevations in blood phosphate levels increase the filtered load of phosphate, increasing the amount of phosphate excreted, and thereby producing a simple feedback system that aids in the control of blood phosphate levels.

The T_m for phosphate is under hormonal control. Parathyroid hormone will lower the T_m for phosphate, as will increased blood cortisone levels.

REABSORPTION OF WATER AND SMALL IONS IN OTHER PORTIONS OF THE NEPHRON

Although approximately two-thirds (64–87%) of the sodium and water of the glomerular filtrate is reabsorbed in the proximal tubules, other portions of the tubular system play significant roles in reabsorption. Their importance in (1) osmoregulation (i.e., production of dilute or concentrated urine), (2) regulation of urine pH, and (3) regulation of extracellular fluid volume will be discussed later, but a few points are appropriate here.

NaCl AND WATER

Reabsorption of NaCl and H_2O from the distal tubules and collecting ducts is an active gradient-limited process like that already described for the intestine (Chapter 30). The rate of Na^+ transport in these portions of the nephron is probably limited not by the vigor of the baso-lateral pump, but by the reduced rate of Na^+ entry across the apical membrane. This is partly because there is no co-transport with organic solutes in the distal nephron, and partly because the apical membrane potential is much less in the distal nephron (Fig. 11), thereby reducing the driving force for Na^+ entry.

ALDOSTERONE

Aldosterone, a major mineralocorticoid hormone from the adrenal cortex, increases the permeability of the apical membrane for Na^+. The Na^+ gains access to the basolateral pump more readily, and the pumping rate goes up. Some investigators believe aldosterone also increases the efficiency of the pump. The effect of aldosterone on apical permeability to Na^+ and on the pump itself are both indirect effects mediated by protein(s) synthesized intracellularly in response to aldosterone. Thus, aldosterone stimulates NaCl and water absorption in the collecting duct. Aldosterone does not have this effect in the proximal tubule.

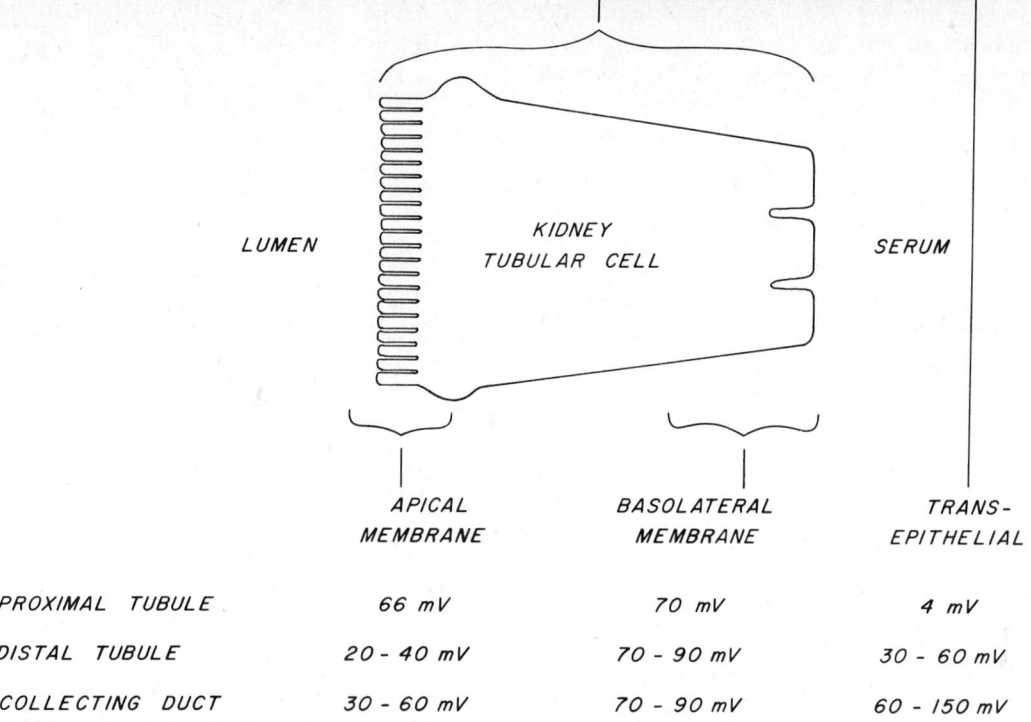

	APICAL MEMBRANE	BASOLATERAL MEMBRANE	TRANS-EPITHELIAL
PROXIMAL TUBULE	66 mV	70 mV	4 mV
DISTAL TUBULE	20 - 40 mV	70 - 90 mV	30 - 60 mV
COLLECTING DUCT	30 - 60 mV	70 - 90 mV	60 - 150 mV

Figure 11 Approximate electrical potential profile of cells located in different parts of the kidney tubules. The precise value of the potentials can vary slightly with the activity of the cells.

POTASSIUM

Almost all filtered K^+ is reabsorbed before the filtrate reaches the distal tubules. Reabsorption of K^+ also can occur all along the distal convoluted tubules and collecting ducts. Nevertheless, urine always contains appreciable amounts of K^+, usually about 10–20% of the filtered load. Thus we must conclude that K^+ is secreted by the distal tubules and collecting ducts at a rate that exceeds its reabsorption in these regions. (Secretion of potassium by the distal tubules will be covered a little later in this chapter.)

CHLORIDE

Reabsorption of Cl^- in the distal tubules and collecting ducts is probably passive following Na^+ reabsorption. In the thick portion of the ascending loop of Henle, Cl^- is actively reabsorbed. This active reabsorption of Cl^- plays a particularly crucial role in the development of medullary hypertonicity, which is necessary for the production of a concentrated urine.

ELECTRICAL POTENTIAL DIFFERENCES

The electrical potential differences across the apical and basal-lateral membranes and across the whole epithelium are compared among various parts of the nephron in Figure 11. Note that, compared with the proximal tubule, the distal tubule has (1) a somewhat greater basolateral membrane potential, (2) a considerably smaller apical membrane potential, and (3) a much greater transepithelial potential difference. All these differences are probably mainly attributable to the paracellular shunts (tight junctions and lateral intercellular spaces) becoming less permeable progressing from the proximal, through the distal tubules, to the collecting duct. (See Chapter 30 for an explanation of this voltage effect.)

There are two exceptions to the above orientation of the transepithelial potential difference. First, in the distal portion of the proximal tubule, the lumen is slightly positive with respect to the blood. This is an indirect result of the fact that HCO_3^- is the anion preferentially reabsorbed in the more proximal region of the proximal tubule. Thus, Cl^- concentration in the tubular fluid increases due to iso-osmotic fluid reabsorption, resulting in a lumen-positive Cl^- diffusion potential developing in the distal portion of the proximal tubule. This luminal positivity can aid in K^+ reabsorption. Second, in the thick portion of the ascending limb of Henle's loop, the lumen is positive with respect to the blood. This orientation of the potential difference is caused by the strong Cl^- pump located there (see above).

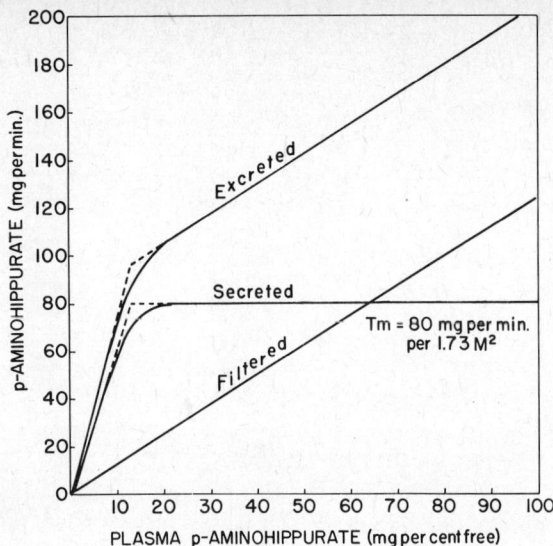

Figure 12 Rates of filtration, secretion, and excretion of PAH by the kidney as a function of plasma concentration. (From Pitts RF: *Physiology of the Kidney and Body Fluids,* ed 3. Chicago, Year Book, 1974.)

TUBULAR SECRETION

In addition to promoting net reabsorption of various materials, the renal tubules generate net fluxes of some substances from peritubular blood to tubular lumen. A net flux in this direction, be it active or passive, is referred to as tubular secretion.

ACTIVE TUBULAR SECRETION

The renal tubules undergo three well-known active secretory processes: (1) secretion of organic acids, (2) secretion of strong organic bases, and (3) secretion of hydrogen ions.

Organic Acids
The proximal tubules possess a rather nonspecific active transport mechanism that can secrete a wide variety of organic acids into the tubular lumen, thereby aiding glomerular filtration in eliminating these acids from the body. Materials handled by this system include hippuric acid and various glucuronides produced mainly in the liver as part of the liver's "detoxifying" function. But because of the low specificity of this system, many foreign compounds that may or may not be toxic to the body are also excreted in this way; these include penicillin, *p*-aminohippuric acid (PAH), chlorothiazide, and various

acetylated sulfonamides. Radiologists take advantage of this transport system when they want to test tubular function, using Diodrast (or something similar), which is transported by this system and is also radiopaque. Urologists take advantage of this system when they want to study renal blood flow—they measure the "clearance" of PAH (don't worry about this now; it will be discussed later).

Figure 12 shows the rates of filtration, secretion, and excretion of PAH as a function of PAH concentration in arterial blood plasma. Note the clear-cut T_m for the tubular secretion of PAH. Since the excretion rate is always equal to the sum of the filtration and secretion rates:

Secretion rate
$$= \text{excretion rate} - \text{filtration rate}$$

or

Secretion rate
$$= (U_{PAH} \times \dot{V}) - (GFR \times P_{PAH}) \quad (3)$$

Strong Organic Bases
The proximal tubules also actively secrete, by a mechanism separate from that for organic acids, various strong organic bases, such as guanidine, thiamine, choline, and

histamine. The physiological importance of this transport system, as well as that for organic acids, is obscure.

Hydrogen Ions

The epithelial cells of proximal tubules, distal tubules, and collecting ducts all actively secrete hydrogen ions into the tubular lumen. The physiological importance of this transport system is tremendous—it provides the means by which the kidneys control the bicarbonate ion concentration in blood plasma, and therefore, help control the pH of body fluids.

Hydrogen ion transport is gradient limited at all levels of the nephron, presumably because H^+ can back diffuse through the paracellular shunts. In the proximal tubules the limiting gradient is rather small. The pH of the proximal tubular fluid can be reduced only to about 6.8. This means that the hydrogen ion concentration in proximal tubular fluid can become, at most, 4 times greater than in arterial blood plasma. Nevertheless, a tremendous quantity of H^+ is secreted by the proximal tubules, resulting in reabsorption of about 90% of the filtered bicarbonate.

In the distal tubules and collecting ducts hydrogen ions continue to be secreted, but at a much reduced rate. (For data on relative rates of H^+ secretion in proximal and distal parts of the nephron, see Chapter 34, Table 2.) But in spite of the much slower rate of H^+ secretion in the distal nephron, a much greater concentration gradient between tubular lumen and blood can be developed, especially in the collecting ducts. A pH of 4.4 can be attained in the collecting ducts; this represents a hydrogen ion concentration 1,000-fold greater than that in blood plasma. The ability to attain such a gradient must be ascribed partly to the "tightness" of the epithelium in this part of the nephron (i.e., the relative impermeability of the paracellular shunts) and partly to the power of the hydrogen ion pump itself.

Hydrogen ions secreted by the tubules can suffer one of four fates: (1) They can combine with filtered bicarbonate and in the overall process accomplish bicarbonate reabsorption. (2) They can combine with nonbicarbonate buffers, such as phosphate and ammonia, and be excreted as such in the urine. (3) They can exist, to a very limited extent, as free hydrogen ions in the tubular fluid (however, since urine cannot attain a pH of less than 4.4, only $4 \times$ 10^{-5} moles/L can be excreted in this form). (4) They can backdiffuse across the tight junctions on into the blood (the extent to which this occurs is unknown). The first process predominates in the proximal tubules, since there the bicarbonate concentration is relatively high. However, by the time the fluid has reached the distal convoluted tubules and collecting ducts, bicarbonate concentration is greatly reduced; this is particularly true when systemic metabolic acidosis has reduced the plasma bicarbonate concentration. These matters will be discussed further in Chapter 34.

Potassium

As mentioned earlier, almost all K^+ is reabsorbed by the proximal tubule, yet an appreciable amount always is present in urine. Thus, a secretory process for K^+ is clearly indicated, but the mechanism or mechanisms are unknown. Under many conditions, the net secretion of K^+ into tubular fluid can be explained by a favorable electrochemical gradient for K^+ between tubular lumen and the capillary. However, there are situations in which the net secretion of potassium can be shown to occur against the electrochemical gradient. Thus, under some circumstances, active transport of K^+ into the renal tubule must be occurring. It seems entirely possible that both passive and active routes of K^+ secretion might be operating.

Regardless of mechanism, it is quite clear that K^+ secretion occurs almost exclusively along the latter half or so of the distal tubule and along the collecting duct. Potassium secretion is an important regulatory process for total body potassium levels, since virtually all K^+ excretion is a result of the secretory process. When aldoesterone stimulates Na^+ reabsorption, an increased rate of K^+ secretion occurs at the same time. This might reflect the fact that increasing the rate of Na^+ pumping out of the tubule renders the lumen electrically more negative via an electrogenic pumping effect (Chapter 1). Thus, K^+ can diffuse passively down its electrochemical gradient.

There is also an interesting relationship between the acid–base balance of an individual and the rate at which the distal tubules secrete K^+. Although there are exceptions, in general K^+ secretion is stimulated during extracellular alkalosis and inhibited during acidosis. The mechanism for this effect is unknown. It was formerly believed that K^+ and H^+ competed for a common secretory carrier in the apical membrane of distal tubular cells. When intracellular pH was low, H^+ secretion would be accompanied by reduced K^+ secretion. The reverse would occur when intracellular pH was high—K^+ would displace H^+ from the carrier, and K^+ secretion would go up at the expense of H^+ secretion. This notion is no longer accepted, partly because it has been found that in many situations H^+ and K^+ secretion rise or fall more or less in parallel, instead of being reciprocally related.

PASSIVE TUBULAR SECRETION

It might seem unreasonable that there could be net passive diffusion of any material into the tubular lumen, since no solutes that can readily diffuse across the epithelium are at higher concentration in peritubular blood than in tubular fluid (except those solutes that have been actively transported there). There are, however, two processes that promote passive secretion into the tubules: (1) electrodiffusion of ions in response to the transepithelial electrical potential difference as mentioned above for potassium, and (2) the phenomenon of "ion trapping" of weak bases and acids, resulting from a transepithelial pH difference (explained below).

Passive Secretion of Ammonia

Ammonia is a base with a pK value of about 9.2. It can bind hydrogen ions to form the ammonium ion as follows:

$$NH_3 + H^+ \rightleftharpoons NH_4^+ \qquad pK = 9.2$$

With this pK value, the Henderson-Hasselbalch equation predicts that at the pH of arterial blood plasma (7.4), the ammonia nitrogen will exist largely in the ionized state (i.e., will exist mostly as NH_4^+). At this pH value, only about one in every 63 molecules is un-ionized. But this un-ionized form (NH_3) can readily diffuse across cell membranes, since it is sufficiently lipid soluble. In fact, it can penetrate cell membranes so rapidly that it is always at nearly the same concentration in tubular fluid as it is in blood plasma and intracellular fluid. By contrast, ammonium ions cannot readily penetrate cell membranes. If the urine has been acidified, as it usually is in the collecting ducts, an even larger fraction of the ammonia nitrogen will be in the ionized form (i.e., the above reaction will be

driven to the right). At a urinary pH of 4.4, only one in every 63,000 molecules is un-ionized. Therefore, it should be clear that if the concentration of NH_3 is the same in collecting duct urine at pH 4.4 as it is in blood plasma at pH 7.4, the concentration of NH_4^+ will be 1,000 times greater in the urine than in the plasma. The base has been trapped in its ionized form in the region of lowest pH. It is trapped there because the ionized form cannot leak back readily across the epithelium. This phenomenon is called ion trapping.

Actually, most of the NH_3 that enters urine comes from the tubular cells (proximal, distal, and collecting duct), rather than from the blood. NH_3 is manufactured there from amino acids, especially glutamine. It tends to diffuse from its site of formation in the cells to the blood and the urine. But because of trapping as the ammonium ion in acidified urine, the total concentration in the urine far exceeds that in the blood. If an alkaline urine is formed, far less NH_4^+ is excreted. From the above discussion, it should be easy to see why.

RENAL CONCENTRATING AND DILUTING PROCESSES

Many animals are capable of disposing of excess water by excreting a dilute, copious urine. But mammals, along with some birds, are unique among vertebrates both in possessing loops of Henle and in their ability to compensate for water deficits by elaborating a urine more concentrated than blood. The efficiency of the concentration process is directly dependent on an anatomical construction of the nephron, which leads to a process called "countercurrent multiplication."

THE RENAL COUNTERCURRENT OSMOTIC MULTIPLIER MECHANISM

Figure 13 summarizes those processes that are most directly involved in determining the final osmolarity of urine.

The rates of glomerular filtration and proximal tubule

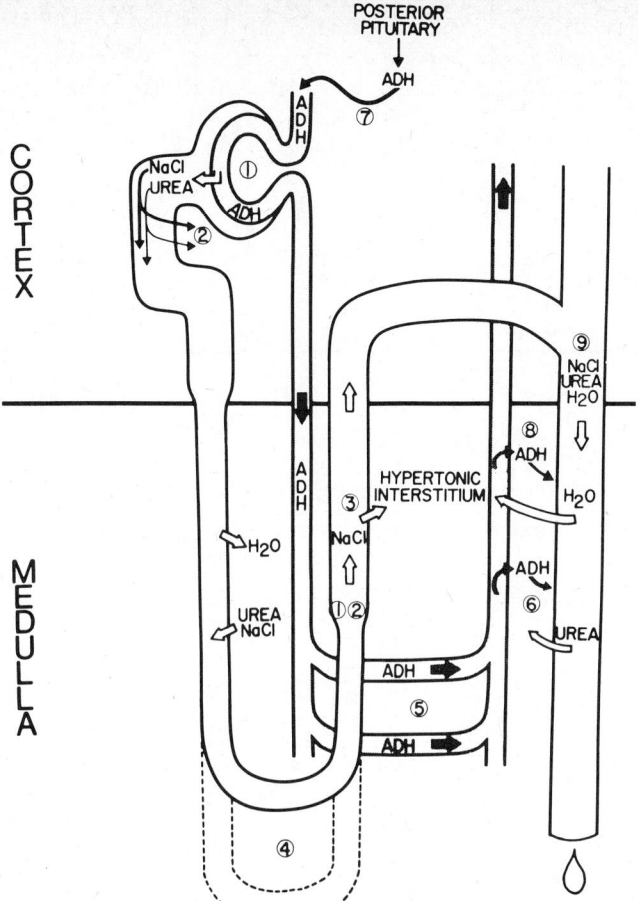

Figure 13 Potential determinants of the renal concentrating mechanism: (1) GFR as a determinant of fluid and solute delivery to ascending limb of loop of Henle. (2) Proximal tubular reabsorption as a determinant of fluid and solute delivery to ascending loop of Henle. (3) Sodium chloride transport in water-impermeable ascending limb. (4) Length and integrity of long loops of Henle in inner medulla and papilla. (5) Rate of medullary blood flow in vasa recta. (6) Urea availability in antidiuretic (ADH)-responsive portion of collecting duct. (7) Presence of ADH. (8) Response of cortical and medullary collecting duct to ADH. (9) Rate of solute and water delivery as a determinant of completeness of osmotic equilibration. (From Schrier RW: *Renal and Electrolyte Disorders,* ed 2. Boston, Little, Brown, 1980.)

reabsorption are important, primarily in determining the rate of Na^+ and water delivery to the more distal portions of the nephron where the concentrating and diluting mechanisms are operative. Although the tubular fluid entering Henle's loop has been reduced to one-third to one-eighth the original filtrate volume, it is virtually isotonic with the cortical interstitial fluid, as discussed previously.

Although GFR and proximal tubular reabsorption of water and electrolytes can influence the final urine osmolarity, they are quantitatively far less important than the processes acting in the more distal portions of the tubule. It turns out that the osmolarity of the interstitial fluid in the tip of the papilla of the medullary region of the kidney is normally quite high, as high as 1,200–1,400 mOsm/L. Thus, in principle, the tubular fluid flowing through that region of kidney could also achieve an osmolarity of 1,200–1,400 mOsm/L. As seen in Figure 13, this can occur when antidiuretic hormone (ADH), secreted from the posterior pituitary, causes the water permeability of the collecting ducts and distal portions of the distal convoluted tubules to increase.

What is the basis for this extraordinarily high interstitial fluid osmolarity? As we shall see, ultimately it depends on the active transport of chloride, and on the electrically obligatory transport of Na, in the thick ascending region of the loop of Henle. This is true even though net NaCl reabsorption by this segment is a small fraction of the total NaCl reabsorbed throughout the nephron (only about half the amount that escapes reabsorption by the proximal tubule). How can the pumping of such a small amount of Cl and Na lead to such a large gradient, especially since the pump can only pump against a 200–300-mOsm/L gradient? The answer lies with the unique anatomical relations among loops of Henle, the site of active Cl pumping, and the collecting ducts.

The juxtamedullary nephrons have long loops of Henle that descend deep into the medulla. The ascending and descending portions of these limbs are quite close to one another and to their parallel vasa recta, permitting water or solute to move easily between these elements.

The functional significance of the loops of Henle was first suggested in 1942 in the form of the hypothesis for "countercurrent multiplication." The hypothesis states that a small difference in osmotic concentration at any point between fluid flowing in opposite directions in two parallel tubes connected in a hairpin manner can be multiplied many times along the length of the tubes. In the kidney, this would result in a high osmolar concentration difference between the corticomedullary junction and the hairpin loop deep in the medulla.

The properties of the loop of Henle that make this mechanism work are as follows:

1. The thick portion of ascending limb actively extrudes Cl into interstitium; Na passively (?) follows. This is the so-called "single effect."

2. The thick portion of the ascending limb is relatively impermeable to water, urea, sodium, and chloride, thereby limiting their passive backdiffusion from interstitium to tubule lumen.

3. The descending limb and the thin ascending limb of the loop of Henle do not actively transport Cl or Na—they are the only tubular segments without active ion transport.

4. The descending limb is permeable to water, somewhat permeable to urea, and nearly impermeable to ions.

5. The thin portion of the ascending loop is permeable to NaCl, somewhat permeable to urea, and impermeable to water.

With these properties in mind, let us imagine the loop of Henle filled with a stationary column of fluid supplied by the proximal tubule.

At the top of Figure 14(a) the loop of Henle is represented digramatically with the descending limb on the left and the ascending limb on the right. The thick portion of the ascending limb is represented by the upper, wider portion of the ascending limb. The numbers represent fluid osmolarity.

We now allow the pump of the thick ascending limb to pump out Cl, and with it, Na. We assume it can only establish a 200-mOsm/L gradient. The conditions shown in the second diagram of Figure 14(a) are thereby established.

Now we know that the descending limb is definitely water permeable, but nearly impermeable to ions. Therefore, the next step involves the movement of water from the descending loop into the medullary interstitium resulting in the conditions shown in the third diagram of Figure 14(a).

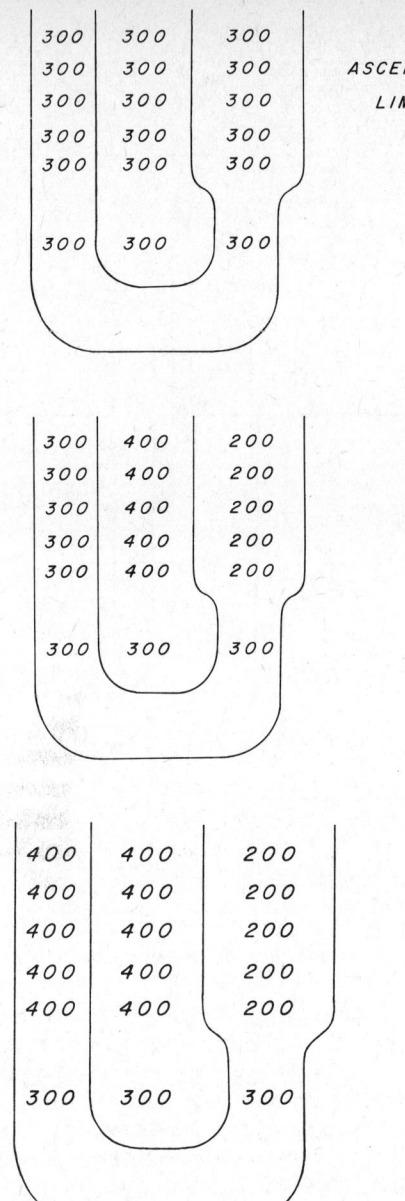

Figure 14(a) Schematic, stepwise demonstration of countercurrent osmotic multiplier. (Adapted from Vander AJ: *Renal Physiology*, ed 2. New York, McGraw-Hill, 1980.)

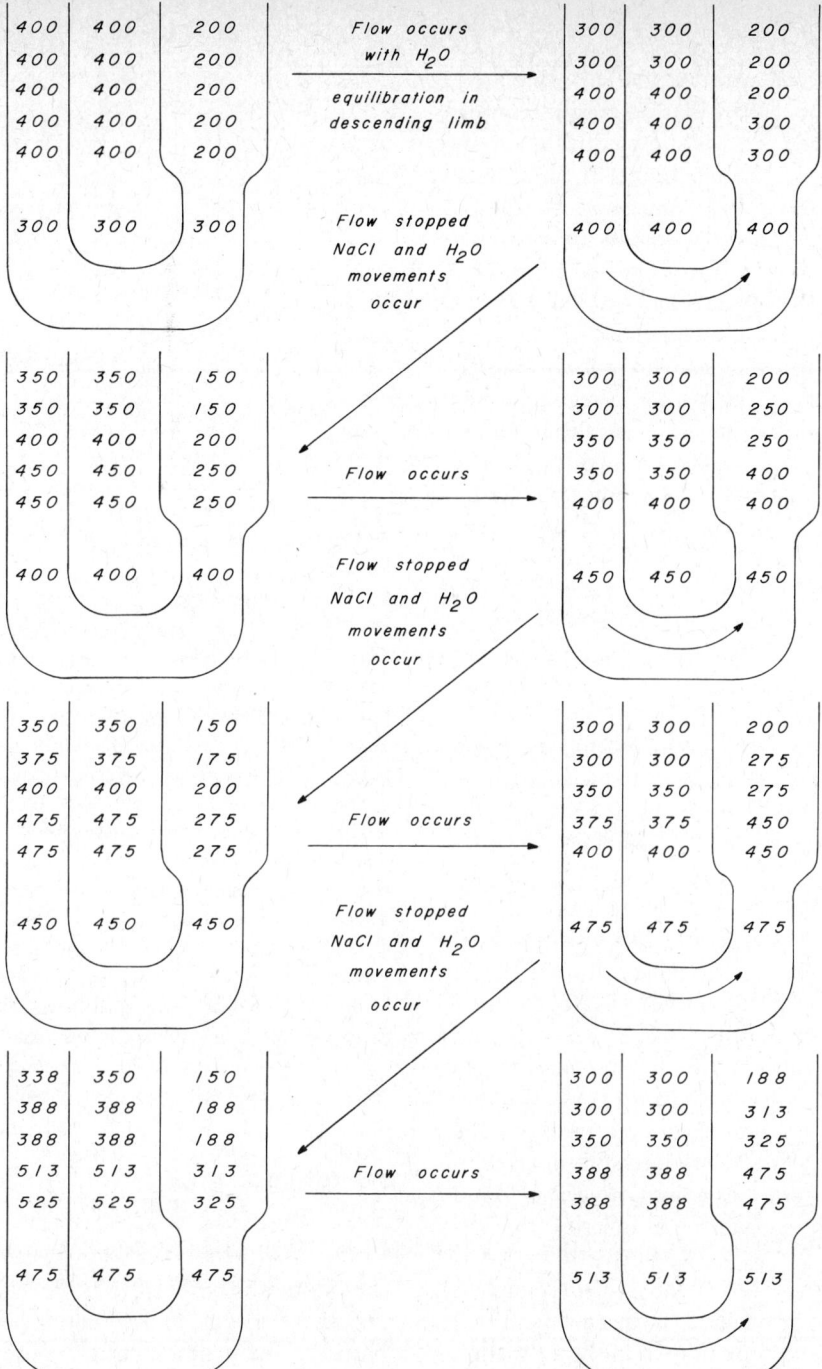

Figure 14(*b*) Schematic, stepwise demonstration of countercurrent osmotic multiplier (continued).

The remainder of Figure 14(*b*) shows schematically how the repetition of these steps can result in the generation of a sizeable osmotic gradient, with the highest osmolarity being at the tip of the loop. Notice that water equilibration occurs at every step. These models show both the movement of tubular fluid and the ion transport and water equilibration processes to be discontinuous, out-of-phase processes. Obviously, this model is a simplification that allows us to follow the buildup of a gradient. In reality, these processes occur simultaneously and continuously.

Figure 14(*b*) shows that the fluid is progressively concentrated as it flows down the descending limb, but becomes progressively diluted in the thick region of the ascending limb. It should be obvious why the latter region of the renal tubule is called "the diluting segment." While only a 200-mOsm/L gradient is maintained across the thick ascending limb at any given horizontal level in the medulla, there is a much larger osmotic gradient from the top of the medulla to the bottom. Thus, the 200-mOsm/L gradient established by the active Cl pump (and passive Na movement) has been multiplied by the countercurrent flow within the loop.

The maximum gradient within the medullary interstitium that can be established depends upon several factors, particularly the lengths of the loops. It has long been known that certain desert animals, such as the kangaroo rat, have very long loops of Henle and can excrete a very concentrated urine. Other important factors include the rate of Cl pumping and the rate of delivery of proximal tubular fluid to the loops (i.e., ultimately, the GFR). In human physiology, the maximal osmolarity that can be attained is about 1,400 mOsm/L.

The preceding presentation of the development of the high medullary osmolarity dealt only with the NaCl content of the interstitial fluids. However, if one measures the NaCl concentration of the medullary interstitium and calculates its contribution to the osmolarity, NaCl is found to account for only perhaps half the osmolarity, the balance being accounted for by urea. Furthermore, it is known that starvation or experimental protein depletion causes most mammals to be unable to form a concentrated urine. However, the precise role of urea in the formation of a concentrated urine of small volume remains obscure.

The large urea concentration found in the medullary

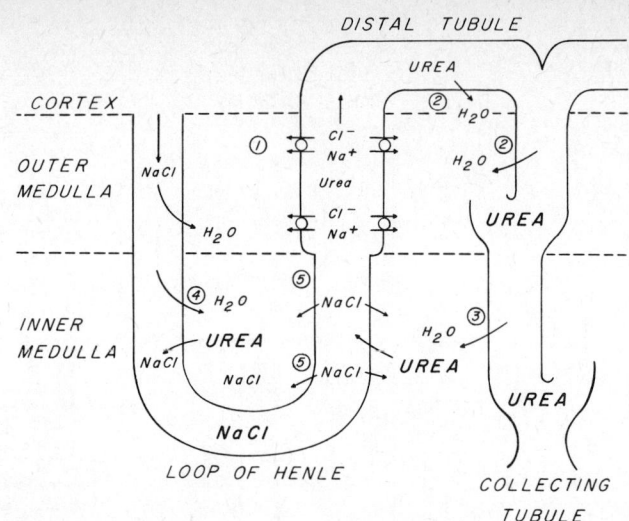

Figure 15 Sequence of events that are believed to be responsible for the high urea concentration found in the medullary interstitium. Both the thin ascending limb of the inner medulla and the thick ascending limb of outer medulla as well as first portion of the distal tubule are always impermeable to water. (1) Active Cl^- transport and passive Na^+ reabsorption dilutes the tubule fluid, but makes interstitial fluid hyperosmotic. (2) Water is reabsorbed down its osmotic gradient, but membrane of distal tubule and outer medullary portion of collecting duct is impermeable to urea, so urea concentration in tubular fluid rises. (3) Both water and urea are reabsorbed in the inner medullary region of the collecting duct. Some urea re-enters the ascending and descending loop and this medullary re-cycling of urea, in addition to trapping of urea by counter-current exchange in the vasa recta (see Figure 17) causes urea to accumulate in large quantities in the medullary interstitium as indicated by large, bold type. (4) This high urea concentration tends to osmotically extract water from descending limb and consequently to concentrate NaCl. (5) When the NaCl-rich fluid reaches the ascending thin limb which is water impermeable but NaCl permeable, NaCl moves passively down its concentration gradient helping to render ascending tubule fluid hypo-osmotic to the surrounding interstitium. (Modified from Jamison RL, Maffly RH: *N Engl J Med* 295:1059–1067, 1976.)

interstitium is believed to be developed by a passive process that depends on differential permeabilities of various parts of the renal tubule. Figure 15 shows the steps thought to account for the concentration of urea in the medullary interstititium.

ROLE OF THE VASA RECTA: COUNTERCURRENT EXCHANGE

Figure 16 illustrates the relative proportions of glomerular filtrate that are reabsorbed in the cortex and medulla respectively from a juxtamedullary nephron. In this figure about 80% of the filtered fluid is reabsorbed in the proximal tubules. Not all this fluid is absorbed into cortical blood vessels. About 25% of the reabsorbed fluid is reabsorbed via medullary blood vessels. The blood flow required for this purpose could have a dissipating effect on the medullary osmotic gradients, because osmotically active solutes would be carried away at the same time. In fact, such an effect is minimized by two factors; (1) medullary blood flow is small compared with that of the cortex, and (2) the close apposition of descending and ascending limbs of the *vasa recta* favors a countercurrent *exchange* of osmotically active solutes from ascending to descending limbs.

In the case of the vasa recta, no active transport of solute occurs nor are there different permeability properties attendant to different portions of the loop. Therefore, the osmolarity of the blood increase as it dives into the medulla and then decreases as it emerges toward the cortex (Fig. 17). This occurs as a result of osmotic equilib-

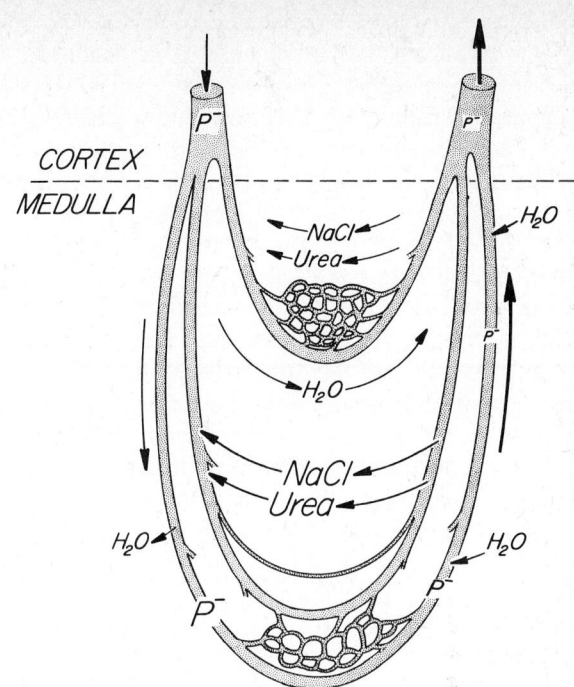

Figure 17 Countercurrent exchange by the vasa recta. The medullary circulation consists of a network of channels with main thoroughfares and branch connections. P^-, plasma protein. The size of the lettering indicates the relative concentrations of each solute with respect to its location in the medulla, but not necessarily with respect to other solutes. The progressive rise of NaCl and urea in the medullary interstitium is due to the loop of Henle and collecting duct (Fig. 15). Since the capillaries are permeable to NaCl and urea, these solutes enter the descending vasa recta and leave the ascending vasa recta. This transcapillary exchange helps "trap" the solutes in the medulla. Conversely, water leaves the descending vasa recta, causing the protein concentration to rise. In the ascending vasa recta, water enters the capillaries. More water enters the ascending vasa recta than is left in the descending vasa recta because of a slight net entry of NaCl and urea in the inner medulla. Thus, the vasa recta serve two functions: to preserve the hyperosmolarity of the medullary interstitium and to return water reabsorbed from the collecting ducts to the systemic circulation.

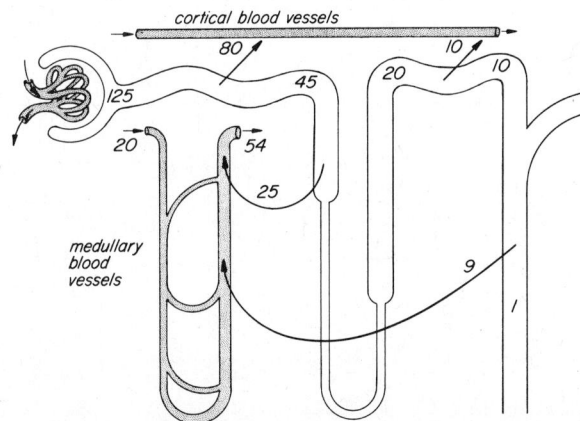

Figure 16 The fluid balance of the kidney, assuming a GFR of 125 ml/min and a medullary blood flow of 20 ml/min. Net water movement is indicated by arrows. Numbers refers to milliliters per minute of water flow for the whole kidney.

rium between the blood and the interstitium at every horizontal level. Such a mechanism depends on the blood flow rate being slow enough to permit equilibration. Obviously, an increase in medullary blood flow rate would result in some loss of osmotically active substances from the medulla, that is, the osmotic gradient could be "washed out."

If the osmolarity of the plasma that enters the vasa recta remains unchanged when it leaves, how can there be a net return of solutes and water to the body during the passage of plasma through the medullary loop? The answer lies in the fact that although the osmolarity is similar, the volume of plasma is larger leaving the vasa recta than entering it. Water and some solutes are drawn into the ascending vessels to increase the volume, returning water to the body.

CONTROL OF URINE OSMOLARITY

The osmolarity of human urine varies from less than 100 mOsm/L to 1,400 mOsm/L, depending on a person's hydration and fluid balance (see Control and Feedback Mechanisms Affecting Renal Function, below). How is this possible? Figure 18 illustrates these two extremes. It can be seen that the key difference in these two states is the presence or absence of ADH. In the absence of ADH, the excreted urine will be dilute, reflecting the activity of the ion pumps in the distal portions of the nephrons. Thus, solute is continually pumped out of tubular fluid, but since these regions of the nephrons are water impermeable in the absence of ADH, the water cannot leave with the solute. Furthermore, the osmotic gradient in the medulla

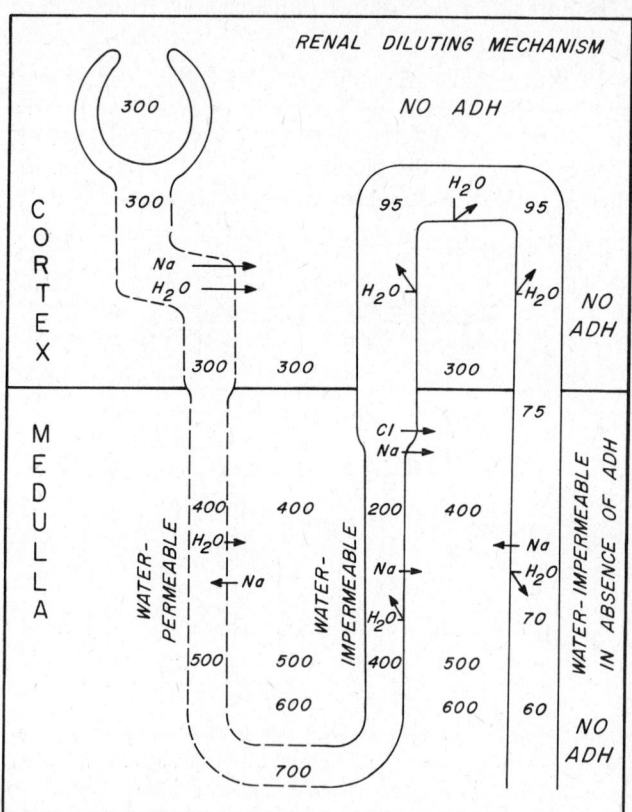

Figure 18 Renal concentrating and diluting mechanisms. Left, when ADH is present. Right, the conditions caused by the absence of ADH. (Modified from Schrier RW: *Renal and Electrolyte Disorders,* Boston, Little, Brown, 1976.)

is somewhat reduced. The reasons for this are not well understood, but could be related to an increased blood flow through the vasa recta, to an increased GFR, and/or to an inhibition of the active transport of Cl^-. At any rate, this reduction in the osmolarity of the medullary interstitium would also serve to maintain a large urine volume.

In contrast, when ADH secretion is maximally stimulated, urine osmolarity can become equal to the osmolarity of the deep medullary interstitial fluid, that is, 1,200–1,400 mOsm/L.

Figure 19 summarizes the changes in tubular fluid volume and osmolarity as it traverses a juxtamedullary nephron. It shows that although a large volume is reabsorbed from the proximal tubule, no change in osmolarity of tubule fluid occurs, that is, proximal tubule reabsorption is iso-osmotic. In the loop of Henle, the fluid osmolarity rises due to the water permeability of the descending limb and its proximity to the hyperosmotic medullary interstitium. It then falls to a level below that of blood plasma because of NaCl reabsorption in the thick ascending limb. Since ADH exerts its effects only on the collecting duct, we see in this portion of the nephron two possible extremes of volume and osmolarity, depending on the presence or absence of ADH. In the absence of ADH, very little water is reabsorbed, and a copious, dilute urine is formed (i.e., diuresis occurs). By contrast, in the presence of ADH, water reabsorption is high, and a small volume of concentrated urine is formed (i.e., antidiuresis occurs).

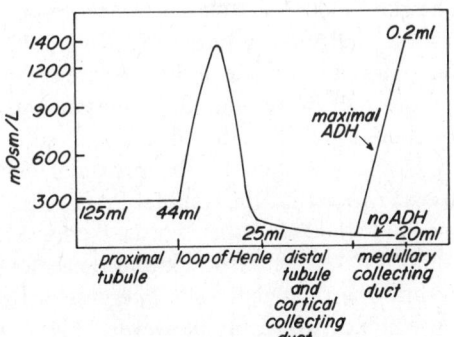

Figure 19 Changes in volume and osmolarity of tubular fluid as it flows along the nephron. (Adapted from *Renal Physiology*, Vander AJ: ed 2. New York, McGraw-Hill, 1980.)

RENAL CLEARANCES

Earlier in this chapter, we discussed the three basic processes of renal function separately. In a clinical environment, it is generally impractical to separate these individual components. However, by applying some of the fundamental ideas previously presented, we can develop a concept of particular clinical importance: the renal clearance of a substance.

Clearance is defined as the volume of plasma that would have to be completely cleared of a substance in a unit of time in order to account for the rate of the substance's appearance in the urine. Mathematically, the clearance of substance X, that is, C_x, is expressed as follows:

$$C_x = \frac{(\text{Urine concentration of } x)\,(\text{urine flow rate})}{\text{plasma concentration of } x} \quad (4)$$

You should be aware that no portion of plasma is actually completely stripped of substance X, while at the same time the remaining plasma is left untouched. Thus, the clearance is termed a virtual value. In addition, it is important to realize that the clearance of a substance reflects the algebraic sum of all the basic factors governing the handling of the substance X, that is, filtration, reabsorption, and secretion.

ESTIMATING THE GLOMERULAR FILTRATION RATE

Clearance measurements are useful when they give us information about specific renal processes. For example, the use of a completely filtered and nonreabsorbed material will give us information concerning the glomerular filtration rate (GFR). Inulin can be used for this determination, since it is a nonmetabolizable carbohydrate that is small enough to pass freely through the glomerular membrane. Consequently, it is found in the same concentration in glomerular filtrate as in plasma. There are no renal transport systems for inulin, and it does not diffuse through the tubular walls. Since it is not transported and does not diffuse out of the tubule, all the inulin filtered is excreted in the urine. Furthermore, the portion excreted in the urine enters the tubule only by filtration. By measuring the urine flow rate and the urine inulin concentration, one can calculate the inulin excretion rate (in milligrams per minute). This rate of inulin excretion is identical with the rate of inulin filtration. The concentration of inulin in the filtrate can be estimated by measuring the plasma inulin concentration. The two are virtually identical. The rate of inulin excretion, which we calculate ($U_{In} \cdot \dot{V}$), equals the filtrate concentration of inulin, which we determine, times the rate of filtrate formation. The GFR can be calculated by this means and averages 125 ml/min.

$$GFR = C_{In} = \frac{U_{In} \cdot \dot{V}}{P_{In}} \tag{5}$$

where U_{In} is the inulin concentration in urine (mg/ml), $\dot{V}$ is the rate of urine flow (ml/min), and P_{In} is the inulin concentration in plasma (mg/ml). The GFR thus calculated is the inulin clearance. Inulin clearance is a reasonable estimate of GFR only because of the characteristics of renal handling of inulin. The clinical relevance is important, however, since we can now use for GFR the value of the inulin clearance to estimate the clinical state of the glomeruli.

A number of analytical difficulties inherent in the measurement of inulin concentration in plasma and urine, have often required the use of alternative markers to measure GFR. One of these markers is creatinine. Urinary creatinine is derived almost entirely from endogenous sources, obviating the need for injecting foreign material. Also, it is very easily measured in the urine. Under normal circumstances, the clearance of creatinine very closely approximates the clearance of inulin. However, the similarity is, to a large part, attributable to compensating errors. Since creatinine is excreted in humans and other primates by secretion as well as filtration, its clerance actually exceeds the simultaneous clearance of inulin by about 20%. But blood plasma contains substances, in addition to creatinine, that react with the normal colorimetric reagents used in creatinine determination. Since these noncreatinine reactants appear in the plasma, but not in the urine, their presence results in an overestimation of plasma creatinine and consequently an underestimation of creatinine clearance by roughly 20%. Thus, the two compensating errors yield a creatinine clearance close to that of inulin. Nevertheless, if this factor is kept in mind, creatinine clearance still represents a relatively easy noninvasive method for estimating GFR.

ESTIMATING RENAL PLASMA FLOW

Another clinically important process is the renal plasma flow (RPF). It should be possible to use the clearance concept to measure this process. All that is needed is a substance that can be filtered by the glomeruli, that cannot be reabsorbed by the tubules, and that can then be completely removed from peritubular capillary plasma by tubular secretion. All the plasma flowing into the kidney would be "cleared" of such a substance. Although it might be hard to believe, such a substance actually exists; it is PAH.

PAH readily crosses the glomerular membrane into the filtrate, but does not diffuse out of the tubules back to the blood. An active transport system exists in the proximal tubular membrane to transport organic acids from the plasma to the tubular fluid. Until the T_m of this system for PAH is exceeded (about 80 mg/min), this process will remove practically all the PAH presented to the proximal tubules by the plasma. Given these facts, it should be clear that the clearance of PAH will be equal to RPF unless the T_m has been exceeded. Thus,

$$RPF = C_{PAH} = \frac{U_{PAH} \cdot \dot{V}}{P_{PAH}} \tag{6}$$

where C_{PAH} is the PAH clearance (ml/min); U_{PAH} is the PAH concentration in the urine (mg/ml); $\dot{V}$ is the flow rate of urine (ml/min), and P_{PAH} is the PAH concentration in plasma (mg/ml) of afferent arteriole. The RPF calculated in this way is about 660 ml/min for the average adult. The dependence of this clearance on the plasma concentration of PAH can be seen in Figure 20.

Critical assumptions made in the interpretation of C_{PAH} being equal to RPF were that all the plasma entering the kidney is filtered and that after filtration it subsequently passes in close apposition to the proximal convoluted tubule in order to permit the residual PAH to be secreted into the tubular fluid. In other words, we have assumed that all the PAH that enters the kidney is removed from the blood into the tubular fluid in one pass of this blood through the kidney. In reality, however, only about 90% of the PAH entering a healthy kidney is removed, with the balance representing blood that supplies nonsecretory parts of the kidney. Thus, what is really calculated is the

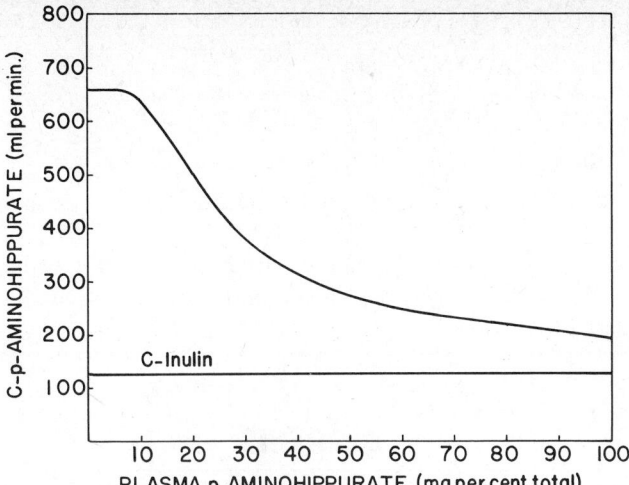

Figure 20 Clearance of PAH as a function of plasma concentration in human physiology: mg% total = mg/100 ml. (From Pitts RF: *Physiology of the Kidney and Body Fluids,* ed 3. Chicago, Year Book, 1974.)

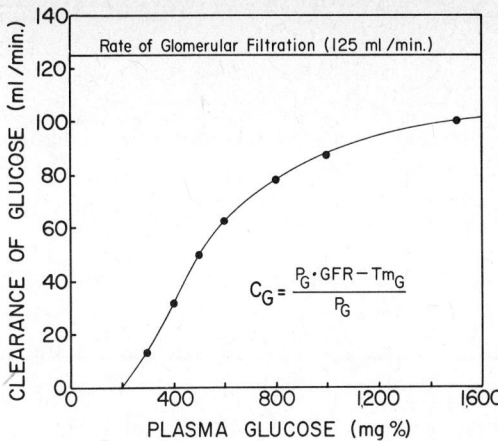

Figure 21 Glucose clearance as a function of plasma concentration of glucose. (From Pitts RF: *Physiology of the Kidney and Body Fluids,* ed 3. Chicago, Year Book, 1974.)

effective renal plasma flow (ERPF). In a diseased kidney the extraction percentage might be even lower if the renal tubules have been damaged, causing an apparently low ERPF. In order to make appropriate corrections, renal arterial and venous blood must be sampled to correct for that percentage of PAH that is neither filtered nor secreted, but that does pass through the kidney.

RENAL CLEARANCE OF GLUCOSE

Clearances for other substances do not bear such a clear correlation to specific aspects of renal function, but several are of interest both clinically and from the standpoint of basic renal physiology. The glucose clearance (shown in Fig. 21, with inulin clearance for comparison) is an interesting example of a T_m system. Below the renal threshold for glucose, the glucose clearance is zero. If the blood glucose is raised beyond the renal threshold, the glucose clearance increases. It will continue to increase and asymptotically approach C_{inulin} (i.e., the GFR).

RENAL CLEARANCE OF WATER

Theoretically, the clearance of water from the kidney could be handled in exactly the same manner as any other substance, but in practice it is difficult to deal with the definition of the "concentration" of water in plasma or in urine. However, the quantitation of water excretion has been facilitated by the concept that urine flow is divisible into two components. One component is the urine volume that would be needed to excrete solutes at the osmotic concentration of solutes in plasma. This iso-osmotic component has been termed osmolar clearance (C_{Osm}). The other component is called free-water clearance (C_{H_2O}) and is the theoretical volume of solute-free water that must be added to (positive C_{H_2O}) or reabsorbed from (negative C_{H_2O}) the iso-osmotic portion of the urine to create either hypo-osmotic or hyperosmotic urine, respectively. These terms are calculated as follows:

$$C_{Osm}(ml/min)$$
$$= \frac{\text{Urine osmolarity } (U_{Osm}) \times \text{urine flow rate } (\dot{V})}{\text{Plasma osmolarity } (P_{Osm})} \quad (7)$$

and

$$\dot{V} = C_{Osm} + C_{H_2O} \quad (8)$$

That is, the volume flow of urine is just equal to the

osmolar flow plus any free water added (or minus any free water reabsorbed). Consequently,

$$C_{H_2O} = \dot{V} - C_{Osm} \qquad (9)$$

Further inspection of these relationships will demonstrate several interesting points:

1. When U_{Osm} equals P_{Osm} (iso-osmotic urine), $\dot{V}$ equals C_{Osm}; therefore, C_{H_2O} is zero.
2. When U_{Osm} is greater than P_{Osm} (hyperosmotic urine), C_{Osm} is greater than $\dot{V}$; therefore, C_{H_2O} will be negative.
3. When U_{Osm} is less than P_{Osm} (hypo-osmotic urine), C_{Osm} is less than $\dot{V}$, and C_{H_2O} is positive.

It is interesting to note that in terms of water clearance, the ability of the normal kidney to reabsorb water is not as great as its ability to excrete water. Consider the example of a person ingesting normal amounts of Na^+ and protein. He will have to dispose of approximately 600 mOsm solute per day per square meter of body surface area. If the renal concentrating ability is intact, this daily solute load could be excreted in 500 ml urine at 1,200 mOsm/L*. With this osmolarity and volume, for a person of 1 sq m surface area, C_{H_2O} can be calculated as follows:

$$C_{Osm} = \frac{U_{Osm} \dot{V}}{P_{Osm}} = \frac{(1,200 \text{ mOsm/L}) (500 \text{ ml/day})}{300 \text{ mOsm/L}}$$

$$C_{Osm} = 2,000 \text{ ml/day}$$

$$C_{H_2O} = 500 \text{ ml/day} - 2,000 \text{ ml/day}$$

$$C_{H_2O} = -1,500 \text{ ml/day}$$

Thus, only 1,500 ml solute-free water will be returned to body fluids during the maximal antidiuresis caused by maximal secretion of antidiuretic hormone (see below). In contrast, with the same daily solute load of 600 mOsm/M², minimal urinary osmolarity of 60 mOsm/L gives a daily urine volume of 10 L; the renal capacity to excrete water is much greater than the capacity to return solute-free water to the body. Specifically,

$$C_{Osm} = \frac{U_{Osm} \dot{V}}{P_{Osm}}$$

*Assume 1200 mOsm/L to be maximal urinary concentration for this individual.

$$= \frac{(60 \text{ mOsm/L}) (10 \text{ L/day})}{300 \text{ mOsm/L}}$$

$$= 2 \text{ L/day}$$

$$C_{H_2O} = \dot{V} - C_{Osm} = 10 - 2 = 8 \text{ L/day}$$

So, with comparable solute loads and relatively maximal and minimal urine osmolarities, the 1.5 L/day water retention is substantially less than that of 8 L/day of excess water elimination. Prevention of total body water deficit is thus largely dependent on water intake as modulated by thirst.

DIURESIS

Diuresis is defined as increased output of urine. This increase can come about from a number of causes, some of which represent homeostatic mechanisms adjusting body fluid and electrolyte composition. Other causes include disease processes and drug effects. The composition of the urine, the volume of which is elevated, will differ widely depending on the causes of the diuresis—there are a number of ways in which an increased urine volume can be attained. A description of several diuretic processes will serve to illustrate the interrelationships among the various components of renal function.

WATER DIURESIS

Water diuresis occurs as a result of a decreased level of plasma ADH. The stimulus for the decreased ADH secretion by the posterior pituitary is a dilution of extracellular fluids (see Osmoreceptor–Antidiuretic Hormone System, below). In the absence of ADH, the walls of the collecting duct become essentially water impermeable (Figs. 18 and 19), resulting in a large urine flow associated with a large free-water clearance (see above).

OSMOTIC DIURESIS

This type of diuresis is caused by substances having the following properties:

1. They are freely filterable across the glomerulus.

2. They undergo limited or no reabsorption by all segments of the renal tubule.

3. They exhibit no direct pharmacological actions on tubular transport systems.

A substance with these properties will retard water (and indirectly, Na) reabsorption merely by its osmotic contribution to the tubular fluid. Normally, the main osmotically active constituents of the glomerular filtrate are Na and its attendant anions (Cl^- and HCO_3^-). Let us administer a substance that has the above-mentioned properties, such as mannitol, such that the plasma concentration is 10 mOsm/L. As the plasma is filtered, mannitol will be present in the filtrate at a concentration of 10 mOsm/L. The reabsorption of Na by the proximal tubule will begin to cause water to be reabsorbed as well. This will result in an increased mannitol concentration, since it cannot be reabsorbed. As its concentration increases it begins to inhibit the reabsorption of water. Remember the proximal tubular epithelium is permeable to water and that tubular water is in osmotic equilibrium with plasma water. Sodium transport continues, however, so that tubular Na concentration falls below that of plasma concentration. Nevertheless, because it is a gradient-limited transport system, not as much Na is removed in absolute amounts by the proximal tubule as would be otherwise. Another aspect is that by creating a lower Na concentration inside the renal tubule, the gradient for passive Na leakage from plasma to tubule has been increased. The result is that osmotic diuretics cause not only a water loss, but a loss of large quantities of Na, K, and Cl as well.

This is a practical problem in the management of the patient with diabetes mellitus. Such patients have unusually high plasma levels of glucose, so high that the filtered load can exceed the glucose T_m. In addition, ketone bodies such as acetoacetate and β-hydroxybutyrate may also be present in excess of their T_m levels. These substances fulfill the requirements listed above for osmotic diuretics. Hence, uncontrolled diabetes mellitus is characterized by large rates of urine flow and loss of electrolytes. The presence of the ketoanions exacerbates the problem of potassium loss. Not only is there reduced reabsorption of K due to the osmotic diuretic effect, but K secretion is actually increased as a result of the presence of these nonreabsorbable anions in the distal tubules.

DIURESIS AS A RESULT OF INHIBITION OF TUBULAR TRANSPORT PROCESSES

When the amount of osmotically active solute in the urine is increased, there is usually an increased urine volume, as we have just seen. The size of this increase is subject to a variety of compensatory effects as, for example, the rate of ADH release. In general, there are two different ways whereby the amount of solute in the urine can be increased: (1) an increase in the filtered load of poorly absorbed substances (i.e., increased GFR), and (2) the inhibition of tubular electrolyte transport processes (mainly Na^+, Cl^-, H^+, and K^+) by specific drugs. A description of the effects of the second method mentioned above will provide an intuitive understanding of the interrelationships among the various tubular transport systems. To this end, two classes of pharmacological agents with differing sites of action will be discussed.

Carbonic Anhydrase Inhibitors

The enzyme carbonic anhydrase catalyzes the reaction

$$H_2O + CO_2 \rightleftharpoons H_2CO_3.$$

The carbonic acid formed is subject to a virtually instantaneous ionic dissociation to hydrogen ion and bicarbonate, a reaction that does not require enzymatic catalyzation. In the kidney this enzyme is found in all renal tubular cells and is also apparently attached to the luminal border of proximal tubular cells. The results of inhibiting this enzyme by an agent such as acetazolamide can be seen in Table 3.

The effects of acetazolamide can be understood in light of the importance of carbonic anhydrase to HCO_3^- reabsorption in the proximal tubule. If the production of HCO_3^- within the luminal cell is inhibited (or, more correctly, the rate is greatly decreased), the ability of the proximal tubule to "reabsorb" luminal HCO_3^- will be decreased. Thus, a higher concentration of HCO_3^- will still be in the luminal fluid as this fluid leaves the proximal tubule. This will require a higher concentration of cations (mainly Na^+) to accompany the increased bicarbonate concentration, which in turn will cause an osmotic equivalent of water to be retained within the tubule. In the distal tubule much of the Na^+ will be reabsorbed. You will recall

Table 3 Effects of Tubular Transport Inhibition on Urinary Composition[a,b]

Substance	Vol (ml/min)	Ion Concentrations (mEg/L)				
		pH	Na$^+$	K$^+$	Cl$^-$	HCO$_3^-$
Control	1	6	50	15	60	1
Acetazolamide	3	8.2	70	60	15	120
Ethacrynic acid	8	6	140	10	155	1
Mannitol	10	6.5	90	20	110	4

[a]Modified from Goodman & Gilman: *Pharmacological Basis of Therapeutics*, ed 6, New York, Macmillan, 1980.
[b]The effects of the osmotic diuretic mannitol are included for comparison.

that in the distal tubule Na$^+$ is reabsorbed by an active transport process, but that a cation (either H$^+$ or K$^+$) must be injected into the tubular fluid to maintain electroneutrality. Since the H$^+$ is formed as a result of the carbonic anhydrase-catalyzed reaction, its availability to replace Na$^+$ is greatly reduced by acetazolamide treatment. Accordingly, K$^+$ is secreted into the tubular fluid, and thus most of the Na$^+$ is reabsorbed, but at the expense of a great loss of K$^+$.

Inhibitors of Active Cl$^-$ Transport in Thick Ascending Limb of Loop of Henle

Furosemide and ethacrynic acid are known to directly block Cl$^-$ transport out of the isolated thick limb segments of the loop of Henle. Such an action will result in the presentation to the distal tubule of much higher than normal concentrations of Cl$^-$ and Na$^+$. The Na$^+$ concentration is high because of the presumed obligatory nature of the movement of Na$^+$ out of the tubule to match the actively extruded Cl$^-$. Most of this NaCl cannot be removed by the Na$^+$ pump of the distal tubule. Since this agent inhibits NaCl reabsorption in the thick ascending limb, it will tend to cause an iso-osmotic urine. That is, the ability of the kidney to produce free-water clearance or reabsorption is impaired. In the case of a patient with a significant positive C_{H_2O}, the ADH levels are low to permit excretion of excess water. In the absence of ADH, the urine osmolarity normally reflects the effects of Cl$^-$ pumping in the ascending limb of Henle's loop (diluting seg-

ment); the more active the pump, the more dilute the fluid that leaves this segment and enters the distal tubule. Thus, inhibiting the pump will result in a less dilute urine. A less dilute urine with respect to solutes means a smaller positive free-water clearance (i.e., less water excreted).

The situation is just the reverse for a patient exhibiting a significant negative C_{H_2O}. In this case, the levels of ADH will be higher, and therefore water will be leaving the collecting ducts drawn by the higher medullary interstitial osmolarity. Since this high osmolarity depends on the Cl$^-$ pump of the ascending limb of the loop of Henle, inhibition of this pump will result in a dissipation of the osmotic gradient, hence less water will be reabsorbed from the collecting duct.

CONTROL AND FEEDBACK MECHANISMS AFFECTING RENAL FUNCTION

THE OSMORECEPTOR– ANTIDIURETIC HORMONE SYSTEM

Besides regulation of water and electrolyte balance, regulation of extracellular osmolarity is also an important renal function. This regulation is vested in the osmoreceptor–antidiuretic hormone system involving the hypothalamus, the posterior pituitary, ADH, and the renal tubules. This system, when activated, promotes water conservation. ADH is secreted into the blood by the posterior pituitary. ADH then promotes increased water reabsorption from the collecting ducts of the kidneys via a cyclic adenosine monophosphate (AMP)-modulated pathway. When no ADH is secreted, the amount of water passing into the urine each day is five to 15 times normal (large positive C_{H_2O}). On the other hand, when large quantities of ADH are secreted, water is reabsorbed to an extreme degree, so that the volume of urine formed each day in a normal person (1.7 sq m surface area) might be as little as 700 ml, or one-half the normal volume. As a result, water is conserved and the extracellular fluids become more dilute.

Antidiuretic hormone would be of no value to the body

without the presence of a concomitant system for regulating the secretion of antidiuretic hormone in proportion to its need. To provide this, special neurons of the anterior hypothalamus, called osmoreceptors, respond to changes in osmolarity of the extracellular fluid. When extracellular osmolarity becomes low, osmosis of water into the osmoreceptors causes them to swell, thereby decreasing their rate of neuronal impulse discharge. Conversely, increased osmolarity in the extracellular fluid pulls water out of the osmoreceptors, causing them to shrink, and thereby increasing their rate of discharge.

The impulses from the osmoreceptors are transmitted from the hypothalamus through the pituitary stalk into the posterior pituitary gland, where they promote the release of antiduiretic hormone.

Thus, ADH is secreted in relationship to the osmolarity of the extracellular fluids—the greater the osmolarity, the greater the rate of ADH secretion, and the lower the osmolarity, the lower the rate of ADH secretion.

The entire mechanism can be synthesized as follows. An increase in osmolarity of the extracellular fluids excites the osmoreceptors, promoting ADH secretion and causing marked reabsorption of water by the renal tubules while solutes continue to be lost into the urine. Consequently, the extracellular fluids become diluted and their osmolarity returns toward normal. By contrast, low osmolarity of the extracellular fluids decreases the activity of the osmoreceptors, thereby decreasing ADH secretion, and large amounts of water are lost into the urine until the extracellular fluid osmolarity returns to normal.

The osmoreceptor system is sufficiently sensitive that increased extracellular osmolarity of only 1–2% above normal causes marked retention of water, whereas a similar decrease causes rapid loss of water.

PLASMA VOLUME REGULATION

The control of plasma volume, and thereby extracellular fluid (ECF) volume, is quite complex and not well understood. The following treatment is an attempt to distill current ideas about this perplexing yet extremely important aspect of renal function.

It should be stated at the outset that the didactic separation of osmotic and volume control employed in this chapter is an artificial device designed to aid the student in understanding the role of the kidneys in the overall economy of body fluids. For example, ADH secretion will result in an increase of ECF volume as well as a decrease in its osmolarity. Hence, it should not be surprising that ADH secretion can be affected by ECF volume changes, as well as osmolarity changes.

Afferent Signals for Volume Regulation

Although we speak of ECF volume regulation, it is important to realize that, physiologically speaking, the function of this renal process is to maintain blood volume, hence the adequacy of the circulation. Realizing this, it is not hard to understand why much of the afferent (i.e., receptor) information is believed to come from baroreceptors located both within and without the kidney. The extrarenal receptors are located in the great veins, the atria, the carotid sinuses, and the aortic arch. The afferent arteriole that is part of the juxtaglomerular apparatus is also believed to have a baroreceptor capability.

In addition to changes in blood pressure, changes in osmolarity and in ECF sodium concentration also serve as signals to be integrated for the control of ECF volume as well as for osmolarity control. These are believed to be located in the cerebrospinal fluid compartments and the liver.

Effector Mechanisms for Volume Regulation

The control of ECF volume is largely effected by controlling the rate of sodium excretion. Thus, as Na and its attendant anion(s) are lost, a certain osmotic equivalent amount of water will be lost as well. Since NaCl is the most abundant osmotically active salt in the extracellular environment, any reduction in its amount must cause a reduction in the amount of extracellular fluid. We should recognize that those changes in the ECF volume that are accompanied by changes in fluid osmolarity will be corrected, at least in part, by the osmoreceptor–ADH system we have already spoken about.

The amount of Na excreted is a result of the filtration process and the reabsorption process, that is,

Na excretion =
Na filtered − Na reabsorbed =
$(GFR \times P_{Na}) - (Na\ pumped - Na\ backleak)$

Adjustment of any of these four variables would result in a change of Na excretion. P_{Na}, the plasma Na concentration, stays within quite narrow limits and therefore does not appear to be a primary variable in controlling Na excretion.

Control of GFR

Although it is quite true that changes in GFR will cause changes in Na excretion rates, it has been clearly demonstrated that natriuresis (i.e., loss of Na via the urine) can occur in subjects in whom ECF volume has been expanded and in whom GFR has remained constant. It is therefore currently believed that changes in GFR alone are not essential for renal regulation of volume and that they are unimportant for control of Na excretion.

We are therefore left with the Na reabsorption process as the key to control of Na excretion, hence of ECF volume. A number of mechanisms appear to operate on the reabsorption process. It is impossible to know which, if any, of these mechanisms is the most important in the overall regulation of ECF. It does not seem unlikely, however, that ECF volume control reflects the interplay of all these factors in a way that would render it impossible to assign primary importance to any one of them.

Nephron Inhomogeneity

We know that juxtamedullary nephrons have somewhat higher rates of glomerular filtration than do cortical nephrons. Therefore, if cortical nephrons also have lower reabsorptive rates, a sodium loss could be caused by shifting a portion of blood flow to them. The problem with this hypothesis is that it does not appear that such a shift in blood flow occurs in ECF volume expansion.

Effects on Na Pumping

Renin–Angiotensin–Aldosterone System

Secretion of renin by the kidney activates a series of physiological mechanisms that have important effects on sodium and potassium balance as well as on arterial blood pressure. Renin is synthesized and stored in the granular juxtaglomerular cells located along the afferent arterioles (Fig. 4). Renin is secreted from the juxtaglomerular cells into the lumen of the afferent arteriole.

Renin substrate, or angiotensinogen, is a protein synthesized by the liver and secreted into the circulation. Its complete structure has not been determined, nor has it

been established as a single substance. Under the enzymatic action of renin in the circulating plasma, angiotensinogen is converted to angiotensin I. Angiotensin I has been found to have little or no physiological action of its own. Rather, it is rapidly converted to an active form, angiotensin II. This reaction is mediated by converting enzyme. The most abundant source of converting enzyme is the lung, and virtually all the angiotensin I is converted to angiotensin II in a single passage through the pulmonary circulation.

The actions of angiotensin II are many and varied. To date, there is still uncertainty as to which of these actions are of physiological importance and which are simply pharmacological effects demonstrable only when the hormone is administered from exogenous sources. The primary effect of concern to us at the moment is the action of angiotensin II on the adrenal cortex, which results in the secretion of aldosterone. The corticosteroid aldosterone is classified as a mineralocorticoid, since its action is to enhance sodium and potassium transport across epithelial membranes. Aldosterone exerts this action on many epithelia, including the salivary ducts and colon, but its most important site of action is the cortical collecting duct of the renal tubule. There it increases sodium reabsorption and enhances secretion of potassium and hydrogen ion into the luminal fluid. In performing these functions, aldosterone plays a key role in regulating fluid and electrolyte balance in the body. By increasing sodium reabsorption at the level of the cortical collecting duct, aldosterone tends to increase the osmolarity of the ECF. This rise in osmolarity provides a stimulus for secretion of ADH, which in turn increases the permeability of the distal tubule and collecting duct to water. Thus, water reabsorption is enhanced and the osmolarity of ECF is returned toward normal. It is through this increase in tubular reabsorption of water that aldosterone indirectly causes an expansion of ECF volume. Not the interplay between control of Na reabsorption and the osmoreceptor ADH system in effecting a change of ECF volume.

The ability of aldosterone to initiate ECF volume expansion is limited, however, in normal subjects. The chronic administration of large doses of the hormone results in expansion of the ECF volume by approximately 1–2 L. After this degree of ECF volume expansion, additional salt and water retention ceases in spite of continued administration of aldosterone. This phenomenon has been referred to as aldosterone or mineralocorticoid "escape." It is unlikely, however, that the kidney truly escapes from the effects of the hormone, since urinary potassium loss persists. It now seems more likely that as a result of volume expansion other factors intervene to suppress tubular sodium reabsorption at an aldosterone-independent site or sites in the nephron. This effect then counterbalances and obscures the sodium-retaining action of aldosterone.

Since aldosterone is involved in the regulation of ECF volume and potassium balance, it is perhaps not surprising that the rate of secretion of aldosterone from adrenal cortex is regulated through these factors. There are actually four known factors involved in stimulating the secretion of aldosterone: (1) angiotensin II, (2) plasma potassium concentration, (3) plasma sodium concentration, and (4) adrenocorticotropic hormone (ACTH). Of these four factors, the first two are probably of most importance in human physiology.

Natriuretic Hormone(s)

Experimental observations have shown that under apparently well-controlled conditions, ECF volume expansion will provoke a small natriuresis (increased rate of Na excretion). In the absence of any other explanation, a new hormone or hormones has been postulated to cause natriuresis. Although many substances that occur in body fluids have been shown to cause natriuresis (e.g., prostaglandins), it has not yet been possible to show that any of them has a physiological function.

Hemodynamic and Oncotic Pressure Factors

Although we have said that GFR and RPF are fairly constant over a range of arterial perfusion pressures, it has been well known for many years that sodium excretion rate has a marked, direct dependence on such pressure. More recently, it was shown that increased renal perfusion pressure or renal vasodilation actually decreased tubular sodium reabsorption with a consequent increase in Na$^+$ and water excretion. This effect was believed to be the result of an increase in the hydrostatic pressure in the peritubular capillary that could be inhibited by increasing the plasma oncotic pressure. Conversely, it is known that

if the oncotic pressure of plasma is lowered, tubular sodium reabsorption is inhibited even at a constant GFR, RPF, and perfusion pressure. Thus, it became obvious that changes in Starling capillary forces were affecting renal tubule sodium reabsorption.

Recall that Starling's hypothesis states that the size and direction of net fluid movement across capillary walls depend on the quantitative relationship among several forces. Thus, the net pressure for fluid movement into a peritubular capillary can be expressed as

$$P_{Int} + \pi_{cap} - P_{cap} - \pi_{Int}$$

where P_{Int} is interstitial fluid hydrostatic pressure, π_{cap} is peritubular capillary fluid oncotic pressure, P_{cap} is peritubular capillary hydrostatic pressure, and π_{Int} is interstitial fluid oncotic pressure. Not that those factors that inhibit sodium reabsorption also tend to decrease the reabsorptive capacity of the peritubular capillary (i.e., increased capillary hydraulic pressure, decreased capillary oncotic pressure).

How is it possible that a decrease in the ability of the capillaries to reabsorb fluid could decrease the ability of the renal tubules to reabsorb sodium? The sodium pump itself is not affected. Remember that sodium is pumped from the cells into the lateral intercellular spaces with chloride following it, to create an osmotic pressure gradient that will permit water to move from the tubule into this intercellular space. Since the intercellular space is relatively small, sufficient NaCl and water enter it to create a hydrostatic pressure (P_{Int}). Normally, this pressure is relieved by fluid movement into the peritubular capillary, but an increase of capillary hydrostatic pressure or a decrease of capillary oncotic pressure makes net movement of fluid into the capillary less favorable. Consequently, the hydrostatic pressure in the intercellular space increases somewhat. The increase of this intercellular fluid pressure is associated with an increase in the leakiness of the tight junctions between the cells on the luminal border. Thus, NaCl and water that has been moved out of the tubule, into the tubular cell, and then to the intercellular space now leaks right back into the tubule lumen, with a resultant decreased net rate of sodium reabsorption.

Experiments have shown that it is the decrease in plasma oncotic pressure that is most important in this mechanism. An iso-osmotic increase in ECF volume brought about by an increase of diffusible electrolytes (e.g., NaCl) and water will actually dilute the plasma protein concentration. Thus, the plasma oncotic pressure will decrease, although total osmotic pressure might remain the same. Conversely, a fall in ECF volume resulting in a loss of protein-free fluid (e.g., sweating, vomiting, diarrhea) would result in an increase of plasma oncotic pressure and thereby facilitate reabsorption. The latter response will actually be facilitated by hemodynamic events in the kidney. A drop in ECF volume will tend to cause a drop in blood pressure. Compensation for this process will involve vasoconstriction, as well as kidney involvement. Thus, renal plasma flow might fall somewhat, but GFR will be only slightly decreased because the resistance of the efferent arteriole as well as the afferent will be increased. This increase in the filtration fraction will result in an even higher peritubular oncotic pressure than normal, further facilitating NaCl and water reabsorption.

Although the details of these events have been most extensively studied in the proximal tubules, it is thought that the principles apply equally well to the late distal convoluted tubule and collecting duct. This mechanism provides a means of rapid response by the kidney to plasma (ECF) volume changes, which, in principle, operates independently of neural or hormonal influences. Obviously, these latter influences will affect this mechanism through changes in systemic blood pressure and renal arteriolar resistance.

Summary

We have seen that a number of factors are involved in the way in which the kidney deals with changes in ECF volume. It is not possible to assign degrees of importance to the various mechanisms discussed. It seems quite likely that the relative importance of the mechanisms will vary depending on the precise nature of the ECF volume change. Figure 22 summarizes the factors we have been discussing and serves to illustrate the complexity of volume control. A further complication is that the response

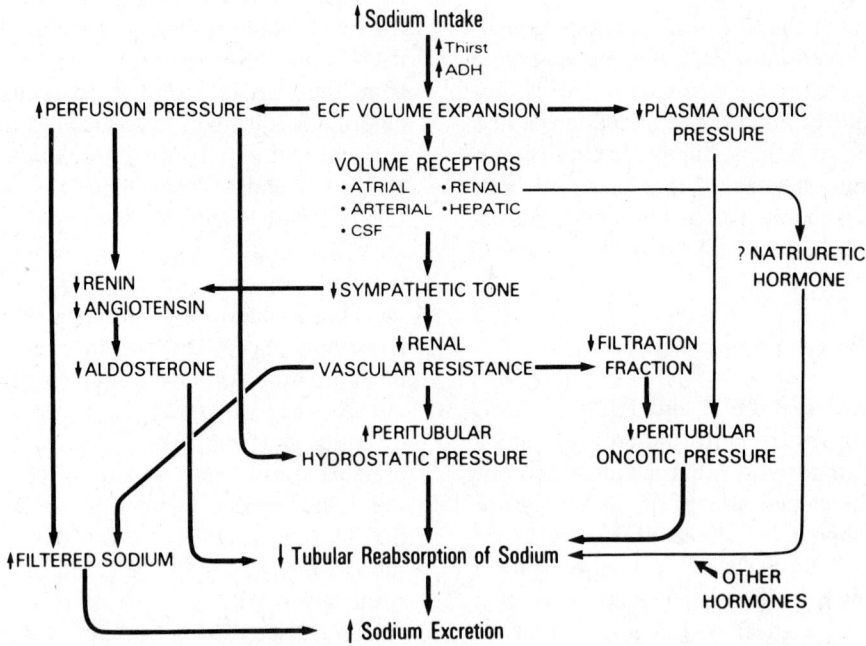

Figure 22 Summary of afferent and efferent mechanisms involved in the response to increases in sodium intake. (From Brenner BM, Rector FC: *The Kidney,* vol. I. Philadelphia, WB Saunders, 1976.)

of the kidney to a volume change can be affected by the prior state of salt and water balance to which it has been exposed. That is, apparently the reactivity of these mechanisms can be reset to alter the degree of response.

body's internal milieu than can the study of any other individual organ. Renal physiology and its correlates, water and electrolyte regulation and acid–base balance, are areas of medicine in which it is difficult to distinguish between basic science and clinical science.

CONCLUSION

With the possible exception of the central nervous system, it seems clear that the kidney is the most complex organ of the body. The responsibility of the kidney for body fluid homeostasis makes an understanding of this organ's functions a necessity for the physician who hopes to deal with, and correct, abnormal or diseased states of the human organism.

Fortunately, very fundamental concepts and theories of operation are directly correlated and useful in understanding clinical aspects of renal dysfunction. Consequently, a thorough understanding of basic renal mechanisms and feedback systems can probably assist more in understanding the complex system of balances that controls the

SELECTED BIBLIOGRAPHY

Brenner BM, Rector FC (eds): *The Kidney,* vol. I. ed. 2 Philadelphia, WB Saunders, 1981.

Deetjen P, Boylan JW, Kramer K: *Physiology of the Kidney and Water Balance.* New York, Springer-Verlag, 1975.

Pitts RF: *Physiology of the Kidney and Body Fluids, ed 3.* Chicago, Year Book, 1974.

Schrier RW: *Renal and Electrolyte Disorders, ed 2.* Boston, Little, Brown, 1980.

Smith H: *Lectures on the Kidney.* Lawrence, University of Kansas, 1943.

Vander AJ: *Renal Physiology, ed 2.* New York, McGraw-Hill, 1980.

SELF-STUDY QUESTIONS

MULTIPLE CHOICE

Select the single best answer.

An autoregulation study in a dog revealed the following: When the systemic arterial blood pressure was decreased, the blood pressure in the renal afferent arteriole fell from 135 to 100 mm Hg, while the rate of plasma flow into the glomerular capillary decreased from 660 to 600 ml/min. At the same time, the blood pressure in the glomerular capillary fell from 47 to 46 mm Hg. The blood pressure in the efferent arteriole remained constant at 20 mm Hg. Finally, the glomerular filtration rate decreased slightly from 120 to 117 ml/min.

1. The filtration fraction
   A. fell by 10%.
   B. increased by 10%.
   C. increased from 0.182 to 0.197.
   D. decreased from 0.200 to 0.179.
   E. was unaffected.

2. How was the rate of plasma flow into the efferent arteriole affected by the change in systemic blood pressure?
   A. Decreased from 660 to 600 ml/min.
   B. Decreased from 540 to 482 ml/min.
   C. Unchanged.
   D. Increased from 540 to 600 ml/min.
   E. Increased from 600 to 660 ml/min.

3. The drop in systemic blood pressure was accompanied by what change in afferent arteriolar resistance?

   A. No change.

   B. 13% increase.

   C. 13% decrease.

   D. 3.5 PRU increase.

   E. 2.6 PRU decrease.

4. The fall in systemic blood pressure was accompanied by what change in efferent arteriolar resistance?

   A. No change.

   B. 0.05 PRU increase.

   C. 0.2 PRU increase.

   D. 0.2 PRU decrease.

   E. Not enough information to calculate.

Select the correct answer(s). (In many instances, more than one answer is correct.)

5. Substance X is found in the urine at the same concentration as it is in the plasma. This proves that

   A. substance X is freely filtered by the glomerulus but that it is neither secreted nor reabsorbed by the tubules.

   B. substance X passively diffuses out of the peritubular capillaries into the tubular lumen.

   C. substance X is actively transported from the lumen to the peritubular space.

   D. the tubular membranes are impermeable to substance X.

   E. none of the above is correct.

6. Substance Y is present in the urine. Does this prove that it is filterable at the glomerulus?

7. Substance Z is filtered, reabsorbed, and secreted. Which changes would lead to an increased renal excretion of Z?

   A. Increase GFR.

   B. Increase plasma concentration of Z.

   C. Inhibit tubular reabsorption of Z.

   D. Enhance tubular secretion of Z.

   E. Any of the above.

8. The following test results were obtained on specimens from a person over a 24-h period:

   Total urine vol   1.4 L
   U_{In} 100 mg/100 ml
   P_{In} 1 mg/100 ml
   U_{Urea} 220 mmole/L
   P_{urea} 5 mmole/L
   U_{PAH} 30 mg/ml
   P_{PAH} 5 mg/100 ml
   Hematocrit   0.40

   A. What are the clearances of inulin? Urea? PAH?

   B. What is the effective renal plasma flow (ERPF)?

   C. What is the effective renal blood flow (ERBF)?

   D. How much urea is reabsorbed?

9. An increase in the plasma concentration of inulin causes which of the following in the clearance of inulin?

   A. Increase.

   B. Decrease.

   C. No change.

   D. Increase, then decrease.

   E. Decrease, then increase.

10. The clearance of substance A is less than that simultaneously determined for inulin. Which of the following might explain this observation?

   A. Substance A is itself a large molecule, and is therefore poorly filtered at the glomerulus.

   B. Substance A is actively reabsorbed by the tubules.

   C. Substance A is bound to a plasma portein.

   D. The tubular membrane is quite permeable to substance A, thereby promoting its rapid passive reabsorption.

11. An increase in the plasma PAH concentration from 5 to 50 mg% would be expected to cause which of the following changes in the renal clearance of PAH?

   A. Increase.

   B. Decrease.

   C. No change.

   D. Increase, then decrease.

   E. Decrease, then increase.

12. If a person with initially normal plasma osmolarity excretes 2 L urine/day having an osmolarity of 525 mOsm/L, then as a result

A. his body fluid osmolarity is unchanged.

B. his body fluid osmolarity will decrease.

C. his body fluid osmolarity will increase.

13. In the preceding problem, the change in body fluid osmolarity would be identical to that of

A. adding 1.5 L pure water.

B. subtracting 1.5 L pure water.

C. adding 1 L 300 mOsm/L fluid.

D. maintaining body water volume constant.

14. The following data were obtained from a patient receiving the carbonic anhydrase inhibitor acetazolamide: Urine flow = 3 ml/min; urine osmolarity = 270 mOsm/L, plasma osmolarity = 286 mOsm/L. What is this patient's free-water clearance?

A. 0.4 ml.

B. 40 ml/min.

C. 0.2 ml/min.

D. 2.8 ml/min.

E. −0.4 ml/min.

15. A patient's plasma sodium concentration is 144 mmole/L. Inulin clearance is 120 ml/min. Urine volume is 36 ml in 30 min, and the urine sodium concentration is 200 mmole/L. What percentage of filtered sodium is this patient excreting?

16. Complete inhibition of active Na^+ transport would cause an increase in excretion of which of the following substances?

A. Water.

B. Urea.

C. Chloride ion.

D. Glucose.

17. A patient is suffering from primary hyperaldosteronism, that is, increased secretion of aldosterone, which is usually caused by an aldosterone-producing adrenal tumor. Is the plasma renin concentration in this patient higher or lower than normal?

18. In a subject with normal RPF and GFR, the plasma concentration of creatinine is 1 mg%, with a urine flow of 2 L/day. The urine concentration of creatinine

is 90 mg%. The plasma and urine osmolarities are equal (approximately 300 mOs/L). Calculate the GFR:

A. 90 L/day.

B. 45 L/day.

C. 180 L/day.

D. 240 L/day.

E. Insufficient data given.

19. What is the osmolar water clearance (C_{osm})?

A. 1 L/day.

B. 2 L/day.

C. 0.5 L/day.

D. 5 L/day.

E. Insufficient data given.

20. What is the free water clearance (C_{H_2O})?

A. −1.5 L/day.

B. 1.0 L/day.

C. −0.5 L/day.

D. 0 L/day.

E. Insufficient data given.

21. When renal plasma flow is 650 ml/ min and the GFR 125 ml/min with a glucose concentration of 90 mg%, the glucose load delivered to the beginning of the proximal tubule will be

A. 90 mg/min.

B. 112.5 mg/min.

C. 468 mg/min.

D. 17.3 mg/min.

E. Insufficient data given.

22. Normal osmolarity of the ECF near peritubular capillaries is 285 mOsm. The osmolarity of the fluid near the tight junctions in the lateral spaces of proximal tubule cells is unknown, but estimates made on the basis of water flow across the tight junctions suggest that it is at least three to four times higher than the normal osmolarity of ECF. Why is there normally a net osmotic movement of water across the tight junctions into the lateral space, yet no net movement from the ECF near the peritubular capillaries?

A. The osmotic pressure of the ECF near the peritubular capillaries balances the osmotic pressure of solutes in the lateral space.

B. The tight junctions resist water movements produced by hydrostatic pressure heads, thus forcing bulk water movements to occur only toward the peritubular capillaries.

C. Expansion of the lateral spaces leads to increased

ease of movement toward the capillaries and decreased permeability of the tight junctions.

D. The reflection coefficient of Na^+ and Cl^- is zero from the lateral spaces to the peritubular ECF but is nonzero across the tight junctions.

ANSWERS

1. C. The filtration fraction is that portion of the plasma entering the glomerulus that is filtered into the renal tubules. Before the blood pressure change, this fraction was $120 \div 660 = 0.182$. After the pressure change, it was $118 \div 600 = 0.197$. Thus, the filtration fraction increased as the blood pressure decreased.

2. B. Remember that the plasma entering the glomerulus has two potential exit points: (1) the efferent arteriole, and (2) the proximal tubule. Thus, in the case of the higher pressure, the plasma flow into the efferent arteriole was $600 - 120 = 540$ ml/min. In the case of the lower pressure, the flow was $600 - 118 = 482$ ml/min.

3. E.

4. C.

The overall system can be represented diagrammatically as follows:

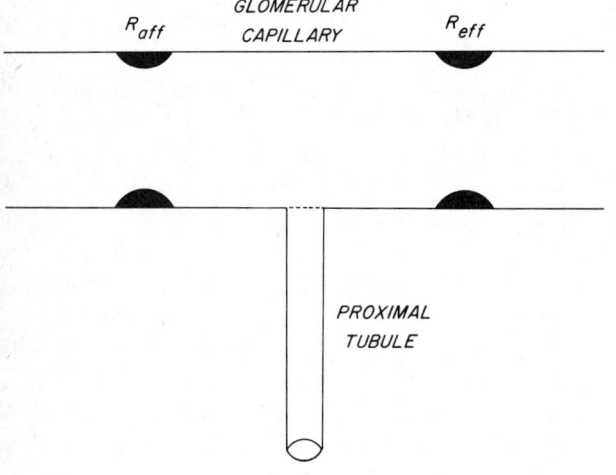

Resistances can be calculated using the relation

$$\text{Resistance (R)} = \frac{\text{change in pressure } (\Delta P)}{\text{flow } (\dot{Q})}$$

The afferent resistance (R_{aff}) was initially:

$$R_{aff} = \frac{135 - 47}{11} = 8.0 \text{ PRU.}$$

Remember that flow $\dot{Q}$ must be in units of milliliters per second to obtain an answer in PRU.
   Finally,

$$R_{aff} = \frac{100 - 45}{10} = 5.5 \text{ PRU.}$$

Thus, R_{aff} fell by $8.0 - 5.5 = 2.5$ PRU.
   The efferent resistance (R_{eff}) was initially

$$R_{eff} = \frac{47 - 20}{9} = 3 \text{ PRU}$$

After the systemic blood pressure fall

$$R_{eff} = \frac{46 - 20}{8.03} = 3.2 \text{ PRU}$$

Thus R_{eff} increased by 0.2 PRU.

5. E. The information given is insufficient to determine the fate of substance X; you should recognize, however, that some responses are nonsensical. For example, if the plasma concentration of X and the urine concentration of X were the same, then in order for A to be true, the urine volume would have to be the same as the volume of the glomerular filtrate (180 L/day). This seems a bit excessive.

6. No. It is possible, of course, but the substance can enter the lumen of the tubules only by secretion.

7. E. A and B are incorrect because they increase the filtered load, while C and D enhance the luminal concentration.

8. A. The general formula for clearance of a substance X is

$$C_x = \frac{(\text{urine concentration of } x)(\text{urine flow rate})}{\text{plasma concentration of } x}$$

The flow rate is 1.4 L/day (1440 min), or about 1 ml/min. Thus, the clearance for inulin is

$$C_{In} = \frac{(100 \text{ mg}/100 \text{ ml})(1 \text{ ml/min})}{1 \text{ mg}/100 \text{ ml}} = 100 \text{ ml/min}$$

For urea,

$$C_u = \frac{(220 \text{ mmole/L})(1 \text{ ml/ min})}{5 \text{ mmole/L}} = 44 \text{ ml/min}$$

For PAH,

$$C_{PAH} = \frac{(30 \text{ mg/ml})(1 \text{ ml/min})}{0.05 \text{ mg/ml}} = 600 \text{ ml/min}$$

B. Effective renal plasma flow is measured by the PAH clearance (if the plasma concentration of PAH is low), so that in this case, ERPF = C_{PAH} = 600 ml/min.

C. To calculate the effective renal blood flow (ERBF), it is important to recognize that a portion of the blood volume is plasma and that a portion is due to the volume of red blood cells (RBCs). The percentage of the blood volume attributable to RBCs is given by the hematocrit and is in this case 0.40, or 40% of the total volume. This means that 60% of the renal blood volume is plasma. But the renal plasma flow is 600 ml/min. Therefore,

0.60 × ERBF = ERPF = 600 ml/min

Thus,

ERBF = 1,000 ml/min

D. The amount of urea reabsorbed is the difference between the amount filtered and the amount ex-creted. Since urea is freely filtered, the urea concentration in the filtrate is the same as that in the plasma, that is, 5 mmole/L and, since the filtration rate is 100 ml/min, the amount of urea filtered per unit time 0.5 mmole/min. The amount of urea excreted per day is 1.4 L urine, with a urea concentration of 220 mmole/L or 308 mmole/day, which is 0.21 mmole/min. So the amount reabsorbed must be the amount filtered (0.5 mmole/min) less the amount excreted (0.21 mmole/min), or 0.29 mmole/min.

9. C. Clearance is the volume of plasma cleared of a substance per unit time. Since inulin is freely filtered, the glomerular filtrate contains the same concentration of inulin as the plasma, regardless of plasma inulin concentration. This means that the same volume of plasma will be cleared of inulin per unit period of time.

10. Any or all of the possibilities. Since clearance is the volume of plasma cleared of substance per unit time, any process that tends either to keep the substance in the plasma or to return it to the plasma after filtration will reduce the clearance of the substance. All the possibilities do this in one way or another. A and C prevent the substance from leaving the plasma in the first place. B and D return the substances to the plasma after it has been filtered.

11. PAH clearance depends to a large extent on active secretion. If the T_m for this system is exceeded, the rate of removal of PAH from the plasma will be reduced, that is, the clearance will decrease. (This should be obvious from Fig. 20.)

12. B. This patient's osmolar water clearance is

$$C_{osm} = \frac{U_{Osm} \dot{V}}{P_{Osm}}$$
$$= \frac{(525 \text{ mOsm/L}) (2,000 \text{ ml/day})}{300 \text{ mOsm/L}}$$
$$= 3500 \text{ ml/day}$$

Thus, free-water clearance is

$$C_{H_2O} = \dot{V} - C_{osm}$$
$$= 2,000 \text{ ml/day} - 3,500 \text{ ml/day}$$
$$= -1,500 \text{ ml}$$

that is, a negative free-water clearance or a return of 1.5 L pure water to plasma. Therefore, this patient's serum osmolarity will decrease.

13. A. See solution above. (Note: Adding as much as you want of a 300 mOsm/L solution will not change plasma osmolarity.)

14. C. Calculate osmolar water clearance

$$C_{Osm} = \frac{U_{Osm} \ \dot{V}}{P_{Osm}}$$
$$= \frac{(270 \ mOsm/L) \ (3 \ ml/min)}{(286 \ mOsm/L)}$$
$$= 2.8 \ ml/min$$

$C_{H_2O} = \dot{V} - C_{Osm} = 3 \ ml/min - 2.8 \ ml/min - 0.2$ ml/min or a positive free-water clearance (loss of plasma water).

15. 1.4%. Filtered Na^+ = (144 mmole/L) (0.12 L/min) = 17.28 mmole/min.

Excreted Na^+ = $(0.036 \ \frac{1}{30} \ min)$ (200 mmole/L) = 0.24 mmole/min.

$$\% = \frac{Excreted}{Filtered} = \frac{0.24}{17.28} \times 100 = 1.4\%$$

16. All are correct. Water and chloride (in most of the tubule) are reabsorbed passively as a result of electrochemical gradients generated by active Na^+ reabsorption. Urea is reabsorbed passively as a result of concentration gradients established by water reabsorption and is thus reabsorbed indirectly by Na^+ reabsorption. Glucose reabsorption depends on the Na^+ gradient from lumen to cell. If this gradient is reduced, glucose reabsorption should be reduced as well.

17. Lower. The increased aldosterone causes positive sodium balance (i.e., increased plasma Na^+), which reflexly inhibits renin secretion. Thus one observes a high plasma aldosterone and a low plasma renin, a strong tipoff to the presence of the disease, since in almost every other situation renin and aldosterone change in the same direction because the renin–angiotensin system is the primary control of aldosterone secretion.

18. C.

$$GFR = C_{creatinine} = \frac{U_{creatinine} \times \dot{V}}{R_{creatinine}}$$
$$= \frac{90 \ mg\% \times 2/day}{1 \ mg\%}$$
$$= 180 \ L/day$$

18. B.

$$C_{Osm} = \frac{U_{Osm}}{P_{Osm}} \times \dot{V}$$
$$= \frac{300 \ mOsm}{300 \ mOsm} \times 2 \ L/day$$
$$= 2 \ L/day$$

20. D. If plasma osmolarity is the same as urine osmolarity, free-water clearance is zero.

21. B. Glucose is completely filtered, that is, the concentration of glucose in the glomerular filtrate is the same as that in the plasma (90 mg%); 125 ml is filtered (with 90 mg/100 ml), representing 112.5 mg glucose presented to the proximal tubule per minute.

22. D. It is important to recognize that osmotic forces are produced only when a barrier is more permeable to water than to solutes. This produces differences in the rates of movement of water and solutes across a barrier with a concomitant osmotic pressure. Consider the various alternatives. A is false because the stated facts contradict the answer. The osmotic pressure near the capillaries is 285 mOsm, while that of the lateral spaces is near 1000 mOsm. B and C are true, but have nothing to do with osmotic water flow, only hydraulic flow.

33

The Physiology of Water and Electrolytes

DONALD W. STUBBS

OBJECTIVES

After completion of this chapter, you should be able to:

1. Understand the significance of water as a solvent for body fluid electrolytes.

2. Comprehend the principle of indicator dilution methods in measuring the volumes of body fluid spaces.

   a. Work problems involving indicator dilution data.

   b. Identify the indicators used to measure specific body fluid spaces.

3. Understand the relationship between solute concentration and osmotic pressure.

   a. Become familiar with milliosmoles/liter (mOsm/L) as the clinical unit of osmolarity.

   b. Distinguish between total osmotic pressure of a solution (e.g., as determined by freezing point) and the effective osmotic pressure (actual osmotic pressure difference developed between two solutions across a membrane).

4. Calculate (a) the normal daily turnover rate of water balance, (b) the daily osmolar waste load, (c) daily urine volume at maximal concentration, and (d) the total daily obligatory water losses.

5. Compare differences in water balance between infant and adult.

6. Describe the avenues of water loss and sources of water gain.

7. Describe the thirst mechanism and its principal stimuli.

8. Describe the antidiuretic hormone mechanism and its principal stimuli.

9. Discuss the pathogenesis of such states of water imbalance as dehydration and overhydration.

10. Calculate the changes in body fluid volumes and contents as a result of intake of different fluids (e.g., distilled water, isotonic saline).

11. Relate renal handling of Na^+ to regulation of extracellular fluid volume.

12. Discuss the pathogenesis of edema in congestive heart failure, cirrhosis, and nephrosis.

13. Recognize and explain Darrow-Yannet diagrams for the different states of water and salt imbalance.

Life presumably began in the sea, which provided a most suitable environment; most marine organisms enjoy a remarkable constancy of environmental pH, electrolyte concentration, and osmotic pressure. In our evolution as land animals we have had to carry our oceans with us in the form of a saline extracellular fluid. Adaptation to living on dry land not only poses the problem of obtaining and conserving water for body fluids, but also that of regulating its electrolyte content, osmolarity, and acid–base balance. The extent to which our regulatory mechanisms occasionally fail us is reflected by the frequency with which physicians are called upon to correct conditions of dehydration, edema, acidosis, alkalosis, and other imbalances.

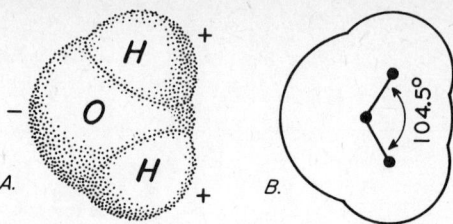

Figure 1 The water molecule. *(a)* "Space-filling" model of the water molecule (surface represented as the van der Waals contours). Note the position of the two hydrogen atoms toward the same side of the oxygen atom. This results in a displacement of the center of positivity toward that side of the molecule and of the center of negativity toward the opposite side. *(b)* "Ball and stick" (crystallographic) model, which better illustrates the covalent bond angle and the distance between nuclei (van der Waals contour is outlined).

PROPERTIES OF WATER

Water is the most common liquid both in living organisms and in the environment; it is truly ubiquitous. However, by comparison with most other liquids, water is found to have a number of uncommon properties that make it rather unique. One of the most important properties, from a physiological point of view, is that water is an unusually good solvent. The uniqueness of water derives from the power of water molecules to attract each other especially strongly, through the formation of "hydrogen bonds" between adjacent molecules. The basis for such strong intermolecular attraction is the electrical asymmetry or dipolar separation of positive and negative charges, which gives the water molecule a considerable ionic character (Fig. 1).

Water dissolves or disperses a wide variety of substances because of the ionic character of its dipolar molecule. Although salts are insoluble in most solvents (e.g., alcohols, ethers, benzene), they readily dissolve in water. Solvation of salts into ion solutes occurs because of the tendency of the water molecules to oppose the electrostatic attraction between ions in a salt crystal and because of a related effect that stabilizes the dissolved ions by forming a shell of hydration (or cage of water molecules) around them. Water also dissolves a number of nonionic (nonelectrolyte) solutes by forming hydrogen bonds with polar functional groups of the solute molecule (e.g., formation of hydrogen bonds with hydroxyl groups of sug-

ars). Nonionic, nonpolar molecules (e.g., hydrocarbons), incapable of forming hydrogen bonds, are fairly insoluble in water.

WATER AND BODY FLUID SPACES

Water is not only the most common liquid of the body, it is the most commonly found of all chemical tissue constituents, with 50–60% of the average adult body weight consisting of water. Because of its ability to dissolve many solutes, it is the most nearly universal solvent of biological systems. Thus, body fluids are aqueous solutions, and water is the principal medium of transport and diffusion of materials within living organisms. In higher animals these body fluids can be divided anatomically into two main fluid spaces: the extracellular fluid (ECF) and the intracellular fluid (ICF). The collective boundary between these two fluid spaces is the cell membrane. The ECF is itself divided into plasma and interstitial fluid (ISF). The boundary between these two subdivisions is the vascular endothelium of the circulatory system.

FLUID SPACE MEASUREMENTS

The first estimates of total body water (TBW) were made by desiccating cadavers, an accurate and direct method, but one of little clinical recommendation. Currently, indirect tracer or indicator methods are used to measure not only TBW, but also ECF and plasma. There is, unfortunately, no single method for measuring ICF or ISF spaces. These must be calculated by obtaining the difference between measurements of two other fluid spaces:

$$ICF = TBW \text{ minus } ECF$$

$$ISF = ECF \text{ minus plasma volume}$$

INDICATOR DILUTION METHODS

Fluid spaces in living subjects are measured by the same general method, that is, by indicator dilution techniques (Fig. 2). Different indicator substances are used according to their abilities to distribute themselves uniformly throughout (equilibration), and be confined within, the specific fluid space to be measured. If a known amount of indicator substance is injected into and distributed throughout a fluid space, and its concentration (after mixing and dilution) is determined from a sample of that fluid, the volume of the fluid space can easily be calculated:

$$\text{Volume} = \frac{\text{amount}}{\text{concentration}}$$

KNOWN AMOUNT OF DYE ADDED   **AFTER EQUILIBRATION OF DYE IN FLUID SPACE**

Figure 2 Determination of fluid volume by indicator dilution method. A known amount of indicator (e.g., dye) is added to the fluid space to be measured. After equilibration, the concentration of the indicator is measured. From the extent of indicator dilution (final concentration), the volume of the fluid space into which the indicator was diluted can be calculated from the equation:
Volume = amount of indicator divided by concentration

TOTAL BODY WATER (TBW)

The heavy isotopes of water, that is, deuterium oxide (D_2O) and tritium oxide (THO), are the indicator substances used most often to determine TBW volume by dilution methods (Table 1). Despite their heavier molecular weights, these isotopes in low concentrations behave like ordinary water in vivo. An advantage to using D_2O is that it is a nonradioactive isotope that poses no radiation hazard whatsoever. Unfortunately, its analysis is not simple because it requires the use of a mass spectrometer. With the advent of convenient liquid scintillation counting, the radioactive isotope THO became the most widely used isotope for measuring TBW. Despite the fact that THO is radioactive, it does have a short biological half-life (half the injected amount is lost from the body in 10 days); it is usually used in low doses to keep total body irradiation below acceptable tolerance levels. Measurements by isotope dilution with D_2O and THO have given values for TBW consistent with those of cadaver desiccation and are considered accurate.

Measurements of total body water have shown two developmental trends: (1) infants have a higher proportion of body weight composed of water than do adults (we gradually become drier with age); and (2) from the time of puberty onward, women have a lower percentage of water than do men. At birth approximately 77% of the body weight is water, but by 2 years of age this proportion has decreased to about 63% (a TBW percentage only a little above that of a male adult). This diminution of TBW

Table 1 Indicator Substances Commonly Used to Determine TBW by Dilution Method

Fluid Space Measured	Useful Indicators
Total body water	Antipyrine, the heavy isotopes of water: deuterium oxide (D_2O) and tritium oxide (THO)
Extracellular fluid	Inulin, mannitol, sucrose, chloride, radiobromide ($^{82}Br^-$), thiocyanate (SCN^-)
Plasma	Evans blue dye (T-1824), radio-iodinated serum albumin (RISA), and red blood cells tagged with radioisotopes ^{51}Cr, ^{32}P, or ^{59}Fe

percentage with age in early life is principally at the expense of ECF volume (i.e., there is a greater decrease in the proportion of TBW as ECF than as ICF). The sexual differences in TBW proportions in adults are simply a matter of adiposity. Adipose cells contain much triglyceride and very little water (only about 10% water); hence women with a greater average proportion of body fat have a lower proportion of water than do men.

EXTRACELLULAR FLUID (ECF)

Whereas TBW can be measured fairly accurately with some assurance, the measurement of ECF volume is equivocal. The principal problem is that the different indicator substances used to determine ECF volume do not give the same results. The ECF is actually a group of spaces, and no single indicator distributes itself precisely throughout the entire group. The ECF is composed of (1) plasma, (2) general interstitial fluid and lymph, (3) interstitial fluid of general connective tissues, (4) bone water, and (5) transcellular fluids consisting of a variety of fluids formed by transport activity of cells (e.g., cerebrospinal fluid, aqueous humor, intraluminal fluids of gut, glandular secretions). Figure 3 illustrates body water compartments in an average male adult.

Small ion indicators, such as chloride or bromide, not only penetrate all five divisions, but also enter some cells (e.g., red blood cells and gastric mucosal cells) and give correspondingly higher values for their volume of distribution. On the other hand, relatively larger molecules, such as the nonutilized saccharides (inulin, mannitol, and sucrose) show incomplete equilbration within the fluids of the extracellular space because of somewhat slow penetration of some connective tissues (e.g., dense connective tissue and cartilage) and an absence of penetration of bone and transcellular fluids. As a result, these nonionic saccharide indicators give values that are undoubtedly too small (i.e., they distribute within a fluid space smaller in volume than the anatomic ECF space). It must be concluded that the ECF is intrinsically not accurately measurable by present indicators and that descriptions of ECF volumes should be defined in terms of the specific indicator substance used.

For clinical purposes in considering disorders of hydration, it is usually sufficient to view the body fluids as a simple two-compartment model (the ECF as the functional ECF of Fig. 3 and the ICF as the remainder). The ECF can be further divided into plasma and ISF (Fig. 4). As a matter of convenience, it is often assumed that the ECF is approximately 20% of the body weight and the ICF 40% (giving a TBW of 60% of body weight).

INTERSTITIAL FLUID

The interstitial fluid (ISF) lies within the extracellular interstices or spaces between the cells. The ISF was once thought to be a free fluid but is now believed to be largely part of an interstitial gel composed primarily of mucopolysaccharides as the solute phase (gel matrix) and containing most of the ISF entrapped within as a nonflowing solvent phase. Normally perhaps no more than 1% of the ISF is a free fluid (the intersititial gel thus prevents

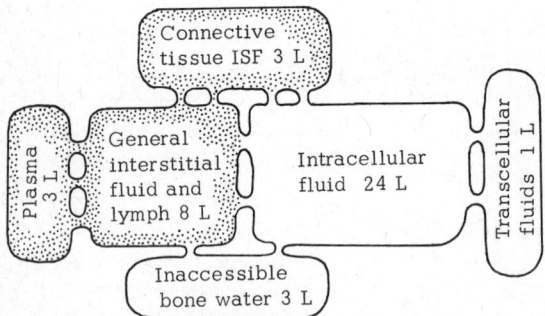

Figure 3 Body water compartments (in liters) in an average 70-kg man. Stippled areas represent ECF compartments in at least moderate equilibration with each other (constitute a functional ECF that is smaller than the true anatomical ECF).

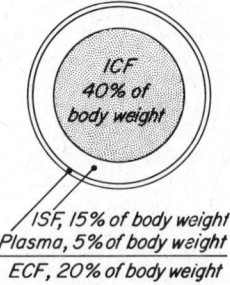

Figure 4 Principal body fluid compartments in an average male adult. Transcellular fluids, which constitute a small percentage of the total body water (TBW), are not shown. TBW, 60% of body weight.

flow of ISF from one level of the body to another). Ordinarily, capillary absorption (Starling mechanism) and lymphatic drainage keep the proportion of free fluid very low. If these mechanisms fail, however, large quantities of free ISF begin to collect, resulting in edema.

PLASMA

Plasma is the noncellular fluid portion of the blood, normally about 57% of the volume of a venous blood sample. The remainder of the blood volume is principally that of the red blood cells (white cells contribute a very small volume), which is described as the hematocrit (Hct), normally about 43% of a venous blood sample. The relative proportions of each are easily determined from centrifugation of a small volume of blood.

However, to determine the absolute total volume of plasma in all the blood, or to determine the total red blood cell mass, indicator dilution methods are used. One method of determining the total plasma volume depends on measurement of the volume of distribution of labeled serum albumin. Albumin can be labeled with the nontoxic Evans blue dye (T-1824), which selectively binds to albumin molecules, or it can be labeled with radioactive isotopes of iodine (^{131}I or ^{125}I), which also combine with albumin (RISA). In either case, corrections must be made for a small amount of albumin leakage from the vascular compartment. Another method of determining the plasma volume involves measuring the total red blood cell volume by means of radioactively tagged red blood cells. From this, the plasma volume can be calculated if the average blood hematocrit, also called the body hematocrit (body Hct), is known—corrections are applied to a venous Hct to obtain an average blood hematocrit.

ELECTROLYTES

The force of attraction of opposite electric charges is inversely proportional to the dielectric constant D of the medium surrounding the charges. Many salts readily dissolve in water as a solvent medium because of its high dielectric constant ($D = 79$). In air or vacuum ($D = 1$)

the crystal lattice of salts, such as NaCl, is held together by strong ionic charges. When placed in water ($D = 79$), salts readily dissociate into individual solute ions (e.g., Na^+ and Cl^-) because their electrostatic attraction for each other is reduced to only $\frac{1}{79}$th of that in air or vaccum. The attraction between the ionic-like water dipoles and the salt ions to form stable hydrated solute ions exceeds the forces tending to attract an ion back to the surface of the salt crystal. Such substances as salts, which yield ions upon being dissolved, are called electrolytes. Electrolytic solutions will conduct electric current.

UNITS OF CONCENTRATION

Two commonly used units in describing concentrations of solutes in body fluids are milliequivalents per liter (mEq/L) and milliosmoles per liter (mOsm/L). Both are derived from the millimole (mmole), which is also used as such. The mEq/L is used exclusively for electrolytes, usually in quantifying the concentration of a specific ion. The mOsm/L is used to describe both electrolytes and nonelectrolytes in terms of the osmotic effect of the total number of solute particles per unit volume of solution, usually without regard for specific identities of individual ion or molecular solute particles, the total number of particles being of principal interest.

MILLIEQUIVALENTS (mEq)

Quantities of electrolytes are described in terms of combining equivalents. For body fluids, the milliequivalent is a more convenient unit that is the equivalent. One mEq H^+ is 6.023×10^{20} separate ions weighing a total of 1.008 mg (i.e., 1 mmole or 1 mg atomic weight of hydrogen). This 1 mEq of H^+ exactly combines with or neutralizes 1 mEq of OH^-, which is also 6.023×10^{20} ions, but weighs a total of 17.008 mg. One mEq of H^+ also combines with 1 mEq of Cl^- (6.023×10^{20} Cl^- ions weighing a total of 35.5 mg) to form 1 mmole of HCl. Thus in acidimetry or alkalimetry, 1 mEq is the amount of a substance that will react with or displace 1 mg atomic weight of H^+ (1.008 mg). Milliequivalents can also be defined in terms of the change in the valence or oxidation state of a substance in an oxidation–reduction reaction.

For clinical purposes in dealing with electrolytes of body fluids, 1 mEq of an ion can be simply defined as its milligram atomic weight divided by its valence. For univalent ions (e.g., Cl^-, atomic weight 35.5) 1 mEq is the mg atomic weight (e.g., 1 mEq of $Cl^- = 35.5$ mg of Cl^-). For divalent ions (e.g., Ca^{2+}, atomic weight 40.1) 1 mEq is the mg atomic weight divided by 2, that is, one-half the mg atomic weight (e.g., 1 mEq of $Ca^{2+} = 20.05$ mg). In other words, the mg atomic weight (or 1 mmole) of a univalent ion is 1 mEq, the mg atomic weight (or 1 mmole) of a divalent ion is 2 mEq, and the mg atomic weight (or 1 mmole) of a trivalent ion is 3 mEq.

Plasma contains 3,265 mg of Na^+ per liter. In terms of mEq of Na^+ (atomic weight 23), this is 142 mEq/L. One L of plasma contains a slightly greater mass of Cl^- (i.e., 3,652 mg of Cl^- per liter). In terms of mEq of Cl^- (atomic weight 35.5), this is 103 mEq/L. Table 2 lists the electrolytic components of plasma expressed in milliequivalents per liter.

Table 2 shows that the total mEq/L of cations equals the total mEq/L of anions. This does not mean that the mass of cations equals the mass of anions—only that the total positive charges equal the total negative charges, according to the principle of macroelectroneutrality. A minor disadvantage in using mEq/L to express electrolytic concentrations is the hidden distortion of absolute amounts or mass of solutes present. This is particularly deceptive for large, massive ions with relatively few charges, such as protein molecules. In terms of mass, the predominant solute of plasma is protein, yet as an anionic

Table 2 Electrolytic Components of Arterial Plasma

Cations		Anions	
Electrolytic Component	*Concentration (mEg/L)*	*Electrolytic Component*	*Concentration (mEq/L)*
Na^+	142	Cl^-	103
K^+	5	HCO_3^-	24
Ca^{2+}	5	HPO_4^{2-}	2
Mg^{2+}	2	SO_4^{2-}	1
Total cations	154	Protein	16
		Organic acid	8
		Total anions	154

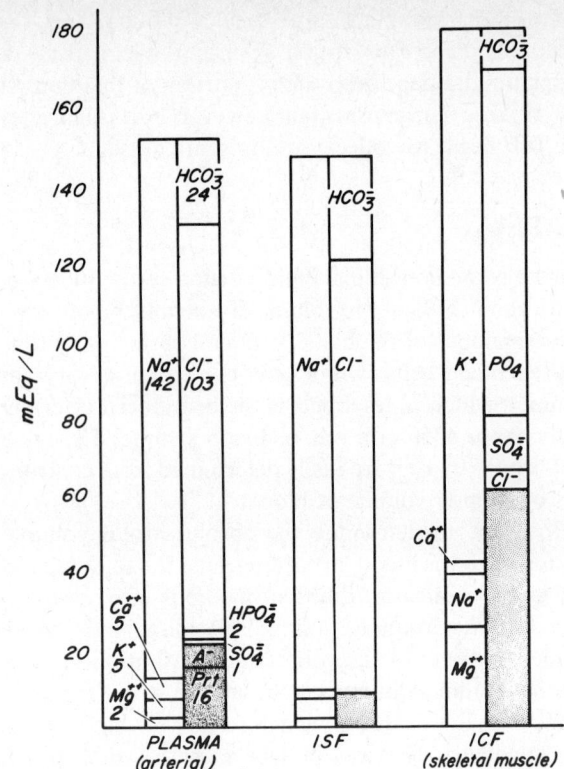

Figure 5 Electrolytes of arterial plasma, ISF, and skeletal muscle ICF compared. Note that the total mEq/L in the ICF is greater than the total mEq/L in the ISF or in plasma. Is this compatible with the statement that the ECF and the ICF are in osmotic equilibrium? (*Hint:* Is osmotic equilibrium determined by mEq/L or by mOsm/L?) A^- represents organic acid anion. This A^- component together with that of protein is stippled.

electrolyte the 70,000 mg of protein in each liter of plasma is represented by a mere 16 mEq/L. Compare this with the 3,265 mg L of sodium, which is represented as 142 mEq/L (Fig. 5).

MILLIOSMOLES (mOsm)

The effects of both electrolytic and nonelectrolytic solutes on the colligative properties of a solution (freezing point, boiling point, vapor pressure, and osmotic pressure) depend on the total number of solute ions or molecules per unit volume. The greater the number of solute ions or molecules, the greater the depression of freezing point, elevation of boiling point, lowering of vapor pressure, and increase in osmotic pressure. A liter of water containing 1 mmole of a nonelectrolyte (e.g., glucose) has 6.023×10^{20} solute particles; if separated from pure water by a semipermeable membrane, it develops an osmotic pressure of about 19 mm Hg (i.e., equal to the pressure exerted by a column of mercury 19 mm in height) at 37°C.

If, instead of glucose, 1 mmole of NaCl is dissolved, the solution should contain twice as many solute particles (12.046×10^{20}), since NaCl dissociates into two ion species in solution (Na^+ and Cl^-). This should result in an osmotic pressure twice as high, since the osmotic pressure is determined by the total number of solute particles. Theoretically, twice as many solute particles should give twice the osmotic pressure. However, while water diminishes interionic attraction between solute ions in solution, it does not eliminate it entirely, and such electrostatic attraction results in a slightly lower effect on osmotic pressure than expected. A salt such as $CaCl_2$ would be expected to dissociate into three ions in solution (Ca^{2+} and 2 Cl^-) and has an even greater effect on the osmotic pressure. A milliosmole can be defined as 6.023×10^{20} solute particles, without regard for specific identities of the solutes. Thus 1 mmole of glucose yields 1 mOsm of solute particles, 1 mmole of NaCl theoretically yields 2 mOsm, and 1 mmole of $CaCl_2$ theoretically yields 3 mOsm. As a rough estimate

$$mOsm = mmoles \times n$$

That is, the number of mOsm in solution is approximately equal to the mmoles of substance dissolved times n, the number of particles or ions released into solution by each molecule of substance dissolved. However, the actual osmolarity observed, as determined by measuring osmotic pressure or other colligative properties of solutions, often does not precisely agree with the expected osmolarity calculated simply on the basis of the number of ions n released by a salt molecule when dissolved. The difference between the observed and the calculated osmolarity can be accounted for by interionic interaction between the oppositely charged ions released. To come up with an accurate calculation of the actual osmolarity to be expected, this ionic interaction must be taken into consideration by means of a correction factor ϕ, which represents the osmotic coefficient (Chapter 1). The osmotic coefficient ϕ

Table 3 Osmotic Coefficients ϕ for Electrolytes at Concentrations Isotonic with Mammalian Extracellular Fluid

Salt	Osmotic Coefficient
NaCl	0.93
KCl	0.92
$CaCl_2$	0.85

must be determined experimentally for each salt. Furthermore, it varies with solute concentration and so must be determined at numerous different concentrations of interest. Such measurements have been made and the osmotic coefficients published tabular form (e.g., Table 3). Thus to predict the mOsm from mmoles of salt solution accurately, the proper equation is

$$mOsm = mmoles \times n \times \phi$$

For example, isotonic saline, which is 154 mmoles of NaCl per liter, should give rise to 308 mmoles of ions per liter, since 1 NaCl dissolves into two ions. However, because of interionic interaction, each NaCl acts as if it gave rise to only 1.86 ions, and the actual osmolarity is only 286 mOsm/L. This effect is taken into account when ϕ is included in the calculation. For example, the mOsm/L of a solution of 154 mmoles NaCl per liter = $154 \times 2 \times 0.93 = 286$ mmOsm/L.

EXTRACELLULAR FLUID ELECTROLYTES

The endothelium of the capillary wall is the anatomical boundary between the two principal divisions of the ECF (between plasma and ISF), and it is readily permeable to crystalloid solutes (small ions and molecules in true solution). It follows that the crystalloid solutes will be present at similar levels of concentration, except for small differences attributable to a simple Donnan equilibrium, and will undergo simultaneous changes in these two divisions. On the other hand, colloidal-size particles (very large solute particles, such as protein molecules in colloidal solution) are largely retained within the vascular system (except for slight albumin leakage), where they exert a small but important effective osmotic pressure difference between plasma and ISF (the plasma colloidal osmotic pressure).

THEORETICAL TOTAL OSMOTIC PRESSURE VERSUS ACTUAL EFFECTIVE OSMOTIC PRESSURE

Plasma has a theoretical total osmotic pressure of about 5,500 mm Hg. This is the osmotic pressure that plasma would exert if it were separated from pure water by a perfect truly semipermeable membrane. Because it is technically difficult to achieve accurate direct measurements of such pressures with semipermeable membranes, and since the effect of solutes on osmotic pressure has as its basis the same principle by which solutes affect the other colligative properties of solutions, the theoretical total osmotic pressure can be calculated from the more easily determined depression of freezing point (e.g., the normal osmolarity of plasma can be determined to be 285 mOsm/L from its depressed freezing point of 0.53°C below zero). When real, live membranes are involved, the actual or effective osmotic pressure exerted approaches the theoretical total osmotic pressure calculated from depression of freezing point only to the extent that the membrane is truly semipermeable*. Solutes that can diffuse across a membrane are ultimately osmotically neutral in terms of actual effective osmotic pressure. Thus in the case of the capillary wall separating plasma from ISF, the only actual or effective osmotic pressure difference between these two fluids is that due to the colloidal osmotic pressure, since all other solutes pass freely through the capillary wall (and therefore interfere with the activity of water molecules equally on both sides of the capillary wall). This colloidal osmotic pressure is an important component of the forces determining fluid exchange across capillaries according to the Starling hypothesis. According to the Starling hypothesis, capillary blood pressure is the pre-

dominant force for moving water out of the capillaries, and the colloidal osmotic pressure is the principal means for water reabsorption back into the capillaries and venules.

INTRACELLULAR FLUID ELECTROLYTES

The relationship between the solutes of the ICF and those of the ISF is quite different from that between the solutes of plasma and of ISF. The cell membrane acts as if it were impermeable to the major ions on each side, although some Na^+ and some K^+ does tend to diffuse across the cell membrane, the Na/K pump transports them back and maintains a constant gradient as if the membrane were impermeable to these ions. Since these ions do not freely pass from the ISF to the ICF, or vice versa, they are osmotically active. Thus the theoretical total osmotic pressure and the actual effective osmotic pressure for fluids separated by the cell membrane are very nearly the same. The total osmotic pressure exerted by the solutes on one side of the cell membrane is balanced by the total osmotic pressure of the solutes on the other side, for a double Donnan equilibrium. In fact, any differences in osmolarity that might occur between ISF and ICF are rapidly corrected by movement of water across the cell membrane to reestablish a balance between these two compartments. Because of this rapidity of equilibration, it may be said that a state of osmotic equilibrium is constantly maintained between the ISF and the ICF; or more generally, it is said that an osmotic equilibrium is maintained between the ECF and the ICF.

WATER BALANCE

TURNOVER

Generally, a normal, healthy person who is not undergoing changes in weight maintains a remarkably precise balance between the amount of water that is gained and lost

---

*With membranes that are somewhat permeable to a solute, the osmotic pressure developed is less than that measured with a membrane completely impermeable to the solute. Calculations of transient actual osmotic pressures with membranes that are not completely impermeable involve the use of a correction factor called a reflection coefficient σ. See Chapter 1.

each day. Regulatory mechanisms normally keep the amount gained equal to the amount lost, that is, the total volume of body water is constant. The amount gained and lost each day is the daily turnover rate. Physicians responsible for supplying hospitalized patients with daily fluid requirements must be able to estimate closely each patient's specific needs. Several rules of thumb have been suggested for estimating the daily turnover rate. Obviously a large adult will require more water each day than will a small infant. One way to estimate the daily turnover rate is on a body weight basis. However, rules based simply on body weight are not uniform except for very narrow weight ranges (from the examples given in Table 4 it can be calculated that the adult will require 37 ml/kg, while the infant will require 125 ml/kg). What might be more convenient is an easily remembered figure that would apply uniformly to patients of different sizes. This is particularly true in pediatrics, where patients vary from tiny 1-kg premature infants to strapping big 70-kg adolescents. A useful way to express the turnover rate is in terms of body surface area. The turnover rate in liters per square meter is the same in infants, older children, and adults, namely, 1.5 L/M^2. This single value can be used to calculate the normal daily turnover rate for children of various sizes and can be used for adults as well, who also come in various shapes and sizes. Although, quantitatively, the rate of turnover per unit area is the same in children and adults, there are some important qualitative differences between children and adults in the physiology of their water balance.

It can be seen from the comparison below that the 2.6 L of water gained and lost each day in the adult represents a very small fraction of his total body water ($\frac{1}{16}$), whereas the 0.375 L of water turned over daily in the infant is a much larger fraction of his total body water ($\frac{1}{6}$). Thus, although the infant has a greater percentage of his body weight as water (77%), a much larger proportion of his total body water is turned over daily. Anything that inter-

feres with either the loss or gain of water will produce a much more rapid and serious imbalance than is the case with adults. Furthermore, the portion of the infant's turnover that is excreted as urine cannot be as concentrated as the urine produced by adults. The infant is sometimes said to have an "immature concentrating ability" (in other words, for each mOsm of excretory solute the infant must obligate or supply a larger volume of water as solvent).

NORMAL ROUTES OF WATER LOSS

The four routes of water loss from the body are (1) expired air, (2) skin evaporation, (3) feces, and (4) urine.

Water Loss in Expired Air

As the usually cooler, drier ambient air is warmed and humidified in its passage through the respiratory tract to the alveoli, it gains water vapor (47 mm Hg) from the moist respiratory epithelium. Upon expiration, this water is lost from the body. This loss is an insensible loss, since one is usually unaware of it (except on cold days when the water vapor in expired breath condenses and can be seen).

Water Loss by Evaporation from Skin

By virtue of its lipid content and keratinization, the skin is a water-resistant, although not waterproof, covering. The evaporative loss of water through the intact skin is less than 3% of that from an equal area of water exposed to air. This continuous loss of water through the skin, independently of any sweat gland activity, is also an insensible loss. The total daily insensible loss from lungs and skin together is somewhat variable (depending on such as variables as temperature, humidity, and metabolic rate), but in round numbers is almost 1 L/day.

Destruction or damage of large areas of this water-resistant covering, as in burns, permits excessive evaporative loss of water. A full-thickness burn (third-degree burn) causes a 20-fold increase in fluid loss through the burned area of skin. A burn involving 50% of the body surface area can cause loss of as much as 6 L of water each day.

Table 4 Comparison of Water Balance in Infant and Adult

Subject	Weight (kg)	Surface Area (M^2)	TBW (L)	Turnover (L)	Proportion of TBW
Infant	3	0.25	2.3 (77%)	0.375	$\frac{1}{6}$
Adult	70	1.75	42 (60%)	2.6	$\frac{1}{16}$

In addition to insensible loss of water through the skin, a variable amount can be lost for cooling purposes by sweating. Under extreme conditions (e.g., physical exertion in the tropics) involving continuous sweating by all skin areas, as much as 15 L/day can be lost as sweat. Under sedentary conditions in temperature-regulated buildings the daily volume of water loss in sweat is only about 100 ml.

Water Loss in Feces

Water loss in feces is normally small (about 100 ml/day), but can be considerable in diarrhea or dysentery. Loss of gastrointestinal (GI) fluid is the most common cause of clinical problems involving loss of water and electrolytes. On the basis of three typical meals a day, the gut must normally cope with a total of about 9 L fluid, of which the dietary intake accounts for only about 2 L or so (Fig. 6). Approximately 7 L of GI fluids are secreted into the gut each day and are essentially recovered completely by intestinal absorption along with the fluids of dietary intake, except the volume of fluid remaining in stools, namely approximately 0.1 L. It might be helpful to think of this large volume of fluid secreted and reabsorbed daily as a

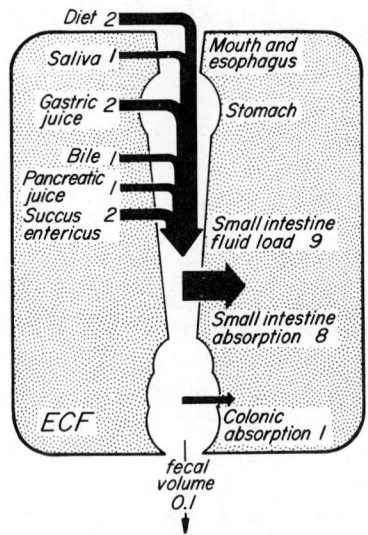

Figure 6 Schematic representation of the average daily loads of fluid (in liters) between the gut and ECF. (Redrawan from Phillips SF: "Fluid and Electrolyte Fluxes in the Gut," *Hosp Pract,* 8:137–146, 1973.)

kind of GI fluid circulation. The conservation or recycling of gastrointestinal fluid can be interrupted by loss through vomiting, diarrhea, or fistulous drainage. One can easily appreciate the serious consequences of failure to reabsorb the large volumes of fluid involved in this circulation. In severe diarrhea (e.g., in cholera), there is not only failure to reabsorb fluid, but also excessive active secretion of fluid into the gut thereby further aggravating the dehydration of diarrhea.

Water Loss in Urine

One of the primary means by which balance between water intake and output is maintained depends on the regulatory mechanisms controlling the daily output of urine. Under conditions of diminished water intake, the urinary load of waste solutes can be excreted in a smaller volume (more concentrated urine), thereby conserving water. However, the renal readjustment of water loss is limited by the maximal ability of the kidneys to concentrate the urine.

The minimal volume of water needed to carry waste solutes (i.e., maximal concentration of urine) represents an obligatory loss of water. This minimal amount of water must be obligated as solvent for the excretion of waste solutes. On an average diet the waste solutes make up a urinary osmolar load of about 600 mOsm/M^2/day. For a 70-kg person (1.7 M^2 surface area), this daily osmolar load of waste solutes is about 1,000 mOsm. Under normal conditions of adequate fluid intake these 1,000 mOsm are excreted in a urine concentration of about 700 mOsm/L and a total volume of about 1.4 L. However, with a need for maximal conservation of water (e.g., in deprivation of water intake), the kidneys can concentrate the urine to 1,400 mOsm/L in a total volume of only 700 ml. The difference between the 700 ml, which is an obligatory volume (smallest possible volume able to carry the osmolar waste load in maximal concentration), and the larger volume normally excreted is a facultative (nonobligatory) volume. This facultative volume of urine represents the regulatory function of the kidneys in balancing water output to water intake under conditions of normal availability of drinking water.

The minimal volume of water able to carry the daily urinary load of waste solutes, the amount of water lost in stools, and the insensible loss of water by vaporization

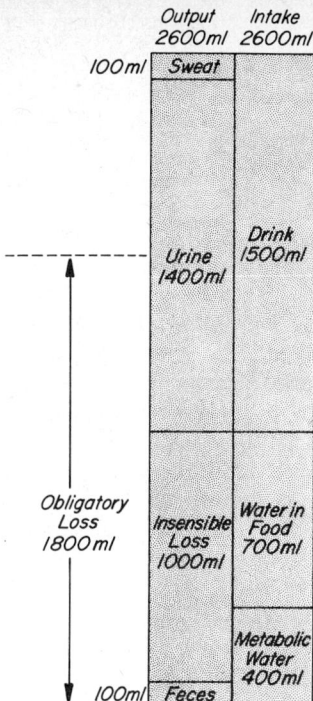

Figure 7 Daily balance between output and input in a normal 70-kg man. Numbers indicate milliliters of water. The obligatory water loss shown includes water loss in feces, insensible water loss, and the minimal, obligatory urine volume (700 ml) necessary to carry a daily load of 1,000 mOsm. The values for water gain are estimated from average ingestion of food and drink, and for the metabolic water from a 3,200-Cal. diet.

from both skin and lungs constitute the total daily obligatory water loss. This loss of about 1.8 L/day is an obligatory loss because it occurs continuously without regard to water intake or to the body's needs for water conservation; that is, the body is obligated to lose this amount each day (Fig. 7).

NORMAL SOURCES OF WATER GAIN

There are three sources of water gain: (1) metabolic water (or water of oxidation), (2) water content of food, and (3) drinking water or water of beverages.

Metabolic Water (Water of Oxidation)

Water is one of the principal products of oxidation of carbohydrates, fats, and proteins. For example, consider the oxidation of glucose and of palmitic acid:

$$C_6H_{12}O_6 + 6O_2 \rightarrow$$
(1 mole of glucose)
$$6CO_2 + 6H_2O + \text{about 680 Calories}$$
(108 ml of water)

$$CH_3(CH_2)_{14}COOH + 23O_2 \rightarrow$$
(1 mole of palmitic acid)
$$16CO_2 + 16H_2O + \text{about 2,400 Calories}$$
(288 ml of water)

For an average diet, the amount of water produced as a product of oxidation of carbohydrate, fat, and protein is about 0.12 ml/Cal (kilocalorie). Thus, a 70-kg man on an average caloric intake of 3,200 Cal/day produces nearly 400 ml of metabolic water each day.

Water Content of Food

Water comprises 70–90% of the mass of the average diet in adults. Even apparently "solid" foods have a high water content. Some of the more succulent vegetables and fruits are more than 90% water. Consider 700 ml/day to be an approximate "ballpark" figure for this source of water.

Drinking Water and Beverages

The physiological regulation of the amount of water taken in as drinking water and beverages depends mainly on the influence of the water on the thirst mechanism. The sensation of thirst is essential to the preservation of life; diseases or conditions that diminish the sensation of thirst can be accompanied by considerable diminution, or even virtual cessation of fluid intake by mouth. A thirst center has been located in the hypothalamus (in close association with the area controlling release of antidiuretic hormone). This hypothalamic center receives information from the periphery, integrates it, and relays it to the cerebral cortex to produce the sensation of thirst. The thirst center is influenced by three kinds of stimuli: (1) peripheral stimuli related to the physical passage of water through the throat into the stomach, (2) internal stimuli related to the state of intracellular hydration, and (3) "volume receptor" activity related to the volume of blood in the vascular system.

Animal experiments have shown that the process of swallowing and the sense of fullness of the stomach provide a coarse control over the volume of water drunk and prevent overhydration before the second category of stimuli (cellular hydration) can exert its influence. That is, the sensation of thirst is modified somewhat by stimuli from the oropharynx (dryness of mouth and throat) and the stomach (fullness). Many dehydrated animals rapidly drink just enough water to make up their deficit and stop before intestinal absorption begins. Such precise "metering" of fluid volume taken in is not quite as well developed in humans.

For more than transient relief of thirst, water must be absorbed into the body fluids. This absorbed water first dilutes the ECF and, in establishing an osmotic equilibrium with the intracellular fluid, eventually dilutes the ICF as well. The hypothalamic thirst center is apprised of the state of cellular hydration (probably from the state of hydration of thirst center neurons themselves), and final adjustment of water intake over a longer period of time (minutes to hours) is mediated by this fine control of water intake.

A third category of stimuli acting on the thirst center is that of a rough measure of the blood volume by various stretch receptors in the vascular system. Thus hypovolemia (e.g., from hemorrhagic shock) produces an intense desire for water. At least a 7–10% decrease in blood volume must occur before thirst is stimulated.

Cerebrocortical activity can readily condition the sensation of thirst and create what might be thought of as a "thirst appetite" (e.g., becoming thirsty from watching a desert movie or a beer commercial on television). This varies from individual to individual of different voluntary habits of drinking (some people were taught as children to drink eight glasses of water a day). An extreme form of excessive voluntary water intake is psychogenic polydipsia (a psychopathological condition of compulsive water drinking).

The relationship between thirst and stimuli such as cellular hydration and blood volume is similar to that between antidiuretic hormone release and these same stimuli (Table 5). The similarities in anatomical location of the centers and in their sensitivities to the same kinds of

Table 5 Stimuli and Inhibitors of Both Thirst and ADH Release

Stimuli	Inhibitors
Hyperosmolar stimuli Decreased hydration of hypothalamic neurons 2° to hyperosmolarity of ECF (e.g., water deprivation, IV infusion of hyperosmolar solutions, eating salty foods.)	Hypo-osmotic inhibitors. Increased hydration of hypothalamic neurons 2° to hypoosmolarity of ECF (e.g., drinking water, IV infusion of 5% glucose solution).
Hypovolemic stimuli Decreased arterial pressure (carotid baroreceptor) 2° to hypovolemia (e.g., hemorrhage) Decreased transmural pressure in thoracic vascular "volume receptors" (stretch receptors in left atrium and great veins) 2° to decreased intrathoracic blood volume (e.g., hemorrhage, quiet standing in upright position.)	Hypervolemic inhibitors. Increased arterial pressure 2° to expanded blood volume. Increased transmural pressure in thoracic vascular "volume receptors" 2° to increased thoracic blood volume (e.g., hypervolemia, reclining position)

stimuli might explain why conditions leading to an increased thirst are also usually accompanied by increased water reabsorption (antidiuresis) by the kidneys. Water balance is achieved by control mechanisms operating both at the intake and output sides of water turnover. Intake, or water gain, is controlled by the thirst mechanism acting on water ingestion, while output, or water loss, is principally controlled by the ADH mechanism acting at the kidneys. These physiological mechanisms operating in the normal everyday control of water intake and output appear to respond primarily to body fluid osmolarity. Regulation of water output is influenced by rate of ADH release in response to changes in body fluid osmolarity. An increase in plasma osmolarity ultimately increases the osmolarity of the ISF, including the ISF surrounding osmoreceptor cells, in the hypothalamus. These osmoreceptor cells, upon experiencing an increase in the osmolarity of surrounding fluid, stimulate other hypothalamic neurons that determine the rate of release of ADH from the posterior pituitary. An increase in ADH secretion conserves water by decreasing water excretion via the kidneys, hence the name antidiuretic hormone (ADH). It is evident that renal conservation of water cannot alone prevent a water deficit; it can only delay it. The continuous loss of water by obligatory routes would result in progressive dehydration if it were not for an additional mechanism acting on the intake side of water balance. The thirst mechanism serves this essential function. The same increase in body fluid osmolarity that stimulates ADH release also stimulates thirst and thereby increases water intake so that one has both increased conservation of water present in the body fluids along with increased intake.

The body water, regulated by these control mechanisms, continually oscillates within a narrow range between slight water excess (hydration with decreased plasma osmolarity) and slight deficit (hydropenia with increased plasma osmolarity), as illustrated in Figure 8. These mechanisms are sufficiently sensitive to limit variations in osmolarity in the normal individual to + 1%.

In the presence of normal posterior pituitary function, the effectiveness of the kidney in conserving water (by decreased water excretion) results from its ability to produce a hypertonic urine under the influence of ADH (i.e., a urine with greater concentration of solutes than that of body fluid). Under this influence of ADH, glomerular filtrate can be converted from an isotonic fluid in the proximal nephron to one that can be maximally concentrated (nearly five times as much as plasma) as the urine passes down the collecting ducts.

Normally we produce a urine osmolarity of about 700 mOsm/L with a daily volume of 1.4 L. We do not ordinarily produce a maximally concentrated urine (in a minimal volume, the obligatory urine volume) unless we have to in order to conserve water. The production of a maximally concentrated urine requires significant amounts of energy (e.g., energy for chloride transport to maintain the medullary osmotic gradient in the kidneys). Theoretically, we could get by on less water ingestion each day and produce a much more concentrated urine, but this would require increased energy costs.

Instead, the thirst mechanism and ADH mechanism operate in a manner such that a maximally concentrated urine is not necessary if drinking water is freely available. If we were to take in only enough water to match our obligatory water loss (1.8 L), we would have to produce a maximally concentrated urine. Instead we take in about 0.8 L of water in excess of that needed to match the obligatory loss (1.8 + 0.8 = 2.6 L/day) and let the kidneys produce a half-maximal concentration of urine (700 mOsm/L).

This conserves energy. Since excretion of a less than maximally concentrated urine conserves energy, why

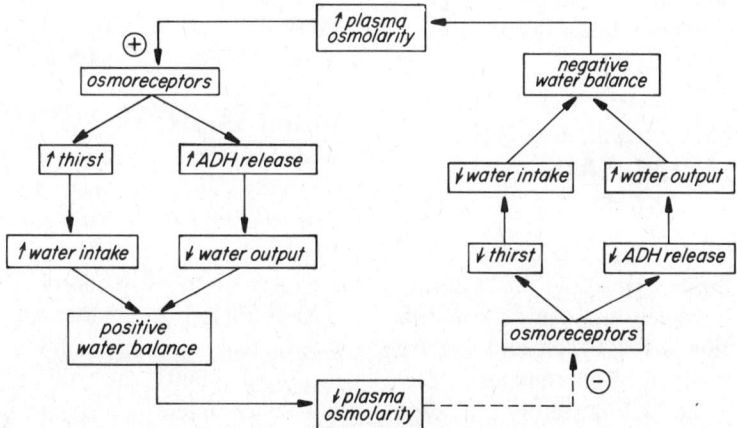

Figure 8 Regulation of water balance. This regulatory mechanism is a function of continuous sensing of plasma osmolarity. In the normal subject it regulates the body water within such narrow limits that osmolarity varies only 1% between hydropenia and hydration.

didn't evolution result in a setting of our thirst centers at a level that would cause an even greater intake of water so that we could excrete an even more dilute urine (say an isotonic urine of 285 mOsm/L)? Part of the answer may be that as terrestrial animals we are adapted to some degree of water conservation. In order to excrete something like an isotonic urine, our primitive ancestors would have had to spend too much time searching for, and drinking water in order to produce a urine this dilute. The setpoints of our hypothalamic thirst center (for water intake) and of our hypothalamic center for control of antidiuretic hormone release (for urine concentration) probably represent a compromise between renal energy conservation and terrestrial water conservation.

We can tolerate disturbances in our ADH release much better than we can disturbances in our thirst mechanism. In the absence of ADH (e.g., in pituitary diabetes insipidus), the kidney excretes large volumes of dilute urine, the daily urinary output usually exceeding 5–10 L (but occasionally even greater, up to 20 L/day). The resultant persistent thirst usually increases intake to equal this loss. In the absence of a normally functioning thirst mechanism, such a rate of loss can lead to very rapid development of a severe water deficit (in only a few hours). Fortunately, primary disturbances of the thirst mechanism are quite rare. However, any disorder that interferes with intake, such as disturbance of consciousness, effectively disrupts the thirst mechanism and removes this most important defense against development of water deficit.

CLINICAL PROBLEMS

Clinical problems in fluid balance can be described either in terms of cause (such as "water depletion" or "salt depletion") or effect (such as "hypertonic contraction" of ECF or "hypotonic contraction" of ECF, respectively). In describing effects, emphasis is placed on changes in the ECF for several reasons: (1) the ECF is readily available (as plasma) for sampling, (2) in most cases the problem begins with some change in ECF volume or osmolarity, and (3) an adequate volume of the ECF (particularly blood) is of considerable significance for proper cardiovascular function. Decreases in ECF volume are dangerous inasmuch as the cardiovascular system is dependent on an adequate blood volume. Development of a shocklike condition is an ominous consequence of the more common types of dehydration (such as in the mixed water and salt depletion of diarrhea).

The following clinical problems will be considered under the broad categories of dehydration and overhydration. Although, strictly speaking, dehydration and overhydration refer specifically to a water deficiency and a water excess, respectively, the terms have come to have a more general meaning among clinicians. Dehydration has become a term applied generally to balance problems resulting in a deficiency of ECF volume, and overhydration simply means excessive ECF expansion.

DEHYDRATION

Basic clinical concepts in dehydration can best be elucidated by consideration of three problems:

1. *Water depletion:* Abnormal loss of water with only small loss of sodium.
2. *Salt depletion:* Abnormal loss of sodium in excess of water loss.
3. *Mixed water and salt depletion:* Loss of abnormal amounts of both water and sodium.

Most dehydrated patients seen by physicians have the mixed depletion of water and salt loss.

Water Depletion

Water depletion occurs as a result of net water loss from the body. This can occur through excessive loss or by normal obligatory loss in the absence of water intake (usually the latter). The effect or result of water depletion is a hypertonic contraction of the ECF.

Even though a person might not be taking anything by mouth (e.g., a shipwreck victim without supplies, or a neglected infant), the daily obligatory water losses continue in complete disregard of the body's inadequate water intake. In the absence of both food and water, the only source of water is metabolic water (from oxidation of endogenous food stores, such as triglycerides of adipose tissue). As obligatory water loss continues, the ECF tends to become hypertonic, just as a bucket of brine in the sun becomes increasingly saline with continued evaporation of water. Both the increase in tonicity and the decrease in volume of the ECF act as stimuli for increased thirst and increased secretion of ADH. The decrease in ECF volume also acts as a powerful stimulus for renal mechanisms augmenting Na$^+$ retention (e.g., increased aldosterone secretion). These mechanisms for Na$^+$ retention add to the hypertonicity of the ECF. The tendency toward hypertonicity of the ECF draws water (by osmosis) from the ICF, thereby adding fluid volume to the contracted ECF space (Fig. 9). This addition of water to the ECF is helpful in maintaining an adequate blood volume. Were it not for

this fluid shift out of the cells, a marked hypovolemia would rapidly develop and produce a shocklike state within 2 or 3 days. However, while the hypernatremia of hypertonic contraction has a favorable effect in delaying the development of hypovolemia, it has an unfavorable effect on the function of the central nervous system (CNS). An early sign of hypernatremia (from any cause) is a peculiar state of lethargy and apathy, on the one hand, and hyperirritability upon stimulation, on the other. Spasticity and seizures or convulsions may also be seen if the hypernatremia is severe. Convulsions and irreversible brain damage have been associated with focal brain hemmorrhages (shrinkage of the brain within the rigid skull tears

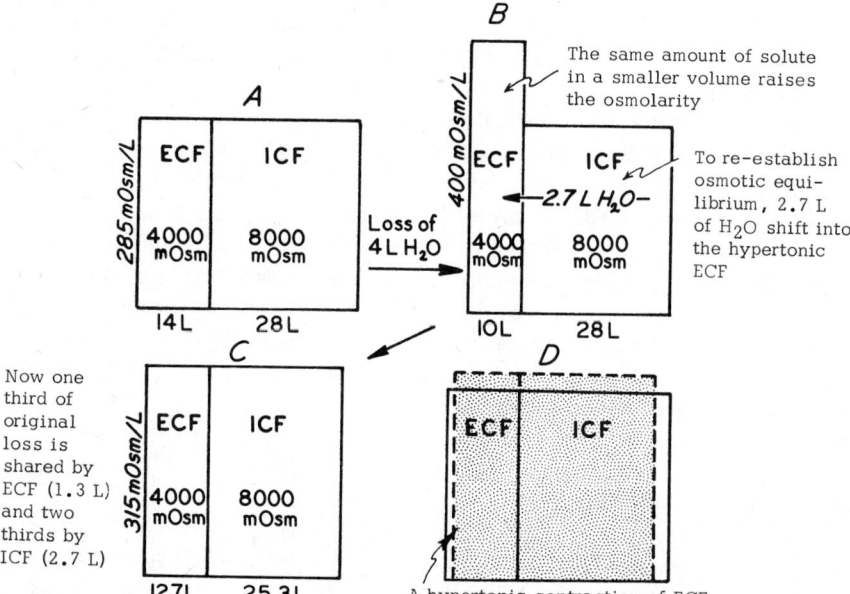

Figure 9 Darrow-Yannet diagram of water depletion. Osmolarity is represented by height, and volume is represented by width. (a) Normal osmolarity and volumes of the ECF and ICF in a 70-kg man. (b) the effect of a 4-L water loss from the ECF (e.g., the obligatory water losses for slightly more than two days). For purposes of instruction, this loss is shown as being completed before any fluid shifts from the ICF to reestablish osmotic equilibrium. In reality, the ECF water loss, the consequent increase in ECF osmolarity, and the osmotic shift of water from the ICF all occur gradually and simultaneously. (c) Final values after reestablishment of osmotic equilibrium (after shift of 2.7 L of water from ICF to ECF). Note that of the original 4 L deficit, two-thirds is shared by the ICF. (d) Typical Darrow-Yannet diagram usually drawn to illustrate water depletion. The initial normal state is indicated by solid lines, the hypothetical intermediate state is omitted, and the final state is drawn with dashed lines, that is, a superimposition of (a) over (c).

small vessels). Thus, development of hypernatremia tends to prevent or delay severe hypovolemia, but poses a threat to life all its own. Hypernatremia is the principal danger in water depletion.

Salt Depletion

Salt depletion is a deficiency of body sodium with little net loss of body water. In the condition of water depletion, just described in the preceding paragraphs, the primary deficit (of water) is usually the result of an inadequate intake (of water). However, in salt depletion the usual cause is excess loss (of salt) rather than inadequate intake (of salt). Renal conservation of sodium is nearly complete in the presence of a salt-free diet. Contrast this ability to conserve sodium with the body's inability to prevent obligatory water loss.

The term salt depletion is applied to a deficit of total body sodium. Occasionally, the term hyponatremia (lowered blood sodium) is used synonymously with salt depletion, perhaps because hyponatremia usually develops in salt depletion. Actually, these two terms are not synonymous. Salt depletion refers to an absolute deficiency of sodium, while hyponatremia describes only a relative decrease in the blood concentration of sodium. Hyponatremia can occur with normal, decreased, or even increased total body sodium. The reason hyponatremia is so often associated with salt depletion is that salt losses are so much in excess of any water loss in pure salt depletion. Because of the disproportionally greater salt loss, a hyponatremia develops.

Salt depletion is sometimes called secondary dehydration because, like water depletion, it causes a contraction of the ECF volume. An important point essential to understanding the relationship between the amount of salt in the body and the volume of the ECF is that sodium is the primary cation of the ECF. Sodium, along with associated anions (particularly Cl^-), is responsible for the major part of the osmolarity of the ECF. A simple estimate of the plasma osmolarity can be made by doubling the plasma sodium concentration (e.g., normal concentration of 142 mEq/L $\times$ 2 = 284 mOsm/L). A physiological rule of thumb is that "as sodium goes, so goes water." The emphasis is placed on sodium, since it is usually handled actively in the body while anions (such as Cl^-) and water usually accompany the sodium passively (except in as-

cending limb of loop of Henle, where it is the Cl^- that is actively pumped).

The most common cause of salt depletion is the loss of sodium-containing body fluids (Fig. 10) followed by replacement of fluid losses with a sodium-free fluid, usually water alone. Of course, replacement of fluid losses with water alone is better than no replacement at all, but it cannot be adequate, since the ECF cannot hold the water if it does not have an adequate amount of solute. In order to balance the osmotic pressure of the ICF, the ECF must have a "skeletal framework" of electrolyte solute to retain the water. Drinking water without replacing the salt deficit results in a shift of most of this water from the diluted ECF into the relatively more hypertonic ICF. For example, patients can replace fluid losses with water alone after sweating profusely or after losing GI fluid from diarrhea or a fistulous drainage. In adrenal insufficiency (e.g., Addison's disease) and in some types of advanced renal insufficiency (with inability to conserve sodium when placed on a low-sodium diet), the sodium loss is a "urinary salt wasting." The convenient availability of urinary sodium measurement affords an easy means of distinguishing between such renal losses of sodium (urinary sodium in excess of 20 mEq/L) and the other, nonrenal losses of sodium associated with avid renal sodium retention (urinary sodium less than 10 mEq/L). In all these circumstances—both renal and nonrenal—the primary event is

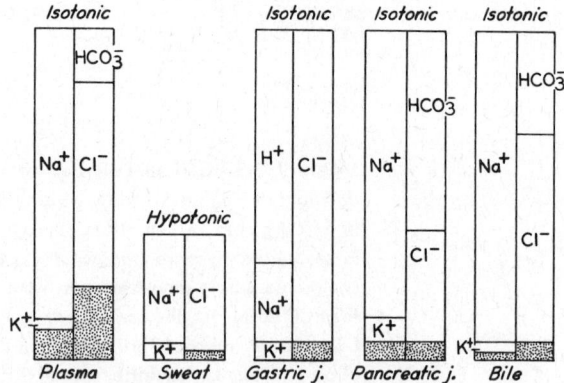

Figure 10 Comparison of osmolarity and ionic composition among plasma, sweat, and several GI secretions. The ionic compositions are only approximations; exact values vary according to method of stimulation of secretion and means of collection.

abnormal loss of the principal ECF electrolyte, sodium, and the accompanying Cl⁻. This results in a severe decrease in the ECF volume.

Under normal conditions, a dilution of the ECF (development of hypo-osmolarity) would be expected to inhibit ADH secretion (see the description of hypo-osmotic inhibitors in Table 5) and result in the excretion of a greater volume of urine of a more dilute nature. However, because of the decrease in ECF volume and its effect on the various volume receptors, this nonosmotic stimulus can take precedence, and ADH secretion can be increased to conserve fluid volume (see the description of hypovolemic stimuli in Table 5). A good point to remember is that the body usually attempts to maintain osmolarity until ECF volume is decreased (to some critical level). From then on it becomes vital to maintain volume (especially blood volume) to prevent circulatory shock. When ADH secretion is increased in response to hyponatremia, osmolarity is permitted to deviate in the interests of conserving water for volume. Renal mechanisms for augmenting Na^+ retention (e.g., increased activity of the renin–angiotensin–aldosterone system, sympathetic-mediated renal hemodynamic changes resulting in an increased filtration fraction) in conditions of salt depletion from nonrenal causes result in almost complete conservation of plasma Na^+ (as in water depletion). However, with continued water ingestion and ADH secretion, hyponatremia develops. In terms of the ECF, a state of hypotonic contraction occurs (Fig. 11).

Because of the more rapid development of a diminished ECF volume, salt depletion is considered a more acutely serious condition than that of water depletion. Not only does the salt-depleted patient have a decreased ECF volume with all the hemodynamic consequences of hypovolemia (hypotension, weak rapid pulse, and eventually deterioration of renal function with oliguria and acidosis), but in addition the patient is threatened by the development of hyponatremia. Hyponatremia (from whatever cause, whether from salt depletion or from water intoxication) produces overhydration of CNS neurons and results in symptoms remarkably similar to those of hypernatremia. Initially hyponatremia causes lightheadedness, headache, and confusion; progresses to lethargy, delirium, and psychotic behavior; and finally predisposes to convulsions as the plasma sodium level becomes lower.

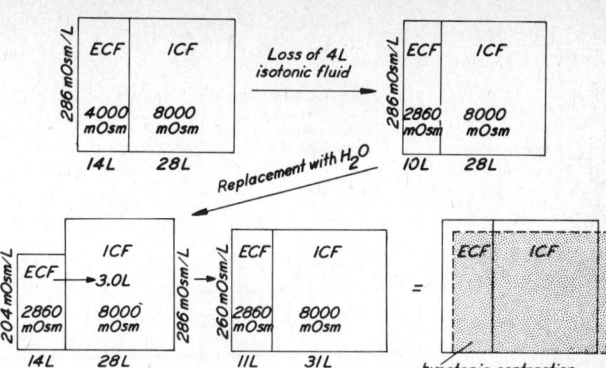

Figure 11 In salt depletion, the osmolarity and volume of the ECF are diminished. ECF hypotonicity results in a shift of water into the ICF. This diminution of ECF volume has serious consequences, namely, hypovolemic shock.

Mixed Water and Salt Depletion

Mixed water and salt depletion can be defined as reduction below normal of both total body water and total body salt, the water reduction usually being slightly in excess of the salt loss. This commonly occurs in patients in whom the same kinds of abnormal fluid losses described in simple salt depletion have occurred (e.g., diarrhea, profuse sweating, and vomiting), but without replacement of fluid loss by any liquids. The water loss slightly outstrips the salt loss because the secretions lost are either isotonic (e.g., GI fluids) and there is continued obligatory loss of water, or the secretions lost may be hypotonic (sweat) in the first place. Nevertheless, because both salt and water are lost, the tonicity of the ECF tends to remain near the upper limits of normal, or only slightly hypertonic, and some clinicians consider this an "isotonic" contraction of the ECF (Fig. 12).

In most instances, the history of onset will reveal a loss of GI fluids (the main route of abnormal electrolyte loss). If replacement of the loss had been attempted by intake of water alone, a simple salt depletion would have developed. But if nothing was taken by mouth (as is often the case with vomiting, and many times in diarrhea as well), a mixed water and salt depletion is the result. About two-thirds of all diarrheas of children in this country result in a mixed water and salt depletion. The hypovolemia forces both ADH and aldosterone secretion to increase in an attempt to conserve what water and salt remain in the ECF.

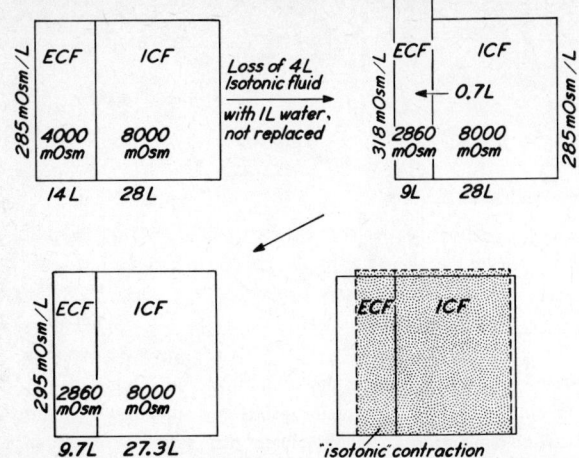

Figure 12 Mixed water and salt depletion. Note the magnitude of shrinkage of the ECF. Since the osmolarity of the ECF is only slightly hypertonic, this is often called an *"isotonic"* contraction.

All three forms of dehydration (water depletion, salt depletion, and mixed water and salt depletion) involve a decrease in ECF volume and portend the danger of hypovolemia. As mentioned above, the danger of hypovolemia is delayed in the hypertonic contraction of water depletion (because of the fluid shift from ICF to ECF). However, the development of hypovolemia is rapid in salt depletion and is particularly ominous in mixed water and salt depletion. Severe hypovolemia and a tendency toward hypotension cause decreased perfusion of most tissues. In an attempt to maintain sufficient blood pressure for perfusion of brain and heart, a widespread vasoconstriction compromises the flow of blood to a large number of different tissues. An important consequence of severe hypovolemia and decreased tissue perfusion is a metabolic acidosis primarily from decreased renal function. Normally, 50–100 mEq of strong acid is produced daily as products of protein metabolism, and this acid is handled indirectly through urinary acidification (see details in Chapter 34). With greatly diminished renal function from severe hypovolemia, these acidic products accumulate and produce a metabolic acidosis.* Furthermore, hypoxemia of

the tissues with diminished perfusion results in an increase in anaerobic metabolism with increased lactic acid production. Thus, lactic acidemia adds to the problems of metabolic acidosis due to decreased renal function. Other renal functions are also diminished, and excretory products are not cleared from the blood at a normal rate. Uremia results from a rise in blood urea nitrogen (BUN).

OVERHYDRATION

With the exception of overhydration from salt and water retention in a few disease states characterized by edema (e.g., congestive heart failure, cirrhosis of the liver, and nephrotic syndrome), overhydration of the ECF (excess volume) is almost always iatrogenic (drug-induced) in origin. Overhydration can be categorized in a manner similar to that for dehydration: (1) water excess, (2) salt excess, and (3) mixed water and salt excess.

Water Excess

Impaired water diuresis, that is, the inability to excrete free water from either endogenous or exogenous sources, is one of the most common metabolic defects observed in hospitalized patients, particularly in postoperative recovery. If a patient can only produce a small volume of urine, what might ordinarily be a proper volume of fluid input becomes a relative excess for that patient under those conditions, for example, forcing fluid intake in patients with oliguria of renal failure. Patients with postoperative water retention are most commonly in this same predicament. The pain and trauma of surgery, pharmacological effects of sedatives, and emotional stress all combine to produce an inappropriate secretion of ADH. Inappropriate ADH secretion, whether from excessive pituitary release (surgical stress, head injury, meningitis) or from tumorous production of ADH-like secretions (bronchogenic carcinoma, adrenocarcinomas of duodenum or pancreas), increases renal reabsorption of water and produces a negative free water clearance (water retention). In addition to the dilutional hyponatremia that results from water retention, sodium excretion eventually increases as

---

*In the case of severe vomiting, the loss of acid gastric fluid more or less compensates for the effect of hypovolemia on renal function, and therefore the ECF of a patient with dehydration from vomiting varies from

nearly normal to quite alkalotic. In the case of diarrhea, however, acidosis is augmented by the loss of considerable intestinal HCO_3^- in diarrheal fluids. The acidosis of diarrheal dehydration can be a severe threat.

a result of ECF volume expansion ("a third-factor" natriuretic effect primarily on proximal tubules, to be discussed later). A dilutional hyponatremia and water intoxication can be produced in almost any patient by the overenthusiastic administration of hypotonic fluids, particularly intravenous infusion of 5% glucose solution (which is initially isotonic to prevent hemolysis during infusion, but gives net gain of water after glucose is taken up and metabolized by cells).

Excess water gain is shared by both the ECF and the ICF in proportion to their solute contents. That is, one-third of the water gain is shared by the ECF, which contains one-third of the total solutes of the body, and two-thirds of the water gain is shared by the ICF, which contains two-thirds of the total solutes. Although there is actually a gradual, essentially simultaneous dilution of both the ECF and ICF, it is helpful to regard the ECF as being diluted first, followed by a shift of water into the ICF. According to this sequential point of view, the ECF becomes hypotonic from the entire water gain, after which some of the water shifts from the ECF to the relatively hypertonic ICF to maintain osmotic equilibrium (Fig. 13).

As in the case of hyponatremia from salt depletion, overhydration of CNS neurons is responsible for most of the signs and symptoms of water intoxication, that is,

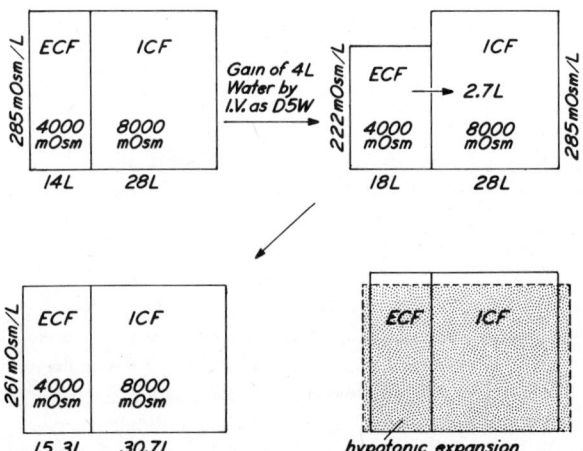

Figure 13 Overhydration from water excess. Dilution of ECF causes a shift of some of the excess water into the ICF (i.e., the ICF shares in the overhydration). Despite a shift of water from ECF to ICF, the ECF is left with a hypotonic expansion.

lethargy and fatigue, mental confusion and delerium, headache, and, in severe cases, convulsions.

Salt Excess

Substantial retention or intake of sodium in excess of water is rare. Circumstances causing increased sodium reabsorption by renal tubules (e.g., increased secretion of aldosterone) also cause increased reabsorption of water (water passively follows sodium). Thus renal sodium retention usually results in an isotonic, or nearly isotonic, expansion of the ECF and hypertension or edema. An increase in salt ingestion stimulates thirst and ADH secretion so that increased ingestion of water and increased renal reabsorption of water in proportion to the salt taken in results in a temporary isotonic expansion of the ECF. Marked hypernatremia due to an excess of salt without a proportional excess of water can only occur iatrogenically. Severe hypernatremia has occurred in accidental substitution of salt for sugar in preparing infants' formula in a hospital nursery. Hypernatremia can also result from excessive administration of hypertonic saline solutions (e.g., saline enemas, infusion of 5% saline intravenously, replacement of amnionic fluid with 20% saline for inducing abortion). The principal danger of hypernatremia is in its effects on the brain. As mentioned earlier, the effects of hypernatremia and hyponatremia are similar. An important difference is that hypernatremia more readily produces permanent brain damage, since it causes multiple hemorrhages and infarcts. (See Fig. 14.)

Mixed Water and Salt Excess

The regulation of body fluid osmotic pressure (at 285 mOsm/L) tends to adjust the volume of water in the body to the amount of solute in the body fluids. By regulating ECF osmolarity, the ratio of volume to solute is kept essentially constant. As solute increases, so does the volume of water (hence osmolarity remains same). This regulation of osmotic pressure by regulation of water content depends on the thirst mechanism (acting on the input side of water balance) and the ADH mechanism (acting on the output side of water balance). As long as this regulation of osmotic pressure remains normal, the volume of the body fluids should depend only on the amount of solute they contain. Since sodium salts constitute the major solutes of the ECF, it can be said that the volume of the ECF

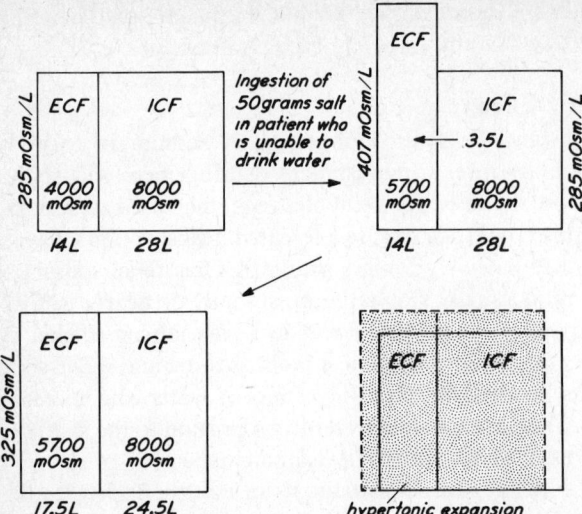

Figure 14 Overhydration from salt excess (e.g., salt poisoning). Hypertonicity of the ECF causes a shift of water from the ICF. This causes expansion of the ECF, but contraction of the ICF (detrimental to brain). The ECF is described as a hypertonic expansion.

is primarily dependent on its content of Na^+ (with accompanying anions). It follows that whatever determines the Na^+ content of the ECF also, in effect, determines its volume (to maintain normal osmolarity). The amount of Na^+ contained in the ECF depends on (1) the dietary intake, and (2) renal handling (retention or excretion of Na^+). Since either increased intake or increased retention of Na^+ is usually accompanied by increased intake of water (via thirst mechanism) and by increased retention of water (via ADH mechanism), most conditions causing excess Na^+ also cause an initial equivalent excess of water, in other words, a mixed water and salt excess with increased volume. Fortunately, the degree to which the ECF volume can expand as a result of increased dietary intake of salt is usually limited by other regulatory mechanisms (e.g., regulation of ECF volume by regulation of Na^+ excretion).

If the daily intake of Na^+ is increased (e.g., taking $NaHCO_3$ for indigestion, sprinkling twice as much salt on food), there is an increase in ECF Na^+, hence an increase in volume (seen as an increase in body weight due to the extra fluid). If Na^+ excretion did not also increase eventually to match the daily increment in Na^+ intake, there

would be a daily increment in H_2O causing an increase in volume (to maintain normal osmolarity) with continuous unlimited expansion. However, in actual fact, there is a limit to the expansion, and it can be assumed that if the ECF volume is to be maintained with any degree of consistancy, a relationship must exist between changes in ECF volume and changes in Na^+ excretion. What happens with increased Na^+ intake and volume expansion is that renal Na^+ excretion gradually increases until it matches intake (Fig. 15), and a new steady state is reached (at a level of expanded ECF that does not increase further).

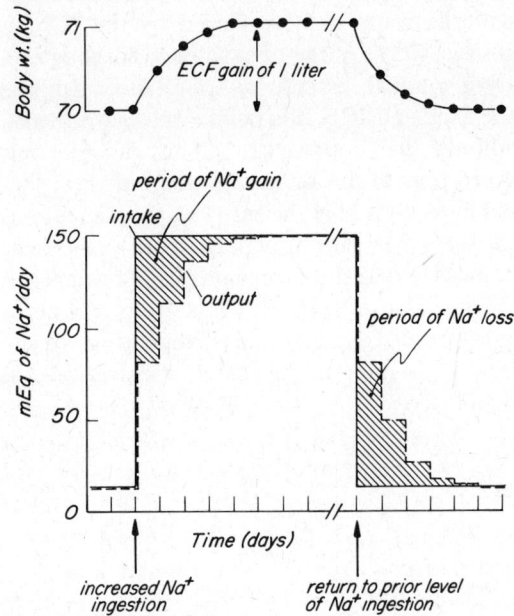

Figure 15 Relationships among sodium intake, sodium excretion, and fluid retention. Initially the subject had a Na^+ intake of 10 mEq/day (and excreted (10 mEq/day). Na^+ intake was increased to 150 mEq/day (at 1st arrow) and maintained at this level until Na^+ excretion once again matched Na^+ intake (after several days). Subsequently, the Na^+ intake was returned to the previous level (at 2nd arrow). Note that the 1 L expansion of the ECF during the period of increased Na^+ intake (as indicated by 1-kg body weight gain) is proportional to the gain of an additional 140 mEq of Na^+ per day, so that isotonicity is maintained (ECF normally contains about 140 mEq/L of Na^+. (Redrawn from Reineck HJ, Stein JH: in Maxwell MH, Kleeman DR (eds): "Regulation of sodium balance," *Clinical Disorders of Fluid and Electrolyte Metabolism*, ed 3. New York, McGraw-Hill, 1980, p 90.)

The kidneys sense and respond to some aspect of ECF volume expansion by increased Na^+ excretion (decreased Na^+ retention), particularly by the proximal tubules. It is probably not volume per se that is monitored, but rather indirect derivative functions such as intravascular or intracardiac pressures, cardiac output, or effectiveness of tissue perfusion (the maintenance of an effective "dynamic circulation"). Two factors previously thought to be primarily responsible for control of Na^+ excretion were glomerular filtration rate (GFR) and aldosterone secretion. It is now clear that these two factors alone do not suffice and that there must be at least a third factor that influences Na^+ excretion in response to ECF volume expansion.

In the case of overexpansion of the ECF, there is a need for increased Na^+ excretion, that is, natriuresis (decreased Na^+ retention). As shown in Figure 15, an increase in ECF volume results in an increase in Na^+ excretion. This increase in Na^+ excretion appears to have come about by the third-factor effect, which decreases Na^+ reabsorption (particularly in the proximal tubule and perhaps other parts of the nephron as well). Increased excretion of Na^+ (along with water) prevents an unlimited expansion of the ECF. On the other hand, a lack of adequate ECF volume (as in dehydration) has the opposite effect: inhibition of the third factor effect and a consequent enhancement of Na^+ retention (along with water).

The precise identity of the third-factor stimulation of Na^+ excretion (natriuresis), which operates primarily by inhibition of proximal tubular reabsorption of Na^+, has not been established. The third factor might simply be an as yet unidentified natriuretic hormone. Unfortunately, evidence for this "salt-losing" hormone is controversial. On the other hand, the third-factor effect might be neural; it could involve sympathetic nervous system (SNS)-mediated changes in the handling of Na^+ by the proximal tubules. Decreased SNS activity to kidneys decreases Na^+ retention (increases Na^+ excretion, i.e., natriuresis), and increased SNS activity increases Na^+ retention. These SNS-mediated effects may operate by altering intrarenal hemodynamics (e.g., changes in the filtration fraction, as discussed in the renal chapter) or by actions on proximal tubular Na^+ transport (e.g., it has been shown that catecholamines can directly enhance active Na^+ transport in vitro). While the specific mechanisms whereby this third-factor effect brings about natriuresis remain to be fully elucidated, it is clear that Na^+ retention and ECF volume are part of a negative feedback system that ultimately matches Na^+ output to Na^+ intake (Fig. 15) and that attempts to stabilize or limit ECF volume changes within reasonable physiological limits.

Unfortunately, conditions that involve either inadequate cardiac output (e.g., heart failure) or a decrease in effective blood volume (cirrhosis of liver with ascites, or nephrosis of the kidneys) act as if there were a decrease in ECF volume (although, in fact, there might be an increase in total ECF volume). These conditions, and others such as hemorrhagic shock and primary and secondary dehydration (which cause a true decrease in blood volume), result in inhibition of the third-factor natriuresis, and thereby stimulate Na^+ and water retention (Fig. 16).

Congestive Heart Failure

The term "congestive heart failure" does not refer to a specific discrete disease (such as USSR influenza), but to a dysfunctional state or condition in which the heart is unable to meet the needs of the peripheral tissues in terms of supply and demand. The principal clinical manifestations of congestive heart failure result from Na^+ and H_2O retention and from increased filling pressure. In either the normal heart or in the failing heart, an increase in diastolic volume stretches myocardial fibers and increases the contractile force (by the Starling mechanism). Although the failing heart operates on a diminished contractility curve, it can increase its stroke work somewhat, up to a point, by such dilation. An increase in filling pressure and an increased blood volume (from Na^+ and water retention) help the heart reach its maximum ability along this curve (although this still might not be an adequate stroke work for the body's needs, if the curve is one of a much lower contractility). The kidneys respond to an inadequate cardiac output from a hypoeffective or failing heart (hypodynamic circulation) in the same way as they would respond to an inadequate supply of blood due to hypotension and/or hypovolemia, namely, by increased retention of Na^+ and H_2O.

The renal circulation is altered by the inadequate cardiac output. GFR is reduced somewhat (less Na^+ filtered), and the fraction of filtered Na^+ that is reabsorbed is increased. Na^+ reabsorption is thought to be increased by

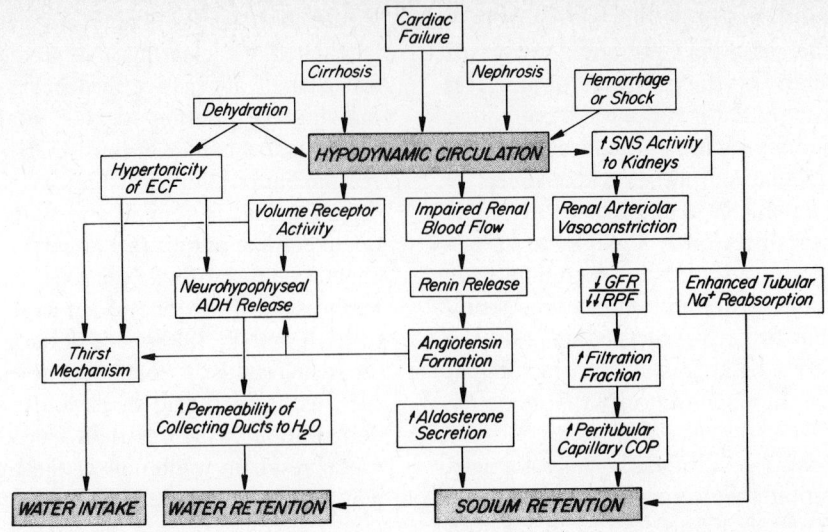

Figure 16. Pathogenesis of sodium and water retention in several clinical conditions. In conditions of dehydration and hemorrhagic shock, the retention of sodium and water is of considerable homeostatic value in restoring or maintaining blood volume, thereby delaying or preventing a more severe hypovolemia. However, in the cases of cirrhosis, cardiac failure, and nephrosis, the retention of sodium and water cannot maintain an adequate blood volume since the balance of hydrostatic and osmotic forces at the capillary level is upset and favors loss of fluid into the ISF (where it produces edema). In these latter conditions the retention of sodium and water is of little or no homeostatic value and only produces a mixed water and salt excess.

an increase in aldosterone secretion acting on the distal nephron (particularly the collecting duct), and by an increase in proximal tubular Na^+ reabsorption from a decrease in third factor natriuretic influence (e.g., by either a decrease in the putative natriuretic hormone, or by an increase in sympathetic activity to the kidneys). With less Na^+ filtered and more Na^+ reabsorbed, the net result is diminished excretion of Na^+. If the intake of Na^+ is unchanged, this diminished excretion results in Na^+ retention along with H_2O and an expansion of the ECF volume. Unfortunately, if the heart has reached a plateau on its diminished contractility curve, additional retention of fluid volume may not improve the ability of the heart to increase its cardiac output to meet tissue needs. In this case, further fluid retention does not correct or remove the original stimulus for Na^+ and H_2O retention (failure of feedback control). The result is edema.

The distribution of the increased ECF volume resulting from Na^+ and water retention is determined by the venous pressure (which affects venous pooling), an effect emphasized by the influence of posture on hydrostatic pressure. A patient able to walk around or remain in an upright position much of the day will accumulate edema fluid in the dependent limbs (ankles and feet). A bedridden patient on his back will have sacral edema.

Cirrhosis with Ascites
In patients with certain forms of liver disease, especially portal cirrhosis (chronic diffuse liver disease with cell loss), fibrosis and distortion of the vascular bed result in portal hypertension. Portal hypertension either from (1) presinusoidal obstruction (obstruction to flow between GI capillaries and hepatic sinusoids) or from (2) sinusoidal or postsinusoidal obstruction (obstruction in or beyond hepatic sinusoids) results in loss of fluid into the abdominal cavity (ascites). This ascites fluid arises from the GI

capillaries in presinusoidal obstruction and is associated with an expansion of the ECF volume due to renal mechanisms for Na^+ and water retention. From a cardiovascular point of view, the most important consequence of portal hypertension is the development of collateral circulation (particularly in formation of esophageal varices) to return the splanchnic circulation to the heart. From the point of view of fluid balance, the most important consequence is renal retention of Na^+ and water. Ascites represents an intraperitoneal expansion of the ECF and is the result of two principal factors: (1) the elevation in portal venous pressure as discussed above, and (2) a diminution in plasma colloidal osmotic pressure due to decreased serum albumin production by the diseased liver. It is the combination of portal hypertension and reduced plasma colloidal osmotic pressure that drives the ISF out of the splanchnic bed into the peritoneum (imbalance of Starling forces).

Loss of a major fraction of the ECF volume into the ascites fluid is thought to result in a hypodynamic circulation which brings about the same renal response found in congestive heart failure, namely an increase in Na^+ and H_2O retention. This would be expected to augment the ECF volume and alleviate the hypodynamic circulation, however, because of diminished plasma colloidal osmotic pressure (due to decreased serum albumin), much of this fluid escapes the vascular system to form more ascites fluid and produce edema elsewhere (again, failure of feedback control).

The Nephrotic Syndrome

Many renal diseases result in proteinuria (loss of protein in urine) so that it may be described as an almost universal manifestation of diffuse renal disease. The severity of clinical manifestations secondary to protein loss from proteinuria is correlated with the degree of loss. For example, secondary manifestations usually do not appear until the protein loss exceeds 2 or 3 grams a day. The continuous urinary drain on the body's stores of protein results in the condition called the nephrotic syndrome. This syndrome is characterized by heavy proteinuria, decreased plasma colloidal osmotic pressure, hypodynamic circulation, Na^+ and water retention, and edema. The result of diminished colloidal osmotic pressure, as was described for cirrhosis above, is transudation of fluid from capillaries into

the ISF. The net result is edema and a diminution of the plasma volume. Edema is frequently the first manifestation of the nephrotic syndrome noted by the patient. In many cases, the edema is bothersome, but not debilitating. However, occasionally it is severe and incapacitating. As the edema becomes generalized, there may be fluid accumulation in the pleural cavity, the peritoneal cavity (ascites), and pericardium. As before, a hypodynamic circulation results in a renal response characterized by Na^+ and water retention. This increased fluid retention would augment plasma volume, except that the low plasma colloidal osmotic pressure favors loss of retained fluid into the ISF (again, loss of feedback control).

In each of the three conditions described above (congestive heart failure, cirrhosis with ascites, and the nephrotic syndrome), the inability of the cardiovascular system to supply the kidneys with an adequate flow of blood (hypodynamic circulation) due to inadequate cardiac output (either from failure of the pump, or from inadequate plasma volume) results in initiation of mechanisms involving renin release, aldosterone secretion, increased Na^+ reabsorption by the proximal tubule (decreased third-factor natriuretic effect), ADH secretion, and thirst, as illustrated in Figure 16. This causes retention of Na^+ and water and is a form of overhydration characterized by an expanded ECF volume (although not necessarily an increased plasma volume) and edema (increased ISF volume).*

As in other conditions of overhydration, iatrogenic causes can also be responsible for ECF expansion. A mixed water and salt excess can be produced by excessive infusion of isotonic saline intravenously (Fig. 17).

---

*Although congestive heart failure, cirrhosis, and nephrosis usually favor retention of both Na^+ and water isotonically, in severe cases with very ineffective or hypodynamic circulation the retention of water can exceed that of Na^+. The result is hyponatremia, which should be taken as a grave sign; that is, the presence and severity of hyponatremia correlate fairly well with the seriousness of the underlying condition. The increased water retention in excess of salt retention in these worsening edematous states may be caused by extensive proximal Na^+ and water reabsorption, which reduces the volume of fluid delivered to the diluting segment (ascending loop) to a "trickle," and by continued secretion of ADH (from nonosmolar stimuli).

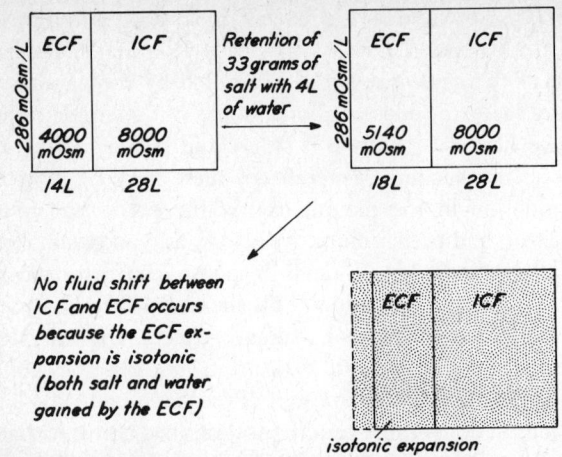

No fluid shift between ICF and ECF occurs because the ECF expansion is isotonic (both salt and water gained by the ECF)

isotonic expansion

Figure 17 Overhydration from mixed water and salt excess (e.g., edema of congestive heart failure). Since this increase in ECF is isotonic, no shifts of water occur to affect the ICF. The ECF undergoes an isotonic expansion.

SELECTED BIBLIOGRAPHY

Brenner BM, Stein JH: *Sodium and Water Homeostasis.* Moravia NY, Churchill Livingstone, 1978.

Goldberger E: *A Primer of Water, Electrolyte and Acid–Base syndromes.* Philadelphia, Lea & Febiger, 1975.

Maxwell MH, Kleeman CR (eds): *Clinical Disorders of Fluid and Electrolyte Metabolism,* ed 3. New York, McGraw-Hill, 1980.

Rose BD: *Clinical Physiology of Acid–Base and Electrolyte Disorders.* New York, McGraw-Hill, 1977.

Schrier RW (ed): *Renal and Electrolyte Disorders.* Boston, Little, Brown, 1980.

SELF-STUDY QUESTIONS

MULTIPLE CHOICE

Select the single best answer.

1. A man weighing 80 kg (176 lb) and 1.8 meters (5 ft, 10 in.) tall has a body surface area of 2m². How many liters of water should he normally gain and lose per day?
   A. 1.5
   B. 2.6
   C. 3.0
   D. 3.4
   E. 4.0

2. The volume of daily turnover is equal to what fraction of total body water (TBW) in the patient described in question 1?
   A. $\frac{1}{6}$
   B. $\frac{1}{10}$
   C. $\frac{1}{16}$
   D. $\frac{1}{20}$
   E. $\frac{1}{25}$

MATCHING

Select the nearest correct answer.

   A. 0.1
   B. 0.12
   C. 1.0
   D. 1.5
   E. 10

3. What is the approximate number of liters of fluid normally presented to gut each day for absorption and reabsorption?

4. What is the normal volume of fluid lost in stools each day (liters)?

5. How many milliliters of water are gained per Calorie from food oxidation (assume average diet)?

6. What is the approximate daily volume of urine (liters) from a normal 70-kg man excreting urine of about 700 mOsm/L?

7. What is the average daily load of waste solutes (in Osmoles) in a 70-kg man on an average diet?

8. What is the average daily insensible loss of water (in liters)?

MULTIPLE CHOICE

Select the single best answer.

9. Given an average daily load of waste solutes (70-kg man on average diet), how many liters of urine would be produced each day to excrete an isotonic urine?

   A. 1
   B. 1.5
   C. 2.6
   D. 3.5
   E. 20

10. Assuming all other routes of water loss to be normal, how many liters of water would this man have to ingest (food and water) to stay in balance?

   A. 2.6
   B. 3.5
   C. 4.3
   D. 4.7
   E. 20

11. The most common cause of water depletion (primary dehydration) is

   A. usual loss of obligatory routes in absence of water intake.
   B. excessive loss by one of the obligatory routes of water loss.
   C. excessive release of ADH.
   D. excessive secretion of aldosterone.
   E. effect of hypovolemia on thirst center.

12. A sign of hypernatremia is
   A. lethargy and apathy.

B. hyperirritability upon stimulation.
C. spasticity.
D. seizures or convulsions.
E. all of the above.

13. A stimulus that ultimately results in aldosterone secretion is
   A. anything that stimulates the thirst center as well.
   B. hypovolemia.
   C. hypernatremia.
   D. hypervolemia.
   E. hypokalemia.

14. When water is lost or gained (in simple water depletion or water excess, respectively), this difference is shared by the ECF and ICF
   A. equally.
   B. ECF, one-fourth; ICF, three-fourths.
   C. ECF, one-third; ICF, two-thirds.
   D. ECF, three-fourths, ICF, one-fourth.
   E. ECF, mainly, ICF, to a very small extent.

15. A patient has returned from vacation south of the border and has the "trots" (diarrhea). If he replaces his fluid loss by drinking an equal amount of sugared tea, he will get
   A. a water depletion.
   B. a salt depletion.
   C. a mixed water and salt depletion.
   D. a worse diarrhea.
   E. tired of sitting.

16. In salt depletion of several days' duration, ADH secretion
   A. will be inhibited by development of a hypoosmolarity of ECF.
   B. will be stimulated.
   C. will be unaffected one way or the other.
   D. will increase Na+ retention by the kidneys.
   E. will stimulate Na+ excretion by the kidneys.

17. Except for a few chronic conditions characterized by edema, overhydration of the ECF is almost always

A. the result of excessive salt intake accompanied by increased water ingestion and retention.

B. iatrogenic in origin.

C. the result of polydipsia (e.g., polydipsia of diabetes, psychogenic polydipsia).

D. the result of ineffective public education on the proper amount of water required each day.

E. only a relative overhydration because of an absolute loss of solute in excess of water loss.

18. Which of the following conditions is associated with Na^+ and H_2O retention (mixed salt and water excess)?

A. Congestive heart failure

B. Cirrhosis with ascites

C. Nephrosis

D. All of the above

E. None of the above

Select the correct answer(s). (In many instances, more than one answer is correct.)

19. A normal man loses 1 L of sweat and then drinks 1 L of distilled water. After the water has been absorbed, increased values relative to the initial conditions (before the loss of sweat) would be noted for

A. plasma sodium concentration.

B. extracellular volume.

C. plasma potassium concentration.

D. intracellular volume.

20. Subject W, a thirsty 70-kg man, drinks 1,000 ml of distilled water. Subject S, also a thirsty 70-kg man, drinks 1,000 ml of isotonic saline solution. Correct statements regarding subject W and subject S include the following:

A. Subject W has the greater change in plasma osmolarity.

B. Subject S has the greater increase in plasma volume.

C. Subject W has the greater change in urine osmolarity.

D. Subject W will have a greater urinary output within 2 h after fluid ingestion.

MATCHING

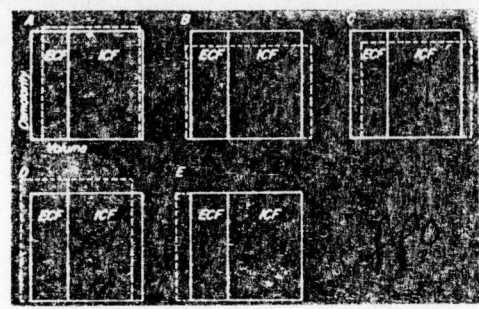

Solid lines = normal
Dashed lines and stippling
= final state

21. Fluid loss from diarrhea replaced with water only.

22. Fluid loss from diarrhea not replaced at all.

23. Excessive administration of isotonic glucose solution (D5W).

24. Substitution of salt for sugar in infant formula.

25. Chronic cirrhosis with ascites formation.

MATCHING

A. $CaCl_2$
B. Urea
C. Albumin
D. NaCl
E. None of these

26. A 155mM solution is isotonic to a red blood cell.

27. Contributes to the plasma colloidal osmotic pressure.

28. A hyperosmotic solution causes red blood cells to shrink first and then to swell and burst (hemolysis).

MULTIPLE CHOICE

Select the single best answer.

29. Among the following, which group of agents is best suited to measure interstitial fluid volume in a laboratory animal by the indicator dilution method?

A. Heavy water, thiocyanate (SCN^-)

B. Urea, ^{51}Cr-labeled RBCs, and ^{131}I-albumin (RISA)

C. Evans blue (T-1824), thiocyanate (SCN$^-$)

D. Evans blue (T-1824), ^{51}Cr-labeled RBCs

E. Evans blue (T-1824), heavy water

30. Changes in response to congestive heart failure include

A. increased release of renin.

B. increased secretion of aldosterone.

C. decreased glomerular filtration rate.

D. increased permeability of collecting ducts to water.

E. all of the above.

MATCHING

A. *Hyponatremia*

B. *Hypernatremia*

C. *Both*

D. *Neither*

31. Lethargy is a symptom.

32. It can cause convulsions.

33. It can cause brain hemorrhages and infarcts.

34. It acts to stimulate release of antidiuretic hormone, but does not stimulate thirst.

MULTIPLE CHOICE

Select the correct answer(s). (In many instances, more than one answer is correct.)

35. The subject ingests a half-mole of salt (about 28 g NaCl), but has no access to drinking water for several hours. During this period

A. antidiuretic hormone secretion is inhibited.

B. plasma colloidal osmotic pressure (colloidal osmotic pressure, or oncotic pressure) decreases.

C. the osmolarity of the renal medullary interstitial fluid decreases.

D. the filtered load of Na$^+$ increases (at glomerulus).

ANSWERS

1. C. The normal daily turnover rate is 1.5 L/m^2, and since he has a body surface area of 2 m^2, this man has a turnover of 3 L/day.

2. C. The TBW of the average man is about 60%. Since 60% of 80 kg is 48 kg, or 48 L, a 3-L turnover is 6.25%, or $\frac{1}{16}$th of 48 L.

3. E. The fluids taken in as beverage and in food in an average diet added to the volume of secreted GI fluids gives a total of about 9 L/day, hence choice E is closest.

4. A. Daily water loss in feces is normally small, viz. 100 ml (0.1 L).

5. B. On an average diet, about 0.12 ml of water is produced as water of oxidation for each Calorie (kilocalorie) of food consumed.

6. D. The normal daily osmolar waste load for a 70-kg man is about 1,000 mOsm. To excrete 1,000 mOsm in urine at a concentration of 700 mOsm/L would take

a daily urine volume of 1.43 L, hence choice D is nearest.

7. C. 1,000 mOsm/day, or 1 Osm/day.

8. C. The actual daily insensible loss is quite variable, but 1 L/day is a good ballpark figure.

9. D. 1,000 mOsm at 285 mOsm/L = 3.5 L.

10. C. Daily losses would be 3.5 L urine + 1 L insensible + 0.1 L in stools + 0.1 L sweat = 4.7 L; however, do not forget to subtract 0.4 L gain from oxidative metabolism, that is, he would have to ingest 4.3 L each day, or about 18 glasses of water daily.

11. A. The most common cause of water depletion (primary dehydration) is via usual amounts of the daily obligatory water losses when there is an inability to obtain or drink water (e.g., desert victims, shipwrecked sailors, neglected infants, feeble elderly people, comatose patients, patients with a neurological defect in the thirst center). B. Less common causes of

water depletion are excessive losses by the obligatory routes: (1) excessive renal excretion of water (e.g., diabetes insipidus), (2) excessive loss of water from lungs (e.g., hyperventilation, a tracheotomy), and (3) excessive loss from skin (e.g., open treatment of burns, fever). C is incorrect. Excessive ADH would cause water retention, not depletion. D is incorrect. Excessive aldosterone favors salt and water retention. E is incorrect. Hypovolemia stimulates thirst, which should promote increased water intake.

12. E. All are correct. Although hypernatremia tends to produce depression of the sensorium with lethargy and apathy, there can also be increased irritability upon stimulation and restlessness. Muscle tremors can progress to tonic spasticity and convulsive seizures similar to those of epilepsy. Sudden (acute) increases in plasma sodium produce more profound neurological effects and carry a greater risk of death than is the case with more gradual (chronic) increases.

13. B. Hypovolemia. A is incorrect because although hypovolemia will stimulate both aldosterone and thirst, other things will not stimulate both. Choices C (hypernatremia) and D (hypervolemia) are incorrect because both indirectly inhibit aldosterone secretion. E is incorrect because hypokalemia directly inhibits aldosterone secretion by the adrenal cortex.

14. C. Losses or gain of pure water are shared between the ECF and ICF according to the ratio of solute in the ECF to solute in the ICF, normally $\frac{1}{3}$ and $\frac{2}{3}$.

15. B. Loss of isotonic GI fluid in diarrhea is a combined loss of solute and water. If such fluid losses were replaced with water alone or with sugared tea the net result would be a net loss of the solute alone (a salt depletion). (Sugared tea is essentially the same as H_2O alone, since sugar usually does not count much as a plasma solute; the glucose and fructose are largely taken up by the liver).

16. B. Stimulation of ADH secrection by severe hypovolemia has precedence over inhibition of ADH secretion by hypo-osmolarity.

17. B. With the exception of overhydration due to salt and water retention in a few disease states characterized by edema (e.g., congestive heart failure, cirrhosis with ascites, and the nephrotic syndrome), overhydration of the ECF is almost always iatrogenic in origin. Voluntary water intoxication is rare, occurring almost exclusively in psychotic compulsive water drinkers, and then usually only if something, such as drug administration, interferes with free-water excretion.

18. D. All of the above (congestive heart failure, cirrhosis with ascites, and nephrosis) are associated with an ineffective, hypodynamic circulation that triggers mechanisms for sodium and water retention (as illustrated in Fig. 16).

19. D. ICF volume is increased. Plasma Na^+ would be diluted; ECF volume would be smaller (the liter of water does not replace the liter of sweat because only one-third of the water remains in the ECF, with the other two-thirds going into the ICF). Sweat contains K^+ (at a slightly higher concentration than plasma), as well as Na^+ and Cl^-. When sweat loss is replaced with water only, there is a net loss of K^+, Na^+, and Cl^-.

20. All are correct. Subject W, who drank the water, has the greater change in plasma osmolarity because the ECF will be diluted by about one-third liter, as in the previous problem (compared with no dilution by the isotonic saline). Subject S, who drank the saline, has the greater increase in plasma volume. The entire liter of isotonic saline remains in the ECF, of which the plasma is one-quarter. Thus, one-fourth of 1 L = 250 ml increase in plasma volume. In the case of the liter of water (subject W), only one-third remains in the ECF, which means that only about 83 ml might be expected to remain in the plasma ($\frac{1}{4} \times \frac{1}{3} = 0.0825$). Actually, the volumes remaining in the plasma are even less than 250 ml for the saline, and 83 ml for water because of dilution of plasma colloid osmotic pressure, but the volume increase in plasma volume will be greater for the saline drinker than for the water drinker. Because subject W has diluted his ECF, he will have a decrease in ADH secretion and therefore put out an increased volume of dilute urine. The urine osmolarity of subject S will not change. Also, subject S will excrete the liter of isotonic saline over a period of about 24 h, while subject W will excrete his water within a couple of hours.

21. C. Salt depletion.

22. A. Mixed water and salt depletion.

23. B. Water excess.

24. D. Salt excess.

25. E. Mixed water and salt retention..

26. D. 155 mM NaCl is isotonic.

27. C. Albumin is the principal plasma protein contributing to plasma colloidal osmotic pressure.

28. B. At first, water moves out of the RBC, but then as urea diffuses into the RBC (due to a low reflection coefficient), the total osmotic pressure of the RBC rises; and water enters, until it swells and bursts.

29. C. ISF cannot be measured directly; it must be calculated by difference after measurement of ECF and plasma volumes. ISF = ECF − plasma. ECF is measured by thiocyanate (SCN⁻), and plasma volume is measured by Evans blue (T-1824).

30. E. All. (Review Fig. 16.) The hypodynamic circulation of congestive heart failure causes impaired perfusion of the kidneys, which in turn causes increased secretion of renin. This also results in increased production of angiotensin II, which stimulates aldosterone secretion. Increased sympathetic activity to renal vasculature decreases renal plasma flow and glomerular filtration. The hypodynamic circulation also increases ADH release from the posterior pituitary. The action of the ADH is to increase permeability of the collecting ducts to water (thereby increasing water reabsorption).

31. C. Both.

32. C. Both.

33. B. Focal brain hemorrhages occur with hypernatremia. These hemorrhages not only interrupt the supply of blood to neurons, but cause damage by pressure of hematomas on adjacent neural tissues, resulting in brain infarcts.

34. D. Neither. Most factors that stimulate release of ADH also stimulate thirst (e.g., hypovolemia, or hypertonicity of ECF).

35. B, D. An increase in the osmolarity of the ECF (e.g., due to NaCl ingestion) stimulates ADH secretion (i.e., ADH secretion is stimulated, not inhibited). The increase in ECF osmolarity also causes a shift of H_2O from the ICF into the ECF (to reestablish osmotic equilibrium). This influx of water into the ECF will cause a moderate dilution of the plasma proteins and thereby decrease the colloidal osmotic pressure. Although ADH secretion increases the permeability of the collecting ducts to water, the increased influx of water from the urine into the medullary ISF does not result in dilution (no decrease in medullary osmolarity) because (1) the volume reabsorbed from urine is small to begin with, and (2) this reabsorbed water is picked up by the vasa recta. The increased Na^+ concentration of the ECF causes the load of Na^+ filtered at the glomerulus to increase.

34

The Physiology of Acid-Base Balance

DONALD W. STUBBS

OBJECTIVES

After completion of this chapter, you should be able to:

1. Understand the relationship between the ion product of water (K_w) and the concentration of H^+ and OH^- in aqueous solutions.

   a. Describe the H^+ concentration in terms of either molarity or pH, and be able to convert from one to the other.

2. Define various substances as acids or bases in terms of whether they are proton donors or acceptors (Brönsted acids and bases).

3. Appreciate the difference between "weak" organic acids, which are only partly dissociated into H^+, and conjugate bases (e.g., carbonic acid) and "strong" mineral acids, which are completely dissociated (e.g., hydrochloric acid). One should also be familiar with the negative logarithm of K, namely the pK, as a measure of weakness or strength of an acid.

4. Appreciate the significance of the combination of a weak acid and its salt as a buffer, and comprehend the mechanism by which buffers act to minimize changes in H^+ concentration.

5. Relate the roles of HCO_3^- and CO_2 in the bicarbonate buffer system and distinguish between "closed" and "open" bicarbonate buffer systems.

   a. Calculate carbonic acid concentration from the Pa_{CO_2}.

   b. Use data on pH, pK, and the concentrations of HCO_3^- and H_2CO_3 in working problems by means of the Henderson-Hasselbalch equation.

6. Describe the role of proteins as buffers and recognize that some amino acid groups are more effective than others as buffers at blood pH. List the properties of hemoglobin that make it the single most effective nonbicarbonate buffer of blood.

7. Review the negative feedback system between arterial PCO_2 (Pa_{CO_2}) and alveolar ventilation ($\dot{V}_A$) and the way in which Pa_{CO_2} affects pH (how primary disturbances in respiratory function can result in dysfunctional changes in ECF pH).

8. Review the negative feedback system between ECF pH and $\dot{V}_A$ (how primary alterations due to nonrespiratory causes can be partly corrected by functional compensatory changes in $\dot{V}_A$).

9. Relate glomerular filtration of HCO_3^- and the secretion of H^+ to HCO_3^- reabsorption and excretion.

10. Understand the indirect mechanism of extant HCO_3^- reabsorption (i.e., conservation of that HCO_3^- that exists in the glomerular filtrate).

11. Describe renal regulation of ECF pH by generation of new HCO_3^- in the process of acidification of the urine

567

(i.e., the reactions involved in acidification of urine that result in the production of new HCO_3^- in addition to that reabsorbed from the glomerular filtrate).

12. Explain the role of urinary conjugate bases in H^+ excretion.

13. Compare maximal effectiveness of HPO_4^{2-} with NH_3 as participants in urinary acidification.

14. List causes and specific examples of respiratory acidosis.

15. Describe the effects of nonbicarbonate buffering on the relationships between PCO_2, HCO_3^-, and pH.

16. List causes and specific examples of respiratory alkalosis.

17. Recognize and use Davenport diagrams (HCO_3^- versus pH with PCO_2 isobars superimposed) in describing different states of acid–base imbalance.

18. List underlying mechanisms and specific examples of metabolic (nonrespiratory) acidosis.

19. List underlying mechanisms and specific examples of metabolic (nonrespiratory) alkalosis.

20. Explain the relationship between K^+ balance and H^+ balance.

21. Describe compensatory mechanisms and their relative effectiveness in correcting the pH back toward normal in the different conditions of acid–base imbalance.

The concentration of hydrogen ions in body fluids (and cells) is closely regulated by homeostatic mechanisms involving buffers, lungs, and kidneys. Despite the fact that H^+ concentration is so much less than that of other important ions (e.g., the concentration of sodium ions in the ECF is millions of times greater than that of the hydrogen ions), it must be carefully regulated within narrow limits. The human body can tolerate neither 60 nEq* more nor 24 nEq less than the 40 nEq normally present in 1 L of ECF. The hydrogen ion is highly reactive. The maintenance of its concentration within the above limits (100–16 nEq/L, or pH 7.0–7.8) is essential for the vital functioning of cellular enzymatic reactions. The rate of combination of enzymes with substrate molecules is dramatically altered by slight changes in hydrogen ion concentration.

H^+ CONCENTRATION AND THE pH SCALE

Water is partially ionized into H^+ and OH^- ions. Since only a very small fraction (0.000000182%) of water exists as ions, that is, about 1 in 550 million molecules, pure

*Nanoequivalent $= 10^{-9}$ equivalent, or 10^{-6} mEq.

water is a very poor electrolyte (i.e., does not readily conduct an electric current). The ionization of water is usually written as

$$H_2O \rightleftharpoons H^+ + OH^-$$

Water exists in a state of equilibrium between the rates of dissociation and association, described by the following equilibrium equation:

$$K_{eq} = \frac{[H^+][OH^-]}{[H_2O]}$$

where K_{eq} is a dissociation or equilibrium constant, and the brackets indicate concentrations (moles/L). Since the concentration of undissociated water is so large in proportion to the very small concentration of the ions, it can be treated as a constant itself and combined with K_{eq} to yield a new constant and a simpler equation:

$$K_w = H^+ \cdot OH^-$$

The new constant K_w is called the ion product of water and is 10^{-14} moles/L (at 25°C). Since the ion product is a constant, the product of H^+ concentration times OH^- concentration at 25°C must always be 10^{-14}. In neutral water the H^+ concentration is 10^{-7}, and the OH^- concentration is 10^{-7} ($10^{-7} \times 10^{-7} = 10^{-14}$). In an acid solution of 10^{-2} M H^+, the OH^- concentration must be 10^{-12} ($10^{-2} \times 10^{-12} = 10^{-14}$).

Table 1 Conversion for pH and H$^+$ Concentration

pH	Eq/L	mEq/L	nEq/L
0	1.0	1000	1 billion
1	0.1	100	100 million
2	0.01	10	10 million
3	1×10^{-3}	1	1 million
4	1×10^{-4}	0.1	100,000
5	1×10^{-5}	0.01	10,000
6	1×10^{-6}	1×10^{-3}	1,000
7	1×10^{-7}	1×10^{-4}	100
7.1	8×10^{-8}	8×10^{-5}	80
7.2	6.3×10^{-8}	6.3×10^{-5}	63
7.3	5×10^{-8}	5×10^{-5}	50
7.4	4×10^{-8}	4×10^{-5}	40
7.5	3.2×10^{-8}	3.2×10^{-5}	32
7.6	2.5×10^{-8}	2.5×10^{-5}	25
7.7	2×10^{-8}	2×10^{-5}	20
7.8	1.6×10^{-8}	1.6×10^{-5}	16
7.9	1.3×10^{-8}	1.3×10^{-5}	13
8	1×10^{-8}	1×10^{-5}	10
9	1×10^{-9}	1×10^{-6}	1
10	1×10^{-10}	1×10^{-7}	0.1
11	1×10^{-11}	1×10^{-8}	0.01
12	1×10^{-12}	1×10^{-9}	1×10^{-3}
13	1×10^{-13}	1×10^{-10}	1×10^{-4}
14	1×10^{-14}	1×10^{-11}	1×10^{-5}

The ion product of water is the basis for the pH scale, a convenient means of expressing a wide range of H$^+$ concentrations from 1.0 to 0.00000000000001 M. Just as it is more convenient to describe a H$^+$ concentration of 0.00000001 M as 10^{-8} M, it is even more convenient to make the exponent positive and call it a pH of 8. Thus 1.0 M (10^0) = pH 0; 0.1 M (10^{-1}) = pH 1; 0.01 M (10^{-2}) = pH 2. To calculate the pH from the H$^+$ concentration, take the negative logarithm of the H$^+$ molarity. For example, the logarithm of 10^{-8} is -8.0; the negative logarithm is $+8.0$, since $-(-8.0) = +8.0$. What is the pH of a solution with 0.0005 M H$^+$? Write 0.0005 as 5×10^{-4}. Take the logarithm of 5, which is 0.7, and add -4.0, which gives -3.3; the pH value is 3.3. Calculate the pH

of a solution with twice as many H$^+$. Note that the pH decreases only 0.3 unit for a doubling of the H$^+$ concentration. A change in pH of 1.0 indicates a 10-fold change in H$^+$. Keep in mind the two most important aspects of pH and you will have no difficulty with it: (1) pH is a logarithmic scale, and (2) as the H$^+$ increases, the pH decreases and vice versa.

There is a trend among some physicians to attempt to deal with H$^+$ concentrations directly, instead of as pH. To make a concentration of 0.00000004 moles/L more convenient to handle, the decimal is shifted to the right nine places to give nanomoles. Since for univalent ions such as H$^+$ and OH$^-$, moles = equivalents, the terms equivalent, milliequivalent, and nanoequivalent are often used in place of moles, millimoles, and nanomoles in acid–base studies. Thus, 0.00000004 moles/L (4×10^{-8}) would be expressed as (40 nEq/L). Table 1 illustrates the pH scale in relationship to H$^+$ concentrations as equivalents, milliequivalents, and nanoequivalents per liter. It can be seen that nanoequivalents are particularly convenient in the pH range of 7.0–8.0.

ACIDS AND BASES

The condition of an aqueous solution is acidic when the concentration of H$^+$ is greater than that of OH$^-$. It is basic when the concentration of H$^+$ is less than the concentration of OH$^-$. A solution is neutral when the two are equal.

J. N. Brönsted originated the definitions of acids and bases, which govern contemporary biochemical usage. An acid is defined as a proton donor, and a base is a proton acceptor. An acid, in the process of giving up a proton, becomes a substance called the conjugate base:

$$\underset{\text{acid}}{HA} \quad \rightleftharpoons \quad \underset{\text{proton}}{H^+} \quad + \quad \underset{\text{conjugate base}}{A^-}$$

The more readily an acid gives up its H$^+$, the stronger it is as an acid, and is said to have a relatively weak conjugate base. The more avidly a base combines with H$^+$ and binds it, the stronger it is as a base and the weaker the acid. Examples include the following:

$$HCl \rightleftharpoons \quad H^+ + Cl^-$$
$$H_2SO_4 \rightleftharpoons \quad H^+ + HSO_4^-$$
$$NH_4^+ \rightleftharpoons \quad H^+ + NH_3$$
$$CH_3COOH \rightleftharpoons H^+ + CH_3COO^-$$
$$H_2CO_3 \rightleftharpoons \quad H^+ + HCO_3^-$$
$$H_2O \rightleftharpoons \quad H^+ + OH^-$$

acid   conjugate base

"Mineral" acids, such as HCl and H_2SO_4, are extremely strong acids (i.e., have very weak conjugate bases) and act as if they were completely dissociated. Organic acids, such as acetic or carbonic acids, are weak acids (i.e., have relatively strong conjugate bases), because they are only partially dissociated. Hydroxides of alkali metals (e.g., NaOH, KOH) are very strong bases, because their OH^- strongly combine with H^+ to form H_2O.

The degree of dissociation of a weak acid, and thus the relative degree of its strength, is described by its dissociation constant K:

$$K = \frac{[H^+]\,[A^-]}{[HA]}$$

Among the different weak acids, those that are relatively stronger have a smaller proportion of the total as undissociated HA and therefore have larger K values than do the relatively weaker ones. Since our interests are most often centered on H^+, a more convenient form of the preceding equation is

$$[H^+] = K\,\frac{[HA]}{[A^-]}$$

Just as a wide range of H^+ concentrations can be handled as negative logarithms (pH), so too can a large variety of K values be expressed as negative logarithms, in this case called pK values. Each acid has its own pK, which provides a convenient measure of that acid's strength: a relatively strong acid has a low pK and a relatively weak one has a high pK. For example, trichloracetic acid (TCA), which is fairly strong for an organic acid, has a pK value of only 0.7. Compare this with trichlorophenol, which is a weak acid with a pK of 6.0. To take the above equation and put it into the form of pH and pK, take the negative logarithms of both sides of the equation

$$-\log [H^+] = -\log K - \log \frac{[HA]}{[A^-]}$$

which is the same as

$$pH = pK + \log \frac{[A^-]}{[HA]}$$

This is the *Henderson-Hasselbalch equation.* It states that the pH of a solution is determined by the strength of dissociation of the acid (represented by its pK) and by the ratio of conjugate base to acid in the solution. For any given acid, the equation has three variables (pH, concentration of conjugate base, and concentration of acid) and one constant (the pK for that acid). If any two of the variables are known, the third can be calculated. For example, HCO_3^- concentration is almost always calculated in this way. The Henderson-Hasselbalch equation is of fundamental importance in understanding body fluid pH, buffer action, and the quantitative assessment of the body's acid–base balance.

BUFFERS

A buffer is a substance (or substances) that tends to diminish the change in pH otherwise produced by the addition of either an exogenous acid or an exogenous base to a solution. A buffer solution commonly consists of a weak acid and a salt of that acid. The salt provides an abundant source of conjugate base. Thus, a buffer is simply a conjugate base/acid pair. The weak acid member acts as a reservoir of H^+ (i.e., is mostly associated, but releases more H^+ if any are used up in neutralization of base). The conjugate base provided by a salt of the acid can react with exogenous H^+ to form the weak acid, thereby taking most of the H^+ out of circulation. Figure 1 illustrates the mechanism of buffer action.

TITRATION CURVES

Suppose one titrates 10 mEq of acetic acid (e.g., 100 ml of 0.1 N acetic acid) with 10 mEq of NaOH and plots the pH against the amount of base added. The result will be a titration curve (Fig. 2). Before any base is added, the pH is that of the weak acid alone (all in the form of acetic acid,

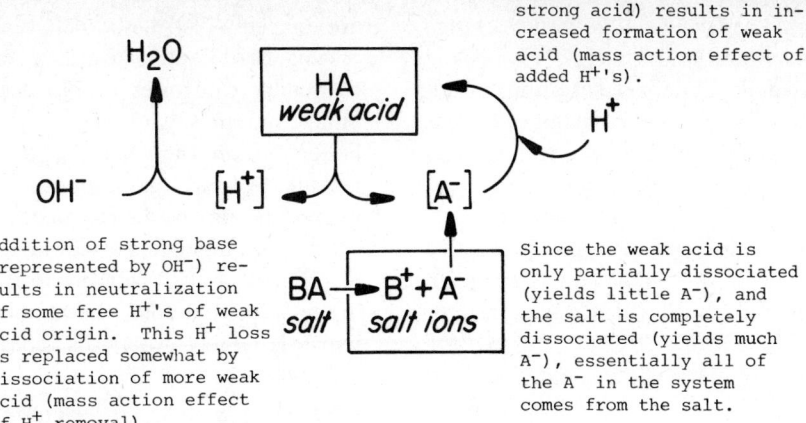

Addition of H^+'s (e.g. from strong acid) results in increased formation of weak acid (mass action effect of added H^+'s).

Addition of strong base (represented by OH^-) results in neutralization of some free H^+'s of weak acid origin. This H^+ loss is replaced somewhat by dissociation of more weak acid (mass action effect of H^+ removal).

Since the weak acid is only partially dissociated (yields little A^-), and the salt is completely dissociated (yields much A^-), essentially all of the A^- in the system comes from the salt.

Figure 1 Mechanism of buffer action.

CH_3COOH). As soon as some NaOH is added it reacts with an equivalent amount of acetic acid to form the salt, sodium acetate ($CH_3COO^-Na^+$) and water. Note that the presence of both acetic acid (a weak acid) and sodium acetate (a salt of the weak acid) constitutes a buffer pair. When 5 mEq of NaOH has been added, the ratio of acetic acid to acetate will be 1:1 (50% acetic acid and 50% acetate), and the pH will be numerically equal to the pK,

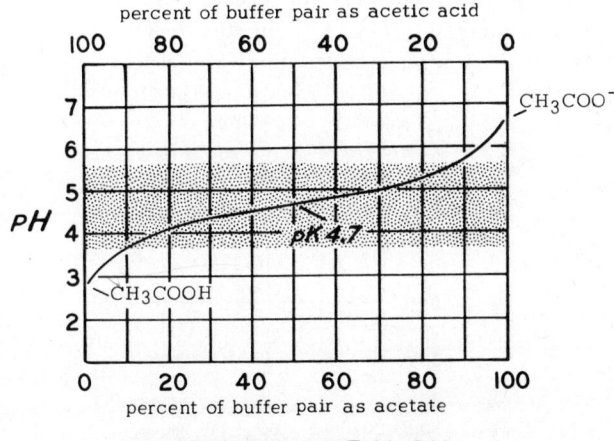

Figure 2 Titration of acetic acid with strong base. The curve describing the change in pH with addition of base is a titration curve. The shaded area one pH unit above and below the pK is the area of effective buffering.

that is, 4.7. After all the base has been added, all of the acetic acid will have been neutralized to acetate, and the pH will be near neutrality. The titration curve, which breaks sharply near both ends (more rapid change in pH), is relatively flat in the middle (slower change in pH). This curve is described mathematically by the Henderson-Hasselbalch equation for buffers as

$$pH = pK + \log \frac{[salt]}{[acid]}$$

or more specifically for acetic acid, which has a pK of 4.7:

$$pH = 4.7 + \log \frac{[acetate]}{[acetic\ acid]}$$

Note that the change in pH per amount of base added is least in the center of the titration curve (flat part of curve). The exact center of the flat part of the curve is the point at which the ratio of acetate to acetic acid is 1:1 (ratio of 1). Since the log of ratio 1 is 0, pH at this point is numerically equal to pK (pH = 4.7 + 0 = 4.7). Maximal buffering ability for most buffers occurs when pH = pK. This is not only true for the addition of base, but is equally true for addition of acid. If one had started with Na-acetate and titrated with a strong acid, the same titration curve would have been plotted (but from right to left). Thus, as a general rule, buffers are most effective for buffering either acid or base when they are present at a pH numerically near their pK value (shaded area of Fig. 2).

Although the best buffering is at a pH exactly equal to pK, there is still adequate buffering ability within a range of about one pH unit above and below the pK. That is, the effective buffering range depends on concentrations of salt and acid being within the same order of magnitude (less than ten times as much of one as the other). In picking a buffer pair to keep a solution near some specific pH, you need to choose a pair with a pK near the pH desired.

An exception to the rule that buffering is only effective at pH values near the pK of the buffer pair is found in so-called "volatile" buffer systems in which one of the pair is a gas or is in equilibrium with a gas, for example, the bicarbonate buffer system of extracellular fluid (ECF), in which the acid member H_2CO_3 is in equilibrium with blood CO_2. The HCO_3^-/H_2CO_3 buffer system used in regulating ECF pH represents vital exploitation of a useful volatile buffer system in maintaining the ECF pH near 7.4.

BLOOD BUFFERS

Because of the presence of buffers in blood, the addition of strong acid in sublethal amounts, for example, results in an increase of H^+ of as little as 1/100,000 or so of what would have occurred in the absence of buffering.

Blood is a complex fluid containing at least five different buffers (Fig. 3). Although several different substances con-

form to the definition of buffer pairs, and although these are unevenly distributed between the red blood cells and plasma, two simplifications are convenient. First, there are two main categories of blood buffers: the bicarbonate buffer system (HCO_3^-/H_2CO_3) and the nonbicarbonate buffer system (hemoglobin, plasma proteins, inorganic phosphate, and organic phosphate). Second, the red cell membrane may be disregarded, that is, blood may be considered as a homogeneous fluid, since both members of the bicarbonate buffer pair penetrate the erythrocyte membrane. In this chapter the nonbicarbonate buffer system, which is largely protein, is represented as $Buf^-/HBuf$.

THE BICARBONATE BUFFER SYSTEM

On the basis of characteristics of typical, nonvolatile buffer pairs, such as the acetate/acetic acid buffer pair illustrated in Figure 2, one might predict that the ratio of HCO_3^-/H_2CO_3 normally present in ECF at 20:1 would not be favorable for effective buffering. As can be seen in Figure 4, the pH of ECF appears to be just outside the effective buffering range of one pH unit on either side of the pK (pH 5.1–7.1) of the bicarbonate buffer pair. However, the bicarbonate buffer pair is not a typical, nonvolatile buffer system; the bicarbonate buffer pair is an example of that special type of "volatile" buffer system in which one member (H_2CO_3) is in equilibrium with a gas (CO_2) in what is often called an "open system." A great

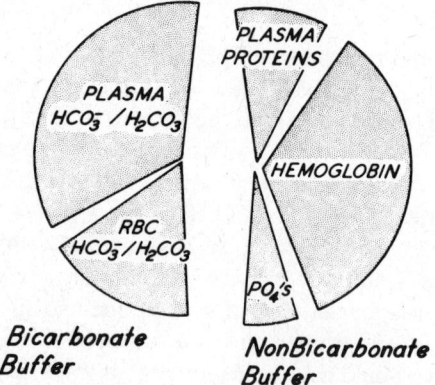

Figure 3 Blood buffers. The two principal categories of blood buffers—bicarbonate and nonbicarbonate—and their approximate proportions.

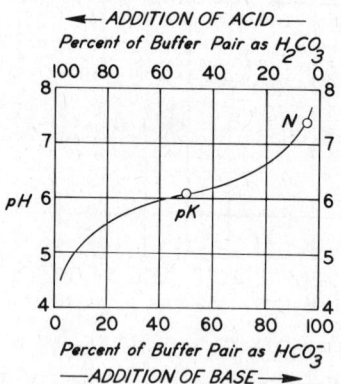

Figure 4 Titration curve for the bicarbonate buffer system. The pK is 6.1. N indicates the normal pH of the blood (for a HCO_3^-/H_2CO_3 ratio of 20:1).

advantage of such a system is that as long as the H_2CO_3 remains in equilibrium with the ECF CO_2 (normally held at a fairly constant level by the respiratory system), the H_2CO_3 member of the buffer pair will remain at a constant concentration despite addition of considerable acid or base. Furthermore, from a regulatory or compensatory point of view, having one of the members of the buffer pair at a much lower concentration than the other permits compensatory mechanisms, which act to correct conditions of acid–base imbalance, to effect relatively large changes in the buffer pair ratio (and therefore to effect change in pH) by fairly small changes in the absolute concentration of the member present in low concentration (i.e., H_2CO_3). For example, in many cases of metabolic acidosis (an acidosis of nonrespiratory origin), the lungs can achieve partial correction of the pH by altering the concentration of H_2CO_3 (a process called respiratory compensation). In time, the HCO_3^- component of the bicarbonate buffer pair can also be adjusted to correct the pH (renal compensation controls HCO_3^- concentration). The ability of the body's compensatory mechanisms to alter both components of the bicarbonate buffer pair is another reason why this is such an important buffer. Finally, the bicarbonate buffer is important in diagnosing conditions of acid–base imbalance, because it is the only system for which quantitative data are available for both members of a buffer pair.

Bicarbonate Buffer Action in an Open versus a Closed System

It is instructive to compare the effects of 10 mEq of strong acid as to (1) 1 L of unbuffered pure water; (2) 1 L of buffered solution with the same concentrations of HCO_3^-/H_2CO_3 as in plasma, but in a closed system (not in equilibrium with a constant partial pressure of CO_2); and (3) 1 L of buffered solution with the same concentrations of HCO_3^-/H_2CO_3 found in plasma, but in an open system (H_2CO_3 in equilibrium with a constant partial pressure of CO_2).

Effect of 10 mEq of Strong Acid on 1 L of Unbuffered Pure Water

Consider first the change in pH and H^+ concentration of 1 L of unbuffered water after addition of 10 mEq in a small volume, so that the total volume is little changed (i.e., after

addition of 10 ml of 1.0 N HCl). Before the acid is added, the liter of pure water has a pH of 7.00 (100 nEq/L). After the addition of 10 mEq of HCl, the pH drops to about 2.00 (10,000,000 nEq/L), or about a 100,000-fold increase in H^+ concentration (a difference of 5 pH units represents a 10^5 or 100,000-fold change in H^+).

Effect of 10 mEq of Strong Acid on 1 L of a Solution Containing 24 mEq of HCO_3^- and 1.2 mEq of H_2CO_3 in a Closed System

Not exposed to a CO_2 gas phase, any H_2CO_3 formed during buffering remains in the system as such, since it cannot escape as CO_2 gas (Fig. 5). Knowing that the pK of H_2CO_3 is 6.1, one can calculate, from the Henderson-Hasselbalch equation, that an initial HCO_3^-/H_2CO_3 ratio of 20:1 will give a pH of 7.40. Thus, before the addition of 10 mEq of HCl, the solution (at pH 7.40) has a H^+ concentration of 40 nEq/L. After the addition of the acid, the 10 mEq of HCl will react with 10 mEq of HCO_3^-, leaving a remainder of only 14 mEq of HCO_3^-/L, and will have formed an additional 10 mEq of H_2CO_3 in the process, giving a total of 11.2 mEq of H_2CO_3/L. This gives a new ratio of HCO_3^-/H_2CO_3 of 14/11.2, or 1.25. Taking the log of the new ratio (log of 1.25 is 0.10) and plugging it into the Henderson-Hasselbalch equation permits calculation of the new pH, which is 6.20 (631 nEq/L):

$$
\begin{aligned}
pH &= 6.10 + \log 1.25 \\
&= 6.10 + 0.10 \\
&= 6.20
\end{aligned}
$$

An increase in H^+ from 40 to 631 nEq/L is only about a 16-fold increase. This is much smaller change than that

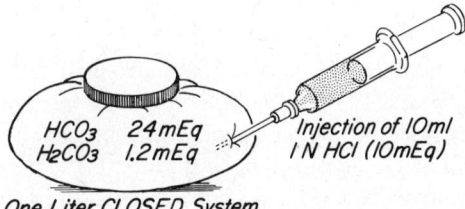

Figure 5 Closed system. Addition of 10 mEq of strong acid to a bicarbonate buffer pair in a closed system (not exposed to a gas phase; any additional H_2CO_3 formed as a result of reaction of HCO_3^- with HCl cannot escape from system as CO_2).

which occurs in unbuffered water, but this closed system would not be sufficient for adequate buffering in vivo.

Effect of 10 mEq of Strong Acid on 1 L of a Solution Containing 24 mEq of HCO_3^- and 1.2 mEq of H_2CO_3 in an Open System (H_2CO_3 is held constant by equilibrium with a constant partial pressure of CO_2).

Consider the change in pH and H^+ following addition of 10 mEq of HCl to a 1 L solution containing the same initial concentrations of HCO_3^- and H_2CO_3 as in the previous example, but in an open system maintained at a P_{CO_2} of 40 mm Hg. As in the previous solution, the initial pH is 7.40 (40 nEq/L), since the HCO_3^-/H_2CO_3 ratio is 20. After the addition of 10 mEq of HCl, 10 mEq of HCO_3^- will have been neutralized, just as before (leaving 14 mEq of HCO_3^-/L), but the additional 10 mEq of H_2CO_3 formed as a result of this reaction will be lost from the system. Since this is an open system in which the H_2CO_3 concentration is determined by the P_{CO_2} (which is held constant), any additional H_2CO_3 formed by neutralization reactions will be in excess of the equilibrium. This excess H_2CO_3 must change to CO_2 bubbles and be washed out of, or "blown off" from, the system. Thus while the HCO_3^- concentration becomes 14 mEq/L as before, the H_2CO_3 must remain 1.2 mEq/L (as determined by the P_{CO_2} of 40 mm Hg). In this case, the ratio becomes 14/1.2 (instead of 14/11.2, as in the closed system); 14/1.2 is a ratio of 11.67, the log of which is 1.07. Plugging this into the Henderson-Hasselbalch equation gives the new pH of 7.17 (68 nEq/L):

$$\begin{aligned} pH &= 6.10 + \log 11.67 \\ &= 6.10 + 1.07 \\ &= 7.17 \end{aligned}$$

The elevation of H^+ concentration from 40 to 68 nEq/L is only about a 1.5-fold increase. It should be appreciated that because the denominator does not change in the HCO_3^-/H_2CO_3 ratio in an open system, this system can be very effective. This open system is illustrated in Figure 6.

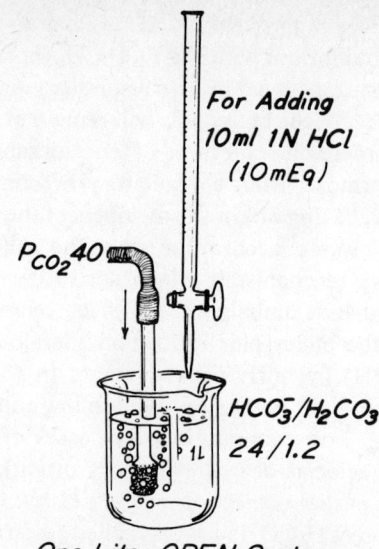

One Liter OPEN System

Figure 6 Open system. Addition of 10 mEq of strong acid to a bicarbonate buffer in an open system. The open system is held to a constant P_{CO_2} (by supply of CO_2 at a given P_{CO_2} of 40 mm Hg). An equilibrium exists between CO_2 and H_2CO_3 ($CO_2 + H_2O \rightleftharpoons H_2CO_3$) such that production of additional H_2CO_3 in buffering results in extra CO_2 formation. This extra CO_2 is washed out or "blown off" from the system. Thus while addition of acid decreases HCO_3^- concentration, the H_2CO_3 concentration remains the same (i.e., is held at 1.2 mEq/L at a P_{CO_2} of 40 mmHg).

NONBICARBONATE BUFFER SYSTEM

The principal constituents of the nonbicarbonate buffer system of blood are blood proteins (plasma proteins to some extent, but primarily hemoglobin of red blood cells). Proteins are not only important in buffering the ECF, but also the ICF (proteins are probably the predominant cellular buffers). In general, proteins are effective buffers because they contain numerous ionic groups that interact reversibly with H^+. The groups responsible for buffering action over a wide range of pH in vitro include the terminal α-amino group at one end of the polypeptide chain and the terminal α-carboxyl group at the other, plus many side groups of internal amino acids. These groups are able to donate or accept protons and are therefore, by definition, acids or bases, respectively.

Within the relatively narrow range of pH in blood (pH 7.0–7.8), many of the different groups are either nearly completely dissociated and unable to accept protons to form weak acids (groups with pK values around 3–4; e.g., the terminal α-carboxyl group), or are highly associated and not willing to donate H^+ (pK of 9–12; e.g., the ϵ-amino group of lysine). The α-amino groups of N-terminal amino acids have a suitable pK for blood buffering (pK of 7.6–8.4) and can reversibly take up and donate protons, but since there is usually only one N-terminal amino acid per protein molecule, they are not very abundant. The only group that is both abundant and of a reasonable pK for blood buffering is the imidazolium group of histidine (pK 5.6–7.0).

Blood proteins, like most proteins, have a net negative charge at ECF pH (a pH above the isoelectric point of most proteins), and can be expressed as Buf^-. Dissociated groups capable of accepting H^+ constitute the conjugate bases that are, in part, responsible for the negative charge of Buf^-. Acceptance of H^+ to form weak acids reduces the overall anionic or negative charge of the protein and, for the sake of convenience, the weak acid form is expressed as HBuf (written as being without charge). The buffer pairs of blood proteins are often represented in this fashion: Buf^-/HBuf (without necessarily identifying the specific ionic groups involved). Hemoglobin, the most abundant protein of blood, not only shares the buffering ability of proteins in general, but is a particularly effective blood buffer in vivo for two reasons: (1) it occurs in high concentration (about 15 g/100 ml blood), and (2) it has a prominent histidine content (about 8% of the amino acids in hemoglobin).

The nonbicarbonate buffer system (Buf^-/HBuf) is of primary importance in dealing with abrupt changes in CO_2 (and H_2CO_3) in acute pulmonary dysfunctions. For example, sudden hypercapnia (increased P_{CO_2}, also called hypercarbia) can occur abruptly in disorders inhibiting pulmonary function. The only immediate defense against this respiratory acidosis is that provided by the nonbicarbonate buffer system. Naturally, one would not expect the bicarbonate buffer system to buffer changes in itself; it cannot buffer against changes in H_2CO_3 due to excess CO_2 retention in respiratory acidosis. Since renal compensatory mechanisms take some time, the nonbicarbonate buffer system makes vital contributions during transitional periods until the kidneys can establish effective control of the blood pH.

RESPIRATORY CONTROL OF pH

Normally the body is in a steady state with respect to CO_2 production and elimination. The rate of CO_2 loss is regulated by the respiratory system on the basis of a negative feedback system (review Chapter 25). The most important factor in regulating the arterial P_{CO_2} is the partial pressure of CO_2 in arterial blood perfusing the central medullary chemoreceptors. Alterations in P_{CO_2} result in reflex changes in pulmonary ventilation so as to counter or reverse these alterations.

The peripheral chemoreceptors (e.g., the carotid body) are sensitive to the concentration of H^+ in arterial blood. Thus changes in arterial pH can also affect ventilation and thereby affect the P_{CO_2} of blood (review the role of peripheral chemoreceptors in Chapter 25).

Hypoventilation can be roughly defined as ventilation initially insufficient to remove CO_2 at the same rate as it is formed and is therefore associated with CO_2 retention and increased H_2CO_3 (since the H_2CO_3 concentration is determined by the P_{CO_2} according to the equation

$$\text{mEq } H_2CO_3 \text{ per liter} = 0.03 \times P_{CO_2} \text{ (in mm Hg)}$$

Hyperventilation can be defined as ventilation initially in excess of that necessary for elimination of CO_2 at the same rate as it is formed and is therefore associated with CO_2 loss and decreased H_2CO_3.

Changes in respiratory ventilation affect the HCO_3^-/H_2CO_3 ratio and therefore affect the pH of the ECF by altering the concentration of the H_2CO_3 component of the bicarbonate buffer pair. Thus, primary disturbances in respiratory function can result in dysfunctional changes in ECF pH (respiratory acidosis and respiratory alkalosis). On the other hand, alterations in pH due to nonrespiratory causes (metabolic acidosis and metabolic alkalosis) can be partially corrected by functional compensatory changes in ventilation.

RENAL CONTROL OF pH

The body is in a steady state with respect to production of nonvolatile ("fixed") acid and the excretion of H^+ by the kidneys. Each day the body produces about 50–100 mEq of nonvolatile acid (mainly from protein metabolism). For example, consider the production of sulfuric acid from methionine oxidation:

$$2 \text{ methionine} + 15O_2 \rightleftharpoons$$
$$2H_2SO_4 + \text{Urea} + 7H_2O + 9CO_2$$

In order to maintain acid–base balance, the body must excrete an equal amount of acidic equivalents daily. The excretion of H^+ from nonvolatile acid is somewhat indirect. Nonvolatile acid (e.g., sulfuric acid from methionine and cysteine metabolism) is considered to be immediately buffered by HCO_3^-, thereby resulting in a loss of HCO_3^- and a gain of an equal amount of H_2CO_3. Since the bicarbonate buffering system of the body is an open system, CO_2 formed from the additional H_2CO_3 produced by this buffering is rapidly disposed of by the lungs. Shortly, the only remaining result of the buffering is a diminished HCO_3^- and, of course, a slightly decreased pH.

$$H_2SO_4 + 2HCO_3^- \rightleftharpoons 2\ H_2CO_3 + SO_4^{2-}$$
$$H_2CO_3 \rightleftharpoons H_2O + CO_2\uparrow$$

Since the original H^+ from the nonvolatile acid are no longer in existence as free H^+, they cannot very well circulate to the kidneys for excretion. The kidneys can, however, restore the normal pH of the ECF by replacing the HCO_3^- lost in buffering. To do this, new HCO_3^- is generated by renal tubule cells from CO_2 and H_2O by way of H_2CO_3 formation. The hydration of CO_2 is catalyzed by the enzyme, carbonic anhydrase (CA). For each HCO_3^- formed in this manner, a H^+ is also formed. The kidneys dispose of these new H^+ by acidification of the urine (mainly distal tubular secretion of H^+) and retain the HCO_3^- for the ECF. The new H^+ secreted are equivalent in amount to those originally present in the nonvolatile acid. Since the new H^+ excreted in the process of HCO_3^- generation are equivalent to those originally buffered, the H^+ from the nonvolatile acids may be considered to be excreted indirectly.

The process of renal tubular regeneration of HCO_3^- is dependent on the hydration of CO_2 within tubular cells:

$$CO_2 + H_2O \underset{CA}{\rightleftharpoons} H_2CO_3 \rightleftharpoons H^+ + HCO_3^-$$

Because the equilibrium is far to the left, in favor of CO_2 + H_2O, appreciable generation of HCO_3^- by this means depends on H^+ secretion out of the tubular cells into the tubular lumen in order to dispose of one of the products and "pull" the reaction toward the right, in favor of HCO_3^- production. This is why the HCO_3^- generating mechanism is dependent on H^+ secretion and on the Na^+–H^+ exchange. This exchange is normally responsible for about one-fifth the total Na^+ reabsorption by the renal tubular cells. It will be seen in the next section that this same system is also involved in reabsorption of extant HCO_3^-, the HCO_3^- that already exists in the glomerular filtrate of plasma. Since the luminal border of the tubular cells is relatively impermeable to HCO_3^- per se, HCO_3^- must be handled indirectly.

BICARBONATE EXCRETION AND REABSORPTION

The amount of HCO_3^- excreted or reabsorbed depends on two factors: (1) the rate of HCO_3^- filtration (from plasma into glomerular filtrate), and (2) the rate of H^+ secretion by tubular cells. Since the kidneys filter more than 4,000 mEq of HCO_3^- each day, the quantitative significance of the filtration rate should be apparent. What might not be so obvious is why HCO_3^- excretion or reabsorption is also in part dependent on the rate of H^+ secretion into the urine. The reason for this dependence has to do with the operation of the HCO_3^--generating mechanism. As described above, in order to produce HCO_3^- from H_2CO_3 in tubular cells, H^+ must also be formed (Fig. 7). The HCO_3^- generated is used in the indirect reabsorption of HCO_3^- (to be described below) and in the production of new HCO_3^- to satisfy the buffering requirements of the ECF (e.g., to replace HCO_3^- lost in neutralization reactions). In order for this HCO_3^--generating mechanism to continue functioning, the H^+ formed must be secreted (and excreted):

$$HCO_3^-\ \text{excreted} =$$
$$HCO_3^-\ \text{filtered} - H^+\ \text{secreted} \quad (1)$$

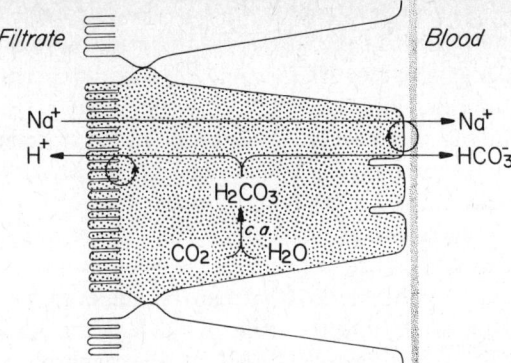

Filtrate

Blood

Na^+ → → Na^+

H^+ ← ← HCO_3^-

H_2CO_3

CO_2 H_2O *c.a.*

Figure 7 Secretion of H^+ into glomerular filtrate in exchange for Na^+. The source of H^+ is intracellular H_2CO_3 produced by hydration of CO_2 (catalyzed by carbonic anhydrase, CA). For each H^+ produced, a HCO_3^- is also formed. The H^+ is secreted into the glomerular filtrate in exchange for a Na^+, which enters the blood along with the HCO_3^- produced.

If the glomerular filtration rate (GFR, or total volume of glomerular filtrate formed per minute) is reasonably constant, the rate of HCO_3^- filtration is directly proportional to the plasma concentration of HCO_3^-. The rate of H^+ secretion in Eq. (1) is determined by a number of factors, including the acid–base status of the body.

1. If the rate of HCO_3^- filtration is greater than the rate of H^+ secretion, HCO_3^- will be excreted. For example, if the filtration of HCO_3^- is high because of a high concentration of HCO_3^- in plasma (as in alkalosis due to HCO_3^- excess), and if the rate of H^+ secretion is less than this (which it would be in alkalosis), then HCO_3^- will be excreted in the urine. This comes about because HCO_3^- reabsorption depends on H^+ secretion. If more HCO_3^- is presented to tubules (via filtration) than can be reabsorbed by the mechanism involving H^+ secretion, the excess HCO_3^- remains in the filtrate and is excreted in urine. This is how the body rids itself of excess bicarbonate.

2. If the rate of HCO_3^- filtration exactly equals the rate of H^+ secretion (rarely the case), then the HCO_3^- excreted in Eq. (1) is zero. That is, in this case there is exactly enough H^+ secretion to provide for reabsorption of all the HCO_3^- presented to the tubules via filtration. In this circumstance all of the HCO_3^- filtered is reabsorbed and none is excreted in the urine.

3. If the rate of HCO_3^- filtration is slightly less than the rate of H^+ secretion (normally the case), then HCO_3^- excreted in Eq. (1) is "negative". That is, this negative excretion is the same thing as net gain of HCO_3^- to the body (not only complete reabsorption of all extant HCO_3^- filtered, but also generation of new additional HCO_3^- by tubular cells). The normal rate of HCO_3^- filtration is about 3.00 mEq/min, and the normal rate of H^+ secretion is about 0.05 more mEq/min (i.e., the H^+ secretion rate is about 3.05 mEq/min). This slight excess of H^+ secretion above the level necessary for participation in reabsorption of filtered HCO_3^- provides for generation of enough new HCO_3^- to exactly replace that lost in buffering nonvolatile acid (e.g., from protein catabolism).

With HCO_3^- present in the glomerular filtrate, H^+ secretion results in the formation of luminal H_2CO_3, which immediately breaks down into H_2O and CO_2 (catalyzed by carbonic anhydrase present in the brush border of tubular cells). The increased H_2O and CO_2 become part of the general local pool of tubular H_2O and CO_2 and can easily diffuse into tubule cells. In the process of forming the H^+ that was secreted for combination with the luminal HCO_3^- in the glomerular filtrate, an HCO_3^- was also generated in the tubule cell. It is this latter HCO_3^-, which diffuses into the peritubular capillary blood (along with a Na^+). That is, the HCO_3^- entering the blood is not necessarily the same HCO_3^- molecule that was present in the glomerular filtrate; reabsorption of bicarbonate is indirect. The bulk of bicarbonate reabsorption (about 90%) is a function of the proximal tubule.

The secretion of H^+ in the presence of luminal HCO_3^- does *not* have an appreciable effect on urinary pH since the H_2CO_3 that forms rapidly breaks down into CO_2 and H_2O. Thus the H^+ that was secreted into the lumen in the presence of HCO_3^- is not excreted in the urine (it has been incorporated into H_2O and reabsorbed). The point is that *any secreted H^+ that combines with luminal HCO_3^- effects indirect HCO_3^- reabsorption, but does not contribute to urinary excretion of acid.* In other words, the mechanism of indirect HCO_3^- reabsorption is an isohydric cycle in which the body undergoes neither net gain nor loss of H^+. Thus, while most of the HCO_3^- that exists in the glomerular filtrate is indirectly reabsorbed, there is

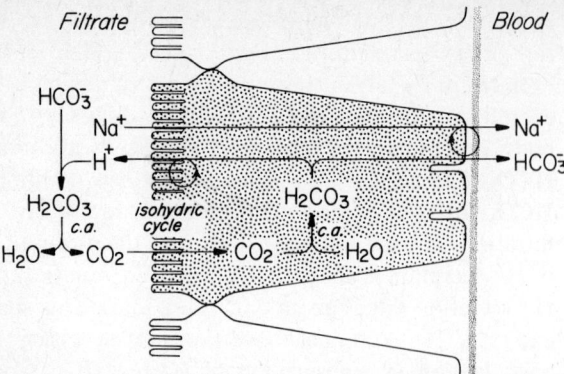

Figure 8 Reabsorption of HCO_3^- is an indirect process dependent on H^+ secretion. Luminal H^+ and HCO_3^- combine to form H_2CO_3, which breaks down into H_2O and CO_2. The H_2O and CO_2 can reenter cells for further formation of cellular HCO_3^- (and further H^+ production). Since this results in no net loss of H^+, it is an isohydric cycle. What is accomplished is not excretion of H^+ (by acidification of the urine), but indirect reabsorption of extant HCO_3^-.

little acidification of the urine by the proximal tubule (Fig. 8).

ACIDIFICATION OF THE URINE

As pointed out previously, the kidneys can prevent imminent development of a state of acidosis despite the body's daily production of nonvolatile acid (we are always poised on the verge of potential acidosis). The kidneys do this by the generation of new HCO_3^- in the process of urinary acidification.

Significant acidification of the urine is principally a function of the distal nephron (distal convoluted tubule and collecting duct). By the time the urine passes into the distal convoluted tubule, most of the isohydric reabsorption of HCO_3^- described in the previous section has already occurred. It is in the distal nephron that a net gain of new HCO_3^- can be generated in the process of urinary acidification. The gain of new HCO_3^- in the process of urinary acidification replaces that lost in buffering nonvolatile acid. In other words, the proximal tubule merely achieves bicarbonate reabsorption (conservation of HCO_3^- already existent in glomerular filtrate), while the distal nephron generates new additional HCO_3^- (in the process of urinary acidification). Urinary acidification by the distal nephron does not require nearly as many H^+ secreted

as does bicarbonate reabsorption in the proximal tubule, but does result in the loss of acidic products (via urine) and net gain of new HCO_3^- (into the blood of peritubular capillaries.)

Generation of new HCO_3^- depends on H^+ secretion and Na^+-H^+ exchange. As described previously, for each HCO_3^- produced, a H^+ is also formed. The H^+ is secreted into the filtrate in exchange for a Na^+. The hydrogen ion gradient (difference between H^+ concentration in urine and that in the blood) that can be achieved by distal secretion of H^+ into the urine is much greater than that possible in the proximal part of the nephron. For a comparison of proximal and distal parts of tubule, see Table 2.

In thinking about the rate of H^+ secretion before the limiting gradient prevents further acidification, it is useful to consider the contents of the glomerular filtrate: the chief cation is Na^+, and the principal anions are HCO_3^-, HPO_4^{2-}, and Cl^-. In other words, glomerular filtrate contains $NaHCO_3$, Na_2HPO_4, and $NaCl$. Since H^+ is exchanged for some of the Na^+, it is instructive to consider the nature of the acids (weak or otherwise) produced from these salts in the filtrate by this exchange. Consider the properties of HCO_3^-, HPO_4^{2-} and Cl^- as conjugate bases. A fourth conjugate base, NH_3, is secreted into the urine by tubular cells.

1. HCO_3^- is a strong conjugate base. That is, HCO_3^- has a strong affinity for combining with H^+ to form the weak acid, H_2CO_3. As described earlier, luminal H_2CO_3

Table 2 H^+ Secretion in Proximal and Distal Segments of Nephron

	Proximal	Distal
Daily H^+ secreted (mEq/day)	3,888	504
Maximum H^+ gradient (urine to blood)	4:1	1,000:1
Maximal acidity of urine (as pH)	6.8	4.4
Acomplishment	Reabsorption of 3,888 mEq HCO_3^- per day	Reabsorption of 432 mEq HCO_3^- per day and generation of 72 mEq new HCO_3^-

breaks down into H_2O and CO_2, and no acidification of urine occurs.

2. HPO_4^{2-} is also a reltively strong conjugate base. It has a strong affinity for combining with H^+ to form the weak acid, $H_2PO_4^-$. Formation of this weak acid, which is excreted, permits considerable secretion of H^+ before the limiting H^+ gradient is reached. The amount of H^+ lost from the body in this manner can be determined by titration of urine with base back to pH 7.4 (physiological neutrality). This $H_2PO_4^-$ in urine is responsible for the "titratable acidity" (Fig. 9) of urine. However, as will be seen below, not all the H^+ secreted into the urine for excretion is in the form of titrable acidity ($H_2PO_4^-$).

3. Cl^- is an extremely weak conjugate base. Cl^- does not bind H^+ to form a weak acid (that is why HCl is such a strong acid). Thus, the exchange of H^+ for the Na^+ in the presence of Cl^- results, in effect, in the formation of the strong mineral acid, HCl. Since the secretion of H^+ is gradient limited, the accumulation of this completely dissociated acid would quickly inhibit further secretion of H^+ into the glomerular filtrate if there were no other conjugate bases to bind H^+. Much of the H^+ can be prevented from accumulating as free hydrogen ion by combining with HPO_4^{2-} to form the weak acid, $H_2PO_4^-$ (as described above). However, there is only a limited amount of HPO_4^{2-} available (all that is available is that which is filtered through the glomerulus into the glomerular

filtrate). To prevent the buildup of free H^+, another strong conjugate base is needed to bind hydrogen ions. No other strong conjugate base is filtered into the glomerular filtrate, but one fortunately is secreted into the filtrate by the tubular cells, namely, ammonia (NH_3).

4. NH_3 is a very strong conjugate base that avidly combines with H^+ to form the extremely weak acid, NH_4^+. Unlike the limited availability of HPO_4^{2-} for combining with free H^+, there is little limitation on the amount of NH_3 available. This NH_3 comes from glutamine (glutaminase reaction) primarily (Fig. 10).

The amount of NH_3 theoretically available for secretion into the urine is equivalent to the total production of nitrogen from protein metabolism, most of which is otherwise converted to urea by the liver. The body's minimal daily dietary protein requirement is about 25 g. Dietary amounts of protein in excess of this yield increments of waste nitrogen. Under normal conditions, only a small percentage of this waste nitrogen becomes NH_3 (most excess nitrogen is excreted as urea). Even maximal requirements for NH_3 secretion into urine (as in severe acidosis) can be fully met without drawing on tissue nitrogen supplies, since considerable excess nitrogen is provided by an

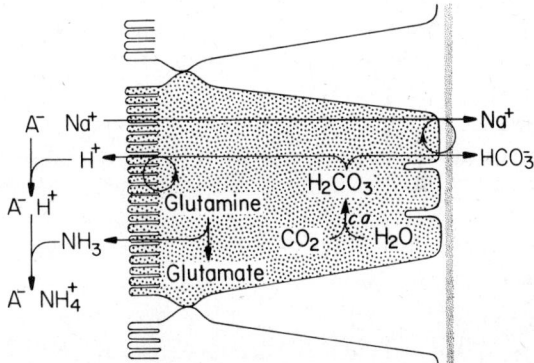

Figure 10 Secretion of ammonia. NH_3 (from deamination of glutamine) is uncharged and lipid soluble, and it diffuses passively through the cell membrane out of the cell into the filtrate, or urine. NH_3 in the urine avidly binds a H^+ to form NH_4^+, which is charged, hydrophilic, and cannot diffuse into the cell. The pK of the NH_3/NH_4^+ buffer pair is 9.3. Thus, essentially all is in the form of NH_4^+ at physiological pH (i.e., NH_3 is an excellent conjugate base for binding H^+).

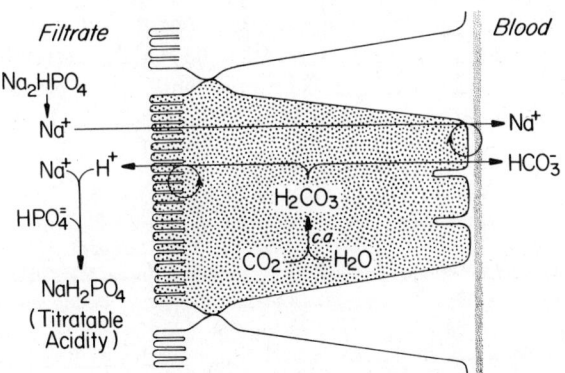

Figure 9 Formation of titratable acidity (NaH_2PO_4) in a distal nephron. For each H^+ secreted and exchanged for Na^+, a new HCO_3^- is generated.

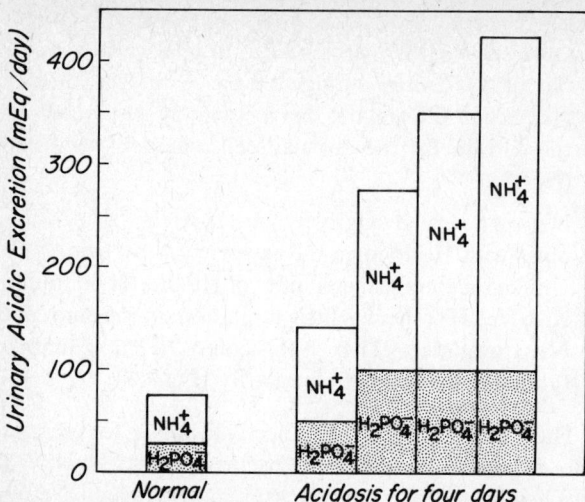

Figure 11 Renal excretory response to an acidosis induced by IV infusion of an acidic solution over a 4-day period. Note that the maximal increase in the excretion of $H_2PO_4^-$ (titratable acidity) is reached by the second day, while NH_4^+ excretion continues to rise throughout the duration of the acidosis.

average daily intake of about 70 g of protein. H^+ secreted by tubular cells and bound with NH_3 cannot be detected by titration of urine back to pH 7.4, since NH_4^+ will not release any H^+ at this pH. Therefore, NH_4^+ is not part of "titratable acidity."

The presence of the conjugate bases NH_3 and HPO_4^{2-} in the luminal fluid permit large amounts of H^+ to be secreted (Fig. 11) before the maximal gradient limits secretion of H^+ (the maximal gradient for free H^+ of 1,000:1 in the distal nephron is equivalent to a minimal urinary pH of 4.4*). Note that the secretion of H^+ that combines with either NH_3 or HPO_4^{2-} results in the generation of a new HCO_3^- (not merely indirect reabsorption of an HCO_3^- that exists in the glomerular filtrate, but the

production of new, additional HCO_3^- by the process of urinary acidification).

DISTURBANCES IN ACID–BASE BALANCE

RESPIRATORY ACIDOSIS

Primary disturbances in respiratory function that impair alveolar ventilation result in increased blood CO_2, hence an increase in H_2CO_3 that raises H^+ concentration (decreases pH).

Causes of Respiratory Acidosis

1. *Depression of medullary respiratory centers:* In severe hypothyroidism, and by overdose of barbiturates, tranquilizers, or morphine.

2. *Reduced airflow:* In severe asthma, emphysema, tracheal obstruction, bilateral pneumothorax, or severe pneumonia.

3. *Restriction of normal ventilatory movements of the chest wall and/or the diaphragm:* In gross obesity (pickwickian syndrome), curare poisoning, paralysis, muscular dystrophy, myasthenia gravis.

Changes in pH Related to Changes in PCO_2

The relationship between PCO_2, H_2CO_3, H^+, and HCO_3^- in a solution containing *only the bicarbonate buffer pair* is simple: doubling the PCO_2 doubles the H_2CO_3 and doubles the H^+ (decreases pH 0.3 units) with no appreciable change in HCO_3^- concentration.* Halving the PCO_2 halves the H_2CO_3 and halves the H^+ (increases pH 0.3 units) with

---

*pH 4.4 = 0.04 mEq of free H^+/L. If the daily volume of urine were 1 L (to use a convenient round figure), then no more than 0.04 mEq of free H^+ could be excreted each day as such. Compare this with the 50–100 mEq of H^+ secreted daily and bound with HPO_4^{2-} and NH_3. In other words, less than 0.1% of the total H^+ excreted exists as free H^+. Because of the limitations of the gradient-limited system of H^+ secretion, the body is dependent on the urinary buffers to accept H^+ in sufficient quantity to achieve the high rate of elimination of 50–100 mEq/day by way of the urine.

*For example, doubling the H^+ concentration from 40 nEq/L (pH 7.40) to 80 nEq/L (pH 7.10) is accompanied by an increase of an additional 40 nEq HCO_3^-/L since one HCO_3^- is produced for each H^+ released from H_2CO_3. If the HCO_3^- were 24 mEq/L to begin with, this would be an increase to 24.00004 mEq/L (an increase of +40 nEq/L). However, this is *not* a measurable increase in HCO_3^-; nanoequivalent (10^{-6} mEq) changes in HCO_3^- are undetectable and have no physiological effect. On the other hand, nanoequivalent changes in H^+ are easily measured (by a pH meter) and have marked physiological effects.

no appreciable change in the HCO_3^- concentration (Fig. 12*a*). However, in blood, changes in either of the bicarbonate buffer pair are exclusively buffered by the nonbicarbonate buffer system. Therefore, in blood, changes in Pco_2 *in the presence of nonbicarbonate buffer* result in smaller changes in pH than if the nonbicarbonate buffer were not present. Doubling the Pco_2 still doubles the final concentration of H_2CO_3, as before, but increases the H^+ to a lesser extent (decreases pH only 0.2 units) because of a simultaneous secondary increase in HCO_3^- that results from interaction of some H_2CO_3 with the nonbicarbonate buffer. Halving the Pco_2 halves the final concentration of H_2CO_3, as before, but decreases the H^+ to a lesser extent (increases pH only 0.2 units) because of a simultaneous secondary decrease in HCO_3^- due to the nonbicarbonate buffer action (Fig. 12*b*). Recall that according to the Henderson-Hasselbalch equation, the pH is determined by the ratio of HCO_3^-/H_2CO_3. Thus the secondary changes in HCO_3^- accompanying the changes in Pco_2 in the presence of nonbicarbonate buffer decrease the pH changes because the ratio does not change as much.

For example, consider a simple bicarbonate solution containing 24 mEq HCO_3^-/L in equilibrium with a Pco_2 of 40 mm Hg giving a H_2CO_3 concentration of 1.2 mEq/L (0.03 × 40 mm Hg = 1.2 mEq/L); this gives a HCO_3^-/H_2CO_3 ratio of 20 (24/1.2) and a pH of 7.40 (7.40 = 6.10 + log 20 = 6.10 + 1.30). If the Pco_2 is doubled to 80 mm Hg in this simple solution, the H_2CO_3 concentration doubles to 2.4 mEq/L, and the HCO_3^-/H_2CO_3 ratio becomes 10 (24/2.4), giving a new pH of 7.10 (7.10 = 6.10 + log 10 = 6.10 + 1.00). However, if instead of a simple bicarbonate solution we had a solution more like blood, that is, one that contained nonbicarbonate buffer in addition to bicarbonate buffer, the doubling of the H_2CO_3 from 1.2 to 2.4 mEq/L would have been accompanied by a secondary increase in HCO_3^- (say, from 24 to 30 mEq/L). Thus the HCO_3^-/H_2CO_3 ratio would not change as much as before (i.e., would be 30/2.4, or ratio of 12.5, instead of 24/2.4, which is a ratio of 10), and the pH would not change as much as before (in this case, the pH would decrease only 0.2 units from 7.40 to 7.20, since 7.20 = 6.10 + log 12.5 = 6.10 + 1.10).

The reaction between H_2CO_3 and the nonbicarbonate system is responsible for the secondary changes in HCO_3^-. For example, when the Pco_2 increases, there is increased H_2CO_3 as before. However, some of this H_2CO_3

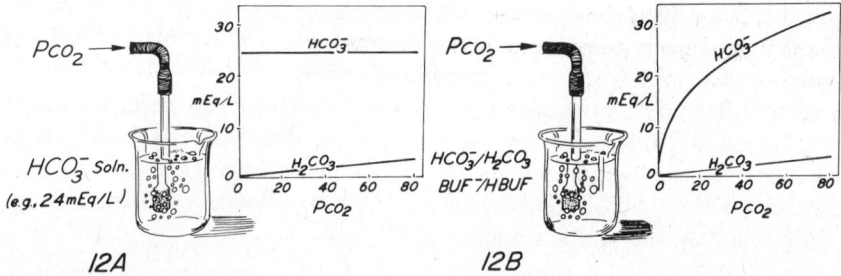

Figure 12 The relationship among Pco_2, H_2CO_3, and HCO_3^- in a solution containing a bicarbonate buffer pair. The Pco_2 can be set at different levels depending on the source (e.g., a gas containing 5% CO_2 has a Pco_2 of about 40 mm Hg). *(a)* The solution contains only the bicarbonate buffer pair. The concentration of HCO_3^- is determined by the amount of bicarbonate salt added in making up the solution (in this case, 24 mEq/L). The concentration of the H_2CO_3 is determined by the Pco_2 (e.g., H_2CO_3 = 0.03 × Pco_2). *(b)* The bicarbonate buffer solution also contains a nonbicarbonate buffer pair (Buf$^-$/HBuf) as in blood. The primary change in H_2CO_3 that results from a change in Pco_2 is now found to be accompanied by a secondary change in HCO_3^- (as a result of the buffer action of the nonbicarbonate buffer system). With the nonbicarbonate system in the same concentration as in vivo, the change in pH is only about two-thirds that which would occur in the absence of the nonbicarbonate system.

reacts with Buf⁻ to form HBuf and HCO_3^-. As some H_2CO_3 is lost by this reaction, more is formed to replace it (it is in equilibrium with P_{CO_2}). Still more buffering by the nonbicarbonate system occurs (with still more HCO_3^- formation) until the buffering capacity of this system has been reached. The amount of H_2CO_3 that one ends up with is just what would be expected from the P_{CO_2} (since any lost in reaction with Buf⁻ is immediately replaced), but in addition there is the secondary increase in HCO_3^-.

Buffer reactions in respiratory acidosis:

$$\boxed{\text{Cause}} \quad \uparrow P_{CO_2} \longrightarrow \uparrow H_2CO_3 \quad \boxed{\text{Primary effect}}$$

$$\text{Some } H_2CO_3 + \text{Buf}^- \longrightarrow \text{HBuf} + HCO_3^- \quad \boxed{\text{Secondary effect}}$$

$$\boxed{\text{Decreases}} \qquad \boxed{\text{Increases}}$$

$$\downarrow \text{pH} = 6.1 + \log \frac{\uparrow HCO_3^-}{\uparrow\uparrow H_2CO_3}$$

Ordinarily, respiratory acidosis is accompanied by this secondary increase in HCO_3^- (from buffering action of nonbicarbonate buffer system). However, if severe hypoxia is present as well, the expected secondary increase in HCO_3^- may not be seen. Tissue hypoxia can cause substantial anaerobic metabolism with resulting lactic acidemia and decreased blood HCO_3^- (from neutralization of lactic acid). The end result is a mixed respiratory and metabolic acidosis.

The reaction of the nonbicarbonate buffer system with H_2CO_3 also occurs under normal circumstances. In passing through the tissues, the P_{CO_2} of blood normally increases from 40 to about 46 mm Hg. It is the buffering of this increased P_{CO_2}, or H_2CO_3, by the nonbicarbonate buffer system (primarily hemoglobin) that is responsible for the increase in HCO_3^- concentration in going from arterial to venous blood (venous blood contains about two more mEq of HCO_3^-/L of blood than does arterial blood). That is, this reaction between H_2CO_3 and Buf⁻ accounts for the fact that the principal form of CO_2 transport from tissues to lungs in venous blood is as HCO_3^-.

RESPIRATORY ALKALOSIS

Primary disturbances of respiratory function resulting in an excessive alveolar ventilation relative to the rate of CO_2 production by the body cause decreased blood CO_2 and thereby diminished H_2CO_3, lowering the H⁺ concentration.

Just as increased P_{CO_2} resulted in a secondary increase in blood HCO_3^- in respiratory acidosis, so too a decreased P_{CO_2} results in a secondary decrease in HCO_3^- in respiratory alkalosis. It is these secondary changes in bicarbonate that serve to emphasize differences between respiratory changes in pH and nonrespiratory (metabolic) changes in pH. Changes in HCO_3^- concentration produced by changes in P_{CO_2} (respiratory acidosis or alkalosis) are inversely related to change in pH, in contrast to nonrespiratory (metabolic) conditions in which HCO_3^- changes (the primary cause of the acid–base imbalance) are the same direction as the pH changes. (See Table 3.)

Causes of Respiratory Alkalosis

1. *Stimulation of medullary respiratory centers:* Hysterical hyperventilation, pain, high fever, initial phase of salicylate toxicity.

2. *Stimulation of peripheral chemoreceptors:* Hypoxemia of cardiopulmonary disease (atelectasis, congenital right to left shunt), hypoxia of high altitude.

3. *Stimulation of pulmonary J receptors (juxtapulmonary–capillary receptors):* Excess stimulation of juxtapulmonary–capillary receptors by pulmonary embolism, pulmonary edema, early pneumonia, or early asthma.

4. *Passive mechanical hyperventilation:* Overventilation with a respirator.

Table 3 Comparison of Changes in pH and HCO_3^- in Respiratory and Metabolic (Nonrespiratory) Imbalances

Acid–Base Imbalance	pH	HCO_3^-
Respiratory acidosis	↓	↑HCO_3^- (2°)
Respiratory alkalosis	↑	↓HCO_3^- (2°)
Metabolic acidosis	↓	↓↓HCO_3^- (1°)
Metabolic alkalosis	↑	↑↑HCO_3^- (1°)

The following equations illustrate the changes in the bicarbonate buffer pair under conditions of hypocapnia ($\downarrow Pa_{CO_2}$) and the resulting respiratory alkalosis:

$$\boxed{\text{Cause}} \quad \downarrow PCO_2 \longrightarrow \downarrow H_2CO_3 \quad \boxed{\text{Primary effect}}$$

$$\text{Some HCO}_3^- + \text{HBuf} \longrightarrow \text{Buf}^- + \text{H}_2\text{CO}_3 \quad \boxed{\text{Secondary effect}}$$

$$\boxed{\text{Decreases}} \qquad \boxed{\text{Increases}}$$

$$\uparrow \text{pH} = 6.1 + \log \frac{\downarrow \text{HCO}_3^-}{\downarrow\downarrow \text{H}_2\text{CO}_3}$$

While some patients with respiratory alkalosis show a grossly obvious hyperpnea (increased rate and depth of respiration), which can be a first clue to an underlying disease state, others maintain a substantial hypocapnia (Pa_{CO_2} of less than 30 mm Hg) with surprisingly little clinically obvious respiratory effort. Symptoms associated with respiratory alkalosis include paresthesias (strange sensations such as numbness or tingling) of extremities and circumoral area, lightheadedness, and occasionally carpopedal spasm (involuntary flexure of wrists and ankles). These are the result of decreased cerebral blood flow and increased pH of body fluids (which decreases the concentration of ionized Ca^{2+}).

Many clinicians find it convenient to visualize changes in acid-base balance by comparing alterations in pH with changes in HCO_3^- concentration (plotted as so-called Davenport diagrams). The changes in pH that result from respiratory acidosis and respiratory alkalosis, and the secondary changes in HCO_3^- that usually accompany them are illustrated in Figure 13. The changes shown in Figure 13 are initial changes that occur before compensatory mechanisms to restore the normal pH can be brought into play, that is, they are those of uncompensated respiratory acidosis and uncompensated respiratory alkalosis. In time, the kidneys will begin to restore acid–base balance (by adjusting ECF HCO_3^- concentration). However, this renal compensatory mechanism—to be discussed further in the section on Compensatory Mechanisms—takes considerable time (several days) to become maximally effective.

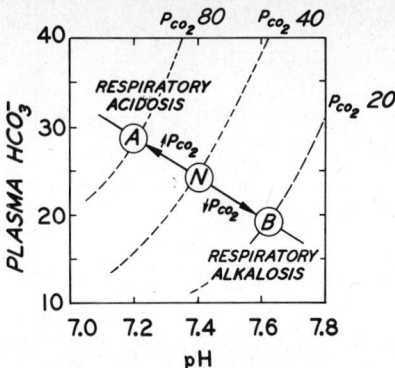

Figure 13 Changes in pH and plasma bicarbonate in respiratory acidosis and alkalosis. The buffer curve (between points A and B) is actually a short segment of a CO_2 titration curve for blood. Point A represents a pure, uncompensated respiratory acidosis; point B illustrates a pure, uncompensated respiratory alkalosis. With increased PCO_2 in respiratory acidosis the pH and plasma HCO_3^- shift to the left (decreased pH) and upward (secondary increase in HCO_3^- along this buffer curve. With decreased PCO_2 in respiratory alkalosis and the pH and plasma HCO_3^- shift to the right (increased pH) and downward (secondary decrease in HCO_3^-) along the curve. Dotted lines are isobars for PCO_2. Each isobar is a single PCO_2 of constant partial pressure.

METABOLIC ACIDOSIS

The adjective "metabolic," as it is used in acid–base balance parlance, means nonrespiratory. That is, the terms metabolic acidosis and metabolic alkalosis are defined as including all disturbances in pH that are nonrespiratory in origin. In only some instances are these disturbances of metabolic origin per se (e.g., the ketoacidosis of diabetes mellitus is caused by excess metabolism of fatty acids to the so-called ketoacids; the production of lactic acid in conditions of increased anaerobic glycolysis). In any case, the term is firmly entrenched in the clinical lingo and should be read as synonymous with nonrespiratory.

Metabolic acidosis is an abnormal state characterized by either gain of strong nonvolatile acid (through excessive production or by inadequate excretion) or by loss of bicarbonate from the ECF (either by loss of bicarbonate-containing diarrheal fluid or by inadequate renal reabsorption of bicarbonate). From whatever cause, the end result of metabolic acidosis is the same: decreased pH of the ECF and decreased HCO_3^- concentration. Bicarbonate loss is indirect in the case of strong acid accumulation

(HCO_3^- lost in neutralization of acid), and direct in the loss of alkaline fluids from the lower intestine or through urinary loss due to failure of the renal reabsorptive mechanism previously described. The concentration of H_2CO_3 and the level of PCO_2 remain normal initially. At first glance, it might seem that H_2CO_3 should increase if there is a gain of strong acid (since $HCO_3^- + acid \rightarrow H_2CO_3$). However, remember that the body is an open system normally regulated to a fixed level of arterial PCO_2. When the H_2CO_3 tends to increase, the concomitant rise in PCO_2 stimulates sufficient additional alveolar ventilation to dispose of the extra CO_2, that is, it is "blown off." This is not a compensatory mechanism, but is merely the normal everyday control of PCO_2 by the negative feedback system. (Later, when the pH decreases by 0.1–0.2 units, a compensatory increase in ventilation stimulated by H^+ effects on chemoreceptors will occur, however, defer consideration of compensatory mechanisms until later.) Thus, with the concentration of HCO_3^- decreased, even though that of H_2CO_3 remains normal, the pH must decrease because of the decreased HCO_3^-/H_2CO_3 ratio.

As an example, consider the metabolic acidosis resulting from a suicide attempt by the drinking of an acidic liquid toilet bowl cleaner (7% HCl):

$$HCl + HCO_3^- \longrightarrow Cl^- + H_2CO_3$$

| Decreases | | Tends to increase |

$$H_2CO_3 \rightarrow H_2O + CO_2 \longrightarrow \boxed{\text{Blown off in lungs}}$$

In buffering the acid by bicarbonate, the HCO_3^- is lost in forming H_2CO_3. This tends to increase the H_2CO_3 concentration, except that as it begins to increase, the concomitant rise in PCO_2 stimulates sufficient additional alveolar ventilation to dispose of it. The net result is a diminished HCO_3^-, but an initially unchanged H_2CO_3.

Causes of Metabolic Acidosis

Acid-Gaining Metabolic Acidosis

1. *Increased anaerobic glycolysis:* Lacticacidemia following impaired tissue oxygenation (as in inadequate cardiopulmonary bypass or in hypoxia of cyanotic heart disease). This is the most common cause of metabolic acidosis.

2. *Incomplete hepatic oxidation of fatty acids:* Ketoacidosis of diabetes mellitus.

3. *Failure to excrete nonvolatile acid equivalents from protein and phospholipid metabolism:* From renal failure (azotemia), or from refractory hypovolemia (hypovolemic shock).

4. *Poisoning by acids or from agents metabolized to acid products:* Acid drinking, the later phase of salicylate toxicity, ingestion of methanol, ethylene glycol, paraldehyde, or NH_4Cl.

Base-Losing Metabolic Acidosis

1. *Loss of GI fluids rich in HCO_3^- content (below stomach):* Diarrhea, prolonged drainage of pancreatic, biliary, or intestinal fluids (either by tube drainage or by devleopment of an external fistula).

2. *Excess urinary excretion of HCO_3^-:* Carbonic anhydrase inhibition, renal tubular acidosis, hyperparathyroidism.

METABOLIC ALKALOSIS

Metabolic alkalosis is an abnormal state characterized by an above-normal concentration of ECF HCO_3^- unattended by a proportionate increase in PCO_2 (initially). Therefore, the HCO_3^-/H_2CO_3 ratio is increased (above 20:1) with a consequent increase in pH (subnormal concentration of H^+ in the ECF). Metabolic alkalosis can be caused by loss of H^+ (e.g., loss of gastric HCl in vomiting), by gain of exogenous base (e.g., overzealous self-medication with baking soda or iatrogenic excessive administration of alkaline intravenous fluids or alkalinizing salts), or by induction of a potassium deficiency (as often occurs with use of potent diuretics, dietary potassium restriction, or by presence of excess mineralocorticoid hormones). From whatever cause, the end result of metabolic alkalosis is the same: increased pH and increased HCO_3^- concentration of the ECF.

Causes of Metabolic Alkalosis

Base-Gaining Alkalosis

1. *Direct gain of exogenous HCO_3^-:* Overzealous self-medication with baking soda as an antacid; excess intravenous infusion of HCO_3^- solutions.

2. *Intake of strong base or alkali:* Accidental ingestion of dilute lye solution.

Acid-Losing Alkalosis

1. *Increased secretion of H+ by renal distal tubules as a result of therapy with most diuretics:* Many diuretics indirectly increase renal excretion of H+ by increasing Na+ presented to distal tubules for exchange with H+ (and with K+).

2. *Loss of gastric acid:* In protracted vomiting.

3. *Hypokalemic alkalosis due to K+ deficiency:* K+ depletion can result from diuretic therapy, dietary K+ restriction, or from mineralocorticoid excess.

That gain of exogenous bicarbonate (e.g., by ingestion or infusion of $NaHCO_3$) causes increased blood HCO_3^- concentration is obvious. It should also be evident that even gain of nonbicarbonate base (e.g., NaOH) increases blood bicarbonate as a result of buffering of the nonbicarbonate base by H_2CO_3:

$$OH^- + H_2CO_3 \rightarrow HCO_3^- + H_2O$$

However, it might not be readily obvious why acid loss causes blood bicarbonate gain. The explanation in the case of acid loss via urine is a fairly simple one. Recall that for each H+ secreted by the nephron a HCO_3^- is generated. Thus, anything increasing H+ secretion (e.g., certain diuretics, hypokalemia) automatically increases HCO_3^- generation. The explanation in the case of acid loss by vomiting acidic chyme is similar. Just as H+ secretion in the kidney involves HCO_3^- production, so too does H+ secretion in the stomach. Normally the acidic chyme is ultimately neutralized by bicarbonate-rich secretions when it reaches the duodenum (the H+ and HCO_3^- produced in gastric secretion are reunited, and acid–base balance is reestablished). However, if the acidic chyme is lost in vomiting, the new HCO_3^- originally produced by gastric secretion never react with the acidic chyme in the duodenum and remain as a net gain of HCO_3^-.

Therapy with highly potent diuretics (e.g., ethacrynic acid, furosemide) or moderately potent diuretics (e.g., the thiazides) results in increased amounts of Na+ and accompanying water presented to the distal nephron. This has at least two important consequences related to the production of a metabolic alkalosis: (1) an increase in Na+ results in increased distal Na+–H+ exchange with subsequent increased excretion of H+ and gain of HCO_3^-, and (2) the increased volume flow that reaches the distal tubule increases K+ secretion, eventually producing a hypokalemia that, in turn, also favors the production of an acidic urine and generation of HCO_3^- (to be discussed below). Although the increased excretion of H+ following diuretic therapy results from two different, independent mechanisms, both usually occur together in such therapy because they have the same common cause: the increased amounts of Na+ and water reaching the distal tubule.

Cause–Effect Relationship Between Potassium Deficiency and Alkalosis

Potassium deficiency is an important cause of metabolic alkalosis (second only to protracted vomiting). Loss of K+ from the ECF produces the potassium deficiency. In an attempt to replace the deficit of K+ in hypokalemic blood, K+ shifts from the ICF of cells (principal location of K+ in the body) to the ECF. As K+ moves out of the cells, H+ (and some Na+) move into the cells. For each H+ that enters a cell, an OH− is left behind in the ECF, thereby causing an extracellular alkalosis with increased ECF HCO_3^- (since $OH^- + H_2CO \rightarrow HCO_3^-$). Like other cells, the renal tubular cells share in this exchange of H+ for K+, but secrete their influx of H+ into the urine (Figure 14b), thereby allowing many more additional H+ from the ECF to enter renal tubular cells.

This acid urine, which results from a potassium deficiency, is called a paradoxical aciduria of hypokalemic alkalosis. It seems paradoxical in the sense that one ordinarily expects the production of alkaline urine as a compensatory effect to help correct the pH in a condition of alkalosis. However, it is not so paradoxical if you consider that the production of acid urine is part of the cause of the alkalosis in this case (by losing too many H+ from the body via the urine).

The potassium-deficiency alkalosis just described involved hypokalemia as cause and alkalosis as effect. The reverse is also seen: alkalosis (from some other cause) can produce hypokalemia. In the absence of potassium deficiency, alkalosis of the ECF (e.g., a respiratory alkalosis) causes shift of H+ (and Na+) out of cells in exchange for K+ shifting into cells (Fig. 14a). This loss of K+ from the ECF into the ICF results in a hypokalemia. Hence, as a general rule, hypokalemia and alkalosis go hand in hand.

2° Movement of H⁺ in Response to I° Movement of K⁺

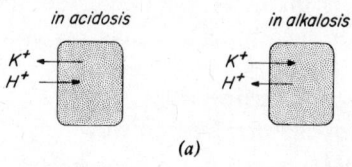

2° Movement of K⁺ in Response to I° Movement of H⁺

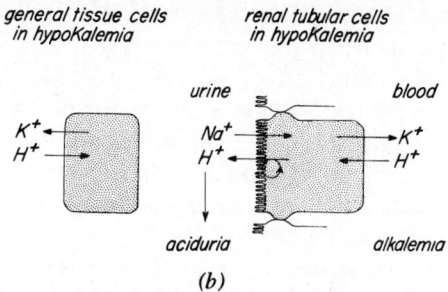

(a)

Metabolic alkalosis from hypoKalemia of K⁺ depletion:

general tissue cells   renal tubular cells
in hypoKalemia   in hypoKalemia

(b)

Figure 14 *(a)* Secondary H⁺ exchange in response to K⁺ movements into or out of cells (as in ECF K⁺ excess or deficiency), and secondary K⁺ exchange into or out of cells in response to primary H⁺ movements (as in acidosis or alkalosis). Arrows indicate direction of ion movement after the disturbance indicated has developed (e.g., after hypokalemia, or after acidosis). *(b)* Metabolic effect of hypokalemia on ECF pH. The shift of K⁺ out of the cells in exchange for H⁺ results in an alkalosis of the ECF. Production of an acid urine (a "paradoxical aciduria") further contributes to the alkalemia.

Not only that, but hyperkalemia and acidosis generally go hand in hand as well. In other words, changes in plasma K⁺ and H⁺ are usually parallel (go in the same direction). (See Fig. 15.) Of course, there are exceptions to almost every general rule of physiology. Any interference with H⁺ secretion by renal tubule cells (e.g., inhibition of carbonic anhydrase by acetazolamide, renal tubular acidosis) will produce exceptions to the parallelism between plasma K⁺ and H⁺ changes.

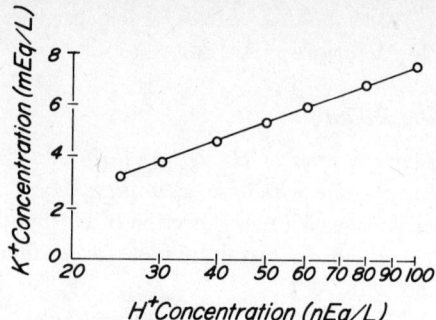

Figure 15 General relationship between plasma H⁺ and K⁺ concentration. This relationship is generally true in conditions not involving interference with H⁺ secreting ability of renal tubule cells.

Ingestion of Exogenous Base

A much simpler but less common form of metabolic alkalosis is that resulting from excess ingestion of exogenous base (not a common cause, since the kidneys have a considerable ability to excrete base). As an example, consider accidental ingestion of a dilute solution of a strong base, such as NaOH (possibly stored in a pop bottle and assumed by a toddler to be a beverage).

$$NaOH + H_2CO_3 \longrightarrow NaHCO_3 + H_2O$$

| Tends to decrease | Increases |

$$CO_2 + H_2O \rightarrow H_2CO_3 \longrightarrow \boxed{\text{Replaces above loss}}$$

When a fixed (nonvolatile) acid is added to the ECF (metabolic acidosis), HCO_3^- is consumed in buffering and the pH falls. When base is added to the ECF, HCO_3^- increases (metabolic alkalosis) and the pH increases. In both cases there is no change in H_2CO_3 initially because of the normal feedback mechanism for respiratory regulation of P_{CO_2}. Thus changes in HCO_3^- and pH resulting from metabolic acidosis or metabolic alkalosis may be plotted along constant P_{CO_2} isobars (Fig. 16). Figure 16 shows the isobar for a P_{CO_2} or 40 mm Hg, since this is the level of arterial P_{CO_2} normally maintained. Additions of fixed acid or base at other P_{CO_2} values would result in similar changes along other isobars.

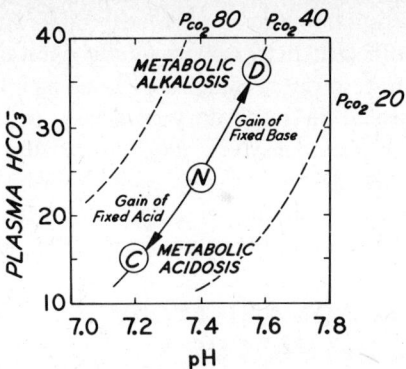

Figure 16 Changes in pH and plasma bicarbonate in metabolic acidosis and alkalosis. The development of a pure uncompensated metabolic acidosis shifts values for pH and HCO_3^- to the left (decreased pH) and downward *(marked decrease in HCO_3^-)* along the normal arterial PCO_2 isobar of 40 mm Hg (to point C). In a pure uncompensated metabolic alkalosis, the values shift to the right (increased pH) and upward *(marked increase in HCO_3^-)* along this same isobar (to point D).

COMPENSATORY MECHANISMS

In order to maintain the H^+ concentration of the ECF within the narrow limits compatible with life (within limits spanning less than one pH unit), the body has three lines of defense: (1) buffers, (2) respiratory compensation, and (3) renal compensation. These are listed in the order of their rapidity of action. Buffering acts to limit H^+ changes almost immediately; respiratory changes begin to correct pH changes in minutes to hours; renal compensation to correct pH is maximal only after several days.

As a generalization one may say that respiratory compensation is never complete (cannot return pH all the way back to normal), while renal compensation can be (although it might not be complete in every case).

RENAL COMPENSATORY RESPONSE TO RESPIRATORY ACIDOSIS AND ALKALOSIS

Since the primary disturbance in respiratory acidosis or alkalosis is a dysfunction of the respiratory system, there

Table 4 Renal and Respiratory Compensatory Responses to Disturbances in Acid–Base Balance

Primary Disturbance	Consequences	Compensatory Response
Hypercapnia ($\uparrow Pa_{CO_2}$)	Respiratory acidosis	Kidney generates additional HCO_3^- for ECF
Hypocapnia ($\downarrow Pa_{CO_2}$)	Respiratory alkalosis	Kidney decreases retention and generation of HCO_3^- for ECF
Positive gain of nonvolatile acid ($\downarrow HCO_3^-$)	Metabolic acidosis	Hyperventilation produces hypocapnia
Positive gain of base ($\uparrow HCO_3^-$)	Metabolic alkalosis	Limited hypoventilation produces some hypercapnia

remain only two of the three lines of defense against acid–base imbalance: buffering and renal compensation. That is, if the respiratory system is part of the problem, it cannot also be part of the solution.

In respiratory acidosis the decreased pH is due to a primary increase in PCO_2 and therefore an increase in H_2CO_3 (and a proportionally smaller secondary increase in HCO_3^- as a result of nonbicarbonate buffering of the H_2CO_3). The basic problem is a decrease in the HCO_3^-/H_2CO_3 ratio. Renal compensation eventually corrects the ratio fairly close to normal by generating greater amounts of HCO_3^-. By the time renal compensation is complete or fairly close to complete (after several days, i.e., chronic respiratory acidosis), there is not only the originally increased H_2CO_3 from the respiratory acidosis, but also a greatly increased HCO_3^- (both components of the bicarbonate buffer pair are in higher concentration than normal, but the ratio is returned to or near 20:1 and the pH is once again 7.4, or close to it).

In respiratory alkalosis the increased pH is due to a primary decrease in PCO_2 and thereby a decrease in the H_2CO_3 component of the bicarbonate buffer pair (along with a proportionally smaller secondary decrease in HCO_3^-). The kidneys can return the increased ratio back toward normal by greatly diminishing the HCO_3^- component. Thus the kidneys decrease ECF HCO_3^- and the HCO_3^-/H_2CO_3 returns to or near the normal ratio of 20:1 (although not normal concentrations).

Changes in HCO_3^- concentration and pH in respiratory acidosis and alkalosis, both before and after renal compensation, are illustrated in Figure 17. Note that renal compensation by addition or depletion of bicarbonate changes the pH along different P_{CO_2} isobars (along a lower than normal isobar in respiratory alkalosis, and along a higher than normal isobar in respiratory acidosis). It is also shown that renal compensation may (potentially) completely correct the pH all the way back toward normal (7.4). Thus patients with chronic respiratory acidosis or alkalosis tend to have ECF pH values not far from normal.

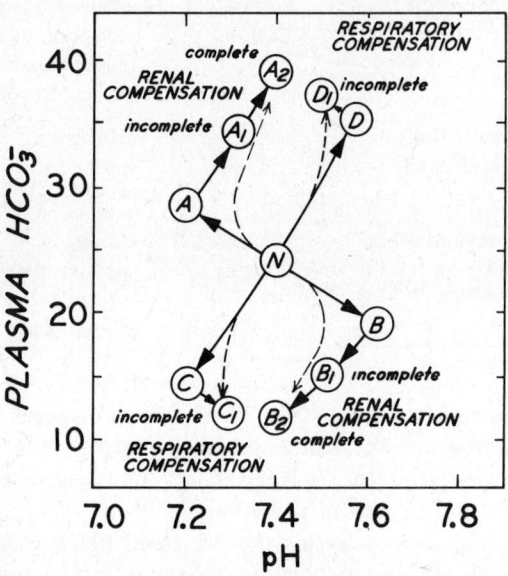

Figure 17 Compensatory changes. As in Figure 16, the arrow from point N to point A illustrates a shift along the normal CO_2 titration curve from the normal P_{CO_2} isobar of 40 mm Hg to a new, higher P_{CO_2} isobar as the result of uncompensated respiratory acidosis. An increase in plasma HCO_3^- from increased renal reabsorption and generation of HCO_3^- brings the pH partly back toward normal along the new isobar to point A_1 (incomplete renal compensation) and eventually all the way back to pH 7.4 at point A_2 (complete renal compensation). Similarly, point B illustrates a shift from the normal P_{CO_2} isobar of 40 mm Hg to a new lower isobar caused by uncompensated respiratory alkalosis. Incomplete (B_1) and complete (B_2) renal compensation by increased HCO_3^- excretion are shown along this new isobar. The arrow from point N to point C represents development of an uncompensated metabolic acidosis, as in Figure 13. Respiratory compensation by increased elimination of CO_2 brings the pH partly back toward normal (C_1). Point D illustrates uncompensated metabolic alkalosis. Incomplete respiratory compensation by hypoventilation shifts the pH a short distance to point D_1. For simplicity, changes have been shown as progressing completely to well-developed states before compensatory mechanisms begin to act. More realist courses of correction are shown by the dashed lines.

RESPIRATORY COMPENSATORY RESPONSE TO METABOLIC ACIDOSIS AND ALKALOSIS

In both metabolic acidosis and metabolic alkalosis, respiratory compensation occurs shortly after a significant change in blood pH (by 0.1–0.2 pH units). Thus it must be emphasized that the early phase of pure, uncompensated metabolic acidosis or alkalosis with normal P_{CO_2} and normal H_2CO_3 concentration is a brief one. Most patients will have some degree of respiratory compensation imposed on their metabolic acidosis or alkalosis by the time they are seen by a physician.

In nonrenal metabolic (nonrespiratory) disturbances of acid–base balance, the body uses all three lines of defense: buffering, respiratory compensation, and, ultimately, renal compensation. However, if the metabolic disturbance in acid–base balance is due to renal causes, then only the first two defenses apply.

In metabolic acidosis the pH is decreased because of a substantive decrease in HCO_3^- with no initial change in H_2CO_3 (i.e., the HCO_3^-/H_2CO_3 ratio decreases). In the initial uncompensated state the H_2CO_3 remains unchanged, since any increase raises the P_{CO_2}, which increases ventilation sufficiently to blow off the additional CO_2 (normal respiratory control of P_{CO_2}). This is not considered a compensatory mechanism. However, as the pH continues to fall, it begins to stimulate peripheral chemoreceptors primarily (aortic and carotid bodies), with the result that ventilation increases even further thereby decreasing P_{CO_2} and H_2CO_3 concentration below normal.* This is a compensatory response to metabolic acidosis and the decrease in H_2CO_3, which will partly raise the buffer ratio back toward normal. In time, if the cause

---

*Respiratory compensation (i.e., hyperventilation) can be very pronounced, leading to Pa_{CO_2} values below 10 mm Hg.

of the metabolic acidosis continues, additional compensation by the kidneys can completely correct the only partially corrected pH achieved by respiratory compensation.

In metabolic alkalosis the increased HCO_3^-/H_2CO_3 ratio, because of increased bicarbonate, can be partly reestablished by a decrease in ventilation, thereby increasing P_{CO_2} and H_2CO_3 concentration. In general, the respiratory compensation for metabolic alkalosis is less effective than is respiratory compensation for metabolic acidosis. The degree of hypoventilation possible in respiratory compensation for metabolic alkalosis is limited, since the stimulatory effect of accompanying hypoxia on the peripheral chemoreceptors and the stimulatory effect of increased Pa_{CO_2} on both peripheral and central chemoreceptors prevents further inhibition of ventilation. Respiratory compensation for metabolic alkalosis in the form of hypoventilation can raise the P_{CO_2} to a maximum of only about 60 mm Hg.

If a partially corrected metabolic alkalosis of nonrenal origin persists, renal compensation eventually helps correct the pH. Normally, functioning kidneys respond to elevated plasma HCO_3^- by an increase in HCO_3^- excretion. That is, bicarbonate normally appears in the urine due to incomplete reabsorption whenever the plasma concentration exceeds the normal level. Decreased or absent formation of titratable acidity and NH_4^+ brings generation of new HCO_3^- to minimal or zero levels. Thus with excretion of HCO_3^- in urine and no new generation of HCO_3^-, there is a net loss of HCO_3^-.

Specific graphic examples of HCO_3^- and pH combinations in simple uncompensated conditions of acid-base imbalance have been illustrated in Figures 13 and 16. Graphic changes in such combinations as a result of compensation for imbalances of a given degree of severity have been illustrated in Figure 17. Although there are a multitude of different possible combinations of HCO_3^- and pH for varying degrees of severity (from mild to severe) and extent of compensation (from slight to maximal), these can be conveniently circumscribed within limits or zones as illustrated in Figure 18.

Table 5 assumes that compensatory states have already been initiated. Note that no respiratory compensation is possible in respiratory acidosis or alkalosis, since the respiratory system is part of the problem, not the solution. Similarly, renal compensation is precluded in renal metabolic acidosis or alkalosis. Indeed, it is the inappropriate

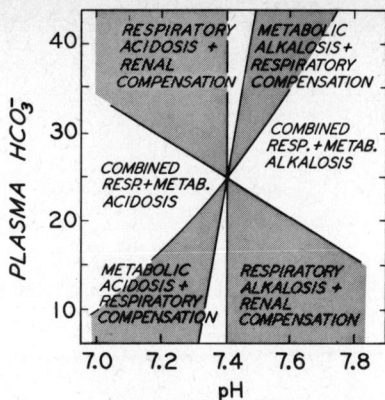

Figure 18 Zones of acid-base imbalance and compensation. The location of the values for pH and plasma HCO_3^- permits identification of the specific state of acid–base imbalance and approximation or the degree of compensation achieved. (Redrawn after the style of Davenport HW: *The ABC of Acid–Base Chemistry*. The University of Chicago Press, 1974.)

production of an alkaline urine that is the principal cause of the renal metabolic acidosis, and it is the inappropriate production of an acidic urine that is the primary cause of the renal metabolic alkalosis.

SUMMARY

To summarize acid–base balance, the following points recapitulate the major points and principles.

Respiratory Imbalance
Carbon dioxide can combine reversibly with water to form carbonic acid. Thus production of CO_2 by cells is considered equivalent to the production of acid and, by the same token, the excretion of CO_2 by the lungs is equivalent to the elimination of acid (elimination of the volatile acid H_2CO_3). An excess of H_2CO_3 constitutes respiratory acidosis, and a deficiency constitutes respiratory alkalosis.

Nonrespiratory ("Metabolic") Imbalance
In addition to producing CO_2, cells also produce nonvolatile acid from the metabolism of amino acids (e.g., the production of H_2SO_4 from methionine and cysteine).

Table 5 Clinical Laboratory Data in Partially Compensated States of Acid–Base Imbalance

	State (blood pH)	Blood HCO_3^-	Blood P_{CO_2}	Urinary pH
Acidosis	Respiratory acidosis	↑ (Compensatory)	↑ (Cause)	Acidic (compensatory)
	Nonrenal metabolic acidosis	↓ (Cause)	↓ (Compensatory)	Acidic (compensatory)
	Renal metabolic acidosis (e.g., carbonic anhydrase inhibition)	↓ (2° cause)	↓ (Compensatory	Alkaline (1° cause)
Alkalosis	Respiratory alkalosis	↓ (Compensatory)	↓ (Cause)	Alkaline (compensatory)
	Nonrenal metabolic alkalosis	↑ (Cause)	Slight↑ (compensatory)	Alkaline (compensatory)
	Renal metabolic alkalosis (e.g., from hypokalemia)	↑ (2° cause)	Slight↑ (compensatory)	Acidic (1° cause)

These nonvolatile acids react with, and are neutralized by, ECF HCO_3^-. A deficiency of ECF HCO_3^- constitutes metabolic acidosis, and an excess constitutes metabolic alkalosis. The kidneys are primarily responsible for replacing HCO_3^- used up in neutralizing nonvolatile acids. Defective handling of HCO_3^- production by the kidneys can result in renal metabolic acidosis or alkalosis. Other causes of metabolic acidosis or alkalosis are nonrenal (e.g., ingestion of exogenous acids or bases).

Lines of Defense Against Acid–Base Imbalance.

The accumulation of acids can produce marked increases in body acidity. To prevent body acidity, the first line of defense is the system of buffers (combinations of substances capable of H^+ acceptance or donation, e.g., a weak acid and a salt of that acid). The most important ECF buffer system is the bicarbonate buffer system. By regulating CO_2 elimination, the respiratory system normally determines the concentration of the ECF H_2CO_3. By regulating HCO_3^- production, the kidneys determine the concentration of ECF HCO_3^-. Thus both components of the bicarbonate buffer system are controlled by physiological mechanisms.

In terms of the Henderson-Hasselbalch equation ap-

plied to the bicarbonate buffer system, the acid–base status of the ECF can be assessed by knowledge of two items of information: Pa_{CO_2} and pH. That is, if Pa_{CO_2} and pH are known, the HCO_3^- is calculated. With calculation of HCO_3^-, both components of the bicarbonate buffer system are known. Primary changes in HCO_3^- indicate metabolic acid–base imbalance (e.g., a primary decrease in HCO_3^- unrelated to a change in CO_2 constitutes metabolic acidosis). Primary changes in H_2CO_3 (as indicated by changes in Pa_{CO_2}) produce respiratory acid–base imbalance (e.g., an excess of Pa_{CO_2} indicates respiratory acidosis).

Compensatory Mechanisms

In respiratory acidosis or alkalosis, the problem is caused by malfunction of the respiratory system. Therefore, to help restore the pH back toward normal (compensation), we cannot count on the respiratory system as the second line of defense (it cannot be part of the solution if it is the primary problem, unless one can correct the underlying clinical condition). The main compensatory mechanism for restoring pH in respiratory acid–base imbalances is renal readjustment of the ECF HCO_3^- concentration (via processes resulting in acidification or alkalinization of the urine). In metabolic acidosis or alkalosis of nonrenal origin there are two possible compensatory mechanisms: respiratory and renal. Of course, if the metabolic imbalance has a renal basis, the kidneys cannot help (again, unless the underlying condition can be corrected). If the cause of the metabolic acid–base imbalance is nonrenal, one could expect early respiratory compensation followed by a later renal compensation (the third line of defense). Unfortunately, respiratory compensation is always only partial (i.e., can never return the pH all the way back to normal). By contrast, renal compensation in acid–base imbalances of nonrenal origin can be very effective provided there is sufficient time for it to develop maximally (takes several days). Sadly, acute severe changes in pH can kill the patient before the kidneys have time to help restore pH. Fortunately, renal compensation in chronic disorders does have time to become fairly effective (e.g., renal compensation in chronic respiratory acidosis or alkalosis is fairly

good although not necessarily totally complete; it is often within one-tenth of a pH unit of normal).

SELECTED BIBLIOGRAPHY

Hills AB: *Acid–Base Balance.* Baltimore, Williams & Wilkins, 1973.

Robinson JR: *Fundamentals of Acid–Base Regulation.* Oxford, Blackwell, 1975.

Rose BD: *Clinical Physiology of Acid–Base and Electrolyte Disorders.* New York, McGraw Hill, 1977.

Winters RW, Engel K, Dell RB: *Acid–Base Physiology in Medicine.* Cleveland, London Company–Copenhagen, Radiometer A/S, 1967.

SELF-STUDY QUESTIONS

MULTIPLE CHOICE

Select the single best answer.

1. The component of the Henderson-Hasselbalch equation that shows the greatest relative change (greatest percentage increase or decrease) as a result of a change in ventilation is
   A. the pK.
   B. the pH.
   C. [HCO_3^-].
   D. the concentration of H_2CO_3.

2. Respiratory ventilation can be increased by
   A. an increase in arterial pH.
   B. an increase in arterial [H^+].
   C. a decrease in arterial [H^+].
   D. a decrease in arterial P_{CO_2}. ($\downarrow Pa_{CO_2}$).
   E. an increase in arterial P_{O_2}. ($\uparrow Pa_{O_2}$).

TRUE OR FALSE

3. The body normally produces no more than about 100 mEq/day of volatile acid.

4. In order to maintain acid–base balance, the kidneys normally excrete an amount of acidic equivalents ($H_2PO_4^-$ and NH_4^+) each day that is equal to the daily gain of nonvolatile acid (from protein metabolism).

5. Carbonic acid is a fixed (nonvolatile) acid participating in CO_2 elimination in an open system.

6. HCO_3^- in the glomerular filtrate must diffuse across the luminal border into the tubular cells of the proximal tubule to be reabsorbed into the blood.

MULTIPLE CHOICE

Select the single best answer.

7. If the normal rate of H^+ secretion is 3.05 mEq/min, and that of bicarbonate filtration is 3.00 mEq/min, how many mEq of new HCO_3^- are generated in a day? (*Hint:* There are 1,440 min/day.)
   A. 4,392
   B. 4,320

C. 72

D. 3

E. 1

8. In the nephron the proximal tubule has the highest rate of H^+ secretion and

   A. is responsible for the major part of acidification of urine.

   B. is responsible for the major part of HCO_3^- reabsorption.

   C. is responsible for the major part of Na^+ secretion into urine.

   D. is able to produce a maximal H^+ gradient between urine and blood of 1,000:1.

   E. the lowest rate of Na^+ reabsorption.

9. Assuming a normal excretion of nonvolatile acid equivalents each day, what percentage in urine is free H^+?

   A. 100%

   B. 50–100%

   C. 10%

   D. Less than 1.0%

   E. None

10. Which of the following is most likely to produce a respiratory acidosis?

   A. Morphine overdose (sublethal)

   B. Salicylate overdose (sublethal)

   C. High fever

   D. Hypoxia of high altitude

11. If the P_{CO_2} of 1 L of bicarbonate solution (in same concentration as in plasma) *containing no nonbicarbonate buffers* is doubled from 40 to 80 mm Hg, and the pH falls from 7.40 (40 nEq of H^+/L) to 7.10 (80 nEq of H^+/L), the amount of HCO_3^- will

   A. decrease as a result of neutralizing additional carbonic acid.

   B. double.

   C. increase a few mEq.

   D. increase 40 nEq.

12. If the P_{CO_2} of blood is doubled from 40 to 80 mm Hg,

and nonbicarbonate buffer base (Buf^-) reacts with the additional H_2CO_3,

   A. the H_2CO_3 concentration cannot double.

   B. the H_2CO_3 concentration will double anyway.

   C. the HCO_3^- concentration will not change appreciably.

   D. the pH will not change appreciably.

13. When ascending to high altitude (e.g., vacation at Lake Titicaca in Peru) the rarefied air causes

   A. depression of respiratory neurons and respiratory acidosis.

   B. hypoventilation with respiratory alkalosis.

   C. hyperventilation (from stimulation by peripheral chemoreceptors), which produces a respiratory alkalosis.

   D. ventilatory adjustments permitting increased P_{O_2} without changing P_{CO_2}.

14. In uncompensated metabolic acidosis or metabolic alkalosis

   A. changes in HCO_3^- occur in the same direction as do changes in H^+.

   B. changes in HCO_3^- are inversely related to changes in pH.

   C. changes in HCO_3^- occur in the same direction as do changes in pH.

   D. changes in HCO_3^- are only secondary to changes in P_{CO_2}.

15. Patients with primary aldosteronism (excess secretion of the mineralocorticoid aldosterone) frequently develop a potassium deficiency and

   A. respiratory acidosis.

   B. respiratory alkalosis.

   C. metabolic acidosis.

   D. metabolic alkalosis.

MATCHING

Each of the following indicates a particular acid–base imbalance with compensation already begun. Select the letter that designates the three conditions (blood bicarbonate, blood P_aCO_2, and urinary pH) that most appropriately describes the acid–base state described.

Answer	Blood HCO$_3^-$	Blood Pa$_{CO_2}$	Urinary pH
A	Increased	Increased	Acidic
B	Decreased	Decreased	Acidic
C	Decreased	Decreased	Alkaline
D	Increased	Slightly increased	Alkaline
E	Increased	Slightly increased	Acidic

16. Sustained hysterical hyperventilation.
17. Excessive intravenous infusion of NaHCO$_3$.
18. Carbonic anhydrase inhibition (e.g., by acetazolamide).
19. Hypokalemia from potassium depletion.
20. Cyanosis from sublethal overdose of morphine.

MULTIPLE CHOICE

Select the single best answer.

21. A patient with severe vomiting caused by complete pyloric obstruction would be expected to develop
    A. a rise in plasma Cl$^-$ concentration.
    B. a rise in plasma HCO$_3^-$ concentration.
    C. increased alveolar ventilation.
    D. increased cerebrospinal fluid pressure.
    E. a drop in arterial pH.

MATCHING

A. *Diabetic acidosis*
B. *Chronic respiratory tract obstruction*
C. *Excessive ingestion of NaHCO$_3$*
D. *Acute hyperventilation*
E. *Diabetes insipidus*

22. Plasma pH elevated, plasma HCO$_3^-$ slightly decreased, urine slightly alkaline.
23. Plasma pH elevated, plasma HCO$_3^-$ elevated, urine alkaline.
24. Plasma pH decreased, plasma HCO$_3^-$ decreased, urine acid.

MULTIPLE CHOICE

Select the correct answer(s). (More than one answer might be correct.)

25. Carbonic anhydrase
    A. occurs in renal tubule cells.
    B. is not present in blood plasma.
    C. occurs in gastric parietal cells.
    D. catalyzes the dissociation of carbonic acid into H$^+$ and HCO$_3^-$.

26. The bicarbonate buffer pair is the most powerful of the acid–base buffers of
    A. interstitial fluid.
    B. tissue cells.
    C. cerebrospinal fluid.
    D. erythrocytes.

MATCHING

Select the single best match for each numbered condition or description in the following table. Each condition or description describes a particular acid–base problem.

	Plasma pH	Pa$_{CO_2}$	Plasma HCO$_3^-$
A	7.0	10 mm Hg	3 mEq/L
B	7.5	30	22
C	7.2	80	30
D	7.5	50	38
E	7.35	45	24

27. A missionary is paddling a canoe along a tributary of the Amazon River. Suddenly, out of the dark, dense foliage along the river bank comes a curare-tipped dart from a blowgun. It strikes him in the thigh; in a few minutes he has difficulty in moving and breathing. In another few minutes he begins to appear cyanotic.

28. A toddler (a young child who has just learned to walk and climb) climbs onto the bathroom sink and opens the medicine cabinet. She finds a bottle (with an insecure cap) of orange-flavored children's aspirin (acetylsalicylic acid). Thinking they are candy, she eats all of them (a sublethal overdose). A few minutes after aspirin absorption has begun in the gut, she has increased rate and depth of respiration.

29. A senile elderly lady who lives by herself has about one-fourth normal glomerular filtration rate. Some years earlier, her husband had died suddenly of severe

hemorrhage from a perforated gastric ulcer. In the belief that her recent symptoms of "heartburn" (reflux esophagitis) might be due to an ulcer, she has been consuming large quantitites of baking soda every day for the past 2 weeks.

MATCHING

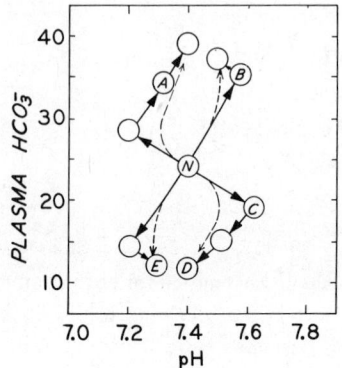

30. Diabetic patient with Kussmaul breathing.

31. Emphysema patient with pH below normal and excreting acidic urine.

32. An otherwise normal woman rapidly ascends to 4,500 m (15,000 ft) in a balloon. She has been at this altitude without an oxygen mask for almost an hour and has experienced an increase in rate and depth of breathing.

33. A vacationer has been visiting at an altitude of 4,500 m (15,000 ft) for 2 weeks.

MULTIPLE CHOICE

Select the single best answer.

34. A middle-aged woman with asthma since childhood has been a heavy smoker since her early teens. During the past few years she has experienced progressive dyspnea and somnolence. Physical examination revealed a cachectic woman with shortness of breath, prolonged expirations with pursing of the lips, and frequent coughing. Her chest was barrel shaped. Laboratory data were as follows: arterial pH, 7.35; arterial HCO_3^-, 32 mEq/L; arterial blood gases, Pa_{CO_2} 60 mm Hg and Pa_{O_2} 60 mm Hg. A likely diagnosis in terms of acid–base balance is

A. acute metabolic acidosis without renal compensation.

B. chronic metabolic acidosis with considerable renal compensation.

C. acute respiratory acidosis without renal compensation.

D. chronic respiratory acidosis with considerable renal compensation.

E. none of the above.

35. A patient with cirrhosis and ascites has been treated aggressively with a potent diuretic (e.g., furosemide). After a few days he experienced symptoms of weakness, muscle cramps, and postural dizziness. His mental state was characterized by a combination of apathy and confusion. After hospitalization the following data were reported: plasma Na^+, 137 mEq/L; plasma K^+, 2.5 mEq/L; arterial blood gases, arterial pH 7.58 and Pa_{CO_2} of 50 mm Hg. A likely diagnosis is

A. respiratory alkalosis without renal compensation.

B. chronic respiratory alkalosis with considerable renal compensation.

C. metabolic alkalosis without any respiratory compensation.

D. metabolic alkalosis with some respiratory compensation.

E. diabetes insipidus.

ANSWERS

1. D. pK is a constant. pH changes a few percent. HCO_3^- can change several percent (secondary to changes in P_{CO_2}). H_2CO_3 can double or halve ($+100\%$ or -50%).

2. B. All other choices would tend to decrease ventilation.

3. False. Normally, active people produce about 23,000 mEq of volatile acid (H_2CO_3) each day.

4. True.

5. False. Carbonic acid is a volatile acid.

6. False. HCO_3^- cannot diffuse as such across luminal border of tubular cells.

7. C. 3.05 mEq of H^+/min $\times$ 1,440 min/day = 4,392 mEq H^+ secreted per day. Of this, 4,320 mEq was secreted to reabsorb 4,320 mEq HCO_3^- (3.00 $\times$ 1440). The remaining 72 mEq of H^+ secreted generated 72 mEq of new HCO_3^- in the process of urinary acidification.

8. B. A is incorrect because the major part of urinary acidification occurs in the distal nephron. C is incorrect because there is no Na^+ secretion into urine. D is incorrect because the maximal gradient between urine and blood at the level of the proximal tubule is 4:1 (the 1,000:1 gradient occurs in the distal nephon).

9. D. Less than 0.1% of the total H^+ excreted exists as free H^+.

10. A. An overdose of morphine, like a number of other narcotics and sedatives, depresses the medullary respiratory centers. Depression of respiration is discernable even with small doses of morphine (producing no disturbance of consciousness or sleep) and increases progressively as the dose is increased. B is incorrect because the effect of salicylate intoxication on respiration is one of stimulation. Thus, the initial phase of toxicity involves a respiratory alkalosis (which later changes to a metabolic acidosis as other effects of salcylate toxicity become manifest). C is incorrect because high fever stimulates respiration and tends to produce a respiratory alkalosis. D is incorrect because hypoxia of high altitude also stimulates respiration (via peripheral chemoreceptors) and produces a respiratory alkalosis.

11. D. The H^+ concentration will increase by +40 nEq/L. Since for each H^+ released from H_2CO_3, a HCO_3^- is also released, the HCO_3^- concentration also increased by +40 nEq/L. A is incorrect because HCO_3^- cannot react with or neutralize H_2CO_3. In that case, what would the products be? One member of a buffer pair cannot buffer the other member of that pair. B is incorrect because the concentration of bicarbonate does not double (it goes from 24 to 24.00004 mEq/L, an increase of +40 nEq/L). C is incorrect for the same reason, that is, the increase in HCO_3^- is very slight in a bicarbonate solution containing no nonbicarbonate buffers.

12. B. The H_2CO_3 will double when the P_{CO_2} doubles (remember that H_2CO_3 in mEq/L = 0.03 $\times$ P_{CO_2} in mm Hg). Thus, A cannot be correct. C is incorrect because, in the presence of nonbicarbonate buffer, a change in P_{CO_2} is accompanied by an appreciable secondary change in HCO_3^-. D is incorrect because a measurable or appreciable change in pH will occur. Buffers minimize changes in pH but cannot altogether or completely prevent any change in pH.

13. C. This is the mechanism by which altitude causes respiratory alkalosis. A is incorrect because the net effect of altitude hypoxia is one of stimulation, not depression. B is incorrect because hypoventilation does not produce the respiratory alkalosis, *hyper*ventilation does. D is incorrect. When breathing air, one cannot change the Pa_{O_2} by altering respiration without changing the Pa_{CO_2} as well. For example, in an attempt to increase the low Pa_{O_2}, the lungs hyperventilate (which helps the Pa_{O_2} slightly) and therefore blow off more CO_2 (decrease Pa_{CO_2}).

14. C. One aspect of nonrespiratory (metabolic) acid–base imbalance of considerable diagnostic significance is that the bicarbonate concentration is markedly changed and in the same direction as the pH. B is incorrect because the bicarbonate change is not inversely related to pH change. D is incorrect because change in P_{CO_2} and accompanying secondary changes in bicarbonate are not part of uncompensated metabolic acid–base imbalance. Even when P_{CO_2} changes occur in respiratory compensation, the secondary changes in bicarbonate concentration are not the only changes in bicarbonate that have occurred.

15. D. Mineralocorticoid excess leads to a K^+ deficiency. K^+ deficiency is the second most common cause of metabolic alkalosis, resulting from a shift of ECF H^+ into cells and the excretion of acid urine.

16. C. The conditions described in A, B, C, D, and E would be easy to recognize if one additional item of information were included, namely the blood pH. As it is, more thought has to go into distinctions between causes and effects. For example, C fits both respiratory alkalosis (which is what hysterical hyperventilation produces) and renal metabolic acidosis (in question 18). In the case of respiratory alkalosis, the decreased blood P_{CO_2} is the cause of the condition. The decreased blood HCO_3^- represents both a secondary decrease in bicarbonate and a more substantial decrease brought about by renal compensation, as reflected in the production of an alkaline urine.

17. D. The conditions described by D are those of metabolic alkalosis of nonrenal origin, produced here by excessive gain of exogenous HCO_3^-. Since the kidneys are not part of the problem, they can participate in compensation (hence the production of an alkaline urine). The cause of the metabolic alkalosis is the increased blood HCO_3^-. The respiratory system makes a weak attempt at compensation; however, there are limits to the degree of hypoventilation that can be achieved, and the blood P_{CO_2} is only slightly increased (not more than a P_{CO_2} of 60 mm Hg).

18. C. The inhibition of carbonic anhydrase inhibits H^+ secretion by the kidneys and results in a renal metabolic acidosis. In this case, the alkaline urine is the cause of the condition and results in decreased blood HCO_3^- because of all the bicarbonate lost in the urine. The decreased blood P_{CO_2} represents respiratory compensation.

19. E. The condition of hypokalemia from K^+ depletion commonly produces a renal metabolic alkalosis. In K^+ depletion, K^+ comes out of cells in exchange for H^+ going in. This loss of H^+ initiates a metabolic alkalosis. Furthermore, gain of H^+ by renal tubule cells (which release K^+ in exchange for H^+ like other cells) increases the acidification of urine. This "paradoxical" production of an acid urine (in a person with alkalosis) helps sustain the alkalosis. This exaggerated acidification of urine greatly increases gain of HCO_3^- (hence increased blood HCO_3^-) and sustains the alkalosis. The slightly increased blood P_{CO_2} represents limited respiratory compensation. (Remember, respiratory compensation for metabolic alkalosis is always poor.)

20. A. This individual has *respiratory acidosis,* which explains the increased blood P_{CO_2}. The increased blood HCO_3^- is generated in the process of urinary acidification (hence the acid urine), which is the normal response of the kidneys to an acidosis of nonrenal origin.

21. B. A rise in plasma HCO_3^- (metabolic alkalosis). A is incorrect because plasma Cl^- would fall for two reasons: (1) there would be a net Cl^- loss because of HCl loss, and (2) whenever HCO_3^- rises, Cl^- falls. C is incorrect because vomiting does not increase ventilation. If any compensatory change were to occur, it would be a slight decrease in ventilation. E is incorrect because it accompanies acidosis, not alkalosis.

22. D. This is a simple respiratory alkalosis due to an acute hyperventilation.

23. C. This is a metabolic alkalosis due to $NaHCO_3$ ingestion.

24. A. This is a metabolic acidosis due to diabetic ketoacidosis.

25. A,B,C. Dissociation into H^+ and HCO_3^- is spontaneous and does not involve enzymatic function.

26. A, C. There is little protein in either ISF or CSF. The main buffer of these fluids is the HCO_3^-/H_2CO_3 buffer pair. The main buffer of tissue cells is probably ICF protein. The main buffer of erythrocytes is hemoglobin.

27. C. This is an acute respiratory acidosis due to paralysis of respiratory muscles by curare. It is characterized by a primary increase of Pa_{CO_2} (80 mm Hg), a secondary rise in HCO_3^- (30 mEq/L), and a lowered pH (7.2). There has not been enough time for renal compensation to occur.

28. B. This is a respiratory alkalosis due to direct stimulation of medullary respiratory centers by salicylate, which results in hyperventilation. This blows off CO_2 (30 mm Hg) and causes a secondary decrease in HCO_3^- (22 mEq/L). There has not been time for renal compensation. There has also not been time for the metabolic acidosis that occurs in the later phase of salicylate toxicity.

29. D. This is a metabolic alkalosis caused by gain of exogenous $NaHCO_3$. Note the high plasma HCO_3^-. The Pa_{CO_2} is increased slightly due to respiratory compensation in the form of hypoventilation (50 mm Hg). Renal compensation in the form of increased HCO_3^- excretion has prevented this from becoming a more severe alkalosis.

30. E. A diabetic patient with Kussmaul breathing has a metabolic acidosis with respiratory compensation.

31. A. The emphysema patient has a respiratory acidosis due to CO_2 retention and a renal compensation in the form of increased acidification of the urine. Point A shows fairly good but incomplete compensation typical of chronic respiratory acidosis.

32. C. The balloonist has an acute respiratory alkalosis due to altitude hypoxia. There has not been sufficient time for renal compensation.

33. D. After undergoing a chronic respiratory alkalosis due to sustained stay at high altitude, this person achieves complete renal compensation (lowers plasma HCO_3^- by increased urinary excretion). Complete or nearly complete renal compensation is characteristic of chronic respiratory alkalosis.

34. D. Choices A and B can be eliminated right away. Although the patient has an acidosis of some sort, it cannot be metabolic (nonrespiratory) because of the higher than normal arterial HCO_3^- (recall that metabolic acidosis is characterized by a decreased HCO_3^-). Therefore, it must be a respiratory acidosis. Now to distinguish between acute or chronic respiratory acidosis. An uncompensated acute respiratory acidosis with a Pa_{CO_2} of 60 mm Hg would have been accompanied by an arterial HCO_3^- of about 26 mEq/L (a small secondary increase in HCO_3^-) and an arterial pH of 7.26. Not the case here. Note that the pH is not far from normal and that the arterial HCO_3^- is 8 mEq more per liter than normal (indicating that there must have been time for nearly complete renal compensation in the form of increased bicarbonate generation). That is, this is chronic respiratory acidosis with plenty of time for renal compensation. Although the renal compensation is not perfectly complete, it is quite good (typical of chronic respiratory acidosis). The history and physical examination suggest chronic pulmonary emphysema; poor alveolar ventilation explains the high Pa_{CO_2} and low Pa_{O_2}.

35. D. Judging from the alkaline arterial blood (pH 7.58), the patient must have some type of alkalosis. Choices A and B can be eliminated; this is not a respiratory alkalosis, since there is an elevated Pa_{CO_2} (it would be below normal in respiratory alkalosis). Choice E can also be eliminated because nothing indicates diabetes insipidus (indeed, patients with severe cirrhosis and ascites frequently have an increased rate of ADH secretion). Therefore, this is a metabolic alkalosis with some respiratory compensation, as indicated by the above-normal Pa_{CO_2}. When plasma K^+ is as low as 2.5 mEq/L or less, muscle weakness and cramps tend to occur. The hypokalemia can also explain the apathetic, stuperous-confusional state. The postural hypotension is a sign of hypovolemia due to overly aggressive diuretic therapy.

35

Hemofluidity and Hemostasis

M. MASON GUEST

OBJECTIVES

After completion of this chapter, you should be able to:

1. Understand the functional features of hemostasis and hemofluidity.

2. Use the nomenclature of blood components involved in coagulation and fibrinolysis.

3. Know and understand how other barriers to hemorrhage assist the coagulation system in promoting hemostasis.

4. Describe the enzyme sequences culminating in the formation of a hemostatic fibrin clot.

5. Describe the enzyme sequences involved in fibrinolysis.

6. Understand the roles of the physiological anticoagulants in maintaining hemofluidity.

7. Describe the most common clinical disorders involving the hemostatic mechanisms.

Blood serves as a medium for transporting essential materials, energy, and products of cellular metabolism to and from the close proximity of fixed tissue cells. Homeostasis, the maintenance of a relatively constant internal environment, would not be possible in vertebrates and other large multicellular animals without a circulating medium. The categories of substances and energy transported include (1) water; (2) electrolytes; (3) amino acids, polypeptides, and proteins; (4) lipids; (5) carbohydrates; (6) respiratory gases (both in cells and in plasma); (7) metabolic products; (8) immune materials (both dissolved in plasma and associated with leukocytes); (9) hormones; (10) enzymes or proenzymes; (11) enzyme inhibitors; and (12) heat. Obviously, this categorization includes some overlap (e.g., proteins and enzymes, proteins, and some hormones). Furthermore, the heat distributed by the blood within the organism is not the only form of energy being carried. The energy transmitted to the blood in the form of pressure and flow by the heart is in part used in the filtration of fluid and dissolved substances from the capillaries into the interstitial space and into the renal tubules.

If the blood is to be an effective transport medium, mechanisms must be available to maintain its volume and composition. In other words, although blood serves the function of maintaining a relative constancy of the internal environment for fixed cells in the organism, it cannot perform this function unless adequate means are provided to insure its own homeostatic state. You have already learned about some of the ways in which blood volume and composition are adjusted and controlled. Obviously, when blood vessels are torn or ruptured, rapid loss of the circulating medium occurs and, if the hemorrhage is not arrested, death occurs as a result of hypovolemia.

This chapter will deal with the maintenance of the blood in a fluid state under normal conditions and with the conversion of blood to a semi-solid state (the blood clot) when vessels are damaged. We will also discuss other means by which hemorrhage can be arrested.

HEMOSTASIS

GENERAL PRINCIPLES

A well-balanced homeostatis is a critical determinant of whether blood can perform its homeostatic functions. If sufficient blood is lost by hemorrhage, its volume is reduced below the critical level required for cardiovascular function. Furthermore, essential materials are removed from the body. On the other hand, overactive hemostatic processes result in blocking of important vascular channels required for maintaining homeostasis in a tissue or organ.

Blood is normally a fluid in vivo, but its inherent properties sometimes cause it to be converted to a semisolid state. This alteration, called clotting or blood coagulation, occurs rapidly when the blood is removed from its containing vessels; but overt clotting within the vessels of the living organism is a relatively rare occurrence, except when severe damage to such vessels is present. The clot prevents the loss of circulating fluid from torn or ruptured blood vessels. In contrast, the formation of a thrombus within an intact vessel is a pathological process. If a thrombus completely occludes the vessel, blood is not delivered to tissues supplied by that vessel and subsequent degenerative changes occur in the tissue it supplies.

Clotting is by no means the only hemostatic mechanism, and it might be secondary to other processes. For example, patients with afibrinogenemia, a congenital absence of fibrinogen in plasma, have less difficulty in maintaining a hemostatic balance than would be predicted from our present knowledge of the clotting system. In some unknown way, other hemostatic mechanisms must compensate for the absence of clot formation. Several hemostatic safeguards have been described: (1) ruptures of small blood vessels ("capillary bleeding") are blocked by platelet plugs; (2) constriction of blood vessels, whether resulting from direct action by the agent causing the injury, or from the release of vasoconstrictor substances (such as the catecholamines or serotonin), reduces the rate of blood loss from ruptured vessels; (3) tissue pressure, when hemorrhage occurs within tissues, can counterbalance the pressure within the torn vessel and reduce blood loss. The detailed biochemistry and physiology of the several hemostatic mechanisms and their apparently exquisite interrelationships can only be inferred at present because there are limited techniques available to investigate these processes in vivo.

Although circulating blood contains all elements necessary for the formation of a clot, manifest clotting in vivo does not ordinarily occur in the absence of injury to a blood vessel. Sufficient injury to a vessel initiates a series of reactions, resulting in the formation of a tenacious and protective clot; in the absence of injury, or if the injury is minor, the clotting sequence is not catalyzed effectively. The balance between normal fluidity and clotting of blood is critical to the survival of the organism. An imbalance that leads to inadequate clotting, such severe injury to a hemophiliac, can result in fatal hemorrhage; an imbalance leading to hypercoagulability of blood can result in thrombosis, embolism, and infarction, and possibly in irreversible changes in tissues and death of the organism. Even minor disarrangement in the fluid–semisolid equilibrium of blood can have pathophysiological significance, since either small hemorrhages or thrombi temporarily affect the functional integrity of tissues.

Clots formed within the blood vessels or in the tissues can remain and become organized through the ingrowth of fibroblasts and other cellular elements, or they might be broken down through the action of proteolytic enzymes. Since plasma contains a precursor of a potent fibrinolytic enzyme, fibrinolysin (plasmin), the potential for clot dissolution is always present. Thus, the fibrinolytic enzyme system is also a factor in the balance between blood fluidity and a semisolid state; excessive fibrinolytic activity can lead to hemorrhages from defective blood vessels, whereas an inadequate fibrinolytic enzyme system can tip the balance in the opposite direction through the absence of clot dissolution.

Localized Hemorrhages

Even though the amount of blood lost from the circulation is insignificant, localized hemorrhages can disrupt the function of affected tissues. Altered function results from partial or complete interruption of blood flow in the vessels that supply the tissue and from altered tissue pressure resulting from hemorrhage into a tissue and the filling of potential cavities with blood. For example, cardiac tamponade from hemorrhage into the pericardium interferes with cardiac filling, hemothorax reduces pulmonary vital capacity, intracranial hemorrhage blocks function of the central nervous system (CNS), and hemorrhage into joint capsules reduces mobility and can result in joint malfunctions. In addition, even though the hemorrhage stops and the blood clot is gradually lysed and reabsorbed, the removal of the residue requires an expenditure of energy and a diversion of functional processes.

Intravascular Coagulation

Thrombosis, or intravascular coagulation, can be as inimical to life as hemorrhage. If a thrombus forms in a vessel that furnishes the major blood supply to a vital organ, such as the heart or medulla oblongata, death results from failure of the ischemic organ to perform its function. Smaller thrombi or thrombi in vessels supplying blood to only part of an organ, although not causing immediate death of the organism, cause morphologic and functional changes.

Embolism

A possible consequence of thrombosis is an embolism. If the thrombus breaks loose from its attachment to the intima of a blood vessel, it is carried in the circulating blood to a new location. Most venous emboli lodge in the pulmonary arterial bed, because this is the first region in the odyssey of the venous clot in which a progressive decrease occurs in the diameter of the blood vessels. Arterial emboli are lodged in the systemic arterial bed, their final destination being determined by their locus of origin and their size with reference to the diameter of the arterial lumen. The consequences of embolization are not basically different from the effects of thrombus formation; however, the effects of thrombi in the more peripheral veins may be relatively benign when compared with the effects from embolization of pulmonary vessels; in the latter case, the blood supply to the left heart is often reduced to a critical level.

COAGULATION

A clot, formed from whole blood, consists of a fibrin network with platelets adhering to the fibrin strands; it contains erythrocytes, leukocytes, and serum (plasma minus fibrinogen) entrapped within the fibrin mesh. The fibrin strands adhere to injured surfaces; thus the clot becomes a barrier to hemorrhage.

To initiate processes culminating in the formation of a clot, procoagulants in blood must be activated. Blood contains a number of substances classified as procoagulants, but it also contains inhibitors of clotting, the anticoagulants. The ratio of activated procoagulants to anticoagulants determines at any instant in a given locus whether or not fibrin is formed. The normal fluid state of blood is partly a reflection of the predominance of the anticoagulants. However, in the event of injury to or rupture of a vessel, initiation of the clotting sequence results in sufficient activation of procoagulants to overwhelm the anticoagulants in the immediate vicinity of the damage. The result is local clot formation, prevention of further blood loss and subsequent repair of the damaged vessel. (The fibrin strands act as a framework for fibroblasts.)

As a convenient basis for discussion, the process of coagulation can be divided into four phases, described in Figure 1.

Phase 1: Reactions Leading to Activation of Factor X

The early stages of phase 1 can be initiated via either of two distinct and perhaps independent routes, the intrinsic and extrinsic pathways. Coagulation processes comprising the intrinsic pathway begin when blood comes in contact with foreign, usually wettable surfaces; coagulation is initiated via the extrinsic pathway when blood is exposed to particulate lipoprotein material (tissue thromboplastin) from damaged cell membranes. (See Fig. 2.)

Although coagulation in vitro can be effectively consummated by the activation of either the intrinsic or extrinsic pathways, it appears that, for effective hemostasis, in vivo, both pathways are usually activated perhaps simultaneously. Involvement of two pathways for the initiation of coagulation may be necessary to activate a sufficient amount of procoagulants in the final common pathway to momentarily overwhelm inhibitors. Thus, if only one pathway is activated in vivo, the sequence of coagulation reactions is usually quenched and little or no fibrin is formed. Evidence for this interpretation is demonstrated by the fact that severe and sometimes fatal bleeding diatheses develop if a deficiency in a procoagulant occurs in either pathway (e.g., factor VIII in the intrinsic pathway or factor VII in the extrinsic pathway) and the other pathway contains normal amounts of its procoagulants.

Intrinsic System

When blood is removed (without contamination by injured tissue) from a blood vessel and placed in a glass test tube, initiation of clotting results from activation of the

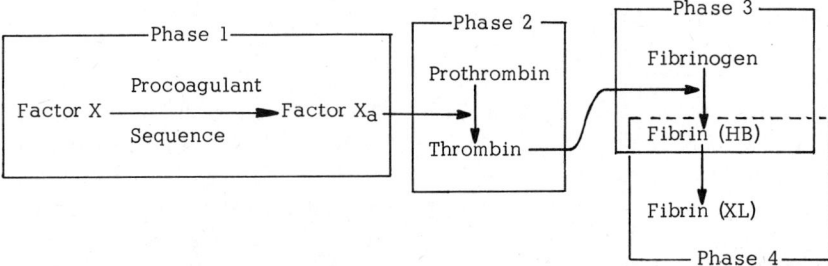

Figure 1 Coagulation. The first phase culminates in the activation of factor X (prothrombin converting enzyme) (see Fig. 2). During the second phase, factor X_a (subscript a designates the activated enzyme) splits prothrombin and releases thrombin (see Fig. 3). The third phase involves the splitting off by thrombin of small molecular-weight-charged peptides from fibrinogen and polymerization of the fibrin monomers (major portion of the fibrinogen molecule) (see Fig. 4). The fourth and final phase is the conversion of the unstable hydrogen-bonded fibrin formed by action of thrombin during third phase to a stable cross-linked fibrin (see Fig. 5). The major features of phases 2, 3, and 4 are agreed on by most investigators. However, the details of phase 1 have not yet been clearly delineated.

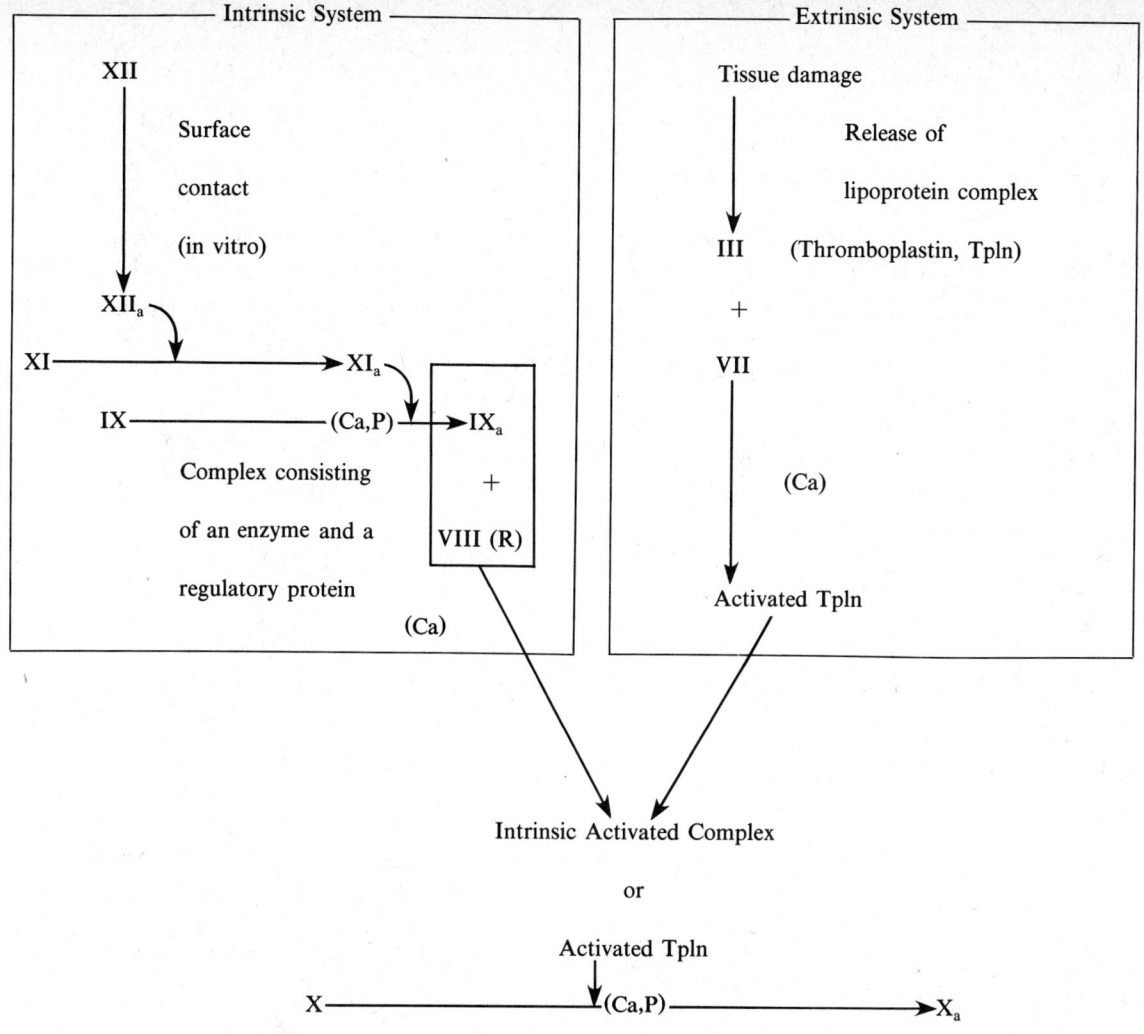

a, where used after Roman numeral, denotes *activated enzyme.*

(P)   denotes *proteolytic step.*

(R)   denotes *regulatory protein.*

(Ca)   denotes *calcium ion dependency.*

Figure 2 The intrinsic and extrinsic pathways that constitute phase 1 of the coagulation process. A lower case letter (a) after a Roman numeral denotes an activated factor, the letter (P) denotes a proteolytic step, and (Ca) indicates calcium dependence.

intrinsic system. Factors XII (glass factor) and XI (plasma thromboplastin antecedent, or PTA) are activated by contact with a nonendothelial surface (such as glass or kaolin). If the test tube has been made nonwettable (e.g., by coating its surface with silicone or with a parafin oil), clotting may be delayed for an hour or more. When it comes into contact with glass, the factor XII proenzyme is converted to its active form, XII_a. However, whether factor XII actually participates in the activation of the intrinsic sequence in vivo is uncertain, since patients lacking this protein do not have bleeding difficulties. Nevertheless, in some way, factor XI is converted to XI_a, and XI_a in turn converts the proenzyeme, factor IX (Christmas factor) to IX_a, a calcium-dependent enzyme. Since factors IX and IXa have different electrophoretic mobilities and different elution profiles following gel filtration, activation by factor XI_a is believed to involve proteolysis. Upon activation of factor IX, factors IX_a and VIII form a complex in which factor IX_a is the enzyme and factor VIII is probably a regulatory protein. The rate of formation of the complex is accelerated by traces of thrombin (autocatalytic reaction). (See Table 1, which lists the nomenclature of each of the coagulation and platelet factors.)

Table 1 Nomenclature of Coagulation and Platelet Factors

Factor No.	Most Commonly Used Name	Investigators
Coagulation factors		
I	Fibrinogen	
II	Prothrombin	
III	Thromboplastins	
	Lung	
	Brain	
	Other tissues	
IV	Calcium	
V	Accelerator globulin (ACG)	Ware, Guest, and Seegers
	Labile factor	Quick
VII	Serum prothrombin conversion accelerator (SPCA)	deVries and Alexander
	Stable factor	Stefanini
	Autoprothrombin I	Seegers
VIII	Antihemophilic factor (AHF)	Brinkhous
	Antihemophilic globulin	Patek and Taylor
	Platelet cofactor	Johnson
IX	Plasma thromboplastin component (PTC)	Aggeler
X	Stuart factor (Stuart-Prower)	Hougie
XI	Plasma thromboplastin antecedent (PTA)	Rosenthal
XII	Hageman factor	Ratnoff
	Glass factor	
	Contact factor	Margolis
XIII	Fibrin stabilizing factor (FSF)	
	Laki-Lorand factor (L-L factor)	
	Fibrinase	Loewy
Platelet factors		
1	Platelet accelerator (equivalent to factor V in plasma)	Seegers
2	Synergistic with thrombin in fibrinogen conversion)	
3	Lipid activator	Seegers

Before considering the role of the intrinsic activated complex formed in the intrinsic sequence, the formation of an activated thromboplastin in the extrinsic sequence is described.

Extrinsic System
Compared with the relatively slow intrinsic pathway, the extrinsic pathway is rapid. The extrinsic system is activated whenever injured cells (e.g., lacerated vessels) release factor III (tissue thromboplastin), a lipoprotein present in the membranes of cells. In the presence of a plasma protein, factor VII (stable factor), and Ca^{2+}, tissue thromboplastin is converted into an activated thromboplastin. Both the intrinsic and the extrinsic pathways culminate in the formation of complexes that have similar activities, but which are not identical in their chemical composition. From this point, the subsequent coagulation sequence of phase 1, whether initiated by the intrinsic or extrinsic system, follows the same course, or final common pathway.

Final Common Pathway of Phase 1
Either, and in the in vivo situation, perhaps both, the intrinsic activated complex and activated thromboplastin convert factor X to X_a. This conversion of factor X to X_a apparently involves a proteolytic step that uncovers the active enzymatic site on this key protein molecule.

Phase 2: Formation of Thrombin
Factor X_a is a proteolytic enzyme that splits the prothrombin molecule; a portion of the prothrombin molecule becomes the enzyme thrombin (Fig. 3). The proteolytic step is calcium ion dependent, and both factor V and lipid appear to be involved in the conversion of prothrombin to thrombin. Factor V is probably a regulatory protein. Platelets can substitute for factor V, and platelet factor 1 could actually be plasma factor V adsorbed onto the surface of platelets. Lipid is also supplied by platelets.

Phase 3: Splitting of Fibrinogen by Thrombin and Polymerization of Fibrin
The significant events of phase 3 are the action of thrombin, a proteolytic enzyme, on fibrinogen, as well as the polymerization of the fibrin monomer formed by this action. When thrombin is formed in phase 2, it then splits

$$X_a, V(R) + phospholipid$$
$$II (prothrombin)—(Ca, P)\longrightarrow thrombin$$

Note that
1. X_a is a proteolytic enzyme.
2. Regulatory protein V is present in plasma, but it is also apparently absorbed onto platelet membranes, and, when supplied by platelets, is termed platelet factor 1.
3. Phospholipid is supplied from platelets and has been termed platelet factor 3.

Figure 3 The conversion of prothrombin to thrombin as phase 2 of the coagulation process. (Ca) indicates calcium dependence, (P) denotes the presence of a proteolytic step, and (R) denotes function of a regulatory protein.

off negatively charged peptides (peptides A and B) from the N-terminal portion of the fibrinogen molecule. The loss of these charged peptides permits the remaining molecule (fibrin monomer), of slightly less molecular weight than fibrinogen, to undergo both end-to-end and side-by-side polymerization. The result of this polymerization is the formation of high-molecular-weight fibrin strands, sometimes called soluble fibrin. The bonding between the fibrin monomers in this stage of polymerization is through relatively weak hydrogen bonds, hence it can be identified as hydrogen-bonded fibrin, or fibrin HB. Fibrin HB forms a relatively weak clot that dissolves in vitro in a 5 M solution of urea and in strong acids or bases. Although polymerization of fibrin monomers to fibrin HB is facilitated by the presence of platelet factor 2, this factor is not essential to the formation of fibrin.

Phase 4: Fibrin Stabilization
The unstable clot formed during phase 3 is not adequate for effective hemostasis, and it is not the normal end product of coagulation in vivo. Persons with normal clotting

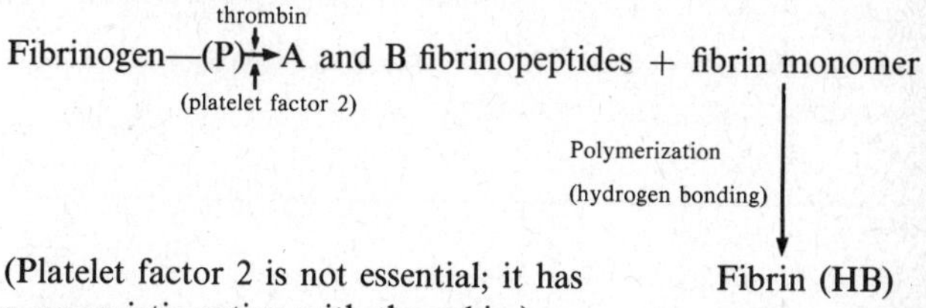

$$\text{Fibrinogen—(P)} \overset{\text{thrombin}}{\underset{\text{(platelet factor 2)}}{\rightarrow}} \text{A and B fibrinopeptides + fibrin monomer}$$

Polymerization
(hydrogen bonding)

(Platelet factor 2 is not essential; it has a synergistic action with thrombin.)   Fibrin (HB)

Figure 4 Fibrin HB denotes hydrogen-bonded fibrin a relatively unstable form of fibrin that is soluble in 5 M urea and at pH values below 4.5 or above 9.

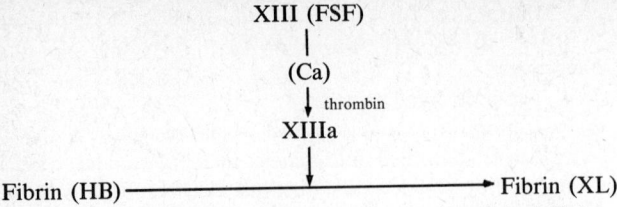

Figure 5 HB, Hydrogen-bonded polymer; XL, cross-linked fibrin.

systems have present in their plasma a proenzyme, called fibrin stabilizing factor (FSF, or factor XIII). Thrombin, in a calcium ion dependent reaction, converts factor XIII to an active enzyme, factor XIII_a. Factor XIII_a is a transglutaminase that brings about the formation of covalent bonds between glutamine and -amino groups of lysine side chains on adjacent fibrin monomers. The resulting stabilized, insoluble fibrin has been called stabilized or insoluble fibrin, but a more precise term is cross-linked fibrin (fibrin XL) (Fig. 5). Whether a clot is composed of fibrin HB or fibrin XL can be tested by determining its solubility in solutions of urea or monochloroacetic acid. Fibrin XL is not soluble in these solutions since its polymerized fibrin molecules are held together by strong covalent bonds.

Some people have a rare congenital deficiency of factor XIII and, as one might expect, they have a hemorrhagic tendency due to the instability of their clots.

Figure 6 depicts the sequential phases of the coagulation process. The process is a relatively complex series of interdependent enzymatic reactions. Several features are remarkable: (1) The process can be initiated via two distinct pathways, the intrinsic and extrinsic; (2) both the intrinsic and extrinsic pathways must be activated for effective in vivo hemostasis; (3) the sequence of clotting reactions is an outstanding example of biochemical amplification (e.g., activation of a minute quantity of an enzyme in the initial reaction activates a larger store of enzyme in the next reaction); (4) many of the reactions are calcium dependent; (5) a number of the activation steps involve proteolysis; and (6) thrombin acts autocatalytically to increase the reactivities of two regulatory proteins: factors V and VIII.

PLATELETS AND HEMOSTASIS

A reasonable understanding of the clotting process requires an insight into the functions of platelets. Platelets

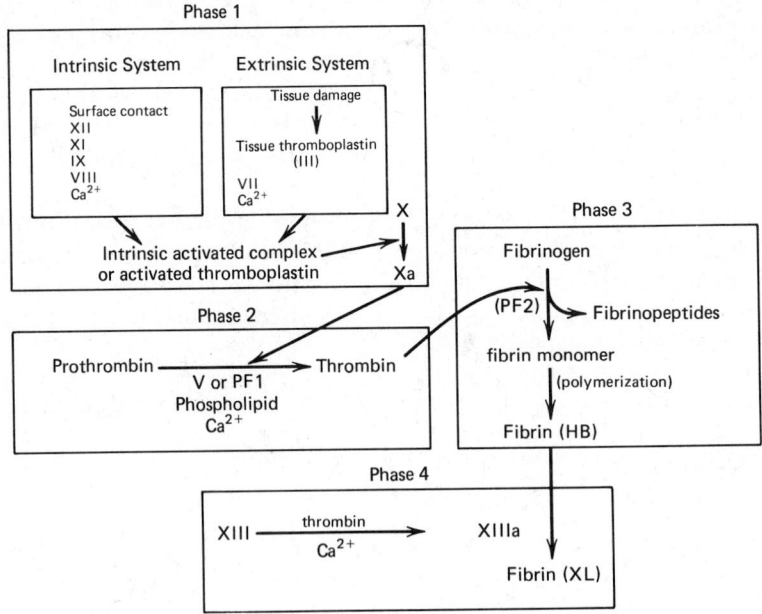

Figure 6 Summarization of Figures 2, 3, 4, and 5, showing relationships.

are cytoplasmic fragments measuring 2–4μm in diameter, derived from megakaryocytes of bone marrow. Their normal concentration in the blood is 150,000–350,000 per cu mm.

Platelets have several functions:

1. *Endothelial supporting function:* Present evidence indicates that platelets are necessary for the repair and maintenance of the endothelial lining of blood vessels.
2. *Platelet plugs:* Platelets form plugs that prevent blood from oozing out of small tears in the endothelium of microvessels. The bleeding time is a measure of the effectiveness of platelets in arresting bleeding from microvessels. In thrombocytopenia—deficient concentration of platelets in the circulating blood—the bleeding time is prolonged, but the clotting time is usually normal.
3. *Clotting factors:* Platelets release clotting factors when they break down (lyse). The platelet clotting factors are given arabic numbers: Platelet factor 1 is equivalent in its action to plasma factor V; platelet factor 2 synergizes with thrombin in the conversion of fibrinogen to fibrin; platelet factor 3 is a phospholipid involved in the interactions of IX$_a$ and VIII in generating the intrinsic activation complex.
4. *Clot retraction:* Platelets are required for clot retraction: after a clot forms, it becomes reduced in size as serum is expressed. In some manner not yet understood, intact platelets exert tension on the fibrin strands. Normal clot retraction is an indication that an adequate number of normal platelets are present.
5. *Serotonin and epinephrine:* When platelets release these substances vasoconstriction occurs, which probably assists in hemostasis.

Platelets adhere to foreign surfaces; they also agglutinate in the presence of certain organic materials. For example, thrombin causes platelets to agglutinate irreversibly and adenosine diphosphate (ADP) induces an agglutination that is usually reversible. When platelets come into contact with collagen, they adhere in clumps (agglutinate).

ANTICOAGULANTS AND ANTICOAGULATION

Physiologic inhibitors of clotting reactions prevent excessive intravascular coagulation. Slow formation of thrombin may be occurring continuously within the body, but the physiologic antithrombins inactivate it as it is formed. Thus, actual clotting occurs only with an explosively rapid thrombin generation when the inhibitors are temporarily overwhelmed. Among the physiologic clotting inhibitors (anticoagulants) heparin, the antithrombins and antithromboplastins appear to be most important.

Heparin
This polysulfuric acid ester is produced and stored in mast cells. With a cofactor in the albumin fraction of plasma, it blocks the conversion of prothrombin to thrombin; and with another plasma factor (antithrombin II), it inactivates thrombin. Heparin is used clinically as an anticoagulant.

Antithrombins
Several proteins in plasma or derived from plasma have the property of inactivating thrombin: (1) Fibrin adsorbs thrombin and thus as fibrin forms, it begins to remove the thrombin responsible for its formation; (2) the cofactor in plasma that acts with heparin is antithrombin II; (3) several substances in plasma act independently of heparin to inactivate thrombin; and (4) when fibrinolysin (plasmin) is present, it partly digests fibrinogen and fibrin. Some of the peptide fragments resulting from the fibrinolytic breakdown of fibrinogen and fibrin are anticoagulants. In addition to acting as antithrombins, some of the larger split products are combined in the clot, but because they are fragments, the clot is of poor quality.

Antithromboplastins
Tissue thromboplastin is slowly inactivated when it comes into contact with plasma. The chemistry of this inactivation is uncertain.

Coumarin-Type Drugs
Dicumarol (bishydroxycoumarin) was found to be the substance in spoiled sweet clover hay that causes fatal hemorrhages in cattle. Dicumarol and synthetic coumarin

drugs, such as Warfarin, have been postulated to be competitive inhibitors of vitamin K, which is required in the process of prothrombin synthesis. When administered orally, coumarin drugs are absorbed and then block the formation of prothrombin in the liver. Since at the start of administration of coumarin drugs, prothrombin is usually in normal concentration in the blood, about 36h is required before the prothrombin level is effectively reduced. Because vitamin K is required for the synthesis of all members of the prothrombin complex (i.e., prothrombin and factors VII, IX and X), the blood concentrations of all these procoagulants are reduced during coumarin therapy. Coumarin drugs have no anticoagulant action when they are added to the blood in vitro.

Substances that Remove or Bind Calcium

To prevent blood from clotting in vitro after it has been removed from the body, substances that remove calcium ions are commonly used. Among such substances are (1) EDTA, a chelating agent; (2) citrate, which prevents the ionization of calcium; and (3) oxalate, which forms insoluble calcium salts. Substances that remove or bind calcium must never be introduced in vivo in attempts to control the coagulation process.

THE FIBRINOLYTIC SYSTEM

As with the blood clotting system, a multiplicity of terms has been used to designate fibrinolytic factors. Two parallel systems of nomenclature are used. The first system has historical precedence and is specific with respect to preferential substrate. Thus, the terms profibrinolysin and plasminogen, fibrinolysin and plasmin, and antifibrinolysin and antiplasmin are sets of synonyms. The profibrinolysin–fibrinolysin–antifibrinolysin system of nomenclature is used in this chapter.

The salient features of the fibrinolytic enzyme system are relatively uncomplicated. Blood clots, when formed in vitro under sterile conditions, usually persist for several weeks, but occasionally break down rapidly. The breakdown results from a proteolytic digestion of fibrin by an enzyme called fibrinolysin. Fibrinolysin is derived from an inactive plasma precursor, profibrinolysin, through activation in various ways. Also present in plasma are inhibitors of fibrinolysin, termed antifibrinolysins. (See Fig. 7.)

ACTIVATION OF THE FIBRINOLYTIC SYSTEM

Activators of profibrinolysin are present in plasma, in the endothelium of blood vessels, and in urine. These activators are proteolytic enzymes. By splitting the proenzyme profibrinolysin, an active site is uncovered; the active site that gives fibrinolysin its proteolytic activity hydrolyzes arginine–lysine C-N linkages in fibrinogen and fibrin. Arginine–lysine C-N bonds in factors V and VIII are also hydrolyzed by fibrinolysin.

The plasma activator of profibrinolysin is probably derived from the endothelium of blood vessels, particularly the endothelium of capillaries and veins. As blood flows through capillaries and veins, it normally washes away some of the endothelial activator. If blood is prevented from flowing for a short period and then flow is again started, the concentration of activator in the plasma flowing out of the vessel is markedly increased, indicating that endothelium is continuously producing the activator. The urine activator, named urokinase, may be derived in part from endothelium of blood vessels, but it is also produced by renal tubular cells.

Apparently small amounts of profibrinolysin are being continuously activated to fibrinolysin within the circulation. However, circulating fibrinolysin is usually absent in detectable quantities because (1) inhibitors of activators are normally present in plasma, and (2) antifibrinolysins in plasma inactivate fibrinolysin. Nevertheless, a fulminating fibrinolytic crisis may develop in shock, in patients undergoing thoracic surgery, in eclampsia during pregnancy, and in some other pathological states. Presumably, under these conditions, large amounts of an activator are released into the circulation. If sufficient amounts of fibrinolysin are formed, fibrinogenolysis (proteolysis of fibrinogen) occurs together with proteolysis of factors V and VIII. Because of the inactivation of clotting factors and because some of the split products of fibrinogen have anticoagulant activity, severe hemorrhage may then occur.

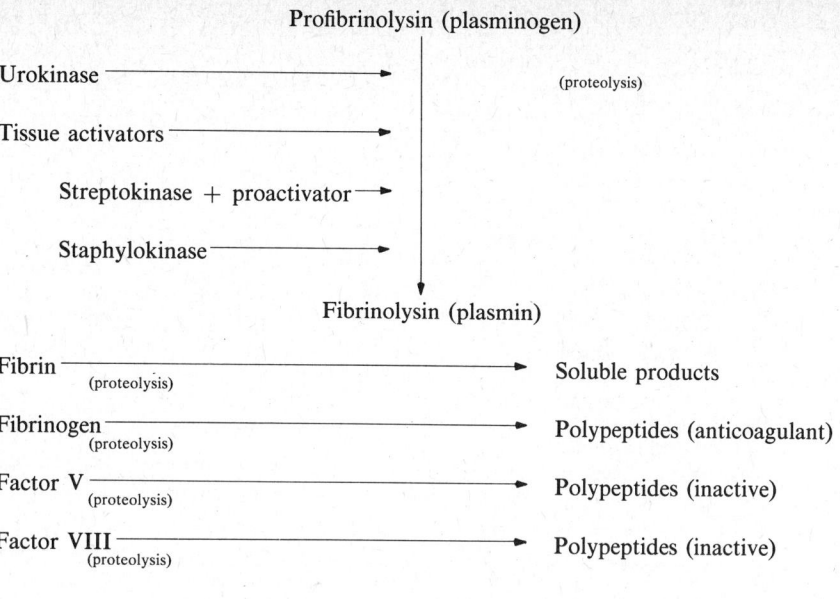

Profibrinolysin (plasminogen)

Urokinase ——————————→   (proteolysis)

Tissue activators ——————————→

Streptokinase + proactivator —→

Staphylokinase ——————————→

↓

Fibrinolysin (plasmin)

Fibrin ————————————————————→ Soluble products
(proteolysis)

Fibrinogen ————————————————→ Polypeptides (anticoagulant)
(proteolysis)

Factor V ——————————————————→ Polypeptides (inactive)
(proteolysis)

Factor VIII ————————————————→ Polypeptides (inactive)
(proteolysis)

Inhibitors
Antikinases inactivate activators of profibrinolysin.
Antifibrinolysins (antiplasmins) inactivate fibrinolysin.
Epsilon aminocaproic acid (EACA) blocks activation of profibrinolysin and inactivates fibrinolysin

Figure 7 Fibrinolytic enzyme flow chart showing the interrelationship between some of the fibrinolytic components.

Urokinase, an activator of profibrinolysin found in the urine of all vertebrate species, has been obtained in a highly purified state by extraction from urine. It is nonantigenic and has been approved for clinical use in dissolving pulmonary thromboemboli.

β-hemolytic streptococci produce an exotoxin, named streptokinase. Streptokinase can be obtained in purified form; when added to human blood, it reacts with profibrinolysin to form a potent activator of profibrinolysin. Streptokinase is also being tested as a clinical thrombolytic agent.

According to currently available evidence, the fibrinolytic enzyme system has three primary functions: (1) It acts as a brake on the clotting system; (2) it causes the dissolution of fibrin, which would otherwise block circulatory flow; and (3) it appears to be responsible for cleaning up fibrin debris as blood passes through capillaries. The latter function is inferred from evidence that erythrocytes tend to adsorb fibrinogen and fibrin, and from the information that capillary endothelium releases a relatively large amount of fibrinolytic activator. Thus, according to this hypothesis, erythrocytes are cleaned of adherent fibrin as a result of a local fibrinolytic activation while they traverse a capillary. The rheological forces present in capillaries may also assist in removing fibrin fragments from erythrocytes.

BLEEDING AND OTHER HEMOSTATIC DISORDERS

Coagulation Defects, Especially Those Involving Factors VIII and IX

All the plasma coagulation factors are subject to genetic or acquired deficiencies, or both (Table 2). However, the genetic features of deficiencies in factors VIII and IX have

Table 2 Nomenclature of Factors

Factor No.	Most Commonly Used Name	Nature of the Factor	Stability	Present in Serum	Genetic Features	Absence or Deficiency Causes
I	Fibrinogen	Euglobulin, between β and γ fractions; MW = 340,000; negatively charged (fibrinopeptides carrying charge are split off by thrombin); normal concentration 200–400 mg	Circulating half-life 3–6 days; relatively stable in stored blood	No	Rare; autosomal recessive	Mild to moderate hemorrhagic condition; excessive menstrual bleeding
II	Prothrombin	α-2 globulin; MW 65,000; a proenzyme or zymogen converted by proteolytic action to thrombin, a highly specific proteolytic enzyme; synthesis is vitamin K dependent. Normal concentration 10–20 mg%.	Circulating half-life 24–48 h; relatively stable in stored blood	$\downarrow$	Rare; most deficiencies are acquired	Acquired deficiency due to vitamin K lack, liver disease, or overtreatment with coumarin drugs; frank hemorrhage can occur if prothrombin falls below 10%
III	Tissue thromboplastin	Particulate lipoprotein material that has not been chemically characterized; lung and brain tissue extracts are most potent thromboplastins; lipid is probably phosphatidyl serine and/or phosphatidyl ethanolamine	Not normally present in blood; extracts of tissues are relatively stable	—	Present in all tissues, regardless of genetic variations	Always present in tissues
IV	Calcium	Calcium ions are required for all phases of coagulation process except conversion of fibrinogen to fibrin-HB	Stable	Yes		Calcium in living animal *always* in sufficient concentration for coagulation; anticoagulation in vitro may be produced by removing calcium or preventing its ionization
V	Accelerator globulin	Globulin; MW > 300,000; appears to act as a regulatory protein in conversion of prothrombin to	Poor stability	No	Automal recessive	Acquired deficiency results from severe liver disease, acute leukemia collagen diseases, DIC, or

610

Table 2 (Continued)

		thrombin; equivalent to platelet factor 1; converted to active form by thrombin				radiation therapy; hemorrhage occurs in various parts of the body; hemarthroses are rare
VII	Serum prothrombin conversion accelerator (SPCA)	β-Globulin; MW, 69,000; accelerates activation of factor X in presence of tissue thromboplastin; member of prothrombin complex; synthesis is vitamin K dependent	Circulating half-life about 5 h; heat labile; stable at 4°C for 2 weeks or longer	Yes	Incompletely autosomal recessive	Acquired: vitamin K deficiency, liver disease; hemorrhagic manifestations are mild to moderately severe; easy bruising and ecchymoses are relatively common; hemarthroses rare
VIII	Antihemophilic factor (AHF)	β-Globulin; MW, 200,000; involved only in formation of intrinsic activated complex; appears to be converted to active form by action of thrombin	Circulating half-life 2–8 h; in vitro half-life about 24 h	No	Sex-linked recessive	Hemorrhagic manifestations may be extremely severe; hemarthroses are characteristic of moderate to severe deficiency
IX	Christmas factor	β-Globulin involved only in formation of intrinsic activated complex; member of prothrombin complex; synthesis is vitamin K dependent	More stable than factor VIII	Yes	Sex-linked recessive	Similar to deficiencies in factor VIII
X	Stuart factor	α-Globulin, which when activated becomes proteolytic enzyme with highly specific action in splitting prothrombin to form thrombin; vitamin K dependent; member of prothrombin complex	Relatively stable in activated form (2 mo at 40°C)	—	Autosomal recessive (congenital deficiency rare)	Acquired: liver disease, vitamin K deficiency; hemorrhagic states that may clinically resemble hemophilia
XI	Plasma thromboplastin antecedent (PTA)	β-Globulin; MW, 100,000–200,000, which apparently interacts with activated factor XII to initiate formation of intrinsic activated complex; has been reported to be vitamin K dependent	Relatively stable in stored blood	Yes	Autosomal recessive (more common among Hebrews)	Spontaneous hemorrhage and hemarthroses are rare; easy bruising is common; severe hemorrhage may follow surgical procedures

Table 2   (Continued)

XII	Hageman factor	γ-Globulin; MW $>$ 100,000; proenzyme activated on contact with foreign surfaces (e.g., glass,); reacts with factor XI to initiate the reaction involved in formation of intrinsic activated complex	Relatively stable	Yes	Autosomal recessive	Prolonged clotting time in glass, but no clinical symptoms associated with its absence
XIII	Fibrin stabilizing factor (FSF)	α-Globulin; MW 320,000; a transglutaminase requiring Ca^{2+} and thrombin for activation; intact sulfhydryl groups required for activity	Relatively heat stable	Yes	Rare; autosomal recessive	Acquired; liver disease; clots are loose and weak; bleeding occurs 24–36 h after injury; wounds heal slowly

received the greatest attention. These defects have had profound effects on the history of the Western world. Queen Victoria was a carrier of hemophilia A (factor VIII deficiency), and many male members of the ruling families in Europe consequently suffered from the disease. A great grandson of Victoria, the son of Nicholas (Tsar of Russia) and Alexandra, was a hemophiliac. Perhaps the Russian Revolution would have been postponed had it not been for the influence of the monk Rasputin, who, through hypnotism, was reputed to have alleviated the suffering of the young Tsarovich.

Both factor VIII deficiency (hemophilia A, classical hemophilia) and factor IX deficiency (hemophilia B) are sex-linked genetic diseases having a similar symptomatology. The diseases are manifest in 50% of male children born to a carrier mother. Hemorrhages in hemophiliacs occur from vessels larger than capillaries in any tissue or organ, but hemorrhages into joints (hemarthroses) are especially debilatating. Extreme pain results from a hemarthrosis, which compromises the mobility of the joint.

When deficiencies in coagulation factors are sufficiently great to require medical intervention, they can be temporarily alleviated by the intravenous administration of fresh plasma or a concentrate of the deficient factor. Many hemophiliacs are currently being given, usually at weekly intervals, prophylactic doses of a plasma concentrate containing the deficient factor.

Platelet Defects

When defects in platelets are the primary cause of bleeding disorders, bleeding is usually capillary bleeding (in contrast to hemorrhages from larger vessels, in which the fibrin clot is involved in hemostasis). When the platelet concentration in blood falls below about 80,000 per cu mm (normal count 150,000–350,000 per cu mm) capillary bleeding may occur because the number of platelets is inadequate to form effective platelet plugs in torn microvessels. The disease in patients who have hemorrhages due to gross deficiencies in platelets is called thrombocytopenic purpura. Such patients frequently have spontaneous bleeding into the skin and mucous membranes, and they usually have a history of abnormal bruising. The hemorrhages from skin capillaries form little reddish or purple blebs beneath the skin; these hemorrhages are called petechiae.

Disseminated Intravascular Coagulation

A relatively slow utilization of coagulation factors, especially fibrinogen and factors V and VIII, with a concomitant decrease in the platelet count has recently been recognized as a serious clinical problem. The syndrome is usually termed disseminated intravascular coagulation (DIC). Although the consumption of clotting factors in DIC is usually slow, in some diseases, such as purpura fulminans, some cases of meningococcemia and some septicemias, it can be rapid. DIC is believed to be caused by the release into the circulation of tissue thromboplastins. Since red cells contain a thromboplastic material, called erythrocytin, excessive hemolysis can also result in DIC. Excessive hemolysis with the release of erythrocytin occurs in (1) incompatible blood transfusions, (2) paroxysmal nocturnal hemoglobinuria, (3) malaria, (4) favism, (5) cold hemoglobinuria, (6) sickle cell disease, and (7) the infusion of hypotonic solutions. Tissue thromboplastins are believed to be released into the circulation in toxemias of pregnancy, in promyelocytic leukemia, in some carcinomas, with severe crush injuries, in severe burns, during manipulation of the lungs, and in a number of infectious diseases.

Disseminated intravascular coagulation can lead to three established clinical consequences: (1) the formation of fibrin, which is usually diffusely deposited on the walls of blood vessels and which may form small thrombi in the microcirculation of several organs and tissues (the kidneys are particularly vulnerable); (2) the formation of some fibrin as strands without occlusion of arterioles and capillaries but which causes destruction to red blood cells (and can lead to a hemolytic anemia); and (3) the most common problem in DIC, which is a bleeding tendency resulting from both a consumption of coagulation factors (and platelets) and antihemostatic properties of split products of fibrinogen (when the fibrinolytic system is activated).

Mammalian organisms have several defenses against DIC. For example, the plasma contains antithromboplastins as well as antithrombins. One of the most effective treatments for DIC is the administration of heparin. Also, both thromboplastin (a particulate material) and fibrin are taken up by the reticuloendothelial system (RES). However, the RES can be overwhelmed. This is believed to occur in the Schwartzman reaction. To induce the Schwartzman reaction under experimental conditions, two doses of endotoxin are given to an animal, such as the rabbit, about 24 h apart. The first dose causes some DIC, but the thromboplastin and fibrin are removed by the RES; hence a severe reaction does not occur. At the time of the second dose of endotoxin, the RES has been saturated with thromboplastin and fibrin (RES blockade) so that it is no longer able to remove the particulate debris. As a result, the second dose of endotoxin causes fibrin thrombi to form in vessels, particularly in renal vessels. This kind of process can occur in several disease states involving DIC.

COAGULATION ASSAYS

In vitro assays have been developed to measure the activities of the coagulation factors. Several are routinely performed in clinical laboratories. Generally, the end point of the assays is the formation of fibrin strands.

PROTHROMBIN TIME

Prothrombin time (PT) is the time required for fibrin to form when calcium and a tissue thromboplastin (usually brain) is added to diluted, citrated, or oxalated plasma. A normal prothrombin time (about 12 s) indicates that prothrombin, factors VII, X, and fibrinogen are present in adequate concentration.

PARTIAL THROMBOPLASTIN TIME

Partial thromboplastin time (PTT) is the time required for fibrin to form when a platelet substitute (cephalin or inosithin) and calcium are added to platelet-poor citrated plasma. The PTT is dependent on plasma factors involved in the intrinsic pathway, as well as factors V, X, and fibrinogen. By mixing the patient's plasma with a plasma that has been made deficient (or is known to be deficient) in a specific factor, the deficient factor in a patient's plasma can be identified and quantitated.

Other, more sophisticated, assays are performed in some clinical laboratories and in research laboratories.

SELECTED BIBILOGRAPHY

Davie EW, Schmer G: Modern concept of concept of blood coagulation, in Schmer G, Strandjord PE (eds): *Coagulation, Current Research and Clinical Applications.* New York, Academic Press, 1973, p 3.

Doolittle RD: Structural aspects of the fibrinogen to fibrin conversion. *Adv Prot Chem* 27:1, 1973.

Guest MM: Circulatory effects of blood clotting, fibrinolysis, and related hemostatic processes, in Dow P(exec ed): *Handbook of Physiology: A Critical Comprehensive Presentation of Physiologic Knowledge and Concepts.* Section 2: *Circulation,* vol 3; WF Hamilton (section ed). Baltimore, Williams & Wilkins, 1965, p 2009.

Guest MM: Functional significance of the fibrinolytic enzyme system. *Fed Proc* 25:73, 1966.

Hougie C: *Fundamentals of Blood Coagulation in Clinical Medicine.* New York, McGraw-Hill, 1963.

Johnson SA, Guest MM: *Dynamics of Thrombus Formation and Dissolution.* Philadelphia, Lippincott, 1969.

Seegers WH: *Blood Clotting Enzymology.* New York, Academic Press, 1967.

Tullis JL: *Clot.* Springfield, Ill, Charles C Thomas, 1976.

Wintrobe MM: *Clinical Hematology,* ed 7. Philadelphia, Lea and Febiger, 1974.

Wolf PL: Disseminated intravascular coagulation: Principles of diagnosis and management, in Schmer G, Strandjord PE (eds): *Coagulation, Current Research and Clinical Applications.* New York, Academic Press, 1973, p 17.

SELF-STUDY QUESTIONS

MULTIPLE CHOICE

Select the single best answer.

1. One of the following substances in blood, which is completely free of tissue thromboplastin, will not have an effect on the clotting time.

   A. Factor VII (SPCA)
   B. Factor X (Stuart factor)
   C. Factor VIII (AHF)
   D. Factor IX (Christmas factor)
   E. Factor XII (Hageman factor)

2. A patient has disseminated intravascular coagulation (DIC). He requires immediate anticoagulation. Which agent would you administer?

   A. Dicumarol
   B. Sodium citrate
   C. Fibrinolysin
   D. Heparin
   E. Sodium oxalate

3. A patient has excessive hemorrhages from his capillaries (petechiae). The cause is most likely that

   A. his fibrinolytic enzyme system has been activated.
   B. his platelet count is less than 50,000/mm.
   C. his blood is deficient in factor VIII (AHF).
   D. he has been given therapeutic doses of Dicumarol.
   E. his ionized plasma calcium concentration is 1.5 mEq/L. (normal about 4.3 mEq/L).

4. The patient's mother is a carrier. None of his eight sisters has clinical symptoms of the disease. He has hemarthrosis. The clotting factor in which he is deficient is labile and is destroyed by fibrinolysin. Choose the most appropriate answer.

   A. Factor V (accelerator globulin)
   B. Factor VII (SPCA)
   C. Factor VIII (AHF)
   D. Factor IX (Christmas factor)
   E. Factor X (Stuart factor)

5. A patient is pathologically fibrinolytic. Which of the following would you administer to control the excessive fibrinolytic activity?

   A. EACA (ϵ-aminocaproic acid)
   B. Streptokinase

C. Fibrinogen

D. Heparin

E. Dicumarol

6. Which of the following when added to freshly drawn blood would not interfere with clot formation?

A. An equal volume of a saturated salt solution ($CaCl_2$)

B. Dicumarol

C. Heparin

D. Na citrate

E. A proteolytic enzyme, such as trypsin or fibrinolysin

7. Which of the following procedures will significantly accelerate the clotting of freshly drawn, nonanticoagulated whole blood in a test tube?

A. Coating the surface that the blood will contact with silicone

B. Cooling the blood to 0°C

C. Rapidly heating the blood to 80°C.

D. Adding ground glass to increase the surface area of glass with which the blood comes into contact

E. Adding a solution of $CaCl_2$ to bring the plasma concentration to calcium to 9 mEq/L (about 2 times the normal concentration of Ca^{2+})

8. Which of the following proteins involved in hemostasis has the highest molecular weight?

A. Prothrombin

B. Fibrinogen

C. Antithrombin II (acts as a cofactor with heparin)

D. Thrombin

E. Fibrin monomer

9. A factor required for blood clotting via the extrinsic pathway, but not via the intrinsic pathway, is

A. factor XI (plasma thromboplastin antecedent).

B. factor IX (Christmas factor).

C. factor VIII (antihemophilic factor).

D. factor VII (serum prothrombin conversion accelerator).

E. factor XI (plasma thromboplastin antecedent).

10. Which of the following would accelerate clotting if administered IV?

A. Profibrinolysin

B. Urokinase

C. Homogenized extract of lung

D. Factor XII

E. Factor XIII

11. Which of the following factors acts in the clotting sequences as a proteolytic enzyme?

A. Fibrin monomer

B. Fibrinolysin

C. Thrombin

D. Tissue thromboplastin

E. Factor XIII

12. The retraction of a clot is most dependent on the

A. leukocyte concentration.

B. erythrocyte concentration.

C. number of distintegrated platelets.

D. number of intact platelets.

E. concentration of factor VIII.

13. Fresh blood was drawn into a clean test tube and failed to clot, even though thrombin was added to it. The missing clotting factor was

A. factor V.

B. factor X.

C. factor XIII.

D. fibrinogen.

E. calcium.

14. A patient was treated with a drug administered IV. You take a fresh blood sample. It clots and then lyses within 15 min. The drug used was

A. heparin.

B. dicumarol.

C. ϵ-Aminocaproic acid.

D. factor VIII.

E. streptokinase.

ANSWERS

1. A. Factor VII acts only with tissue thromboplastin via the extrinsic pathway. Factor X is common to both the intrinsic and extrinsic pathways (final common path). Factors VIII, IX, and XII are involved in the intrinsic pathway, which does not require tissue thromboplastin.

2. D. Heparin is the only immediately acting anticoagulant, except for sodium oxalate and sodium citrate. Sodium oxalate should never be given in vivo. It is extremely toxic (precipitates the plasma calcium). Sodium citrate also cannot be given IV in large enough concentrations to act as an anticoagulant because it prevents plasma calcium from being ionized. Dicumarol acts as an anticoagulant only because it reduces prothrombin production by the liver (in vivo). Fibrinolysin per se does not prevent clots. It lyses them after they are formed.

3. B. Capillary and other microvessel hemorrhages are controlled only by formation of platelet aggregates. When the count is below 50,000/mm³ there are too few platelets for effective hemostasis.

4. C. The patient is a hemophiliac. Although either a deficiency in factor VIII or IX can be the cause of hemophilia, it cannot be IX because factor IX is relatively stable. Both factors VIII and V are labile and are destroyed by fibrinolysin, but a deficiency in factor V does not cause hemophilia.

5. A. EACA is micromolecular inhibitor of both fibrinolytic activation and the enzyme fibrinolysin (it is an antiactivator and an antifibrinolysin).

6. B. Only Dicumarol of the substances listed will not interfere with clot formation when added to freshly drawn blood. (Dicumarol acts only in vivo to prevent the formation of prothrombin by the liver.) A is incorrect because half-saturation of plasma by most salts will precipitate out most of the protein enzymes.

7. D. Ground glass by increasing the wettable surface area will accelerate the activation of factor XII (Hageman factor). With respect to C, heating blood to 80 °C inactivates (denatures) the enzymes in blood, and with respect to E the normal blood concentration of $CaCl_2$ appears to be about optimal for blood clotting.

8. B. Fibrinogen has a molecular weight of about 325,000. The other listed proteins are smaller.

9. D. Factor VII is the only extrinsic pathway factor listed.

10. C. Lung is an excellent source of tissue thromboplastin. Clotting occurs extremely rapidly in the presence of significant amounts of membranes from cells. Although factors VII and XIII are procoagulants, neither factor initiates the clotting sequences.

11. C. Only thrombin meets the specified requirements. Fibrinolysin, the only other listed proteolytic enzyme, does not act in the clotting sequences.

12. D. Intact platelets are responsible for clot retraction. Disintegrated platelets have no influence on clot retraction.

13. D. Thrombin can act directly on fibrinogen to initiate its conversion to fibrin. Therefore, when thrombin is added to blood, a clot will form if fibrinogen is present.

14. E. Streptokinase reacts with a factor in human plasma to form an activator of profibrinolysin. With the conversion of profibrinolysin to fibrinolysin, this proteolytic enzyme is made available to hydrolyze the fibrin clot. The other listed substances do not cause the breakdown of a clot after its formation.

36

Energy Exchange and Temperature Regulation

M. MASON GUEST

OBJECTIVES

After completion of this chapter, you should be able to:

1. Explain how heat is transferred within organisms, and especially between organisms and their environment.

2. Describe the cutaneous circulation.

3. Explain the ways in which physiological modifications augment or decrease heat loss from the body.

4. List the salient features of a hypothesis that attempts to explain the central control of heat loss, heat conservation, and heat generation.

5. Describe the clinical disabilities relating to temperature regulation (e.g., fever, heat stroke, and hypothermia).

Living cells function only within a limited temperature range. When a cell freezes it is no longer metabolically active; at temperatures above 43°–45°C cellular constituents can be damaged, and if the cell remains at an abnormally high temperature for several minutes, especially at temperatures above 45°C, the injury can be irreversible. Neurons in the central nervous system (CNS) are especially sensitive to abnormally high temperatures. Most cold-blooded animals (poikilotherms) can survive cooling to just above the freezing point of their cells but, like homothermic (warm blooded) animals, they might be unable to survive if their body temperatures go above 43°–45°C.

Poikilothermic animals are largely at the mercy of their environment while homothermic animals have acquired the ability to control the temperature of the fluids and cells within the core of the body as well as the composition and chemistry of the fluids bathing internal cells. The temperature of the interior of the body, the core temperature, remains remarkably constant in homotherms under the usual environmental conditions encountered in the temperate zones.

This chapter describes mechanisms through which a relatively uniform core temperature is maintained. Also described and discussed are abnormal core temperatures and the effects of these extremes on survival of the organism.

PHYSICAL PROCESSES BY WHICH HEAT IS TRANSFERRED

CONDUCTION

Conduction is transfer of thermal energy by collision between molecules in motion. Molecules traveling at high velocities transfer some of their energy by collision to molecules moving at slower velocities. Heat is not only transferred within homogeneous materials by conduction, but by conduction between different materials with which they come in contact. Heat is transferred by conduction to or from the air that comes in contact with the animal body; however, heat transfer by conduction ceases when air in contact with the skin reaches the skin temperature; for conduction to continue, the shell of air in contact with the skin must be removed by convection (see below).

CONVECTION

Convection is movement of a fluid due to thermal currents. Since a heated fluid expands and is then less dense than a cool fluid (law of Gay-Lussac), it rises, setting up a circulation (convection current). Animals lose heat by conduction to air in contact with the skin; the heated air rises allowing cooler air to contact the skin, etc. When an animal is immersed in water, conduction and convection occur, but since water has a much higher thermal capacity than does air, heat loss (or heat gain) is more rapid. Forced convection (sometimes termed forced conduction) in gases and liquids results from wind, movement of air by fans, movement of water by stirrers, and movement of an object surrounded by air or water.

EVAPORATION

Since change of state from liquid to gas requires an increase in mean velocity of molecular thermal motion, energy must be provided. The energy for the change of state comes from surfaces where evaporation occurs. Thermal motion of molecules comprising the surface layer is slowed and the surface is cooled as energy is transferred to the molecules of water that thereby gain sufficient velocity to become a gas (water vapor). The heat lost from the surface for each gram of water that evaporates is 0.58 kcal.

Water continuously passes through the skin by insensible perspiration and through the epithelium of the respiratory passageways. At ambient temperatures below 30°C about 900 ml is lost from the body in these ways each day. Most of the insensible perspiration fluid evaporates and removes heat from the body. That which is absorbed by clothing or removed in other ways contributes little or not at all to heat loss from the body.

Sweating occurs when environmental temperatures are high and/or when the body temperature rises. In nonacclimated persons subjected to high environmental temperatures, 1–2 L of sweat may be lost per hour. It has been reported that persons acclimatized to high temperatures can lose as much as 4 L of sweat per hour. However, these rates of sweating cannot be maintained for periods of several hours, even though the fluid lost is replaced. Notwithstanding the importance of sweating in cooling the body, you should be aware that the amount of heat lost from the body is not necessarily a function of the quantity of sweat produced. Sweat that does not evaporate removes little or no heat.

The amount of evaporation from a surface is dependent on the relative humidity; this is a ratio of two absolute humidities. Absolute humidity is defined as the mass of water vapor present in a unit volume of the atmosphere. The amount that a unit volume of air can hold depends on the temperature; at higher temperatures more water can be suspended in the air in the form of vapor than at low temperatures. The term relative humidity refers to the ratio of the existing absolute humidity to the absolute humidity required to produce saturation at the ambient temperature. If the air is saturated (100% relative humidity), no more water vapor can be taken up by the air and evaporation ceases. When this condition exists, sweating is not effective in losing heat. Furthermore, if there is no convection, 100% relative humidity could occur in the immediate vicinity of a moist surface even though the ambient air at a distance from the surface is not saturated.

RADIATION

Radiation is transfer of heat energy by electromagnetic waves (or by photons). Heat is transferred primarily by electromagnetic waves in the infrared portion of the spectrum (6,000–200,000 Å). The sun's radiant energy is in the near infrared (6,000 to about 30,000 Å) while radiation from the human body reaches a peak at about 90,000 Å and extends to 200,000 Å. Radiant energy is transferred from warmer to cooler bodies; it can be emitted or absorbed. An efficient emitter is also an efficient absorber. An object with the emissivity of 1, the ideal "black body," absorbs all the energy incident upon it and emits the maximum amount of energy consistent with its surface temperature. The human skin, white or black, is a 97% perfect black body. The infrared wavelengths are beyond the visible range of the spectrum, and as DuBois has stated: "There is certainly no difference in color between the skin of a white man and a black man in a dark room in the middle of a moonless night."

TRANSFER OF HEAT WITHIN AND FROM ANIMAL BODIES

RELATIONSHIP BETWEEN CORE AND SURFACE TEMPERATURE

The core temperature (interior temperature) in homothermic animals, when compared to poikilothermic (cold blooded) animals, is regulated and relatively constant. In healthy human beings under normal conditions it usually varies less than 1°C. On the other hand, surface temperature can fluctuate widely, depending on physiological activity and external conditions.

The ambient temperature is usually below the core temperature of homothermic animals and heat is consequently transferred from the animal to its surrounding environment. The following discussion pertains to this situation (core temperature > environmental temperature).

Maintenance of core temperature depends on balance between heat production and heat loss. Heat is primarily produced by oxidative exothermic chemical reactions. Heat loss from the core occurs when heat is transferred to pulmonary airway surfaces and to the external surfaces of the body. Five percent of heat transfer occurs by conduction through tissue, while 95% is carried to the surface by circulating blood. Actively metabolizing tissues transfer their heat to the blood. The temperature of the arterial blood delivered to somatic and visceral metabolizing tissues is slightly below that of the tissues. Heat flows from the hotter tissue to the cooler blood. Each calorie produced by a tissue raises the temperature of 1 ml of blood approximately 1°C (since blood is mostly water and water has a specific heat of 1).

Blood returning to the right atrium from the surface of the body is usually cooler than arterial blood that enters the skin circulation. If some combination of radiation, conduction, convection, and evaporation permits heat transfer to the environment, blood flowing through skin vessels (especially subpapillary venous plexuses) gives up heat. The resultant cooler venous blood from the skin and from the upper respiratory tract mixes with hot venous blood from highly metabolizing tissues, such as contracting skeletal muscle, and the blood entering the right heart is thereby brought to a temperature between the two extremes.

The skin of a person having a surface area of 1.8 m² comprises about 3% of the weight of the body. It is only 1–1.5 mm thick. This thin layer is an effective radiator of heat in homothermic animals that have no fur, such as humans. On the other hand, the skin together with its underlying adipose tissue becomes a heat conserving insulator when blood flow in these tissues is reduced to a minimum. The conversion from radiator to insulator or from insulator to radiator depends on the adjustment in blood flow to the skin.

CUTANEOUS BLOOD FLOW

The cutaneous circulation serves two functions: it supplies the metabolic requirements of the skin and transfers heat to the surface of the body. Through neural control, blood flow to the skin of an adult human being can be as much as 4,000 ml/min or as little as 20 ml/min. With maximal flow, heat loss from the body surface is promoted; with minimal flow, heat loss from the surface is reduced.

The skin is overly vascularized for a tissue with low metabolic requirements. (It is the most poikilothermic of the tissues in the body; its temperature is usually 2°–10°C below that of the core.) Although variations in vascular patterns of the skin occur in different regions of the body, a common feature is the presence of "preferential" throughfare metarterioles and extensive venous plexuses in the dermis and hypodermis. When both the muscular arterioles and metarterioles are dilated, the venous plexuses are engorged with blood and act as radiators of heat to the environment.

In the face (especially ears), hands, and feet (especially apices of fingers and toes), arteriovenous shunts are a common feature. These are coiled channels with muscular walls and an interior diameter which, when the vessel is dilated, varies between 20 and 70 μm in diameter. When the shunts are dilated, large volumes of warm arterial

blood are poured directly into the venous plexuses and heat loss is thereby greatly augmented (Fig. 1).

Vasoconstriction of arterioles and metarterioles in the skin as well as the arteriovenous (AV) shunting mechanism is under control of a sympathetic center in the posterior hypothalamus (Figs. 6 and 7). When center is active, sympathetic impulses cause constriction of arterioles, metarterioles, and AV shunts, with consequent reduction or absence of flow and therefore, a reduction in heat loss from the surface. Contrariwise, when the heat-losing mechanisms function, the vasoconstrictor–sympathetic center in the posterior hypothalamus is inhibited and skin arterioles, metarterioles and AV shunts dilate. No vasodilator nerves to blood vessels have been identified in human skin, although cholinergic sympathetic vasodilator fibers have been reported to be present in skin of several other mammalian species. It is postulated that, with inhibition of the sympathetic center in the posterior hypothalamus, the circularly arranged smooth muscles in the

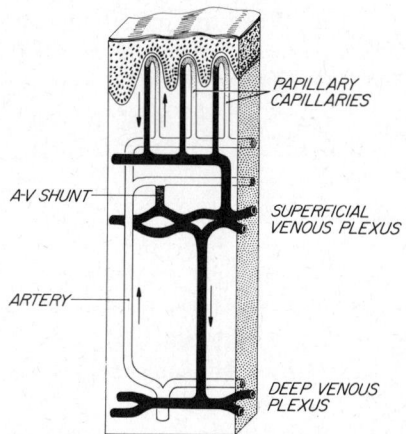

PAPILLARY CAPILLARIES

A-V SHUNT

SUPERFICIAL VENOUS PLEXUS

ARTERY

DEEP VENOUS PLEXUS

Figure 1 Schematic representation of blood vessels in skin. Capillary loops arise in the papillae of the dermis and return to enter a relatively superficial subpapillary venous plexus. Venous blood may drain from the superficial plexus into shallow surface veins for return to the heart or may enter a deeper plexus before returning via more internal veins (depending on requirements of temperature regulation). In the skin of exposed apical areas a large number of arteriovenous shunts directly connect arterioles and venules in the superficial layer of the dermis. Opening of the AV shunts permits large volumes of blood to enter the superficial plexus in hands, feet and face.

resistance vessels and AV shunts of human skin spontaneously relax (reduce tone).

Some areas of the skin in the human species appear to be particularly sensitive to circulating pressor and vasodilator agents. These are the flush areas (upper thorax, neck, and head). After the inhalation of amyl nitrite, these areas have a deep flush while the extremities show strong vasoconstriction. The mechanism responsible for, and the functional significance of, the flush are unknown. It appears to be unrelated to either skin nutrition or temperature control.

REACTIVE HYPEREMIA

Although metabolism of skin is relatively low compared with most other organs, when it is deprived of blood flow for a period of 10–20 min, smooth muscles of its blood vessels relax and the skin consequently reddens due to dilation of its vessels and the increase in flow. This reaction, called reactive hyperemia, indicates that skin vessels are capable of local vascular response that prevent nutritional damage to skin cells. However, time required for response is much longer than in other tissues in which local control of blood flow occurs.

TRIPLE RESPONSE

Injury to the skin also produces hyperemia. When the skin is stroked firmly with a blunt, pointed object, the triple response occurs. Immediately following the stroke, the stroked area becomes white; this is believed to be due to blood being mechanically forced out of the microvessels and is not part of the triple response. A red line resulting from damage to the microvessels and possibly involving release of histamine then appears. Twenty to 40 s later a red flare is seen. The flare involves a much broader area of skin than the point at which contact was made and is believed to result from arteriolar vasodilation brought about by the axon reflex. The axon reflex is a vascular response in the neighborhood of the injury. An afferent nerve fiber normally carrying sensory impulses to the spinal cord from a pain ending in the skin has several branches to arterioles supplying blood to the skin in the region of the sensory terminus. The nerve endings of the branch fibers release a vasodilator substance into or onto smooth muscle cells in an arteriole. Thus, when a painful

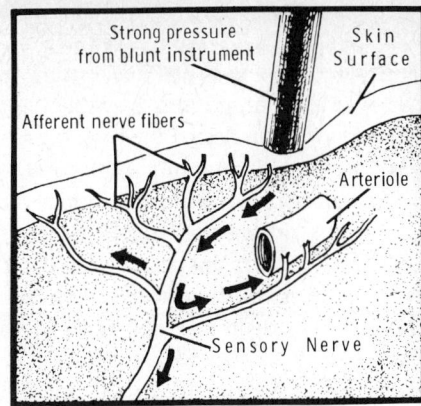

Figure 2 Schematic representation of the axon reflex in response to strong pressure along the skin surface with a blunt instrument. Arrows indicate the pathways of impulses in a sensory nerve from the site of stimulation to adjacent arterioles to produce local vasodilation (flare).

stimulus is applied, the smooth muscles of the arteriole relax in that region of the skin causing an increased blood flow into capillaries and venules supplied by the arteriole, hence the flare (Fig. 2). The third feature of the triple response is the wheal that develops in people with sensitive skin. It is a raised area under and on both sides of the line of the stroke; it is believed to be caused by the release of histamine. Histamine brings about arteriolar dilation, venular constriction, and perhaps direct damage to capillary endothelium. Regardless of whether histamine causes direct damage to capillary endothelium, increased pressure within microvessels can result in transudation of fluid into the tissues (highly localized edema).

The responses of skin blood vessels to direct heating or cooling of localized areas of the skin are, in general, vasodilation during warming and vasoconstriction during cooling. However, the degree of vasodilation is greater if the subject's core temperature is higher than normal, indicating that the local response is in part dependent upon control exerted via the hypothalamus.

HUNTING REACTION

When a part of the body is immersed in ice water, marked vasoconstriction occurs. However, if a finger, toe, or ear is maintained in the cold for more than 10 min, an alternation of vasoconstriction and vasodilation occurs. The cyclic change in blood flow is called a "hunting reaction." The response is of local origin and is probably caused by cooling of the smooth muscles of the blood vessels to the point at which they lose their ability to respond to sympathetic nerve vasoconstrictor impulses. When they relax, blood flow increases raising their temperature, and then they again respond to vasoconstrictor stimuli. The axon reflex, initiated by pain resulting from severe chilling, can also contribute to the oscillatory responses of the resistance vessels.

Control systems for skin circulation involve higher CNS centers, spinal cord reflexes, and local control. Within reasonable limits of temperature variations, the hypothalamic control system determines the general pattern of response. However, extremes of temperature and other traumatic influences, including relatively long-term ischemia, can modify or even reverse reactions dictated from the hypothalamus through local mechanisms in the skin.

COUNTERCURRECT HEAT EXCHANGE

An important feature of heat conservation, which is dependent on the anatomy of the circulation, is the counter-

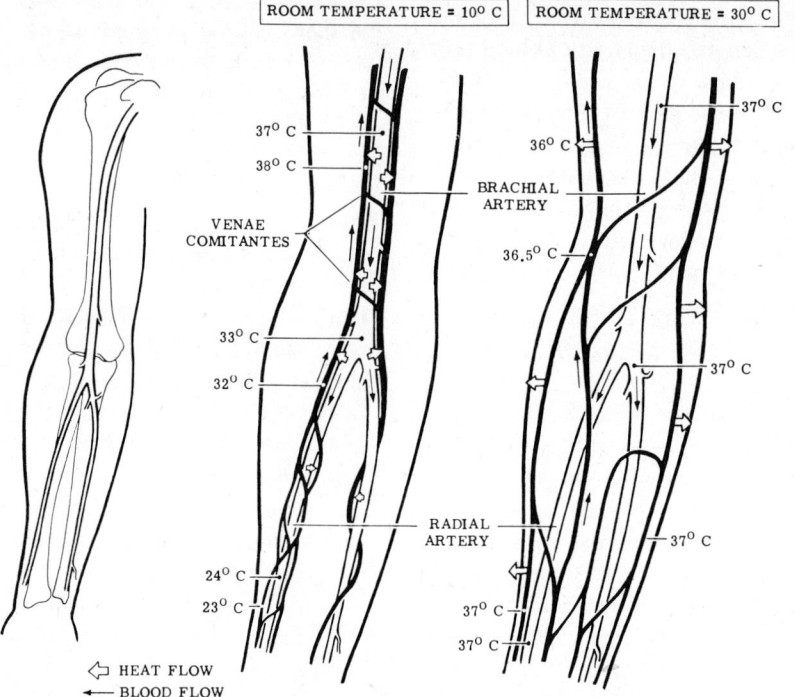

Figure 3 Countercurrent heat exchange in human arm under conditions of cold environment (10°C). The temperature of arterial blood leaving axillary region is normally 37°C. When the body is exposed to cold, venous flow shifts from surface veins to deep veins (venae comitantes) running parallel with the arterial supply. As a result of the closeness of the parallel vessels, heat is transferred from the warm arterial blood to the cool venous blood (countercurrent exchange). The result of arterial warming of deep veins is that the venous blood, when returned to the chest, has been warmed back to 37°C, thereby conserving core temperature. In a warm environment (e.g., room temperature of 30°C), venous blood drains into the surface veins permitting more heat loss to the environment. (Modified from Selkurt EE: *Physiology,* ed 4. Little, Brown and Co., Boston 1971.)

current exchange mechanism. In response to cold, blood flow in the extremities shifts from superficial veins to deep veins paralleling arteries. Cool blood returning to the core of the body via a (deep) vein is warmed by heat given off from the adjacent artery. Concurrently, temperature of the arterial blood going to the extremity is reduced so that less heat is lost from the extremity to the environment (Fig. 3). The consequence of the countercurrent heat exchange is that core temperature is maintained, but the temperature of the extremities and especially their surfaces is significantly reduced. Thus less heat is lost from the body as a whole because the temperature of the surface of the extremities is not greatly different from the temperature of the air and surrounding objects. This mechanism of heat conservation is especially effective in wading birds and mammals.

SUMMARY OF MECHANISMS INVOLVED IN HEAT CONSERVATION AND HEAT DISSIPATION

Heat is produced in the body by exothermic chemical reactions and friction. In homotherms the central core is maintained at a relatively constant temperature while the surface temperature can fluctuate widely. When heat needs to be conserved, vasoconstriction of skin arterioles is brought about by sympathetic discharges. Hence less blood, and consequently less heat, is transported to the skin and its surface becomes relatively cool. In addition, countercurrent heat exchange reduces the temperature of the arterial blood on its way to the extremities and warms the venous blood returning to the core.

When heat production is greater than heat loss, core temperature rises. Continuation of the rise in core temperature is prevented by increased flow of blood to the skin. Arterioles and arteriovenous shunts in the skin become dilated and the venous plexus becomes engorged with warm blood. With rise in surface temperature of the skin, conduction to cooler air in the shell of air around the body occurs. The heated air rises (convection) and is replaced by cooler air. Concurrently, more heat is lost by radiation because of a greater temperature differential between the surface of the body and relatively cool surrounding objects. Also, sympathetic nerve activity brings about the secretion of sweat. Increase in the amount of water evaporated effects a heat loss from the body surface of 0.58 kcal for every gram of sweat vaporized.

NEURAL CONTROL OF HEAT LOSS, HEAT CONSERVATION, AND HEAT GENERATION

The core temperature of the human body is ordinarily maintained within 1°C. However, severe exercise, exposure to very low temperatures, and so forth result in shifts from the normal 37°C core temperature (Fig. 4).

THE HYPOTHALAMIC CENTERS CONTROLLING THE CORE TEMPERATURE

To understand present concepts of how homothermic organisms maintain a relatively constant core temperature one should comprehend the principle involved in the operation of a thermostat. Thermostats are on and off switches used to automatically keep rooms within a narrow temperature range. The thermostatic control is the key part of a negative feedback mechanism. The temperature sensor of the thermostat is usually a thermocouple. The two dissimilar metals in contact expand at different rates and, as

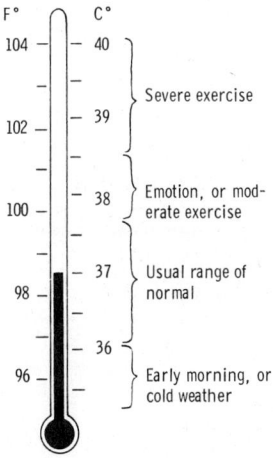

Figure 4 Range of core temperatures in normal persons.

the temperature rises, open an electrical switch. The valve, permitting gas to flow into the furnace, has been held in the open position by an electrical current activating an electromagnet. Without the current to keep the magnetically activated valve open, gas flow into the furnace stops. The fire in the furnace goes out, and the room begins to cool. When the temperature of the room falls a few tenths of a degree, the thermostat's electrical switch closes; the current then activates the electromagnet, which opens the valve, and gas again flows into the furnace. The flowing gas is ignited by a pilot light. The temperature at which the thermostat is set is called the setpoint.

The principal thermostatic control in homothermic animals is located within the preoptic nucleus and closely adjacent regions of the anterior hypothalamus (AH-POA). Other neurons in the CNS are sensitive to temperature changes as well, but they exert only minor influences on temperature control of the body core. The temperature control mechanisms are depicted in Figures 5–7.

The information presented here is a simplification of a complex system. It summarizes concepts that have evolved from many years of experimental work in numerous laboratories. The most effective experimental techniques for evaluating the role of the hypothalamus in temperature regulation include ablation of discrete areas of the brain, electrical stimulation of groups of neurons or individual neurons, and especially direct heating or cooling of specific regions by means of thermodes (devices that can be either cooled or heated by electrical means or by passing hot or cold water through them).

Until recently, it was believed that the anterior hypothalamus, functioning as a heat losing center, responded primarily to temperatures above the setpoint (normally 37°C), while the posterior hypothalamus, functioning as

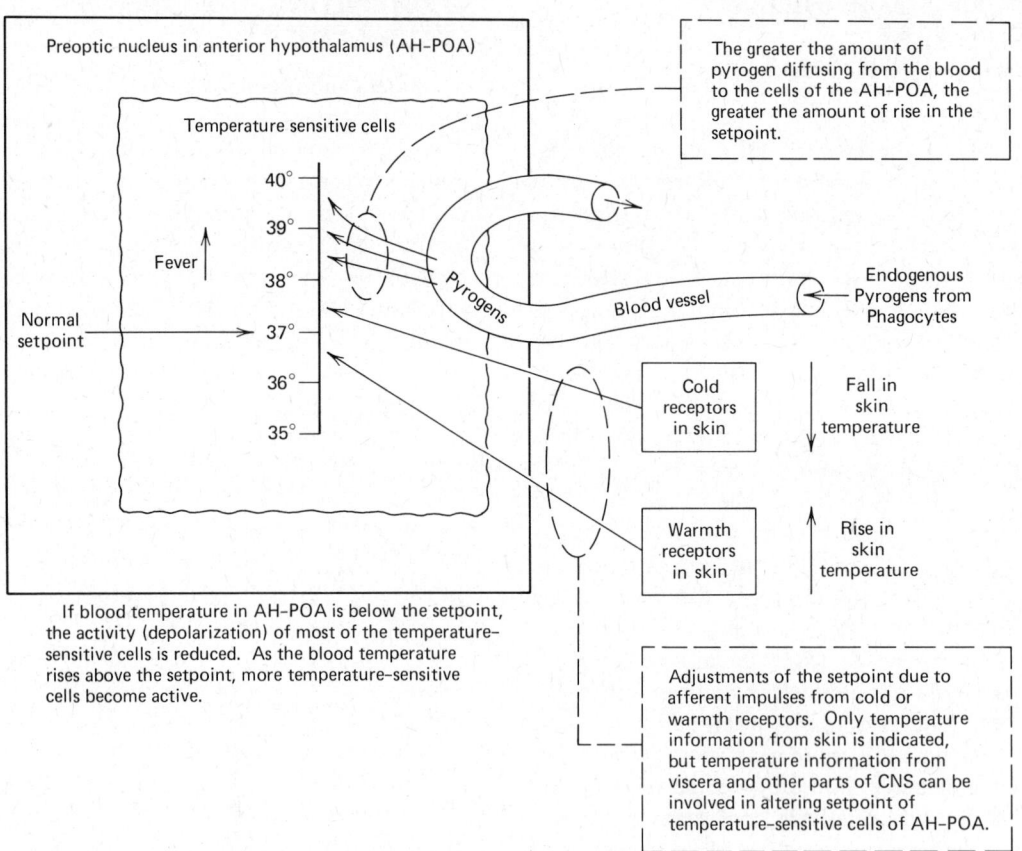

Figure 5 Central temperature control. The hypothalamic thermostat and the control of its setpoint.

the heat conservation center, was activated by temperatures below the setpoint. However, more recent evidence indicates that the preoptic nucleus of the anterior hypothalamus (AH-POA) contains the primary temperature-sensitive cells and that these cells (neurons) act like a thermostat. When the temperature of the blood perfusing the AH-POA is below the setpoint, most of the temperature-sensitive neurons are quiescent (hyperpolarized). When the temperature of blood in the AH-POA is above the setpoint, the temperature-sensitive cells become active and discharge cyclically.

Referring to Figure 5, note that the normal setpoint of the physiological thermostat in healthy persons is 37°C—or within a few tenths of a degree above or below this temperature—and that the setpoint can be altered. First, it is altered by changes in skin temperature; if the warmth receptors in the skin are stimulated, the setpoint is reduced by a small increment (one- to three-tenths of a degree); if the cold receptors in the skin are stimulated, the setpoint is raised by a correspondingly small increment. Second, it is raised by pyrogens diffusing from blood ves-

sels into the cells of the AH-POA; pyrogens are the agents responsible for fever (see below). Third, it may be raised by exercise; however, raising of the setpoint by exercise has not been definitely proved. Fourth, the setpoint is raised by severe dehydration (see below). Fifth, it can be raised by certain neurological disorders, such as tumors of the third ventricle or the hypothalamus.

The temperature control process involves negative feedback, which maintains the core temperature at, or close to, the physiologic setpoint in the AH-POA. When the temperature of the blood perfusing the AH-POA is below the setpoint, both heat production and heat conservation mechanisms are activated. When the temperature of the perfusing blood is above the setpoint, heat-losing mechanisms are activated. These responses are depicted in Figure 6 (perfusing blood temperature above the setpoint) and in Figure 7 (perfusing blood temperature below the setpoint).

Neuronal interconnections with the temperature-sensitive neurons of the AH-POA are extremely complex; these interconnections have not yet been adequately described.

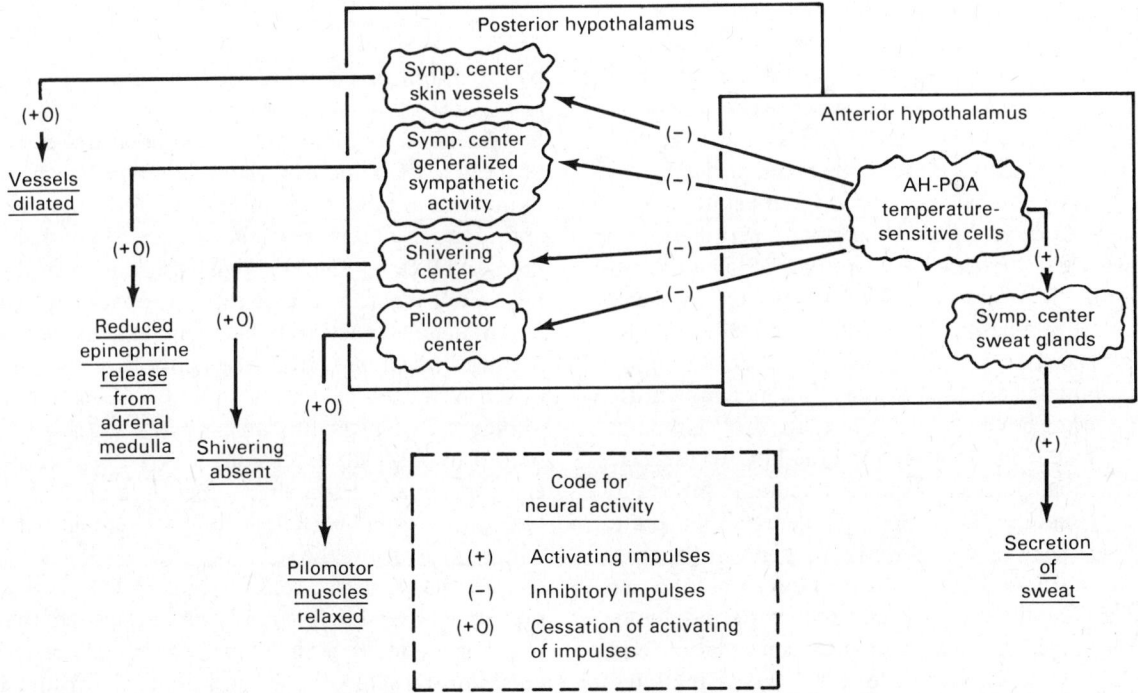

Figure 6 Central temperature control. When temperature of blood perfusing AH-POA is above the setpoint.

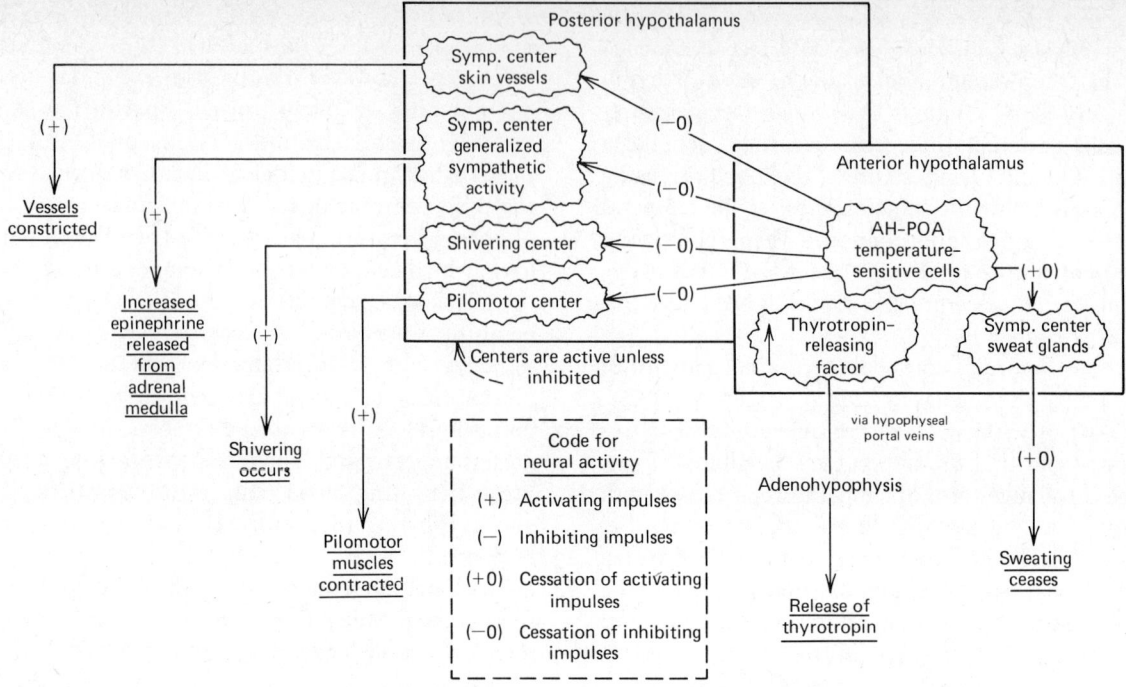

Figure 7 Central temperature control. When temperature of blood perfusing AH-POA is below the setpoint.

Nevertheless, we have information about activation or inhibition of CNS centers modifying peripheral responses (those involving heat conservation or loss and heat production) when the temperature of the blood perfusing the AH-POA is either raised above or lowered below the setpoint. We also have information about peripheral responses involving secretion of sweat, contraction, or relaxation of smooth muscles in blood vessels of skin, and contraction of pilomotor muscles and of skeletal muscles involved in the shivering response. The schematic representations show the effects of changes in core temperature above (Fig. 6) or below the setpoint (Fig. 7) on temperature-sensitive cells in the AH-POA, the kind of message these cells probably send via complex neural pathways to CNS effector centers and the peripheral effects that result.

With reference to a rise above the setpoint in the temperature of the blood perfusing the AH-POA (Fig. 6) we hypothesize that the centers in the posterior hypothalamus involved in control of heat conservation and heat-production mechanisms are active unless inhibited

by impulses originating in the temperature sensitive cells of the AH-POA. When the temperature of the perfusing blood is above the setpoint, the temperature-sensitive cells of the AH-POA increase their activity; consequently, the number of inhibitory impulses to the heat conservation and production centers in the posterior hypothalamus increases, blocking neural activity originating in these centers. Concurrently, the sweating center in the anterior hypothalamus is activated by an increase in the number of impulses reaching it from the temperature-sensitive cells of the AH-POA.

Figure 7 depicts probable relationships between the temperature-sensitive cells of the AH-POA and the CNS effectors centers when the temperature of the blood perfusing the AH-POA falls below the setpoint; the fall in temperature below the setpoint reduces the activity of the temperature sensitive cells in the AH-POA. As a result, reduction or cessation of inhibitory impulses to the centers in the posterior hypothalamus occurs and activating impulses to the sympathetic sweating center in the anterior hypothalamus also cease. Consequently, heat production and conservation centers become active and the sympathetic sweating center becomes inactive.

Keep in mind that sensory information about a temperature change of the skin surface, when it reaches the cerebral cortex, can also result in a behavioral response. Thus we put on more clothing when our skin is cooled or we move out of the direct sunlight when we are overheated.

SUMMARY OF HEAT-CONSERVING AND HEAT-PRODUCING MECHANISMS

AUGMENTATION OF HEAT-CONSERVING PROCESSES

1. *Vasoconstriction occurs in arterioles and metarterioles of the skin, and AV shunts are closed.* When temperature of blood perfusing the AH-POA is below the setpoint, the posterior hypothalamus is not inhibited by neural impulses from the temperature-sensitive cells of the AH-POA; in consequence, the activated, when not inhibited, sympathetic center in the posterior hypothalamus then sends neural impulses to alpha receptors in smooth muscle of the skin vessels, which thereby constrict, reducing blood flow to the skin.

2. *The countercurrent heat-exchange mechanism is activated.* Venous blood return is shifted from superficial to deeper veins; blood in the deeper veins gains heat from adjacent arteries. Consequent cooling of arterial blood on its way to the extremities results in less heat loss to environment.

3. *In furry mammals the hair is fluffed due to contraction of pilomotor muscles.* Erection of hair traps air, which acts as an insulator. As in #1 (above), the pilomotor center in the posterior hypothalamus is not inhibited by the AH-POA when the temperature of the blood perfusing the AH-POA is below its setpoint; consequently, the pilomotor center becomes active. In humans, the remnant of the pilomotor response is goose flesh, and goose flesh does not conserve heat.

4. *Sweating does not occur.* Activating impulses from the temperature-sensitive cells of the AH-POA to the sweating center in the anterior hypothalamus are absent when the temperature of the perfusing blood is below the setpoint of the AH-POA.

AUGMENTATION OF HEAT-PRODUCING PROCESSES

1. *Shivering occurs.* When temperature in the POA falls below the setpoint, inhibition of the shivering center ceases and it becomes active. Rhythmic coordinated contractions of skeletal muscle through involuntary neural control increase blood flow to these contracting muscles and increase their metabolic rate; consequently, increased heat production occurs.

2. *An increase in intermediary metabolism occurs if an animal is chronically exposed to low temperatures.* In rats and many other mammals shivering stops with continued exposure to cold; however, metabolism continues at an increased rate. The increased metabolism may be due to increased production and release of thyroid hormone (Fig. 7) and/or to increased epinephrine production and release (calorigenic effect of epinephrine) and/or to uncoupling of oxidative phosphorylation.

SUMMARY OF HEAT-LOSS MECHANISMS

1. *Skin blood vessels dilate and AV shunts in skin open.* When temperature of the blood perfusing the AH-POA is above its setpoint, the vasoconstrictor sympathetic center in the posterior hypothalamus is inhibited by impulses from the temperature-sensitive cells of the AH-POA. In the absence of sympathetic stimulation, the skin blood vessels dilate. (Possibly, the vasodilation is mediated by bradykinin released during activation of the sweat glands.)

2. *The sweat glands secrete sweat.* When the temperature of perfusing blood in the AH-POA is above its setpoint, neural impulses from the temperature-sensitive cells

cause the sweating center in the anterior hypothalamus to become active and it then sends sympathetic neural impulses to cholinergic endings in the sweat glands.

3. *Panting occurs in some mammals, such as the dog, which has sweat glands on only the pads of feet.* Increased movement of air in the respiratory dead space augments evaporation from the respiratory tract and consequently, heat loss.

4. *The animal consciously or unconsciously exposes as much skin as possible.* Such maneuvers as arms away from trunk of body and legs spread apart result in increased heat loss by radiation.

EFFECTIVE TEMPERATURE AND COMFORT

To serve as a measure of the rate of heat loss from the human body, Houghton and Yaglow in the 1920s devised a scale termed the effective temperature. Based on three variables, that is (1) temperature (dry bulb), (2) relative humidity (wet bulb temperature), and (3) velocity of air movement, effective temperature relates the sensation of warmth or cold in human subjects when exposed at various combinations of temperature, relative humidity, and air velocity to the equivalent sensation when exposed to a temperature of air saturated with water vapor and not moving. Effective temperature charts are used by heating and air conditioning engineers. These charts give the effective temperature corresponding to any combination of the three variables.

In recent years the Weather Bureau in the winter months has been reporting the wind chill temperature. This temperature scale is an attempt to describe exposures at low temperatures and at different wind velocities in terms of the equivalent discomfort experienced from an exposure at a reference temperature when no wind is blowing. The wind causes forced convection, thereby increasing the rate at which heat is lost from the body. Wind chill temperatures are not as precise as the effective temperatures in assessing discomfort, because humidity is ignored.

Individual assessment of thermal comfort is based on numerous variables. Among these temperature, relative humidity, and air movement are important. In addition, the temperature of nearby objects and direct exposure to a source of heat contribute significantly to feelings of comfort or discomfort. The temperature of nearby objects determines whether heat is lost or gained by the body through radiation; line of sight exposure to a heat source such as the sun or a hot stove permits heat transfer to the skin by radiation; line-of-sight exposure to an object at a lower temperature than the skin removes heat by radiation. Physiologists report that subjective impressions of the thermal environment are highly dependent on the mean skin temperature. Thus cold discomfort resulting from a low effective temperature can be reduced by simultaneous exposure to direct radiation from the sun or from a bonfire and other sources.

EXTREMES OF BODY TEMPERATURE

The setting of the hypothalamic thermostat can be altered by disease and by excessive heating or cooling of the body. The extremes of body temperature are depicted in Figure 8.

FEVER (PYREXIA, HYPERPYREXIA)

When pathogenic organisms invade the body, toxins, such as endotoxins, are released from the cell walls of the pathogens. These toxins stimulate phagocytic cells (monocytes, granulocyts and Kupffer cells) to produce and release proteins called endogenous pyrogens. (Until relatively recently it was believed that pyrogens were produced only when leukocytes, primarily granulocytes, degenerated. It is now known that all phagocytic cells produce endogenous pyrogens and that these cells do not need to die to release the pyrogens. Furthermore, monocytes produce many times more pyrogen than granulocytes.)

The pyrogens produced by the body's phagocytic cells are carried in the blood to the AH-POA. Here, some of the pyrogen molecules diffuse out of the microcirculation

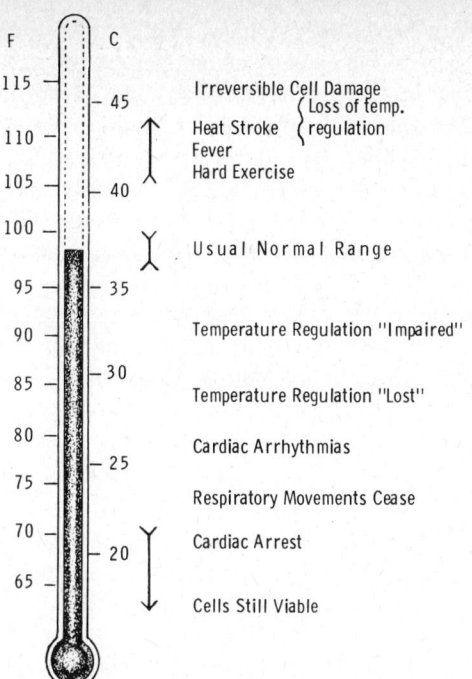

Figure 8. Extremes of body temperature. Note especially that a body temperature rise of only 7 to 10°C above the normal range can cause irreversible damage to cells with possible consequent death, while a temperature drop of 37°C below the normal temperature range can occur without irreversible damage to cells. If the cells freeze slowly, large ice crystals are formed. These usually destroy the cells in which they form.

and stimulate formation of prostaglandins of the E series. Prostaglandins, and possibly also cyclic AMP, are believed to be the mediators directly responsible for raising the thermostatic setting (setpoint) of the neurons of AH-POA.

Fever can result from causes other than the release of pyrogens. The setpoint of the AH-POA is raised in some neurological disorders, in dehydration, and possibly during severe exercise.

When the hypothalamic thermostat is reset at a higher temperature, for example, 40°C, the body responds as if it had been cooled. Heat conservation mechanisms are activated, shivering occurs, and sweating is inhibited. The skin becomes cool and the patient experiences the sensation of being cold. This phase is the "chill." Heat conservation and production mechanisms continue to be overactive until the core blood temperature reaches the new higher setting of the hypothalamic thermostat. When the core temperature reaches the new setting, it then is maintained at or near the new setpoint (Fig. 9).

With remission of the disease process, pyrogens are no longer released by phagocytic cells; hence, the factor causing the high setting of the thermostat is removed and the thermostatic setting reverts back to 37°C. Heat loss mechanisms then dominate. The point in time when the body temperature starts to fall is called the "crisis." Activation

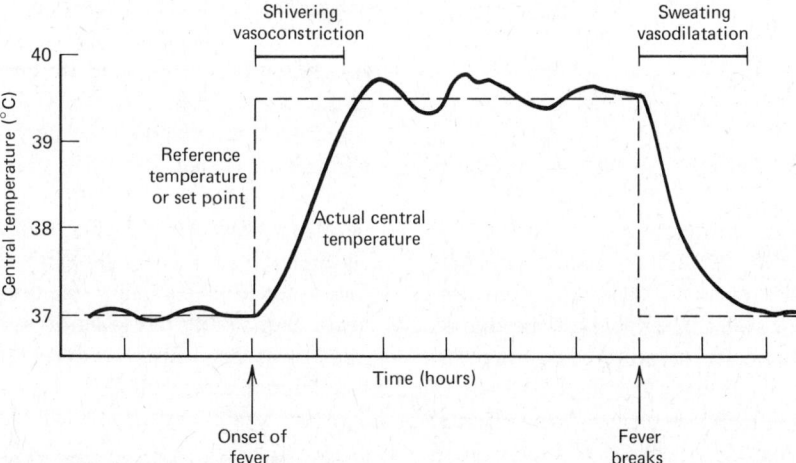

Figure 9 Time course of typical febrile episode. The actual body temperature lags behind the rapid shifts in the setpoint. Note that regulation is maintained during the fever but is less precise, so that temperature fluctuations are generally greater than normal.

of the heat-loss mechanisms and reduction in heat production cause the core temperature to fall (Fig. 6). The patient feels uncomfortably hot, his subpapillary venous plexuses are engorged with blood, his skin is red and warm to the touch, and he sweats profusely.

DEHYDRATION

Severe dehydration causes an upward resetting of the hypothalamic thermostat (dehydration fever). The mechanism responsible for this change is uncertain. However, the higher setting helps to conserve water because sweating will not occur until a higher than normal temperature is reached.

HEAT EXHAUSTION AND HEAT SYNCOPE

Heat exhaustion is a subjective feeling of extreme fatigue during or following an acute exposure to high temperature. *Heat syncope* is loss of consciousness as a result of an acute exposure to high temperature. Both heat exhaustion and heat syncope are primarily due to cardiovascular deficits. Hypovolemia results from loss of water and electrolytes through excessive sweating. In addition, vasodilation in skin blood vessels increases the size of the vascular compartment. Hypovolemia added to peripheral vasodilation results in inadequate venous return, inadequate cardiac output, and arterial hypotension.

HEAT STROKE

This extremely serious disability results from breakdown in functioning of the CNS, especially the hypothalamic thermostatic neurons. If body temperature rises to between 40° and 43°C, CNS function is impaired. Severe disturbances in consciousness occur together with a failure to sweat. Absence of sweating appears to be due to malfunction of the preoptic temperature-control center. When a breakdown in function of AH-POA occurs, a vicious circle (positive feedback) results. As the temperature rises, metabolism increases because of increases in chemical reaction rates, Q_{10}. As body temperature rises, more heat may be produced than can be removed from the body by the poorly functioning heat loss mechanisms, and

temperature continues to rise. Death occurs because of cell damage if the cycle is not broken. Antipyretics such as aspirin are often administered, and barbiturates are sometimes given to help reduce core temperature, but not uncommonly the only effective way to break the vicious cycle is to actively cool the body. Ice baths and ice cold enemas are used, but medical personnel must realize that the neurons of the AH-POA have been damaged (the patient has become poikilothermic). Therefore, he must be removed from the ice bath shortly before his temperature reaches 37°C, and his body temperature must then be externally controlled until his hypothalamus is again functional.

HYPOTHERMIA

A sufficiently reduced core temperature resulting from rapid, profound cooling of the body also leads to failure of the anterior hypothalamus to function in controlling body temperature. The normal response to cooling, as has been discussed, is constriction of skin arteriovenous shunts, vasoconstriction of skin arterioles, inhibition of sweating and activation of the shivering mechanism. These adjustments are usually sufficient to prevent a significant fall in core temperature. However, if the subject, without adequate clothing, is exposed for a period of time to extreme cold, or if he is immersed in water at a low temperature, the heat conservation and production mechanisms may not be sufficient to prevent a progressive fall in core temperature. When the core temperature has fallen to a temperature of about 29°C, the hypothalamic control centers are inactivated, and normal responses to cold are eliminated; hence, shivering stops and circulation to the skin is no longer under central control. Also, as the temperature falls, metabolism decreases (decreased rate of chemical reactions, Q_{10}). With the fall in temperature, heart rate and respiratory rate decrease. When the temperature of the body has fallen to a temperature between 25° and 15°C, respiratory movements cease and the heart stops contracting.

EXERCISE HYPERTHERMIA

The temperature rise during exercise deserves brief comment. Depending on the severity of exercise, highly metabolizing skeletal muscles produce more heat than can be released from the body surface, even though sweating and vasodilation of blood vessels of the skin are present. As a result, core temperature rises above 37°C. It may go as high as 41°C.

Some physiologists argue that in exercise there is an upward resetting of the hypothalamic thermostat, because neither sweating nor skin vasodilation during exercise is as great as when core temperature is raised by other means. (If sweating were maximal during exercise, severe dehydration would occur and, if skin vasodilation were maximal, an excessive amount of blood could be diverted from the active skeletal muscles.) However, there is no direct evidence that the thermostatic setting is altered during exercise.

SELECTED BIBLIOGRAPHY

Carlson LD, Hsieh ACL: *Control of Energy Exchange.* New York, Macmillan, 1970.

Dill DB, Adolph EF, Wilber CG: *Handbook of Physiology.* Section 4: *Adaptation to the Environment.* Baltimore, Williams & Wilkins, 1964.

DuBois EF: *Fever and the Regulation of Body Temperature,* publication 13. American Lecture Series, Springfield, Ill, Charles C Thomas, 1948.

Hardy JD: Thermal comfort and health. *ASHRAE J* 13:43, 1971.

Montagna S, Parakkal PF: *The Structure and Function of the Skin.* New York, Academic Press, 1974.

SELF-STUDY QUESTIONS

MATCHING

A. *Convection*
B. *Conduction*
C. *Evaporation*
D. *Forced convection*
E. *Radiation*

Select the best answer(s). (In many instances, more than one answer is correct.)

1. The method of heat transfer between the sun and the earth.

2. A method of heat transfer that depends on the principle stated in the law of Gay-Lussac.

3. The only available method of heat transfer when the environmental temperature is greater than body temperature.

4. The method by which heat is transferred from the skin to the air.

5. The method of heat transfer that is usually more effective in West Texas than on the Gulf Coast.

6. The major method of heat loss from the body of a swimmer when swimming in water below skin temperature.

7. The principal method by which the body gains heat from surrounding solid objects at a temperature higher than body temperature.

8. The principal method by which heat is lost from the lungs and respiratory passages.

TRUE OR FALSE

9. Core temperature is always higher than body surface temperature.

10. Only 5% or less of the heat transferred from the core to the surface of the body occurs by conduction.

11. Only homothermic animals have a core temperature.

12. Blood flow to the skin may vary from 20 to 4,000 ml/min.

13. Vasodilator fibers to the skin blood vessels of man have been identified.

14. Countercurrent heat exchange protects the feet of arctic birds from damage when exposed to severe cold.

15. Countercurrent heat exchange is an effective mechanism for increasing blood flow to the limbs.

16. When blood flow to the skin has been reduced or absent for 10 or more minutes, reactive hyperemia results.

17. In the triple response the red line is believed to result from the release of histamine.

18. The flare is believed to be the manifestation of the axon reflex.

19. In the triple response the wheal is due to fluid loss from blood vessels.

20. One of the effects of histamine is constriction of veins.

MULTIPLE CHOICE

Select the best answer(s). (In many instances, more than one answer is correct.)

21. Hypothalamic centers bring about an increase in metabolic rate when the core temperature is
   A. above a setpoint of 37°C.
   B. above 37°C, but below a setpoint of 40°C.
   C. 40°C, and this is the setpoint.
   D. 2° below the normal setpoint of 37°C.

22. During the induction phase of a fever
   A. one can be certain that the cause of the fever is the release of pyrogens from degenerating leukocytes.
   B. essentially no water is lost through sweating.
   C. the patient subjectively feels hot because the core temperature is rising.
   D. radiant heat loss is at a minimum.

23. When the core temperature rises above 43°C (109.4° F)
   A. central nervous system functions become abnormal.
   B. heat loss mechanisms are probably depressed.
   C. heat production is increased.
   D. the animal usually becomes poikilothermic.

24. In a hypothermic homotherm with a core temperature
   A. of 20° C, the blood is more viscous than it was at 37°C.
   B. below 20°C, the heart usually ceases to contract.
   C. below 20°C, respiration usually has ceased.
   D. between 37° and 30°C, blood vessels in the skin are constricted.

ANSWERS

1. E.

2. A. The law of Gay-Lussac states that the volume of a gas is directly proportional to its absolute temperature. Convection involves heating of a gas (or liquid), expansion (to a larger volume), and then displacement by colder gas (or liquid) that is heavier than the heated and expanded fluid substance.

3. C. Evaporation will remove heat from a wet surface if the air is not already saturated.

4. B.

5. C. Relative humidity is low in West Texas; therefore, rate of evaporation is increased.

6. D. Best answer appears to be forced convection, although the first step in the heat loss is conduction from skin to water.

7. E. You might argue that conduction (B) is a good answer. However, the body is not usually in contact at any one time with many solid objects.

8. C. Moist surfaces of lungs and respiratory passages promote evaporation. However, if inhaled air is saturated, no heat is lost from respiratory passages by evaporation.

9. False. Although core temperature is usually higher than skin temperature, if the environmental temperature is considerably higher than the core temperature

and air is saturated with water vapor, skin temperature can be higher than core temperature.

10. True. More than 95% of the heat is transferred by the circulating blood.

11. False. All animals have a core temperature, but it varies less in homothermic animals than in poikilothermic animals.

12. True. The skin circulation varies widely depending upon the need to conserve or lose heat.

13. False. Vasodilator fibers to blood vessels of skin of some animals, but not humans, have been identified.

14. False. Countercurrent exchange helps maintain core temperature, but it prevents heat from reaching the apical tissues.

15. False. Countercurrent exchange redistributes flow in limbs but does not increase flow.

16. True.

17. True.

18. True.

19. True.

20. True.

21. B, D. See Figure 7 for explanation.

22. B, D. B is correct because during the induction phase, the setpoint is above the core temperature; hence the sympathetic sweating center in the anterior hypothalamus is not activated. D is correct because blood flow to the skin is greatly reduced; hence the skin tends to be relatively cool during the induction phase and radiant heat loss is thereby reduced. A is incorrect because (1) present evidence indicates that leukocytes do not need to degenerate to release pyrogens; (2) fever also can be caused by neurological disturbances, dehydration, and possibly exercise; and (3) although not mentioned in the text, some patients become hyperthermic during anesthesia. We have no evidence that pyrogens are involved in (2) or (3). C is incorrect because the patient feels cold and often shivers during the induction phase.

23. All are correct. A is correct because above 43°C (109.4°F), neurons are damaged. The damage appears to include alterations in some proteins and injury to cell membranes. B is correct because of damage and lack of function of the hypothalamic centers involved in the control of peripheral mechanisms, promoting heat loss; sweating usually ceases and the peripheral circulation is no longer under the control of the CNS. C is correct because heat production is increased through the increased rate of chemical reactions at the higher temperature. D is another way of stating that the animal no longer is able to control its core temperature. It should be noted, however, that most poikilothermic (cold-blooded) animals have some control over the core temperature. In the heat-damaged normally homothermic animal, all intrinsic control of body temperature is lost.

24. All are correct. A is correct because viscosity increases as the temperature decreases; at 20°C (68°F), both the plasma and cells become more viscous than at 37°C, and the red blood cells become too rigid to squeeze through the true capillaries; consequently, oxygen delivery to tissue cells is essentially eliminated, most of the metabolism is anaerobic and metabolic acidosis develops. B is correct because heart contractions usually cease at some temperature between 20° and 15°C. Fibrillation frequently occurs during the rewarming, but usually not during the cooling phase. C is correct because as the temperature is lowered, respiration ceases before the heart stops beating. D is correct because when the core temperature is below the setpoint, the sympathetic vasoconstrictor center in the posterior hypothalamus controlling skin vessels is not inhibited by the AH-POA; hence this center is active and causes vasoconstriction of skin blood vessels, resulting in heat conservation.

37

Receptors

BURGESS CHRISTENSEN

OBJECTIVES

After completion of this chapter, you should be able to:

1. Describe differences in structure between receptor types.
2. Classify according to function different receptor types and primary afferent fibers.
3. Describe how receptors can signal intensity and quality.
4. Describe receptive field properties, generator potential, adaptation, and adequate stimulus.

RECEPTORS

Every animal has structures that respond to specific forms of energy. These energy transducers are called sensory receptors, and their excitation provides the first step of any sensory process. The degree of sophistication to which any animal reacts to specific stimuli depends on the number and variety of receptors, as well as on the diversity of their central connections. This chapter discusses the way in which sensory receptors, particularly those in the body wall, change the form of energy to which they are exposed into electrical impulses to be carried by nerve axons. Chapter 38 deals specifically with the organized system of sensory receptors located on the body surface, the somatosensory system. We will consider the nature of the input to the nervous system, the anatomical and physiological organization of the sensory pathways, and the types of modulation that occur at different levels along these pathways.

MORPHOLOGY

Although the information carried by axons is electrical in nature, this form of energy is not usually directly applied to the body surface. Stimuli from the environment that may be applied to the body surface are normally mechanical or thermal in nature. The primary function of the receptor is to transform the different forms of energy into action potentials that may be transmitted along the primary afferent axons to the CNS. Receptors are, therefore, transducers. There are three major somatosensory receptor categories: (1) mechanoreceptors responsive to light mechanical stimulation of the body surface; 2) nociceptors responsive to damaging mechanical and thermal stimuli; and 3) thermoreceptors responsive to changes in temperature occurring in the environment. Structurally receptors may be classified into two main types. One type consists of simple free nerve endings. The primary afferent axon loses its myelin sheath and ends within the dermis or epidermis. It is not clear how transduction occurs in such structures, and even more difficult to understand is how they are specifically activated by certain stimuli to the exclusion of others. However, many high threshold mechanoreceptors, those believed to be involved in the mediation of painful sensations, are of this type. A second type

consists of an axon terminal associated with a separate receptor structure. Specificity and the functional form of the response arise from this relationship. The association between axon terminal and receptor takes on several forms including simple close apposition of the axon terminal with the receptor structure as is found in a hair follicle or Pacinian corpuscle. However, this association may be more intimate taking the form of a synaptic contact between the receptor and associated terminal. Although this latter type of association has no physiological basis in this example, electron micrographs indicate that structures similar to the synaptic elements seen in the CNS may be found at these receptor sites. Figure 1a shows the relationship of several receptor types located in skin. Figure 1b shows a Merkel's touch corpuscle that may be one receptor type coupled synaptically to its nerve fiber.

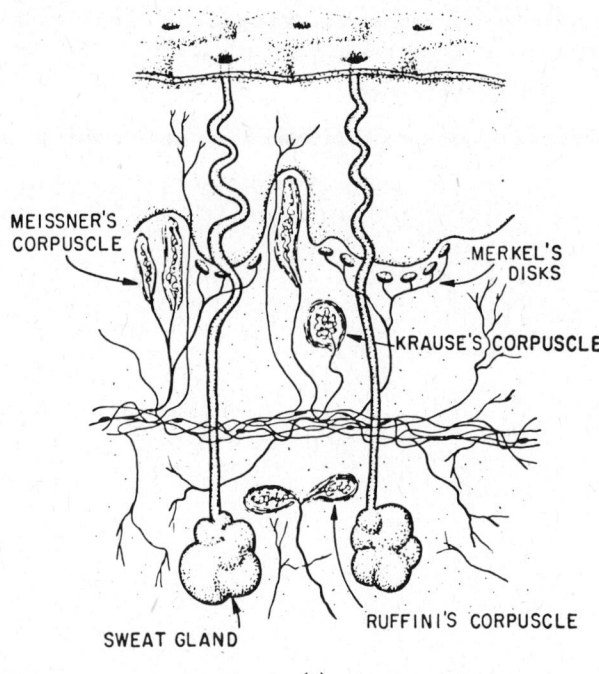

(a)

Figure 1 *(a)* Some receptor types found in the skin of the human fingertip. Two sweat glands are shown with their canals extending to the skin surface. A nerve plexus runs horizontally giving off branches, some of which innervate specific receptor types, some terminating as free nerve endings. (Adapted from Eyzagmire C, Fidono SJ: "Physiology of the nervous system, cutaneous and joint receptors, 2nd edition, Year Book Medical Publishers, Inc., 1975.)

PHYSIOLOGY

Generator Potential

Just how mammalian receptors change the form of energy is not entirely clear, except for a few well-studied examples, such as the Pacinian corpuscle or photoreceptors of the retina. A stimulus applied to the receptor must cause a conductance change either directly or through the influences of a neurotransmitter allowing current to flow across the membrane, which gives rise to a change in the potential. This potential difference has many of the characteristics of a synaptic potential, and, in fact in some instances, it may be produced by the liberation of a neurotransmitter from a receptor structure such as the one shown in Figure 1b.

The potential change produced in the axon terminal innervating a receptor is called the generator potential. Its function is to generate action potentials in the primary afferent fiber just like the synaptic potential following the release of neurotransmitter. Some features in common with synaptic potentials are the following. The generator

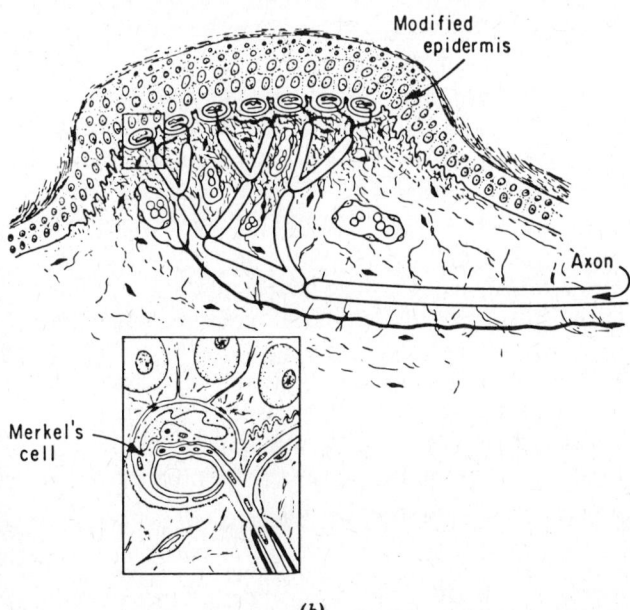

Figure 1. (con't) *(b)* Diagram of a type-I touch corpuscle. The receptor consists of a modified dome of thickened epidermis that overlies a series of receptor cells called Merkel's cells (inset), which are innervated by fine axon branches. The association between axon terminal and receptor cell may be similar to a synaptic contact.

potential is passively or electrotonically conducted in the nerve fiber just like the synaptic potential. This means that unlike the action potential, which is actively propagated, its amplitude will decrease as a function of distance from the site of generation. The conductance change that underlies the potential change is likely to be associated with entry of sodium ions into the axon terminal. This is reasonable since sodium is one of the biological ions present whose equilibrium potential is in the depolarizing direction. It may be that in some cases the conductance change occurring at the nerve terminal is nonspecific. Such a conductance change would allow sodium, potassium, and chloride ions to diffuse across the membrane according to their individual concentration gradients. The result would be a decrease in the membrane potential to zero. In order to arrive at zero, the membrane potential would necessarily have to cross the threshold level for the action potential and hence, an action potential would be generated in the axon.

The amplitude and rate of change of the generator potential is directly proportional to the intensity and rate of application of the stimulus applied to the receptor. This probably means that as the stimulus intensity is increased more ion channels are opened, which results in more current flow and a larger potential change. The net effect of a larger potential change is an increase in the output of the axon in terms of numbers of action potentials. Ultimately this may be interpreted in specific regions of the cerebral cortex as a change in the intensity of the stimulus but not necessarily as a change in the quality of the stimulus. This is an important distinction. Intensity may also be coded by the number of activated receptors. The relationship between the strength or magnitude of the stimulus and the frequency of firing of the primary afferent fiber is not linear, but is rather a power function of the form:

$$\text{Response} = \text{constant} \times (\text{stimulus strength})^n$$

where n is less than 1. This relationship, shown in Figure 2, indicates that initially as the stimulus strength is increased the output rises linearly but ultimately reaches a plateau. Therefore, during the early rise equal increments of stimulus intensity produces similar increments in output whereas at higher intensities the output falls off rapidly with large increments in stimulus strength. Interestingly, such a relationship has been found between stimulus strength and the resulting sensation actually perceived.

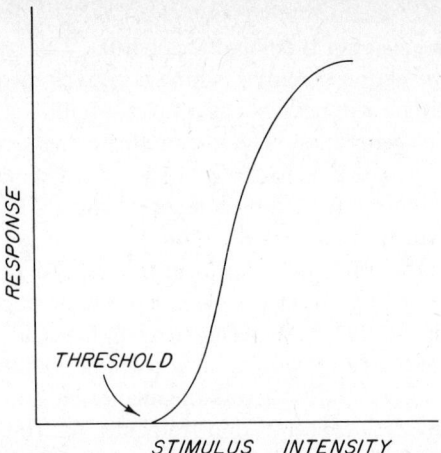

Figure 2 Shows the nonlinear relationship between stimulus intensity and perceived sensation. In the midrange, a small increase in stimulus produces a large increase in sensation. At the two extremes, a larger increase in stimulus is needed to produce a noticeable change in the intensity of the sensation.

Human dorsal root fibers range from about 20 μm in diameter to less than 1 μm. Cutaneous fibers have been divided into three size categories: A α = (6–17 μm diameter, 36–102 m/s), A δ (1–5 μm diameter, 3.5–20 m/s), C (less than 0.3–1.5 μm diameter, less than 1 m/s). Figure 3 shows the size distribution of cutaneous nerve fibers.

It should be remembered that large fibers have faster

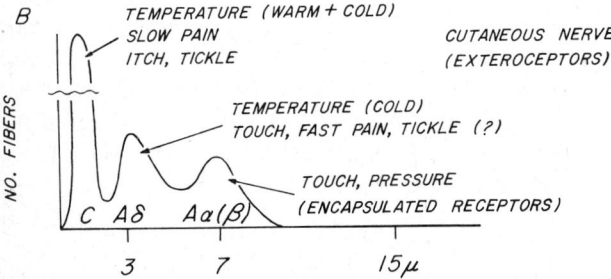

Figure 3 Histograms showing the distribution of cutaneous nerve fibers according to receptor function and axon diameter. There are two major fiber groups, the A group, all of which are myelinated, and the C group, all of which are unmyelinated. The A group contains subgroupings alpha, beta, and delta. Note that the larger diameter fibers conduct most rapidly. (Adapted from Eyzagmire C, Fidono SJ: "Physiology of the nervous system, cutaneous and joint receptors, 2nd edition, Year Book Medical Publishers, Inc., 1975.)

conduction velocities than smaller fibers. A useful generalization may be made that concerns the information carried over these fibers. Pain and temperature information (protopathic sensibilities) are carried over the small myelinated and unmyelinated fiber population and light touch (epicritic sensibilities) over the large myelinated fiber population. More specifically, the A α, β-fiber group innervates receptors activated only during application or removal of a stimulus. Examples are transient detectors with little directional sensitivity such as Pacinian corpuscles and some types of hair receptors. Pacinian corpuscles are particularly sensitive to vibration. Large fibers also innervate velocity detectors such as some hair and nonhair mechanoreceptors located in hairy and non-hairy (glabrous) skin. These mechanoreceptors are concerned both with stimulus position and velocity.

C fibers are unmyelinated and have the smallest diameter. They innervate receptors that also respond to stimulus position and velocity. Response properties of C fibers are unusual in two respects. They require long contact time before they are activated and often there is some discharge following removal of the stimulus. The receptors innervated by C fibers are concerned with tactile sensibility and sensations of itch and tickle.

The A δ and C fiber groups also innervate receptors that sense changes in temperature. These receptors usually give a maintained response to a constant temperature. Nociceptors are a special class of receptors whose activation gives rise to sensations of pain. These receptors are innervated by A δ and C fibers. Either mechanical or temperature stimuli will excite them but the intensity of the stimulus must be elevated. Polymodal receptors are innervated only by C fibers. These receptors respond well to both mechanical and temperature stimuli. The responses of different receptor types are shown in Figure 4.

Note that gentle stroking of the moderate pressure receptor (Fig. 4*b*) barely activates the fiber but a needle thrust produces a vigorous response.

Once the electrical information arrives in the CNS it is transmitted across synaptic junctions to postsynaptic neurons that are either tract cells or relay cells. The former are distributed over large areas, for example, within the dorsal horn of the spinal cord, whereas the latter are located within discrete well-defined areas of the CNS called nuclei.

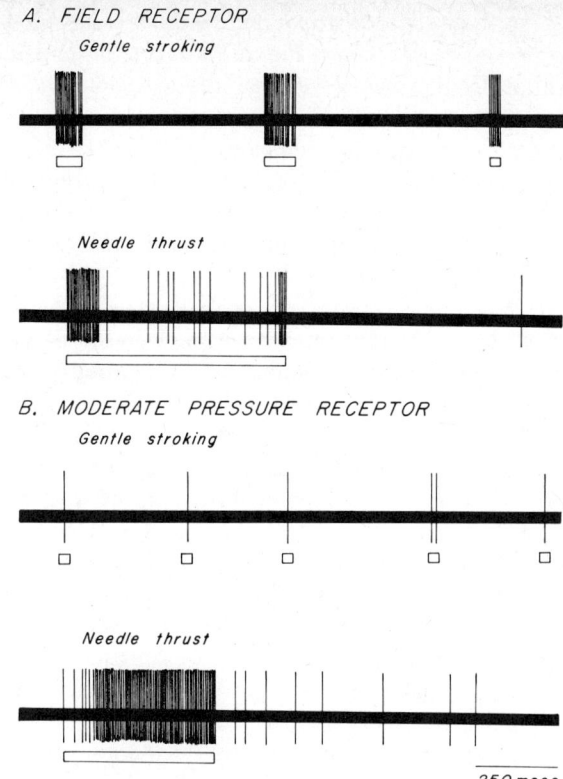

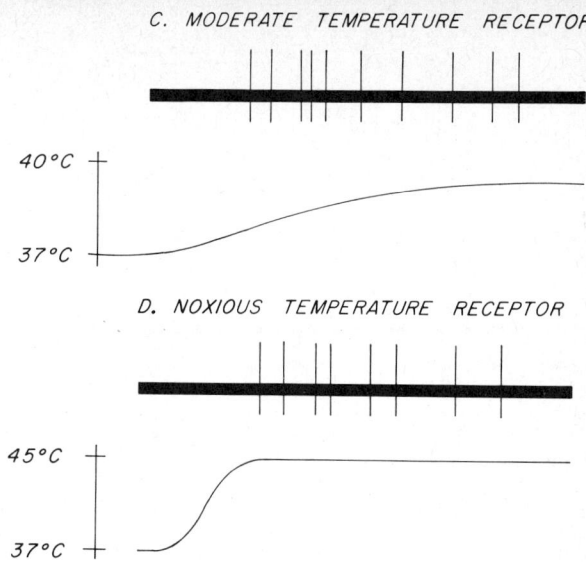

Figure 4. (con't) *(c,d)*. Response properties to moderate and noxious temperature receptors. The moderate temperature receptor begins firing below 40°C in a range that does not produce painful sensations. The noxious temperature receptor begins firing near 45°C near the temperature threshold for pain. (Adapted from Eyzagmire C, Fidono SJ: "Physiology of the nervous system, cutaneous and joint receptors, 2nd edition, Year Book Medical Publishers, Inc., 1975.)

Figure 4. *(a,b)* Response properties of two receptor types found in skin to two different stimuli. The field receptor is a low-threshold mechanoreceptor and responds best to gentle stroking. The moderate pressure receptor may be considered to belong to the low end of the nociceptor group. Its response to needle thrust is quite vigorous. The response to gentle stroking of this receptor type, although present, might never reach cortical levels and therefore, no perceived sensation would occur.

Adaptation

Primary afferent fibers when activated by applying a continuous steady stimulus to their receptor have three types of signal responses. Combinations of these three types may also occur. They may respond by a continuous burst of action potentials throughout the duration of the applied stimulus. This is called a tonic or static response and is probably associated with sensations that may be continuously perceived, such as touch. On the other hand, primary afferent fibers may respond with an initial burst of action potentials that slowly or rapidly declines to a lower frequency tonic response or may even decline to zero. This

initial burst of activity is called a phasic response and is probably important in signalling velocity or changes in direction of an applied stimulus. For example, one might imagine that every time the stimulus direction is changed, a brief burst of action potentials follows signalling the change in direction. Finally, the afferent fiber might respond initially with only one or two action potentials, followed by a period of no activity during the maintained stimulus. This is termed a transient response and is probably important in signalling the sensations of vibration or flutter. How these three types of signals are produced may be determined in part by the receptor's structure or by the time course of the generator potential, as shown in Fig. 5 *a, b*.

When the stimulus is first applied and an initial burst of action potentials occurs (the phasic response) followed by a decline in activity, we say that the receptor is adapting to the applied stimulus. If the decline in activity is rapid then this is a rapidly adapting receptor, if slow then

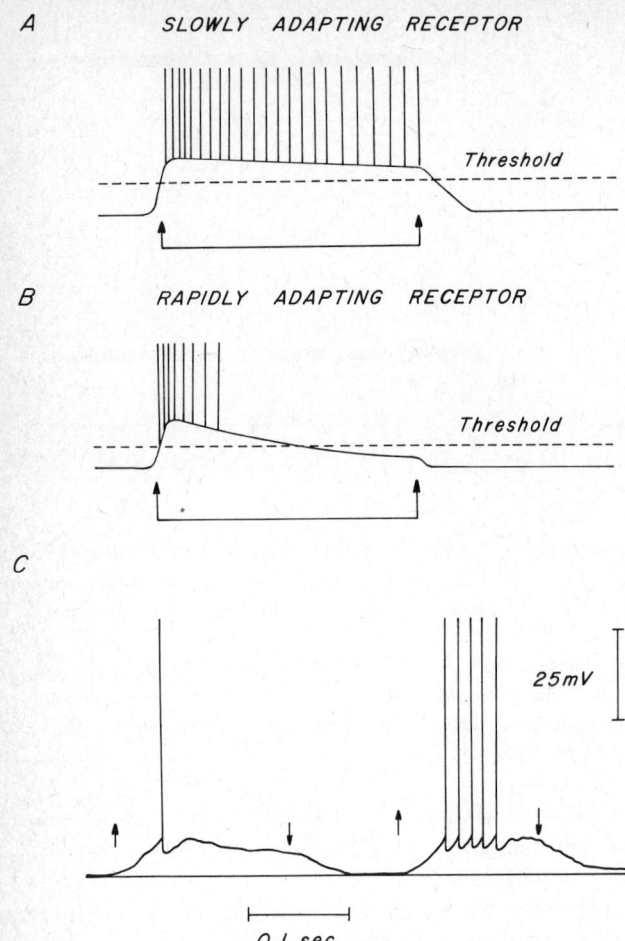

A   SLOWLY   ADAPTING   RECEPTOR

Threshold

B   RAPIDLY   ADAPTING   RECEPTOR

Threshold

C

25 mV

0.1 sec

Figure 5. Response properties of slowly and rapidly adapting receptor types. *(a)* Spike activity is maintained as long as the underlying dc (generator) potential is above threshold. *(b)* Spike activity decreases and ultimately stops when the generator potential is near threshold. The stimulus duration is indicated by the length of the arrows. *(c)* An actual recording from a receptor with the underlying variation in the generator potential. The up arrow is stimulus on and down arrow stimulus off. *(c.* Redrawn from Eyzagmire L, and Kuffler SW: *J. Gen. Physiol.* 39:98, 1955.)

a slowly adapting receptor. Receptors that adapt are thus able to signal both rate of change of the applied stimulus as well as the steady state. Figure 5a–c shows schematically the discharge pattern from slowly adapting and rapidly adapting receptors.

The spike discharges are superimposed on the generator potential. As can be seen in Figure 5b, when the generator potential declines to a level near the threshold the discharge is turned off. In Figure 5a, where the generator potential is maintained well above threshold, the discharge is maintained. Figure 5c shows an intracellular recording from a slowly adapting mechanoreceptor found in crayfish muscle. Here the fluctuations in generator potential are quite apparent.

The Pacinian corpuscle will serve as an example of a transient receptor. This receptor is found in the skin and in the mesentery of the abdominal cavity. Stimulation of the Pacinian corpuscle gives rise to the sensation of vibration. When a maintained dc mechanical stimulus is applied to a Pacinian corpuscle the primary afferent response recorded is a single action potential. When the stimulus is removed another action potential is recorded. During the steady state when the stimulus is maintained, however, no activity is recorded. This receptor consists of a specific morphological structure that results in its rapidly adapting or transient behavior. The structure consists of several fluid filled compartments surrounded by lamellae derived from connective tissue (Fig. 6a).

The nerve terminal is located in the center of this structure. When the stimulus is applied, the mechanical disturbance is transferred through the fluid filled compartments to the nerve terminal resulting in a depolarizing potential change across the nerve terminal membrane. Figure 6B shows the current flow across the nerve terminal membrane during stimulus application. If this depolarization is great enough to reach threshold an action potential is produced in the afferent fiber. The mechanical disturbance, however, is rapidly absorbed by the fluid filled compartments of the receptor structure. The nerve terminal is no longer deformed by the stimulus and the membrane potential declines to its resting level. When the stimulus is removed, the disturbance again is transferred to the nerve terminal resulting in a second action potential. One can readily see that if the mechanical disturbance were to oscillate rapidly, for example, as when stimulated by a tuning fork, it would be transmitted with a high degree of fidelity to the nerve terminal. Pacinian corpuscles can follow frequencies that are really limited only by the ability of the nerve to depolarize and generate an action potential. These frequencies may be as high as 1,000 Hz.

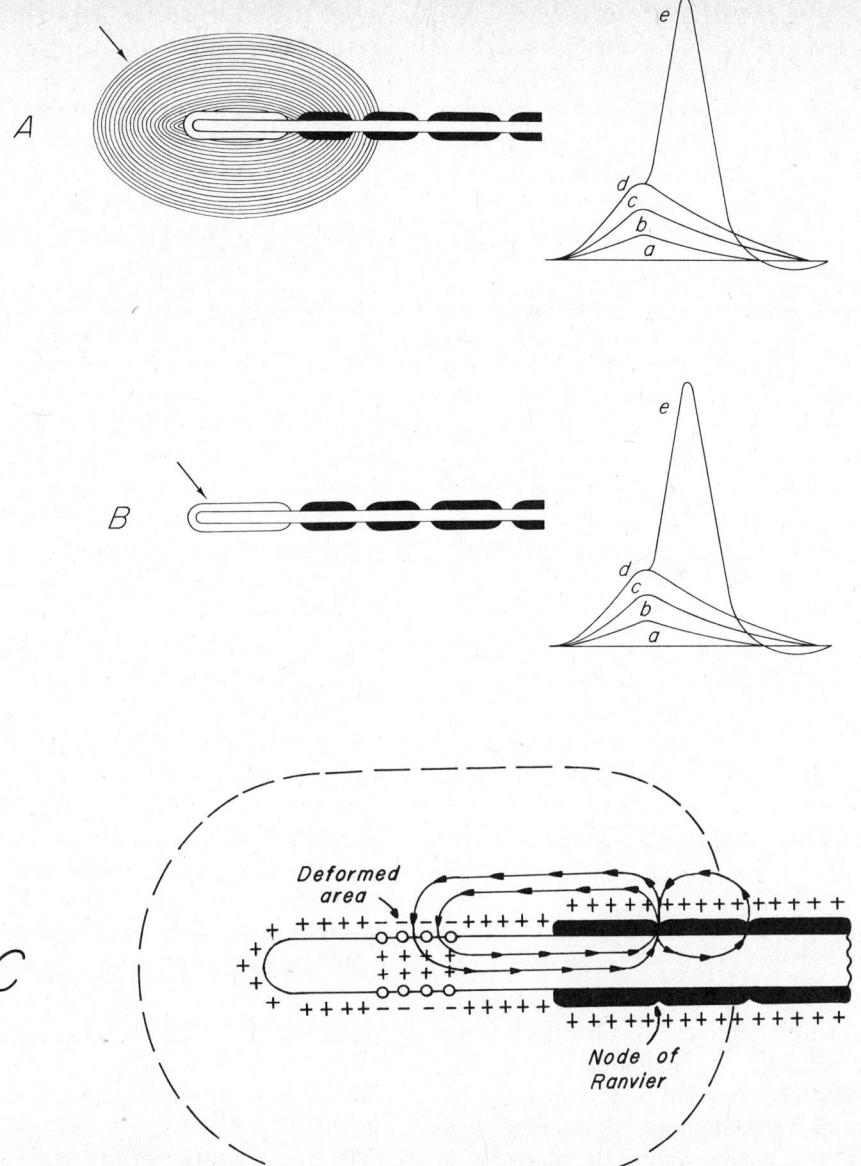

Figure 6. *(a)* Layered structure of the Pacinian corpuscle capsule innervated by a single myelinated fiber. To the right is the receptor generator potential produced by increasing (a–d) the stimulus intensity indicated by the arrows. When of sufficient strength the generator potential produces an action potential (e). *(b)* A similar sequence occurs when the outer lamellae are removed. *(c)* The path taken by current flowing in the axon terminal and across the membrane. (A-B redrawn from Lowenstein WR: Biological transducers, *Sci. Am.* 203:98, 1960) (C. From: Guyton AC, Medical Physiology, 4th ed., ch 48, Saunders, 1971 as modified from Loewenstein: Ann. N.Y. Acad. Sci., 94:510, 1961)

The hair follicle receptor is another example of a rapidly adapting receptor type. If the hair is bent and held in that position the sensation disappears. This is because no depolarizing potential is being generated in the primary afferent fiber. If the hair is bent rapidly a transient discharge results. This may be interpreted as faster motion. Hence, in this sense we are detecting changes in velocity. Such receptors give information about how fast stimuli are moving across the surface of the body.

Other types of receptors are slowly adapting and tell the nervous system about long term changes in the environment. An obvious example is temperature receptors. As the temperature of the environment changes we are continuously aware of the changes. These receptors may produce activity in the primary afferent fibers for long periods of time. There are also slowly adapting touch receptors. It is a curious phenomenon that although these slowly adapting receptors are continuously sending signals to the CNS we may become totally unaware of the sensations. This is because we are able to block this information from our conscious perception based on our level of attention. Thus, a student reading in a noisy environment may not be distracted. One can readily test this with a self-experiment. Place a coin on the back of the hand and see how the initial sensation decreases with time. This is in part due to the adaptation of the receptors, however, if occupied with another task one would soon "forget" that the coin is there.

ADEQUATE STIMULUS

One way to classify a primary afferent fiber is to record its activity to different forms of stimuli. This has been done for a variety of mammals including human subjects. Often a primary afferent fiber will respond to several types of stimuli. Sometimes when one stimulus is presented that does not activate the primary afferent fiber, it will respond to this same stimulus if it has received previous intense stimulation. This introduces two new concepts, the first being that of the adequate stimulus and the second that of sensitization.

In spite of the fact that receptors may respond to several types of stimuli there is usually one to which it responds best. This is the receptor's adequate or best stimulus. An example of a stimulus applied to a specific receptor and one which readily demonstrates the principle of the adequate stimulus is the visual response produced by applying pressure to the retina through the eyelid with a finger. Although the perception of light occurs, the necessary mechanical deformation measured in terms of energy to produce the sensation of light is several orders of magnitude greater than the amount of light-hitting photoreceptors, which will also produce the same sensation of light. In the latter case it is one or two photons. Hence, we can conclude that the proper stimulus for the retina is photons and not mechanical deformation. It is important to note, however, that although in the first case mechanical stimulus is applied, the sensation of light is perceived. This is a result of the precise central connections at the cortical level and not a function of the receptors or primary afferent fibers stimulated. In other words, these same fibers stimulated by electrical means would give rise to the perception of flashes of light because of their central connections. The function of the receptor, therefore, is to respond to the proper stimulus. Some cutaneous receptors respond not only to mechanical stimuli, but also to a change in temperature. However, some of these are exquisitely more sensitive to the mechanical stimuli than to temperature changes and are rightly classified as low threshold mechanoreceptors. On the other hand, there are also true polymodal cutaneous receptors, ones that respond equally well to mechanical and temperature stimuli. It is not clear what types of sensations result from activity in these fibers, although it is known that the primary afferent fiber belongs to the unmyelinated C fiber group. They are probably involved in the protopathic type of sensibility.

It should be readily obvious by now that the primary afferent fiber, the first link of the environment with the CNS, is usually precisely tuned in terms of its response to specific stimuli. This fact has suggested to physiologists that the wiring of sensory systems is specific with the tendency to keep information as to type, separate and discrete. This means, for example, that when a hair on the skin is touched or bent, we perceive a specific sensation, which we recognize as a hair being touched because a specific receptor has been activated, and the information travels over a precise pathway to a specific area of the brain where the activity is interpreted and the appropriate sensation results. This hypothesis suggests that there is specificity at the receptor level, as well as in the central connections, and that the discrimination of sensory experience is a combination of activity in specific fibers and in specific cortical neurons that give rise to the sensation.

As noted above it is possible to activate receptors with other than their proper stimulus. The converse is also true for certain receptor types, that is, it is possible by intense stimulation to lower the threshold of a receptor. This is particularly so for nociceptors. For example, a heat nociceptor will normally start to discharge when the skin temperature reaches about 43°–45°C. If the temperature is elevated for several minutes, to say 48°C, and then permitted to return to normal, this receptor can then respond to temperature changes below 43°C, in the non-noxious range. This phenomenon is termed sensitization. The sensation produced by the lower temperature is still pain. Sensitization is accompanied by reddening of the area probably due to release of irritant chemical substances from damaged tissue in the surrounding area. Possibly these substances reduce the threshold for activation. Sensitization explains why an injured area becomes sensitive to non-noxious stimuli following injury or damage to the tissue.

RECEPTIVE FIELD

Primary afferent fibers usually do not innervate a single receptor. However, each receptor is usually innervated by a single fiber. One can imagine, therefore, that a recording from a single primary afferent fiber might be expected to respond from several stimulated points on the body surface. These points will be close together but the total area will vary in size depending on how many receptors are innervated by the fiber and the interreceptor spacing. The area on the body surface that will produce activity in a single afferent fiber when the adequate stimulus is applied is called the receptive field. The conscious perception of that area of the body being stimulated depends on the central connections within the cerebral cortex to which the primary afferent fiber projects. If the receptive field for a single afferent fiber included an entire finger then we would not be able to distinguish where on the finger we were being touched. However, if the finger were supplied by several afferent fibers with distinct but contiguous re-

ceptive fields, then as the stimulus moves from one field to the next the sensation that a stimulus were moving across the finger in a spatially oriented direction would be perceived, since each different fiber projects to a slightly different area of the cortex.

The ability to distinguish between two points touched simultaneously on the body surface, called two point discrimination, is related to this concept of receptive field. If the receptive fields are large, two stimuli simultaneously applied to the skin will have to be relatively far apart. As the receptive field decreases in size the two points can be moved closer together. A test on your own body with a pencil will quickly convince you that the receptive fields on the parts of our body that are used for highly discriminiative purposes, such as the finger tips, must be very small for the two-point threshold is also small. The two point threshold or limen is the minimal distance between two points on the body surface that may be perceived as being distinct and separate. A consequence of this concept is that the density of innervation of the distal body parts is much greater than the proximal body parts. In the CNS this simply means that many more neurons are devoted to information processing for the distal body parts than for the proximal body parts.

SELECTED BIBLIOGRAPHY

Burgess PR: Sensory receptors and primary afferent fibers, in Eyzaguirre C, Fidone S (eds): *Physiology of the Nervous System.* Chicago, Year Book, 1975.

Burgess PR: Cutaneous and joint receptors, in Eyzaguirre C, Fidone S (eds): *Physiology of the Nervous System,* Chicago, Year Book, 1975.

Burgess PR, Perl ER: Cutaneous mechanoreceptors and nociceptors, in Iggo A (ed): *Handbook of Sensory Physiology,* vol 2, *Somatosensory System.* Berlin, Springer-Verlag, 1973.

Iggo A: Cutaneous receptors with a high sensitivity to mechanical displacement, in DeReuck AVS, Knight J (eds): *Touch, Heat and Pain.* Boston, Little, Brown, 1966.

Mountcastle VB: Sensory receptors and neural encoding: Introduction to sensory processes, in Mountcastle VB (ed): *Medical Physiology,* ed 13, St Louis, CV Mosby, 1974.

SELF-STUDY QUESTIONS

MATCHING

A. *Velocity detector*
B. *Flutter vibration*
C. *Slowly adapting*
D. *Thermal receptors*
E. *Nociceptors*

1. Pacinian corpuscle
2. Hair follicle
3. Merkel's touch corpuscle
4. Free nerve endings

MULTIPLE CHOICE

Select the correct answer(s). (In many instances, more than one answer is correct.)

5. The generator potential has all the following features in common with synaptic potentials *except* being
   A. passively conducted.
   B. graded in amplitude.
   C. hyperpolarizing.
   D. a property of summation.

6. All the following can be used by the nervous system to code intensity of sensation *except*
   A. magnitude of the stimulus.
   B. number of activated receptors.
   C. number of active primary afferent fibers.
   D. increasing the duration of the stimulus.

7. Cutaneous receptive fields at a peripheral area are smaller than those located on proximal body parts because
   A. less cortical area is allocated to peripheral body parts.
   B. different receptors are found on peripheral body parts.
   C. the density of receptors is greater at peripheral body parts.

   D. smaller receptors are located in peripheral body parts.

8. All the following are true about areas of the body with large receptive fields *except*
   A. small cortical representation.
   B. poor localization of sensory stimulus.
   C. proximal body areas have larger receptive fields than peripheral body areas.
   D. precise localization of sensory stimulus.

9. The amplitude of the generator potential
   A. increases linearly with stimulus strength.
   B. can be influenced by inhibitory neurons.
   C. regulates the amplitude of the action potential.
   D. varies with the log of the stimulus intensity.

10. During a maintained stimulus, afferent fibers innervating slowly adapting receptors
    A. give a maintained response.
    B. turn off immediately.
    C. vary their response sinusoidally.
    D. do not respond.

11. A receptor's adequate stimulus is
    A. the one that elicits the largest response.
    B. the one best suited to activate the receptor.
    C. the one that elicits the least number of action potentials.
    D. always mechanical.

12. Students found the two-point limen to be 2 mm on the fingertip and 5 mm on the shoulder. Which of the following is correlated with this observation?
    A. The finger representation is larger in cortex than is the shoulder.
    B. The finger has a greater density of innervation than does the shoulder.
    C. The receptive fields on the finger are smaller than on the shoulder.
    D. All of the above.

13. The frequency of firing of an afferent fiber in response to sensory stimulation depends primarily on
    A. duration and amplitude of the generation potential.
    B. duration of stimulation.
    C. the diameter of the axon.
    D. the number of ion channels in the axon membrane.

14. All the following are true about nociceptors *except*
    A. they give rise to sensations of pain.
    B. they can respond to mechanical or thermal stimuli.
    C. they are free nerve endings.
    D. they are excited by light touch.

MULTIPLE CHOICE

Select the single best answer.

15. Which receptors are rapidly adapting (no maintained response)?
    A. Warm receptors
    B. Merkel cells (type I)
    C. Meissner's corpuscles
    D. Ruffini endings (type II)
    E. Polymodal nociceptors

16. Which receptors would respond to a light touch on the palm?
    A. Mechanical nociceptors
    B. Cold and warm receptors
    C. Receptors around hair follicles
    D. Meissner's and Ruffini (type II) corpuscles and Merkel cells (type I)
    E. Meissner's corpuscles only

17. A receptor potential
    A. is conducted to the CNS.
    B. can cause an action potential in the afferent nerve fiber.
    C. is abolished by local anesthetics.
    D. is an all-or-none event.
    E. inhibits the afferent nerve fiber.

18. The term receptive field of a neuron refers to
    A. none of the following.
    B. the firing of rate of the presynaptic cell.
    C. the area of the sensory periphery that affects the discharge.
    D. where the neuron projects.
    E. the location of the neuron.

19. Receptive fields
    A. of primary afferent fibers from sense organs can be either excitatory or inhibitory.
    B. tend to be larger for primary afferent fibers than for central sensory neurons.
    C. can be inhibitory for central sensory neurons.
    D. refer only to cutaneous sensation.
    E. remain normal following peripheral nerve injury.

20. Which one of the following receptors is rapidly adapting?
    A. Muscle spindle
    B. Pacinian corpuscle
    C. Golgi tendon organ
    D. Ruffini ending
    E. Touch corpuscle (Merkel's)

21. Which of the following cutaneous receptors responds selectively to a gentle mechanical stimulus applied within its receptive field?
    A. Receptors C and D only
    B. Receptors C, D, and E.
    C. Polymodal nociceptors.
    D. Mechanical nociceptors.
    E. Meissner's corpuscle.

22. All the following statements about sensory receptors are correct, *except*
    A. they are particularly sensitive to a single form of stimulation.
    B. they convert stimulus energy to changes in a membrane conductance or voltage.
    C. a single sensory fiber innervates a restricted portion of the sensory surface.
    D. primary afferent fibers can code stimulus intensity by their discharge rate.
    E. afferent fibers from the skin have excitatory and inhibitory receptive fields.

23. The following properties are characteristic of a primary afferent fiber from a sensory receptor *except*

A. discharge is due to a receptor potential.

B. discharge can fatigue with repetitive stimulation.

C. it is an excitatory receptive field.

D. it is an inhibitory receptive field.

E. it can display spontaneous discharge.

MULTIPLE CHOICE

Select the correct answer(s). (In some cases, more than one answer is correct.)

24. Sensory receptors

A. are usually specialized to respond best to one type of stimulus.

B. transduce a stimulus by changes in membrane conductance.

C. can be classified as exteroceptors, interoceptors, or proprioceptors.

D. usually make monosynaptic inhibitory connections in the spinal cord.

25. The determinant of receptor specificity is

A. position of the receptor on the body.

B. composition of the receptor surface membrane.

C. connective tissue surrounding the receptor.

D. composition of the axonal membrane.

26. In general, receptors or afferent fibers code equal steps in stimulus intensity by

A. equal increments in firing frequency of afferent fibers at all stimulus intensities.

B. larger increments in firing frequency of afferent fibers in the low intensity range.

C. equal increments in the magnitude of the receptor potential at all stimulus intensities.

D. smaller increments in the magnitude of the receptor potential in the high-intensity range.

27. A mechanical stimulus to the skin could activate

A. Type I (Merkel) and type II (Ruffini) endings.

B. hair follicle receptors.

C. Meissner's corpuscles.

D. pacinian corpuscles.

ANSWERS

1. B.

2. A.

3. C.

4. D,E.

5. C. Generator potentials initiated at receptors are always excitatory (depolarizing) in nature. Their function is to drive the membrane potential to threshold.

6. D. Increasing the duration will either give rise to the same sensation or the perception of the sensation may simply disappear. This latter can occur if the receptor is rapidly adapting. In general, anything that will increase the firing rate of the afferent fiber may give a perceived increase in intensity. Duration will not increase firing rate.

7. C. Because the density is greater at peripheral body parts each afferent fiber innervates a smaller area. A is incorrect because the situation is reversed, that is, more cortical areas are devoted to peripheral body parts because these areas are more densely innervated.

8. D. When the receptive fields are large, precise localization is lost. If, for example, a receptive field were to consist of the entire upper arm, one would be unable to distinguish different localities on this body part, since wherever one touched the same receptor would be activated.

9. D. Only answers A and D are reasonable. Most if not all biological phenomena show saturation. As one function (stimulus) increases, the response ultimately slows and reaches a maximum. In this case, the relationship is logarithmic.

10. A. Only reasonable answer.

11. B. Many different kinds of stimuli can activate receptors, but each receptor is best tuned to respond to a particular stimulus type, the adequate stimulus. In any case, the same sensation is always perceived.

12. D. All the answers are associated.

13. A. It is the generator potential-causing threshold that produces the fiber activity. B and D can contribute, but are not primarily responsible.

14. D. By definition, nociceptors require a strong stimulus.

15. C. All receptors listed, except that Meissner's corpuscle can give a maintained response. This is an important velocity detector found in nonhairy (glabrous) skin, especially dermal papillae of fingertips.

16. D. These receptor types are low threshold mechanoreceptors found on nonhairy skin. C. Found on hairy skin.

17. B. Local anesthetics block action potentials by blocking sodium channels.

18. C. The neuron referred to here could be a primary afferent fiber, or a cortical neuron, or any in between.

19. B. This is because for central neurons the receptive field organization can be altered by inhibitory mechanisms. See Chapter 38.

20. B.

21. E. Nociceptors require strong stimuli.

22. E. The afferent fibers do not have inhibitory receptive fields. However, central neurons can.

23. D.

24. A. Only A is incorrect. Both excitatory and inhibitory connections may be made.

25. A.

26. C. This indicates the nonlinear relationship between stimulus and response.

27. E. All are mechanoreceptors.

38

The Somatosensory System

BURGESS CHRISTENSEN

OBJECTIVES

After completion of this chapter, you should be able to:

1. Describe the anatomical features of the ascending dorsal column and spinothalamic sensory pathways.

2. Describe the functional properties of each pathway.

3. Compare the properties of sensory receptive fields on different body areas.

4. Consider the functional significance of convergence and divergence as it applies to the nuclei integrating sensory information.

5. Describe the role of inhibitory and excitatory synaptic mechanisms in cerebral sensory integration.

6. Locate the areas and cytoarchitecture of the cortical representations for somatic sensation.

7. Describe the columnar organization of sensory cortex and its functional significance.

Exteroceptors include those receptors located in the body wall. Their activation by stimuli from the environment produce sensations that reach consciousness. The pathways over which this information travels in part makes up the somatosensory system. Proprioceptors are located in the deeper tissues such as the skeletal muscle and joints and when activated, also send information over somatosensory pathways about the position of our body in space. However, activation of these receptors probably does not produce conscious sensations. It is the accurate interpretation of the electrical impulses traveling over these pathways that enables the organism to continue to exist in its environment. The network involved in distributing this information can be succinctly divided into two, both with respect to sensory modality represented (i.e., kind of sensation) and to the anatomical pathways over which the information travels. These are the dorsal column system over which information about discriminative touch (that which is well localized), vibration, and joint position travels, and the spinothalamic system over which travels information about pain, temperature, and crude touch.

The most primitive type of sensibility is that carried by pathways involved with protopathic sensations. This is the sensibility to sensations of pain and temperature that may be poorly localized, but absolutely necessary for the survival of the organism in its environment. The information that gives rise to these sensations are carried over the spinothalamic pathways. Phylogenetically, this is the oldest pathway and is present in all vertebrates. Fibers innervating cutaneous or deep organ receptors or that end as free nerve endings and are specifically sensitive to damaging kinds of mechanical stimuli or to extreme temperature changes, enter the spinal cord via the dorsal roots. The receptors are called nociceptors because they respond to noxious stimuli, that which evokes a painful sensation. They synapse with second-order neurons within the dorsal horn of the spinal cord. These second-order neurons send their axons across the spinal cord and ascend in the ventral and lateral white columns called the ventral and lateral spinothalamic tracts. As their name suggests, these fibers are destined for the posterior nuclei of the sensory thalamus. The important point, however, is that these fibers carry information about pain and temperature, and the fibers are crossed at the level of the spinal cord.

A more highly developed and phylogenetically newer system is that which carries information about epicritic sensibilities. These are sensations resulting from light touch and gentle stimulation of the body surface. Such sensations can be precisely localized by the individual on the body surface. Again, as for the pathway described above, information is carried by the discharging primary afferent fibers to the spinal cord where, instead of making synaptic contact directly with second-order neurons that cross the cord to ascend the neuraxis, they stay on the same side and ascend in the posterior white matter (dorsal columns) where they synapse with their second-order neurons in the dorsal column nuclei located at the junction of the spinal cord with the medulla. As these axons ascend in the dorsal columns, they can give off axon collaterals that synapse on cells within the dorsal horn. However, the second-order relay neurons are located in the dorsal column nuclei.

Figure 1 *a, b* shows the locations of fiber tracts and the distributions of the cells for these two pathways. To contrast the two pathways, the second-order neurons of the dorsal column system are located in nuclei at the rostral end of the spinal cord. Axons from these cells cross the neuraxis to form the medial lemniscus and project along with the already crossed spinothalamic tract fibers to the posterior nuclei of sensory thalamus. The dorsal column system is most highly developed in primates, in particular the human, and gives us our fine discriminative capabilities especially with the distal portions of our extremities, such as the fingertips, where the density of innervation of the primary afferent fibers mediating these sensations is the highest.

The primary afferent fibers that innervate the body surface are collected into nerve trunks and finally into the dorsal roots. Each dorsal root will innervate a specific area of the body. This body area is called a dermatome. The dermatomes can be directly related to spinal segments. Hence, if spinal injury were to occur, the level of the injury can be readily determined by noting the dermatome on which the sensory deficit occurs. Clinically important dermatomal regions are shown in Figure 2.

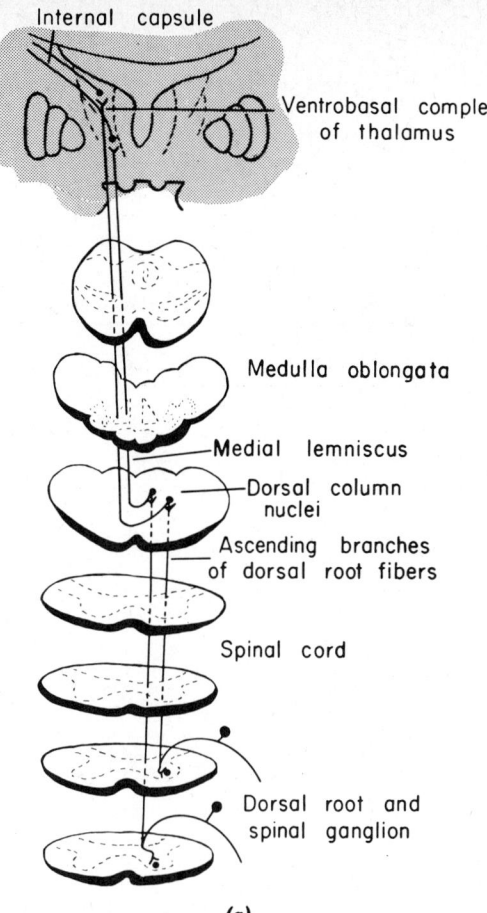

Figure 1 The two major afferent pathways for somatic sensations. *(a)* Dorsal column pathway for epicritic sensations. Note crossing of pathway at level of dorsal column nuclei.

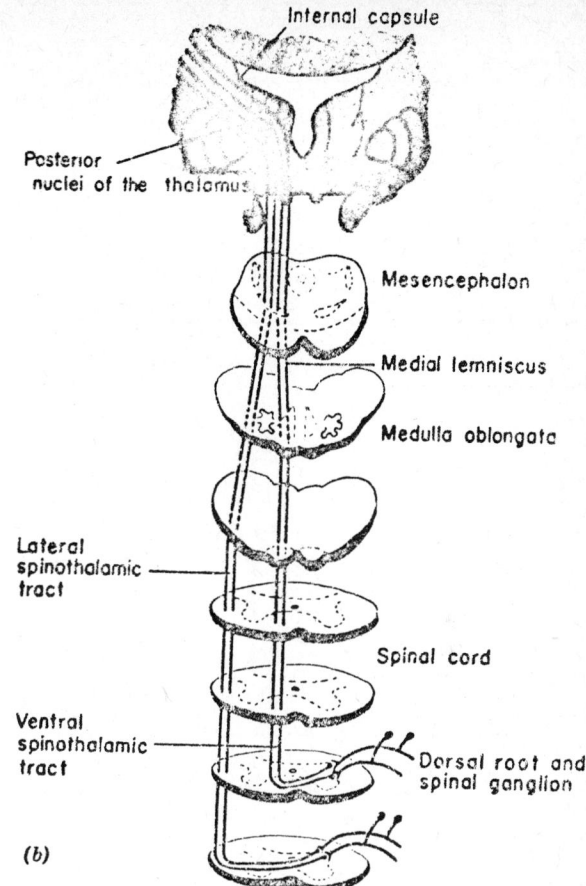

Figure 1 (con't) *(b)* Spinothalamic pathway serving protopathic sensation. Note crossing of pathway at or near level of entrance of primary afferent fibers. (From Guyton AC, Textbook of Medical Physiology, 4th ed. Chap. 49, Saunders, 1971).

THE DORSAL COLUMNS

The dorsal columns are really fiber tracts or fasciculi. There are two fasciculi located on each side of the spinal cord. The most lateral fasciculus is called the fasciculus cuneatus and the most medial the fasciculus gracilis. As the dorsal roots enter the spinal cord and the ascending axons begin to form the dorsal columns, fibers from the most caudal part of the spinal cord get pushed medially as they travel toward the medulla. This is simply because,

as one ascends the neuraxis, more fibers are being added. Thus, the dorsal columns are organized topographically with the hindlimbs of mammals or the lower extremeties of primates represented medially in the fasciculus gracilis and the forelimbs or upper extremities laterally in the fasciculus cuneatus. This topographical organization is important for it is maintained all the way to cortex and is a useful concept for localizing lesions subsequent to injury or vascular accident.

As the axons of the primary afferent fibers ascend in the dorsal columns, they give off axon collaterals that synapse with cells within the dorsal horn. In this way information

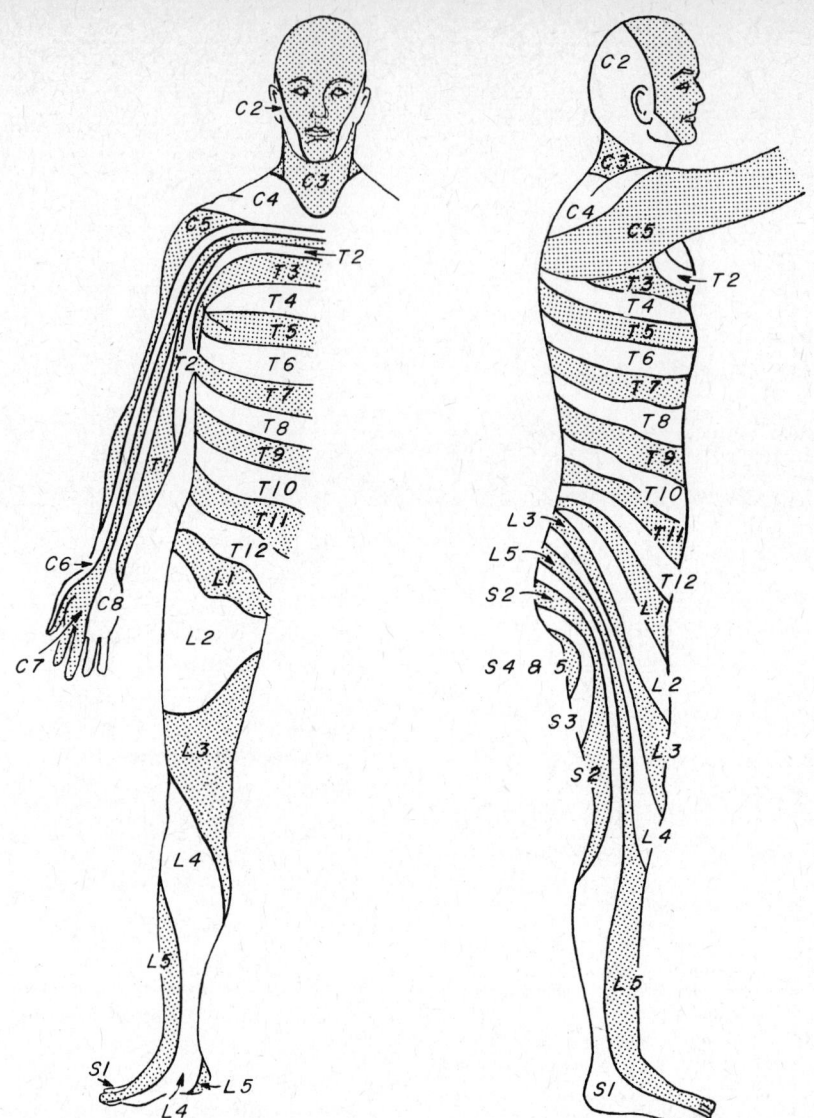

Figure 2 Major dermatomal organization of the body surface. Of particular importance for identifying levels are T4 at the level of the nipples and T10 the umbilicus.

about an event occurring at peripheral receptor sites is transmitted in a local fashion to spinal cord pathways important for reflex action. Each time an axon gives off a collateral, the diameter of the axon decreases. Consequently, the conduction velocity also decreases. This slowing of conduction velocity may be as much as 80% for

some receptor types. Some fibers do not make it all the way to the dorsal column nuclei but instead exit the dorsal columns and synapse on cells that give rise to Clarke's column. The cells that form Clarke's column are destined for the cerebellum and a synonym is, therefore, the dorsal spinocerebellar tract.

DORSAL COLUMN NUCLEI

Anatomical Organization

The first relay for sensory information destined for sensory cortex and traveling via the dorsal column system is the dorsal column nuclei located at about the level of the obex, the junction between spinal cord and medulla. Here again there are two nuclei on each side of the spinal cord. Each nucleus corresponds to one of the above-named fasciculi, the nucleus cuneatus and nucleus gracilis.

The dorsal column nuclei are divided into several cellular regions based on their function. The most important is that region that gives rise to the second order neurons that project to sensory thalamus. As in any nuclear area, divergence and convergence of the incoming fibers can take place. However, it is important to realize that specificity is maintained and the mixing of sensory modalities does not occur. It appears, therefore, that what occurs in the case of the direct projection fibers (i.e., those that go to the thalamus) is one of an increase in gain that results in a more powerful input to the higher brain centers. Other cellular areas within these nuclei are probably involved in local integration and inhibition. Many of the interneurons form axoaxonic contacts that are probably involved in presynaptic inhibition. Finally, as in most nuclei, a feedback system comes from cortical areas. These also are able to modify the feed-forward activity by the dorsal column fiber system.

Not only is there good anatomical evidence that cells in specific regions of the nuclei project to the thalamus, but there is good physiological evidence as well. Electrophysiological recordings from single cells coupled with antidromic stimulation of their axons in the medial lemniscus has confirmed that cells project to the thalamus. Stimulation experiments of this type have also shown that activity of cells within the nucleus is capable of being modified by cortical activity.

The receptive field size of cells within the dorsal column nuclei is larger than the peripheral receptive field of the primary afferent fibers. This simply means that single fibers forming adjacent fields in the periphery converge on the same cell in the nuclei. In spite of this increase in receptive field size, the primary modalities are conserved just as they were for the primary afferent fiber.

SPINOTHALAMIC SYSTEM

We must now return to the level of the spinal cord to describe the anatomy and physiology of the second ascending sensory system, the spinothalamic system. This sensory pathway is primarily concerned with the transmission of information dealing with gross localization of input to the body surface, temperature, and pain. As you now know, the primary afferent fibers activated by these kinds of stimuli are among the slowest conducting.

ANATOMY

There are three divisions of the spinothalamic pathway: the spinobulbar and paleospinothalamic, the fibers of which project to the reticular formation located in the medulla, and the neospinothalamic, the fibers of which project directly to the thalamus along with the fibers of the medial lemniscus.

The origin of the spinothalamic tracts, unlike the dorsal column system, is from the cells located within the dorsal horn of the spinal cord. Remember that the dorsal column system fibers come from the cells located in dorsal root ganglia. In the monkey, approximately 80% of the second-order cells are located in the deeper layers of the dorsal horn. However, cells are also located in superficial layers. These latter cells are particularly responsive to noxious stimuli and are therefore believed to be the cells that carry information about pain. The cells in the dorsal horn of the spinal cord are then the second-order neurons that relay information to higher brain centers. Unlike the second-order neurons of the dorsal column system, they are not located in a discrete nucleus; instead, they are distributed along the length of the spinal cord. The majority of the axons from these second-order neurons then cross the spinal cord and ascend in the ventral and ventrolateral white columns. At this point it is valuable to compare two aspects of the spinothalamic and dorsal column systems, because these differences are important for clinical diagnosis. First, remember that the dorsal column system is organized topographically with the medial fasciculus gracilis containing the fibers from the lower extremities and the lateral fasciculus cuneatus from the upper extremities. Since the spinothalamic system is a crossed system at the level of the spinal cord, the lower extremities are pushed laterally and the upper extremities medially since they are the last to enter the spinal cord as one ascends the neuraxis. The second important point is that if a spinal injury occurs involving only one half the spinal cord, then certain sensory modalities below the lesion will be lost on the ipsilateral half of the body surface and others lost on the contralateral half. In the case wherein the cord is hemisected, there will be loss of crude touch, pain, and temperature on the contralateral half of the body surface and loss of two point discrimination and vibration sensibilities on the ipsilateral half of the body surface below the lesion site. The anatomical organization that gives rise to this clinical syndrome, also called the Brown-Séquard syndrome, is shown in Figure 3.

PHYSIOLOGY

It is relatively easy to understand that primary afferent fibers innervating receptors that are excited by changes in temperature or a strong mechanical stimulus will evoke sensations of heat or cold and pain. However, it is reasonable to ask what is meant by crude touch. Such terminology simply suggests absence of resolving power or ability to localize to a small area. The way in which this occurs can be determined by examining the size of receptive fields. The physiological properties of single neurons in the dorsal horn has been studied in great detail in many different animals. In the monkey the receptive fields range from a whole digit to half a limb. As with the dorsal column system, the smaller fields are located distally. By comparison, receptive fields on the digits represented in the dorsal column system may be only a few millimeters. Four categories of responses have been recorded from single cells within the dorsal horn: (1) responses to gentle stimulation of hair with fields extending onto glabrous skin (these cells are also excited by noxious pressure, heat, and cold); (2) responses to gentle mechanical stimulation of the skin, but not to hair (these cells also respond to noxious heat and cold); (3) cells responding to strong mechanical stimulation, noxious heat, and cold; and (4) cells that respond to pressure applied to deep structures. Interestingly, the activity in many of these cells can be inhibited by applying stimuli to the contralateral limb. These results indicate that

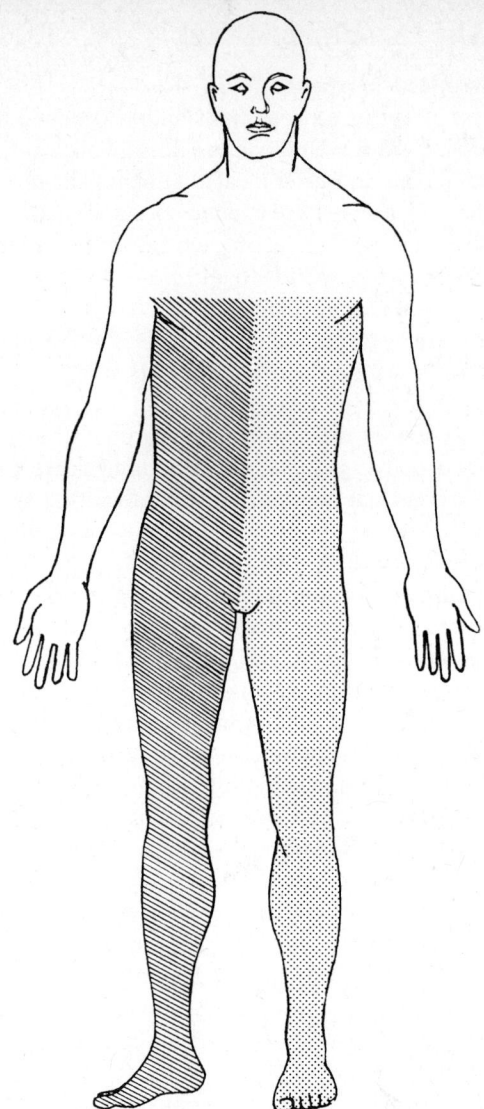

Figure 3 Representation of Brown-Séquard syndrome following left hemisection of the spinal cord at about T4. Loss of pain and temperature represented by slanted lines and two-point discrimination and vibrations sensation on ipsilateral side represented by the dotted area.

there is a good deal of divergence and convergence on the part of the primary afferent fibers arriving in the spinal cord and synapsing on the cells giving rise to the spinothalamic system. This results in a loss of specificity, as well as localization, because of the increase in size of the receptive fields.

THE THALAMUS

The thalamus is the last relay station for most sensory systems before the sensory information attains cortical levels. This includes not only somatosensory systems, but vision, audition, and taste as well. For the somatosensory system, all pathways converge on a few thalamic nuclei. At this level both spinothalamic and dorsal column systems converge. We have not mentioned sensory input from the face; they are similar, except that the second-order neurons are located in the divisions of the trigeminal nucleus located in the pons, medulla, and upper spinal cord.

ANATOMY

All sensory systems converge on the dorsal thalamus. The specific nuclei concerned with the somatosensory system include the ventral nuclear group of the dorsal thalamus. The neurons of the dorsal column nuclei that project to the thalamus cross the midline forming the medial lemniscus. Their main target is the ventral posterior lateral nucleus of the thalamus (VPL). There is somatotopic organization within the thalamus, but since the axons in the medial lemniscus have now traversed the midline, the gracile (lower extremities) nucleus is now represented in a lateral position, and the fibers from the cuneate in a medial position. Figure 4 is a diagram showing the somatotopic organization of the thalamus.

Single cell recording from the thalamus has shown that its topographical organization is a mirror image of the organization seen in the dorsal column nuclei. The receptive field organization is similar, that is, receptive fields on the digits are smaller than on proximal body parts. For any thalamic neuron, the peripheral drive is maximal when the stimulus is at the center of the receptive field and minimal at the edge. Therefore, there is a gradient across the receptive field. The represented sensory modalities appear to be the same as those found in the dorsal column nuclei, with the added observation that there is some evidence for a spatial segregation (e.g., deep structures in dorsal part of VPL and skin in the ventral part). Neurons in the thalamus may be spontaneously active and respond with multiple discharges after a single synchronous volley applied to a peripheral receptive field or nerve. The frequency of action potentials is a function of the strength of the stimulus. Following an electrical stimulus to the afferent input, cells in the thalamus are less responsive. This observation is correlated with an afterhyperpolarization of the thalamic neurons, as well as with presynaptic inhibition, probably occurring via thalamic interneurons. The spinothalamic input to VPL also shows some of the same properties. The cells have restricted receptive fields, are modality specific, and have almost pure contralateral input.

The posterior thalamic nucleus (PO) receives most of the spinothalamic input carrying information about pain and temperature, although this nucleus also receives an input from the dorsal column nuclei. The receptive fields are larger than those found for cells in VPL, tend to be located on proximal body parts, and are frequently bilateral. This organization is consistent with the concept that information arriving over the spinothalamic system is of the nondiscriminative type.

SOMATOSENSORY CORTEX

By comparison with lower mammals, the higher mammals and especially primates have developed cortical areas specifically designed both anatomically and functionally for a particular purpose. The sensory cortices are a particularly good example of this development. In the primate, for example, there are two cortical areas that are functionally important for somatic sensations. These are called somatosensory areas I and II (SI and SII) and are the primary receiving areas for information arriving at the cortical level. From this point, the information is disseminated to other areas of the cortex as well as fed back to the sensory relay nuclei, as discussed above. One of the most important functions of the sensory cortex is the interpretation of the incoming information so that an appropriate motor response may occur. How this is achieved remains one of the unsolved mysteries of the nervous system. It is, however, at the level of the cortex that conscious sensory perception occurs. The primary sensory cortex (SI) is located in the sensory strip or postcentral gyrus, the area of cortex just posterior to the Sylvian sulcus. When this area is removed many of the neurons in sensory thalamus degenerate. When this area is electrically stimulated in awake humans sensory illusions may occur. Physically

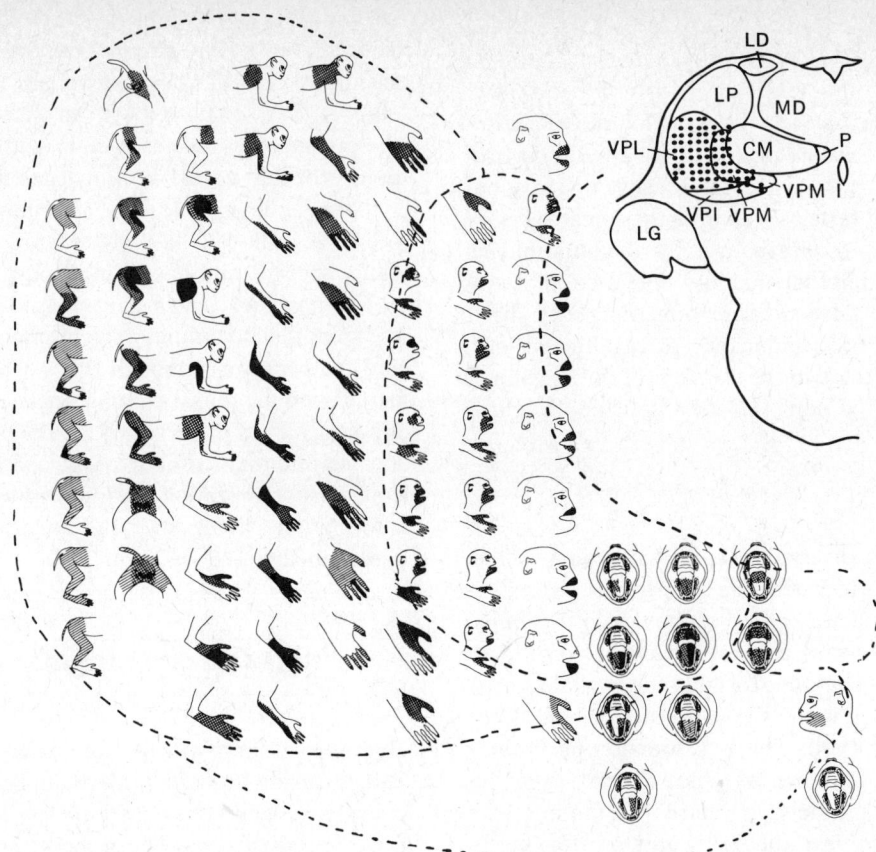

Figure 4 Somatotopic organization in a single frontal plane through the thalamus. Shaded area indicates density of projection as determined by the evoked potential method. With the exception of midline structures all responses were produced by contralateral body stimulation, the location of which is indicated on each figurine. The position of the figurine within the thalamus indicates approximate location of the electrode tip. (From: Medical Physiology, 13th ed., Mountcastle VB, ed., Neural mechanisms in somesthesia, ch. 10, Fig. 10–8, 1979, C.V. Mosby)

this area is quite separate but is adjacent to the motor area located in the precentral gyrus.

ANATOMY

The sensory cortex probably has the most complicated cellular organization of any cortical area. Organization of CNS tissue on the basis of cellular characteristics is called cytoarchitecture. The cerebral cortex has been divided into several areas on the basis of cellular architecture. These are called Brodmann's areas after the famous neurologist who first described them. SI is located in Brodmann's areas 3, 2, and 1. Area 3 is the transitional area between sensory and motor cortex and has been further subdivided into areas 3a and 3b. The cortical areas involved in somatic sensation and the somatotopic organization of the body are shown in Figure 5. Small lesions in VPL produce anterograde degenerative changes in fibers

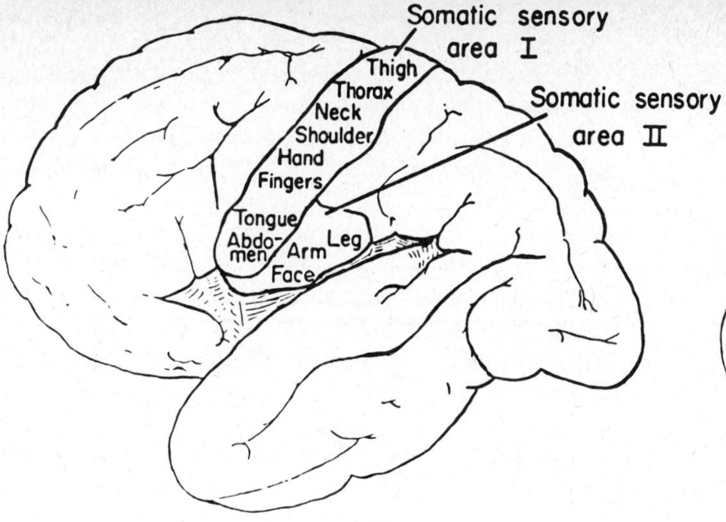

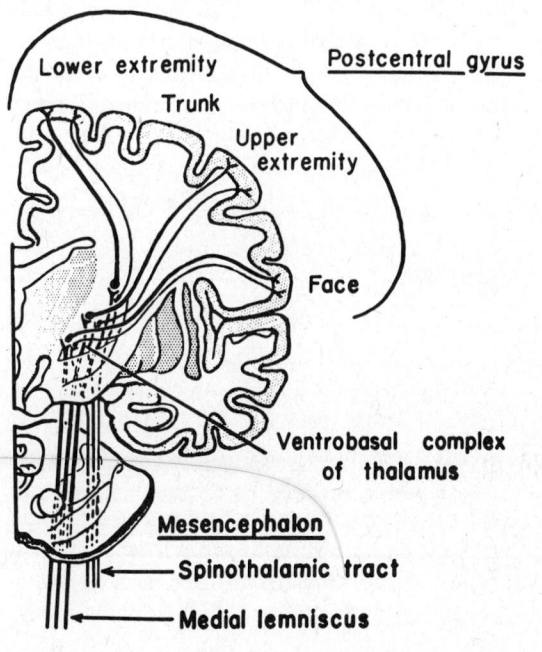

(a)

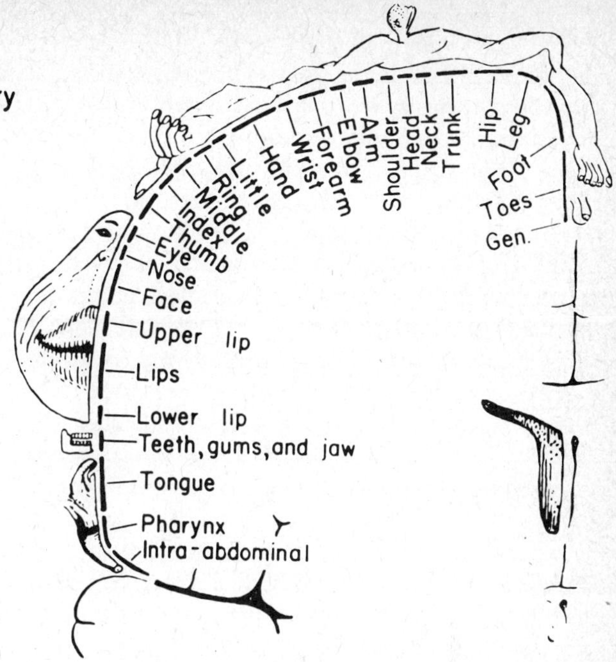

Somatic sensory area I

Thigh
Thorax
Neck
Shoulder
Hand
Fingers
Tongue
Abdomen
Arm
Leg
Face

Somatic sensory area II

(b)

Hip
Leg
Little
Ring
Middle
Index
Thumb
Eye
Nose
Face
Upper lip
Lips
Lower lip
Teeth, gums, and jaw
Tongue
Pharynx
Intra-abdominal
Hand
Wrist
Forearm
Elbow
Arm
Shoulder
Head
Neck
Trunk
Foot
Toes
Gen.

Postcentral gyrus

Lower extremity

Trunk

Upper extremity

Face

Ventrobasal complex of thalamus

Mesencephalon

Spinothalamic tract

Medial lemniscus

(c)

p. 657

Figure 5 *(a)* Somatotopic organization and areas of cortex denoted SI and SII, as well as their relationship to the motor area on the percentral gyrus. *(b)* Representation of the body in SI viewed in a frontal plane. Note the disproportionate allocation of cortex to peripheral body parts. *(c)* Projection onto SI of dorsal column and spinothalamic systems via thalamic relay nuclei in the ventrobasal complex. (A and C. From: Guyton AC, *Medical Physiology,* 4th ed., Saunders, 1971, ch. 49) (B. Adapted from: Penfield W, and Rasmussen J, The cerebral Cortex of Man, Macmillan, 1950).

projecting to these cortical areas. Lesions in the rostral portion of VPL show changes in the transitional cortical sensory area. This is the area that receives information from muscle sensory receptors and possibly a better term for this area is sensorimotor cortex.

One of the most distinctive features of the cellular organization of the sensory cortex is its layered structure. Sensory cortex contains six layers of cells. Most of these cells fall into one of two categories: pyramidal or stellate cells. Pyramidal cells have triangular-shaped cell bodies with a dendritic tree polarized in its orientation. One major dendrite (apical dendrite) emanates from the soma and is directed towards the pial surface of the cortex. As this dendrite traverses the cortex it gives off branches located within the other cortical layers. These branches receive many synaptic inputs from cells located within cortex, as well as from cells from other regions of the CNS. From the base of the triangle arise the basal dendrites of the pyramidal cell. These dendrites are directed ventrally and laterally and receive both extrinsic and intrinsic input from other cells. Pyramidal cells can be found in all layers of sensory cortex, except layer I, which is essentially acellular. Layer IV contains relatively few pyramidal cells. Another unusual feature of pyramidal cells is the small membranous extrusion located on the dendrites, called a dendritic spine. This apparatus appears to be a specialized structure for making synaptic contacts with axon terminals.

Stellate cells, as their name suggests, are star-shaped with a round soma and dendrites emanating from around the soma. The cell bodies of stellate cells are small, whereas pyramidal cells are considerably larger, ranging from 10 μm to as large as 50 μm, with the larger ones located in the deeper layers of the cortex. The stellate cells are found almost exclusively in layer IV of the cortex. This layer is also called the granule cell layer, and for this reason sensory cortex is sometimes called granular cortex. Stellate cells usually do not contain spines, although some stellate cells do have spines. Figure 6 illustrates several examples of Golgi-stained pyramidal and stellate cells found in sensory cortex.

Lesions within the sensory thalamus have shown that the afferent input zone based on terminal degeneration is located within layers IV, III, and I. It was previously thought that stellate cells were the primary cortical recipient cells for the incoming afferent terminals because they are located in the thalamic projection zone. Another point of view, however, is that the middle layers within sensory cortex contain both the apical dendrites of more ventrally located pyramidal neurons and the basal dendrites of dorsally located pyramidal neurons. Thus thalamic terminals in layers III and IV are in a position to access cortical neurons of all layers. Experimental studies designed to investigate these possibilities have shown without doubt that cells from most if not all layers can receive synaptic contacts from thalamic fibers.

The synaptic arrangements on the two prevalent cortical cell types are also different. Morphologically one can distinguish two types of synaptic contacts in most areas of the CNS. One is a terminal containing round vesicles and is believed to be functionally excitatory. The second contains flat vesicles and is believed to be functionally inhibitory. Pyramidal neurons receive contacts on their dendritic spines from terminals containing round vesicles and making an asymmetrical membrane differentiation. On the somas of pyramidal cells the predominant synaptic type is from terminals containing flattened vesicles making a symmetrical membrane differentiation. Stellate neurons on the other hand, receive both round–asymmetrical as well as flat–symmetrical synapses on their dendrites. Following lesions within the thalamus most of the degenerating terminals are the round–asymmetrical type ending on spines. With the exception of those stellate cells that have spines this suggests that most of the primary thalamic recipient neurons are pyramidal cells.

PHYSIOLOGY

Single cell recordings from cortex following a peripheral electrical shock reveal that these cells respond with an initial short burst followed by a later response. The cells are sensitive to gentle mechanical stimuli and are isolated in layers II–VI with most in the middle layers. These cells have clearly defined receptive fields that are small distally and larger proximally on the body surface. The receptive fields are smaller than in both the dorsal column nuclei and thalamus. Recordings from three basic types of cells have been made: (1) those responding to cutaneous stimulation; (2) those responding to stimulation of deep structures; and (3) those unresponsive to gentle stimulation. Of

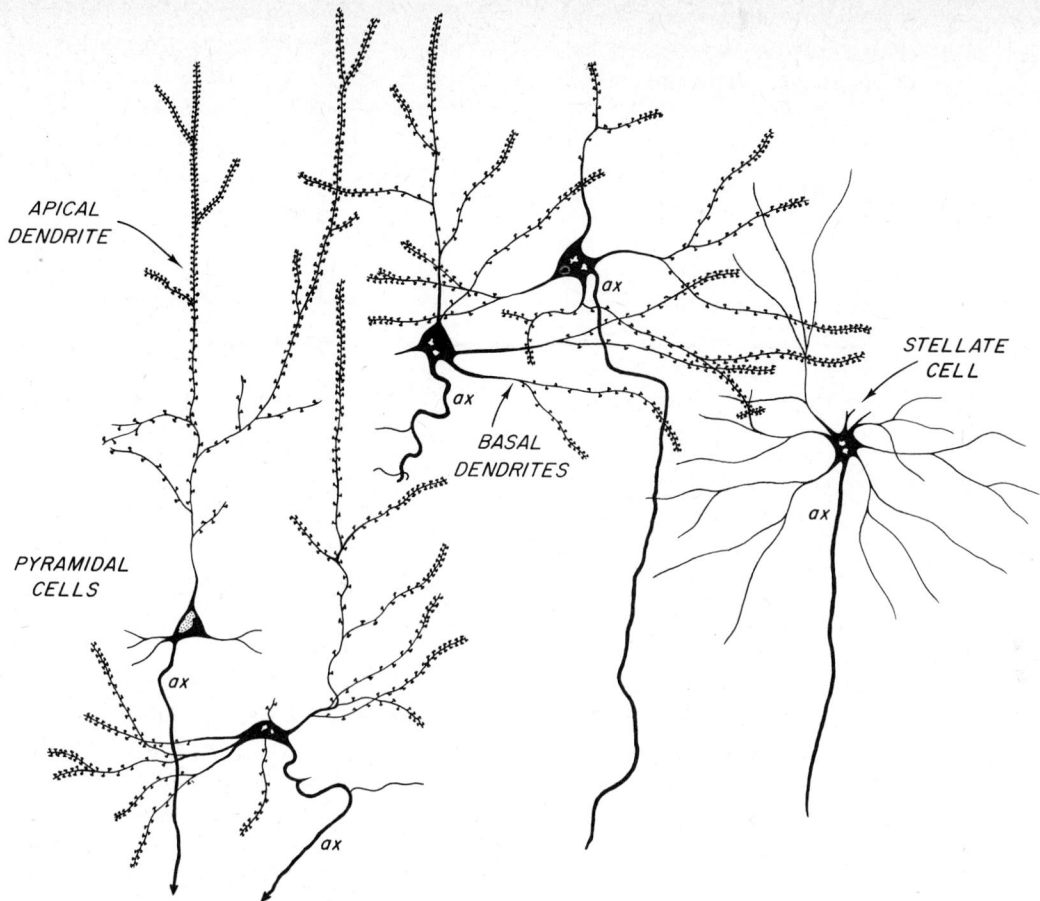

APICAL
DENDRITE

STELLATE
CELL

BASAL
DENDRITES

PYRAMIDAL
CELLS

ax

ax

ax

ax

ax

Figure 6 Examples of pyramidal and stellate neurons and their distribution within the layers of sensory cortex. Note the organization of the dendritic trees. Pyramidal cell dendrites are polarized with the apical dendrite pointing to the cortical surface. Dendrites from stellate cells emanate from the soma in a random fashion.

those responding to cutaneous stimulation, more than 70% are rapidly adapting.

The postcentral gyrus can be divided into three physiological areas that coincide with the cytoarchitectural boundaries described above. Cutaneous input is located in area 3, input from receptors in deep structures in area 2, and a mixed population of cells in area 1. Area 3 also contains some slowly adapting cells having a cutaneous input.

One of the most interesting physiological properties of neurons in somatosensory cortex is their arrangement into functional columns. Within a column, cells have the same

modality, receptive field, and same temporal behavior. For example, an electrode passing through the cortex in a direction orthogonal to the surface of the cortex may encounter a cell in a superficial layer that responds to movement of hair on a contralateral body part. If the electrode is advanced it will record from another cell responding to hair and with a similar receptive field. The physiological estimate of the width of each column is approximately 500 μm, in agreement with the anatomical evidence for a column. A small percentage of cells within SI responds preferentially to the direction of movement of the stimulus across the receptive field. In a small proportion of these there is also a velocity component. This columnar organization is present in other sensory cortical areas. It has reached a high degree of complexity, especially in the primary visual areas.

MODULATION OF SENSORY TRANSMISSION

The function of relay nuclei is twofold. They allow for the transfer of information to the next higher level within the CNS, but the nucleus is an area in which sensory processing can occur as well. As we have already seen, this involves synaptic arrangements via interneurons located within the nuclei that can receive input from intrinsic neurons, or it can involve direct feedback from the next receiving station in the pathway. In this section we will examine the physiological properties of these interactions and how they influence the processing of sensory information.

Surround or Afferent Inhibition

We must now begin to use our knowledge of neuronal interaction including the concepts of divergence, convergence, synaptic excitation, and inhibition to perform functional tasks such as the perception of a sensory stimulus. Let us suppose for the sake of discussion that a single primary afferent fiber innervates a number of hairs covering a 1-sq cm area. If two primary afferent fibers innervating adjacent receptive fields were to make synpatic contact on a single dorsal column nuclei cell, and activity in that cell were to represent conscious perception of hair movement, we would perceive a stimulated area of 2 sq cm. If this convergence of information on single neurons were to proceed all the way to the cortex, we would soon lose our fine discriminatory abilities. From single cell recordings in the dorsal column nuclei we know, however, that the activity of a cell is reduced by gentle stimulation to an area surrounding the receptive field for that cell. This phenomenon is called surround inhibition. Its importance in visual processing at the level of the retina has been well documented. How surround inhibition occurs has not been directly established, but rather, has been inferred from both physiological as well as anatomical evidence. The physiological evidence is that just described, that is, when a single cell is activated by stimulation of its receptive field

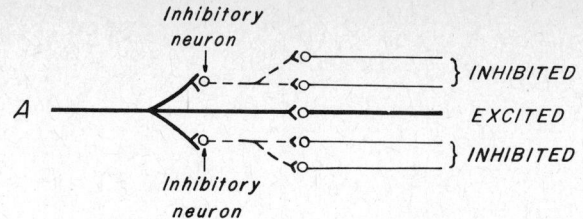

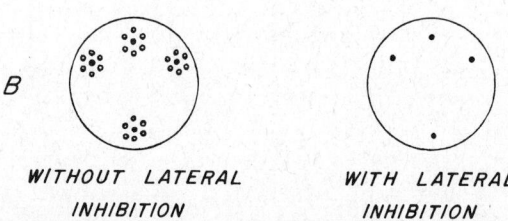

Figure 7 *(a)* The influence of diverging afferent inputs can alter the excitability of postsynaptic neurons through inhibitor interneurons. *(b)* The altered size of cutaneous receptive fields resulting from this inhibitory mechanism. (Adapted from: Guyton AC, *Medical Physiology,* 6th ed., ch. 47, Saunders, 1981)

center the response may be inhibited by stimulation of the surrounding area. The morphological correlate of this result can be found in the synaptic relationships that occur in the dorsal column nuclei, especially the number of axoaxonal synaptic contacts. This type of inhibition can occur either by presynaptic or postsynaptic mechanisms, and the phenomenon is believed to be prevalent in the dorsal column nuclei, but less so at thalamic and cortical levels. Spinothalamic tract cells also show this phenomenon, but the inhibitory phenomenon is produced by stimulation of contralateral body parts rather than the area surrounding the receptive field. Figure 7 is a diagram showing neuronal connections that could result in surround inhibition.

SELECTED BIBLIOGRAPHY

Brodal A: The somatic afferent pathways, in Brodal A (ed): *Neurological Anatomy,* ed 3. New York–Oxford, Oxford University Press, 1981.

Darian-Smith I: Somatic sensation. *Annu Rev Physiol* 31:417, 1969.

Mountcastle VB: Neural mechanisms in somesthesia, in Mountcastle VB (ed): *Medical Physiology,* ed 13. St Louis, CV Mosby, 1974.

Wall PD: The sensory and motor role of impulses traveling in the dorsal columns towards cerebral cortex. *Brain* 93:505, 1970.

Werner G, Whitsel BL: The somatic sensory cortex: Functional organization, in Iggo A (ed): *Handbook of Sensory Physiology,* Vol 2, *Somatosensory System.* Berlin, Springer-Verlag, 1973.

Woolsey CN: Organization of somatic sensory and motor areas of the cerebral cortex, in Halrow HF, Woolsey CN (eds): *Biological and Biochemical Bases of Behavior.* Madison, University of Wisconsin Press, 1958.

SELF-STUDY QUESTIONS

MULTIPLE CHOICE

Select the single *incorrect* answer.

1. In the spinothalamic tract
   A. many spinothalamic tract cells respond to noxious stimuli.
   B. the axons of spinothalamic tract cells decussate within a few segments of the cell body.
   C. some spinothalamic tract cells respond to tactile stimuli.
   D. thermoreception is thought to play a role.
   E. spinothalamic tract cells are activated by application of a vibratory stimulus (e.g., tuning fork) to bony prominences.

2. Primary afferent fibers
   A. generally synapse on both dorsal root ganglion cells and dorsal horn cells.
   B. can send a branch into the dorsal columns.
   C. can exceed 1 m in length.
   D. can synapse directly with α-motor neurons.
   E. can establish synaptic contacts with neurons located rostral or caudal to the segment of entry.

3. An example of place coding in the somatosensory system includes
   A. sensory homunculus in postcentral gyrus.
   B. nucleus gracilis transmits tactile information originating from foot.
   C. pacinian corpuscles signal vibration and nociceptors signal pain.
   D. graded stimulus intensity causes graded response in type II (Ruffini) receptors.
   E. the spinothalamic tract carries different information than the dorsal column pathway.

4. In the spinothalamic tract
   A. many spinothalamic tract cells are activated by nociceptors.
   B. most spinothalamic tract axons decussate within the spinal cord.
   C. there is a characteristic somatotropic organization, with the arm representation lateral and the leg medial at C2.
   D. the site of termination is in several thalamic nuclei, including ventral posterior lateral (VPL) and the intralaminar complex.
   E. the cord is accompanied by the spinoreticular tract.

5. The Gate theory of pain has been used to explain
   A. relief of pain by dorsal column stimulation.
   B. acupuncture analgesia.
   C. effectiveness of scratching or rubbing to reduce itch
   D. ability of descending pathways to interfere with nociceptive responses.
   E. observation that some spinothalamic tract cells that are excited only by nociceptors.

6. Sensory modalities represented by the dorsal column pathway include
   A. appreciation of changes in position.
   B. sense of texture and shape.
   C. pain.
   D. dorsolateral part of nucleus ventralis posterolateralis.
   E. postcentral gyrus adjacent to lateral (Sylvian) fissure.

Select the single best answer.

7. The dorsal column–medial lemniscus pathway synapses in which thalamic nucleus?
   A. Ventrolateral (VL).
   B. Ventral posterior lateral (VPL).
   C. Ventral anterior (VA).
   D. Ventral posterior medial (VPM).
   E. Medial dorsal (MD).

8. An elderly person suddenly develops a hemiplegia on the right and a reduction in somatic sensibility on the right side of the body and face. Which is the most likely lesion?
   A. Brown-Séquard (hemisected spinal cord).
   B. Central cord lesion.
   C. Wallenberg (syndrome of the lateral medulla).
   D. Posterior internal capsule stroke.
   E. Lesion of postcentral gyrus.

9. Of the following, which would you judge to be the most suitable place to make a lesion to relieve pain due to a malignancy in the left leg?
   A. Just dorsal to the denticulate ligament at T6 on the right.
   B. Just ventral to the denticulate ligament at T6 on the right.
   C. Just dorsal to the denticulate ligament at T6 on the left.
   D. Just ventral to the denticulate ligament at T6 on the left.
   E. In the midline at T6.

10. An example of a lemniscal fiber is an axon in the
    A. fiber tract arising from the nucleus gracilis.
    B. internal capsule.
    C. spinal tract of trigeminal.
    D. spinothalamic tract.
    E. solitary tract.

11. Which of the following is an example of pattern coding of sensory information?
    A. Lower body represented in fasciculus gracilis.
    B. Stimulus intensity represented by greater discharge rate.
    C. Face represented in postcentral gyrus near lateral fissure.
    D. Vibratory stimulus activates Pacinian corpuscles.
    E. Pain depends on activity in spinothalamic tract.

12. Information about the intensity of a stimulus applied to the skin can be transmitted to the CNS by
    A. C, D, and E.
    B. C and E only.
    C. the discharge rate of the receptor activated.
    D. the number of receptors activated.
    E. the receptor types activated.

13. The central projections of primary afferent fibers (dorsal root fibers) may
    A. C, D only.
    B. D, E only.
    C. synapse in the dorsal root ganglion.
    D. ascend in the dorsal columns to terminate in supraspinal nuclei.
    E. have collaterals ending on motor neurons and interneurons.

14. The umbilicus is within which dermatome?
    A. C2
    B. C8
    C. T4
    D. T10
    E. L1

15. The dorsal column pathway has a highly somatotopic organization. Which of the following subdivisions of the dorsal column pathway carries information

concerning the lower part of the body and the lower extremity?

A. Fasciculus cuneatus.

B. Fasciculus gracilis.

C. Dorsal part of the medial lemniscus at midolivary level.

D. Dorsolateral part of nucleus ventralis posterolateralis.

E. Postcentral gyrus adjacent to lateral (Sylvian) fissure.

16. For the face, what brainstem nucleus appears to be comparable to a dorsal column nucleus?

A. Superior (main) sensory nucleus of the trigeminal.

B. Mesencephalic nucleus of the trigeminal.

C. Spinal nucleus of the trigeminal.

D. Solitary nucleus.

E. Nucleus ambiguus.

MULTIPLE CHOICE

Select the best answer(s). (In many instances, more than one answer is correct.)

17. Two-point discrimination is better in areas of the skin

A. with primary afferents that have small receptive fields.

B. with primary afferents that have low thresholds.

C. with a high density of afferent innervation with partially overlapping receptive fields.

D. on the fingertip as compared with the arm.

18. A patient having a loss of pain and temperature sensation in both arms and both legs with preservation of light touch, vibration, and limb position is likely to have a disease involving

A. large myelinated fibers in the peripheral nerves.

B. the dorsal half of the spinal cord.

C. the central region of the spinal cord (central cord syndrome).

D. small myelinated and unmyelinated fibers in the peripheral nerves.

19. Axons forming the fasciculus cuneatus

A. include ascending collaterals of primary afferents supplying the hand.

B. synapse in the nucleus cuneatus.

C. signal touch, vibration, and position.

D. mediate pain.

ANSWERS

1. E. Spinothalamic tract cells include those activated by strong mechanical stimuli, temperature, or mechanical stimuli requiring long contact time with the receptive field.

2. A. Synaptic contacts are not found in dorsal root ganglia.

3. B. The nucleus cuneatus (medial nucleus) receives information from the foot. Since the dorsal column fibers do not cross the spinal cord, they get pushed medially as they ascend the neuraxis. The lower body is therefore represented in the medial nucleus and upper body the lateral nucleus.

4. C. Just the opposite is true. See previous answer for

details of why this is so. Remember spinothalamic system is corssed at spinal cord level.

5. E. The fact that some spinothalamic tract cells are excited by noxious stimulation has nothing to do with the Gate theory of pain. This is a red herring.

6. C. Pain information travels in the spinothalamic tracts.

7. B. VL and part of VA are associated with ascending motor systems. VPM projects to sensory cortex but receives fibers from the trigerminal nuclei. Hence, it is the analogous system for the dorsal column–medial temniscus, but represents the face. MD projects to frontal cortex.

8. D. This stroke involves both sensory and motor systems over the entire right half of the body. This eliminates answers A, B, and C because lesions in this area would involve both sides of the body. A lesion of the postcentral gyrus would result in a sensory loss only. Thus answer E cannot be correct.

9. B. Interruption of the spinothalamic tract is important. These tracts are located in the ventral white matter. This eliminates answers A, C, and E. Since this tract is crossed soon after the entering dorsal root fibers synapse with the dorsal horn cells that project into the spinothalamic tract, interruption of the ascending fibers on the right side will interrupt sensory information arriving from the left leg.

10. A. Lemniscal fibers arrive over the dorsal column pathway. Origin of fibers is in the dorsal column nuclei (gracile and cuneate).

11. B. Pattern coding as opposed to place coding suggests firing activity of the involved neurons. A, C, and E are examples of place coding. D refers to the adequate stimulus for Pacinian corpuscles.

12. A. All can contribute to the perception of intensity.

13. B. No synapses are made in dorsal root ganglion.

14. D.

15. B. Medial lower extremities and lateral upper extremities. The opposite is true for the spinothalamic system because it is crossed in the spinal cord.

16. A. The primary afferent fibers innervating sensory receptors of the face run in the trigeminal nerve. They synapse with second-order neurons in the two divisions of the trigeminal nucleus within the spinal cord and medulla. These two divisions include a main sensory nucleus and a spinal nucleus. The latter receives information about protopathic sensibilities similar to the spinothalamic system. The main sensory nucleus receives information about epicritic sensibilities and is analogous to the dorsal column–medial lemniscal system. The mesencephalic nucleus of the trigeminal system is involved with muscle receptor afferents. The solitary and ambiguous nucleus are not part of the trigeminal system.

17. All are correct.

18. D. A and B would involve other sensory deficits over a circumscribed area. C could be true except that it would probably not involve the entire body below the face.

19. A, B, C.

39

The Motor System

BURGESS CHRISTENSEN

OBJECTIVES

After completion of this chapter, you should be able to:

1. Describe the structure and function of muscle sensory receptors.

2. Describe the static and dynamic responses of the muscle spindle receptor and the functional importance of these responses.

3. Discuss the mechanisms involved in sensorimotor integration: divergence and convergence.

4. Describe the connectional and functional organization of simple reflex pathways.

5. Describe the cellular organization, input, and output pathways to the cerebellum.

6. Describe information processing in the cerebellum.

Sensory systems tell the organism about environmental changes, and the motor system produces appropriate responses to these sensory stimuli in the form of changes in posture and movement. The parts of the CNS involved in motor functions receive information from other sensory systems as well as from receptors located in the individual muscle groups. Many of the pathways carrying motor information parallel those of the different sensory systems. Figure 1 shows the areas of the peripheral and CNS involved in motor control and where they are connected.

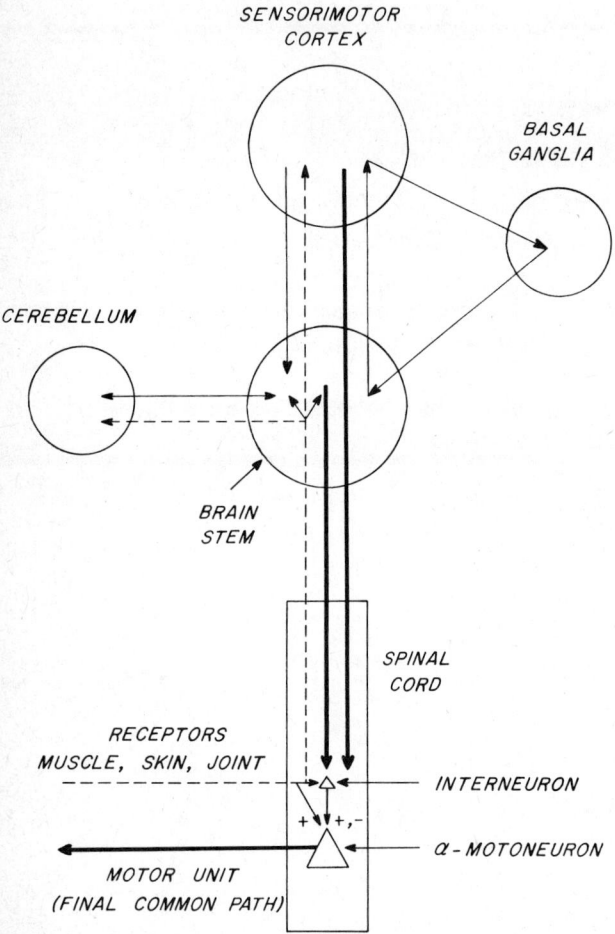

SENSORIMOTOR CORTEX

BASAL GANGLIA

CEREBELLUM

BRAIN STEM

SPINAL CORD

RECEPTORS MUSCLE, SKIN, JOINT

INTERNEURON

α - MOTONEURON

MOTOR UNIT (FINAL COMMON PATH)

Figure 1 Organization of the motor system, showing the major components and their areas of interaction and integration. (Adapted from: Eyzaguirre C, and Fidone SW, *Physiology of the Nervous System,* 2nd ed. chapt. 12, Year Book Med., 1975)

Each level progressing toward the motor cortex adds a degree of complexity to the control of movement. When the spinal cord is isolated from the higher motor areas, many reflexes remain. These reflexes represent the foundation upon which our motor behavior is built. Integration of the motor centers above the level of the spinal cord results in smooth, coordinated, voluntary movements.

The output of the CNS as it relates to the motor system ultimately results in contraction of different muscles or muscle groups. The motor neurons located in the spinal cord and medulla and their axons in the ventral roots and cranial nerves that innervate these muscles are called the final common path. Although many anatomical names are given to the different muscles, we will refer to them here according to their physiological function as flexors or extensors. Extensors are those muscles that are used to maintain body posture against the force of gravity. Physiological flexors are the muscles not falling into this group.

MUSCLE RECEPTORS

Two sensory receptors are found in muscle that relay information to the CNS about changes in tension, length, and rate of change in length. Each has particular structural characteristics as well as physiological properties important to proprioception. Proprioception means the reception of sensory information that tells the body about movement and its position in space.

The muscle spindle is located within the muscle mass in parallel with the other muscle fibers. A portion of the spindle contains striated muscle, and the name intrafusal fiber is often used to describe the spindle. Nonspindle muscle is referred to as extrafusal fibers.

Spindles are 4–7 mm in length and 80–200 μm in diameter. They consist of a polar region that is muscular and an equatorial region that is almost devoid of contractile tissue. Two kinds of intrafusal fibers are associated with the spindle: the nuclear bag and the nuclear chain. Figure 2 shows a schematic diagram of the spindle. Usually two nuclear bag fibers and four to five nuclear chain fibers form each spindle. Figure 2 shows the nuclear bag fiber to consist of a nonmuscle structure, the equatorial region and

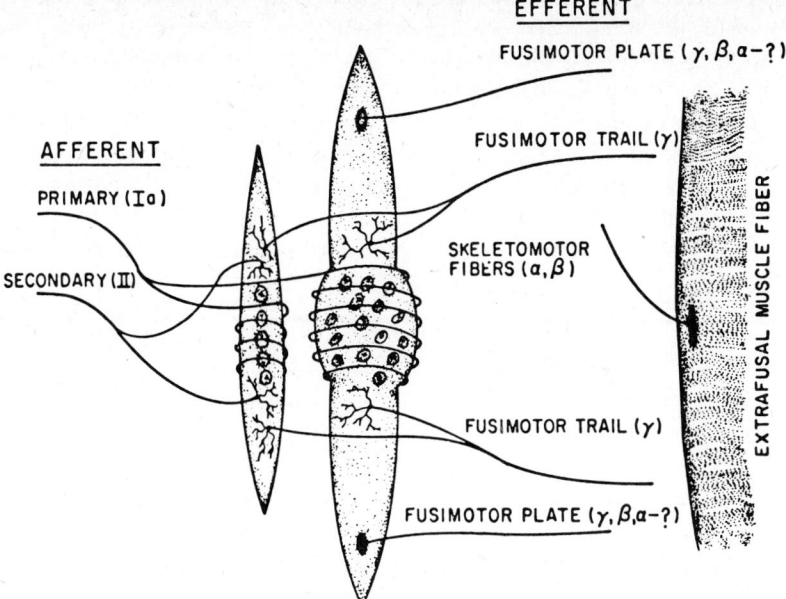

Figure 2 Schematic diagrams showing the muscle spindle receptor and its innervation. Both nuclear bag and nuclear chain fibers are innervated by group 1a fibers, whereas the group II fibers innervate mostly nuclear chain fibers. Two kinds of motor innervation occurs. Although there is considerable overlap, fusimotor plate endings are found mostly on bag fibers, and trail endings (supplied only by gamma fibers) are found mostly on chain fibers. α and β fibers supply the innervation of the extrafusal muscle fibers. (Adapted from: Eyzaguirre C, and Fidone SW, *Physiology of the Nervous System,* 2nd ed. chapt. 12, Year Book Med., 1975)

muscle (contractile) tissue of the polar regions, while the nuclear chain fiber consists of only muscle tissue. This difference, along with the organization of its innervation, gives rise to the physiological properties of the spindle.

The muscle spindle is innervated by both afferent as well as efferent fibers. There are two types of afferent fiber that innervate different regions of the spindle. The group Ia fiber innervates the equatorial region of both the nuclear bag and nuclear chain fibers forming the annulospiral or primary ending. These fibers are among the largest in the mammalian nervous system and therefore conduct very rapidly. The group II fiber innervates the polar region of the nuclear chain forming the flower spray or secondary ending. These fibers send information directly to motor neurons within the spinal cord, as well as relaying information toward higher structures. This direct pathway to

motor neurons is involved in producing basic reflex movements.

Intrafusal fibers are also innervated by neurons located in the spinal cord called γ motor neurons. The axons from these neurons are smaller than those of the α motor neurons and therefore have a slower conduction velocity. Recent evidence indicates that α motor neurons may also innervate intrafusal fibers but that γ motor neurons apparently do not innervate extrafusal fibers. The junction between the motor axon and the intrafusal fiber is similar to the endplate of the extrafusal fiber and functionally it is the same. The liberation of transmitter results in contraction of the intrafusal fiber. The way in which the sensory input to the spinal cord is connected to the γ motor outflow results in an exquisitely well-tuned feedback system necessary for motor control.

Let us now examine four physiological situations that will help clarify the function of the spindle and what it must be telling the CNS about the state of the muscle. We will record the activity from the group Ia fiber during passive muscle stretch and active contraction. Figure 3 diagrammatically depicts each experimental situation. In the first case, the muscle is at its resting length, and some tonic activity is generated in the afferent fiber because of

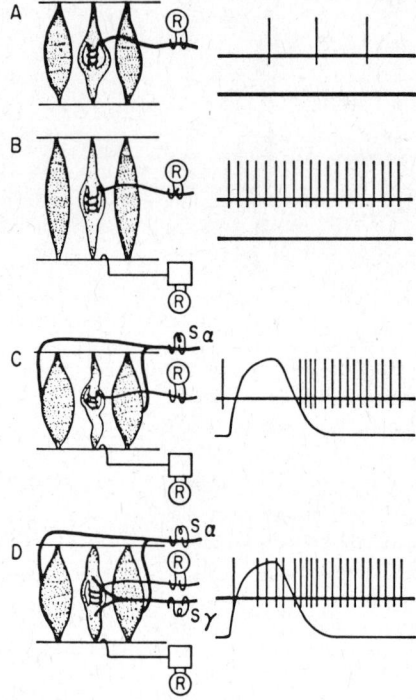

Figure 3 Diagram showing the effects of stretch and contraction of the intra- and extrafusal fibers on the discharge rate of the group Ia afferent fiber. *(a)* The muscle mass is at resting tension, producing a tonic low rate discharge from the afferent fiber. *(b)* The stretch is increased, resulting in an increased tonic discharge rate. *(c)* An electrical stimulus is applied to the α fibers innervating the extrafusal muscle, resulting in contraction (depicted in the lower trace) and a temporary pause in the discharge rate of the spindle afferent. *(d)* Both the α and γ fibers are stimulated resulting in contraction of the intra- and extrafusal fibers. The lower trace shows the contraction of the extrafusal fibers. During this contraction phase, the spindle afferent discharge rate is slightly decreased, but does not shut off entirely because the intrafusal fiber contracts. (Adapted from: Eyzaguirre C, and Fidone SW, *Physiology of the Nervous System,* 2nd ed. chapt. 12, Year Book Med., 1975)

mechanical deformation of the receptor ending. When the muscle is passively stretched, the activity in the Ia fiber increases reflecting further changes in the mechanical coupling between receptor organ and receptor ending around the equatorial regions of the nuclear bag and nuclear chain fibers. A similar result would be seen with the secondary receptor ending. In the third case, an electrical stimulus is applied to the nerve fibers, which innervate only the extrafusal muscle fibers, that is, the α motor neuron axons. Thus, the muscle fibers contract and, because the spindle is in parallel with the extrafusal muscle fiber, the spindle is unloaded. During muscle contraction, activity in the primary receptor ending pauses, then starts again following the contraction. If we now combine stimulation of both α as well as γ nerve fibers both the extra- and intrafusal muscle fibers will contract and the unloading will be partially overcome.

We must now recall the fact that the nuclear bag fiber consists of both muscle and nonmuscle tissue whereas, the nuclear chain fiber is composed entirely of muscle. Furthermore, we must remember that the primary receptor ending innervates both kinds of tissue, whereas the secondary ending innervates only muscle. These differences give rise to two response properties of group Ia and group II fibers. These responses are called static and dynamic. A static response is a constant or maintained response to steady input; a dynamic response is one that increases rapidly when a stimulus is applied or decreases rapidly when a stimulus is removed. Both primary and secondary spindle afferents give static responses but only the primary fiber gives a dynamic response. When the muscle is passively stretched both fibers respond with a static response. However, if the stretch is, for example, sinusoidal, the primary spindle afferent will respond with an oscillatory or dynamic response whereas, the secondary afferent will produce a well-maintained or static response (Fig. 4). What is it about the structure of the spindle or the relationship of the receptor endings on the intrafusal fiber that could give rise to these very different physiological responses? The answer can be found in the physical properties of the polar and equatorial regions of the spindle. The equatorial regions of both the nuclear bag and nuclear chain fibers contain few myofibrils and are, therefore, very elastic. The polar regions are essentially muscle tissue that is more viscous than elastic. The equatorial regions are in

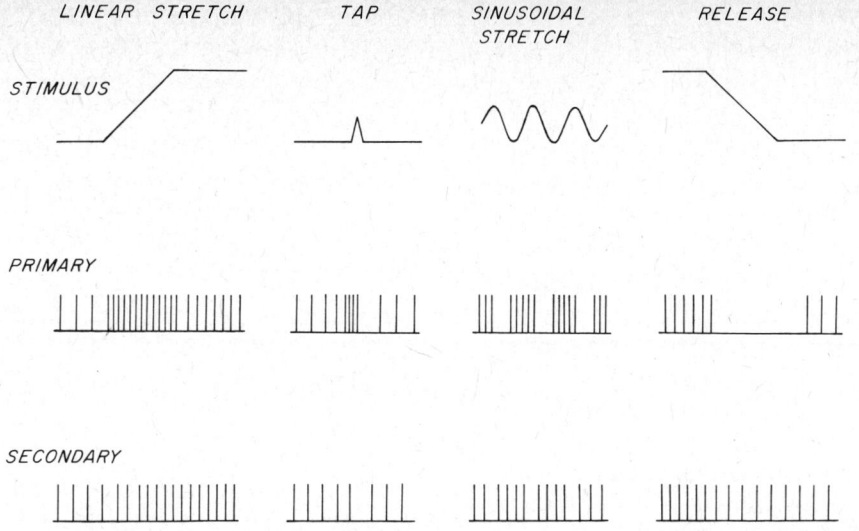

Figure 4 Diagram showing the response properties of the primary and secondary fibers during a variety of stimulus input functions. During the ramp function the primary fiber shows its dynamic properties by increasing its discharge rate rapidly whereas the secondary ending only slowly increases its discharge rate. During the plateau phase of the ramp function the primary endings discharge at a slower rate than during the dynamic part of the stretch, whereas the secondary ending discharges at the same rate. The dynamic versus phasic properties are more evident during a transient stimulus shown by the tap or sinusoidal stretch. The primary fiber "sees" the transient stimulus, but the secondary endings are unable to do so. During a linear decrease in stretch shown by the decreasing ramp function, the primary ending shuts off completely whereas the secondary ending shows a linear decrease in stretch. (Adapted from: Eyzaguirre C, and Fidone SW, *Physiology of the Nervous System,* 2nd ed. chapt. 12, Year Book Med., 1975)

series with this viscous element. When the muscle is stretched, the viscous element acts as a dashpot and initially imparts a force proportional to the velocity of the stretch to the primary ending located in the elastic region. During the maintained portion of the stretch, the viscous element will tend to give slowly but the slack is taken up by the elastic element in the equatorial regions and thus a maintained or static response is produced. The secondary endings, on the other hand, innervate the viscous polar regions of the nuclear chain fiber and their static response is maintained by the elastic element of the equatorial region. The importance of these two physiological responses with respect to the information they send to the CNS must be determined. It appears that the dynamic response must

be telling the CNS about the velocity with which a muscle changes length, whereas the static response is important for relaying the average muscle length at any time.

The fusimotor fibers that innervate the bag or chain intrafusal fibers are able to selectively increase either the dynamic or static responses of the group Ia or group II afferent fibers. Obviously this enables even greater control over the quality and quantity of information and the balance that may exist between the two types of responses. These selective influences are believed to be produced by separate activation of either the bag or chain fibers since one type of fusimotor fiber innervates only bag fibers, while another type can innervate both chain and bag fibers.

GOLGI TENDON ORGAN

The Golgi tendon organ (GTO) is the second type of muscle receptor, and as its name suggests it is not located within the muscle mass but rather within the connective tissue of the muscle tendons. Because of its unique position in series with the extrafusal muscle fibers, it can be activated both during passive stretch of the muscle as well as active contraction. In other words, unlike the spindle no unloading can occur on the GTO during contraction of extrafusal fibers.

Structurally the GTO is simpler than the spindle receptor. Group Ib fibers branch as they near the tendon, finally becoming unmyelinated and surrounded by a capsule of connective tissue. It is probably the mechanical association with the surrounding tissue that results in the deformation of the unmyelinated nerve endings during stretch or contraction of the muscle that activates the GTO.

It is much more difficult to activate the GTO by passive stretch of the muscle. This is probably because the elastic components of the muscle mass absorb much of the tension. For this reason it was believed that the GTO was important for signaling the CNS about excessive stretch or contraction of the muscle. This idea was reinforced because of the reciprocal inhibition that group Ib fibers have on the muscle it innervates and it was thought that this was a protective reflex action to avoid damaging the muscle itself. However, subsequently it has been shown that the GTO is exquisitely sensitive to muscle contraction responding to less than 0.1 g of force applied directly to the receptor endings. Therefore, under isometric conditions in which the muscle is not permitted to shorten, the GTO should be readily activated; under isotonic conditions the activation of the GTO may be much more subtle. Figure 5 illustrates the response patterns of the GTO.

INTEGRATIVE MECHANISMS IN THE CNS

It is appropriate at this point to discuss some of the mechanisms used by neuronal circuits that allow for alterations

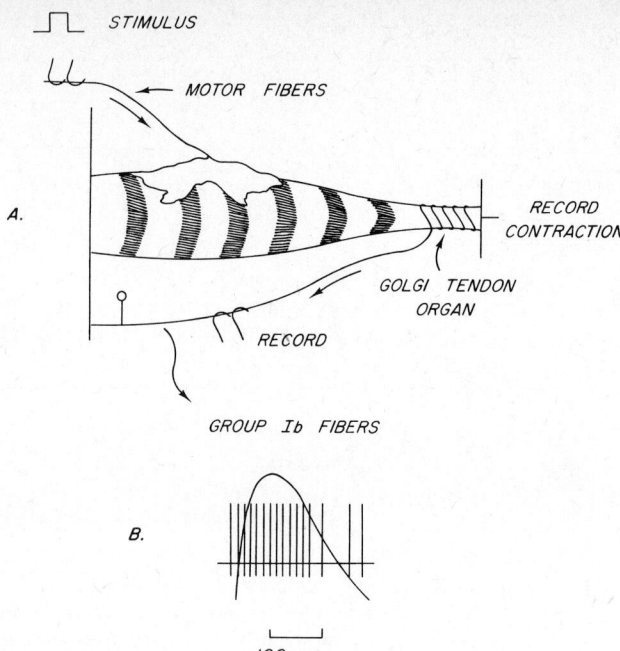

Figure 5 Diagram showing the response properties of the GTO. The motor fibers are simulated *(a)* and the response from Ib fiber is rounded *(b).* Because of its position in series with the extrafusal muscle fibers, it responds with an increase in discharge rate during contraction (smooth curve in *b*) of these muscle fibers. This should be contrasted with the response properties of the spindle receptor. (Adapted from: Eyzaguirre C, and Fidone SW, *Physiology of the Nervous System,* 2nd ed. chapt. 12, Year Book Med., 1975)

and dissemination of information through the various pathways. Some of these have already been discussed under the topic of synaptic transmission and the properties of the synaptic potential.

The first important concept is that when an afferent fiber enters the CNS it branches. Some of these branches serve to send information to very local neuron populations (within the same spinal segment for example). Other branches can ascend or descend for several segments or can even travel to nuclei long distances from the entry point, as occurs for the dorsal column system. One purpose of interconnecting different areas is to permit coordination between spatially separated parts such as lower and upper extremities. When, for example, a group Ia fiber enters the cord, it will branch locally and synapse with several alpha motor neurons. Each motor neuron may receive several synaptic contacts from one or several branches of this single axon. This mechanism for distributing information to a pool of cells is called divergence. This process can be viewed from the postsynaptic side, where each neuron receives input from a large number of presynaptic axons originating from many different cells. This is called convergence. Why are these arrangements important for integrative processes? When activity arrives almost simultaneously from more than one input source to a postsynaptic neuron the synaptic potentials will summate. This process is known as spatial summation. It originates from spatially separated input sources. We have already discussed the possibility that a single fiber can give rise to a train of action potentials. As each action potential arrives at a branch point, each daughter branch will be activated and will conduct action potentials to the next branch point until it finally arrives at the axon terminal. The arrival of a train of action potentials at the termination will result in discrete amounts of transmitter being released over a period of time. The resulting synaptic potentials occurring close to each other in time will summate. This process is called temporal summation, since it occurs from a single input source being activated for a short period of time. Obviously both spatial and temporal summation can occur simultaneously through divergent and convergent sources. Figure 6 illustrates all these important concepts.

Each motor neuron can receive only a few synaptic contacts from a single Ia fiber. However, the total synaptic input for a single motor neuron might be more than 5,000. These inputs come from a variety of sources, including Ia, local interneurons, and descending fibers from cortex. As each fiber is activated the effect of the neurotransmitter on the postsynaptic motor neuron will be to change its state of excitability. If the transmitter is excitatory in nature the cell's excitability will be enhanced. If the transmitter is inhibitory, the excitability will be decreased. The reference point, of course, for the state of excitability is the threshold level for action potential generation.

We must now change our point of view from the single neuron to pools or groups of neurons and imagine that excitability states are fluctuating continuously depending on the variety of signals arriving over spatially separated inputs, as well as the temporal pattern of the input.

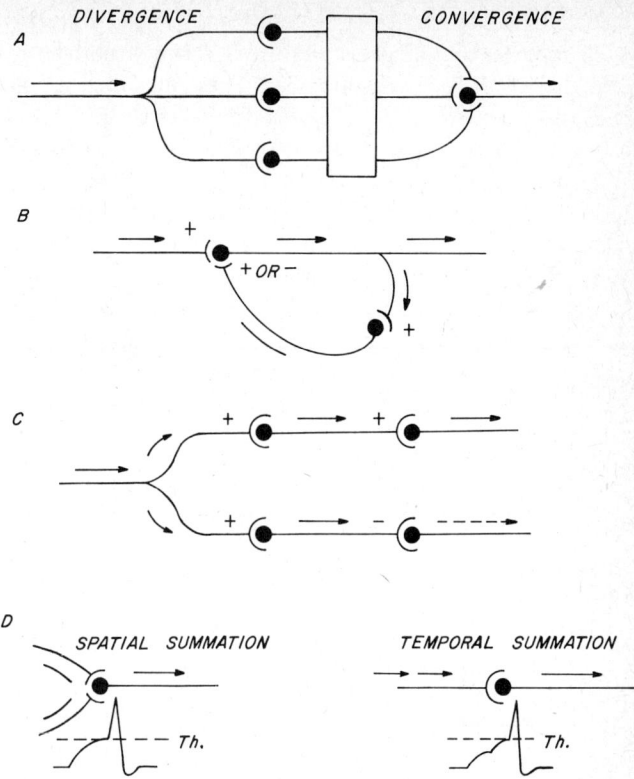

Figure 6 Schematic diagram showing the concepts of divergence and convergence that occur in the CNS and how these anatomical arrangements can result in the summation (both temporal and spatial) of synaptic potentials that alter the discharge rate of postsynaptic neurons. *(a)* A fiber input forms several terminal branches that diverge and synapse with three different postsynaptic neurons. The outputs from these neurons then converge onto a single neuron. *(b)* The convergence takes the form of feedback synapses through recurrent collaterals. *(c)* The divergence results in the separation of information into two separate pathways. *(d)* Convergence of inputs onto a single neuron can produce spatial summation (information arriving over spatially separated inputs) when discharges arrive almost simultaneously. Temporal summation is depicted as discharges arriving over one input at sufficiently high discharge rates to cause the synaptic potentials to summate. Note that both spatial and temporal summation can occur on the same postsynaptic neuron. (Adapted from: Eyzaguirre C, and Fidone SW, *Physiology of the Nervous System,* 2nd ed. chapt. 14, Year Book Med., 1975)

Furthermore, we must not forget the importance of feedback interneurons that are activated through collateral branches of the postsynaptic fiber. These may also influence the intensity, duration, and state of excitability of the postsynaptic cell. If a group of Ia fibers is activated by stretch of an entire muscle, for example, the input to spinal motor neurons will in effect activate some of these neurons. These neurons are said to be brought into the firing zone. The excitability of other neurons in the group will be enhanced but they will not discharge impulses. These cells are subliminally excited and are said to comprise the subliminal fringe around the firing zone. These cells are not brought into the firing zone because all spindles are not equally activated and motor neurons do not all receive equal numbers of synaptic input.

Now suppose we activate two synergistic muscles and record the efferent activity from the ventral root, which contains axons returning to the same muscle. Activation of each group of Ia fibers by stretching each muscle individually results in a certain amplitude potential in the ventral root, which represents those cells belonging to the firing zone. Activation of both groups of Ia fibers by simultaneously stretching both muscles produces a potential greater than the simple algebraic sum of the individual responses. This is because the subliminal fringe of both motor neuron pools overlaps, and some cells that were not activated by stretch of either muscle alone are now brought into the firing zone. This process is called facilitation. Now suppose we activate nearly all the motor neurons in each individual pool by a very strong stretch to the muscle. The individual responses are large, but when the two muscles are simultaneously stretched, the response is smaller than the algebraic sum of the two individual responses. This is because the firing zones now overlap, since they share input by the motor neurons. This decrease in the response is called occlusion. Figures 7 and 8 diagrammatically illustrate the concepts of facilitation and occlusion. Facilitation arises from the overlap of subliminal fringes, and occlusion arises from the overlap of firing zones.

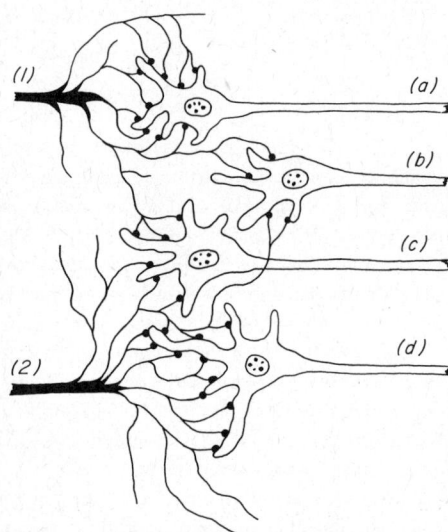

Figure 7 Diagram showing how the properties of divergence and convergence can alter the output from a pool of neurons. Postsynaptic neurons b and c receive common inputs from input terminals 1 and 2. When terminal 1 or 2 is active, the excitability of neurons b and c is raised, but not to threshold. Neurons a and d will both fire because of the numerous synaptic contacts they receive. When terminal groups 1 and 2 are both active, neurons b and c are then brought into the firing zone. (Adapted from: Guyton AL, *Medical Physiology*, 4th ed., chap. 47, Saunders, 1971)

SENSORIMOTOR INTEGRATION

We are now going to discuss the connections between the sensory or input side to the CNS and the motor or output side. The connections between the cellular elements within the CNS, specifically the spinal cord, form what is called a reflex arc. We have already discussed in some detail the different sensory receptors, both cutaneous and muscle, that may be involved. The most conceptually simple reflex would be one consisting of only two cells. This is the monosynaptic reflex arc; such a combination between two cells exists between the Ia spindle afferent fiber and motor neurons, as shown in Figure 9. Let us go through the processing steps of this simple reflex arc to examine how the physiological concepts of receptor activation, action potential conduction, and synaptic transmission come together to produce a motor response. The muscle sensory receptor, the spindle, can be activated by

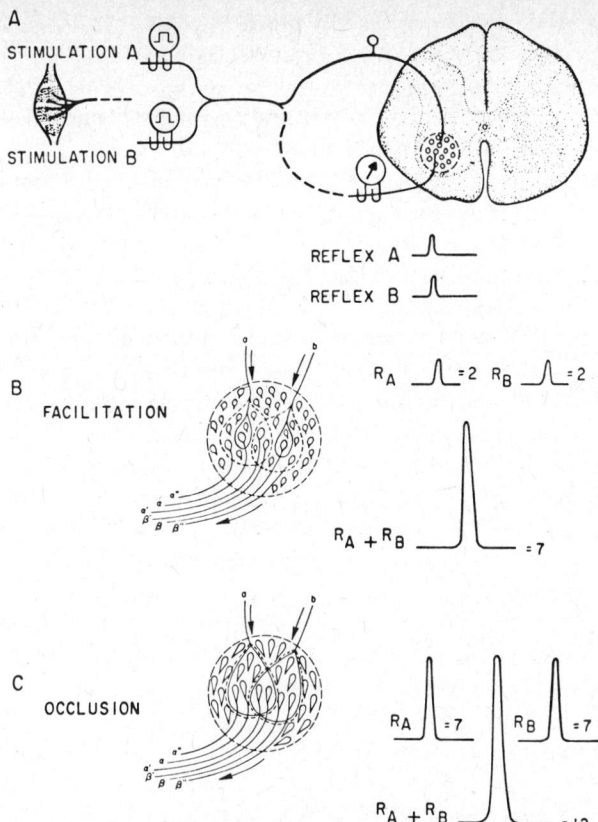

A

STIMULATION A

STIMULATION B

REFLEX A

REFLEX B

B

FACILITATION

R_A = 2 R_B = 2

$R_A + R_B$ = 7

C

OCCLUSION

R_A = 7 R_B = 7

$R_A + R_B$ = 12

Figure 8 Diagram showing the properties of facilitation and occlusion as they relate to reflexes. *(a)* The experimental setup. A muscle nerve is dissected and separated into two branches, each receiving a pair of stimulating electrodes. The reflex activity is recorded on the ventral root. The monosynaptic reflex is evoked by low-intensity stimulation of the afferent fibers. When each branch is stimulated separately, reflexes A and B are elicited. *(b)* When both branches are stimulated simultaneously, the reflex produced (depicted by A + B) is greater than the algebraic summation of the individual responses. This is because neurons in the motoneuron pool brought into the subliminal fringe by each individual stimulus are now brought into the firing zone by the summation of both inputs. *(c)* The results of occlusion. When the individual stimulus is increased so as to bring all motor neurons into the firing zone from each stimulus, the algebraic sum is greater than when both stimuli are applied simultaneously. This is because of the common input shared by each group of input fibers that are brought into the firing zone by the individual stimulus. (Adapted from: Eyzaguirre C, and Fidone SW, *Physiology of the Nervous System,* 2nd ed. chapt. 14, Year Book Med., 1975)

pulling on the tendon. Activation presumably occurs by mechanical deformation of the primary fiber ending. A series of action potentials is conducted along the afferent Ia fiber. At first the response represents the dynamic process, which is proportional to the rate of stretch of the muscle, followed by a static or maintained response. These action potentials travel centrally, finally reaching the terminal branching of the central end of the Ia afferent fiber. The fine branches give rise to synaptic endings, many of which end near the cell body of the motor neuron pool on proximal dendrites. Upon arrival of the action potentials, the process of neurotransmitter release occurs. Each action potential is involved in releasing several packages or quanta of transmitter from the nerve terminal. It should

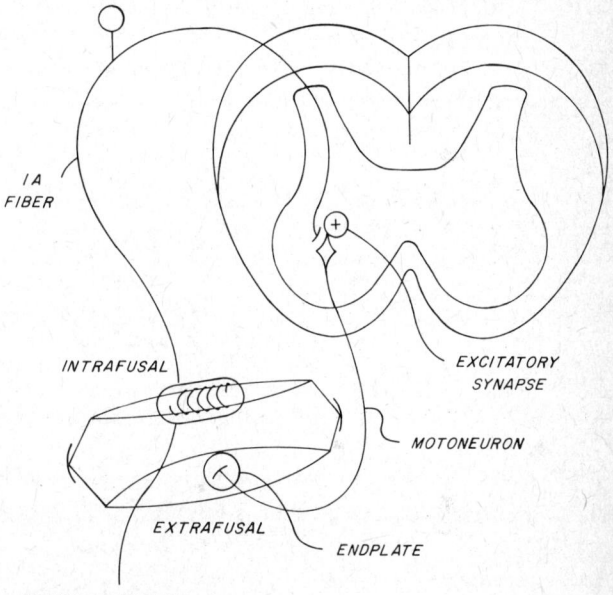

IA FIBER

INTRAFUSAL

EXTRAFUSAL

PRIMARY ENDING

EXCITATORY SYNAPSE

MOTONEURON

ENDPLATE

Figure 9 Schematic diagram showing the organization of the monosynaptic reflex arc. The Ia afferent fiber innervates the intrafusal muscle stretch receptor (spindle). Its cell body is located in the dorsal root ganglion; the central branch of the axon enters the spinal cord through the dorsal horn. This fiber makes an excitatory synaptic contact with α motor neurons. Axons from these cells exit the ventral horn and travel back to the same muscle, where they form the endplate on the extrafusal muscle fibers. Stretch of the muscle activates the spindle receptor, which in turn produces discharges in the α motor neuron, leading to contraction of the extrafusal muscle fibers.

be made clear here that the number of quanta released is probably less than five from a single axon terminal on any one motor neuron. Furthermore, Ia afferents can innervate several motor neurons, but they probably make only a few synaptic contacts (1–3 terminal boutons on the average) with any one motor neuron. Therefore, unlike the synaptic relationship between the motor nerve and a muscle cell, a single action potential in a Ia afferent fiber will elicit only a subthreshold synaptic potential on the motor neuron. Through the mechanisms of temporal summation and spatial summation an action potential or train of action potentials can be elicited from the postsynaptic motor neuron. This is the process of integration that occurs at synaptic junctions within the CNS. When summation of the synaptic potential crosses threshold to generate an action potential in the postsynaptic motor neuron, these action potentials travel down the axon, exit the spinal cord via the ventral roots, and finally invade the motor nerve terminal branches on the same muscle. Again, transmitter is released, and this time muscle contraction results. The muscle fibers innervated by the single motor nerve contract almost simultaneously. This group of muscle fibers is called the motor unit. If several motor nerve axons are recruited into the firing zone, several motor units will contract. The strength of the contraction will obviously depend on the number of motor units activated. The size of the motor unit varies. It is smaller in muscles located in the distal extremities, such as hands and fingers, and larger on proximal body parts. This corresponds to the fact that we use the fingers for fine motor control, and the smaller motor unit allows for smaller, and hence, finer movements. Smooth-graded contractions can be produced by finely tuning the output from the motor neuron pool. The fine tuning is a result of controlling the outflow of activity from the motor neuron pool by the descending activity from motor cortex acting directly or indirectly on γ and α motor neurons.

The GTO is involved in the inverse stretch reflex. This receptor is connected centrally in just the opposite manner of the spindle receptor. When the muscle is stretched passively or when it contracts under isometric conditions, the GTO is activated and centrally inhibits the contraction of the homonymous muscle group. This inhibition is disynaptic involving an interneuron within the spinal cord. Antagonist muscle groups are reflexly facilitated just as

with the stretch reflex. The precise function of the GTO has been difficult to determine. It might simply act as another proprioceptor. It might also be involved in the relaxation of different motor units during sustained muscle contraction to avoid muscle fatigue.

This description completes the monosynaptic circuit analysis. It is important to understand how this reflex circuit works not only because it is a simple but elucidating analysis of basic physiological processes, but also because this is the circuit that comes into play during neurological testing. When a tendon is tapped with a reflex hammer, the muscle is transiently stretched, activating in a synchronous fashion many of the muscle spindle receptors. Reflex contraction of the same muscle occurs following the appropriate delay for axonal conduction to and from the CNS plus a delay for transmission across a single synapse. By measuring the strength of the contraction, it is possible to localize specific lesions affecting the motor system.

Muscles act against each other in an antagonistic manner. The integrity of the agonist–antagonist relationship between the muscle groups must be maintained. When one group contracts, the other group must not contract at the same time, if coordinated movement is to occur. The CNS has been wired to obviate this possibility by supplying inhibitory inputs to the motor neurons of the antagonist muscle group. Dale's law states that different neurotransmitters cannot be released from axon terminals of the same axon. Therefore, we would not expect inhibition of the motor neurons that innervate the antagonist muscle group to occur directly by the Ia afferent fibers. Inhibition occurs through interneuron pathways. Collaterals of the Ia afferent fibers synapse on neurons within the ventral horn, which in turn synapse on the motor neurons innervating the antagonist muscle group. The Ia afferents excite the interneuron pool, which in turn liberate an inhibitory transmitter substance on the motor neurons of the antagonist muscle group.

The relationship between agonist–antagonist muscle groups is important in the next reflex to be analyzed. The withdrawal reflex involves pathways that activate motor neuron pools on the opposite side of the spinal cord. This reflex is initiated by cutaneous afferents that synapse on the motor neuron pool. The classical example of this reflex is illustrated by a person who steps barefoot onto a sharp

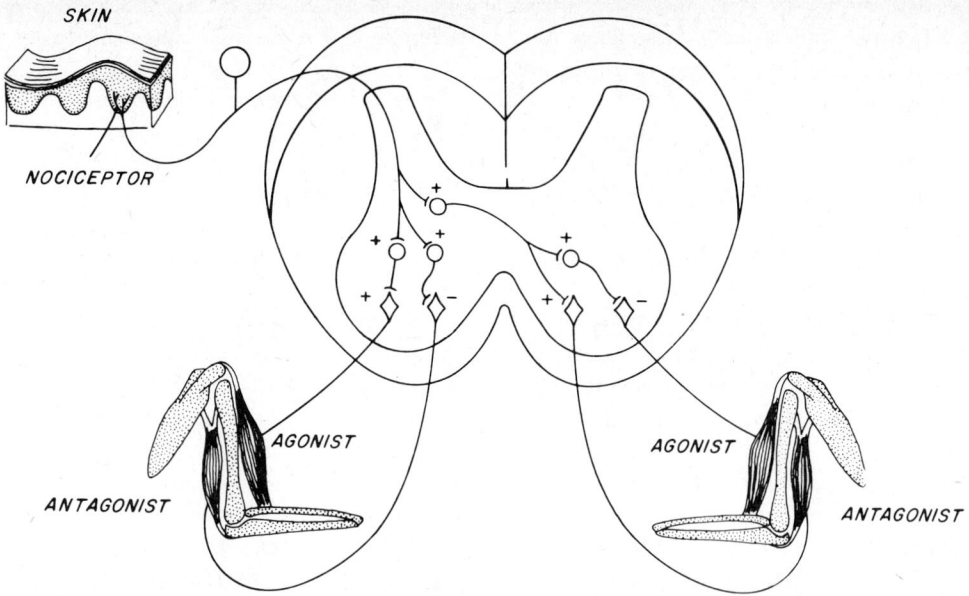

Figure 10 Schematic diagram showing the anatomical organization that results in the withdrawal reflex. Noxious input from the skin elicits the reflex via excitatory synaptic contacts with agonist muscles from the injured limb and produces concomitant inhibition of antagonist muscles. Through polysynaptic pathways, the converse occurs in the contralateral limb.

object. The immediate reflex action is withdrawal of the injured foot. Concomitant with the ipsilateral response is the extension of the contralateral leg. This enables a person to support the body weight and is an integral part of the entire reflex. Let us analyze in detail how this circuit works. The afferent activity is initiated in the primary afferent fibers by stimulation of cutaneous nociceptors in the foot. As their name suggests, these are high threshold mechanoreceptors, probably free nerve endings. These fibers enter the spinal cord and synapse on interneurons located in the ventral horn. As already stated in Chapter 38, you know that they also synapse on cells in the dorsal horn whose fibers give rise to the spinothalamic tract. These interneurons in turn form excitatory synaptic contacts on motor neurons, which return to physiological flexors. Other interneurons form inhibitory synaptic contacts on physiological extensors. This enables the injured limb to flex, thus removing the noxious stimulus from the skin. Simultaneously, interneurons project to the opposite side of the spinal cord forming excitatory synaptic con-

tacts on physiological extensors and inhibitory synaptic contacts on physiological flexors of the contralateral leg. Figure 10 is a diagram showing the circuitry involved. This withdrawal reflex with crossed extension shows how a slightly more intricate circuitry can produce a more complicated and coordinated response. However, the same basic mechanisms of axonal conduction, synaptic transmission, and CNS integration that produced the simple stretch reflex are involved here.

MOTOR CORTEX

Although electrical stimulation of many areas of the cortical surface can produce movement, only that area directly in front of the central sulcus is considered the primary motor cortex. This region is Brodmann's area 4, and the cells within the deeper layers give rise to fibers that descend the neuraxis, forming the pyramidal or corticospinal tract and the corticobulbar tract. These fibers will terminate on interneurons and motor neurons within the ventral gray matter of the spinal cord.

The cytoarchitecture of motor cortex is quite different from the sensory cortex previously discussed. It lacks the large numbers of small stellate cells (also called granule cells) found in layer IV, and for this reason this cortex is called agranular cortex. The motor cortex also contains some of the largest neurons found in the CNS. These neurons are called the gigantopyramidal cells of Betz. These cells were originally thought to be the sole source of corticospinal fibers, but it is now known that this tract receives input from other cortical areas (Brodmann's areas 6, 3, 2, 1, 5, and 7).

There is some functional importance to be attached to the layering of the motor cortex as was previously pointed out for sensory cortex. There are a variety of other subcortical structures that receive information from motor cortex besides the lower motor neurons supplied by the pyramidal system. Most of the pyramidal tract fibers are derived from cells in the deeper portion of layer V. Corticothalamic fibers arise from cells in layer VI. These project primarily back to the ventral anterior lateral nucleus of the thalamus (motor thalamus). Corticostriate fibers arise from cells in superficial layer V, and fibers in midlayer V project to the red nucleus, the pontine nuclei, and the reticular formation. Originally, most if not all cells in layer V were believed to provide the axons of the pyramidal system and other subcortical structures were to be supplied by collaterals from these axons. New data point out clearly that activity in the separate subpopulations of cells located in the output layers V and VI of motor cortex can selectively activate different subcortical structures.

Much of the afferent input to the motor cortex comes from the thalamus, in particular the ventral anterior lateral nucleus (motor thalamus). This link supplies information primarily from the cerebellum. The thalamic projection is somatotopically organized and is a mirror image of the map found in the primary somatosensory cortical area in the postcentral gyrus. However, it is important to be aware that cortical influences may arrive via other pathways including other thalamic nuclei. Thus the pallidum relays back to motor cortex via the thalamus, and there are various intrinsic cortical inputs as well as input arriving from other cortical areas.

As stated above, corticospinal fibers originating from deep layer V pyramidal cells synapse directly onto α and γ motor neurons. There is now good physiological evidence for differences in the activity of motor neurons supplying distal musculature as compared with those supplying proximal musculature. As suggested by the somatotopic organization of the motor cortex more cells are devoted to supplying the distal musculature. More than one pyramidal tract fiber supplies a single lower motor neuron, and the excitatory postsynaptic potentials (EPSPs) of the lower motor neurons supplying distal musculature are larger when activated by the descending corticospinal input than those EPSPs recorded from motor neurons supplying proximal musculature. The thresholds for stimulation are also lower for fingers, toes, and face, and the synaptic responses appear to be mediated by the largest fibers in the pyramidal tracts. These findings are in accord with our present concepts of corticospinal organization in human physiology, such as the well-defined control that exists over the distal musculature. The transition area between somatosensory and motor cortices (Brodmann's area 3a) is interesting because they receive the information arising in the group I fibers innervating the primary spindle ending. The group II fibers innervating the secondary spindle ending end in area 4. The somatotopic organization and the juxtaposition of similar body parts point to the possibility for direct access to the motor cortex by group I fibers.

PYRAMIDAL AND EXTRAPYRAMIDAL MOTOR SYSTEMS

The terms pyramidal and extrapyramidal system deserve some mention because of clinical usage. Strictly speaking, the pyramidal tract consists of all those fibers located or traversing the pyramids located in the medulla oblongata regardless of their site of origin. There is uniform agreement, however, that the pyramidal system refers strictly to those fibers originating in the cortex and most of which are destined to form the corticospinal tract. These fibers are strictly related to motor function. There are, however, many fibers of similar origin and function that end on brainstem nuclei, but that never traverse the pyramids. These fibers belong to the corticobulbar system. Since their function is the same, they are often included clinically under the term pyramidal system.

The term extrapyramidal encompasses properly all those fibers that do not belong to the pyramidal tract. However, usage has restricted this term to reference to motor systems involving fibers that are neither pyramidal nor corticobulbar as defined above. In its most strict usage this term refers to the basal ganglia system described below; however, some investigators include the cerebellum and pontine nuclei, since they are not pyramidal, but are motor structures. These terms appear to have some use clinically since distinctly separate and recognizable diseases are associated with the two systems. Diseases involving the pyramidal system are also sometimes called upper motor neuron diseases in reference to the fact that the cells of origin are located in higher centers (i.e., cortex). These diseases are characterized by increased muscle tone, hyperreflexia, and spasticity. By contrast, the extrapyramidal system is generally involved with the coordination of movement. Diseases involving areas of this system are characterized by often involuntary movement disorders and tremors, as well as increased muscle tone. Because of the elaborate connections within the CNS, however, it is unlikely that strict lesion of one system does not involve the other; thus a word of caution should be given regarding the use of these terms.

BASAL GANGLIA

The basal ganglia consist of several structures below the cortex that are involved in motor function. There is some disagreement as to all the structures to be included; however, the caudate nucleus, putamen, globus pallidus, substantia nigra, and the subthalamic nucleus are generally included. The basal ganglia system is foremost involved in the integration of information traveling through the thalamus. This is evident from the fact that the basal ganglia receive well organized projections from the cortex and in turn supply most of their projections to the thalamus, in particular the ventral anterior lateral nucleus. However, in recent years it has become apparent that the basal ganglia must be more intimately involved with other functional roles, as the complex organization of their projections suggests.

The known connections of the basal ganglia are shown in Figure 11. Both the caudate and putamen receive directly from cortex. These structures in turn project to the pallidum and the substantia nigra. The pallidum also projects to the subthalamic nucleus and ventral anterior lateral thalamic nucleus. Reciprocal connections are also apparent as shown in the figure.

The basal ganglia cannot be considered centers for motor function, but rather are mainly concerned with the integration of motor activity. Lesions involving the basal ganglia result in an increase of motor activity, and therefore it is thought that these areas have an inhibitory influence on the motor pathways with result in voluntary movements. Major diseases associated with lesions of the basal ganglia include Parkinsonism, Huntington's chorea, and ballism.

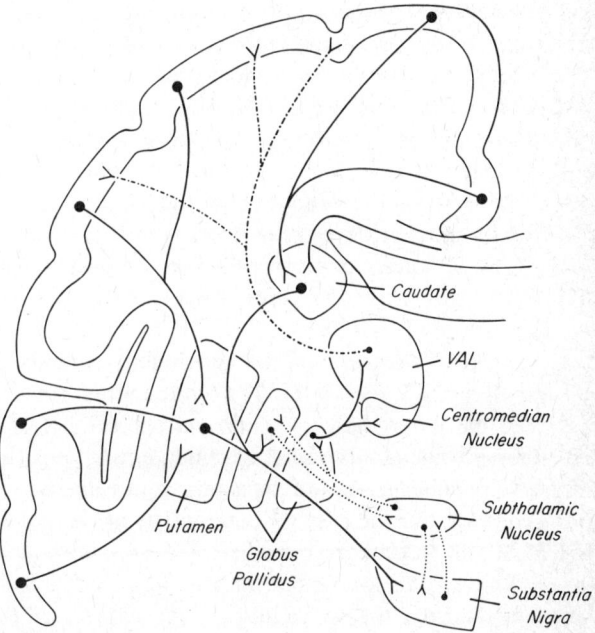

Figure 11 Connections of the basal ganglia. (Adapted from: Brodal A, *Neurological Anatomy,* chap. 4, Oxford Univ. Press, 1981)

THE CEREBELLUM

The cerebellum sits astride the brainstem in a unique position perfectly suited to monitor activity initiated at the periphery and ascending the spinal cord as well as the activity initiated at cortical levels and descending to the spinal cord. The function of the cerebellum is to add quality to motor function. It is involved in predicting, judging, and correcting the motor act in order to achieve the highest level of motor control. Since most, if not all, motor acts involving skeletal muscle are initiated either from some sensory input or from voluntary action through the cerebral cortex, it is not surprising that the cerebellum receives inputs not only from the proprioceptive afferents derived from muscle, but from cutaneous receptors, vestibular apparatus, and the visual system as well. The cerebellum is apparently not involved in conscious sensation. If removed, there appears to be no deficit in cognitive functions, although there is marked alteration in the ability to produce smooth motor acts.

The cerebellum can be divided into three main phylogenetic areas: the archicerebellum, paleocerebellum, and the neocerebellum. The latter also can be grouped into the corpus cerebelli. The older portion includes the floccular nodular lobe, which is linked to the vestibular system. The corpus cerebelli also contains older and newer regions. The older structures are located on the midline and, as the cerebellum developed, newer regions pushed out laterally. As the cerebellum developed, so did the quality of the motor act. It is in the primate that the cerebellum has achieved its greatest development, and this is associated with a large repertoire of motor acts that the primate can achieve. In this sense the cerebellum has imitated the development of neocortex.

The internal structure of the cerebellum consists of white matter and gray matter. The white matter consists of axons that enter the cerebellum via one of the three cerebellar peduncles—the inferior, middle, and superior cerebellar peduncles—as well as axons from cells located in the cerebellum. The gray matter consists of those cells that form the cerebellar cortex and the deep cerebellar nuclei. These latter structures number four and are the main output from the cerebellum. Their names are the fastigial, globose, emboliform, and dentate nuclei.

Inputs to the cerebellum from the spinal cord and periphery arrive by the dorsal and ventral spinocerebellar tracts (DSCT, VSCT), the spinal olivary tract (SOT), and the dorsal columns. Most of this input is uncrossed and arrives via the inferior cerebellar peduncle. These inputs include the muscle afferents (Ia, Ib, II fibers) and cutaneous and joint afferents coming up the dorsal column system. The cells that project to the cerebellum are second-order neurons located in Clarke's column (DSCT), the intermediate zone in the spinal cord (VSCT, SOT), and the dorsal column nuclei. The VSCT afferents cross the cord and enter the cerebellum via the superior cerebellar peduncle. The dorsal column input is mainly uncrossed, but there is also a crossed input. The SOT projects contralaterally to the inferior olive. The neurons in the olive then cross back to the other side and enter the cerebellum via the inferior cerebellar peduncle.

The input to the cerebellum from the brainstem is concerned primarily with posture and equilibrium and thus arrives from the vestibular system. This area projects to the archicerebellum, consisting of the flocculonodular lobe. There is also input from the tectal areas of the brainstem, which is involved in visuomotor coordination. Input from the cerebral cortex is mainly from motor and premotor areas in the frontal cortex. These afferents descend to the pontine nuclei. The neurons in the pons send axons across the midline to enter the cerebellum via the middle cerebellar peduncle. The dentate nucleus within the cerebellum forms the feedback loop to the motor and premotor cortex. Cells within this nucleus send axons to the red nucleus as well as directly to the motor thalamus, that is, the ventral lateral nucleus (VL). These axons form a crossed projection back to motor cortical areas.

There are five neuronal types within the cerebellum: Purkinje cells, granule cells, Golgi cells, basket cells, and stellate cells. The organization of this organ is quite unique because of the obvious geometrical relationship between the cell types. Two types of axon terminations are made by afferent fibers arriving in the cerebellum. One type is the mossy fiber, which includes almost all the afferent input. These fibers terminate primarily on granule cells. The climbing fiber is the other axon terminal type and arises almost exclusively from olivary cells. All input to the cerebellum from extrinsic fibers form excitatory synaptic connections. Mossy fibers project to granule cells, forming the parallel fiber network within the outer portion of cerebellar cortex. Climbing fibers, on the other hand, project directly to Purkinje cells. The Purkinje cell axons project directly to the deep cerebellar nuclei. Curiously, all synaptic connections within the cerebellum are inhibitory, except for extrinsic input and the parallel fiber system formed by the granule cells. The parallel fiber system projects to all the other cell types, including Purkinje cells, basket cells, Golgi cells, and stellate cells, forming excitatory connections with them. Basket and stellate cells form inhibitory contacts with Purkinje cells, and Golgi cells form inhibitory feedback contacts with granule cells. These complicated circuits are shown in Figure 12*b*.

The cerebellum represents a unique structure in the CNS that has been extremely well studied both from the point of view of its neuronal organization as well as the precise way in which the cells are synaptically coupled. Just like the retina, the cerebellum is a small processor of information. Electrical activity arrives over extrinsic inputs and is switched directly to the cortical processing units—those cells discussed above—and simultaneously to the deep cerebellar nuclei, the output cells. The function of the intrinsic cellular material is to regulate the output from the deep cerebellar nuclei. Let us examine how this is done.

First, there is a simple geometrical relationship that exists between Purkinje cell dendrites and parallel fibers. The elaborate dendritic trees of the Purkinje cells are oriented orthogonal to the long axis of the folia (Figure 12 *a*). The granule cells onto which mossy fibers form excitatory synaptic contacts, give rise to a system of parallel fibers that run parallel to the long axis of the folia and, therefore, othogonal to the plane of the Purkinje cell dendrites. Activation of a circumscribed population of granule cells will then drive a narrow band of Purkinje cells. In addition, stellate and basket cells are excited by the parallel fiber network, and these cells feed back inhibitory synapses onto Purkinje cells. One would expect, therefore, that this combined synaptic drive onto the Purkinje cells would result in no output from these cells.

However, within the area of active parallel fibers, the synaptic drive to the Purkinje cells overcomes the negative input of the basket and stellate cells, whereas the Purkinje cells outside this narrow band will receive a net inhibitory influence. In this way, the border of this band of Purkinje

cells is sharpened by this process, called lateral inhibition. The result is that a localized mossy fiber input from, for example, DSCT or VSCT neurons, results in a narrow band of Purkinje cell excitation surrounded by a steep, wide gradient of Purkinje cell inhibition. The negative or inhibitory input by the Purkinje cell axons onto the deep cerebellar nuclei functions to regulate the output of the cerebellum for specific areas. It is precisely those cells in the cerebellar nuclei that are excited by input collaterals inhibited by Purkinje cell feedback. Climbing fibers form a very powerful input to Purkinje cells. They also contact via collaterals Golgi and basket cells that inhibit Purkinje cell activity. Thus, this circuit acts in a similar fashion in producing lateral inhibition of Purkinje cell activity for those cells lying outside the sphere of influence of the climbing fibers.

Functionally, the cerebellum acts like a servomechanism. Input states are compared continuously with output states and appropriate corrections made to ensure the desired power and range of motion. The actual movement is supplied by information derived from muscle, joint receptors, and vestibular apparatus. The intent of the movement comes from the voluntary motor areas of the cerebral cortex. The information comes together in the cerebellum, where it can be compared.

SELECTED BIBLIOGRAPHY

Brodal A: Pathways mediating supraspinal influences on the spinal cord: The basal ganglia, in Brodal A (ed): *Neurological Anatomy.* New York–Oxford, Oxford University Press, 1981.

Brodal A: The cerebellum, in Brodal A (ed): *Neurological Anatomy.* New York–Oxford, Oxford University Press, 1981.

Henneman E: Organization of the motor system—a preview, in Mountcastle VB (ed): *Medical Physiology,* ed 13. St Louis, CV Mosby, 1974.

Henneman E: Feedback control of muscle: Introductory concepts, in Mountcastle VB (ed): *Medical Physiology,* ed 13. St Louis, CV Mosby, 1974.

Henneman E: Peripheral mechanisms involved in the control of muscle, in Mountcastle VB (ed): *Medical Physiology,* ed 13. St Louis, CV Mosby, 1974.

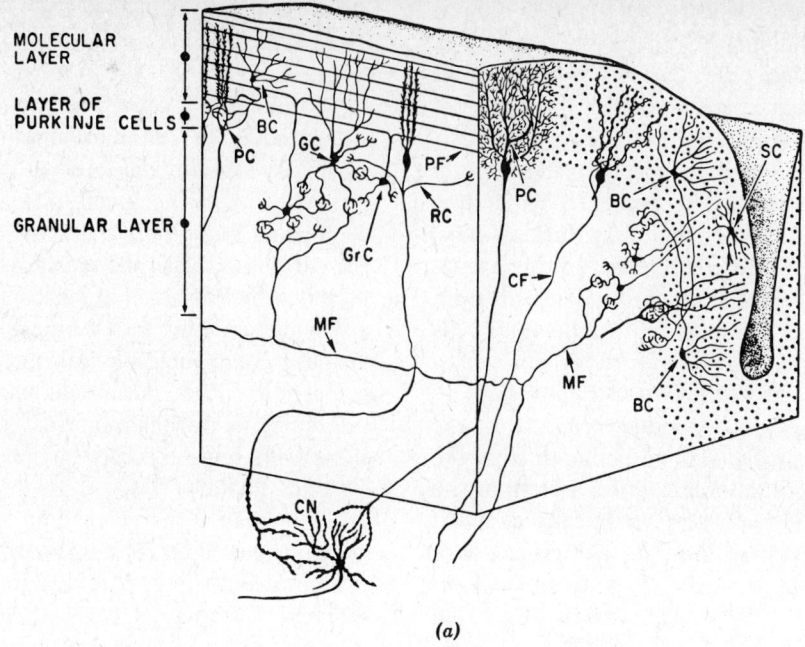

(a)

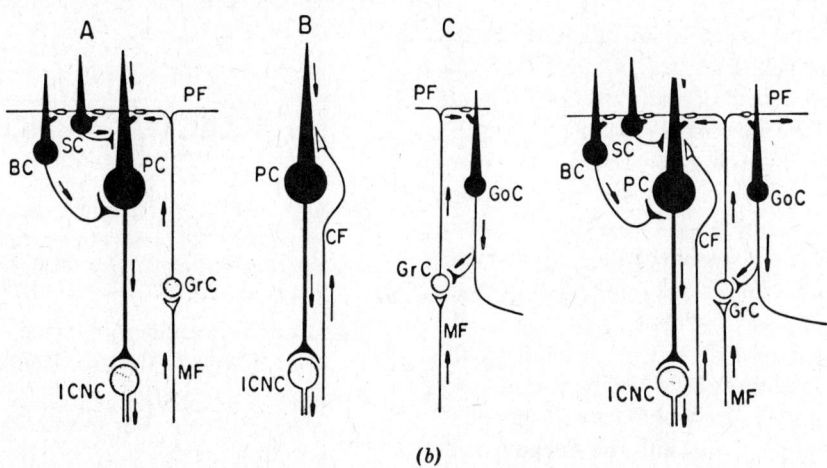

(b)

Figure 12 Diagram showing the neuronal organization of the cerebellar cortex. PC, Purkinje cell; BC, basket cell; GC, Golgi cell; GrC, granule cell; PF, parallel fiber; RC, recurrent collateral; MF, mossy fiber; CF, climbing fiber; CN, deep cerebellar nuclear cell; SC, stellate cell. The bottom diagram shows in greater detail the synaptic organization, without reference to cell position within the cortex. (Adapted from: Eyzaguirre C, and Fidone SW, *Physiology of the Nervous System,* 2nd ed., Year Book, 1975, chap. 19)

Henneman E: Signal reflexes and the control of movement, in Mountcastle VB (ed): *Medical Physiology,* ed 13. St Louis, CV Mosby, 1974.

Henneman E: Motor functions of the brainstem and basal ganglia, in Mountcastle VB (ed): *Medical Physiology,* ed 13. St Louis, CV Mosby, 1974.

Henneman E: The cerebellum, in Mountcastle VB (ed): *Medical Physiology,* ed 13. St Louis, CV Mosby, 1974.

Henneman E: Motor functions of the cerebral cortex, in Mountcastle VB (ed): *Medical Physiology,* ed 13. St Louis, CV Mosby, 1974.

SELF-STUDY QUESTIONS

MULTIPLE CHOICE

Select the single best answer.

1. Which of the following form input to the cerebellum?
   A. α motor neurons
   B. Granule cells
   C. Climbing fibers and mossy fibers
   D. Group Ia afferents
   E. Purkinje fibers

2. Which cell type forms the efferent pathway from cerebellar cortex?
   A. Purkinje cells
   B. Granule cells
   C. Stellate cells
   D. Spinothalamic cells
   E. Basket cells

MATCHING

A. *Output from cerebellar cortex*
B. *Inhibit(s) granule cells.*
C. *Arise(s) from neurons of the inferior olive.*
D. *Terminate(s) in granule cell layer as mossy fibers*
E. *Give(s) rise to parallel fibers.*

3. Dorsal spinocerebellar tract
4. Purkinje cell fibers
5. Granule cells
6. Climbing fibers

7. Golgi cell
8. Activation of the GTO produces
   A. facilitation of the α motor neurons innervating the same muscle.
   B. contraction of antagonistic muscles.
   C. inhibition of the α motor neurons innervating the same muscle.
   D. reciprocal inhibition.
   E. none of the above.

9. The branching of afferent fibers within the CNS that can synapse with many neurons is called
   A. divergence.
   B. convergence.
   C. reciprocal innervation.
   D. negative feedback.
   E. positive feedback.

10. The group II fiber is particularly responsive to
    A. rate of change in length of the spindle.
    B. stretch of the spindle.
    C. activation of alpha motoneurons.
    D. activity in corticospinal tracts.
    E. changes in temperature.

11. The GTO is located within
    A. the intrafusal muscle fibers.
    B. the tendon.
    C. the joints.
    D. the skin.
    E. none of the above.

TRUE OR FALSE

12. Muscle spindles respond to stretch.
13. Active γ motor neurons increase the output from muscle spindles.
14. GTOs can be excited during muscle contraction.
15. Excitation of α motor neurons increases spindle output.
16. The physical relationship between intra- and extrafusal fibers determines in part their functional properties.
17. Ia fibers are the primary afferent axons innervating spindles and Ib fibers innervate GTOs.

MULTIPLE CHOICE

Select the single best answer.

18. The monosynaptic reflex requires all of the following, *except*
    A. Golgi tendon organs.
    B. group Ia muscle afferents.
    C. primary endings on muscle spindles.
    D. α motor neurons
    E. acetylcholine.

19. Excitation of γ motor neurons to an extensor muscle results in all of the following, *except*
    A. contraction of intrafusal muscle fibers in the extensor muscle.
    B. decreased activity in group Ib afferents from the extensor muscle.
    C. increased extensor group Ia afferent activity.
    D. excitation of extensor α motor neurons.
    E. inhibition of α motor neurons innervating antagonistic flexor muscles.

20. All the following characterize the pathway for the "claspknife" reflex, *except*
    A. functions as a servomechanism that helps modulate muscle tone.
    B. monosynaptically inhibits α motor neurons.
    C. mediates a reflex involving stretch receptors.
    D. receives input from Golgi tendon organs.
    E. protects muscles from injurious levels of tension.

21. A motor unit is characterized by all of the following, *except*
    A. consists of an α motor neuron and the muscle fibers it innervates.
    B. produces a fasciculation when active.
    C. produces a fibrillation when active.
    D. acts in concert with other motor units to maintain muscle tone.
    E. has more muscle fibers in the quadriceps than in extraocular eye muscles.

22. All the following characterize activity in the flexion reflex afferents, *except*
    A. reaches consciousness.
    B. leads to polysynaptic excitation of flexor α motor neurons and inhibition of extensor α motor neurons ipsilaterally.
    C. monosynaptically excites flexor α motor neurons.
    D. initiates reflexes that exhibit local sign.
    E. arises from nociceptors, touch receptors, and pressure receptors.

23. All the following statements about motor function are correct, *except*
    A. the basal ganglia modify thalamocortical interactions.
    B. the lateral corticospinal tract is primarily involved with fine, skilled movements of the distal extremities.
    C. cerebellar influence over α motor neuron activity is mainly via direct fibers from the Purkinje cells to the spinal cord.
    D. convergence of input from the basal ganglia and the cerebellum occurs in the ventrolateral (VL) thalamic nucleus.
    E. the recticulospinal and vestibulospinal tracts function in the maintenance of a balanced posture.

24. All the following statements about muscle spindles are true, *except*
    A. spindles are attached at their ends.
    B. tapping a tendon will activate the primary endings on spindles.
    C. Ia fibers have both a dynamic and static response.

D. γ motor neurons synapse on extrafusal muscle fibers.

E. γ motor neurons affect either dynamic or static responses.

25. All the following statements about muscle spindles are true, *except*

A. they signal the length of the muscle.

B. they signal the rate of change of muscle length.

C. they are activated by contraction of the extrafusal muscle fibers with which they are associated.

D. their sensitivity is controlled by the CNS.

E. their Ia afferent fibers make monosynaptic connections with α motor neurons.

MATCHED PAIRS

Select the single incorrect answer.

26. The following pairs are correctly matched, *except*

A. Phasic stretch reflex,–group Ia muscle afferents.

B. Tonic stretch reflex,–group Ia muscle afferents.

C. Claspknife reflex,–group Ib afferents.

D. Phasic stretch reflex,–monosynaptic pathway.

E. Deep tendon reflex (DTR),–Golgi tendon organ.

27. Which receptor type could respond to these muscle manipulations?

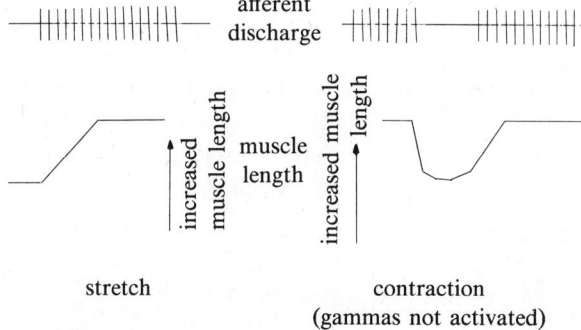

A. Muscle spindle primary ending

B. Golgi tendon organ

C. Muscle spindle secondary ending

D. Pressure—pain ending

E. Pacinian corpuscles

MULTIPLE CHOICE

Select the single best answer.

28. The phasic stretch reflex

A. C, D only.

B. C, D, and E.

C. results from the dynamic response in group Ia muscle afferents.

D. can become hyperactive following lesions of the internal capsule.

E. is unaffected by peripheral nerve lesions.

29. Motor neurons

A. are located in the dorsal and intermediate spinal gray matter.

B. are inactive when a person is at rest.

C. are excited by pathways descending from supraspinal structures.

D. influence only the tonic discharge of group Ia muscle afferents.

E. innervate extrafusal muscle fibers.

30. Spinal cord reflex pathways

A. C and D only.

B. C, D, and E.

C. are affected by activity descending from the brainstem.

D. are not activated by afferent activity that is perceived.

E. are not used in voluntary or willed postural changes.

31. γ Motor neurons

A. C and D only.

B. C, D, and E.

C. receive input from supraspinal structures.

D. innervate intrafusal muscle fibers.

E. can selectively alter muscle spindle sensitivity to a change in muscle length and/or rate of change in muscle length.

32. The phasic stretch reflex

A. C and D only.

B. C, D, and E.

C. is commonly termed the "deep tendon reflex" because it is elicited by excitation of receptors in tendons.

D. will generally become hyperactive following an "upper motor neuron" lesion.

E. can be elicited by tapping a tendon, but not by rapidly extending or flexing a joint.

33. Choose the neural mechanism illustrated in the diagram.

(+) indicates excitation.   (−) indicates inhibition

A. Recurrent inhibition

B. Recurrent facilitation

C. Stretch reflex

D. Flexion reflex

E. Claspknife reflex

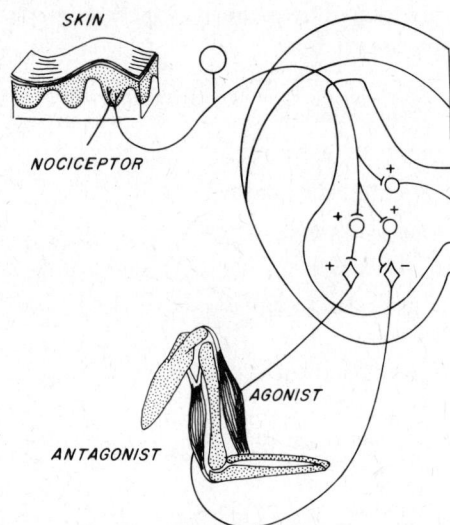

34. Lesions involving the basal ganglia can betray themselves by which of the following signs?

A. All of the following.

B. Tremors.

C. Disorders of muscle tone.

D. Involuntary movements other than tremors.

E. Disturbances of gait or posture.

35. The function of the basal ganglia is best described by which of the following statements?

A. Serve as major intermediate relay nuclei for sensory information ascending from the thalamus to the cerebral cortex.

B. Involved with thalamocortical interactions and, on the basis of disordered function, appear to have an inhibitory role.

C. Major inhibitory relay nuclei for corticodescending influences on the brainstem and spinal cord.

D. Predominantly involved with hypothalamic and autonomic nervous system function.

E. Serve to integrate motor activity via large-fiber tracts descending from the globus pallidus to the spinal cord.

36. Which of the mechanisms listed below can be used by a descending pathway to effect a change in motor activity?

A. C, D, and E.

B. C and D.

C. Direct changes in the excitability of α motor neurons.

D. Changes in the level of activity in γ motor neurons.

E. Primary afferent depolarization.

37. The claspknife reflex

A. C, D only.

B. C, E only.

C. is elicited by Golgi tendon organ activity.

D. utilizes a monosynaptic pathway to α motor neurons.

E. effectively shortens a muscle to reduce the tension on individual muscle fibers.

38. The excitability of α motor neurons is a function of the activity in

A. all of the following.

B. group Ia muscle afferents.

C. group Ib afferents.

D. interneurons.

E. γ motor neurons.

39. Spinal cord reflexes

A. C, D only.

B. C, E only.

C. are not significantly influenced by descending pathways.

D. play a role in the maintenance of posture.

E. generally become active only during rapid or extreme movements.

40. Select the *one* argument that is *unsupportable* on the basis of fact.

A. The cerebellum serves to coordinate sensorimotor function; hence disorders of this structure would be expected to result in asynergy during movement.

B. A number of descending pathways influence γ motor neuron activity; therefore, the central effects of group Ia afferent activity play a major role in motor function.

C. The dyskinesias and disorders of tone associated with lesions of the basal ganglia imply that this deep nuclear complex exercises an inhibitory influence over cortical activity.

D. Due to the vast array of sensory input to the cerebellum, cerebellar lesions would be expected to result in generalized losses of sensibility.

E. Since the reticulospinal pathways provide a major route for effecting changes in activity of autonomic preganglionic neurons, lesions of the brainstem can result in autonomic dysfunction.

MULTIPLE CHOICE

Select the best answer(s). (In many instances, more than one answer is correct.)

41. Muscle spindles

A. are length sensitive.

B. can be activated by tapping a muscle tendon.

C. have afferents that make monosynaptic connections with α motor neurons.

D. are tension sensitive.

42. Reflex withdrawal of an extremity from a noxious stimulus

A. results from the excitation of polysynaptic spinal pathways.

B. can involve afterdischarge.

C. results from activity in afferent fibers with diameters less than that of group I afferents.

D. can be exaggerated in patients with spinal cord lesions.

43. The γ-loop mechanism may be active in

A. maintaining muscle tone.

B. postural adjustments.

C. initiation of movement.

D. the regulation of phasic and tonic stretch reflexes.

44. One could differentiate a Golgi tendon organ (GTO) afferent from a muscle spindle afferent because only the GTO afferent would

A. increase firing during stretch of the muscle.

B. decrease firing during stretch of the muscle.

C. decrease firing during contraction of the muscle.

D. increase firing during contraction of the muscle.

45. Which statement is *not* correct with respect to muscle spindles?

A. Muscle spindles contain two types of modified muscle fiber.

B. Muscle spindles receive two types of sensory nerve endings.

C. Muscle spindles receive a motor innervation.

D. Muscle spindle afferents discharge when the muscle is stretched.

E. Muscle spindle afferents discharge when the muscle contracts.

Select the single best answer.

46. Which of the following belongs to the anterior lobe of the cerebellum?

A. Tonsil

B. Nodule

C. Flocculus

D. Lingula (lobule I)

E. Lobule IX

47. The cerebellar cortex

A. is basically six layered, like the cerebral cortex.

B. sends descending fibers into the spinal cord to terminate on α-motor neurons.

C. inhibits the deep cerebellar nuclei.

D. is not somatotopically organized.

E. contains pyramidal cells that send fibers to join the corticospinal tracts.

48. Purkinje cells

A. are inhibited by granule cells.

B. are inhibited by climbing fibers.

C. are inhibited by parallel fibers.

D. inhibit neurons in the deep cerebrellar nuclei.

E. excite neurons in the vestibular nuclei.

49. The established synaptic arrangement between granule cells and Purkinje cells is (1) granule cell, (2) granule cell axon, (3) basket cell, (4) Purkinje cell.

A. All of the above

B. 1 and 2 only

C. 1 and 3 only

D. 2 and 3 only

E. None of the above

50. Afferent information from the spinal cord reaches the cerebellar cortex of the anterior lobe and part of the posterior vermis via the

A. inferior cerebellar peduncle only.

B. inferior and middle cerebellar peduncles only.

C. inferior and superior cerebellar peduncles only.

D. all three of the cerebellar peduncles.

E. middle and superior cerebellar peduncles only.

51. Suppose that a disease process were to destroy the cerebellum. What sensory modality would be lost?

A. None

B. Pain

C. Temperature

D. Touch

E. Position sense

MATCHED PAIRS

Select the single incorrect response.

52. The following pairs are correctly matched, *except*

A. Dentate nucleus–neocerebellum

B. Flocculonodular lobe–fastigial nucleus

C. Granule cells–climbing fibers

D. Purkinje cells–inhibition of deep cerebellar nuclei

E. Dentate nucleus–ventrolateral thalamic nucleus (VL)

MULTIPLE CHOICE

Select the single best answer.

53. Granule cells of the cerebellum

A. C and D only

B. C, D, and E

C. constitute the majority of cerebellar cortical neurons.

D. send axons into the molecular layer.

E. inhibit Purkinje cells.

54. The neocerebellum (corticocerebellum)

A. C and D

B. C, D, and E

C. receives fibers from the pontine nuclei.

D. sends axons to the ventral posterior thalamic nucleus.

E. sends axons into the dorsal cerebellospinal tract.

ANSWERS

1. C. Both climbing fibers (from inferior olive and accessory olivary nuclei in the medulla) and mossy fibers (from neurons in the spinal cord and brainstem) form inputs to the cerebellum. Granule cells are intrinsic to cerebellar cortex, and Purkinje cells form the efferent system from cerebellar cortex.

2. A. Only Purkinje cells are efferent from cerebellar cortex. Granule, stellate, and basket cells are intrinsic to cerebellar cortex.

3. D.

4. A.

5. E.

6. C.

7. B.

8. C. Activation of the GTOs will produce at least inhibition of the same muscle. This is the characteristic claspknife reflex.

9. A.

10. B. The group II fibers do not have the phasic response that the group I fibers seem to have.

11. B. As the name suggests, within the tendon in series with the extrafusal muscle fibers.

12. True.

13. True.

14. True.

15. False.

16. True.

17. True.

18. A. GTOs are not involved in the monosynaptic reflex arc. Acetylcholine is, of course, the transmitter at the motor endplate.

19. B. GTOs will not be affected.

20. A. Not involved in muscle tone.

21. C. Fibrillation is the uncoordinated contraction of muscle fibers due to muscle disease. The motor unit is activated only when the motor axons are active.

22. C. Flexion reflex is a synonym for the withdrawal reflex. Activity is carried over polysynaptic pathways. Since it involves nociceptors, the afferent activity certainly reaches conscious perception. This reflex, although not as brisk, can be produced by stimulation of nonnociceptor afferents.

23. C. The only direct Purkinje cell output from the cerebellum goes to the vestibular nuclei. All other output comes from the deep cerebellar nuclei. See Figure 13 for basal ganglia interaction in the motor system.

24. D. γ Motor neurons synapse on intrafusal muscle fibers.

25. C. Contraction of the extrafusal muscle fibers in the absence of γ motor control reduces activity in the spindle afferents.

26. E. Deep tendon reflex is a clinical synonym for the monosynaptic or phasic stretch reflex.

27. A, C on the left and B on the right.

28. A. The reflex is produced by spindle afferents that can be damaged in peripheral nerve lesions. Lesions of the internal capsule can affect descending motor pathways involving input to the γ and α motor neurons.

29. C. Both γ and α motor neurons are excited by descending pathways from supraspinal structures.

30. C. Spinal cord reflex pathways are affected by descending pathways. D and E are false statements.

31. B.

32. D. It becomes hyperactive because of removal of descending inhibitory influences. This is used as a clinical test.

33. D. This is a diagram for one-half the withdrawal reflex.

34. A. The basal ganglia are involved in regulating information flow to and from the motor cortex.

35. B. Disease of the basal ganglia can result in gross movement disorders that are not easily controlled.

36. A. Descending influences can affect both α and γ motor neurons. Primary afferent depolarization is involved in presynaptic inhibition, which is believed to be present within the spinal cord, both in the motor and sensory areas of integration.

37. C. A polysynaptic pathway is involved, but no active shortening (contraction) of the muscle occurs.

38. A.

39. D.

40. D. No sensory deficits are seen following removal of the cerebellum. However, motor deficits appear as the loss of ability to produce smooth motor acts.

41. All are correct.

42. All are correct. After discharge is the continuing discharge of neurons following stimulus removal. This probably involves recurrent collaterals. Activity is carried over nociceptors innervated by the small myelinated and unmyelinated fiber groups. These reflexes can be exaggerated because of removal of descending inhibitory influences.

43. All are correct.

44. D. The GTO increases its firing rate during contraction.

45. E. Muscle spindle afferents do not discharge when the muscle contracts, except when the γ motor neurons are active.

46. D. All others are part of the posterior cerebellum.

47. C. The Purkinje cells inhibit the deep cerebellar nuclei and thus control their output.

48. D. See previous answer. Granule cells give rise to the parallel fiber system, exciting Purkinje cells. Climbing fibers are extrinsic fibers from the olive, which excite Purkinje cells.

49. A. This is an established synaptic arrangement between granule cells and Purkinje cells.

50. A. Afferents from the spinal cord enter the cerebellum via the inferior cerebellar peduncle (restiform body).

51. A. No sensory modality would be lost.

52. C. Mossy fibers input to granule cells, which in turn give rise to parallel fibers.

53. C. Sends output to ventrolateral thalamic nucleus.

54. C.

40

The Limbic System

BURGESS CHRISTENSEN

OBJECTIVES

After completion of this chapter, you should be able to:

1. Describe the anatomical structures that form the limbic system.

2. Describe the functional aspects of limbic structures and their importance in the role of behavior.

FUNCTIONS OF THE LIMBIC SYSTEM

The limbic system is closely associated with the regulation of behavior. The anatomical areas that make up this system include the amygdala, hippocampus, and several of the midline structures, such as the mamillary and septal nuclei, as well as parts of the thalamus (Fig. 1). The interconnections of these areas are numerous and complex, and the interested student should refer to a neuroanatomical text for more detailed information.

The time frame that encompasses changes in behavior or emotions can be on the order of seconds, minutes, and even days. For the cellular physiologist who normally deals with events that can occur at most over several seconds the limbic system is very difficult to study. First, the physiologist interested in electrical events that occur between neurons attempts to apply some input or stimulus to the system and record some type of output usually in the form of electrical activity from a single cell or small number of cells. This is a difficult approach for the limbic system because the appropriate stimulus is hard to define. Furthermore, because of the diffuse interconnections, one is hard put to decide where to record the output of the system. Questions as to how one defines emotional or behavioral changes have been best dealt with by behavioral scientists working with the intact animal, rather than by the physiologist. For these reasons, research in this area at the single-cell level has been restricted. On the other hand, approaches designed to examine the interconnections and behavioral alterations in the animal following stimulation of specific limbic system structures or following lesions in this area have yielded a good deal of information about function. We will discuss some of the important functions of the limbic system without reference to details of mechanisms, since they are largely unknown.

The hypothalamus represents one of the main structures to which many of the other limbic areas project. Experiments on primates involving chronically implanted electrodes have shown that specific hypothalamic structures are involved in both reward and punishment behavior. Electrodes are implanted in specific hypothalamic areas. These electrodes are connected to a stimulator activated by a lever. The animal has access to this lever and presses it almost continuously if the electrodes are located in a reward area. In a corollary of this type of experiment, the electrodes can be placed in an area that produces painful or unpleasant sensations. The circuit is arranged such that the stimulator is turned off by pressing the lever. The animal soon learns to keep the lever pressed. There are secondary centers that produce similar but less intense sensations in the hypothalamus and other limbic structures including the amygdala, septum, and basal ganglia.

The hypothalamus is also intimately involved in producing the behavioral pattern characterized by increased alertness, defensive postures, piloerection, dilated pupils, and vocalizations, all of which can be described by the term rage. Many of these reactions are closely associated with wakefulness, alertness, and sleep. This suggests both excitatory and inhibitory connections with the reticular activating system. Neuroanatomical research has suggested that this is the case. There is also an important relationship of the hypothalamus with the autonomic system, which has been discussed previously. Activation of the sympathetic nervous system usually results in widespread effects on several body functions, including increased heart rate and arterial pressure and increased metabolic rate. On the other hand, activation of the parasympathetic system can result in much more specific action, such as increased or decreased heart rate, esophageal spasm, and increased peristalsis of the upper gastrointestinal tract. Finally, the relationship between hypothalamic structures and the anterior pituitary gland is important for hormonal control of this organ. These have also been discussed previously.

The amygdala consists of several nuclei located in the anterior pole of the temporal lobe just underlying the cortical gray matter. This structure is closely associated with olfaction, as one of the major divisions of the olfactory tract ends here. Human olfaction is not as highly developed as it is in other mammals, however. The amygdala receives inputs from most other areas of the limbic structures, and, as indicated above, projects to the hypothalamus where its activity will produce many of the behavioral responses discussed above. Stimulation of the amygdala can produce certain involuntary movements of the head or body. These include movements associated with eating, such as licking, chewing, and swallowing, circling movements, and occasionally clonic and rhythmic

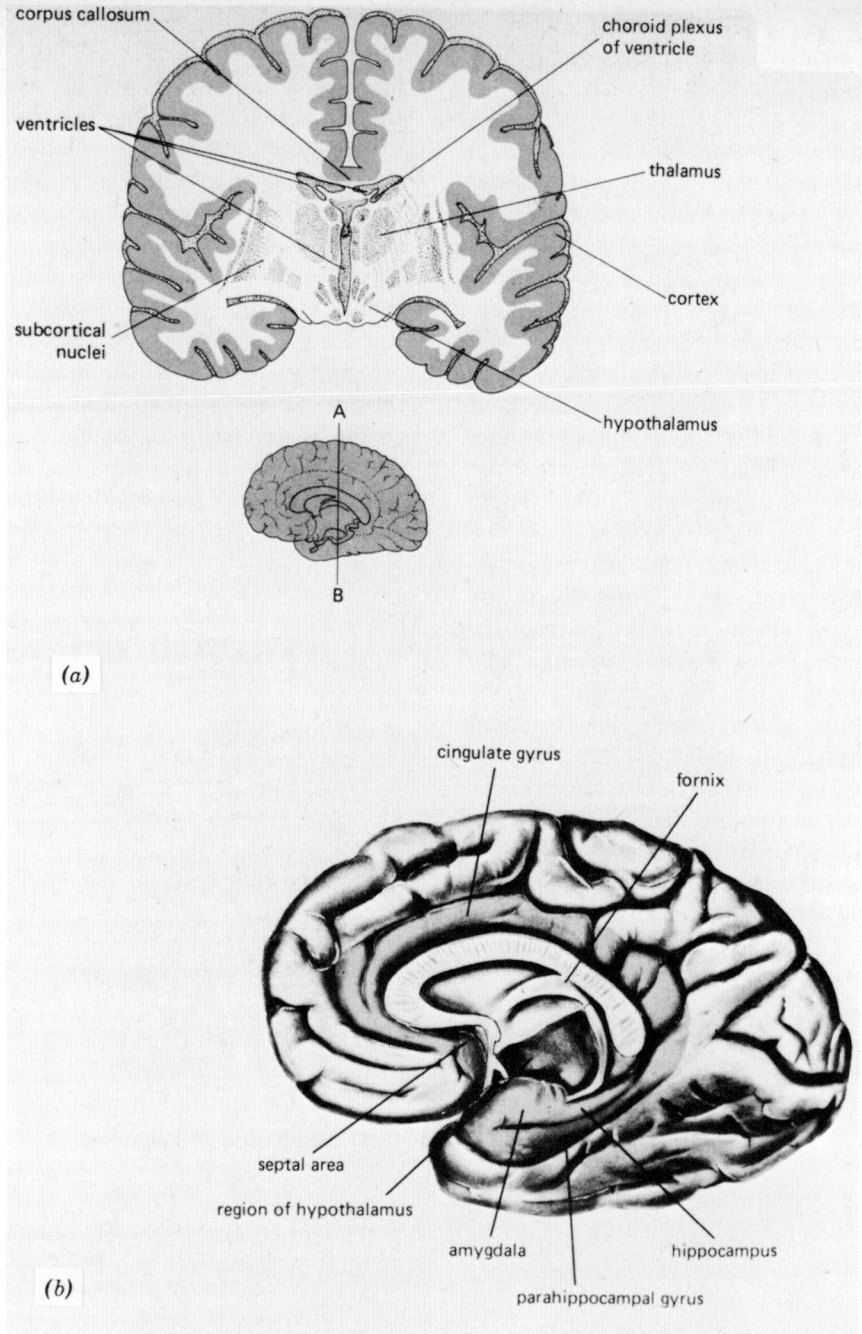

corpus callosum

choroid plexus
of ventricle

ventricles

thalamus

cortex

subcortical
nuclei

A

B

hypothalamus

(a)

cingulate gyrus

fornix

septal area

region of hypothalamus

amygdala

hippocampus

parahippocampal gyrus

(b)

Figure 1 *(a)* Midfrontal plane section through the brain showing structures of the limbic system. *(b)* View of a whole brain in the midsagittal plane showing the structures of the limbic system. (Adapted from: Vander AJ, Sherman JH, and Luciano DS, *Human Physiology The Mechanisms of Body Function,* 3rd ed., McGraw Hill, chap. 8, 1980)

movements, erection, copulatory movements, ejaculation, ovulation, uterine activity, and premature labor.

The function of the hippocampus has received a great deal of interested attention recently for two reasons: (1) the hippocampus is associated with memory and learning, topics of long interest to physiologists and behavioral scientists; and (2) it has been possible to produce isolated preparations of hippocampal tissue amenable to well-controlled electrophysiological manipulations with microelectrodes. Although these isolated preparations will probably yield little information about the function of the hippocampus, they should give detailed information about the kinds of transmitters used at specific synaptic junctions, as well as information about the synaptic organization of extrinsic and intrinsic inputs to hippocampal neurons.

The hippocampus is located just caudal to the amygdala with which it makes connections. It forms one of the major inputs to the hypothalamus and other parts of the diencephalon through the fornix. As with the amygdala, stimulation of the hippocampus produces many of the same behavioral patterns as stimulation of the associated hypothalamic structures. Of special interest is the fact that electrical stimulation of the hippocampus can produce focal epileptic seizures that result in visual, olfactory, auditory, and tactile hallucinations. As pointed out earlier, stimulation of certain hypothalamic structures produce

behavioral patterns associated with reward and punishment. It has been well established that when such affective emotional states are associated with learning paradigms that learning occurs more readily. The hippocampus is believed to play an important role in this ability to learn. How the hippocampus and its associated structures play this role remains, unfortunately, an unsolved mystery.

Associated with the limbic structures discussed so far is a ring of cortex about which we know very little. These cortical structures include hippocampal, cingulate, and orbitofrontal cortex. Ablation and stimulation studies of these areas have yielded little information because little change is seen following such procedures. These cortical areas are likely to be involved in associating a variety of cortical areas with the underlying subcortical limbic structures.

SELECTED BIBLIOGRAPHY

Brooks McC, Koizumi K: The hypothalamus and control of integrative processes, in Mountcastle VB (ed): *Medical Physiology,* ed 13. St Louis, CV Mosby, 1974.

Eyzaguirre C: The hypothalamus and the limbic system, in Eyzaguirre C, Fidone S (eds): *Physiology of the Nervous System,* ed 2. Chicago, Year Book, 1975.

SELF-STUDY QUESTIONS

1. All the following structures form part of the limbic system *except*

   A. hypothalamus.

   B. cinglulate gyrus.

   C. amygdala.

   D. hippocampus.

   E. thalamus.

2. All the following are important functions of the limbic system *except*

   A. memory.

   B. regulation of emotion.

   C. motor control of posture.

   D. regulation of behavior.

   E. olfaction.

3. The structure of the limbic system that has been shown to be most directly associated with memory and learning is

   A. hippocampus.

   B. septal nuclei.

   C. hypothalamus.

   D. basal ganglia.

   E. postcentral gyrus.

ANSWERS

1. E. The thalamus does not form directly part of the limbic system although there are direct connections between limbic structures and the thalamus. One example is the projections from the mamillary bodies to the anterior nucleus in the thalamus, which then project to cingulate cortex. This is to emphasize that limbic structures connect to many wide areas of the brain.

2. C. Although stimulation of certain limbic structures will produce movement, especially of the head and mouth region, these structures do not play an important role in the control of posture. Again, it is difficult to dissociate many of the behavioral patterns produced by limbic stimulation from structures that receive direct projections from the limbic system.

3. A. The hippocampus is one of the most well-studied structures of the limbic system. In part this is due to advanced techniques in removing the hippocampus from the animal and being able to work on it as an in vitro preparation. However, its role in memory and learning has been known for a long time.

41

The Autonomic Nervous System

ARTHUR M. BROWN

OBJECTIVES

After completion of this chapter, you should be able to:

1. Describe the difference between the autonomic and somatic outflows from a functional point of view.

2. Present the differences and similarities between the sympathetic and parasympathetic divisions of the autonomic nervous system.

3. Discuss the nature of synaptic transmission in these two divisions at ganglion and effector cell levels.

4. Briefly discuss denervation supersensitivity.

5. Describe central control mechanisms.

The autonomic nervous system (ANS), or visceral efferent nervous system, is that part of the motor outflow of the central nervous system (CNS) that is directed toward cardiac muscle, smooth muscle, and glandular cells. The latter are effector cells concerned with visceral function and homeostasis. The ANS is thus distinguished from the other motor outflow of the CNS, the somatic efferent nervous system that innervates skeletal muscle and is involved in somatic interactions with the environment. However, the two efferent systems are not mutually exclusive, and such body functions as exercise, eating, sleeping, and sexual intercourse are performed by patterns of nervous activity involving both systems. Whereas the somatic efferent nervous system is often directed by conscious activity, its activity can be performed subconsciously, as in postural adjustments. On the other hand, the ANS usually operates below the level of consciousness, although it is possible to learn to control heart rate by means of biofeedback techniques. Conscious control of other visceral function, such as blood pressure, is also possible.

The anatomy and physiology of the somatic efferent nervous system is relatively simple. The neurons that innervate the somatic muscles release an excitatory transmitter, acetylcholine (ACh), at well-developed neuromuscular synapses. The effector response is an excitatory one leading to contraction; inhibitory responses do not occur. The complex behavior of the system resides in the huge numbers of complex patterns of neural activity that are possible. By contrast, the anatomy and physiology of the ANS at the effector level is more complicated, but there are fewer patterns of nervous activity because function is restricted to homeostasis. These patterns may however be very specific or discrete, and they discount the earlier notion of mass or diffuse discharge of the ANS.

The differences between ANS and somatic efferent nervous system are shown in Figure 1 and can be summarized

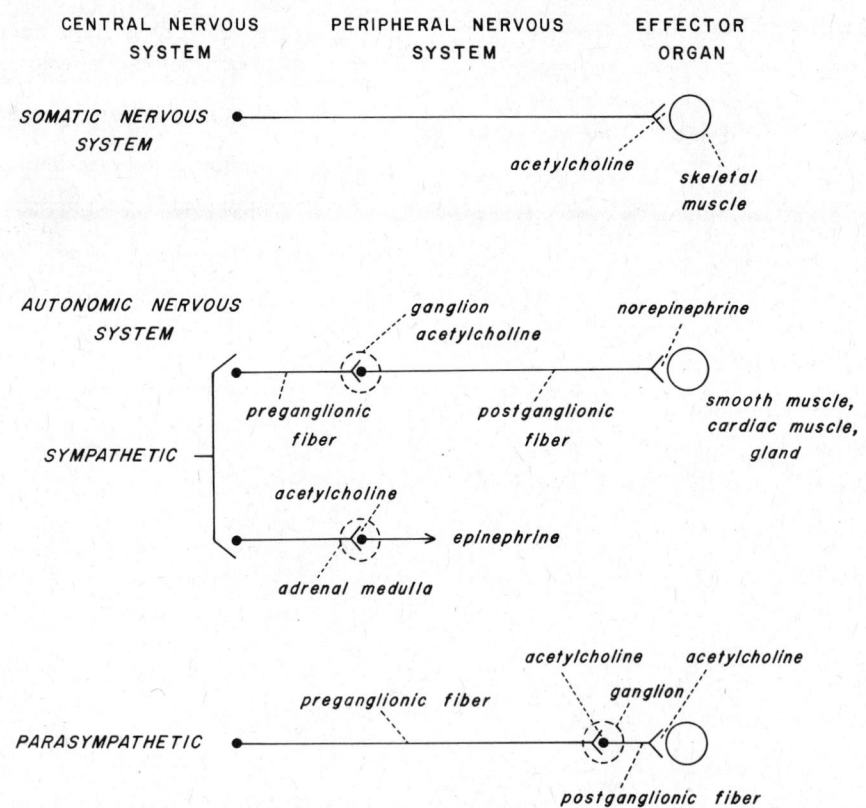

Figure 1 Efferent divisions of the peripheral nervous system.

as follows. Fibers of the somatic nervous system innervate skeletal muscle, cause it to contract, and do not make synapses with other nerve cells, once they have left the CNS. Fibers of the ANS synapse in ganglia with other nerve cells after they have left the CNS, innervate smooth or cardiac muscle or gland cells, and can produce excitation or inhibition of effector cells.

ORGANIZATION OF THE ANS

The ANS has two components: sympathetic and parasympathetic (Fig. 2). The cell bodies of the sympathetic component are located in the lateral horns of the thoracolumbar segments of the spinal cord, and the cell bodies of the parasympathetic component are located in the brain and sacral regions of the spinal cord. The preganglionic neurons of the sympathetic component make synapses with

neurons in the lateral (paravertebral), ganglia (sympathetic trunks), and collateral (prevertebral) ganglia (coeliac, aorticorenal, superior and inferior mesenteric ganglia). The axons of the ganglion cells then run to the organs containing the effector cells. The arrangement for the parasympathetic component is different. The preganglionic neurons make contact with neurons that are already located in the organ containing the effector cells of interest, and the postganglionic axons are restricted to that particular organ.

Many effector cells receive a dual innervation from the sympathetic and parasympathetic components. In the vast majority of cases the effects are opposite in nature and synergistic, since the two divisions are usually activated reciprocally. The result is a finer degree of control and a greater range of options than can be achieved by one component alone.

The effector cells and the ways in which the sympathetic and parasympathetic components of the ANS act on them are outlined in Table 1. These effects and the associated reflexes are discussed in chapters that deal with specific organ systems; in the present chapter only the broad principles involved in ANS function are presented.

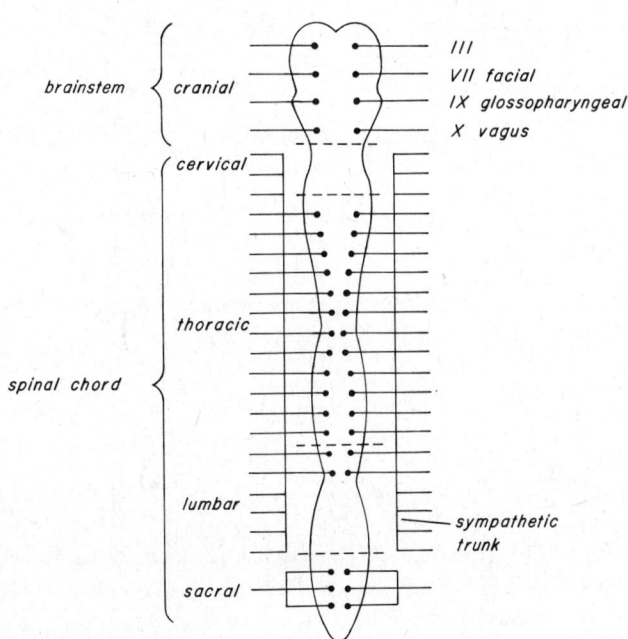

- ● parasympathetic
- ● sympathetic

Figure 2 Neutral control mechanisms. Origins of the sympathetic and parasympathetic divisions of the autonomic nervous system.

SYNAPTIC TRANSMISSION IN THE ANS

Synapses occur at two sites, the autonomic ganglia and the effector cells. The pre- and postsynaptic neurons are covered by glia or Schwann cells, except where the synapses occur. As it courses near the effector cells, the axon has varicosities bereft of supporting cells, and the postsynaptic membrane, although more distant than the endplate, is also bare. In the ganglia, one presynaptic neuron can innervate several postsynaptic neurons, so that spatial as well as temporal summation can occur. Spatiotemporal summation can also occur at the effector cells as a result of their dual innervation. The ganglia may also contain interneurons that have an inhibitory postsynaptic action on ganglion cells.

Synaptic action at autonomic synapses is similar in broad outline to that found in synapses located elsewhere in the nervous system. The presynaptic neuron releases a transmitter that diffuses to the postsynaptic membrane,

Table 1 Effects of Autonomic Nervous System Activity

Effectors	Sympathetic Nervous System	Parasympathetic Nervous System
Eye		
Muscles of the iris	Contract radial muscle (widens pupil)	Contract sphincter muscle (makes pupil smaller)
Ciliary muscle	Relaxation (tightens suspensory ligaments, thus flattening for lens far vision)	Contraction (relaxes ligament, enabling lens to become more convex for near vision)
Heart		
SA node	Increases heart rate	Decreases heart rate
Atria	Increases contractility	Decreases contractility
AV node	Increases conduction velocity	Decreases conduction velocity
Ventricles	Increase contractility	—
Arterioles		
Coronary	Constriction (direct effect only)	Dilation
Cerebral	Constriction	Dilation
Skin and mucous membrane	Constriction	—
Skeletal muscle	Constriction or dilation	—
Abdominal viscera and kidneys	Constriction	—
Salivary glands	Constriction	—
Penis or clitoris	Constriction	Dilation (causing erection)
Veins	Constriction	—
Lung		
Bronchial muscle	Relaxation	Contraction
Bronchial glands	Inhibit secretion	Stimulate secretion
Salivary glands	Stimulate secretion	Stimulate secretion
Fat cells	Stimulate lipid breakdown	—
Urinary bladder	Relaxation	Contraction
Uterus	Pregnant: contraction Nonpregnant: relaxation	Variable —
Stomach		
Motility	Decreases	Increases
Sphincters	Contraction	Relaxation
Secretion	Possibly inhibition	Stimulation
Intestine		
Motility	Decreases	Increases
Sphincters	Contraction	Relaxation
Secretion	Possibly inhibition	Stimulation
Gallbladder and ducts	Relaxation	Contraction
Liver	Glycogenolysis, gluconeogenesis	—
Pancreas		
Exocrine glands	Decreased secretion	Stimulate secretion
Endocrine glands (islets)	Inhibit insulin secretion, stimulate glucagon secretion	Stimulate insulin secretion
Reproductive tract (male)	Ejaculation	—
Skin		
Muscles causing hair to stand erect	Contraction	—
Sweat glands	Stimulates secretion	Stimulate secretion
Lacrimal glands	—	Stimulate secretion

*Adapted from Goodman LS, Gilman A: *The Pharmacological Basis of Therapeutics*, ed. 5. New York, Macmillan, 1975.

where it bombards a specific receptor. The receptor is coupled to an ionophore, and when the receptor is hit by the appropriate transmitter, a transient change in excitability of the post-synaptic membrane is produced.

The main transmitter in autonomic ganglia is acetylcholine (ACh); there are two different types of ACh receptor. These are the so-called muscarinic and nicotinic receptors. ACh produces a fast excitatory postsynaptic potential (EPSP) in the ganglion cells that can in turn produce an action potential. This ganglionic ACh receptor, which is excited by nicotine and blocked by hexamethonium, is called a nicotinic receptor. Presynaptic stimulation also produces an inhibitory postsynaptic potential (IPSP) as well as a late EPSP (Fig. 3). The late EPSP can be produced by the mushroom poison muscarine and is blocked by atropine. For this reason, the receptor is termed a muscarinic receptor. The time course of the late EPSP could be related to the remoteness of the receptors from the released ACh, making the diffusional delay significantly greater. Other mechanisms might be involved as well. For example, if the transmitter is not quickly removed, it can reactivate receptors from which previous

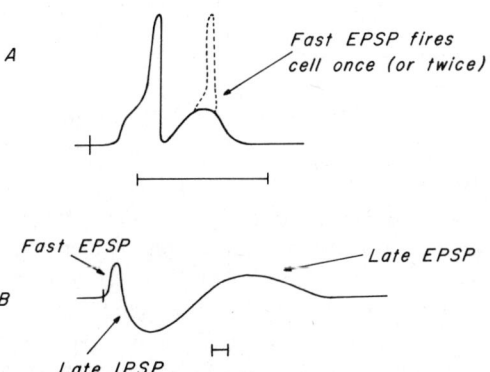

Figure 3 Responses of a postganglionic neuron to a preganglionic volley. *(a)* Response of a postganglionic neuron to a volley in preganglionic axons. The preparation is bathed with a physiological electrolyte solution. The chief response seen is an excitatory postsynaptic potential (EPSP) that elicits one or two action potentials. The latency of the EPSP is brief, as indicated by the time mark. *(b)* Response of the same neuron on a slower sweep. The time mark represents the same interval as in *a*. In this case, synaptic transmission is depressed by curare. The short latency EPSP follows the shock artifact. This is followed by an inhibitory postsynaptic potential (IPSP) and a late excitatory postsynaptic potential.

transmitter molecules have escaped. The IPSP is probably produced by a different transmitter, either norepinephrine (NE) or dopamine. It may result from synaptic action of the interneurons restricted to the ganglia. These interneurons have cytoplasmic granules that fluoresce when they contain NE or dopamine; they are sometimes called small intensely fluorescent (SIF) neurons.

ACh is the transmitter for the synaptic action of sympathetic preganglionic neurons on the chromaffin cells of the adrenal medulla. These cells are depolarized by preganglionic action potentials and release epinephrine and NE as a result.

ACh is also the transmitter at the effector cells for the parasympathetic postganglionic neurons; it is for this reason that the parasympathetic division of the ANS is sometimes referred to as cholinergic. At effector cells, the agonistic action of ACh is opposed by atropine, hence the receptors are muscarinic. The effects on excitability are diverse (Table 1) because similar agonist–receptor combinations can be coupled to ionophores that may differ greatly. For sinoatrial (SA) nodal cells in the heart, an increase in potassium permeability is produced. This hyperpolarizes the membrane and slows the heart rate, that is, the action is inhibitory. A reduction in membrane calcium current may also occur. For the radial smooth muscle cells of the iris, an EPSP is produced, indicating that an increased permeability to potassium cannot have occurred; contraction ensues and the pupil is constricted.

Most of the postganglionic neurons of the sympathetic outflow release NE as the chemical transmitter, and for this reason the sympathetic division has often been referred to as adrenergic. There is one exception, namely sympathetic neurons innervating the proximal limb vessels of certain vertebrates that release ACh. These are discussed in the cardiovascular section. In adrenergic neurons, NE is synthesized in the nerve from tyrosine via dopamine, as shown in Figure 4. NE is stored in dense-core vesicles, to be released when the axon is invaded by an action potential. The NE diffuses to the receptor site; its subsequent fate is more complicated than that of ACh. Part of the NE diffuses away in the bloodstream, where a small amount is enzymatically degraded by catechol-*o*-methyltransferase (COMT) present in the postsynaptic cell, but the greatest amount is taken up again by the presynaptic axon for subsequent storage in vesicles. An equilibrium reaction takes place between the NE in the

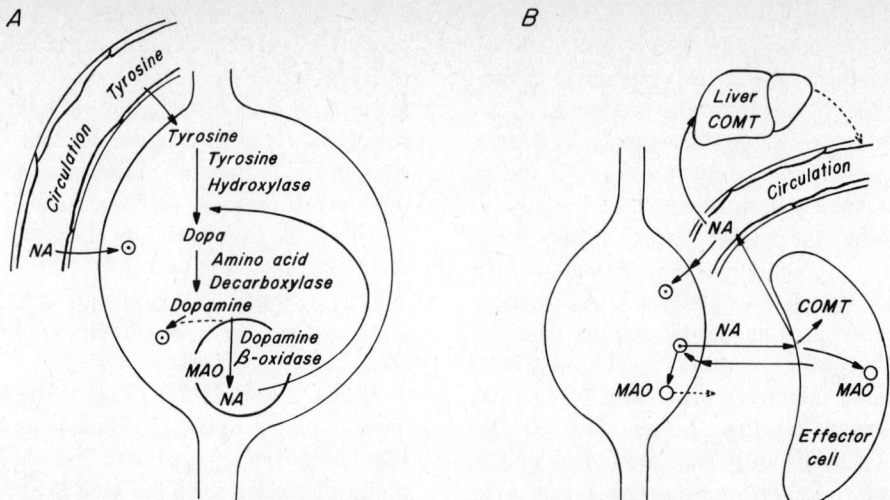

Figure 4 Biosynthesis of norepinephrine (noradrenalin) in sympathetic nerve. NA, noradrenalin; MAO, monoamine oxidase. Note storage granules represented by small circle with central dot, enlarged in lower portion of figure to show processes occurring therein. *(a)* Note feedback action of NA on formation of dopa and transport of NA between blood vessels and neuron. *(b)* Fate of noradrenalin released in sympathetic nerves and tissues. COMT, catechol-o-methyltransferase. Dotted line represents inactivated NA. Note multiple pathways of movement of NA. (From Axelrod J, Kopin IJ: The uptake, storage, release and metabolism of noradrenaline in sympathetic nerves, *Prog Brain Res* 31:21, 1969.)

cytoplasm and that stored in the vesicles, and excess cytoplasmic NE is inactivated by deamination. This is done by the enzyme monoamine oxidase or MAO. The NE that spills over into the circulation is degraded in the liver by COMT.

The synaptic action of NE on smooth muscle and gland cells is complicated and a simplified version is presented here. For a more complete account, a recent review by Wiener is suggested. On the basis of the rank order of potency of effector cell responses, the three catecholamine agonists norepinephrine (NE), epinephrine (E), and isoproterenol (I), activate two different receptors, α- and β- receptors. Each category has two receptor subsets: α_1, α_2 and β_1, β_2. The potency sequences for the different receptors are as follows:

$$\alpha_1, \alpha_2, \text{NE} = \text{E} >> \text{I}$$
$$\beta_1, \text{I} >> \text{E} \cong \text{NE}$$
$$\beta_2, \text{I} >> \text{E} \cong \text{NE}$$

α_1-receptors are located on smooth muscle cells and produce contraction (Table 2). Phenylephrine is a strong α_1-agonist, and phenoxybenzylamine is a potent α_1-antagonist. α_2-Receptors are located pre- and postsynaptically. The presynaptic α_2-receptors act to inhibit the release of NE from the presynaptic endings. Certain drugs such as clonidine are more potent as α_2-agonists than they are as α_1-agonists, and the antihypertensive action of this compound might be related to this effect. One α_2-antagonist is the drug yohimbine. β_2-Receptors are also located on smooth muscle, but they produce muscular relaxation.

Table 2 Norepinephrine Receptor Pharmacology

	Agonist	*Antagonist*
α_1	Amidephrine or phenylephrine	Prazosin or phentolamine
α_2	Amidephrine or clonidine	Yohimbine
β_1	Isoproterenol	Practolol
β_2	Salbutinol	

The effects of NE on blood flow may therefore be complicated. When NE is perfused into a vascular bed vasoconstriction results. When α_1-receptors are blocked, β_2-receptor effects are unmasked and a small vasodilation effect occurs. The reversal of the response in the presence of α_1-antagonist is greater for E than for NE, as the above response sequence indicates. β_1-Receptors are located in the heart and certain endocrine organs. Activation of β_1-receptors is accompanied by a stimulatory or secretory response; for example, in the heart contractility and heart rate are increased, and in the pancreatic β cells insulin secretion is increased.

The mechanisms whereby adrenergic receptors alter the state of effector cells are unknown. In heart muscle, the increase in cardiac contractility (β_1 effect) might be attributable to increased Ca entry during an action potential. The increased heart rate, also a β_1 effect, may be due to an increase in the rate of inactivation of the outward membrane current. In both cases the effects might not be limited to the sarcolemma, but could also involve changes in intracellular cyclic adenosine monophosphate (cAMP). In smooth muscle cells the α_1 effects might be mediated by increases in Ca current, but again this has not been proved. Moreover, the increase in Ca current could result from membrane depolarization produced by effects of NE on noncalcium membrane channels. The mechanisms for α_2 and β_2 effects are even less certain, although reductions in membrane Ca currents could be involved.

DENERVATION SUPERSENSITIVITY

After denervation, the receptors on ganglion cells and effector cells become supersensitive to application of the normal transmitter. There are several possible mechanisms for this effect. For ganglion cells, it appears that an increase in the number of ACh receptors occurs. For adrenergic receptors, the absence of presynaptic endings that take up the greater part of the released NE is probably a major factor. Other mechanisms would include alterations in receptor–agonist affinity and receptor–ionophore coupling.

AUTONOMIC REFLEXES

There are two classes of autonomic reflexes: those that are restricted mainly to homeostasis of the internal environment and those associated with somatic responses that involve a change in the state of internal body function. The first category includes reflexes involving blood pressure, airway size, intestinal movements, and pupillary size. The input for the reflexes comes almost exclusively from enteroreceptors. The second category includes respiration, micturition, coitus, responses to exercise, and noxious stimuli. The input for the reflexes comes from somatic and visceral afferents. The responses directed toward homeostasis are simple negative feedback operations directed toward maintaining resting setpoints. The autonomic reflexes involved in somatic responses also operate in a negative feedback manner, but generally the setpoint or state of the system has been altered or reset, probably as a result of changes in the CNS. A peripheral mechanism whereby resetting can occur is the possibility that the autonomic outflow might reset the receptors for visceral reflexes; that is, the baroreceptors may be reset by the sympathetic nerves to the carotid sinus or aortic arch.

CENTRAL CONTROL OF AUTONOMIC FUNCTIONS

The autonomic outflow can be altered by afferents from visceral or somatic receptors and by input from the brain. However, in the absence of these inputs, preganglionic neurons might still be actively discharging. It appears that tonic activity might be an intrinsic property of autonomic neurons.

Regulation of this tonic activity is possible at all levels of the CNS. The simplest form occurs in the spinal cord. Acute transection leads to suppression of all reflexes. This is followed by a chronic stage in which organ system reflexes persist, but regulation is incomplete. Large swings in blood pressure occur, and "mass" reflexes involving somatic withdrawal responses, defecation, and micturition occur. Bladder emptying is incomplete. The mechanism for these large transients, sometimes referred to as hyperreflexia, is not known, but must involve a reduction of an inhibitory input upon autonomic neurons from supraspinal neurons.

Many of the major reflex actions of the autonomic system are carried out in the medulla oblongata. This is because a large amount of efferent information carried in the cranial nerves, particularly 1X and X is integrated at this site. A bulbospinal animal exhibits blood pressure and heart rate control characterized by reciprocal actions of the sympathetic and parasympathetic division and respiration is maintained in a reasonable fashion. Much of the tonic activity of autonomic neurons results from excitatory input from medullary neuronal pools, such as the pressor and depressor center. However, the bulbospinal preparation does not exhibit the full range of autonomic responses seen as part of the total response involving somatic reflexes. For this an intact brain is required.

The hypothalamus is sometimes referred to as the head ganglion of the autonomic nervous system. This is because it is essential for thermal regulation and those autonomic reflexes, such as sweating, skin blood flow, and cardiac output important for maintaining body temperature. The hypothalamus also plays an essential role in stress, in the so-called fight-or-flight responses, in rage, and in various emotional states. The hypothalamus is also involved in hormonal regulation, an important mechanism for homeostasis. The role of the hypothalamus in these functions is more fully discussed in Chapter 45.

The limbic system, cerebellum, and cerebral cortex also project onto the autonomic system via brain centers and descending pathways. However, the significance of the projections from these areas is not well understood.

SELECTED BIBLIOGRAPHY

Axelrod J, Koplin IJ: The uptake, storage, release and metabolism of noradrenaline in sympathetic nerves. *Prog Brain Res* 31:21, 1969.

Weiner N: Norepinephrine, epinephrine, and the sympathomimetic amines, in Gilman AG, Goodman LS, Gilman A (eds): *The Pharmacological Basis of Therapeutics,* ed 6. New York, MacMillan, 1980, p 138.

SELF-STUDY QUESTIONS

MULTIPLE CHOICE

Select the correct answer(s). (In many instances, more than one answer is correct.)

1. Synaptic transmission at sympathetic ganglia
   A. is electrical.
   B. involves release of ACh presympathetically.
   C. is mediated by adrenergic receptors exclusively.
   D. involves nicotonic or muscarinic receptors, or both.

2. The autonomic outflow
   A. has at least two synapses outside the spinal cord.
   B. is identical to the somatic outflow with respect to peripheral synapses.
   C. does not originate in the spinal cord exclusively.
   D. has only a single chemical transmitter.

3. The cardiac motor vagus
   A. has its target cell action on muscarinic receptors.
   B. has its ultimate peripheral synaptic action on nicotonic receptors.
   C. increases K conductance in cardiac pacemaker cells.
   D. arises in the hypothalamus.

4. Adrenergic receptors
   A. can be activated by at least four types of agonists.
   B. are probably coupled to a number of different membrane proteins ionophores.
   C. (if α_1) can have amidephrine or phenylephrine as agonists.
   D. (if β_1) can have isoproteronol as an agonist.

5. Some characteristics of the autonomic nervous system are that

A. it operates mainly at conscious levels.

B. it has one major subdivision originating in cells at C_1–C_5 of the spinal cord.

C. its components usually act antagonistically.

D. it receives a large input from the hypothalamus.

6. Denervation supersensitivity could be related to

A. increased numbers of receptors.

B. enhanced receptor–ionophore coupling.

C. reduced uptake mechanism, because of absence of adrenergic endings.

D. decreased receptor affinity for transmitters.

7. The sympathetic nervous system

A. increases stomach motility.

B. relaxes intestinal secretion

C. produces constriction of the gallbladder and ducts.

D. produces hepatic glycogenolysis.

8. The parasympathetic outflow

A. causes contraction of bronchial muscle.

B. increases stomach motility.

C. decreases atrial contractility.

D. causes contraction of the urinary bladder.

ANSWERS

1. B, D. Electrical transmission at synapses occurs rarely, and never between sympathetic preganglionic neurons and sympathetic ganglion cells. Transmission in chemical, cholinergic, and postsynaptic receptors can be nicotinic or muscarinic, or both.

2. A, C. The autonomic outflow has synapses in the peripheral ganglion and at the effector or target organ cells, whereas the somatic system has a single peripheral synapse at the target cell. The parasympathetic division has a cranial component, as well as a spinal component.

3. A, C. The vagal outflow is blocked by atropine, which acts at the postsynaptic receptor, making it muscarinic. The K conductance in SA nodal cells is increased by ACh, and the Ca conductance may be reduced. The cells of origin for cranial nerve X are located in the medulla.

4. All are correct. The agonists are presently classified as α or β, with subscripts 1 and 2 for each. The effects of membrane currents vary widely with α_1-agonists probably increasing Ca and Na conductance and α_2-agonists possibly reducing Ca conductance. β_1-agonists enhance Ca current in heart muscle and shift I–V relationships for K conductances in pacemaker cells, so

that the rate of turnoff is faster at a given potential.

5. D. The ANS is mainly concerned with with automatic or homeostatic regulation that occurs at subconscious levels (e.g., they work during sleep). The components act synergistically; for example, subthreshold effects can be summed to produce a response that might not occur if only one component were acting. The hypothalamus is sometimes called the head ganglion of the ANS.

6. A, B, C. There is evidence that innervation reduces the number of receptors and localizes them to the region of the synapse. Another possibility, enhanced receptor–ionophore coupling, could involve a "membrane particle" that activates the ionophore. This is possible for β-receptor activation of Ca channels, which can be antagonized by ACh. β-agonistic action might involve adenylate cyclase activation. Apparently the cholinergic receptor can prevent the membrane particle from activating adenylate cyclase. There is no evidence that receptor affinity is altered by denervation, although the proportions of different receptor types (i.e., muscarinic nicotinic ratio) may be changed.

7. D. See Table 1.

8. All are correct. See Table 1.

42

The Human Eye

KEN-ICHI NAKA
RAYMOND Y. CHAN

OBJECTIVES

After completion of this chapter, you should be able to:

1. Illustrate the anatomy of the eye.
2. Describe the process of photochemical reactions in the receptors.
3. Describe the functional morphology of the retina and visual pathway.
4. Provide an understanding of psychophysical phenomena related to vision.
5. Explain visual optics and visual fields.
6. Describe functions of extraocular muscles and control of the eye movement.

The eye shaped in the form of a globe of about 3 cm in diameter sits in a conical cavity in the skull, the orbit. The eyeball, as seen in Figure 1, has three coats or shells. The sclera, the outermost shell, covers the posterior two-thirds of the eye; it is a dense fibrous membrane, part of which is exposed through the conjunctiva as the white of the eye. The innermost shell is the retina that contains various types of nerve cells including the photoreceptors. Between the sclera and the retina is the choroid, the middle layer, that contains blood vessels that supply nutrients to the receptors, and a sheet of cells filled with dark pigment, the pigment epithelium that absorbs extra light.

Entering the eye through the cornea, light travels across the anterior chamber, the lens, the vitreous, and the retina before reaching its final destination, the receptors, the photon-catching cells (Fig. 1). The anterior chamber is the space between the cornea and the iris; the posterior chamber is the space between the iris and the lens. The aqueous humor fills the two chambers. The posterior segment cavity, the vitreal chamber, makes up the bulk of the eyeball and is filled with the vitreous humor. In the back of the eyeball is a hole through which the optic nerve, a bundle of axons from the retinal ganglion cells, exits, and through which the blood vessels enter and leave the eyeball. The exit–entry area lacks retinal photoreceptors and is called the optic disc, or blind spot. The area centralis, a depression having a diameter of about 0.5 mm (or 5 degrees), is located about two disc diameters temporal to the temporal edge of the optic disc. This area has the most dense population of cones and provides an area of very high visual acuity; it is also used for direct gaze. The fovea centralis, which surrounds the area centralis, is about 1.5 mm in diameter and is covered with a yellow pigment, the macular lutea. The rest of the retina is loosely called the periphery in which rods, the receptors for night or scotopic vision, are predominant.

OPTICS OF THE EYE

The eye is equipped with optics to form an image of the outside world, an iris to control the amount of light entering into the eye, as well as the degree of the depth of focus and an array of receptors to capture photons. Optically the

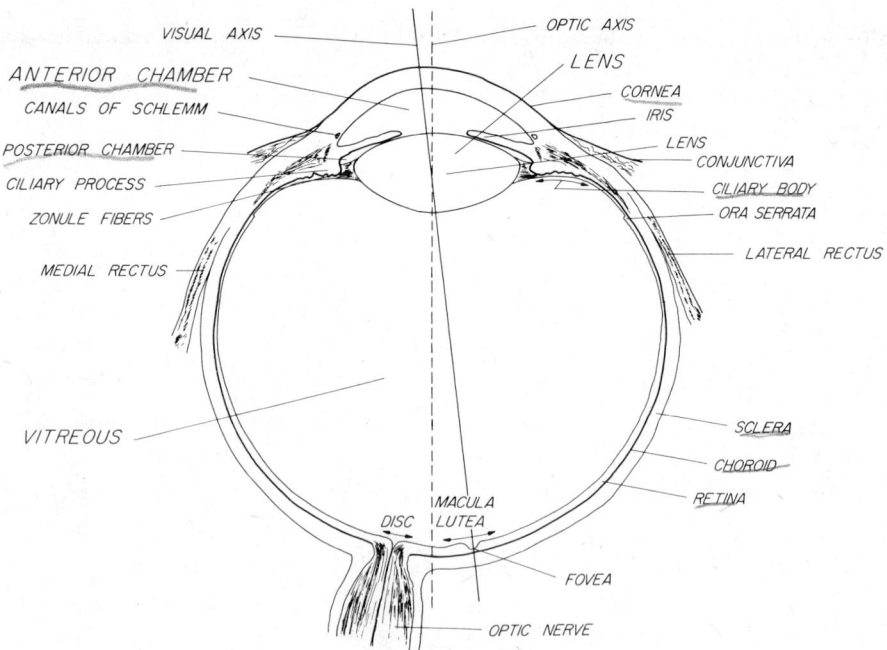

Figure 1 Horizontal section of right human eye.

average human eye is equivalent to a curved surface with a radius of 5.6 mm placed 22.2 mm in front of the optical center of the retina, the fovea. The simplified optics focuses an object located at infinity onto the retina and is called the reduced eye. In the human eye the cornea is largely responsible for light refraction (or convergence), and the lens plays a minor part in it. The human cornea has a refractive power of +45.0 diopters (1 per focal length in meters), and the average human lens has only a maximal refractive power of +15.0 diopters. The primary function of the lens is accommodation, an adjustment of the eye's dioptric power to focus clearly objects at various distances from the eye.

CORNEA

The cornea is a transparent window in front of the eye and is a structural continuation of the sclera. The corneal stroma, a lamellar structure, consists of submicroscopic collagen fibrils arranged in an organized fashion to make it transparent to light. The cornea is avascular; its oxygen and nutrient supplies, CO_2 loss, and waste product removal are by means of diffusion. The outer surface of the cornea, the corneal epithelium, is the interface between the air and eye, and is protected by lids and lubricated by tears. The epithelium is continuous with the epithelial layers of the conjunctiva. The corneal transparency is maintained by intact corneal endothelium and epithelium, by a proper intraocular pressure as well as by a proper osmotic equilibrium with the surrounding fluids.

LENS

The crystalline lens, an avascular structure, is made of viscous protein fibers that are transparent and continuously renewed. However, the volume of the lens remains little changed throughout life. The lens is encapsulated in a strong elastic outer membrane that controls its shape; when no tension is applied, the lens tends to form a spherical shape. The outer edge of the membrane is attached to

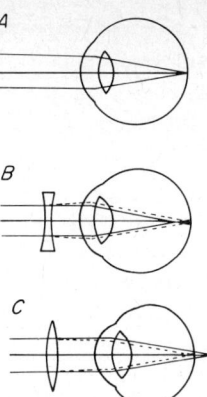

Figure 2 Schematic representation of optics of emmetropic (normal) eye *(a)*, myopic (nearsighted) eye *(b)*, and hyperopic (farsighted) eye *(c)*. Dotted lines show corrections of refractive errors with lenses of negative *(b)* and positive *(c)* powers.

about 70 suspensory ligaments (zonules) that, in turn, are attached to two sets of ciliary smooth muscles: the meridional and circular fibers. Their contraction will relax the ligament to make the lens more ball-like (more dioptric, shorter focal distance), and their relaxation will lead to a flattening of the lens to make it less dioptric (Fig. 2). As in the cornea, intake and removal of gases, nutrients, and waste occur by means of diffusion.

VITREOUS

The vitreous that fills the posterior segment cavity is a viscous, gel-like fluid. Collagen and hyaluronic acid are its main chemical constituents. It also contains water, a high level of Na^+ ions, and a low level of K^+ ions, and is therefore considered an extracellular fluid.

AQUEOUS

The aqueous humor fills the anterior and posterior chambers and performs two functions: (1) it provides, together with the vitreous, an adequate intraocular pressure (12–20 mm Hg) to keep the spherical shape of the eyeball; and (2) by diffusion, it provides nutrients to the lens and cornea, and removes waste. The formation and drainage of the aqueous are important means of keeping the integrity of the optics of the eye.

The aqueous is formed through active processes from the blood plasma and is secreted into the space between the lens and iris (i.e., posterior chamber). Circulation of the aqueous is a passive process depending on the temperature gradient within the eye. The aqueous drains through a channel located near the junction between the sclera and the cornea into the canal of Schlemm.

dional and the circular fibers. The fibers are controlled by the ciliary ganglion, a para-sympathetic nerve. Contraction of the two fibers slackens the tension on the zonular fibers, so that the lens becomes balloon-like by its own elasticity, a condition of the maximal dioptric strength of the lens and of near vision. A relaxation of the fibers gives rise to a tension on the zonular fibers to produce flatter surfaces, especially the anterior surface, a condition of the minimal dioptric strength of the lens and, therefore, of a farsighted vision (Fig. 2). The process of controlling the shape of the lens, and resulting changes in the eye's dioptric strength are called accommodation.

ACCOMMODATION

The crystalline lens is suspended around its edge by means of about 70 ligaments, the zonule of Zinn, from the ciliary body that contains two sets of smooth muscles, the meri-

PUPILLARY APERTURE

The aperture of the iris, the pupil, is controlled by the two sets of muscles in the iris, the sphincter and the dilator. The innervation of the former is parasympathetic and of

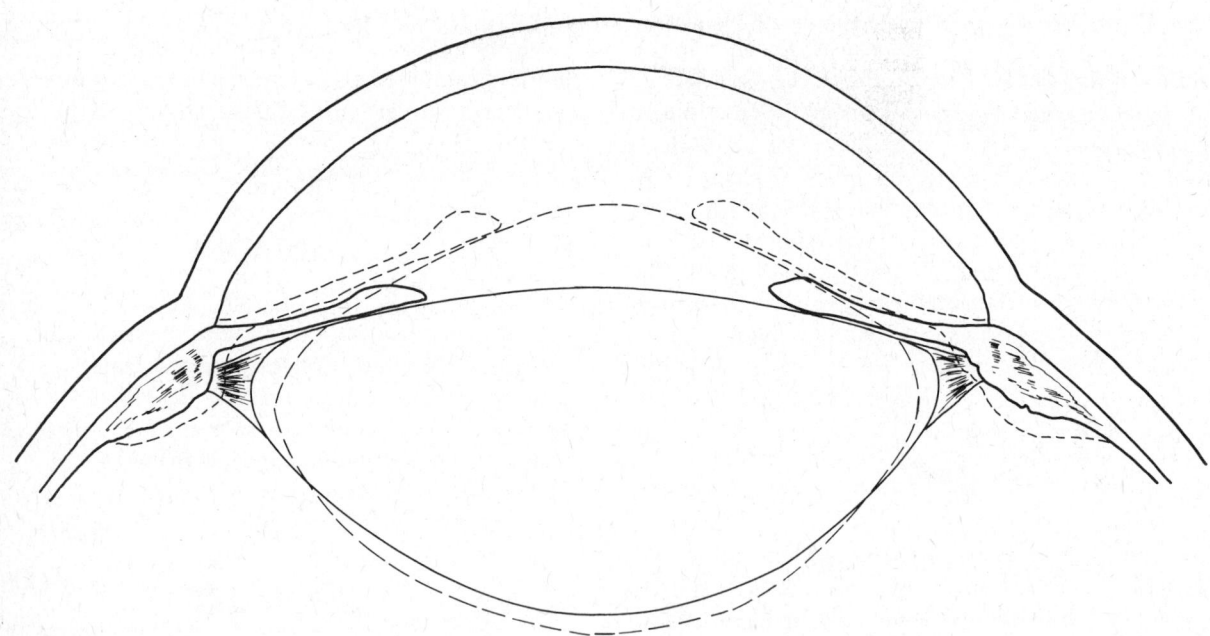

Figure 3 Schematic representation of changes in shape and position of the lens during accommodation. Continuous lines represent the relaxed condition, in which an object at the infinity is in focus. Broken lines represent accommodated condition, in which a near object is in focus.

the latter, sympathetic. Contraction of the sphincter and/ or relaxation of the dilator results in the pupil's constriction or miosis; the opposing processes result in the pupil's dilatation or mydriasis. As in the case of a camera, a small aperture of the lens (miosis) increases the lens' depth of focus, and a large aperture (mydriasis) produces a very shallow depth of focus. The normal diameter of the pupil varies from 8 mm in mydriasis, a condition that can normally be produced when a person stays in darkness, to 2 mm in miosis, a condition achieved when a person stays in very bright surroundings, as on the beach. The range corresponds to a 4-F-stop change in a lens' aperture or roughly to a change in illuminance when one wears a pair of sunglasses.

Pharmacological agents that produce miosis are called miotics and can be either cholinergic agonist or adrenergic blocker. Those that produce mydriasis are mydriatics and are either adrenergic agonist or cholinergic blocker.

ERRORS OF REFRACTION

A normal eye with a relaxed accommodation (i.e., with the most tension applied to the lens) will focus a distant object on the photoreceptor array, located about 22.2 mm behind the corneal vertex. This condition is called the emmetropia, an absence of refractive error (Fig. 2a). Often there will be a mismatch between the focal length of the optics and the axial length of the eye: emmetropia no longer exists and gives way to a condition referred to as refractive error or as ametropia. There are two kinds of error: (1) myopia, in which images form in front of the retina (Fig. 2b), and (2) hyperopia, in which images form behind the retina (Fig. 2c). Myopia results either when the optical refraction of the eye is more powerful (i.e., a shorter focal length for a given eyeball), or the axial length of the eyeball is too long. In myopia a sharp image is formed some distance in front of the retina, and images on the retina are blurred. The nearest point that a myopic eye can form a sharp image on its retina is called the far point. In the normal eye the point is at infinity. The degree of myopia in diopter is the reciprocal of the distance in meters between the far point and the eye. Errors of refraction

in a myope can be corrected by placing a divergent or minus lens in front of the eye so that an object at infinity looks to the eye as if the object is at the far point. When the axial length of an eye is too short, relative to the focal length of its optics, hyperopia results (i.e., the dioptic power of the optic is too small). A hyperope can be corrected by adding more refractive power to the eye—an addition of a convergent or plus lens.

The third kind of refractive error is astigmatism in which the corneal surface is not perfectly symmetrical. The corneal curvature might be different for the two meridians to produce different dioptric strengths for the horizontal and vertical optical planes. In such a case, either a horizontal or vertical image will be sharply focused on the retina, and the other image will be blurred. A cylindrical lens is used to correct this imbalance in dioptric powers.

VISUAL ACUITY

FIELD OF VISION

All objects having a similar size on the retina are contained in a cone, or ray formed by the eye and the objects in space; the size of the image on the retina is uniquely defined by the cone. This cone of light is to be distinguished from the class of retinal photoreceptors known as cones, described in a subsequent section. It is usual practice, therefore, to indicate a distance on the retina in terms of the aperture of the cone in degrees. An object that produces a 0.3-mm retinal image subtends in a cone with an aperture of 1 degree. The width of the optic nerve head or blind spot is about 1.5 mm on the retina, but it is usually referred to as 5 degrees.

When a subject is asked to gaze at a small point, the subject sees it through the optic pathway, which joins the point and the fovea; the pathway is called the visual axis and the point the fixation point. The line that joins the center of the pupil perpendicular to the cornea is called the geometrical (or optical) axis of the eye and usually does not coincide with the visual axis (Fig. 1). The visual field is defined as the portion of a space within which objects are visible simultaneously to an eye steadily fixating on one direction, a fixation point. (Both central and peripheral visions are included in the visual field.)

Any optics, including the eye, do not produce a point image of a point source, but rather produces its somewhat blurred image. Even when the optics are perfect, the blurring of a point image results, due to a diffraction pattern called the point-spread function. When the eye's pupil is small, its optical aberration is the diffraction pattern, and an eye with a small pupil is almost a perfect optical instrument. When a pupil is larger than 2 mm in diameter, the diffraction is no longer the optics' limitation, but other aberrations become a major source of blurred images.

The smallest separation of two point sources that an eye can discriminate is called its visual acuity, and is 1 min or one-sixtieth of an arc, or 5 μm on the retina. The fovea has the best visual acuity.

NEURAL RETINA
(SENSORY RETINA)

VISUAL PIGMENT

The eye's optics form a sharp image of an object on the photoreceptor array. For the object to be seen, photons radiating from it have to be caught by the photoreceptors (rods and cones). Photoncatching pigments or photosensitive pigments are located in the outer segments of the receptors and form a stack of minute discs enclosed in rods (like a stack of pancakes) and are infoldings of the outer-segment membrane in cones. It is estimated that more than 80% of an outer segment's protein forms the pigment.

The human retina contains four kinds of photosensitive pigments, one for the rod or scotopic vision, and the remaining three for color or photopic vision. Human rods are the most sensitive to light of 500 nm (blue-green), and the three cones are most sensitive to lights of 450 nm (blue), 535 nm (yellow-green), and 580 nm (orange-red). The three cones correspond to the colors blue, green, and red—and are commonly referred to as the blue, green, and red cones; however, they are more properly the blue-, green-, and red-photon-catching cones or pigments.

Little is known about photosensitive pigments in human cones, but the pigment of the rod has been studied extensively; rod pigments in many animals are identical to that in the human and are called rhodopsin, or visual purple. An unbleached solution of rhodopsin appears purple because rhodopsin catches blue-green light; when the solution is exposed to light, it becomes colorless—a process referred to as bleaching of rhodopsin.

Rhodopsin is composed of retinal, an aldehyde of vitamin A, and opsin, a lipoprotein. Opsin combines only with the 11-cis isomer of retinal. When a rhodopsin molecule absorbs a photon, the 11-cis isomer is photoisomerized into an all-trans isomer, a bathorhodopsin (formally known as prelumirhodopsin). The isomerization is followed by a chain of many spontaneous reactions, ending in the production of all-*trans*-retinal and free opsin. Most of these reactions take only a fraction of milliseconds to complete. Somewhere between the isomerization of rhodopsin and the final stage, a reaction occurs that triggers the neural events of vision; however, the reaction is not as yet known.

In order for us to be able to keep seeing, all-*trans*-retinal has to be isomerized back to the 11-cis form, or a new supply of 11-*cis*-retinal has to be brought in. A mixture of all-*trans*-retinal and opsin turns into rhodopsin through an intermediate stage of a mixture of 11-*cis*-retinal and opsin by means of retinal isomerase, an enzyme known to exist in the pigment epithelium (Fig. 4). Vitamin A is stored as its ester; it is converted into a 11-cis form in the liver and is carried to the pigment epithelium through the blood supply. The chemical circuits involved in the for-

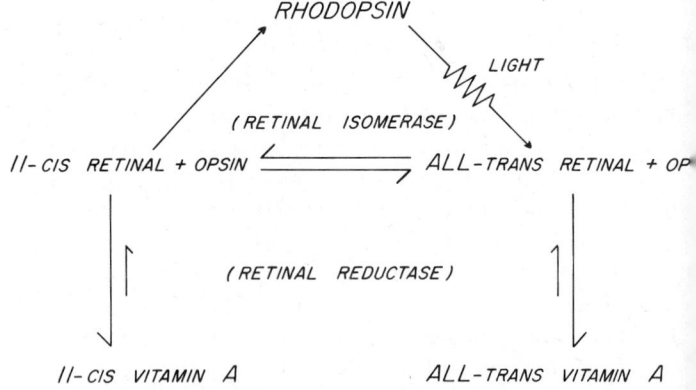

Figure 4 Simplified chemical circuit for production and degeneration of rhodopsin.

mation and decomposition of rhodopsin are shown in Figure 4.

NEURAL RETINA

The neural retina is about 250 μm in thickness, one-quarter the thickness of a piece of paper, and designed to transform information contained in the optical image cast on the receptor array into a series of spike discharges carried by the optic nerves to the lateral geniculate body in the brain.

The retina has a layered structure that is composed of three nuclear layers and two plexiform (synaptic) layers. The outer nuclear layer contains the cell bodies of photoreceptors, the inner nuclear layer the cell bodies of bipolar, horizontal, and amacrine cells, and the third layer the cell bodies of ganglion cells. The outer plexiform interconnects the outer and inner nuclear layers, and the inner plexiform layer interconnects the inner nuclear layer with the layer of ganglion cells. The general layout of cellular organization of the retina is shown in Figure 5. The bipolar and ganglion cells provide a direct pathway of information, and the horizontal and amacrine cells spread information laterally within the retina.

NEURAL PROCESSES IN THE RETINA

Absorption of photons produces a hyperpolarization of receptors, that is, potential inside a receptor becomes more negative, and such a polarization is maintained as long as light shines. The receptor's polarization is transmitted to the two second-order neurons, the horizontal and bipolar cells, to produce changes in their membrane potential.

When only one kind of photosensitive pigment is stimulated, the amplitude of the resulting polarizations in the receptor and horizontal cells fit the Michaelis-Menten equation:

$$V/V_{max} = I/(I + I_{1/2})$$

where V is the amplitude of polarization, V_{max} the maximal amplitude of a cell's response, I the illuminance, and $I_{1/2}$, the illuminance required to produce the half-maximal response.

The second-order cells either hyperpolarize or depolarize, depending on how many kinds of receptors are stimulated (a function of the spectral composition of light) and on the stimulating light's spatial distribution. Most bipolar cells form a spatial organization known as a concentric receptive field in which stimulation of the field's center and periphery produces opposing changes in the cells' membrane potential (Fig. 6). This antagonistic influence arises either from the center or the periphery of a cell's concentric receptive field. The concentric receptive fields have been found in the ganglion cells as well as in the cells in the lateral geniculate; they are the basic building blocks of information transmission in the visual system.

The proximal layer of retina consists of two kinds of cells, the amacrine and ganglion cells. The ganglion cells have axons that carry the retina's output (information) to the lateral geniculate as a series of spike discharges. Little is known about the amacrine (literally nonaxon) cells; they may or may not produce spike discharges. It is likely that the cells are involved in complex signal transformations within the retina.

ELECTRICAL ACTIVITY IN THE RETINA

The distal cells (i.e., photoreceptors, horizontal cells, and bipolar cells) in the retina do not produce spike discharges and communicate with each other by means of relatively slow, non-explosive changes in their membrane potential. The ganglion cells produce spike discharges that can be recorded from the retina, the optic tract, or the lateral geniculate nuclei (LGNs). Discharges from the cells can be classified into three main categories: on-, off-, and on–off discharges. For example, a small spot of light placed at a concentric receptive field's center will produce an on-discharge from an on-center ganglion cell and will produce an off-discharge from an off-center ganglion cell; an annulus of light will produce opposite discharge patterns from the cells. In animals, some ganglion cells are known to signal specific events, such as a motion of a spot of light to a given direction. Such cells are called directionally sensitive ganglion cells. We do not know, however, whether there are specific, "task-performing" ganglion cells in the human retina.

The mass electrical potential from the retina or eyeball

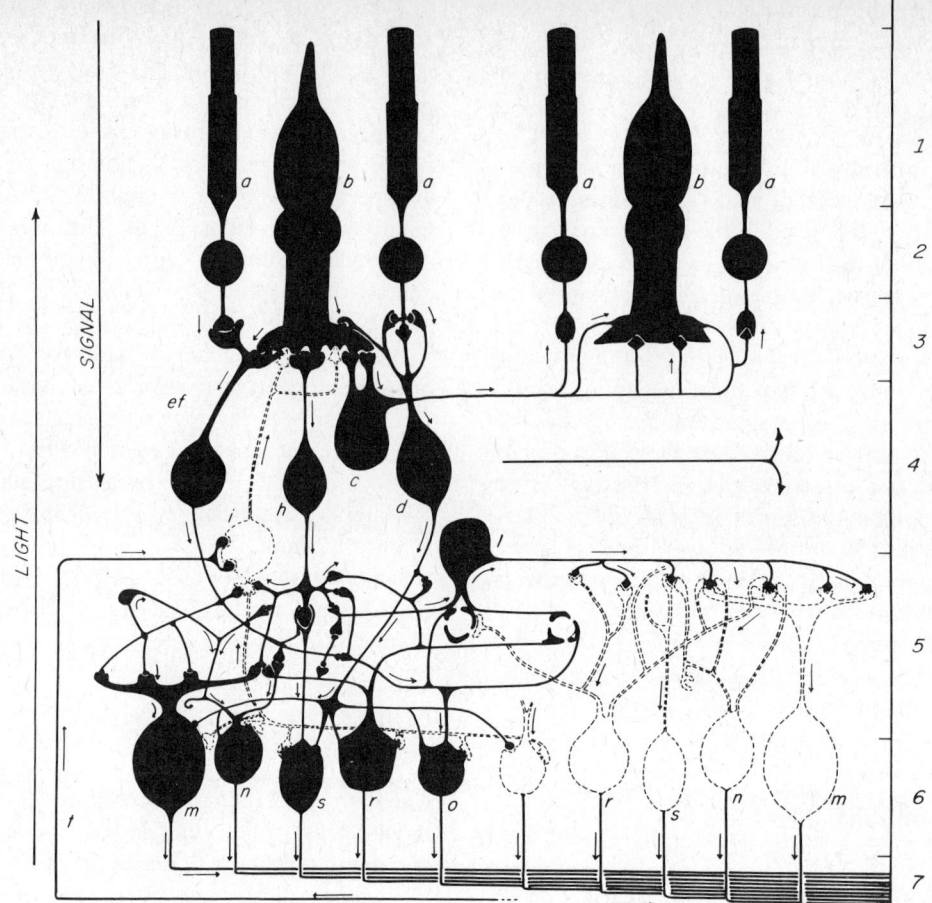

Figure 5 Highly schematized neural circuitry of the primate retina. *(a,b)* Rods and cones. *(c)* Horizontal cells. *(d,e,f,h)* Bipolar cells *l,* Amacrine cell. *(m,n,o,s,r)* Ganglion cells, *(i)* Interplexiform cell, a specialized cell type. Layers are as follows: (1) outer segment of receptor (site of photochemical reaction); (2) outer (receptor) nuclear layer; (3) outer plexiform (synaptic) layer; (4) inner nuclear layer; (5) inner plexiform (synaptic) layer; (6) ganglion cell layer; (7) optic nerve fiber layer. The group of receptors on the left is stimulated by light; flow of information within the retina are shown by small arrows. Note that light reaches the outer segment of receptors by going through the retina.

in situ is called the electroretinogram (ERG), which was discovered more than 100 years ago. The ERG can be reliably recorded from the human cornea using a contact-lens electrode. The ERG is a compound potential having its source in the neural elements, such as the receptor, horizontal, and bipolar cells, and in the glial element, such as the Müller fiber.

A typical mammalian ERG is shown in Figure 7; the

ERG has several waves usually referred to as a, b, c, and d waves. Considerable effort has been expended to dissect the ERG into several components having sources in the specific cell types or specific retinal layers. So far no definitive answer has been advanced. It is generally accepted that the a wave represents the photoreceptor potential, that the b wave originates in the bipolar cell layer and may reflect activities of Müller cells, that the c wave originates

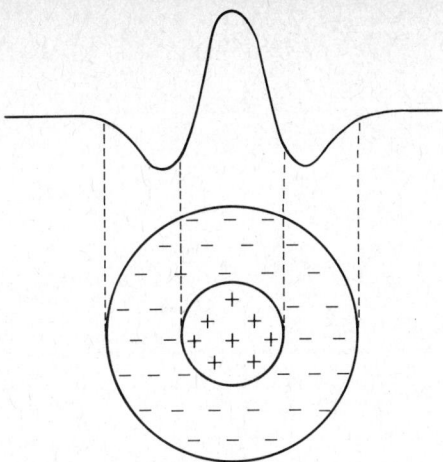

Figure 6 Concentric receptive field of the on-center type. Upper portion of illustration shows side view of the field; lower portion shows top view of the field. In bipolar cells the positive signs are for a depolarization and the negative signs are for a hyperpolarization of the cells' membrane potential. In off-center bipolar cells the polarities are reversed. In ganglion cell the positive signs represent initiation of spike discharges (on-discharges) and the negative signs represent depression of discharges followed by rebound discharges (off-discharges).

in the pigment epithelium, and that the d wave is an off response.

THE VISUAL PATHWAY

The visual event begins when photons travel through the cornea, anterior chamber, the lens, the vitreous, and the anterior portion of the retina and are finally absorbed by

the photon-catching pigments located in the outer segments of the photoreceptors. Isomerization of the visual pigment, as a result of photon absorption, leads to changes in membrane potential of the photoreceptors, the first-order neuron in the visual pathway. This variation in membrane potential is then transmitted to the second-order neurons, namely, the horizontal cells and bipolar cells. From the second-order neurons, the variation of membrane potential is transmitted to ganglion cells, with amacrine cells probably serving a modulatory function through complex interactions among them. The ganglion cells are the third order neurons and produce spike discharges (action potentials). The axons of the ganglion cells run toward the disc (optic nerve head) to form the optic nerve. The nerve fibers from both nasal halves of the retina crossover at the optic chiasm to join the nerve fibers from the respective contralateral temporal halves of the retina on the way toward the lateral geniculate bodies (LGN) of the thalamus (Fig. 8) superior colliculus, and pretectal area. Most optic fibers (about 80%) enter the LGN, which give rise to fibers forming the optic radiations toward the visual cortex of the occipital lobes. Within the visual cortex Brodmann's area 17 is the visual receptive area, while Brodmann's area 18 and 19 are regions for visual perception and fixation.

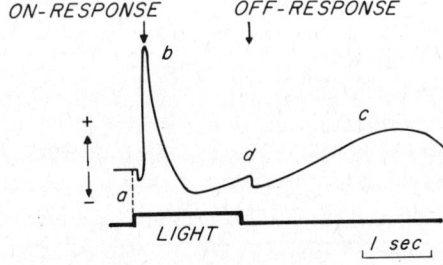

Figure 7 Schematized mammalian electroretinagram (ERG) showing a, b, c, and d waves. Electrical signs show an electrode (active) on the cornea relative to a reference (ground) electrode.

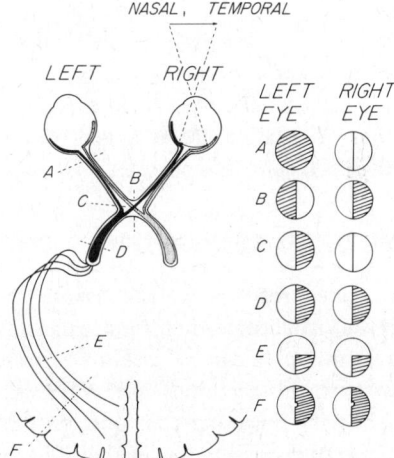

Figure 8 Schematized diagram of the central visual pathways, indicating the types of visual defects that will result from lesions located at various points in the pathway. The central unhatched areas in F are represented in the ipsilateral lobes.

Lesions that interrupt the visual pathway at various levels produce characteristic deficiencies in the appropriate visual fields (Fig. 8).

ELECTRICAL ACTIVITIES IN THE LGN AND VISUAL CORTEX

The lateral geniculate nuclei (LGNs), or bodies, are the relay or stopover stations for the optic fibers from the retinas to their continued journey toward the visual cortex at the occipital lobes. Here, the third-order visual neurons (i.e., ganglion cells) make their last synapse before the visual signals reach the central processing unit, namely the visual cortex. The electrical activities recorded from the LGNs are exclusively spike discharges (action potentials). Neurons in the LGN assume the same concentric receptive field organization as the neurons in the retina (bipolar and ganglion cells). The cells in the visual cortex respond more to lines, bars, and edges, rather than to a flashing spot of light and the stimulus orientation is critical to evoke the cells' response. Based on their response properties, the cells in the visual cortex are classified into simple and complex cells. Simple cells have elongated receptive fields in which an excitatory area is flanked by inhibitory areas. A moving line or bar of light with the same orientation as that of a field produces the most vigorous response from the cells. Complex cells respond to stimuli with more complex parameters such as the length of a bar of light or the direction of a bar's motion.

The cells in the visual cortex form a column in which cells of similar orientation are arranged roughly perpendicular to the surface of the cortex.

VISUAL EVOKED RESPONSE

The visual evoked response (VER) is an electrical response from the visual cortex in response to a flash of light. It can be measured by means of a skin electrode attached to the scalp over the occipital lobe. VER is a massive response, primarily reflecting the function of the foveal cones. It is a useful clinical tool in evaluation of the unilateral optic nerve or macula disease. As withERG, VER is an objective test and is particularly useful in pediatric patients, from whom subjective feedback of normal visual function is not possible.

PSYCHOPHYSICS OF VISION

ROD AND CONE VISIONS

The human retina has two kinds of photoreceptors: the rod and cones. The former are for night or scotopic vision and are based on a single photosensitive pigment, rhodopsin. The latter are for daylight or photopic vision and are based on the three different photosensitive pigments or cone pigments. The cones are concentrated in the fovea, and the rods are more numerous in the periphery of the retina. When we gaze at an object, we are using our cones, and when we see an object obliquely, we are using our rods. Scotopic vision is lower in spatial and temporal discrimination, that is, this vision is low in acuity and cannot follow fast changes, a phenomenon often referred to as a low-flicker fusion frequency. In return for these features, scotopic vision achieves a very high sensitivity; it is so sensitive when fully dark adapted, that an absorption of a few photons within 1 ms, among hundreds of rods, is enough to give a sensation of seeing. Based on such an observation, it is argued that individual rods respond to an absorption of single photons. As photons cannot be divided into smaller units, this is the maximal achievable sensitivity of any photon-sensing instrument.

In scotopic vision only the single pigment, rhodopsin, is involved. In such a system, light of various colors (or spectral composition) can be matched or made indistinguishable from light of a given color by simply changing the given color's intensity; for example, on a moonless night, everything looks gray. The interchangeability of color and intensity in a single pigment system is known as the *principle of univariance*. The spectral sensitivity of scotopic vision, as one would expect, is identical to the absorption characteristic of rhodopsin.

Photopic vision has a better spatiotemporal resolution and can discriminate color. The human phototopic spectral sensitivity curve has its sole peak at 560 nm.

A person, more likely male than female, who lacks cones totally, is a rod monochromat and has to wear very dark sunglasses during the daytime. He sees the world as a very bright or dazzling moonless night, since rods are much more sensitive to light than cones. Persons who lack one or two of the three cone pigments are color defective, commonly called color blind. Persons who lack two cone pigments are extremely rare, but those who lack one cone pigment are not uncommon. This condition of absence of one cone pigment is called dichroism and refers to the fact that any color can be matched by a combination of two colors. In dichroism, the case in which blue pigment is lacking, is very rare; more common is the absence of the red, more precisely, orange-red cones, and the yellow or, more precisely, yellow-green-catching cones. People with these conditions are called protanopes and tenteranopes, respectively. The color defects can be screened by the Ishihara color chart.

ADAPTATION

The sensitivity of the eye is not fixed, but varies according to the environmental conditions, a phenomenon known as adaptation. There are two kinds of adaptation: dark (or bleaching) and light (or field) adaptations. Dark adaptation is an increase in the eye's sensitivity when the eye is kept in darkness—it is our common experience that when we first enter a cinema theater, we do not see vacant seats, but when we are inside for awhile, we begin to see them.

Increase in the concentration of photosensitive pigment in the receptors underlies the process of dark adaptation. For example, when an eye is fully dark adapted, all the opsin is combined with 11-*cis*-retinal, so that receptors reach their highest efficiency to catch photons.

Dark adaptation has two phases: the initial rapid phase reflecting the increase in cone sensitivity, and the slower phase, reflecting the increase in rod sensitivity.

Field or light adaptation is an adjustment of the eye's sensitivity to the average illumination of its environment; under a dim light or when light is scarce the eye's sensitivity is high, and under a bright light or when light is abundant, the eye's sensitivity is low. Several explanations have been advanced to account for such changes in sensitivity.

EYE MOVEMENT SYSTEMS

Six pairs of extraocular muscles (EOM) control the various eye movements: the superior and inferior recti, the medial and lateral recti, and the superior and inferior oblique muscles (Fig. 9). The primary and secondary ac-

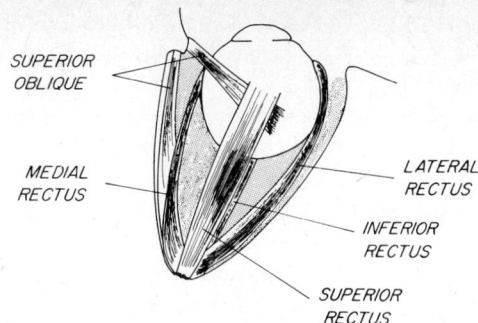

Figure 9 Extraocular muscles of right eyeball. Space between the eyeball and the orbit is filled with fat. Inferior oblique is situated below the eyeball and runs parallel to superior oblique.

tions of the extraocular muscles while the eyes are at the primary position are listed in Table 1.

The primary objective of eye movement is to aim the eye in such a position so that the target of interest in the environment can be projected directly onto the most sensitive portion of the retina, namely the fovea. There are primarily five specific type of supranuclear eye movements: saccadic, smooth pursuit, vergence, position maintenance, and nonoptic reflex.

Saccadic eye movement is the fast, voluntary eye movement executed when the eyes shift from one object to another. This type of eye movement is usually very precise, although slight overshoot or undershoot may be normal. Saccadic eye movement is mediated by the frontal lobes; in particular the horizontal conjugated left saccade is controlled by the right frontal lobe while the horizontal

Table 1 Primary and Secondary Action of Extraocular Muscles

Extraocular Muscles	Primary Action	Secondary Action
Medial rectus	Adduction	Insignificant
Lateral rectus	Abduction	Insignificant
Superior rectus	Elevation	Adduction Incycloduction
Inferior rectus	Depression	Adduction Excycloduction
Superior oblique	Incycloduction	Depression Abduction
Inferior oblique	Excycloduction	Elevation Abduction

conjugated right saccade is controlled by the left frontal lobe. The vertical saccade has no laterality.

The smooth pursuit system is utilized when a slow tracking movement of an object is required as in watching horse racing. This eye movement is mediated by the occipitoparietal lobes. Contralateral responsibilities (i.e., the right occipitoparietal lobe controls the smooth pursuit movement to the left, and the left occipitoparietal lobe controls the smooth pursuit movement to the right) also exist for horizontal pursuit movements, but unilateral occipitoparietal lobe may have bilateral influence on horizontal pursuit movements. The vertical smooth pursuit movement is the responsibility of both occipitoparietal lobes.

The vergence movement is the very slow and disconjugated eye movements used when an observed object is moving toward the eye (which stimulates the convergence movement), or moving away from the eye (which stimulates the divergence movement). This eye movement is necessary so that images of the observed object falls on the corresponding areas of the retinas. The occipitoparietal lobes are again responsible for this eye movement.

Ocular motility is a fascinating subject. Strabismus is defined as the inability of maintaining straight visual axes of both eyes in the primary position. This is usually a manifest deviation of an eye. If the eye is deviated nasally (or inward), it is termed esotropia, a condition frequently encountered in pediatric patients. If the eye is deviated temporarily (or outward), it is termed exotropia. The clinical diagnosis and management of strabismic conditions are beyond the scope of this chapter.

SELECTED BIBLIOGRAPHY

Armington JC: *The Electronretiongram.* New York, Academic Press, 1974.

Brindley GS: *Physiology of the Retina and Visual Pathway,* ed 2. Baltimore, Williams & Wilkins, 1970.

Crescitelli F, Dartnall HJA: Human visual purple. *Nature* 172:195–196, 1953.

Davson H: *The Physiology of the Eye,* ed 3. New York, Academic Press, 1972.

Davson H, Graham LT Jr: *The Eye,* vol 1–6. New York, Academic Press, 1969–1974.

Denton EJ, Pirenne MH: The absolute sensitivity and functional stability of the human eye. *J Physiol (Lond)* 123:417–442, 1954.

Fine BS, Yanoff M: *Ocular Histology.* Hagerstown MD, Harper & Row, 1972.

Gregory, RL: *Eye and Brain: The Psychology of Seeing.* London, Weidenfeld & Nicolson, 1967.

Hartline HK: The response of single optic nerve fibers of the vertebrate eye to illumination of the retina. *Am J Physiol,* 121:400–415, 1938.

Keidel WD, Neff, WD: *Handbook of Sensory Physiology,* vol 1–3: *Auditory System.* Berlin, Springer-Verlag, 1974–1976.

Pirenne MH: *Vision and the Eye.* London, Chapman & Hall, 1967.

Polyak SL: *The Retina.* Chicago, University of Chicago Press, 1941.

Rushton WAH: Pigments and signals in color vision. *J Physiol (Lond)* 220:1–31p, 1972.

SELF-STUDY QUESTIONS

MULTIPLE CHOICE

Select the correct answer(s). (In many instances, more than one answer is correct.)

1. Rod photoreceptors
   A. synapse in the inner plexiform layer.
   B. make one-to-one connections with bipolar cells in the fovea.
   C. are more sensitive than cones only in the green region of the spectrum.
   D. are more concentrated in the parafoveal area than they are in the retinal periphery.

2. An individual neuron in the lateral geniculate body

A. may project to the calcarine cortex.

B. may receive inputs from either eye, but not both.

C. has an antagonistic center-surround receptive field.

D. makes synaptic connections in the outer plexiform layer.

3. During visual field testing, a patient was found to have a left homonymous hemianopsia. The lesion producing the field defect

A. is in the left visual radiation.

B. could be in the right lateral geniculate body.

C. could be in the optic chiasm.

D. cannot be localized specifically without more information.

4. Photochemical dark adaption

A. occurs more rapidly in cones than in rods.

B. results when pigment epithelial cells lose their melanin.

C. occurs to a greater extent in rods than in cones.

D. does not occur in the fovea, since inner retinal layers are absent there.

5. In the cells making up the retina,

A. the conversion of light energy to electrical energy by photoreceptors is dependent on conversion of a vitamin A-containing compound to its isomer.

B. rods function at low light intensity, whereas cones function best in bright light.

C. many photoreceptors converge on a single ganglion cell in the retinal periphery, resulting in high sensitivity, but low visual acuity.

D. regeneration of rhodopsin requires light.

6. A function or characteristic of the retinal pigment epithelium is

A. absorption of stray light.

B. phagocytosis of rod outer segments.

C. that it is a site of blood–vitreous barrier.

D. secretion of aqueous humor.

7. Regarding rhodopsin,

A. the outer segment of a rod contains rhodopsin stored in the walls of membranous discs.

B. dark adaptation refers to the increase in light sensitivity as more rhodopsin is regenerated.

C. vitamin A is required for rhodopsin synthesis.

D. it absorbs light optimally at three wavelengths, corresponding to the three primary colors.

8. Visual acuity

A. is measured by the angle of separation at which two points can be distinguished.

B. can be decreased by opacities in the vitreous or the lens.

C. is aided by contrast enhancing mechanisms in the central nervous system.

D. can be measured accurately with the visual evoked response.

9. Concerning color vision,

A. color vision defects occur only as a result of an inherited disorder.

B. color-defective vision occurs when one of three cone pigments is abnormal or absent.

C. the electro-oculogram can be used to evaluate cone function.

D. the light absorption curves of the cone pigments overlap, allowing for perception of intermediate colors.

10. Concerning retinal organization,

A. the cones of the fovea make approximately one-to-one connections, via bipolar cells, with ganglion cells.

B. retinal interneurons may synapse with photoreceptors and bipolar cells.

C. the rod ganglion cell ratio in the peripheral retina allows for better function in this area at low light intensities, as compared with the macula.

D. ganglion cells outnumber photoreceptors by 100:1.

11. The structure(s) in the eye responsible for the refraction of light is/are

A. anterior chamber.

B. lens.

C. retinal pigment epithelium.

D. cornea.

Select the single best answer.

A. *Rod*
B. *Cone*
C. *Retinal pigment epithelium*
D. *Ganglion cell*
E. *Horizontal cell*

12. This cell is most highly concentrated in the part of the retina just outside the macula.

13. The choroid lies on one side of this cell layer.

14. Axons from these cells synapse in the lateral geniculate body.

Select the single best answer.

15. There are _____ pigments in the human retina.
    A. One
    B. Two
    C. Three
    D. Four
    E. Five

16. Rod (scotopic) vision is for
    A. day (photopic) vision.
    B. twilight vision.
    C. night vision.
    D. morning.
    E. none of the above.

17. In human eye, converging power is provided mainly by
    A. humor.
    B. cornea.
    C. lens.
    D. vitreous.
    E. photoreceptors.

18. Visual acuity is a measure of
    A. rod function.
    B. primary cone function.
    C. both rod and cone function.
    D. peripheral vision.
    E. ganglion cell function.

19. The most typical receptive fields in the visual pathway is
    A. circular.
    B. concentric.
    C. triangular.
    D. square.
    E. none of the above.

20. In the human visual pathway the third-order neurons are
    A. lateral geniculate body.
    B. ganglion cell.
    C. bipolar cell.
    D. cortical cell.
    E. amacrine cell.

21. Cells in the visual cortex differ from those in the lateral geniculate body and of the retina in that they
    A. have concentric receptive fields.
    B. can receive input from both eyes.
    C. all have simple receptive field.
    D. all have complex receptive field.
    E. none of the above.

22. In abducted position, the primary function of superior oblique muscle is
    A. elevation.
    B. depression.
    C. incycloduction.
    D. excycloduction.
    E. none of the above.

23. Saccadian eye movement is mediated by the
    A. occipitoparietal lobe.
    B. frontal lobes.
    C. cerebellum.
    D. brainstem.
    E. lateral geniculate body.

24. The retinal neurons that give rise to fibers of the optic nerve are the
    A. rods and cones.
    B. bipolar cells.
    C. amacrine cells.

D. ganglion cells.

E. horizontal cells.

25. Integration of signals to produce binocular vision occurs at the level of

    A. retina.

    B. optic chiasm.

    C. lateral geniculate body.

    D. visual cortex.

    E. none of the above.

26. Images stabilized on the retina appear

    A. brighter than usual.

    B. larger than usual.

    C. more richly colored than usual.

    D. more dense than usual.

    E. to disappear.

27. All of the following are true *except*

    A. visual acuity depends significantly on the diameter of the cone cells.

    B. cone cells dark adapt more rapidly than do rods.

    C. cones are inherently slightly more light sensitive than are rods.

    D. rods show greater convergence on ganglion cells than do cones.

    E. not all photoreceptor cells have the same photopigment.

28. Myopia is associated with all of the following *except*

    A. an increase in the axial length of the eyeball.

    B. an image forming in front of the retina.

    C. shorter focal length.

    D. requiring a converging lens to see clearly.

    E. none of the above.

ANSWERS

1. C. Rods synapse directly with bipolar cells, but not in the fovea. Rods are concentrated outside the fovea; they absorb light maximally in the green region of the spectrum (500 μm), and at this wavelength are more sensitive than cones.

2. A, B, C. The lateral geniculate body is the thalamic relay nucleus for retinal ganglion cells transferring information to the visual cortex. An individual cell in the LGN may receive input from only one eye (i.e., binocular vision is not represented at the single cell level in the LGN). Their receptive field properties are similar to those found for ganglion cells in the retina.

3. B, D. Left homonymous hemianopsia is blindness of half the retina in the left visual field. The left visual field projects to the nasal half of the left retina and the temporal half of the right retina. Ganglion cells from this part of the retina all project to the right LGN and from there to the right visual cortex. This rules out the left visual radiations. Lesions in the optic chiasm are likely to involve fibers coming from both eyes;

thus visual deficits will appear in opposite visual fields (i.e., bitemporal hemianopsia). Lesions posterior to the chiasm will involve fields from both eyes.

4. A, C. Photochemical dark adaptation occurs more rapidly in cones, but not to the same extent as in rods. This is why rods function primarily at night and cones in daylight.

5. A, B, C. Regeneration of rhodopsin takes place in the absence of light.

6. A, B, C.

7. A, B, C. Rhodopsin is found in rods. The cones contain the three pigments necessary for color vision.

8. A, B, C.

9. B, D. Inherited disorders are not the only reason for color vision deficiencies.

10. All are correct.

11. B, D.

12. A.

13. C.

14. D.

15. D. Three in cones and one in rods.

16. C.

17. C. It changes shape by contraction of the ciliary muscle.

18. B. The cones are involved in acuity. Primarily around the fovea, where they connect 1:1 with ganglion cells.

19. B. An excitatory or inhibitory center surrounded by an area of opposite polarity.

20. B. Photoreceptor, bipolar, ganglion cell. Horizontal and amacrine cells are interneurons and are important for processing visual information within the retina.

21. B. Cells in the LGN receive from one or the other eye.

22. C. The upper part of the eye will turn toward the midline.

23. B.

24. D. Ganglion cells are the output cells of the retina.

25. D. Although the lateral geniculate nuclei in the thalamus receive retinal ganglion fibers from both eyes, individual cells within the LGN receive input from one eye or the other, but not both. It is only when the projections reach cortical levels that individual cells receive inputs from both eyes.

26. E. This is why the eye is continuously moving (small saccadic movements), thus avoiding stabilizing images on the retina.

27. Acuity does not depend on the size of the cell, but rather on how they are connected. Cones are connected to fewer ganglion cells than rods and therefore have smaller receptive fields.

28. D. A diverging lens is required.

43

The Auditory and Vestibular System

KEN-ICHI NAKA
RAYMOND Y. CHAN

OBJECTIVES

After completion of this chapter, you should be able to:

1. Know the anatomy of the human ear and vestibular apparatus.

2. Understand the functional morphology of the cochlea.

3. Describe the central auditory pathway.

4. Illustrate the functional morphology of the semicircular apparatus.

5. Describe the central pathway important in maintaining posture.

HUMAN EAR

The human ear contains the transducers for the perception of sound and for the maintenance of posture and equilibrium. The auditory receptors are located in the cochlea, which is a spiral-shaped apparatus situated within the inner ear. The vestibular apparatus includes three semicircular canals, saccule, and utricle, all located in the inner ear. Both the auditory and vestibulary apparatus are innervated by cranial nerve VIII.

ANATOMY

The human outer ear is made up of two principal structures: the auricle (or pinna) and the external auditory meatus (Fig. 1). The auricles collect sound waves and

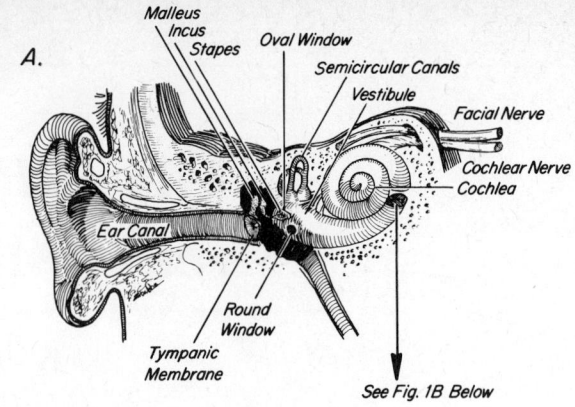

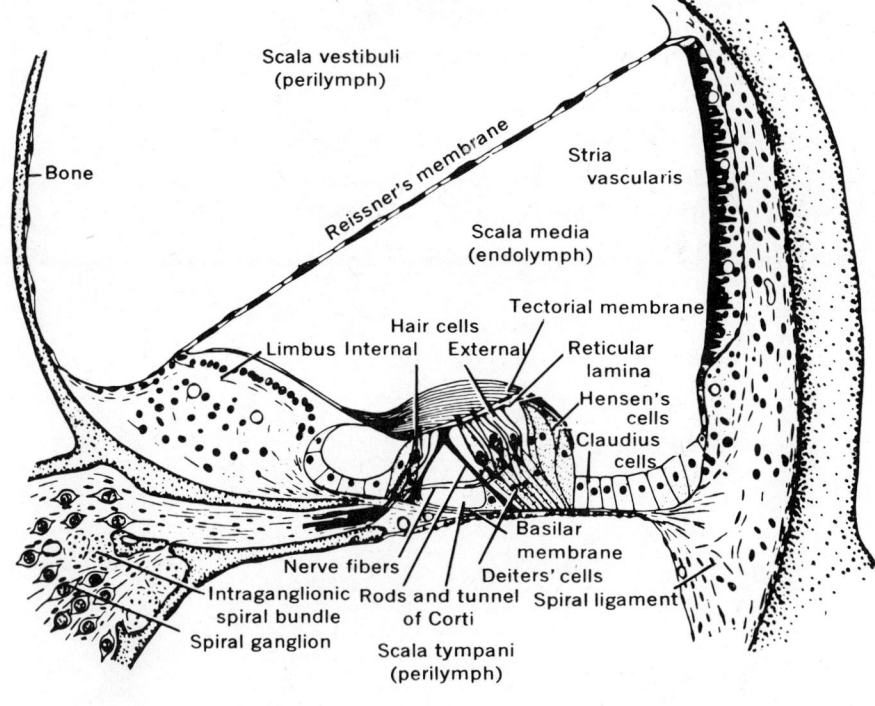

(b)

Figure 1 *(a)* Gross structures of the ear and vestibular system. Note the relationship of the bony structures in the middle ear. Also note the relationship between the cochlea and the three semicircular canals of the vestibular system. *(b)* A section of the cochlea has been removed and is shown in greater detail.

direct them to the auditory meatus. The external auditory meatus spans about 2.5 cm and leads inward to the tympanic membrane (or the eardrum). The meatus is lined by short hairs and cerumen-producing glands to keep foreign objects out of it.

The middle ear includes the tympanic membrane; three bony auditory ossicles, that is, the malleus, incus, and stapes and two muscles, the tensor tympani and stapedius. The middle ear cavity is connected to the pharynx through the Eustachian (auditory) tube.

The tympanic membrane separating the outer and middle ear is concave; it is formed by a very thin fibrous membrane covered with a thin layer of skin on the outer side and a mucous membrane on the inner side (Fig. 2). The tympanic membrane has two parts: the pars tensa and pars flaccida. The former occupies the major portion of the membrane and is tightly stretched as a vibrating membrane. The latter, as its name implies, is less stiff because it lacks the layer of fibrous connective tissue that is found in the pars tensa.

The manubrium (handle) of the malleus is attached to the upper surface of the tympanic membrane pointing at the 1 o'clock position in the right ear and at the 11 o'clock position in the left ear. The rounded head of the malleus fits (or articulates) into a socket in the incus, the intermediate bone in the series. The incus and stapes make a ball-and-socket joint, providing a freely rotating motion between them. The base (or the foot plate) of the stapes fits into the oval window of the vestibule.

The impact of sound waves produces vibration of the tympanic membrane that is carried across the air-filled

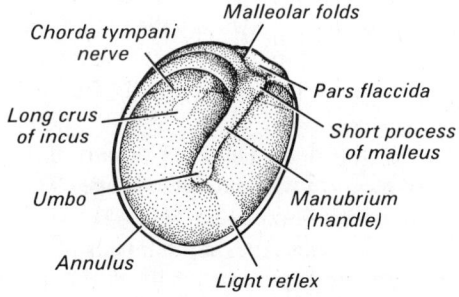

RIGHT TYMPANIC MEMBRANE

Figure 2 Detail of the tympanic membrane.

cavity of the middle ear by a lever system formed by the three ear bones. The rocking motion of the stapes produces waves in the fluid in the inner ear. The three ear ossicles suspended by various ligaments are passive devices to transform the vibration of the tympanic membrane in the air into the corresponding vibration of the round window, which interfaces the air in the middle ear with the perilymph. Thus sound energy carried by the air is (impedance) matched to that in the perilymph.

The middle ear contains two skeletal ossicular muscles: the tensor tympani and stapedius. The former attaches by a tendon to the manubrium of the malleus and the latter's tendon is attached to the neck of the stapes. Contraction of the tensor tympani tightens the tympanic membrane and that of the stapedius suppresses the movement of the lever system. Contractions of both muscles in response to a loud sound would have a dampening effect on the vibration of the three ossicles to protect the delicate inner ear from excessive vibration.

The auditory tube interconnecting the middle ear with the nasopharynx is normally closed so as to screen out all the interfering sounds, such as speech and breathing, from entering the middle ear. During eating, chewing, yawning, and similar motions the tube opens so as to equalize the pressure at both sides of the tympanic membrane. The tube is kept open by performing a Valsalva maneuver, that is, forced expiration against a closed glottis, a maneuver tried when a sudden change in the environmental pressure occurs. Although it is essential to equalize pressures, the auditory tube may also become a liability, as it offers a passageway by which infections may invade the middle ear.

The inner ear is contained within the bony labyrinth of the temporal bone and consists of the cochlear and the vestibular apparatus. The portion of the cochlea adjoining the oval window is the vestibule, which is also continuous with the vestibular apparatus. The cochlea is in the shape of a helix with $2\frac{1}{2}$ turns. In cross section the cochlea has three ducts, separated by the thin vestibular membrane (of Reissner) and the basilar membrane (Fig. 1b). The scala vestibuli forms the top duct, the scala media in the middle, and the scala tympani at the bottom, to form a three-duct helix. One end of the scala vestibuli is the oval window and the other end meets with the scala tympani at the apex of the helix, the helicotrema. The other end of the scala tympani terminates at the round window. Thus the scala vestibuli and tympani form a continuous duct from the oval to the round windows and filled with perilymph. The scala media (or the cochlear duct) is a triangular duct interposed between the scala vestibuli and tympani, and is formed by the vestibular membrane (to be separated from the scale vestibuli) and the stria vascularis, a vascularized membrane lining the external wall of the bony labyrinth. The cochlear duct is filled with endolymph.

The sensory epithelium of the cochlea is the organ of Corti, which lies on the basilar membrane and is composed of supporting and hair cells. The supporting cells are tall epithelial cells of various types and are attached to the basilar membrane. Their free surfaces form a reticular membrane over the spiral organ. The hair cells are the receptor cells, with hairlike processes that extend through the openings in the reticular membrane into the endolymph of the cochlear duct. The hair cells are grouped into two types: the internal and external hair cells. The former are located nearest the bony spiral lamina and arranged in a single row, each receiving a single nerve fiber from the cranial nerve VIII; the latter are located more distally and arranged in several rows, each receiving multiple innervation. The distal end of the hair cell is embedded in an overlying thin elastic tectorial membrane.

The movements of the foot plates of the stapes produce a series of traveling waves in the perilymph of the scala vestibuli. The wave produces distortion of the basilar membrane and the site along the cochlea at which this distortion is maximal is determined by the frequency of the sound wave. High-pitched sounds produce distortion of the membrane near the base of the cochlea, and low-pitched sounds produce distortions near the apex. The tips of the hair cells are held by the tectorial membrane, whereas their base vibrates with the basilar membrane. The mechanical distortion of the hair produces cochlear microphonics, a generator potential that is instantaneous and reproduces almost exactly the waveform of the original sound. The cochlear microphonic generates the action potential of the cochlear nerve.

VESTIBULAR SYSTEM

The vestibular apparatus consists of three semicircular canals: the anterior (superior), the posterior, and the lat-eral (horizontal), which are deeply embedded in the ducts hollowed out of the temporal bone and lined with connective tissue. The three canals are located in planes approximately at right angles to each other (Fig. 3). The lateral duct is in a plane 30° inclined from the horizontal head plane. The anterior and posterior ducts lie in the vertical head planes and are called the vertical semicircular ducts. The canals communicate with the vestibule by five openings rather than six because of the fusion of the two ducts. Each duct contains two canals, the osseous canal and the membranous canal. The former with its larger diameter contains the perilymph, and the latter with its smaller diameter contains the endolymph. Each semicircular duct widens at one end before it merges with the utricle of the vestibule because the diameter of the ducts is about 0.25 mm, whereas that of the utricle is 2–4 mm. The dilation, the ampulla, is about 2 mm in diameter and contains the neuroepithelium of the semicircular duct. The ampulla contains a crest of hair cells, the crista ampullaris, which are the end organs of the vestibular nerve (Fig. 4). From the apical end of these hair cells, cilia extend upward into the cupula, a gelatinous mass that forms a fluid-tight partition, or flap, across the ampulla.

The function of the semicircular duct is to detect angular acceleration of the head. When the head begins to rotate, the inertia of the endolymph within one or more pairs of ducts produces a relative movement of the endolymph in the direction opposite to that of the head. The movement of the endolymph causes the cupula to bend in the direction of the endolymph flow. The mechanical distortion will depolarize the hair cells, leading to discharges of the vestibular nerve.

The vestibule contains two membraneous sacs, the larger utricle and the smaller saccule. The former lies in the superior–posterior portion of the vestibule in close approximity to the semicircular canal. The utricle communicates with the canal as well as with the saccule by means of a very fine duct. The anterior part of the utricles widens into a recess in which a spade-shaped sensory area, the macula utriculi, is located (Fig. 4). The utricular macula has a complex shape, but its major plane is nearly horizontal with the head erect. The saccular macula is hook shaped and its major axis is perpendicular to the plane of the ultricular macula. The macula contains cells of two types: supporting cells and hair cells. On the surface of the macula is a gelatinous substance that contains

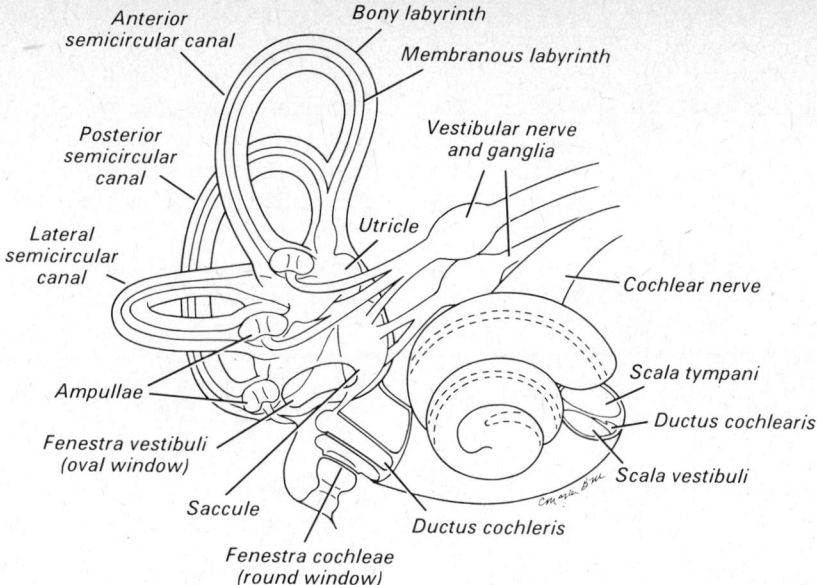

Figure 3 — Anterior semicircular canal; Bony labyrinth; Membranous labyrinth; Posterior semicircular canal; Vestibular nerve and ganglia; Lateral semicircular canal; Utricle; Cochlear nerve; Ampullae; Scala tympani; Fenestra vestibuli (oval window); Ductus cochlearis; Saccule; Scala vestibuli; Ductus cochleris; Fenestra cochleae (round window)

Figure 3 Detail of the semicircular canals and their relationship to the cochlea. Note the innervation by cranial nerve VIII and the position of the ganglia. These contain the cell bodies of the nerves and are analogous to dorsal root ganglia found along the spinal cord.

minute calcareous concretions called statoconia or otoconia. These are similar to the larger ear stones or otolith found in certain invertebrates and fish. The utricles are thought to be affected by gravity; when the head is tipped, the statoconia slides onto the underlying macula and distorts hair cells. Thus utricle and saccule might function as an accelerometer.

AUDITORY PATHWAYS AND MECHANISM

The sequence of events leading to sound perception begins with vibration of the tympanic membrane set off by the sound waves. The corresponding oscillation is transmitted to the oval window by an impedance-matching mechanism consisting of three auditory ossicles. Vibration of the oval window creates a corresponding pressure wave transmitted through the perilymph in the cochlea, resulting in deformation of the cochlear partitions (namely, the Reissner and basilar membranes), which in turn distort

the tectorial membrane of the organ of Corti, stimulating the hair cells and converting the mechanical energy into corresponding electrical potentials. This electrical potential is then conducted by the dendritic processes of the spiral ganglion cells, which innervate the hair cells. When the potential crosses threshold action potentials are generated in the cochlear fibers and conducted centrally. The axons of the spiral ganglion cells form the cochlear nerve, which enters the brainstem near the junction of the pons and medulla. The nerve then bifurcates and travels to both anterior and posterior cochlear nuclei. Through various uncrossed and crossed pathways, fibers from both cochlear nuclei reach both the ipsilateral and contralateral inferior colliculi. From the inferior colliculi, fibers travel through medial geniculate bodies to terminate at the auditory cortex located at the transverse temporal gyrus (Brodmann's area 41). The sound is presumably perceived when the electrical potential reaches the auditory cortex. However, intelligent interpretation of the perceived sound probably involves the auditory association areas adjacent to the auditory cortex.

Most cochlear nerve fibers are afferent fibers. There is good evidence to support the presence of efferent fibers,

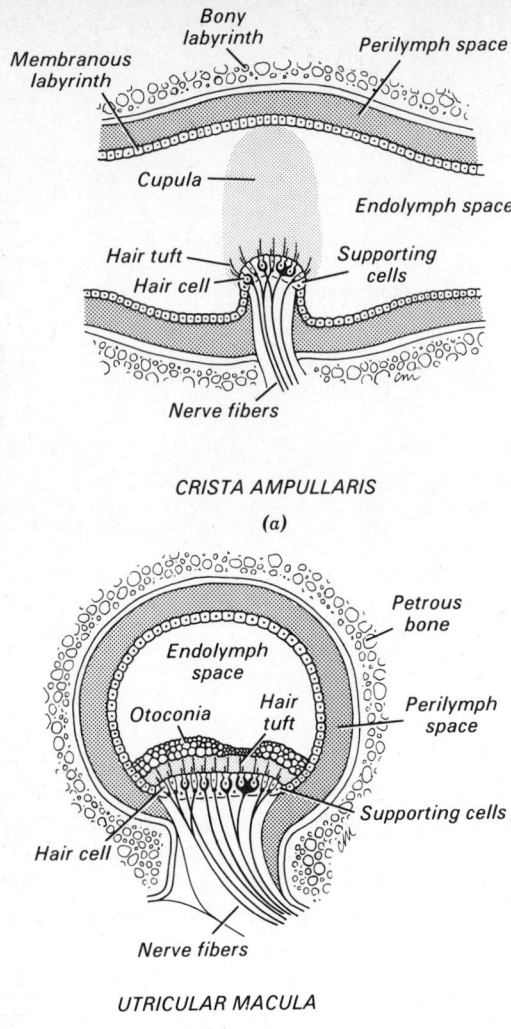

CRISTA AMPULLARIS

(a)

UTRICULAR MACULA

(b)

Figure 4 *(a)* The sensory end organ, which lies in the ampulla at the opening of the semicircular canals. These hair cells are stimulated by rotation of the head. *(b)* The sensory end organ, which lies in the utricle and the saccule. In each compartment the end organ lies in a different plane, each perpendicular to the next. These end organs are believed to be stimulated by gravity when the head is tilted.

which are believed to be involved in feedback regulation and in modification of the neural activity at the receptor level. From the electrophysiological standpoint, two types of cochlear potential can be obtained: the endocochlear positive potential recorded from the cochlear duct with respect to scala tympani, and the cochlear microphonic potential recorded from the vicinity of the round window, representing in a large part the oscillatory vibration of the cochlear partition. The so-called summating potential results from shifts in the DC level of the cochlear microphonic potential; its origin and function are uncertain. It is interesting to note that the action potentials recorded from each of the afferent auditory nerve fibers is frequency dependent, offering further verification of the anatomical selectivity of the human cochlea as a superb audiofrequency analyzer.

AUDITORY RESPONSES AND FUNCTIONAL ASSESSMENT

Sound waves are generated by vibratory devices, such as expiratory air passing through vocal folds or striking a bell with a hammer. In either case, the vibratory waves reach the tympanic membrane after propagating through air as cycles of alternating compression and rarefaction of the air molecules. The amplitude of the sound wave determines the sound intensity, that is, loudness (Fig. 5). The amplitude is ordinarily expressed in decibels, which is a logarithmic scale defined as

$$\text{Decibel (dB)} = 20 \log_{10} \frac{\text{pressure of sound applied}}{\text{reference level pressure}}$$

The reference level pressure is a constant at 0.000204 dyn/cm^2, which corresponds to the lowest threshold of hearing in an average individual. (As will be discussed shortly, the auditory threshold is frequency dependent.) The frequency of sound determines pitch. A high pitch sound implies high frequency. Addition of harmonics (i.e., integral multiples of the primary frequency of a particular sound) to a single-frequency sound wave creates musical quality, or timbre. It is generally believed that amplitude and frequency of a sound wave are independent variables. However, the physiological perception of sound (i.e., auditory threshold) is frequency dependent (Fig. 5). The audible frequency range for humans is approximately 20–20,000 cycles per second (Hz), with the best sensitivity around 1,000–4,000 Hz. The ability to hear very high frequencies gradually declines with age. The best listening frequency range is 250–750 Hz. Included within this range

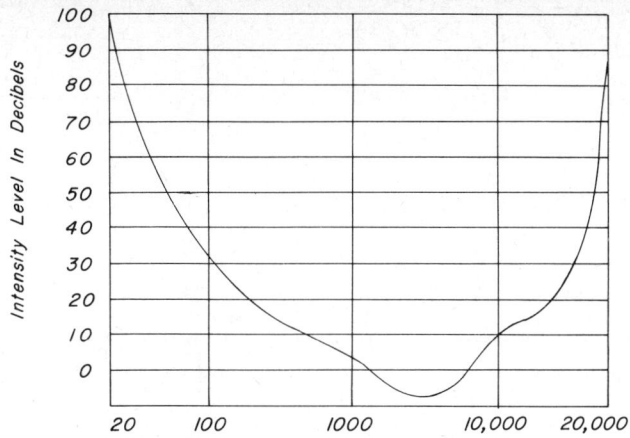

Figure 5 Diagram showing the auditory threshold. The intensity necessary to reach threshold is minimal at around 1000–4000 Hz, but increases rapidly away from these frequencies in order to maintain perception of the sound.

is the average female voice. The pitch of the male voice is usually 100–200 Hz lower.

Deafness can be classified into three groups.

1. *Conduction deafness:* Attributable to interruption of sound transmission to the inner ear. This can occur as a result of an obstruction of the auditory canal by wax, exudates, or foreign bodies; thickening of the tympanic membrane after recurrent otitis media; otosclerosis; or abnormal attachment of stapes to the oval window. This condition can be readily diagnosed by performing tuning fork tests to be described shortly.

2. *Nerve deafness:* Attributable to damage to the auditory neural pathway, which may be at the level of the sensory hair cells or anywhere along the path of the cochlear nerve. Frequent exposure to loud noise can eventually damage the sensory hair cells, while ischemia, tumor, and toxic chemicals are frequent insults to the cochlear nerve pathways.

3. *Cortical deafness:* Attributable to damage to, or atrophy of, the auditory cortex secondary to pressure or ischemic events, or both. However, the diagnosis is usually by exclusion; it is a rarely encountered situation.

VESTIBULAR PATHWAYS AND MECHANISM

The initiation of the vestibular response is a mechanical event, namely head rotation, which agitates the endolymph within the membranous labyrinth, bending the hair cells cilia within the cupula of the crista or pushing the calcium carbonate crystals against the hair tufts within the macula, with the resulting generation of action potentials by the mechanoreceptors, the hair cells. These electrical potentials are then transmitted to the innervating nerve fiber bundles whose cell bodies are located at the vestibular (Scarpa's) ganglion in the proximal end of the internal auditory canal. Nerve fibers leaving the vestibular ganglion form the vestibular portion of cranial nerve VIII. The vestibular nerve is accompanied by the cochlear nerve coursing to the brain stem. At the upper medulla, the vestibular nerve bifurcates into ascending and descending tracts. Those fibers that arise in the cristae of the semicircular canals enter the superior and medial vestibular nuclei, while those that arise in the maculae of the utricle and saccule enter the inferior and lateral nuclei (Fig. 6), and a small portion of vestibular fibers project to the cerebellum. Nerve fibers enter the ascending medial longitudinal fasciculus (MLF) projecting to the oculomotor nuclei from both the superior and medial vestibular nuclei. Fibers from the superior vestibular nucleus are ipsilateral, whereas those from the medial vestibular nucleus are bilateral. From the medial vestibular nucleus, nerve fibers also project bilaterally through the descending MLF to reach the cervical musculature. Anatomical communication between stimulation of semicircular canals to the elicitation of head rotation and nystagmus is thus established. From the lateral vestibular nucleus, nerve fibers project ipsilaterally down the lateral vestibulospinal tract extending through the length of the spinal cord, eventually influencing the antigravity muscle responses to maintain posture. The primary efferent nerve fiber projection from the inferior vestibular nucleus is to the cerebellum. Direct cortical projection from the vestibular system is uncertain. However, it appears that the ventroposteroinferior (VPI) nucleus might serve as the relay station for vestibulocortical communications.

Electrical potentials measured from the vestibular sys-

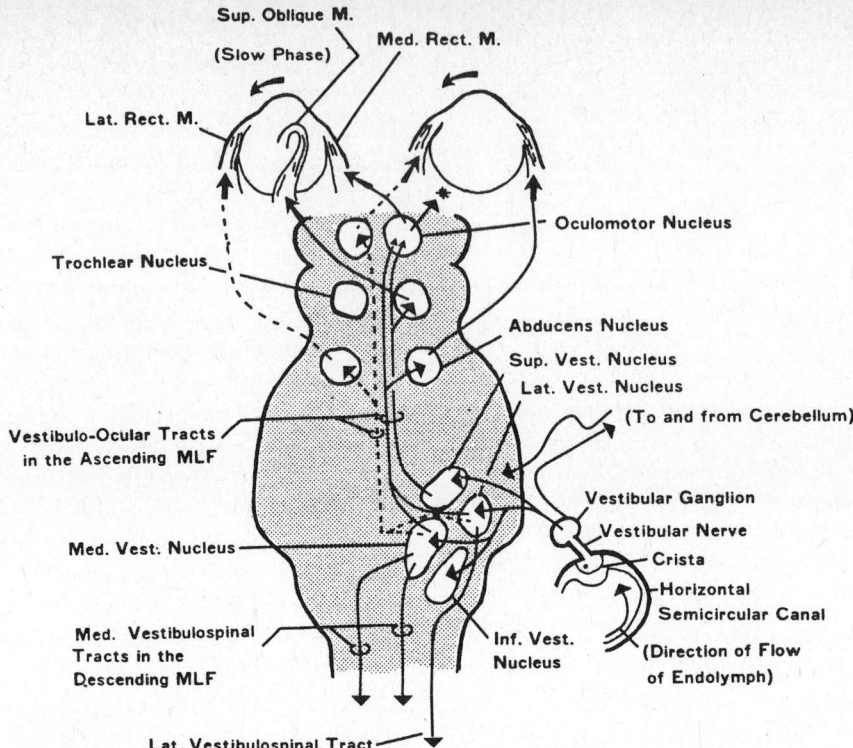

Figure 6 Diagram showing the anatomical organization of the auditory and vestibular pathways in the brainstem. Nystagmus induced by centripetal flow of endolymph in a horizontal semicircular canal is indicated. The dotted lines show the probable route of activating stimuli. MLF, Medial longitudinal fasciculus.

tem include primarily the action potentials recorded from the nerve fibers innervating the five neuromechanoreceptors (i.e., vestibular nerve); although potential differences between the hair cells and the endolymph can be recorded. Bending of the cupula in the semicircular canal could either increase or decrease the firing rate of individual fibers depending on the deflection direction (i.e., away or toward the utricle). Such changes in firing patterns are then transmitted to higher centers to be decoded.

VESTIBULAR RESPONSES AND FUNCTIONAL ASSESSMENT

Vestibular responses begin to express themselves when the inputs to the two vestibular end organs are unbalanced.

Such responses include head rotation, swaying, nystagmus, and vertigo. Vestibular nystagmus is described as regular oscillatory eye movements, which include a slow and a fast phase. The slow phase of nystagmus is in the same direction as the endolymph movement within the semicircular canal, and the fast phase is the compensatory jerk-back eye movement presumably originated by the brain. The direction of nystagmus, however, is defined as the direction of the fast phase of eye movement. Vestibular nystagmus can be elicited by overstimulation of the semicircular canal system such as by continuous rotation or by caloric stimulation as discussed in the next section. True vertigo is a sensation of rotation of the environment accompanied with vestibular nystagmus. The mechanism of vertigo depends on the smooth pursuit (i.e., tracking movement) and saccadic (fast jump) eye movements, and the fact that vision is preserved during smooth pursuit but not during saccadic movements. The slow phase of vestibular nystagmus (generated by the ocular smooth pursuit mechanism) creates a rotational sensation of the environment opposite to its direction of movement. Since there is no vision during the fast phase of vestibular nystagmus, which is generated by the ocular saccadic mechanism, a continuous rotational sensation of the environment may be perceived.

Assessment of vestibular system function can be accomplished both clinically and quantitatively. Clinical assessment provides visual observation, whereas vestibulometry, or electronystagmography (ENG) ensures careful calibration and recording of the visually observed phenomena. Commonly used techniques include the bithermal caloric, rotation, and optokinetic (OKN) tests.

The *bithermal caloric test* consists of irrigating the ear canal with water that is colder or warmer than body temperature (e.g., 30° or 44°C) for 30–40 s while the patient is lying supine with head elevated at 30° so as to bring the horizontal semicircular canals to vertical position. For a normal person, caloric-induced nystagmus is in the direction opposite the stimulating ear for cold water and in the same direction for warm water, hence the abbreviation COWS, cold, Opposite–Warm, same. An important aspect is to observe any symmetry or asymmetry between the two ears. Unilateral weakness suggests the presence of a lesion at the peripheral vestibular apparatus.

The *rotation test* can be performed in a variety of ways, all centered to elicit nystagmus opposite to the direction of constant acceleration. For example, the Barany chair is motor driven to provide constant acceleration. Its main advantages are easier stimulation to vertical semicircular canals and less discomfort (therefore more suitable for pediatric patients). Diminished or absence of rotational nystagmus in both directions is an indication of peripheral vestibular apparatus pathology.

The *optokinetic nystagmus test* is accomplished by asking the patient to fixate at a slowly moving repetitive stimulus so as to create sequence of ocular smooth pursuit followed by saccadic eye movement, i.e., nystagmus. Both horizontal and vertical nystagmus can be induced by either presenting the moving pattern horizontally or vertically. Asymmetrical OKN is seen both in brainstem lesions and in cerebellar lesions. However, an isolated vertical OKN asymmetry seems to suggest only a high brainstem lesion, either midline or bilateral.

SELECTED BIBLIOGRAPHY

Ballenger JJ: *Diseases of the Nose, Throat and Ear.* Philadelphia, Lea & Febiger, 1977.

Bateman HG: *A Clinical Approach to Speech Anatomy and Physiology.* Springfield, IL, Charles C Thomas, 1977.

Bradford LJ: *Physiological Measures of the Audio-Vestibular System.* New York, Academic Press, 1975.

Clark RG: *Manter and Gatz's Essentials of Clinical Neuroanatomy and Neurophysiology,* ed 5. Philadelphia, FA Davis, 1975.

Eyzaguirre C, Fidone S: *Physiology of the Nervous System,* ed 2. Chicago, Year Book, 1975.

Ganong WF: *Review of Medical Physiology,* ed 7. Los Altos, CA, Lange Medical Publications, 1975.

Norther JL: *Hearing Disorders.* Boston, Little, Brown, 1976.

Sach MB: *Physiology of the Auditory System—A Workshop.* National Educational Consultants, 1971.

SELF-STUDY QUESTIONS

MULTIPLE CHOICE

Select the single best answer.

1. All of the following are common to the auditory and vestibular systems, *except*
   A. cranial nerve VIII.
   B. endolymphatic space.
   C. hair cells.
   D. efferent innervation of the end organ.
   E. topographical representation of the end organ in the temporal cortex.

2. All of the following are true concerning efferent fibers to the cochlea, *except*
   A. they arise, in part, from a division of the superior olivary complex.
   B. they innervate the cochlear hair cells.
   C. they have an inhibitory effect on cochlear hair cells.
   D. they innervate the stapedius muscle.
   E. they hyperpolarize cochlear hair cells.

3. All of the following are true when a person's head is rotated to the right, *except*
   A. the cupula of the left lateral semicircular duct is deflected away from the utricle (utriculofugal).
   B. the ciliary processes of the hair cells in the crista are bent in the direction of endolymph flow.
   C. activity in the afferent neurons innervating the left lateral semicircular duct is increased.
   D. the vestibulo-ocular reflex causes the eyes to deviate to the left.
   E. there is a compensatory postural adjustment to prevent falling to the left.

4. All of the following are true concerning the middle ear, *except*
   A. middle ear air pressure is normally equalized to atmospheric pressure.
   B. vibration of the tympanic membrane by sound is transmitted to the perilymph of the vestibule by the ossicles.
   C. the tympanic membrane "protects" the round window by phase-delaying sound waves.
   D. the stapes transmits mechanical energy directly to the basilar membrane of the cochlea.
   E. middle ear muscles contract in a reflex response to loud sounds.

5. A warm caloric irrigation of the right ear will result in all of the following, *except*
   A. the direction of nystagmus (fast phase) is toward the right.
   B. the patient senses that he is spinning toward the right.
   C. the cupula in the right lateral semicircular duct is deflected away from the utricle.

D. activity of the right lateral ampullary nerve is relatively greater than that of the left lateral ampullary nerve.
   E. the patient has a tendency to fall toward the left.

6. All of the following are true concerning the semicircular ducts of the membranous labyrinth, *except*
   A. they are sensitive primarily to angular acceleration.
   B. they consist of three ducts in each inner ear.
   C. they contain fluid that has a high concentration of potassium ions.
   D. each is connected to the lumen of the utricle.
   E. each lies in a plane that aligns exactly with the cardinal (vertical and horizontal) axes of the head.

7. After caloric irrigation of the right external auditory canal with ice water, all of the following occur, *except*
   A. a left-beating (fast phase left) nystagmus.
   B. a decrease in firing frequency of primary afferents innervating the right lateral semicircular duct.
   C. depolarization of hair cells in the crista of the right lateral semicircular duct ampulla.
   D. deflection of the cupula in the right lateral semicircular duct ampulla away from the utricle.
   E. deflection of the stereocilia away from the kinocilium of each hair cell in the crista of the right lateral semicircular duct ampulla.

8. All of the following statements about the lateral semicircular ducts are true, *except*
   A. are analogous to angular accelerometers.
   B. have a receptor structure (the cupula) having different specific gravity than that of the surrounding endolymph.
   C. have primary afferents that project to the vestibular nuclei and the cerebellum.
   D. are easily tested by caloric irrigation of the ear.
   E. are involved with horizontal eye movements in response to horizontal head motions.

9. All of the following are true about slow movement of the eyes during nystagmus, *except*
   A. used to define its direction clinically.
   B. in the direction of the endolymph movement.

C. in the direction of fast pointing or falling.

D. in the direction of a preceding rotation.

E. in the direction opposite to that of vertigo.

10. All of the following statements about endolymph are true, *except*

A. is a special extracellular fluid contained within the membranous labyrinth.

B. has a high potassium and low sodium content.

C. is found in the cochlea.

D. surrounds the membranous labyrinth.

E. the specific gravity is much lower than the statoconial membrane of the otolith organs.

11. All of the following concerning the vestibular apparatus and the cochlea are true, *except*

A. have continuous endolymphatic and perilymphatic spaces.

B. have hair cells in their sensory neuroepithelia.

C. have different cortical projections.

D. both are always involved when peripheral disease occurs.

E. have afferent and efferent innervation of their sensory receptor cells.

12. The cochlear microphonic can be characterized by each of the following statements, *except*

A. it is an instantaneous electrical signal generated in the hair cells.

B. it faithfully mimics the auditory stimulus waveform.

C. it is dependent on and proportional to basilar membrane displacement.

D. it is transmitted unchanged to the dorsal cochlear nucleus.

E. it uses the high transmembrane endocochlear potential set up by the stria vascularis.

13. All of the following regarding the vestibular system are true, *except*

A. the membraneous labyrinth contains head linear acceleration receptors and head angular acceleration receptors.

B. like the cochlea, the sensory neuroepithelium contains only type II hair cells.

C. vestibular primary afferent fibers project to the vestibular nuclei and to the cerebellum.

D. the vestibular system and appropriate motor systems attempt to stabilize images on the retina during rapid head motion and head tilts.

E. nonphysiological stimuli or pathology of the vestibular system can produce nausea, vertigo, spontaneous nystagmus, or disorientation.

Select the single best answer.

14. Ciliary processes of the sensory cells of the vestibular apparatus

A. are oriented randomly in the crista ampullaris.

B. are bathed in perilymph.

C. are functionally (physiologically) polarized.

D. are of the same type(s) as those in the cochlea.

E. consist of a single kinocilium encircled by many thin processes called stereocilia.

15. The otolith organs

A. sense both linear and angular acceleration of the head.

B. are routinely tested by caloric irrigation of the ear.

C. contain sensory hair cells that lack a kinocilium.

D. are primarily associated with the ascending vestibulo-ocular pathway.

E. are important for postural adjustments of the limbs and trunk when the head is tilted.

16. The otolith organs are thought to be sensitive to linear acceleration because

A. the ciliary processes of the sensory cells are graded in height.

B. the sensory cells are a single type specialized to sense shearing forces.

C. the specific gravity (density) of the statoconia (otoliths) is greater than that of the surrounding fluid.

D. the hair cells are morphologically polarized along a single axis.

E. gravity normally produces compression of the statoconial membrane that activates the sensory cells.

17. Functional polarization of the vestibular hair cells means

    A. the direction in which the ciliary processes are bent is the determinant of the response (change in membrane potential) of the sensory cell.

    B. the responses of the sensory cells are related to the shape of the cell body (pear-shaped type I or cylindrical type II).

    C. the semicircular ducts are sensitive to the position of the head in space.

    D. the otolith organs are primarily associated with the descending projections of the vestibular nuclei.

    E. the uniform height of the ciliary processes allows them to bend in one direction only.

18. The cochlear hair cells

    A. are divided into two types, as are vestibular hair cells.

    B. are bathed in perilymph.

    C. produce an action potential.

    D. have no kinocilium.

    E. make synaptic contact only with efferent fibers.

19. The electrical potential of the cochlea or central auditory system that is directly proportional to the magnitude of basilar membrane displacement is the

    A. endocochlear potential.

    B. cochlear microphonic.

    C. summating potential.

    D. cochlear nerve action potential.

    E. brainstem evoked response (BSER).

20. A nucleus in the medulla that receives primary afferent fibers from the cochlea is the

    A. dorsal cochlear nucleus.

    B. spiral ganglion.

    C. trapezoid body.

    D. inferior colliculus.

    E. restiform body.

21. Which of the following is a constant potential in the cochlea and is necessary for normal hearing?

    A. Endocochlear potential

    B. Cochlear microphonic

    C. Summating potential

    D. Primary afferent action potential

    E. Hair cell generator potential

22. In contrast to vestibular hair cells, cochlear hair cells

    A. are flask shaped.

    B. synapse with afferent and efferent terminals.

    C. are bathed in endolymph.

    D. have one kinocilium and many stereocilia.

    E. have only stereocilia.

23. Which of the following theories best explains the fact that, although the upper frequency limit of hearing exceeds the maximum firing rate of cochlear primary afferents, the lower frequency limit is as low as 18 Hz?

    A. Place theory

    B. Frequency theory

    C. Duplex theory

    D. Torsion pendulum theory

    E. Impedance matching theory

24. The receiving station in the medulla for impulses from spiral ganglion neurons is

    A. ipsilateral superior olive.

    B. nuclei of the lateral lemniscus.

    C. inferior colliculus.

    D. medial geniculate.

    E. ventral cochlear nucleus.

25. The cochlear microphonic

    A. is a "minicochlear" topographic representation in the transverse temporal gyrus.

    B. is an electrical signal generated in the hair cells and mimicks the auditory stimulus waveform.

    C. is a compound action potential recorded in the cochlear nerve.

    D. sets up the basilar membrane traveling wave.

    E. is a clinical test of middle ear function.

26. The sensory receptor cells of the cochlea,

    A. are tall cylindrical cells, similar to the type II vestibular hair cells.

    B. have no kinocilium.

C. have stereocilia embedded in the tectorial membrane.

D. are bathed in the endolymph of the scala media.

E. all of the above.

27. According to the place theory, high-frequency tones

A. selectively stimulate only outer hair cells.

B. cause the basilar membrane to vibrate uniformly like a microphone.

C. stimulate few neurons in the cochlear nucleus and most neurons of the inferior colliculus.

D. produce maximum displacement of the basilar membrane near the base of the cochlea.

E. none of the above.

28. The auditory ossicles act as

A. amplifier.

B. damping device.

C. impedance matching device.

D. sharpening device.

E. none of the above.

29. One of the four structures that does not form the cochlea duct is the

A. scala macula.

B. scala media.

C. scala vestibuli.

D. scala tympani.

E. basilar membrane.

30. High-pitched (high-frequency) sound excites hair cells that are

A. close to the oval window.

B. farthest from the oval window.

C. near the helicotrema.

D. midway between the oval window and the helicotrema.

E. in the cupula.

31. Vibration of the oval window is transmitted to the hair cells

A. through perilymph in the scala vestibuli.

B. through spiral lamina.

C. through cochlear nerve.

D. through endolymph in the scala media.

E. by the organ of Corti.

32. Mechanical distortion of the auditory hair cells produces

A. microphonic potential.

B. chemical transmitter.

C. inhibitory potential.

D. reflex eye movement.

E. none of these.

33. The perilymph has a much higher concentration of _____ than the endolymph.

A. potassium

B. sodium

C. chloride

D. calcium

E. none of these

34. Sound waves are transmitted to a series of bones connected in the following order:

A. malleus to incus to stapes.

B. incus to malleus to stapes.

C. stapes to malleus to incus.

D. stapes to incus to malleus.

E. malleus to stapes to incus.

35. Microphonic potentials are produced by

A. electrical stimulation of the hair cell.

B. chemical stimulation of the hair cell.

C. mechanical stimulation of the hair cell.

D. nerve stimulation of the hair cell.

E. none of the above.

36. The cochlear duct

A. is continuous with the scala vestibuli.

B. is continuous with the scala tympani.

C. is continuous with the helicotrema.

D. ends as a blind sac.

E. opens into the middle ear.

37. Sound perception originates in

A. the organ of Corti.

B. the vestibular apparatus.

C. semicircular duct.

D. vestibule.

E. otolith organs.

38. The semicircular canals are essential

    A. for detecting a sudden change in the body posture.

    B. for detecting sound.

    C. for maintaining a proper auditory reflex.

    D. for tracking objects in space.

    E. none of these.

39. The organ most affected by changes in gravitational force is

    A. the scala media.

    B. the utricle and saccule

    C. the vestibule.

    D. the ampulla.

    E. organ of Corti.

40. The most peripheral point in the auditory pathway at which the central nervous system applies feedback control of sensory input if the

A. organ of Corti.

B. oval window.

C. otolith organ.

D. tympanic membrane.

E. round window.

41. Unilateral deafness would be produced by unilateral destruction of the

    A. auditory cortex.

    B. medial geniculate nucleus.

    C. cochlear nucleus.

    D. lateral lemniscus.

    E. inferior colliculus.

42. The structure that improves sound energy transfer from air to cochlear fluids (perilymph) is

    A. the auditory tube.

    B. the ossicular chain.

    C. the tectorial membrane.

    D. the organ of Corti.

    E. the otolith organs.

ANSWERS

1. E. The vestibular end organs probably do not have a topographical representation in the cortex. The auditory end organs do have an organized pattern of tonal localization in the temporal lobe.

2. D. They do not innervate the stapedius muscle. Inhibition and hyperpolarization are synonyms here.

3. C. When the head is rotated to the right, the fluid in the lateral semicircular canals moves to the left. The cupula and hair cells will move in the same direction as the fluid, namely to the left. In the left ear this will be away from the utricle (located on the midline) and in the right ear toward the utricle. Deflection away from the kinocilium (and in this case the utricle) results in a decrease in afferent discharge and toward an increase in afferent discharge. When the body turns to the right the world appears to move to the left, and the eyes (slow phase) will attempt to follow or track

the movement of the world. When the eyes reach a maximum deflection they are rapidly (saccade) returned in the opposite direction (fast phase).

4. D. The stapes transmits vibration directly to the perilymph via the oval window.

5. C. The direction of nystagmus (fast phase) is in the same direction as the sensation of spinning (COWS). If spinning is towards the right the fluid moves in the opposite direction, and along with the fluid the hairs are bent in this direction. In the right canal this will be toward the utricle, which will result in greater discharge of the afferent fibers. There is a tendency to fall toward the left.

6. E. The lateral duct is inclined about 30° from the horizontal.

7. C. Again, Cold, opposite–warm, same. In the right

canals the direction of nystagmus is to the left (opposite), which means the fluid is moving toward the right. The sensation of spinning is in the direction of the fast phase (left) and the eyes move slowly toward the right in order to watch the world go by. If the fluid is moving toward the right the cupula and hairs move in the same direction, which is away from the utricle and kinocilium. This will decrease the afferent discharge. If the discharge is discreased then the membrane potential must not be moving in the depolarizing direction.

8. B. The otolith organ has a receptor structure of a higher specific gravity than that of the surrounding endolymph.

9. A. The fast phase is used to define the direction on nystagmus although it is the slow phase that is physiologically most important. However, it is easier to see the fast phase during eye examination. The fast phase is in the direction of rotation, the slow phase in the direction of fluid movement. After one stops rotating, the fluid also changes direction. This is why the slow phase changes to the opposite direction.

10. D. Endolymph is found within the membranous labyrinth. Perilymph surrounds the membranous labyrinth.

11. D. These areas can be disease specific, affecting each system individually. In addition, the middle ear is susceptible to infections that will not affect the vestibular apparatus because of anatomical separation.

12. D. The receptor or generator potential produced by the hair cells in the cochlea become action potentials, which then travel to nuclei in the brainstem.

13. B. The vestibular neuroepithelium contains both type I and II hair cells.

14. C. Ciliary processes are morphologically polarized, are bathed in endolymph, are not the same as in the cochlea, and the kinocilium is to one side.

15. E. The otolith organs primarily sense changes in linear acceleration, rather than in angular acceleration, which is the primary function of the semicircular canals. It is the latter that are tested by caloric irrigation. Sensory hair cells of both the otolith organs and semicircular ducts contain a kinocilium, and again it is the latter that are primarily involved in the vestibulo-ocular reflex.

16. C. It is the sliding motion of the statoconial membrane following head tilt that excites the sensory receptors. Since they are heavier than the surrounding fluid, they move under the influence of gravity in the fluid. It is bending of these hair cells, not compression by gravity, that results in excitation of the afferent fibers. The hair cells are actually divided into opposite areas of morphological polarization in the macula.

17. A. Because of their morphological polarization, some hair cells will increase their firing rate, and others will decrease their firing rate, depending on the direction in which they are bent.

18. D. Unlike the vestibular hair cells the cochlear hair cells have no kinocilium.

19. B.

20. A. The spiral ganglion contains the cell bodies of the primary afferent fibers coming from the cochlea. These primary afferents end in both the dorsal and vental cochlear nuclei of the medulla. Second order fibers from the cochlear nuclei cross the midline forming the trapezoid body. Most of these ascend in the lateral lemniscus and end in the inferior colliculus. Fibers from the inferior colliculus project to the lateral geniculate or cross to the opposite inferior colliculus. The lateral geniculate then projects to auditory cortex in the temporal lobe.

21. C. The cochlear microphonic is superimposed on the summating potential.

22. E. The cochlear hair cells have only stereocilia. The vestibular hair cells have both a kinocilium and stereocilia.

23. C. The duplex theory states that the frequency theory applies principally to the detection of low frequencies, the place theory to high frequencies, and that they overlap broadly for midrange coverage.

24. E. The ventral and dorsal cochlear nuclei receive afferent input fibers arising in the spiral ganglion.

25. B.

26. E.

27. D. High frequencies do not travel as far as low frequencies. Low frequencies excite the basilar membrane near the apex of the cochlea.

28. C.

29. A. The macula is part of the vestibular system.

30. A. High frequencies travel only short distances.

31. A.

32. A. Inhibitory potential can be produced by efferent innervation of sensory cells.

33. B. It has only a slightly higher concentration of chloride.

34. A.

35. C.

36. D. The cochlear duct is a synonym for scala media, which ends as a blind sac at the apex of the cochlea.

37. A. The organ of Corti is the sensory organ of hearing.

38. A. The vestibular apparatus tells the brain about the position of the body in space.

39. B.

40. A. The organ of Corti is the sensory organ of hearing and receives efferent innervation. The otolith organs are sensory organs for the vestibular apparatus and also receive efferent innervation.

41. C. Only the cochlear nuclei receive afferent innervation from one ear. Fibers from both ears send information to all the other structures listed.

42. B. The ossicular chain serves as an impedance matching device.

44

Taste and Olfaction

ARTHUR M. BROWN

OBJECTIVES

After completion of this chapter, you should be able to:

1. Describe the functional morphology of taste organs.

2. Describe the functional morphology of olfactory organs.

SENSE OF TASTE

The receptors for taste are the taste buds, which are located on the tongue, pharnyx, larynx, and roof of the mouth. The taste bud is made up of receptor cells, supporting cells, and afferent nerve fibers arranged segmentally much like the segments of an orange. A taste bud is onion shaped, measures 60–80 μm in length and 40 μm in diameter, and contains about 40 taste cells, which are modified epithelial cells. The human taste organs contain about 9,000 taste buds. Distal ends of taste cells form microvilli arranged around a minute pore on the top of the taste buds. The microvilli are thought to be the site of taste reception. Access of chemical stimuli is obtained via the pore (Fig. 1). Near the pore the taste cells are interconnected through tight junctions so that the inside of taste buds are insulated from the outside. A first-order sensory neuron can send afferent projections to several taste buds, and a single taste bud can be innervated by several primary afferent neurons. A simple afferent fiber connected to a single taste bud can be stimulated by any of the four traditional taste stimuli, or chemicals: sweet, sour, salty, and bitter. There is no apparent specificity at the level of the taste buds, but there are clear differences among taste buds in the pattern of impulse discharge provoked by the same stimulus and by different stimuli. Through the use of appropriate stimuli, it appears that sweet and salty tastes come from the front of the tongue, bitter from the back, and sour from the sides. The perceptual processes are thus likely to be embedded in subsequent or second-order neurons that process the afferent input.

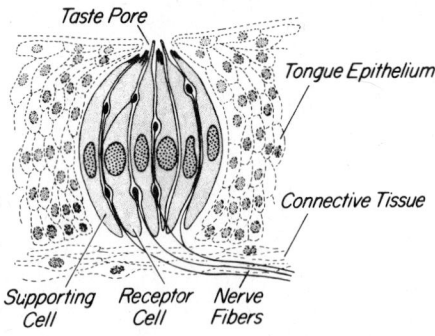

Taste Pore

Tongue Epithelium

Connective Tissue

Supporting Cell   Receptor Cell   Nerve Fibers

Figure 1 Morphology and nerve supply of a taste bud.

The transduction mechanism for taste is unknown. The stimulating chemicals are thought to interact with specific receptor sites on the receptor cell membrane and produce a depolarizing receptor potential. The receptor sites must be specific for many taste stimuli, such that sugars or H^+ in the concentrations used do not have specific effects on other nerve membranes. The coupling between receptor cell and afferent fiber is also unknown.

The properties of the chemical stimuli necessary to evoke the four primary categories of taste have been examined. Sourness requires the presence of H^+, although the degree of sourness may differ at constant pH. Other factors may be involved, including ionic strength and surface charge at the receptor.

A salty taste is given by NaCl. Most other salts produce a mixed taste of saltiness and bitterness. Bitterness is not produced by any single structural component, and many bitter substances may be converted into sweet-tasting by small changes in molecular structure. It has been suggested that all sweet-tasting substances of which sugar is the prototype bind to the receptor membrane nonionically via hydrogen bonds.

Cranial nerves VII, IX, and X carry the afferent information and have cell bodies located in the geniculate, superior petrosal, and nodose ganglia, respectively. The anterior two-thirds of the tongue is innervated by the lingual nerve. The taste afferent fibers branch from the nerve as the chorda tympani and enter the brainstem as cranial nerve VII. The posterior third of the tongue is innervated by cranial nerve IX, and the pharyngeal taste areas are innervated by cranial nerve X. The afferent fibers in IX or X (first-order neurons) have their synapse in the nucleus tractus solitarius. Fibers pass from neurons in this area (second-order neurons) to the thalamus, to another synapse. From the thalamus there are projections via the internal capsule to cortical areas near the somatosensory cortex. The central pathways for taste are closely associated with the somatosensory system of the face.

OLFACTION

The sense of smell is initiated by receptors located in the mucus-secreting membrane, called the olfactory mucosa, situated in the upper part of the nasal cavity (Fig. 2). The cell bodies of the afferent neurons are located in the olfactory epithelium; they send short processes out to the mucosal surface. A longer axon projects to the brain in the olfactory nerve, or cranial nerve I. The primary olfactory neurons are separated from each other by supporting cells that have a prominent endoplasmic reticulum, suggesting a secretory function in addition to the supporting func-

tion. (The function of supporting cells in sensory receptors is presumably a metabolic one, much as glial cells and Schwann cells apparently provide nutrition to axons and nerve cell bodies elsewhere in the nervous system.)

The central axons enter the olfactory bulb, where they synapse with dendrites of second-order cells called mitral cells. Horizontal interneurons are present that intercommunicate; centrifugal fibers also synapse here. The usual convergence and divergence produce a seemingly uninterpretable wiring maze here as is common elsewhere in the nervous system. The axon of the initial cells project to the cortex and from there information is relayed to the thalamus and hypothalamus, where they are involved in controlling feeding and reproduction.

The transduction process requires that the odorous stimulus permeate the mucous layer and make contact with the receptor cells. Whether there are receptor cells that respond to only one specific stimulus, or whether one receptor cell may respond to a variety of stimuli with different patterns of discharge is unknown. The mechanisms of transduction also appear to be unknown. In a general sense the lock-and-key hypothesis of Ehrlich, which applies to receptor theory in general, is thought to hold, but the details are totally unknown. The best model for olfaction and taste is another well-studied chemically gated change in membrane excitability, the end-plate potential.

The basis for odor discrimination is also unknown. Again, as for all the special senses, there are two main possibilities: (1) specific receptors coupled to specific pathways, and (2) nonspecific receptors coupled to specific pathways with the discriminatory capabilities embedded in recognition among different discharge patterns.

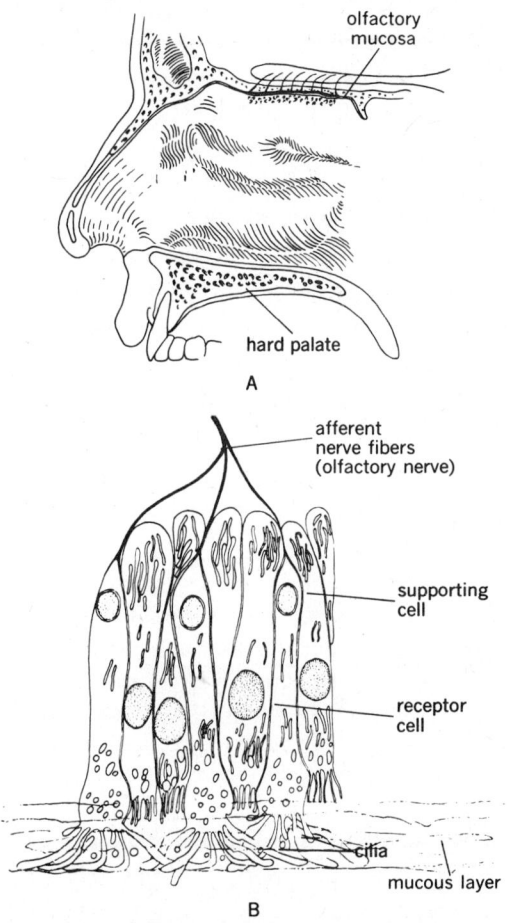

Figure 2 Morphology of olfactory receptors.

SELECTED BIBLIOGRAPHY

Vander AJ, Sherman JH, Luciano DS: *Human Physiology.* New York, McGraw-Hill, 1980, p. 584.

SELF-STUDY QUESTIONS

MULTIPLE CHOICE

Select the single best answer.

1. Taste receptors respond to
   A. only one kind of stimulant.
   B. all kinds of stimulant.
   C. one kind of stimulant predominantly, although it can respond to others as well.
   D. two kinds of stimulant applied simultaneously.

2. Taste receptors show
   A. adaptation.
   B. no adaptation.
   C. an increase in sensitivity with constant stimulation.
   D. none of these.

3. The initial reaction of gustatory reception occurs
   A. in the gustatory nerve fiber.
   B. in the gustatory cell.
   C. in the sustentacular cell.
   D. in the cells of the mucous membrane.

4. The olfactory cell
   A. has a nerve fiber (axon).
   B. has no nerve fiber (axon).
   C. is a second-order neuron.
   D. is none of these.

5. A wine connoisseur can recognize several hundred varieties of wine by sampling them. Given that there are only four types of taste receptors, how is this possible?

ANSWERS

1. C. There is some specificity among taste buds regarding stimulus, but it is not absolute.

2. A.

3. D. The epithelial cells respond to the olfactory stimulus and are coupled to the first order neuron by a transmitter. Similar arrangements occur in hair cells of the cochlea and chemoreceptors in the carotid and aortic bodies.

4. D.

5. Most substances stimulate two or more types of receptors, and most receptors are stimulated by more than one substance. Hence the discriminatory ability of the wine connoisseur rests on the great number of effects possible by combination of the four basic tastes in different relative concentrations. Olfaction also has an important role in taste.

45

Neuroendocrinology

AILEEN K. RITCHIE

OBJECTIVES

After completion of this chapter, you should be able to:

1. State the anatomical relationship of the hypothalamus to the adenohypophysis and neurohypophysis.

2. State the origin of the neurohypophyseal hormones.

3. List the elements involved in the neural reflex and central regulation of oxytocin and antidiuretic hormone secretion.

4. Explain the importance of the median eminence and pituitary portal system to adenohypophyseal function.

5. List the functions of the hypophysiotropic substances.

6. Describe the major physiological effects, the cellular origin, the target organs, and the general chemical families of the seven adenohypophyseal hormones.

7. Explain the positive and negative feedback control of adenohypophyseal function.

8. List the factors that control growth hormone secretion and the physiological effects of growth hormone, as well as describe the roles of somatotropin-releasing factor, somatostatin, and somatomedin in regulating the release or mediating the effects of growth hormone.

9. State the role of the hypothalamic–pituitary axis in regulating the menstrual cycle, ovulation, and puberty.

10. Explain the regulation of prolactin, thyroid stimulating hormone, adrenocorticotropic hormone, and melanocyte stimulating hormone secretion.

Once you have an appreciation of the above objectives you should be able to explain, predict, and diagnose the various pathological or adaptive conditions that occur when there is a hormonal disturbance involving the hypothalamic–pituitary axis.

The major objective of this chapter is to introduce the pituitary hormones and to provide information regarding the hypothalamic regulation of pituitary function and the integrating role of the hypothalamic–pituitary unit within the endocrine system.

The physiological role of most of the pituitary hormones have been, or will be, covered in those chapters concerned with the function of the individual target organs affected by the pituitary hormones. The physiological role of growth hormone (or somatotropic hormone), which affects nearly all tissues of the body, is described here in greater detail.

ANATOMY OF THE HYPOTHALAMIC-PITUITARY UNIT

This chapter discusses the interaction of the nervous system with the endocrine glands. Integration of endocrine function is achieved primarily at the level of the hypothalamus through its control over the pituitary gland. To understand the functional relationship between these two organs, it is necessary to examine their anatomical relationship.

The pituitary gland (hypophysis) lies in a concavity of the sphenoid bone called the sella turcica and is connected to the base of the brain by a thin pituitary stalk. The gland is composed of two divisions that are distinct in embryonic origin, in histological composition, and in their functional relationship to the hypothalamus. The *neurohypophysis* is composed primarily of neural tissue; it originates embryonically as a ventral evagination from the floor of the third ventricle. The *adenohypophysis* is the more typical glandular portion of the pituitary and arises from a dorsal outgrowth of the buccal cavity in the developing embryo.

The neurohypophysis consists of the *pars nervosa* (also known as the neural lobe or infundibular process); the *median eminence,* which is a specialized neurovascular region located at the base of the hypothalamus; and the *infundibular stem* (Fig. 1). The pars nervosa is composed mainly of densely packed nerve endings in association with columns of fenestrated capillaries. Two neurohypo-

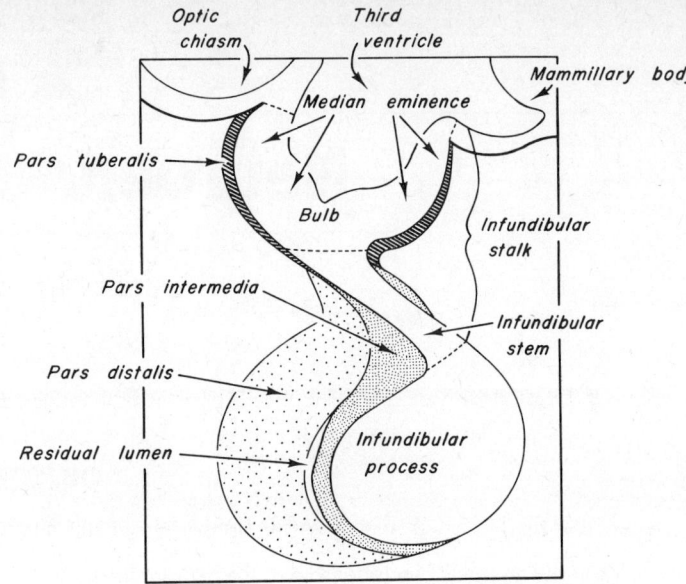

Figure 1 Diagrammatic outline of the structure and standard nomenclature of the hypothalamic–pituitary unit of the hypophysis of a macaque monkey *(Macaca mulatta).* (Redrawn from Rioch DM, Wislocki GB, O'Leary JL: A Précis of Preoptic, Hypothalamic and Hypophysial Terminology with Atlas, *Res Publ Assoc Nerv Ment Dis* 20:3, 1940.

physeal hormones synthesized elsewhere are released from these nerve endings and carried into the general circulation by the capillaries. The infundibular stem contains the axons of neurons, which originate in the supraoptic and paraventricular nuclei of the hypothalamus, traverse the median eminence and terminate in the pars nervosa. Since the median eminence plays no role in neurohypophyseal function, but is extremely important in the control of adenohypophyseal function (see below), the classical designation of the median eminence as a part of the neurohypophysis is unfortunate.

The adenohypophysis is also comprised of several anatomical regions, the major one of which is the *pars distalis.* A small segment, which extends partially up the pituitary stalk, is called the *pars tuberalis,* and a third segment, which adjoins the pars nervosa, is called the *pars intermedia.* The pars distalis, frequently referred to as the anterior lobe of the pituitary, is the source of the seven anterior pituitary hormones. One of these hormones is also produced in the pars intermedia, a segment that is prominent in many vertebrates but absent in the human. The glandular cells of the pars distalis receive no innervation from the hypothalamus.

The superior hypophyseal artery supplies blood to a rich capillary plexus (primary plexus) located in the median eminence. The capillaries drain into the long hypophyseal portal vessels of the pituitary stalk, then ramify again to form the sinusoidal capillary loops (secondary plexus) that supply blood to the pars distalis. The inferior hypophyseal artery distributes blood primarily to the pars nervosa; however, a branch also supplies blood to a primary capillary plexus located in the lower stalk. This lower stalk blood is then delivered to the pars distalis by the short hypophyseal portal vessels. Therefore, the adenohypophysis is mainly supplied by portal blood from the median eminence and lower stalk, while the neurohypophysis receives only arterial blood (Fig. 2).

The primary capillary plexus of the median eminence and lower stalk is densely populated by nerve endings of axons that arise primarily in the ventral hypothalamus. These neurons are collectively referred to as the tuberohypophyseal tract. Neurohumoral substances are released by the nerve endings, collected by the fenestrated capillaries, then carried by the long and short hypophyseal portal veins to the adenohypophysis where the substances regulate the secretion of hormones produced by the glandular cells of the adenohypophysis. The glandular cells of the anterior pituitary are not innervated by the hypothalamus, thus the secretion by these cells of anterior pituitary hormones is regulated entirely by hypothalamic neurohumoral substances. In contrast, the hormones of the neurohypophysis (posterior pituitary) are secreted in the pars nervosa directly from nerve terminals of neurons that originate in the hypothalamus (Figure 2).

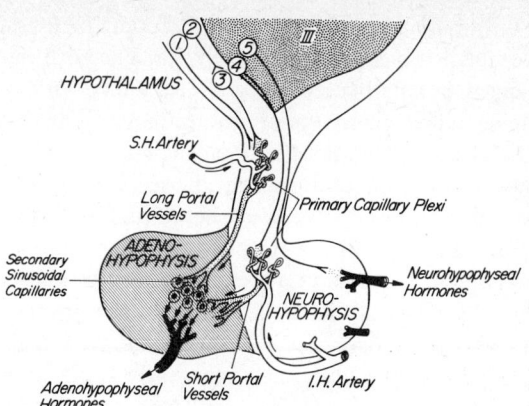

Figure 2 Neural and vascular connections between the hypothalamus and hypophysis. Neuron 5 represents the neurons of the supraopticohypophyseal and paraventriculohypophyseal tracts, with hormone-producing (ADH, oxytocin) cell bodies in the hypothalamus and secretory nerve terminals in the neurohypophysis. Neurons 4 and 3 are the hypothalamic neurons of the tuberohypophyseal tract, which secrete the hypophysiotropic hormones into the substance of the median eminence or stalk in anatomical relationship to the primary capillary plexi (which in humans form spiral loops called gomitoli). The hypophysiotropic hormones enter the primary capillary plexi and are delivered to the pituitary via long and short portal vessels that eventually ramify to become the sinusoidal capillaries (secondary plexus) of the adenohypophysis. Neurons 1 and 2 are the functional links between the remainder of the brain and the hypothalamic hypophysiotropic hormone secreting neurons. Neuron 2 represents a possible axoaxonic transmission resulting from a neuron which ends on terminals of a hypophysiotropic hormone secreting neuron.

ytocin causes milk letdown and uterine smooth muscle contraction.

NEUROHYPOPHYSIS

The two hormones secreted from the neural lobe of the pituitary are *antidiuretic hormone* (ADH), also called vasopressin, and *oxytocin.* ADH plays an important role in maintaining fluid and electrolyte homeostasis, and ox-

HORMONE SYNTHESIS, TRANSPORT, AND SECRETION

ADH and oxytocin are both octapeptides (molecular weight 1,100) that differ in only 2 amino acids; however, oxytocin is 100-fold more potent than ADH in causing milk letdown in humans, and ADH is several hundredfold more potent than oxytocin in causing antidiuresis in dogs. The hormones are secreted from neurons having cell bodies located in the paraventricular (PV) and supraoptic (SO) nuclei of the hypothalamus. Although some types of stimuli can elicit the secretion of both ADH and oxytocin, the hormones are synthesized in different neurons, and their release is independently regulated. The hormones are packaged in small secretory granules along with carrier proteins called *neurophysins* (molecular weight of about 10,000–12,000). Synthesis and packaging occur in the cell bodies before axonal transport of the secretory granules to their release sites in the pars nervosa. Nerve terminal depolarization, resulting from an action potential, activates voltage dependent Ca^{2+} channels and the subsequent entry of Ca^{2+} ions causes discharge of the granule contents via exocytosis. Before or during exocytosis, the neurohypophyseal hormones apparently split from their carrier protein and enter the capillary system as the free hormone. Neurophysin I is released with ADH and neurophysin II is released with oxytocin.

ANTIDIURETIC HORMONE

Regulation of fluid and electrolyte balance involves the complex interaction of ADH, the renin–angiotensin–aldosterone system, thirst, drinking behavior, and kidney function. This section is simplified to emphasize the role of ADH. ADH promotes water retention by increasing the water permeability of the collecting ducts of the kidney, and at very high concentrations ADH also has a pressor effect due to its action on blood vessels (hence the name vasopressin).

Injection of hypertonic saline into the carotid artery accelerates electrical activity recorded from neurons located in the SO and PV nuclei and is accompanied by release of ADH. This response is evidence of the existence of central osmoreceptors, which regulate the discharge of the posterior pituitary hormone. The receptors are very sensitive since only a 1% change in plasma osmolality affects the quantity of ADH secreted. Dehydration stimulates and water loading inhibits ADH secretion by this mechanism (see Fig. 8, chap. 33).

Another important regulator of ADH secretion is a neural reflex involving peripheral volume receptors. These volume receptors, primarily the baroreceptors located in the atrium and juxtaglomerular apparatus of the kidney, send afferent signals to the brain that are eventually received by the neurons in the PV and SO nuclei. A 7–15% drop in blood volume is required to trigger ADH secretion by this mechanism; however, 10 to 100 times higher levels of ADH are elicited than are released by osmoregulation.

The levels of ADH released by hypovolemia are high enough to elicit a pressor response via activation of vasopressin receptors located in the walls of certain blood vessels. Therefore, hemorrhage elicits release of ADH in sufficient quantity to help restore blood pressure in addition to its role in water retention.

Water conservation is mediated not only by decreased excretion but also by increased intake of water. It is therefore not surprising that the levels of hyperosmolality and hypovolemia that evoke ADH release also stimulate the thirst mechanism.

Other neurogenic stimuli that affect ADH release and that can override the responses evoked by osmolar or volume changes include a wide variety of physical and emotional stresses (e.g., cold stress, trauma, anxiety, and pain). Adrenergic and cholinergic neurotransmitters and several neuropeptides may play a role in mediating these responses, since nicotinic cholinergic agonists, β-endorphin, and angiotensin stimulate ADH release and β-adrenergic agonists inhibit release. Heavy cigarette smokers tend to have higher levels of ADH due to nicotinic stimulation of ADH release; the diuretic effect of alcohol is due to inhibition of ADH release.

Diabetes insipidus (DI) is a disease caused by a deficiency of ADH (neurogenic DI) or by failure of the collecting duct cells of the kidney to respond to ADH (nephrogenic DI). These defects lead to the excretion of large volumes (5–10 L/day) of dilute urine (polyuria). The increased thirst (polydipsia) and drinking, also characteristic features of this disease, are caused by the fluid loss and increase in plasma osmolaity. Neurogenic DI has a familial or idiopathic origin or occurs as a result of irreversible damage to the hypothalamic neurons that synthesize ADH. Damage to the posterior lobe of the pituitary causes only temporary neurogenic DI since function is eventually restored by ADH secretion from nerve terminals at sites proximal to the pituitary.

The clinical symptoms of polyuria and polydipsia are not only characteristic of DI, but are also present in cases in which the primary disturbance is a persistent, excessive intake of water (psychogenic water drinking). The chronic intake of water inhibits the secretion of ADH and in addition diminishes the intrarenal concentration gradient on which water reabsorption by the collecting ducts ultimately depends.

Differential diagnosis is accomplished by assessing the

response of the kidneys to cautious restriction of water intake or to administration of ADH. Persons with neurogenic DI excrete a concentrated urine in response to ADH administration but continue to excrete a dilute urine upon moderate dehydration. Patients with a psychogenic water drinking problem excrete a concentrated urine most effectively upon restriction of water intake and to a lesser extent in response to ADH administration. Patients with nephrogenic DI fail to elaborate a concentrated urine in response to ADH administration or to the osmotic stimulus caused by temporary water restriction.

Overproduction of ADH, known as syndrome of inappropriate antidiuretic hormone (SIADH), can be caused by ADH secretory tumors or by malfunction of CNS regulation of ADH release, or as a side effect of many drugs. The syndrome is characterized by hyponatremia and low output of urine of inappropriately high osmolality.

OXYTOCIN

Oxytocin is released in response to suckling through an afferent reflex pathway that ultimately evokes impulses in the PV and SO nuclei of the hypothalamus. The liberated oxytocin causes contractions of the myoepithelial cells of the mammary gland, thus causing milk ejection (milk letdown) from mammary glands prepared for lactation. Oxytocin secretion is also evoked by anticipation of nursing and inhibited by emotional stress.

Oxytocin causes contraction of the uterine smooth muscle, and the responsiveness of the muscle varies according to the phase of the menstrual cycle or stage of pregnancy. Although the physiological role of oxytocin on myometrial contractility is not completely known, the peptide is used clinically to induce labor and to control postpartum hemorrhage.

ADENOHYPOPHYSIS

The adenohypophysis secretes seven hormones of known function. Four hormones, that is, adrenocorticotropic hormone (ACTH), thyroid stimulating hormone (TSH), follicle stimulating hormone (FSH), and luteinizing hormone (LH), have other endocrine glands as their target organs. Growth hormone (GH), also called somatotropic hormone (STH), and prolactin affect many nonendocrine cell types. Melanocyte stimulating hormone (MSH) has specific effects on the pigment-containing melanocytes of epidermis. Recently beta endorphin, a peptide with potent opiate activity, has been found in pituitary extracts but its hormonal function is not yet known.

In view of the numerous endocrine functions of the adenohypophysis it is not surprising that the anterior pituitary was once considered the master gland that controls and integrates the activities of the endocrine system. Although numerous clinical and laboratory experiments have long implicated the hypothalamus as an important link between the pituitary and central nervous system (CNS), the role of the anterior pituitary as an antonomous organ persisted, largely due to failure to locate neural connections between the brain and anterior pituitary.

During the 1930s, the missing link was provided by the discovery of the portal system of blood vessels, which supplies venous blood from the hypothalamus directly to the anterior pituitary. In 1937 several investigators realized the functional significance of the portal system by hypothesizing the existence of molecules arising in the hypothalamus that could be carried by the portal system to the anterior pituitary.

Most substances released from the hypothalamus are peptide in nature; thus, the development in the 1960s of the highly specific and sensitive radioimmunoassay techniques greatly facilitated the isolation and identification of these substances. The direct proof of the portal vessel–chemotransmitter hypothesis followed with the purification and chemical identification of a thyrotropin releasing hormone from median eminence extracts that was shown to cause the release of thyroid stimulating hormone when added to cultured organ transplants of the anterior pituitary.

HYPOTHALAMIC CONTROL OF PITUITARY FUNCTION

Neural Input

The CNS receives input from a wide variety of environmental and internal stimuli. A partial list of internal stimuli includes enteroceptors, which sense blood volume, temperature, osmolality, reproductive tract stimulation, and glucose concentration. The CNS is also the source of such ill-defined stimuli as stress, anxiety, sleep–wake rhythms, and excitement. Knowledge of the mechanisms and pathways by which these neural stimuli affect hormone secretion is incomplete but must ultimately include conveyance to the hypothalamic neurons that control pituitary function. Some of the regions of the hypothalamus that have been tentatively implicated in the control of adenohypophyseal function are illustrated in Figure 3. These regions are referred to as the hypophysiotropic area of the hypothalamus.

The Hypophysiotropic Hormones

The neurohumoral substances released by the hypothalamus into the pituitary portal system are called the hypophysiotropic hormones. By convention, the substances are called hormones if their chemical structure is known, and factors when their structure is unknown. Except for the prolactin inhibitory factor, which is probably dopamine, all the hypophysiotropic substances appear to be peptides. The effects of the hypophysiotropic hormones are probably mediated by binding to specific receptors located in the plasma membrane of the pituitary cell and possibly by internalization of the hormones via endocytosis. Specific high-affinity membrane binding sites have been demonstrated in the pituitary for the four chemically identified hypophysiotropic hormones, that is, gonadotropin releasing hormone, thyrotropin releasing hormone, somatostatin, and dopamine.

The hypophysiotropic substances affect both synthesis and secretion of the pituitary hormones and the differentiation of specific pituitary cell types. Although the effect on protein synthesis is distinct from the immediate stimu-

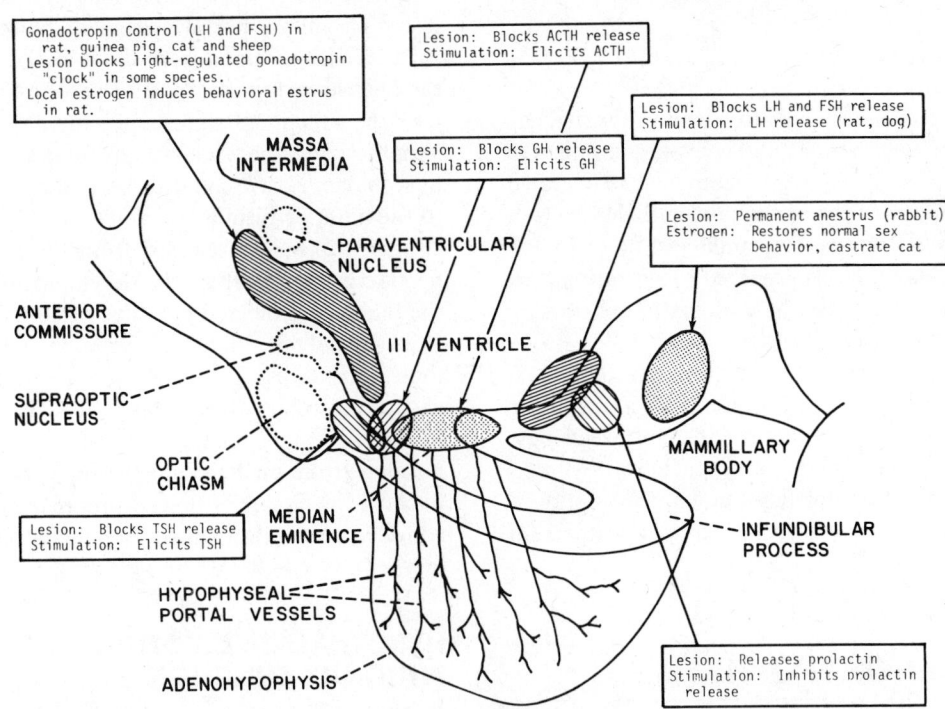

Figure 3 Composite diagram of certain hypothalamic regions that have been implicated in adenohypophyseal function. Regions indicated are to be regarded as tentative and are by no means as clearly demarcated as they appear to be in the diagram. Note that hypothalamic organization with respect to gonadotropin control and sex behavior shows interspecies variation. (Modified with permission from Tepperman J:*Metabolic and Endocrine Physiology,* Chicago, Year Book, 1980.)

lation of secretion, with prolonged stimulation secretion can become dependent on synthesis of hormone or of some other protein involved in hormone release. The effects on protein synthesis and cellular differentiation may be mediated through intracellular messengers via activation of the adenylate cyclase system or possibly by an intracellular effect from internalization of the hypophysiotropic peptide.

Pituitary hormone secretion occurs by exocytosis and is mediated by Ca^{2+} ions. The releasing factors bind to membrane receptors, then stimulate or modify secretion by mobilizing internal stores of Ca^{2+} through an affect on the adenylate cyclase system or by causing influx of Ca^{2+} into the cell through voltage dependent Ca^{2+} channels. Thyrotropin releasing hormone has recently been shown to activate voltage-dependent Ca channels by increasing the frequency of spontaneous action potentials recorded from rat anterior pituitary cells.

Feedback Control Mechanisms

Negative feedback mechanisms play a fundamental role in preserving stability within the neuroendocrine system. Increases in the blood level of hormones secreted by the target gland (thyroxine, corticosteroids, and sex steroids) can inhibit release of the pituitary tropic hormone (TSH, ACTH, or gonadotrophin) that initiates the secretion from the target gland. Similarly, decreases in the blood level of a target hormone increase the output of the pituitary tropic hormone. This feedback may occur at the level of the pituitary, the hypothalamus, or at central extrahypothalamic sites. Positive feedback also serves as an important control mechanism, the most notable example being the midcycle rise in estrogen that initiates ovulation by stimulating secretion of luteinizing hormone. Feedback by the target organ hormone is referred to as long loop feedback.

Three of the adenohypophyseal hormones—growth hormone, prolactin, and MSH—have no known target gland hormones to take part in feedback control. Significantly, additional hypophysiotropic substances are present that exert tonic inhibitory influence over the secretions of these particular hormones. That is, these three adenohypophyseal hormones are subject to a dual control by hypothalamic agents (releasing factors and inhibitory factors).

Feedback by pituitary hormones at the level of the pituitary, hypothalamus, or higher levels is called short loop feedback, and when the hypothalamic hormones influence their own secretion, it is referred to as ultrashort loop feedback. Short loop negative feedback is generally not as effective as long loop negative feedback.

Finally, there is feedback interaction from different hypothalamic–pituitary–target hormone systems. For example, estrogen sensitizes the pituitary to the stimulatory effects of thyrotropin releasing hormone and directly stimulates the synthesis and secretion of pituitary prolactin.

Endocrine Rhythms

Large sleep-related surges in the secretion of ACTH, TSH, growth hormone, prolactin, and, under certain circumstances, the gonadotropins (during puberty) occur every 24 h in what are called circadian rhythms. These hormones also exhibit episodic secretions, which are regularly repeated bursts of secretion of lower amplitude with a 1–3-h interval. The time of onset and duration of the surges are not necessarily coincident for the different hormones, and the rhythms are influenced by various humoral, cerebrospinal fluid (CSF), and neural inputs from the internal and external environment. It is not known whether these rhythms are generated by an intrinsic oscillator and merely entrained by the inputs or whether the rhythms are dependent on the inputs. The adaptive value of these endocrine rhythms to the individual is also unknown.

ORIGIN AND CHEMISTRY OF THE ADENOHYPOPHYSEAL HORMONES

The adenohypophysis is a composite of several distinct glandular cell types with each of the seven major hormones of established function being the product of a particular cell type. All the pituitary hormones are stored in membrane-bound secretory granules which, upon appropriate signal, release their contents via Ca^{2+}-dependent exocytosis.

The cell types are distinguishable by their characteristic ultastructure and by immunohistochemical identification of their hormonal content. These cell types (and the hormones they contain) are the somatotropes (growth hormone, or somatotropic hormone), lactotropes (prolactin), thyrotropes (TSH), corticotropes (ACTH and MSH), and gonadotropes (LH and FSH). The gonadotropes include cells that store only LH or FSH and cells that store both. All the cells are located in the pars distalis, with the exception of the corticotropes, which are also found in a region of the neurohypophysis, possibly the human counterpart to the pars intermedia.

Before the development of electron microscopic and immunochemical labeling techniques the cells were classified by their histochemical staining properties. The glycoprotein hormones (LH, FSH, and TSH) and the peptide hormones derived from glycoprotein precursors (ACTH and MSH) belong to the basophilic group. The acidophilic group contains growth hormone and prolactin, and the chromophobes contain no characteristic secretory granules.

Approximately 85% of pituitary adenomas are of the chromophobe type. Because of their destructive invasiveness, they tend to cause primary hypopituitarism by damaging the pituitary secretory cells and secondary hypopituitarism by impairing production or transport of the hypophysiotropic hormones. Approximately 10–15% of pituitary adenomas are of the acidophilic type and secrete growth hormone and/or prolactin, whereas basophilic tumors are rare.

The anterior pituitary hormones and two related placental hormones have been divided into three general chemical classes.

1. *Corticotropin-related peptides:* ACTH, two melanocyte stimulating peptides (α- and β-MSH), β-lipotropin (β-LPH), a peptide with lipid mobilizing activity, and β-endorphin, a peptide with potent opiate activity. Although none of these is a glycopeptide, all are products of a common high-molecular-weight glycoprotein precursor found in corticotropes of the anterior and neural lobes of the pituitary and at a number of other brain sites. Proteolytic cleavage of the glycoprotein precursor gives rise to ACTH (39 amino acids; molecular weight of 4,500) and β-LPH (91 amino acids; molecular weight of 9,500). The first 13 amino acids of ACTH are identical to α-MSH, and the sequences of β-MSH (a 22 amino acid peptide) and β-endorphin (a 31-

amino acid peptide) are contained within the structure of β-LPH. ACTH, β-LPH, and MSH all exhibit varying degrees of corticotrophic, lipotropic, and melanotropic activity. In view of their overlapping biological actions and common precursor, it is not surprising that patients with diseases associated with overproduction of ACTH also show signs of hyperpigmentation (melanotropic activity). The hormonal roles (if any) of β-LPH and β-endorphin remain to be established.

2. *Somatomammotropin hormones:* Growth hormone (molecular weight 21,700), prolactin (molecular weight, 22,500), and chorionic somatomammotropin (molecular weight, 21,600) are classified by virtue of their amino acid sequence homologies and overlapping biological actions. They are single-peptide chains with intrachain disulfide bridges and have probably evolved from a common molecular ancestor.

3. *Glycoprotein hormones:* TSH (molecular weight, 26,600), FSH (molecular weight, 32,000), LH (molecular weight, 30,000), and chorionic gonadotropic hormone (molecular weight, 57,000) are each composed of two peptide chains, α and β. The amino acid sequence of the α chain is nearly identical and is interchangeable for all hormones in this group. The distinctive sequences and hormonal specificity are provided by the β chain, but both chains are necessary for biological activity.

THYROID STIMULATING HORMONE

Thyroid stimulating hormone (TSH), also known as thyrotropin, is responsible for the maintenance of the thyroid gland and stimulation of thyroid hormone secretion. The thyroid hormones in turn participate in regulation of basal metabolic rate, CNS and bone development, and, in some animals, adaptation to cold.

Regulation of TSH Secretion

The hypothalamus exerts positive control over pituitary TSH by secreting thyrotropin releasing hormone (TRH) into the hypophyseal portal system. TRH is a tripeptide (pyro-Glu–His–Pro–NH$_2$) that stimulates both the synthesis and secretion of TSH; however, the latter response does not require protein synthesis. There is also an indication that somatostatin (see section on Growth Hormone) may function as a thyrotropin release-inhibiting hormone, since administration of antibodies to somatostatin prevent stress-induced inhibition of TSH secretion.

The most potent inhibitor and the primary factor in control of the rate of TSH secretion is the plasma thyroid hormone level. When the thyroid hormone level is low, TSH secretion is stimulated; when the thyroid hormone level increases, it is inhibited. The primary site of the feedback inhibition is the pituitary thyrotrope; however, some positive and negative feedback can occur as well at

hypothalamic or extrahypothalamic sites (Fig. 4a). If hypothalamic influence is removed by pituitary stalk lesion, the baseline levels of TSH secretion are not only lower, but the pituitary becomes sensitive to feedback inhibition by even lower levels of thyroid hormone. This indicates that the amount of TRH secreted from the hypothalamus determines the "set point" for feedback control and thus the tonic level of TSH secretion (Fig. 4b). Upon exposure to cold stress in neonates and some animals, TSH secretion is stimulated by release of sufficient TRH to break through the feedback inhibition (Chapter 46).

Thyroid hormone inhibits TSH secretion by rendering the pituitary thyrotrope less sensitive to stimulation by TRH. The effect takes an hour to develop and requires the synthesis of new protein. Thyroid hormone also has marked effects on the metabolic activity of the thyrotropic cells, since hyperthyroidism causes the cells to regress and hypothyroidism causes hypertrophy. Other hormones may also inhibit (glucocorticoids) or enhance (estrogen) TSH secretion by affecting the pituitary response to TRH.

The major physiologic stimulus to TSH secretion in many animals is exposure to cold. Thermoreceptor neurons located in a region of the anterior hypothalamus are sensitive to changes in core body temperature. This region controls neural input into an adjacent thyrotropic area that mediates TRH secretion into the median eminence.

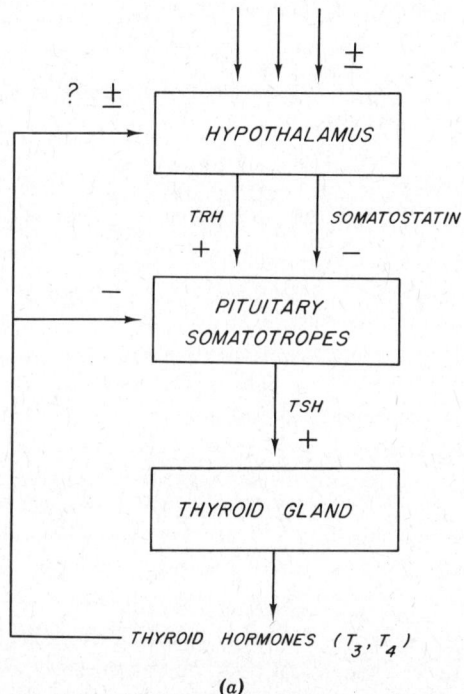

(a)

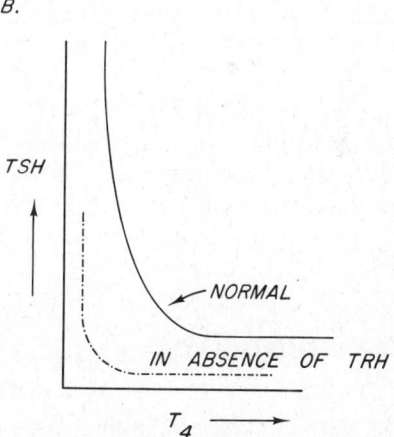

Figure 4 Hypothalamic–pituitary–thyroid regulation of TSH secretion. *(a)* The circulatory level of free thyroid hormone regulates TSH secretion by negative feedback. The controlled variable in this closed loop system is the plasma thyroid hormone level. Open loop neural input into TSH secretion is mediated by the stimulatory and inhibitory effects of the hypothalamic hormones TRH and somatostatin.

Figure 4 (con't) *(b)* The hypothalamic secretion of TRH also determines the set point for feedback control of TSH secretion. (Figure 4 *b* is adapted from Williams RH (ed): *Textbook of Endocrinology.* Philadelphia, WB Saunders, 1974.)

The thermoreceptor neurons also influence the other behavioral (e.g., food intake), autonomic (e.g., shivering, perspiration, and regulation of blood flow), and endocrine mechanisms that control heat production and loss.

Disorders

Hyperthyroidism

Hyperthyroidism as a consequence of excessive TSH production is rare. The usual cause of thyrotoxicosis is autonomous secretion of thyroid hormone or excessive stimulation of the thyroid gland by TSH-like immunoglobulins. Since pituitary function is intact, feedback inhibition by the high plasma thyroid hormone levels causes low plasma TSH levels and insensitivity to administration of TRH.

Primary Hypothyroidism

Low thyroid hormone levels resulting from ablation (surgical thyroidectomy or radioiodine therapy) or associated with diseases of the thyroid gland are accompanied by marked changes in the function of the pituitary thyrotropic cells. Because of low thyroid hormone feedback, the pituitary thyrotropes undergo hypertrophy, the TSH content increases, plasma TSH levels are high, and patients show an exaggerated TSH response to TRH administration.

Secondary and Tertiary Hypothyroidism

Hypothyroidism caused by impairment of pituitary (secondary) or hypothalamic (tertiary) function is accompanied by low plasma TSH levels and partial atrophy of the thyroid gland. If the administration of TRH stimulates TSH secretion, the disorder is of hypothalamic rather than pituitary origin. Since selective loss of pituitary TSH is rare, patients with secondary or tertiary hypothyroidism nearly always show signs of adrenocortical and gonadal insufficiency.

ADRENOCORTICOTROPIC HORMONE

ACTH plays an important role in the body's adaptation to various types of physiological, pathological, and psychological stress. The primary target of ACTH is the adrenal cortex where ACTH controls the growth of certain adrenocortical cells and the synthesis and secretion of glucocorticoid hormones. The latter affect the metabolism of glucose, proteins and fats. ACTH also regulates to some extent the synthesis of the adrenal mineralocorticoids and adrenal androgens.

Regulation of ACTH Secretion

Trauma, cold exposure, burns, surgery, exercise, hypoglycemia, infection, anxiety, and excitement are some of the stimuli that evoke ACTH secretion. Portions of the visceral brain, such as the amygdaloid nuclei and reticular activating system, are particularly involved in the psychic stimulation and circadian rhythm of ACTH output.

Control of ACTH secretion is similar to TSH output in that it is regulated by both a neuroendocrine mechanism and by feedback from the target organ hormone. The hypothalamus exerts positive control by releasing corticotropin releasing factor (CRF) into the median eminence. Feedback inhibition of ACTH secretion by circulating glucocorticoid hormone also provides a potent level of control. High glucocorticoid levels inhibit secretion and low levels stimulate secretion. One site of the glucocorticoid inhibition of ACTH secretion occurs at the level of pituitary corticotropes. Since glucocorticoids have been shown in vitro to decrease mRNA levels for the high molecular weight precursor of ACTH, part of the long term inhibitory effect of the glucocorticoids may be mediated by inhibition of ACTH synthesis. Feedback inhibition also occurs at the hypothalamic level with subsequent reduction of CRF release. The degree and time course of inhibition are complicated and may include rate-sensitive as well as level-sensitive feedback mechanisms and CRF mediated alterations in the set point for feedback inhibition.

Disorders

Cushing's syndrome describes four different disorders with common clinical features caused by excessive levels of plasma glucocorticoids. Only one of the diseases is of pituitary or hypothalamic origin, but their differential diagnosis depends on a knowledge of the pituitary–adrenal regulatory mechanisms described earlier.

Cushing's syndrome as a result of Cushing's disease is due to hypersecretion of ACTH from the pituitary. Because the pituitary is still sensitive to negative feedback, injection of the synthetic steroid dexamethasone inhibits

ACTH secretion. The primary defect in Cushing's disease may be overproduction of CRF by the hypothalamus. Excessive ACTH levels are also caused by ectopic ACTH-secreting neoplasms. These neoplasms are insensitive to inhibition of ACTH secretion upon diagnostic administration of dexamethasone. Cushing's Syndrome can also be the direct result of an adrenal neoplasm that secretes glucocorticoids autonomously. In this case ACTH levels are low because of feedback inhibition by high plasma levels of endogenous glucocorticoids. Diagnostic administration of dexamethasone will cause no further decrease in the plasma ACTH level. The fourth variant is due to excessive treatment with glucocorticoids (iatrogenic Cushing's syndrome).

Hyperpigmentation is frequently observed in persons with ectopic ACTH neoplasms or Addison's disease (primary adrenal insufficiency), and in patients adrenalectomized for treatment of Cushing's disease. In each instance there is hypersecretion of ACTH that is unchecked by glucocorticoid feedback inhibition. In addition to high levels of ACTH, which itself has some MSH activity, these persons have high circulating levels of β-MSH.

Hyposecretion of ACTH occurs in persons with damage to the pituitary, to the hypothalamus, or to the pituitary portal system. The urinary steroid levels of these persons are low but increase upon ACTH administration.

GROWTH HORMONE

Growth hormone affects numerous metabolic processes and promotes growth in nearly all tissues of the body. This growth is a result of increase in cell size (hypertrophy) and, in some tissues, in cell number (hyperplasia). Although the deficiency of growth hormone during childhood causes dwarfism, the actions of growth hormone are not confined to juveniles. In adults growth hormone continues to help the body cope with physical (e.g., acute trauma, surgery) and emotional (e.g., sudden loud noises, excitement, and anxiety) stress and adapt to both short- and long-term nutritional imbalances such as occur during hypoglycemia, exercise, and starvation.

Actions of Growth Hormone

The major physiological actions of growth hormone are growth of bone and cartilage and metabolic effects, which include increased protein synthesis, decreased carbohydrate utilization, and increased lipolysis by adipose tissue.

Growth hormone causes an increase in protein deposition by influencing a number of metabolic processes. In liver and muscle cells growth hormone stimulates components of the protein-synthesizing machinery by increasing total RNA synthesis and also enhances protein synthesis by independent stimulation of amino acid uptake into cells. The effects of growth hormone are not only important for growth and general tissue maintenance, but for their contribution to wound repair, compensatory hypertrophy, and organ regeneration mechanisms.

The effects of growth hormone on carbohydrate metabolism are complicated. Initially and transiently growth hormone has an insulin-like effect, which is to enhance glucose uptake into muscle cells. After a latent period, growth hormone renders the tissue resistant to the action of insulin, thus decreasing glucose utilization and increasing the plasma glucose level; this is in part responsible for the diabetogenic or contrainsulin actions sometimes ascribed to growth hormone. Conversely, growth hormone deficiency causes abnormal sensitivity to insulin.

In adipose tissue growth hormone stimulates lypolysis and raises the plasma unesterified fatty acid level. Fatty acids can then be used as a source of energy in preference to glucose and protein. Since high plasma unesterified fatty acid levels inhibit glucose utilization by cells, this could also contribute to the contrainsulin effect of growth hormone.

The most striking effects of growth hormone are on stimulation of cartilage and bone growth, since the absence of growth hormone in childhood causes dwarfism, and high levels cause bone overgrowth and bone deformation. The overall metabolic effects observed in cartilage are increased sulfate incorporation into chondroitin sulfate, increased uptake of amino acids, increased RNA and protein synthesis, and mitogenesis and cell replication. It is now known that these effects are mediated by a polypeptide (around 63 amino acids) called somatomedin, since growth hormone itself has no direct effect on cartilage tissue.

Several lines of evidence led to the discovery of somatomedin. Serum from hypophysectomized rats had no stimulatory effect on cartilage explants in vitro, but serum from hypophysectomized rats treated in vivo with growth hormone was as effective as normal serum. Since addition of purified growth hormone to cartilage explants had no stimulatory action, a growth hormone-dependent factor was postulated to be the mediator of the action of growth hormone on cartilage. The factor was subsequently purified and named somatomedin. When purified somatomedin was added to cartilage extracts it demonstrated all the metabolic stimulatory effects on cartilage formerly ascribed to growth hormone.

With the exception of some pathological states (severe starvation and Laron dwarfism) an increase or decrease in growth hormone is paralleled by changes in serum somatomedin levels. The liver appears to be the site of somatomedin production, however, the role of growth hormone in the control of somatomedin synthesis is uncertain since stimulation requires unphysiologically high concentrations of growth hormone. Once in the blood, somatomedin binds specifically to a serum carrier protein that greatly prolongs the biological half life of circulating somatomedin to 2–4 h. Since the biological half life of somatomedin in hypophysectomized animals is very short (8 min), some investigators have suggested that growth hormone may be involved in controlling the synthesis of the carrier protein. Somatomedin unequivocally mediates all the stimulatory effects of pituitary growth hormone on cartilage tissue. In tissue other than cartilage it is not yet clear whether growth hormone or its intermediary is responsible for the physiological action. Hepatic tissue, for example, has been shown to have specific, saturable growth hormone receptors; moreover, the metabolic effects of growth hormone in liver, muscle and adipose tissue can be demonstrated in the absence of somatomedin.

Regulation of Growth Hormone Secretion

Physiological stimuli for growth hormone secretion include hypoglycemia, increases in plasma amino acid levels (mainly arginine), decreases in plasma unesterified fatty acid levels, exercise, stress (physical and emotional), and starvation. Control of growth hormone release is under dual control from the hypothalamus by a somatotropic-releasing factor (SRF) and a release-inhibiting substance called somatostatin. There is good evidence for the existence of SRF, probably a polypeptide, in median eminence extracts; however, its chemical structure is not yet known.

Somatostatin, which has been chemically identified, is a 14-amino acid polypeptide.

The mechanisms by which the various metabolic and neural signals are received and integrated in the hypothalamus are largely unknown. There is, however, good evidence for the presence of glucose receptors in certain regions of the hypothalamus. It is the glucose receptor that probably mediates the response to exercise (and to hypoglycemia), since infusion of glucose inhibits the rise in plasma growth hormone that occurs during exercise. The response to stress is mediated independently, since glucose infusion does not inhibit the response to stressful stimuli.

In addition to its presence in hypophysiotropic neurons, somatostatin is also found in other regions of the CNS, in pancreas, gastrointestinal epithelium, and thyroid gland. This compound not only inhibits growth hormone secretion, but also the secretion of numerous other hormones (e.g., TSH, gastrin, secretin, parathyroid hormone, renin, vasoactive peptide, glucagon, insulin), suppresses neuronal activity, and inhibits gastrointestinal motility and splanchnic blood flow. Since the general inhibitory effects on secretion, neuronal activity, and smooth muscle all involve Ca^{2+} ions, it is believed that somatostatin might in some way affect Ca^{2+} levels in cells that have receptors for somatostatin.

Abnormalities of Growth Hormone Secretion

Hypersecretion

A pituitary adenoma involving the somatotropic cells will cause excessive production of growth hormone. If the tumor appears during childhood (before closure of the epiphyseal plates) the disproportionate growth leads to gigantism. Affected persons have been known to exceed a height of 270 cm (9 ft). After adolescence there is no further bone lengthening; however, the bones can continue to increase in thickness. Thus a feature of excessive production of growth hormone in adults is marked deformation of the small bones of the hands and feet and the membranous bones. This condition, known as acromegaly, is characterized by protruding jaws and forehead, pronounced supraorbital ridges, enlarged nose, hands, and feet, and hunched back. The soft tissues (e.g., kidney, liver, tongue) hypertrophy as well.

High plasma levels of growth hormone occur during starvation; however, in this instance somatomedin levels are unaccountably low, so that neither giantism nor acromegaly occurs. High levels of growth hormone in association with low levels of somatomedin are also present in the inherited disease of Laron dwarfism. Laron dwarfs are somehow insensitive to the effects of growth hormone.

Hyposecretion

Hyposecretion of growth hormone can occur as a result of pituitary or hypothalamic destruction, from a genetic disorder causing selective hyposomatotropism, and in some idiopathic cases involves no recognizable pituitary disease. Hyposecretion of growth hormone in adults produces no obvious clinical manisfestations; however, occurrence in early childhood causes dwarfism. Since numerous other defects can lead to dwarfism, diagnosis relies on measurement of plasma growth hormone levels after administering a challenge to growth hormone secretion (e.g., by injection of arginine).

PROLACTIN

The major physiological function of prolactin (formerly called luteotropic hormone after its action on the corpus luteum in some experimental animals) is to prepare the mammary gland for lactation. Its scope of action also includes less well characterized effects on other reproductive processes (in both male and female), on maternal behavior, osmoregulation, and cell metabolism. The physiological role of these other functions is not known. Some of the types of stimuli that evoke prolactin secretion are suckling, pregnancy, estrogen, sleep, and hypoglycemia.

Until 1970, when the chemical structure of prolactin was determined, the lactogenic functions of the hormone were mistakenly attributed to growth hormone. Although growth hormone has lactogenic activity, not surprising in view of its considerable amino acid sequence homology with prolactin, it is clearly not the physiological hormone responsible for lactation, since unlike prolactin, blood levels of growth hormone do not correlate with stimulation of milk secretion.

Control of Prolactin Secretion

A unique aspect in the control of prolactin secretion is that the hypothalamus normally exerts a tonic inhibitory influence on the lactotrope cells of the anterior pituitary. Certain hypothalamic neurons continually secrete a prolactin inhibitory factor (PIF) that inhibits both prolactin synthesis and secretion. Lesions to the pituitary stalk, therefore, cause an increase in prolactin secretion, while the secretion of all other anterior pituitary hormones decreases. There is compelling pharmacological evidence that PIF is dopamine (DA). Technical difficulties, however, have hampered the unequivocal demonstration of an inverse relationship between DA levels in stalk blood and levels of prolactin secretion. Even less certain is the identity of prolactin releasing factor (PRF). While TRH is capable of stimulating prolactin secretion from pituitary lactotropes, there is skepticism that TRH acts as the physiological releasing factor for prolactin.

Prolactin is itself involved in short loop feedback inhibition at the level of the hypothalamus and possibly acts by activating dopaminergic terminals in the median eminence. Finally, estrogen secretion from the ovary and placenta increases prolactin secretion by sensitizing the pituitary lactotropes to the effects of PRF. Estrogen also stimulates hypertrophy and hyperplasia of the pituitary lactotropes; thus, the high persistent levels of estrogen that occur during pregnancy cause the pituitary to enlarge up to 10 times its original weight.

Physiological Role of Prolactin

During pregnancy, high estrogen levels induce secretion of pituitary prolactin, and the placenta secretes a similar lactogen called chorionic somatomammotropin. These prolactins, along with estrogen, progesterone, the glucocorticoids, and growth hormone, participate in the growth and development of the maternal breast for lactation. Lactation, however, does not occur because high levels of placental estrogen and progesterone inhibit this particular action of prolactin. After child delivery, prolactin is able to stimulate milk secretion because of the sudden withdrawal of estrogen and progesterone.

After parturition, prolactin is released from the pituitary in response to suckling. Suckling stimulates sensory afferents that inhibit the secretion of PIF, and possibly stimulate the secretion of PRF. Suckling also stimulates milk ejection by causing oxytocin secretion and inhibits gonadotropic hormone secretion. This latter effect is responsible for the amenorrhea and infertility of nursing mothers and is possibly mediated by prolactin acting at the hypothalamic level to inhibit gonadotropic hormone secretion.

Disorders

Hypersecretion of prolactin in nonpregnant women or non-nursing mothers frequently causes galactorrhea (persistent lactation) and amenorrhea. A number of disorders can contribute to this hyperprolactemia, i.e., pituitary or ectopic prolactin secreting tumors, hypothalamic damage or interruption of the pituitary portal system, drugs that interfere with dopaminergic transmission (tranquilizers, α-methyldopa, reserpine), and irritative lesions of the anterior chest wall.

GONADOTROPIC HORMONES

The pituitary gonadotropic hormones are luteinizing hormone (LH) and follicle stimulating hormone (FSH). In females FSH stimulates ovarian follicle cell development and LH is important for ovulation and maintenance of the corpus luteum. The FSH-primed ovary secretes estrogen, and the LH-primed ovary secretes estrogen and progesterone. In males LH, also called interstitial cell stimulating hormone (ICSH), acts on the Leydig cells of the testes, where it is a regulator of testosterone secretion. Testosterone in turn is a major mediator of spermatogenesis. FSH acts together with LH to promote spermatogenesis.

Control of Gonadotropin Secretion

Secretion of the pituitary gonodotropic hormones is regulated by hypothalamic releasing hormone(s) and by complex and still poorly understood feedback interactions by the gonadal steroid hormones. A decapeptide (GLU-HIS-TRP-SER-TYR-GLY-LEU-ARG-PRO-GLY-NH$_2$) called gonadotropic releasing hormone (GnRH) is released from the hypothalamus and is capable of stimulating the secretion of both FSH and LH.

Basal gonadotropic hormone secretion is regulated principally by negative feedback from estrogen in females and from testosterone in males. This feedback inhibition occurs at both the hypothalamic and pituitary levels. In addition, the release of FSH in males is inhibited at the pituitary level by a peptide released from the testes called inhibin. In females a positive feedback mechanism involving estrogen is of major importance in causing the cyclic preovulatory release of LH. Positive feedback involves stimulation of hypothalamic GnRH release and an increase in the sensitivity of the pituitary to GnRH. Progesterone also exerts positive and negative feedback effects primarily at the hypothalamic level by complex interactions that depend on the estrogen level.

Although only a single GnRH seems to exist in regulating both FSH and LH secretion, there is considerable variability in the plasma ratios of FSH and LH, depending on sex, age, and stage of the ovarian cycle. This variability is partly explained by the differential response of the FSH and LH producing cells to GnRH depending on the circulating levels of steroidal hormones and inhibin.

Changes in internal and external environment also in-

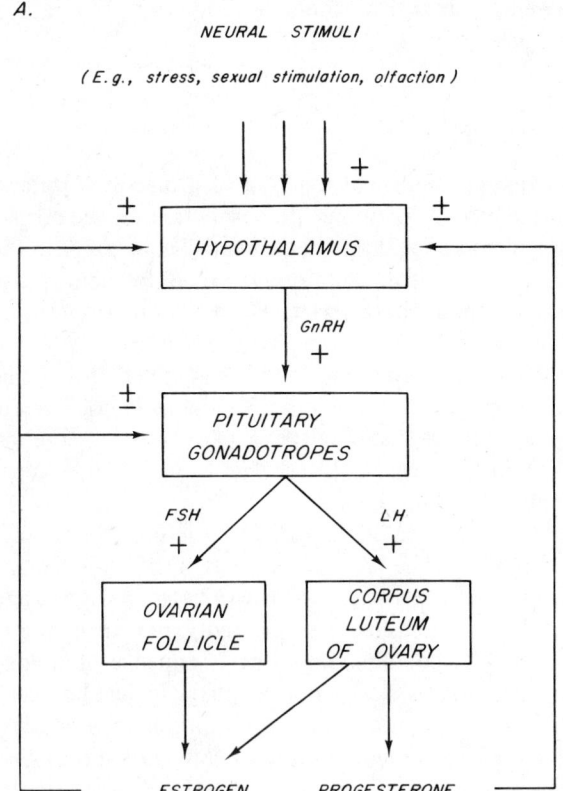

Figure 5 *(a)* The hypothalamic–pituitary–gonad axis in females.

fluence neural input into the hypothalamic neurons that mediate GnRH release. A partial list of the factors that influence reproductive activity include olfactory, visual, and tactile stimuli; physical and psychological stress; nutritional state; and the light and dark cycle.

Neuroendocrine Control of Ovulation and the Menstrual Cycle

It was previously thought that the hypothalamus in females, but not in males, contained an inherent rhythm that led to cyclic stimulation of pituitary gonadotropic hormone secretion. It is now apparent that the cyclic na-

ture in plasma levels of pituitary gonadotropins in females is determined by the changing pattern of ovarian hormone secretions that feed back at the level of the hypothalamus and pituitary. These events are briefly outlined below and are illustrated in Figure 5b.

1. A few days before menstruation, the pituitary secretion of FSH causes induction and growth of follicular cells around the oocyte. The developing follicle begins to secrete estrogen, inhibiting further FSH secretion.

2. Around midcycle, the estrogen levels begin rising very steeply. A preovulatory surge in LH secretion is brought about by positive feedback from estrogen. The LH surge causes follicle rupture, followed by ovulation.

3. The cavity of the ruptured follicle becomes filled with luteal cells that proliferate under the influence of LH. This new mass of cells, called the corpus luteum, secretes large quantities of both estrogen and progesterone. If fertilization does not occur, the corpus luteum degenerates, progesterone and estrogen levels decline, and menstruation ensues. New follicle cell growth for the following cycle already begins in this late luteal stage as the declining levels of estrogen relieve feedback inhibition of FSH secretion.

If fertilization occurs, the life of the corpus luteum is prolonged by placental hormones. As a result, high levels of estrogen and progesterone persist and menstruation does not occur. Since high levels of progesterone antagonize the positive feedback effect of estrogen on LH secretion, this ensures that ovulation will not occur when the uterus contains a fertilized egg. It is also the basis for the use of progesterone, in the company of smaller concentrations of estrogen, as an oral contraceptive.

Menopause, the cessation of the ovarian cycle, occurs when the ovary becomes depleted of primordial follicles. Since the production of estrogen by the ovaries falls to very low levels, the decline in feedback inhibition gives rise to high persistent secretion of the gonadotropic hormones.

Onset of Puberty

GnRH, the gonadotropins, and the capacity for feedback inhibition by gonadal steroid are already present at birth. Long before puberty the gonads and secondary sexual structures are also capable of being stimulated by exogenous hormones. The prepuberal state is therefore not a

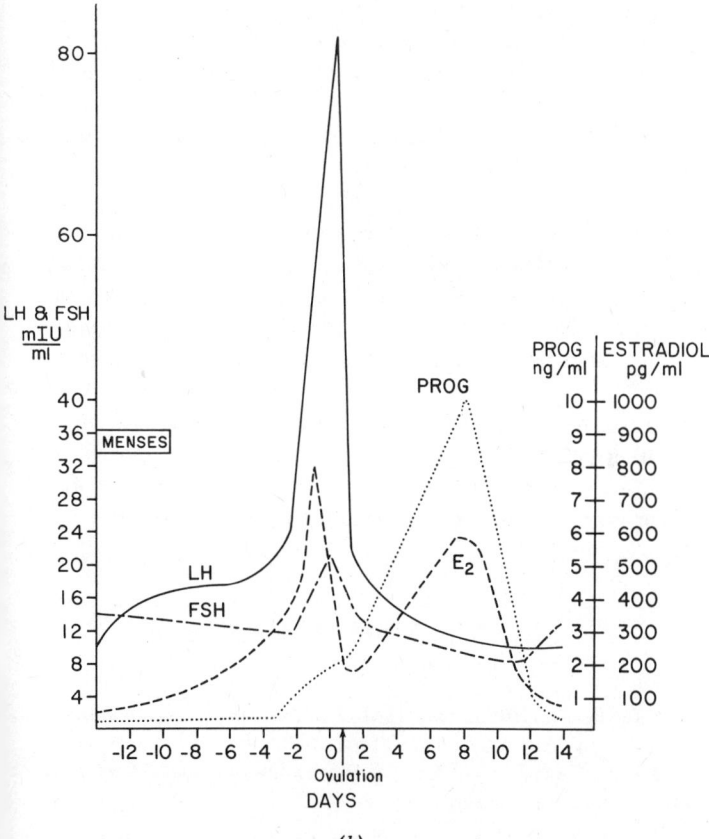

Figure 5 (con't) *(b)* Schematic representation of plasma levels of gonadotropins and gonadal steroids during the human menstrual cycle. The cycle is centered on day 0, the day of midcycle LH peak. (Figure 5b is from Speroff L, and RL Vande Wiele: Regulation of the Human Menstrual Cycle, Am J Obstet Gynecol 109:234, 1971.)

consequence of inability to respond to hormone stimulation but due to lack of stimulation of GnRH secretion. Before puberty the hypothalamus is extremely sensitive to the negative feedback effect of gonadal steroids on GnRH secretion. Thus, gonadotropin levels remain low despite low levels of circulating gonadal steroids.

The onset of puberty is associated with an increase in gonadotropin secretion and appears to be triggered by a gradual decrease in hypothalamic sensitivity to the negative feedback effects of the gonadal steroids. In females the positive feedback effect of estrogen also occurs around puberty, thus providing the mechanism for cyclic ovarian activity.

Disorders

Gonadal atrophy and hypogonadism due to hyposecretion of the gonadotropins can result from congenital or destructive lesions of the pituitary or hypothalamus and frequently occur in association with ACTH, TSH, and growth hormone deficiency. Damage to the hypothalamus can also cause precocious puberty if it results in a premature rise in gonadotropic hormone levels by interfering with feedback inhibition of GnRH secretion. Isolated gonadotropic hormone deficiency occurs as a genetic disorder in males and is caused by insufficient production of GnRH. This is confirmed by diagnostic administration of GnRH. In females, hypothalamic–pituitary dysfunction is a frequent cause of amenorrhea. In this case, diagnostic administration of GnRH to distinguish between a defect of hypothalamic or pituitary origin is not very useful, since variable results are caused by the numerous factors that modulate the stimulatory influence of GnRH on LH and FSH release. Amenorrhea may also be caused by inhibition of cyclic LH secretion from psychological stress or hyperprolactemia.

MELANOCYTE STIMULATING HORMONE

Melanocyte stimulating hormone, also known as intermedin, is normally secreted in small quantities by cells in the pars intermedia. MSH induces pigmentation by stimulating melanocytes and is extremely important for color adaptation in a wide variety of vertebrates. Its effects in humans are more subtle. In fact, the pars intermedia merges with the neural lobe of the pituitary and assumes a more vestigial nature in humans. No acute effects of MSH administration have been observed in humans, but long-term exposure to MSH (8–10 days) will eventually affect the depth of pigmentation.

Regulation of Secretion

Melanotrope-inhibitory (MIF) and releasing (MRF) factors acting on the cells of the pars intermedia have been isolated from stalk median eminence extracts. MIF is believed to be the tripeptide Pro-leu-Gly-NH_2. Like prolactin secretion from the anterior pituitary, MSH secretion appears to be under predominantly inhibitory control since lesions to the pituitary stalk result in a marked increase in MSH release. Stimuli (stress, suckling, sound, touch, hypertonicity) that induce MSH secretion do so primarily by suppressing the hypothalamic secretion of MIF and permitting the intrinsic capacity of the pars intermedia to secrete high levels of MSH. Positive control, by release of MRF, may also be involved.

MSH secretion is further influenced by estrogens acting directly on the pituitary gland, or indirectly by an action on hypothalamic release of MIF. The immediate effect of estrogen is stimulation of MSH secretion, followed by both increased synthesis and release of MSH. Progesterone potentiates the effect of estrogen. The hyperpigmentation changes that occur during pregnancy are most likely caused by the effect of these steroids on MSH secretion. Other instances of hyperpigmentation occur in association with disease states involving overproduction of ACTH (primary Addison's disease and Cushing's disease).

SELECTED BIBLIOGRAPHY

Jeffcoate SL, Hutchinson JSM: *The Endocrine Hypothalamus.* New York, Academic Press, 1978.

Tepperman J: *Metabolic and Endocrine Physiology.* Chicago, Year Book, 1980.

Tolis G, Labrie F, Martin JB, et al: *Clinical Neuroendocrinology —A Pathophysiological Approach.* New York, Raven Press, 1979.

Williams RH: *Textbook of Endocrinology.* 6 ed., Philadelphia, WB Saunders, 1981.

SELF-STUDY QUESTIONS

MULTIPLE CHOICE

Select the correct answer(s). (In many instances, more than one answer is correct.)

1. A patient complaining of lethargy, mental sluggishness, and inability to tolerate cold is found to have low plasma levels of total thyroid hormone and TSH.
   A. Primary thyroid dysfunction is indicted.
   B. A hypothalamic or pituitary disorder is indicated.
   C. TSH secretion is normally not regulated by feedback inhibition from thyroid hormone.
   D. Administration of TRH would be of diagnostic value in establishing the location of the defect.

2. Immunoassay of the peptide and protein hormones is a valuable tool in the clinical diagnosis of endocrine disorders. Unless special precautions are taken, all the following groups exhibit immunological cross-reactivity because of considerable sequence homology. The groups that exhibit overlapping biological actions as well are
   A. prolactin and growth hormone.
   B. ADH and oxytocin.
   C. ACTH and MSH.
   D. TSH, LH, and FSH.

3. Conditions that could lead to high serum levels of prolactin are
   A. acidophilic pituitary adenoma.
   B. hypothalamic lesion.
   C. pregnancy.
   D. hypertension treated by α-methyldopa.

4. A 22-year-old patient experiencing polydipsia and excreting large volumes (5 L/day) of dilute (50–200 mOsms/kg of water) urine is diagnosed as having neurogenic diabetes insipidus (inadequate ADH secretion). Which of the following observations are consistent with this diagnosis?

A. Kidney elaborates a concentrated urine in response to an injection of hypertonic saline.
B. Medical history indicates that the patient had once received a head injury in an automobile accident.
C. Kidney elaborates a concentrated urine upon temporary restriction (less than 24 h) of fluid intake.
D. Kidney elaborates a concentrated urine, and urine flow rate diminishes in response to administration of ADH.

5. Long loop feedback inhibition plays a major role in regulating the tonic levels of which of the following adenohypophyseal hormone(s)?
   A. TSH
   B. Prolactin
   C. LH and FSH
   D. ACTH

6. The pituitary portal system
   A. carries the hypothalamic hypophysiotropic hormones to the anterior pituitary.
   B. is important for regulation of oxytocin secretion.
   C. is important for regulation of ACTH secretion.
   D. carries the neurohypophyseal hormones to the neural lobe of the pituitary.

7. Primary adrenocortical insufficiency is associated with
   A. low levels of glucocorticoid hormone.
   B. hyperpigmentation.
   C. high levels of ACTH.
   D. low levels of ACTH.

8. Growth hormone secretion
   A. in excessive quantities causes acromegaly because of concomitant high levels of somatomedin.
   B. is inhibited by somatostatin.
   C. is stimulated by hypoglycemia.
   D. is inhibited by somatomedin.

TRUE OR FALSE

9. The ovarian cycle in women is determined by an inherent rhythm in the brain.

10. Neurons from the paraventricular and supraoptic nuclei terminate in the neurohypophysis, where they evoke the secretion of ADH and oxytocin from glandular cells of the pars nervosa.

11. Hormone from both the hypothalamic neurosecretory neurons and the pituitary glandular cells is stored in secretory granules and released by Ca^{2+}-dependent exocytosis.

12. Each hypophysiotropic hormone is highly selective for one particular pituitary cell type.

13. Onset of puberty reflects the ability of the gonads to respond to the gonadotropic hormones.

ANSWERS

1. **B,D.** Feedback inhibition is an important regulatory mechanism in controlling TSH secretion. When the hypothalamic–pituitary system is functioning normally, the free thyroid hormone level and TSH levels are inversely correlated. Thus primary thyroid dysfunction would lead to abnormally high plasma levels of plasma TSH. If administration of TRH restores thyroid hormone levels, a hypothalamic lesion or some disturbance of transfer of TRH to the pituitary is involved. If TRH has no effect, the defect is of pituitary origin.

2. **A,B,C.** The biological specificity of TSH, LH, and FSH resides in their distinctive β chains. Their immunological cross-reactivity is due to identical α chains.

3. **All are correct.** Acidophilic tumors frequently secrete prolactin. The high estrogen level during pregnancy stimulates secretion of prolactin and hypertrophy and hyperplasia of pituitary lactotropes. Hypothalamic lesions and α-methyldopa interfere with the tonic dopamine inhibition of the pituitary lactotropes.

4. **B,D.** The patient cannot respond to a hyperosmolar stimulus (injection of hypertonic saline or water deprivation) because of a defect in ADH production. Damage to the neurohypophyseal neurons is suspected from the medical history. The patient's kidney function is not impaired, as evidenced by the response to ADH administration.

5. **A,C,D.** Answer B is incorrect because prolactin does not cause secretion of target organ hormones and therefore is not regulated by long loop feedback.

6. **A,C.** Answers B and D are incorrect because neurons carry the neurohypophyseal hormones to the neural lobe of the pituitary.

7. **A,B,C.** Primary adrenal cortical deficiency causes low levels of glucocorticoid hormone. Consequently, the low level of feedback inhibition causes hypersecretion of pituitary ACTH and MSH.

8. **A,B,C.**

9. **False**

10. **False**

11. **True**

12. **False**

13. **False**

46

The Thyroid Gland

DONALD W. STUBBS

OBJECTIVES

After completion of this chapter, you should be able to:

1. Describe the embryological development of the thyroid gland.

2. Relate follicular cell height to thyroid secretory activity in response to thyroid stimulating hormone.

3. Distinguish between inorganic iodide and organic iodine, and state the relationship of each to thyroid gland function.

4. Describe the steps of hormonogenesis in correct order (including the order of: amino acid incorporation into thyroglobulin protein, iodination, and coupling of iodinated tyrosines to form iodothyroninies.

5. Compare the different types of inhibitors of hormonogenesis and identify the specific sites of actions.

6. Interrelate structural and functional relationships.

7. Describe the equilibrium between free thyroid hormones and reversibly bound thyroid hormones.

8. Distinguish between the physiological significance of bound and free hormone, and know the effect of changes in the amount of carrier protein (in the absence of long-term changes in thyroid function).

9. Compare thyroxine with triiodothyronine in terms of the relative proportion of bound to free hormone and free hormone concentration and potency.

10. Describe the relationship between hypothalamic thyrotropin-releasing hormone (TRH), pituitary thyrotropin (also called thyroid stimulating hormone and abbreviated TSH), and thyroid hormone secretion in the normal control of thyroid function.

11. List the actions of thyroid hormones.

12. Relate the significance of inadequate thyroid function to the development of cretinism.

13. Describe the signs and symptoms of either hypo- or hyper-secretion of thyroid hormones.

14. Recognize the principal cause of endemic goiter.

15. Relate the concentration of TSH to different states of thyroid function (e.g., hyperthyroidism, euthyroidism, hypothyroidism of thyroidal failure, hypothyroidism of pituitary failure.).

The thyroid gland maintains the level of metabolism in the tissues optimal for their normal function. Thyroid hormone stimulates the oxygen consumption of most tissues, and it is necessary for normal growth and maturation. The thyroid gland is not essential for life, but its absence is associated with poor resistance to cold temperatures, mental and physical slowing, and mental retardation and stunting of growth in children (dwarfism). Conversely, excess thyroid hormone causes body wasting, nervousness and irritability, tachycardia, tremor, and excess heat production. Thyroid function is controlled by the thyroid stimulating hormone (TSH) from the anterior pituitary gland. The secretion of this trophic hormone is, in turn, regulated by a negative feedback system in which thyroid hormone exerts an inhibitory influence on the pituitary. In this way, changes in the internal and external environment bring about appropriate adjustments in the rate of thyroid secretion.

The unique features of the thyroid gland are the following.

1. The thyroid is the largest endocrine gland in the body.
2. Its hormonal secretion is stored extracellularly as part of the thyroglobulin molecule in the follicular colloid.
3. It has greater capacity for hormonal storage than any other endocrine gland, normally storing about a 2 month supply.
4. Its superficial location permits direct palpation and evaluation in physical examination. It is the most superficial endocrine gland in females and the second most superficial endocrine gland in males.
5. Thyroid hormone is unique among endocrine secretions in its iodine content.

PHYLOGENY AND ONTOGENY

Born at the dawn of chordate evolution in the Ordovician waters of the Paleozoic era, the thyroid gland is considered one of the more ancient vertebrate endocrine glands. Evolved from the pharyngeal endostyle (a ciliated mucus-secreting groove on the floor of the pharynx) of lower chordates, clearly recognizable follicular thyroid tissue is found only in vertebrates (and normally present in all species thereof).

The human thyroid anlage first appears just before the end of the first month after conception as a ventral diverticulum off the floor of the pharynx ("ontogeny recapitulates phylogeny"). This diverticulum undergoes a caudal descent, and the primitive hollow stalk connecting the primordium with its site of orgin undergoes elongation (the thyroglossal duct). Later in gestation the thyroglossal duct undergoes solidification and dissolution, leaving at its point of origin a small dimple, the foramen cecum. Remnants of this duct may persist in some individuals, giving rise to thyroglossal cysts. During the third month the thyroid acquires a follicular arrangement of cells capable of iodide accumulation and colloid formation (about the same time as the adenohypophyseal development of thyrotroph cells capable of TSH secretion). In addition to the follicular cells originating from the thyroidal diverticulum, the thyroid also receives a small contribution from the fourth (last) pharyngeal pouches (the ultimobranchial bodies). These latter cells, which are incorporated into the follicles but do not produce or come into contact with colloid, are called parafollicular cells (or C cells). The parafollicular cells are also endocrine cells, but they have a distinct function (calcitonin formation) unrelated to that of the follicular cells (formation of thyroxine and triiodothyronine). These cells, originally of neural crest origin, constitute only a very small percentage of the total thyroid cells. For discussion of parafollicular C cell function, see the Chapter 47.

ANATOMY

The normal thyroid is a bilobed gland weighing about 20–30 g. The upper lobes of the gland lie in apposition to the lateral surfaces of the lower part of the thyroid cartilage. The two lobes are connected by an isthmus of thyroid tissue that crosses the front of the trachea (Fig. 1). The recurrent laryngeal nerves lie sandwiched between the thyroid lobes and the trachea (a proximity offering a constant hazard during thyroid surgery).

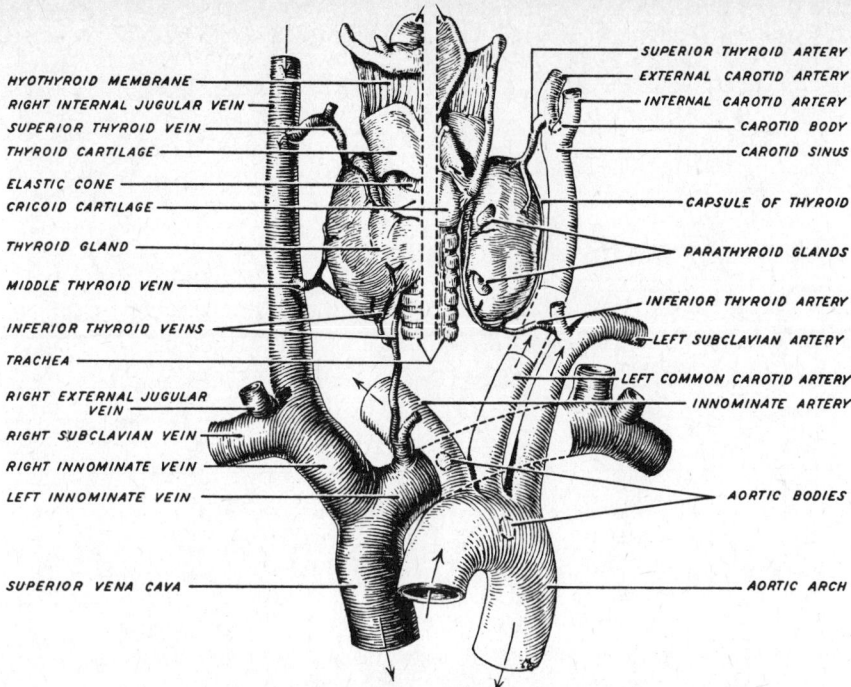

HYOTHYROID MEMBRANE
RIGHT INTERNAL JUGULAR VEIN
SUPERIOR THYROID VEIN
THYROID CARTILAGE
ELASTIC CONE
CRICOID CARTILAGE
THYROID GLAND
MIDDLE THYROID VEIN
INFERIOR THYROID VEINS
TRACHEA
RIGHT EXTERNAL JUGULAR VEIN
RIGHT SUBCLAVIAN VEIN
RIGHT INNOMINATE VEIN
LEFT INNOMINATE VEIN
SUPERIOR VENA CAVA

SUPERIOR THYROID ARTERY
EXTERNAL CAROTID ARTERY
INTERNAL CAROTID ARTERY
CAROTID BODY
CAROTID SINUS
CAPSULE OF THYROID
PARATHYROID GLANDS
INFERIOR THYROID ARTERY
LEFT SUBCLAVIAN ARTERY
LEFT COMMON CAROTID ARTERY
INNOMINATE ARTERY
AORTIC BODIES
AORTIC ARCH

Figure 1 Human thyroid and parathyroid glands and relations to adjacent structures. Anterior surfaces of the thyroid, larynx, and trachea plus the venous system are on the left; posterior surfaces and the arterial system are on the right. (From Turner CD, Bagnara JT: *General Endocrinology,* ed 6. Philadelphia, WB Saunders, 1976.)

Besides being a large gland, the thyroid has a tremendous potential for enlargement, goiterous glands weighing up to many hundreds of grams. Glandular enlargement can result in tracheal or esophageal compression and may produce varying degrees of respiratory distress and/or dysphagia. Compression of the recurrent laryngeal nerves can result in dysphonia and/or dyspnea.

The blood supply to the thyroid is provided by two pairs of arteries and is drained by three pairs of veins that anastomose freely on the surface of the gland. The thyroid gland is exceptionally well vascularized and has a high relative blood flow rate (600 ml/100 g/min) in excess of that to the kidneys (400 ml/100 g/min).

By light microscopy the gland is seen to be made up of numerous spherical follicles containing a clear, structureless colloid (Fig. 2). Each follicle is lined with a single layer of closely packed cuboidal cells (the follicular cells). Normally these follicular cells are about 15μm high, but increase in height with increased secretory activity and decrease in height with decreased secretory activity (Fig. 3). Each follicle is surrounded by a basement membrane and is invested with a rich capillary network. Occasionally parafollicular C cells may be seen. These are larger paler cells, which rest on the same basement membrane as the follicular cells but do not come into contact with the colloid of the follicular lumen.

IODINE METABOLISM

The subjects of iodine metabolism and thyroid function are intimately bound together. Although iodine itself was not discovered as an element until 1812 (as a direct result of the Napoleonic Wars), the therapeutic properties of

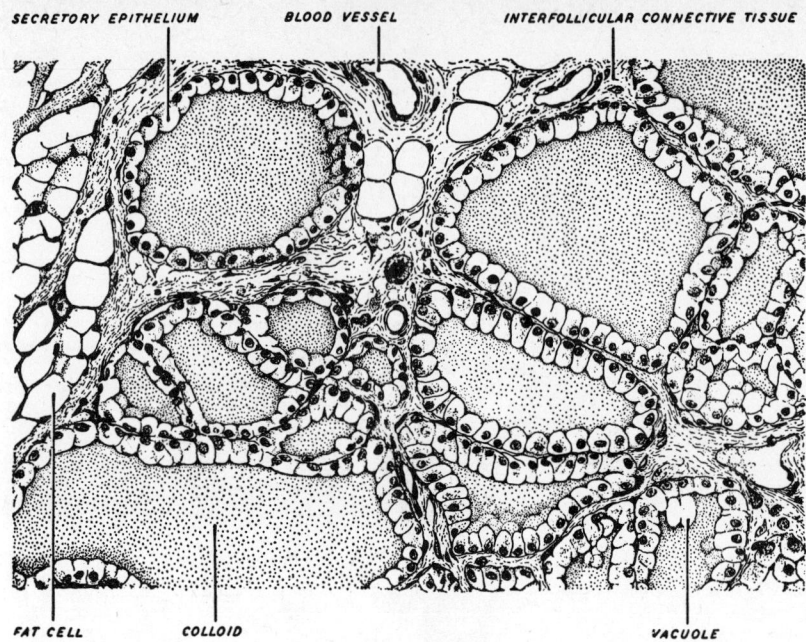

SECRETORY EPITHELIUM   BLOOD VESSEL   INTERFOLLICULAR CONNECTIVE TISSUE

FAT CELL   COLLOID   VACUOLE

Figure 2 Histologic features of normal thyroid gland. (From Turner CD, Bagnara JT: *General Endocrinology,* ed 6. Philadelphia, WB Saunders, 1976.)

substances now known to contain iodine may have been discovered and rediscovered many times as being successful in curing endemic goiter (possibly as far back as the reign of the Chinese emperor, Shen-Nung, about 2900 BC).

IODINE KINETICS

The thyroid has been described as an efficient collector of a rare element, iodine. Iodine is not an abundant substance and, strictly speaking, must be considered to be a trace element (it is only 0.000007% of the earth's crust). Iodine is absorbed from the gut in the form of iodide. Iodide is accumulated from the blood into the thyroid, where some of it is oxidized to a nonionic form and then incorporated into thyroid hormone, which is stored and ultimately secreted. Thus, iodine exists in the body in one of two principal forms: either as inorganic iodide (iodide anion, I^-), which constitutes about 3% of the total iodine in the body; and as organic iodine (nonionic iodine, I^o, in covalent bond with carbon), which is the remaining 97% of the total.

Many studies have been carried out on kinetics of iodine metabolism in humans, and a number of models have been proposed. The most common model is the three compartment one, in which iodide in plasma is taken up by the thyroid and primarily released into circulation as thyroid hormone. This thyroid hormone, in turn, gets degraded by the tissues, and the released iodide recycles again. Excretion pathways from plasma iodide into urine and from plasma hormone into feces are also included. While this simple model is not adequate to explain in detail all the data from many kinetic studies, it does provide a good general approximation sufficient for understanding first principles of thyroid function.

It is difficult to state the average daily dietary intake of iodine, since this varies widely throughout the world (depending on iodine content of soil and water and on culturally established dietary preferences). As an approximation, 500 μg is taken as an average daily intake for most parts of the United States. Dietary iodine is ingested in both the nonionic and ionic forms. Iodine is rapidly absorbed (mainly small intestine) as the ionic iodide (I^-).

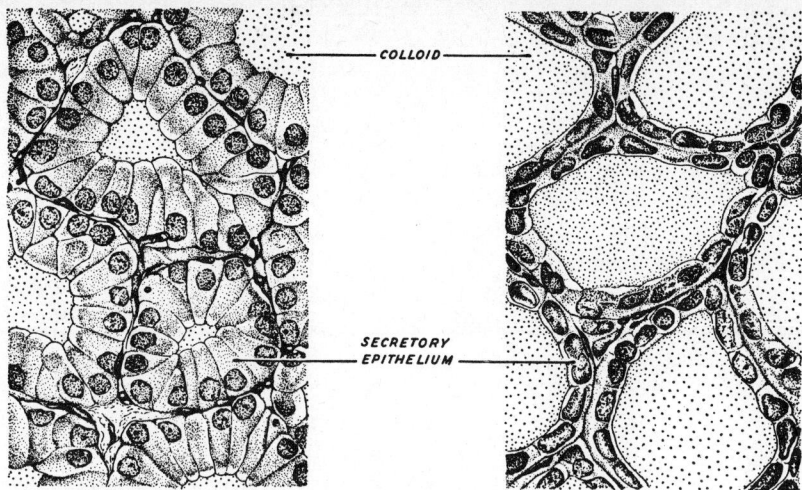

COLLOID

SECRETORY
EPITHELIUM

Figure 3 (Left) Thyroid from rat receiving TSH injections. (Right) Thyroid from rat six months after complete removal of the pituitary gland. Both glands drawn to scale. The height of the epithelium is inversely proportional to the diameter of the lumen. (From Turner CD, Bagnara JT: *General Endocrinology,* ed 6. Philadelphia, WB Saunders, 1976.)

Nonionic iodine (I^o) is first reduced to ionic iodide (I^-), the form that is absorbed.

Iodide enters a fluid space slightly larger than the volume of the extracellular fluid (ECF) space. That is, iodide is largely confined to the ECF plus the cells of some tissues and/or their secretions (I^- is found in erythrocytes and follicular cells, and is concentrated into the intraluminal fluids of the gastrointestinal tract, particularly saliva and gastric juice).

There are two main sites for the clearance (removal) of I^- from the ECF: (1) clearance by the thyroid, and (2) clearance by the kidneys. Of the two, the only one that ultimately results in a substantial loss of iodine from the body is that of the kidneys. Most of the net removal of I^- from the ECF by the thyroid is eventually recycled back into the ECF iodide pool, and only a small amount is actually lost from the body each day as a result of thyroid function (loss or organic iodine as conjugated hormone secreted into the gut as bile).

Thyroidal Clearance of Iodide

The thyroid both takes up iodide from and leaks iodide into the ECF. Since more is taken up each day than leaks back, the thyroid has a net daily gain of iodide (has a net clearance of iodide from the ECF). Some of the iodide leaking out is simply I^- that was transported into the gland but not immediately utilized, and which then leaked right back out. However, most of the iodide leaking back into the ECF has at one time been a constituent of organic compounds (organic iodine) of the thyroid. Some substances such as iodotyrosines released from follicular colloid are subject to deiodination by follicular cells and give up their iodine in the form of iodide, some of which reenters the ECF pool of I^-. Indeed, most of the net gain of iodide taken up by the thyroid does become organic iodine. The inorganic I^- within the thyroid is oxidized to nonionic I^o, which becomes organic iodine in the process of thyroid hormone synthesis (discussed below). Some of this organic iodine leaves the thyroid in the form of thyroid hormone secretion. Circulating thyroid hormone is picked up by the various tissues for utilization and, in the process, is metabolized by reactions that include deiodination. As a result of the tissue metabolism of thyroid hormone, about four-fifths of the daily secretion of organic or hormonal iodine is recycled back into the ECF pool as I^- (from deiodination of thyroid hormone in peripheral tissues). The remaining one-fifth of the hormonal iodine is lost from the body because some of the thyroid hormone is picked up by the liver and secreted in bile.

Renal Clearance of Iodide

Most of us are in a steady-state condition in which the amount of iodine taken in each day equals the amount lost. Most of the iodine lost each day is by way of urinary iodide (renal clearance). Since, as mentioned above, only a small amount of iodine is lost each day in biliary secretion (Fig. 4), it follows that renal clearance is responsible for the major amount of daily iodine loss. Of the 500 μg taken in each day, approximately 98% appears in the urine.

HORMONOGENESIS

Thyroid hormones are synthesized as a part of thyroglobulin (TG), a large storage protein with a molecular weight of 660,000. Thyroglobulin synthesis occurs on the

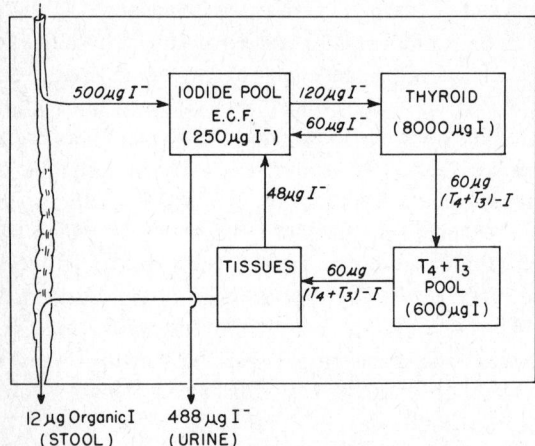

Figure 4 Normal pathways of iodine metabolism in a state of iodine balance. Note that most (approximately 90%) of body iodine store is present in the thyroid (chiefly in the organic nonionic form). Arrows indicate daily flux of iodine from one compartment to another. (From Ingbar SH, Woeber KA: in Williams RH (ed): The Thyroid Gland, *Textbook of Endocrinology*, ed 6. Philadelphia, WB Saunders, 1981.)

endoplasmic reticulum and is independent of iodination, which occurs after the large molecule is formed.

Iodine incorporation into TG and into the thyroid hormones, thyroxine (T_4) and triiodothyronine (T_3), follows the following scheme: (1) concentration of iodide by the thyroid gland; (2) organification (also called iodination), which involves oxidation of iodide to iodine with subsequent bonding to certain carbons of the phenol ring in tyrosyl units in the thyroglobulin molecule; (3) coupling of iodotyrosyl units within thyroglobulin to form iodothyronines (T_4 and T_3), which are also part of the thyroglobulin molecule; and (4) release of free iodothyronine hormones following proteolysis of thyroglobulin. All four of these steps are stimulated by thyrotropin, or thyroid stimulating hormone (TSH), from the anterior pituitary gland. All TSH actions on the thyroid appear to involve increased formation of cyclic AMP. The participation of iodide is illustrated in Figure 5.

CONCENTRATION OF IODIDE

Removal of inorganic iodide from the blood is almost exclusively the result of renal and thyroidal clearance. The thyroid gland actively transports iodide against a concentration gradient into the cells and colloid of the follicles. The normal ratio of thyroidal I^- to serum I^- (T/S ratio) is about 25:1, that is, the concentration of I^- in the gland is about 25 times higher than in serum or plasma. Following prolonged inhibition of organification of iodide to tyrosyl iodine by certain antithyroid drugs containing the thiourea group (e.g., propylthiouracil, methimazole, carbimazole), the I^- concentration in the gland can reach a level several hundred times greater than in serum. On the other hand, other inhibitors of thyroid function, which act by blocking I^- transport, can decrease the T/S ratio to 1:1, that is, I^- enters and leaves the gland by simple diffusion in this case. The best known of these I^- pump blockers are perchlorate (ClO_4^-) and thiocyanate (SCN^-).

ORGANIFICATION OF IODIDE (IODINATION)

Although some of the newly accumulated iodide leaks back out of the gland, most is immediately oxidized from iodide (I^-) to iodine (I°). It is generally accepted that this

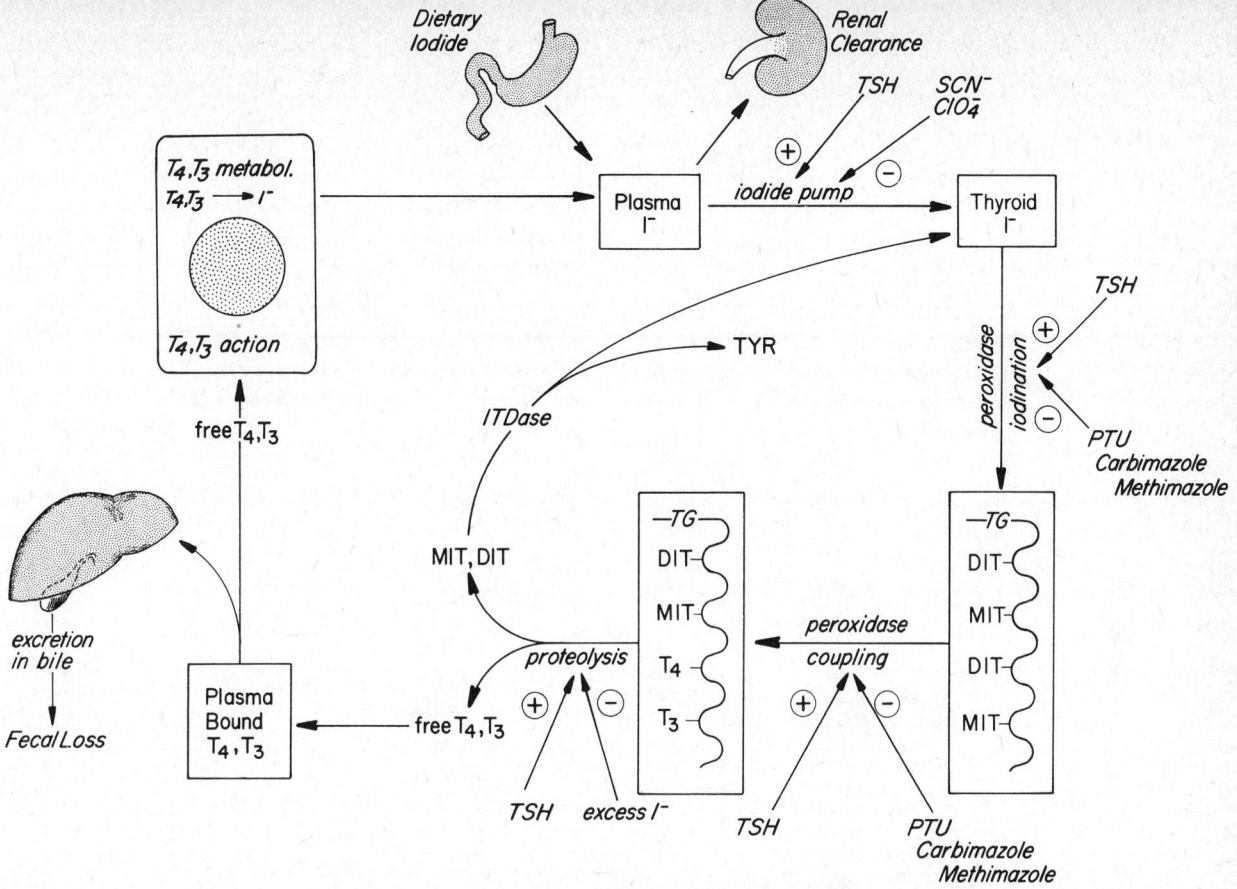

Figure 5 Iodide cycle. ECF I⁻ is trapped in the thyroid, oxidized, and bound to iodotyrosines in thyroglobulin (TG): coupling of iodotyrosines forms T_4 and T_3. The hormones are released into plasma and transported largely bound to transport proteins. Some T_4 is deiodinated to T_3. Ultimately, both T_4 and T_3 are completely deiodinated by tissue utilization. The released I⁻ reenters the ECF iodide pool. (Redrawn from DeGroot LJ, Stanbury JB: *The Thyroid and Its Diseases,* ed 4. New York, John Wiley & Sons, 1975, p 38.)

oxidation involves a reaction catalyzed by a peroxidase enzyme: hydrogen peroxide acts as an electron acceptor (is reduced to water) and iodide acts as an electron donor (is oxidized to iodine).

$$H_2O_2 + 2HI \xrightarrow[\text{peroxidase}]{} 2H_2O + I_2$$

Once oxidized, the iodine combines with some of the tyrosyl units of the thyroglobulin molecule. Substitution of one iodine into the phenol ring of tyrosine results in the formation of monoiodotyrosine (MIT). Substitution of a second iodine into an MIT gives a diiodotyrosine (DIT). Thiourea drugs (propylthiouracil, methimazole, and carbimazole) block this step of hormone synthesis.

COUPLING

Soon after tyrosyl units of thyroglobulin are iodinated to MITs and DITs, hormonal iodothyronines T_4 and T_3 are formed by the coupling reaction. In the coupling reaction the iodinated phenol group of one iodotyrosyl molecule is thought to be oxidized to an unbound, free radical, which

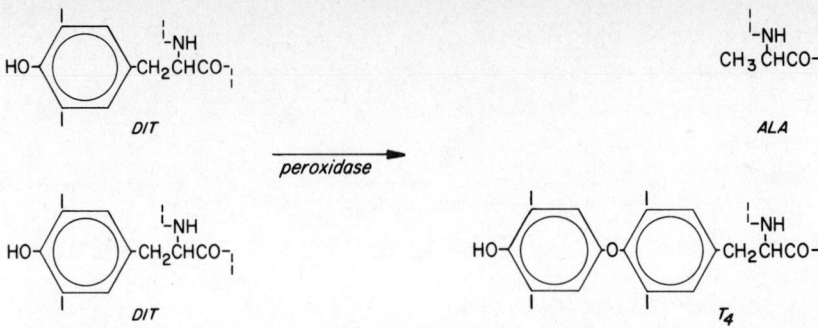

Figure 6 Conversion of two diiodotyrosyl units (as part of thyroglobulin chain) to thyroxine (T$_4$) and alanine, both of which remain as part of thyroglobulin chain. The dotted lines on the α-amino and carboxyl groups indicate continuation of the peptide chain backbone of thyroglobulin.

immediately couples with a nearby bound iodotyrosyl to produce a bound iodothyronine. The net reaction involves, in effect, the splitting out of an alanine unit and the joining of the iodinated rings by an ether bridge. The coupling of one MIT and one DIT to form a T$_3$ or the coupling of two DITs to form a T$_4$ involves an oxidative reaction that might use the same peroxidase enzyme involved in organification (Fig. 6).

1. Active transport of iodide into follicular cell (I$^-$ pump at base of cell).

2. Synthesis of noniodinated, nonglycoprotein by rough endoplasmic reticulum. Addition of carbohydrate occurs primarily at Golgi apparatus.

3. Iodination of tyrosyl units of thyroglobulin near surface of apical microvilli and incorporation of iodinated thyroglobulin into colloid.

4. Formation of colloid droplets by pinocytosis.

5. Merging of colloid droplets with lysosomes to form phagolysosomes.

6. Proteolysis of thyroglobulin in phagolysosomes with release of hydrolytic products. MIT and DIT are deiodinated by ITDase and iodide is recycled. T$_4$ and T$_3$ are released into blood for circulation.

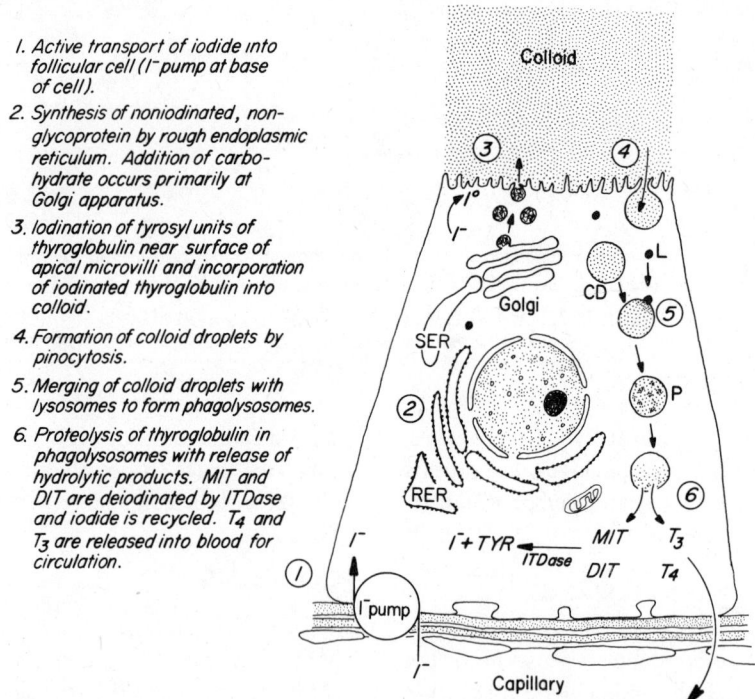

Figure 7 Schematic representation of thyroidal hormonogenesis. CD, Colloid droplet; L, lysosome; RER, rough endoplasmic reticulum; G, Golgi apparatus; P, phagolysosome.

PROTEOLYSIS

Thyroid hormones are stored in peptide linkage in thyroglobulin. This large-molecular-weight protein is stored extracellularly in the colloid space. For secretion of T_4 and T_3, thyroglobulin is drawn into the follicular cell by the process of pinocytosis (Fig. 7) as colloid droplets. These fuse with lysosomes containing proteolytic enzymes, and thyroglobulin digestion occurs. This complex glycoprotein molecule, containing approximately 6,000 amino acids, is digested to release its amino acid constituents including T_4, T_3, DIT, and MIT. T_4 and T_3 are released to the circulation. Very little MIT and DIT leave the follicular cell, but rather, are deiodinated by iodotyrosine deiodinase (ITDase), thereby conserving iodide for reuse by the thyroid. The importance of this conservation mechanism is demonstrated by a rare condition in which ITDase is deficient; this deficiency leads to iodine deficiency and goiter due to the loss of iodotyrosines from the cell and subsequent renal excretion.

STRUCTURAL–FUNCTIONAL CORRELATIONS

Various kinds of biochemical and histological methods have provided evidence of how the thyroid accomplished hormonal synthesis and secretion. Figure 7 summarizes the events of hormonogenesis.

TRANSPORT OF THYROID HORMONES

Upon entering the blood, T_4 and T_3 are bound to certain plasma proteins in a firm but reversible bond: 99.97% of the total blood T_4 is bound and 99.70% of the total T_3 is bound. The remainder is free hormone in reversible equilibrium with the bound hormone. Much of what is known about the specific binding of thyroid hormones has been derived from study of labeled hormone by the technique of zonal electrophoresis (Fig. 8). These studies have disclosed two plasma proteins with which T_4 is primarily associated: thyroxine-binding globulin (TBG) and thyroxine-binding prealbumin (TBPA). To a small extent, some

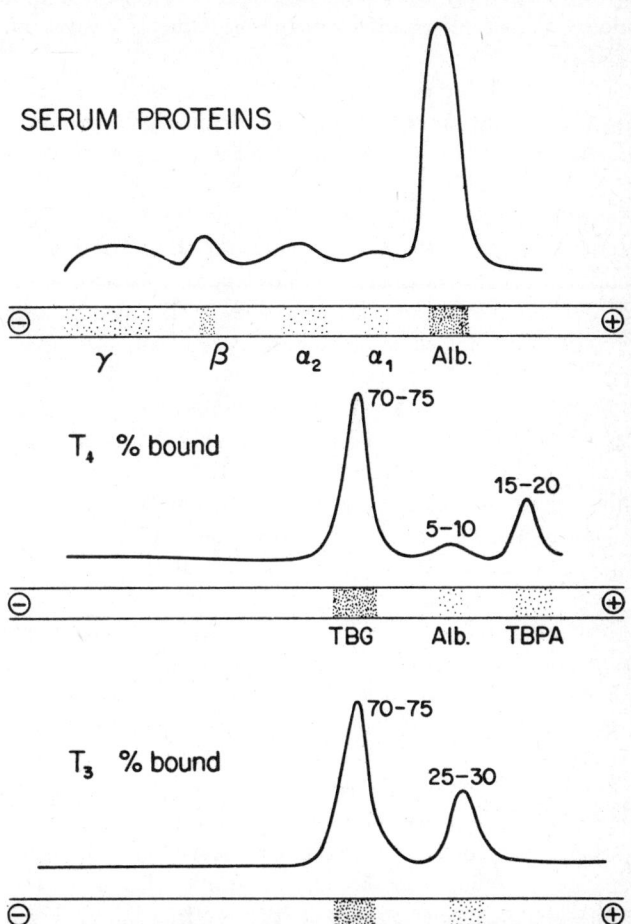

Figure 8 Diagrammatic representation of the distribution of T_4 and T_3 among serum thyroid hormone binding proteins. (Upper curve) Distribution and relative amounts (by height) of serum proteins on paper electrophoresis. (Middle curve) Distribution of T_4 among TBG(T_4-binding globulin); Alb(albumin); TBPA(thyroxine-binding prealbumin). T_4 is bound predominantly by TBG, to a lesser extent by TBPA, and slightly by Alb. (Lower curve) Distribution of T_3 among T_3-binding proteins; T_3 is bound predominantly by TBG. Some T_3 is bound by Alb, but essentially none by TBPA. The relative percentages bound by each protein for a normal adult are given in numbers above each electrophoretic peak. (From Rosenfield RL, Refetoff S, Hoffer PB, et al: in James AE, Wagner HN Jr, Cook ED (eds): "Diagnosis of Thyroid Diseases in Pediatrics", *Pediatric Nuclear Medicine.* Philadelphia, WB Saunders, 1974, pp 376–399).

T_4 is bound to serum albumin. T_3, on the other hand, is bound mainly by TBG and, to a small extent, by serum albumin. For all practical purposes, none is bound to TBPA.

Tissues take up the free hormones. That is, the proportion of total hormone that is free (unbound) is the primary determinant of hormonal effects. Since the free portion is the mediator of hormonal action, it is logical that it is this fraction that is regulated by the negative feedback control system (acting on the pituitary–thyroid axis). The principal role of the transport proteins appears to be one of conferring macromolecular properties to the thyroid hormones that alter their rate of utilization and loss. The normally negligible urinary excretion of T_4 and T_3 is due to their limited filtration at the glomerulus. Also, the transport proteins serve as a reservoir able to supply additional hormone as needed. Thus, as tissues take up free hormone, it is replaced by release from bound form. While such functions of the transport proteins are helpful, they are not essential. Individuals with no TBG, grossly elevated TBG, or who are devoid of serum albumin have normal thyroid hormone function. Certain hormones can diminish the amount of TBG (e.g., androgens, glucocorticoids) and others can increase TBG content (e.g., estrogens, as in some birth control pills), which affects the total amount of circulating thyroid hormone (bound hormone plus free hormone), but does not change the concentration of free hormone (except briefly). For example, suppose the concentration of TBG were doubled. The additional availability of binding sites (because of additional carrier protein) would cause an immediate drop in free T_4 and T_3 as some of the free hormone becomes bound to the new binding sites on the additional TBG. This would then result in an increase in T_4 and T_3 secretion by the thyroid (via the negative feedback control system: a drop in free hormone concentration removes feedback inhibition on TSH secretion, TSH secretion increases and stimulates additional hormone synthesis and release) until the free hormone concentration is restored to normal. The negative feedback control system is a function of the free hormone concentration. At this point in time a new equilibrium will have been established between free and bound hormone as before, except that the amount of bound hormone, and therefore the amount of total hormone (sum of bound plus free), will be greater than before. If one were measuring only total hormone (e.g., as protein-bound iodine, or as total T_4 by radioimmunoassay) it would appear that an excess of hormone were present, although the individual would be euthyroid (because the free hormone level would be normal), not hyperthyroid (Fig. 9). Remember that the free thyroid hormone concentration (which determines the level of hormonal activity) is maintained at the normal level (by the negative feedback control system) despite changes in total hormone amount due to changes in the amount of carrier protein.

METABOLISM OF THYROID HORMONES

The plasma half-life of thyroid hormones is largely determined by binding to the transport proteins. Thyroxine is 99.97% bound (0.03% free or soluble hormone) and has a very long half-life (7–10 days) in the human. Triiodothyronine is less firmly bound to the transport protein, but 99.7% is protein bound (0.3% free) and has a half-life of 1–2 days.

The most important pathway for T_4 metabolism is

Figure 9 (at right) Graphic representation of the sequence of events following an acute change in serum TBG capacity (e.g., a change in TBG concentration) in a subject with normally controlled thyroid hormone secretion and metabolism. The communicating reservoir principle is used for analogy. The width of the two large reservoirs represents available T_4-binding site capacity in serum (TBG) and in peripheral cells (tissue), which are partially saturated by T_4 (gray areas). The fluid represents thyroid hormone (T_4 in this example, although on analogous diagram can be drawn for T_3). The height of fluid in the small central reservoir represents free T_4 concentration in equilibrium with bound T_4 in each of the large reservoirs. Free T_4 is proportional to the level of saturation of the binding sites in serum (TBG) and in cells (tissue). Thyroidal secretion (supply) of hormone is represented by the input of fluid through the faucet, and hormone metabolism (disposal) by the overspill of the tissue reservoir. (Redrawn from Refetoff S: "Thyroid Hormone Transport" in DeGroot LJ, et al (eds): *Endocrinology.* New York, Grune & Stratton, 1979, vol 1, p 351.)

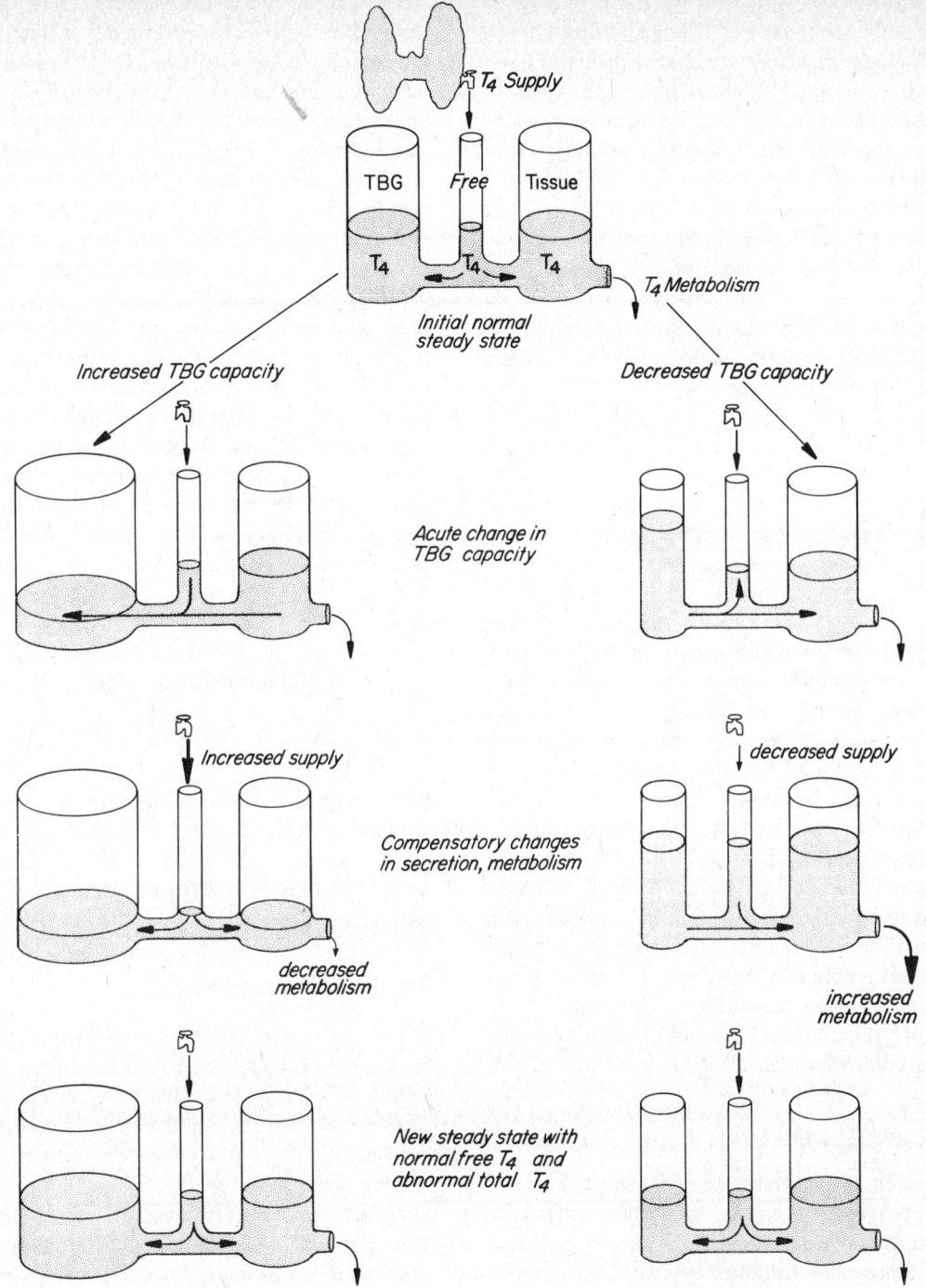

T4 Supply

TBG   Free   Tissue

T4   T4   T4

T4 Metabolism

Initial normal
steady state

Increased TBG capacity

Decreased TBG capacity

Acute change in
TBG capacity

Increased supply

decreased supply

Compensatory changes
in secretion, metabolism

decreased
metabolism

increased
metabolism

New steady state with
normal free T4 and
abnormal total T4

5'-monodeiodination to T_3. At least one-fourth of the T_4 secreted each day is converted to T_3, accounting for 60–80% of the T_3 used each day. This occurs in the liver, kidney, pituitary, and possibly all tissues of the body; it has led to the speculation that T_4 is a prohormone and that only T_3 has metabolic activity. Currently, the bulk of information favors biological potency for T_4 as well as T_3. Thus, the thyroid secretes two thyroid hormones: (1) T_4 predominantly, and (2) T_3 to a slight extent (most T_3 in circulation is derived from monodeiodination of secreted T_4).

Further metabolism of the thyroid hormones T_4 and T_3 involves deiodination, decarboxylation, and deamination with conjugation with glucuronic acid or sulfate and renal excretion.

CONTROL OF THYROID FUNCTION

Follicular cells are intrinsically capable of thyroid hormonogenesis and even have a small degree of autoregulation of secretion. However, in the absence of TSH, the biosynthesis and release of thyroid hormone occur at rates much too slow for adequate function. Each of the steps in hormonogenesis is accelerated by TSH.

Although the dependence of normal thyroid secretion upon adequate secretion of TSH has been recognized for over 50 years, only recently have the influences affecting TSH secretion come under careful scrutiny. Two principal factors in regulating the release of TSH from thyrotrophic cells of the anterior pituitary have been shown to be important: (1) hypothalamic secretion of TRH, and (2) the participation of circulating free thyroid hormone in a negative feedback control system.

HYPOTHALAMIC CONTROL

The hypothalamus is required for normal function of the pituitary. Hypothyroidism can be induced by section of the pituitary stalk if regeneration of a physical connection is prevented. Since neural connection with the anterior pituitary is scant, it had been presumed, initially without

proof, that hypothalamic control was mediated by humoral factors thought to pass down the hypophyseal portal system to the pituitary. That proof was forthcoming and now the factor controlling TSH secretion has been extracted, purified, identified, and synthesized: thyrotropin-releasing hormone (TRH) is a modified tripeptide. TRH is bound at specific binding sites (receptors) on the cell membrane of TSH-secreting cells (thyrotroph cells of anterior pituitary) and modulates the response of these cells to the negative feedback effects of circulating free thyroid hormone acting at other sites in TSH-secreting cells (primarily the nucleus). The inhibitory influence of the thyroid hormones is diminished and the production and secretion of TSH is increased in response to increased release of TRH. The output of TSH is a result of a balance struck between the stimulatory effects of TRH and the inhibitory effects of the thyroid hormones. Thus, when there is a large excess of T_4 and/or T_3, as in Graves' disease, the response of the anterior pituitary to increased TRH is abolished. Furthermore, in hypothyroidism when basal levels of T_4 and T_3 are much reduced, the response of the anterior pituitary to TRH is greatly increased.

It has been suggested that the hypothalamus could also participate in the negative feedback control system (by sensing the blood levels of thyroid hormones and altering the rate of release of TRH). However, there is no unequivocal evidence to support this. Instead, the hypothalamic neurons seem to be involved in the interpretation of stimuli from other areas of the CNS, stress, and thermal changes and provide tonic control of TSH release. Direct negative feedback of T_4 and T_3 appears to occur only at the pituitary (Fig. 10).

PITUITARY CONTROL

While the hypothalamus can affect the rate of TSH release by means of TRH, the usual day-to-day control of TSH secretion is accomplished by the negative feedback control system between the anterior pituitary and the concentration of free thyroid hormone in circulation.

The interactions among hypothalamus, anterior pituitary, and thyroid are analogous to the relationship between a person regulating room temperature, a thermostat, and a furnace. For a particular setting of the thermostat the furnace maintains the room temperature at

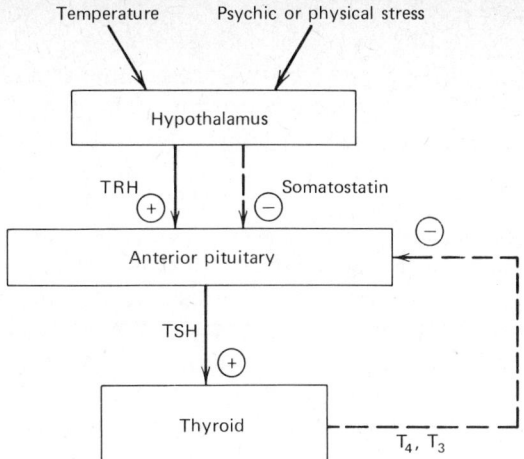

Figure 10 Diagrammatic representation of negative feedback control system regulating thyroid function. Free T₄ and T₃ inhibit secretion of TSH (increases in thyroid hormone decrease TSH secretion; while decreases in thyroid hormone increase TSH secretion). TRH from the hypothalamus determines the set point at which feedback operates. Somatostatin, an inhibitory releasing factor for growth hormone, may also participate as an inhibitor of TSH release (i.e., there may be dual hypothalamic control for TSH regulation similar to that for prolactin and growth hormone).

a relatively constant level by means of the negative feedback control system acting between the thermostat and the furnace. The person in the room acts as a modulator and may choose any particular set point to suit his own needs or comfort. In the body the thyroid gland (like the furnace) acts in a negative feedback control system with the anterior pituitary thyrotrophs (like the thermostat). The hypothalamus (like the person or modulator who sets the thermostat) sets the level of thyroid hormone concentration around which the system operates (the set point) to suit the body's needs. That is, the hypothalamus functions to modulate the system to operate around a specific set point.

For example, acute exposure of experimental anmials to cold results in a prompt increase in TRH release, followed by a rapid increase in TSH output by anterior pituitary thyrotroph cells and a subsequent increase in thyroid hormone secretion by the thyroid gland. The increased thyroid hormone increases the animals' metabolic rate (see under Calorigenic Effect), thereby helping maintain body temperature. Certain hypothalamic lesions prevent this response. To some extent, humans also respond to cold exposure with an increase in TRH release. A prompt and dramatic secretion of TRH and TSH occurs upon cooling in the newborn. This response operates well between birth and the first year of age, but diminishes with age beyond the first year and becomes minimal in adults.

ACTIONS OF THYROID HORMONES

Although the thyroid is not absolutely essential for life, it must be present and functional if an adequate or normal rate of metabolism is to be maintained, and for normal growth and development. The actions of thyroid hormones may be conveniently classified into two categories: (1) metabolic effects, and (2) developmental effects.

METABOLIC EFFECTS

The thyroid hormones have a multiplicity of metabolic effects on many tissues. Since no one primary effect on which all the other actions depend has yet been discovered, discussion of these various effects must remain largely descriptive.

Calorigenic Effect

The most prominent metabolic effect of thyroid hormones on adult warm-blooded animals is their calorigenic effect. This effect, which is due to an increase in the metabolic rate, can be measured directly by calorimetry (measurement of body heat loss), or indirectly by measurement of oxygen consumption (on the basis of 4.825 Calories being produced from each liter of oxygen consumed). That thyroid function affects metabolic rate as shown by its influence on the rate of O_2 consumption has been known since first reported by Magnus-Levy in 1895. In the past, determination of the metabolic rate under resting conditions, that is, the basal metabolic rate (BMR), has been used as a clinical test for possible abnormalities of thyroid function. However, since it lacks specificity (other conditions can also alter the BMR), and since it is inaccurate in all but the most experienced hands, it has gradually been replaced by more specific biochemical tests.

Thyroid hormone can increase the metabolic rate and oxygen consumption in all tissues studied except a few (spleen, brain and testes).

T_3 causes a more prompt but short-lived effect than does T_4. Also, T_3 is about five times more potent than T_4. The precise mechanism by which thyroid hormones produce an increased metabolic rate is uncertain. The calorigenic response can be inhibited or reversed by inhibitors of protein synthesis. An important implication of this finding is that the increase in metabolic rate may depend on the de novo synthesis of a specific enzyme such as Na^+K^+-ATPase, the action of which serves to accelerate the metabolic rate.

Effects on Protein Metabolism

One might intuitively predict that, since thyroid hormones are necessary for normal growth, these hormones (in physiological amounts) must have a widespread stimulatory effect on protein anabolism in general. T_4 and T_3, along with growth hormone and insulin, do have an overall stimulatory effect on protein synthesis in general. However, the effects of thyroid hormones on growth are biphasic in the sense that they are stimulatory at one dose and inhibitory at a higher dose. Consider the condition of hypothyroidism in immature animals and young patients: replacement doses stimulate or restore protein synthesis and growth, but excessive doses inhibit growth.

In addition, thyroid hormones stimulate the synthesis of certain specific enzymes to an even greater extent than protein synthesis in general. It is possible that specific induction of certain individual enzymatic activities may be the primary effect from which the numerous secondary effects arise (e.g., growth, maturation, calorigenesis). Several agents that block protein synthesis (e.g., actinomycin D, puromycin) inhibit or reverse both the metabolic and developmental effects of thyroid hormones.

Effects on Carbohydrate Metabolism

Thyroid hormones enhance the rate of absorption of glucose and glactose by the duodenum. They also potentiate insulin action on glucose uptake by tissues. Consequently, ingestion of carbohydrates by a hyperthyroid individual results in a rapid rise in blood glucose (due to enhanced absorption) with glucosuria (increased blood glucose increases the rate of filtration of glucose at the glomerulus to a level above the tubular maximum (T_m) for glucose reabsorption by the proximal tubules) followed by a rapid fall in blood glucose. The blood glucose level falls because of increased insulin secretion in response to the rapid rise in blood glucose, and because the thyroid hormones potentiate the effect of insulin to increase uptake of glucose by peripheral tissues (by insulin-stimulated facilitated diffusion). Although the peripheral action of insulin is augmented, the overall requirements for insulin are increased, probably as a result of increased metabolism of insulin. Thus, excess thyroid function may activate or intensify diabetes mellitus.

Effects on Lipid Metabolism

Thyroid hormones affect most areas of lipid metabolism, that is, synthesis, mobilization, and degradation. In general, thyroid hormone administration or secretion favors a decrease in body lipids because of a predominance of effect on degradation. That is, although synthesis is also stimulated by the action of thyroid hormone, degradation is increased to a greater extent. Since the amount of lipid present at any given instant is a balance between synthesis and degradation, the net result of a disproportionately greater increase in degradation is a decrease in lipid (lipolysis). Plasma and adipose tissue levels of triglycerides, phospholipids, and cholesterol decline. The decrease in blood cholesterol produced by thyroid hormone action has been recognized for years and is described as a "classic" effect. Most, if not all the lipolytic effects of thyroid hormones are mediated synergistically in conjunction with the direct lipolytic actions of other hormones (catecholamines, growth hormone, glucocorticoids, and glucagon).

In the case of a deficiency or lack of thyroid hormone, both synthesis and degradation of lipids are depressed. Again, the effect on degradation is greater, that is, the depression of degradation by thyroid hormone deficiency is greater than the depression of synthesis. Thus, the overall effect of thyroid hormone is one of net lipogenesis resulting in an accumulation of plasma cholesterol and other lipids.

DEVELOPMENTAL EFFECTS

The most conspicuous demonstration of thyroidal participation in the regulation of maturation is in amphibian metamorphosis. Although mammalian development is not as abrupt as the conversion of a tadpole into a frog, man and other higher vertebrates also require thyroid hormone for proper development. Experimentally, administration of antithyroid drugs to pregnant animals inevitably results in offspring with signs of cretinism. In humans, spontaneous cretinism occurs from (1) inadequate iodine intake, (2) congenital absence of a thyroid, or (3) congenital deficiency in any of the four steps in hormonogenesis. Developmental retardations in untreated cretins include delayed bone ossification, delayed union of epiphyses, delayed dentition, and hypoplasia of the brain. Typical results are dwarfism and imbecility. Diagnosis at the earliest possible date is of utmost importance. A successful outcome depends on prompt diagnosis and early treatment. Delay increases the chances of permanent irreversible mental retardation.

MECHANISM OF ACTION OF THYROID HORMONES

Although much is known about the effects of thyroid hormones, it has not been possible to find a fundamental mechanism of action that unifies the many biological effects observed.

EFFECTS ON OXYGEN CONSUMPTION AND MITOCHONDRIA

One of the oldest observations is that T_4 (and T_3) increase the oxygen consumption of the whole animal and many tissues in vitro. Since mitochondria account for 90% of the body's O_2 consumption, attention was turned to these organelles as the site of action. Swelling of mitochondria and uncoupling of oxidative phosphorylation (increased substrate and O_2 consumption with decreased ATP production) were early observations but required more than physiological amounts of thyroid hormone. Uncoupling is attractive in that it would explain the effect of excess T_4 in decreased efficiency of substrate utilization and excess heat production. Theories have been advanced for altered mitochondrial function with increased substrate delivery, increased mitochondrial numbers, or increased mitochondrial enzyme protein.

EFFECTS ON THE NUCLEUS AND PROTEIN SYNTHESIS

Current interest centers on this area as an explanation for the lag time in thyroid hormone effects (12–18 h for T_3 and 24–48 h for T_4), and the finding that thyroid hormone receptors reside in the nucleus (similar to steroid hormones). Increased protein synthesis as required in the effects of thyroid hormone on growth, specific increases in multiple specific enzymes, and the observation that inhibitors of protein synthesis block many of the effects of thyroid hormone would be explained. Effects of increased RNA polymerase and specific mRNA are observed as the earliest effects of thyroid hormone in vitro.

EFFECTS ON HEAT PRODUCTION AND Na+K+ ATPASE

Increased production of heat in tissues is an attractive hypothesis for thyroid hormone action. This would explain extremes of deficiency or excess thyroid hormone, but it is more difficult to explain the normal physiological function of these hormones. One proposed mechanism is by increasing the Na^+ pump, and thus ATP consumption, leading to increased substrate utilization and O_2 consumption.

At present it is impossible to determine a single mechanism of action for thyroid hormone and to determine which of the above effects are secondary. This may be because thyroid hormone has multiple effects, or it may just be that the right question has not been asked in the right manner.

THYROID DYSFUNCTION

HYPERTHYROIDISM (THYROTOXICOSIS)

Hyperthyroidism is a disease of adults, particularly females; it occurs rarely in children until just before, or during, adolescence. Hyperthyroidism can result from (1) excess secretion of TSH by a hypophyseal tumor (a rare cause of hyperthyroidism), (2) autonomous hypersecretory nodules or adenomas of the thyroid (not an uncommon cause among middle-aged females), or (3) most commonly from Graves' disease, a condition of diffuse enlargement and hypersecretion of the entire thyroid (diffuse toxic goiter) which, in about half the patients, is accompanied by exophthalmos (protrusion of the eyes).

The cause of Graves' disease has puzzled clinical endocrinologists for many years. It was once thought to be the result of excess secretion of TSH. Since pituitary extracts have thyrotrophic activity and can be shown to produce protrusion of eyeballs in a few experimental animals (mainly fish), it seemed likely that excess TSH or TSH and an exophtalmic producing substance (EPS) could explain both the hyperplasia of the thyroid and the exophtalmos of Graves' disease. However, a variety of experiments make it almost certain that the pituitary is not involved in Graves' disease. In fact, because of the negative feedback inhibition of the thyrotropic cells of the pituitary by excess thyroid hormone, TSH secretion is suppressed in Graves' disease (Fig. 11).

Instead of excess TSH as the cause of the hyperthyroidism of Graves' disease, the thyroid stimulating agent responsible is an unusual antibody that, in reacting with thyroid antigens, causes effects not unlike those of TSH. When tested in bioassay, this unique gamma globulin (an IgG) has a more delayed onset of action and longer duration of action in stimulating the secretion of thyroid hormones than does TSH. Therefore, this thyroid-stimulating agent has been named long-acting thyroid stimulator (LATS). LATS duplicates many of the effects of TSH both in vivo and in vitro. This circulating antibody, which is present in the plasma of most patients with Graves' dis-

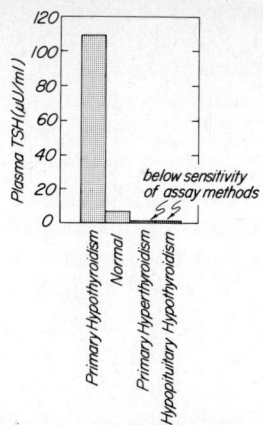

Figure 11. Plasma thyroid stimulating hormone (TSH) levels in various conditions. The plasma TSH concentration is quite high in primary hypothyroidism (failure to produce adequate amounts of thyroid hormones due to a deficiency of the thyroid gland) because of absence of the normal negative feedback of thyroid hormones on pituitary TSH-producing cells. Plasma TSH is quite low in primary hyperthyroidism (excessive secretion of thyroid hormones due to an autonomous hyperactivity of the the thyroid gland) for the opposite reason: excessive negative feedback inhibition of the TSH-producing cells. In hypopituitary hypothyroidism (inadequate production of thyroid hormones by the thyroid because of lack of adequate TSH production by the pituitary), which is a rare form of hypothyroidism, there is also a very low level of plasma TSH. In both primary hyperthyroidism and hypopituitary hypothyroidism the levels of TSH are so low that they are below the ability of the assay methods to detect.

ease, is undoubtedly a significant contributor to the cause of the hypresecretory goiter. However, LATS is not inevitably found in every single patient with Graves' disease. LATS is now considered to be only one of several thyroid-stimulating immunoglobulins implicated in Graves' disease. Those patients lacking in LATS will usually have at least one of the others present. Since some euthyroid relatives of patients with Graves' disease show LATS or LATS-like immunoglobulins present but no clinical signs, it must be concluded that other factors might also be operative in the incompletely understood pathogenesis of this disease.

The etiology of exophthalmos is even more obscure; it is unknown. It has been suggested that there might be an EPS from the pituitary (a glycopeptide fragment of TSH has been implicated). However, this seems unlikely since exophthalmos has persisted in some patients who have undergone hypophysectomy. That EPS could be derived from TSH is difficult to reconcile with the apparent inhibition of TSH secretion in thyrotoxic ophthalmopathic patients with Graves' disease and with the absence of possible TSH fragments (that would have been detected by radioimmunoassay) in sera of patients with ophthalmopathy. Another possibility is a second antibody distinct from LATS that has been demonstrated in the IgG fraction of sera from some patients with exophthalmos.

Thyrotoxicosis, or excess thyroid hormone, whether from a hypersecretory gland or from overzealous treatment of hypothyroid patients with exogenous thyroid hormone, produces effects that are largely the result of overly accelerated rates of tissue metabolism. A good way to remember many of the signs and symptoms of thyrotoxicosis is to think of your own signs and symptoms after overexercising on a hot summer day.

The need for elimination of excess heat generated by increased metabolism results in a hot, moist skin due to sweating and vasodilation—hyperthyroidism is a disease that has occasionally been diagnosed by a handshake. Hyperthyroid patients are intolerant of heat (subjectively feel overheated) and prefer cool environments. Alterations of the cardiovascular system are among the most prominent features of hyperthyroidism, for example, tachycardia and palpitation (often accompanied by dyspnea). There is an increase in stroke volume that, along with increased heart rate, causes increased cardiac output. In persons over age 40, this increased cardiac output can lead to a kind of heart failure called high output failure. Arterial blood pressure shows an increased systolic pressure (from the increased stroke volume) and decreased diastolic pressure (due to a more rapid runoff from peripheral vasodilation).

A patient's story of increased appetite but with continued weight loss (negative caloric balance) is characteristic of hyperthyroidism. A common complaint is "loose bowels" (anything from increased frequency of stools to diarrhea). Another prominent group of symptoms in hyperthyroidism arise from alterations in central nervous system function. Hyperthyroid patients tend to be hyperactive, "nervous," and fidgety to the point that it seems impossible for them to be still an instant. Thyrotoxic patients are often emotionally unstable and easily lose their tempers, or may have fits of crying for no apparent reason. Examination may reveal a fine rhythmic tremor of the extended hands, extended tongue, or tightly closed eyes. Since the nervous system does not respond to the caloriginic effects of thyroid hormones, it might seem paradoxical that symptoms of hyperactivity should occur. Part of the explanation may lie in increased sensitivity to catecholamines. Although catecholamine secretion is normal in hyperthyroidism, the excess thyroid hormone seems to enhance catecholamine effects. Adrenergic blocking drugs (particularly beta blockers like propranolol) have been found to decrease many of the nervous and cardiovascular symptoms (they reduce anxiety, tremulousness, tachycardia, and palpitations). Figure 12 summarizes the signs and symptoms of hyperthyroidism.

Thyroid Storm

Thyroid storm or crisis is an abrupt extreme accentuation of thyrotoxicosis. This serious, life-threatening complication in patients with hyperthyroidism can occur following thyroid surgery (due to release of stored hormone by manipulation of the gland) or may be triggered by independent illnesses, trauma, or infection in nonsurgical patients. The clinical picture is dominated by manifestations of hypermetabolism. Fever is present and may be extreme; profuse sweating occurs. Marked tachycardia (often with arrhythmias) may be accompanied by high output failure and pulmonary edema. Increased tremulousness and restlessness occur at the onset and progress to delirium and frank psychosis, and eventually coma. Hypotension may develop due to a shocklike collapse of peripheral vascular resistance (extensive vasodilation).

HYPOTHYROIDISM

Hypothyroidism can occur at any age. It is more common in females than in males. Growth is stunted in both cretinism (congenital hypothyroidism) and in juvenile hypothyroidism (hypothyroidism acquired by a previously normal child); however, mental retardation is more severe in cretinism.

Myxedema is a term coined nearly a century ago by Dr. William Ord to describe what he considered to be the most

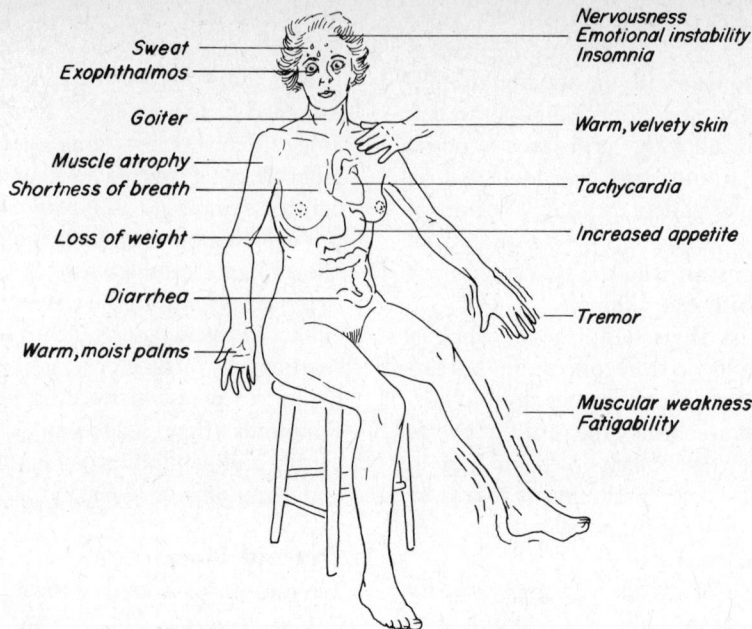

Sweat
Exophthalmos
Goiter
Muscle atrophy
Shortness of breath
Loss of weight
Diarrhea
Warm, moist palms

Nervousness
Emotional instability
Insomnia
Warm, velvety skin
Tachycardia
Increased appetite
Tremor
Muscular weakness
Fatigability

Figure 12 Drawing of a hypothetical patient illustrating many of the signs and symptoms of hyperthyroidism of Graves' disease. (Redrawn from Netter FH; in *The CIBA Collection of Medical Illustrations.* Vol 4: *Endocrine System and Selected Metabolic Diseases.* CIBA, 1965, p 49).

characteristic feature of adult hypothyroidism. The term myxedema means "mucous edema": a generalized nonpitting edema resulting from increased ISF colloid osmotic pressure due to accumulation of mucoprotein in the interstitial fluid of the dermis and other sites. Many clinicians have inappropriately used the terms myxedema and hypothyroidism synonymously. Hypothyroidism is the condition of deficiency of circulating thyroid hormone and may exist for many months or years before the appearance of clinically detectable myxedema.

Common causes of hypothyroidism are (1) spontaneous atrophy (formation of small fibrous thyroid with no previous history of goiter), (2) iatrogenic causes (e.g., surgical removal, excessive ^{131}I therapy, overenthusiastic treatment of hyperthyroidism with antithyroid drugs), (3) chronic lymphocytic thyroiditis (CLT, or Hashimoto's disease), and (4) the development of hypothyroidism from severe dietary deficiency of iodine (endemic goiter). A decrease in thyroid hormone production due to lack of sufficient iodine in the diet results in loss of negative feedback on

TSH secretion; as a result, TSH secretion increases and produces an enlarged thyroid (goiter). If the iodine deficiency is not severe, the increased TSH secretion may stimulate the thyroid to maximal utilization of what little iodine there is, with near normal production of thyroid hormone (but only as long as TSH secretion remains high, that is, a new set point in the negative feedback relationship). If the iodine lack is severe (e.g., nearly absent), no amount of TSH secretion can bring about adequate hormonogenesis (there just is not enough iodine available), and the only result of increased TSH secretion is thyroid enlargement. In both cases the thyroidal enlargement is called a nontoxic goiter.

To remember some of the signs and symptoms of hypothyroidism, think of how you would feel upon being partly awakened from a deep sleep in the middle of a very cold winter night.

The symptoms of hypothyroidism are generally those of diminished function due to a decreased metabolic rate. In contrast to the need for heat elimination seen in hyperthyroidism, the hypothyroid patient has quite the opposite problem: need for heat conservation. These patients often have the sensation of being cold or chilled. Vasoconstriction causes the skin to be cool and dry. The dryness is due to decreased secretory function of both sebaceous glands and sweat glands. With extreme dryness the skin may become scaly and, if severe, may resemble ichthyosis. Myxedema is most apparent in the suborbital area around the eyes (puffy eyes with drooping upper lids) and on the dorsa of hands and feet. Edematous changes in the vocal cords result in a hoarse, deep "froggy" voice (some clinical endocrinologists claim to have diagnosed hypothyroidism over the telephone).

Cardiovascular changes include both a decreased stroke volume and a diminished heart rate (i.e., a decreased cardiac output). The heart is enlarged and flabby. Its contractility is subnormal and may be a direct cause of the dilation (increased stretch to meet requirements of force of contraction). The large heart with reduced cardiac output is reminiscent of, or mimics, congestive failure (and occasionally the hypothyroid heart actually is in failure). Despite diminished cardiac output, the blood pressure tends to be either normal or hypertensive due to increased peripheral resistance (i.e., a greater increase in the total peripheral resistance than the decrease in cardiac output). Atherosclerosis is commonly present and thought to be related to increased plasma cholesterol.

The symptoms related to alimentary function are also additional examples of low idling speed of the living machinery. Although anorexia is common, most patients show a gain in weight. This is due both to the additional fluid of myxedema and the increase in adipose stores due to net lipogenesis. However, gross obesity is not common in hypothyroidism, contrary to popular opinion. Constipation, a frequent complaint, is the result of diminished propulsive activity and can lead to fecal impaction.

The mental picture in hypothyroidism of adults is one of slow-witted, good-natured complacency (sometimes with amusing congeniality). Mentation tends to be slow and memory lapses are common. Lethargy and somnolence are characteristic. Dozing in the sun or at the fireside may be favorite pastimes. These patients tend to sleep at the lightest excuse and may occasionally fall asleep under inappropriate circumstances. In profound hypothyroidism there may be psychotic disorders. Some hypothyroid patients become depressed and withdrawn. Later frank psychotic behavior characterized by maniacal or paranoid behavior may supervene ("myxedema madness"). Figure 13 summarizes the signs and symptoms of hypothyroidism.

ENDEMIC GOITER

Endemic goiter is said to exist when an enlargement of the thyroid gland can be detected in over 10% of a population. The principal cause of endemic goiter in most areas of the world where it is found (particularly mountainous regions) is a deficiency of dietery iodine due to the presence of little or no iodine in soil and water. Genetic factors may accentuate sensitivity to iodine deficiency, but the lack of sufficient iodine for normal hormonogenesis is the principal defect. The striking disappearance of endemic goiter in those areas where iodized salt (1 mg iodine per 10 g NaCl) has proved the validity and adequacy of the iodine deficiency theory. Because of the addition of iodine to salt

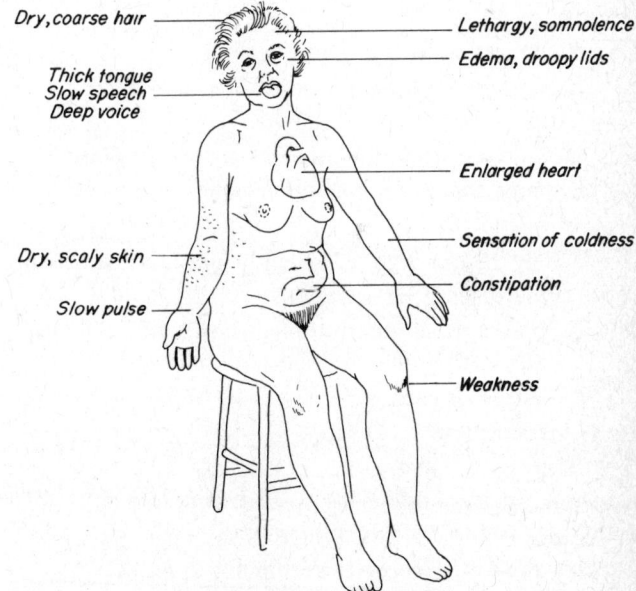

Figure 13 A hypothetical patient, illustrating many of the signs and symptoms of hypothyroidism. (Redrawn from Netter FH: in *The CIBA Collection of Medical Illustrations.* Vol 4: *Endocrine System and Selected Metabolic Diseases.* CIBA, 1965, p 49).

and to enriched bread, iodine deficiency goiter is almost never seen in this country.

Apart from gradual enlargement of the thyroid, there may be few or no symptoms. Apparently the stimulating action of increased TSH is able to compensate for a low iodide intake (which at first causes a low rate of hormonogenesis) by bringing about maximal utilization of what little iodide exists in the diet. Only in the most severe endemic districts where the iodine deficiency is almost absolute, do patients with goiter exhibit signs of hypothyroidism. In such areas endemic cretinism is also seen.

SELECTED BIBLIOGRAPHY

DeGroot LJ, Stanbury JB (eds): *The Thyroid and Its Diseases.* New York, John Wiley & Sons, 1975.

DeGroot LJ, et al (ed): *Endocrinology,* Vol 1: *Thyroid Gland.* New York, Grune & Stratton, 1979, p 305.

Ingbar SH, Woeber KA: in Williams RH (ed): *Textbook of Endocrinology.* Philadelphia, WB Saunders, 1981.

Werner SC, Ingbar SH (eds): *The Thyroid, A Fundamental and Clinical Text,* ed 4. Hagerstown MD, Harper & Row, 1978.

SELF-STUDY QUESTIONS

MULTIPLE CHOICE

Select the single best answer.

1. All of the following characterize the thyroid gland, except:
   A. largest endocrine gland in the body.
   B. hormonal product is stored extracellularly.
   C. great capacity for hormonal storage.
   D. secretes triiodothyronine (T_3) predominantly.
   E. most superficial endocrine gland in females.

2. Normally the follicular cells are cuboidal, approximately 15 μm high. This cell height increases with
   A. increased TSH secretion.
   B. prolonged administration of antithyroid drug that blocks the organification of iodine.
   C. iodine deficiency.
   D. deficiency of ITDase.
   E. all of the above.

3. Which of the following thyroid/serum (T/S) iodide ratios would you be expected to have after taking 500 mg of sodium perchlorate/day for 2 weeks?
   A. 2,000:1
   B. 200:1
   C. 20:1
   D. 1:1
   E. 1:100

4. Normally approximately 100 μg of thyroid hormone (containing 60 μg of organic iodine) is lost or used up each day. Which of the following statements about this is true?
   A. About 20% of this thyroid hormone is deiodinated, thereby returning approximately one-fifth of the daily secretion of organic iodine back into the ECF pool of inorganic iodide.
   B. All of this organic iodine is recycled back into the ECF pool of inorganic iodide.
   C. Most of the daily secretion of organic iodine is cleared from the plasma by renal excretion of thyroid hormone.
   D. About one-fifth of the daily secretion of organic iodine is lost from the body in bile.
   E. If no new thyroid hormone were synthesized, the normal store of hormone in the colloid would provide for about 2 weeks' worth of normal secretion.

5. A principal site of action of antithyroid drugs of the thiourea type (e.g., propylthiouracil, or PTU) is
   A. inhibition of iodide pumping.
   B. inhibition of peroxidase enzymatic activity.
   C. inhibition of iodotyrosine deiodinase (ITDase).
   D. inhibitin of lysosomal proteolysis in phagolysosomes.
   E. inhibition of hypothalamic secretion of TRH.

6. In order for the thyroid gland to synthesize and secrete hormonally active compounds, the thyroid

 A. requires only passive diffusion of I^- from plasma into follicular cells.

 B. must couple iodotyrosines to form iodothyronines.

 C. must inactivate iodotyrosine deiodinase (ITDase), thereby liberating iodoproteins and, ultimately T_4 and T_3.

 D. produce a large intrathyroidal inorganic iodide pool by inhibiting organification of iodide.

GREATER THAN, LESS THAN

Mark A if item in column A is greater; mark B if item in column B is greater, and mark C if both items are essentially the same.

Column A	*Column B*
7. Percentage of total T_4 bound to carrier proteins.	7. Percentage of total T_3 bound to carrier proteins.
8. The total amount of carrier protein available to bind thyroid hormone in women with increased estrogen levels.	8. The normal total amount of carrier protein available to bind thyroid hormone in normal women.
9. The concentration of total thyroid hormone in women with increased estrogen levels.	9. The concentration of total thyroid hormone in normal women.
10. The concentration of free thyroid hormone in women with increased estrogen levels.	10. The concentration of free thyroid hormone in normal women.

MULTIPLE CHOICE

Select the single best answer.

11. A patient secreting an excess amount of thyroid hormone (hyperthyroidism) would be expected to have

 A. positive nitrogen balance.

 B. zero nitrogen balance (neither positive nor negative).

 C. negative nitrogen balance.

 D. diminished urinary excretion of nitrogen.

 E. diminished protein catabolism.

12. Excess thyroid hormone favors

 A. enhanced rate of glucose and galactose absorption from gut.

 B. hyperglycemia followed by hypoglycemia after a high carbohydrate meal.

 C. glycosuria at some time following a high carbohydrate meal.

 D. increased requirement for insulin.

 E. all of the above.

13. Which of the following is characteristic of excess thyroid hormone?

 A. Decreased synthesis of lipids

 B. Decreased degradation of lipids

 C. Increased concentration of blood cholesterol

 D. Net lipogenesis

 E. None of the above

MATCHING

Match letters representing points on the graph with numbered statements below.

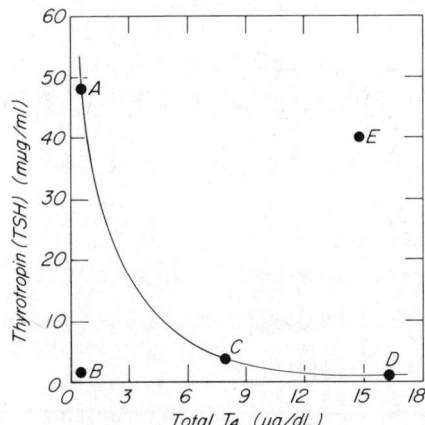

14. Hypophysectomy

15. Administration of excess exogenous thyroxine

16. Chronic administration of propylthiouracil (PTU)

MULTIPLE CHOICE

Select the single best answer.

17. Endemic goiter is most often characterized by
    A. profound hypothyroidism with generalized myxedema.
    B. thyrotoxicosis.
    C. congenital absence of TSH.
    D. exophthalmos in about half the people with goiter.
    E. euthyroidism.

18. Chronic administration of a therapeutic agent to combat bacterial infection results in thyroid enlargement (hyperplasia and hypertrophy) associated with an increase in free T_4 as a side effect. This drug could have its effect on thyroid function by
    A. inhibiting the binding of TSH to thyroid receptor sites.
    B. stimulating deiodination of iodotyrosines.
    C. interfering with the ability of thyroid hormone to inhibit TSH secretion by the anterior pituitary (loss of negative feedback control).
    D. inhibiting release of hypothalamic TRH into the hypophyseal portal system.
    E. causing competitive inhibition of iodide uptake by follicular cells.

19. A 9-year-old white female, who weighs 29 kg (64 lb) and is 132 cm (4 ft, 4 in.) tall, complains of being cold all the time, is slow to answer questions, has a dry, pale, scaly skin and puffy eyelids, and is often sleepy. Her neck has a normal appearance. Your diagnosis should be
    A. cretinism.
    B. acquired juvenile hypothyroidism.
    C. endemic goiter.
    D. Graves' disease.
    E. toxic nodular goiter.

MULTIPLE CHOICE

Select the correct answer(s). (In many instances, more than one answer is correct.)

20. Congenital absence of a thyroid gland (athyreosis) is associated with
    A. delayed skeletal maturation.
    B. elevated blood levels of growth hormone.
    C. elevated blood levels of thyrotropin (TSH).
    D. elevated blood levels of inorganic iodide (I^-).

21. Thyroxine (T_4) administration in a slightly hypothyroid patient will result in
    A. an increase in the basal metabolic rate (BMR).
    B. a decrease in the secretion of triiodothyronine (T_3).
    C. inhibition of TSH secretion.
    D. an increase in blood cholesterol levels.

22. The thyroids are removed from a group of dogs. Four to 6 weeks elapse before the animals show the first signs of hypothyroidism. This prolonged period before symptoms appear is due to
    A. the low metabolic rate of dogs.
    B. the large amount of thyroid hormone stored as thyroglobulin.
    C. the complete reabsorption of all thyroid hormone excreted in bile.
    D. plasma protein binding of thyroid hormone.

23. An excessive secretion of thyroxine is likely to cause
    A. decreased propulsive motility of the gut and constipation.
    B. increased heart rate and increased stroke volume.
    C. somnolence.
    D. the skin to become warm and moist.

24. Patients with Graves' disease
    A. invariably have exophthalmos.
    B. have a few autonomous hypersecretory adenomas of the thyroid.
    C. have a profound hypersecretion of TSH by the anterior pituitary.
    D. often have LATS or other thyroid-stimulating immunoglobulin.

25. An increase in thyroid gland secretion brought about by an increase in TSH (e.g., by administration of TSH) is likely to cause
    A. an increase in the rate of iodide uptake by the thyroid gland.
    B. an increased height of the follicular cells.
    C. a decrease in the diameter of the follicular colloid.
    D. a decrease in systolic blood pressure.

ANSWERS

1. D. All the statements characterize the thyroid except D. T_3 is not the predominant secretion of the thyroid, T_4 is. Some T_3 is secreted, but not nearly as much as T_4. Some of the T_4 that is secreted is ultimately converted to T_3 in a few tissues, but little T_3 is actually secreted by the normal thyroid.

2. E. Increased TSH secretion results in increased cellular activity with increased cell height and decreased colloid (colloid thyroglobulin is hydrolyzed into thyroid hormones at a greater rate and used up faster). (See Fig. 3.) All the situations would lead to increased TSH because thyroid hormone synthesis and release are decreased.

3. D. Perchlorate (ClO_4^-) inhibits the active transport of iodide into the follicular cells and colloid. Thus, the only iodide that gets in does so by passive diffusion, and the thyroid is unable to accumulate iodide; the T/S ratio is near unity.

4. D. One-fifth of the daily secretion of thyroid hormone is lost in bile, and the remaining four-fifths is completely deiodinated, thereby returning 80% of the organic iodine back into the ECF pool of inorganic iodide (recycled).

5. B. Drugs of the thiourea type (e.g., PTU) inhibit peroxidase activity, thereby inhibiting organification and coupling.

6. B. Iodotyrosines must be coupled to form the hormonally active iodothyronines, T_4 and T_3, as part of the thyroglobulin molecule. A could also be correct, but only under very special circumstances. When there is a congenital defect in the iodide trapping mechanism, large oral doses of iodide can greatly increase the plasma concentration of iodide. This increases the amount of iodide entering the thyroid (although still by passive diffusion only, and with a low T/S ratio). This can supply enough intrathyroidal iodide to maintain normal hormonogenesis. The therapeutic administration of up to 1 g/day has converted individuals with this trapping defect from hypothyroidism to euthyroidism.

7. A. 99.97% of total T_4 is bound while 99.70% of total T_3 is bound. At first glance these percentages might seem to be essentially the same (you may have picked C as the answer), however, the differences are significant. For example, although the total amount of T_4 is 50 times as great as the total amount of T_3, the free amount of T_4 is only five times greater because of the greater proportion of T_4 bound (T_3 less bound).

8. A. Increased estrogen (e.g., the use of birth control pills containing estrogen) results in the production of increased carrier protein.

9. A. An equilibrium exists between the free hormone and that bound to carrier protein (some of the bound hormone becomes free and an equal amount of free hormone becomes bound each instant). An increase in carrier protein makes available additional binding sites and increases the amount of hormone bound (at the expense of the amount free). As more free hormone becomes bound, the concentration of free hormone briefly decreases. Feedback control mechanisms regulating the concentration of free hormone cause a temporary increase in thyroid hormone production to restore free hormone concentration to normal. The amount of free hormone is normal again, the amount bound is still increased (with a new equilibrium between free and bound), and the total amount (bound plus free) is increased.

10. C. As described above, the feedback control mechanisms attempt to keep the concentration of free hormone normal. Thus, although the total amount of hormone is increased, the amount free is normal.

11. C. Negative nitrogen balance results from inhibition of protein synthesis and/or stimulation of protein catabolism by excess thyroid hormone. There is a net loss of body protein with greater nitrogen excretion than nitrogen intake as dietary protein (negative nitrogen balance).

12. E. All of the choices are correct. The increased rate of glucose absorption results in a sharp rise in blood glucose (hyperglycemia) followed by enhanced tissue

uptake of the glucose (resulting in a "reactive" hypoglycemia). At the time of the hyperglycemic peak the renal threshold is exceeded and glucosuria results. Although thyroid hormone enhances or facilitates the action of insulin on glucose uptake, it also increases the requirements for insulin secretion (perhaps because of increased metabolic breakdown of insulin).

13. E. None of the choices is correct. Thyroid hormone excess would cause increased lipid synthesis, an even greater increase in lipid degradation, decreased blood lipids (particularly decreased blood cholesterol), and net lipolysis (not net lipogenesis).

14. B. Removal of the pituitary removes the source of TSH and greatly diminishes thyroid function (as evidenced by near absence of T_4).

15. D. The T_4 level is high because of the exogenous T_4 given. The TSH level is low because all this T_4 inhibits TSH secretion by the negative feedback control system.

16. A. The PTU inhibits hormonogenesis, resulting in a low T_4. This, in turn, releases the anterior pituitary from negative feedback inhibition and results in a large increase in TSH secretion.

17. E. The goiter is caused by increased TSH that is usually able to bring about enough increase in thyroid function, despite the presence of low dietary iodine or the presence of other factors tending to diminish hormonogenesis, that the individual is able to remain euthyroid. Rarely are the conditions that produce endemic goiter so severe as to cause hypothyroidism and/or endemic cretinism.

18. C. Inhibition of the negative feedback system is the only one of these choices that would result in an increase of T_4 secretion and also in an increase in thyroid size.

19. B. She has hypothyroidism, but is too tall to be a cretin. She has acquired the hypothyroidism after doing a fair amount of growing. Spontaneous atrophy of the thyroid would explain the absence of thyroid enlargement. The other choices (C, D, and E) all include goiters of the neck.

20. A, C. Absence of the thyroid does not cause an elevation of growth hormone secretion (if anything, it is decreased). Since the thyroid is responsible for only a small fraction of the clearance of iodide from the plasma, the absence of a thyroid does not have much effect on iodide concentration in the blood.

21. A, B, C. The calorigenic effect of T_4 will increase BMR. The feedback inhibitory effect of the exogenous T_4 on the pituitary will decrease TSH secretion and, thus, decrease endogenous secretion of T_4 and T_3 by the patient's thyroid (which is already secreting at a slightly subnormal rate to begin with). The blood cholesterol will not increase further, but will probably fall as a result of the additional T_4 provided exogenously.

22. D. As kcal/m², metabolic rates of mammals are very similar. The storage function of thyroglobulin is irrelevant since these animals have had their thyroids removed. If thyroid hormone in bile were completely reabsorbed in the gut, it would diminish somewhat the loss of circulating plasma thyroid hormone; however, there is not complete reabsorption and some is lost each day by this route. The decline in circulating plasma thyroid hormone is fairly prolonged, mainly because of binding to carrier proteins.

23. B, D. Propulsive activity would be increased, not decreased. Insomnia rather than somnolence is caused by excessive thyroxine.

24. D. Patients with Graves' disease do not invariably have exophthalmos (only about half do). There is diffuse enlargement of the gland, not nodular formation. TSH secretion is depressed by the high levels of thyroid hormone secreted (see Fig. 11). Graves' disease appears to be an autoimmune disease characterized by immunoglobulins that have TSH-like activity.

25. A, B, C. The systolic pressure would increase as a result of an increase in stroke volume.

47

Parathyroid Hormone, Vitamin D, and Calcitonin in the Homeostatic Regulation of Calcium Metabolism

DONALD W. STUBBS

OBJECTIVES

After completion of this chapter, you should be able to:

1. Describe the flow of calcium into and out of the extracellular fluid space and identify which aspects of calcium kinectics are under endocrine control.

2. Distinguish between the nonhormonally regulated calcium ion buffer exchange function of bone and the endocrine-controlled process of bone resorption and deposition.

3. Describe the three states of calcium in plasma and identify which one is of primary physiological significance.

4. Diagram the interrelationships among 1,25-diOH D, parathyroid hormone, and calcitonin, including feedback regulation by plasma $[Ca^{2+}]$.

5. Discuss the role of 1,25-diOH D in the indirect influence of PTH on intestinal absorption of calcium and how this brings about intestinal adaptation to different levels of dietary calcium.

6. Relate the evolution of the parathyroid glands to tetrapod adaptation to a terrestrial existence.

7. Describe the embryological developement of the para-

thyroids and the ultimobranchial bodies and their ultimate association with the thyroid gland after birth.

8. List the sequence of steps involved in the biosynthesis of PTH and the specific role played by the 25-amino acid signal sequence in the formation of PreproPTH. This should include discussion of translocation of intermediates and products from one part of the chief cell to another.

9. Explain the influence of changes in ECF $[Ca^{2+}]$ on the biosynthesis and release of PTH.

10. Describe the effects of PTH on intestine, bone, and kidney and discuss how these actions relate to the overall homeostatic control of plasma calcium concentration.

11. Distinguish between rapid and late effects of PTH on bone resorptive activity.

12. Explain the influence of the phosphaturic effect of PTH on calcium regulation.

13. Describe the various clinical conditions of hyper- and hyposecretion of PTH.

14. Discuss the interrelationship between CT and gastrin and what influence this might have in diminishing alimentary hypercalcemia.

15. Describe the effect of CT on bone resorption.

16. Relate vitamin D requirement to rickets.

Calcium is one of the most abundant minerals of the body. The total amount of calcium is almost 2% of the body weight (about 1.3 kg in a 70-kg person). Disregarding the oxygen and hydrogen in body water, calcium ranks third in abundance of elements after carbon and nitrogen. That an inorganic element would comprise such a relatively substantial portion of the body might seem surprising at first; however, in considering the role of calcium as a structural component (along with phosphorus) of bone, its abundance should make more sense (more than 99% of the body's calcium is in bone). However, the calcium ion came to have numerous important, although perhaps less obvious, physiological functions long before the evolution of bone, which made its first appearance in the form of bony plates and scales in ostracoderm fish of the Ordovician waters of the Paleozoic era. Our earlier marine ancestors lived in an environment rich in calcium (seawater contains 40 mg calcium per deciliter), and this readily available ion was long ago incorporated into the function of numerous organ and cell systems to the point that it eventually became absolutely essential. Some of these nonstructural, functional roles are (1) in excitation–contraction coupling in muscle, (2) in excitation–secretion coupling in many secretory cells, (3) participation as factor IV in blood coagulation, (4) cofactor function in various enzyme systems, and (5) most importantly, from a clinical point of view, calcium has a membrane-stabilizing effect that prevents excessive excitability of the neuromuscular system, that is, the ability of excitable cells to generate an action potential is greatly enhanced and can even become spontaneous if ECF calcium levels are below normal. This latter function in the neuromuscular system demands precise regulatory systems to maintain the extracellular fluid (ECF) calcium concentration within narrow limits. Although the other nonstructural functions also have an essential requirement for minimal levels of Ca^{2+}, the levels required are much lower than those necessary for normal neuromuscular excitability, and therefore they do not ap-

pear to be much affected by even the lowest hypocalcemic levels possible in vivo, which would have a greatly adverse or fatal effect on neuromuscular function.

That the requirement of a proper level of calcium ion in the ECF for maintaining normal neuromuscular excitability is of greater importance than its more obvious role as a structural component in bone can be seen in the extent to which the control mechanisms regulating plasma calcium concentration permit demineralization of bone, when necessary, in order to keep the plasma Ca^{2+} level within fairly narrow limits. The level of ECF calcium is regulated by gastrointestinal (GI), renal, and skeletal factors, which, under the influence of endocrine controls, normally function together to maintain ECF Ca^{2+} levels within these remarkably constant and narrow limits, in spite of the large and variable amounts of calcium in the diet, demands of increased skeletal formation during growth, and losses occurring in pregnancy and lactation. The plasma calcium concentration is vigorously defended; changes as small as $\pm 1\%$ result in activation of compensatory homeostatic mechanisms that counteract such perturbations and restore the calcium concentration to normal. This vitally important homeostatic control of ECF Ca^{2+} involves the coordinated effects of parathyroid hormone (PTH) and a hydroxylated derivative of vitamin D (1,25-diOH D) for the most part, and perhaps calcitonin (CT) to a much lesser extent.

OVERVIEW OF CALCIUM KINETICS AND COMPARTMENTALIZATION

The dietary intake of calcium is quite variable depending on customs and preferences. Of the amount taken in with the diet, approximately one-third is absorbed. However, the net gain is much less than this, since a considerable amount of calcium is secreted into, and excreted from, the gastrointestinal tract. For example, if a diet provides 1,000 mg of calcium daily, as much as 800 mg may appear in the feces, giving a net gain of only 200 mg/day (Fig. 1). In a person in calcium balance, the dietary gain is matched by renal excretion. Furthermore, the exchange of calcium between ECF and bone is normally balanced (as much is deposited into bone each day as is liberated or resorbed from bone), if a person is neither forming more bone as growth nor losing bone due to pathology.

INTESTINAL ABSORPTION AND SECRETION

The absorption of calcium occurs in the small intestine. The average daily intake of dietary calcium in the United States is between 600 and 1,600 mg, but this varies from day to day between far greater extremes. To accommodate the needs of the body and prevent wide fluctuations, since neuromuscular excitability is dangerously high in hypocalcemia and cardiac arrhythmias can develop in hypercalcemia, the efficiency of absorption is increased when dietary calcium is low and decreases when intake is high. The proportion of dietary calcium absorbed varies as a function of PTH influence mediated by a dihydroxy derivative of vitamin D. That is, intestinal absorption becomes adapted to calcium requirements.

In considering intestinal absorption of calcium, a distinction should be made between total absorption, an estimated quantity, and net absorption, which can be measured. Total absorption is the amount of calcium absorbed from the total load of calcium that enters the small intestine. This total amount presented includes calcium in the diet and calcium in intestinal secretions, bile, and shed epithelial cells—how much is secreted cannot be known with certainty. On the other hand, net absorption is the difference between total absorption and calcium excretion and is relatively easy to measure: it is simply the difference between calcium in the diet and that in feces. At very low levels of dietary calcium, the amount of calcium secreted into the gut and lost in feces can exceed the amount absorbed, that is, there can be a net loss. At higher levels of intake, net absorption increases to give a net gain (Fig. 2). The amount of calcium secreted into the gut and lost in the feces does not vary much and can be considered to be an obligatory loss. It is not regulated nor controlled.

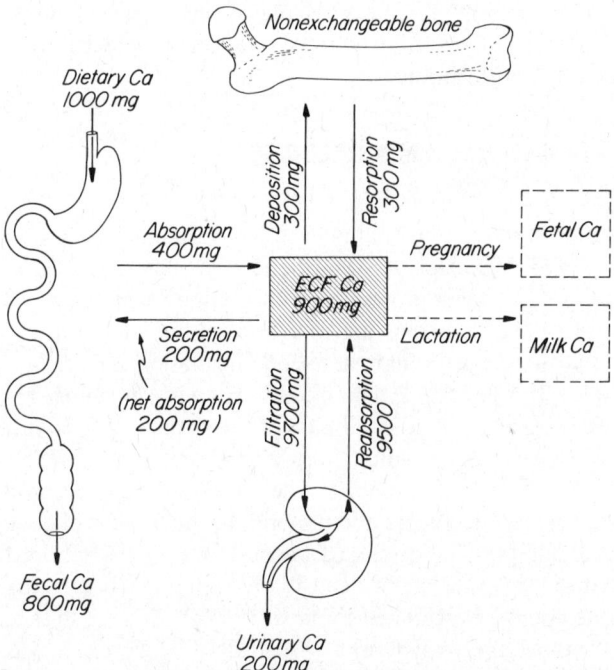

Figure 1 Diagrammatic schema of calcium kinetics and compartmentalization for a normal man on an average calcium intake for a diet including substantial amounts of dairy products (e.g., milk and cheese), and in a steady-state condition of calcium balance. All values shown are mg/day, except for the ECF; the 900-mg indicated for the ECF is the total ECF pool of Ca.

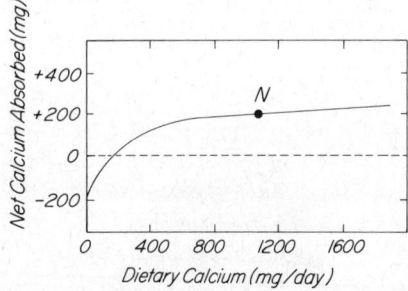

Figure 2 The relationship between net calcium absorbed and dietary intake of calcium in a normal 70-kg subject. Point N indicates normal Ca intake and normal net absorption for a diet that includes a substantial amount of dairy products.

Homeostatic regulation of net calcium absorption by the gut depends solely on the PTH influence on absorption, and not at all on control of the secretion of calcium.

RENAL FILTRATION AND REABSORPTION

In a normal steady-state condition of calcium balance, the amount of calcium excreted in urine each day is equal to the daily net absorption from the intestine. This excreted amount of urinary calcium represents that fraction of the filtered load (through the glomerulus) that fails to be reabsorbed (by the renal tubules). There is no evidence of tubular secretion of calcium into the urine. Normally in a person with a total plasma calcium level of 10 mg/dl, about 98% of the filtered calcium is reabsorbed. Over a 24-h period this amounts to 9,500 mg of calcium reabsorbed out of a total of 9,700 mg of calcium filtered. This leaves 200 mg to be excreted in the urine (Fig. 3). If the plasma level of calcium concentration is just below 9 mg/dl, renal absorption is increased to about 99%. At hypocalcemic levels below 7.5 mg/dl the efficiency of renal tubular reabsorption can increase to the point of essentially 100% recovery (complete conservation of calcium by kidneys).

Calcium reabsorption is similar to that for sodium, that is, about two-thirds of the filtered load is reabsorbed by the proximal tubules, with the remainder being handled by the loop of Henle, distal tubule, and collecting duct—

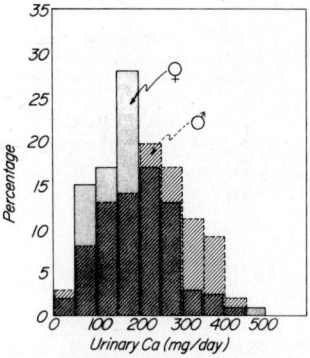

Figure 3 Frequency distributions (as percentages of persons excreting different amounts) of urinary calcium excreted per day in a population of healthy men and women.

there might even be a loose coupling between calcium and sodium reabsorption in the proximal tubule. Similarly, just as aldosterone regulates the reabsorption of sodium in the distal part of the nephron, so too does PTH have its primary influence on calcium reabsorption in this same part of the nephron. The influence of PTH on tubular reabsorption of calcium was first suspected when an acute increase in urinary calcium was observed following parathyroidectomy. The other side of the coin is the severalfold decrease in calcium clearance brought about by administration of PTH. This conservatory influence of endogenous PTH secretion is of considerable homeostatic influence (see Physiological Actions of PTH below).

There is normally some linkage between the net GI absorption of calcium each day and the daily urinary excretion of calcium. This connection is based, in part, on the influence of dietary absorption on the plasma concentration and thereby on the filtered load at the glomerulus and in part on the influence of plasma calcium concentration on PTH secretion and thereby on the efficiency of tubular reabsorption.

DEPOSITION AND RELEASE FROM BONE

It should be obvious that new bone is formed in the growing child. However, it might be less obvious that even in the mature skeleton new bone formation takes place despite a lack of further increase in total bone mass in the nongrowing adult. This new bone formation is part of a cellularly mediated dynamic equilibrium involving resorptive breakdown of a small part of the skeletal mass (with loss of both the organic matrix and the mineral) and formation of an equal amount of new bone (with production of both organic matrix and mineral) in the process of bone remodeling (to be discussed later). It was initially hoped that the rate of mineral deposition in such newly forming bone could be estimated by measuring the rate of uptake of radioactive phosphorus or calcium. However, it was found that radioisotopes were taken up by bone far too rapidly to account for deposition of additional mineral in new bone formation. Instead, it was realized that most of the initial uptake by bone must be in a rapid, freely reversible exchange mechanism between soluble mineral ions in the ECF and ions in part of already extant bone mineral.

The radioisotope content of bone after administration of labeled calcium and/or phosphorus is considered to consist of at least two distinct compartments: (1) that compartment of mineral that is part of newly formed "exchangeable" bone and freely exchanges ions with those in the ECF, and (2) that mineral that has been deposited and stabilized in the essentially irreversible process of mature bone formation ("nonexchangeable" bone). It is the formation of additional mineral in the new bone formation (deposition) and the release of older stable mineral from bone breakdown (resorption of nonexchangeable bone) that is under endocrine control and essential to the precise regulation of ECF calcium concentration within narrow limits. The freely reversible, rapid exchange system is not regulated by hormones but helps buffer changes in ECF calcium.

Buffering of Changes in ECF Ca²⁺ by the Nonhormonally Regulated Exchange System

The buffering of changes in ECF Ca^{2+} concentration depends on a direct interaction between Ca^{2+} in the ECF and that in newly formed "exchangeable" bone mineral. This exchange of ions and its buffering action on the concentration of ECF Ca^{2+} can easily be observed in vitro with newborn mouse bone kept alive in culture media. This isolated bone takes up or releases calcium in response to an increase or a decrease in the calcium concentration of the culture medium bathing the bone (Fig. 4). The ratio of exchangeable bone calcium to ECF calcium in vivo is even more favorable for buffering of calcium levels than is the in vitro situation illustrated. Thus abrupt changes in plasma calcium are initially buffered by this direct, nonhormonally regulated, free exchange system, thereby diminishing the magnitude of the change in concentration that would have otherwise occurred. The exchangeable bone participating in this buffer system represents only a small part (about 1%) of the total bone mass of the skeleton; however, the exchange rate is extremely rapid and can involve as much as 100 g of calcium taken up and released each day (Fig. 5). Although this exchange mechanism provides a fairly effective means of minimizing large changes in ECF Ca^{2+} concentration, it cannot, by itself, keep the Ca^{2+} concentration at the precise level best suited to optimal neuromuscular function.

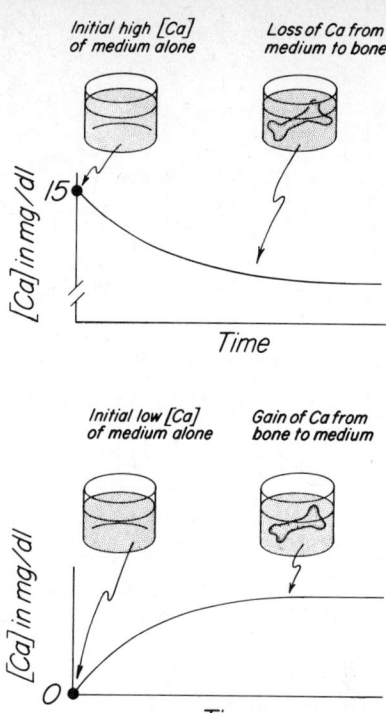

Figure 4. Buffering of changes in extracellular calcium by a nonhormonally regulated exchange system of bone. When neonatal bone from experimental animals is placed in a tissue culture medium containing a high concentration of calcium, the bone gradually takes up calcium causing the level in the medium to fall. If the bone is placed in a medium with little or no calcium, however, the calcium level in the medium will gradually rise as calcium is released from exchangeable bone stores.

An analogy can be drawn between buffering of H^+ concentration changes and the buffering of Ca^{2+} concentration changes. In both instances the buffering effect is rapid and constitutes the first line of defense against changes in ECF ion concentration. In each case the buffer mechanism cannot totally prevent a change in ion concentration nor can it regulate the concentration to a certain optimal value —it can only minimize changes. In order to return the ionic concentration to, or at least near to, normal, other mechanisms must operate. In the case of H^+, the regulatory mechanisms for returning the pH to normal are compensatory actions of the lungs and kidneys. In the case of Ca^{2+}, the regulatory mechanisms are compensatory

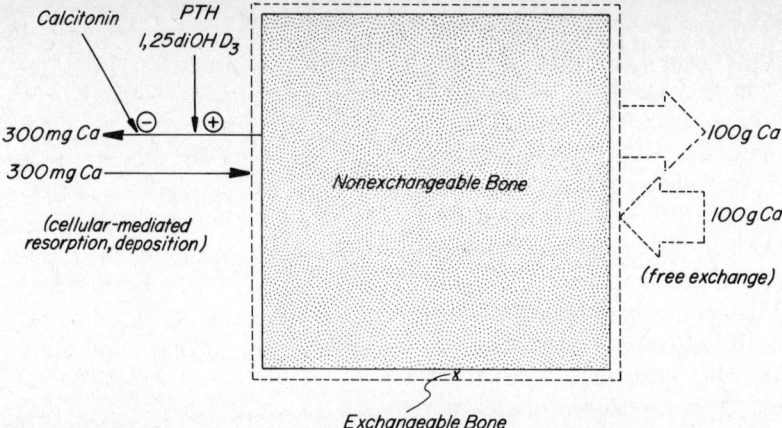

Figure 5 Turnover of mineral ions in newly formed exchangeable bone and older, more stable, nonexchangeable bone. The amounts indicated are amounts of calcium taken up and released per day. Note that the resorption of mineral from nonexchangeable bone is regulated by PTH, calcitonin, and 1,25-diOH D.

actions brought about by secretory changes in PTH, calcitonin, and 1,25-diOH D.

Endocrine Regulation of Bone Mineral Deposition and Resorption

In the free exchange system described above and illustrated in Figure 5, there is a physicochemical equilibrium in which enormous amounts of calcium are taken up and released each day (high turnover rate). This chemical equilibrium provides for the coarse buffering of perturbations in ECF Ca^{2+} concentration. However, the fine control of ECF Ca^{2+} concentration depends on the hormonally regulated system of bone deposition and resorption, which involves the turnover of much smaller amounts of calcium. Although the amounts taken up for deposition in new bone and released from resorption of older bone are about three orders of magnitude smaller (measured in milligrams per day instead of grams per day) than in the free exchange system, a shift in the bone deposition–resorption system to a greater predominance of either deposition or resorption can nevertheless have a substantial effect in regulating the plasma Ca^{2+} concentration, particularly if maintained for a few days. In the absence of hormonal regulation of this fine control (e.g., in patients

who experience severe hypoparathyroidism following thyroidectomy), the equilibrium level around which the free exchange system operates involves a plasma concentration of calcium that is often less than 7 mg/dl (total plasma calcium). Such low levels of plasma calcium are a serious threat and can bring about tetany and occasionally death by asphyxiation. Thus the hormonally regulated fine control system is essential for the maintenance of the plasma calcium at its normal level of 10 mg/dl (total plasma calcium). This system also helps to prevent excessively high concentrations of plasma calcium (which can also be dangerous in that they tend to precipitate cardiac arrhythmias). In the absence of endocrine control of calcium homeostasis, changes in calcium concentration of blood are slow to return toward normal (can take days). To compare rapidity of changes with and without endocrine control, see Figure 6.

Further difficulties in the use of radioisotopes to measure calcium kinetics arise from attempts to calculate the rate of endocrine-controlled bone resorption. Studies measuring release of previously incorporated isotopes from bone and studies estimating resorption from rate of release of collagen breakdown products (e.g., urinary excretion of hydroxylysine and hydroxyproline; amino acids found exclusively in collagen, the principal organic component of

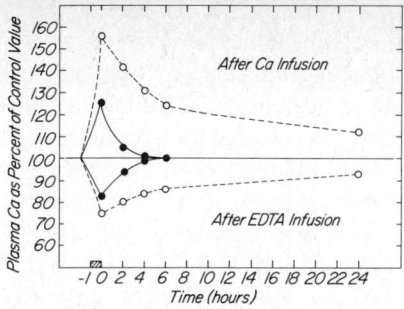

Figure 6 The change in plasma calcium concentration in response to an abrupt increase (calcium infusion) or decrease (chelation of calcium by EDTA) in plasma calcium in normal intact dogs (●-●) and in thyroparathyroidectomized dogs (o-o). Note that the normal intact animal is able to respond quickly to either hypo- or hypercalcemia and restore the plasma calcium level to normal within 4–6 h. Because of the rapid exchange mechanism between calcium in plasma and in exchangeable bone, the thyroparathyroidectomized dogs have some return of the plasma calcium toward normal (buffering mechanism), but in the absence of endocrine control cannot compensate by rapidly returning the calcium level all the way to normal, as can the intact dogs. (Redrawn from Sanderson PH, Marshall F, and Wilson RE, et al: "Calcium and Phosphorus homeostasis in the parathyroidectomized dog; evaluation by means of ethylenediamine tetraacetate and calcium tolerance tests." *J Clin Invest* 39:661, 1960.)

bone) give different values for the rate of resorption. The estimates based on calcium radioisotope release indicate that about 500 mg of calcium are released daily as a result of resorptive breakdown of stable bone. This value is about three times higher than that based on collagen degradation. If the true value is assumed to be somewhere, between these two different estimates, the daily turnover rate (of resorption and deposition) can be taken to be close to 300 mg/day (Fig. 5).

STATE OF CALCIUM IN BLOOD

As mentioned above, the plasma calcium concentration is held within narrow limits by homeostatic mechanisms. The total amount of calcium (all forms) in plasma is normally 10.0 ± 0.5 mg/dl (5.0 mEq/L or 2.5 moles/L). This consists of three fractions: (1) ionic Ca^{2+}, (2) chelated complexes of calcium with small ions (e.g., with phosphate and citrate), and (3) protein-bound calcium (primarily bound with albumin):

1. ionic Ca^{2+}		4.8 mg/dl
2. calcium complexed with small ions		0.6
3. protein-bound calcium		4.6
Total plasma calcium		10.0 mg/dl

The ionic Ca^{2+} and the small ion complexes easily pass through semipermeable membranes (e.g., dialysis tubing) and therefore constitute the filterable fraction of plasma calcium. On the other hand, protein-bound calcium does not pass through such membranes (because of the large size of protein molecules) and is called the nonfilterable fraction. When the concentration of plasma proteins is normal (about 7 g/dl), the ionic fraction is approximately one-half the total plasma calcium. This ionic fraction is the biologically active form of calcium and thus is most important for neuromuscular and secretory functions. It is also this ionic fraction that participates in a negative feedback relationship with the endocrine glands that regulate it.

Since the ionic fraction is the principal functional form of calcium, it should be of primary diagnostic significance. Unfortunately, methods utilized to determine the ionic fraction alone (e.g., the use of a Ca^{2+}-specific electrode) are not yet widespread in clinical laboratories. Until such methods become more common, many physicians must be satisfied with more routine laboratory methods that measure the total plasma calcium. In patients with abnormally high concentrations of plasma proteins, the total plasma calcium level is also higher than normal (although the ionic calcium concentration remains normal). This comes about by an increase in the amount of calcium bound to protein (similar to the effect of excess protein on total T_4). To avoid misdiagnosis, this influence of abnormal concentration of plasma protein must be taken into account. As a rule of thumb, one can assume a change in total calcium of 0.8 mg/dl for each gram per deciliter difference in protein concentration from normal. For example, if a patient with cirrhosis or renal nephrosis has a plasma protein concentration of only 4 g/dl (i.e., 3 g/dl less than normal), his total plasma calcium concentration of 7.6 mg/dl would not be considered hypocalcemic (is normal for the circumstances, i.e., would have a normal concentration of ionic Ca^{2+}).

OVERVIEW OF ENDOCRINE CONTROL OF Ca²⁺ HOMEOSTASIS

The endocrine control of calcium homeostasis involves three hormones, i.e., parathyroid hormone, calcitonin, and a dihydroxy derivative of vitamin D, and three target organs, i.e., intestine, bone, and kidney (Table 1).

The best-known endocrine regulator of plasma calcium is parathyroid hormone. A fall in the level of plasma calcium ion below normal has a stimulatory effect on the release and synthesis of PTH. PTH has a primary, direct action on bone and kidney, and has a secondary, indirect action on the small intestine. By stimulating the rate of bone resorption, PTH increases the release of calcium and phosphate from bone mineral into the ECF. At the same time PTH also acts on renal tubular reabsorption to conserve calcium and excrete much of the phosphate. These actions of PTH on both bone and kidney favor increased levels of plasma calcium and explain the principal role of PTH as a hypercalcemic hormone—a hormone that raises only the plasma calcium concentration.

Bone is only a reservoir of mineral, and the kidney can only conserve what is already present; neither can provide new, additional calcium for the body. To ensure a continued supply of new calcium (e.g., to replace that withdrawn from the reservoir and to replace any lost in urine or feces), there must be some regulation of dietary absorption of new exogenous calcium. PTH plays an important, albeit indirect, role here; PTH exerts its influence on the intestine via an additional action on kidney (other than its action on calcium and phosphate reabsorption): PTH acts

to increase the formation of 1,25-diOH D. This dihydroxy derivative of vitamin D is often considered to be a renal hormone, although derived from what has traditionally been thought of as a vitamin. 1,25-diOH D acts to increase intestinal absorption of calcium by a mechanism that remains to be fully established. It appears to involve the formation of an intestinal calcium-binding protein (CaBP) thought to act on the luminal side of intestinal cells to increase entry of calcium into the cell. 1,25-diOH D also facilitates the direct actions of PTH on bone resorption and renal tubular reabsorption. Thus by means of its multiple actions, both direct and indirect, PTH has a hypercalcemic effect, that is, it raises plasma Ca²⁺ concentration, elevating the lowered plasma Ca²⁺ back to normal. As the plasma Ca²⁺ level returns, this removes the hypocalcemic stimulus on the parathyroid glands and the increased secretory activity declines, constituting a negative feedback relationship between ionic calcium and PTH secretion.

Another hormone thought to have a possible negative feedback relationship with ionic calcium is calcitonin. In this case it was a rise in plasma Ca²⁺ that was considered to act as a stimulus for hormonal secretion. A rise in plasma calcium can have a stimulatory effect on the secretion of CT by parafollicular C cells of the mammalian thyroid gland. CT is a hypocalcemic hormone, that is, it lowers the concentration of calcium ion in plasma, which is widespread throughout vertebrates (from the ultimobranchial glands of fish to the parafollicular C cells of the thyroid in mammals). If the level of plasma calcium is increased orally or by intravenous infusion, a rapid rise in CT secretion can be produced (how much the plasma Ca²⁺ has to rise to do this varies among different species). When calcium is infused into an intact animal, the rise in plasma calcium is smaller and returns to normal sooner than is the case in thyroidectomized animals (which lack CT).

The role of CT and its negative feedback regulation of plasma Ca²⁺ in humans is not nearly as significant as that of PTH. Two problems in assigning a primary role to CT in homeostatic regulation of plasma Ca²⁺ are (1) plasma levels of Ca²⁺ remain normal long after thyroidectomy, and (2) chronic excessive levels of CT (e.g., from medullary carcinoma of the thyroid) do not produce hypocalcemia.

Table 1 Principal hormones regulating ECF [Ca²⁺] and their sites of action

Hormone	Target Organ(s)
Parathyroid hormone (PTH)	Bone and kidney directly (intestine indirectly via 1,25-diOH D)
1,25-Dihydroxy vitamin D (1,25-diOH D)	Intestine and bone.
Calcitonin (CT)	Bone (specific physiological significance remains to be established)

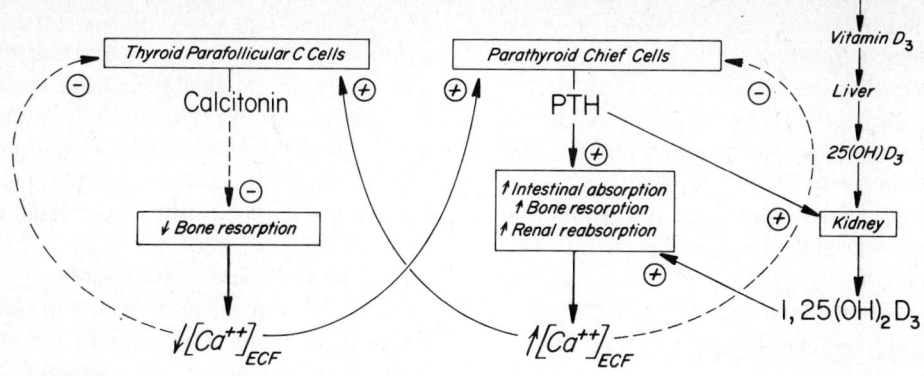

Hormone that may decrease ECF $[Ca^{++}]$ Hormones that increase ECF $[Ca^{++}]$

Figure 7. Current concept of endocrine control of extracellular fluid calcium concentration. Inhibitory effects are indicated by a negative sign and a dashed arrow; stimulatory effects are indicated by a plus sign and solid arrow. The principal control of ECF Ca^{2+} is thought to be exerted by PTH along with $1,25(OH)_2D$. The role of calcitonin in the minute-to-minute regulation of ECF Ca^{2+} is thought to be less important than PTH and $1,25(OH)_2D$ in humans.

Figure 7 summarizes the current concept of hormonal regulation of ECF Ca^{2+} concentration by means of negative feedback relationships between the hormones and their secretory cells. Since 1,25-diOH D formation is stimulated by PTH, and since it both increases Ca^{2+} absorption and augments PTH action elsewhere, 1,25-diOH D is also part of the PTH feedback loop.

The role of the kidney regulation of tubular reabsorption of mineral ions and as an endocrine organ in its own right (secretion of 1,25-diOH D) emphasizes the pivotal role that the kidney plays in the homeostatic regulation of plasma Ca^{2+} concentration.

THE PARATHYROID GLANDS AND PTH

The parathyroid glands were first recognized as distinct anatomical entities by the English zoologist Richard Owen during the dissection of an Indian rhinoceros in 1862. However, in the absence of any knowledge of their physiological significance, this anatomical discovery in a single animal species created little interest, and the parathyroid glands had to be rediscovered almost 20 years later. Hence discovery of the parathyroid glands as anatomical entities is usually accredited to Sandstrom, who in 1880 described them in humans and other animals. For many years they were called Sandstrom's glands.

Knowledge of the essential nature of the parathyroids was not forthcoming until the very end of the nineteenth century. It had long been known that surgical removal of the thyroid gland occasionally resulted in a fatal tetany (in severe tetany death follows laryngospasm and convulsions, which produce asphyxiation). Between 1891 and 1900 it was conclusively demonstrated that tetany was not a consequence of thyroidectomy itself, but results from inadvertent removal, ischemia, or trauma of the parathyroids as a result of thyroidectomy (even in the enlightened twentieth century this is still the most common cause of hypoparathyroidism). Several years later, in 1909, the symptoms of parathyroidectomy were compared with those of experimental hypocalcemic tetany, and it was shown that an infusion of calcium salts prevented hypoparathyroid tetany. The conclusion was correctly drawn that the tiny parathyroid glands, the smallest of the endocrine glands, are involved in the essential regulation of blood calcium concentration.

The proof of the endocrine nature of the parathyroids was furnished by Collip in 1952 when he obtained a physiologically active extract of the parathyroid glands. This parathyroid extract not only restored normal function to parathyroidectomized animals when administered in therapeutic doses, but could also produce hyperparathyroidism in large doses.

PHYLOGENY AND ONTOGENY

The parathyroid glands are the only endocrine glands not found in all classes of jawed vertebrates.* They first appear in the amphibia and occur in these and all higher vertebrates. It is interesting to note that although piscine ancestors undoubtedly did not possess parathyroid glands, they might well have had the means to make and store vitamin D (e.g., a rich source of vitamin used therapeutically for many years is cod liver oil). Fish also have ultimobranchial glands, separate distinct endocrine glands that secrete CT (salmon CT is used therapeutically in man to inhibit excessive bone resorption, e.g., in Paget's disease). It is quite possible that piscine CT and vitamin D have aided fish in their regulation of ECF Ca^{2+}.

Since tetrapods are the only animals that have parathyroid glands, it has been suggested that the evolution of these glands was closely associated with terrestrial adaptation—an adaptation to an environment relatively lower in availability of calcium. This is an intriguing idea, since (1) in the ontogeny of the parathyroid glands they are derived from embryonic gill pouch epithelium, and (2) fish gills not only carry out exchange of respiratory gases, but also regulate ion exchanges. In higher animals the gill-derived parathyroid glands continue to regulate an important ion, namely, calcium, an ion that is indirectly essential to normal respiratory activity (the usual cause of death in hypocalcemic tetany is asphyxiation).

In the mammalian embryo the third and fourth pairs of pharyngeal gill pouches give rise to outgrowths that are involved in the formation of the parathyroids, thymus,

*Agnathans lack parathyroid glands and ultimobranchial glands. Sharks and bony fish lack parathyroid glands only.

and ultimobranchial bodies (which give rise to calcitonin-secreting parafollicular C cells of the thyroid) (Fig. 8). As is the case with the ontogeny of the thyroid, the parathyroid and ultimobranchial primordia break away from their branchial points of origin and migrate to the ventral portion of the neck. Not only do the ultimobranchial cells become incorporated into the thyroid gland in mammals, but occasionally parathyroid glands also tend to become closely adherent to or even partially embedded within the substance of the thyroid.

It would appear that in tetrapods the parathyroid glands have evolved, along with the already present ultimobranchial glands and the vitamin D system, to aid in the homeostatic regulation of calcium; for example, to bring about control of calcium ion mobilization from a bony skeleton. The mineralized skeleton, originally evolved in fish as a structural protective system, has been exploited to serve an additional role in terrestial vertebrates—that of a calcium reservoir that can be drawn upon in times of calcium scarcity or deprivation.

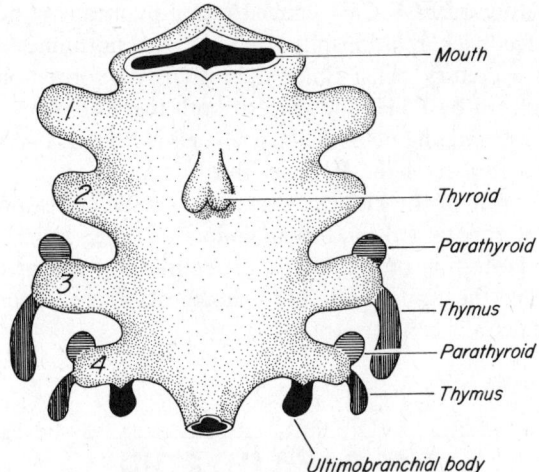

Figure 8 Schematic diagram of a 4-week-old human embryonic pharynx illustrating branchial gill pouch origins of several glandular derivatives. The parathyroid glands are derived from the third and fourth gill pouches. The ultimobranchial bodies are found on the fourth gill pouch, but are thought to have migrated there from a neural crest origin. The cells of the ultimobranchial bodies later become incorporated in the thyroid gland.

GROSS AND MICROANATOMY OF PARATHYROIDS

The parathyroid glands in humans usually consist of a superior pair located on the upper dorsal surfaces of the lateral thyroid lobes and an inferior pair that can be attached to, or even embedded within, the lower part of the thyroid. The inferior pair can even lie free anywhere between the thyroid and the mediastinum. Thus the usual condition is one of four parathyroid glands. However, there can be as few as two or as many as six. Variability in number and location are considerations of much importance in surgical treatment of primary hyperparathyroidism, since failure to locate all hypersecretory glands can preclude the efficacy of surgical correction. Their usual close association with the thyroid gland, with which they share several blood vessels in common (both pairs of parathyroids are supplied by the inferior thyroid arteries and drained by the lateral and inferior thyroid veins), occasionally leads to destructive trauma and/or ischemia in the course of thyroid surgery.

The parathyroid glands are brownish ovoid bodies weighing about 40 mg each. Their anatomic relationship to the posterior portion of the thyroid gland is illustrated in Figure 9.

Histologically, there are three recognized categories of cell types identified: (1) chief cells (the principal secretory cells), (2) oxyphil cells, and (3) transitional cells. The main feature of chief cells by light microscopy is their apparently empty cytoplasm. Electron microscopy shows, however, that the cytoplasm is not at all empty, but contains the usual cytoplasmic organelles, such as mitochondria, Golgi apparatus, and rough endoplasmic reticulum. Oxyphil cells are larger and have an acidophilic cytoplasm and smaller nuclei. Transitional cells have properties between the first two. The chief cells are the only kind found until about puberty when transitional and oxyphil cells begin to appear. Electron microscopic observations tend to favor the idea that there is only one type of secretory cell, the chief cell, and that these undergo secretory cycles—not all chief cells are in the secretory phase at one time. The oxyphil cell may be only an involutionary form of the chief cell that is no longer functional. The transitional cell is probably an intermediate stage in this involution.

SYNTHESIS AND SECRETION

Studies of PTH synthesis, like those of several other polypeptide or protein hormones (e.g., insulin), have contributed much to the understanding of the biosynthetic process whereby secretory polypeptides are assembled and packaged for export. A feature in common among the several biosynthetic systems is that of sequential proteolytic cleavage from larger inactive precursors to smaller active hormones and inactive fragments. This sequence of successive cleavages is characteristic of the biosynthesis of most, and perhaps all, peptide and protein hormones (and also many other kinds of secretory proteins).

The first indication that PTH was derived from a larger precursor protein came from chromatographic and pulse-labeling (brief incorporation of radioactive constituents resulting in incorporation of the label into the product) studies of parathyroid extracts. Analysis revealed the presence of a 90-amino acid protein that was six amino acids longer than the established structure of PTH. Cleavage of a hexapeptide off the amino-terminal and of this precursor, called ProPTH, yielded the 84-amino acid PTH molecule (Fig. 10).

Further studies using parathyroid mRNA in heterologous cell-free systems capable of in vitro peptide synthesis demonstrated that an even larger precursor (of 115 amino acids in length) exists transiently. This larger precursor,

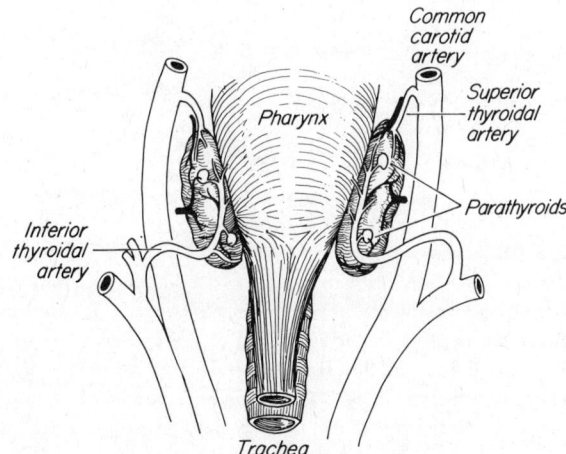

Figure 9 Usual positions of superior and inferior pairs of parathyroid glands on dorsal (posterior) aspect of thyroid gland in human.

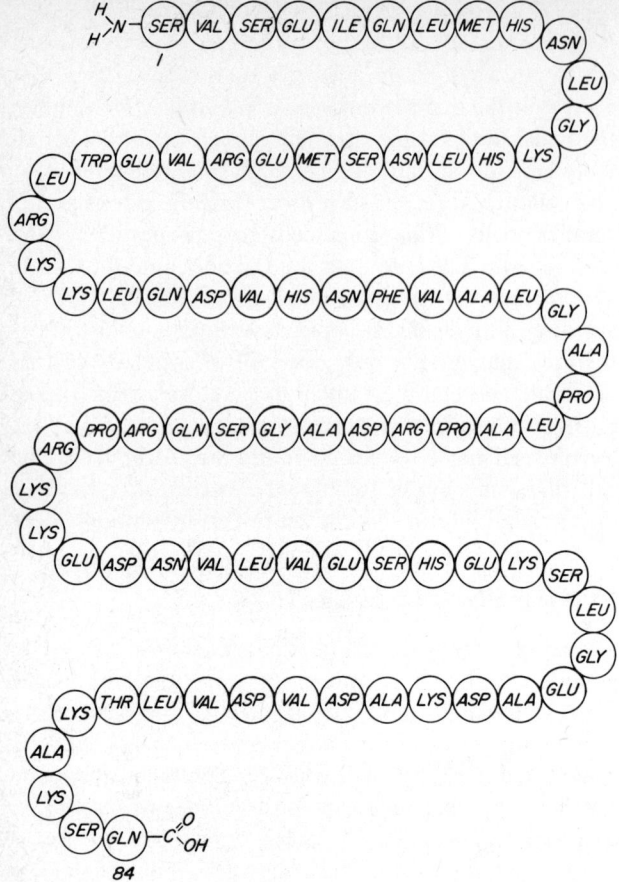

Figure 10. Amino acid sequence of human parathyroid hormone (PTH).

vesicles or granules. According to the "signal hypothesis," mRNAs coding for these secretory proteins destined to cross into the cisternae contain at their 5' end (immediately following the initiating codon) a specialized "signal sequence" of codons. The "signal sequence" of codons cause ribosomal translation of the proper amino acids into a corresponding initial signal sequence peptide that is the amino terminal end of the chain. As this specialized, very lipophilic chain emerges from the ribosome, it interacts with the endoplasmic reticulum membrane to attach the ribosome and to vectorially direct the newly assembled chain through the membrane. This gradually enlarging peptide chain, now on a membrane-bound polyribosome enters the cisternal space. As the chain emerges, the signal sequence peptide is enzymatically split off on the cisternal side by an enzyme sometimes referred to as a *signalase* (Fig. 11). In the case of PreproPTH, the signal sequence is 25 amino acids long. Within 1 min of the beginning of translation, the signal sequence peptide is clipped off; the PreproPTH precursor is very short-lived or transient (Fig. 12).

The hexapeptide portion of ProPTH is thought to aid somehow in the translocation of the newly formed

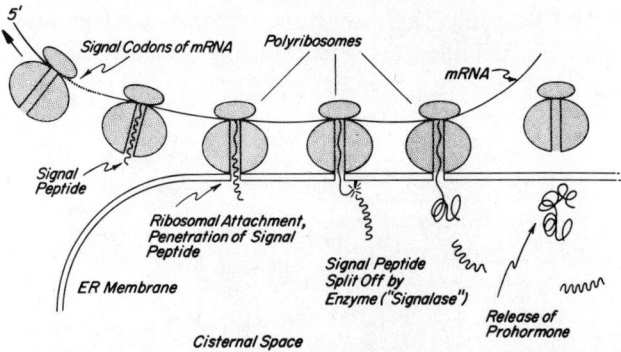

Figure 11. Schematic representation of the signal hypothesis for synthesis of secretory proteins on membrane-bound polyribosomes. The initial signal sequence of codons on mRNA results in translation of the signal peptide at the beginning of the synthesized peptide chain. The purpose of this signal peptide appears to be one of attachment of the ribosome to the endoplasmic reticulum membrane for vectorial discharge of the forming peptide chain through the membrane into the cisternal space. Shortly after emerging through the membrane, the signal peptide is split off by a membrane-bound "signalase" enzyme.

designated PreproPTH, is synthesized by polyribosomes of the RER and almost immediately shortened to ProPTH by the hydrolytic splitting out of 25 amino acids. The need for a "pre" form that is many amino acids longer than ProPTH is explained by the *signal hypothesis*. It has been known for some time that proteins destined to be exported (secreted) are synthesized by polyribosomes that attach to the membrane of the rough endoplasmic reticulum. As the secretory protein chain begins to emerge from the ribosome, it burrows unidirectionally through the endoplasmic reticulum membrane into the cisterna. When completed the secretory protein is translocated via the cisternal space to the Golgi apparatus for packaging in secretory

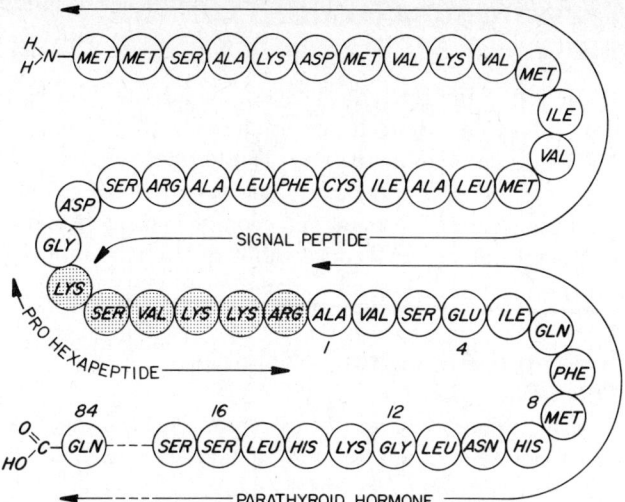

Figure 12. Amino acid sequence of bovine preproparathyroid hormone. PreProPTH as such exists only transiently since the signal peptide is split off during or shortly after formation. After removal of the signal peptide, the remaining sequence of amino acids constitutes the precursor of PTH, ProPTH. ProPTH is converted to the 84 amino acid PTH by splitting off a hexapeptide. Only a few of the 84 amino acid units of PTH are illustrated here.

ProPTH to the Golgi apparatus where this 6-amino acid sequence is then enzymatically split off to form PTH. The PTH becomes enclosed within a vesicular membrane prior to secretion by exocytosis at the cell membrane. Neither PreproPTH nor ProPTH is secreted. This sequence is illustrated in Figure 13.

REGULATION OF SYNTHESIS AND SECRETION BY Ca²⁺

Unlike the thyroid gland (the largest endocrine gland), which can store enough hormone to supply normal amounts for months, the parathyroids (the smallest endocrines) store no more than a few hours worth. That is, the parathyroids depend to a much greater extent on rates of biosynthesis and degradation than on release of preformed stored hormone. Since the parathyroids do not store much PTH, it is fortunate that an increase in biosynthesis requires only a few hours to respond to a fall in ECF [Ca²⁺]. Because it does take at least this long to be stimulated, it is assumed that the decrease in Ca²⁺ acts on

transcription (DNA-directed synthesis of mRNA) rather than on translation (ribosomal response to extant mRNA). Ca²⁺ also affects the rate of intracellular degradative pathways that catabolize some of the PTH and its precursors. When ECF [Ca²⁺] is decreased, the rate of degradation of ProPTH and PTH also decreases making more available for secretion. An increase in ECF calcium concentration increases degradation.

The influence of calcium ion on PTH secretion involves the production of cAMP (activation of adenyl cyclase is an intermediate step in the secretion of many polypeptide hormones). Parathyroid adenylcyclase is sensentive to inhibition by calcium. At hypercalcemic levels, cyclic adenosine monophosphate (cyclic AMP) production is minimal. As plasma calcium concentrations drop below normal, this inhibition is diminished and cyclic AMP levels in the gland rise, stimulating PTH secretion. Other agents that stimulate PTH secretion, such as β-agonists

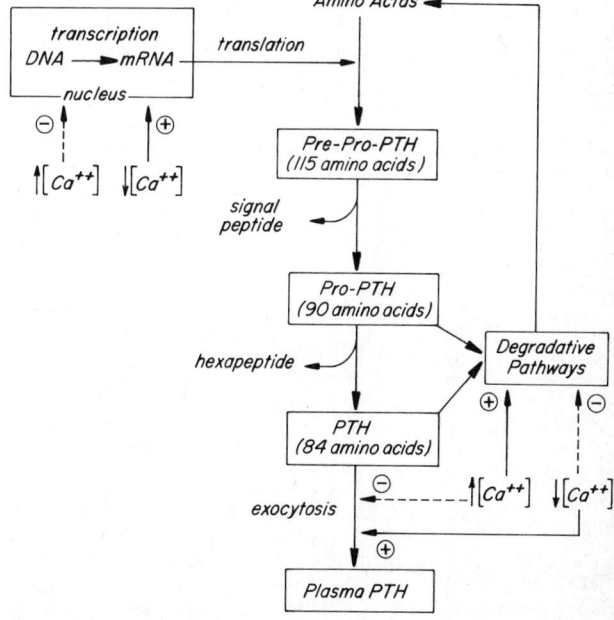

Figure 13 The biosynthesis and release of PTH. The formation of precursors and their sequential cleavages to form PTH are shown along with the effects of changes in ECF [Ca²⁺] on synthesis, degradation, and release (by exocytosis). Stimulatory effects are indicated by a plus sign; inhibitory effects are indicated by a dashed arrow and a negative sign.

(epinephrine, isoproterenol) also increase parathyroid cAMP. Although epinephrine or adrenergic nerve activity is not known to play a physiological role in the regulation of PTH secretion, the influence of catecholamines on parathyroid β receptors may explain some of the hyperparathyroidism associated with pheochromocytomas, particularly in those cases in which surgical removal of the catecholamine-secreting tumor is followed by a return of PTH secretion to normal.

In summary, the principal stimulus for the biosynthesis and secretion of PTH is a drop in ECF calcium ion concentration. This results in (1) increased release of PTH by a mechanism involving increased cyclicAMP production, (2) decreased degradation of preformed ProPTH and PTH, and (3) increased biosynthesis of PTH from increased production of mRNA (transcription). These actions result in an inverse sigmoid relationship between ECF calcium ion concentration and PTH secretion (Fig. 14). Although epinephrine has been shown to increase PTH secretion by β stimulation of cyclic AMP formation, it is not known to what extent β activation plays a role in the physiological regulation of parathyroid secretion.

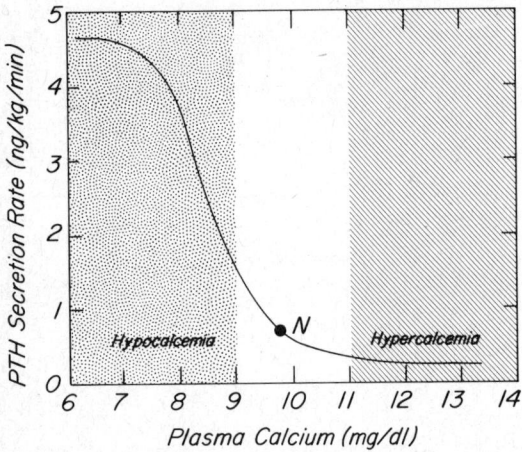

Figure 14 Secretory response of parathyroid glands to change in plasma calcium. The rate of PTH secretion was measured in calves subjected to various different levels of plasma calcium. It can be seen that the greatest changes in PTH secretion occur in the hypocalcemic range. N indicates the normal rate of secretion at normal levels of total plasma calcium. (After Mayer GP: in Talmage RV, Owen M, and Parsons JA, (eds): "Effect of Calcium and Magnesium on Parathyroid Secretion Rate in Calves," *Calcium Regulating Hormones.* Amsterdam, Excerpta Medica, 1975.)

PHYSIOLOGICAL ACTIONS OF PTH

PTH, acting indirectly on intestine and directly on bone and kidney, has four major established effects, that is, PTH stimulates an increase in (1) intestinal absorption of calcium (and phosphate), (2) bone resorption of calcium (and phosphate), (3) renal excretion of phosphate (phosphaturia), and (4) renal conservation of calcium (anticalciuria).

Intestinal Absorption of Calcium (and Phosphate)

It has long been known (for over four decades) that man and other animals adapt to low dietary calcium intake by increasing their efficiency of intestinal absorption (i.e., by increasing the fraction of calcium absorbed). Similarly, when placed on a high calcium intake, there is a marked diminution in efficiency of absorption. The ability of the intestine to adapt to the body's calcium needs led to the postulation of some humoral factor (originally thought to be produced by bone) that directs the small intestine to alter its calcium absorption as necessary. Actually two humoral factors are involved (neither one of which is from bone), and their concentrations are determined by the level of plasma calcium ion. One is PTH and the other is 1,25-diOH D. Until recent years the influence of PTH on calcium absorption has been somewhat controversial, however, it is now well accepted that PTH acts primarily by enhancing the formation of 1,25-diOH D in the kidney. Just how PTH does this is not known for sure but it appears to involve cAMP formation. The result is an induction of increased 1-hydroxylase activity of mitochondria of renal cortical cells, which increases conversion of 25-OH D (from liver) to 1,25-diOH D.

Although 1,25-diOH D is known to increase intestinal absorption of calcium (and phosphate), the precise mechanism is not completely understood. Initially 1,25-diOH D enters intestinal mucosal cells and combines with a cytosol receptor (a protein of about 100,000 molecular weight). The 1,25-diOH D-receptor complex enters the nucleus and reacts with a different (nuclear) receptor, as is typical of all steroid hormones. The subsequent transcription/translation sequence initiated by 1,25-diOH D-chromatin combination results in the formation of a calcium-binding protein (CaBP). That this protein appears in increased amounts in response to increased amounts of 1,25-diOH D, that it binds calcium, and that physiological mechanisms that modify calcium absorption also alter the level of calcium-binding protein, all argue convincingly for its participation in intestinal adaptation to dietary calcium. Other factors are also thought to participate, that is, an alkaline phosphatase enzyme and a calcium-dependent ATPase enzyme. Just how the calcium binding protein and the associated enzyme activities bring about increased calcium absorption remains to be elucidated. The calcium-binding protein has been shown to reside in the apical microvilli in at least one species and facilitates entry of calcium into the mucosal cell. There is evidence to indicate that the calcium is actively taken up by the mitochondria, which migrate to the base of the cell. The calcium released from the basal mitochondria is hypothesized to be actively transported out of the cell in exchange for sodium (Fig. 15).

In summary, the entire sequence is as follows. If dietary calcium is deficient, (1) the plasma calcium levels tend to decrease, (2) a drop in plasma calcium stimulates increased PTH secretion, (3) an increase in PTH stimulates increased renal formation of 1,25-diOH D, (4) increased 1,25-diOH D stimulates increased formation of intestinal mucosal calcium-binding protein, and (5) calcium-binding protein facilitates transport of dietary calcium from intestinal lumen into the blood, helping to raise the plasma calcium level back to normal.

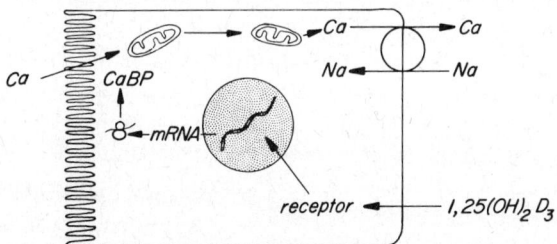

Figure 15 Proposed scheme of 1,25-diOH D_3 action on intestinal absorption of calcium. 1,25-diOH D_3 enters mucosal cells and reacts with a cytosol receptor. The cytosol receptor–1,25-diOH D_3 complex enters the nucleus and directs formation of messenger RNA (mRNA) for the biosynthesis of a calcium-binding protein (CaBP). The CaBP allows entry of calcium across the luminal border. The calcium is thought to be transported to the serosal border by mitochondria. This calcium is actively transported out of the cell in exchange for sodium.

Bone Resorption of Calcium (and Phosphate):

PTH has a biphasic action on bone, depending on the level of circulating PTH. At minimal levels, PTH has a predominantly anabolic action, that is, it increases formation of organic matrix and deposition of minerals to increase bone mass. Thus at low concentrations PTH is a regulator of bone formation. At the usual levels found in healthy normocalcemic persons, both bone formation and bone resorption are stimulated. This is the basis of the normal ongoing process of bone remodeling in which older bone is resorbed and then replaced with new bone. Remodeling is a continuous process from the earliest stages of embryonic bone until death. The remodeling involves all bone surfaces (periosteal, haversian, and endosteal) and is apparently essential to maintenance of the mechanical integrity of bone as an effective structural material. At any given time in normal adults about 2–4% of the total bone surface is involved in remodeling. Since there is no net gain or loss of skeletal mass at this level of PTH secretion, formation and resorption must be equal. This balance represents a coupling of the activity of units consisting of resorptive osteoclasts tunneling through bone and closely following bone-forming osteoblasts, which replace bone lost. At higher levels of PTH secretion the number and activity of osteoclasts greatly increases and causes a predominance of catabolic resorptive activity. At these higher levels, PTH also has a dual temporal effect on catabolic bone resorption depending on the duration of exposure to the elevated levels of the hormone. The initial, immediate response depends upon the effect of PTH on activity of extant osteocytes and osteoclasts. Within minutes, increased levels of PTH can increase osteocytic and osteoclastic osteolysis by these bone cells. This acceleration of resorptive processes increases the breakdown of organic matrix (with increased released of hydroxyproline from collagen hydrolysis) and the release of calcium and phosphate into the ECF.

In addition to this early effect, which is prompt in onset (within a few minutes) but of transient duration, a later, more prolonged elevation of PTH produces an even greater amount of osteoclastic activity, but only after 12–24 h. This later response to chronically elevated levels of PTH stimulates the production of increased numbers of new osteoclast cells (as characterized by waves of mitotic divisions leading to osteoclasts from osteoprogenitor cells). This increased resorptive activity of increased numbers of osteoclasts is sustained in duration (as long as PTH is secreted in increased concentration and often long after PTH levels have returned to normal).

The mechanism of action of PTH on bone cells involves two early events: (1) activation of bone cell adenylate cyclase with increased production of cyclicAMP, and (2) an increased uptake of ECF Ca^{2+} into bone cells (paradoxically, while the ultimate effect of increased PTH is to raise ECF levels of calcium, the initial effect on bone cells causes a brief drop in ECF calcium levels). The increase in bone cell cyclicAMP is thought to increase the efflux of calcium from mitochondria into the cytosol. This combined effect of PTH on calcium entry into the cell from the ECF and upon the increased release of calcium from mitochondria leads to a substantial increase in cytosol calcium levels. This cytosolic calcium acts as a second messenger to mediate the PTH effects on bone.

The resorptive activity stimulated by elevated levels of PTH, like the increased efficiency of intestinal absorption, increases the release of both calcium and phosphate into the ECF space. Adequate quantities of 1,25-diOH D are an absolute requirement for the bone resorptive effect of PTH. Since a predominance of catabolic resorptive activity over anabolic osteoblastic bone-forming activity causes a loss of bone mass, this mechanism should be regarded as an emergency withdrawal of mineral from the bone reservoir as necessary to maintain ECF calcium concentration (i.e., the maintenance of adequate plasma calcium for proper neuromuscular function takes precedence over the structural integrity of bone).

Renal Phosphaturic Effect

The phosphaturic effect of parathyroid gland extracts has been known since 1911. Since PTH increases urinary phosphate excretion within 10–15 min, it should be counted as one of the earliest effects of the hormone. PTH causes an increased phosphaturia by inhibiting phosphate reabsorption. This is brought about by a diminution in the tubular maximum T_m for phosphate reabsorption in the proximal tubule, the major site for phosphate reabsorption. At the same time, PTH also decreases the proximal reabsorption of several other urinary solutes: Na^+, Ca^{2+}, and HCO_3^-. Most of the Na^+ and Ca^{2+} rejected by PTH action proximally is reabsorbed distally with the result

that the PTH influence at the proximal site has little effect on the overall excretion of these two cations. Indeed, as pointed out below, the principal influence of PTH on calcium reabsorption is to produce an increase in distal tubular reabsorption, an effect that brings about a net overall increase in renal reabsorption of calcium. The proximal decrease in bicarbonate reabsorption, which results from a decrease in proximal H^+ secretion, causes increased urinary excretion of bicarbonate and can ultimately produce a base-losing metabolic acidosis if maintained for long.

Although an increase in phosphate excretion is one of the most rapid effects of PTH, this effect is preceded by an increased renal production and urinary excretion of cyclicAMP. The PTH inhibition of phosphate reabsorption has been shown to be mediated by cyclicAMP. The specific mechanism by which an increase in cyclic AMP inhibits phosphate reabsorption is not known, but is thought to involve activation of a protein kinase and subsequent phosphorylation of a protein involved in inhibition of phosphate transport.

Since PTH is primarily concerned with the maintenance of adequate levels of ECF calcium concentration, one might wonder at the need for a phosphaturic action. The following rationale may explain the significance of this PTH effect. As a result of both the indirect action of PTH on the intestinal absorption of calcium and phosphate and the direct effect on bone resorption of calcium and phosphate (described above), the levels of both of these ions tend to increase. Increasing both concentrations would substantially exceed the solubility product of calcium times phosphate in the equilibrium between these ions and calcium phosphate in exchangeable bone. Increasing the levels above the solubility product results in increased deposition of both ions (to reestablish equilibrium), which detracts from the gain of calcium from intestine and bone. Furthermore, calcium phosphate could precipitate in nonosseous soft tissues. Thus from a teleological point of view it seems reasonable that the homeostatic mechanism for elevating ECF calcium from intestine and bone would also include a means of disposing of the accompanying phosphate by increasing its urinary excretion.

Renal Anticalciuric Effect

There were originally two schools of thought about the principal action of PTH on plasma calcium: (1) that PTH acted primarily on bone and that changes in plasma calcium were secondary to bone changes, and (2) that the kidney, in handling phosphate, could produce secondary changes in bone deposition or release. It is now known that PTH has direct effects on both bone and kidney. In addition to the well-known renal phosphaturic effect described above, PTH has also been shown to have a net stimulatory effect on renal reabsorption of calcium. Parathyroidectomy is followed by an increase in urinary calcium excretion. Administration of physiological amounts of PTH to parathyroidectomized animals is associated with a fall in urinary excretion of calcium. However, in the presence of excessive amounts of PTH, as in primary hyperparathyroidism, urinary calcium excretion may actually increase despite an increase in calcium reabsorption (due to an even greater increase in the filtered load as a result of severe hypercalcemia). Even though a hyperparathyroid patient might have a greater than normal excretion of urinary calcium (despite increased renal reabsorption) at any given level of hypercalcemia, this excretion is much less than that in a normal individual if his blood calcium were raised to that same level (Fig. 16).

PTH has a dual action on renal tubular calcium transport: (1) inhibiting proximal reabsorption of Ca^{2+} (along with proximal inhibition of Na^+, PO_4, and HCO_3^-), and (2) enhancing distal reabsorption of Ca^{2+}. The distal site of action (apparently distal convoluted tubule and part of collecting ducts) involves the tubular maximum for calcium reabsorption T_{mCa} and has a greater effect than the proximal site inasmuch as the overall net effect is one of increased reabsorption. The proximal effect has been shown to be mediated by cyclicAMP and there is some evidence that this is also true for the distal action. The role of vitamin D or its hydroxylated metabolites on either the renal phosphaturic or anticalciuric actions of PTH has not been established and remains controversial.

CLINICAL PROBLEMS

Hyperparathyroidism

Hyperparathyroidism is the most common disorder in parathyroid function. It is a chronic disease that can result

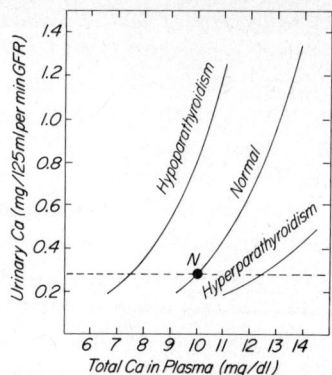

Figure 16. Relationship between total plasma calcium concentration and urinary excretion of calcium. This is a comparison of the renal responses among three different categories of subjects to varying levels of plasma calcium. Hypoparathyroid patients (with insufficient PTH secretion) will lose large quantities of urinary calcium despite subnormal plasma calcium concentrations (and reduced filtered load of calcium at the glomerulus). Hyperparathyroid patients (with excessive PTH secretion) retain considerable amounts of calcium even in the face of greatly elevated plasma calcium levels. At very high levels of plasma calcium the greatly elevated filtered load overwhelms the ability of the kidneys to reabsorb calcium, and urinary levels rise above normal despite increased tubular reabsorption. However, it can be seen that, for any given level of plasma calcium, hyperparathyroid patients excrete much less calcium than would normal subjects at that same level. Point N indicates the normal plasma calcium concentration and the normal level of urinary calcium (all urinary clacium levels expressed as milligrams per normal glomerular filtration rate of 125 ml/min).

from either neoplasia or hyperplasia. Most persons with primary hyperparathyroidism have a single neoplastic parathyroid gland (usually an adenoma, rarely a carcinoma). Some individuals with parathyroid adenomas have adenomas of other endocrine glands as well; these patients belong to one of two categories of genetic conditions of multiple endocrine neoplasia, called type I and type II. There is a certain amount of overlap between these conditions of multiple endocrine adenomas in that both types can include parathyroid adenomas.

A few patients with primary hyperparathyroidism have a generalized, nonneoplastic increase in the size and number of parathyroid cells in all four glands. This generalized hyperplasia and hypertrophy of the parathyroids (of unknown etiology) can result in an increase in gland size up to 100 times normal. Occasionally excessive secretion of PTH or a PTH-like agent is produced from ectopic non-parathyroid tumors (pseudohyperparathyroidism).

Primary hyperparathyroidism is most commonly seen in middle-aged patients, between the third and fifth decades, and much more often in women than in men.

An excessive secretion of PTH initially causes changes in urinary electrolyte excretion (increased excretion of phosphate, decreased excretion of Ca^{2+} and H^+) and increased bone resorption by osteocytes (osteocytic osteolysis) and by extant osteoclasts. These changes result in an early increase in plasma calcium (hypercalcemia) and a decrease in plasma phosphate (hypophosphatemia). The diminution in renal tubular secretion of H^+ produces a fall in HCO_3^- reabsorption accompanied by a reciprocal rise in Cl^- reabsorption (since some anion must accompany Na^+ in its reabsorption); these renal changes produce a wasting of urinary bicarbonate and, as a consequence, a kind of renal tubular acidosis of endocrine origin. Because of the reciprocal relationship between Cl^- and HCO_3^-, there is a rise in plasma Cl^-; the systemic acid–base imbalance, often described as a hyperchloremic metabolic acidosis, is usually neither severe nor life threatening.

Increased PTH secretion also causes increased renal synthesis of 1,25-diOH D from 25-OH D. The increased formation of this active form of vitamin D, often referred to as a renal hormone, stimulates increased intestinal absorption of calcium, which adds to and sustains the hypercalcemia. PTH and 1,25-diOH D act together synergistically to further enhance bone resorption by an increase in osteoclast formation and activity.

The most common complaints presented by patients with primary hyperparathyroidism are of renal origin. Recognizable renal signs and symptoms are present in as many as 80%. Of particular importance are nephrolithiasis (formation of renal calculi, or kidney stones) and nephrocalcinosis (deposition of calcium deposits in and around renal tubules). Characteristic symptoms are painful renal colic and hematuria with passage of urinary "sand and gravel."

The use of specialized radioisotope and microradiographic techniques can reveal bone involvement in about two-thirds of hyperparathyroid patients; however, osseous involvement detectable by more common clinical methods (e.g., ordinary x-ray films and bone biopsy) reveals only about one-half of these persons (about 30% of hyperparathyroid patients). For many years the classical pathognomonic bone lesion of excess PTH secretion was considered to be "osteitis fibrosa cystica" (von Recklinghausen's disease of bone): a severe stage of bone loss marked by loss of marrow and some of bone with replacement by fibrous tissue and by the presence of various-sized cysts. However, the frequency with which osteitis fibrosa cystica has been seen is gradually decreasing. Instead the most common osseous involvement seen today is a simple diffuse osteopenia (e.g., diffuse spinal rarefaction) indistinguishable from that of senile osteoporosis. In addition to diffuse demineralization, localized subperiosteal resorption is of diagnostic significance. A prominent symptom of bone demineralization is bone pain, probably owing to microfractures.

The reason for a decline in the incidence of osteitis fibrosa cystica and an increase in observation of simple osteopenia is not known for certain. It may be only that hyperparathyroidism is being diagnosed much earlier and that severe end stage destruction of bone with fibrosis and cysts is often prevented by effective therapy (usually surgical removal of the overactive glands).

Peptic ulcer disease is fairly commonly seen in patients with hyperparathyroidism (to a much greater extent than in the population as a whole). A small fraction of these patients have a hereditary association of gastrin-producing tumors of the pancreatic islets along with their hyperparathyroidism (multiple endocrine neoplasia type I). In these patients the excess gastrin results from autonomous excessive secretion of the hyperactive pancreatic tumor cells. However, in most hyperparathyroid patients an excess of gastrin results from increased secretion by nontumorous gastric antral cells. These cells are known to be stimulated by an increase in calcium and therefore might have enhanced release of gastrin as a consequence of hypercalcemia. Hypercalcemia also appears to predispose hyperparathyroid patients to pancreatitis, however, the mechanism by which this is brought about has not been discovered. Anorexia, nausea, and vomiting are less serious gastrointestinal problems brought about by the hypercalcemia of hyperparathyroidism.

A profound influence of hyperparathyroidism on the neuromuscular system has been known from some of the earliest descriptions of the disease. An increased appreciation for influence of hypercalcemia and/or excess PTH on muscle function has raised awareness of such symptoms as fatigability and weakness to the point where they are perhaps second only to renal manifestations as common complaints in hyperparathyroidism. Central nervous system effects of hypercalcemia may involve poor memory, lethargy, somnolence, or personality changes.

Secondary hyperparathyroidism is a condition of parathyroid gland hypertrophy and hyperplasia and a greater than normal secretion of PTH as an adaptive respose to conditions tending to produce hypocalcemia (e.g., resistance to PTH, hyperphosphatemia of chronic renal failure, vitamin D deficiency). The tendency toward hypocalcemia may be difficult to establish from analysis of plasma calcium concentration since the compensatory secretion of PTH may result in partial correction of the lowered blood calcium (except in PTH resistance). That is, secondary hyperparathyroidism is an adaptive response that usually serves to protect calcium homeostasis. This is in contrast to primary hyperparathyroidism, which is the result of an intrinsic abnormality of the parathyroid glands leading to excessive, often autonomous, secretion. The homeostatic maintenance of plasma calcium levels in secondary hyperparathyroidism draws on the bone reservoir of mineral ions; plasma calcium regulation takes precedence over the structural function of bone, and loss of mineral content continues despite detrimental loss of bone integrity.

Hypoparathyroidism

Hypoparathyroidism can result from (1) surgical damage, (2) idiopathic failure, and (3) lack of target organ response to PTH (pseudohypoparathyroidism). All are fairly rare disorders of parathyroid function, but among the three categories that of surgical damage is the most frequent. The major signs and symptoms of hypoparathyroidism are directly attributable to hypocalcemia. The decreased level of plasma calcium is a consequence of the loss of PTH influence on kidney, bone, and intestine.

A deficiency of PTH influence on renal function results in a fall in calcium reabsorption and an increase in that of phosphate. Urinary calcium excretion increases at first but then diminishes as hypocalcemia develops. An additional renal influence that becomes decreased in hypoparathyroidism is that of 1,25-diOH D formation. A diminution of 1,25-diOH D decreases intestinal absorption of calcium,

and, along with decreased PTH, decreases the resorptive activity of bone (bone density becomes slightly increased).

The most prominent feature of hypocalcemia is an increase in neuromuscular excitability expressed as latent or overt tetany (depending on the degree of hypocalcemia). Although tetany is usually defined as an excess in neuromuscular motor function, the condition is often accompanied by sensory phenomena as well (e.g., paresthesias felt as a tingling around mouth and fingertips). The basis for the hyperexcitability of the neuromuscular system resides in the influence of calcium on membrane ionic permeabilities (particularly that of Na^+). A deficiency of calcium leads to increased Na^+ entry, which causes depolarization. This can result in increased excitability and spontaneous firing. Various muscles become tense and progress to spasm, particularly those of hands and feet (carpopedial spasm). Rarely such spontaneous contracture of skeletal muscle can produce asphyxiation by laryngospasm. Less severe states of hypocalcemic excitability may produce only a latent tetany in which various diagnostic manipulations may be necessary to uncover the condition (by tapping over a nerve, producing transient ischemia with a blood pressure cuff, or by measuring the response to galvanic stimulation). Central manifestations of hypocalcemia can vary from confusion to grand mal-like seizures.

Pseudohypoparathyroidism is a rare genetic disorder that has many of the same signs and symptoms as simple hypoparathyroidism (e.g., hypocalcemia, tetany) and, in addition, is marked by distinctive skeletal and developmental defects (e.g., round face, short neck, thick-set stocky body build, bradydactylly) (Fig. 17). The hypoparathyroid-like conditions are the result of target organ resistance to PTH. PTH is secreted (often in much higher amounts than normal), but bone, kidney, and intestine have greatly reduced response. Often other kindred have the same skeletal and developmental defects, but with normal PTH function. Since they look like persons with pseudohypoparathyroidism, but do not have hypoparathyroid symptoms, they are said to have pseudopseudohypoparathyroidism.

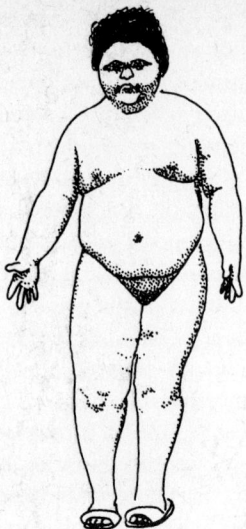

Figure 17 Pseudohypoparathyroidism in a young girl with short stature, stocky body build, rounded face, and short fingers. Her mother has a similar appearance but no chemical abnormalities of hypoparathyroidism (the mother has pseudopseudohypoparathyroidism). (Drawn from Potts VT: in Stanbury VB, Wyngaarden JB, and Fredrickson DS (eds): "Pseudohypoparathyroidism," *The Metabolic Basis of Inherited Disease,* ed 4, New York, McGraw-Hill, 1978.)

CALCITONIN-SECRETING CELLS OF THE THYROID

The thyroid gland contains two separate types of epithelial cells: (1) the abundant follicular cells, and (2) sparse parafolicular C cells (Chapter 46). The embryonic origins of these two cell types are quite different: the follicular cells are derived from an entodermal pouch on the floor of the embryonic pharynx. The parafollicular C cells migrate to the thyroid from the ultimobranchial bodies of the fourth gill pouch body. These cells arrived at the ultimobranchial bodies after an earlier migration from the neural crest, and these C cells are considered part of the APUD series of Pearse, the so-called *A*mine *P*recursor *U*ptake, *D*ecarboxylation cells.

On the basis of studies of calcium homeostasis in dogs, Copp postulated the existence of a parathyroid hypocalcemic hormone that he named calcitonin. Subsequent studies in other species have confirmed the existence of CT, but have shown that it is predominantly found in the parafollicular C cells of the mammalian thyroid, although occasionally in parathyroids and thymus. At one time the name thyrocalcitonin was suggested to reflect this thyroidal origin, but since this hormone is found in the ultimobranchial bodies of submammalian vertebrates from fish to birds, it is now simply called calcitonin. Within a surprisingly short span of only 6 years (1962–1968), the hormone was discovered, isolated, sequenced, and synthesized. The calcitonins from all species studied, including several mammalian and submammalian species, are all 32-amino acid polypeptides with a seven-member ring at the amino terminal end, formed by a cysteine-to-cysteine disulfide bond. Oddly, salmon CT is 10 times more potent in humans than is human CT, a fact made use of in the administration of salmon CT for therapy in Paget's bone disease. It has been suggested the biological activity of salmon CT is little different from that of human CT, but only appears to be more active because of a slower rate of inactivation by degradation. The structure of human CT is illustrated in Figure 18. Radioimmunoassay of CT is a valuable diagnostic tool in identification of thyroid medullary carcinoma (an adenocarcinoma of thyroidal parafollicular C cells), the only known cause of continued hypersecretion of CT.

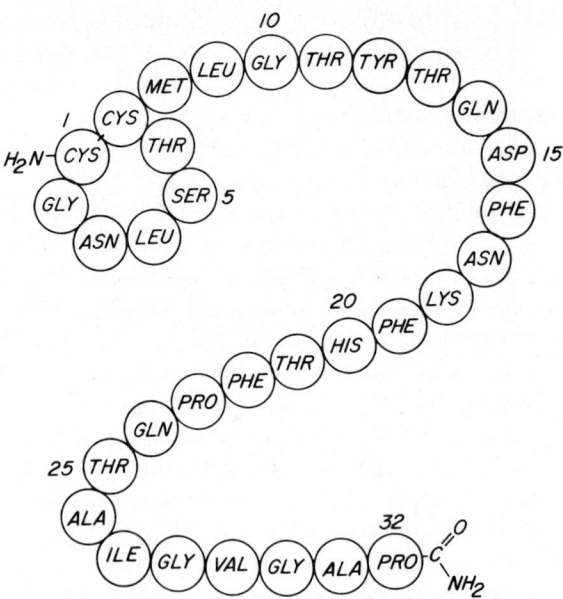

Figure 18 Amino acid sequence of human calcitonin.

SYNTHESIS AND SECRETION OF CT

Some of the same types of studies that established the sequence of cleavages of large precursors in the formation of PTH (e.g., pulse-chase studies, translation of C cell mRNA in cell-free systems) have indicated that CT is first translated by polyribosomes as a large polypeptide of about 15,000 molecular weight. Proteolytic activity associated with the endoplasmic reticulum splits off what appears to be a 25- to 30-amino acid signal sequence, leaving a slightly smaller precursor to CT. Even so, this smaller precursor is quite a bit larger (a molecular weight of 12,000 or more than 100 amino acids in length) than CT. Why such a much larger precursor is necessary to form a small 3,500-molecular-weight molecule of CT is unknown, but it could be that more than one product is formed, in analogy to the large precursor of ACTH, which splits into several biologically active products such as adrenocorticotropic hormone (ACTH), melanocyte-stimulating hormone (MSH), and endorphans. The naming of these precursors as PreproCT and ProCT awaits their isolation and amino acid sequence identification.

Animal experiments have shown that parafollicular C cells have a minimal or basal level of secretion that is increased by a rise in plasma calcium ion concentration. The situation in normal, healthy humans is less clear, the amounts of calcium necessary to stimulate the secretion in man are larger than the kinds of changes that occur physiologically. It is generally agreed that the parathyroid glands have the predominant role in maintaining plasma calcium ion within narrow limits. However, that is not to say that CT does not participate to some extent. It may have a physiological role in blunting or diminishing alimentary hypercalcemia following a high calcium intake. Animal experiments have shown changes in CT secretion to be more closely associated with feeding than anything else. CT can be shown to increase after a calcium-containing meal before any change in plasma calcium concentration takes place. This is similar to the anticipatory ("early warning") rise in insulin secretion brought about by glucose in the intestine prior to any increase in blood glucose levels. The mediator in the case of insulin is gastric inhibitory peptide (GIP); that mediating the increase in CT is thought to be gastrin. Gastrin injection (or more commonly the synthetic peptide, pentagastrin) can be used as a rapid and convenient provocative test for CT release in suspected subclinical cases of medullary carcinoma of the thyroid.

CT can diminish the magnitude of alimentary hypercalcemia by inhibiting release of calcium into the ECF from bone (antiresorptive action) at the same time that increased amounts of calcium are entering the ECF from the intestine. The possibility of a functional relationship between gastrin and CT is further emphasized by the finding that CT can inhibit gastrin release (which raises the possibility of a feedback relationship not unlike that between gastrin and gastric H^+ secretion).

MECHANISMS OF ACTION

The best-known effect of CT, demonstrated both in vivo (experimental animals and people) and in vitro (bone preparations in culture medium), is inhibition of bone resorption. This should have a tendency to lower plasma calcium concentration in vivo. However, in the normal adult only about 300 mg of calcium is resorbed each day. Inhibition of this small amount should not have a marked effect on the plasma calcium concentration. It might be that CT has a more pronounced effect only on bone that undergoes a higher rate of turnover, such as in growing children or in patients with Paget's disease. Furthermore, any tendency for hypocalcemia could result in an overwhelming compensatory response by the parathyroid glands. This could explain why there are no long-term alterations in plasma calcium concentration after total thyroidectomy and after chronic hypersecretion of CT from medullary carcinoma of the thyroid.

There are two principal means by which CT inhibits bone resorption: (1) a rapid decrease in osteocytic and osteoclastic osteolysis by extant osteocytes and osteoclasts, and (2) a more prolonged inhibition of formation of osteoclasts from their precursor osteoprogenitor cells.

CT could have effects on renal reabsorption of phosphate. There have been some indications of an inhibition of phosphate reabsorption leading to an increase in urinary phosphate (phosphaturia). Renal CT receptors capable of binding the hormone have been demonstrated. However, there is no general acceptance that a renal action of CT has any physiological significance.

VITAMIN D

CHEMISTRY

Vitamin D is a fat-soluble vitamin derived from a typical perhydrocyclopentanophenanthrene steroid precursor; however, it is itself a so-called secosterol in that there has been an opening of the B ring and a rotation of the A ring. The natural endogenously synthesized form of vitamin D in animals is cholecalciferol, or vitamin D_3. Vitamin D_3 is formed by the action of ultraviolet light (e.g., from sunlight) on 7-dehydrocholesterol in skin (Fig. 19). Irradiation of plant sterols (e.g., from yeast or fungi) produces a synthetic vitamin D called ergocalciferol, or vitamin D_2, which is often commercially added to processed food (fortification). A considerable amount of vitamin D_3 is also formed by a nonphotochemical process and stored by fish (e.g., cod liver oil is rich in vitamin D_3).

ABSORPTION AND METABOLISM

Being a fat-soluble vitamin, vitamin D is absorbed like most other lipids: first it is solubilized with bile salts along with other fats to form micelles, which facilitate transfer to the apical borders of mucosal cells in the duodenum. Vitamin D is incorporated into chylomicrons within the absorptive cells; these chylomicrons are discharged into the interstitial fluid on the serosal side and carried into the circulatory system via lymphatics.

After entering the circulation, vitamin D must undergo two successive hydroxylations to be transformed into its active form, 1,25-diOH D. Both vitamins D_2 and D_3 undergo the same metabolic transformations, and both dihydroxy derivatives are equally active in mammals (1,25-diOH D_2 is less active than 1,25-diOH D_3 in birds). Once vitamin D from any source—skin or diet—gains entrance into the blood, it is rapidly taken up by the liver and hydroxylated at the C-25 position, to give the intermediate, 25-hydroxy D, which is then released back into the blood. The hepatic enzyme, vitamin D 25-OH, is weakly inhibited by its product (feedback inhibition); however, if large amounts of vitamin D are present, they are apparently hydroxylated and produce increased amounts of 25-OH D (perhaps by a second nonspecific hydroxylase). Hence, the feedback regulation may be only effective at normal physiological concentrations of vitamin D. 25-Hydroxy D is taken up by kidneys and is hydroxylated at the C-1 position, to give the final active form, 1,25-diOH D. The renal C-1 hydroxylase is subject to regulation by PTH and by phosphate. Either an increse in PTH or a decrease in phosphate stimulates this renal enzymic activity (Fig. 20).

Kidney cells also have a second hydroxylase that can act on 25-OH D. This second renal enzyme acts on the C-24 position of 25-OH D, to form 24,25-diOH D, which

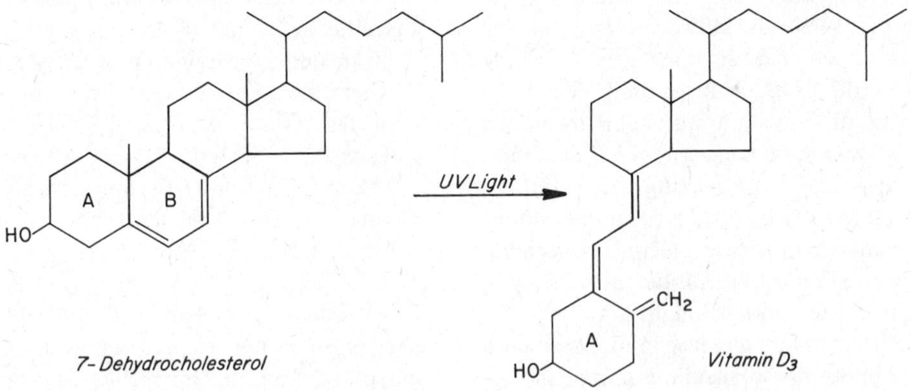

Figure 19 Conversion of the skin precursor, 7-dehydrocholesterol, to cholecalciferol (vitamin D_3) by a photochemical reaction involving ultraviolet light.

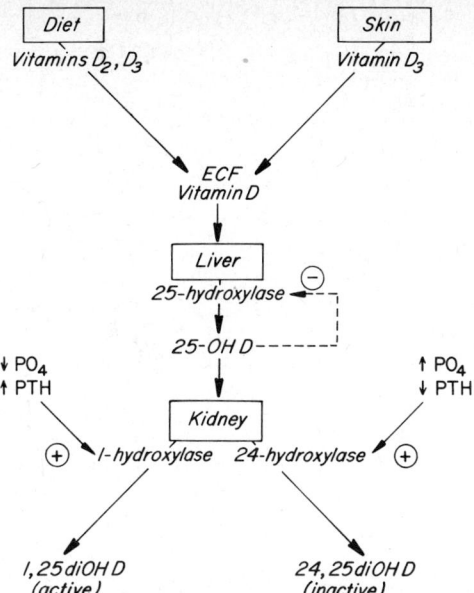

Figure 20 Metabolism of vitamin D. The activity of renal 1-hydroxylase, which causes the formation of the active form of vitamin D, 1,25-diOH D, is stimulated either by a decrease in ECF phosphate (PO_4) or by an increase in parathyroid hormone (PTH). There is a reciprocal relationship between renal 1-hydroxylase and renal 24-hydroxylase; that is, when one goes up, the other goes down.

appears to lack much biological activity. There is a reciprocal relationship between the 1-hydroxylase and the 24-hydroxylase. When plasma calcium levels are below normal, more PTH is secreted and the activity of the 1-hydroxylase increases. When plasma calcium levels are at or above normal, less PTH is released and 1-hydroxylase activity is diminished, while that of the 24-hydroxylase is increased. Changes in plasma phosphate concentration also produce reciprocal changes in these two renal hydroxylases.

PHYSIOLOGICAL ACTIONS OF VITAMIN D

The two most important physiological roles of vitamin D are (1) it is necessary for proper mineralization of bone (this is its oldest known action and was the basis for considering it a vitamin that prevents a deficiency disease, rickets), and (2) more recently it has been found to be an active participant in the homeostatic regulation of both calcium and phosphate. In some respects both PTH and vitamin D have parallel actions that tend to raise both plasma calcium and phosphate, except that PTH has a powerful renal phosphaturic effect that negates the other actions of PTH tending to raise plasma phosphate (giving a net gain of calcium alone), while vitamin D may favor renal retention of phosphate. Under some circumstances vitamin D participates in mechanisms (previously discussed) to raise plasma calcium levels, but at other times (discussed below) can help to raise phosphate.

Vitamin D, in its active form as 1,25-diOH D, acts on intestine, bone, and kidney. The effect of 1,25-diOH D on intestinal absorption of calcium (and phosphate) as the mediator of PTH influence on the gut has already been described under PTH actions.

1,25-diOH D has two principal actions on bone: (1) an indirect, permissive action that permits adequate mineralization of bone (by making calcium and phosphate available from the intestine), and (2) a direct PTH-like effect that enhances osteoclastic resorption. In physiological concentrations, PTH and 1,25-diOH D require each other for their actions on bone mineral mobilization; however, at high pharmacological concentrations, 1,25-diOH D can act in the absence of PTH. This pharmacological effect is the basis for the therapeutic use of massive amounts of vitamin D and/or dihydrotachysterol (a reduced analog of vitamin D) in preventing hypocalcemia of hypoparathyroidism. Although vitamin D has been implicated in control of phosphate reabsorption by renal tubules since 1941, there has not been enough definitive work since then to deny or establish the point.

Dual Role of 1,25-diOH D as Both a Phosphate-Mobilizing Hormone and a Calcium-Mobilizing Hormone

Low plasma levels of inorganic phosphate (hypophosphatemia) increase formation of 1,25-diOH D (by stimulating renal 1-hydroxylase), in turn resulting in mobilization of phosphate from bone, intestine, and perhaps kidney as well. This helps return plasma levels of phosphate back to normal. Similarly, low levels of plasma calcium (hypocalcemia) increase formation of 1,25-diOH D (from increased stimulation of PTH secretion). This helps restore plasma levels of calcium. It may well be disturbing to consider that this single renal hormone (1,25-diOH D) could have two different functions stimulated by two different signals for its production. This may seem all the more puzzling, since plasma calcium and phosphate tend to have reciprocal effects on each other (due to the solutility product relationship). Thus one might ask how 1,25-diOH D can increase plasma phosphate and yet increase plasma calcium under different circumstances.

The reason that 1,25-diOH D can act to restore plasma phosphate without increasing calcium under conditions of hypophosphatemia, as well as to restore plasma calcium under conditions of hypocalcemia without increasing phosphate, is that in the former condition there is a decrease in PTH secretion, while in the latter there is an increase in PTH. The increase or decrease in PTH, and therefore the increase or decrease of its action at the kidney, is the key difference (see Fig. 21.).

Under conditions of diminished plasma phosphate, there is a stimulatory effect on renal 1-hydroxylase, resulting in increased formation of 1,25-diOH D. At the same time there is an increase in plasma calcium due to the

solubility product effect (whenever phosphate drops, calcium rises, and vice versa). This increase in plasma calcium will, in turn, decrease PTH secretion to minimal levels. As long as some minimal PTH is present, a physiological increase in 1,25-diOH D brought about by the hypophosphatemic stimulation of 1-hydroxylase can stimulate osteoclastic resorption of bone and release calcium and phosphate into the ECF. More importantly, the increase in 1,25-diOH D will increase the absorption of calcium and phosphate from intestine. Thus there is a tendency for both calcium and phosphate to become elevated; however, with only minimal PTH present there will be less reabsorption of calcium by the distal tubules and collecting ducts, and urinary calcium excretion will increase. Also, there will be a diminution in phosphaturia. The net result of all these actions is to cause an increase in plasma phosphate along with little or no increase in calcium.

Under conditions of diminished plasma calcium ion, a different sequence of actions takes place. As described earlier, hypocalcemia stimulates PTH secretion, which, in turn, increases the formation of 1,25-diOH D. That is, this time there is an increase in both PTH and 1,25-diOH D.

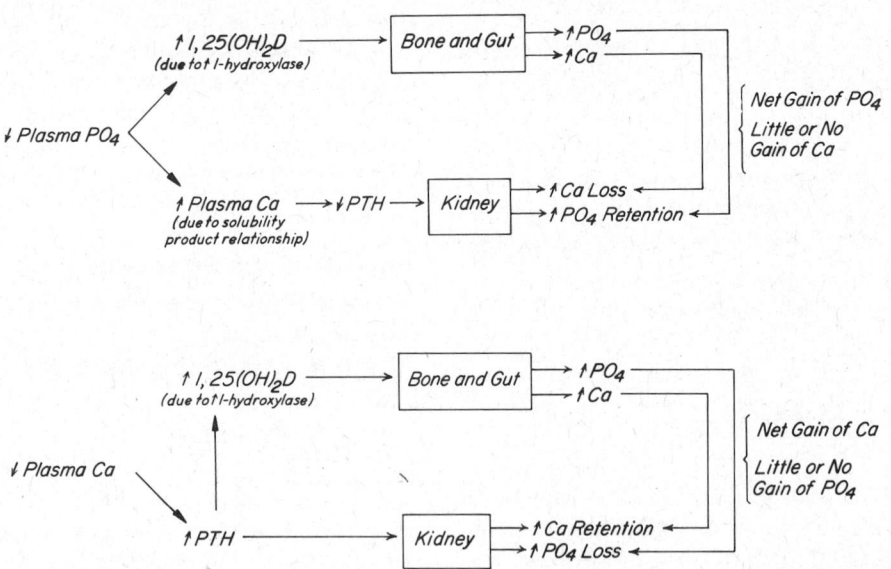

Figure 21. Dual role of 1,25(OH)$_2$D as both a phosphate-mobilizing hormone and a calcium-mobilizing hormone.

The increase in 1,25-diOH D increases absorption of calcium and phosphate from intestine and (along with PTH action) increases osteoclastic bone resorbtion (as before). However, because PTH has a powerful phosphaturic effect and a calcium reabsorbing action, the additional phosphate obtained from intestine and bone is excreted while calcium is retained. The net result is an increase in calcium without an increase in phosphate.

VITAMIN D AND RICKETS

The substance cholecalciferol came to be named as a form of vitamin D as a result of the early history of its discovery and the conception of its role in preventing or curing the disease rickets (in children) or osteomalacia (in adults). In rickets, also called rachitis, there is defective mineralization of the newly formed organic matrix of the growing skeleton; in osteomalacia there is decreased mineralization in remodeling of "nongrowing" bone, leading to decreased bone density. If rickets of children is untreated, there is progressive development of deformities of the pelvis and extremities (characterized by bowing of tibias, femurs, and ulnas). Fractures are common, as is delayed eruption of the teeth.

Although rickets has been known from ancient times, the first clear description of the disease did not appear until 1650. Up until the time of the industrial revolution rickets was of minor significance. However, during the last century and the early part of the present one, rickets became epidemic in industrialized cities, particularly in northern Europe, primarily the result of the shifting of increased numbers of people from rural existence to an urban one. Increased air pollution was blocking the preventive or antirachitic action of ultraviolet radiation on the skin, while greater time was spent indoors (e.g., in homes, schools, and factories). Although rickets was not usually fatal, it did have crippling effects persisting into adulthood. One such deformity was that of a contracted pelvis that, in women, contributed to a high incidence of difficult labor (dystocia). Dystocia, in turn, contributed to high maternal and neonatal morbidity and mortality. Understandably, there was a considerable urgency to discover the means for a prevention or cure, or both.

The first step in the path leading to the conquest of rickets came in 1919 with the success of Sir Edward Mellanby in producing experimental rickets in puppies. Mellanby showed that cod liver oil feeding (long a folk remedy in Scandinavian countries) cured or prevented rickets. At the same time other workers discovered and emphasized the equally effective procedure of increased exposure of skin to ultraviolet radiation (e.g., sunlight). Mellanby had originally concluded that the active ingredient in cod liver oil was vitamin A (which had been recently discovered by McCollum); however, McCollum held that vitamin A was not responsible and demonstrated that the antirachitic agent acted differently. McCollum concluded that this must be a new vitamin, which he named vitamin D. If the antirachitic agent could only be obtained in the diet, the perception of it as a vitamin would not be unusual. However, the role of sunlight or other UV sources on skin biosynthesis, which was once ignored, complicates the picture somewhat. Many regard the 1,25-diOH derivative of vitamin D as a hormone.

An important key to the prevention of rickets was the discovery by Steenbock and associates that UV radiation can act on certain plant sterols to convert them to antirachitic agents. This important finding led to the implementation of UV irradiation of foods and the addition of commercially produced vitamin D, which has virtually eliminated rickets as a major medical problem in affluent Western nations.

Thus, the first part of the vitamin D story is that of the conquest of a disease. The second part has been the more recent discovery of vitamin D participation (as 1,25-diOH D) in homeostatic regulation of calcium and phosphate.

SELECTED BIBLIOGRAPHY

Deftos LJ: III. Calcitonin, in Gray CH, James VHT (eds): *Hormones in Blood,* vol 2. New York, Academic Press, 1979, p 97.

DeGroot LJ, Cahill GF Jr, Martin L, et al (eds): *Endocrinology.* Section 1: *Disorders of Bone and Mineral Metabolism: Relation to Parathyroid Hormone, Calcitonin, and Vitamin D,* vol 2. New York, Grune & Stratton, 1979, p 551.

Harrison HE, Harrison HC: *Disorders of Calcium and Phosphate Metabolism in Childhood and Adolescence.* Philadelphia, WB Saunders, 1979.

Hendy GN, Papapoulos SE, Lewin IG, et al: I. Parathyroid hormone, in Gray CH, James VHT (eds): *Hormones in Blood,* vol 2. New York, Academic Press, 1979, p 1.

Massry SG, Goldstein DA: Parathyroid Hormone, Calcium, and the Kidney, in Suki WN, Eknoyan G (eds): *The Kidney in Systemic Disease.* New York, John Wiley & Sons, 1981, p 417.

Papapoulos SE, Lewin IG, Clemens TL, et al: II. Vitamin D, in Gray CH, James, VHT (eds): *Hormones in Blood,* vol 2. New York, Academic Press, 1979, p 53.

Aurbach GD, Marx SJ and Spiegel AM: Parathyroid hormone, calcitonin, and the calciferols, in Williams RH (ed): *Textbook of Endocrinology.* Philadelphia, WB Saunders, 1981, p 922.

SELF-STUDY QUESTIONS

MULTIPLE CHOICE

Select the single best answer.

1. A hypocalcemia of 7 mg/dl (total plasma calcium) should have a significant effect on
   A. blood coagulation.
   B. excitation–contraction coupling in skeletal muscle.
   C. excitation–secretion coupling in insulin secretion.
   D. neuromuscular excitability.
   E. all of the above.

2. The daily intake of dietary calcium in a 70-kg person who includes a substantial amount of dairy products in his diet should be about
   A. 100 mg.
   B. 200 mg.
   C. 400 mg.
   D. 600 mg.
   E. 1,000 mg.

3. In a person with normally functioning parathyroid glands, a decrease in plasma calcium concentration to 7 mg/dl (e.g., by infusion of a chelating agent) should result in what level of renal reabsorption of filtered calcium?
   A. 50%
   B. 75%
   C. 90%
   D. 95%
   E. 100%

Select the correct answer(s). (In many instances, more than one answer is correct.)

4. A deficiency of parathyroid gland secretion results in a decrease in
   A. plasma calcium ion concentration.
   B. bone osteoclastic activity.
   C. intestinal absorption of calcium.
   D. phosphaturia.

5. Intestinal absorption of calcium
   A. is enhanced following binding of 1,25-diOH D_3 to a mucosal cell cytosolic receptor.
   B. is not influenced by PTH secretion.
   C. becomes adapted to a low intake of dietary calcium by an increase in efficiency.
   D. is primarily accomplished by diffusion across the colonic mucosa into the blood.

6. The renal tubular reabsorption of calcium
   A. for a given level of plasma calcium ion concentration is greater in primary hyperparathyroid patients than in normal subjects.
   B. for a given level of plasma calcium ion concentration is smaller in hypoparathyroid patients than in normal individuals.
   C. is influenced by the effect of PTH on distal tubular cells and collecting duct cells.
   D. is always so high in hyperparathyroid patients with hypercalcemia that they never excrete more

urinary calcium than do normal subjects having a total plasma calcium of 10 mg/dl.

7. As a consequence of trauma to, and ischemia of, the parathyroids during a difficult thyroidectomy, severe primary hypoparathyroidism develops in a patient. The following signs and symptoms are likely:

A. Hypocalcemia

B. Hyperphosphatemia

C. Carpopedal spasm

D. Profuse phosphaturia

8. PTH is known to inhibit

A. proximal tubular reabsorption of phosphate.

B. proximal tubular reabsorption of calcium.

C. proximal tubular reabsorption of bicarbonate.

D. distal and collecting duct reabsorption of calcium.

9. Calcitonin

A. secretion is elevated in patients with medullary carcinoma of the thyroid.

B. from salmon ultimobranchial glands is ineffective in influencing bone resorption in humans.

C. has fewer amino acids (is a shorter polypeptide) than does PTH.

D. secretion is inhibited by administration of pentagastrin.

10. In the case of a patient with an autonomous hypersecretory adenoma of one of the parathyroid glands

A. there is danger of death from asphyxiation due to laryngospasm and convulsions.

B. the total plasma calcium concentration must be elevated above 10 mg/dl.

C. the other three nontumorous parathyroids will have hyperplasia and hypertrophy.

D. the likelihood of kidney stone formation is greater than in the general population.

11. The proposed function of the 25-amino acid signal peptide at the amino terminal of the PreproPTH chain as it is being translated by polyribosomes is to

A. prevent proteolytic attack on the nascent precursor polypeptide as it formed.

B. provide for attachment of the translating polyribosomes to the endoplasmic reticulum.

C. provide a ready source of polypeptides to the cell as energy-yielding substrates.

D. vectorially direct the nascent chain being formed through the endoplasmic reticular membrane into the cisternal space.

MATCHING

A. *PTH*
B. *1,25-diOH D$_3$*
C. *Both*
D. *Neither*

12. can increase osteoclastic resorption of bone mineral.

13. causes decreased renal reabsorption of phosphate.

14. is derived from a larger precursor molecule.

15. is present in deficient amounts in rickets.

16. formation increases as a result of hypocalcemia in otherwise normal animals or persons.

17. formation increases by a decrease in plasma phosphate.

ANSWERS

1. D. Although calcium does participate in each of the four systems listed, a total of plasma calcium level of 7 ml/dl is not low enough to interfere significantly with the first three. It will, however, greatly increase neuromuscular excitation and can cause tetany.

2. E. The daily intake of calcium in a 70-kg subject with substantial amounts of dairy products in his diet (i.e., as in many people in more affluent countries) is 1,000 mg.

3. E. At a normal level of total plasma calcium of 10 mg/dl, the percentage of filtered calcium that is reabsorbed is about 98%. As the plasma level decreases,

the secretion of PTH increases from normally functioning glands and increases the rate of renal tubular reabsorprion of calcium. At a blood level of less than 7.5 mg/dl, this reabsorption becomes essentially complete (100%).

4. All are correct. A. PTH is a hypercalcemic hormone (increases plasma calcium) that acts by increasing intestinal absorption of calcium, by increasing release of calcium from bone mineral, by increasing renal reabsorption of calcium thereby conserving it, and by decreasing renal reabsorption of phosphate (which tends to increase plasma calcium since whenever plasma phosphate tends to decrease, plasma calcium tends to increase). When there is a deficiency of PTH, these actions are diminished and less calcium enters the ECF thereby producing a hypocalcemia. B. PTH deficiency results in decreased osteoclastic activity by decreasing the activity of extant osteoclasts and by decreasing the formation of osteoclasts from osteoprogenitor cells. C. PTH augments renal formation of 1,25-diOH D; therefore, a decrease in PTH causes a decreased intesinal absorption of calcium due to a deficiency of calcium-binding protein. D. With less PTH there is less of a phosphaturia.

5. A, C. Like most steroid hormones, 1,25-diOH D_3 binds to a cytosolic receptor within its target organ's cells; the hormone–receptor complex enters the nucleus and initiates transcription/translation of a protein or proteins important in intestinal absorption of calcium (Fig. 15). Also, the intestine does become adapted; when plasma calcium levels tend to fall because of decreased dietary intake, PTH secretion increases and therefore intestinal efficiency increases because of increased 1,25diOH D_3 formation. B is incorrect because intestinal absorption of calcium *is* influenced by PTH secretion (via formation of 1,25-diOH D_3). D is incorrect because intestinal absorption occurs in the small intestine by an active transport process.

6. A, B, C. Choices A and B are correct and are illustrated by Figure 16. C. The site of action of PTH on calcium reabsorption is predominantly in the distal tubules and collecting ducts. D is incorrect because although PTH does increase the rate of reabsorption,

if at the same time plasma calcium concentration becomes quite high, so much more calcium is filtered at the glomerulus that the reabsorptive process is overwhelmed and urinary calcium excretion increases anyway (see hyperparathyroid curve of Fig. 16).

7. A, B, C. Hypocalcemia develops because of decreased entry of calcium into the ECF from intestine, bone, and kidney. Hyperphosphatemia results from a decrease in the phosphaturic effect due to PTH deficiency. Carpopedal spasm is one of the first spasms to result from hypocalcemic hyperexcitability of the neuromuscular system. D is incorrect because instead of an increase in urinary phosphate, there is a decrease.

8. A, B, C. PTH has an inhibitory effect on the proximal reabsorption of calcium, phosphate, and bicarbonate. The net effect of PTH on calcium reabsorption, however, is to increase it by stimulating calcium reabsorption to a greater extent in the distal tubules and collecting ducts than to inhibit it proximally.

9. A, C. Calcitonin secretion is increased by tumorous C cells of medullary carcinoma of the thyroid. B is incorrect because salmon calcitonin is even more active on bone resorption (it inhibits it) in humans than is human calcitonin. C. Yes, calcitonin is only 32 amino acids long, whereas PTH is 84 amino acids long. D is incorrect; just the opposite is true. Calcitonin secretion is stimulated by gastrin and the synthetic five amino acid peptide, pentagastrin.

10. D. With hypersecretion of PTH, the concentration of phosphate in the urine increases. If there is also a considerable increase in plasma calcium ion, there can be an increase in urinary calcium as well, despite increased reabsorption of calcium (see hyperparathyroid curve of Fig. 16). The combination of increased phosphate and calcium in the urine can greatly exceed the solubility product for these two ions and result in precipitation of insoluble calcium phosphate, which leads to stone formation (nephrolithiasis). A is incorrect because the danger of death from aphyxiation due to laryngospasm and convulsions is true of hypocalcemia of hypoparathyroidism, not the hypercalcemia of hyperparathyroidism. B is incorrect because although the total plasma calcium is usually greater

than 10 mg/dl, this is not always true. If there were a subnormal concentration of plasma proteins, there would be a decrease in the protein-bound fraction of plasma calcium, although the ionic calcium level could be increased from hyperparathyroidism. Thus the patient could have a serious excess of ionic calcium without the total plasma calcium being over 10 mg/dl. C is incorrect. When only one parathyroid is hypersecretory, due to a single adenoma, the remaining glands become surpressed by the hypercalcemia that develops. There are conditions in which all four parathyroid glands can become hyperplastic and hypersecretory, but these are nonneoplastic in nature.

11. B, D. According to the signal hypothesis, the two functions of the signal sequence of amino acids at the amino terminal head of the forming PreproPTH are to provide for attachment of the ployribosomes to the endoplasmic reticular membrane and to direct the chain vectorially through the membrane into the cisternal space for subsequent translocation to the Golgi apparatus.

12. C. Both PTH and 1,25-diOH D can increase osteoclastic resorption of bone mineral. PTH cannot act in the absence of 1,25-diOH D, but pharmacological amounts of 1,25-diOH D can act in the absence of PTH.

13. A. PTH has a rapid phosphaturic effect (increased phosphate in urine due to decreased renal reabsorption). 1,25-diOH D does not cause phosphaturia.

14. A. PTH (84 amino acids) is derived from ProPTH (90 amino acids), which is derived from PreproPTH (115 amino acids). 1,25-diOH D is derived from 25-OH D (slightly smaller by one hydroxyl group), which is derived from vitamin D (smaller by another hydroxyl group).

15. B. There is a deficiency of 1,25-diOH D because of a deficiency of vitamin D (from skin or diet) in rickets. PTH is usually increased as a result of secondary hyperparathyroidism in rickets.

16. C. As a result of hypocalcemia there is a direct stimulation of biosynthesis and release of PTH. PTH, in turn, stimulates renal 1-hydroxylase activity and increases 1,25-diOH D formation. So both increase as a result of hypocalcemia.

17. B. A decrease in plasma phosphate increases 1,25-diOH D formation in kidney by stimulating 1-hydroxylase activity. A decrease in plasma phosphate has no direct effect on PTH formation. It has an indirect effect; decreased plasma phosphate favors increased plasma calcium (solubility product effect), which tends to suppress PTH secretion.

48

The Endocrine Pancreas

DONALD W. STUBBS

OBJECTIVES

After completion of this chapter, you should be able to:

1. Compare the phylogenic and embryonic relationships between islet hormones and gut hormones.

2. Distinguish between A, B, and D islet cells in terms of their specific hormones secreted and the topographical distribution of these cells within human islets.

3. Describe the mechanism by which insulin, as a molecule composed of two peptide chains, is synthesized in vivo from a single-chain precursor.

4. Relate the structural and functional correlations between initiation of insulin formation as preproinsulin, its migrations, packaging, and ultimate storage as insulin within β-granules at the cell membrane before secretion by exocytosis.

5. List agents known to stimulate or inhibit insulin secretion.

6. Compare the roles of Ca^{2+} and cyclicAMP in the glucose-stimulated release of insulin from B cells.

7. Compare similarities between insulin and glucagon biosynthesis.

8. Compare similarities and differences between agents acting on insulin release and those acting on glucagon release.

9. Describe islet cell interactions and how these may come about.

10. Compare the actions of insulin and other hormones on various aspects of carbohydrate, protein, and fat metabolism under endocrine control.

11. Describe the metabolic changes that comprise adaptation to fasting, as well as emphasize the central role of diminished insulin secretion in relationship to continued actions of contrainsulin hormones.

12. Compare the acute metabolic derangements of insulin deficiency in diabetes with the metabolic changes in adaptation to fasting.

13. Describe the mechanism of ketogenesis and contrast its beneficial effects in fasting with its deleterious effects in type I ketosis-prone, juvenile-onset diabetics.

14. Compare similarities and differences of acute metabolic complications between type I (juvenile-onset) and type II (maturity-onset) diabetes.

15. Diagram the interrelationships between different areas of metabolism in diabetes and show how these lead to the more prominent symptoms of this endocrine disease.

This chapter is concerned with a fairly small but extremely important mass of endocrine tissue: the endocrine pancreas, made up of the islets of Langerhans. These islets, numbering more than one million in normal adults, are scattered throughout the lobules of the exocrine pancreas. Summed together, they collectively amount to 1 or 2 g of endocrine tissue, including three major cells types (A, B, and D cells), a network of autonomic fibers and extensive vascularity. Each individual islet is a minute multihormonal organ that senses and responds to bloodborne substances (e.g., nutrient metabolic fuels, other hormones) and to neural impulses in a manner resulting in homeostatic regulation of metabolic fuels (either as reserve stores or as mobilized fuel for immediate usage) in relationship to feeding and fasting.

Some acute disorders of islet function can result in serious life-threatening conditions calling for rapid and heroic countermeasures by clinicians. Fortunately, most islet disorders are either very rare (e.g., tumorous glucagonomas secreting excess glucagon) or at least uncommon (e.g., excessive autonomous secretion of insulin from a tumorous insulinoma). Unfortunately, one disorder, the absolute or relative deficiency of insulin secretion, which results in the disease, diabetes mellitus* (sugar diabetes), is a common malady in the human and several other animal species. Approximately four to five million people in this country (i.e., about 2% of the population) are known to have diabetes (Fig. 1). It is further estimated that at least an equal number have undiagnosed diabetes. Thus, the total prevalence of diabetes in the United States is probably about 4% of the population. In 1975, the National Committee on Diabetes estimated that as many as 10 million Americans have diabetes.

PHYLOGENY AND ONTOGENY

Insulin and glucagon, long regarded as typical vertebrate hormones from pancreatic islet tissue, are now known to

*Diabetes mellitus is characterized by a polyuria of sweet-tasting urine, whereas diabetes insipidus has a polyuria with a tasteless or insipid urine. Hereafter in this chapter the term *diabetes* will refer to diabetes mellitus.

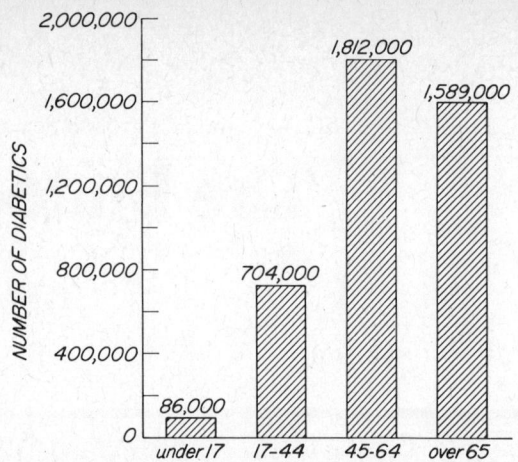

Figure 1 Prevalence (number of cases in existence at a point in time) of diabetes mellitus in the United States as reported in 1973. (From the National Center for Health Statistics, 1973.)

be limited neither to vertebrates nor to the pancreas. It has clearly been demonstrated that not only will mollusks respond to mammalian insulin (by a lowering of blood glucose) but that some (e.g., *Mytilus edulis,* the common edible mussel) have intestinal cells that secrete molluscan insulin. Insulin has also been found in the gut of protochordate invertebrates such as the amphioxus. The gut of amphioxus also has gastrin- and glucagon-producing cells. The direct association of insulin-secreting cells with the gastrointestinal (GI) tract continues into primitive vertebrates such as Agnathan cyclostomes (e.g., lamprey eels and hagfish) where islet tissue is embedded in the intestinal wall.

The close association of islet cells with the gut can also be seen in the embryogenesis of pancreatic islet tissue in higher vertebrates. The embryonic mammalian pancreas develops from two primitive foregut diverticula (dorsal and ventral). Islet endocrine cells have been shown to be present in these epithelial buds of the gut. The classical assumption has been that these endocrine cells, and the exocrine cells that will form the pancreatic acini, arise from the same undifferentiated gut endoderm. Recent evidence has suggested (although not conclusively) that these endocrine cells, as well as endocrine cells of the gut (and many other endocrine cells that produce peptide hormones), may share a number of morphological and biochemical properties because of a possible common origin from neuroectoderm. Cytochemical studies have suggested that several different kinds of neuroendocrine cells have common metabolic patterns, such as the uptake of certain amine precursors and the decarboxylation of these to form biogenic amines. This possible system of cells has been called the APUD system (an acronym for *A*mine *P*recursor *U*ptake and *D*ecarboxylation). It is further suggested that gut and islet cells are part of the APUD system and that these particular members of the system arise from cells of neural crest origin that have migrated to the gut. Although there is some evidence against the idea of neural crest origin of insulin-secreting cells, it seems clear that endocrine cells of the gastrointestinal mucosa and those of pancreatic islets are very closely related. All these islet and mucosal cells are, if not part of a category of APUD cells, at least part of a system of gastroenteric–pancreatic cells (GEP system) that has its orgins both phylogenetically (in evolution) and ontogenetically (in embryogenesis) in the gut.

This commonality of origin goes a long way toward explaining the presence in islet tumors of cells that can secrete what are commonly accepted as typical gastrointestinal hormones (e.g., the secretion of gastrin by pancreatic islet tumors in the Zollinger-Ellison syndrome) and the normal presence of hormones typically considered to be islet hormones in GI musosa (e.g., the secretion of glucagon by certain gut endocrine cells).

FUNCTIONAL ANATOMY

The pancreatic islets were first described by Langerhans in 1869. Islets are compact, ovoid groups of polygonal epithelial cells scattered throughout the pancreas (with a greater concentration of islets in the tail of the pancreas than in its body and head). The islet cells are smaller than those of the surrounding exocrine pancreas (Fig. 2). The total mass of endocrine tissue as islets comprises only about 2% of the wet weight of the pancreas. As is true of many endocrine tissues, islets have an unusually rich vascular supply (in comparison with most nonendocrine tissues, e.g., the exocrine acinar tissue). Islet capillaries,

which form glomerulus-like networks within the islets, are of the fenestrated variety.

The presence of nerve fibers and ganglion cells in the vicinity of islet cells was recognized by Langerhans. Islets are now known to receive both adrenergic sympathetic and cholinergic parasympathetic innervation. Although islet cells continue to function in the absence of such innervation (e.g., in a pancreas transplanted to another region of the body), the autonomic nervous system does exert a small modifying influence in the normal pancreas. Since only a small fraction of islet cells (fewer than 10%) receive autonomic axon terminals, it could be that the gap junctions between islet cells serve as a system of intercellular connections (a functional syncytium), which could electronically disperse neural signals throughout all the islet cells.

DIFFERENCES IN ISLET CELLS

That islets might have more than one kind of cell was recognized as long ago as the turn of the century (by Diamare in 1899). A few years later, Lane described differences in staining between islet cells after alcohol fixation. Since a minority of the islet cells would stain with certain dyes after fixation with alcohol, while the majority wouldn't, he called these stained cells A cells (for alcohol fixation). Bensley added the agranular C cell in 1911, but it wasn't until 20 years later that Bloom described another granular cell (in addition to the A and B granular cells), which was named the D cell. The three principal secretory cells of interest to most endocrinologists are the A, B, and D cells. In routine histologic preparations for light microscopy, the islets are easily identified as groups of cells smaller and less dense than those of the surrounding exocrine acinar cells (Fig. 2); however, specialized staining techniques are required to demonstrate the different cell types. Distinction between the different cell types can also be seen on the basis of secretory granule morphology as revealed by electromicroscopy.

B Cells and Insulin

B cells comprise more than two-thirds of all islet cells and are therefore the most numerous of islet cell types. Perhaps for this reason their function in the synthesis and release of insulin was elucidated much earlier than func-

INTERLOBULAR SEPTUM   PANCREATIC ACINUS

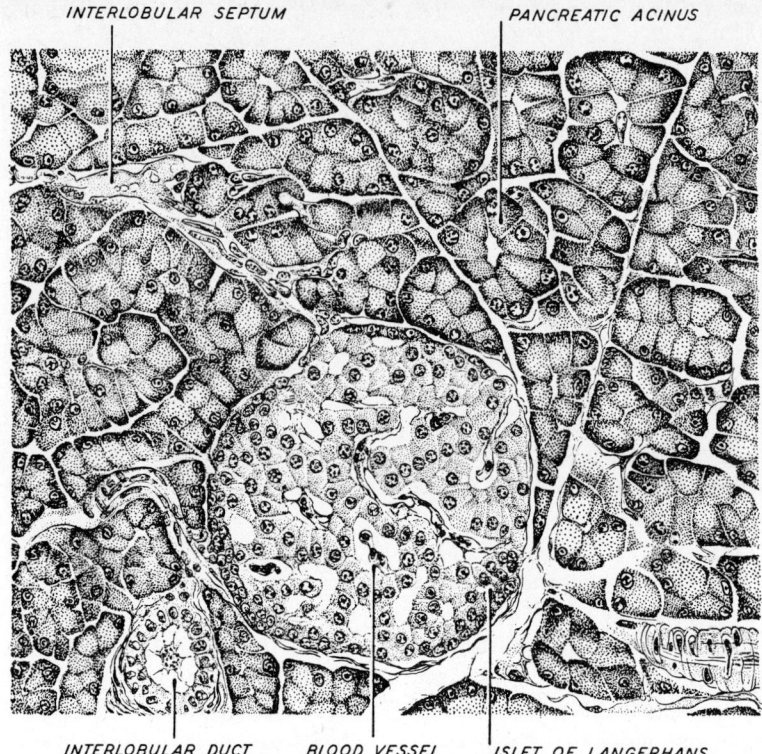

INTERLOBULAR DUCT   BLOOD VESSEL   ISLET OF LANGERHANS

Figure 2 Appearance of section of pancreas including an islet by light microscopy. (From Turner CD, Bagnara JT: *General Endocrinology,* ed 6. Philadelphia,
WB Saunders, 1976.)

tions of the other two major cell types. In 1921, Banting
and Best (while Best was a medical student) successfully
extracted islet hypoglycemic preparations (would lower
blood glucose in experimental animals and humans) from
dog pancreases, thereby verifying earlier conjectures that
pancreatic islets were involved in endocrine control of
carbohydrate metabolism.

B cells were later suggested (in 1938) as the specific cell
type from which insulin was produced and released. This
conclusion was reached after examining pancreases from
dogs made experimentally diabetic by chronic administration of anterior pituitary extracts. B cell changes observed
were interpreted as representing functional exhaustion.
More conclusive evidence that B cells were the source of
insulin was obtained after the availability of drugs with a
highly specific toxicity for B cells (e.g., alloxan and strep-
tozotocin). Selective destruction of B cells results in loss

of insulin secretion and results in a severe experimental
diabetes. More recently, direct evidence for B cell origin
of insulin has been obtained by fluorescent antibody techniques and by bioassay of isolated B cells (obtained by
microdissection).

A Cells and Glucagon

Many early pancreatic extracts and insulin preparations
used in the treatment of diabetics produced an unexpected
early transient increase in blood glucose before the expected hypoglycemic effect. As early as 1923 it was suggested that such preparations contained a contaminating
second hormone, and the name *glucagon* was coined.
However, glucagon itself was not extracted and identified
until many years later. Direct evidence for identification
of A cells as the source of glucagon has been recently
obtained by immunofluorescent techniques.

D Cells and Somatostatin

Until as recently as 1975, the functional role of the D cell was a complete enigma. The first useful clue leading to the correct solution of the mystery came from a completely unexpected source: studies of hypothalamic releasing and inhibiting factors. A hypothalamic factor inhibiting the secretion of growth hormo..e (somatotropin) was discovered in the late 1960s and early 1970s. It was chemically identified as a 14-amino acid peptide and named somatotropin-release-inhibitory hormone, or somatostatin. In 1973, it was found that somatostatin injection could lower blood insulin content. Perfusion experiments on isolated dog pancreases showed this effect to be a direct inhibition of B cell insulin secretion (i.e., completely independently of any effect on somatotropin inhibition). Further investigations showed that somatostatin also inhibits glucagon release from A cells. That a hypophyseal hormone might directly act on islet cells was unexpected, but what was even more surprising was the discovery that this hormone was also produced by certain islet cells (in addition to discovery of somatostatin secretion found in some cells of the stomach, duodenum, and jejunum). Distribution of somatostatin-immunofluorescent cells in islets was found to be identical to that of D cells.

Distribution of Islet Cell Types

Immunofluorescence techniques using antibodies to glucagon, insulin, and somatostatin have permitted mapping of the specific topographical pattern of A, B, and D cells, respectively. In humans, the large vascular channels that penetrate the islets tend to divide the islet mass into several smaller subunits. Within each subunit, the A and D cells occupy the outer periphery, whereas B cells are located centrally (Fig. 3).

In the drug-induced experimental diabetes of streptozotocin-treated rats and in the spontaneous clinical diabetes of many patients with chronic type I insulin-requiring juvenile-onset diabetes, the number of insulin-containing B cells is greatly reduced (essentially zero in some chronic type I juvenile-onset diabetics). Since there are fewer B cells, the A and D cells that remain become the predominant cell types (Fig. 4).

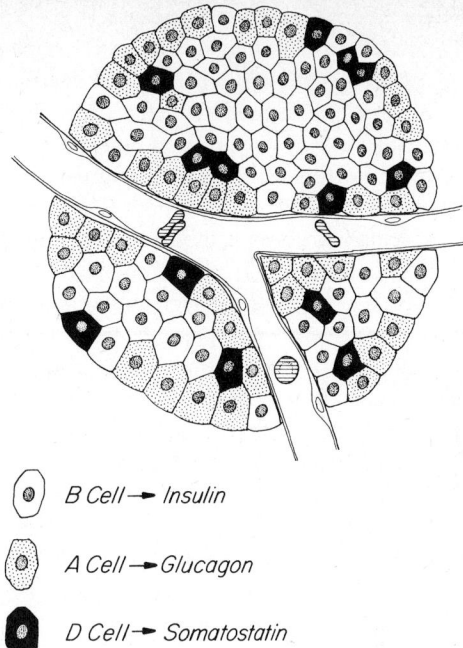

B Cell → Insulin

A Cell → Glucagon

D Cell → Somatostatin

Figure 3 Schematic representation of the number and distribution of glucagon-, insulin-, and somatostatin-secreting cells in islets of humans. (After Bajaj VS: *Insulin and Metabolism.* Amsterdam, Excerpta Medica, 1977.)

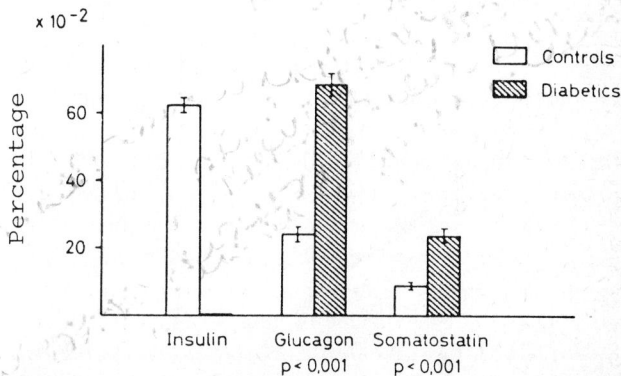

Figure 4. Graphic representation of percentage of total islet cell volume occupied by insulin, glucagon- and somatostatin-secreting cells in normal subjects and in some patients with chronic type I insulin-dependent juvenile-onset diabetes. In normal persons, about two-thirds of the islet cells are B type, and one-fourth are A type, and approximately one-tenth are D cells. Chronic type I juvenile-onset diabetes in a number of patients clearly is the consequence of a nearly complete absence of B cells and the resulting absolute deficiency of insulin. Others have some B cells remaining, and the basis for their insulin deficiency is not as clear. (From Bajaj VS: *Insulin and Metabolism.* Amsterdam, Excerta Medica, 1977.)

Other Pancreatic Islet Hormones

Glucagon, insulin, and somatostatin are not the only hormones released by islet cells; however, the specific cells of origin for other islet hormones have not been clearly identified. Gastrin is known to be present in small amounts in normal human islets and in large amounts in neoplastic cells of pancreatic ulcerogenic tumors (Zollinger-Ellison syndrome). It has not been established whether this gastrin comes from specific G cells, similar to the gastrin-secreting G cells of the antral gastric mucosa, or from one of the typical islet cells (e.g., possibly from D cells).

A powerful gastric-stimulating polypeptide hormone that increases secretion of both HCl and pepsin was reported in 1971 to be isolated from the pancreas. Pancreatic polypeptide appears to be produced in special cells located in the islet periphery with A and D cells (and to a lesser extent scattered throughout the exocrine parenchyma). These are sometimes called PP cells or F cells.

B CELL FUNCTION: BIOSYNTHESIS AND SECRETION OF INSULIN

More than 60 years have passed since the discovery of insulin by Banting and Best. By the end of the first half of this period, all that was known about the structure of insulin was that it is a small protein made up of two peptide chains attached together by disulfide bridges. Not until 1955 was the complete structure (with identification of the amino acid sequences) defined. During the next decade, three different research groups independently accomplished a rather remarkable achievement, the first in vitro synthesis of a specific protein: insulin.

Until 1967, most biochemists believed the synthesis of insulin by B cells in vivo to involve production of the A and B peptide chains individually (by ribosomes) followed by a joining together via two disulfide bonds (at a post-ribosomal site). However, in 1967, Steiner discovered the true process by which B cells form insulin. He incubated pieces of an insulin-producing islet cell tumor (an insulinoma obtained at surgery) with tritium-labeled leucine. The tumor cells incorporated the radioactively labeled leucine, and Steiner was able to isolate a labeled protein, formed early in the incubation, that was larger than insulin. This larger protein (molecular weight of about 9,000) could be converted to typical insulin (molecular weight of about 6,000) by further incubation with a proteolytic enzyme. Thus, it appeared that the tumor cells made a single long-chain precursor (called proinsulin) folded upon itself like a lowercase *e* (Fig. 5), presumably to facilitate formation of the disulfide bridges by proper alignment of the sulfhydryl amino acid units (cysteine units). A central part of the proinsulin protein chain not containing disulfide bonds (called the *connecting peptide*) was then split out enzymatically, thereby providing the final insulin molecule in a highly efficient yield. It is now known that this same process produces insulin by nontumorous B cells.

In the process of insulin biosynthesis and release, not all proinsulin is converted to insulin and connecting peptide, also called C peptide. In addition to equimolar amounts of insulin and connecting peptide released by B cell secretion (amounting together to about 94% of total secretion), a small amount of proinsulin (about 6%) is also released and enters the circulation.

Proinsulin is cross-reactive with insulin in the insulin radioimmunoassay (RIA). Insulin-developed antisera cannot readily distinguish between the two, whereas connecting peptide is not cross-reactive with insulin antibodies. The cross-reactivity of proinsulin with anti-insulin serum poses two problems: (1) to obtain a true measure of insulin alone, some pains must be taken to separate insulin and proinsulin before assay (e.g., with gel filtration and column fractionation); and (2) neither insulin nor proinsulin (or even both together) can be accurately measured by RIA in patients previously treated with exogenous animal insulin (because of development of antibodies against bovine or porcine insulin used in insulin therapy). Since the connecting peptide does not cross-react with such antibodies against therapeutic insulin, and since it is released from B cells in equimolar quantities with insulin, it is possible to obtain an accurate idea of B cell secretory activity, even in patients previously treated with animal insulins, by an RIA for connecting peptide in the blood.

Experiments involving cell-free protein synthesis by isolated ribosomal systems utilizing islet cell nucleic acids have given evidence of early production of a very transient

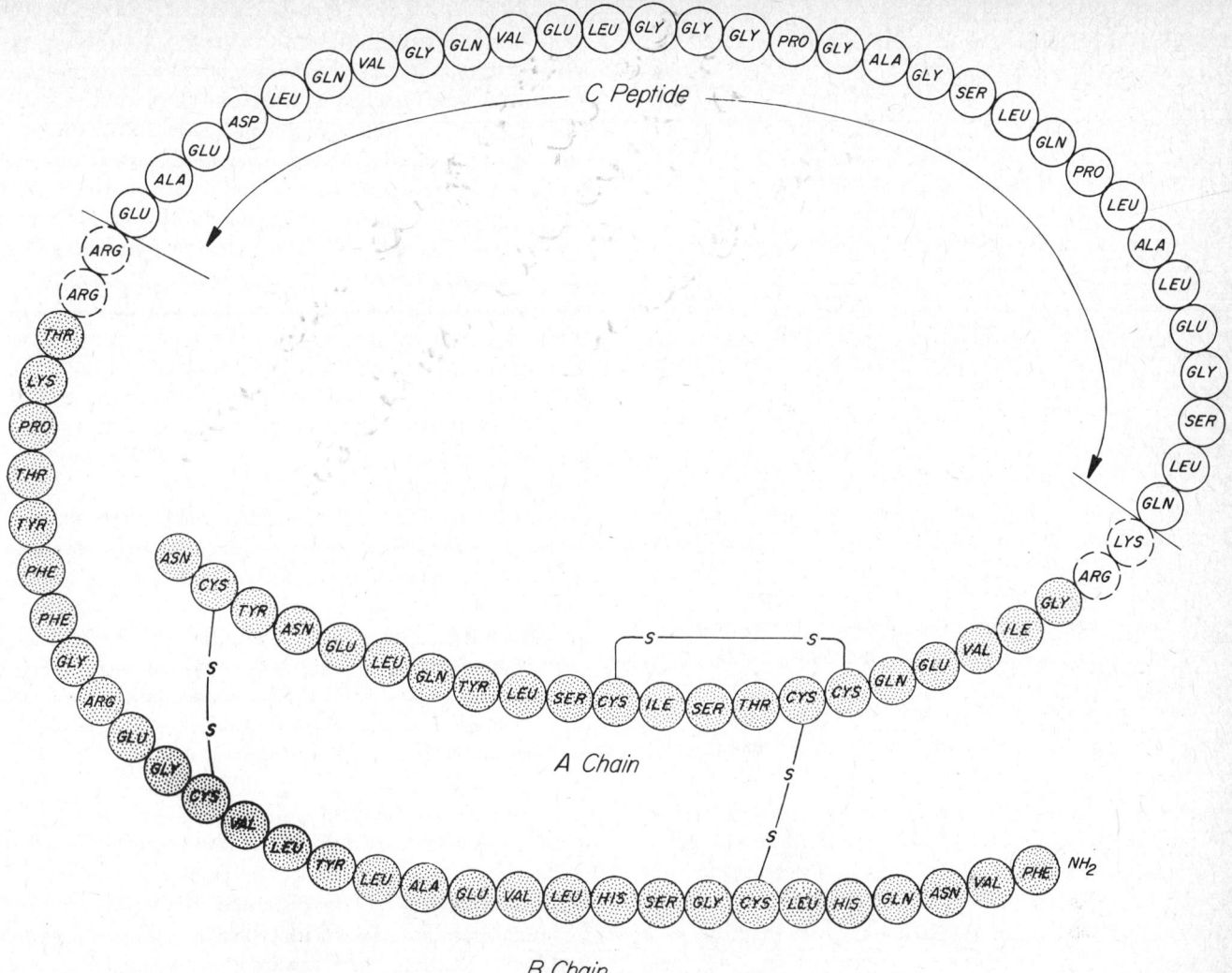

Figure 5 The primary structure of human proinsulin. The insulin moiety is depicted by the stippled circles. The dashed circles represent the pairs of basic amino acids split out by proteolytic cleavage in the conversion of proinsulin to insulin, resulting in the removal of 31-amino acid residues representing the C peptide.

precursor to proinsulin, namely preproinsulin. Preproinsulin is some 20-odd amino acids longer than proinsulin. According to the so-called signal hypothesis, a number of polypeptide hormones (e.g., parathyroid hormone) and other secretory proteins are first formed with a 20–30-amino acid signal peptide as the initial segment of the forming chain of amino acids. In the translation process, as the signal peptide begins to emerge from the ribosomes forming the peptide to be secreted, it facilitates attachment of these ribosomes to the endoplasmic reticulum and permits penetration of the peptide through the membrane into the cisternal space.

STRUCTURAL–FUNCTIONAL CORRELATIONS

Considerably more is known about structural and functional correlations in insulin synthesis than about any other islet cell hormone. Knowledge of events in B cell secretion is extensive, because these cells have long been subjected to continuous intensive investigation at physiological, biochemical, and morphological levels (perhaps to a greater extent than any other endocrine tissue). Ultrastructural studies of B cells show a Golgi complex of elongated cisternae in close topographical relationship with a rough endoplasmic reticulum (RER) on one side and numerous secretory granule vesicles on the other. Transitional elements called microvesicles appear to bud off of the RER and move toward the Golgi complex, where they fuse with the outer cisternae (contribute to the outer stacks). The Golgi cisternae on the other side also participate in producing vesicles by budding. These vesicles are larger and gradually begin to accumulate deposits of secretory granules. Eventually, the secretory granule vesicles line up perpendicular to the plasma membrane of the cell surface for sequential discharge of their contents by a process of exocytosis (Fig. 6).

The progressive sequence involved in insulin biosynthesis (including conversion of proinsulin to insulin), its packaging into secretory granules, and their migration to the cell surface for discharge, involves correlation between static and dynamic studies. The ultrastructural changes, which can only be visualized from static electron micrographs, must be matched with dynamic temporal changes determined by autoradiography at different times (involving deposition of silver grains over radioactively labeled cell structures) and analyses of radioactive label in products from islet cell extracts at different periods of time. Since insulin is the principal protein manufactured by active B cells, it is possible to introduce radioactively labeled amino acids into the molecule at an early stage of its biosynthesis and then follow the progress of that label as it moves from one part of the cell to another (Fig. 6). To label the protein chain forming in the RER, a brief "pulse" (exposure for only a few minutes) of tritiated leucine is incubated with active B cells. The radioactive leucine is taken up and incorporated into forming protein and can be visualized by autoradiography. Preproinsulin is synthesized by polyribosomes attached to endoplasmic reticulum. During or shortly after the formation of preproinsulin and its emergence into the cisternal space, the initial signal peptide is split off, and the remaining proinsulin chain folds in on itself and forms the two disulfide bridges. This proinsulin is then translocated (in the form of transitional microvesicles that split off of the endoplasmic reticulum) to the Golgi apparatus. The hydrolytic cleavage of the connecting peptide and four basic amino acids) out of the proinsulin molecule to give insulin occurs within the Golgi apparatus. Immature secretory vesicles containing insulin and connecting peptide molecules are budded off from the Golgi cisternae and become mature secretory granule vesicles with typical crystallized β granules inside as these move toward the cell surface.

All that remains for secretion to occur is the lining up of mature secretory granule vesicles perpendicular to the cell membrane in preparation for exocytosis. Exocytosis appears to involve interaction between cellular microfilaments and microtubules. The cell web of numerous microfilaments that exists just beneath the B cell plasma membrane may be of particular importance. The principal protein of this cell web is chemically similar to actin. Myosin has also been detected. These contractile proteins are thought to be responsible for active movements of the cell membrane (e.g., ruffling) and for such transfer activity as exocytosis.

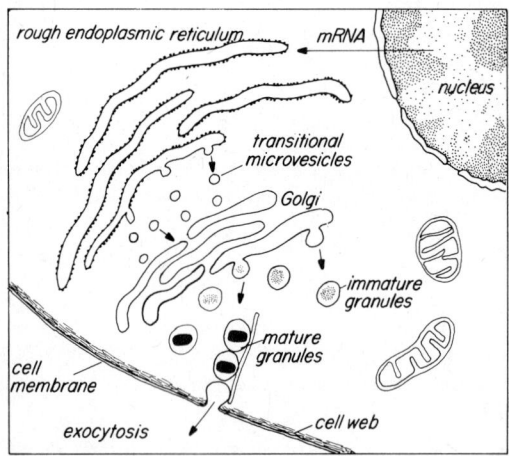

Figure 6 Schematic summary of the proinsulin-to-insulin biosynthetic mechanism of B cells.

STIMULI FOR INSULIN SECRETION

A few animals are able to feed continuously (e.g., marine filter feeders, blood parasites, fetuses), but most postnatal animals, including humans, are intermittent feeders. In nature, this might be a consequence of time needed to locate food or catch prey, or both, or in the case of civilized humankind, it is usually a matter of social convenience. Throughout our vertebrate evolution we must have had to deal with the problem of "feast or famine" (although our present three meals a day constitute moderate feasts and the between-meal periods rather short famines, the problem is the same). During the absorptive period immediately following digestion, more nutrient fuels enter the bloodstream than can be taken up by the tissues using them. This surplus must either be wasted or placed in reserve stores. Nature efficiently chose the latter solution in order to provide fuel during periods of fasting and evolved a system of hormonal controls to regulate it. To maintain a reasonably constant internal milieu, there are mechanisms for sensing entry of exogenous nutrient fuels and for dealing with any surplus after feeding. There are also mechanisms for mobilizing storage depots to supply needs when food is not being absorbed from the gut. A rise in blood glucose during the absorptive period appears to be the principal metabolic signal by which the system is informed that we are in a fed state, and the release of increased amounts of insulin is the primary endocrine signal that the surplus of incoming fuel is to be stored. Similarly, a drop in blood glucose and a fall in insulin secretion are the principal signals initiating mechanisms for fuel store mobilization in the fasting periods between meals. The fuel stores consist of relatively large molecules (e.g., triglycerides, glycogen, and protein) synthesized from smaller subunits (e.g., from fatty acids and glycerol, glucose, and amino acids, respectively). The formation of the reserves is an anabolic process; their later mobilization constitutes a catabolic process (Fig. 7).

Since glucose is not the only nutrient fuel involved, it makes sense that other fuels (e.g., amino acids, ketones, and to some extent fatty acids) also influence insulin secretion, although glucose is of greatest significance. The B cells have been compared to the behavior of a thermostat, continuously monitoring the levels of circulating glucose and other nutrients, and responding in an appropriate

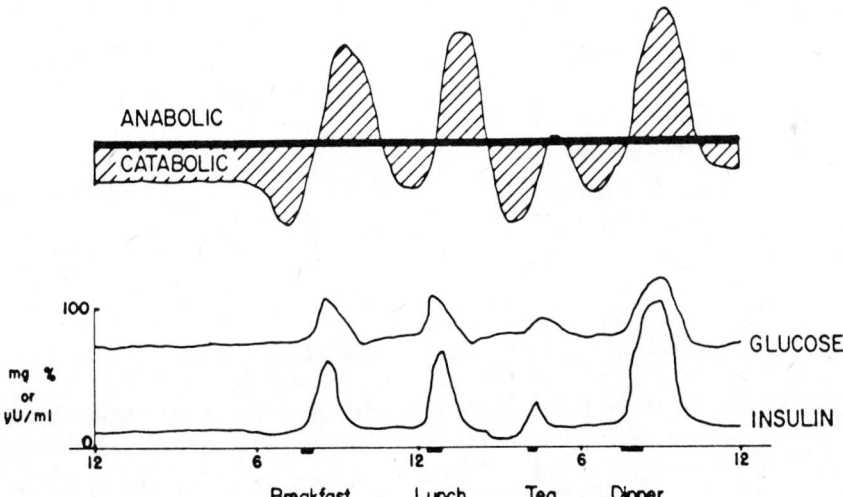

Figure 7 Record of continuous serial measurements of blood glucose and insulin concentrations in a healthy individual eating 3 meals a day plus a midafternoon snack at tea time. (From Cahill GF Jr in Ezrin C et al. (eds) *Clinical Endocrinology*. New York, Appleton-Century-Crofts, 1977.)

manner to initiate either anabolic or catabolic phases. The fundamental importance of glucose derives from its essential role in brain metabolism. The brain has obligatory requirements for two substrates: oxygen and glucose. Thus, in the fasting periods between meals, it is imperative that a continuous supply of blood glucose be maintained to satisfy brain requirements.

Agents Stimulating or Inhibiting Insulin Secretion

The secretion of insulin by B cells can be either stimulated or inhibited by a variety of physiological and pharmacological agents. Among the principal physiological agents stimulating insulin release are increases in blood levels of D-glucose and a number of L-amino acids, in addition to localized increases in islet acetylcholine from vagal cholinergic parasympathetic fibers. Ketones also have a stimulatory effect. Fatty acids are of minor significance as stimuli. The stimulatory effects of the amino acids ketones and acetylcholine are dependent on the presence of at least substimulatory levels of glucose (which is said to have a permissive effect). Thus, glucose has a predominant role in the stimulation of B cell function, even for nonglucose stimuli (Table 1).

Table 1 Physiological and Pharmacological Agents Acting on B Cells

Stimulates Insulin Secretion	Inhibits Insulin Secretion
D-Glucose	Hypoglycemia
D-Glucosamine	2-Desoxyglucose
Many typical L-amino acids	Epinephrine ($_\alpha$effect)
$_\alpha$Aminoisobutyric acid (nonmetabolized amino acid)	Norepinephrine ($_\alpha$effect)
	Somatostatin
Ketones	Starvation
Fatty acids (slightly)	Exercise
Sulfonylureas	
Intracellular cyclic AMP	
Intracellular Ca^{2+}	
Glucagon	
GIP	
Isoproterenol (β effect)	
Acetylcholine	

Glucose

On an equimolar basis (i.e., molecule for molecule), D-glucose is the single most potent stimulator of insulin secretion in normal subjects. It is also the most studied. Despite considerable acquisition of data, the precise mechanism of action remains unknown; however, from the mass of accumulated information, a few generalizations are possible. Glucose appears to influence insulin release in two ways: (1) by a hormone-like interaction of glucose with specific glucoreceptors on the surface of B cells, and (2) by a sustaining influence that depends on metabolism of glucose taken up by the B cells.

If, instead of a single bolus the blood glucose is suddenly raised experimentally and maintained for a half-hour or more, a biphasic release of insulin is seen (Fig. 8). There is an abrupt initial peak (similar to that following a single bolus of glucose), followed by a brief fall and then a second rise to a plateau maintained for the duration of the glucose infusion. Normally, changes in blood glucose are more gradual (note that the abscissa in Figure 7 is in hours, not minutes).

Amino Acids

Amino acids were late in being recognized as nutrient fuel stimulators of insulin release. One of the first suggestions of a causal relationship came in 1955 as a result of in vitro experiments. A few years later, a striking decrease in

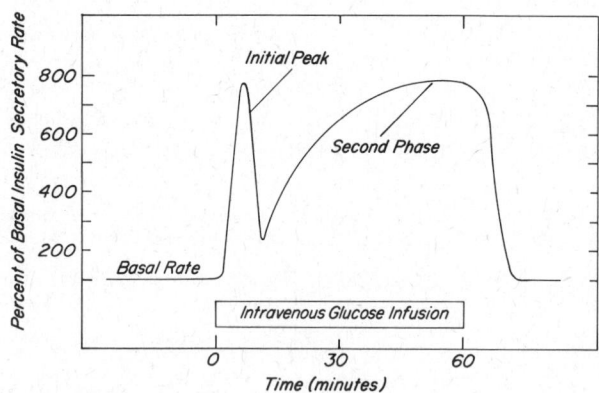

Figure 8 The biphasic pattern of insulin release in response to an abruptly increased glucose concentration (by infusion) that was maintained for an hour and then terminated (a square-wave stimulus).

blood glucose in vivo was observed in patients with insulinomas after L-leucine administration. During the years that followed almost all essential amino acids were shown to stimulate insulin secretion, both in vitro and in vivo. In normal subjects, insulin secretion following protein meals results in levels exceeding those expected on the basis of leucine content. This led to the assumption that other amino acids must have a strong stimulatory effect. In a healthy subject, L-arginine is the most potent amino acid stimulator of insulin secretion. Mixtures of amino acids act synergistically (the sum of the estimated contributions of each individually is not nearly as great as the effect of the mixture). Little is known of the mechanism by which amino acids act, but since several nonmetabolizable amino acid analogs are effective, metabolism must not be required.

The Role of Ca^{2+} and CyclicAMP

Two related events occur approximately simultaneously immediately after glucose stimulation of B cells: (1) an increase in intracellular cyclicAMP, and (2) an increase in intracellular Ca^{2+}. The increase in free calcium ion is primarily the result of an increased cellular uptake of extracellular Ca^{2+}. The presence of extracellular Ca^{2+} is absolutely essential to glucose stimulation of insulin release. The increase in cyclicAMP appears to modify or modulate the increase in intracellular Ca^{2+}, probably by causing the release of ionic calcium from intracellular organelles or membranes. In this case cyclicAMP is not acting as a second messenger. If the increase in B cell cyclicAMP were a second messenger, anything increasing cyclicAMP in these cells would trigger the sequence of events, culminating in insulin release. However, agents (e.g., glucagon) that can increase cyclicAMP in the absence of glucose are ineffective. On the other hand, agents that increase intracellular Ca^{2+} do trigger insulin release even in the absence of glucose (intracellular Ca^{2+} can be increased by administration of a special antibiotic that is a specific membrane carrier ionophore for divalent cations).

It has been suggested that Ca^{2+} acts in stimulus–secretion coupling in analogy with the role of Ca^{2+} in excitation–contraction coupling in muscle. Intracellular Ca^{2+} in B cells is thought to interact with the contractile proteins of cell web microfilaments. Thus, the analogy of the role

of Ca^{2+} might be even more appropriate than originally realized. Numerous microtubules, which are linear hollow rods, are oriented perpendicular to the cell membrane. In tissue culture preparations of B cells, long columns of microtubules can be seen separating linear columns of stacked β granules (Fig. 9). A similar though shorter

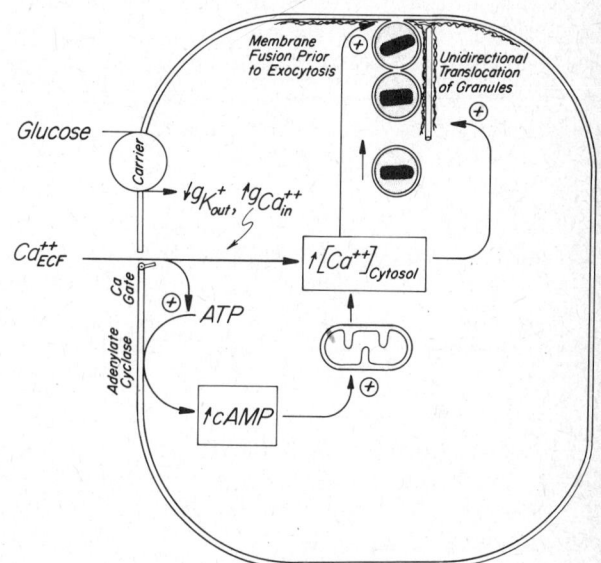

Figure 9. Model of insulin release by B cell as a result of glucose stimulation. Glucose enters the B cell by carrier-mediated facilitated diffusion. Either the combination of glucose with a glucose receptor or the formation of an intermediary metabolite triggers a depolarization of the resting membrane potential (by a decrease in outward potassium conductance) with spiking Ca^{2+} action potentials (from increased inward calcium conductance). The resulting increase in cytosolic Ca^{2+} acts on several intercellular enzymic systems (via calmodulin) including adenylate cyclase. This results in increased formation of cAMP that, in turn, increases release of intracellular Ca^{2+} from cellular organelles (e.g., mitochondria). Since increased cytosolic Ca^{2+} (from whatever source) increases cAMP, this constitutes an amplification system with positive feedback. The increase in cytosolic Ca^{2+} is essential for increased insulin release. The Ca^{2+} may favor fusion of storage granule membranes with the cell plasma membrane (by neutralization of surface charge) and may bring about vectorial (unidirectional) translocation of the storage granules to the plasma membrane by reacting with microtubules and microfilaments. Thus cytosolic Ca^{2+} is of principal importance as the primary mediator of excitation-secretion coupling that results in the release of granular insulin from the B cell by exocytosis.

linear orientation of rows of β-granules exists in B cells in situ. Agents that alter the integrity of these microtubules (e.g., colchicine, vincristine, and vinblastine) inhibit glucose-induced release of insulin. In perfusion studies having abrupt changes in glucose concentration, these agents affect both phases of the biphasic response. The microtubules seem to provide the proper orientation of the β-granules, and the microfilaments probably provide the appropriate contractile force, triggered by Ca²⁺, to move these granules to the cell surface for secretion via exocytosis.

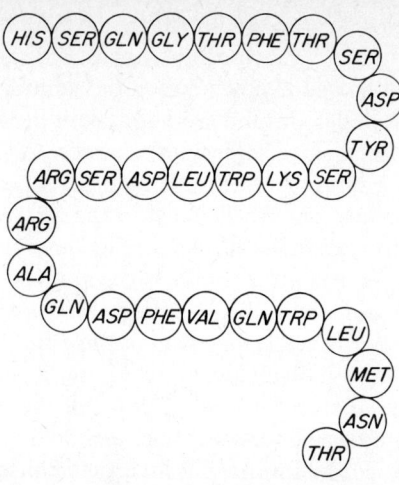

Figure 10 Structure of human glucagon. Glucagon is a single-chain polypeptide of 29-amino acids (molecular weight 3,485).

A CELL FUNCTION: BIOSYNTHESIS AND SECRETION OF GLUCAGON

Interest in A cell function has lagged far behind that for B cells, since for many years neither the lack nor the excess of glucagon secretion had been perceived to be directly involved with any major syndromes. There has been little of the sense of urgency or compelling need for information such as that pervading insulin research. More recent ideas about possible participation of excess glucagon along with a deficiency of insulin in nonobese diabetics may change this and result in increased priorities for studies of A cell function.

Evidence that glucagon, like insulin, is derived from a larger precursor molecule has accumulated from two different lines of study: (1) detection of radioactively labeled higher-molecular-weight biosynthetic products in systems with glucagon synthesis in vitro, using tritiated amino acids; and (2) analysis of plasma and pancreatic extracts containing peptides larger than glucagon that cross-react with antiglucagon sera. These data, along with theoretical considerations about peptide biosynthesis in general, leave little doubt that glucagon is made from a precursor (proglucagon) that is probably at least three times as large as typical pancreatic glucagon. Hydrolysis of proglucagon yields glucagon, as shown by the structure of glucagon (Fig. 10).

Although data on the sequential events in A cells are scarce, it is generally believed that the same steps followed in the biosynthesis of insulin by B cells are also followed in the biosynthesis of glucagon by A cells: (1) synthesis in the RER, (2) transfer to the Golgi complex, and (3) packaging for export as secretory granule vesicles. Furthermore, there is evidence that the mode of secretion is the same for insulin and glucagon: exocytosis. This is becoming recognized as a widespread process of secretion in a variety of cell types, including fibroblasts, granulocytes, macrophages, plasma cells, pancreatic acinar cells, and many peptide hormone-secreting cells (secretion of glucagon, insulin, PTH, STH, TSH, prolactin, oxytocin, ADH, and release of thyroglobulin into colloid space). Catecholamines are also released by exocytosis.

The secretion of glucagon by A cells does not vary much over a 24-hour period, unlike the secretion of insulin by B cells. The principal stimulus for glucagon release is a substantial drop in the blood glucose level (i.e., D-glucose is a physiological inhibitor of A cells in normal subjects). A cells are also stimulated to secrete glucagon in response to an increase in L-amino acids. Either a sudden decrease in D-glucose or a rapid increase in L-amino acids are strong stimuli for A cell secretion and result in a biphasic response. The principal physiological role of glucagon is to prevent hypoglycemia by increasing hepatic glycogenolysis and gluconeogenesis. Thus, when there is

Table 2 Physiological and Pharmacological Agents Acting on A Cells

Stimulates Glucagon Secretion	Inhibits Glucagon Secretion
Strenuous exercise	—
Hypoglycemia	Hyperglycemia
Many L-amino acids	Somatostatin
Cholecystokinin (CCK)	Insulin
SNS or isoproterenol (β effect)[a]	Ketones
Starvation	Fatty acids
Prostaglandins	—

[a]Both insulin and glucagon secretion are increased by β stimulation.

a drop in blood glucose to hypoglycemic levels (i.e., below 50 mg%), it is very helpful to have stimulation of glucagon from A cells. This makes obvious teleological sense. What may not be so obvious is why both insulin and glucagon secretion are stimulated by amino acids. When a high-protein meal is eaten (e.g., broiled fish), the increase in amino acids stimulates insulin secretion, which facilitates the uptake of these amino acids by peripheral tissues, particularly muscle. That glucagon is stimulated at the same time protects the cerebral cortex from a potential hypoglycemia due to the increased insulin secretion. Glucagon secretion is also stimulated by starvation (Table 2).

ISLET CELL INTERACTIONS

Interesting possibilities for islet cell-to-cell interactions have been raised by the recent discovery of specialized junctional complexes between adjacent islet cells. These junctions represent regions in which the plasma membranes of neighboring cells come into well-defined relationships with each other: (1) fusion of the outer layers of two adjacent trilaminar plasma membranes (tight junctions); (2) close but not fused apposition of the outer membrane layers along a narrow gap (about 20 Å wide) with interconnecting channels that bridge the cytoplasms of the two cells (gap junctions); and (3) localized aggregations of intracellular tonofilaments and extracellular material between neighboring membranes (desmosomes).

The structure of desmosomes, well defined in electron micrographs of thin sections of islet cells, appears to serve a simple mechanical function: cell-to-cell adhesion. The other two junctional complexes, whose ultrastructures have been studied by transmission electron microscopy of shadow-cast replicas of freeze-fractured membranes, appear to have a more physiological function in cell-to-cell communication or interaction. It has been shown for certain in other tissues that gap junctions permit low-molecular-weight molecules (< 500 MW) to pass from one cell to the next. Although this could permit interaction via as yet unknown low molecular weight metabolites, this would rule out communication between islet cells by means of their known polypeptide secretions (of high molecular weight). On the other hand, tight junctions may permit local interaction by such hormones released into limited intercellular channels or compartments. Tight junctions in cavitary epithelia (surrounding a cavity of lumen) appear to segregate subcompartments of the spaces around cells (e.g., separation of basolateral spaces of renal tubule cells from the luminal glomerular filtrate). Although islet tight junctions do not completely surround the cells as in cavitary epithelia, they do appear to delimit networks or channels between adjacent islet cells. By these limited routes, islet cell secretions (e.g., polypeptide hormones) could act locally without entering the circulatory system.

The demonstration that certain islet cells can be influenced by secretory products of other islet cells raises the possibility of complex paracrine interactions (paracrine secretions are not intended for distribution in the cirulatory system, but act locally on nearby cells in the vicinity). Such local paracrine effects include glucagon secretion by A cells stimulating B and D cells, insulin secretion by B cells suppressing A cells, and somatostatin secretion from D cells inhibiting all the other islet cells. An interesting clinical aspect of these interactions arises from indications that a deficiency of insulin in B cells in diabetes appears to result in inappropriately excessive glucagon secretion by A cells. This relative hyperglucagonemia combined with insulinopenia could make the acute metabolic derangements of diabetes the result of a dual hormonal abnormality (an excessively high glucagon/insulin ratio), as proposed by Unger.

THE ROLES OF INSULIN AND CONTRAINSULIN HORMONES IN FEEDING AND FASTING

The importance of regulating blood glucose to maintain adequate brain function has been mentioned above. However, blood glucose regulation and carbohydrate metabolism do not exist alone in a vacuum, independent of other blood-borne nutrient fuels and their metabolism. Endocrine control mechanisms operating in anabolic or catabolic phases in relationship to feeding or fasting involve not only insulin, but a number of other hormones—the so-called contrainsulin hormones, the actions of which are, for the most part, opposite to those of insulin. Before considering the changes in endocrine secretion and resulting alterations of metabolism in fasting, and the similarities between these changes and those of diabetes, it is necessary to consider the balance of forces regulated.

CARBOHYDRATE METABOLISM

In affluent Western societies about 50% of caloric intake is carbohydrate. The principal carbohydrate fuel absorbed from the gut is D-glucose. This absorbed glucose enters the hepatic portal system, where most of it (about three-fourths) is taken up by the liver (the remaining one-fourth enters the general circulation, transiently raising the blood glucose somewhat before entering tissue cells). Some of the glucose taken up by the liver is synthesized into hepatic glycogen and stored, but most of the glucose taken up by liver cells is metabolized to pyruvate and acetyl-CoA. Little of this acetyl-CoA from glucose is oxidized completely to CO_2 and H_2O (most of the energy requirements of liver are met by oxidation of fatty acids). Thus, liver differs from many other tissues that use glycolysis as a significant source of energy (e.g., skeletal muscle). The major role of glycolysis in liver is to provide pyruvate for conversion to acetyl-CoA, which can be used in a variety of reactions (e.g., acetyl-CoA is the principal substrate for hepatic fatty acid biosynthesis). Fatty acids synthesized in liver are incorporated into triglycerides for export to adipose tissues in the form of very low density lipoproteins (VLDL). In this manner, surplus calories from carbohydrate are stored peripherally as fat.

Glycogenesis and Glycogenolysis

Of the glucose taken up by the liver during the postprandial absorptive period (the period following ingestion when digestive products are absorbed into the blood), about one-fourth is converted to hepatic glycogen. The metabolic pathway leading to glycogen formation (glycogenesis) is different from that of glycogen mobilization or breakdown (glycogenolysis). That is, one is not simply the reverse of the other. The formation of glycogen depends on the availability of glucose and the activity of glycogen synthetase. Glycogenolysis depends primarily on the activity of glycogen phosphorylase (Fig. 11). The amount of liver glycogen at any time depends on the extent to which the balance is tipped to favor glucose polymerization into glycogen by the anabolic glycogenic pathway, or the mobilization of glycogen by the catabolic glycogenolytic pathway. The flow of glucose and its phosphate intermediates is determined to some extent by intrinsic metabolic controls (e.g., allosteric effects on certain key enzymes reflecting the availability of fuels and the state of energy available as ATP). Superimposed on these intrinsic controls is an endocrine set of controls (mainly by insulin, glucagon, and epinephrine) whose actions are mediated by cyclicAMP.

The hepatic enzyme, glucose-6-phosphatase (G6Pase) has an important role in mechanisms regulating blood glucose in that it controls the release of free glucose from hepatic glucose-6-phosphate (G6P) into the blood. In a sense, it is an enzymatic valve or faucet, the activity of which controls the flow of glucose from the liver. A specific G6Pase is necessary, since removal of phosphate from G6P does not occur simply by reversal of hepatic glucoki-

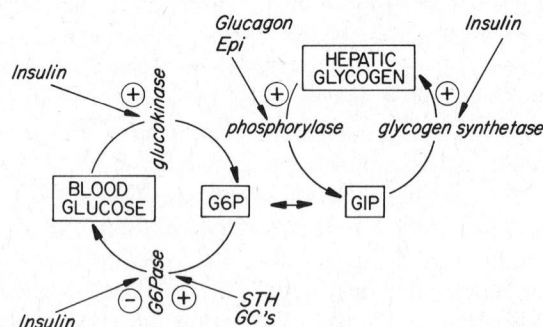

Figure 11 Metabolic pathways and points of endocrine control between glucose and hepatic glycogen.

nase phosphorylation of glucose. The absence of G6Pase in muscle prevents muscle glycogen from contributing glucose directly to blood (muscle participates indirectly via the Cori cycle). Thus, hepatic G6Pase activity is an important part of the catabolic pathway from liver glycogen to blood glucose. The principal hormones that enhance G6Pase activity are growth hormone (somatotropic hormone, or STH) and the glucocorticoids (GCs). Insulin decreases G6Pase activity (i.e., insulin tends to retard the flow of glucose from liver to blood).

Glycolysis and Gluconeogenesis

Although liver glycogen is an important source of blood glucose for brief fasting periods (e.g., the usual intermeal fasting periods such as from supper to breakfast), it is quantitatively minute (representing about 200–300 Calories in adult). Thus if blood glucose is to be maintained for more than several hours (e.g., a fast of a day or more), there has to be some additional and sustained mechanism for producing glucose. Such a source is hepatic gluconeogensis; the formation of "new" glucose from noncarbohydrate sources, particularly from certain amino acids (the gluconeogenic amino acids).

The gluconeogenic pathway from pyruvate allows net synthesis of glucose from various precursors of pyruvate and oxaloacetate (OxA). The amino acids alanine, glycine, serine, and cysteine are converted to pyruvate, which in turn forms OxA by means of the enzyme pyruvate carboxylase. Other gluconeogenic amino acids are convertible to intermediates of the Krebs tricarboxylic acid cycle (TCA cycle), which eventually forms OxA. These gluconeogenic amino acids are asparatic acid, arginine, histidine, glutamic acid, proline, valine, methionine, and threonine. OxA from either source is converted to phosphoenolpyruvate (PEP) by the enzyme PEP carboxykinase and enters gluconeogenesis by this route (Fig. 12).

Although any of the above gluconeogenic amino acids arising from catabolism of muscle protein (the principal reservoir of body protein) can provide substrate for new glucose synthesis in liver, the amino acids predominantly exported for this do not resemble hydrolysis of muscle protein. Instead, a very high proportion (about one-half) of the amino acids made available by muscle is made up of only two amino acids: alanine and glutamate. The liver has a predilection for these same two amino acids as preferred sources for gluconeogenesis.

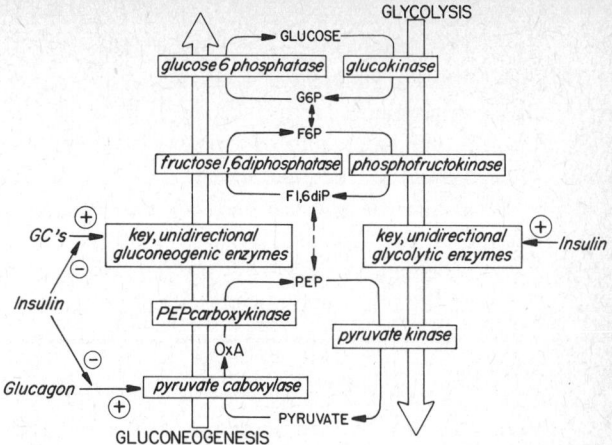

Figure 12 Metabolic control by hormonal regulation of enzymic activities. Insulin favors glycolysis by induction of glycolytic enzymes and by inhibition of glucagon and glucocorticoid effects on gluconeogenesis. The glucocorticoids (GC's) favor gluconeogenesis by induction of gluconeogenic enzymes. Glucagon is thought to favor gluconeogenesis by increasing the activity of pyruvate carboxylase.

The energy-requiring gluconeogenic pathway from pyruvate to glucose is opposite in direction to that of the energy-releasing glycolytic pathway from glucose to pyruvate. These two pathways are reciprocally related; when one is increased, the other is decreased, a fortunate circumstance that prevents the establishment of futile continuous cycling or recycling between glucose and pyruvate. The predominance of traffic in one direction or the other is regulated in part by intrinsic allosteric controls operating at specific control points, characterized by pairs of unidirectional reactions in opposite directions. In addition to this intrinsic control, the glycolytic–gluconeogenic system is under endocrine control with respect to the activities of these key unidirectional enzymes. The key unidirectional enzymes in the glycolytic pathway (glucokinase, phosphofructokinase, and pyruvate kinase) are increased by increased insulin secretion, favoring glycolysis. The key unidirectional enzymes in the gluconeogenic pathway (pyruvate carboxylase, PEP carboxykinase, fructose 1–6-diphosphatase, and G6Pase) are enhanced by increases in glucocorticoid (GC) secretion. Glucagon also stimulates hepatic gluconeogenesis apparently by increasing the activity of pyruvate carboxylase. The actions of both glucagon and the GCs are inhibited by insulin. Thus, the endocrine control superimposed on intrinsic control reinforces the reciprocal relationship between the two pathways (Fig. 12).

During periods of increased availability of glucose, more than enough glucose is at one's disposal for hepatic energy needs (via entry into glycolysis and the TCA cycle), indeed much of the glucose taken by liver is only oxidized as far as acetyl-CoA, which is then synthesized into other substances, such as fatty acids. However, when glucose is not being absorbed from the gut, the energy-supplying pathway of hepatic glycolysis is decreased and the energy-requiring pathway of hepatic gluconeogenesis is increased (see above). The source of energy for hepatic gluconeogenesis comes from hepatic fatty acid oxidation. The primary source of fatty acids for this comes from mobilization of adipose fat stores (see Fat Metabolism; below).

PROTEIN METABOLISM

The principal reservoir of tissue protein is in skeletal muscle—skeletal muscle, which is nearly one-half the body weight, is itself about 20% protein. Thus, it is mainly from the breakdown of this large reservoir of muscle protein that amino acids are mobilized and supplied to the liver for gluconeogenesis. The control of gluconeogenesis exists on two levels: (1) regulation of hepatic gluconeogenic enzymic activities as mentioned above, and (2) alteration of supply of amino acids from muscle protein (the hepatic capacity for gluconeogenesis is not saturated by concentrations of plasma amino acid normally present; increases in the supply of precursors result in increases in the rate of gluconeogenesis).

Regulation of protein turnover in muscle involves changes in either rate of protein synthesis or rate of its degradation. These rates can be estimated experimentally by monitoring the flow of radioactive amino acids into or out of muscle protein. Both protein synthesis and degradation are influenced by hormones. In terms of response in relation to fasting and feeding, the principal hormones involved are insulin and the GCs. Insulin primarily favors protein anabolism by increasing protein synthesis and, to a lesser extent, by decreasing protein breakdown (i.e., insulin has both an anabolic effect and an anticatabolic effect). An insulin-induced anabolic increase in protein synthesis involves increased uptake of amino acids and increased ribosomal assembly of these amino acids into protein. The means by which insulin inhibits catabolic protein breakdown is obscure, but may involve decreased lysosomal activity. Glucocorticoids have effects opposite to those of insulin and other muscle protein anabolic hormones. The principal effect of GCs is to increase protein breakdown into amino acids (a primary catabolic influence). However, the mechanism for enhancing protein catabolism is not well understood. It does not appear to involve increased lysosomal activity. At least one non-lysosomal protease participating in intracellular proteolysis is known to be located in the myofibrillar fraction of muscle cells and is increased in activity by GCs. GCs also favor decreased muscle protein by antianabolic effects: decreased uptake and incorporation of amino acids into muscle protein. Endocrine regulation of protein metabolism is illustrated in Figure 13.

FAT METABOLISM

Evolutionary adaption to intermittent feeding has favored selection of the most efficient storage, namely triglycerides, for the excess calories obtained at mealtime. Triglycerides not only have twice as many calories per gram as carbohydrate and protein, they are stored almost anhydrously (adipose cells are only about 10% water), thereby diminishing the weight of non-energy-storing tissue constituents. Thus, a gram of adipose tissue has more than 10 times as much available energy as an equal weight of any other tissue. If we had to depend only on other tissues for energy stores, we would have to weigh considerably more than we do. Triglycerides for storage are primarily formed

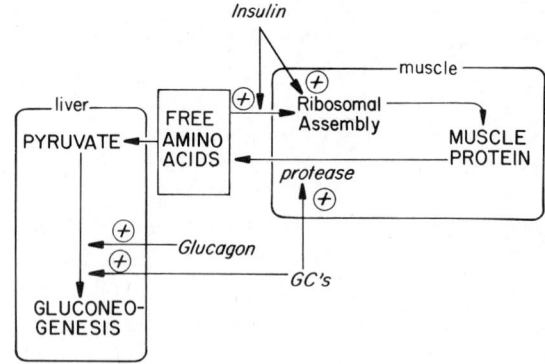

Figure 13 Endocrine regulation of protein metabolism.

in the liver (mainly from glucose and, to a less extent, from some amino acids) and then exported to adipose tissue. Adipose cells are also capable of limited triglyceride synthesis from glucose, but in humans, the liver is the principal organ for this conversion.

During absorption of food from the gut, more than enough glucose to meet the immediate energy needs of the body enters the hepatic portal system. Most is extracted by the liver and, of this, about three-fourths is oxidized to pyruvate by glycolysis. Only a small fraction is completely oxidized to CO_2 and H_2O (via the tricarboxylic acid cycle). Most of the acetyl-CoA obtained from pyruvate (via pyruvate dehydrogenase) is available for synthesis into fatty acids and triglycerides (lipogenesis). Although some amino acids (leucine, isoleucine, lysine, phenylalanine, and tyrosine) can contribute, glucose is the principal source of substrate for fatty acid synthesis. The metabolism of glucose not only provides acetyl-CoA, but also glycero-3-phosphate (a derivative from an intermediate in glycolysis), the precursor of the glycerol moiety of triglyceride. Another essential product of glucose metabolism (via the phosphogluconate, or hexosemonophosphate shunt pathway) is NADPH, which is required as a source of reducing equivalents in fatty acid synthesis.

The triglycerides synthesized in liver and those arising from absorption in the gut, arrive at adipose cells by different carriers. Triglycerides from liver are incorporated into very low density lipoproteins (VLDL). Triglycerides from the gut enter the blood from the thoracic duct as chylomicrons. However, in both cases, fatty acids are released from blood triglycerides for incorporation into adipose cells (for reesterification into adipose triglyceride) by the same mechanism: hydrolysis catalyzed by extracellular lipoprotein lipase (located in capillary endothelium of adipose tissue). Although hydrolysis releases both fatty acid and glycerol, only the fatty acid is taken up by adipose cells. Reesterification in adipose cells requires the formation of intracellular glycero-3-phosphate from adipose glycolysis.

When there is need for mobilization of adipose triglyceride stores (lipolysis), these stored intracellular triglycerides are acted upon by an intracellular hormone-sensitive lipase to release both fatty acids and glycerol into the blood. There is a reciprocal relationship between the activities of lipoprotein lipase (necessary for formation of adipose triglyceride from VLDL and chylomicrons) and hormone sensitive lipase (necessary for hydrolysis of adipose triglyceride in order to release fatty acids and glycerol). Lipoprotein lipase can be increased only by insulin or some prostaglandins.

The hormone-sensitive lipase, however, is activated by a great number of hormones: epinephrine and norepinephrine, growth hormone (STH), ACTH, glucagon, GCs, and thyroid hormones. GCs are also necessary to provide a permissive action for hormone-sensitive lipase stimulation by the other lipolytic hormones (Fig. 14). An increase in

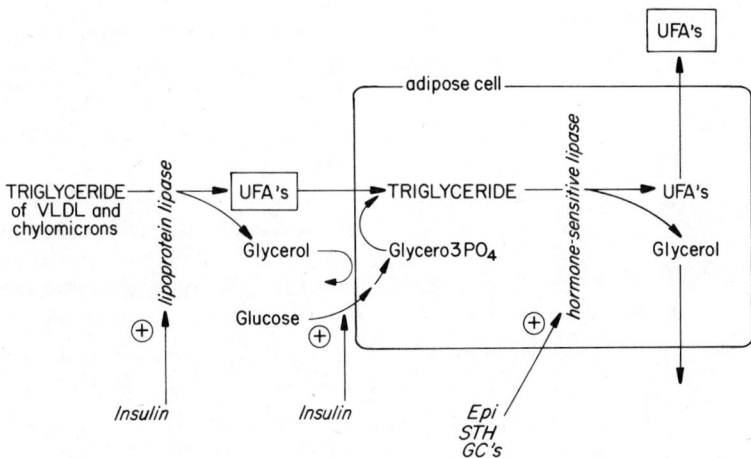

Figure 14 Endocrine regulation of fat metabolism.

hormone-sensitive lipase results in increased release of unesterified fatty acids (UFAs) into the blood, where they reversibly bind to serum albumin. UFAs can be used directly by many tissues (e.g., by striated muscle fibers of heart and skeletal muscle). UFAs can also be used indirectly after being converted to ketones by hepatic metabolism (ketogenesis). UFAs are partially oxidized by liver to acetyl-CoA, which can then enter several hepatic pathways: fatty acid synthesis, TCA cycle, cholesterol synthesis, and ketogenesis (Fig. 15). At such times as those normally resulting in increased mobilization of adipose triglycerides (e.g., in fasting), there is little hepatic fatty acid or cholesterol synthesis. Therefore, most of the acetyl-CoA is divided between the TCA cycle and ketogenesis. When the level of UFA concentration in the blood increases, hepatic uptake and oxidation increase, resulting in a moderate increase in formation of ketones.* The formation of ketones involves two important events: (1) an increase in available fatty acids for substrate (supplied by

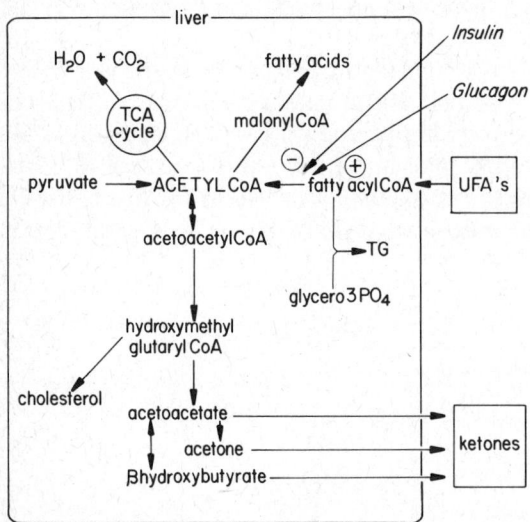

Figure 15 Pathways of acetyl-CoA in liver.

---

*In clinical usage, the terms *ketones, ketone bodies,* and *ketoacids* are used interchangeably to indicate acetoacetate (which is both a ketone and a strong organic acid), β-hydroxybutyrate (which is a strong organic acid, but not a ketone), and acetone (which is a ketone, but not an acid). No one commonly used term accurately describes all three to the satisfaction of chemists.

lipolysis of adipose stores), and (2) a change in the "set" of the liver toward the increased conversion of fatty acids to ketones (and a diminution in the reesterification of fatty acids to triglycerides). Glucagon and insulin have opposite effects on the ketogenic "set" (as if they operated a ketogenic switch that can be switched "on" by glucagon and switched "off" by insulin). The insulin influence is the more powerful of the two; glucagon has little ketogenic influence in the presence of normal amounts of insulin. Only when insulin levels are greatly diminished is glucagon able to increase formation of ketones.

EFFECTS OF FASTING

The surplus fuels provided by feeding result in initiation of anabolic mechanisms for storage (by glycogenesis, protein synthesis, and lipogenesis), while the need for mobilization of these stores during fasting requires initiation of catabolic mechanisms (via glycogenolysis, gluconeogenesis, lipolysis, and ketogenesis). Of all the endocrine events involved in bringing about such changes, *the principal signal for switching from one state of metabolism (anabolism or catabolism) to the other is the level of blood insulin.* Within a few hours after feeding, when absorption has diminished, the insulin level drops from its meal-related peak to a lower, basal level. This decrease in insulin (the primary signal that the body has entered a fasting period) inhibits anabolic processes and initiates catabolic responses necessary to ensure the central nervous system of an adequate supply of glucose. If fasting is prolonged beyond a simple intermeal hiatus, the basal level of insulin decreases and the sensitivity of islet B cells to stimuli for insulin secretion also diminishes. Several contrainsulin hormones with effects opposite to those of insulin are either transiently increased (e.g., glucagon, STH) or remain at near-normal levels. In either event, their actions are less opposed by those of insulin. Thus, the most important endocrine adaptation to fasting and starvation that permits survival over prolonged periods of food deprivation (up to weeks and months) is the persistent decrease in B cell secretion of insulin.

With a decrease in insulin secretion there is little or no synthesis of liver glycogen (since there is a decrease in hepatic glycogen synthetase activity), and there is an increase in glycogenolysis (since there is an increase in glycogen phosphorylase activity). Thus, less G6P moves in the direction of glycogen formation and more is formed from glycogen breakdown. The G6P level in hepatocytes would tend to increase, except that increased G6Pase activity releases more free glucose from the liver into the blood. During an overnight fast, about 80% of the hepatic glucose released into the blood comes from hepatic glycogenolysis. However, the other 20% comes from a different hepatic pathway leading to G6P, namely gluconeogenesis. This not only involves increased activity of key hepatic enzymes for gluconeogenesis, but also arises from an increased supply of amino acids from peripheral sources (mainly from muscle). Recall that insulin has both anabolic and anticatabolic effects on peripheral protein metabolism. With a decrease in insulin, the catabolic and antianabolic effects of GCs predominate and increase the release of amino acids. If fasting is prolonged, most of the glucose produced by liver and released into blood must eventually come from increased gluconeogenesis. At first, the rate of muscle protein breakdown and the rate of gluconeogenesis gradually increase to a peak level and then begin to decline. That gluconeogenesis is not maintained at a high level for more than several days is fortunate, since this would result in loss of muscle function. Instead, demands for glucose begin to decrease. This permits gluconeogenesis to diminish to a less devastating level.

The body's demands for glucose become decreased by several mechanisms. Many tissues able to use fatty acids or ketones, or both, as a source of energy decrease their glucose uptake and utilization. This is due to both a decrease in insulin-dependent facilitated diffusion of ECF glucose down its concentration gradient into peripheral cells (less insulin, less uptake) and to an inhibitory effect of increased ECF fatty acids and ketones on peripheral glucose utilization. The increase in fatty acids that occurs is also the result of diminished insulin secretion. Insulin favors the storage of fat and inhibits fat mobilization. Thus, when insulin decreases, there is decreased lipogenesis and increased mobilization of adipose TGs with increased release of UFAs into blood. Many tissues can directly use UFAs as a source of energy, thereby sparing glucose for the brain. In addition, many UFAs are taken up by the liver, where they are partially oxidized to ketones. A number of tissues can use ketones as sources of energy (e.g., heart and skeletal muscle). Another mechanism that helps reduce demands for blood glucose involves an adaptation of the brain. After a few days of fasting, the brain decreases its obligatory requirements for glucose by one-half. The brain gradually increases in ability to utilize ketones, which results in the glucose-sparing effect.

With fasting, the secretion of glucagon increases (while that of insulin decreases) and raises the ketogenic capability of the liver. This, along with the increased supply of fatty acids from adipose mobilization, results in a gradual increase in ketones. Although the increase in blood UFAs is partly responsible for increased ketone formation, UFAs also inhibit the use of ketone by muscle, and muscle consumption of UFAs increases while that of ketones decreases in prolonged starvation. With a decrease in utilization of ketones by muscle and the aforementioned increased production of ketones by liver, the net result is an increased level of ketones in blood. Teleologically, this may be beneficial in starvation in that it makes more ketones available for use by the brain. In the normal (nondiabetic) subject who is fasting, this rise in ketone levels is self-limited due to negative feedback. Ketones are a stimulus for B cell secretion of insulin (Table 1). Thus, a rise in ketone levels begins to stimulate B cells and puts a brake on any further decreases in insulin secretion in the normal individual. This limits the extent to which ketones can increase and prevents a fatal buildup of strong organic acids in blood in the starving but otherwise normal subject.

In summary, the decrease in insulin that is the principal endocrinological change in fasting results in increased hepatic production of glucose from glycogenolysis and gluconeogenesis, increased amino acid availability from muscle protein, and increased lipolysis and ketogenesis. The decreased use of glucose by muscle and brain eventually reduce demands for glucose. The remaining glucose requirements are met by the liver without calling upon excessive mobilization of muscle protein. The principal energy requirements of peripheral tissues are met by switching to fatty acid oxidation as the primary source of calories. These changes are illustrated in Figure 16.

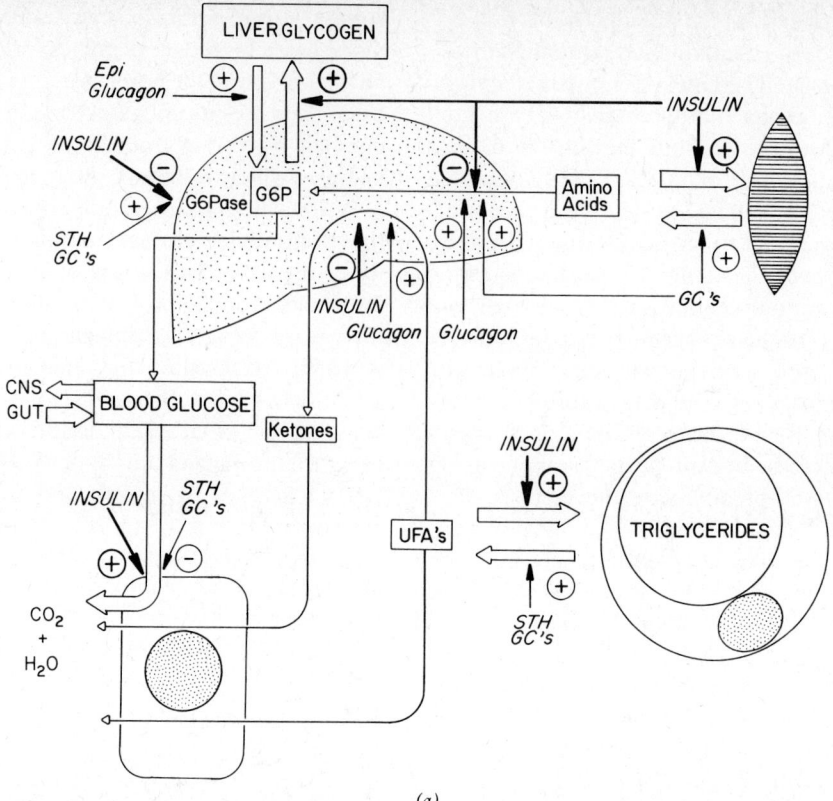

Figure 16 Summary of metabolic effects. *(a)* Normal flow of metabolites in a subject having a predominance of insulin anabolism in the insulin–contrainsulin balance, as in a normal postprandial subject. +, Stimulation; −, inhibition.

DIABETES

Primary idiopathic diabetes is a genetically influenced disorder of metabolic regulation characterized by either a relative or absolute deficiency of insulin. In its fully developed clinical form, diabetes displays acute metabolic disturbances marked by fasting hyperglycemia, muscle wasting, hyperlipemia, and ketosis. Chronic complications include large vessel disease (e.g., arteries) in the form of atherosclerosis, small vessel disease (e.g., capillaries) in the form of renal glomerulosclerosis, and retinal neovascularization and neuropathy of both the somatic and autonomic nervous systems (Fig. 17). Diabetes mellitus is not

a single disease, but a genetically and clinically heterogeneous group of diseases that share glucose intolerance in common. Diabetes mellitus is certainly at least two diseases distinguishable from each other on the basis of severity and insulin need: type I, insulin-dependent diabetes mellitus (formerly juvenile-onset diabetes), and type II, non-insulin-dependent diabetes mellitus (formerly maturity or adult onset diabetes mellitus). Type I, or insulin-dependent diabetes mellitus (IDDM) is usually characterized clinically by abrupt onset of symptoms, insulinopenia, dependence on injected insulin to sustain life, and susceptibility to ketosis and frank ketoacidosis if left untreated. Classically, this type of disease occurs in juveniles, and it was formerly termed juvenile diabetes. However, it can be recognized and become symptomatic for the first

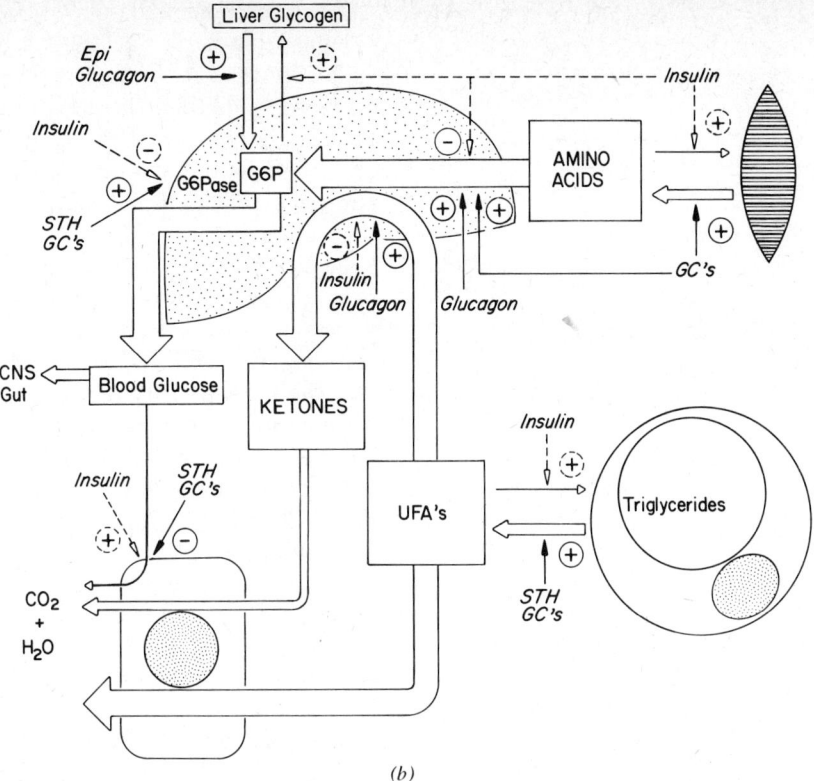

Figure 16 (con't) *(b)* Altered flow of metabolites resulting from loss of insulin–contrainsulin balance due to a relative deficiency of insulin and a predominance of catabolism. These same qualitative changes are seen in both starvation and diabetes, except that those in diabetes are quantitatively greater; for example, much higher blood glucose concentration (hyperglycemia) and higher ketone concentrations (ketosis). Dotted lines illustrate insulin actions that are diminished or absent. The relative rate of handling of metabolites in a given direction is indicated by the width of the arrow for flow in that direction.

time at any age. When left untreated by insulin replacement, IDDM is associated with the acute metabolic disturbances mentioned above and with secondary symptoms that these disturbances cause, such as polyuria, dehydration, polydipsia, polyphagia, cachectic emaciation, fatigue, and ketoacidosis (which can lead to a fatal coma). Fortunately, this more severe form of diabetes accounts for less than 10% of all diabetes.

The great majority of diabetics have the second subclass of diabetes, type II, or non-insulin-dependent diabetes mellitus (NIDDM), formerly called maturity- or adult-

onset diabetes. About 60–90% of NIDDM subjects are obese. Although in most patients in whom NIDDM develops onset is after age 40, the NIDDM type also occurs in a few young persons who do not require insulin and who are not ketotic. Patients with NIDDM are not dependent on insulin for prevention of ketosis. However, they might require some exogenous insulin for correction of symptomatic, or persistent, fasting hyperglycemia if this cannot be achieved with the use of diet or oral agents. In many obese NIDDM subjects, dietary control and weight loss might be sufficient to prevent metabolic derangements.

Specific organ changes

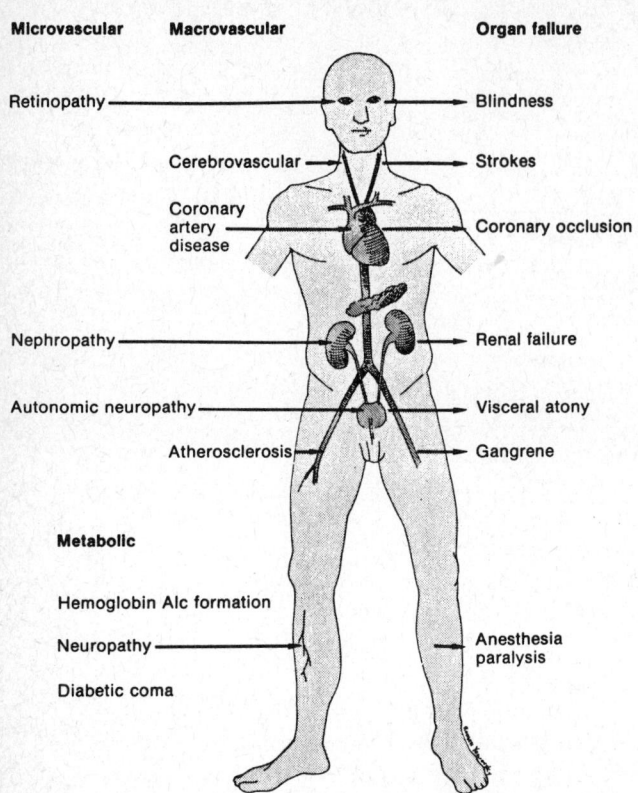

Microvascular   Macrovascular   **Organ failure**

Retinopathy ─────────────────────────── Blindness

          Cerebrovascular ─────── Strokes

          Coronary
          artery ─────────────── Coronary occlusion
          disease

Nephropathy ─────────────────────────── Renal failure

Autonomic neuropathy ──────────────── Visceral atony

         Atherosclerosis ──────── Gangrene

Metabolic

Hemoglobin Alc formation

Neuropathy ──────────────────────── Anesthesia
                                      paralysis

Diabetic coma

Figure 17 Specific organ changes in diabetes. (From Gorman CK in Ezrin C, Godden JO, Volpe R: *Systemic Endocrinology,* ed 2. Hagerstown MD, Harper & Row, 1979.)

TYPE I: INSULIN-DEPENDENT DIABETES MELLITUS

Since type I, (IDDM, insulin-dependent diabetes mellitus) most often appears to be simply the result of an absolute deficiency of insulin, as a consequence of B cell loss or damage (Fig. 4), it is perhaps the easiest to understand in terms of the characteristic acute metabolic disturbances that occur. These metabolic changes appear to be nothing more than a gross exaggeration of the adaptive mechanism for fasting. Diabetes has been described by Tepperman as a ghastly caricature of what occurs in starvation. Recall that in fasting and starvation there is a decrease in insulin, which permits the opposing actions of contrainsulin

effects to become predominant. In IDDM there is severe insulinopenia (insulin deficiency), which permits contrainsulin effects to operate freely without any opposition whatsoever. In addition, there is considerable evidence that glucagon levels are increased. Whether this is due to increased secretion by A cells (possibly as a result of loss of paracrine inhibition of A cells by insulin as mentioned in the section on islet cell interactions), or to decreased metabolic clearance (by liver and kidney), or both, has not yet been clearly established. Although Unger has proposed an interesting theory that diabetes is a bihormonal defect resulting from both decreased insulin and increased glucagon, there is some evidence refuting the significance of the elevated glucagon, and one must consider the issue unresolved at the present time.

It appears that glucagon does have definite contrainsulin effects expressed only in the absence or considerable diminution of insulin secretion. However, since these effects are largely corrected by exogenous insulin replacement, it seems more logical to consider the primary defect to be one of insulin deficiency. When insulin is absent, all the contrainsulin effects of all of the contrainsulin hormones, including glucagon, are freely and fully expressed. Further perusal of Figure 16 should provide the basis for explaining diabetic hyperglycemia, muscle wasting, hyperlipemia, and ketosis as metabolic derangements of IDDM.

Hyperglycemia is the result of liver production of glucose in excess of its peripheral utilization. Liver production is greatly increased by diabetic insulinopenia because of an upset of the balance between glycogen synthesis and glycogen breakdown (resulting in greater breakdown). Furthermore, the stimulatory effect of insulin on glycolysis is lost, as is the inhibitory effect on hormones favoring gluconeogenesis (e.g., GCs and glucagon). Hence, the key enzymes regulating traffic through these pathways are set for accelerated gluconeogenesis. In addition, there is an abundant supply of substrate for gluconeogenesis in the form of increased release of amino acids from muscle protein. In the absence of the retarding effect of insulin on G6Pase, the glucose comes flowing out the liver in great abundance. At the peripheral level, there is decreased uptake and utilization of glucose by the insulin-dependent facilitated transfer system in many tissues (in muscle and adipose but not in liver and brain). As a result of decreased peripheral utilization and increased hepatic production, the blood glucose must increase, hence the development of hyperglycemia, the sine qua non of diabetes.

In the absence of anabolic and anticatabolic actions of insulin on muscle protein, protein catabolism is rampant and results in elevated release of amino acids. These are mainly converted to alanine and glutamate, which are avidly extracted by the liver to supply substrate for accelerated gluconeogenesis. Thus, effects of decreased availability of insulin on muscle protein metabolism are responsible for sustaining the excessive hepatic production of blood glucose. Most of this excess glucose is wasted because it is lost from the body in urine. Excessive levels of glucose in the plasma result in equally excessive levels in the glomerular filtrate. When the rate of filtered glucose exceeds the capacity of the proximal renal tubules for reabsorption (when the filtered load is greater than the T_m, or tubular maximum for reabsorption), glucose spills over into the excreted urine. This loss of valuable nutrient fuel is the least of problems related to glucosuria. The principal problem is the osmotic diuresis that is produced by the presence of all that glucose in the urine. This causes loss of fluid and electrolytes and results in dehydration and hypovolemia.

Whereas muscle wasting and dehydration are serious, unfortunate circumstances that add considerably to patient deterioration, they are not nearly the acute threat to life that ketoacidosis is. A prominent consequence of insulinopenia is lipolysis of adipose triglycerides. This raises the concentration of UFA's and increases their uptake by the liver. Simply raising the concentration of UFA's favors increased hepatic oxidation of fatty acids to ketones. In addition, increased levels of glucagon (along with decreased insulin) may enhance this process by increasing the transport of fatty acids across hepatocyte mitochondrial membranes. In any case, there is increased production of ketones. In a normal nondiabetic who has increased production of ketones (as in fasting), there is a negative feedback mechanism that prevents excessive ketogenesis (via ketone stimulation of B cell secretion of insulin). This negative feedback inhibition of ketogenesis by insulin is inoperative in the type I juvenile diabetic who has no B cells to secrete insulin. As fat mobilization continues to supply more fatty acids, the capacity for ketogenesis continues to increase, resulting in increased keto-

nemia. Furthermore, the diabetic situation with ketones in blood is very similar to that of glucose in blood: not only is there excess production, but also decreased utilization. Since two of the so-called ketones are strong organic acids, an accumulation in blood produces a metabolic acidosis (ketoacidosis). Before the availability of insulin, ketoacidosis was the primary cause of death in IDDM.

Not all the UFAs mobilized from adipose stores are converted to ketones by the liver. Despite a prominent oxidation by hepatocyte mitochondria, many escape this fate and instead are reesterified to form triglycerides. These are partly stored in the liver and might explain why fatty livers are not uncommon in diabetes. Many are released into the blood as VLDLs and cause hyperlipidemia. Of the fatty acids oxidized by hepatocyte mitochondria, the predominant product is acetyl-CoA. Much goes into the formation of ketones as described above and illustrated in Figure 15. However, some of this excess acetyl-CoA production ends up in cholesterol synthesis, resulting in increased hepatic release of cholesterol into the blood as low-density lipoprotein (LDL). This hypercholesterolemia contributes further to the hyperlipidemia of diabetes. Whereas the development of ketosis is the most serious acute metabolic complication of fat metabolism in diabetes, the hyperlipidemia is thought to be of considerable significance in the chronic complication of increased incidence of atherosclerosis.

TYPE II: NON-INSULIN-DEPENDENT DIABETES MELLITUS

This form of diabetes usually begins in middle-aged or older people; however, it is seen in a few young persons who do not require insulin and who are not ketotic. Some of these cases represent maturity-onset diabetes of the young (MODY), inherited as an autosomal dominant. Symptoms tend to begin more gradually than in the typical juvenile form (type I), and often the diagnosis is made when an asymptomatic patient is found to have an elevated plasma glucose on routine laboratory examination.

Although type I (IDDM) is characterized by a grossly subnormal or absent insulin secretory response to glucose, insulin responses in type II (NIDDM) are quite variable. There may be normal levels of insulin, mild insulinopenia, or above-normal levels of insulin associated with insulin resistance. That resistance to insulin action may be the result of target cell hyporesponsiveness to insulin as a result of a receptor defect, has been indicated by measuring the binding of radioactively labeled insulin molecules on target cells from type II (NIDDM) subjects. The number of insulin receptors in such patients has been found to be less than that for normal subjects. Thus, to obtain adequate insulin action, greater than normal amounts of insulin are required. Interestingly, when type II diabetics reduce their weight by diet control, the number of insulin receptors increases towards normal along with improvement in glucose tolerance. Furthermore, in addition to receptor defects, postreceptor defects have recently been implicated in the insulin resistance seen in type II. Further knowledge of insulin action is necessary to delineate the specific defects in postreceptor events that can contribute to the insulin resistance and glucose intolerance of NIDDM.

Thus, despite ability to secrete insulin in response to a glucose load (either ingested or infused intravenously), these obese type II insulin-resistant patients develop many of the same acute metabolic derangements as do type I juvenile-onset diabetics, except that they are not prone to

develop ketoacidosis. A hyperglycemic dehydration syndrome, termed nonketotic hyperosmolar coma, is the most threatening acute complication in some very hyperglycemic patients with nonketosisprone diabetes (and is most frequently observed in the middle-aged). These patients characteristically develop the classic manifestations of symptomatic hyperglycemia, polyuria, and polydipsia that increase in intensity and persist for several days to weeks before they seek medical attention or before the significance of their syndrome is appreciated. Patients with hyperglycemic dehydration are an extremely heterogeneous group. However, they do share certain features in common—that impaired consciousness (including coma) and shock can occur in association with marked hyperglycemia and dehydration, but in the absence of marked hyperketonemia and metabolic acidosis. The mortality in such patients with this syndrome occurs most commonly in type II adult-onset diabetics. There is general agreement that in these patients hypergiycemia results in a persistent osmotic diuresis that eventually causes a marked dehydration with significant deficits of Na⁺ and K⁺ and marked contraction of total body water and circulating extracellular fluid volume.

Two possible explanations for the absence of marked

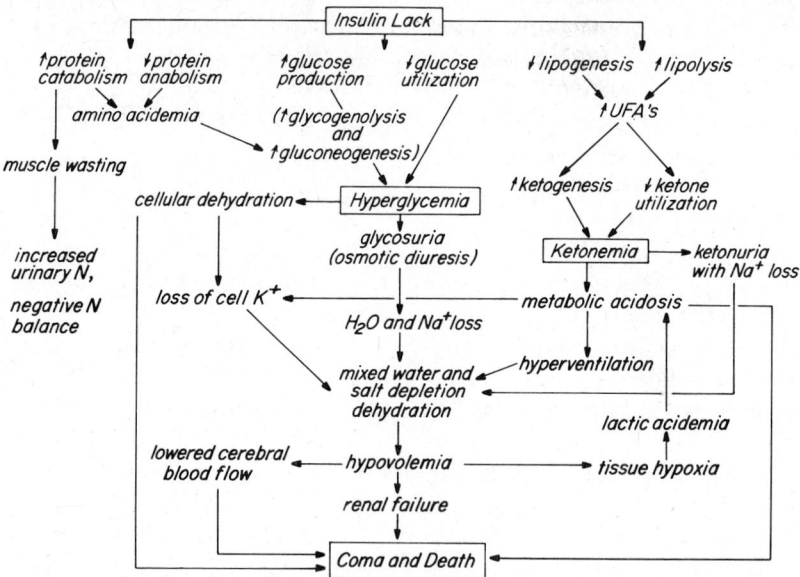

Figure 18 Summary of acute metabolic derangements in diabetes. (Modified and reproduced with permission from Tepperman, J: *Metabolic and Endocrine Physiology*, ed 4. Chicago, Year Book Medical Publishers, 1980.)

ketonemia in type II (maturity-onset) diabetics are (1) lower concentrations of blood UFAs (UFA levels are two or three times higher in ketosis prone patients), and (2) the portal blood of type II diabetics has a higher concentration of insulin, which can prevent full activation of the hepatic "set" for ketogenesis (via carnitine acyltransferase activity).

ETIOLOGY

Diabetes of unproved origin (most diabetics) is termed idiopathic. Idiopathic diabetes is apparently caused by some combination of (1) genetic predisposition, (2) destruction of B cell function and/or, (3) receptor-mediated insulin resistance. There is either an absolute or relative deficiency in insulin. That is, in type I (IDDM) there is usually an absolute deficiency of insulin due to inadequate or even total absence of insulin secretion; in obese, type II (NIDDM) subjects insulin is secreted (sometimes in greater than normal amounts), but is a relative deficiency due to increased tissue resistance to insulin action (primarily from a decrease in the number of receptors).

Idiopathic diabetes is distinguished from secondary types of diabetes resulting from (1) destruction of the pancreas by disease (pancreatitis, hemochromatosis, pancreatic cancer), (2) surgical excision, or (3) endocrine disease involving excess secretion of contrainsulin hormones (e.g., excess growth hormone, corticosteroids, glucagon, catecholamines, or somatostatin).

The precise cause of idiopathic diabetes is unknown, but several theories are currently popular. Much interest has recently been generated by evidence for destruction of islet tissue as a result of viral infection, particularly as a possible cause of type I (IDDM).

It has long been known that diabetes tends to run in families. The likelihood that a person will develop diabetes is greatly increased if one or both parents are diabetic, particularly for the development of type II (NIDDM).

SUMMARY

A diagram illustrating the acute metabolic derangements of protein, carbohydrate, and fat metabolism, and their interrelationships in diabetes is shown in Figure 18.

SELECTED BIBLIOGRAPHY

Burman D, Holton JB, Pennock CA(eds): *Inherited Disorders of Carbohydrate Metabolism.* Baltimore, University Park Press, 1980.

DeGroot LJ, Cahill GF Jr, Martin L, et al (eds): *Endocrinology,* Section on Pancreatic Islets. New York, Grune & Stratton, 1979, p 907.

Ezrin C, Godden JO, Walfish PG (eds): *Clinical Endocrinology, A Survey of Current Practice.* New York, Appleton-Century-Crofts, 1977.

Erzin C, Godden JO, Volpe R: *Systematic Endocrinology* ed 2. Hagerstown MD, Harper & Row, 1979, p 18.

Foster DW: Diabetes mellitus, in Isselbacher KJ, Adams RD, Braunwald E, et al (eds): *Harrison's Principles of Internal Medicine.* Baltimore, University Park Press, 1980, p 1741.

Katzen HM, Mahler RJ (eds): *Diabetes, Obesity and Vascular Disease. Advances in Modern Nutrition,* vol 2. New York, John Wiley & Sons, 1978.

Renold AE, et al: in Stanbury JB, Wyngaarden JB, Fredrickson DS (eds): *The Metabolic Basis of Inherited Disease.* New York, McGraw-Hill, 1978.

Tepperman J: *Metabolic and Endocrine Physiology,* ed 4. Chicago, Year Book, 1980, p 225.

Volk BW, Wellman KF (eds): *The Diabetic Pancreas.* New York, Plenum Press, 1977.

West KM: *Epidemiology of Diabetes and Its Vascular Lesions.* New York, Elsevier-North-Holland, 1978.

Porte D Jr and Halter JB in Williams RH (ed): *Textbook of Endocrinology,* ed 6. Philadelphia, WB Saunders, 1981.

SELF-STUDY QUESTIONS

MULTIPLE CHOICE

Select the single best answer.

1. The number of reported, known diabetics in this country is four to five million. What is the estimated number of unknown, undiagnosed diabetics?
   A. None. Essentially all diabetics have been diagnosed.
   B. Very few. Most diabetics are easily recognized by their symptoms.
   C. Another four to five million. There are about as many undiagnosed diabetics as diagnosed.
   D. About 20 million in all. There are many more undiagnosed diabetics than those known to have the disease.
   E. More than 200 million. Essentially everyone gets it sooner or later.

2. Which of the following stimulate insulin release from B cells?
   A. D-Arginine
   B. D-Leucine
   C. α-Aminoisobutyric acid (α-AIB)
   D. All of the above
   E. None of the above

3. An abrupt marked increase in blood glucose maintained for about a half-hour results in
   A. a minute, almost imperceptible rise in insulin release unless the increase in blood glucose is maintained for a much longer period of time.
   B. a brief monophasic (single peak) release of insulin.
   C. a biphasic release of insulin (a brief peak followed by a second increased and sustained release).
   D. a triphasic release of insulin (a brief peak followed by a second sustained release and then a sharp decrease below the base level of insulin secretion).
   E. a biphasic release of glucagon (a brief peak followed by a second increased and sustained release).

4. A junctional complex found between islet cells that appears to form isolated channels within the interstitial space (possibly for limited distribution of paracrine secretions from one islet cell to another) is the
   A. desmosome.
   B. gap junction.
   C. tight junction.
   D. nexus (gap junction).
   E. adnexus.

MATCHING

A. *A cells of islets*
B. *B cells of islets*
C. *C cells of islets*
D. *D cells of islets*

5. Cells that secrete a small protein (large polypeptide) hormone of 51-amino acids and a molecular weight of about 6,000.

6. Cells that secrete a large polypeptide hormone of 29-amino acids and about 3,500 molecular weight.

7. Cells that secrete a 14-amino acid peptide originally discovered as a hypothalamic hormone affecting secretion of growth hormone by the anterior pituitary gland.

MATCHING

In normal, healthy adult B cells

A. *Transitional microvesicles (MV)*
B. *Rough endoplasmic reticulum (RER)*
C. *Golgi cisternae*
D. *Early (immature) secretory granules*
E. *Mature secretory granules*

8. Site of disulfide bridge formation within proinsulin molecule.

9. Site of preproinsulin formation.

10. Primary site of conversion of proinsulin to insulin (by membrane-bound proteases).

MATCHING

A. *Increased secretion*
B. *Decreased secretion*
C. *No effect on secretion*

11. Effect of D-glucose on A cell function.
12. Effect of D-glucose on B cell function.
13. Effect of endogenous catecholamines on B cell function (acting on α-adrenergic receptors).
14. Effect of somatostatin on both A and B cell function.
15. Effect of insulin on A cell function.
16. Effect of L-arginine on A cell function.

MATCHING

In a normal, healthy subject

A. *Glucagon only*
B. *Insulin only*
C. *Both glucagon and insulin*
D. *Neither glucagon nor insulin*

17. Derived from a larger molecular-weight precursor.
18. Secretion increases following a high protein meal.
19. Increased secretion stimulated by somatostatin.
20. Secreted by exocytosis (emiocytosis).
21. Normally goes up and down several times daily in relation to intake of average meals (containing typical proportions of carbohydrate, fat and protein).

MULTIPLE CHOICE

Select the correct answer(s). (In many instances, more than one answer is correct.)

22. Agents that alter the integrity of B cell microtubules and thereby inhibit glucose-induced release of insulin are
    A. colchicine.
    B. vincristine.
    C. vinblastine.
    D. tolbutamide.
23. In the stimulation of insulin release from B cells in an isolated, perfused pancreas,

A. any agents increasing intracellular cyclic 3'5' AMP (e.g., glucagon) will trigger the secretory events leading to insulin release, even in the absence of glucose.
B. any agents increasing intracellular Ca^{2+} (e.g., the administration of a membrane carrier ionophore for divalent ions) will trigger the secretory events leading to insulin release even in the absence of glucose.
C. the presence of Ca^{2+} is not essential.
D. agents that alter the integrity of the cellular microtubular system (e.g., colchicine) inhibit glucose-induced secretion of insulin.

24. Which of the following pharmacological agents should be expected to increase B cell release of insulin?
    A. Isoproterenol (a β-adrenergic agent)
    B. Parasympathomimetic agents (agents with same effects as acetylcholine)
    C. Norepinephrine plus an α-adrenergic blocking agent
    D. Norepinephrine plus a β-adrenergic blocking agent

MATCHING

A. *Activity increased by insulin action*
B. *Activity decreased by insulin action*
C. *Activity not directly influenced by insulin action*

25. Key unidirectional glycolytic enzymes of liver cells (glucokinase, phosphofructokinase, and pyruvate kinase).
26. Entry of D-glucose into muscle and adipose cells by facilitated diffusion.
27. Ribosomal assembly of amino acids into protein by muscle cells.
28. Lipoprotein lipase of adipose capillary endothelia.
29. Hepatic mitochondrial uptake of fatty acyl-CoA (via carnitine acyltransferase activity) for oxidation to acetyl-CoA and ketones.

MULTIPLE CHOICE

Select the single *incorrect* choice.

30. Of the following hormonal effects, which one is *not* opposite to the effect produced by insulin?

A. Glucocorticoids have a protein catabolic effect on skeletal muscle.

B. Glucocorticoids favor increased activity of glycogen synthetase.

C. Glucagon favors hepatic gluconeogenesis.

D. Growth hormone increases the activity of adipose hormone-sensitive lipase.

E. Glucocorticoids increase the activity of key unidirectional gluconeogenic enzymes of liver cells.

31. Which of the following statements about fasting is *incorrect?*

A. The principal signal for switching from anabolism to catabolism is a decrease in insulin secretion.

B. Several contrainsulin hormones with effects generally opposite to those of insulin are either transiently increased (e.g., glucagon, growth hormone) or remain near normal levels in starvation.

C. In starvation, the body's demands for glucose become decreased.

D. In starvation, there is both an increased supply of unestrerified fatty acids to the liver plus an increased hepatic ketonegic ability (resulting in a gradual increase in the output of ketones).

E. The consumption of ketones by skeletal muscle continues to increase throughout the duration of starvation, even if prolonged.

32. Which one of the following statements about type I (IDDM) (juvenile-onset diabetes) is *incorrect?*

A. It is more common in younger people than in middle-aged or elderly people.

B. Ketoacidosis is a common acute metabolic complication of insufficient insulin replacement therapy.

C. It constitutes about 90% of all diabetic cases in the United States today.

D. In most cases, it cannot be managed by diet alone.

E. It is usually the result of deficiency of insulin secretion.

ANSWERS

1. C. Another four to five million. It is estimated that the number of unknown diabetics is approximately equal to that of the known.

2. C. α-Aminoisobutyric acid (AIB) is a nonmetabolizable L-amino acid that nevertheless stimulates insulin secretion. Several other nonmetabolizable amino acids have been found to stimulate insulin secretion. That these nonmetabolized L-amino acid analogs can stimulate insulin release without being converted to intermediary metabolites is a point in favor of the idea that islet cells recognize certain stimulant molecules through specific receptors. A and B are incorrect because B cells are not stimulated to secrete insulin by D-amino acids.

3. C. The normal response to an abrupt marked increase in blood glucose that is maintained at a high level for a half-hour or more is a biphasic release of insulin (Fig. 8). As soon as the blood glucose returns to normal, the insulin secretion returns to its previous level (the basal level).

4. C. Tight junctions (zona occludens in cavitary epithelia) between islet cells have been shown (by freeze-fracture technique) to fence off specific intercellular channels. A is incorrect. Desmosomes (maculae adhaerentes) are found in many epithelia. They serve in adhesion of cells to one another. B is incorrect. Gap junctions contain subunits within the narrow gap between adjacent cell membranes and appear to mediate electrotonic coupling between cells. They can also permit cellular interchange of small molecular weight

compounds. D is incorrect. Nexus is just another name for gap junction. E is incorrect. Adnexus means appendage and is not a junctional complex—the most common appendages referred to as adnexa are the uterine appendages: ovaries, fallopian tubes, and uterine ligaments.

5. B. Insulin is the hormone described (51 amino acids, molecular weight of about 6,000); it is secreted by the B islet cells.

6. A. Glucagon is the hormone described (29 amino acids, molecular weight of about 3,500); it is secreted by the A islet cells.

7. D. Somatostatin is the hormone described (14 amino acids; originally discovered in hypothalamus); it is found in D islet cells.

8. B. The RER is the site at which disulfide bridge formation occurs within the proinsulin molecule.

9. B. The ribosomes of the RER are the site of preproinsulin formation.

10. C. Most of the conversion of proinsulin to insulin occurs in the Golgi complex (by membrane-bound proteases).

11. B. Decreased secretion. The secretion of glucagon by A cells in inhibited by D-glucose (just the opposite of what D-glucose does to B islet cell secretion).

12. A. The primary stimulus for insulin secretion by B islet cells is D-glucose. On an equimolar basis (molecule for molecule), D-glucose is the single most potent stimulus for B cell secretion.

13. B. Endogenous catecholamines (epinephrine and norepinephrine) act on B cell α-adrenergic receptors to inhibit insulin secretion (Table 1).

14. B. Somatostatin inhibits many functions, including glucagon secretion by A islet cells and insulin by B islet cells (i.e., both are inhibited).

15. B. Insulin has an inhibitory effect on A cell secretion of glucagon. Thus, in diabetes due to a deficiency of insulin there tends to be an excess of glucagon.

16. A. L-Amino acids stimulate glucagon secretion by A cells. L-Amino acids also stimulate insulin secretion by B cells. The sense of it all is that a high-protein meal that only stimulated insulin secretion would possibly result in a severe hypoglycemia. The concomi-

tant stimulation of glucagon secretion counters the hypoglycemic action of an increase in insulin secretion in the absence of significant glucose absorption from the gut (i.e., with a high-protein meal). Although insulin might increase glucose loss from the blood (into peripheral cells), glucagon would increase glucose addition to the blood (from hepatic glycogenolysis and gluconeogenesis).

17. C. Both insulin and glucagon are derived from larger-molecular-weight precursors (insulin from proinsulin and glucagon from proglucagon).

18. C. Both insulin and glucagon secretions are stimulated by a high-protein meal, as discussed above in the answer to question 16.

19. D. Neither is stimulated by somatostatin (they are inhibited).

20. C. Both glucagon and insulin (also a number of other hormones) are secreted by exocytosis (emicytosis).

21. B. With a normal intake of average meals, there is little or no fluctuation of glucagon secretion, but there are considerable increases and decreases in insulin.

22. A, B, C. Colchicine, vincristine, and vinblastine all interfere with microtubular function and, thereby, inhibit B cell secretion of insulin. D is incorrect. Tolbutamide is a sulfonylurea that stimulates insulin secretion.

23. B,D. Increasing intracellular Ca^{2+} does trigger secretion, even in the absence of glucose. Agents interfering with microtubular integrity inhibit glucose-induced secretion of insulin (see answer to question 22 above). A would be correct if cyclic AMP were to act as a second messenger (as it does in many cells). However, if this were the case, glucagon stimulation of B cell cyclic AMP would initiate the events that culminate in insulin secretion whether glucose were present or not. But this is not the case, and glucagon does not stimulate insulin secretion in the absence of glucose. C is incorrect. Ca^{2+} is essential (this is true for many endocrine glands).

24. A, B, C. Isoproterenol, an exogenous catecholamine with only β-adrenergic effects, stimulates insulin release (Table 1). Parasympathomimetic agents (mimic acetylcholine) stimulate insulin release (an action

which can be blocked by atropine). The principal endogenous catecholamines (epinephrine and norepinephrine) have both α- and β-adrenergic effects. When administered by themselves, the α effects on B cells predominate and inhibit insulin secretion. However, when given with an α-blocking agent, the β-adrenergic effects are unmasked and cause stimulation of insulin secretion. D is correct. The combination of norepinephrine and a β-adrenergic blocking agent would further enhance the already predominant α-inhibitory effect on B cells and decrease insulin secretion.

25. A. Insulin favors hepatic glycolysis by increasing the activities of the key unidirectional glycolytic enzymes (glucokinase, phosphofructokinase, and pyruvate kinase) (Fig. 12).

26. A. One of the earliest discovered roles of insulin is stimulation of glucose uptake (by facilitated diffusion) into adipose and skeletal muscle tissues.

27. A. Insulin not only increases the uptake of amino acids by muscle fibers, but also increases ribosomal assembly of the amino acids into protein (Fig. 13).

28. A. Adipose tissue lipoprotein lipase activity (located in the endothelial cells of adipose tissue capillaries), which is necessary for transfer of triglyceride from blood into adipose cells, is increased by insulin. That is, insulin favors adipose fat storage. At the same time, tissue hormone-sensitive lipase activity, which is necessary for mobilization of stored fat, decreases in activity. In other words, these two enzymes, which affect opposite aspects of the handling of triglyceride, are reciprocally related.

29. B. Insulin suppresses hepatic ketogenesis (is said to turn off the "switch" for ketogenesis). Rapidly acting insulin preparations suitable for intravenous injection are used for treatment of diabetic ketoacidosis. The markedly increased hepatic ketogenesis, the primary cause of ketonemia, is a consequence of both increased uptake of unesterified fatty acids by the liver and an increased transport of coenzyme A derivatives of these fatty acids (fatty acyl-CoA) across the inner mitochondrial membrane (via carnitine acyltransferase activity). A greatly enhanced rate of hepatic oxidation of these long-chain acyl-CoA derivatives now appears to be a major factor in the increased rate of ketogenesis. Insulin therapy not only decreases peripheral mobilization of UFAs from adipose stores, but also decreases this hepatic mitochondrial uptake and oxidation of the fatty acyl-CoA to ketones (possibly by decreasing the heaptic concentration of carnitine).

30. B. Various contrainsulin hormones do not necessarily act opposite to insulin in all aspects of insulin action. Some may even have one or two actions that are similar to those of insulin. Glucocorticoids are an example of a contrainsulin hormone that has many actions opposite to the insulin, but one that is similar: increased activity of glycogen synthetase in liver cells.

31. E. As fasting is prolonged and UFA use by skeletal muscle increases, the use of ketones by muscle decreases. Although excess hepatic ketogenesis is the principal cause of increased concentration of blood ketones, diminished use by muscle contributes as well.

32. C. The traditional classification of diabetes as type I (IDDM or juvenile-onset) and type II (NIDDM or maturity-onset) is probably acceptable as a clinical convenience; however, it should be recognized that the descriptions are generalizations and that overlap exists. For example, a relatively mild insulin-dependent maturity-onset type of diabetes can occur in the very young, and a severe, insulin-dependent ketosis-prone diabetes can begin in middle-aged persons. Nevertheless, from an overall statistical point of view, about 10% of all diabetics are those with the type I juvenile-onset, insulin-dependent, ketosis-prone diabetes. The remainder (90%) have the type II maturity-onset, insulin-independent, nonketosis-prone type of diabetes.

49

Adrenal Glands

CHARLES E. HALL

OBJECTIVES

After completion of this chapter, you should be able to:

1. Explain how the catecholamines are synthesized, stored, transported, and inactivated.

2. Identify the types of activity displayed by epinephrine and norepinephrine.

3. Understand the essential differences between glucocorticoid and mineralocorticoid hormones as regards their location within the gland, as well as their control mechanisms, target organs, transport, and physiological effects.

4. Have an appreciation of the various clinical syndromes that develop because of adrenocortical hyperactivity or deficiency, associate the abnormalities with specific hormone activity, and identify the nature of the target organ dysfunction.

The purpose of this chapter is to (1) convey an understanding of the location and formation of epinephrine and norepinephrine within the adrenal medullary cells; (2) describe the synthe-sis, storage, release, and inactivation of the two catecholamines; (3) describe the stimuli that elicit release of the respective catecholamines and consider the physiological and metabolic responses brought about by the release of these hormones; (4) describe the physiological functions subserved by mineralocorticoid and glucocorticoid steroid hormones from the adrenal cortex; (5) consider the control mechanisms that govern mineralocorticoid and glucocorticoid hormone secretion, and how they are brought into play; (6) depict the effects of glucocorticoid hormones, or their lack, on fat, protein, and carbohydrate metabolism; (7) identify the action of glucocorticoid hormones on the course of various disease processes; (8) consider the interplay among aldosterone, Na^+ and K^+ metabolism, and the renin–angiotensin system; (9) consider the formation and role of adrenocortical sex hormones; and (10) illustrate the hormonal effects, or their lack, by a consideration of various clinical syndromes.

The paired adrenal glands consist of an inner medullary core of modified ganglion cells that are derived embryologically from the neural crest and termed pheochromocytes, and a surrounding shell of epithelioid cells constituting the cortex. In humans, the paired adrenal glands lie at the superior pole of the kidneys, the right adrenal being adherent to the liver and inferior vena cava and the left lying entirely free. Hormones of the medulla (i.e., epinephrine, norepinephrine, and perhaps dopamine) are water-soluble catecholamines, whereas those of the cortex are lipid-soluble steroids. Medullary cells have an affinity for the chromate ion and are therefore said to be chromaffin. A variety of evidence indicates that there are two types of cells present. Both possess the enzymes that convert tyrosine to norepinephrine, but only one contains the enzyme phenylethanolamine-N-transferase (PNMT), which methylates the terminal amine of norepinephrine to yield epinephrine, that is, epinephrine can be synthesized only by these cells (Fig. 1).

Not only are there two cell types, and hence at least two hormones, but there are believed to be at least two receptors, α-receptors and β-receptors, for them in responding tissues. Some tissues lack either or possess only one, whereas others have both. These adrenergic receptors vary in abundance in different tissues—and in the same tissue from different animals—and also in their affinity for and sensitivity to epinephrine and norepinephrine.

Each of the hormones is confined to granules contained within cytoplasmic vesicles that can be differentially stained; upon appropriate nerve stimulation the vesicles rupture, discharging their contents into cellular cytoplasm, from whence it diffuses into the adjacent capillaries. Evidently the type of nerve stimulation that elicits norepinephrine secretion differs from that inducing epinephrine secretion. Hemorrhage, which is best combatted by vasoconstriction, causes predominantly norepinephrine secretion, whereas hypoglycemia, which is corrected by glycogenolysis and insulin suppression, causes chiefly the secretion of epinephrine. Thus the two hormones may cause similar or dissimilar responses, have different effects at different doses, or contrary effects in different animals or on different tissues in the same animal, which complicates an understanding of the physiological role of the catecholamines.

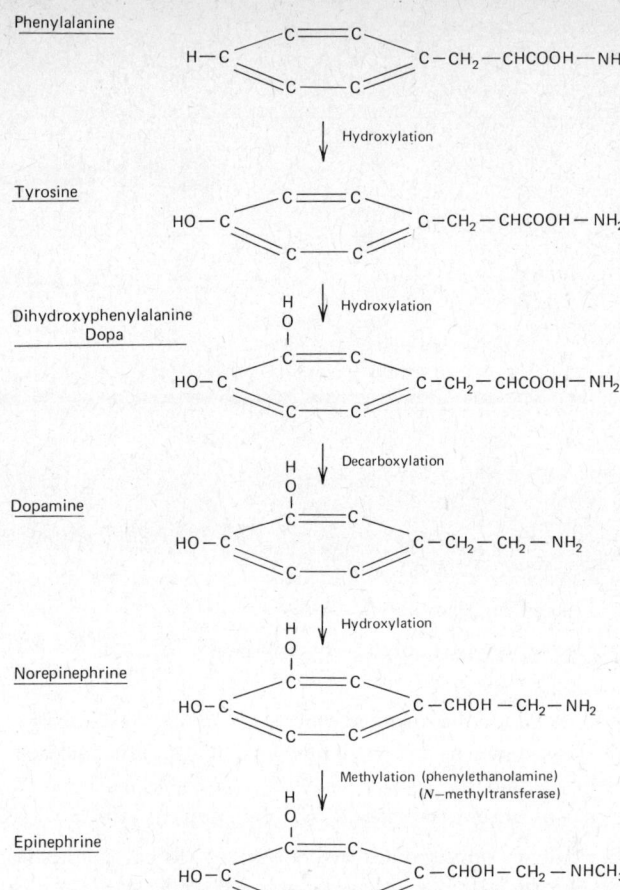

Figure 1 The synthesis of adrenal catecholamines.

PHYSIOLOGICAL ROLE OF CATECHOLAMINES

Denervation of the adrenal medulla causes atrophy of the chromaffin cells, which cease to function. Conversely, stimulation of the sympathetic nerves causes the medullae to secrete both epinephrine and norepinephrine. The former constitutes about 80% and the latter about 20% of the catecholamines secreted, under resting conditions, but these percentages can vary. Medullary denervation causes no catastrophic physiological changes. This is because the medulla acts as a central ganglion of the autonomic system and as a reserve or backup for the sympathetic nervous system. As long as the rest of the system is functioning well, the reserves are rarely necessary, and certainly not essential, to normal function. Some studies have shown that demedullated animals show a delayed recovery from hypoglycemia, but recovery does occur. When hormone-secreting tumors develop in these glands, however, hyperfunctional syndromes develop, with conspicuous signs and symptoms.

The sympathetic nervous system is activated by environmental changes that dictate "fight or flight" strategies; epinephrine is unusual among hormones in that both physiological and emotional changes are immediate consequences of its intravenous injection. An injection of epinephrine provokes a feeling of apprehension. Thus when confronted by danger—as our ancestors might have when they chanced unannounced upon a saber-toothed tiger—flight is an appropriate response. But while the sight of the tiger or the sound of its approach might provoke feelings of fear and facilitate speedy departure, once one's back is turned, and the threat is no longer visible, fear could conceivably diminish prematurely. However, epinephrine secreted at the initial encounter serves to reinforce the desire to be elsewhere without undue delay. Clearly, this would have survival value. The ability to flee or fight is enhanced by the catecholamines, which cause vasoconstriction in the skin and gastrointestinal arterioles and dilation in those supplying skeletal muscle (at least transiently), thus rerouting blood from areas in which its presence is not immediately crucial, to those where great need exists. The overall effect of epinephrine is vasodilation, which would cause blood pressure to fall were it not for the increased heart rate (chronotropic effect) and force of ventricular contraction (inotropic effect), which tend to cause it to rise. Dilation of coronary vessels assures adequate perfusion of the heart muscle, promoting maximal delivery of oxygen and metabolic substrates. Catecholamines also cause contraction of the spleen, which has an exceptionally high hematocrit value, forcing stored erythrocytes into the circulation. Increased muscular work increases heat production, which, if the skin vessels were constricted, would cause elevation of body temperature; however, the accumulation of local metabolites, due both to decreased perfusion and to the direct effect of rising temperature, causes dilation, thereby offsetting the vasoconstrictive hormonal effects.

Simultaneously, the catecholamines promote glycogenolysis in liver and muscle, thereby raising blood glucose and affording increased energy supplies. They also dilate the pupils, ensuring a full field of view, and the bronchiolar musculature, facilitating respiration.

METABOLIC EFFECTS OF THE CATECHOLAMINES

Both epinephrine and norepinephrine have significant metabolic effects, that of the former being many times the greater. Increased levels of epinephrine from pheochromocytomas may double the basal metabolic rate.

Epinephrine promotes hepatic glycogenolysis by activating the enzyme phosphorylase. Both catecholamines also increase intestinal absorption of glucose, and both inhibit insulin secretion (an α effect). Thus, in a time of crisis, adequate glucose is made available to the central nervous system (CNS), heart, and skeletal musculature. At the same time, oxygen consumption is increased. Both catecholamines also activate hormone-sensitive lipase within adipose cells, increasing lipolysis and the release of unesterified fatty acids (UFAs). This substrate is actively metabolized by both skeletal and heart muscle, and shortly after UFAs begin to rise, oxygen consumption increases because of a rising metabolic rate. This is the calorigenic effect.

HORMONE BIOSYNTHESIS

Hormone production begins with L-tyrosine, which is either synthesized by medullary cells or brought to them via the bloodstream. Tyrosine hydroxylase (TH) effects metahydroxylation, yielding dihydroxyphenylalanine (DOPA), which is then decarboxylated to yield dopamine. The latter is attacked by dopamine β-oxidase (DBA), giving rise to norepinephrine, which is β-hydroxylated dopamine (Fig. 1). Two of these steps, principally the first, are rate limiting points at which hormone synthesis can be naturally or therapeutically regulated. The subpopulation of cells that contain PNMT then methylate the terminal amino group, yielding epinephrine. Several alternative biosynthetic pathways have been postulated and might be operant in infrahuman mammals; however, there is no good evidence of their presence in humans.

The norepinephrine formed by tissues other than the adrenal medulla, and by all the medullary hormone-secreting cells (in contrast to epinephrine that is formed to all intents and purposes under normal circumstances only by certain specific cells in the medulla) constitutes the best candidate for feedback control. Rising concentrations of free norepinephrine within cells inhibits the formation of both tyrosine and tyrosine hydroxylase, and falling concentrations have the opposite effect.

Both catecholamines exert prompt responses within seconds of intravenous administration. The duration is equally brief, since upon reaching the liver, both are rapidly destroyed by the enzyme catechol-o-methyl transferase (COMT).

PHEOCHROMOCYTOMAS

These are tumors of the adrenal medullary cells. As such they may secrete excessive quantities of either or both of the catecholamines. When primarily epinephrine is secreted the result has been described as a metabolic syndrome, characterized by tachycardia, flushing, hyperglycemia, systolic hypertension, pupillary dilation, sweating, and other symptoms. On the other hand, when only norepinephrine is secreted excessively, a "hypertensive syndrome" characterized by elevation of both systolic and diastolic blood pressures ensues. Most commonly, both hormones are secreted excessively, although one or the other can predominate. The symptomatology is therefore variable. Hormone secretion is generally not a continuous process, but is rather sporadic, brought about by the same stimuli that would increase normal catecholamine secretion. Paroxysmal attacks are the rule. The disorder is amenable to surgical correction.

THE ADRENAL CORTEX

The adrenal glands oocupy a unique position in the endocrine system, being the only glands necessary for life in all vertebrate species. The medulla is dispensable, but the cortex is not. The outer essential part of the adrenals (the cortex) consists of glandular epithelium of mesodermal origin, which synthesizes and secretes steroid hormones. Unlike the adrenal medulla, the cortex is primarily under hormonal control that continues after denervation.

Adrenocortical hormones are either C_{18} estrogens, C_{19} androgens, or C_{21} corticoids. The sex hormones contribute to the growth spurt in females at puberty, and the adrenal androgens to the development of axillary and pubic hair in them. Adrenal androgens exert important metabolic effects at puberty in females (adrenarche), and aid in tissue protein synthesis. Waning secretion (adrenopause) at menopause may be an element in the prevalence of senile osteoporosis, perhaps due to deficient deposition of the protein ossein. In adult males, adrenal androgens are less important in normal physiology, since the more potent testicular androgen testosterone dominates the picture.

During development in utero, the fetal adrenal androgen dehydroepiandrosterone plays an important role in both fetal and maternal physiology, as described in Chapter 52. Similarly, in certain adrenocortical hyperfunctional disease syndromes, excessive androgen or estrogen secretion can elicit a wide variety of signs and symptoms.

The true corticoids fall into two fairly well-defined subclasses according to principal activity. The mineralocorticoids exert their primary action on sodium, potassium, and water metabolism, whereas the glucocorticoids act chiefly on carbohydrate, fat, and protein metabolism. Mineralocorticoid activity appears to be the simpler, since the most potent hormones of this class are essentially free of glucocorticoid potency. By contrast, glucocorticoid hormones also possess weak mineralocorticoid activity. When large quantities of the natural glucocorticoids are administered therapeutically or secreted in excess, their intrinsically weak mineralocorticoid activity is thus multiplied and can give rise to serious electrolyte disturbances.

Glucocorticoid therapy has proved a most useful treatment for a wide variety of diseases. The naturally occurring adrenal steroid, cortisone (which must be endogenously converted to cortisol before it acts), or cortisol itself when given in effective doses, had excessive mineralocorticoid side effects, causing sodium retention, edema, and alkalosis. Synthetic analogs, such as prednisone and prednisolone, largely overcame that problem.

The principal and most potent mineralocorticoid secreted by the adrenal cortex is aldosterone; cortisol (hydrocortisone) fills that niche in the glucocorticoid series. It will suffice to consider these as subserving the respective functions, although again, in hyperadrenocorticism, other mineralo -or glucocorticoids may underlie the physiological disturbances. Similarly, cost or other considerations may dictate the use of other natural or synthetic steroids to obtain a desired therapeutic effect. Aldosterone is rarely used as a mineralocorticoid, either clinically or experimentally. The same effects can be obtained with less expensive deoxycorticosterone, the inherently weaker activity being overcome by giving larger quantities: certain of the fluorinated synthetic steroids are more potent than aldosterone, and others are more potent than cortisol.

SYNTHESIS AND SECRETION OF ADRENAL CORTICOIDS

Adrenal corticoids can be elaborated by: (1) progressive cleavage of the C-17 side chain of cholesterol, and (2) building the molecule from acetate. Since the adrenal cortex contains cholesterol in abundance, and is able to entrap it from the circulation as needed, the first is the preferred method, but the second is available if required.

The adrenal cortex is comprised of three anatomically distinct zones. The outer zona glomerulosa is the source of aldosterone, the middle zona fasciculata synthesizes the glucocorticoids and the mineralocorticoid deoxycorticosterone. The function of the inner zona reticularis in human physiology is obscure. The sequence of steps by which cholesterol is degraded first to pregnenolone, the immediate precursor of all of the hormonal steroids, and then either to dehydroepiandrosterone or to progesterone (which is the parent hormone of testosterone), estradiol, cortisol or aldosterone is shown in Figure 2.

Since enzymes are required for the various interconversions, a deficiency or absence of any of them can impair the formation of successive steroids in the sequence. This has the effect of creating one or more deficiencies of hormones on the one hand and an excess of the precursor, which cannot be effectively processed. If it too is a hormone, there will be excessive effects of the now superabundant steroid. Furthermore, if cortisol is the deficient hormone, the lack of inhibitory feedback on the hypothalamico–hypophyseal axis will cause excessive ACTH to be secreted, and the adrenal cortices will enlarge and become hyperactive, causing even more of the undesirable precursor hormone to be formed.

Aldosterone and cortisol are secreted by two different zones of the adrenal and are controlled by separate mechanisms. The ability of the adrenal cortex to remove cholesterol from the circulation, or to synthesize it de novo, and to carry out the subsequent steps leading to hormone production, depends largely on ACTH from the adenohypophysis. If ACTH is lacking, the gland involutes and becomes extremely atrophic. Glucocorticoid production does not cease entirely, but drops to about 3% of its original value, with severe glucocorticoid insufficiency ensuing. Furthermore, glucocorticoid secretion is no longer adjustable, but continues at a fixed low level.

Aldosterone secretion drops by about one-third, probably due to the general diminution of adrenal metabolic activity. The hormone is so potent, however, that mineralocorticoid deficiency does not develop. Because ACTH has only slight control over aldosterone, the level of secretion is still fully controllable by other means in either direction as needed.

Excessive ACTH secretion leads to enlargement and hyperactivity of the glands. Cortisol secretion is greatly and permanently augmented, but aldosterone secretion is only slightly and transiently, increased; therefore, only hyperglucocorticism is manifest.

ACTION OF ACTH ON THE ADRENAL CORTEX

The surfaces of ACTH-sensitive adrenal cortical cells bear receptors that bind the hormone. In the process, adenyl cyclase is activated and adenosine triphosphate (ATP) is converted to $3',5'$-adenosine monophosphate (cyclic AMP), which is instrumental in activating various protein enzymes. The latter enhance the phosphorylation of cellular proteins, which in turn promote steroidogenesis and glandular growth.

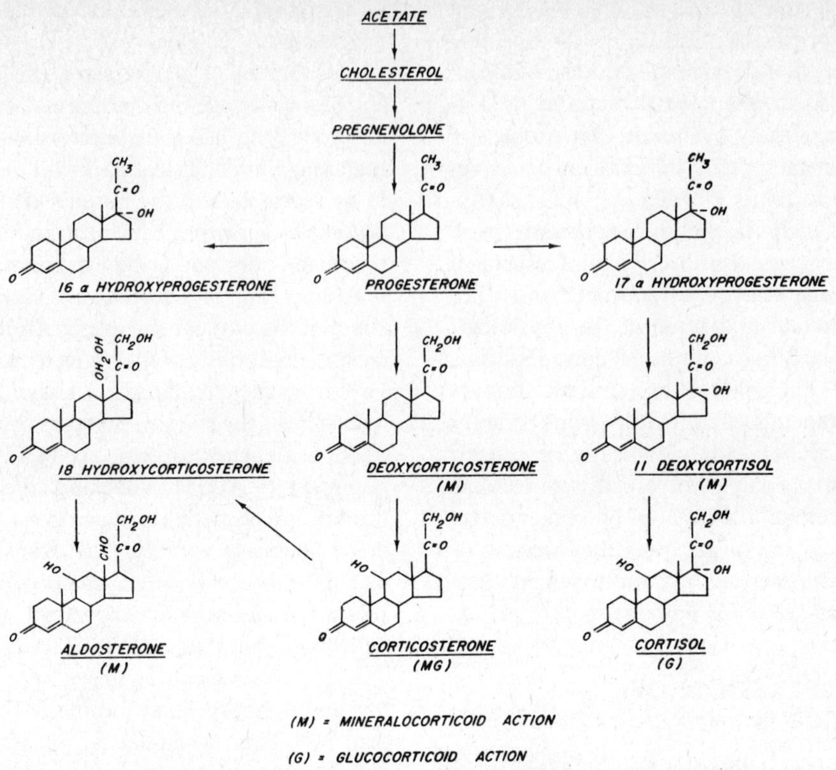

Figure 2 Principal biosynthetic sequence of adrenocorticoids.

The magnitude and duration of adrenal responsiveness to ACTH are governed by a number of factors. A minimum quantity engenders detectable steroidogenesis in humans, namely, a plasma concentration of 10 pg/ml:30 times that dosage produces an almost maximal response. Repeated equivalent small doses of ACTH daily can eventually double the level of cortisol responsiveness. Furthermore, the response to small doses given slowly by intravenous drip is greater than when it is given rapidly as a bolus injection.

REGULATION OF ACTH SECRETION

In health, only the adenohypophysis secretes ACTH, although a wide variety of diverse neoplasms acquire the ability. The normal ACTH-secreting cells synthesize and release the hormone when stimulated by corticotropin-releasing hormone (CRH) from the hypothalamus. Because the CNS has phasic activity, manifested by a circadian rhythm in many physiological processes, there is a diurnal variation in ACTH secretion, paralleled by plasma cortisol concentrations. Both decline in parallel during the late afternoon and evening in day workers, reach a nadir after 2–3 h of sleep, and rise to a maximum an hour or two after awakening. The ACTH cycle is present even in patients lacking adrenal glands and probably reflects a natural hypothalamic neural rhythm. Any perturbation created by the injection of ACTH or glucocorticoids, by adrenal insufficiency or excess, or by pituitary malfunction, will alter the basic pattern, as will stressful stimuli of any kind.

The pituitary–adrenal interaction constitutes a typical negative feedback, or servomechanism. In the normal state, falling blood cortisol concentrations cause increased ACTH secretion, and vice versa. The endogenous hypersecretion of cortisol by autonomous adrenocortical tumors, or the administration of synthetic glucocorticoids such as dexamethasone can entirely suppress ACTH secretion, which, conversely, can be persistently elevated in states of adrenal insufficiency or stress. The question then arises as to whether the feedback from the adrenal glands inhibits the pituitary or the hypothalamus. It appears that direct pituitary inhibition, which both reduces the rate at which ACTH is synthesized and diminishes pituitary sensitivity to CRH, is the more important, although suppression of hypothalamic secretion of corticotropin-releasing hormone is also possible. Figure 3 presents a synopsis of ACTH–adrenocortical relationships.

STRESS AND THE ADRENAL CORTEX

A characteristic of adrenal insufficiency is the reduced ability to withstand stress. Relatively mild degrees of stress, well tolerated by intact animals or humans, can be fatal when the adrenal glands are absent or hypofunctional. Also, the glucocorticoid requirement of patients being treated for adrenal insufficiency rises markedly if stressful circumstances are experienced. It is not surprising then that the imposition of stress, given normal-functioning adrenals, produces a prompt increase in ACTH secretion, followed by a parallel rise of cortisol. This occurs regardless of the time of day and independent of concurrent plasma cortisol levels. Chronic stress can give rise to typical Cushing's syndrome, discussed later.

TRANSPORT OF GLUCOCORTICOIDS

Blood contains a glycoprotein, corticosteroid-binding protein (CBG), or transcortin, a high-affinity, low-capacity glucocorticoid carrier. Albumin acts as a secondary, low-affinity, high-capacity carrier. Normally, about 75% of circulating cortisol is strongly but reversibly bound to CBG, some 15% is loosely bound to albumin, and 10% is free. The latter is the physiologically active component, since bound cortisol cannot enter cells. However, the entry of cortisol into cells upsets the delicate equilibrium between bound and free cortisol, which then causes dissociation of cortisol from CBG. Obviously, the bound cortisol is protected from hepatic inactivation or renal excretion and serves as a readily availble reservoir of active hormone.

Plasma CBG concentration does not normally fluctuate and is unaffected by the actual levels of circulating adrenal hormones. However, high circulating estrogens, as in pregnancy, can increase the concentration two- or threefold. This increases the percentage of bound cortisol and the decline in free cortisol, reducing the feedback inhibition on the hypothalamico–adenophyophyseal axis, causing increased ACTH secretion, and the total cortisol levels rise until the normal ratio of bound to free cortisol is reestablished. The result is hypercortisolism without physiological consequences, since the active free component is at normal levels. When total plasma glucocorticoid levels rise, as in Cushing's syndrome or high-dosage hormone therapy, the carrier capacity may be inadequate, and larger proportions of free cortisol can result. Since this increases feedback suppression of cortisol secretion, ACTH secretion is reduced, thus lowering adrenocortical activity.

PHYSIOLOGICAL ACTIONS OF GLUCOCORTICOIDS

The principal actions of glucocorticoids are as follows:

1. Protein metabolism
2. Fat metabolism
3. Carbohydrate metabolism
4. Antistress
5. Anti-inflammatory
6. Antiallergic

PROTEIN METABOLISM

Glucocorticoids have antianabolic effects on protein metabolism. Apparently the transport of amino acids into the liver is increased, but is decreased in most other tissues. Increased availability of substrates to the liver, together with enhanced hepatic enzyme activity then has secondary effects: (1) increased synthesis of plasma proteins, and (2) increased deamination of amino acids and increased glucose formation (gluconeogenesis).

The reduced entry of amino acids into nonhepatic tissues diminishes the availability of substrate for protein synthesis. At the same time, intracellular protein breakdown continues at a normal or perhaps accelerated rate. In any event, the net effect is protein catabolism. Lymphoid tissues are particularly sensitive, and widespread involution of this system ensues, along with the possibility of great muscular wasting, marked muscular weakness, thinning of the skin, and of capillary walls, the latter leading to easy bruising and petechial hemorrhages.

High blood cortisol levels decrease growth hormone synthesis, which undoubtedly contributes to impaired cellular anabolism. Restoration of growth hormone, however, will not completely repair the defect, which can even be demonstrated in vitro. The intracellular amino acid pool of extrahepatic tissues, being heightened by continued endogenous catabolism and impaired ability to synthesize proteins, provides a drive for outward diffusion, contributing to elevated plasma amino acid levels.

FAT METABOLISM

Cortisol tends to increase both the intracellular oxidation of fatty acids and their mobilization from adipose tissue. Although the net result is a reduction in depot fat, particularly in the legs and arms, in certain specific target areas, such as the face, neck, and upper torso, fat deposition is increased (buffalo obesity). In part, this may be due to an increase of food intake, providing fatty acids at rates that exceed their elevated rate of mobilization and oxidation. Another possibility is that although the qualitative effects of cortisol are the same, the quantitative effects on different fat depots can vary. Thus in certain depots fatty acid oxidation and mobilization can be less stimulated than in others, enabling the latter to accumulate fat because of the prevailing general increase in circulating fatty acids.

Glucocorticoids exert a contrainsulin or "diabetogenic" effect, because of two inherent properties. First, they inhibit the entry of glucose into many cells, which tends to elevate its plasma concentration, and second, they enhance gluconeogenesis, the production of glucose from noncarbohydrate precursors, further exacerbating the response. The fasting hypoglycemia characteristic of adrenal insufficiency reflects the fact that the two processes are subnormally operative. Death in hypoglycemic convulsions is a common fate of untreated adrenalectomized animals, and was, before the advent of glucocorticoid therapy, true also of human patients with Addison's disease. Cortisol depresses the oxidation of intracellular NADH, an essential step in the glycolytic process. Since the overall effects of cortisol, particularly as they apply to the first process, are directly opposite to those of insulin, steroid diabetes is less responsive to insulin control than is pancreatic diabetes, in which only insulin deficiency requires correction.

OTHER GLUCOCORTICOID EFFECTS

In addition to their metabolic properties, glucocorticoids possess a number of other interesting and therapeutically important activities. In a nonspecific way, they are antiinflammatory, antiallergic, and defervescent. Synthetic glucocorticoid steroids, such as dexamethasone, are among the most widely prescribed drugs and are often successful in treating a wide variety of diseases of unknown cause.

Although inflammation is generally a useful process, used by the body to combat a variety of assaults, it can be detrimental and even life threatening. For example, whereas the replacement of air by fluid in the lungs can have useful features, (e.g., diluting chemical irritants or pathogens), alveoli filled with fluid are relatively useless for gaseous exchange, and a patient can "drown" in alveolar fluid. Similarly, the loss of epithelial surface in exfoliative dermatitis can lead to a fatal outcome as a result of dehydration and electrolyte disturbances. The anti-inflammatory action of glucocorticoids can be used to combat these processes without knowing their cause.

The anti-inflammatory effect can be used to block the development of inflammation or to resolve an established process. When cells are damaged, regardless of how this is accomplished, the cytoplasmic lysosomes tend to disrupt. The resulting release of digestive enzymes and other lysosomal contents initiates the inflammatory processes. Glucocorticoids tend to inhibit lysosomal rupture, thereby preventing release of the irritants. Capillary permeability is also reduced, thereby minimizing the formation of edema fluid and retarding the escape of inflammatory leukocytes into the affected area. The production of leukocytes by the lymphoid apparatus and bone marrow is also reduced, and the antibody–antigen reaction is either impaired or blocked. Finally, the defervescent effect of glucocorticoids reduces fever and thereby erythema.

Obviously the beneficial effects of inflammation are also impaired. The waxing and waning of fever, which can be used as a guide to the progress of disease or to the success of its treatment, is no longer a useful index of either, if fever is prevented. Similarly, depression of leukocyte activity, of antibody production, and of the antibody–antigen reaction can deprive the body of some of its most useful defenses. Interference with normal transcellular amino acid transport with inhibition of protein synthesis also impairs reparative processes and wound healing. When glucocorticoids are used for their nonspecific therapeutic properties, special attention must be paid to supportive therapy. Bacterial peritonitis leads to pain, fever, and rigidity of abdominal muscles, each of which can be prevented by treatment with glucocorticoids; however, the infection per se accelerates because so many of the body's defenses are impaired. The danger is that the amelioration of signs and symptoms tends to foster optimism regarding recovery, while the basic infection is becoming ever more life threatening.

HEMATOLOGICAL EFFECTS

Within minutes of cortisol injection and quickly reaching a maximum within hours, eosinopenia and lymphopenia occur. During chronic treatment (or endogenous hypersecretion), there is widespread involution of lymphatic tissue, lymph nodes, spleen, thymus, and other tissue. Curiously, and inexplicably, erythrocyte production is increased by cortisol. Erythropenia is commonly associated with Addison's disease and erythrocytosis with Cushing's syndrome.

The diminution of circulating lymphocytes reduces the availability of antibodies. This has a favorable influence on the rejection phenomena associated with tissue or organ transplantation from one person to another of dissimilar genetic constitution. Glucocorticoids are among the most therapeutically valuable agents in the physician's armamentarium. Nevertheless, antibodies against dangerous internal (e.g., latent tuberculosis) or external pathogens (e.g., bacteria, pathogenic fungi, viruses) are also

diminished, increasing vulnerability to infection, but masking many of the usual signs and symptoms of illness.

PERMISSIVE EFFECTS

Finally, cortisol participates in the control of many functions for which it is not directly responsible, but that do not occur unless it is present—these are called "permissive" or "conditional" actions. As examples, both epinephrine and somatotrophic hormone (STH, or growth hormone) promote lipolysis, but only in the presence of adequate glucocorticoid secretion.

DISPOSITION OF CORTISOL

The continuous secretion of glucocorticoids by the adrenal cortex is required to counteract their disposal by the liver. The native steroids are relatively insoluble in plasma water, but are solubilized by bile and then excreted into the intestines by the liver. Various relatively water-soluble esters, mainly sulfates and glucuronides, are also manufactured by the liver, and these are excreted in urine.

ALDOSTERONE AND ITS CONTROL

The principal adrenocortical mineralocorticoid is aldosterone. As noted earlier it differs from cortisol and other glucocorticoids structurally, and is secreted by the adrenal glomerulosa rather than the zona fasciculata. It has quite different control mechanisms. Four different factors are known to exert control over aldosterone secretion. These are, in the likely order of importance,

1. The renin–angiotensin system
2. Extracellular K^+
3. Total body Na^+
4. ACTH

THE RENIN–ANGIOTENSIN CONTROL SYSTEM

Inability to secrete renin has the same effect on aldosterone secretion as does the absence of ACTH on cortisol secretion, namely, a reduction to extremely low levels, although not complete disappearance. The ability to synthesize steroids is an inherent capability of the adrenal cortex, although the capacity is minimal without hormonal stimulation.

Autonomous renin hypersecretion is accompanied by secondary hyperaldosteronism. Primary hyperaldosteronism diminishes renin secretion and hence angiotensin II formation. Aldosterone secretion is enhanced by stimuli that increase renin secretion, such as assumption of the upright posture, reduced plasma volume, and hypotension, and is reduced by those having the opposite effect (e.g., recumbency hypervolemia, high salt intake). Figure 3 depicts this control system.

CONTROL BY POTASSIUM

Aldosterone secretion is increased by infusions of potassium too small to elevate plasma K^+. In a quantitative sense aldosterone secretion is more sensitive to K^+ than to angiotensin II. However, in man, at least, severe hyperkalemia fails to elevate plasma aldosterone in the absence of angiotensin, whereas angiotensin II will do so even if plasma K^+ is reduced. As yet, it is uncertain as to whether plasma K^+, extracellular K^+, or total body K^+ is the important component.

The effect of aldosterone is to increase K^+ secretion by the kidney, the intestines, and the salivary and sweat glands, thus tending, under hyperkalemic conditions, to restore normal electrolyte balance. The physiologically acceptable range of plasma K^+ is relatively narrow, hyperkalemia can quickly lead to cardiac arrest, and many influences can elevate plasma K^+. It is therefore important that an effective means of controlling the range be present. If K^+ is infused, plasma aldosterone will remain elevated indefinitely. Hypokalemia is a hallmark of primary hyperaldosteronism; the resulting muscular weakness is a common presenting sign.

SODIUM BALANCE AND ALDOSTERONE SECRETION

Inadequate aldosterone secretion leads to excessive loss of Na^+ by the kidneys, and also by the sweat, intestinal, salivary, and lacrimal glands; when it is excessive, the reverse obtains. If aldosterone is administered over an extended period, there is renal sodium retention for a day or two, after which sodium excretion returns to normal. This is the "escape" phenomenon. A slight positive sodium balance remains, both because not all the initially retained sodium is lost, and because the other sodium-handling glands do not exhibit escape. Recent studies have shown that the distal nephron continues to resorb increased sodium under continuous aldosterone treatment. After a day or so of sodium retention, however, progressively larger quantities of sodium are delivered to it; since maximal resorption never reaches 100%, larger quantities are excreted, accounting for the escape phenomenon.

Sodium does not appear to be related reciprocally to aldosterone output directly. Although changes in body sodium are potent regulators, they appear to act indirectly. Losses and increases of body sodium are reflected by parallel changes in body water, increasing extracellular fluid volume, which causes reciprocal changes in angiotensin II, and consequently changes in aldosterone secretion. Body sodium changes modulate the response of aldosterone to changes in potassium concentration. Increased aldosterone secretion due to hyperkalemia is blunted by hypernatremia and augmented by hyponatremia.

It has been suggested that the adrenal zona glomerulosa can be directly stimulated by sodium deficiency and depressed by sodium excess. Alternatively, these electrolyte changes might stimulate or depress the secretion of an adenohypophyseal hormone other than ACTH (an adrenoglomerulotropin), which then influences aldosterone secretion. Neither postulate has been proved.

TRANSPORT OF ALDOSTERONE

In blood, aldosterone binds principally to albumin and has only slight affinity for CBG. Under in vitro conditions at 4°C, plasma binds 2.8 times as much aldosterone as an albumin solution containing the same concentration of albumin, but at 37°C the binding is the same in both.

ADRENAL SEX HORMONES

Three male and two female sex hormones are normal products of adrenal steroidogenesis. These are, respectively, androstenedione, dehydroepiandrosterone, and testosterone (androgens); progesterone (progestagen) and 17

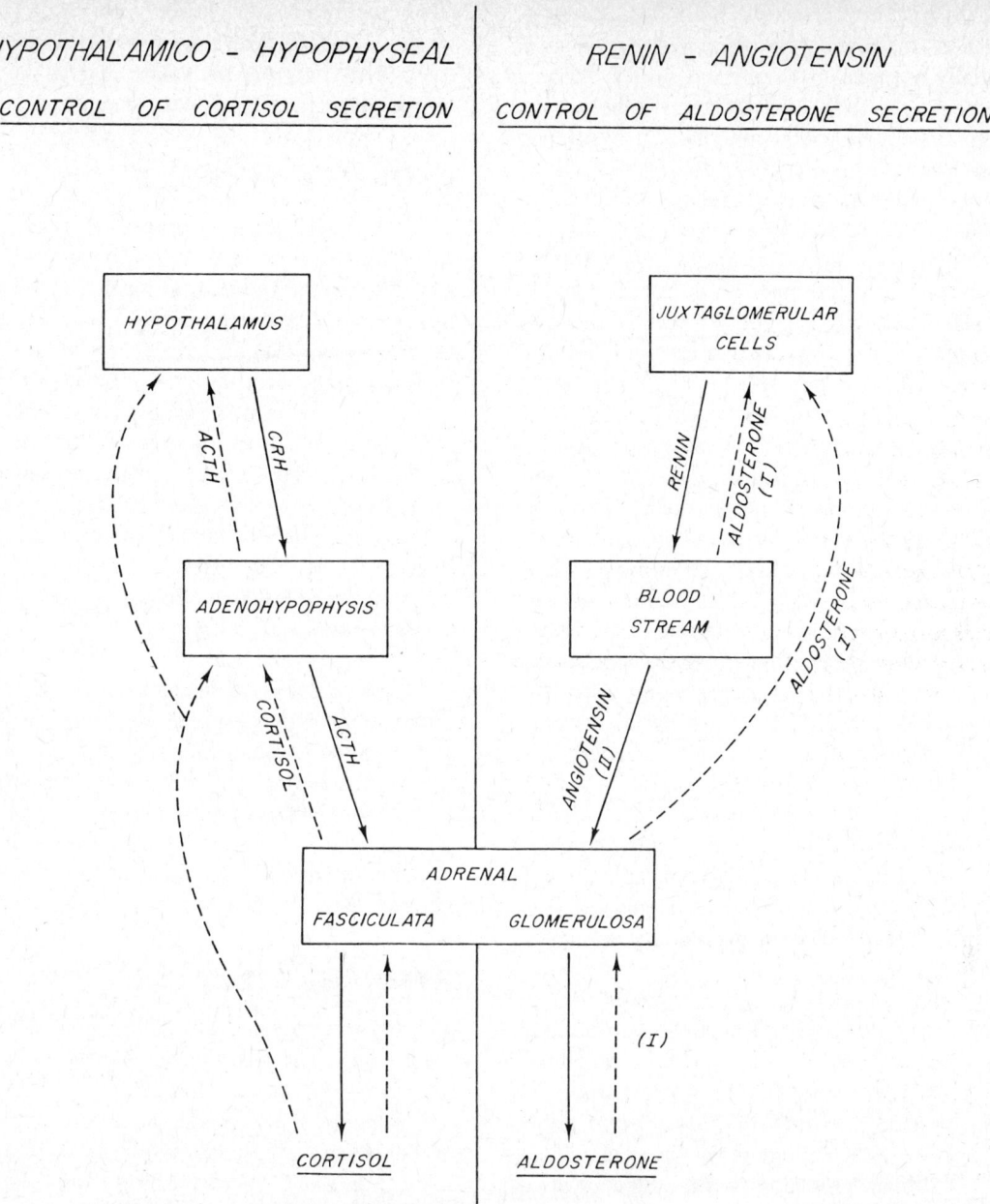

Figure 3 Schematic representation of hypothalmico–hypophyseal–adrenocortical and renin–angiotensin adrenocortical hormonal interrelationships. Solid arrows indicate stimulatory, dotted arrows, inhibitory activity. Indirect effects are indicated by (I), although some of the other feedback inhibitory effects may also involve indirectly acting components.

β-estradiol (estrogen). None of these hormones, with the exception of dehydroepiandrosterone (by the fetal adrenal), is normally secreted in notable quantities. Animal studies have demonstrated that combined adrenalectomy and gonadectomy causes a more pronounced involution of accessory reproductive organs than does gonadectomy alone, indicating that in gonadectomized animals, at least, there is significant secretion of adrenal androgens and estrogens.

Under pathological conditions, as in adrenal hyperplasia or neoplasia, excess sex hormone secretion gives rise to the adrenogenital syndrome. Increased isosexual hormone secretion is difficult to detect in the male adult because of great variability in the "normal" manifestations of sexuality; by contrast, in the female adult disturbances in the menstrual cycle are common. In children, sexual precocity and premature development of genitalia are prominent features. Testicular development is inhibited, since the increased level of adrenal androgens suppress gonadotrophic hormone secretion by the adenohypophysis. Heterosexual hormone secretion leads to the signs and symptoms of masculinity in females (e.g., hirsutism, increased muscular strength, deep voice) and feminism in males (breast enlargement, female fat distribution), and pseudohermaphroditism when it occurs in a fetus.

The contribution of adrenal androgens to the development of males following birth is insignificant, owing to the dominance of testicular androgens. In females, however, they contribute significantly to the pubertal growth spurt (adrenarche) and the development of axillary and pubic hair.

With the background that has been presented, we can now undertake a consideration of clinical hyperfunctional and hypofunctional syndromes. These syndromes fall into three main classes, with the exception of hyperaldosteronism, which seldom occurs in a "pure" form. Often, mixed deficiencies or excesses are manifested. The three classes are those exhibiting problems with glucocorticoid, mineralocorticoid, or sex hormone secretion.

HYPERFUNCTIONAL STATES

Tumors or hyperplasia of the adrenal cortex that give rise to excessive secretion of glucocorticoids are classified as Cushing's syndrome. Similar pathology confined to the adrenal glomerulosa causes hyperaldosteronism, hence electrolyte disturbances. More rarely, the sex hormones are involved (adrenogenital syndrome).

CUSHING'S SYNDROME

Cushing's syndrome occurs more frequently in females than in males (4:1). The naturally occurring form is diagnosed in about 0.001% of autopsies. Similar physiological disturbances are encountered in patients treated overenthusiastically with glucocorticoids. About 10% of patients with Cushing's syndrome of adrenal origin show no discernible adrenal pathology.

Because the adrenal glands secrete such a broad spectrum of hormones, a wide variety of signs and symptoms may be present, with great variation among patients. Although Cushing's syndrome is commonly equated with hypercortisolism, it must be remembered that other glucocorticoids with greater effects on electrolyte metabolism can be excessively secreted as well, along with true mineralocorticoids (e.g., deoxycorticosterone) and androgens, or estrogens, or both.

The signs and symptoms attributable to glucocorticoid excess include centripetal obesity, thinning of the skin and bones, muscle wasting, latent or overt diabetes, easy bruisability, abnormal pigmentation (if there is excessive ACTH secretion) hypokalemia, metabolic alkalosis, and atherosclerosis with systolic, and sometimes diastolic, hypertension.

HYPERALDOSTERONISM

Hyperaldosteronism occurs in three defined forms: (1) primary hyperaldosteronism (Conn's disease), caused by an adrenal tumor and corrected by its surgical removal; (2) pseudoprimary hyperaldosteronism, caused by bilateral adrenal hyperplasia and less amenable to surgical correction; and (3) secondary hyperaldosteronism, in which the adrenal stimulation reflects either a response to some physiological disturbances (e.g., nephrosis, cirrhosis, heart failure, pregnancy) usually involving edematous states, or conditions of excessive angiotensin stimulation (e.g., malignant hypertension, juxtaglomerular hyperplasia or tumor, and Bartter's syndrome).

The symptomatology of primary aldosteronism can be predicted from an appreciation of aldosterone activity. It is associated with mild sodium retention, hypokalemia accompanied by muscular weakness and occasionally paralysis, metabolic alkalosis, expanded extracellular fluid volume with hypertension and edema, and diminished renin-angiotensin activity.

Secondary hyperaldosteronism is not a clinical entity, but is rather a common component of a wide variety of pathophysiological disorders. In contrast to primary disease, there is a combined excess of both renin–angiotensin and aldosterone activity. Usually some circulatory disorder underlies the aldosteronism, and it is this underlying condition, rather than the hormonal disorder, that requires treatment. Neither hypokalemia nor hypertension is a problem, since the elevated aldosterone levels, although high, are not greatly excessive, and are compensatory.

In one form, primary juxtaglomerular disease, the blood pressure is elevated because both angiotension II and aldosterone are elevated. In another form, Bartter's syndrome, there is juxtaglomerular cell hyperplasia, hyperreninemia, hyperaldosteronism, and hypokalemia, yet the blood pressure is depressed rather than elevated. Vascular responsiveness to injected angiotensin II is low, suggesting that a refractoriness of resistance vessels to the vasoconstrictor action of angiotensin may be the fundamental disturbance.

ADRENAL DEFICIENCY

Although adrenocortical hyperfunction can be manifested as a pure excess of either glucocorticoids or mineralocorticoids, adrenal deficiency usually involves some loss of both, although one or the other can predominate. In the naturally occurring diseases, as distinct from therapeutic total or subtotal adrenalectomy that has the same effect, the distinction is between acute (e.g., the Waterhouse-Friderichsen syndrome) or chronic (Addison's disease) insufficiency.

In the acute form, usually the consequence of massive intraadrenal hemorrhage, the loss of adrenal hormones is relatively sudden and severe, and the signs and symptoms life threatening, particularly if the adrenal hemorrhages are precipitated by active infection. Until the advent of corticosteroid therapy, the acute form was almost uniformly fatal.

The chronic form, Addison's disease, is usually a slowly developing condition due to progressive destruction of the adrenal glands by some underlying disorder (e.g., tuberculosis, syphilis, autoimmune disease, metastatic carcinoma). Since the body has time to adapt to the deficiency state, the signs and symptoms are less severe than in acute, fulminating, adrenal insufficiency. In fact, the patient who is maintained on an adequate diet and who is not under stress will show only slight evidence of diminished hormone secretion. However, with essentially complete adrenocortical destruction, wherein there may be an absolute hormone deficiency, or in the milder forms, in which the sudden onset of stress creates an immediate relative deficiency, an acute and always potentially fatal Addisonian crisis with cardiovascular collapse can ensue. In such cases, the usual accompaniments of adrenal insufficiency appear in an exaggerated form.

MINERALOCORTICOID (ALDOSTERONE) DEFICIENCY

The primary defect in aldosterone deficiency is a diminished ability of the kidneys to resorb sodium, which, in the renal tubules and collecting ducts, is normally exchanged for potassium and hydrogen ion. Unless the increased sodium loss is compensated for by increased sodium intake—many untreated Addisonian's exhibit a greatly increased salt appetite, consuming vast quantities—sodium depletion will occur, hence loss of extracellular fluid (ECF) volume. The concomitant retention of K^+ and H^+ leads to hyperkalemia and metabolic acidosis. The falling plasma volume compromises renal function and promotes hypotension. The kidneys secrete increased renin, and plasma angiotensin II rises, but blood pressure remains low. Among the reasons for the vascular collapse are (1) low plasma volume, (2) reduced vascular responsiveness to pressor agents, and (3) impaired glucocorticoid-dependent hepatic synthesis of angiotensinogen, the substrate upon which renin acts to release angiotensin I, which might therefore be relatively or absolutely deficient. Cardiac output falls, and death occurs as the culmination of a shocklike state.

GLUCOCORTICOID (CORTISOL) DEFICIENCY

In this condition, peripheral tissues tend to consume glucose extravagantly, but to use it poorly; since anorexia is the rule, gluconeogenesis is impaired and liver glycogen reduced. This combination of effects encourages hypoglycemia, the cerebral effects of which are apathy, confusion, and occasionally frank psychosis. Mobilization and utilization of fat are decreased since they depend, in part, on efficient glucose utilization and because adipose hormone-sensitive lipase requires the presence of GCs. Feedback inhibition of ACTH is diminished, and the consequent hypersecretion causes increased mucocutaneous pigmentation. This is brought about because the sequence of amino acids comprising melanocyte-stimulating hormone (MSH) is a component of ACTH, hence the latter also has that activity.

Glucocorticoid deficiency is characterized by decreased tolerance to stressors of any kind, with important connotations. First, stresses that would be well tolerated by normal persons might precipitate a sudden addisonian crisis. Second, in a patient being treated with steroids, the hormone requirement during stress is vastly augmented and, to prevent a crisis, steroid administration must be increased.

The manifestations of addisonian crisis are those of sudden cardiovascular collapse, brought about by simultaneously impaired function of the heart, peripheral resistance vessels, and the nervous system.

SECONDARY ADRENAL INSUFFICIENCY

Secondary adrenal insufficiency is caused by reduced ACTH stimulation of the adrenal cortex, a complication of panhypopituitarism. Since aldosterone secretion is normal, the symptomatology is that of cortisol deficiency, and electrolyte disturbances are rare. Abnormal pigmentation does not occur, since ACTH secretion is low.

SELECTED BIBLIOGRAPHY

Eisenstein AB: *The Adrenal Cortex*. Philadelphia, WB Saunders, 1974, pp 283–322.

Leung K, Munck A: Peripheral actions of glucocorticoids. *Annu Rev Physiol* 37:245, 1975.

Melby JC: Clinical pharmacology of systemic corticosteroids. *Annu Rev Pharmacol* 17:511, 1977.

Mulrow PJ: The adrenal cortex. *Annu Rev Physiol* 34:409, 1972.

Williams RH: *Textbook of Endocrinology*. Philadelphia, WB Saunders, 1974, pp 283–322.

SELF-STUDY QUESTIONS

MATCHING

A. *Epinephrine*
B. *Norepinephrine*
C. *Both*
D. *Neither*

1. Predominant secretion of the resting adrenal medulla.

2. Synthesis requires the enzyme dopamine β-hydroxylase.

3. Disappears following glandular denervation.

4. Secretion reduced by rising concentration of norepinephrine.

5. IV infusion promotes tachycardia.

6. When injected intravenously causes emotional perturbation.

7. Destroyed by catechol-o-methyl transferase (COMT).

8. Can cause marked vasodilation.

9. Causes marked vasoconstriction.

10. Secreted by pheochromocytomas.

MULTIPLE CHOICE

Select the single best answer.

11. Adrenal steroid, which must be chemically altered in order to act at the cellular level, is
   A. cortisol.
   B. cortisone.
   C. aldosterone.
   D. deoxycortiosterone.
   E. progesterone.

12. The adrenal zona glomerulosa is
   A. the innermost zone.
   B. highly responsive to ACTH.
   C. the primary source of aldosterone.
   D. the thickest zone.
   E. absent in Addison's disease.

13. Cortisol secretion
   A. reaches peak at about 5 PM in day workers.
   B. responds uniformly to a fixed ACTH dosage.
   C. is secreted only in response to ACTH stimulation.
   D. will increase in response to rising levels of CBG.
   E. is inhibited by aldosterone.

14. Mild stress, in contradistinction to severe stress,
   A. does not affect the level of adrenal hormone secretion.
   B. is inocuous in adrenal-deficient animals.
   C. does not increase the therapeutic cortisol requirements of adrenally insufficient patients.
   D. operates only through the hypothalamus.
   E. causes quantitatively less ACTH secretion.

15. Corticosteroid binding protein (transcortin) is
   A. a high-affinity, high-capacity cortisol carrier.
   B. a high-affinity, low-capacity cortisol carrier.
   C. a low-affinity, high-capacity cortisol carrier.
   D. a low-affinity, low-capacity cortisol carrier.
   E. none of the above.

16. The plasma concentration of transcortin
   A. follows a circadian rhythm.
   B. is subject to marked fluctuations.

   C. is increased by estrogens.
   D. carries about 50% of the total cortisol.
   E. rises in Cushing's syndrome.

17. High cortisol concentrations in plasma, if sustained, would cause
   A. impaired tissue growth.
   B. lympholysis.
   C. inhibition of ACTH.
   D. decreased growth hormone secretion.
   E. all of the above.

18. A condition accompanied by hyperaldosteronism, hypotension, hyperreninemia, and hypokalemia is known as
   A. Conn's disease (primary hyperaldosteronism).
   B. Secondary hyperaldosteronism.
   C. Bartter's syndrome.
   D. Cushing's syndrome.
   E. None of the above.

19. Which of the following is *not* a sign of chronic ACTH hypersecretion?
   A. Elevated cortisol secretion
   B. Hyperpigmentation
   C. Elevated aldosterone secretion
   D. Osteoporosis
   E. Hyperglycemia

20. The adrenal cortex is most responsive to a small dose of ACTH when it is given
   A. over an extended, but limited, period.
   B. rapidly as a bolus injection.
   C. to a patient with Cushing's syndrome.
   D. as the first of a series of injections.
   E. together with aldosterone.

21. Which of the following factors is known to increase the circulating concentration of transcortin?
   A. Progesterone
   B. ACTH
   C. Cortisol
   D. Estradiol
   E. None of the above

Select the best answer(s). (In many instances, more than one answer is correct.)

22. An excessive ACTH secretion would be expected to cause
    A. persistently increased aldosterone secretion.
    B. transiently increased aldosterone secretion.
    C. transiently elevated cortisol secretion.
    D. persistently elevated cortisol secretion.

23. Mild hyponatremia would have the effect of
    A. increasing aldosterone secretion.
    B. increasing cortisol secretion.
    C. increasing the responsiveness of the adrenal cortex to hyperkalemia.
    D. decreasing the responsiveness of the adrenal cortex to hyperkalemia.

24. Among the classical features of Conn's syndrome are
    A. muscular paralysis.
    B. hypokalemia.
    C. sodium retention.
    D. ACTH hypersecretion.

25. Of the following hormones, a primary secretion of the adrenal fasciculata is
    A. progesterone.
    B. deoxycorticosterone.
    C. aldosterone.
    D. cortisol.

26. Acute, nonspecific stress has the effect of
    A. increasing cortisol requirements.
    B. increasing delivery of cortisol to peripheral tissues.
    C. decreasing the sensitivity of the hypothalamus to cortisol inhibition.
    D. increasing aldosterone requirements.

27. Glucocorticoid hormones are synthesized
    A. by the degradation of cholesterol.
    B. by being built up from acetate.
    C. with 17-hydroxyprogesterone as an intermediate.
    D. when the adrenal cortex is stimulated by ACTH.

28. Aldosterone secretion is stimulated by
    A. ACTH.
    B. hypokalemia.
    C. angiotensin II.
    D. hypernatremia.

29. Acute and chronic adrenal insufficiency differ in that
    A. chronic insufficiency is the more life threatening.
    B. acute insufficiency is more often the result of fulminating infection.
    C. cortisol deficiency is greater in the acute form.
    D. the chronic form more easily escapes detection.

30. Alterations in adrenocortical function in the female are
    A. approximately coincidental with the onset of puberty.
    B. approximately coincidental with the menopause.
    C. important in the development of secondary sexual characteristics.
    D. involved in aspects of bone development.

ANSWERS

1. A. About 80% of the resting catecholamine secretion is epinephrine.
2. C. Norepinephrine and epinephrine, since the former is the precursor of the latter.
3. C. All hormonal secretion of the medulla ceases after denervation of the gland.
4. C. Norepinephrine blocks the conversion of tyrosine and therefore diminishes all catecholamine secretion.
5. A. Epinephrine increases cardiac excitability and excitation causing tachycardia. In the intact preparation, norepinephrine raises arterial pressure, causing reflex cardiac slowing.

6. A. Epinephrine has a marked effect on the CNS, causing a feeling of apprehension.

7. C. Both hormones are inactivated by this enzyme.

8. A. Epinephrine produces vasodilation in skeletal muscle arterioles. The predominant effect of norepinephrine in most vascular beds (but not coronary) is vasoconstriction.

9. B. See above.

10. C. Pheochromocytomas can secrete predominantly one or the other, but usually both are elevated.

11. B. Cortisone must first be converted to cortisol before it can act at the cellular level.

12. C. The adrenal glomerulosa is the primary source of aldosterone, none of which is elaborated elsewhere.

13. D. Cortical secretion peaks at around the time of awakening. The response to ACTH is conditioned by previous exposure, and even after hypophysectomy continues to be secreted in small amounts. Aldosterone has no effect on its secretion.

14. E. Only this statement is unqualifiedly true.

15. B. Transcortin has a high affinity for cortisol, but it carries other steroids and is in far less amount than albumin, which has a low affinity.

16. C. Transcortin varies little from day to day, within the day, or during illness. It carries about 75% of circulating cortisol and is markedly increased by estrogenic hormones.

17. E. All of the statements are true.

18. C. Conn's disease (primary hyperaldosteronism) includes hyporeninemia due to aldosterone excess. In secondary hyperaldosteronism, normotension is the rule. Cushing's syndrome does not include hyper-

aldosteronism. Only Bartter's syndrome embodies the given features.

19. C. All the given features, except aldosterone hypersecretion, are signs of chronic excessive ACTH activity. Aldosterone hypersecretion is only a transient response to ACTH stimulation.

20. A. Slow administration is the most effective way of giving ACTH. B, D, and E are incorrect. It is possible, although unlikely, for a patient with Cushing's syndrome to respond excessively to a small dose of ACTH.

21. D. Only estrogens are known to increase transcortin synthesis.

22. B, D. ACTH increases aldosterone secretion only transiently, but cortisol permanently.

23. A, C. Hyperkalemia is a most effective stimulus, but the magnitude of the aldosterone response is buffered by the sodium status, with hyponatremia effectively increasing it.

24. A, B, C. Primary hyperaldosteronism does not affect ACTH secretion per se.

25. B, D. Little if any progesterone is secreted, and aldosterone is secreted by the glomerulosa.

26. A, B, C. The aldosterone requirement is not affected by non specific stress.

27. All are correct.

28. A, C. Both hypokalemia and hypernatremia negatively affect aldosterone secretion.

29. B, D. Chronic insufficiency is usually better tolerated because it develops slowly, giving the body time to adapt. For the same reason, it has milder symptoms, except for pigmentary changes.

30. E. All are correct.

50

Male Endocrine Function

CHARLES E. HALL

OBJECTIVES

After reading this chapter, you should be able to:

1. Describe the waxing and waning hormonal relationships that control testicular testosterone secretion.

2. Appreciate the many different areas of physiological control in which androgens participate.

3. Understand the essential elements of male gametogenesis.

4. Comprehend the role of the various accessory sex organs in spermatogenesis, sperm viability, and transport.

5. Identify factors affecting the sperm that diminish their viability and efficacy in achieving fertilization.

6. Explain the effects of hypogonadism and hypergonadism on body growth and habitus.

7. Identify the principal target organs controlled by testosterone.

8. Understand how testosterone is secreted, transported, and inactivated.

The purpose of this Chapter is to (1) illustrate the structural composition of the testes and its relationship to their gametogenic and endocrine functions; (2) provide a description of the hormonal interrelationships among the brain (hypothalamus), the adenohyophysis, and the testes, which will facilitate appreciation of endocrine control of the male gonads; (3) describe the process of steroidogenesis within the Leydig cells; (4) identify various metabolic, physiological, and anatomical changes influenced by androgenic hormones, thus facilitating an appreciation of their raison d'être; and (5) consider the waxing and waning of testosterone secretion during life and the effect thereof on physiological and anatomical changes.

The testes have a dual function: (1) to produce viable gametes (sperm) to fertilize the female ova, together with the fluids necessary for their preservation, sustenance, and transport to the exterior; and (2) to synthesize and secrete steroids having a C_{19} configuration and classed as androgens. The steroids influence a wide range of anatomical, physiological, and biochemical phenomena, most prominently including the development and maintenance of the accessory sexual organs and of those physical attributes to which the term "maleness" is applied.

ANATOMY OF THE MALE REPRODUCTIVE SYSTEM

From a functional standpoint, the testes contain two important components: the seminiferous tubules, comprised of the sperm-forming germinal epithelium, together with the sperm-protecting Sertoli (sustentacular) cells, which can also elaborate and secrete a hormone or hormones; and the hormone-secreting Leydig (interstitial) cells, which cluster in pockets formed by contiguous seminiferous tubules. The latter constitute about one-fifth of the testicular bulk.

Figure 1 shows numerous, coiled, seminiferous tubules, which generate sperm, massed together. Their contents discharge into another coiled tube, the epididymis, a conduit to the vas deferens, which through its ampulla conveys the material to the prostate gland. The prostatic end of the ampulla has openings leading from the paired seminal vesicles, which adds its secretions to the stream. The contents are then emptied into the ejaculatory duct, which passes through the prostate, receiving contributions via the prostatic ducts before reaching the internal urethra. The urethra, the jumping off point for spermatic fluids, adds others, chiefly mucus, secreted by the glands of Littré that line it, and by the adjacent Cowper's (bulbourethral) glands.

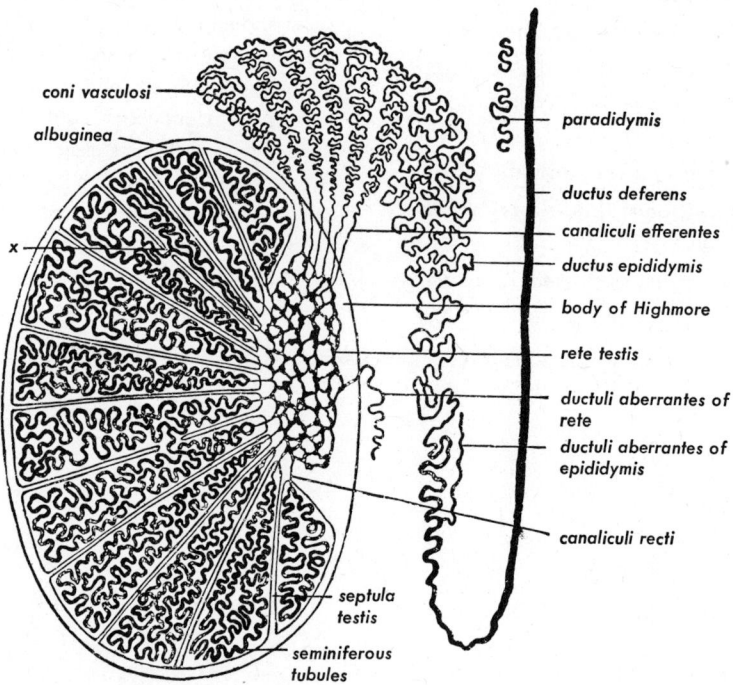

Figure 1 Diagram of arrangement of seminiferous tubules and excretory ducts in the testis and epididymis. x, Communication between seminiferous tubules of different lobules.

SPERMATOGENESIS

The seminiferous tubules in adult males (Fig. 2) are lined by a few layers of germinal epithelial cells, spermatogonia, arranged at the periphery. Constant proliferation assures replacement of those lost through attrition, or maturation. Some of them undergo differentiation as they pursue the pathway to becoming mature sperm, a process requiring almost 10 weeks. Three phases can be distinguished: sper-

matocytogenesis, in which spermatogonia replicate themselves or give rise, by mitotic division, to enlarged primary spermatocytes. These, by meiotic division, generate two secondary spermatocytes, each of which then contains half the number of chromosomes, which, by mitotic division become spermatids, still having only 23 unpaired chromosomes. If fertilization occurs, these associate with a similar complement in the ovum, to provide the requisite 23 pairs (46) of chromosomes. The spermatids then undergo a series of cellular transformations charac-

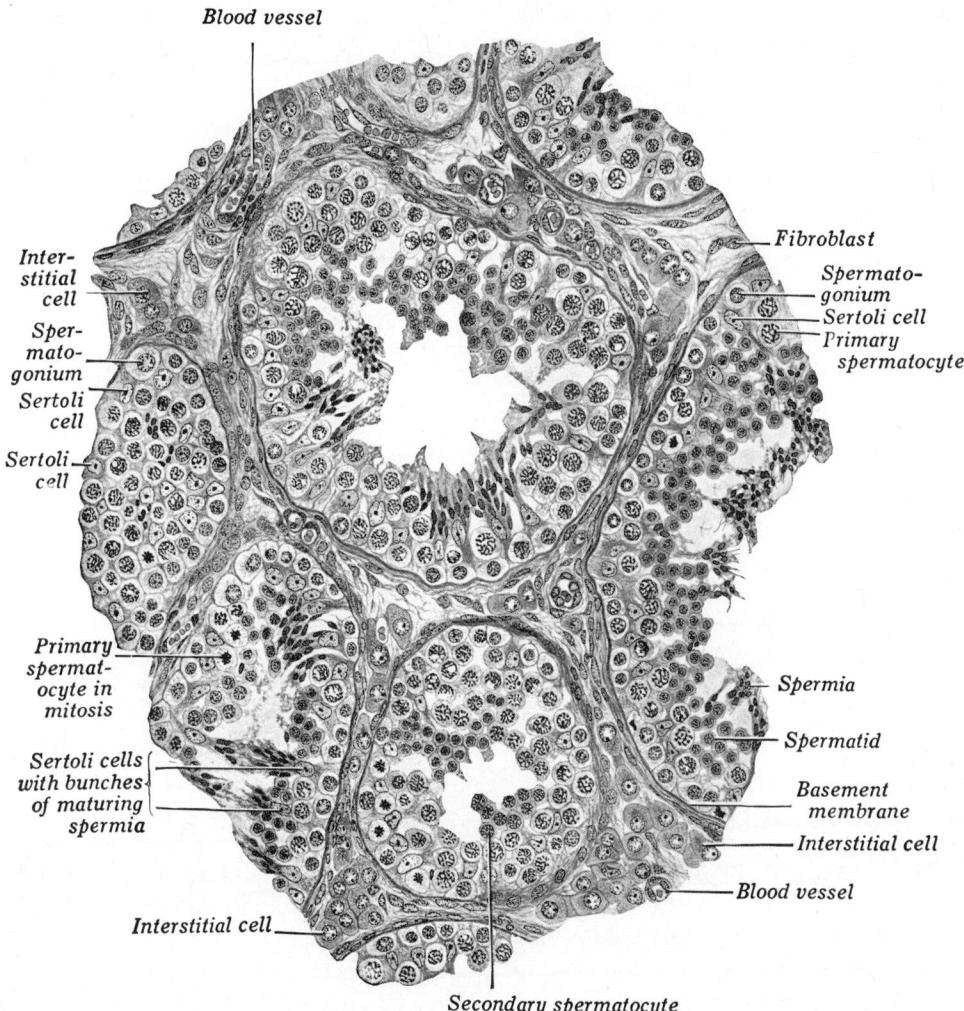

Figure 2 Human testis (from operation). The transections of the tubles show various stages of spermatogenesis. (170 X.) (From Maximow AA, Bloom W: *A Textbook of Histology,* ed 7. Philadelphia, WB Saunders, 1957, p 479.)

terized by reduction in volume and elongation, at the completion of which they become spermatozoa.

Each spermatozoon consists of a head, neck, body, and tail. The head is tipped by a structure, the acrosome, rich in hyaluronidase and proteases, enabling it to penetrate the membrane of the ovum. The nuclear material, which fertilizes the ovum, is compacted into a dense cephalad mass. The tail segment is rich in adenosine triphosphate (ATP), doubtless a source of energy for the 20-cm/h spurt with which, in the female genital tract, the sperm begins its quest for an ovum. There, the spermatozoa undergo further alterations, improving their ability to penetrate the zona pellucida and fusing with the ovum.

These phenomena are collectively termed *capacitation,* although, as evidenced by recent fertilizations achieved in vitro and subsequently successfully implanted into the uterus—developing normally into fetuses delivered alive at term—they are not indispensable.

Spermatozoa removed from the testes lack motility or the capacity to fertilize ova. Both are acquired after a few days sojourn in the epididymis, a process termed *maturation.* It is uncertain whether maturation is simply an aging phenomenon of a specific duration, or whether the milieu of the epididymis, rich in hormones, enzymes, and nutrients, contributes importantly to the change. Within the genital ducts, sperm remain viable for up to 6 weeks. Once they leave, through a process called *ejaculation,* they survive only 1–3 days at best. Their motility is enhanced by an alkaline and retarded by an acid pH: indeed, highly acid media are quickly fatal. Longevity can be prolonged by cooling; the semen of cattle and other animals has been preserved for a year or more by freezing at $-100°C$ without significant loss of potency.

SEMINAL VESICLE AND PROSTATIC FUNCTIONS

Seminal vesicle and prostatic functions add volume and nutrients to the ejaculate; they also provide vitamins, enzymes, and amino acids, as well as adjusting the pH. Stored in the epididymis, sperm motility is low, perhaps because CO_2 is generated by their metabolic activity. Fe-

male vaginal secretions are also acidic. Thus, in both areas, motility and fertility are compromised. Secretions from the seminal vesicles appear to provide primarily nutrients, whereas those from the prostate, being highly alkaline, overcome the detrimental acidic conditions that would otherwise obtain.

SEMEN

The sperm-containing seminal fluid, derived from the testes and associated sex organs, has a pH of about 7.4 and an average volume of 3–4 ml. The sperm count is extremely variable within limits. Normal males can ejaculate semen with sperm counts varying from $35 \times 10^6/ml$ to $400 \times 10^6/ml$. An average of $200 \times 10^6/ml$ is usually cited; with values below $20 \times 10^6/ml$, infertility is common. Although only a single sperm is necessary for fertilization—and once the ovum is fertilized it is impermeable to other sperm—large numbers are needed. Many fall by the wayside, abandon the chase, or get lost en route.

MALE FERTILITY

Inflammation of the testes, orchitis, can produce temporary or permanent sterility. Certain specific infectious diseases, such as mumps and typhus, are especially apt to cause transient orchitis, which often leads to permanent sterility. Acute orchitis from a variety of other causes might have no permanent sequelae.

High temperature is inimical to sperm viability. In many mammals, including humans, the testes are carried in a thermoregulatory scrotal sac, which maintains the temperature at a lower level by a few degrees than the general core temperature. If, as in cryptorchism, the testes are retained within the abdominal cavity, or inguinal canal, the spermatogenic epithelium survives until it becomes active at puberty. It then undergoes degeneration and fails to produce sperm, although androgen production by the Leydig cells proceeds relatively normally. In some smaller mammals, the testes descend into the scrotal sac only during the breeding season and are withdrawn into the abdominal cavity, where they become inactive, at other times. No damage occurs as long as they are inactive. Very large mammals, such as elephants and whales, have permanently abdominal testes, but also have internal body temperatures a few degrees lower than those of smaller mammals.

Even short exposure to very high temperatures (e.g., sauna baths, turkish baths) can produce temporary sterility and thus have marginal, if unreliable, contraceptive value. A breed of Merino sheep was developed in which the rams were found to be relatively infertile. Shaving their heavily wool-coated scrotal sacs was found to improve fertility enormously.

ABNORMAL SPERM

Infertility, or low fertility, in a male with a normal sperm count, is often associated with functional or structural abnormalities of sperm. The tails, heads, or both might be abnormal in number, size, or shape. Alternatively, there might be no visible identifiably aberrant morphology. The sperm might have a normal appearance, but be nonmotile or weakly motile, swim in circles, or otherwise display impaired mobility. If a large proportion of sperm display abnormal structure or behavior, infertility is a likely consequence.

The testes of male infants at birth are poorly developed. Semiferous tubules are small, lack a definite lumen, and contain only spermatogonia. The testosterone-secreting Leydig cells are identifiable in the fetus at 8 weeks; they secrete androgens, under the influence of maternal chorionic gonadotropic hormone, which stimulate the growth and development of male secondary sex characteristics in utero. At birth, the maternal hormone is withdrawn; since the hypothalamico–hypophysial axis of the newborn is not functional with respect to gonadotropins, the Leydig cells involute. They are not recognizable a few months after birth and remain in this state until the onset of puberty, when they attain the adult structure.

Infantile testes are fully capable of secreting androgen if stimulated by pituitary interstitial cell-stimulating hor-

mone (ICSH), as witness the occasional appearance of precocious puberty. Juvenile adenohypophysis is capable of synthesizing ICSH, for if the pituitary gland from a young, sexually immature animal is transplanted into the sella turcica of an hypophysectomized adult male whose testes have become atrophic, full testicular repair occurs. The hypothalamus does not acquire control over the sexual apparatus until the onset of puberty, which signals its maturation. It then begins to secrete the releasing hormones—which also stimulate hormone synthesis—goading the adenohypophysis into action, followed by the testes.

At puberty, the increasing titer of gonadotropin-releasing hormone (Gn-RH) causes the adenohypophysis to release follicle-stimulating hormone (FSH) and ICSH. FSH acts on Sertoli cells of the seminiferous tubules and ICSH on interstitial (Leydig) cells. FSH enlarges the testes by causing growth of the seminiferous tubules and stimulating spermatogenesis. The Sertoli cells are believed to secrete an inhibitory hormone (inhibin), a water-soluble nonsteroidal hormone that exerts negative feedback control over FSH secretion. ICSH acts primarily on the Leydig cells which enlarge and begin to secrete testosterone. Testosterone enters the bloodstream to exert its many functions. One of these functions, insofar as the testes are concerned, is to establish servocontrol over ICSH secretion; another is to encourage sperm maturation.

Human puberty begins at about the 8th year, and ends in the middle or late teens. During this period, semen begins to appear and contains mature sperm. Subsequently, the physiological effects of androgen become evident. Until puberty, the growth rate is comparable in boys and girls, but with the advent of testosterone, male growth rate accelerates. The long bones elongate rapidly, because testosterone, like growth hormone, stimulates proliferation of the cartilagenous endplates. They also thicken, because periosteal deposition of protein matrix (ossein) increases, as does the deposition of calcium salts. Testosterone accelerates the rate at which long bones elongate, but shortens the duration by speeding the union between the epiphyses and shafts. Hypogonad males grow more slowly than normal, but continue over a longer period. Thus, as children and young adolescents, they are of shorter stature than normal, but as adults they might be taller than they otherwise would have been. Conversely, excess androgen secretion (or treatment) hastens growth rate but reduces its duration. Persons so affected might outgrow their playmates as children, but will be smaller as adults.

HORMONE DEFICIENCIES

The spermatogenic epithelium is matured and sustained by FSH from the pituitary, but sperm maturation also requires adequate androgen, produced by the Leydig cells under stimulation by hypophyseal ICSH, identical with luteinizing hormone (LH) of females. If FSH is lacking or insufficient, the spermatogenic epithelium degenerates, with unavoidable consequences to sperm production. Various pituitary disorders have that effect. If ICSH is deficient, then even if sperm are produced, their maturation and development are impaired because testosterone, essential to these processes, is inadequate.

Nature had reasons for placing certain hormone-secreting cells exactly where they are. The cells of the adrenal medulla would produce norepinephrine regardless of their location, but only in immediate contiguity with the adrenal cortex would they receive optimal levels of the cortisol so essential to epinephrine synthesis. Similarly, Leydig cells could produce testosterone at other sites in the body, but would not then be as intimately associated with the spermatogenic epithelium, which is so dependent on testosterone.

The androgens in general—and to a lesser extent the estrogens—are protein anabolic hormones; that is, they cause nitrogen retention and net protein deposition. Testosterone, the most potent of the naturally occurring testicular androgens, is also the most anabolically effective of these. This action is largely responsible for the growth spurt just described and on the related effects on bone growth. The spurt in adolescent girls is fostered both by the weaker adrenal androgens and by ovarian estrogens, but the latter are particularly efficient at bringing about epiphyseal closure, hence, the spurt is of shorter duration than in males. Among the most dramatic effects is that on skeletal muscle.

EFFECTS ON MUSCLE

The bulk and strength of skeletal muscle are characteristic of virile males, and are notably less developed before puberty or in hypogonad persons. In the latter, androgen treatment can effect dramatic changes, but it is doubtful —despite the belief of many athletes and athletic coaches to the contrary—that in a male with normal androgen production, treatment with additional exogenous testosterone will have any appreciable effect, other than psychological. Once muscles are exposed to as much androgen as they can use, more is unlikely to be effective. Testosterone production in many heavily muscled males may be less than that of asthenic males. Muscles of the former are either used more or have greater inherent sensitivity to adrogen stimulation. In children the androgen effect is very dramatic, and because of the greater muscular development often associated with precocious puberty or precocious pseudopuberty, such a child is often termed an "infant Hercules."

The anabolic action of androgens deviates a relatively higher proportion of amino acids into protein synthesis than would be the case if they were absent. The skeletal muscles and bone together constitute by far the largest repository for protein deposition, hence each is strongly influenced by androgens.

THE INTEGUMENT

Androgens cause a thickening and coarsening of the skin and subcutaneous tissues. Males have much thicker skins than do females, unless the latter are exposed to excess androgen, in which case masculinization of the skin occurs. Hypogonad persons often experience difficulty in acquiring a tan on exposure to sunlight. This condition is amenable to androgen therapy, which increases melanin deposition in skin.

Testosterone stimulates hair growth in the axillae, on the face and chest, and over the pubis. In other areas the effect is less evident, but the growth rate of hair and its abundance are both stimulated. The absence of testes prevents the development of baldness in men who would have a genetic propensity toward that change and, concomitantly, females from such lineages become bald if hyperandrogenicity develops (e.g., from adrenogenital syndrome). Body and facial hair increases in women exposed to excess androgen. Cosmetic problems are less evident in blondes and more conspicuous in dark-skinned brunettes. The luxuriant development of beards and moustaches in women has distressing psychological consequences.

Growth and secretory activity of sebaceous glands is stimulated by androgens, an effect that is most conspicuous at puberty, when the secretion of adrenal androgens increases in both sexes and testicular androgen secretion begins in the male. The resulting hyperactivity of sebaceous glands causes acne, a condition that eventually disappears spontaneously as the sebaceous glands become accustomed to androgen, but is susceptible to corrective therapy if it is unduly embarrassing psychologically.

LARYNX

Androgens cause a thickening of the laryngeal mucosa and a lengthening of the vocal cords, making them resonate at a lower frequency. This effect gives the characteristic deep voice of the adult male. During the Middle Ages, castration was often resorted to in order to preserve particularly fine soprano voices in young boys.

SECONDARY SEX ORGANS

The organs most sensitive to androgens, and therefore commonly used in bioassaying them, are the seminal vesicles and prostate. These atrophy to a negligible size after male castration. Enlargement of normal glands from excess androgen is less conspicuous because, as in the case of skeletal muscle, once the maximal normal growth has been achieved, further growth is minimal. Enlargement of the penis and scrotum also depends on androgens, but these organs do not atrophy after castration. Similarly, their size is largely genetically determined. They can be large in men who have relatively low androgen secretion, and vice versa. The female homolog, the clitoris, is also enlarged when androgen production is increased, as in certain disorders of the ovary or adrenal cortex. Sensitivity of the penis and clitoris to androgen decreases with age. Enlargement can be extreme when androgen hyperstimulation of a fetus occurs in utero, or develops in early childhood, but can be negligible in an adult.

MISCELLANEOUS EFFECTS

Hypogonad males suffer from mild anemia; the erythrocyte count of the normal man is higher by about a million per cubic millimeter than that of females, both effects indicating a stimulatory influence of androgens on red cell mass. It is not known whether this is a direct effect on bone marrow or merely an indirect one due to heightened metabolism.

The basal metabolism of male adults is estimated to be almost 10% higher than it would be without testicular androgens; it can be further augmented by testosterone treatment.

The androgen anabolic effect, stimulating as it does protein synthesis in the three largest organs of the body—the skeleton, striated muscles, and skin—involves increased formation of protein (both enzymes and structural elements). Consequently, basal metabolism is increased and protein mass is augmented.

TESTOSTERONE SYNTHESIS

Testosterone is synthesized within Leydig cells under the influence of adenohypophyseal ICSH. Cholesterol, the ubiquitous steroid hormone precursor, is transported to the mitochondria, where the 17-C side chain is removed, giving rise to Δ^5 pregnenolone, which via several intermediates ultimately yields testosterone (Fig. 3).

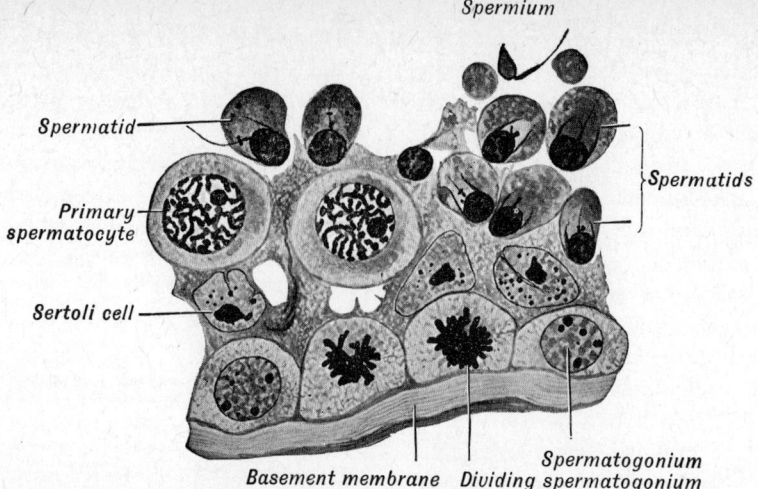

Figure 3 Human testis, from young adult; seminiferous epithelium with mitoses of spermatogonia. The spermatids show a caudal sheath. Iron-hematoxylin stain. (750X.) (Maximow AA, Bloom W: *A Textbook of Histology,* ed 7. Philadelphia, WB Saunders, 1957, p 479.)

SECRETION, TRANSPORT, AND CELLULAR ACTION

Free testosterone is secreted from the cell surface into the surrounding capillaries. There, 97–99% of free testosterone binds to a β-globulin testosterone-binding globulin, the form in which it is transported. The remainder (free) is presumably the hormonally effective form. The bound fraction cannot enter cells and is therefore metabolically inert.

Testosterone interacts with two cell types. One cell type is typified by the prostate, containing the enzyme 5α-reductase. Within these cells, testosterone acts as a precursor, undergoing enzymatic conversion to dihydrotestosterone, the active hormone (Fig. 4). In other cells, such as those of bone and muscle, the reductase is lacking, and testosterone acts directly.

Cells may lack cytosol receptors for testosterone or dihydrotestosterone (DHT), or both, in which case neither steroid exerts metabolic activity. This is an example of target-organ insensitivity to a hormone, in which appropriate alterations are not initiated in spite of adequate plasma levels of the required hormone. Hormone secretion has a diurnal rhythm, with several peaks. The purpose of this phenomenon is unknown, but in effect it has a result similar to that achieved by incorporating an active drug in time-release capsules.

The testes produce androgens weaker than testosterone, such as androstenedione and dehydroepiandrosterone, and nonandrogenic steroids such as etiochalanolone. Similarly, tissues other than the testes, such as the seminal vesicles, both synthesize and metabolize testosterone. The quantities thus produced are small, but may serve to provide locally higher concentrations than otherwise would be available. Apparently the Sertoli cells secrete an androgen-binding protein (ABP), which is controlled by FSH. This may serve to provide locally high intratubular concentrations of testosterone.

Androgens act within tissues by binding to a cytoplasmic receptor protein; the complex then enters the nucleus and adheres to a nuclear protein. DNA transcription is then activated. Large amounts of RNA appear, soon followed by increases in cellular protein and eventually by cell proliferation in certain tissues (e.g., the prostate); testosterone is transformed intracellularly into DHT. In tissues of that type, only DHT acts hormonally.

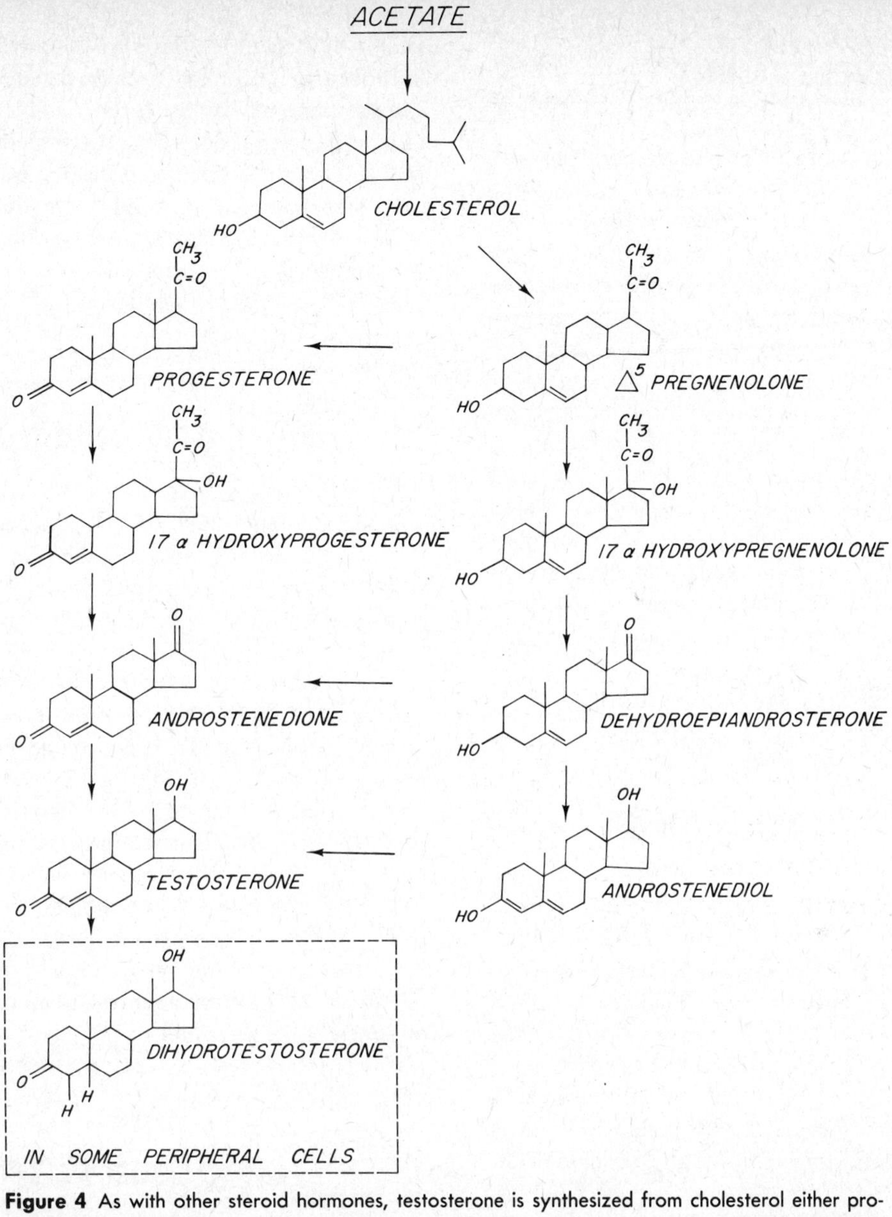

Figure 4 As with other steroid hormones, testosterone is synthesized from cholesterol either produced within the gland proper from acetate, or extracted from the circulation passing through. In many somatic cells testosterone acts directly, but in others it is a precursor for the active steroid dihydrotestosterone.

ANDROGENS IN PERINATAL DEVELOPMENT

Leydig cell development of the fetal testes parallels both the circulating levels of chorionic gonadotropin and the synthesis and secretion of testosterone. Substantial levels of testosterone are present in fetal testes and plasma by the eighth week and rise to a maximum at the eleventh to fifteenth week, correlating well with both Leydig cell proliferation and differentiation of secondary sex characteristics. Thereafter a precipitous drop occurs, basal values, which persist until birth being reached by about the eighteenth week. Within a few hours of birth, male infants demonstrate a markedly higher plasma concentration of testosterone than do females, 242 ng/L as compared with 29 ng/L. Both sexes show a marked reduction in circulating testosterone during the first postnatal week, followed by a steep rise in males until the second month. It is noteworthy that at birth the level of globulin-bound testosterone is only one-third that of adults, and that of free testosterone correspondingly greater.

SEXUAL DEVELOPMENT

In the female fetus sexual differentiation is a passive process, that is, the female habitus develops unless it is suppressed by androgens. In the absence of testicular function, the müllerian ducts become fallopian tubes, uterus and wolffian ducts involute, and feminization of external genitalia takes place. The presence or absence of ovaries does not affect the changes. If there is androgen exposure before the twelfth week of gestation, the labia minora fuse, accompanied by clitoral hypertrophy and virilization of external genitalia.

Testosterone appears to be directly responsible for the differentiation of wolffian ducts into epididymis, vas deferens, and seminal vesicles, whereas development of external genitalia seemingly is dependent on prior conversion to DHT at the intracellular level.

The hypersecretion of adrenal androgens will thus masculinize a female fetus, the external genitalia and habitus being those of a male. Testes, of course, are absent. Testicular dysgenesis similarly leads to the syndrome of male pseudohermaphrodism, because deficient testicular androgen production permits expression of female external genitalia and internal accessory sex structures. Similarly, exposure of a male fetus to excessive estrogenic hormones may permit persistence of müllerian duct structures along with male duct development. There may be hypospadias. Interference by estrogens with the normal production of testicular androgens is believed to be causative.

ANDROGEN INACTIVATION

Testosterone and DHT are metabolized by the liver, and to a lesser extent by muscles, the prostate, and other tissues to neutral 17-ketosteroids, which are then conjugated with glucuronic or sulfuric acid to form glucuronides or sulfates. These are highly water-soluble and are excreted in the urine. The testes contribute 30% to the total 17-ketosteroid excretion, 70% representing degradation of adrenal steroids. Total excretion in adults ranges 7–20 mg/day; female levels being about two-thirds that of males and consisting only of adrenocorticosteroid metabolites. Some of the 17-ketosteroids, such as androstenedione and dehydroepiandrosterone, have androgenic activity, whereas others, such as etiocholanolone, do not. Because only one-third of the 17-ketosteroids of males represents testicular activity, only one-half of which is derived from testosterone, these metabolites are an inaccurate reflexion of testosterone secretion.

LIBIDO AND POTENTIA

Sexual desire in males does not begin until testosterone secretion is initiated and the external genitalia begin to respond. Failure of the testes to develop (testicular agenesis), or to secrete testosterone, or castration in childhood, inhibits the development of libido, and sexual interest in females does not develop. In lower mammals, sex drive disappears following castration of adults. In man, however, sex drive becomes encephalized, and to a large extent becomes a mental function. Castration can then diminish sex drive somewhat, but does not eradicate it. Conversely, the intensity of sex drive does not correlate well with either circulating androgen levels or habitus. Strongly masculine types may have relatively little interest in the opposite sex, whereas frail types may be obsessed.

Because of this, the value of testosterone administration, as a treatment for impotence in males not suffering from hypogonadism, is dubious. Since so much of the drive is cerebrally dominated, psychological problems are often of more importance than hormonal disturbances. However, an individual who believes that a certain substance is of value in this cerebrally dominated function will probably benefit from taking it, whether it be testosterone or one of the myriad of materials of purported aphrodisiacal merit, such as rhinoceros horn, turtle shell, oysters, or whatever. It appears likely that there is a certain amount of drive applied to the psychic sex appetite by adequate amounts of testosterone, but its magnitude is difficult to estimate.

Further manifestations of the cerebral component of libido are the occurrence of nocturnal emissions in males during erotic dreams, or the arousal of sexual desire by sights, or even thoughts, with sexual content. Some members of either sex may display homosexual preferences and have little or no interest in heterosexual contacts. This sex drive is also cerebrally dominated, for such persons do not usually exhibit endocrine dysfunction.

Androgen levels are highest, and drive greatest, during the late teens and early twenties. Often it declines somewhat thereafter, but the individual variations, even within the same person at different times, are such that there is no reliable pattern. Normal production of both sperm and androgen may continue into old age, and males, as a rule, do not experience the relatively abrupt cessation of gonadal function characteristic of menopausal females. Some individuals do experience a marked diminution in Leydig cell function and testosterone levels in midlife. Symptoms similar to those experienced by women during the waning of ovarian activity in late adult life, such as irritability, depression, and hot flashes, can appear. The term "male climacteric" has been applied to this pattern.

THE SEXUAL ACT

Performance of the sexual act necessitates erection of the glans penis if introitus is to be achieved; this is accomplished by activity of parasympathetic nerves that simultaneously constrict the venous outflow of blood and dilate the arterial inflow, thus expanding erectile tissue. The latter is composed of large venous sinusoids, encapsulated by thick fibrous coats. When the sinusoids are filled under high pressure, the penis swells, elongates, and hardens.

This sequence can be initiated by psychic stimuli, by certain drugs that irritate the bladder and urethra (e.g., catharides), or by tactile stimuli applied to the penis, scrotal sac, perineal region, anus, or even more remote areas of the body. Erection is maintained by continuance of these stimuli until ejaculation occurs, upon which, as a rule but not inevitably, there is a waning of desire, and the stimuli become less effective. Blood then leaves the erectile tissues via the veins and the penis collapses.

At the height of sexual stimulation, nerves travelling in the lumbar sympathetic outflow carry impulses to the hypogastric plexus and there to seminal reservoirs and their associated glands. Contraction of the ampulla, epididymis, and vas deferens, causes sperm to be extruded along with the accompanying glandular secretions from the seminal vesicles and prostate. This act of emission propels the semen toward the urethra, along with mucous added by the bulbourethral glands. The material enters the internal urethra, initiating impulses from the pudendal nerve to the sacral region of the spinal cord. Centers there are stimulated to send rhythmic impulses outward. These impinge on the striated muscle surrounding the base of the erectile tissue, and elicit fluctuating pressure changes within it, causing the urethral contents to be emitted in short bursts during the act of ejaculation. A feeling of wellbeing usually follows.

HYPOGONADISM

Hypogonadism can be complete, as from testicular agenesis or castration, or partial. Incomplete hypogonadism can be reflected in decreased testosterone secretion; in some impairment of spermatogenesis, or a deficiency of both. Although normal Leydig cell function is possible without spermatogenesis, normal spermatogenesis is not possible in the complete absence of Leydig cell function. Testicular failure may be a primary disorder, or it may be secondary to inadequate stimulation by FSH and ICSH (hypogonadotropic hypogonadism). The latter, in turn, can be due to a dysfunction of the pituitary gland, which is unable to produce or secrete the gonadotropins adequately, or to failure of the hypothalamus to provide the releasing hormones necessary to activate the adenohypophysis.

Boys may be thought of as having natural hypogonadotropic eunuchoidism until the advent of adolescence. Then the hypothalamus begins to secrete the decapeptide, gonadotropin-releasing hormone (GnRH), since it has attained secretory maturity, the pituitary secretes gonadotropins and testicular hormones are insufficient in amount to inhibit the release of either. The result is a rising titer of testosterone and presumably "inhibin" until these attain levels adequate to exert feedback suppression on the hypothalamico-hypophysial axis. Subsequently, falling blood levels of testosterone and inhibin are countered by rising levels of gonadotropin, which are in turn kept from becoming excessive by rising levels of testicular hormones. Cryptorchid testes never display normal spermatogenesis, although testosterone production may be normal.

Hypogonadotrophic hypogonadism occurs commonly when anterior pituitary function is subnormal. The pituitary can be destroyed therapeutically by x-irradiation or surgery, or by a disease process such as tumor, inflammation, or occasionally after a massive hemorrhage. Certain diseases of the hypothalamus also reduce the ability of the hypophysis to secrete gonadotropins. Sometimes there is an associated impairment of the hypothalamic satiety center, which is manifested as hyperphagia. The result is hypothalamic eunuchoidism and obesity, an association to which the term adiposogenital dystrophy (Fröhlich's syndrome) is applied.

Inflammatory or other diseases may so damage the testes that they are unable to respond to gonadotropin. The germinal epithelium is much more sensitive than the Leydig cells, so that testosterone secretion may be essentially normal, although sperm production is not. This is primary hypogonadism.

HYPERGONADISM

In childhood this is easy to recognize, since a young boy who grows a beard and moustache and becomes heavily muscular is not apt to go unnoticed. In an adult the signs and symptoms are less conspicuous.

Primary hypergonadism is usually caused by a Leydig cell tumor, which produces large quantities of testosterone. In children, these are quickly identified and treated, in adults the problem may be recognized only at autopsy.

Hypergonadism secondary to excessive production of gonadotropins is termed precocious puberty when it occurs in a child and results in both increased testosterone production and the premature appearance of spermatozoa. It is referred to as precocious pseudopuberty when testosterone secretion is abnormally elevated, but spermatozoa are absent. The first is caused by prematurity of hypothalamic function with respect to both FSH and ICSH, whereas, in the second, only ICSH is involved. Of course, precocious pseudopuberty can also be due to an androgen producing adrenocortical lesion, or to premature sensitivity of Leydig cells to normal gonadotropin levels.

Adult males seldom seek treatment for this type of hypergonadism; consequently its status as a disorder is uncertain.

SELECTED BIBLIOGRAPHY

Gomes WR, Van Denmark NL: The male reproductive system. *Annu Rev Physiol* 36:307, 1974.

James VHT, Serio M, Martini L: *The Endocrine Function of the Human Testis,* 2 vols. New York, Academic Press, 1973.

Johnson AD, Gomes WR, Van Denmark NL: *The Testes,* 4 vols. New York, Academic Press, 1970.

Williams RH: *Textbook of Endocrinology,* ed. 6, Philadelphia, WB Saunders, Co 1981, p 293.

SELF-STUDY QUESTIONS

MULTIPLE CHOICE

Select the single best answer.

1. Reduced male fertility would likely result from
   A. a sperm count below 50×10^6/ml.
   B. exposure of sperm to a pH of 7.4.
   C. exposure of sperm to temperatures of 0°C.
   D. severe testosterone deficiency.

2. Infantile testes secrete little testosterone because
   A. they are functionally incapable of reacting to hormones.
   B. Leydig cells have not yet developed.
   C. Leydig cells have developed, but lack the necessary enzymes.
   D. of hypothalamic immaturity.

3. To be classified as androgenic hormones, steroids must possess
   A. a C_{18} configuration.
   B. a C_{21} configuration.
   C. a C_{16} configuration.
   D. a C_{19} configuration.

4. The onset of adult testicular function depends primarily on maturation of
   A. the testes.
   B. the anterior pituitary.
   C. a combination of the above.
   D. the hypothalamus.

5. Testosterone is believed to act indirectly, after conversion to dihydrotestosterone, on
   A. muscles.
   B. skin.
   C. bone.
   D. the prostate gland.

6. Normal female sexual differentiation depends on
   A. estrogens.
   B. pituitary FSH.
   C. hypothalamic activity.
   D. relatively low androgen exposure.

7. Sexual impotence in males is most commonly due to
   A. psychologic disturbance.
   B. aging.
   C. hypogonadism.
   D. precocious puberty.

MULTIPLE CHOICE

Select the correct answer(s). (In many instances, more than one answer is correct.)

8. The urinary excretion of neutral 17-ketosteroids in male adults is
   A. a reliable quantitative index of testosterone secretion.
   B. a reliable index of total androgen secretion.
   C. generally lower than in female adults.
   D. a reflection of hepatic inactivation of androgens.

9. Testosterone has the following effects on the development of long bones of the extremities:
   A. proliferation of cartilagenous endplates.
   B. periosteal deposition of bone matrix increases.
   C. increased deposition of calcium salts.
   D. accelerates closure of epiphyseal endplates.

10. Adult Leydig cell function depends directly on
    A. pituitary FSH secretion.
    B. pituitary ICSH secretion.
    C. attainment of puberty.
    D. hypothalamic maturity.

11. Hypogonadotrophic eunuchoidism is
    A. characteristic of preadolescent boys.
    B. a condition restricted to males.
    C. a sign of subnormal hypophyseal activity.
    D. endemic in certain geographic regions.

12. Precocious pseudopuberty in males is usually characterized by

A. elevated secretion of testosterone.

B. premature appearance of spermatozoa.

C. elevated secretion of ICSH.

D. elevated secretion of FSH.

13. Secretions of the seminal vesicles are

A. nutritive for spermatozoa.

B. concerned with pH adjustment of seminal fluid.

C. importantly controlled by testicular androgens.

D. add volume to the ejaculate.

ANSWERS

1. D. The sperm count is within the normal range, and sperm can be frozen without loss of activity. An acid pH is deleterious, and normal spermatogenesis requires the participation of testosterone.

2. D. Immature testes will produce viable sperm if they are hormonally stimulated. It is generally believed that hypothalamic maturity is the requisite normal initiator of spermatogenesis.

3. D. All hormones with primary androgenic activity are C_{19} steroids. While other steroids might possess weak activity, their much greater primary potency causes them to be classified under whichever primary category they may fall.

4. D. The onset of normal hypophyseal and testicular function appears to await the attainment of hypothalamic maturity.

5. D. Of the organ systems indicated, only the prostate contains the enzyme required to convert testosterone to DHT, and is the one in which testosterone acts indirectly.

6. D. Sexual differentiation in females, as opposed to sexual development, is a neutral phenomenon, the direction taken unless androgenic inhibition is imposed.

7. A. All the situations described, except the last, may be accompanied by impotence, but the first is generally believed to be the most important.

8. B, D. Male adults average higher 17-ketosteroid excretion than do women, but because many precursors are of adrenocortical origin, the contribution of testosterone is not accurately reflected in the measurement. It is, however, an index of total androgen production and reflects their hepatic inactivation.

9. All are correct.

10. B. Normal Leydig cell function depends indirectly on puberty and the hypothalamus, and directly on ICSH.

11. A, C. Hypogonadotrophic eunuchoidism occurs in both sexes. It is characteristic of prepubertal individuals and reflects hypothalamic–pituitary deficiency.

12. A, C. Precocious pseudopuberty is a disorder characterized by premature gonadal hormone production without spermatogenesis. The testosterone is usually secreted in response to ICSH stimulation.

13. A, B, C, D. All are correct.

51

The Ovaries and Female Endocrine Function

CHARLES E. HALL

OBJECTIVES

After completion of this unit, you should be able to:

1. Describe the essential functional structure of the ovary.

2. Discuss the cell types involved in hormone synthesis.

3. Describe the changes that occur in the ovary and its hormones after birth, during adolescence and reproductive life, and at the menopause.

4. List the hormonal steroids secreted by the ovary and display an understanding of their purpose, site of action, and feedback control over their tropic hormones.

5. Display an understanding of the intracellular action of estrogens and their inactivation.

6. Discuss the normal menstrual cycle, its hormonal characteristics, and the sequential uterine changes that characterize the different stages.

7. Understand the hormonal function of the mature ovarian follicle and of the corpus luteum.

8. Appreciate and describe the anabolic activity of sex steroids and its role in growth, maturation, and development.

Female endocrinology is concerned with the phenomena of individual growth and development, the functions of the ovary, and its interrelationships with the hypothalamico–hypophyseal axis on the one hand; with the primary target organs of ovarian hormones on the other; and with the characteristics of the menstrual cycle, pregnancy, and lactation.

THE SEX ORGANS

The paired ovaries, as with the testes, contain germinal epithelium, which forms ova, the germ cells that carry one-half of a female's genetic material. However, unlike the testes, the ovarian hormones are not secreted by isolated pockets of specialized cells located remotely from germinal epithelium. Instead, they are formed by cellular components of the ovarian follicles; two chemically distinct and physiologically different types of steroid hormone are secreted—both by the same cells, but at different stages of their secretory activity and under the influence of quite different tropic stimulation.

Each ovum is surrounded by a layer of epithelioid granulosa cells, forming a primordial follicle. About 500,000 are present at birth in each ovary, but by puberty 20–30% have undergone atresia, and only a total of about 300,000 remain. Of these, fewer than 500 will actually develop to the stage of ovulation, the remainder also becoming atretic. Reproductive life ceases at menopause, when only a few primordial follicles, fated to disappear quickly, remain.

OVARIAN FUNCTION

A state of normal physiological hypogonadotropic hypogonadism exists in the female until puberty, and the ovaries are relatively inactive. A growing body of evidence indicates that small amounts of gonadotropins, mainly follicle stimulating hormone (FSH), are secreted even in young children, and that even then there is probably some type of feedback control between the ovary and pituitary. The ovary is relatively quiescent however, and the full-blown reciprocal interaction between the two glands does not appear until puberty, when the hypothalamus attains functional maturity.

With the onset of puberty, a number of specific sexual characteristics appear. The age at which puberty begins is highly variable, being influenced by racial, geographic, nutritional, psychic, socio-economic, and other factors. The first occurrence of menstrual bleeding, the menarche, at any time between the ages of 9 and 16 may be regarded as normal. Earlier or later appearance suggests an abnormal process. Several months to a year will elapse before the menses become regular, and a similar period until they become painful. A large number of the cycles occurring during this span of time will be anovulatory.

Further indications of hormone activity coincident with these changes are the growth spurt, appearance of axillary and pubic hair, and breast budding. The last undoubtedly reflects the activity of ovarian hormones, but the first two are probably ascribable to the rising titers of adrenal androgens due to maturation of that adrenal function. This is known as the adrenarche.

FOLLICULAR DEVELOPMENT AND OVULATION

These processes necessitate the proper interaction between the adenohypophysis and its gonadotropins, and the ovary and its steroids. The following schema illustrates the essential features.

The immature pituitary has been secreting small amounts of FSH and LH, and the ovary small amounts of estrogen, inadequate to elicit detectable changes in the target organs. The hypothalamus, meanwhile, has been undergoing functional maturation. There is, even prior to this time, some feedback control over the hypothalamic "gonadostat," since the administration of small quantities of androgens or estrogens will suppress the low levels of circulating gonadotropin in the serum of children. Maturation of the gonadostat function confers increased sensitivity upon it, and its heightened activity initiates a surge in circulating FSH. This, in turn, causes a burst of activity in a certain population, probably about 20%, of the ovarian follicles. In these there occurs proliferation and growth of granulosa and theca cells, which secrete a fluid rich in estrogens, and leading to the formation of an antrum or pocket, within them (Fig. 1).

As FSH acts on the ovary, many follicles and the contained ova enlarge. Additional layers of granulosa cells surround the ova, and, in turn, are enveloped by several layers of theca cells. The latter are the ovarian homologues of the Leydig cells and secrete the ovarian steroid hormones. They become epithelioid in character, constitute the theca interna, and are themselves encircled by a connective tissue capsule, the theca externa.

The ovaries are attached by means of the ovarian ligament to the fallopian tubes, into which ova are discharged.

FOLLICULAR GROWTH AND DEVELOPMENT

Until puberty, the ovaries are essentially inactive, producing little estrogen, no progesterone, and containing quiescent follicles. Maturation of the hypothalamus leads to significant secretion of GnRH, increasing the elaboration and secretion of FSH and LH from the anterior pituitary gland. FSH, as in the case of the testes, is concerned with the development of germ cells, that is, it is a gametogenic hormone. LH, also as in the testes, is concerned with ovarian hormone secretion and is an hormonogenic hormone.

Testosterone production in the human male testes is a continuous process during reproductive life, cyclic fluctuations of secretion occurring within a 24-h period, but with scant variation in day-to-day levels. Estrogen production in the human female during reproductive life is a cyclic phenomenon, waxing and waning over a monthly period, the minima being attained at the menses, when the products of a degenerating endometrium are discharged and a new cycle begins. Significant progesterone secretion occurs only after ovulation, or, as sometimes occurs, the ovum becomes entrapped and is not released. Granulosa and theca cells continue to proliferate after antrum forma-

tion, and the very large population of theca cells, stimulated by even the relatively low levels of circulating LH, secrete increasing amounts of estrogen into the circulation. Continued growth leads to the development of vesicular follicles: thereafter, cellular proliferation is concentrated at that pole of the follicle bearing the ovum.

THE DOMINANT FOLLICLE AND OVULATION

A number of ova will ovulate in species in which the females bear litters, but in humans and in the larger mammals, as a rule, a single ovum is released. Exceptions can lead to multiple births. A single follicle gains ascendency, and the remainder of the growing follicles undergo regressive or involutionary changes (atresia). It is believed that the dominant follicle is slightly more advanced than the others. Together the population of growing follicles secrete enough estrogen to inhibit the hypothalamico–hypophyseal axis, and FSH levels fall, removing growth support from the follicles. The dominant follicle requires less hormonal support, either because it is more mature and its cells are therefore more responsive to estrogen, or because the local estrogen concentration provided by its own secretory activity synergizes with FSH, permitting continued growth despite declining FSH levels.

The dominant follicle continues to grow, ultimately reaching 1–1.5 cm in diameter. The later stages of this growth phase depend on LH, as does ovulation. It is presumed that at this stage of the cycle, the rising estrogen levels, which through negative feedback have been inhibiting FSH and LH release, now exert a positive stimulatory effect, resulting in a preovulatory rise of both, but the LH increase is three- to fivefold greater than that of FSH. This surge of stimulatory hormones, occurring about 48 hours preceding ovulation, induces rapid follicular swelling, and a portion of the follicular wall thins as the volume expands. Fluid begins to leak through this weakened area; soon after it ruptures, it extrudes the ovum. The rising levels of estrogen also cause cyclic changes in the mucosa and musculature of the fallopian tubes. Prior to ovulation, rhythmic contractions of the fallopian tube musculature

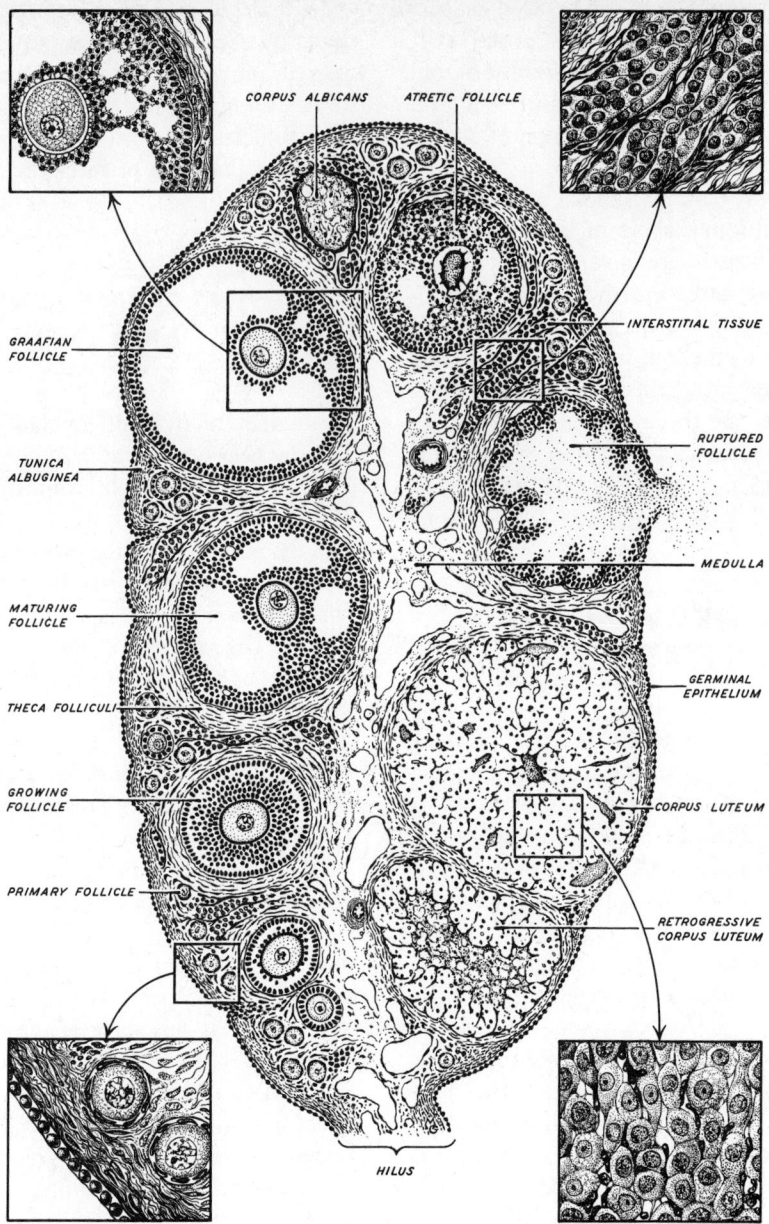

Figure 1 Diagram of a composite mammalian ovary. Progressive stages in the differentiation of a graafian follicle are indicated on the left. The mature follicle may become atretic (top) or ovulate and undergo luteinization (right). (Turner CD: *General Endocrinology*, W.B. Saunders Co., Phila., 1966, p 400.)

begin, bringing its fimbriated orifice into apposition with the ovarian surface. Ciliary action increases within the tubular lumen, moving the ovum and follicular fluid toward the uterus.

Ovulation in women can take a few seconds or as much as 15 min and will occur in four out of five women exactly 14 days before menstruation begins, even though each may have cycles of different duration (Table 1). In some mammals (e.g., the rabbit), ovulation can be initiated in reproductively active females at almost any time by forms of sexual excitement, particularly by copulation. Nevertheless, in many, mating does not occur unless the ovum is at a stage where ovulation is possible, one signalled by female receptivity. There has been speculation as to whether or not ovulation may be induced by sexual excitement in some women, once the follicle has reached a critical stage of development, but no decision has been reached.

Once the ovum has been extruded, or even when, as sometimes happens, it has not (entrapped ovum), the follicular theca cells change both their morphology, and their capacity for hormonal biosynthesis, under the influence of LH. They become fat-laden, yellowish bodies (corpora lutea), which now, in addition to the enzymatic complement necessary for estrogen synthesis, contain those yielding progesterone. The two hormones have actions on a number of structures, including the uterus and its appendages, the breast, vaginal mucosa, and functions such as the psyche, as we shall see.

In the rat and other lower mammals, LH causes transformation of the follicle into a corpus luteum, but the action of another pituitary hormone, prolactin or luteotropic hormone (LTH) is required for progesterone synthesis. This does not appear to be the case in women. Apparently progesterone synthesis begins with the LH surge, even before the morphological change (luteiniza-

tion) is visibly evident. It can even participate in the ovulation process, which is preceded by follicular swelling and the appearance of proteolytic enzymes that dissolve the capsular wall.

THE CORPUS LUTEUM AND ITS FUNCTION

Transformation of the follicle into a corpus luteum proceeds over several days in a normal menstrual cycle, during which it progressively enlarges and incorporates an increasing amount of lipid. This may aid in the accommodation of larger amounts of lipid-soluble steroids than would otherwise be possible. In about 1 week it reaches maximal development, and the quantities of estrogen and progesterone secreted suffice to suppress the pituitary gonadotropins (FSH and LH) providing its support. Withdrawal of FSH and LH causes it to lose gradually much of its ability to secrete hormones, or to maintain its physical structure. The process of involution proceeds, until by about the 26th day of the cycle it has become a whitish corpus albicans, which eventually disappears entirely, being replaced by reparative connective tissue (Fig. 1).

THE NEW CYCLE

Very large quantities of estrogen and progesterone are secreted by the follicle after ovulation, particularly when it has become a fully formed corpus luteum. Consequently both the hypothalamus and pituitary gland are greatly inhibited, and the circulating levels of both FSH and LH are very low. Follicular growth is therefore suppressed. Except for the dominant follicle, the ovary is functionally quiescent.

Involution of the corpus luteum on the 26th day of the cycle removes feedback inhibition of the hypothalamico–hypophyseal axis, restoring normal function to it. Both FSH, which stimulates growth of the follicles, and LH, which controls their secretory capacity, now increase, the

Table 1 Relationship between Length of Menstrual Cycle and Day of Ovulation

Cycle Length (days)	Ovulation Day	% of Cycle Length
21	7 ± 1	$\frac{1}{3}$
28	14 ± 1	$\frac{1}{2}$
40	26 ± 1	$\frac{2}{3}$

former proportionately more than the latter, and a new cycle is initiated. This process is repeated monthly throughout reproductive life. The hormonal changes during the cycle are shown in Figure 2.

It should be noted, however, that at both ends of reproductive life, during the year or so following menarche, or preceding menopause, many cycles are anovulatory. In the intervening years also, there may be few or many of these. Since the hypothalamus is both a part of the central nervous system, as well as the source of releasing hormones (which also control synthesis of hypophyseal gonadotropins), it is obviously susceptible to inputs from other centers in the central nervous system (CNS). Psychic factors, therefore, very importantly influence female reproductive physiology. Normalcy of the cycle depends on the proper sequencing and concentration of gonadotropins and the ovarian hormones. Suitability of the uterus for implantation of a fertilized ovum depends on the proper sequencing and concentrations of steroid hormones from the ovary. It follows, therefore, that any disruption of the pattern will adversely affect reproduction.

For these reasons, and because in the male timing is not critical, female reproductive function is much more vulnerable to psychological disturbances than that of the

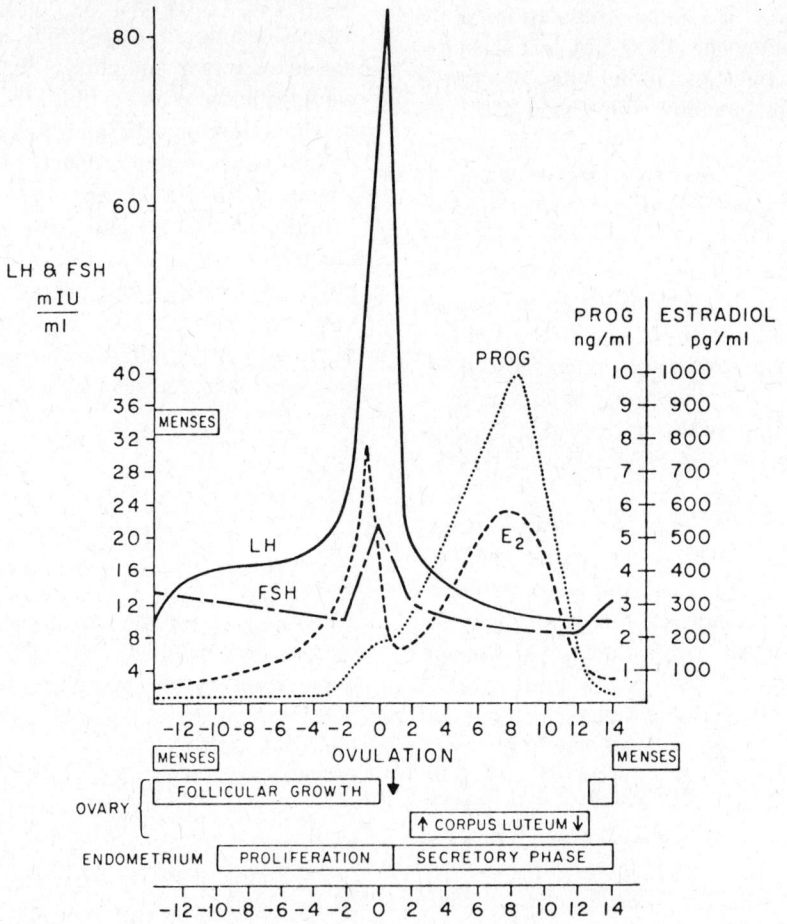

Figure 2 Circulating hormone levels during ovarian and endometrial events in the human. (Modified, with permission, from Speroff L, Vande Wiele RL: "Regulation of the Human Menstrual Cycle," *Am J Obstet Gynecol* 109:234, 1971.)

male. Many women experience abnormalities of various kinds in the cycles. These may be of irregular length, or of longer or shorter duration; there may be anovulatory cycles, menstrual flow may be scanty, exuberant, shortened, or prolonged. This may have organic, nutritional, or other causes, but is traceable in many instances to disturbed emotional or mental equilibrium. The hypothalamic centers controlling the sympathetic nervous system, and therefore involved in "fight or flight" adaptations, are contiguous to those modulating gonadal function. In the female, prolonged sieges of rage, fear, worry, pain, and the like, are inimical to normal physiological operations of the hypothalamus, pituitary, ovaries, and uterus. Consequently, menses may be irregular or otherwise abnormal. Constitutionally temperamental women are more susceptible to this consequence of emotional turmoil than are those of a phlegmatic nature.

OVARIAN HORMONES

Two types of ovarian steroids are normally secreted. Those classed as estrogens have 18 carbon atoms and an unsaturated A ring in the nucleus. Progesterone, on the other hand, classed as a luteoid hormone, or a progestagen, has 21 carbon atoms and is structurally related to the adrenal cortical hormones. The structural formulas of these steroids are given in Figure 3.

ESTROGENS

These steroids differ from others secreted by gonads and the adrenal cortices in having a completely unsaturated, phenolic, A ring. Small amounts are secreted by the adrenal cortices and human testes; in some species (e.g., the horse) the testes secrete large quantities, but extraovarian estrogen is of little physiological significance in humans, except in certain pathologic states. During pregnancy, the placenta, which becomes a sort of combined pituitary–ovary and adrenal cortex, secretes huge quantities.

Steroid hormones are classed as estrogens (folliculoids),

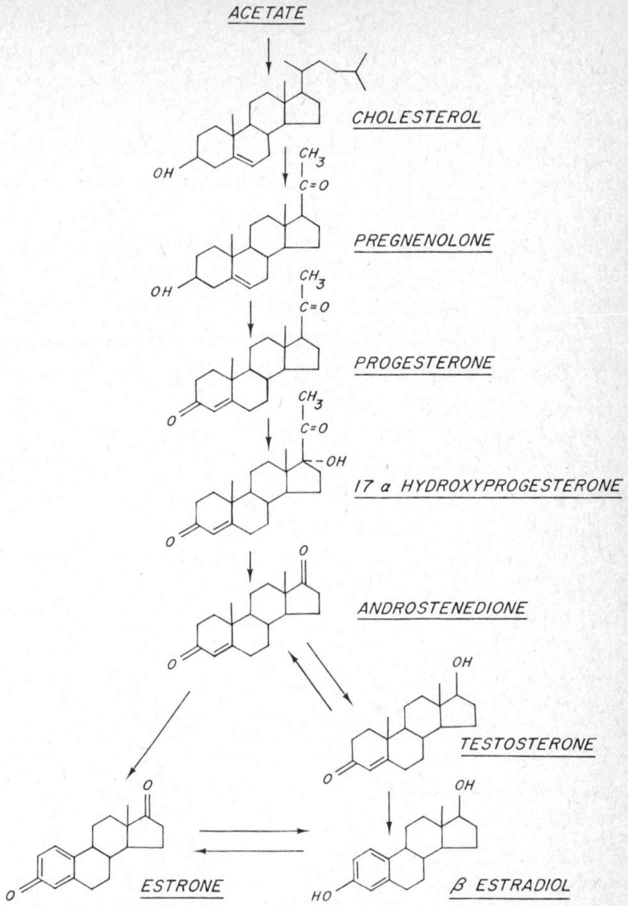

Figure 3 Biosynthesis of ovarian hormones. Typically, the ovarian hormones are derived from cholesterol by the steps shown. Prior to corpus luteum formation, the small quantities of progesterone produced are largely converted to estrogens. The corpus luteum produces much larger total quantities of progesterone, so that despite the secretion of large amounts sufficient quantities remain to cause even larger secretions of estrogen.

progestagens (luteoids), corticoids, or androgens (testoids), and the first classification is the only one exhibited to the exclusion of any of the others. That is, while members of the three other classes might fall into more than one of the categories, estrogens do not. Certain androgens have progesterone-like actions; progesterone can act as a weak mineralocorticoid, and the corticoids may have androgenic properties, but estrogens exhibit no overlapping hormonal properties. Furthermore, estrogenic activity does not even require a steroid configuration. Diethylstilbestrol (DES), used extensively to fatten cattle, is a more potent estrogen than is β-estradiol and is not a steroid.

Three estrogenic substances are present in female plasma in concentrations high enough to be physiologically important. Two of them, β-estradiol and estrone, are ovarian secretions that are metabolized by peripheral tissues into the third, estriol. Estrone is a 17-ketosteroid, but since the A ring is unsaturated not a neutral 17-ketosteroid, and therefore does not interfere in the determination of androgen metabolites. Estradiol has a potency about 12 times that of estrone, which is, in turn, much more active than estriol and is therefore considered to be the principal ovarian hormone (Fig. 3). Estrone is also secreted by the ovary, but the great preponderance of estrone in the female blood stream is derived from the degradation of androstenedione, synthesized by the ovary and by the adrenal cortex, by peripheral tissues.

INTRACELLULAR ACTION OF ESTROGEN

The action of estrogen on and in cells is strikingly reminiscent of what has already been described for androgen. Upon entering the blood stream most of the estrogen promptly couples to a carrier β-globulin, and is transported to tissues in the bound form. Bound estrogen cannot enter cells. Therefore, either there is uncoupling at the cell surface, or free, unbound estrogen enters cells, thus disturbing the equilibrium of the bound-free relationship and causing uncoupling of estrogen from the estrogen carrier–protein complex to correct the resulting disequilibrium.

Penetrating the cell membrane, estrogen enters the cytosol, where it combines with a specific cytosolic receptor protein, a reaction that alters the configuration of the protein. The complex then enters the nucleus, where it reacts with certain constituents to stimulate by the process of transcription (via DNA-directed synthesis of mRNA), the synthesis of specific proteins with structural or functional attributes.

METABOLISM OF ESTROGENS

The levels of estrogen in the bloodstream reflect the balance struck between continued hormone synthesis by the ovary and/or placenta, and hormonal destruction by peripheral tissues. The principal organ effecting removal is the liver. Hepatic failure, as from cirrhosis, etc. impairs the ability to inactivate estrogens. Among the results of this, in males, is impaired testicular function, (from suppression of gonadotropins), and gynaecomastia from the direct estrogenic effects on the mammary glands.

Three separate hepatic functions are involved in estrogen metabolism. Active estrogens, such as β-estradiol and estrone, are oxidized to minimally active estriol: estrogens are conjugated to form esters, chiefly sulfates and glucuronides, which being more water soluble, can effectively be excreted, and finally, about 20% are excreted along with bile, into the intestines.

THE BIOLOGICAL ACTION OF ESTROGENS

Quantitatively, the estrogens are by far the most potent of steroid hormones. The housing of animals used to bioassay estrogens cannot be completely freed from hormone contamination by normal washing procedures, and the remaining minute traces can exert hormonal effects.

The small quantities of estrogen secreted by prepuberal females undoubtedly has biological activity, but it is difficult to delineate precisely. At puberty a large surge in estrogen production by the developing ovarian follicles occurs, with concomitant physical changes. Until this time, the general habitus of females and males is not dramatically different, but thereafter distinctive differences begin to appear.

The externally visible changes are impressive, but the internal structural modifications are equally distinctive. The estrogens have anabolic activity, albeit less potent than androgens in this respect, but the close temporal relationship between menarche and adrenarche assures adequate levels of both to encourage the puberal growth spurt. Because estrogens more effectively than androgens bring about union between bone shafts and epiphyses, estrogen-induced increase in osteoblastic activity is curtailed sooner than in males, final height being achieved earlier and at a smaller stature. Females with delayed puberty, or hypogonadism, as in similarly affected males, become taller then they otherwise would have been, because of an extended growth period. The reduced effectiveness of estrogens, as contrasted with androgens, on protein anabolism and bone growth, is paralleled by a less prominent retention of calcium and phosphate, and a smaller muscle mass. In females as compared with males, estrogens are also anabolic with respect to fat, with the result that female adults, as compared with young girls or adult males, have a greater percentage of adipose tissue. It is deposited subcutaneously, in the breasts, and in the buttocks and hips, giving the typical female contours notably lacking before puberty or in females with hypogonadism.

There is some thickening of the skin by estrogens, but being accompanied by increased vascularity, it is softer and more delicate than in males. The effect of estrogens on epithelial growth is much more conspicuous in the vagina, to be described later.

ESTROGENS AND THE FEMALE REPRODUCTIVE ORGANS

The pubertal estrogen surge has profound effects on the internal and external genitalia and accessory organs. In the uterus, this causes proliferation and thickening of the endometrium, the accessory glands and its muscular coat, the myometrium, which develops increased tonus. This is accompanied by increases of alkaline phosphatase, endometrial glycogen content, and tissue fluids. Growth of spiral arterioles in the endometrium is induced.

Analogous changes occur in the structure and function of the fallopian tubes. The number and activity of ciliated epithelial cells increase: rhythmic contractions of the musculature are stimulated, and there is an expansion of the glandular component of the endothelium.

The labia minora, labia majora, and clitoris all increase in size and vascularity, and are more edematous, under the influence of estrogens. Growth of the clitoris, however, is dominated by adrenal androgens.

CERVIX AND VAGINA

Estrogens cause lengthening of the columnar cells lining the endocervical glands and enhance their metabolic activity. The result is a greater volume of cervical mucus of decreased viscosity, which facilitates the survival and movement of sperm cells.

The vaginal epithelium is quite thin in prepuberal and postmenopausal women, making it vulnerable to injury and them susceptible to various forms of vaginitis, but is greatly thickened under the influence of estrogens. As the proliferating superficial cells increasingly separate from their blood supply, they undergo loss of cytoplasm and cornify. At the same time, the glycogen content is increased, and the pH of vaginal fluid decreased.

MAMMARY GLANDS

Estrogens increase the adiposity of mammary glands, accompanied by duct proliferation, increasing pigmentation, and nipple growth. These preparatory steps set the stage for milk production. But before that occurs, development of the alveoli under progesterone, secretory capacity through the influence of prolactin, and a large number of other hormones will be required.

PROGESTERONE

The corpus luteum secretes both estrogens and progesterone. Many of the functional and structural changes initiated or dependent on progesterone require the prior action of estrogen. Obviously, alveolar development in mammary glands would be purposeless, without the development of ducts to convey the milk. In some respects the two hormones act synergistically, whereas in others they are antagonistic. Synthetic progestogens have androgenic effects, causing masculinization of external genitalia in female fetuses and hypospadias in males.

EFFECTS ON THE UTERUS

Estrogens cause the uterus to undergo what is termed the "proliferative" phase of development, whereas progesterone initiates and maintains the "secretory" phase. The latter is essential to successful implantation and nidation of the ovum, either of which is unsuccessful in a proliferative endometrium. Endometrial stromal cells swell and come to contain large quantities of nutrient materials on which nidation depends, including protein, glycogen, lipids, and electrolytes. Uterine motility is inhibited. Further growth of the endometrium and of its coiled, spiral arteries is encouraged.

EFFECTS ON THE FALLOPIAN TUBES

Progesterone decreases ciliary activity and motility in the fallopian tubes. The lining cells contain increased quantities of nutrient materials to aid in survival of the ovum until it arrives at uterine sites suitable for implantation.

EFFECTS ON THE MAMMARY GLANDS

Progesterone promotes lobular–alveolar development in the breast tissues and increases their fluid contents. The glands are thus equipped with the structural components essential to milk secretion and transport, but will require activation by still other hormones before lactation occurs.

EFFECTS ON THE CERVIX

Progesterone antagonizes the stimulatory effect of estrogens on the cervical epithelium, the cells of which become smaller. Less mucus is formed, and its viscosity decreases.

EFFECTS ON THE VAGINA

Superficial layers of the vagina are lost, and the epithelial layer becomes thinner and shows leukocyte invasion. The shed surface cells show a curious folding, and the thin cytoplasm is less acidophilic than it is under estrogen stimulation.

EFFECTS ON TISSUE METABOLISM

Progesterone has a catabolic effect, believed to account for the rise in basal body temperature (about 1°C) that occurs immediately following ovulation. Perhaps, during pregnancy, the property is of value in providing nutrients to the developing fetus. Non-ovulatory cycles are not, except for those in which the ovum is entrapped, accompanied by progesterone secretion, or by a rise in basal body temperature. The latter is a reliable index of ovulation, when intercurrent infections and the like can be excluded, but is difficult to estimate unless special precautions are taken to avoid activities (e.g., eating, smoking, exercise) that interfere with reliable measurements.

Progesterone resembles certain of the adrenocortical steroids rather closely, and may act as a weak mineralocorticoid (a cortical hormone affecting mineral metabolism). In adrenalectomized animals, which lose sodium excessively and hence lose ECF volume but fail to excrete potassium normally in the urine, progesterone will enhance the resorption of sodium, chloride, and water by the kidney and promote potassium excretion. When given to intact animals, however, sodium and water excretion actually increase. This is because the weak mineralocorticoid progesterone competes with the strong one, aldosterone, for renal receptor binding sites. To the extent that aldosterone is replaced then, there is sodium and water loss. To the adrenalectomized animal that has no mineralocorticoid receptors occupied, even a weak mineralocorticoid is better than none, leading to sodium and water retention by renal tubular cells.

HORMONAL INTEGRATION AND THE MENSTRUAL CYCLE

It is convenient to begin with the hypothalamus, because although realizing that other centers in the CNS might be entitled to primacy under some circumstances, initiating activity that activates the hypothalamus, we know little about their role or location.

In an earlier section, it was emphasized that although there is some gonadal function in the prepubertal phase, it is neither physiologically significant nor cyclic. Both features depend on intrinsic maturation of the hypothalamus, which occurs at puberty. It is presumed that it secretes the releasing hormones because feedback inhibition from gonadal hormones is inadequately supressive. The resulting gonadotropic hormone secretion stimulates the production of estrogens, and later in the cycle, progesterone, until they reach levels that cause feedback inhibition. Gonadotropin-releasing hormone (GnRH) is synthesized in cell bodies of neurons in the median eminence of the hypothalamus and is transmitted by axoplasmic flow to nerve terminals, where it is released and transported via the portal vessels to the anterior pituitary cells that produce gonadotropins. It is a decapeptide, having the formula   GLU-HIS-TRP-SER-TYR-GLY-LEU-ARG-PRO-GLY-NH$_2$.

Until recently it was believed that each of the gonadotropins FSH and LH responded to a specific releasing hormone. The current view is that GnRH releases both. Which one is secreted depends on intrinsic pituitary responsiveness. Clearly, GnRH can and does release FSH, making it unnecessary to presume a specific FSH-RH; however, a few investigators do not believe that the latter neurohormone has been completely excluded by the available evidence.

Hypothalamic hypophysiotropic control can be visualized as follows. Because of inadequate feedback suppression by estrogens, the mature hypothalamus begins to secrete GnRH, which causes the pituitary gland to secrete FSH and LH. The first, which has been shown to bind only to granulosa cells (in the rat ovary), stimulates growth of the ovarian follicles and the second causes secretion of estrogens. Estrogens then exert feedback suppression on both the pituitary (to limit the sensitivity of gonadotropin-secreting cells to LRH), and on the hypothalamus (to reduce its LRH secretion). The second of these is the more important.

The relationships between the hypothalamus–pituitary–ovary axis are complicated. They are imperfectly understood at present and can not be appreciated at all if one presupposes that endocrine regulation depends only on quantitative relationships between the various hormones involved.

Once hypothalamic releasing hormone has effected gonadotropin release and thus ovarian hormone secretion, several types of feedback, both positive and negative, have been demonstrated. Pituitary hormones control their own secretion by two internal feedback mechanisms. The first, autofeedback, is a process of direct interaction between the gonadotropins and the pituitary proper; the second, a short-loop feedback, involves gonadotropic inhibition of hypothalamic activity. Added to this is steroid feedback, both on the hypothalamus and on the pituitary. Estrogen and progesterone tend to inhibit both the pituitary and the hypothalamus, but Gn-RH causes the pituitary to release relatively more FSH when estrogens are low, and to release relatively more LH when they are high. The balance of evidence indicates that the same cells secrete FSH and LH, and that only one releasing hormone, GnRH, acts on them. The type of hormone actually elaborated therefore depends on the quantity of stimulation that acts on them, their receptivity to it, and other factors affecting their enzymatic machinery and substrate availability.

In castrate female animals, where removal of feedback inhibition results in elevated blood levels of both FSH and LH, small doses of estrogen cause a prompt fall, but the pituitary content of both is further elevated. Thus, gonadotropin release is more inhibited than synthesis. Furthermore, in castrate females, where the usual rise in circulating gonadotropins has been suppressed by estrogen treatment, a single additional dose of estrogen causes a prompt surge in circulating LH. These findings demonstrate that a positive feedback on LH release can be superimposed upon, and is therefore independent of, its negative feedback impact on gonadotropin secretion. This type of response accounts for the surge in LH that precedes and initiates ovulation and occurs in spite of a general feedback suppression of gonadotropic hormone secretion.

Another interesting observation is that in hormone-responsive cells, there is often a reciprocal relationship between the prevailing hormone levels and the available surface binding sites. Thus, when hormone stimulation is interpreted by the cell as excessive, it protects itself by reducing the reactive sites made available, and, conversely, when hormone levels are low, by increasing them commensurately. This phenomenon helps explain a fundamental tenet in endocrinology, that when a hormone is absent, or essentially so, an exquisite sensitivity to it results.

UTERINE CHANGES IN THE MENSTRUAL CYCLE

Sloughing and growth of the endometrium during the menstrual cycle are easily visualized if it is recognized that two different layers are involved. One of them, the basalis, does not participate in the sloughing, but serves as the regenerative base for growth in the succeeding cycle. It has its own straight arterioles that do not lengthen under hormonal influences, and hence its blood supply is not compromised by the events leading to desquamation of the second layer, the functionalis. The integrity of the functionalis, which thickens during the first 26 days or so of a 28-day cycle, depends on a second set of arterioles that

lengthen and coil as it grows—these ultimately rupture—withdrawing nutritive support and dissecting its layers with blood lacunae.

The usual 28-day menstrual cycle can be divided into four recognized stages: (1) the proliferative, (2) the secretory, (3) the ovulatory, and (4) the menstrual. Menstrual cycles may vary in length from 20 to 45 days and oscillate from one month to another during irregular cycles.

PROLIFERATIVE (FOLLICULAR) PHASE

The proliferative phase has a duration of 14 days in a typical cycle. Mitotic activity in the stroma increases, its cells assume a spindle shape and it becomes more edematous. The edema wanes and has almost disappeared by the time ovulation occurs. The covering is a single layer of columnar epithelial cells. At intervals, there is an infolding of the surface to form crypts or glands that penetrate to the basalis. These are generally straight in the proliferative phase, but become increasingly tortuous toward the end, and highly so during the following phase. An arteriolar system grows along with the stroma. Since both the glandular epithelium and the spiral arterioles grow more rapidly than does the stroma, they become progressively more tortuous, especially during the secretory phase.

SECRETORY (LUTEAL) PHASE

The endometrium now continues under estrogen stimulation and comes under the influence of increasing quantities of progesterone. Maximal endometrial hypertrophy occurs in this phase and it becomes about 5 mm thick. Under progesterone stimulation the epithelial cells become laden with glycogen and glycoprotein, present in only trace amounts during the proliferative phase. Both the glands and arterioles of the functionalis become coiled and increasingly tortuous as their growth proceeds more rapidly than that of the stroma, forcing accommodation to a progressively restricted space. The stroma becomes quite edematous and highly vascular.

OVULATORY PHASE

The ovulatory phase refers to the period when ovulation occurs, and is not characterized by uterine changes.

MENSTRUAL PHASE

Toward the end of the luteal phase, the high levels of estrogen and progesterone exert increasing feedback inhibition on the secretion of the pituitary gonadotropins, which progressively diminish. This removes structural and functional support from the corpus luteum, which then begins to involute. The declining levels of estrogen and progesterone become inadequate to support the lush uterine stroma, which begins to lose fluid. This causes the endometrium to shrink. The coiled spiral arterioles begin to collapse, accommodating to the lesser thickness, and become progressively kinked. Ruptures eventually occur at these kinks and the escape of blood produces lacunae, which dissect the layers. These areas become necrotic and slough away. Eventually, the endometrial functionalis sloughs away, over several days, back to the basalis with its independent vascular supply, and the stage is set for the next cycle.

THE CLIMACTERIC AND THE MENOPAUSE

The end of reproductive life is signaled by a period of irregular ovarian activity, declining ovarian hormone secretion, increased gonadotropin secretion, and increasing numbers of anovulatory cycles. This extends over a period of several years, known as the climacteric, and terminates in the cessation of ovarian activity at the menopause. In most women this will occur before the age of 50, and is signaled by cessation of menses.

As the aging ovaries become less proficient at secreting estrogens, the negative feedback inhibition of pituitary gonadotropin secretion becomes less effective. Both FSH and LH rise, the latter less than the former, which should stimulate the ovary into increased activity, but are unable to effect that purpose.

The reduction in circulating estrogens withdraws support of the vaginal epithelium, which becomes thinner, drier, and more vulnerable to irritation and infection. The breasts also undergo involutional changes. Some estrogen secretion persists, probably from the adrenal glands. Because of this, and because women differ in their response to declining estrogen levels, some women experience more difficulty than do others.

SELECTED BIBLIOGRAPHY

Cowell CA, Wilson R: The ovary, in C Ezrin, JO Godden, R Volpe, Wilson R (eds): *Systematic Endocrinology*. Hagerstown MD, Harper & Row, 1973, p 213.

Eskins BA: Female endocrinology—The ovary and menstrual disorders, in Schneeberg NG (ed): *Essentials of Clinical Endocrinology*. St Louis, CV Mosby, 1970, p 359.

Williams RH: *Textbook of Endocrinology,* ed. 6, Philadelphia, WB Saunders, 1981, p. 355.

SELF-STUDY QUESTIONS

MULTIPLE CHOICE

Select the single best answer.

1. A hormone arising principally from peripheral tissues is
   A. estrone.
   B. estradiol.
   C. estriol.
   D. dehydroepiondrosterone.

2. Estrogen in the bloodstream binds chiefly to
   A. albumen.
   B. an α-globulin.
   C. a β-globulin.
   D. an α_2-globulin.

3. The menarche is
   A. the beginning of female reproductive life.
   B. the beginning of menses.

C. the inception of ovarian activity.

D. abnormally early if it occurs at the age of 9.

4. Ovulation is a process that

    A. is triggered by a surge of FSH.

    B. is triggered by a surge of LH.

    C. is facilitated by high estrogen levels.

    D. requires sexual excitation.

5. Estrogen activity

    A. is confined to C_{18} steroids.

    B. is greater in estriol than in estrone.

    C. is transported by β-globulin in blood.

    D. exerts its primary cellular action in the cytosol.

MULTIPLE CHOICE

Select the best answer(s). (In many instances, more than one answer is correct.)

6. Of the original complement of primordial follicles, the number that will actually ovulate during the typical reproductive period is about

    A. 500.

B. 5,000.

C. 1%.

D. 25%.

7. Estrogenic hormones are secreted chiefly

    A. by granulosa cells.

    B. by theca cells.

    C. in response to FSH.

    D. in response to LH.

8. Estrogenic activity on the uterus includes

    A. proliferation of the endometrium.

    B. hypertrophy of the myometrium.

    C. increased muscle tone.

    D. increased coiling of the arterioles.

9. Increasing levels of circulating estrogen in the preovulatory phase of a normal menstrual cycle

    A. reflects beginning activity of a corpus luteum.

    B. will cause thickening of the uterine myometrium.

    C. reflects FSH stimulation of ovarian steroidogenesis.

    D. reflects increased activity of theca interna.

ANSWERS

1. C. Of the steroids listed, estriol is the only one largely formed by peripheral tissues. Estradiol and estrone are secreted primarily by the ovaries and dehydroepiandrosterone mainly from the adrenal cortex.

2. C. Estrogen-binding protein is a β-globulin.

3. B. The onset of menses signals the beginning of menarche. Ovarian activity begins in the fetus, and reproductive life does not begin until ovulation occurs, some time after the menarche. Age 9 is not too early.

4. B. A surge of LH triggers ovulation. FSH is usually falling, and high estrogen levels are not facilitatory. Sexual excitation is required only in a few species.

5. C. Estradiol is more potent than estrone, and diethylstilbestrol, a nonsteroid, is even more so. Estrogens are

transported in blood by a β-globulin and act principally within the cellular nucleus.

6. A, C. About 500 (or 1%) of the original follicular complement will ovulate.

7. B, D. Estrogens are secreted primarily by theca interna cells under stimulation by LH. Granulosa cells secrete some estrogen and FSH determines their level of activity.

8. All are correct.

9. B, D. There is no corpus luteum before ovulation, and FSH has no direct effect on estrogen secretion. Rising estrogen levels reflect stimulation of the theca interna by LH, and cause thickening of the myometrium.

52

Pregnancy, Parturition, and Lactation

CHARLES E. HALL

OBJECTIVES

After completion of this chapter, you should be able to:

1. List the steps incident to fertilization, as well as state its usual timing and location.

2. The hormonal and nutritive functions of the placenta, the timing of its formation, and functional decline.

3. The essential role subserved by the corpus luteum during pregnancy, as well as its hormonal functions.

4. The various organs, tissues, and physiological processes affected, respectively, by estrogens and progesterone, and how and where these hormones are synthesized.

5. The hormonal and mechanical factors involved in parturition.

6. The modus operandi of certain contraceptive methods.

7. The fetal contribution to the hormones of pregnancy.

8. The anatomical and hormonal requirements for lactation.

The purpose of this chapter is to (1) present the salient features of fertilization, as well as how, when, and where it usually occurs; (2) consider the function and life of the placenta; (3) compare and contrast the contributions of the ovary and the placenta with the maintenance of pregnancy; (4) consider the hormonal changes that initiate, support, and terminate pregnancy; (5) consider the contribution of the fetus to the hormonal support of pregnancy and its own physiological requirements; (6) describe the act of parturition; (7) consider various contraceptive methods; and (8) describe lactation and its hormonal and anatomical features.

If everything goes the way nature intended, which is by no means always the case, the encounter between an active spermatozoon and a mature ovum will result in fertilization. Unless this is followed by successful implantation somewhere in the genital apparatus, ideally on the surface of the uterine endometrium, the process ends there. If, however, implantation is successful, a state of pregnancy or gestation ensues, during which the fertilized ovum develops into a fetus.

Sperm, aided by propulsive movements of the uterus and fallopian tubes, reach the entrance of the latter within about 5 minutes of their start on the journey. Fertilization usually occurs at about the midpoint of the tube. The fertilized ovum, or blastocyst, makes its way to the uterus, where it becomes implanted. A placenta then develops and pregnancy is under way.

The placenta immediately begins to secrete chorionic gonadotropin, a glycoprotein of about 30,000 MW that structurally and functionally resembles LH. This sustains the corpus luteum and its hormones. The latter, in turn, support the uterus and prevent endometrial sloughing and vaginal bleeding. A menstrual period is then missed and will fail to recur until the fetus is delivered. Chorionic gonadotropin, which can be detected within 8 days of ovulation, rises to a maximum at 7 weeks and declines thereafter. It accounts for "positive" pregnancy tests, its function is to support corpus luteum function for the first trimester of pregnancy, which will terminate otherwise. Thereafter, the placenta takes over, producing sufficient estrogen and progesterone so that the ovaries are no longer needed to maintain pregnancy.

Placental estrogen is largely estriol, a negligible component in nonpregnant females. It arises mainly from transformation of the substrate dehydroepiandrosterone, a steroid secreted by fetal adrenal glands and supplied to the maternal circulation. In addition to supporting the uterus, placental estrogens stimulate glandular development of the mammary glands, enlarge the female genitalia, and relax the pelvic ligaments, thus facilitating fetal delivery.

Maternal progesterone levels, largely derived from the placenta, rise tenfold during pregnancy, during which the hormone has several functions.

1. Encouraging the development of decidual cells in the uterine endometrium, nutritionally important to the fetus.

2. Increasing uterine secretions (e.g., glycogen), which nourish the developing fetus.

3. Inhibition of uterine contractions, thus forestalling spontaneous abortion.

4. Stimulating lobular–alveolar development of the mammary glands.

It will be evident from the foregoing that the placenta assumes, in effect, the functions of both the adenohypophysis and ovaries insofar as female reproductive hormones are concerned. Initially, chorionic gonadotropin is needed to support ovarian secretions during the first trimester of pregnancy, since the placenta has not yet become capable of adequately secreting steroids. Once this capacity has been attained, however, both adenohypophysis and ovaries are dispensable to gestation.

Additional adenohypophyseal-like hormones are secreted by the placenta, chiefly, beginning at about the fifth week of pregnancy, and in increasing quantities until term, placental somatomammotropin, a protein of about 38,000 MW. Among its important functions is that it mimics prolactin, causing milk secretion by the developed breast. In addition, it has growth hormone-like effects on deposition of protein, which may be of help to the developing fetus, and on carbohydrate metabolism. Under experimental conditions it favors the development of a diabetic glucose-tolerance curve. It may reduce glucose utilization by the mother, thus conserving it for the fetus.

The metabolic demands of pregnancy also are accompanied by changes in the secretion of other hormones. The high estrogen and progesterone levels inhibit pituitary FSH and LH secretion, which become vanishingly low. Corticotropin and thyrotropin secretion increase. Growth hormone probably also increases, but the similar structure and function of placental lactogen make it difficult to quantify the two.

The secretion of mineralocorticoids and glucocorticoids by the adrenal cortex also increases. The rising aldosterone secretion leads to sodium and water retention, contributing to the weight gain during pregnancy. Increased thyrotropin secretion leads to enlargement of and increased hormone secretion by the thyroid glands. Fetal demand for calcium and phosphorus is met by increased secretion of parathyroid hormone, which, especially if dietary intake is suboptimal, removes the minerals from maternal bone. The hormone relaxin, which is produced by both the corpus luteum and the placenta, also rises during pregnancy. In many infrahuman species this hormone may be essential to proper relaxation of pelvic ligaments during parturition, but its role in the human is obscure, since estrogens and progesterone, both of which are abundant at this time, have that same effect.

PARTURITION

This process, at term, culminates pregnancy by expelling the fetus. High levels of circulating and locally acting progesterone prevailing during pregnancy tend to maintain the uterus in a quiescent state. In fact, the end of the first trimester of pregnancy is a crucial period, because at this time function of the corpus luteum is failing and the placenta is taking over progesterone secretion. In some pregnancies, and particularly in some women, this integration is suboptimal, placental maturation not quite compensating for waning ovarian function, and progesterone levels fall precipitously, leading to spontaneous abortion provoked by inadequate uterine hormonal support and increased uterine irritability.

The placenta has a built-in life span, which defines the term duration. If the fetus is removed in monkeys without damaging the placenta, the latter will deliver at term. A few days beforehand, the placenta begins to degenerate, and falling progesterone levels progressively diminish the uterine inhibition that has prevailed. Uterine irritability, therefore, increases. At the same time, the inhibitory effect of placental hormones on oxytocin secretion by the hypothalamus also weakens. The resultant surge of oxytocin, aided apparently by the fetal hypothalamus, helps initiate the uterine contractions that cause labor pains. Another powerful contractile hormonal stimulus is provided by the prostaglandins, secreted by both uterine and fetal membranes, which also rise concomitantly.

The mechanical stretch placed on uterine muscle by the fetus and its associated structures, aided by phasic stretching initiated by strong fetal movements, also favors contraction. Thus, parturition is initiated by a combination of hormonal and mechanical events.

CONTRACEPTIVE METHODS

As we have seen, the hormonal changes that bring about maturation of the ovum, its release, transport and implantation, involve relatively precise timing, quantitation, and sequencing of reproductive hormones. The hormones themselves initiate changes in vaginal and uterine pH, secretions, structure, and motility favorable to fertilization and implantation. Obviously any or all of these, or various combinations of them, can be altered so as to inhibit or prevent conception.

THE RHYTHM METHOD

In women, the ovum is fertilizable only for about 24 h after ovulation, and most sperm cannot survive for longer within the female, although a few may live for up to 72 h. Optimal fertility will occur if coitus takes place immediately following ovulation, a process signalled by an abrupt rise of 1°C or so in the basal body temperature. Until a day or so preceding ovulation, and about a day thereafter, fertilization usually does not occur. The difficulty in attaining or obviating fertilization is accurate timing of ovulation. To be on the safe side, it has been recommended that avoidance of intercourse for 3 days before and after ovulation, or other precautions taken, can prevent pregnancy. There is a lingering suspicion that some women at least might ovulate in response to the stimuli arising during coitus, which, along with irregular cycles, obviates such planning.

HORMONAL METHODS

Various synthetic estrogens or progestagens, singly or in combination, have been used to prevent conception. As we have seen, ovulation is triggered by a sudden surge of LH. If this is prevented, ovulation will not occur. Either estrogens or progesterone, if given in sufficient quantity, may have that effect, and if given early enough in the cycle can even prevent maturation of ova by inhibiting the hypothalamus and pituitary. The difficulty is in avoiding unwanted side effects, particularly abnormal menstrual bleeding. By combining relatively small amounts of estrogens and progestins in various forms of "the pill," it has been possible to achieve that end.

As a practical matter, synthetic forms of the two steroids are preferred, since they are more resistant than the naturally occurring hormones to hepatic inactivation. Although the formulations available do not promote abnormal menstrual patterns if properly used, there is a higher incidence of hypertension among women who take the pill, as compared with those who do not. This appears to be due to an increased formation of renin substrate by the liver and to increased renin secretion by the kidneys.

The pill is usually taken daily from the first few days of the menses, until after the expected ovulation date. Discontinuance then allows for normal regression of the uterine endometrium and a relatively normal menstrual period, ovulation having been prevented.

Other formulations have been devised, with the design not of preventing ovulation, but of maintaining the uterus in the proliferative phase until the ovum arrives: to be met with a surface unsuitable for implantation and nidation. This is attainable with lower hormone dosage than those designed to inhibit ovulation. This form of therapy also alters motility of the fallopian tubes and the quality of cervical mucus in ways inimical to successful fertilization and implantation of the ovum.

THE PLACENTAL UNIT

The placenta subserves a number of vital functions during pregnancy. In the final analysis its efficacy depends on a very large and quite permeable membrane that attains maximal efficiency at about 34 weeks of pregnancy. In the early months, it functions also as a storage organ, to supply fetal needs during the later stages, accumulating the basic materials, including protein, glycogen, and minerals. Later in pregnancy this function is assumed by the fetal liver and becomes less requisite. Large quantities of hormones, including chorionic gonadotropin, chorionic somatomammotropin, estrogen, and progesterone are secreted, the relative quantity of each depending on the stage of pregnancy.

DIFFUSION ACROSS THE PLACENTAL MEMBRANES

Since the fetus lacks direct access to the exterior, its nutritional requirements, gaseous exchange, and excretory functions must be supplied by the mother via the placenta.

Nutritional needs are subserved by the ability of the placenta to remove and store essential materials from the maternal circulation on the one hand, and to synthesize others from substrates therein (e.g., glycogen from glucose). These are then made available to the fetus.

FETAL EXCRETION

Certain substances formed during fetal metabolism, such as urea, diffuse easily and are only slightly higher in fetal than in maternal blood. Others, such as creatinine, which diffuse more slowly, have a much higher concentration gradient, diffusion being driven thereby. In either case the placental membranes act passively, diffusion being accomplished by differences in the concentration gradients between fetal and maternal circulations. Active absorption as well as active secretion have been known to occur.

GASEOUS EXCHANGE

The fetus requires abundant oxygen to support the active metabolism incident to growth. A corollary is the generation of much CO_2, which cannot be allowed to accumulate. An oxygen pressure gradient of about 20 mm Hg exists between maternal (50 mm Hg) and fetal (30 mm Hg) circulations, allowing for an effective passive O_2 diffusion. Also fetal hemoglobin is present at approximately a 50% greater concentration than maternal Hb and, since its oxygen dissociation curve is shifted to the left, it has about a 25% greater O_2 transport capacity than the maternal pigment. The actively metabolizing fetus produces large amounts of CO_2, which, as it traverses the placenta, tends to transfer to the maternal circulation. The consequence is that the pH of fetal blood rises while that of maternal blood falls. The oxygen binding capacity of fetal blood rises and of maternal blood falls in parallel in accordance with the Bohr effect. These factors together provide adequate oxygen to the fetus.

Carbon dioxide is disposed of in a similar manner. Although the gradient is only about 3 mm Hg, this driving force is enhanced by the extremely high solubility of CO_2 in placental tissue, thus facilitating transfer from fetal to maternal circulation. As pregnancy advances and the placenta ages, breaks in the membrane may occur, allowing passages of red-blood cells in one or both directions. The fetal blood volume is small, and the cardiovascular consequences of relatively slight hemorrhages may be severe. Then too, if the fetal red blood cells are genetically different from those of the mother, and enter the maternal circulation, antibodies against them will develop giving rise to erythroblastosis fetalis, with grave consequences to the fetus.

LACTATION

The anatomical features, glandular elements for milk synthesis, and ducts necessary to conduct secretion to the exterior, have been fully developed under the influence of estrogens and progesterone during pregnancy. Under the influence of somatomammotropin and prolactin, milk synthesis has also occurred, but its secretion has been prevented by steroid suppression of the hypothalamus. Milk secretion may be initiated at any time in the later stages of pregnancy, if abortion occurs, thereby removing placental inhibition. Full secretory potential is not attained until a few days after the birth of the infant, since until the placenta is discharged a degree of suppression persists.

As soon as placental inhibition is withdrawn, the hypothalamus begins to secrete oxytocin. The hormone acts on the muscular basket cells that surround mammary alveoli. These are thus compressed, and milk is discharged into the ducts and thence to the nipples. Such milk discharge is an active process, hence milk does not continuously drip from the breast between nursings.

Suckling is, itself, the most effective stimulus to oxytocin release. Because oxytocin also powerfully contracts uterine musculature, uterine contractions tend to occur during suckling. Normally the basal level of prolactin and oxytocin secretion quickly return to the prepregnant level within a few weeks, and milk synthesis and discharge are sustained only by constant suckling. In certain primitive cultures breast feeding is continued for a variable number of years, because other food is less readily available. Under such conditions milk secretion may persist for years. In former years, some women earned a more or less continual living by becoming "wet nurses" to a clientelle of mothers who did not wish to, or could not, suckle their own offspring. If suckling ceases, milk secretion quickly stops.

If the nipples are denervated, suckling at those outlets will not maintain milk secretion, but if adjacent innervated nipples are suckled, the denervated members will lactate. The stimulus by suckling provided to milk secretion is therefore mediated by nerve impulses, but the secretory impulse is clearly hormonal. This constitutes a neurohormonal reflex, with an afferent neuronal limb and an humoral efferent limb; mediated by the hypothalamus, which is a component of both the nervous and endocrine systems. Each time that suckling occurs, there is a spurt of prolactin secretion that repletes the milk supply for the next feeding. A synopsis of the hormonal stimuli is given in Figure 1.

During breastfeeding, the preoccupation of the hypothalamus with the production and secretion of prolactin and oxytocin preempts the normal sequencing of LH-releasing hormone secretion. The adenohypophysis therefore does not properly release gonadotropins. FSH and LH, and hormone secretion by the ovaries is therefore vanishingly low. The normal sexual cycle usually fails to recur until breastfeeding ceases. However, because of the complexity of neurohormonal factors involved in the sexual cycle, the many environmental and emotional facets that play on it, and the considerable differences in the susceptibility of women to them, there is great variability in the degree of suppression achieved by nursing. Nevertheless, it is a period during which, for a variety of reasons, fecundity is exceedingly low.

Milk is a watery fluid, containing variable proportions of fat, carbohydrate, and protein; its composition varying with the species, time of lactation, and nutritional status.

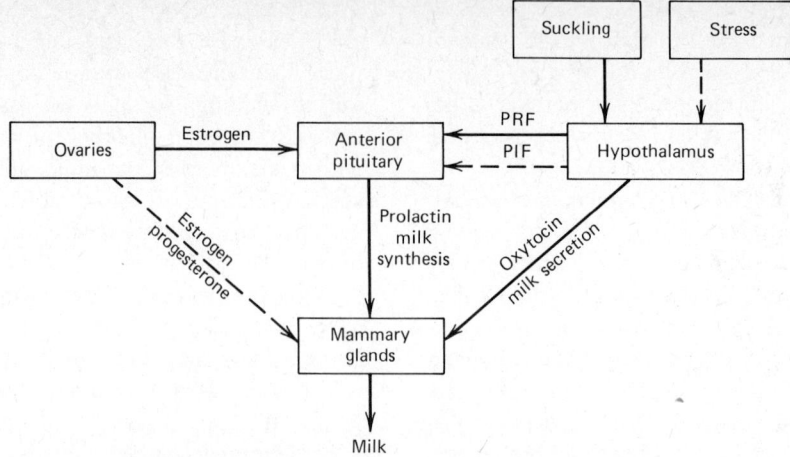

Figure 1. Synopsis of hormonal stimuli during suckling.

The quantity and quality both benefit from raising either protein or fat content of the diet and are impaired by increasing carbohydrate intake. Raising dietary fat also raises the level of milk fat. The milk of "lean" species, such as the horse, has a low-lipid, high-CHO content, whereas that of "fat" animals, such as swine or the marine cetacea, which require large blubber depots to minimize heat loss, have the reverse pattern and a high protein content (Table 1).

During lactation, the daily volume of milk secretion in women can reach 1–3 L. This represents a considerable metabolic drain in terms of water, fat, carbohydrate, and protein, which, if it is to be sustained without detriment to the mother, requires increased intake of an adequate diet.

Lactation is an extremely complicated phenomenon that requires the interaction of a variety of hormones. In addition to prolactin and oxytocin, growth hormone, insulin, thyroid hormones, adrenocorticoids, and parathyroid hormones are required. Because of this, and because prolactin secretion is a labile component, lactation may fail, or be inadequate, for a variety of endocrinological reasons. Psychological factors also have an important role, and highly excitable women tend to have more problems with breastfeeding than do those of a more placid temperament.

Table 1 Percentage Composition of Mammalian Milk

	Water	Ash	Fat	Carbohydrate	Protein
Human	86.4–88.6	0.2–0.3	2.9– 4.6	5.3–7.5	1.2– 2.7
Horse	89.9–90.2	0.3–0.4	0.6– 1.3	6.3–7.9	2.1– 3.8
Cow	85.6–88.2	0.7	3.5– 5.3	4.6	3.2– 3.8
Swine	81.0–84.0	1.0–1.1	3.7– 7.1	3.1–4.3	7.0–15.0
Dolphin	41.1–75.5	0.5–0.6	11.0–45.8	0.4–1.3	9.4–11.2

SELECTED BIBLIOGRAPHY

Nathanielsz PW: Endocrine mechanisms of parturition. *Annu Rev Physiol* 40:411, 1978.

Simpson ER, MacDonald PC: Endocrine physiology of the placenta. *Annu Rev Physiol* 43:163, 1981.

Thorburn GD, Challis JRG: Endocrine control of parturition. *Physiol Rev* 59:863, 1979.

SELF-STUDY QUESTIONS

MULTIPLE CHOICE

Select the best answer.

1. Fertilization of an ovum in women is optimal within
   A. 24 h of ovulation.
   B. 48 h of ovulation.
   C. 72 h of ovulation.
   D. any of these periods.

2. Chorionic gonadotropin
   A. is a glycoprotein.
   B. has an LH-like action.
   C. stimulates corpus luteum function.
   D. all of the above.

3. Estriol is an estrogen that
   A. is abundant in the latter half of the menstrual cycle.
   B. is enhanced by the presence of a fetus.
   C. is more potent than estrone.
   D. indicates pregnancy.

4. Placental somatomammotropin
   A. has an anti-insulin effect.
   B. increases glucose utilization.
   C. has prolactin-like effect.
   D. has a molecular weight of about 95,000.

5. Uterine contractions leading to parturition are favored by
   A. increasing oxytocin secretion.
   B. increasing prostaglandin secretion.
   C. decreased progesterone secretion.
   D. all of the foregoing.

6. In women who take "the pill" for contraceptive purposes
   A. abnormal menstrual patterns are common.
   B. Basal body temperatures rise more precipitously at the midmenstrual period.
   C. the ovum dies soon after ovulation.
   D. the incidence of hypertension is increased.

7. Milk secretion is inhibited by
   A. oxytocin.
   B. prolactin.
   C. progesterone.
   D. growth hormone.

8. Lactation activates the secretion of
   A. the parathyroid glands.
   B. LH.
   C. aldosterone.
   D. oxytocin.

9. Parturition is encouraged by
   A. falling progesterone secretion.
   B. increasing oxytocin secretion
   C. prostaglandins.
   D. all of the above.

ANSWERS

1. A. Fertilization is optimal for only a brief period (in women) and is unlikely to occur beyond 24 h after ovulation.

2. D. All the given characteristics apply to chorionic gonadotropin.

3. A. Estriol arises largely from conversion of estrone and estradiol by peripheral tissues, and is therefore abundant when substrates are high (i.e., in the latter half of the menstrual cycle). It is less potent than estrone or estradiol. The mere presence of an early fetus would not necessarily affect it, although advanced placental development would enhance the quantity circulating, although not the inherent activity.

4. C. Placental somatomammotropin, MW 75,000, has

insulin-like properties and inhibits maternal glucose utilization. The hormonal activity most closely mimics that of prolactin.

5. D. Progesterone inhibits uterine irritability and hence a decrease would enhance it. Both oxytocin and prostaglandins directly stimulate myometrial activity.

6. D. If the "pills" are taken properly, abnormal menstrual activity is rare. A rising basal body temperature in the midmenstrual period would indicate ovulation, which is suppressed by many oral contraceptives and would not, in any case, rise more precipitously because of them. There is no direct effect on an ovulated ovum, although its implantation might be prevented. An increased incidence of hypertension is found in women who take oral contraceptives.

7. C. Milk secretion is inhibited by progesterone, the withdrawal of which, at parturition, permits prolactin and oxytocin to cause lactation. Growth hormone is not inhibitory.

8. D. Oxytocin is the best answer, although A is a possibility. B and C are not directly affected.

9. D. A, B, and C are involved in promoting parturition.

Index